Proteins and Proteomics

A LABORATORY MANUAL

Proteins and Proteomics

A LABORATORY MANUAL

www.proteinsandproteomics.org

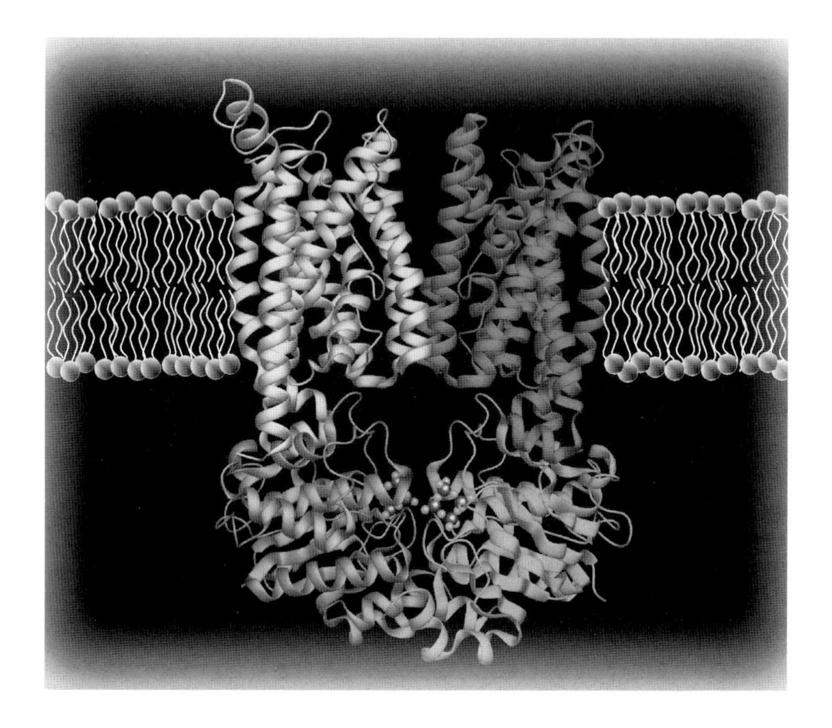

Richard J. Simpson

Joint ProteomicS Laboratory (JPSL) of the Ludwig Institute for Cancer Research and the Walter and Eliza Hall Institute of Medical Research, Melbourne, Australia

CSHL PRESS

COLD SPRING HARBOR LABORATORY PRESS
Cold Spring Harbor, New York

Proteins and Proteomics

A LABORATORY MANUAL

Printed in the United States of America

Publisher and Acquisition Editor	John Inglis
Managing Editor	Jan Argentine
Project Manager and Developmental Editor	Michael Zierler
Developmental Editor	Judy Cuddihy
Project Coordinator	Inez Sialiano
Production Editor	Dotty Brown
Desktop Editor	Susan Schaefer
Production Manager	Denise Weiss
Cover Designer	Ed Atkeson

Front cover artwork (paperback edition): Ribbon diagram of the *E. coli* vitamin B12 importer BtuCD, an integral membrane protein of the ABC transporter superfamily. The two membrane-spanning BtuC subunits that form the translocation pathway are colored yellow and red, while the cytoplasmic BtuD subunits that couple ATP hydrolysis to ligand transport are colored green and blue. Lipids forming the membrane bilayer are schematically indicated. (Image provided by Kaspar P. Locher and Douglas C. Rees, California Institute of Technology/ Howard Hughes Medical Institute; background chromatographs courtesy of Leemor Joshua-Tor, Cold Spring Harbor Laboratory.)

Library of Congress Cataloging-in-Publication Data

Proteins and proteomics : a laboratory manual / Richard J. Simpson
p. cm.
Includes bibliographical references and index.
ISBN 0-87969-553-6 (cloth : alk. paper) -- ISBN 0-87969-554-4 (pbk. : alk. paper)
1. Proteomics--Laboratory manuals. 2. Proteins--Laboratory manuals.
I. Simpson, Richard J.
QP551 .P697776 ~~2002~~ 2003
572.6--dc21

2002031580

1 0 9 8 7 6 5 4 3 2

All Cold Spring Harbor Laboratory Press publications may be ordered directly from Cold Spring Harbor Laboratory Press, 500 Sunnyside Blvd., Woodbury, N.Y. 11797-2924. Phone: 1-800-843-4388 in Continental U.S. and Canada. All other locations: (516) 422-4100. FAX: (516) 422-4097. E-mail: cshpress@cshl.edu. For a complete catalog of all Cold Spring Harbor Laboratory Press publications, visit our World Wide Web Site http://www.cshlpress.com.

Contents

Proteins and Proteomics
Companion Web Site

A COMPANION WEB SITE (www.proteinsandproteomics.org) to *Proteins and Proteomics: A Laboratory Manual* provides supplemental information about this fast-moving field of research. The site will include:

- References linked to Medline.
- Links to other databases of value to working scientists.
- Selected figures from the book for use in troubleshooting.
- Assistance in troubleshooting HPLC and 2D electrophoresis problems.
- A chapter on the analysis of carbohydrate from proteins (Oxley et al.; see abstract below).

Analysis of Carbohydrate from Glycoproteins

David Oxley,* Graeme Currie,† and Antony Bacic†

**The Babraham Institute, Babraham, Cambridge CB2 4AT, United Kingdom; †Plant Cell Biology Research Centre, School of Botany; University of Melbourne, Victoria 3010, Australia*

ABSTRACT

The identification and analysis of glycans associated with proteins is a formidable challenge due to the complexity and diversity of these carbohydrate structures. This chapter attempts to guide the "nonexpert" through the strategies used for the detection and analysis of carbohydrates on proteins. The "best" approach is very much dictated by the availability of material and the extent of information required. Many of the techniques can be used in most biochemistry/molecular biology laboratories, whereas others are very much in the bailiwick of the specialist laboratories that have access to highly sophisticated instrumentation. This chapter will answer at the very least the question most asked: Is my protein glycosylated? For many investigators, this level of information may suffice. For those inspired to ask the next logical question—What is the structure of this carbohydrate?—this chapter will provide sufficient background to proceed with confidence.

In addition to a brief history of carbohydrate research and a presentation of carbohydrate nomenclature, the chapter explains the fundamental steps involved in the analysis of glycoprotein-derived glycans. Each section within the chapter is designed to provide a concise understanding of the methods available to obtain the goal of the section, some of the limitations of each method, and references to some excellent texts on practical matters. Four widely used protocols are included, which detail the removal of glycans from glycoproteins and the preparation of monosaccharides for analysis by gas chromatography coupled with mass spectrometry.

- A protocol on the use of a multicompartment electrolyzer for the isoelectric fractionation of samples prior to 2D gel electrophoresis (Herbert et al.; see abstract below).

Sample Preparation for High-resolution Two-dimensional Electrophoresis by Isoelectric Fractionation in a Multicompartment Electrolyzer

Ben R. Herbert,* Pier Giorgio Righetti,† John McCarthy,* Jasmine Grinyer,* Annalisa Castagna,† Matthew Laver,* Matthew Durack,* Gerard Rummery,* Rebecca Harcourt,*and Keith L. Williams*

**Proteome Systems, North Ryde, Sydney, NSW, 1670, Australia; †University of Verona, Department of Agricultural and Industrial Biotechnologies, Strada Le Grazie No. 15, 37134 Verona, Italy*

ABSTRACT

Two common problems when using a broad pH gradient with two-dimensional gel electrophoresis for the separation of proteins are low resolution of hydrophobic, highly acidic or basic proteins and poor detection of low-abundance proteins. Increasing resolution and enhancing detection on 2D gels is possible with the use of narrow and ultra-narrow range (1–3 pH units and <1 pH unit, respectively), immobilized pH gradients (IPGs). However, when narrow pH gradients are loaded with an entire cell lysate, a large proportion of the protein sample is not isoelectric within the separation range of the pH gradient. These "extraneous" proteins severely disturb the separation, because they have pIs outside the pH range of the IPG. This phenomenon is aggravated at high protein loads. Although it can be almost eliminated by loading a small amount of protein, this unfortunately is of very limited use for proteomics.

One solution to these problems is to employ a multicompartment electrolyzer (MCE), an instrument that fractionates protein samples isoelectrically prior to the creation of 2D maps. The resulting protein fractions match the pH intervals to be adopted as the first dimensions of the subsequent 2D maps. The fractionated protein mixture, devoid of proteins with isoelectric points outside of the range of the IPGs, can be loaded in a 2D map at much higher levels, thus ensuring greater sensitivity and detection of low-abundance proteins. Isoelectric fractionation using the MCE is fully compatible with subsequent 2D protocols, because it is based on a focusing technique that yields highly concentrated samples devoid of salts and buffers. The current MCE instrument uses commercially available, amphoteric, buffered membranes that are matched to the pH endpoints of commonly used IPGs.

Additional information will be added after the book is published. To access the Web Site:

1. Open the home page of the site.
2. Follow the simple registration procedure that begins on that page (no unique access code is required, since the site is open to anyone who completes the registration process).
3. Your e-mail address and password (selected during the registration process) become your log-in information for subsequent visits to the site.

The FAQ section of the site contains answers about the registration procedure. For additional assistance with registration, to inform us of other Web address changes, and for all other inquiries about the proteinsandproteomics.org Web Site, please e-mail support@proteinsand proteomics.org or call 1-800-843-4388 (in the continental U.S. and Canada) or 516-422-4100 (all other locations) between 8:00 A.M. and 5:00 P.M. Eastern U.S. time.

Preface

NOW THAT THE FIRST DRAFT OF THE HUMAN GENOME SEQUENCE is in the public domain, the primary focus of biologists is rapidly shifting toward gaining an understanding of how genes function, i.e., the functional roles of the full complement of encoded proteins. As well as defining structural characteristics of proteins, this task requires an understanding of the temporal and spatial location of proteins within the cell, including the intricate nature of how proteins interact with one another. Analytical protein chemistry, or proteomics as it is now commonly known, has a vital role in this daunting task. As information began to flow from the various genome projects, it became apparent to Cold Spring Harbor Laboratory Press that there was a growing need to provide researchers with a source of reliable proteomics protocols. Not long after, at the urging of my colleague Joe Sambrook, author of the enormously successful manual *Molecular Cloning*, I was invited to tackle the challenge of writing a laboratory manual of analytical methods and protocols for proteomics studies. Thus, *Proteins and Proteomics: A Laboratory Manual* was conceived.

Proteins and Proteomics is aimed at those who wish to isolate proteins and peptides for subsequent proteomic analysis. It is written for an audience ranging from early graduate students to experienced investigators. *Proteins and Proteomics* is not an encyclopedic book covering all possible proteomics methods. Rather, the book covers only those proteomics methods and technologies that are in current use in my laboratory or those of trusted colleagues. In each chapter, I have endeavored to provide sufficient background knowledge to underpin the accompanying protocols. In areas outside the immediate ken of the protein chemist, such as glycobiology and informatics, I have sought the contributions of experts to cover these specific fields of experimentation. Accordingly, I thank Antony Bacic and Parag Mallick and their colleagues for valued contributions in glycobiology and informatics, respectively.

A work such as this does not see the light of day without much outside help and support. My first thanks go to my friends and colleagues in the Parkville precinct in Melbourne, who have responded generously to my sometimes intemperate requests to provide illustration material or technical review. I am greatly indebted to the editorial and production staff at Cold Spring Harbor Laboratory Press for their dedication and tireless efforts in checking references, facts, faulty constructions, and stylistic abominations and keeping me on schedule (almost). I owe a special debt of gratitude to Judy Cuddihy for her cheerful optimism that raised my spirits, Kaaren Janssen and Maryliz Dickerson for their guidance in the beginning of the project, Tamara Howard for diligent fact checking, Inez Sialiano for coordinating the project, Dorothy Brown for editorial assistance, Susan Schaefer for page layout, Denise Weiss for her elegant design of the book, and most of all, Michael Zierler for his unstinting support as Senior Developmental Editor in steering the book to completion. I also acknowledge the generous support of Jan Argentine, my Managing Editor, and John Inglis, the Director of Cold Spring Harbor Laboratory Press, for overseeing the project.

Finally, I owe a special debt of gratitude to Mary Whitham and Pamela Jones for their secretarial assistance. I must also thank Simone Pakin especially for her superb information gathering skills and for her dedication and shockproof resilience that helped us both survive the process. My thanks also go to the members of the Joint ProteomicS Laboratory (JPSL), especially Robert Moritz, Hong Ji, David Frecklington, Lisa Connolly, James Eddes, Eugene Kapp, and Gavin Reid, for the support that they have given me in more ways that I can list here. I also acknowledge the rich intellectual and collegial environment at the Ludwig Institute for Cancer Research and especially the ungrudging support given to me by its director, Tony Burgess. Without his sustaining enthusiasm, this book would not have been possible.

To this long list of creditors I must finally add my partner, Donna Dorow, who has had to suffer with increasing patience the growing and seemingly never-ending demands that the "book" came to make on our time (the never-ending weekend and late night writing episodes), my attention, and temper.

Richard J. Simpson

Acknowledgments

THE AUTHOR WISHES TO THANK THE FOLLOWING COLLEAGUES for their valuable assistance:

Ruedi Aebersold
Alastair Aitken
Ron D. Appel
Manuel Baca
Antony Bacic
Tomas Bergman
Tom Berkelman
Willy Bienvenut
Reinhard I. Boysen
Edward J. Bures
Annalisa Castagna
Ella Cederlund
Andrea Cinnamon
Lisa Connolly
Patrick W. Cooley
Garry L. Corthals
Graeme Currie
Jenny M. Cutalo
Marc Damelin
Catherine Déon
Leesa J. Deterding
Sam Donohoe
Janice L. Duff
Matthew Durack
Richard H. Ebright
James S. Eddes
David A. Fancy
David Frecklington
Ernesto Freire
Parag S. Ghandi
Robert Goode
David R. Goodlett
Andrew A. Gooley
Robin Gras
Timothy J. Griffin
Jasmine Grinyer
Melanie P. Gygi
Steven P. Gygi
Rebecca Harcourt
Lara G. Hays
Milton T.W. Hearn
Thomas P. Hennessy
Ben R. Herbert
Cameron J. Hill
Denis F. Hochstrasser
Wendy L. Holstein
Femia G. Hopwood
Geoff Howlett
Marion I. Huber
Toshiaki Isobe
Ole N. Jensen
Hong Ji
Hans Jörnvall
Eugene A. Kapp
Hooi Hong Keah
Rosalind Kim
Nancy Laird
Martin R. Larsen
Matthew Laver
Larry J. Licklider
Gavin MacBeath
Gregory S. Makowski
Parag Mallick
Matthias Mann
Edward M. Marcotte
John McCarthy
Scott A. McLuckey
Helmut E. Meyer
Robert L. Moritz
Philippe Mottay
Markus Müeller
Nikolai Naryshkin
Richard A.J. O'Hair
Yoshiya Oda
David Oxley
Sang-Hyun Park
Scott Patterson
Junmin Peng
Ronald T. Raines
Melinda L. Ramsby
Juri Rappsilber
Gavin E. Reid
Andrey Revyakin
Pier G. Righetti
Gerard Rummery
Michael T. Ryan
Jean-Charles Sanchez
David M. Schieltz
Albert Sickmann
Pamela A. Silver
Andrew J. Sloane
Paul E. Smith
Christopher S. Spahr
Hanno Steen
Allan Stensballe
Wayne R. Stochaj
Nobuhiro Takahashi
Masato Taoka
Kenneth B. Tomer
Klaus K. Unger
Adrian Velazquez-Campoy
Anne Verhagen
David B. Wallace
Michael P. Washburn
Valerie C. Wasinger
Keith L. Williams
Yoshio Yamauchi
Eugene C. Yi
Jian-Guo Zhang
Lynn R. Zieske

Foreword

STUDENTS AND EXPERIENCED RESEARCHERS FACE SIMILAR challenges when attempting to delve into a new field of research. They need to learn new terminology, concepts, and theories, define current research topics in the field, and master a new set of methods and techniques. Finding suitable resources may be as difficult as mastering the subject matter itself, especially when the topic is an emerging and rapidly evolving field, such as proteomics. Often, textbooks, glossaries, and reference manuals are sparse—if they exist at all—and the available information must be gleaned from numerous articles in the primary literature.

In this comprehensive book, Richard Simpson and a group of leading proteomics experts attempt the impossible: to condense theory, background information, protocols, and information resources into a single volume. They succeed. This manual contains virtually everything one would hope to find in both a textbook and laboratory manual: overviews, introductory materials, and a theoretical base for proteomics. Detailed protocols for common proteomics experiments, complete source lists for the tested materials, and web links to crucial reagent resources are also thoughtfully provided.

Proteins and Proteomics: A Laboratory Manual is an invaluable information tool both for the experienced protein chemist who bravely ventures into the new world of proteomics and for the novice to proteins and proteomes. By focusing on what is currently considered the bedrock of proteomics technologies, Professor Simpson ensures that—in spite of the rapid advances that characterize contemporary proteomics research—this volume will remain relevant and current for years to come.

Ruedi Aebersold
Professor, Institute for Systems Biology

CHAPTER 1

Introduction to Proteomics

Biological macromolecules are the main actors in the makeup of life.... To understand biology and medicine at a molecular level...we need to visualize the activity and interplay of large macromolecules such as proteins. To study protein molecules, principles for their separation and determination of their individual characteristics had to be developed. One of the most important chemical techniques used today for the analysis of biomolecules is mass spectrometry (MS), one of the subjects of the 2002 Nobel prize award.

The 2002 Nobel prize in Chemistry was awarded "for the development of methods for identification and structure analyses of biological macromolecules" with one half going jointly to John B. Fenn (Virginia Commonwealth University, Richmond, USA) and Koichi Tanaka (Shimadzu Corporation, Kyoto, Japan) "for their development of soft desorption ionization methods for mass spectrometric analyses of biological macromolecules."

This is a revolutionary breakthrough. Chemists and biologists can now rapidly and reliably identify what proteins a sample contains. Hence, scientists can both "see" the proteins and understand how they function within cells.

NOW THAT MORE THAN 40 GENOMES, INCLUDING THE HUMAN GENOME, have been fully sequenced and are in the public domain, the next challenge for biologists will be to connect gene to function, genotype to phenotype, to find out what the genes really do! The rapid pace of genome sequencing efforts during the past several years has resulted in many newly discovered genes that have been ascribed no function or a function that at best has been poorly described. For an up-to-date monitor of complete and ongoing sequencing projects, see the GOLD Web Site (Genomes OnLine Databases at http://wit.integratedgenomics.com/GOLD). This impetus to understand the function of newly discovered genes is leading biologists toward the systematic analysis of the expression levels of the components that constitute a biological system, chiefly, mRNA (transcriptomics), proteins (proteomics), and metabolites (metabolomics) (see Figure 1.1). Because proteins are central to biological function and obvious candidates for drug targeting, proteomics is enjoying a rapidly increasing level of attention.

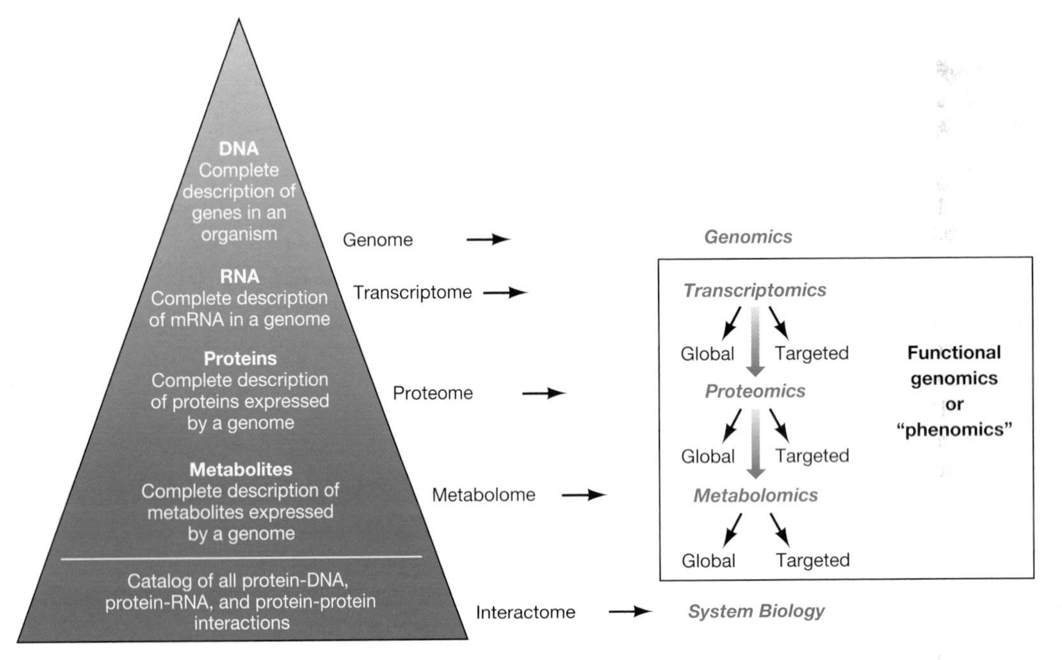

FIGURE 1.1. Functional genomics or phenomics. Genomics provides an overall description of the complete set of genetic instructions (the genes) contained within the genome that are available to a cell, i.e., the "blueprint" of a cell. Functional genomics, on the other hand, represents a systematic approach to elucidating the function of the novel genes revealed by complete gene sequences. (Phenomics has been suggested as an all-embracing term to describe functional genomics.) Functional genomics adopts a hierarchical strategy aimed at gaining a comprehensive and integrative view of the workings of living cells. There are a number of different approaches for studying the functional analysis of novel genes. These can be grouped into four domains: genome (the complete set of genes for an organism and its organelles), transcriptome (the complete set of mRNA molecules), proteome (the complete set of proteins), and metabolome (the complete set of metabolites, the low-molecular-weight intermediates). Researchers have now added the suffix "*ics*" to describe the utility for analyzing these domains. For example, the task of comparing the mRNA profiles using DNA arrays is now referred to as cellular (or tissue) transcriptomics, and the task of separating the cell's proteins and comparing their expression profiles is referred to as expression proteomics. Studying all of the proteins encoded by a genome (the proteome) without focusing on a particular cell type, growth conditions, and subcellular localization is the domain of global proteomics. Focus on protein expression within a particular cell type and/or subcellular organelle is the domain of targeted proteomics. It is now clear that these domains are not an end in themselves, but a vehicle to understanding an organism's entire metabolism, now referred to as metabolomics (Raamsdonk et al. 2001; Oliver 2002). Thus far, the fully sequenced genome studies have yielded many insights into the functional properties of proteins, especially the emergence of networks of interacting proteins (the term "interactome" has been coined to describe protein-protein networks). Understanding interactions between encoded proteins of a given genome is a critical first step in functional genomic analysis (Xenarios and Eisenberg 2001; Gerstein et al. 2002). To understand biology at the system level, and to develop models that explain the dynamics of cellular and organismal function (rather than the characteristics of isolated parts), all of these domains must be integrated (the "legome," total systems biology can be likened to assembling all of the component parts of a "Lego" set) in a quantitative and temporal manner (for a review on systems biology, see Brenner 1999b; Kitano 2002).

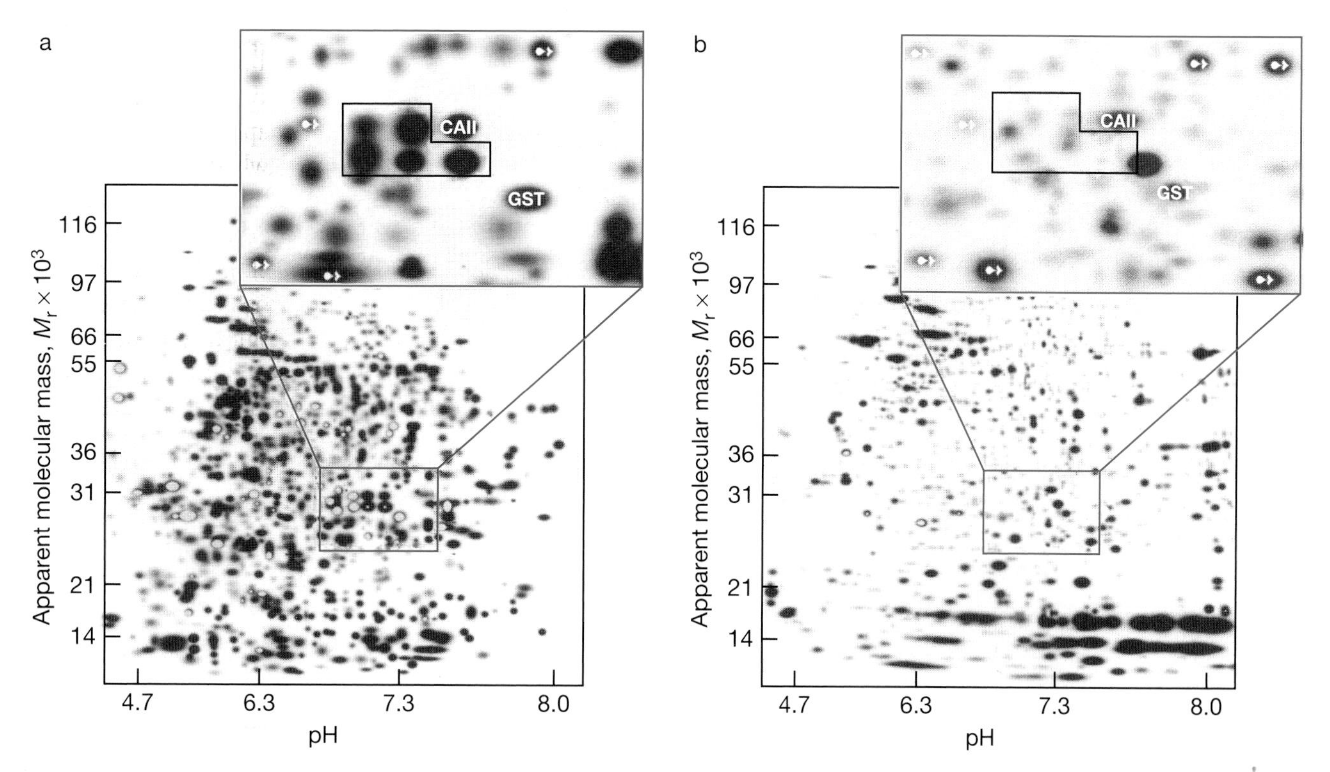

FIGURE 1.2. 2D gel electrophoresis of proteins from a whole-cell extract stained with Coomassie Brilliant Blue. (*a*) Wild-type C57/Black/6J murine colonic crypts; (*b*) polyps from multiple intestinal neoplasia (MIN) mice (Cole et al. 2000). The synthetic gel images were generated using PDQuest software. Differentially expressed proteins are marked in light gray. The insets show that carbonic anhydrase (CAII) and GST are both highly expressed in wild-type crypts and MIN polypts, whereas expression of several CAII isoforms (*outlined box*) is dramatically reduced in MIN polypts. (Reproduced, with permission, from Simpson and Dorow 2001 [© Elsevier Science].)

DEFINING PROTEOMICS

Proteomics or, more appropriately, functional proteomics refers to the branch of discovery science focusing on proteins. Initially, the term was used to describe the study of the expressed proteins of a genome using two-dimensional (2D) gel electrophoresis, and mass spectrometry (MS) to separate and identify proteins and sophisticated informatics approaches for deconvoluting and interrogating data. This approach is now referred to as "expression" or "global profiling" proteomics (Figures 1.2 and 1.3). The scope of proteomics has now broadened to embrace the study of "protein-protein" interactions (protein complexes), referred to as cell-mapping proteomics (Blackstock and Weir 1999) (see panel below on THE MANY FACES OF PROTEOMICS).

The term proteome, coined in 1994 as a linguistic equivalent to the concept of genome, is used to describe the complete set of proteins that is expressed, and modified following expression, by the entire genome in the lifetime of a cell. It is also used in a less universal sense to describe the complement of proteins expressed by a cell at any one time (from *Nature* 1999). Today, proteomics is a scientific discipline that promises to bridge the gap between our understanding of genome sequence and cellular behavior; it can be viewed as more of a biological assay or tool for determining gene function.

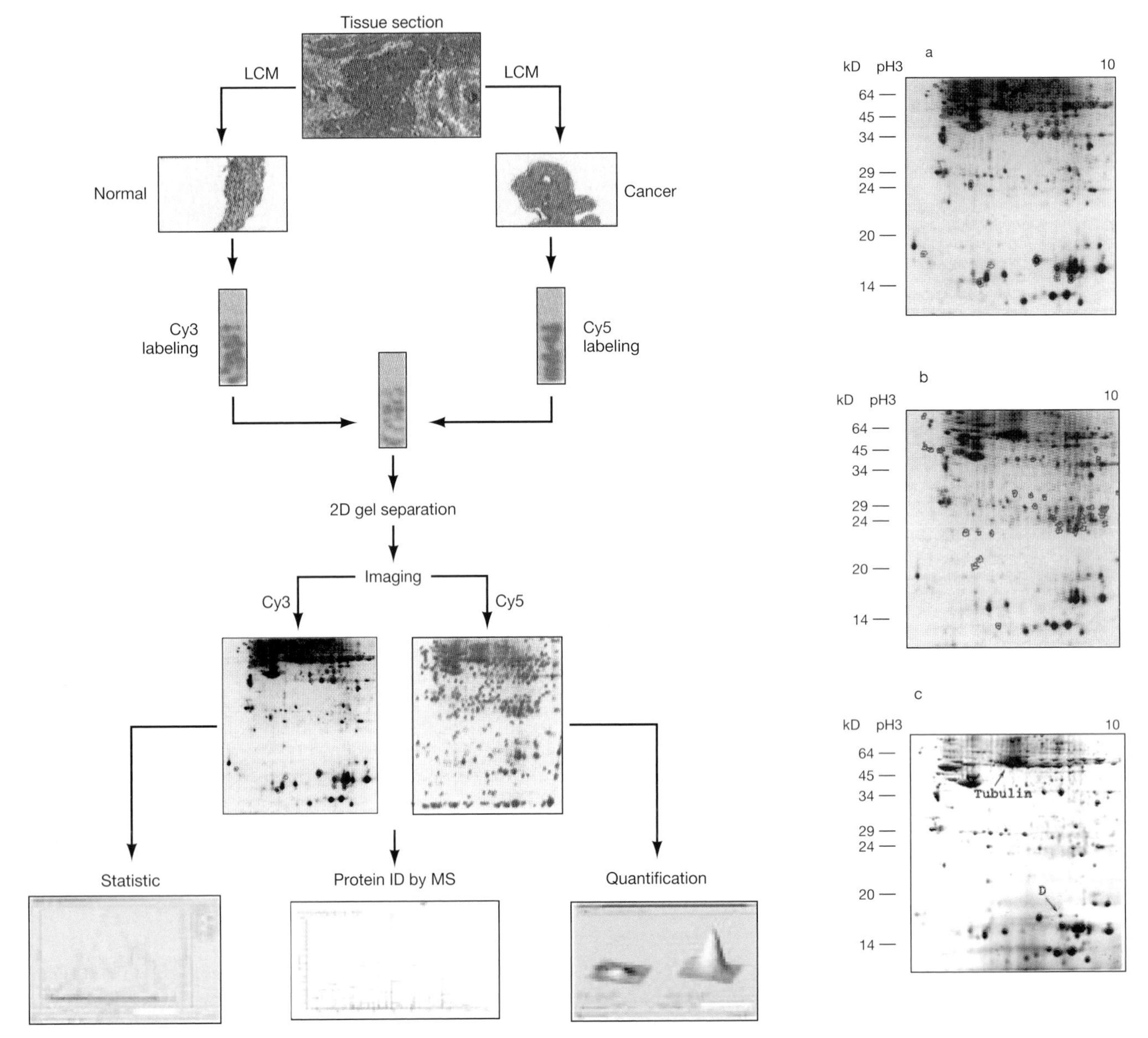

FIGURE 1.3. Differential in-gel electrophoresis (DIGE) for the identification of cancer markers. (*Top, left*) In DIGE, an emerging technology for proteome analysis (Unlu et al. 1997), two pools of proteins (e.g., normal cells and cancer cells procured from the same tumor sample using laser-capture microdissection) (Emmert-Bock et al. 1996) are labeled with 1-(5-carboxypentyl)-1′-propylindocarbocyanine halide (Cy3), *N*-hydroxy-succinimidyl ester and 1-(5-carboxypentyl)-1′-methylindodi-carbocyanine halide (Cy5) *N*-hydroxysuccinimidyl ester fluorescent dyes, respectively. The Cy3- and Cy5-labeled proteins are mixed and then separated in the same 2D gel. The 2D gel protein profiles can be rapidly imaged by the fluorescent excitation of either the Cy3 or Cy5 dye. (Cy3-labeled gel images are collected at an excitation wavelength of 540 nm and at an emission wavelength of 590 nm, whereas the Cy5-labeled gel images are collected at an excitation wavelength of 620 nm and an emission wavelength of 680 nm.) A comparison of the resulting images allows quantitation of each protein spot. (*Right, a*) Cy3 image of proteins from normal cells; (*b*) Cy5 image of proteins from tumor cells. Because both protein pools are electrophoretically separated in the same gel, those proteins existing in both pools will migrate to the same location in the 2D gel, thereby minimizing the inherent reproducibility problem associated with 2D gels. Quantitation of the protein profile can be rapidly and accurately achieved over a wide dynamic range (e.g., four orders of magnitude) based on fluorescent intensity (Patton 2000). The separated proteins in the 2D gel are next visualized by SYPRO Ruby staining (*right, panel c*). (The SYPRO Ruby-stained image is scanned at an excitation wavelength of 400 nm and an emission wavelength of 630 nm.) Protein spots of interest are excised and in-gel-digested with trypsin, and the peptides extracted for the purpose of identification by mass spectrometry (MS) methods described in Chapters 7 and 8. (ID) Identification. (Adapted, with permission from Zhou et al. 2002.)

THE MANY FACES OF PROTEOMICS

- ***Proteomic analysis (or analytical protein chemistry).*** The large-scale identification and characterization of proteins, including their posttranslational modifications, such as phosphorylation and glycosylation. Analysis is done with the aid of mass spectrometry or Edman degradation. For analysis of protein phosphorylation, see Chapter 9; and for amino-terminal sequence analysis using the Edman degradation procedure, see Chapter 6.
- ***Expression proteomics (or differential display proteomics).*** Two-dimensional gels are used for global profiling of expressed proteins in cell lysates and tissues. This conventional approach is being challenged by non-2D gel methods, such as liquid-based isoelectric focusing (IEF) or ion-exchange chromatography/reversed-phase high-performance liquid chromatography (RP-HPLC). Proteins are typically identified by mass spectrometry (MS). In many situations, these methods are complemented by DNA-based array methods. Includes *quantitative proteomics* (for a review of proteomics strategies for the quantitative analysis of paired protein samples [e.g., normal vs. diseased] utilizing stable isotope labeling combined with chromatographic separations, see Chapter 8 and Patterson 2000a,b).
- ***Cell-mapping proteomics (or cataloging of protein-protein interactions).*** Protein-protein interactions and intracellular signaling circuitry are determined by the identification of protein complexes (obtained by affinity purification and protein identifications by MS) or by direct DNA readout (e.g., yeast two-hybrid, phage display, ribosome display, and RNA-peptide fusions). For reviews on protein networks, see Legrain (2002) and Mayer and Hieter (2000); on cell-mapping proteomics, see Blackstock and Weir (1999), Lakey and Raggett (1998), and Duan et al. (2002); and on mapping protein-protein interactions with combinatorial biology methods that rely on direct DNA readout, see Pelletier and Sidhu (2001).

WHY PROTEOMICS IN ADDITION TO GENOMICS?

Large-scale Genome Sequencing: What Have We Learned?

One of the most exciting biological achievements to emerge during the past 40 years has been the completion of draft DNA sequences of the human genome, published by the International Human Genome Sequencing Consortium (a publicly funded project) (Lander et al. 2001) and by Celera Genomics (a commercial effort) (Venter et al. 2001). These Herculean efforts provide a blueprint of the information needed to create a human being and reveal, for the first time, the organization of a vertebrate's DNA (for an overview of this project, see Baltimore 2001). One of the interesting findings about the human genome is the number of genes found. The public project estimates that there are 31,000 protein-encoding genes, whereas Celera finds ~26,000, with many more still to be found. (A current estimate is that the number of protein-encoding genes may be on the order of 60,000.)

Interestingly, the number of coding genes in the human sequence is not dramatically different from the numbers reported for phylogenetically remote organisms: 6,000 for a yeast cell, 13,000 for a fly, 18,000 for a worm, and 26,000 for a plant (Genomes OnLine Databases at http://wit.integratedgenomics.com/GOLD). The number of genes reported for multicellular organisms is not highly accurate because of the limitations of existing ab initio gene prediction methods used to identify genes (Dunham et al. 1999). The existence of an open reading frame (ORF) in genomic data does not necessarily imply the existence of a functional gene. In human DNA, gene prediction by ab initio methods is notoriously difficult because of the extensive alternative splicing (Black 2000), lower density of exons, and high proportion of interspersed repetitive sequences. Given the unreliability of ab initio gene prediction software, all genes will need to be experimentally identified and annotated. For example, the error rate in the annotations for 340 genes from the *Mycoplasma genitalium* genome was ~8% (Brenner 1999a). Hence, verification of a gene product by proteomic analysis is an important first step in annotating the genome.

Disparity between mRNA Profiling and Protein Profiling

No simple correlation exists between changes in mRNA expression levels (transcriptomics) and those in protein levels (proteomics). Indeed, the link between transcript levels and protein levels in a given cell or tissue is tenuous, to say the least, and it is clear that array-based gene expression monitoring or other gene expression methods for measuring mRNA abundances, alone, are insufficient for analyzing the cell's protein complement (for a review of global gene expression methodologies, see Lockhart and Winzeler 2000). Recent studies show a marked disparity between the relative expression levels of mRNAs and those of their corresponding proteins (Anderson and Seilhamer 1997; Gygi et al. 1999a). A further complication arises when considering the complementarity of genomics and proteomics. Despite the adage that one gene gives rise to one protein, the situation in eukaryotic cells is more likely six to eight proteins per gene (Strohman 1994). Thus, there may be several hundred thousand human proteins after splice variants and essential posttranslational modifications are included. For example, 22 different forms of human α-1-antitrypsin have been observed in human plasma (Hoogland et al. 1999). Fortunately, such biological complexity can be unraveled using proteomic studies to understand how cells modulate and integrate signals.

Origins of Cellular Complexity

From the genome sequencing efforts to date, it is clear that the physiological complexity of organisms is not merely a consequence of gene numbers. For instance, humans (although composed of ~10,000,000,000,000 cells) have fewer than twice as many genes as the 959-cell nematode, *Caenorhabditis elegans*. Rather, evolution of the increased complexity of higher-order organisms is due to a number of other mechanisms, such as alternative splicing (Mironov et al. 1999; Black 2000), diversification of gene regulatory networks, and the ability of intracellular signaling pathways to interact with one another (Weng et al. 1999; Davidson et al. 2002). Biological signaling pathways can interact to form complex networks comprising a large number of components. Such complexity arises from the overlapping functions of components, from the connections among components, and from the spatial relationship between components in the cell. Additionally, many cellular processes are performed and regulated not by individual proteins but by proteins acting in large protein assemblies or macromolecular complexes. For instance, the eukaryotic ribosome, which translates RNA into protein, consists of ~80 unique proteins (Wool et al. 1995), and the RNA polymerase II transcription complexes in eukaryotic cells, which is involved in DNA replication, comprises at least 50 different proteins (Pugh 1996). For a further discussion of protein-protein interactions, see Chapter 10.

The traditional view of protein function tends to focus on the biochemical activity of a single protein molecule such as the catalysis of a given reaction or the binding of a ligand to its cognate receptor. This local function is often referred to as the "molecular function" of a protein. However, an expanded view of protein function is beginning to emerge in the post-genome era, with a protein being defined as an element in its network of interactions. This notion of expanded function has been variously referred to as "contextual function" or "cellular function" (see Kim 2000). The contemporary view of function is that each protein in living matter operates as an integral component of an intricate web of interacting molecules. For excellent reviews on this subject, see Weng et al. (1999) and Eisenberg et al. (2000).

Integrated Biology (Total Systems Biology)

To achieve a full understanding of how a complex organism works, biologists must develop an integrated (or global) view of a cell's mRNA and protein complements and a detailed knowledge of how these complements change with development and the environment (especially in disease). Mathematically, expression profiles from both mRNA and protein are required to fully understand how a gene network operates (see Hatzimanikatis and Lee 1999). For example, an integrated genomic and proteomic analysis of a systematically perturbed glucose/galactose-utilizable pathway in yeast concluded that an analysis of both mRNAs and proteins is crucial for understanding biological systems (see Ideker et al. 2001).

Protein-protein interactions are a crucial component of this integrated biology. Already, a large proportion of known protein-protein interactions in yeast have been identified by genome-scale yeast two-hybrid assays (Legrain and Selig 2000; Schwikowski et al. 2000; Uetz et al. 2000; Hazbun and Fields 2001; Ito et al. 2001a,b; Legrain et al. 2001) and direct affinity capture methods (Gavin et al. 2002; Ho et al. 2002). However, the interactions detected by these physical methods may include nonspecific interactions of no biological significance. Biologically important protein-protein interactions require that the interacting partners be in specific *protein states* (interactions may result in the transition of one protein state to another), but physical methods, like two-hybrid assays and affinity capture, do not distinguish between protein states of a given protein molecule. The following is a list of attributes that define protein states (also see Figure 1.4).

- ***Covalent modification*** (e.g., phosphorylation, glycosylation, lipidation, nitrosylation, acetylation, and ubiquitination). A protein may occur in its "active" or "inactive" form depending on the state of covalent modifications, such as phosphorylation (Hunter 2000a,b).
- ***Cellular localization.*** Depending on the biological status of the cell, a protein molecule may reside in one or several cellular locations, such as the nucleus, cytosol, plasma membrane, mitochondria, and endoplasmic reticulum.
- ***Presence of ligands.*** The binding of small molecules and ions (e.g., heme, metal ion, glucose, ATP, ADP, GTP, and GDP) to proteins alters protein states, affecting properties such as rates of enzyme catalysis and allostery.
- ***Alternate splicing.*** Different forms of a protein molecule may result from alternate splicing of the gene product.
- ***Proteolytic cleavage.*** Truncated forms of a protein molecule may result from specific amino- or carboxy-terminal cleavage or internal cleavage. These truncations alter the

state and activity of the protein. For example, certain proteolytic enzymes are produced as inactive precursors (zymogens), which must be cleaved to generate an active enzyme. By restricting the synthesis, location, or activity of the necessary proteases, cells have a means of regulating proteolytic activity with respect to time and cellular localization (Khan and James 1998; Kobe and Kemp 1999). Examples include proteases involved in blood clotting, catabolic digestion (e.g., pepsin, rennin, trypsin, chymotrypsin, and carboxypeptidase are secreted as inactive precursors), apoptosis (e.g., the key effector molecules, the caspases, are present as inactive zymogens [Shi 2002]), cleavage of viral precursor proteins to functional units, and pattern formation in multicellular organisms (Khan and James 1998). Proteolytic activity is further controlled by specialized proteins that specifically inhibit the active proteases (Bode and Huber 1992; Khan and James 1998).

- ***Oligomeric state.*** A protein molecule may exist in a multiprotein complex or as a homodimer or homo-oligomer. From yeast interactome studies, it is estimated that at least 78% of yeast proteins occur in complexes (Gavin et al. 2002).
- ***Protein conformation.*** Three-dimensional structure information on different protein states is important for understanding their biological behavior. For example, protein function is often regulated by allosteric mechanisms (Monod et al. 1963), in which effector molecules bind to regulatory sites distinct from the active site, usually inducing conformational changes that alter the activity. Allosteric effectors usually bear no structural resemblance to the substrates of their target protein, the classic example being end products of metabolic pathways acting at early steps of the pathway to exert feedback control. Intrasteric regulation (Kemp and Pearson 1991), on the other hand, includes autoregulation of protein kinases and phophatases by internal amino acid sequences that resemble the substrate (such internal amino acids are often referred to as pseudosubstrates). This type of regulation is considered the counterpart of allosteric control (see Figure 1.5). Examples of protein kinases whose regulation is mediated by intrasteric autoregulatory sequences include Twitchen kinase, Titin kinase, CaMK-1, insulin receptor kinase, and MAP kinase ERK2 (for reviews, see Kobe and Kemp 1999; Huse and Kuriyan 2002).

For a descriptive database of biological protein interactions organized in terms of protein states and state transitions, see LiveDIP (http:www.dip.doe-mbi.ucla.edu/) (Duan et al. 2002). Additional information on DIP (the Database of Interacting Proteins) can be found in Chapter 11 and in Table 1.1, which provides a list of Web-accessible databases containing information on protein-protein interactions.

One of the difficulties (and challenges) in studying protein-protein interactions is that at any given time, the pool of molecules of a protein inside a cell most likely represents one or several of the protein states of that particular protein, depending on the cellular context. This is a major impetus of proteomics, especially cell-mapping proteomics, which aims to describe all protein-protein interactions (both spatially and temporally) within a given cell. The challenge of proteomics is to utilize existing technologies (and to develop new technologies) to define all of the protein states for a given protein molecule. Such information is of crucial importance in post-genome biology, especially total systems biology, since it will shed light on the molecular mechanisms underlying biological processes. For reviews, see Xenarios and Eisenberg (2001) and Gerstein et al. (2002) for protein-protein interactions, and for systems biology, see Brenner (1999b) and Kitano (2002).

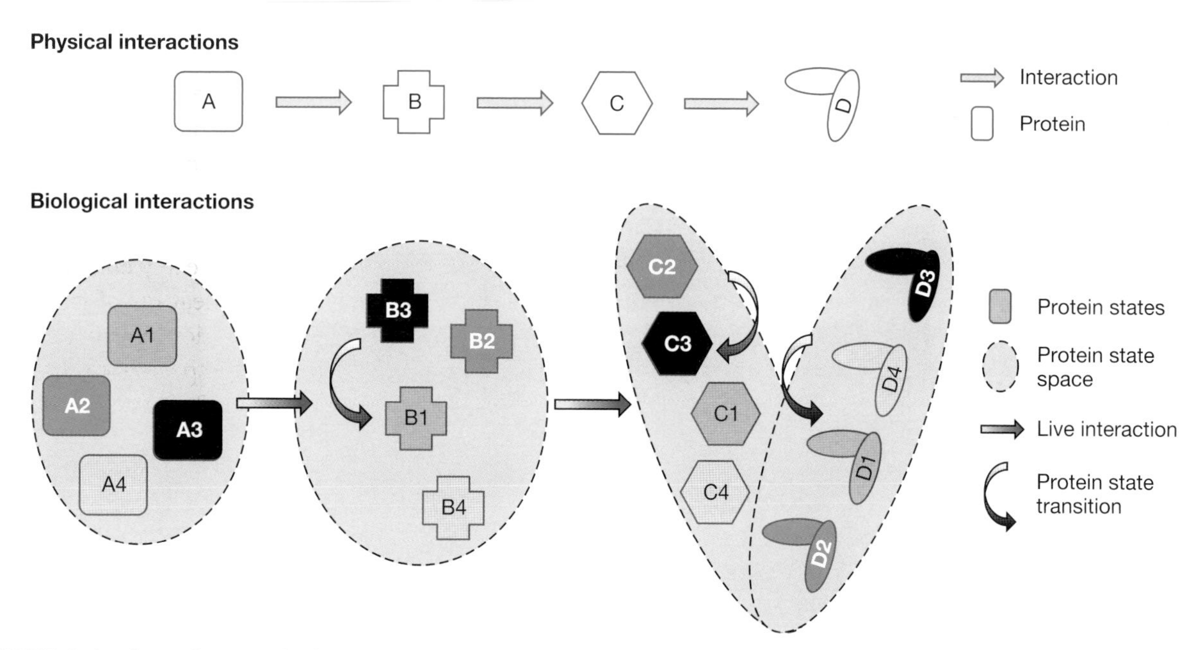

FIGURE 1.4. Physical versus biological protein-protein interactions. Physical interactions are shown as binary relationships between pairs of proteins. Each protein exists as a collection of states (forms) in the protein state space. For a summary of protein states and their attributes, see the panel on INTEGRATED BIOLOGY. For example, protein X exists in any of the protein states X1, X2, etc., which may represent different posttranslationally modified forms of protein X, conformational state, or different cellular localizations of this protein; protein Y can exist in multiple protein states as well. A given protein state, e.g., X1, interacts only with a given protein state of its interacting partner. A database, which describes protein interactions by protein states and state transitions, designated LiveDIP, has been described recently (http:www.dip.doe-mbi.ucla.edu) (Duan et al. 2002). (Adapted, with permission, from Duan et al. 2002.)

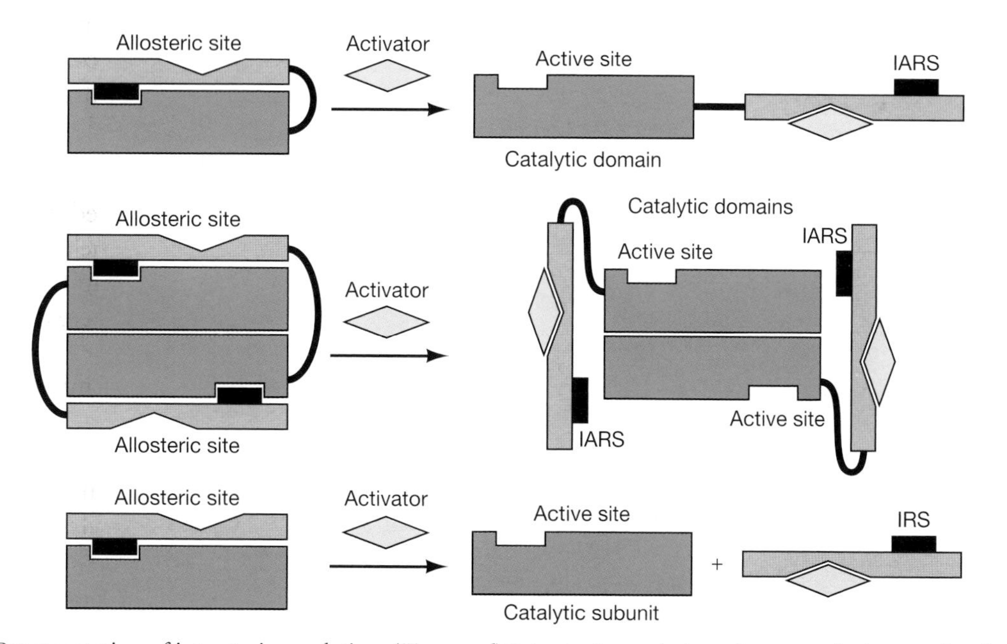

FIGURE 1.5. Representation of intrasteric regulation. (*Top panel*) Intrasteric regulation of enzymes by intramolecular interactions through an intrasteric autoregulatory sequence (IARS) that is contiguous with the catalytic domain containing the active site. The enzyme is maintained in an inactive state through the binding of the IARS in an intramolecular fashion that masks the active site. Allosteric activation by an activatory ligand or protein results in the release of the IARS from the active site. (*Middle panel*) Intrasteric regulation of homodimeric enymes by intermolecular interactions, through an IARS in *trans*. The IARS interacting with one subunit is a part of another subunit. (*Bottom panel*) Intrasteric interaction of heteromeric enzymes by intermolecular interactions, through an intrasteric regulatory sequence (IRS) on a distinct subunit. For a review of active-site-directed protein regulation, see Kobe and Kemp (1999). (Reproduced, with permission, from Kobe and Kemp 1999.)

TABLE 1.1. Protein interaction databases and data sets

Database	Acronym	URL	Content	Number of interactions	References
Alliance for Cellular Signaling/Nature Publishing Group	AFCS	www.afcs.org	The "Mini Molecule Pages" contain information from journals about interactions between various signaling molecules.	3214	Abbott (1999)
Amaze	AMAZE	www.ebi.ac.uk/research/amaze	Graphical information about signaling pathways.	3000 (via KEGG)	van Helden et al. (2000)
Biomolecular Interaction Network Database	BIND	www.bind.ca	Designed to store full descriptions of interactions, molecular complexes, and pathways.	6171	Bader and Hogue (2000)
Database of Interacting Proteins	DIP	dip.doe-mbi.ucla.edu	Protein-protein interactions, including identities of the interacting proteins, their interacting regions, binding affinity, and experimental methods.	11,000	Xenarios et al. (2002)
Database of Ligand-Receptor Partners	DLRP	dip.doe-mbi.ucla.edu	Annotated ligand-receptor interactions.	181	Graeber and Eisenberg (2001)
Live Database of Interacting Proteins	LiveDIP	dip.doe-mbi.ucla.edu	Protein-protein interactions annotated like DIP but also containing data on biological interactions, which are described in terms of protein states and state transitions.	Continuously updated subset of DIP	Duan et al. (2002)
Encyclopedia of *Escherichia coli* Genes and Metabolism	EcoCyc	ecocyc.org	Describes the genome and the biochemical machinery of *E. coli*. The goal of the project is to describe the functions of each of its molecular parts, to facilitate a system-level understanding of *E. coli*.	Genes from 13 species	Karp et al. (2002)
Enzymes and Metabolic Pathways Database	EMP	www.empproject.com/	EMP contains experimental journal publications compiled into a structured, indexed, and easily searchable form. Available commercially as WIT2 (igweb.integratedgenomics.com/WIT2)	3000	

Kyoto Encyclopedia of Genes and Genomes	KEGG	kegg.genome.ad.jp/	Signaling pathway information can be browsed via the PATHWAY database.	3000	Kanchisa and Goto (2000)
Molecular INTeraction Database	MINT	cbm.bio.uniroma2.it/mint/	Focuses on experimentally verified protein-protein interactions. Both direct and indirect relationships are considered.	3786	Zanzoni et al. (2002)
Munich Information Center for Protein Sequences	MIPS	mips.gsf.de/proj/yeast/CYGD/interaction/	Searchable index of physical and genetic interactions from high-throughput (2-hybrid, coprecipitation) yeast experiments.	9750	Mewes et al. (2002)
PathCalling Yeast Interaction Database	PathCalling	portal.curagen.com/extpc/com.curagen.portal.servlet.Yeast	Index of interactions from high-throughput 2-hybrid yeast experiment.	957	Uetz et al. (2000)
Predictome	PREDICTOME	predictome.bu.edu	Database of predicted links between the proteins of 44 genomes based on three computational methods (chromosomal proximity, phylogenetic profiling, and domain fusion) and large-scale experimental screenings of protein-protein interaction data.	300,000	Mellor et al. (2002)
Protein Interaction on the Web	PRONET	www.myriad-pronet.com/	Protein-protein interactions from high-throughput two-hybrid experiments.	NR	
Science's signal transduction knowledge environment	STKE	stke.sciencemag.org	Canonical pathways document basic properties of signaling modules, and specific pathways describe specific components and relations in a particular organism, tissue, or cell type.	NR	Ray and Gough (2002)
Signaling pathway database	SPAD	www.grt.kyushu-u.ac. jp/spad/	Browseable maps of canonical extra-cellular signaling pathways.	NR (~1000)	
Transpath		transpath.gbf.de/index.html	Browseable map of signaling pathways.	NR	Heinemeyer et al. (1999); Wingender et al. (2000)

These databases and data sets are available on the World Wide Web as of July 22, 2002. For additional resources, see Chapter 11.

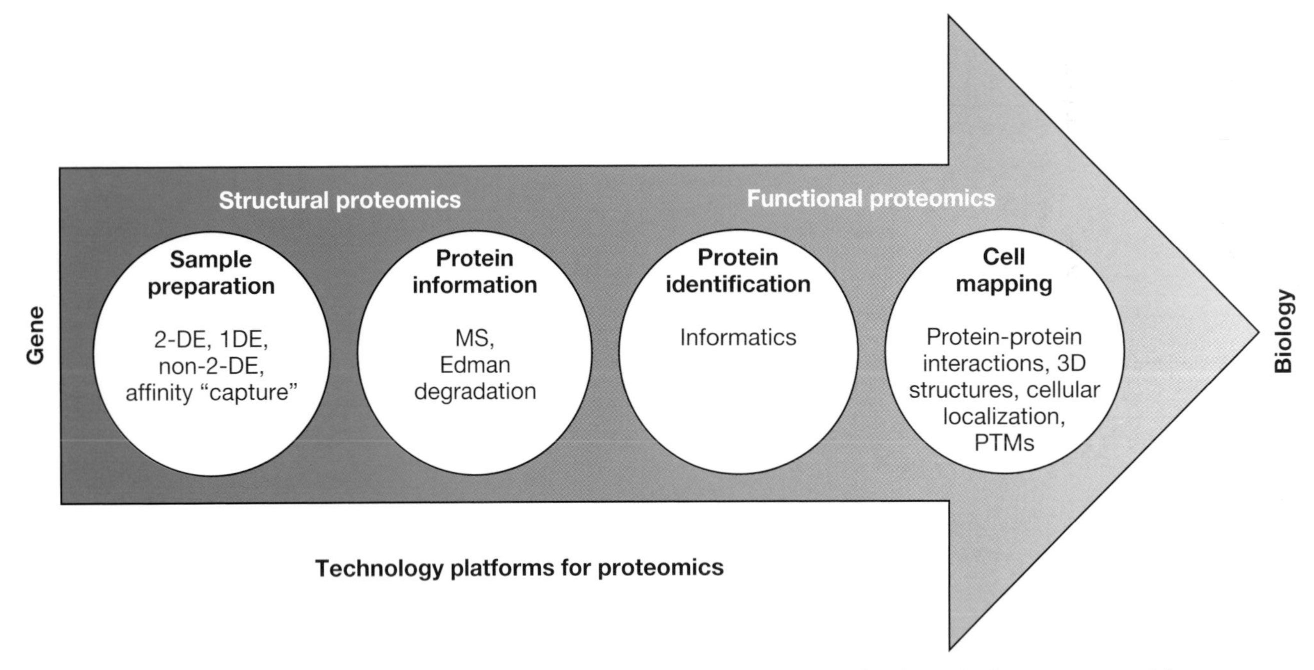

FIGURE 1.6. Proteomics: Bridging the gap between gene and product. Four key technology platforms are essential components for most proteomic strategies: sample preparation, obtaining protein information (albeit peptide mass/sequence information), protein information (use of informatics), and targeted proteomics (study of protein-protein interactions, posttranslational modifications (PTMs), cellular localization, etc.). For a description of the Edman degradation procedures, see Chapter 6. An outline of mass spectrometry (MS) methods for obtaining protein information is given in Chapter 8.

IDENTIFICATION AND ANALYSIS OF PROTEINS

Four key platform technologies are crucial to any proteomics strategy aimed at elucidating the function of an unknown gene (Figure 1.6):

- Sample preparation and handling (see Chapters 2, 3, 4, 8, and 10).
- Determination of partial amino acid sequence information (see Chapters 6, 7, and 8).
- Protein identification and quantification (see Chapters 7, 8, and 11).
- Cell mapping (see Chapter 10).

These platform strategies require different, yet complementary, types of expertise. Figure 1.7 provides an outline of the various identification and analysis strategies for solving any proteomics problem.

Protein Separation Strategies

One of the rate-limiting steps in any proteomic analysis study is obtaining, and then handling, sufficient quantities of a target protein(s) from its original biological source. Before embarking on such a study, it is very important to consider dynamic range considerations. For example, the dynamic range of protein abundance in a biological sample can be as high as 10^6 (i.e., protein abundance may range from 10 copies/cell for transcription factors up to 1,000,000 copies/cell for the more abundant molecules). The task of purifying trace-abundant proteins (e.g., growth factors and their receptors, or transcription factors) from natural biological sources is often extremely difficult, involving extremely large quantities of starting

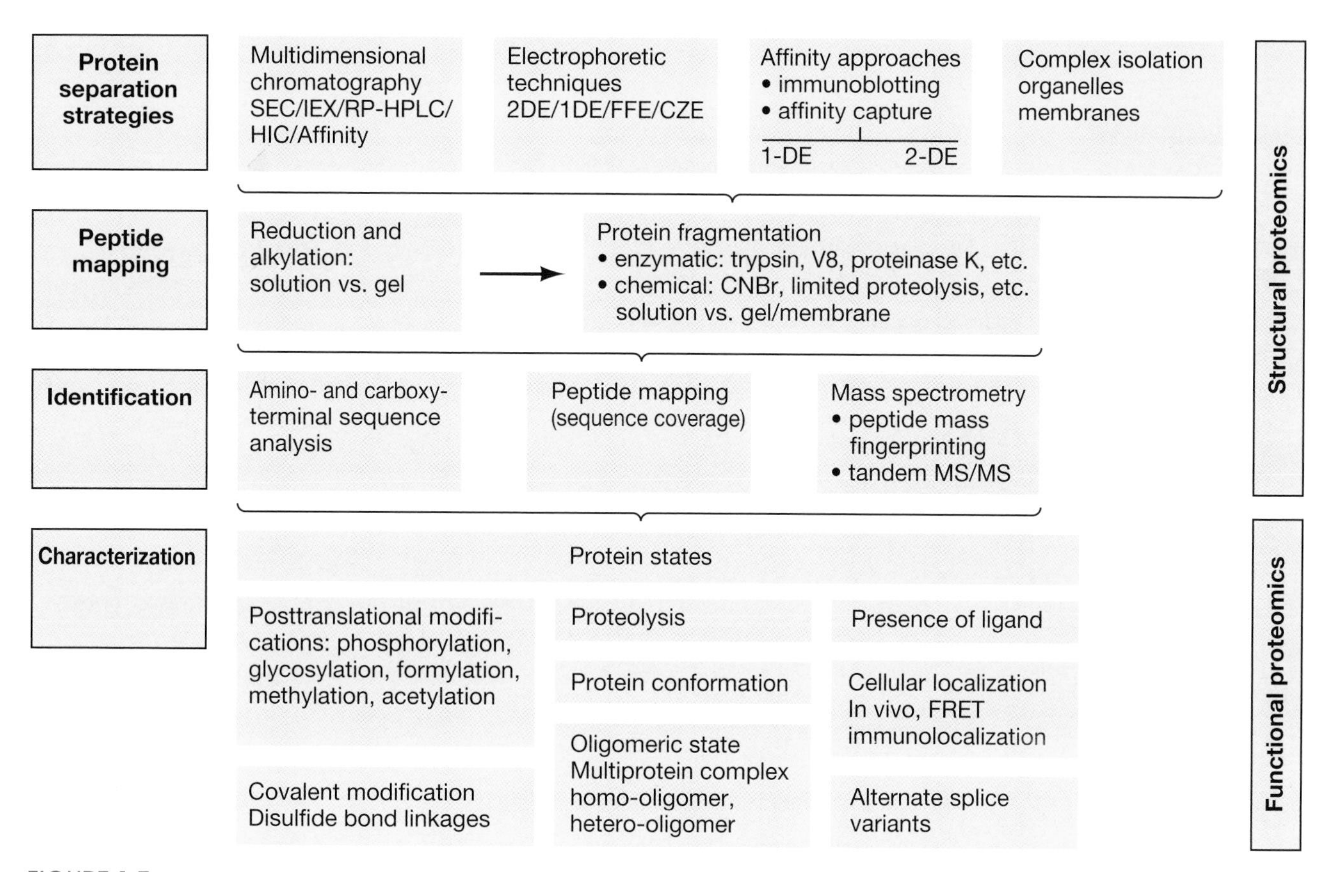

FIGURE 1.7. Proteomic strategies used for the identification and analysis of proteins. (SEC) Size-exclusion chromatography; (IEX) ion-exchange chromatography; (RP-HPLC) reversed-phase high-performance liquid chromatograpy (see Chapter 5); (HIC) hydrophobic interaction chromatography; (2DE) two-dimensional gel electrophoresis (see Chapter 4); (1DE) one-dimensional gel electrophoresis (see Chapter 2); (FFE) free-flow electrophoresis; (CZE) capillary zone electrophoresis; (FRET) fluorescence resonance energy transfer.

material (Table 1.2). Typically, a 1–2-millionfold purification is required to acquire a homogeneous target protein if identification is by amino-terminal sequence analysis using Edman chemistry. Considerably less material is required and purification to homogeneity is not essential if identification is by MS-based methods.

The classical method for quantitative and qualitative expression proteomics combines protein separation by high-resolution 2D gel electrophoresis (Klose 1975; O'Farrell 1975; Görg et al. 1988; also see Chapter 4) with MS or MS/MS identification of selected protein spots (see Chapter 8 and reviews by Yates 1998a,b; Patterson 2000a,b; Griffin et al. 2001). Because even the best 2D gels can routinely separate no more than 1500 proteins, this technique is limited to the most abundant proteins if a crude protein mixture (e.g., whole-cell lysate) is used (see Figure 1.8) (Gygi et al. 2000). Unlike other separation methods, such as RP-HPLC (see Chapter 5) and free-flow electrophoresis, which can tolerate large amounts of sample, 2D electrophoresis is limited by the amount of material that can be applied to the first-dimension immobilized pH gradient gel (~150 μg to low milligram quantities; see Chapter 4). Hence, 2D gels have limited "scale up" capability. For this reason, it is often desirable to "trace enrich" for a particular subclass of proteins. By analyzing proteins in a cellular compartment or organelle, it is possible to reduce the complexity and differences in abundance of a subset of proteins within a cell (for a list of various cellular organelles and compartments that have been subjected to detailed proteomic analysis, see Table 3.2).

TABLE 1.2. Examples of low-abundance proteins and peptides isolated from natural biological sources

Protein	Source	Yield (μg)	Reference
Multipotential colony-stimulating factor	Pokeweed mitogen-stimulated mouse spleen-cell-conditioned medium (10 liters)	1	Cutler et al. (1985)
Human A33 antigen	Human colon cancer cell lines (10^{10} cells)	2.5	Catimel et al. (1996)
Platelet-derived growth factor (PDGF)	Human serum (200 liters)	180	Heldin et al. (1981)
Granulocyte colony-stimulating growth factor (G-CSF)	Mouse lung-conditioned medium (3 liters)	40	Nicola et al. (1983)
Granulocyte-macrophage colony-stimulating growth factor (GM-CSF)	Mouse lung-conditioned medium (18 liters)	12	Burgess et al. (1986)
Coelenterate morphogen	Sea anemone (200 kg)	20	Schaller and Bodenmuller (1981)
Peptide YY (PYY)	Porcine intestine (4000 kg)	600	Tatemoto (1982)
Tumor necrosis factor (TNF)	HL60 tissue culture medium (18 liters)	20	Wang and Creasey (1985)
Murine transferrin receptor	NS-1 myeloma cells (10^{10} cells)	20	van Driel et al. (1984)
Fibroblast growth factor (FGF)	Bovine brain (4 kg)	33	Gospodarowicz et al.(1984)
Transforming growth factor-β (TGF-β)	Human placenta (8.8 kg)	47	Frolik et al. (1983)
Human interferon	Human leukocyte-conditioned medium (10 liters)	21	Rubinstein et al. (1979)
Muscarinic acetylcholine receptor	Porcine cerebrum (600 g)	6	Haga and Haga (1985)
$β_2$-Adrenergic receptor	Rat liver (400 g)	2	Graziano et al. (1985)

Adapted, with permission, from Simpson and Nice (1989).

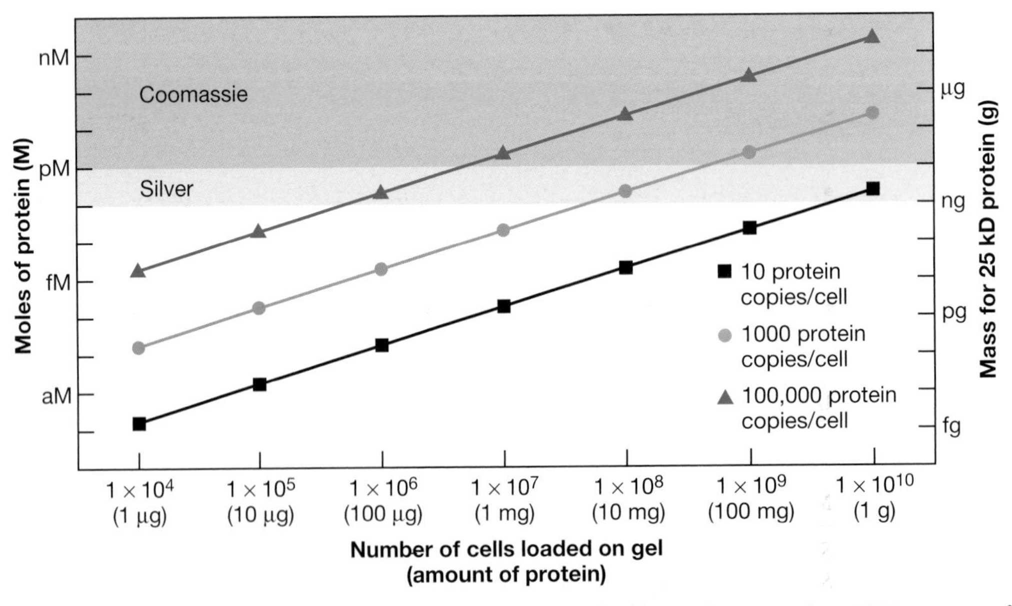

FIGURE 1.8. Relationship between protein copy number and cell quantity. Assuming 100% recovery of protein from cells. For 10^6 cells containing 1000 copies of a specific protein molecule per cell, there are 1.6 pg of this protein present (this value is calculated using Avogadro's number, which dictates that there are 6.02252 x 10^{23} molecules in 1 mole or gram-molecular weight of a given substance). For a 25-kD protein, this translates to 40 pg of protein present. (Courtesy of Ben Herbert, Proteome Systems Ltd., Sydney, Australia.)

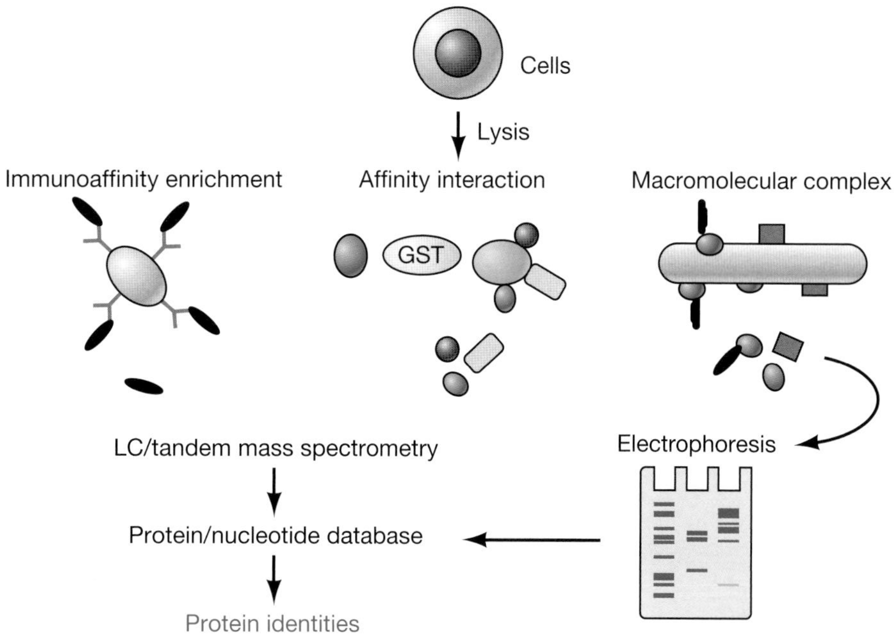

FIGURE 1.9. Cell mapping: Affinity capture methods.

Cell Mapping and Identification of Proteins in Complexes

One way to observe interacting proteins involved in a given biological process is to specifically enrich for these proteins (see Chapter 10). Typically, this requires knowledge of the activity of at least one protein in the multiprotein complex. Under nondenaturing conditions, interacting proteins can be enriched from complex protein mixtures (e.g., cell lysates) using methods (see Figure 1.9) such as:

- ***Coimmunoprecipitation*** or "pull-down" techniques using antibodies directed against one of the component proteins (Adams et al. 2002).
- ***Coprecipitation*** using affinity-tagged recombinant proteins and antibodies directed against the "tag" epitope (see Chapter 10, Protocol 1) (Séraphin et al. 2002).
- ***Protein-affinity-interaction chromatography*** (e.g., using recombinant glutathione *S*-transferase (GST)-fusion proteins and glutathione-affinity chromatography) (see Chapter 10, Protocol 2) (Einarson and Orlinick 2002)
- ***Isolation of intact multiprotein complexes*** (e.g., nuclear pore complexes, ribosome complexes, and spliceosomes).

TABLE 1.3. Representative macromolecular complexes identified by proteomics methods

Multiprotein complex	Number of protein components identified in complex	References
S. cerevisiae nuclear pore complex	~30	Rout et al. (2000); Allen et al. (2001)
S. cerevisiae 80S ribosome	75	Link et al. (1999)
Human spliceosome complex	46	Neubauer et al. (1998)
S. cerevisiae U1 snRNP complex[a]	20	Neubauer et al. (1997)
E. coli chaperonin GroEL	~300	Houry et al. (1999)
S. cerevisiae 26S proteosome	>24	Verma et al. (2000)

[a]snRNP = small nuclear ribonucleoprotein.

Determination of Partial Amino Acid Sequence

Usually, the final step of most proteomic studies, independent of the purification method employed, utilizes either SDS-PAGE (see Chapter 2) or 2D acrylamide gels (see Chapter 4) to separate the proteins for identification and characterization. Following electroblot transfer to an inert membrane, such as polyvinylidine difluoride (PVDF), intact proteins can be identified directly by amino- or carboxy-terminal amino acid sequence analysis or indirectly from peptides generated by *in-gel* or *on-membrane* digestion of the protein with a protease, usually trypsin (see Figure 6.12 and associated text). MS-based methods usually identify a protein, not by analyzing it directly, but by analyzing the peptides derived from proteolytic digestion. The main advantage of this approach is the ease with which proteolytically generated peptides can be recovered from gel slices (and inert membranes), compared with the difficulty of recovering intact proteins from acrylamide gels. Moreover, a small number of peptides usually yield sufficient information to permit protein identification (by peptide mass fingerprinting [PMF] and/or MS/MS of individual peptides). (For a detailed description of methods and strategies for the mass spectrometric identification of proteins, see Chapter 8.) The steps typically involved in the MS-based identification of a protein are illustrated in Figure 1.10.

Essential Elements of a Mass Spectrometer

Mass spectrometers consist of three essential components:

- An ionization source, which converts molecules in either solution or solid form into gas-phase ions.
- A mass analyzer, which separates the gas-phase ions according to their individual mass-to-charge ratios (m/z).
- An ion detector, which measures the m/z of each ion.

A mass analyzer uses a physical property (e.g., time-of-flight [TOF]) or electric or magnetic field (quadrupole or ion trap) to separate ions of a particular m/z that are subsequently measured by striking the ion detector. An important contribution of MS to proteomics has been the development in the mid 1980s of two soft ionization techniques that create ions of proteins and peptides.

- ***Matrix-assisted laser desorption ionization (MALDI)*** creates ions by excitation with a laser of a sample that is mixed or dissolved with an excess amount of a matrix component. The laser energy strikes the crystalline matrix (which has an absorption wavelength that matches closely with the laser wavelength) and causes rapid excitation of the matrix and subsequent plume of matrix and analyte (protein/peptide) ions into the gas phase. The singly charged ions are then guided to the mass analyzer and the detector by electrostatic lenses. MALDI is generally suitable for high-throughput analysis of complex mixtures of analytes and is used typically in conjunction with TOF analyzers to produce accurate measurement of molecular weight in the low ppm level (see Figures 1.10b and 1.11a,b).
- ***Electrospray ionization (ESI)*** creates gas-phase ions by applying a potential to a flowing liquid that contains the analyte and solvent molecules. A fine spray of microdroplets is generated upon application of a high electrical tension through a needle. Solvent is removed as the droplets enter the mass spectrometer by heat or some other form of energy such as energetic collisions with an inert gas (in some instruments, a heated capillary is placed following the electrospray needle to facilitate solvent evaporation). In contrast to MALDI, ESI yields multiply charged ions that require mass spectral deconvolution. The detection limits that can be achieved with ESI have improved dramatically by reducing the flow rate to nanoliter/minute levels using capillary chromatography (see Chapters 7 and 8). ESI is used typically in conjunction with quadrupole or ion-trap mass analyzers to produce information by tandem mass spectrometers (MS/MS) which is diagnostic of amino acid sequence (see Figures 1.10c and 1.11c,d). For a more detailed description of MS, see Chapter 8.

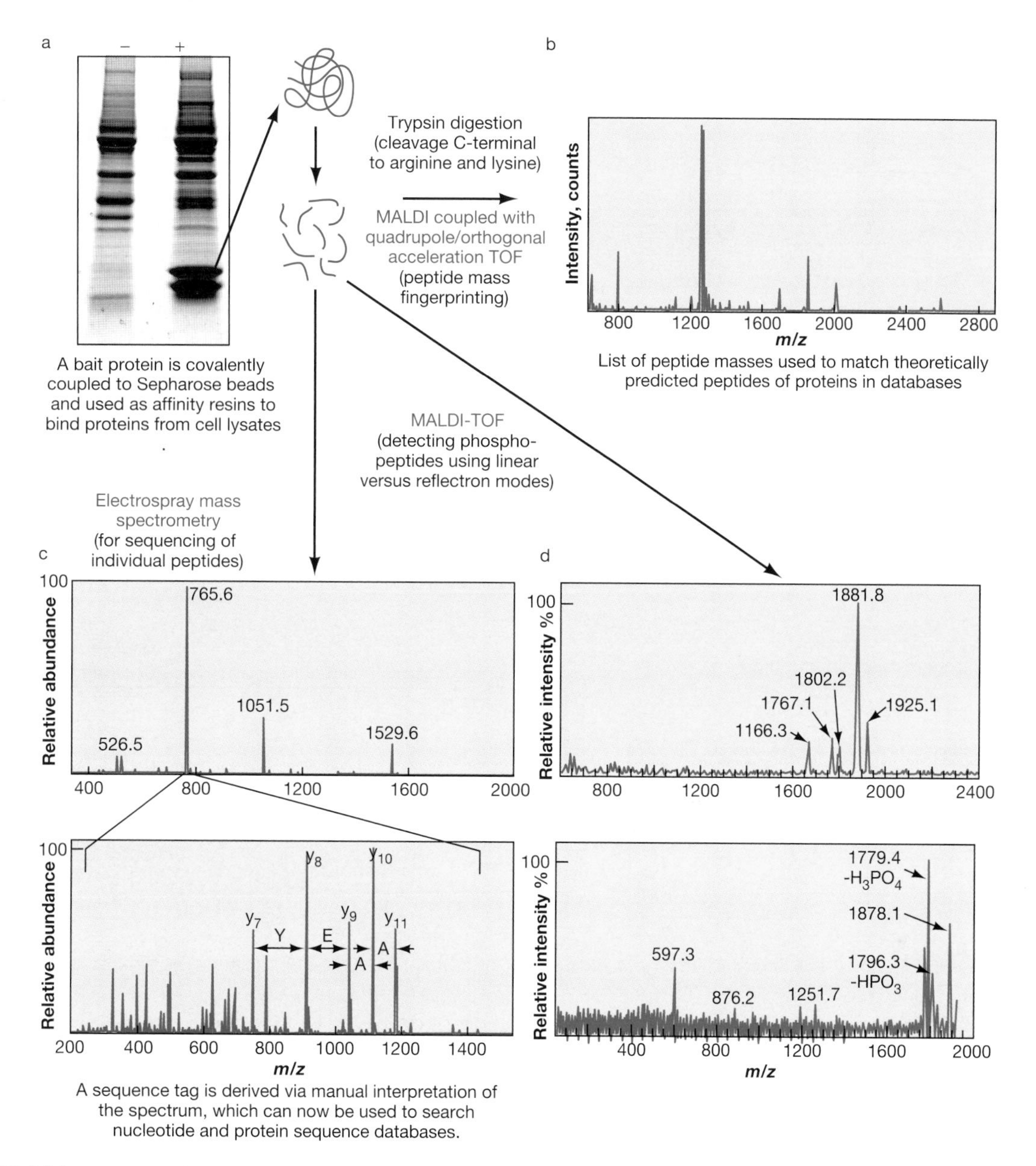

FIGURE 1.10. A strategy for mass spectrometric identification of proteins and posttranslational modifications. (*a*) Affinity capture using a bait protein to isolate protein-binding partners from a cell lysate. A GST fusion protein containing the SOCS-1 SOCS box sequence (Zhang et al. 1999) is used to illustrate the principle of affinity capture. SDS-PAGE analysis of affinity column eluates from GST-control (–), and GST-SOCS-1-SOCS-box glutathione (+) are shown. After 1D gel electrophoresis, the gel was stained with Coomassie and the protein bands of interest excised and subjected to trypsin digestion (see Chapters 2 and 7). (*b*) Analysis of an aliquot of the tryptic peptide mixture using a MALDI quadrupole/orthogonal acceleration TOF mass spectrometer (see Chapter 8). The resultant spectrum represents a peptide mass fingerprint (PMF) of a protein. The peptide masses can be entered into an algorithm, which matches them against theoretically predicted peptides of proteins in publicly available databases (see Chapter 8). (*c*) Analysis of another aliquot of the peptide mixture using an electrospray ion-trap mass spectrometer, which is coupled on-line to a capillary RP-HPLC. (*Top panel*) Peptide masses at a given time. A typical experiment entails isolating the most intense ion in the spectrum (i.e., *m/z* 765.6), and performing collision-induced dissociation to generate sequence ions. (*Bottom panel*) From the resultant MS/MS spectrum, amino acid sequence information can be derived via manual interpretation or by using an algorithm that correlates the experimental spectrum with those in a database (see Chapter 8). (*d*) For phosphopeptide analysis, peptide-containing fractions from a capillary RP-HPLC separation are subjected to MALDI-TOF analysis operated in "linear mode" (*top panel*) and also in "reflectron mode" (*bottom panel*). It can be seen that the major peptide ion detected in "linear mode" (*m/z* 1881.8) is metastable, resulting in a major ion loss of 98 daltons (–H_3PO_4) and a lesser ion loss of 80 daltons (–HPO_3) that are observable in "reflectron mode." Ion losses of 98 and 80 daltons, or multiples thereof, indicate that a parent ion is phosphorylated (see Chapter 9) (Zugaro et al 1998).

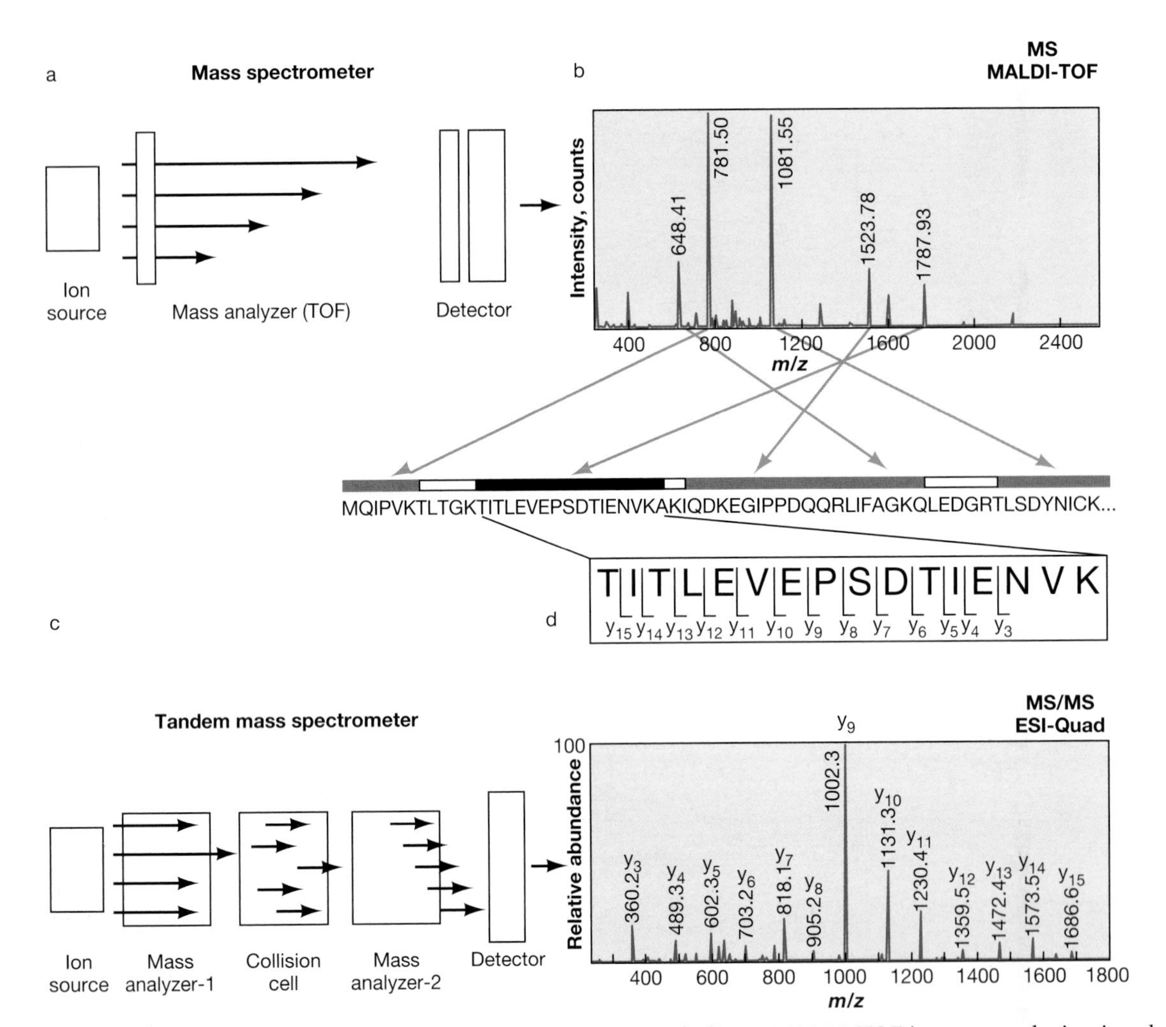

FIGURE 1.11. Basic elements of MS and tandem MS. (*a*) Schematic of a linear MALDI-TOF instrument depicts ions leaving the ion source and entering the mass analyzer, the TOF tube (see Chapter 8). Ions of different masses travel along the TOF tube at different speeds. The time they take to hit the detector is correlated with their mass. (*b*) Mass spectrum of a peptide mixture using a MALDI-TOF instrument. The peptide ions comprise a "peptide mass fingerprint" (PMF) for a specific protein. This information permits the identification of the protein from which the peptides were derived (see Chapter 8). Tryptic peptides derived from human ubiquitin (sequence below) are shown as an example. (*c*) Schematic of a triple-quadrupole tandem mass spectrometer (see Chapter 8). Ions enter the first quadrupole (Mass analyzer-1) where they are analyzed. One of these ions is selected to enter the collision cell whereupon it is fragmented upon collision with helium gas. The fragment ions leave the collision cell and are separated in the next quadrupole (Mass analyzer-2) before being scanned out to the detector. (*d*) MS/MS spectrum of a human ubiquitin peptide. The fragment ions in this spectrum correspond to amino acid residues TITLEVEPSDTIENVK. Proteins can be identified from such peptide sequence information.

In contrast to peptides, the molecular mass of intact proteins is usually insufficient to allow database identification. (This problem is compounded by the fact that very little is known about the extent of posttranslational modifications of most proteins.) However, the task of identifying intact proteins directly by MS will most likely improve with the development of "top-down" sequencing strategies (McLafferty 1999; Sze et al. 2002; see the information panels on FRAGMENTATION MECHANISMS OF PROTONATED PEPTIDES IN THE GAS PHASE and "TOP-DOWN" PROTEIN SEQUENCE ANALYSIS USING MS/MS in Chapter 8) where intact proteins are fragmented within a specialized Fourier transformed-ion cyclotron resonance (FT-ICR) mass spectrometer, and the derived fragments are selected for further fragmentation by collision with an inert gas. This produces sufficient information for identification (see the information panel on "TOP-DOWN" PROTEIN SEQUENCE ANALYSIS USING MS/MS in Chapter 8).

Mass Spectrometric Identification Methods Used in Proteomics

- ***MALDI and peptide mass fingerprinting (PMF).*** MALDI-MS is used to determine the accurate mass of a group of peptides derived from a protein by digestion with a sequence-specific protease, usually trypsin, thus generating a peptide mass map or peptide mass fingerprint. Because trypsin cleaves proteins at the amino acids arginine and lysine, the masses of tryptic peptides can be predicted theoretically for any entry in a protein sequence database. Experimentally obtained peptide masses are compared with those obtained theoretically, and the protein can be identified correctly if there are a sufficient number of peptide matches for a protein in the database. PMF is thus popular for identifying proteins from species for which complete genome sequences have been determined. The method is not suited for searches of EST databases, because the ESTs represent only portions of gene-coding sequences, which may be too short to cover a sufficient number of peptides observed in the experimentally obtained PMF. Digests of complex protein mixtures are not suited for PMF, because it is unclear which peptides in the complex peptide mixture originate from the same protein. Rather, PMF is better suited for identification of proteins separated by 2D gels, where ancillary information about protein molecular weights and isoelectric points can be used to aid in identification. High mass accuracy is critical for unambiguous identifications.
- ***Electrospray ionization and tandem mass spectrometry.*** Individual peptides from a peptide mixture are isolated in the first step in the mass spectrometer and fragmented by collision-induced dissociation (CID, i.e., by collision with an inert gas in a collision cell) during the second step in order to sequence the peptide (hence, the term tandem mass spectrometry). The fragments obtained by this method are derived from the amino or carboxyl terminus of the peptide and are designated b or y ions, respectively (for a discussion on the principles of gas-phase fragmentation, see the information panel on FRAGMENTATION MECHANISMS OF PROTONATED PEPTIDES IN THE GAS PHASE in Chapter 8). CID spectra contain redundant pieces of information, such as overlapping b- and y-series ions, multiple internal ions from the same peptide, and immonium ions. This redundancy makes CID spectra a rich source of sequence information that is highly specific to an individual peptide. These data also contain information that is diagnostic of the amino acid sequence of peptides. Because peptides fragment in the CID process in a predictable manner, sequences in the database can be used to predict an expected pattern to that observed in the experimentally observed MS/MS spectrum (see Figure 1.11d). Advantages of this approach, compared to PMF, are that a protein in a complex mixture can be identified from the CID spectrum of a single peptide, and matching one or more tandem mass spectra to peptide sequences in the same protein provides a high level of confidence in the identification process. Hence, the MS/MS identification approach is amenable to searching EST databases (Verhagen et al. 2000). For a full description of MS-based approaches for indentifying proteins, see Chapter 8.

DIFFERENTIAL DISPLAY PROTEOMICS (COMPARATIVE PROTEOMICS)

A fundamental aspect of proteomics research is the determination of protein expression levels between two different states of a biological system (i.e., relative quantification of protein levels), such as that encountered between a normal and diseased cells or tissues. This is often referred to as differential display or comparative proteomics. Although using image analysis to contrast the position and intensity of stained spots in a 2D electrophoresis gel has been used for comparative proteomics (Görg et al. 2000; Zhou et al. 2002b), this approach lacks adequate dynamic range and is more qualitative than quantitative. To add a quantitative dimension to proteomics, especially non-2D electrophoresis gel-based proteomic strategies, an alternative technique based on stable isotope dilution and MS has emerged. The stable isotope dilution approach has been used extensively in the pharmaceutical industry to quantitatively measure small molecules. The method relies on the premise that the relative signal intensity in a mass spectrometer of two analytes that are chemically identical but of different stable isotope compositions (e.g., ^{2}H, ^{13}C, ^{15}N, ^{18}O) can be resolved in a mass spectrometer and hence yield a true measure of the relative abundance of the two analytes in the sample (Browne et al. 1981; De Leenheer and Thienpont 1992). Although MS is not quantitative per se due to variances in detector response and ionization yields of different analytes, observed peak ratios of isotopic analogs are highly accurate because the isotopic species are chemically identical and they are analyzed in the same experiment using the same instrument. For comparative proteomic analyses, proteins from a reference sample are compared with a second sample that is isotopically labeled with a heavy stable isotope. Following isotopic labeling, which can be performed either in vivo or in vitro, proteins are proteolytically digested and the generated peptides are analyzed by liquid chromatography [LC]–MS/MS to determine both the identity of proteins and the relative ratio of proteins expressed (see Figure 1.12). Because all of the peptides in the sample exist in pairs of identical amino acid sequence (i.e., with the same physicobiochemical properties) but different isotope patterns (different masses), they are expected to behave identically during isolation, fractionation, and ionization. By determining the ratio of intensities of the lower- and higher-mass components of a peptide, the relative abundance of that peptide (and hence protein) can be established. For a summary of various in vitro and in vivo isotopic labeling approaches, see Table 1.4 and Chapter 8, Protocols 9 and 10.

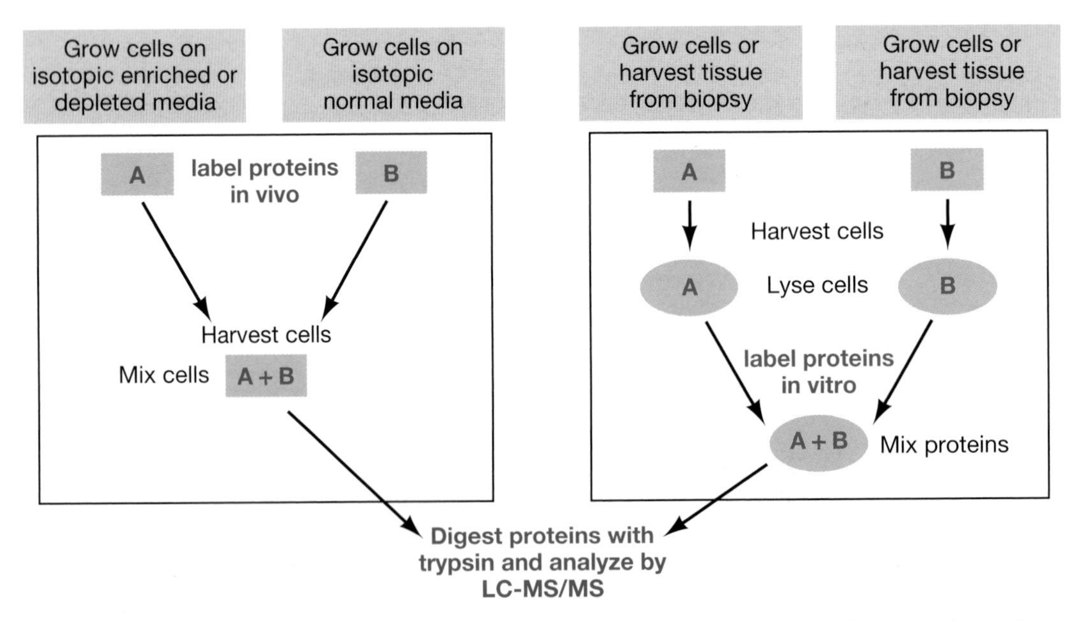

FIGURE 1.12. Scheme for determining protein expression (comparative proteomics). MS can be used as a universal detector to determine relative protein expression between two samples of interest. In vivo and in vitro isotopic labeling of proteins is followed by protein purification, proteolytic digestion, and analysis by LC-MS/MS to determine both the identity of peptides (and hence, proteins) and relative ratio of proteins expressed (by comparison of isotope ratios).

TABLE 1.4. Summary of various in vitro and in vivo isotopic labeling approaches

Approach	References
In vivo labeling (incorporation of stable isotopes into metabolic products)	
Microorganisms: All ^{14}N atoms are replaced by ^{15}N using ^{15}N-substituted media. Because ^{15}N-substituted media are difficult and expensive to make for mammalian systems, this method has been restricted to microorganisms that can be grown in this medium.	Oda et al. (1999); Lahm and Langen (2000); Conrads et al. (2001)
Mammalian cell lines: Mammalian cell lines are grown in media lacking a standard amino acid but supplemented with an isotopically labeled form of that amino acid (e.g., deuterated leucine).	Ong et al. (2002)
In vitro labeling (incorporation of stable isotopes performed postbiosynthesis by chemical derivatization)	
^{18}O labeling: In this method, ^{18}O is incorporated from $H_2{}^{18}O$ into carboxyl groups of peptides during proteolysis or deglycosylation. Although this method is robust, after amide bond cleavage, trypsin continues to exchange ^{18}O into carboxyl groups of peptides with basic amino carboxy-terminal amino acids (the rate of exchange can be structure-specific). Accurate isotope ratio measurements require careful deconvolution of mass spectral data.	Kuster and Mann (1999); Mirgorodskaya et al. (2000); Yao et al. (2001)
Esterification: Carboxyl groups of aspartic and glutamic residues and the carboxyl terminus of peptides are differentially methylated with either $^{1}H3$-or $^{2}H3$-methanol.	Goodlett et al. (2001)
Isotopically coded affinity tags (ICAT): An isotopically labeled affinity reagent is attached to cysteine residues in all proteins in the population. (The ICAT reagent is a sulfhydryl-directed alkylating reagent composed of iodoacetic acid attached to biotin through a short, oligomeric coupling arm.) Following tryptic digestion, the labeled peptides are affinity-purified using the incorporated affinity tag (biotin was used in the first iteration of the method) to achieve a pronounced enrichment (and simplification) of the peptide mixture. In a second iteration of the method, considerable improvements have been made by attaching the cysteine peptides to solid beads and photo-releasing the peptides afterward.	Gygi et al. (1999b,c); Zhou et al. (2002a)
Alternative isotopically coded affinity tags: A number of variations on the ICAT method have been proposed. Like ICAT, these methods share the requirement of in vitro chemical modification of the peptides or proteins and a peptide selection step. The method of Regnier differs from ICAT in that it does not include an affinity capture step.	Munchbach et al. (2000); Goodlett et al. (2001); Cagney and Emili (2002); Regnier et al. (2002)

POSTTRANSLATIONAL MODIFICATIONS

The vast majority of all eukaryotic proteins are posttranslationally modified and more than 200 posttranslational modifications (PTMs) of amino acids have been reported thus far (Krishna and Wold 1997). Practically all PTMs are associated with either an increase or a decrease in molecular mass. Hence, MS is an ideal tool in proteomics studies of PTM identification and characterization. It is generally acknowledged that knowing the nature of PTMs will facilitate our understanding of protein function. For a list of reported PTMs and their molecular masses, see Appendix 1.

In general, the MS methods used for protein identification are also applicable to the analysis of PTMs. However, PTM analyses are inherently more difficult than simple protein/peptide identification for the following reasons:

- ***Dynamic range is low.*** PTMs of proteins are typically at a low stoichiometry; hence, high sensitivity of detection is required. For example, if only 5–10% of a protein kinase substrate is phosphorylated, then a further 90–95% of the modified protein may be required for detection at the levels used for the nonphosphorylated form of the substrate.
- ***Isolation of the modified peptide containing the PTM is required*** to identify the PTM and the position (i.e., the site) of that modification within the peptide sequence. In contrast, proteins can be identified by the amino acid sequence of a single peptide by their CID spectrum.
- ***PTMs are frequently labile.*** Because the covalent bond between the PTM and amino acid side chain in the peptide is typically labile, it is sometimes difficult to maintain the peptide in its modified state during sample preparation and subsequent ionization in the mass spectrometer.
- ***PTMs are frequently transient in nature.*** This makes their "capture" for the purposes of analysis extremely difficult.

The two major PTMs of proteins are phosphorylation and glycosylation. Of these, the reversible phosphorylation of proteins ranks among the most important PTM that occurs in the cell. For a detailed discussion and protocols currently used to identify and characterize phosphorylation, see Chapter 9.

Phosphorylation of Proteins

Phosphorylation of proteins is a ubiquitous regulatory mechanism in both eukaryotes and prokaryotes. Intracellular phosphorylation is regulated by protein kinases (dephosphorylation is controlled by protein phosphatases), which are activated in response to extracellular signals and trigger cells to switch on or off many diverse processes such as metabolic pathways, kinase cascade activation, membrane transport, gene transcription, and motor mechanisms. The human genome contains 575 protein kinase domains, representing 2% of the total genome, and is the third most populous domain in proteins (Lander et al. 2001). For reviews on the structural roles of protein phosphorylation and the roles of phosphorylation in signal transduction, see Hunter (2000a,b), Pawson and Nash (2000), Adams (2001), and Johnson and Lewis (2001).

Protein phosphorylation can be examined in several ways. Table 1.5 provides a comparison of mass spectrometric, Edman degradation, and ^{32}P-labeling and phosphopeptide-mapping methods. For a detailed discussion and protocols for phosphoprotein and phosphopep-

TABLE 1.5. Comparison of widely used methods for phosphoprotein and phosphopeptide analysis

	Phosphopeptide mapping of ^{32}P-labeled proteins and peptides	Aminio-terminal sequencing using the Edman degradation procedure	Mass spectrometry
Sensitivity	most sensitive method	less sensitive (low pmole levels)	highly sensitive (low fmole levels)
Radioactivity requirement	yes	in some methods	no
Localization of phosphorylation sites	mutagenesis studies usually required to detect sites of phosphorylation	yes (however, tyrosine phosphorylation sites can be difficult)	yes
Homogeneous protein required	yes	yes	no
Sample throughput	very slow (this approach is labor-intensive)	slow (very labor-intensive)	can be high if automated LC-MS/MS methodologies utilized

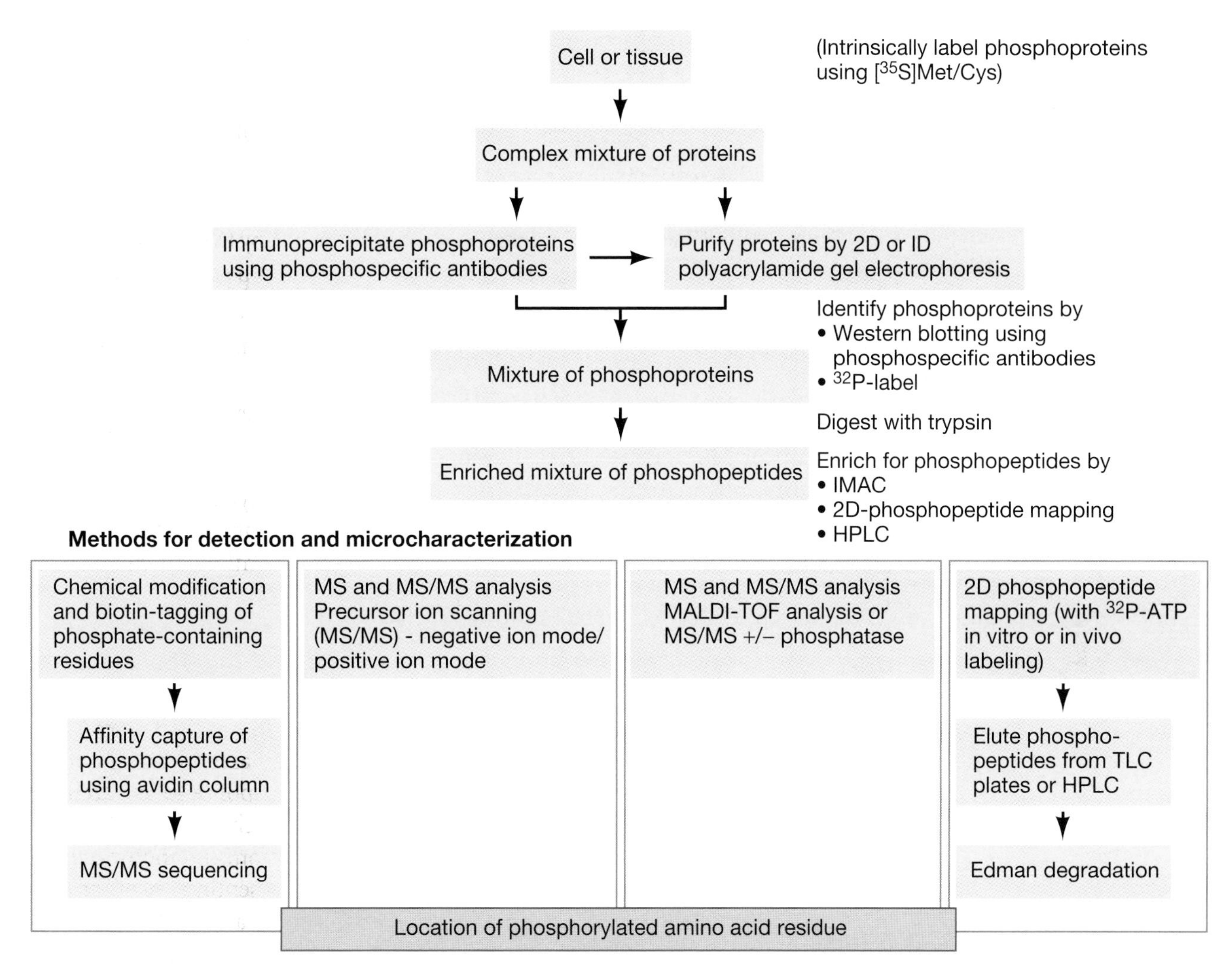

FIGURE 1.13. Summary of the various analytical methods for phosphopeptide/phosphoprotein enrichment, detection, and micro-analysis. (TLC) Thin-layer chromatography; (IMAC) immobilized metal affinity chromatography.

tide analysis, see Chapter 9 and recent reviews covering various aspects of this topic (see, e.g., Resing and Ahn 1997; Yan et al. 1998; Aebersold and Goodlett 2001; Sickmann and Meyer 2001; Mann et al. 2002).

Because only a portion of the proteins in a proteome are phosphorylated at any given time, it is essential to employ an enrichment step for phosphoproteins/phosphopeptides prior to their detection and subsequent microcharacterization. Figure 1.13 provides a summary of the various analytical methods for phosphoprotein and phosphopeptide enrichment, detection, and microanalysis.

Enrichment Methods for Phosphoproteins and Phosphopeptides

Several antibodies are commercially available that recognize low-abundance tyrosine-phosphorylated proteins and that can be used for immunoprecipitation. However, these antibodies are not very effective for enriching phosphopeptides. Currently, no commercially available antibodies can be used for immunoprecipitating proteins that are phosphorylated on serine or threonine residues. Phosphopeptides can be enriched using other methods employing various chromatography separation modalities. For example, phosphopeptides can be concentrated and desalted using reversed-phase C_{18} columns prior to ESI-MS (see Chapter 9, Protocol 5). However, because of the hydrophilic nature of many phosphopeptides, significant losses can occur since these peptides are not retained on the chromatographic support (Verma et al. 1997). Other chromatographic supports that can be used to enrich phosphopeptides include Oligo R3, a perfusion chromatography resin originally used to purify oligonucleotides (Matsumoto et al. 1997; Neubauer and Mann 1999), and porous graphitic carbon (Chin and Papac 1999, see Chapter 9).

Affinity selection of phosphopeptides using Fe(III)-, Al(III)-, and Ga(III)-loaded immobilized metal affinity chromatography (IMAC) columns provides a rapid means for enriching phosphopeptides. IMAC exploits the high affinity of phosphate groups to bind to metal-chelated resins, especially Fe(III)- and Ga(III)-loaded resins (Porath 1992; Posewitz and Tempst 1999; Stensballe et al. 2000; Xhou et al. 2000). Although IMAC has been successfully used in on-line and off-line formats to analyze phosphopeptides using MS (Mann et al. 2002), a major drawback is that these columns also bind acidic peptides (i.e., those rich in aspartic acid and glutamic acid). One way of attaining a much higher specificity for phosphopeptides is to esterify acidic residues (methyl esterification) prior to loading the peptides onto the IMAC column (Ficarro et al. 2002). For protocols outlining the use of IMAC for phosphopeptide enrichment, see Chapter 9, Protocols 1, 3, and 4.

Chemical Modifications

Another way to enrich for phosphopeptides from complex digests is by replacing the phosphate group with an affinity moiety such as biotin (Meyer et al. 1993; Oda et al. 2001) and then using affinity capture (e.g., avidin) to isolate the derivatized peptides. This can be accomplished by β-elimination of phosphoserine and phosphothreonine residues under strongly alkaline conditions to yield dehydroalanine or dehydroaminobutyric acid residues, respectively. This approach allows a simple means for derivatizing serine or threonine with any desired R-group (Adamczyk et al. 2001; Goshe et al. 2001; Oda et al. 2001). Reagents with a sulfyhdryl group (e.g., 1,2-ethanedithiol) add readily to these α,β-unsaturated amides, and the free sulfhydryl group in the derivatized peptide is available for reaction with a biotinylated affinity tag (Goshe et al. 2001; Oda et al. 2001). An additional attractive feature of this approach is that peptides can be isotopically labeled at their initial phosphorylation sites; thus, in addition to affinity selection with avidin, the phosphopeptides can be quantitated. A schematic depiction of this chemical modification method is shown in Figure 1.14. However, the β-elimination strategy does have some serious drawbacks:

- O-phosphorylation sites suffer from a lack of specificity because O-glycosidic linkages are also affected by β-elimination under identical conditions (Greis et al. 1996). In some regulatory proteins, the same serine residue may be either O-phosphorylated or O-glycosylated, and the β-elimination strategy will not distinguish between them (Wells et al. 2001).
- The strategy is inherently compromised by lower reactivity of phosphothreonine and non-reactivity of phosphotyrosine.
- Yields from β-elimination reactions tend to be substoichiometric.

FIGURE 1.14. Alternative methods for chemical derivatization and solid-state purification of phosphopeptides. The protocol of Zhou at al. (2001) (*left*) modifies phosphopeptides with free sulfhydryls that are then trapped by covalent attachment to iodoacetic-acid-linked glass beads. Acid elution regenerates phosphopeptides, which are then analyzed by MS. In the protocol of Oda et al. (2001) (*right*) phosphoproteins are β-eliminated, biotinylated, and purified by avidin affinity chromatography. Biotinylated peptides are further purified in a second avidin binding step and analyzed by MS. (Reproduced, with permission, from Ahn and Resing 2001.)

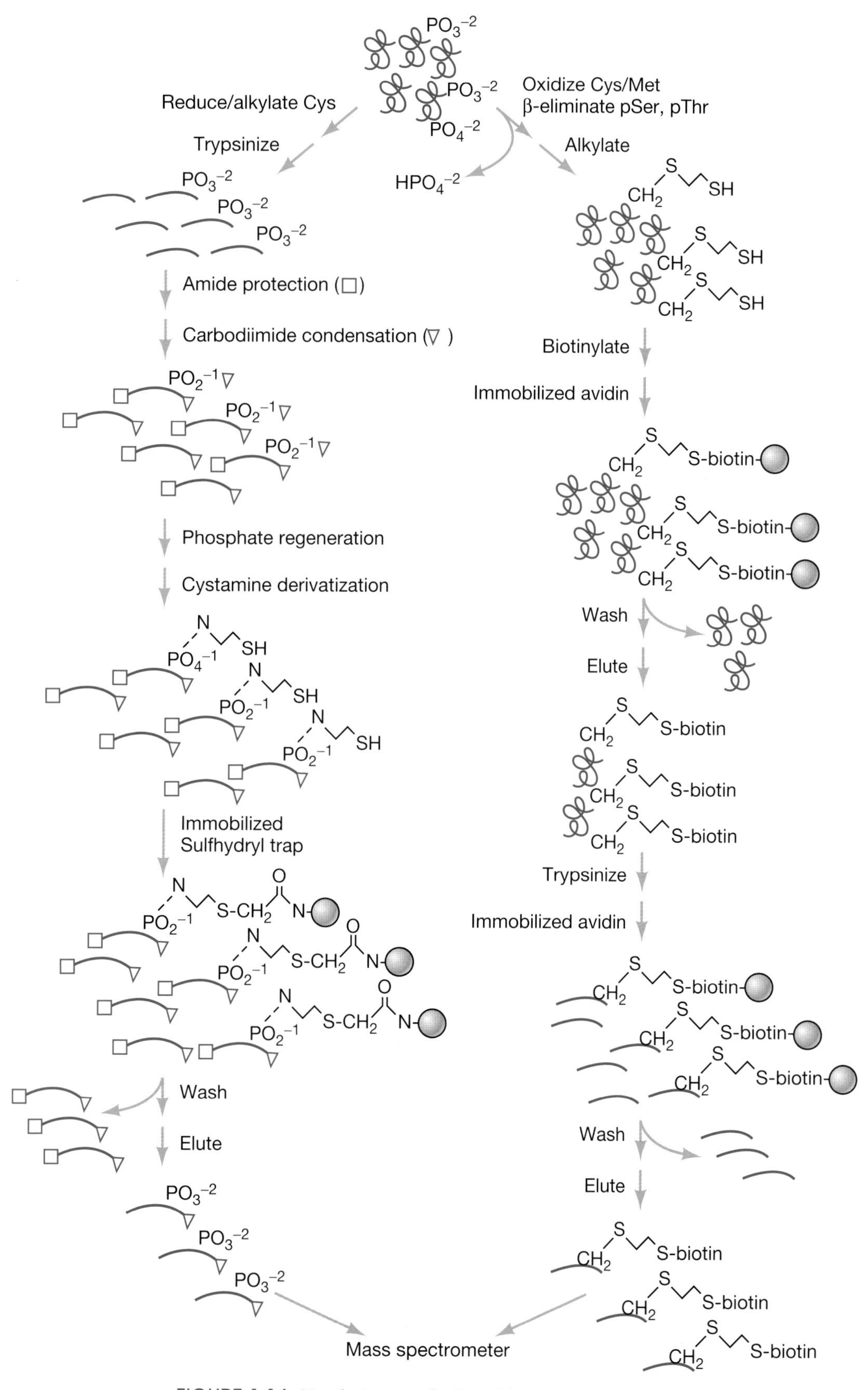

FIGURE 1.14. (*See facing page for legend.*)

Another chemical method for analyzing the phosphoproteome, which is applicable to phosphotyrosine-containing peptides as well as those containing phosphoserine and phosphothreonine residues, has recently been described by Zhou et al. (2001). This approach begins with a proteolytic digest that has been reduced and alkylated to eliminate cross-reactivity from cysteine residues. Both the amino and carboxyl termini of each peptide are chemically blocked, and phosphoroamidate adducts at the phosphorylated residues are formed by carbodiimide condensation with cystamine (see Figure 1.14). Free sulfhydryl groups produced from this step are covalently captured onto glass beads via coupling to iodoacetic acid. Peptides are then eluted from the glass beads using trifluoroacetic acid, and regenerated phosphopeptides are analyzed by MS. Although there are multiple reaction steps, this method may be amenable to automation and high-throughput.

Analysis of Phosphorylation Sites

Several methods detect phosphorylation, including traditional methods such as thin-layer chromatography (TLC) and Edman sequencing, and methods that utilize MS. In contrast to the MS-based methods, the traditional methods typically require the use of radioactively labeled samples and some prior knowledge of the protein sequence. Proteins can be radiolabeled in vivo using [^{32}P]orthophosphate or in vitro using purified protein kinase and [γ-^{32}P]ATP. ^{32}P-phosphorylated proteins are proteolytically digested (typically using trypsin), and the generated tryptic peptides are fractionated using 2D TLC or HPLC. Isolated ^{32}P-labeled tryptic peptides are detected by scintillation counting or autoradiography and typically subjected to phosphoamino acid analysis and Edman sequencing, monitoring for a loss of radioactivity at each cycle of the Edman degradation (see Chapter 6, Protocol 5). Because there is usually insufficient peptide to detect the nonlabeled amino acids at each cycle of the Edman degradation, phosphorylation sites are deduced by aligning the cycles at which radioactivity appears with a list of predicted peptides to ascertain those that contain serine, threonine, or tyrosine at the same position. MS-based methods for phosphorylation site mapping are summarized in Table 1.6 (for more details, see Chapter 9, Protocols 6–8).

PROTEIN MICROARRAYS

A daunting challenge after a genome has been fully sequenced is to understand the functional roles of all the encoded proteins. Although mRNA (and DNA) microarrays allow investigators to track the activity of thousands of genes (and by inference, protein function) on a global scale (Bowtell and Sambrook 2002), these studies are limited by the poor correlation between mRNA and protein expression levels. They also fail to take into consideration the significant roles of PTMs in protein-protein interactions. The success of mRNA and DNA microarrays has triggered efforts to create similar arrays of proteins, which could be used to study enzyme-substrate, DNA-protein, and protein-protein interactions on a proteomic scale. Early efforts to create protein arrays were hindered by the problem of positioning proteins on a solid surface in a correctly folded, fully active form, as well as the logistics of synthesizing large numbers of proteins in a high-throughput fashion. The first of these problems was successfully overcome with the report by MacBeath and Schreiber (2000) of protein arrays containing more than 10,000 proteins on a piece of glass just over half the size of a microscope slide. To make their arrays, these authors used a robot that was originally

TABLE 1.6. A selection of MS methods for detecting phosphorylation

Method	Reference
MALDI	
Post-source decay: Phosphopeptides can lose phosphate under conditions of post-source decay (PSD). MALDI-PSD is performed on reflectron-equipped MALDI-TOF instruments. The molecular ion of interest is selected and undergoes PSD in the first field-free region of the instrument; the reflectron component of the instrument then focuses the fragments such that they are detected at their correct mass (loss of 80 or 98 daltons from the parent molecular ion is diagnostic for phosphorylated peptides). Serine and threonine tend to show a predominant neutral loss of 98 daltons (owing to H_3PO_4 loss) as compared with a loss of 80 daltons (owing to loss of HPO_3) which can be used to differentiate them from tyrosine phosphopeptides, which generally show only a loss of 80 daltons. This approach can be valuable if used on peptide mixtures that are first enriched for phosphopeptides using IMAC resins.	Annan and Carr (1996)
Alkaline phosphatase treatment: MALDI analysis of peptides before and after treatment with alkaline phosphatase can be used to detect phosphopeptides (a characteristic mass shift occurs owing to the loss of phosphate [80 daltons or multiples] after treatment with phosphatase). See protocol(s) in Chapter 9.	Liao et al. (1994); Zhang et al. (1998); Kussman et al. (1999); Zhou et al. (2000); Larsen et al. (2001)
Electrospray ionization	
In-source collision-induced dissociation (CID): If phosphopeptide ions are fragmented in negative-ion mode, $H_2PO_4^-$ (97 daltons), PO_3^- (79 daltons), and PO_2^- (63 daltons) are detected as phosphate-specific diagnostic ions (under low-energy CID, phosphotyrosine will be observed to generate the 79- and 63-dalton ions, but not the 97-dalton ion).	Huddelston et al. (1993); Hunter and Games (1994)
Precursor ion scanning: Peptides carrying a phosphate group can be readily detected by precursor ion scanning because of the loss of phosphate (79 daltons) under alkaline conditions. A triple-quadrupole mass spectrometer operating in negative mode is generally used for this application. This approach is highly selective and sensitive and applicable for serine, threonine, and tyrosine-phosphorylated residues. One shortcoming is that sequencing must be conducted in positive-ion mode immediately after detecting the loss of phosphate in negative-ion mode (this requires a change in polarity of the instrument and rebuffering of the sample).	Carr et al. (1996); Wilm et al. (1996); Neubauer and Mann (1999)

designed to synthesize DNA arrays. The high-precision contact-printing robot dips a quill-like tip into a well containing a single purified protein and then turns to a glass microscope slide, where it spots a tiny 1-nl drop (150–200 μm in diameter) onto a glass slide. Ultimately, the robot is able to place 1600 spots/cm^2 on the slide. As proof of principle, MacBeath and Schreiber demonstrated that members of the protein array were able to specifically bind to other well-characterized binding partners, small drug-like molecules, and protein kinases, which chemically modify protein targets. For a set of detailed protocols on establishing and using protein microarrays, see Chapter 10, Protocols 10–14. An alternative way of arraying functionally active proteins, using microfabricated polyacrylamide gel pads to capture protein samples and microelectrophoresis to accelerate diffusion, was reported recently by Arenkov et al. (2000).

The second obstacle to protein arrays, the generation of proteome-scale proteins, is currently being addressed with significant developments in high-throughput large-scale recombinant protein expression and isolation (see, e.g., Cahill 2000; Braun et al. 2002).

TABLE 1.7. Protein arrays: Classes of capture molecules

Capture molecules	Source	Technique	References
mAb	mouse	hybridoma	Goldman (2000)
scFv/Fab diabodies	antibody libraries	phage display, in vitro evolution	Gao et al. (1999); Ryu and Nam (2000); Krebs et al. (2001); Lecerf et al. (2001); Raum el al. (2001)
Affinity binding agents	recombinant fibronectin structures	in vitro evolution	Kreider (2000)
Affibodies	microorganism	heterologous expression	Gunneriussion et al. (1999a,b)
Aptamers (DNA/RNA/ peptide)	library	SELEX/mRNA display, in vitro evolution	Jayasena (1999); Brody and Gold (2000); Hoppe-Seyler and Butz (2000); Lee and Walt (2000); Lohse and Wright (2001); Wilson et al. (2001)
Receptor ligands	synthetic	combinatorial chemistry	MacBeath et al. (1999); Lee and Walt (2000)
Substrates of enzymes	synthetic; pro- and eukaryotic organisms	protein purification, recombinant protein technology (bacterial fusion proteins, baculovirus, peptide synthesis)	Arenkov et al. (2000); MacBeath and Schreiber (2000); Zhu et al.(2000)

This table summarizes classes of molecules that have the potential to be used or are actually used as capture molecules in protein microarray systems. Abbreviations: (Fab) Antigen-binding fragment; (scFv) single-chain variable region fragment; (mAb) monoclonal antibody.
Reproduced, with permission, from Templin et al. (2002).

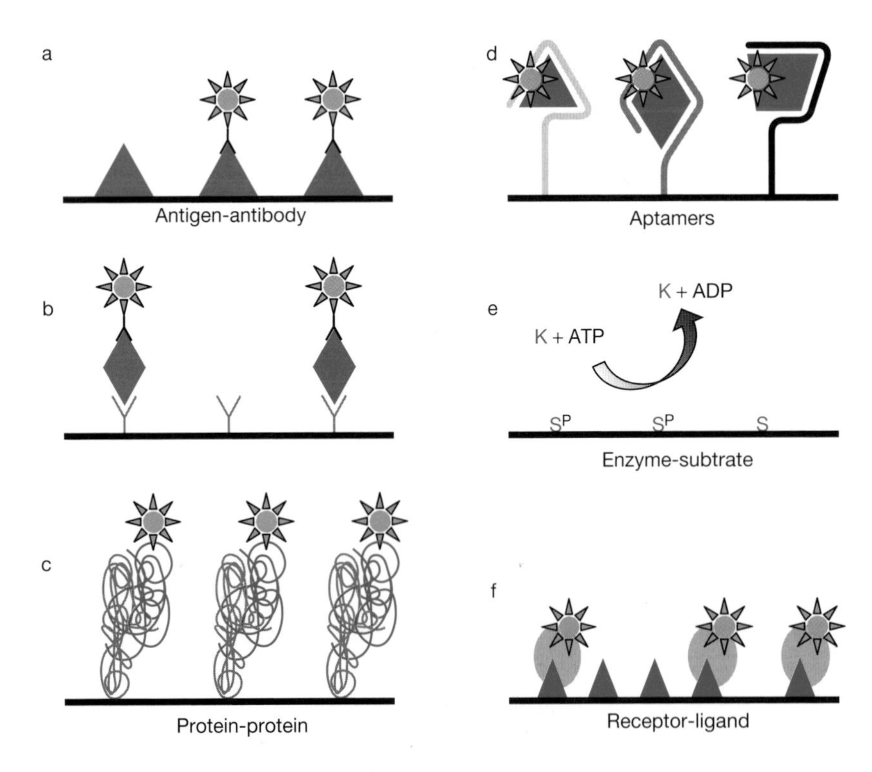

FIGURE 1.15. Classes of capture molecules for protein microarrays. For specific interaction analysis, different classes of molecules can be immobilized on a planar surface to act as capture molecules in a microarray assay. (*a*) Antigen-antibody interaction; (*b*) scheme of a Sandwich immunoassay; (*c*) a specific protein-protein interaction. (*d*) A different class of binders, where synthetic molecules referred to as aptamers act as capture molecules. They can be composed of nucleotides, ribonucleotides, or peptides. (*e*) Interactions of enzymes with their specific substrates, where a substrate (S) for kinases is immobilized and phosphorylated (P) by the repective kinase. (*f*) A typical example for a receptor-ligand interaction, where synthetic low-molecular-mass compounds are immobilized as capture molecules. (Reproduced, with permission, from Templin et al. 2002 [© Elsevier Science].)

CAPTURE MOLECULES AND TARGET MOLECULES

In principle, any type of ligand-binding assay that relies on the formation of an immobilized "capture molecule" product and a "target molecule" present in a biological matrix can be miniaturized to run in parallel in a protein microarray format. Although antibodies are the most prominent capture molecules used to identify target molecules, the labor-intensive nature (and cost) of monoclonal antibodies has led to the development of alternatives. Of these, the most promising approach has been phage-display methodology (Ryu and Nam 2000; Lecerf et al. 2001) combined with highly diverse, fully synthetic libraries ($\sim 10^{11}$ independent clones) to generate synthetic antibodies (Gao et al. 1999; Knappik et al. 2000). Another class of capture molecule are aptamers (i.e., synthetic molecules such as DNA/RNA/peptides that act as highly specific capture molecules with nM-pM affinities) (see Table 1.7 and Figure 1.15) (Jayasena 1999; Brody and Gold 2000; Hoppe-Seyler and Butz 2000; Green et al. 2001; Wilson et al. 2001). For recent reviews of protein microarray technology, see Templin et al. (2002) and Li (2000).

REFERENCES

Abbott A. 1999. Alliance of US labs plans to build map of cell signalling pathways. *Nature* **402:** 219–220.

Adams J.A. 2001. Kinetic and catalytic mechanisms of protein kinases. *Chem. Rev.* **101:** 2271–2290.

Adams P.D., Seeholzer S., and Ohh M. 2002. Identification of associated proteins by coimmunoprecipitation. In *Protein–protein interactions: A molecular cloning manual* (ed. E. Golemis), pp. 59–74. Cold Spring Harbor Laboratory Press, Cold Spring Harbor, New York.

Adamczyk M., Gebler J.C., and Wu J. 2001. Selective analysis of phosphopeptides within a protein mixture by chemical modification, reversible biotinylation and mass spectrometry. *Rapid Commun. Mass Spectrom.* **15:** 1481–1488.

Aebersold R. and Goodlett D.R. 2001. Mass spectrometry in proteomics. *Chem. Rev.* **101:** 269–295.

Ahn N.G. and Resing K.A. 2001. Toward the phosphoproteome. *Nat. Biotechnol.* **19:** 317–318.

Allen N.P., Huang L., Burlingame A., and Rexach M. 2001. Proteomic analysis of nucleoporin interacting proteins. *J. Biol. Chem.* **276:** 29268–29274.

Anderson L. and Seilhamer J. 1997. A comparison of selected mRNA and protein abundances in human liver. *Electrophoresis* **18:** 533–537.

Annan R.S. and Carr S.A. 1996. Phosphopeptide analysis by matrix-assisted desorption time-of-flight mass spectrometry. *Anal. Chem.* **68:** 3413–3421.

Arenkov P., Kukhtin A., Gemmell A., Voloshchuk S., Chupeeva V., and Mirzabekov A. 2000. Protein microchips: Use for immunoassay and enzymatic reactions. *Anal. Biochem.* **278:** 123–131.

Bader G.D. and Hogue C.W. 2000. BIND—A data specification for storing and describing biomolecular interactions, molecular complexes and pathways. *Bioinformatics* **16:** 465–477.

Baltimore D. 2001. Our genome unveiled. *Nature* **409:** 814–816.

Black D.L. 2000. Protein diversity from alternative splicing: A challenge for bioinformatics and post-genome biology. *Cell* **103:** 367–370.

Blackstock W.P. and Weir M.P. 1999. Proteomics: Quantitative and physical mapping of cellular proteins. *Trends Biotechnol.* **17:** 121–127.

Bode W. and Huber R. 1992. Natural protein proteinase inhibitors and their interactions with proteinases. *Eur. J. Biochem.* **204:** 433–452.

Bowtell D. and Sambrook J. 2003. *DNA microarrays: A molecular cloning manual.* Cold Spring Harbor Laboratory Press, Cold Spring Harbor, New York.

Braun P., Hu Y., Shen B., Halleck A., Koundinya M., Harlow E., and LaBaer J. 2002. Proteome-scale purification of human proteins from bacteria. *Proc. Natl. Acad. Sci.* **99:** 2654–2659.

Brenner S.E. 1999a. Errors in genome annotation. *Trends Genet.* **15:** 132–133.

———. 1999b. Theoretical biology in the third millennium. *Philos. Trans. R. Soc. Lond. B Biol. Sci.* **354:** 1963–1965.

Brody E.N. and Gold L. 2000. Aptamers as therapeutic and diagnostic agents *J. Biotechnol.* **74:** 5–13

Browne T.R., Van Langenhove A., Costello C.E., Biemann K., and Greenblatt D.J. 1981. Kinetic equivalence of stable-isotope-labeled and unlabeled phenytoin. *Clin. Pharmacol. Ther.* **29:** 511–515.

Burgess A.W., Metcalf D., Sparrow L.G., Simpson R.J., and Nice E.C. 1986. Granulocyte/macrophage colony-stimulating factor from mouse lung conditioned medium: Purification of multiple forms and radioiodination. *Biochem. J.* **235:** 805–814.

Cagney G. and Emili A. 2002. De novo peptide sequencing and quantitative profiling of complex protein mixtures using mass-coded abundance tagging. *Nat. Biotechnol.* **20:** 163–170.

Cahill D.J. 2000. Protein arrays: A high-throughput solution for proteomics research? *Proteomics: A Trends Guide* **1:** 47–51.

Carr S.A., Huddleston M.J., and Annan R.S. 1996. Selective detection and sequencing of phosphopeptides at the femtomole level by mass spectrometry. *Anal. Biochem.* **239:** 180–192.

Catimel B., Ritter G., Welt S., Old L.J., Cohen L., Nerrie M.A., White S.J., Heath J.K., Demediuk B., Domagala T., Lee F.T., Scott A.M., Tu G.-F., Ji H., Moritz R.L., Simpson R.J., Burgess A.W., and Nice E.C. 1996. Purification and characterization of a novel restricted antigen expressed by normal and transformed human colonic epithelium. *J. Biol. Chem.* **271:** 25664–25670.

Chin E.T. and Papac D.I. 1999. The use of porous graphitic carbon column for desalting hydrophilic peptides prior to matrix-assisted laser desorption/ionization time-of-flight mass spectrometry. *Anal. Biochem.* **273:** 179–185.

Cole A.R., Ji H., and Simpson R.J. 2000. Proteomic analysis of colonic crypts from normal, multiple intestinal neoplasia and p53-null mice: A comparison with colonic polyps. *Electrophoresis* **21:**1772–1781.

Conrads T.P., Alving K., Veenstra T.D., Belov M.E., Anderson G.A., Anderson D.J., Lipton M.S., Pasa-Tolic L., Udseth H.R., Chrisler W.B., and Smith R.D. 2001. Quantitative analysis of bacterial and mammalian proteomes using a combination of cysteine affinity tags and 15N-metabolic labeling. *Anal. Chem.* **73:** 2132–2139.

Cutler R.L., Metcalf D., Nicola N.A., and Johnson G.R. 1985. Purification of a multipotential colony-stimulating factor from pokeweed mitogen-stimulated mouse spleen cell conditioned medium. *J. Biol. Chem.* **260:** 6579–6587.

Davidson E.H., Rast J.P., Oliveri P., Ransick A., Calestani C., Yuh C.H., Minokawa T., Amore G., Hinman V., Arenas-Mena C., Otim O., Brown C.T., Livi C.B., Lee P.Y., Revilla R., Rust A.G., Pan Z., Schilstra M.J., Clarke P.J., Arnone M.I., Rowen L., Cameron R.A., McClay D.R., Hood L., and Bolouri H. 2002. A genomic regulatory network for development. *Science* **295:** 1669–1678.

De Leenheer A.P. and Thienpont L.M. 1992. Applications of isotope-dilution mass-spectrometry in clinical-chemistry, pharmacokinetics, and toxicology. *Mass Spectrom. Rev.* **11:** 249–307.

Duan X.J., Xenarios I., and Eisenberg D. 2002. Describing biological protein interactions in terms of protein states and state transitions: The LiveDIP Database. *Mol. Cell. Proteomics* **1:** 104–116.

Dunham I., Shimizu N., Roe B.A., Chissoe S., Dunham I., Hunt A.R., Collins J.E., Bruskiewich R., Beare D.M., Clamp M., et al. 1999. The DNA sequence of human chromosome 22. *Nature* **402:** 489–495.

Einarson M.B. and Orlinick J.R. 2002. Identification of protein–protein interactions with glutathione-*S*-transferase fusion proteins. In *Protein–protein interactions: A molecular cloning manual* (ed. E. Golemis), pp. 37–57. Cold Spring Harbor Laboratory Press, Cold Spring Harbor, New York.

Eisenberg D., Marcotte E.M., Xenarios I., and Yeates T.O. 2000. Protein function in the post-genomic era. *Nature* **405:** 823–826.

Emmert-Buck M.R., Bonner R.F., Smith P.D., Chuqui R.F., Zhuang Z., Goldstein S.R., Weiss R.A., and Liotta L.A. 1996. Laser capture microdissection. *Science* **274:** 998–1001.

Ficarro S.B., McCleland M.L., Stukenberg P.T., Burke D.J., Ross M.M., Shabanowitz J., Hunt D.F., and White F.M. 2002. Phosphoproteome analysis by mass spectrometry and its application to *Saccharomyces cerevisiae. Nat. Biotechnol.* **20:** 301–305.

Frolik C.A., Dart L.L, Meyers C.A., Smith D.M., and Sporn M.B. 1983. Purification and initial characterization of a type β transforming growth factor from human placenta. *Proc. Natl. Acad. Sci.* **80:** 3676–3680.

Gao C., Mao S., Lo C.H., Wirsching P., Lerner R.A., and Janda K.D. 1999. Making artificial antibodies: A format for phage display of combinatorial heterodimeric arrays. *Proc. Natl. Acad. Sci.* **96:** 6025–6030.

Gavin A.C., Bosche M., Krause R., Grandi P., Marzioch M., Bauer A., Schultz J., Rick J.M., Michon A.M., Cruciat C.M., et al. 2002. Functional organization of the yeast proteome by systematic analysis of protein complexes. *Nature* **415:** 141–147.

Gerstein M., Lan N., and Jansen R. 2002. Proteomics. Integrating interactomes. *Science* **295:** 284–287.

Goldman R.D. 2000. Antibodies: Indispensable tools for biomedicial research. *Trends Biochem. Sci.* **25:** 593–595.

Goodlett D.R., Keller A., Watts J.D., Newitt R., Yi E.C., Purvine S., Eng J.K., von Haller P., Aebersold R., and Kolker E. 2001. Differential stable isotope labeling of peptides for quantitation and de novo sequence derivation. *Rapid Commun. Mass Spectrom.* **15:** 1214–1221.

Görg A., Postel W., Domscheit A., and Gunther S. 1988. Two-dimensional electrophoresis with immobilized pH gradients of leaf proteins from barley (*Hordeum vulgare*): Method, reproducibility and genetic aspects. *Electrophoresis* **9:** 681–692.

Görg A., Obermaier C., Boguth G., Harder A., Scheibe B., Wildgruber R., and Weiss W. 2000. The current state of two-dimensional electrophoresis with immobilized pH gradients. *Electrophoresis* **21:** 1037–1053.

Goshe M.B., Conrads T.P., Panisko E.A., Angell N.H., Veenstra T.D., and Smith R.D. 2001. Phosphoprotein isotope-coded affinity tag approach for isolating and quantitating phosphopeptides in proteome-wide analyses. *Anal. Chem.* **73:** 2578–2586.

Gospodarowicz D., Cheng J., Lui G.-M., Baird A., and Bohlent P. 1984. Isolation of brain fibroblast growth factor by heparin-Sepharose affinity chromatography: Identity with pituitary fibroblast growth factor.

Proc. Natl. Acad. Sci. **81:** 6963–6967.

Graeber T.G. and Eisenberg D. 2001. Bioinformatic identification of potential autocrine signaling loops in cancers from gene expression profiles. *Nat. Genet.* **29:** 295–300.

Graziano M.P., Moxham C.P., and Malbon C.C. 1985. Purified rat hepatic β_2-adrenergic receptor. *J. Biol. Chem.* **260:** 7665–7674.

Green L.S., Bell C., and Janjic N. 2001. Aptamers as reagents for high-throughput screening. *BioTechniques* **30:** 1094–1096.

Greis K.D., Hayes B.K., Comer F.I., Kirk M., Barnes S., Lowary T.L., and Hart G.W. 1996. Selective detection and site-analysis of O-GlcNAc-modified glycopeptides by beta-elimination and tandem electrospray mass spectrometry. *Anal. Biochem.* **234:** 38–49.

Griffin T.J., Goodlett D.R., and Aebersold R. 2001. Advances in proteome analysis by mass spectrometry. *Curr. Opin. Biotechnol.* **12:** 607–612.

Gunneriussion E., Nord K., Uhlen M., and Nygren P. 1999. Affinity maturation of a *Taq* DNA polymerase specific affibody by helix shuffling. *Protein Eng.* **12:** 873–878.

Gunneriusson E., Samuelson P., Ringdahl J., Gronlund H., Nygren P.A., and Stahl S. 1999. Staphylococcal surface display of immunoglobulin A (IgA)- and IgE-specific in vitro-selected binding proteins (affibodies) based on *Staphylococcus aureus* protein A. *Appl. Environ. Microbiol.* **65:** 4134–4140.

Gygi S.P., Rochon Y., Franza B.R., and Aebersold R. 1999a. Correlation between protein and mRNA abundance in yeast. *Mol. Cell. Biol.* **19:** 1720–1730.

Gygi S.P., Corthals G.L., Zhang Y., Rochon Y., and Aebersold R. 2000. Evaluation of two-dimensional gel electrophoresis-based proteome analysis technology. *Proc. Natl. Acad. Sci.* **97:** 9390–9395.

Gygi S.P., Han D.K., Gingras A.C., Sonenberg N., and Aebersold R. 1999b. Protein analysis by mass spectrometry and sequence database searching: Tools for cancer research in the post-genomic era. *Electrophoresis* **20:** 310–319.

Gygi S.P., Rist B., Gerber S.A., Turecek F., Gelb M.H., and Aebersold R. 1999c. Quantitative analysis of complex protein mixtures using isotope-coded affinity tags. *Nat. Biotechnol.* **17:** 994–999.

Haga K. and Haga T. 1985. Purification of the muscarinic acetylcholine receptor from porcine brain. *J. Biol. Chem.* **260:** 7927–7935.

Hatzimanikatis V. and Lee K.H. 1999. Dynamical analysis of gene networks requires both mRNA and protein expression information. *Metab. Eng.* **1:** 275–281.

Hazbun T.R. and Fields S. 2001. Networking proteins in yeast. *Proc. Natl. Acad. Sci.* **98:** 4277–4278.

Heinemeyer T., Chen X., Karas H., Kel A.E., Kel O.V., Liebich I., Meinhardt T., Reuter I., Schacherer F., and Wingender E. 1999. Expanding the TRANSFAC database towards an expert system of regulatory molecular mechanisms. *Nucleic Acids Res.* **27:** 318–322.

Heldin C.-H., Westermark B., and Wasteson A. 1981. Platelet-derived growth factor. *Biochem. J.* **193:** 907–913.

Ho Y., Gruhler A., Heilbut A., Bader G.D., Moore L., Adams S.L., Millar A., Taylor P., Bennett K., Boutilier K., Yang L., Wolting C., Donaldson I., Schandorff S., Shewnarane J., Vo M., Taggart J., Goudreault M., Muskat B., Alfarano C., Dewar D., Lin Z., Michalickova K., Willems A.R., Sassi H., Nielsen P.A., Rasmussen K.J., Andersen J.R., Johansen L.E., Hansen L.H., Jespersen H., Podtelejnikov A., Nielsen E., Crawford J., Poulsen V., Sorensen B.D., Matthiesen J., Hendrickson R.C., Gleeson F., Pawson T., Moran M.F., Durocher D., Mann M., Hogue C.W., Figeys D., and Tyers M. 2002. Systematic identification of protein complexes in *Saccharomyces cerevisiae* by mass spectrometry. *Nature* **415:** 180–183.

Hoogland C., Sanchez J.C., Walther D., Baujard V., Baujard O., Tonella L., Hochstrasser D.F., and Appel R.D. 1999. Two-dimensional electrophoresis resources available from ExPASy. *Electrophoresis* **20:** 3568–3571.

Hoppe-Seyler F. and Butz K. 2000. Peptide aptamers: Powerful new tools for molecular medicine. *J. Mol. Med.* **78:** 426–430.

Houry W.A., Frishman D., Eckerskorn C., Lottspeich F., and Hartl F.U. 1999. Identification of in vivo substrates of the chaperonin GroEL. *Nature* **402:** 147–154.

Huddleston M.J., Annan R.S., and Carr S.A. 1993. Selective detection of phosphopeptides in complex mixtures by electrospray liquid chromatography. *J. Am. Soc. Mass Spectrom.* **4:** 710–717.

Hunter A.P. and Games D.E. 1994. Chromatographic and mass spectrometric methods for the identification of phosphorylation sites in phosphoproteins. *Rapid Commun. Mass Spectrom.* **8:** 559–570.

Hunter T. 2000a. The role of tyrosine phosphorylation in cell growth and disease. *Harvey Lect.* **94:** 81–119.

——. 2000b. Signaling—2000 and beyond. *Cell* **100:** 113–127.

Huse M. and Kuriyan J. 2002. The conformational plasticity of protein kinases. *Cell* **109:** 275–282.

Ideker T., Thorsson V., Ranish J.A., Christmas R., Buhler J., Eng J.K., Bumgarner R., Goodlett D.R., Aebersold R., and Hood L. 2001. Integrated genomic and proteomic analyses of a systematically perturbed metabolic network. *Science* **292:** 929–934.

Ito T., Chiba T., and Yoshida M. 2001a. Exploring the protein interactome using comprehensive two-hybrid projects. *Trends Biotechnol.* (suppl.) **19:** S23–S27.

Ito T., Chiba T., Ozawa R., Yoshida M., Hattori M., and Sakaki Y. 2001b. A comprehensive two-hybrid analysis to explore the yeast protein interactome. *Proc. Natl. Acad. Sci.* **98:** 4569–4574.

Jayasena S.D. 1999. Aptamers: An emerging class of molecules that rival antibodies in diagnostics. *Clin. Chem.* **45:** 1628–1650.

Johnson L.N. and Lewis R.J. 2001. Structural basis for control by phosphorylation. *Chem. Rev.* **101:** 2209–2242.

Kanchisa M. and Goto S. 2000. KEGG: Kyoto encyclopedia of genes and genomes. *Nucleic Acids Res.* **28:** 27–30.

Karp P.D., Riley M., Saier M., Paulsen I.T., Paley S., Pellegrini-Toole A., Bonavides C., and Gama-Castro S. 2002. The EcoCyc Database. *Nucleic Acids Res.* **30:** 56–58.

Kemp B.E. and Pearson R.B. 1991. Intrasteric regulation of protein kinases and phosphatases. *Biochim. Biophys. Acta* **1094:** 67–76.

Khan A.R. and James M.N. 1998. Molecular mechanisms for the conversion of zymogens to active proteolytic enzymes. *Protein Sci.* **7:** 815–836.

Kim S.H. 2000. Structural genomics of microbes: An objective. *Curr. Opin. Struct. Biol.* **10:** 380–383.

Kitano H. 2002. Systems biology: A brief overview. *Science* **295:** 1662–1664.

Klose J. 1975. Protein mapping by combined isoelectric focusing and electrophoresis of mouse tissues. A novel approach to testing for induced point mutations in mammals. *Humangenetik.* **26:** 231–243.

Knappik A., Ge L., Honegger A., Pack P., Fischer M., Wellnhofer G., Hoess A., Wolle J., Pluckthun A., and Virnekas B. 2000. Fully synthetic human combinatorial antibody libraries (HuCAL) based on modular concensus frameworks and CDRs randomized with trinucleotides. *J. Mol. Biol.* **296:** 57–86.

Kobe B. and Kemp B.E. 1999. Active site-directed protein regulation. *Nature* **402:** 373–376.

Krebs B., Rauchenberger R., Reiffert S., Rothe C., Tesar M., Thomassen E., Cao M., Dreier T., Fischer D., Hoss A., et al. 2001. High-throughput generation and engineering of recombinant human antibodies. *J. Immunol. Methods* **254:** 67–84.

Kreider B.L. 2000. PROfusion: Genetically tagged proteins for functional proteomics and beyond. *Med. Res. Rev.* **20:** 212–215.

Krishna R. and Wold F. 1997. Identification of common post-translational modifications. In *Protein structure—A practical approach,* 2nd edition (ed. T.E. Creighton), pp. 91–116. Oxford University Press, New York.

Kussmann M., Hauser K., Kissmehl R., Breed J., Plattner H., and Roepstorff P. 1999. Comparison of in vivo and in vitro phosphorylation of the exocytosis-sensitive protein PP63/parafusin by differential MALDI mass spectrometric peptide mapping. *Biochemistry* **38:** 7780–7790.

Kuster B. and Mann M. 1999. 18O-labeling of N-glycosylation sites to improve the identification of gel-separated glycoproteins using peptide mass mapping and database searching. *Anal. Chem.* **71:** 1431–1440.

Lakey J.H. and Raggett E.M. 1998. Measuring protein-protein interactions. *Curr. Opin. Struct. Biol.* **8:** 119–123.

Lander E.S., Linton L.M., Birren B., Nusbaum C., Zody M.C., Baldwin J., Devon K., Dewar K., Doyle M., FitzHugh W., et al. 2001. Initial sequencing and analysis of the human genome. *Nature* **409:** 860–921.

Lahm H.W. and Langen H. 2000. Mass Spectrometry: A tool for the identification of proteins separated by gels. *Electrophoresis* **21:** 2105–2114.

Larsen M.R., Sorensen G.L., Fey S.J., Larsen P.M., and Roepstorff P. 2001. Phospho-proteomics: Evaluation of the use of enzymatic de-phosphorylation and differential mass spectrometric peptide mass mapping for site specific phosphorylation assignment in proteins separated by gel electrophoresis. *Proteomics* **1:** 223–238.

Lecerf J.M., Shirley T.L., Zhu Q., Kazantsev A., Amersdorfer P., Housman D.E., Messer D.E., and Huston J.S. 2001. Human single-chain Fv intrabodies counteract in situ huntingtin aggregation in cellular models of Huntington's disease. *Proc. Natl. Acad. Sci.* **98:** 4764–4769.

Lee M. and Walt D.R. 2000. A fiber-optic microarray biosensor using aptamers as receptors. *Anal. Biochem.*

282: 142–146.
Legrain P. 2002. Protein domain networking. *Nat. Biotechnol.* **20:** 128–129.
Legrain P. and Selig L. 2000. Genome-wide protein interaction maps using two-hybrid systems. *FEBS Lett.* **480:** 32–36.
Legrain P., Wojcik J., and Gauthier J.M. 2001. Protein–protein interaction maps: A lead towards cellular functions. *Trends Genet.* **17:** 346–352.
Li M. 2000. Applications of display technology in protein analysis. *Nat. Biotechnol.* **18:** 1251–1256.
Liao P.C., Leykam J., Andrew P.C., Gage D.A., and Allison J. 1994. An approach to locate phosphorylation sites in a phosphoprotein: Mass mapping by combining specific enzymatic degradation with matrix-assisted laser desorption/ionization mass spectrometry. *Anal. Biochem.* **219:** 9–20.
Link A.J., Eng J., Schieltz D.M., Carmack E., Mize G.J., Morris D.R., Garvik B.M., and Yates J.R. III. 1999. Direct analysis of protein complexes using mass spectrometry. *Nat. Biotechnol.* **17:** 676–682.
Lockhart D.J. and Winzeler E.A. 2000. Genomics, gene expression and DNA arrays. *Nature* **405:** 827–836.
Lohse P.A. and Wright M.C. 2001. In vitro protein display in drug discovery. *Curr. Opin. Drug Disc. Dev.* **4:** 198–204.
MacBeath G. and Schreiber S.L. 2000. Printing proteins as microarrays for high-throughput function determination. *Science* **289:** 1760–1763.
MacBeath G., Koehler A.N., and Schreiber S.L. 1999. Printing small molecules as microarrays and detecting protein-ligand interactions en masse. *J. Am. Chem Soc.* **121:** 7967–7968.
Mann M., Ong S.E., Gronborg M., Steen H., Jensen O.N., and Pandey A. 2002. Analysis of protein phosphorylation using mass spectrometry: Deciphering the phosphoproteome. *Trends Biotechnol.* **20:** 261–268.
Matsumoto H., Kahn E.S., and Komori N. 1997. Separation of phosphopeptides from their non-phosphorylated forms by reversed-phase POROS perfusion chromatography at alkaline pH. *Anal. Biochem.* **251:** 116–119.
Mayer M.L. and Hieter P. 2000. Protein networks—Built by association. *Nat. Biotechnol.* **18:** 1242–1243.
McLafferty F.W., Fridriksson E.K., Horn D.M., Lewis M.A., and Zubarev R.A. 1999. Techview: Biochemistry. Biomolecule mass spectrometry. *Science* **284:** 1289–1290.
Mellor J.C., Yanai I., Clodfelter K.H., Mintseris J., and DeLisi C. 2002. Predictome: A database of putative functional links between proteins. *Nucleic Acids Res.* **30:** 306–309.
Mewes H.W., Frishman D., Guldener U., Mannhaupt G., Mayer K., Mokrejs M., Morgenstern B., Munsterkotter M., Rudd S., and Weil B. 2002. MIPS: A database for genomes and protein sequences. *Nucleic Acids Res.* **30:** 31–34.
Meyer H.E., Eisermann B., Heber M., Hoffmann-Posorke E., Korte H., Weigt C., Wegner A., Hutton T., Donella-Deana A., and Perich J.W. 1993. Strategies for nonradioactive methods in the localization of phosphorylated amino acids in proteins. *FASEB J.* **7:** 776–782.
Mirgorodskaya O.A., Kozmin Y.P., Titov M.L., Korner R., Sonsken C.P., and Roepstorff P. 2000. Quantitation of peptides and proteins by matrix-assisted laser desorption/ionization mass spectrometry using (18)O-labeled internal standards. *Rapid Commun. Mass Spectrom.* **14:** 1226–1232.
Mironov A.A., Fickett J.W., and Gelfand M.S. 1999. Frequent alternative splicing of human genes. *Genome Res.* **9:** 1288–1293.
Monod J., Changeux J.P., and Jacob F. 1963. Allosteric proteins and cellular control systems. *J. Mol. Biol.* **6:** 306–329.
Munchbach M., Quadroni M., Miotto G., and James P. 2000. Quantitation and facilitated de novo sequencing of proteins by isotopic N-terminal labeling of peptides with a fragmentation-directing moiety. *Anal. Chem.* **72:** 4047–4057
Nature 1999. Proteomics, transcriptomics: What's in a name? *Nature* **402:** 715.
Neubauer G. and Mann M. 1999. Mapping of phosphorylation sites of gel-isolated proteins by nanoelectrospray tandem mass spectrometry: Potentials and limitations. *Anal. Chem.* **71:** 235–242.
Neubauer G., Gottschalk A., Fabrizio P., Séraphin B., Luhrmann R., and Mann M. 1997. Identification of the proteins of the yeast U1 small nuclear ribonucleoprotein complex by mass spectrometry. *Proc. Natl. Acad. Sci.* **94:** 385–390.
Neubauer G., King A., Rappsilber J., Calvio C., Watson M., Ajuh P., Sleeman J., Lamond A., and Mann M. 1998. Mass spectrometry and EST-database searching allows characterization of the multi-protein spliceosome complex. *Nat. Genet.* **20:** 46–50.

Nicola A., Metcalf D., Matsumoto M., and Johnson G.R. 1983. Purification of a factor inducing differentiation in murine myelomonocytic leukemia cells—Identification as granulocyte colony stimulating factor. *J. Biol. Chem.* **258:** 9017–9023.

Oda Y., Nagasu T., and Chait B.T. 2001. Enrichment analysis of phosphorylated proteins as a tool for probing the phosphoproteome. *Nat. Biotechnol.* **19:** 379–382.

Oda Y., Huang K., Cross F.R., Cowburn D., and Chait B.T. 1999. Accurate quantitation of protein expression and site-specific phosphorylation. *Proc. Natl. Acad. Sci.* **96:** 6591–6596.

O'Farrell P.H. 1975. High resolution two-dimensional electrophoresis of proteins. *J. Biol. Chem.* **250:** 4007–4021.

Oliver S.G. 2002. Functional genomics: Lessons from yeast. *Philos. Trans. R. Soc. Lond. B Biol. Sci.* **357:** 17–23.

Ong S.-E., Blagoev B., Kratchmarova I., Kristensen D.B., Steen H., Pandey A., and Mann M. 2002. Stable isotope labeling by amino acids in cell culture, SILAC, as a simple and accurate approach to expression proteomics. *Mol. Cell. Proteomics* **1:** 376–386.

Patterson S.D. 2000a. Proteomics: The industrialization of protein chemistry. *Curr. Opin. Biotechnol.* **11:** 413–418.

———. 2000b. Mass spectrometry and proteomics. *Physiol. Genomics* **2:** 59–65.

Patton W.F. 2000. A thousand points of light: The application of fluorescence detection technologies to two-dimensional gel electrophoresis and proteomics. *Electrophoresis* **21:** 1123–1144.

Pawson T. and Nash P. 2000. Protein-protein interactions define specificity in signal transduction. *Genes Dev.* **14:** 1027–1047.

Pelletier J. and Sidhu S. 2001. Mapping protein-protein interactions with combinatorial biology methods. *Curr. Opin. Biotechnol.* **12:** 340–347.

Porath J. 1992. Immobilized metal ion affinity chromatography. *Protein Expr. Purif.* **3:** 263–281.

Posewitz M.C. and Tempst P. 1999. Immobilized gallium (III) affinity chromatography of phosphopeptides. *Anal. Chem.* **71:** 2883–2892.

Pugh B.F. 1996. Mechanisms of transcription complex assembly. *Curr. Opin. Cell. Biol.* **8:** 303–311.

Raamsdonk L.M., Teusink B., Broadhurst D., Zhang N., Hayes A., Walsh M.C., Berden J.A., Brindle K.M., Kell D.B., Rowland J.J., Westerhoff H.V., van Dam K., and Oliver S.G. 2001. A functional genomics strategy that uses metabolome data to reveal the phenotype of silent mutations. *Nat. Biotechnol.* **19:** 45–50.

Raum T., Gruber R., Riethmuller G., and Kufer P. 2001. Anti-self antibodies selected from a human IgD heavy chain repertoire: A novel approach to generate therapeutic human antibodies against tumor-associated differentiation antigens. *Cancer Immunol. Immunother.* **50:** 141–150.

Ray L.B. and Gough N.R. 2002. Orienteering strategies for a signaling maze. *Science* **296:** 1632–1633.

Regnier F.E., Riggs L., Zhang R., Xiong L., Liu P., Chakraborty A., Seeley E., Stoma C., and Thompson R.A. 2002. Comparative proteomics based on stable isotope labeling and affinity selection. *J. Mass Spectrom.* **37:** 133–145.

Resing K.A. and Ahn N.G. 1997. Protein phosphorylation analysis by electrospray ionization-mass spectrometry. *Methods Enzymol.* **283:** 29–44.

Rout M.P., Aitchison J.D., Suprapto A., Hjertaas K., Zhao Y., and Chait B.T. 2000. The yeast nuclear pore complex: Composition, architecture, and transport mechanism. *J. Cell Biol.* **148:** 635–651.

Rubinstein M., Rubinstein S., Familetti P.C., Miller R.S., Waldman A.A., and Pestka S. 1979. Human leukocyte interferon: Production, purification to homogeneity, and initial characterization. *Proc. Natl. Acad. Sci.* **76:** 640–644.

Ryu D.D. and Nam D.H. 2000. Recent progress in biomolecular engineering. *Biotechnol. Prog.* **16:** 2–16.

Schaller H.C. and Bodenmuller H. 1981. Isolation and amino acid sequence of a morphogenetic peptide from hydra. *Proc. Natl. Acad. Sci.* **78:** 7000–7004.

Schwikowski B., Uetz P., and Fields S. 2000. A network of protein-protein interactions in yeast. *Nat. Biotechnol.* **18:** 1257–1261.

Séraphin B., Puig O., Bouveret E., Rutz B., and Caspary F. 2002. Tandem affinity purification to enhance interacting protein identification. In *Protein–protein interactions: A molecular cloning manual* (ed. E. Golemis), pp. 313–328. Cold Spring Harbor Laboratory Press, Cold Spring Harbor, New York.

Shi Y. 2002. Mechanisms of caspase activation and inhibition during apoptosis. *Mol. Cell* **9:** 459–470.

Sickmann A. and Meyer H.E. 2001. Phosphoamino acid analysis. *Proteomics* **1:** 200–206.

Simpson R.J. and Dorow D.S. 2001. Cancer proteomics: From signaling networks to tumor markers. *Trends*

Biotechnol. **19:** S40–S48.

Simpson R.J. and Nice E.C. 1989. Strategies for the purification of subnanomole amounts of protein and polypeptides for microsequence analysis. In *The use of HPLC in receptor biochemistry,* pp. 201–244. A.R Liss, New York.

Stensballe A., Andersen S., and Jensen O.N. 2000. Characterization of phosphoproteins from electrophoretic gels by nanoscale Fe(III) affinity chromatography with off-line mass spectrometry analysis. *Proteomics* **1:** 207–222.

Strohman R. 1994. Epigenesis: The missing beat in biotechnology? *Bio/Technology* **12:** 156–164.

Sze S.K., Ge Y., Oh H., and McLafferty F.W. 2002. Top-down mass spectrometry of a 29-kDa protein for characterization of any posttranslational modification to within one residue. *Proc. Natl. Acad. Sci.* **99:** 1774–1779.

Tatemoto K. 1982. Isolation and characterization of peptide YY (PYY), a candidate gut hormone that inhibits pancreatic exocrine secretion. *Proc. Natl. Acad. Sci.* **79:** 2514–2518.

Templin M.F., Stoll D., Schrenk M., Traub P.C., Vohringer C.F., and Joos T.O. 2002. Protein microarray technology. *Trends Biotechnol.* **20:** 160–166.

Uetz P., Giot L., Cagney G., Mansfield T.A., Judson R.S., Knight J.R., Lockshon D., Narayan V., Srinivasan M., Pochart P., Qureshi-Emili A., Li Y., Godwin B., Conover D., Kalbfleisch T., Vijayadamodar G., Yang M., Johnston M., Fields S., and Rothberg J.M. 2000. A comprehensive analysis of protein-protein interactions in *Saccharomyces cerevisiae. Nature* **403:** 623–627.

Unlu M., Morgan M.E., and Minden J.S. 1997. Difference gel electrophoresis: A single gel method for detecting changes in protein extracts. *Electrophoresis* **18:** 2071–2077.

van Driel I.R., Stearn P.A., Grego B., Simpson R.J., and Goding J.W. 1984. The receptor for transferrin on murine myeloma cells one step purification based on its physiology and partial amino acid sequence. *J. Immunol.* **133:** 3220–3224.

van Helden J., Naim A., Mancuso R., Eldridge M., Wernisch L., Gilbert D., and Wodak S.J. 2000. Representing and analysing molecular and cellular function using the computer. *Biol. Chem.* **381:** 921–935.

Venter J.C., Adams M.D., Myers E.W., Li P.W., Mural R.J., Sutton G.G., Smith H.O., Yandell M., Evans C.A., Holt R.A, et al. 2001. The sequence of the human genome. *Science* **291:** 1304–1351.

Verhagen A.M., Ekert P.G., Pakusch M., Silke J., Connolly L.M., Reid G.E., Moritz R.L., Simpson R.J., and Vaux D.L. 2000. Identification of DIABLO, a mammalian protein that promotes apoptosis by binding to and antagonizing IAP proteins. *Cell* **102:** 43–53.

Verma R., Annan R.S., Huddleston M.J., Carr S., Reynard G., and Deshaies R.J. 1997. Phosphorylation of Sic1p by G1 Cdk required for its degradation and entry into S phase. *Science* **278:** 455–460.

Verma R., Chen S., Feldman R., Schieltz D., Yates J., Dohmen J., and Deshaies R.J. 2000. Proteasomal proteomics: Identification of nucleotide-sensitive proteasome-interacting proteins by mass spectrometric analysis of affinity-purified proteasomes. *Mol. Biol. Cell* **11:** 3425–3439.

Wang A.M. and Creasey A.A. 1985. Molecular cloning of the complementary DNA for human tumor necrosis factor. *Science* **228:** 149–154.

Wells L., Vosseller K., and Hart G.W. 2001. Glycosylation of nucleocytoplasmic proteins: Signal transduction and O-GlcNAc. *Science* **291:** 2376–2378.

Weng G., Bhalla U.S., and Iyengar R. 1999. Complexity in biological signaling systems. *Science* **284:** 92–96.

Wilson D.S., Keefe A.D., and Szostak J.W. 2001. The use of mRNA display to select high-affinity protein-binding peptides. *Proc. Natl. Acad. Sci.* **98:** 3750–3755.

Wingender E., Chen X., Hehl R., Karas H., Liebich I., Matys V., Meinhardt T., Prüss M., Reuter I., and Schacherer F. 2000. TRANSFAC: An integrated system for gene expression regulation *Nucleic Acids Res.* **28:** 316–319.

Wool I.G., Chan Y.L., and Gluck A. 1995. Structure and evolution of mammalian ribosomal proteins. *Biochem. Cell Biol.* **73:** 933–947.

Xenarios L. and Eisenberg D. 2001. Protein interaction databases. *Curr. Opin. Biotechnol.* **12:** 334–339.

Xenarios I., Salwínski L., Duan X.J., Higney P., Kim S.-M., and Eisenberg D. 2002. DIP, the Database of Interacting Proteins: A research tool for studying cellular networks of protein interactions. *Nucleic Acids Res.* **30:** 303–305.

Xhou W., Merrick B.A., Khaledi M.G., and Tomer K.B. 2000. Detection and sequencing of phosphopeptides affinity bound to immobilized metal ion beads by matrix-assisted laser desorption mass spectrometry.

J. Am. Soc. Mass Spectrom. **11:** 273–282.

Yan J.X., Packer N.H., Gooley A.A., and Williams K.L. 1998 Protein phosphorylation: Technologies for the identification of phosphoamino acids. *J. Chromatogr. A* **808:** 23–41.

Yao X., Freas A., Ramirez J., Demirev P.A., and Fenselau C. 2001. Proteolytic 18O labeling for comparative proteomics: Model studies with two serotypes of adenovirus. *Anal. Chem.* **73:** 2836–2842.

Yates J.R., III. 1998a. Database searching using mass spectrometry data. *Electrophoresis* **19:** 893–900.

———. 1998b. Mass spectrometry and the age of the proteome. *J. Mass Spectrom.* **33:** 1–19.

Zanzoni A., Montecchi-Palazzi L., Quondam M., Ausiello G., Helmer-Citterich M., and Cesareni G. 2002. MINT: A Molecular INTeraction database. *FEBS Lett.* **513:** 135–140.

Zhang J.-G., Farley A., Nicholson S.E., Willson T.A., Zugaro L.M., Simpson R.J., Moritz R.L., Cary D., Richardson R., Hausmann G., Kile B.J., Kent S.B.H., Alexander W.S., Metcalf D., Hilton D.J., Nicola N.A., and Baca M. 1999. The conserved SOCS box motif in suppressors of cytokine signaling binds to elongins B and C and may couple bound proteins to proteasomal degradation. *Proc. Natl. Acad. Sci.* **96:** 2071–2076.

Zhang X., Herring C.J., Romano P.R., Szczepanowska J., Brzeska H., Hinnebusch A.G., and Qin J. 1998. Identification of phosphorylation sites in proteins separated by polyacrylamide gel electrophoresis. *Anal Chem.* **70:** 2050–2059.

Zhou H., Watts J.D., and Aebersold R. 2001. A systematic approach to the analysis of protein phosphorylation. *Nat. Biotechnol.* **19:** 375–378.

Zhou W., Merrick B.A., Khaledi M.G., and Tomer K.B. 2000. Detection and sequencing of phosphopeptides affinity bound to immobilized metal ion beads by matrix-assisted laser desorption/ionization mass spectrometry. *J. Am. Soc. Mass Spectr.* **11:** 273–282.

Zhou H., Ranish J.A., Watts J.D., and Aebersold R. 2002a. Quantitative proteome analysis by solid-phase isotope tagging and mass spectrometry. *Nat. Biotechnol.* **20:** 512–515.

Zhou G., Li H., DeCamp D., Chen S., Shu H., Gong Y., Flaig M., Gillespie J.W., Hu N., Taylor P.R., et al. 2002b. 2D differential in-gel electrophoresis for the identification of esophageal scans cell cancer-specific protein markers. *Mol. Cell. Proteomics* **1:** 117–124.

Zhu H., Klemic J.F., Chang S., Bertone P., Casamayor A., Klemic K.G., Smith D., Gerstein M., Reed M.A., and Snyder M. 2000. Analysis of yeast protein kinases using protein chips. *Nat. Genet.* **26:** 283–289.

Zugaro L.M., Reid G.E., Ji H., Eddes J.S., Murphy A.C., Burgess A.W., and Simpson R.J. 1998. Characterization of rat brain stathmin isoforms by two-dimensional gel electrophoresis-matrix assisted laser desorption/ionization and electrospray ionization-ion trap mass spectrometry. *Electrophoresis* **19:** 867–876.

WWW RESOURCES

For additional Web addresses, see Table 1.1.

http:dip.doe-mbi.ucla.edu The DIP Database (Database of Interacting Proteins)
http://wit.integratedgenomics.com/GOLD GOLD: Genomes OnLine Database homepage

CHAPTER 2

One-dimensional Polyacrylamide Gel Electrophoresis

POLYACRYLAMIDE GEL ELECTROPHORESIS (PAGE) is one of the most reliable methods available for the separation of proteins in complex mixtures and for assessing protein purity. Other applications of PAGE in protein structure and proteomics research include the following:

- Determination of protein relative molecular mass (M_r).
- Verification of protein concentration.
- Detection of protein modifications.
- Global protein expression profiling using a combination of isoelectric focusing (IEF) and SDS-PAGE, i.e., two-dimensional gel electrophoresis (see Chapter 4).
- Peptide mapping using partial proteolytic digestion within the stacking gel (Chapter 7).
- First stage of immunoblotting, i.e., electophoretic transfer to inert immobilizing matrices (Chapter 6).
- High-throughput proteomic approaches that entail either direct amino- or carboxy-terminal sequence analysis (Chapter 6) or a combination of in-gel proteolytic digestion (Chapter 7) and mass spectrometry for protein identification (Chapter 8).

Because the electrophoretic mobilities of individual proteins change disproportionately as either the pH (Figure 2.1) or the concentration of acrylamide (Figure 2.2) is changed, the possibility that the single component observed under one set of conditions results from the accidental coelectrophoresis of two or more proteins can be minimized by running electrophoresis at different pH values or more than one concentration of acrylamide.

Although the majority of investigators still rely on the original Coomassie Brilliant Blue and silver staining techniques, as well as on radiolabeling, recent advances in the use of sensitive protein visualization procedures, especially fluorescent staining methods (see Table 2.1) have greatly increased the overall sensitivity of protein detection in polyacrylamide gels (for a recent review of fluorescent staining techniques, see Patton 2000).

This chapter presents some basic PAGE separation methods that have been found to be useful for a variety of proteomic purposes. A number of staining procedures are also described,

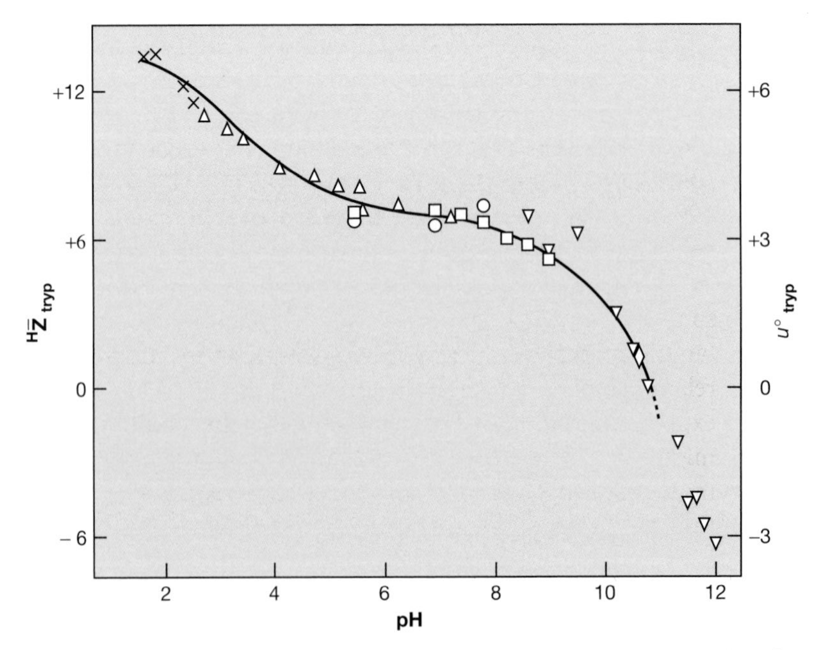

FIGURE 2.1. Comparison of electrophoretic mobilities of trypsin at 0°C (u^{o}_{tryp} points) and acid-base titration curve of trypsin determined at 20°C ($^{H}\bar{Z}_{tryp}$, continuous curve). (Reprinted, with permission, from Duke et al. 1952.)

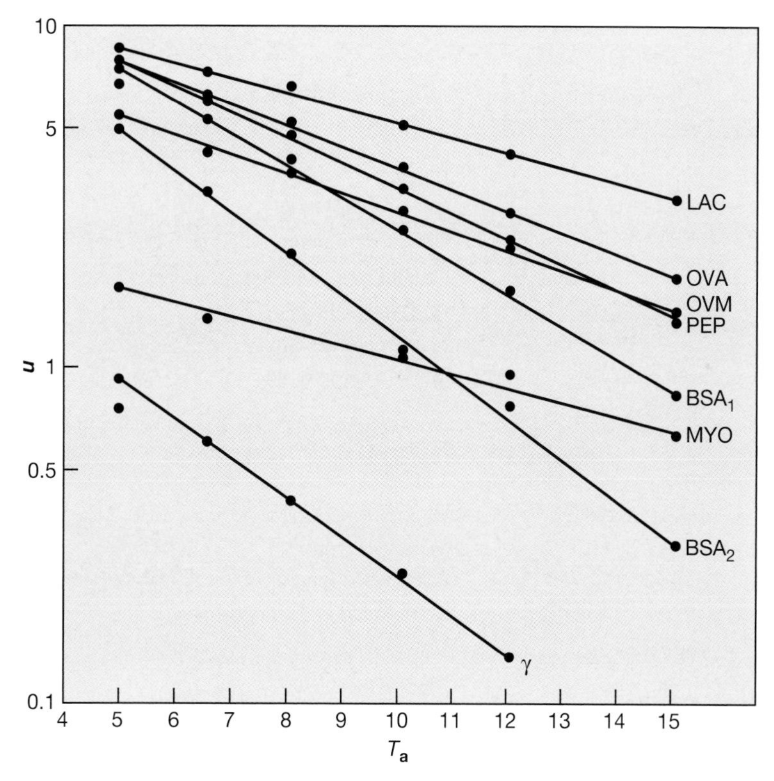

FIGURE 2.2. Effect of various concentrations of polyacrylamide (T_a) on the electrophoretic mobility (u) of various proteins. (LAC) β-lactoglobulin; (OVA) ovalbumin; (OVM) ovomucoid; (PEP) pepsin; (BSA_1) bovine serum albumin monomer; (BSA_2) bovine serum albumin dimer; (MYO) myoglobulin; (γ) immunoglobulin G. (Reprinted from Kyte 1995 [originally from Morris 1966].)

including an improved Coomassie staining technique (Wong et al. 2000) that is faster and more sensitive than the colloidal blue method (Neuhoff et al. 1990) and a silver staining method that is compatible with mass spectrometry (Sinha et al. 2001).

GENERAL PRINCIPLES OF POLYACRYLAMIDE GEL ELECTROPHORESIS

Most electrophoretic procedures for resolving proteins in use today are based on zonal or discontinuous electrophoresis in polyacrylamide gels (Raymond and Weintraub 1959; Davis 1964; Ornstein 1964). Zone electrophoresis in polyacrylamide gels, known as polyacrylamide gel electrophoresis (PAGE), simultaneously exploits differences in molecular size and charge for the purposes of fractionation. This technique, which is based on the mobility of charged solutes in an applied electrical field, is influenced not only by charge, but also by voltage, distance between the electrodes, size and shape of the molecule, temperature, and time. The relationship of the mobility of a particle to the electric field, to a first approximation, is expressed as follows: $M = v/E$, where M is the mobility of the particle ($cm^2 \, sec^{-1} \, V^{-1}$), v is the migration velocity the particle ($cm \, sec^{-1}$), and E is the electric field strength ($V \, cm^{-1}$). For a more detailed description of electrophoresis theory, see Cantor and Schimmel (1980) and Kyte (1995).

Polyacrylamide Gels May Be Continuous or Discontinuous

Polyacrylamide gels may be cast with continuous or discontinuous buffer systems and acrylamide concentrations. Continuous gels, originally described by Weber and Osborn (1969), contain a constant acrylamide concentration, whereas Laemmli discontinuous gels (Laemmli

TABLE 2.1. Fluorescence staining methods for PAGE of proteins

Dye name	Excitation maxima/ Emission maxima (nm)	Principal applications	Features
Gel Stains			
Nile Red dye	300, 540/640	1D gels IEF gels Lipoprotein detection Blotting applications	Fair sensitivity (5–50 ng/band) Similar performance as Coomassie Blue staining methods
SYPRO Ruby protein gel stain	280, 450/610	1D PAGE, 2D gels Mass spectrometry Edman sequencing	Highest sensitivity (1–2 ng/band) Better performance than the best silver staining methods
SYPRO Ruby IEF protein gel stain	280, 450/610	IEF gels Mass spectrometry	Highest sensitivity (1–2 ng/band) Better performance than the best silver staining methods
SYPRO Orange protein gel stain	300, 470/570	1D SDS-PAGE Mass spectrometry Edman sequencing Capillary gel electrophoresis	Good sensitivity (4–10 ng/band) Little protein-to-protein variability Better performance than colloidal Coomassie Blue staining methods
SYPRO Red protein gel stain	300, 550/630	1D SDS-PAGE Mass spectrometry Edman sequencing Capillary gel electrophoresis	Good sensitivity (4–10 ng/band) Little protein-to-protein variability Better performance than colloidal Coomassie Blue staining methods
SYPRO Tangerine protein gel stain	300, 490/640	1D SDS-PAGE Blotting applications Zymography Electroelution Mass spectrometry Edman sequencing	Good sensitivity (4–10 ng/band) Little protein-to-protein variability No organic solvents or acids Better performance than zinc-imidazole reverse staining methods
Blot Stains			
SYPRO Ruby protein blot stain	280, 450/618	Blotting membranes[a] Immunodetection (western blotting) Mass spectrometry Edman sequencing	Highest sensitivity (1–2 ng/band) Better performance than colloidal gold staining methods
SYPRO Rose Plus protein blot stain	350[b]/610	Blotting membranes[a] Immunodetection (western blotting) Mass spectrometry Edman sequencing	Highest sensitivity (1–2 ng/band) Readily reversible Better performance than colloidal gold staining methods
SYPRO Rose protein blot stain	350[b]/590, 615	Blotting membranes[a] Immunodetection (western blotting) Mass spectrometry Edman sequencing	Good sensitivity (15–30 ng/band) Readily reversible Better performance than Amido Black staining methods

Reprinted, with permission, from Patton (2000).
[a]Nitrocellulose or PVDF membranes.
[b]UV epi-illumination only.

1970) are a composite of a short wide-pore "stacking gel" (3–4 %T) (for a definition of T, see page 45) layered on top of a long, small-pore "resolving gel" (7–25 %T). The stacking and resolving gels in the Laemmli system are also discontinuous with respect to buffer composition (i.e., multiphasic buffers are employed that differ with respect to both pH and ionic composition). These two parameters, i.e., different buffer and acrylamide compositions, allow samples of relatively large volumes to be concentrated in the upper stacking gel prior to their separation upon entering the lower resolving gel. In contrast, continuous gels utilizing monophasic buffers are limited by the requirement that samples be both highly concentrated and low in volume.

Electrophoretic Separations Are Designed to Maximize the Resolution of Proteins

The electrophoretic separation of proteins and peptides on polyacrylamide gels must have as high a resolution as possible. This resolution is achieved by using a stacking gel with multiphasic buffer systems (see Figure 2.3) (Chrambach and Rodbard 1971). An important advantage of the stacking gel is the ability to concentrate proteins from very dilute solutions. This process significantly improves resolution of the subsequent separation by shrinking the original sample into very thin, highly concentrated starting zones so that all of the protein molecules begin the electrophoretic separation at very nearly the same point. Stacking results from the formation of a limited high-voltage gradient or *Kohlrausch boundary* (Kohlrausch 1897) in which proteins are confined to a thin and highly concentrated zone of intermediate mobility between leading chloride and trailing glycine ions. Briefly, at the pH of the stacking gel, the sample ion (typically, a mixture of proteins or their detergent derivatives) is initially sandwiched between an upper solution and a lower solution containing ions of the same sign as the sample ion. When a current is passed through the mixture, one ion (e.g., chloride) migrates faster than the other slower ion (e.g., glycine). The sample ion, being of intermediate mobility, is compressed between the chloride and glycine ions, forming a narrow zone or band of high concentration. Thus, the stacking process is able to sandwich the proteins into a *disc* or narrow band. For electrophoretic separation to occur, the proteins must be released from the boundary after they have been stacked. Unstacking (resolution) is accomplished by abruptly increasing the pH and decreasing the pore size of the separating gel (relative to the

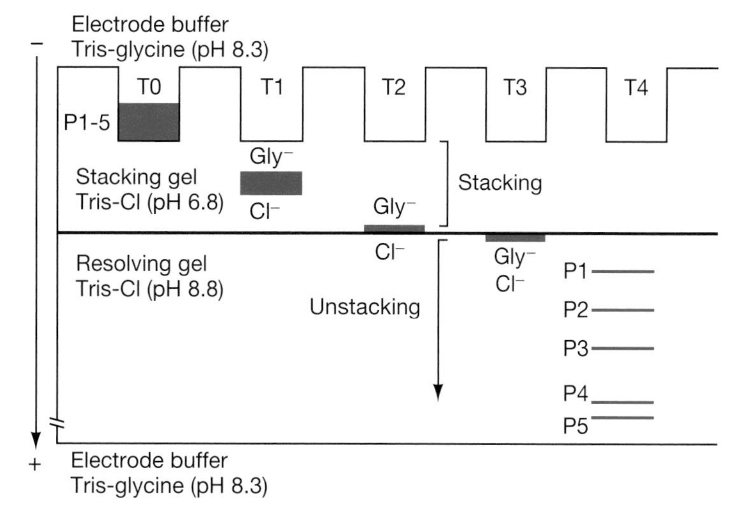

FIGURE 2.3. Discontinuous multiphasic gel-buffer electrophoresis. Electrophoresis through the stacking (pH 6.8) and resolving (pH 8.8) gels of a hypothetical sample containing five proteins (P1–5) is shown as a function of time. At the start (T0), the proteins in the original sample are in a large volume and at a low pH (pH 6.8). As the voltage is applied, they are compressed to a small volume, or disc, as they migrate through the spacer (*large pore*, ~3% acrylamide) by being sandwiched in the boundary between the upper solution (Tris-glycine buffer, pH 8.3) and the solution of the original sample and the spacer (Tris-chloride, pH 6.8) (T1). At some time after stacking is complete (T2), the proteins reach the *small-pore* gel (~5–7% acrylamide) boundary where changes in their mobilities occur when they encounter both decreased pore size and a higher pH (Tris-chloride, pH 8.8) (T3). Because of the special viscous properties of the gel, proteins of equal free mobility but of appreciable different molecular weights (different diffusion constants) will migrate with markedly different mobilities and will be resolved (T4). (Modified, with permission, from Makowski and Ramsby 1997 [©Oxford University Press].)

stacking gel). Under these conditions, the sample ions (proteins) migrate into the separating gel and the trailing (glycine) ions continuously overtake and pass the sample ions. Thus, a comparatively uniform voltage gradient is established in which electrophoretic separation of samples occurs (Davis 1964).

Since the initial description of the Laemmli procedure, considerable effort has gone into trying to optimize these electrophoretic conditions. Foremost has been the development of the multiphasic zone electrophoretic (MZE) theory by Chrambach and Jovin (1983), which predicted more than 4200 potentially useful buffer systems (this has now been simplified to 19) for optimizing SDS-PAGE. For a practical description of PAGE, see Shi and Jackowski (1998), Makowski and Ramsby (1997), and Hames (1990).

The Polyacrylamide Gel Matrix Can Be Formed with a Wide Range of Pore Sizes

Polyacrylamide gels are formed from the polymerization of monomeric acrylamide (CH_2=$CHCONH_2$) and the bifunctional cross-linking monomer *N,N′*-methylenebisacrylamide (CH_2=$CHCONHCH_2NHCOCH$=CH_2) as shown in Figure 2.4. Chemical polymerization of acrylamide and bisacrylamide is initiated by either chemical (e.g., a combination of TEMED [tetramethylethylenediamine] and ammonium persulfate) or photochemical free-

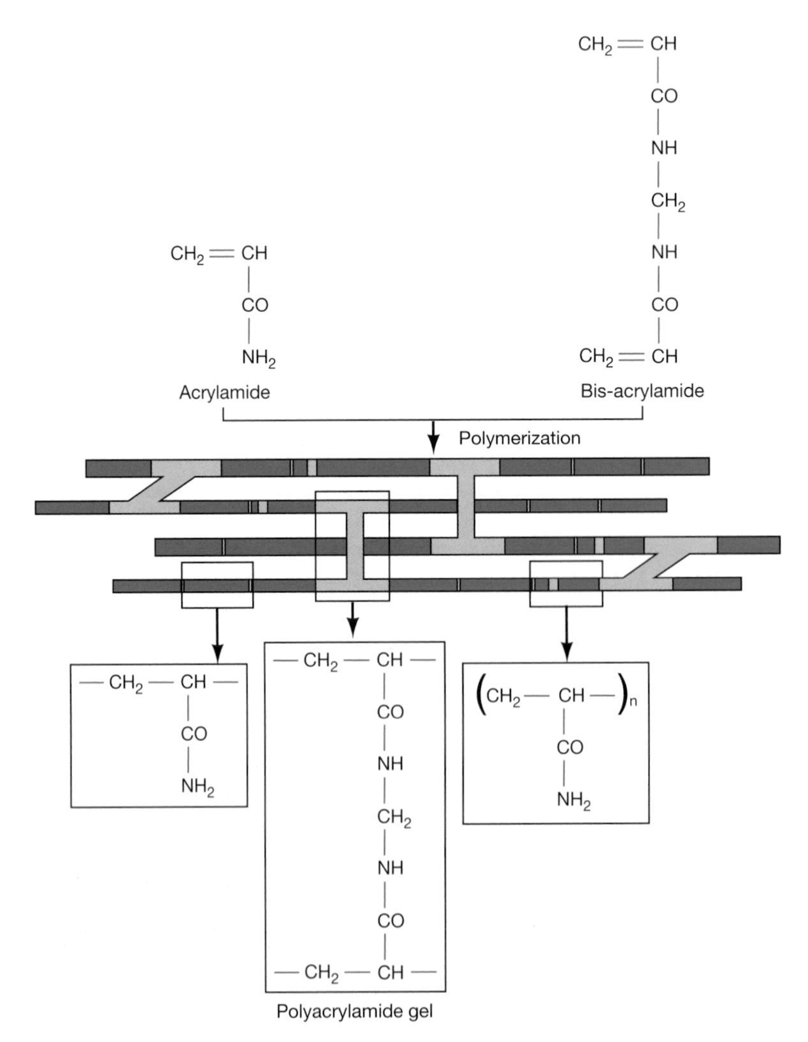

FIGURE 2.4. Chemical structures of acrylamide, bisacrylamide, and polyacrylamide. (Adapted, with permission, from Hames 1998 [©Oxford University Press].)

radical-generating systems such as riboflavin-5′-phosphate or methylene blue. Persulfate free radicals, which form when ammonium persulfate is dissolved in H_2O, activate the acrylamide monomer, and the electron carrier TEMED catalyzes this reaction. Riboflavin can also generate free radicals in the presence of oxygen and UV light as a result of photodecomposition. Riboflavin and ammonium persulfate are sometimes used in combination to generate free radicals. Polyacrylamide polymers can be formed with a wide range of pore sizes by varying the amount of total acrylamide used per unit volume and the degree of cross-linkage. For instance, as the proportion of cross-linker is increased, the pore size decreases. It should be noted that the rate of polymerization also affects the gel pore size. Increasing the rate of polymerization (e.g., by increasing the temperature of polymerization) can produce a gel with a small-pore structure, whereas lowering the polymerization temperature can produce large-pore gels. Factors that influence the efficiency of monomer-polymer conversion can also affect gel pore size. For example, the use of poor-quality acrylamide and extremes of pH can result in greater gel porosity due to incomplete conversion of monomer to polymer. Gel additives can also markedly influence gel pore structure. On the one hand, urea (e.g., 8 M urea is often added to gels to fractionate oligopeptides in the range of 1,200–10,000 daltons) (Swank and Munkres 1971) causes the formation of small-pore-size gels by accelerating persulfate-driven polymerization of monomeric acrylamide to near completion. On the other hand, inclusion of the polymer polyethylene glycol in the gel solution results in macroporous gel formation due to lateral acrylamide chain formation (Righetti 1995).

Gel Pore Size Is the Major Factor in Determining Protein and Polypeptide Resolution in Polyacrylamide Gels

Polyacrylamide gels range from soft (low polyacrylamide concentration) to hard (high polyacrylamide concentration), with the properties of acrylamide mixtures being defined by the letters T and C (Hjertén 1962). T denotes the total percentage concentration of the monomers (acrylamide plus *N,N*′-methylenebisacrylamide) in grams per 100 ml, and C denotes the percentage (by weight) of the cross-linker *N,N*′-methylenebisacrylamide relative to the total amount of monomers. In general, the pore size is inversely proportional to %T.

$$\%T = \{[\text{acrylamide (grams)} + \text{bisacrylamide (grams)}]/100 \text{ ml}\} \times 100$$

$$\%C = \{\text{bisacrylamide (grams)}/[\text{acrylamide (grams)} + \text{bisacrylamide (grams)}]\} \times 100$$

Judicious selection of acrylamide gel concentration is critical for optimal separation of proteins by zone electrophoresis. For a Laemmli gel (Laemmli 1970), T is equal to 10% and C is equal to 2.6%. A guide for choosing the appropriate acrylamide concentration for a uniform gel for separating a particular molecular-weight range of sample proteins is given in Table 2.2. Soft gels with concentrations of acrylamide <3% are almost liquid and very difficult to handle unless they are strengthened by the inclusion of 0.5% agarose. At the other end of the spectrum, high-concentration acrylamide gels (polyacrylamide gels will form up to ~35% acrylamide) are also very difficult to handle due to their brittleness and are therefore of limited usefulness.

A major limitation of uniform acrylamide concentration gels is their restricted molecular separation range. For example, gels with a uniform pore size large enough to resolve large proteins are unlikely to resolve small polypeptides as well. This limitation can be circumvented by using pore gradient gels whose pore size changes from the top to the bottom of the gel, thus enabling the separation of proteins and polypeptides over a larger molecular-weight range and with higher resolution than a uniform concentration gel. Although more difficult to prepare, gradient gels resolve wider ranges of protein relative molecular weights. In practice, the effective range of a polyacrylamide gel is 5–20% for a uniform gel concentration and

TABLE 2.2. Protein and peptide molecular-weight separation guidelines for polyacrylamide gels

Acrylamide concentration		
%T	%C	M_r range of sample polypeptides
Uniform acrylamide concentration gels		
5	2.6	25,000 to 300,000
10	2.6	15,000 to 100,000
10	3.0[a]	1,000 to 100,000
15	2.6	12,000 to 50,000
Gradient acrylamide gels[b]		
3.3–12	2.0[c]	14,500 to 2,800,000
3–30	8.4	13,000 to 1,000,000
5–20	2.6	14,000 to 210,000
8–15	1.0	14,000 to 330,000

Reproduced. with permission, from Shi and Jackowski (1998).
[a]When using Tricine instead of glycine in running buffer (Schägger and von Jagow 1987).
[b]This list is by no means exhaustive and other gel ranges may also be used.
[c]Granzier and Wang (1993).

4–20% for a gradient concentration gel. Table 2.2 provides a guide for choosing a gradient gel range appropriate for the size of the proteins being fractionated. Although the effective resolving range of a polyacrylamide gel is primarily determined by the pore size of the gel, other factors can have important roles, including:

- ***Buffer composition:*** Tricine (pK 8.15), for example, instead of glycine (pK 9.6) in the running buffer facilitates the resolution of small peptides at high acrylamide concentrations (Schägger and von Jagow 1987).
- ***pH:*** The sieving properties of SDS-PAGE (see below), for example, can be altered by a subtle change of pH of the resolving gel, which does not require extensive knowledge of MZE theory (Makowski and Ramsby 1993, 1997). For instance, by increasing the pH from 8.9 to 9.2, the linear calibration range of a 10% gel can be extended to include low-M_r proteins of 10–20K without sacrificing the resolution of high-M_r proteins as occurs with high-percentage gels.
- ***Gel additives:*** For example, the detergents cetyltrimethyl ammonium bromide (CTAB), sulfobetaine 3-12, 3[(3-chloramidopropyl)dimethyl-ammonio]-1-propane sulfonate (CHAPS), Triton X-100 (Caglio et al. 1994), and SDS (Caglio et al. 1994; Sobieszek 1994) and the polymers dextran, ficoll, and polyvinylpyrrolidone (PVP) (Gersten and Bijwaard 1992a,b) inhibit polymerization of acrylamide to varying extents, resulting in large pore sizes. The inclusion of such additives provides an additional parameter to PAGE, which can be useful in resolving complex mixtures of proteins, especially large-M_r proteins.

SDS IS USED TO CREATE DENATURING POLYACRYLAMIDE GELS

An important refinement of PAGE that has gained widespread acceptance is the use of SDS in polyacrylamide electrophoresis as a simple and powerful way to dissociate proteins into individual chains and separate them according to their molecular weights (Shapiro et al. 1967). The following are probably the most widely cited SDS-PAGE procedures:

- Continuous buffer system (normally a phosphate-phosphate system) with a uniform gel (Weber and Osborn 1969).
- Laemmli (Tris-glycine-SDS system; Laemmli 1970) and Neville (borate-sulfate-SDS system; Neville 1971) modifications of the discontinuous buffer system (Davis 1964; Ornstein 1964) with a uniform gel.

In recent years, continuous and discontinuous buffer systems with gradient gels have been developed to improve protein and polypeptide resolution, and precast gels are now commercially available. Protocol 1 provides details for preparing, running, and storing discontinuous SDS-PAGE gels. Several protein visualization procedures are described here, including a rapid, high-temperature Coomassie Blue method (Protocol 3) and a modified silver staining method that is compatible with mass spectrometry-based protein identification (Protocol 7). The classic Coomassie Blue protein-staining procedure and a rapid variant using a commercially available stain-impregnated paper are detailed in Protocols 2 and 4, respectively. Finally, two protocols using the fluorescent stains SYPRO Ruby and SYPRO Orange are described in Protocols 5 and 6. Additional protocols for staining gels and visualizing proteins are described in Chapters 4 and 7. For a brief description of the history and chemical structure of Coomassie Brilliant Blue, see the panel on COOMASSIE BRILLIANT BLUE. For a description of possible problems that may arise during the preparation and use of SDS-PAGE, see Table 2.6 (Protocol 1).

Denaturation and binding of the anionic detergent SDS to proteins and peptides generally result in a relatively uniform negative charge, because most proteins bind similar amounts of SDS (~1.4 g of SDS per gram of polypeptide; Reynolds and Tanford 1970) and become highly negatively charged by the addition of strongly acidic sulfonic acid groups. By these means, possible changes in electrophoretic behavior of different tertiary protein structures are negated since complete unfolding of proteins occurs by SDS binding. Polypeptide chains studied by PAGE in the presence of SDS have been shown to migrate according to their molecular weights (Weber and Osborn 1969). Because SDS-protein complexes have electrophoretic mobilities in polyacrylamide gels that are generally directly proportional to the logarithm of the length of the polypeptide chain (see Figure 2.5), the molecular weight of an unknown protein can be determined to within 10% of its true value by comparing its electrophoretic mobility with that of a set of standard proteins. A procedure for determining the apparent molecular mass of a protein using SDS-PAGE is given in Protocol 8.

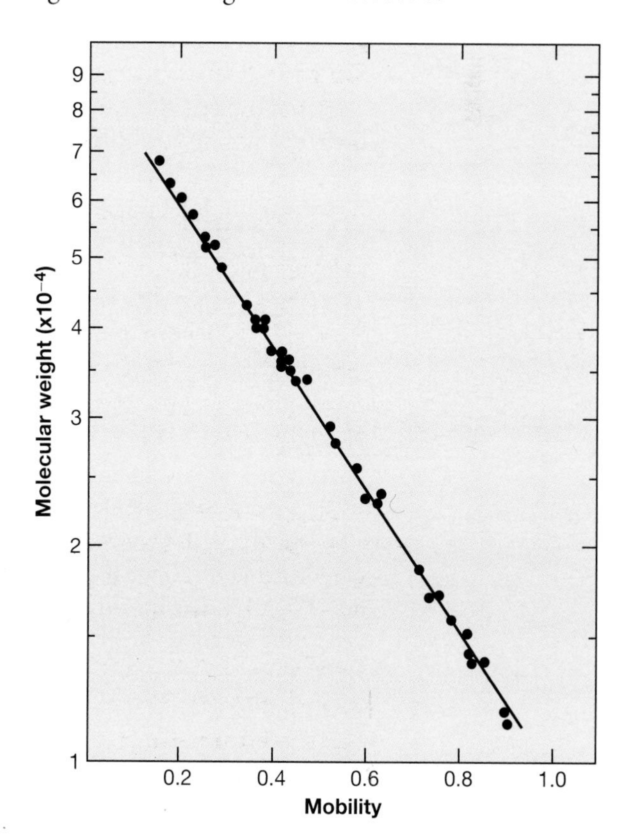

FIGURE 2.5. Comparison of the molecular weights of 37 different polypeptide chains in the molecular range from 11,000 to 70,000 with their electrophoretic mobilities on gels. (Reprinted, with permission, from Weber and Osborn 1969.)

Coomassie Brilliant Blue

Coomassie Blue R250 and G250 are still the most popular chemicals used for the staining of protein bands in polyacrylamide gels. Coomassie Blue is an aminotriarylmethane dye that forms strong but not covalent complexes with proteins, most probably by a combination of van der Waals forces and electrostatic interactions with NH_3^+ groups. Despite its name, Coomassie Brilliant Blue is classified as a magenta dye and is used to stain proteins on dot blots and after electrophoresis through polyacrylamide gels. At the pH used for detection and assay of proteins (~pH 1), Coomassie Blue is mostly in its highly protonated reddish form, with smaller proportions of green and blue forms, which are all in equilibrium with each other. Only the neutral (Blue) form of the dye binds strongly to proteins. The uptake of dye is approximately proportional to the amount of protein, following the Beer-Lambert law. Under normal conditions of staining, Coomassie Blue binds strongly to arginine and lysine residues and with lower affinity to aromatic side chains. The number of strong binding sites varies widely from protein to protein (from 13 to >100) and correlates with the content of basic amino acids, rather than the total mass of the protein. The molar absorbance of the bound dye varies widely between proteins, from ~34,000 $mole^{-1}\ cm^{-1}$ (β-lactoglobulin) to ~58,000 $mole^{-1}\ cm^{-1}$ (glutamate dehydrogenase).

Historical Footnote: Coomassie dyes date back to early last century, originally being developed for the staining of woolen jumpers. They were named to commemorate the 1896 British occupation of the capital Kumasie (Coomassie) of the Ashanti tribe, in what has now become modern-day Ghana (presumably, the name was given when colonization was still in fashion!). Coomassie Brilliant Blue R250 (R stands for red and G for greenish blue) was first used as a laboratory reagent to stain proteins in 1963. (Chemically, Coomassie Blue G250 is the same as Coomassie Brilliant Blue R250, but with extra methyl groups.) Robert Webster, a graduate student in the laboratory of Stephen Fazekas de St. Groth at the Australian National University in Canberra, was searching for a way to locate influenza virus proteins that had been separated by electrophoresis on cellulose acetate strips. At that time, Australia had a thriving wool industry and government laboratories were intensively investigating the mechanism of action of various classes of dyes used for wool-dying. Fazekas and Webster reasoned that these dyes must have a high affinity for proteins and they obtained samples of a great many dyes from the Commonwealth Scientific and Industrial Organization. Included among them was Coomassie Brilliant Blue R250, which had been used since the turn of the century in the textile dying industry. Webster soon found that Coomassie Brilliant Blue R250 was a very sensitive stain for proteins, but he was frustrated by extreme day-to-day variation in the intensity of the staining. At home one night, he suddenly realized that the answer to the problem was to fix the protein before staining. He went back to the laboratory and fixed the separated influenza virus proteins with sulfosalicylic acid. After these results were published (Fazekas de St. Groth et al. 1963), the method was rapidly adapted to stain proteins separated by electrophoresis through polyacrylamide gels (Meyer and Lamberts 1965). To this day, staining with Coomassie Brilliant Blue remains a standard method to detect, visualize, and quantitate fixed proteins.

Because Coomassie Brilliant Blue R250 is now a trademark of Imperial Chemical Industries PLC, the dye is generally listed in biochemical catalogs as Brilliant Blue. Two forms of the dye are available: Brilliant Blue G and Brilliant Blue R, which are given different numbers (42655 and 42660) in the Colour Index, a kind of dyer's Bible, in which dyes are classified and arranged according to color. Brilliant Blue G is less soluble in water and alcohols than Coomassie Blue R250. Both dyes stain fixed proteins efficiently. They are used at a concentration of 0.05–0.1% in methanol–glacial acetic acid–water (50:10:40 v/v). The dyes should be dissolved in methanol before acetic acid and water are added.

Adapted, in part, from GRADIPRINT™ issue No. 6, August 1988 (with permission) and background provided by Joe Sambrook.

Some Proteins Behave Anomalously on SDS-PAGE

There are many reports of proteins, especially glycoproteins, behaving anomalously on SDS-PAGE gels even when SDS and thiol reagent are in excess. Artifactually high M_r estimates often result, probably due to these proteins not binding the normal amount of SDS (Segrest and Jackson 1972; Hames 1990), which results in reduced net charge and lower mobilities during electrophoresis. However, this problem can be alleviated by determining the M_r of glycoproteins at a number of polyacrylamide gel concentrations to yield an asymptotic M_r that approximates the real molecular weight (Segrest and Jackson 1972) or, alternatively, by using polyacrylamide concentration gradient gels (Hames 1990). Proteins with high negative charge (Kaufman et al. 1984) and positive charge (e.g., histones) or rich in proline residues (e.g., collagenous proteins) all yield abnormally high M_r values using SDS-PAGE (Hames 1990).

CTAB Can Replace SDS in Some Denaturing PAGE Applications

Although the anionic detergent SDS is usually the choice for denaturing gel electrophoresis, other detergents have been developed for more specialized purposes. For example, neutral detergents such as Triton X-100 and Nonidet P-40, as well as some zwitterionic detergents (e.g., CHAPS) have been utilized in isoelectric focusing and two-dimensional gel electrophoresis because they do not alter the charge of proteins and facilitate solubilization (see Chapter 4).

SDS has drawbacks in some situations. For example, SDS can cause protein aggregation and precipitation as well as abnormal protein migration leading to poor resolution, and it crystallizes at low temperatures (Shi and Jackowski 1998). These limitations of SDS have led to the development of cationic detergent solubilization and PAGE (Mócz and Bálint 1984; Akin et al. 1985; MacFarlane 1989; Atkins et al. 1992). A method for cationic detergent PAGE using CTAB (Atkins et al. 1992) is given in Protocol 9.

ACETIC ACID–UREA PAGE SEPARATES PROTEINS ON THE BASIS OF THEIR SIZE AND CHARGE

In SDS-PAGE, proteins are fractionated on the basis of their molecular masses by the sieving effect of the polyacrylamide gel matrix. However, this system is of limited value for separating proteins of very similar molecular masses that contain posttranslational modifications such as acetylation, methylation, and phosphorylation (e.g., histone variants). To analyze such modified proteins, PAGE systems have been used in which both molecular size and charge of the protein are involved in the separation process. Although the prevailing pH in such a system may be any value, a value of 3 is typically used. At this pH, the protein will be positively charged and will migrate toward the cathode in an electric field. (The pK_a values of the carboxyl side chains of aspartic acid and glutamic acid are 3.8 and 4.2, respectively, so the negative charge contribution from these amino acids to the overall charge of a protein at pH 3 will be minimal.) For this reason, acid-urea PAGE is commonly employed to separate proteins of similar size but different charge (Panyim and Chalkley 1969; Harwig et al. 1993). A simple acid-urea PAGE procedure is given in Protocol 10.

TRICINE SDS-PAGE PROVIDES EXCELLENT SEPARATION OF SMALL POLYPEPTIDES

Using the Laemmli discontinuous system (Laemmli 1970), the mobilities of small proteins (e.g., M_r <14K) in SDS-PAGE may no longer be proportional to their M_r when the protein charge properties become significant relative to their size. Although gradient gels in the 3–30% range improve resolution over a broader M_r range, they still cannot fully resolve oligopeptides with M_r <10K. This has led to the development of SDS-PAGE systems capable of extending the resolution of oligopeptides into the low-M_r range. Swank and Munkres (1971) modified the SDS-PAGE method by including 8 M urea in a continuous 0.1 M Tris-phosphate (pH 6.8) buffer system, resulting in the resolution of oligopeptides with M_r as low as 2.5K. In a separate approach, a discontinuous buffer system in which the trailing ion in the Laemmli system (glycine) is replaced by tricine has been developed by Schägger and von Jagow (1987). Because tricine migrates much faster than glycine at the usual pH values used in a stacking gel, the stacking limit is shifted toward the low-M_r range so that small SDS-oligopeptide complexes are well separated from SDS. By these means, oligopeptides in the range M_r 1–100K can be resolved using acrylamide concentrations as low as 10%. The tricine SDS-PAGE method is described in Protocol 11.

NONDENATURING PAGE (NATIVE GEL ELECTROPHORESIS) SEPARATES NATIVE PROTEINS AND PROTEIN COMPLEXES

Although SDS-PAGE is probably the most widely used electrophoretic system for analyzing proteins, it should be emphasized that this method separates denatured proteins. Thus, SDS-PAGE cannot be used to separate intact protein complexes or native proteins whose biological activity must be retained in the gel for subsequent functional testing such as enzyme activity, receptor binding, and antibody binding. In these situations, it is necessary to use nondenaturing gel electrophoresis, also called native gel electrophoresis, which separates proteins on the basis of their size and charge properties. Proteins that are more highly charged at the pH of the separating gel have a greater migration rate in the gel, whereas the acrylamide pore size (dictated by the acrylamide concentration used and the acrylamide:bisacrylamide ratio) serves to sieve proteins of different sizes. Variables that affect nondenaturing gel electrophoresis include:

- ***Charge:*** The charges of proteins differ widely at different pH values. Therefore, choosing a buffer in a native gel system that will provide optimal resolution depends on the specific proteins under investigation. The further the electrophoresis buffer pH is from the pI values of the proteins to be analyzed, the higher the charge on the proteins. This leads to shorter times required for electrophoretic separation as well as reduced band diffusion. However, the closer the pH of the electrophoresis buffer to the pI values of the proteins under investigation, the greater the charge differences between proteins and the greater the likelihood of accomplishing separation of the proteins being analyzed. Hence, conditions of optimal resolution of proteins in native gels must be determined by trial and error. Most proteins are negatively charged at pH 8.8, which is the common pH used for native gel electrophoresis. Alternatively, nondenaturing gel electrophoresis may be carried out at low pH, in which case the anode and cathode should be reversed. Although the continuous and discontinuous buffer systems described for SDS-PAGE can be applied to nondenaturing gel electrophoresis, the discontinuous systems yield the higher resolution for most native proteins. It should be noted that some native proteins aggregate and may precipitate at the high protein concentrations accomplished in the stacking gel of the discontinuous gel system. Under these conditions, such proteins either fail to migrate into the resolving gel or cause "streaking." Using a continuous buffer system may circumvent this problem. In general, almost any buffer with a pH between 3 and 10 and a concentration between 0.01 and 0.1 M may be used for native gel electrophoresis. Caution must be exercised in choosing an appropriate buffer system because extreme pH conditions may denature the protein of interest with subsequent loss of biological activity. Similarly, some protein complexes may require the presence of a metal ion or reducing agent to retain activity, which must be added to the electrophoresis buffer. Typical buffer systems in the pH range of 3.8–10.2 used for native gel electrophoresis include Tris-glycine (pH 8.3–9.5), Tris-borate (pH 7.0–8.5), Tris-acetate (pH 7.2–8.5), Tris-citrate (pH 7.0–8.5), and β-alanine-acetate (pH 4–5); for other useful continuous buffer systems, see McLellan (1982).

- ***Ionic strength:*** Buffer ionic strength has a crucial role in electrophoresis, and values typically used are in the range of 0.01–0.1 M. If the ionic strength is too low in native gel systems, proteins may aggregate nonspecifically, whereas high ionic strengths will result in increased heat generation during electrophoresis.
- ***Acrylamide gel porosity:*** Acrylamide concentrations and acrylamide:bisacrylamide ratios may be varied to achieve different sieving effects and to optimize protein separation (see Table 2.2).
- ***Temperature:*** Nondenaturing gel electrophoresis is typically performed at 0–4°C in order to minimize loss of biological activity due to protein denaturation and/or proteolytic degradation.

A high-pH (pH 8.8) discontinuous native gel PAGE method is described in Protocol 12. For a low-pH (pH 3.8) method, see Shi and Jackowski (1998).

PROTOCOL 1

SDS-PAGE of Proteins

PROBABLY THE MOST WIDELY USED METHOD FOR ANALYZING MIXTURES of proteins is SDS-PAGE. In this technique, proteins are reacted with the anionic detergent SDS (or sodium lauryl sulfate) to form negatively charged complexes. Initial heating of the protein sample at 95°C in the presence of excess SDS and a thiol reagent (e.g., 2-mercaptoethanol or dithiothreitol) denatures the protein mixture and disrupts any disulfide bonds. Under these conditions, all reduced polypeptides bind the same amount of SDS on a weight basis (1.4 g of SDS per gram of polypeptide) independent of amino acid composition and sequence. The resolving power of SDS-PAGE is greatly enhanced by preceding the protein separation phase with a "stacking gel," which employs the principles of isotachophoresis to concentrate samples from relatively large volumes into very small bands (or zones), but does not separate them. In the separating gel, the negatively charged SDS-protein complexes migrate through the sieve-like polyacrylamide matrix and are separated solely on molecular-weight differences. SDS-PAGE of proteins can be performed in either "tube" or "slab gel" systems. The slab gel system is more widely used for the following reasons:

- Multiple samples can be resolved on the same gel, thereby allowing a better comparison of test and control samples.
- Molecular weights can be more accurately determined.
- Slab gels lend themselves easily to manipulations such as scanning, drying, autoradiography, electroblotting, and photography.

Electrophoresis equipment for preparing and using slab gels comes in a fantastic array of sizes and styles to meet all budgets and tastes. This protocol describes the preparation and use of SDS-PAGE employing a full-size slab gel and a discontinuous buffer system.

This protocol was contributed by Hong Ji and Robert Goode (Joint ProteomicS Laboratory of the Ludwig Institute for Cancer Research and Walter and Eliza Hall Institute of Medical Research, Melbourne, Australia).

MATERIALS

CAUTION: See Appendix 3 for appropriate handling of materials marked with <!>.

IMPORTANT: All reagents should be electrophoresis grade or better.

Reagents

Buffers for discontinuous buffer system (Laemmli 1970)

4x Stacking gel buffer (0.5 M Tris-HCl at pH 6.8)

Dissolve 60.5 g of Tris base in 850 ml of H_2O and adjust pH to 6.8 with 6 M HCl <!>. Readjust pH to 6.8 at room temperature. Add H_2O to 1000 ml and store the buffer at 4°C.

4x Resolving gel buffer (1.5 M Tris-HCl at pH 8.8)

Dissolve 181.5 g of Tris base in 850 ml of H_2O and adjust pH to 8.8 with 6 M HCl. Cool the solution to room temperature and readjust pH to 8.8. Add H_2O to 1000 ml and store at 4°C.

10x Running buffer

Dissolve 30.0 g of Tris base, 144.0 g of glycine <!>, and 10.0 g of SDS in 1000 ml of H_2O. The pH of the buffer should be 8.3 and no pH adjustment is required. Store the running buffer at room temperature and dilute to 1x before use.

2x SDS-PAGE sample buffer

Mix:

2.0 ml of 4x stacking gel buffer
1.6 ml of glycerol
3.2 ml of 10% SDS
0.8 ml of 2-mercaptoethanol <!>
0.4 ml of 1.0% bromophenol blue <!>

n-butanol (H_2O-saturated) <!>

Combine 50 ml of *n*-butanol or isobutanol <!> (or isopropanol <!>) and 50 ml of H_2O in a glass bottle and shake. Use the top phase to overlay gels. Store at room temperature indefinitely.

Methanol <!>

Polymerization initiator solutions

The three polymerization systems are photopolymerization, methylene blue polymerization, and, the most commonly used, chemical polymerization. The two chemical polymerization initiators in this system are ammonium persulfate and TEMED (*N,N,N′,N′*-tetramethylethylenediamine).

Ammonium persulfate solution (10%) <!>

Dissolve 1.0 g of ammonium persulfate in 10 ml of H_2O.

Because ammonium persulfate is very hygroscopic and begins to degrade immediately after being dissolved in H_2O, this solution should be made fresh daily.

TEMED <!>

TEMED can be used as supplied.

Although this chemical is subject to oxidation and is very hygroscopic, it can be stored in a tightly closed, dark glass bottle for at least 6 months at room temperature.

SDS (10%) stock solution <!>

Dissolve 10 g of SDS in 80 ml of H_2O, and then add H_2O to 100 ml. This stock solution is stable for 6 months at room temperature.

30%T (2.6%C) Acrylamide stock solution <!>

Add 29.22 g of acrylamide and 0.78 g of bisacrylamide <!> to 100 ml of H_2O. Filter the stock solution through Whatman filter paper and store at 4°C. Prepare fresh stock acrylamide solution every few weeks.

Equipment

Centrifuge

Electrophoresis apparatus

For a list of commercially available slab gel systems and supplier details, see Table 2.3.

Gel-loading pipette tips

Glass plates

Heat block or water bath preset to 95°C

PCC-54 detergent (Pierce) or equivalent

This detergent is a cleaning solution designed for washing glass plates.

Power supply

A power supply capable of providing ~200 V and 500 mA (either constant voltage or constant current) is sufficient for SDS-PAGE. A number of power supplies are available commercially, e.g., Bio-Rad Power Pac 3000 (3000 V and 400 mA).

Biological Sample

Protein solution(s) or pellets of cellular proteins

For instance, as prepared in Chapter 3.

TABLE 2.3. Some commercially available slab gel systems for performing SDS-PAGE

System name	Supplier	Gel size (cm x cm)
Xcell II Mini-Cell	Novex	10 x 10
Hoefer SE 600 Ruby Standard Dual Cooled Vertical Unit	Amersham Bioscience	18 x 16
Hoefer SE 600 Standard Dual Cooled Vertical Unit	Amersham Bioscience	18 x 16
Hoefer SE 660 Standard Dual Cooled Vertical Unit	Amersham Bioscience	18 x 24
Hoefer MiniVE Vertical Electrophoresis System	Amersham Bioscience	10 x 8 to 10 x 10.5
Hoefer SE 250 Mighty Small II Mini-Vertical Unit	Amersham Bioscience	10 x 8
Hoefer SE 260 Mighty Small II Mini-Vertical Unit	Amersham Bioscience	10 x 10.5
Protean II xi Cell (16 cm)	Bio-Rad	16 x 16
Protean II xi Cell (20 cm)	Bio-Rad	16 x 20
Protean II XL Cell	Bio-Rad	18.5 x 20
Criterion Mini Vertical Cell	Bio-Rad	13.3 x 8.7
Mini-PROTEAN 3 Electrophoresis Module	Bio-Rad	8.3 x 7.3

URL addresses: Amersham Bioscience (http://www.apbiotech.com), Bio-Rad (http://www.bio-rad.com), Novex (http://www.invitrogen.com).

Additional Reagents

The final step of this protocol requires the reagents listed in any one of the protein visualization protocols, Protocols 2–7.

METHOD

Pouring a Slab Gel

1. Clean the glass plates.

   a. Soak the glass plates in 2% PCC-54 cleaning solution for 3–24 hours.

   b. Rinse the plates with tap water thoroughly and then once with distilled H_2O.

   c. Dry the glass plates with clean tissue paper and then clean them with Kimwipes soaked in methanol. Dry the plates in the air.

2. Assemble the gel-casting unit.

   a. Form the gel sandwich by assembling the spacers and two glass plates in the clamps.

   b. Align the bottom part of the spacers and two glass plates at the same level and then tighten the clamp.

   c. Place the gel sandwich onto the casting stand.

   d. Insert the Teflon sample application comb and mark the glass plate at a level ~1.0–1.5 cm below the bottom of the comb teeth.

      The Protean II xi electrophoresis unit (Bio-Rad) is used as an example to illustrate the assembling of the gel casting unit. For photographs of the assembled casting unit, see Figure 2.6a–f).

3. Pour the resolving gel.

   a. Using Table 2.2 (see introduction to this chapter) as a guide, decide on an acrylamide concentration for the gel.

   b. Prepare the appropriate resolving gel mixture using the recipes in Table 2.4. Make sure that the solution is well mixed before adding the TEMED.

   c. Use a 10-ml pipette to transfer the mixture to the glass-plate sandwich up to the marked level (marked in Step 2d) (see Figure 2.6).

   d. Carefully overlay the gel with an ~2-mm-deep layer of H_2O or H_2O-saturated *n*-butanol or isopropanol solution.

      This prevents air from reaching the gel, which inhibits polymerization of the acrylamide, and ensures that the gel surface is flat.

   e. After polymerization is complete (~30 minutes), pour off the overlaying H_2O and carefully remove any remaining liquid with filter paper without damaging the gel surface. If the gel is overlaid with *n*-butanol (or isopropanol), drain the overlay liquid, and then wash the gel surface with H_2O.

      Polymerization of the gel is evidenced by a clear refractive index change that can be seen between the gel and the overlay liquid.

4. Pour the stacking gel.

   a. Select an acrylamide concentration for the stacking gel and make the appropriate mixture, using the recipes in Table 2.5. Make sure that the solution is well mixed.

   b. Carefully overlay the resolving gel with the stacking gel solution until the height of the stacking gel is ~2.0–3.0 cm.

TABLE 2.4. Resolving gel recipes

For making 100 ml of resolving gel mixture[a]

%T	Acrylamide (30.0%) (ml)	Resolving gel buffer (ml)	H_2O (ml)	10% SDS (ml)	10% Ammonium persulfate[b] (ml)	TEMED[c] (ml)
8.0	26.67	25.0	46.53	1.0	0.75	0.05
9.0	30.00	25.0	43.20	1.0	0.75	0.05
10.0	33.33	25.0	39.87	1.0	0.75	0.05
11.0	36.67	25.0	36.53	1.0	0.75	0.05
12.0	40.00	25.0	33.20	1.0	0.75	0.05
13.0	43.33	25.0	29.87	1.0	0.75	0.05
14.0	46.67	25.0	26.53	1.0	0.75	0.05
15.0	50.00	25.0	23.20	1.0	0.75	0.05
16.0	53.33	25.0	19.87	1.0	0.75	0.05
17.0	56.67	25.0	16.53	1.0	0.75	0.05
18.0	60.00	25.0	13.20	1.0	0.75	0.05
19.0	63.33	25.0	9.87	1.0	0.75	0.05
20.0	66.67	25.0	6.53	1.0	0.75	0.05

[a]Enough for four 1.0-mm thick Bio-Rad Protean II xi gels (160 x 150 x 1 mm^3).
[b]The 10% ammonium persulfate solution must be made fresh.
[c]Add TEMED just before pouring the gel.

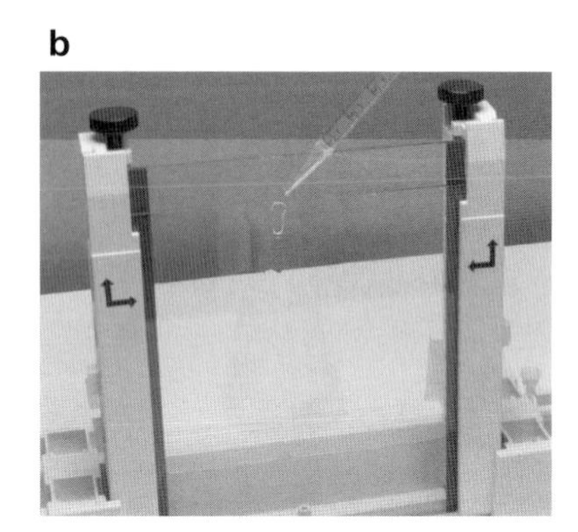

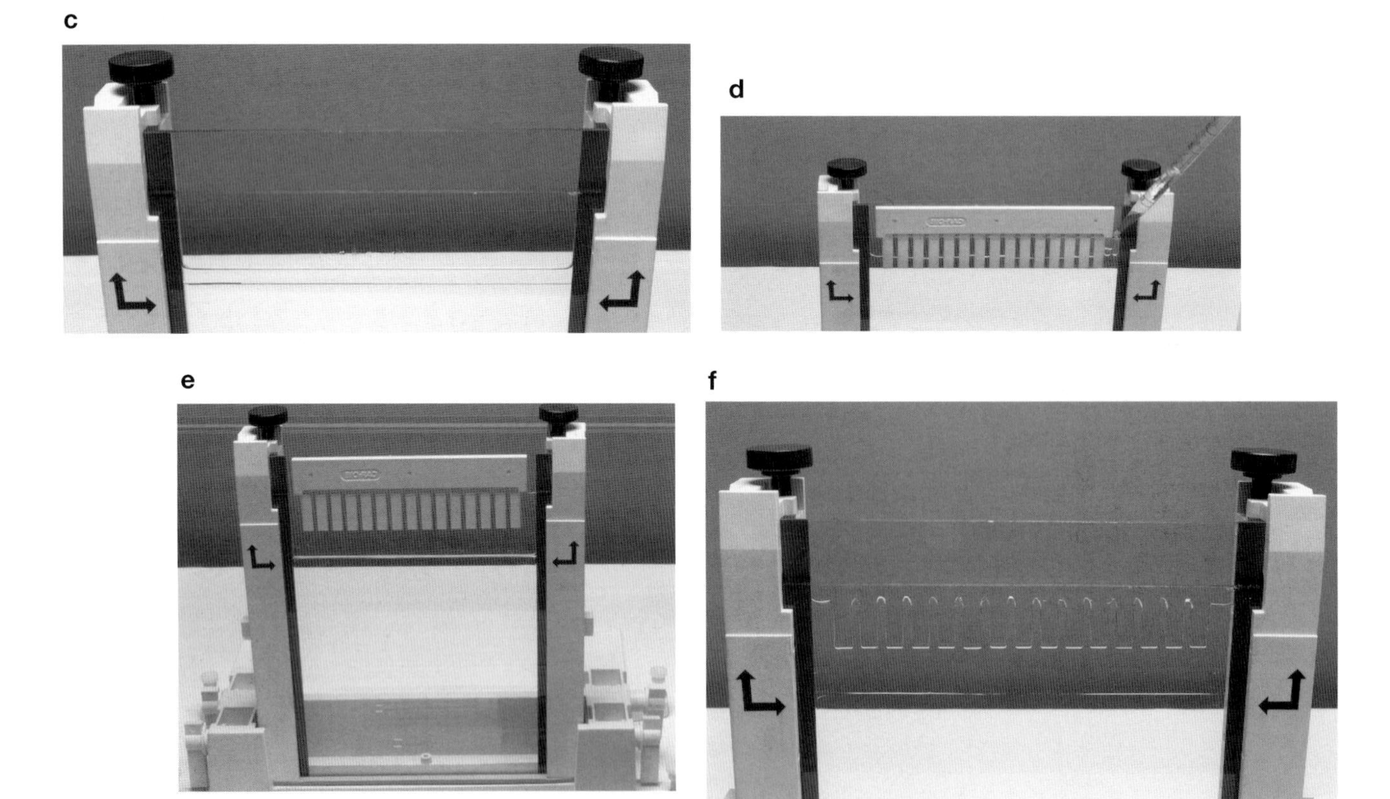

FIGURE 2.6. Preparation of a vertical SDS-polyacrylamide gel for the separation of proteins.

TABLE 2.5. Stacking gel mixture

For making 40 ml of stacking gel mixture[a]						
%T	Acrylamide 30.0% stock (ml)	Stacking gel buffer (ml)	H_2O (ml)	10% SDS stock (ml)	10% Ammonium persulfate[b] (ml)	TEMED[c] (ml)
3.0	4.00	10.0	25.36	0.4	0.2	0.04
4.0	5.36	10.0	24.00	0.4	0.2	0.04
5.0	6.66	10.0	22.70	0.4	0.2	0.04

[a]Sufficient for four 1.0-mm thick Bio-Rad Protean II xi gels (160 x 150 x 1 mm^3).
[b]The 10% ammonium persulfate solution must be made fresh.
[c]Add TEMED just before pouring the gel.

c. Insert the Teflon comb into this solution, leaving 1.0–1.5 cm between the top of the resolving gel and the bottom of the comb. Make sure that no air bubbles are trapped beneath the teeth of the comb.

Insert the comb into the stacking gel at an angle to reduce the chance of trapping air bubbles under the comb's teeth. Trapped bubbles can be released by gently tapping on the glass plate near the trapped bubbles.

d. Allow the stacking gel mixture to polymerize for ~2 hours. Refractive index changes around the comb indicate that the gel has set. It is useful at this stage to mark the positions of the bottoms of the sample wells on the glass plates with a marker pen.

5. Carefully remove the sample comb from the stacking gel, and assemble the cassette in the electrophoresis apparatus according to the manufacturer's instructions.

6. Fill the top reservoir with running buffer ensuring that the buffer fully fills the sample loading wells, and look for any leaks from the top tank. If there are no leaks, fill the bottom tank with running buffer, then tilt the apparatus to dispel any bubbles caught under the gel.

Alternatively, remove trapped air bubbles by squirting running buffer across the bottom edge of the gel through a long bent needle or hooked Pasteur pipette. The gel is now ready to receive protein samples.

Preparation of Samples

Carry out either Step 7 or Step 8.

7. Prepare protein solutions:

a. Mix the protein solution with 2x SDS-PAGE sample buffer in a 1:1 ratio. Although under ideal conditions the binding ratio for SDS and polypeptide is 1.4 g of SDS per gram of polypeptide, to ensure enough SDS is present, the concentration of protein in the final solution should not be higher than 10 μg/μl.

EXPERIMENTAL TIP: To load the entire protein sample onto the gel, bear in mind that the sample volume (i.e., protein solution and sample buffer) should not exceed the volume of the wells.

b. Heat the samples in a heat block or water bath for 2 minutes at 95°C to denature the proteins and ensure the maximum amount of SDS binding to the proteins. Allow the samples to cool to room temperature. Remove insoluble materials by centrifugation.

8. Prepare total cellular protein samples:

   a. Loosen the prepared cell pellet by vortexing the pellet briefly.

   b. Add 2x SDS-PAGE sample buffer directly to the cell pellet and vortex.

      The resultant cellular protein lysate is highly viscous. The amount of sample buffer to add varies depending on the cell line under study. However, a good amount to use is 100 μl of 2x sample buffer per 1 x 10^5 cells. The optimal amount must be determined empirically.

      EXPERIMENTAL TIP: To load the entire protein sample onto the gel, bear in mind that the sample volume (i.e., protein pellet and sample buffer) should not exceed the volume of the wells.

   c. Centrifuge the cellular protein lysate at 100,000*g* for 20 minutes and collect the supernatant.

      Protein concentration for a certain number of cells differs slightly for different cell lines. If the protein concentration exceeds 10 μg/μl, add SDS powder. Addition of SDS to 5% final concentration will not interfere with the electrophoretic separation.

   d. Heat the samples in a heat block or water bath for 2 minutes at 95°C to denature the proteins and ensure the maximum amount of SDS binding to the proteins. Allow the samples to cool to room temperature. Remove insoluble materials by centrifugation.

Running a Discontinuous Slab Gel

9. Use a pipette and gel-loading pipette tips to load the samples into the sample well.

10. Connect the power supply to the electrophoresis apparatus with the anode (+) linked with the bottom reservoir and the cathode (–) connected to the upper reservoir.

11. Pass a constant voltage of 200 V at 8°C (with cooling system connected), or 90 V at room temperature, through the gel until the bromophenol blue dye front reaches the bottom of the gel. This will take about 6–8 hours (at constant 200 V) and ~16–18 hours at constant 90 V.

12. Turn off the power supply and disconnect the electrodes. Remove the gel plates from the apparatus and carefully remove a spacer. Use the spacer to gently pry the gel plates apart, leaving the gel stuck to one plate.

13. Visualize proteins using an appropriately sensitive staining method.

    For commonly used protocols that are compatible with mass-spectrometry-based protein identification procedures, see Protocols 2 (Coomassie Blue) and 7 (silver stain).

Alternative Protocol: Pouring Linear Gradient Gels with the Bio-Rad 385 Gradient Former

The preparation of fixed-concentration SDS-PAGE gels is described in Protocol 1. However, the use of linear gradient gels, which are polyacrylamide gels having a gradient of increasing acrylamide concentration (and linearly decreasing pore size), can have advantages over fixed-concentration gels. First, a much greater range of protein M_r values can be separated on a linear gradient gel than on a fixed-concentration gel. Second, there is a greater likelihood of resolving proteins with very similar M_r values on gradient gels than on fixed-concentration gels. The most common gradient gel contains 4–20% acrylamide; however, the range of acrylamide concentration, should be chosen on the basis of the size of the proteins being separated (see Tables 2.2 and 2.4).

Additional Materials

CAUTION: See Appendix 3 for appropriate handling of materials marked with <!>.

Gel casting chamber with <100 ml capacity
Gradient former (Model 385, Bio-Rad)
Isobutanol <!>
Magnetic stirrer
Peristaltic pump capable of delivering 10 ml/min
Stir bar (1-inch magnetic flea)
30.8%T (2.6%C) Acrylamide stock solution <!>

Add 60 g of acrylamide and 1.6 g of bisacrylamide <!> to 200 ml of H_2O. Filter the stock solution through Whatman filter paper and store at 4°C. Prepare fresh stock acrylamide solution every few weeks.

Method

1. Calculate the volume of solution required for each chamber using either method a or b.
   a. Multiply spacer thickness by length, width, and number of gels to be poured and divide by two.
   b. Set up the casting chamber ready for pouring gels and fill with H_2O. Measure the amount required to fill the cassettes and divide by two. Be sure to dry the apparatus before pouring gels.
2. Deliver the gel solution to the casting chamber either from the top or bottom.

   For pouring the gel from the top:
   a. Set up the gel sandwich as in Figure 2.6.
   b. Cut the tubing to minimize the distance between the stopcock opening and the pump and the pump to the gel sandwich. Attach the tubing to the gradient former and the peristaltic pump. Tape the end of the tubing to the center of the glass plate, above the final level of the gel.
   c. Place the stirring bar into the mixing chamber.
   d. Make up appropriate light and heavy acrylamide solutions for suitable sample separation, without the polymerization initiator solutions (see Tables 2.2 and 2.4). De-gas these if desired.
   e. Add the polymerization initiators to the light solution and quickly mix and pour it into the reservoir chamber. Make sure that the valve stem is closed.
   f. Add the polymerization initiators to the heavy solution, mix, and add to the mixing chamber.
   g. Start the magnetic stirring motor and adjust the speed so that the bottom of the vortex caused by the stirring motion is of equal height to the light solution in the reservoir chamber.
   h. Open the valve stem and stopcock and turn on the peristaltic pump with the speed set to pump the entire gradient solution into the casting chamber in <10 minutes (e.g., 10 ml/minute).
   i. Run the pump until all of the solution is pumped into the sandwich.
   j. Remove the tubing from the gel casting chamber and place it in a waste receptacle.
   k. Overlay the gel with H_2O-saturated *n*-butanol.
   l. Fill each chamber with H_2O and turn on the pump at maximum speed to flush out the system.

   For pouring the gel from the bottom:
   a. If using a gel casting chamber with a gasket at the bottom, attach the tube from the pump to this gasket.
   b. Follow the directions above up to Step h, except place the light solution in the mixing chamber and the heavy solution in the reservoir.
   c. When all of the acrylamide solution has passed into the casting chamber, stop the pump, being careful not to allow air bubbles to enter the casting chamber.
   d. Complete Steps j–l above.

TABLE 2.6. Problems that may arise during the preparation and use of SDS gels

Problem	Cause	Remedy
Decreased rate of gel polymerization	Oxygen is present.	Degas the solutions.
	Stock solutions (especially acrylamide and persulfate) are aged.	Renew the stock solution.
Formation of a sticky top to the gel	Penetration of the gel by *n*-butanol.	Overlay the gel solution with *n*-butanol without mixing them. Do not leave *n*-butanol standing on a polymerized gel.
Poor sample wells Distorted or broken wells	Stacking gel resists the removal of the comb.	Remove the comb carefully or use a gel of lower %T.
Wells with a loose webbing of polyacrylamide	Comb fits loosely.	Replace the comb with a tighter-fitting one.
Unsatisfactory staining Weak staining	Dye is bound inefficiently	Use a more concentrated dye solution, a longer staining time, or a more sensitive stain. The stain solution should contain organic solvent (e.g., methanol), which strips the SDS from the protein to which the dye may then bind.
Uneven staining	Dye penetration or destaining is uneven.	Agitate the gel during staining and destaining. Increase the staining and destaining times.
Stained bands decolorized	Dye has been removed from the protein.	Restain the gel. Reduce the destaining time or use a dye that stains proteins indelibly, e.g., Procion navy MXRB.
Gel marked nonspecifically by the dye	Solid dye is present in the staining solution.	Ensure full dissolution of the dye, or filter the solution before using it.
Contaminants apparent	Apparatus and/or stock solutions are contaminated.	Clean or renew them as required.
	Nonproteinaceous material in the sample (e.g., nucleic acid) has been stained.	Try another stain that will not stain the contaminants.
	Samples have cross-contaminated each other because of overloading or sideways seepage between the gel layers.	Do not overfill sample wells. Ensure good adherence of the gel layers to each other by thorough washing of the polymerized gel before application of subsequent layers.

Protein bands not sufficiently resolved	Insufficient electrophoresis.	Prolong the run.
	Separating gel's pore size is incorrect.	Alter the %T and/or %C of the separating gel.
	Amounts loaded differ greatly.	Keep the loadings roughly similar in size each time.
Small changes in electrophoretic mobilities of standard proteins from time to time	Constituents of the gel vary in quality from batch to batch or with age.	Use one batch of a chemical for as long as possible. Replace aged stock solutions and reagents.
Distortion of bands: Bands smeared or streaked	Proteins in the sample are insoluble or remain aggregated in the sample solvent.	Use fresh sample solvent and/or extra SDS and reducing agent in it (especially for concentrated sample solutions).
	Insoluble matter or a bubble in the gel has interfered with protein band migration.	Filter the stock solutions before use and remove any bubbles from the gel mixtures.
Protein migration uneven (bands bent)	Pore size of the gel is inconsistent.	Ensure that the gel solutions are well mixed and that polymerization is not too rapid (to slow it down, reduce the amount of persulfate added).
	Part of the gel has been insulated.	Remove any bubbles adhering to the gel before electrophoresis.
	Electrical leakage.	Ensure that side spacers are in place.
	Cooling of the gel is uneven (allowing one part of the gel to run more quickly than another).	Improve the cooling of the gel, or reduce the heating by reducing the voltage or ionic strength of buffers.
	Band and/or its neighbors are overloaded.	Repeat the electrophoresis, but with smaller loadings. Leave gaps (i.e., unloaded sample wells) between neighboring, heavily loaded samples. If necessary, modify the gel and/or electrophoresis parameters (e.g., see above Protein bands not sufficiently resolved) so that the relative mobilities of the protein bands change, and they do not interfere with each other.
Bands not of uniform thickness	Sample well was at the very end of the row of wells (the "end well effect").	Avoid using the end wells.
	The sample was loaded unevenly.	Check that the sample well bottoms are straight and horizontal (see above Poor Sample wells)

Adapted from Smith (1994).

Protein Visualization Procedures

THE MOST WIDELY USED METHODS FOR VISUALIZING PROTEINS in polyacrylamide gels and on electroblot membranes use labels that are fixed to the protein molecules after gel electrophoresis, but rely on noncovalent interactions. Although the contrast of these methods is theoretically lower than could be accomplished with covalent binding, the former approach is readily compensated by the availability of a much higher number of interaction sites (for a review, see Rabilloud 2000). Because noncovalent interactions are generally reversible, these staining methods are compatible with a variety of proteomic applications, including chemical amino- and carboxy-terminal sequencing methods (see Chapter 6) and mass spectrometry (see Chapter 8). Detection of proteins after gel electrophoresis is typically carried out directly in the polyacrylamide gel or on the electroblot membrane. The visualization labels commonly used for noncovalent binding to proteins can be categorized as follows:

- ***Organic dyes.*** Examples are Coomassie Brilliant Blue and Amido Black. Detection is by light absorption.
- ***Fluorescent probes.*** Examples are Sypro dyes and Nile red. Detection is by fluorescence.
- ***Metal ion binding.*** This category can be further divided into those that are visualized by differential salt binding (e.g., negative or reverse staining with zinc-imidazole) and those detected by metal ion reduction, such as silver stains.

DETECTION WITH ORGANIC DYES

Conventional Coomassie Brilliant Blue R250 staining of polyacrylamide gels in an aqueous solution containing methanol and acetic acid remains the most popular approach for detecting proteins separated by acrylamide gels (Protocol 2). Due to the high extinction coefficient and high affinity of Coomassie Blue dyes for proteins, the detection limits are ~30–300 ng. A number of improvements in the Coomassie Blue staining procedure have led to an increased sensitivity of detection of proteins in polyacrylamide gels. Perhaps the most significant improvements are the rapid Coomassie Blue staining method described in Protocol 3 and the development of colloidal Coomassie Blue staining for background-free detection of proteins in polyacrylamide gels (Neuhoff et al. 1988, 1990).

FLUORESCENT PROBES

Organic fluorophore probes are virtually nonfluorescent in aqueous solutions, but they become highly fluorescent in nonpolar solvents or upon association with SDS-protein complexes, such as those encountered in SDS-PAGE. Typical fluorophore probes of this type, which are gaining popularity in integrated proteomics platforms for global analysis of protein expression (Patton 2000), include the SYPRO probes SYPRO Ruby (Protocol 5) and SYPRO Orange (Protocol 6). Both of these dyes are commercially available electrophoresis stains that can detect proteins in SDS-PAGE gels using a simple, one-step staining procedure. The limits of sensitivity of the SYPRO dyes are in the 2–10 ng range, rivaling the sensitivity of standard silver staining methods (Patton 2001).

DETECTION BY METAL ION BINDING

Metal ions permit protein visualization either by differential salt binding or by metal ion reduction. The zinc-imidazole reverse staining method is based on the slower precipitation of protein-salt complexes as compared to the precipitation of zinc-dodecyl salts free within the gel. The result is translucent protein bands within a darker background (see Chapter 7, Protocol 10). The most common protein detection method that utilizes metal ion reduction is silver staining. The sensitivity of silver staining with current protocols (see Chapter 4, Protocol 11 and Chapter 7, Protocol 11) is in the low-nanogram range, which is ~50–100 times more sensitive than conventional Coomassie Brilliant Blue staining, 10 times more sensitive than colloidal Coomassie Brilliant Blue staining, and approximately twice as sensitive as zinc-imidazole staining (Rabilloud 2000). It should be emphasized that conventional silver staining procedures are not compatible with proteomic identification methods involving spectrometry procedures, owing to protein cross-linking by formaldehyde during the protein fixation process. Although the silver staining method can be modified to omit aldehydes that preclude further analysis (see Protocol 7 and Chapter 4, Protocol 12), the modifications significantly reduce the sensitivity compared to the conventional silver staining methods. For additional information on the visualization of gel-separated proteins, see Chapters 4 and 7.

PROTOCOL 2

Conventional Coomassie Blue Staining

COOMASSIE BRILLANT BLUE R250 (CBR-250) AND SILVER STAINING are the most widely used methods for the routine visualization of proteins separated by SDS-PAGE. Proteins stained with CBR-250 are compatible with subsequent amino- and carboxy-terminal sequence determination (Chapter 6) and mass spectrometric analysis (Chapter 8). CBR-250 is an organic dye that complexes with basic amino acids, such as arginine, lysine, and histidine, as well as tyrosine (de Moreno et al. 1986; for review of CBR-250 staining and other staining procedures, see Rabilloud [2000] and Patton [2000, 2001]). Conventional CBR-250 staining is capable of detecting as little as 30–100 ng of protein, but sensitivity can be improved by performing the staining and destaining at elevated temperatures (Protocol 3) or by using colloidal Coomassie Blue staining (Chapter 4, Protocol 10). For additional information about Coomassie Blue, see the panel on COOMASSIE BRILLIANT BLUE in the introduction to this chapter.

MATERIALS

CAUTION: See Appendix 3 for appropriate handling of materials marked with <!>.

Reagents

Acetic acid (5%) <!>
See Step 5.

Coomassie Brilliant Blue (CBR-250) stain
To 400 ml of H_2O, add 500 ml of methanol <!>, 100 ml of acetic acid, and 1 g of CBR-250. Stir this stock for 4 hours to overnight to dissolve the Coomassie powder <!>, and then filter with Whatman filter paper.

Destain
To 810 ml of H_2O, add 120 ml of methanol and 70 ml of acetic acid.

Equipment

Plastic container (standard Tupperware sized to fit gel)

Biological Sample

Protein samples separated on one- and two-dimensional gels by electrophoresis

METHOD

EXPERIMENTAL TIP: Reuse of the reagents is not recommended because evaporation and contamination can reduce their effectiveness.

1. After electrophoresis is complete, place the gel in the plastic container containing enough CBR-250 stain to cover the gel. For example, use 100 ml of CBR-250 stain for a typical minigel (8 x 10 cm or 8 x 8 cm).

2. Agitate the gel for 5–20 minutes at room temperature.

3. Discard the used CBR-250 stain.

4. Add an appropriate amount of destain and agitate with a piece of Kimwipe in the solution to absorb excess dye until a suitable background is achieved.

5. Store the gel in 5% acetic acid or dry in the air. To air-dry the gel, place it between two wetted sheets of cellophane, pull the cellophane taut, and allow the gel to dry undisturbed. Air-dried gels can be stored indefinitely.

PROTOCOL 3

Rapid Coomassie Blue Staining

THIS METHOD, WHICH IS A MODIFIED VERSION OF THE CONVENTIONAL COOMASSIE protocol (Protocol 2), speeds up the destaining process for faster results with increased sensitivity and is compatible with mass-spectrometry-based methods for identifying proteins. This protocol is adapted from Wong et al. (2000).

MATERIALS

CAUTION: See Appendix 3 for appropriate handling of materials marked with <!>.

Reagents

Reagent A
Dissolve 0.5 g of CBR-250 <!> in 250 ml of isopropanol <!>, 100 ml of acetic acid <!>, and 650 ml of H_2O.

Reagent B
Dissolve 0.05 g of CBR-250 in 100 ml of isopropanol, 100 ml of acetic acid, and 800 ml of H_2O.

Reagent C
Dissolve 0.02 g of CBR-250 in 100 ml of acetic acid and 900 ml of H_2O.

Reagent D
Mix 100 ml of acetic acid with 900 ml of H_2O.

Equipment

Microwave oven with 1000 W output

Microwave-safe plastic container (standard Tupperware with a hole punched in the lid to vent gases)

Biological Sample

Protein samples separated on a one-dimensional gel by electrophoresis

METHOD

EXPERIMENTAL TIP: Reuse of the reagents is not recommended because evaporation and contamination can reduce their effectiveness.

Staining the Gel

1. After electrophoresis is complete, place the gel in a microwave-safe plastic container containing enough Reagent A to cover the gel. For example, use 100 ml of Reagent A for a typical minigel (8 x 10 cm or 8 x 8 cm).
2. Heat the gel in a microwave oven on full power until the solution is boiling (~2 minutes).
3. Cool the gel for 5 minutes at room temperature with gentle shaking.
4. Discard the used Reagent A and rinse the gel briefly with H_2O. At this stage, bands containing more than 100 ng of proteins can be visualized despite the blue background.
5. Add 100 ml of Reagent B and heat the gel in a microwave oven on full power until the solution is boiling (~80 seconds).
6. Discard the hot Reagent B and rinse the gel with H_2O. After this step, bands containing more than 50 ng of proteins can be visualized.
7. Add 100 ml of Reagent C and heat the gel in a microwave oven on full power until the solution reaches the boiling point (~80 seconds).
8. Discard the hot Reagent C and rinse the gel with H_2O. After this step, bands containing more than 25 ng of proteins can be visualized.

Destaining the Gel

9. Add 100 ml of Reagent D and heat the gel in a microwave oven on full power until the solution reaches the boiling point (~80 seconds).
10. Place a piece of Kimwipe in the solution to absorb excess dye.
11. Cool the gel for 5 minutes at room temperature. At this stage, bands containing 5 ng of proteins or more can be visualized.
12. Repeat Steps 9–11 twice more, or shake the gel in Reagent D for 15 minutes or more at room temperature. At this point, some protein bands containing as little as 2.5 ng of proteins can be visualized.

PROTOCOL 4

InstaStain Blue Gel Paper Stain

THIS COMMERCIALLY AVAILABLE PROCEDURE IS YET ANOTHER MODIFICATION of the standard Coomassie protocol. The method uses Coomassie dye dried onto paper to apply the stain to the gel. It obviates the problems associated with handling liquid stains and can reduce staining background. However, in our experience, the paper is prone to sticking to the gel if the system is overheated. This staining method is adapted from the instructions included in the Pierce InstaStain Blue Gel Stain Paper and is compatible with mass-spectrometry-based protein identification procedures.

MATERIALS

CAUTION: See Appendix 3 for appropriate handling of materials marked with <!>.

Reagents

Reagent A
Mix 400 ml of methanol <!> with 500 ml of H_2O, and add 100 ml of acetic acid <!>.

Equipment

Filter paper (3) cut to gel size (e.g., 3MM paper, Whatman)
Glass plates (2)
Glass rod
InstaStain Blue Gel Stain Paper (Pierce) cut to the size of the gel
Microwave oven with 1000 W output
Plastic container (standard Tupperware sized to fit gel)

Biological Sample

Protein samples separated on a one-dimensional gel by electrophoresis

METHOD

1. After electrophoresis is complete, place the gel in a plastic container containing enough H_2O to cover the gel. For example, use 100 ml of H_2O for a typical minigel (8 x 10 cm or 8 x 8 cm). Gently shake the gel for 5 minutes.

2. Repeat the H_2O wash two more times (the total wash time is 15 minutes).
3. Remove the H_2O and add enough Reagent A to cover the gel. Shake the gel for 10 minutes at room temperature.
4. Remove the gel from Reagent A and place it on a glass plate. Make sure that it remains moist.
5. Place the InstaStain Blue Gel Stain Paper onto the top of the gel, with the blue side of the paper facing down. Roll out any air bubbles with a glass rod.
6. Moisten the three pieces of filter paper with Reagent A and place them on top of the InstaStain Blue Gel Stain Paper. Roll out any air bubbles with a glass rod.
7. Place the second glass plate on top of the filter papers.
8. Place the gel/paper/glass sandwich into the microwave oven and heat on high until it is lukewarm (~8 seconds). Make sure that the gel is not cooked, as it will stick to the InstaStain Blue Gel Stain Paper.
9. Remove the top glass plate, the filter papers, and the InstaStain Blue Gel Stain Paper, and place the gel into a plastic container in enough Reagent A to cover the gel.

   EXPERIMENTAL TIP: If the gel sticks to the Stain Paper, place the gel and Stain Paper into Reagent A and shake until the paper can be peeled off the gel, and then place the gel into fresh Reagent A.

10. Shake the gel for 1 hour at room temperature.
11. Discard Reagent A and wash the gel in H_2O for at least 1 hour at room temperature, or until the background is as desired. This should allow visualization of bands containing ~20 ng of protein.

    The gel may be stored overnight in H_2O without loss of sensitivity or band intensity.

PROTOCOL 5

SYPRO Ruby Fluorescent Staining

BY FAR SYPRO RUBY IS PERHAPS THE MOST SENSITIVE FLUORESCENT staining technique in common use for detecting polyacrylamide-gel-resolved proteins (1–2 ng protein/band). Unfortunately, at this low level of detection, stained protein bands are invisible to the naked eye, and a fluorescence scanner is required for their detection (the excitation maximum and emission maxima for SYPRO Ruby are 280 nm and 450/610 nm, respectively). Bands containing ~50 ng of protein can be seen with the naked eye with the aid of a UV transilluminator. This protocol has been adapted from the SYPRO Ruby staining protocol provided by Genomic Solutions. The method is also mass-spectrometer-compatible. For a review of fluorescent detection technologies, including imaging devices and detection instrumentation, see Patton (2000).

MATERIALS

CAUTION: See Appendix 3 for appropriate handling of materials marked with <!>.

Reagents

Glycerol (2% v/v)
  Mix 20 ml of glycerol into 980 ml of H_2O.

Reagent A
  Mix 400 ml of methanol <!> with 500 ml of H_2O, and add 100 ml of acetic acid. <!>

Reagent B
  Mix 100 ml of methanol with 840 ml of H_2O, and add 60 ml of acetic acid.

SYPRO Ruby protein gel stain <!>

Equipment

Aluminum foil

Fluorescence scanner (Typhoon 8600, Molecular Dynamics or equivalent)

Plastic container (standard Tupperware sized to fit gel)

Biological Sample

Protein samples separated on a one-dimensional gel by electrophoresis

METHOD

EXPERIMENTAL TIP: Reuse of the reagents is not recommended because evaporation and contamination can reduce their effectiveness.

IMPORTANT: At all stages of this protocol, when handling the gel, use powder-free gloves because powder from the gloves fluoresces.

1. After electrophoresis is complete, place the gel in a plastic container containing enough Reagent A to cover the gel. For example, use 100 ml of Reagent A for a typical minigel (8 x 10 cm or 8 x 8 cm).
2. Shake the gel for 30 minutes at room temperature.
3. Discard the used Reagent A, and add enough SYPRO Ruby stain to fully cover the gel, cover it with aluminum foil, and shake the gel for between 90 minutes and overnight at room temperature.

   New SYPRO Ruby stain requires ~2 hours staining to detect 1-ng bands on a fluorescence scanner (Molecular Dynamics Typhoon 8600 or equivalent). Completing Step 3 with used stain for 2 hours reduces the detection sensitivity to bands containing ~8 ng of protein.
4. Discard the SYPRO Ruby stain (or store for later use, see note to Step 3).
5. Add an appropriate amount of Reagent B to the gel and shake it for 30–60 minutes at room temperature.
6. Use a fluorescence scanner set at 532 nm to scan the gel, or store the gel covered in an appropriate amount of 2% glycerol.

   The gel may be stored in 2% glycerol for up to 6 months without any fading of the fluorescent signal.
7. For longer-term storage, preserve the gel by drying using a gel dryer or dry in the air. To air-dry the gel, place it between two wetted sheets of cellophane, pull the cellophane taut, and allow the gel to dry undisturbed. Air-dried gels can be stored indefinitely.

PROTOCOL 6

SYPRO Orange Fluorescent Staining

THIS METHOD USES A FLUORESCENT DYE that is very sensitive to protein (4–10 ng protein/band). SYPRO Orange cannot be visualized with the naked eye and thus a fluorescence scanner is required for detection of protein bands (e.g., Molecular Dynamics Typhoon 8600 or equivalent; the excitation maximum [Ex] and emission maxima [Em] for SYPRO Orange is 300 nm and 470/570 nm, respectively). This same protocol can be used for SYPRO Red stain (Ex and Em = 300 nm and 550/630 nm). For a review of fluorescent detection technologies, including imaging devices and detection instrumentation, see Patton (2000). The method was adapted from the SYPRO Orange staining protocol provided by Amersham Biosciences. This staining method is mass-spectrometer-compatible.

MATERIALS

CAUTION: See Appendix 3 for appropriate handling of materials marked with <!>.

Reagents

Reagent A (100-ml stock)

Mix 92.5 ml of H_2O with 7.5 ml of acetic acid <!>. Add 20 µl of SYPRO Orange gel stain concentrate <!>.

SYPRO Red fluorescent protein stain uses the same binding principle as SYPRO Orange, and hence this protocol can be used with SYPRO Red protein gel stain.

Reagent B (1000-ml stock)

Mix 925 ml of H_2O with 75 ml of acetic acid.

Equipment

Aluminum foil

Fluorescence scanner (Typhoon 8600, Molecular Dynamics or equivalent)

Plastic container (standard Tupperware sized to fit gel)

Biological Sample

Protein samples separated on a one-dimensional gel by electrophoresis

METHOD

EXPERIMENTAL TIP: Reuse of the reagents is not recommended because evaporation and contamination can reduce effectiveness.

IMPORTANT: At all stages of this protocol, when handling the gel, use powder-free gloves because powder from the gloves fluoresces.

1. After electrophoresis is complete, place the gel in a plastic container containing enough Reagent A to cover the gel. For example, use 50–100 ml of Reagent A for a typical minigel (8 x 10 cm or 8 x 8 cm).

2. Cover with aluminum foil to protect the gel from bright light.

3. Shake the gel for 60 minutes at room temperature.

4. Remove Reagent A and set it aside for use in Step 5. Add an appropriate amount of Reagent B to the gel and shake it for 1 minute at room temperature.

   Longer rinsing times can lower the background, although it also reduces sensitivity.

5. Use a fluorescence scanner to scan the gel (SYPRO Orange has an excitation maximum of 472 nm and an emission maximum of 570 nm). After scanning the gel, return it to Reagent A.

6. The gel may be stored in Reagent A for several weeks, although signal intensity decreases with time. Alternatively, air-drying the gel between two sheets of cellophane, although it slightly reduces sensitivity, may preserve the gel. To air-dry the gel, place it between two wetted sheets of cellophane, pull the cellophane taut, and allow the gel to dry undisturbed. Air-dried gels can be stored indefinitely.

PROTOCOL 7

Mass-spectrometry-compatible Silver Staining

THIS ALTERNATIVE TO THE CONVENTIONAL STANDARD SILVER STAINING method uses the same principle, but it omits some reagents to permit mass-spectrometry-based protein identification. It is more sensitive than conventional Coomassie Blue staining procedures, allowing visualization of as little as 1–2 ng of protein. For a detailed review of silver staining procedures from which this protocol has been adapted, see Sinha et al. (2001).

MATERIALS

CAUTION: See Appendix 3 for appropriate handling of materials marked with <!>.

Reagents

Reagent A
Mix 600 ml of ethanol <!> with 200 ml of acetic acid <!> and 1200 ml of H_2O.

Reagent B
Dissolve 6 g of potassium tetrathionate and 98 g of potassium acetate in 600 ml of ethanol and 1400 ml of H_2O.

Reagent C (0.2% $AgNO_3$)
Dissolve 2 g of silver nitrate <!> in 1000 ml of H_2O.

Reagent D (Developer)
Combine 60 g of potassium carbonate <!>, 600 µl of formaldehyde <!>, and 250 µl of 10% sodium thiosulfate pentahydrate. Adjust the volume to 2000 ml with H_2O.

Reagent E (Stop solution)
Combine 80 g of Tris and 40 ml of acetic acid. Adjust the volume to 2000 ml with H_2O.

Equipment

Plastic container (standard Tupperware sized to fit gel)

Biological Sample

Protein samples separated on one- and two-dimensional gels by electrophoresis

METHOD

EXPERIMENTAL TIP: Reuse of reagents is not recommended because evaporation and contamination can reduce their effectiveness.

1. After electrophoresis is complete, place the gel in a plastic container containing enough Reagent A to cover the gel. For example, use 100 ml of Reagent A for a typical minigel (8 x 10 cm or 8 x 8 cm).
2. Shake the gel for 30 minutes at room temperature, and then discard the used Reagent A.
3. Repeat Steps 1 and 2 three more times.
4. Add an appropriate amount of Reagent B to the gel, and shake it for 45 minutes at room temperature.
5. Discard the Reagent B.
6. Wash the gel with H_2O for 10 minutes at room temperature, and then discard the H_2O.
7. Repeat Step 6 five more times (the total wash time is 60 minutes).
8. Add an appropriate amount (e.g., 200–250 ml for a 20 x 20-cm gel) of Reagent C to the gel and incubate it with continuous shaking for 1–2 hours at room temperature.
9. Discard Reagent C in a suitable waste container and quickly rinse the gel with H_2O for 5–15 seconds.
10. Add an appropriate amount of Reagent D to the gel. Continuously shake the gel at room temperature.
11. Observe the gel during development of the protein bands. Once an appropriate sensitivity is reached (i.e., clearly visible bands without a strong background), discard Reagent D.
12. Immediately cover the gel with Reagent E and shake it for 30 minutes at room temperature.
13. Discard Reagent E and store the gel for up to 6 months in H_2O. Alternatively, the gel can be air-dried and stored indefinitely.

PROTOCOL 8

Apparent Molecular Mass Determination

BECAUSE MIGRATION OF SDS-PROTEIN COMPLEXES in polyacrylamide gels is generally proportional to the apparent molecular mass (M_r) of the protein, SDS-PAGE is a widely used procedure for determining the M_r of a protein.

MATERIALS

Biological Molecules

Protein sample of interest
Protein molecular-weight standards

Additional Reagents and Equipment

Step 1 requires the reagents and equipment listed in Protocol 1.
Step 2 requires the reagents and equipment listed in one of the protein visualization protocols, Protocols 2–7.

METHOD

1. Run sample protein and protein molecular-weight standards on the same SDS-PAGE slab gel following the steps in Protocol 1.

   Running samples and protein standards on the same slab gel eliminates any variability due to acrylamide concentration and electrophoresis conditions. Although kits of protein standards are commercially available from several sources, detailed lists of protein standards and their M_r values can be found in Weber and Osborn (1969), Neville (1971), and Hames (1990). For careful M_r determinations using this method, it is advisable to run gels of more than one acrylamide concentration (Ferguson 1964; Neville 1971). For a detailed discussion on the effect of both polyacrylamide and cross-linker concentrations on protein migration using various commercially available calibration proteins for SDS-PAGE, see Makowski and Ramsby (1997).

2. After electrophoresis is complete, carry out one of the protein visualization protocols (Protocols 2–7).

3. (*Optional*) Dry the gels prior to measurement of R_F values.

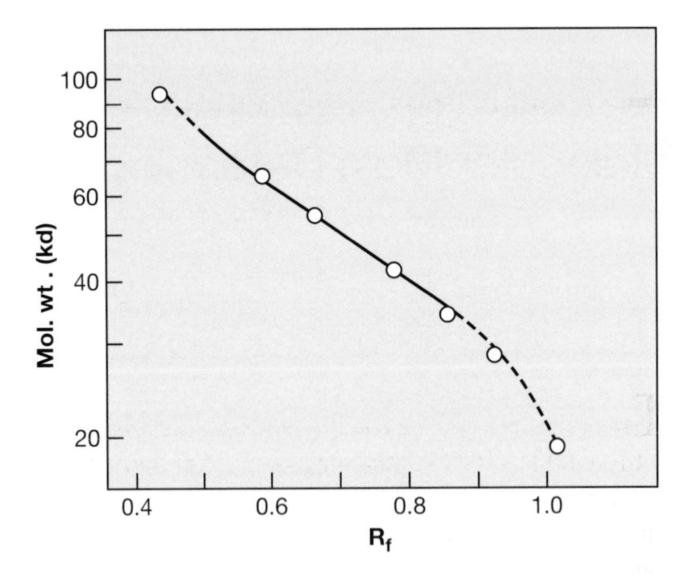

FIGURE 2.7. Semilogarithmic graph of molecular weight versus relative mobility (R_F). (Reprinted, with permission, from Bollag et al. 1996.)

4. Measure the distance of protein migration and of the tracking dye (which is typically bromophenol blue).
5. Calculate relative mobility (R_F) values

   R_F = distance of protein migration/distance of tracking dye migration

6. Plot the mobilities against the known M_r expressed on a semilogarithmic scale. The line covering the middle of the gel should be almost linear (see Figure 2.7).
7. Read the M_r of the unknown protein from the graph based on its R_F value.

   KEY POINT: Generate a new standard curve for each gel.

PROTOCOL 9

CTAB-PAGE

THIS PROTOCOL IS A VARIATION ON THE STANDARD SDS-PAGE METHOD described in Protocol 1. Although SDS-PAGE is the method of choice for most denaturing gel electrophoresis procedures, the anionic detergent SDS still presents some drawbacks. For example, SDS forms crystals at low temperatures and, in some cases, causes proteins to aggregate or precipitate. In addition, some proteins are not well-resolved in SDS gels or may migrate anomalously. In these situations, the use of a cationic detergent for PAGE offers an alternative approach. The discontinuous, cationic detergent PAGE method described here is based on the method of Atkins et al. (1992). This system uses the cationic detergent CTAB and includes a stacking gel based on the zwitterion arginine (used as a stacking agent) and tricine (*N*-tris[hydroxymethyl]-methylglycine) used as a counterion and buffer. Some proteins separated on the CTAB electrophoresis system retain their native enzymatic activity, provided the samples are prepared without boiling and without the addition of a reducing agent.

MATERIALS

CAUTION: See Appendix 3 for appropriate handling of materials marked with <!>.

IMPORTANT: All reagents should be electrophoresis grade or better.

Reagents

Acrylamide stock solution (40%T, 2.67%C) <!>

Mix 38.96 g of acrylamide and 1.04 g of bisacrylamide <!> to 100 ml of H_2O. Filter the stock solution using Whatman filter paper and store at 4°C. Prepare fresh stock acrylamide solution every few weeks.

CTAB sample buffer

- 10 mM tricine-NaOH (pH 8.8) <!>
- 1% CTAB
- 10% glycerol
- 10 µl/ml of a saturated aqueous solution of crystal violet <!>

2-Mercaptoethanol (2%) <!>

Optional, see Step 3.

Resolving gel mixture (6% acrylamide, 375 mM tricine-NaOH at pH 8)

Mix:

- 9.4 ml of H_2O
- 2.4 ml of acrylamide:bisacrylamide (40%T, 2.67%C)
- 4 ml of 1.5 M tricine-NaOH (pH 8)
- 0.16 ml of 10% ammonium persulfate <!>
- 16 µl of TEMED <!>

Add the TEMED just before pouring the gel. This makes enough gel mixture to pour two minigels, 1.5-mm thick.

Stacking gel mixture (0.7% agarose, 0.1% CTAB, 125 mM tricine-NaOH at pH 10)

Add 35 mg of agarose to 1.25 ml of 0.5 M tricine-NaOH at pH 10 and 50 µl of 10% CTAB. Add H_2O to 5 ml. Melt the agarose by boiling or microwaving the mixture before pouring the gel. This makes enough gel mixture to pour stacking gels for two minigels, 1.5-mm thick.

5x Tricine running buffer

Dissolve:

22.4 g of tricine
5 g of CTAB
75 ml of 1 M arginine (free base) in H_2O

Add H_2O to a final volume of 1 liter. The pH of the buffer should be ~8.2 and no pH adjustment is required. Store the running buffer at room temperature and dilute to 1x before use.

Biological Sample

Protein samples, either lyophilized or as pellets

Additional Reagents and Equipment

Boiling water bath

Optional, see Step 3.

Step 1 requires some of the reagents and all of the equipment listed in Protocol 1.

METHOD

1. Prepare the gels as described in Protocol 1, but substitute the resolving gel mixture and stacking gel mixture for those in Protocol 1.
2. Fill the upper and lower buffer reservoirs with 1x tricine running buffer.
3. Prepare the protein samples by dissolving them in 50 µl of CTAB sample buffer at room temperature, to a final concentration of 5 mg/ml. If retaining biological activity is not a concern, heat the samples for 3 minutes in a boiling water bath in the presence of 2% 2-mercaptoethanol.
4. Load the samples into the sample wells.
5. Start electrophoresis from the anode (+) to the cathode (–) at 100 V as the samples migrate through the stacking gel.
6. Increase the voltage to 150 V as the samples begin to migrate through the resolving gel.
7. Turn off the power supply and stop the electrophoresis when the dye front reaches the bottom of the gel.

PROTOCOL **10**

Acid-Urea Continuous PAGE

ACID-UREA CONTINUOUS PAGE IS IDEALLY SUITED for studying minor structure variants (of slightly different charge) or modified forms of the same protein (e.g., histones). Because SDS-PAGE systems are usually unable to separate two proteins of similar size but different charge from each other, these two electrophoresis systems complement each other. This protocol is based on the methods of Panyim and Chalkley (1969) and Rabilloud et al. (1996).

MATERIALS

CAUTION: See Appendix 3 for appropriate handling of materials marked with <!>.

Reagents

Acrylamide:bisacrylamide stock solution (50%T, 1.5%C) <!>
Mix 48.7 g of acrylamide and 1.3 g of bisacrylamide per 100 ml of H_2O. Filter the stock solution using Whatman filter paper and store at 4°C. Prepare fresh stock acrylamide solution every few weeks.

Diphenyliodonium chloride stock (1 mM) <!>
Prepare in H_2O. This solution can be stored for 1 week at 4°C in the dark.

Glacial acetic acid <!>

Methylene blue stock (2 mM) <!>
Prepare in H_2O. This solution can be stored for 1 year at room temperature in the dark.

Reservoir buffer (0.9 M acetic acid)
Prepare as 5.4% glacial acetic acid in H_2O.

Sample buffer (0.9 M acetic acid containing 2.5 M urea <!> and 0.1% pyronin Y <!>)

Sodium *p*-toluenesulfinate (20 mM) <!>
Prepare in H_2O. This solution can be stored for 1 week at 4°C in the dark.

Urea <!>

Equipment

Light source for photopolymerization
A daylight fluorescent lamp (15 W) positioned ~10 cm from the gel can be utilized to catalyze the methylene blue photopolymerization. Alternatively, two light boxes (each with 12-W neon tubes) also positioned 10 cm from the gel on both sides can be used for photopolymerization.

Biological Sample

Protein samples, either lyophilized or as pellets

Additional Equipment

Step 2 requires all of the equipment listed in Protocol 1, although a minigel apparatus will be sufficient.

METHOD

1. Prepare the gel mixture (15% acrylamide, 2.5 M urea, 0.9 M acetic acid at pH 2.7) by mixing:

acrylamide:bisacrylamide (50%T, 1.5%C)	3.0 ml
glacial acetic acid	0.54 ml
urea	1.5 g
2 mM methylene blue	0.15 ml
20 mM sodium toluenesulfinate	0.25 ml
1 mM diphenyliodonium chloride	0.2 ml
H_2O	to 10 ml

2. Pour the gel into the minigel mold.
3. Photopolymerize for 2 hours with constant illumination. Pre-electrophoresis is not necessary.
4. Fill the buffer reservoirs with the reservoir buffer.
5. Dissolve the protein samples in sample buffer, up to 35 mg of protein/ml.
6. Load the protein samples into the sample wells of the gel.
7. Start the electrophoresis from the anode (+) to the cathode (–). Follow the manufacturer's recommendations for the optimal voltage to use. For example, use 130 V constant voltage if the Bio-Rad Mini-Protean II Cell minigel is used.
8. Turn off the power supply and stop the electrophoresis when the dye front reaches the bottom of the gel.

PROTOCOL 11

Electrophoresis of Peptides (Tricine–SDS-PAGE)

THE MOST WIDELY USED PROTEOMICS METHOD for quantitatively analyzing protein mixtures is SDS-PAGE. It is particularly useful for monitoring protein purification, and because the method separates proteins on the basis of their size, it can also be utilized to determine the relative molecular mass (M_r) of proteins (Protocol 8). Parameters affecting the resolution of proteins or peptides separated by SDS-PAGE include the following:

- The ratio of acrylamide to the cross-linker, bisacrylamide.
- The percentage of acrylamide/bisacrylamide used to prepare the stacking and separating gels.
- The pH of the stacking and separating gel buffers and the components of these buffers.
- The method by which the sample is prepared.

Typical Laemmli gel systems (Laemmli 1970), which utilize glycine in the running buffer, are capable of resolving proteins in the molecular mass range of ~200,000 daltons down to ~3000 daltons. In this protocol, the tricine gel system of Schägger and von Jagow (1987) is described, in which tricine is substituted for glycine. This system permits the resolution of peptides as small as 500 daltons, making it suitable for SDS-PAGE peptide mapping and for preparing samples for amino- and carboxy-terminal sequence analysis (see Chapters 7 and 6, respectively).

MATERIALS

CAUTION: See Appendix 3 for appropriate handling of materials marked with <!>.

Reagents

Lower reservoir buffer (0.2 M Tris-HCl at pH 8.9)

Resolving gel mixture <!>

Prepare a 10% acrylamide mixture following the recipe given in Table 2.4 (Protocol 1).

Stacking gel mixture <!>

Prepare a 4% acrylamide stacking gel mixture following the recipe given in Table 2.5 (Protocol 1).

Upper reservoir buffer

0.1 M Tris
0.1 M tricine
0.1% (w/v) SDS <!>

The pH of the buffer will be ~8.3.

Biological Sample

Protein solution(s) or pellets of cellular proteins

Additional Reagents and Equipment

Steps 1 and 3 require some of the reagents and all of the equipment listed in Protocol 1. Step 5 requires the reagents listed in either Protocol 2 or 7.

METHOD

1. Cast the discontinuous gel following the steps in Protocol 1.
2. Fill the upper and the lower buffer reservoirs with the appropriate buffers.
3. Prepare the protein samples for loading onto the gel according to Protocol 1.
4. Start electrophoresis from the cathode (–) to the anode (+) at a constant current of 25 mA/gel.
5. Visualize the proteins by staining with Coomassie Blue or silver as described in Protocol 2 or 7.

PROTOCOL 12

Nondenaturing PAGE of Proteins

NONDENATURING GEL ELECTROPHORESIS, also referred to as native gel electrophoresis, is commonly run at pH 8.8. At this pH, most proteins are negatively charged and migrate toward the anode. All stock solutions used in this system are similar to those used for the Laemmli system (Laemmli 1970) but lack SDS (see Protocol 1). Native gels can also be run at low pH (for a low pH, discontinuous system, where proteins are stacked at pH 5 and resolved at pH 3.8, see Shi and Jackowski 1998).

MATERIALS

CAUTION: See Appendix 3 for appropriate handling of materials marked with <!>.

Reagents

Acrylamide stock (30%T, 2.6%C) <!>

Add 29.22 g of acrylamide and 0.78 g of bisacrylamide <!> to 100 ml of H_2O. Filter the stock solution through Whatman filter paper and store at 4°C. Prepare fresh stock acrylamide solution every few weeks.

Ammonium persulfate (10%) <!>

Electrophoresis buffer

Dissolve 3.0 g of Tris base and 14.4 g of glycine <!> in H_2O and adjust the volume to 1 liter. The final pH should be 8.3.

5x Sample buffer

Mix:

- 15.5 ml of 1 M Tris-HC1 (pH 6.8)
- 2.5 ml of a 1% solution of bromophenol blue <!>
- 7 ml of H_2O
- 25 ml of glycerol

Solid samples can be dissolved directly in 1x sample buffer. Samples already in solution should be diluted accordingly with 5x sample buffer to give a solution that is 1x sample buffer. Do not use protein solutions that are in a strong buffer which is not near to pH 6.8 as it is important that the sample is at the correct pH. For these samples, it will be necessary to dialyze against 1x sample buffer. The desired protein concentration will depend on the sensitivity of the protein detection (visualization) method to be used (see the Introduction to Protein Visualization Procedures, and Protocols 2–7).

Separating gel buffer (1.5 M Tris-HCl at pH 8.8)

Stacking gel buffer (0.5 M Tris-HCl at pH 6.8)

TEMED <!>

Equipment

A microsyringe, standard disposable pipette tips, or gel-loading tips for loading samples onto the gel

Biological Sample

Protein solution(s) or pellets of cellular proteins

Additional Reagents and Equipment

Step 1 requires all of the equipment listed in Protocol 1.
Step 2 requires the reagents listed in either Protocol 2 or 3 (optional).

METHOD

1. The procedures for setting up the gel cassette, pouring the gel, performing the electrophoresis, and handling, storing, and staining the gel are the same as those described in Protocol 1.

2. If proteins are to be detected by their biological activity, run duplicate samples. Stain one set of samples for protein and the other set for activity. Most commonly, look for enzyme activity in the gel. This is achieved by washing the gel in an appropriate enzyme substrate solution that results in a colored product appearing in the gel at the site of the enzyme activity. Alternatively, visualize the proteins using CBR-250 stain, as described in Protocol 2 or 3.

REFERENCES

Akin D.T., Shapira R., and Kincade J.M. 1985. The determination of molecular weights of biologically active proteins by cetyltrimethylammonium bromide-polacrylamide gel electrophoresis. *Anal. Biochem.* **145:** 170–176.

Atkins R.E., Levin P.M., and Taun R.S. 1992. Cetyltrimethylammonium bromide discontinuous gel electrophoresis: M_r-based separation of proteins with retention of enzymatic activity. *Anal. Biochem.* **202:** 172–178.

Bollag D.M, Rozycki M.D., and Edelstein S.J., eds. 1996. *Protein methods,* 2nd edition. Wiley, New York.

Caglio S., Chiari M., and Righetti P.G. 1994. Gel polymerization in detergents: Conversion efficiency of methylene blue vs. persulfate catalysis, as investigated by capillary zone electrophoresis. *Electrophoresis* **15:** 209–214.

Cantor C.R. and Schimmel P.R. 1980. *Biophysical chemistry.* II. *Techniques for the study of biological structure and function,* pp. 676–680. W.H. Freeman, San Francisco.

Chrambach A. and Rodbard D. 1971. Polyacrylamide gel electrophoresis. *Science* **172:** 440–451.

Chrambach A. and Jovin T.M. 1983. Selected buffer system for moving boundary electrophoresis on gels at various pH values, presented in a simplified manner. *Electrophoresis* **4:** 190–204.

Davis B.J. 1964. Disc electrophoresis. II. Method and application to human serum proteins. *Ann. N.Y. Acad. Sci.* **121:** 401–349.

de Moreno M.R., Smith J.F., and Smith R.V. 1986. Mechanism studies of Coomassie blue and silver staining of proteins. *J. Pharm. Sci.* **75:** 907–911.

Duke J.A., Bier M., and Nord F.F. 1952. On the mechanism of enzyme action. LIII. The amphoteric properties of trypsin. *Arch. Biochem. Biophys.* **40:** 424–436.

Fazekas de St. Groth S., Webster R.G., and Datyner A. 1963. Two new staining procedures for quantitative estimation of proteins on electrophoretic strips. *Biochim. Biophys. Acta* **71:** 377–391.

Ferguson K.A. 1964. Starch-gel electrophoresis: Application to the classification of pituitary proteins and polypeptides. *Metabl. Clin. Exp.* **13:** 985–1002.

Gersten D.M. and Bijwaard K.E. 1992a. Separation of proteins by sodium dodecyl sulfate-polyacrylamide gel electrophoresis in the presence of soluble, aqueous polymers: Ficoll and polyvinylpyrrolidone. *Electrophoresis* **13:** 399–401.

——. 1992b. Polyacrylamide gel electrophoresis in vertical, inverse and double-crossing gradients of soluble polymers. *Electrophoresis* **13:** 282–286.

Granzier H.L.M. and Wang K. 1993. Gel electrophoresis of giant proteins: Solubilization and silver-staining of titin and nebulin from single muscle fiber segments. *Electrophoresis* **14:** 56–64.

Hames B.D., ed. 1998. *Gel electrophoresis of proteins: A practical approach,* 3rd edition. Oxford University Press, New York.

Hames B.D. 1990. One-dimensional polyacrylamide gel electrophoresis. In *Gel electrophoresis of proteins: A practical approach,* 2nd edition (ed. B.D. Hames and D. Rickwood). Oxford University Press, New York.

Harwig S.S., Chen N.P., Park A.S., and Lehrer R.I. 1993. Purification of cysteine-rich bioactive peptides from leukocytes by continuous acid-urea-polyacrylamide gel electrophoresis. *Anal. Biochem.* **208:** 382–386.

Hjertén S. 1962. "Molecular Sieve" chromatography on polyacrylamide gels, prepared according to a simplified method. *Arch. Biochem. Biophys.* (suppl. 1) 147–151.

Kaufmann E., Geisler N., and Weber K. 1984. SDS-PAGE strongly overestimates the molecular masses of the neurofilament proteins. *FEBS Lett.* **170:** 81–84.

Kohlrausch F. 1897. Ueber concentrations-verschiebungen durch elecrolyse im innern von lösungen und lösungsgemischen. *Ann. Phys. Chem.* **62:** 210–239.

Kyte J. 1995. *Structure in protein chemistry,* pp. 27–33. Garland Publishing, New York.

Laemmli U.K. 1970. Cleavage of structural proteins during the assembly of the head of bacteriophage T4. *Nature* **227:** 680–685.

MacFarlane D.E. 1989. Two dimensional benzyldimethyl-n-hexadecylammonium chloride–sodium dodecyl sulfate preparative polyacrylamide gel electrophoresis: A high capacity high resolution technique for the purification of proteins from complex mixtures. *Anal. Biochem.* **176:** 457–463.

Makowski G.S. and Ramsby M.L. 1993. pH modification to enhance the molecular sieving properties of sodium dodecyl sulfate-10% polyacrylamide gels. *Anal. Biochem.* **212:** 283–285.

——. 1997. Protein molecular weight determination by sodium dodecyl sulfate polyacrylamide gel electrophoresis. In *Protein structure: A practical approach,* 2nd edition (ed. T.E. Creighton), pp. 1–27. IRL Press/Oxford University Press, New York.

McLellan T. 1982. Electrophoresis buffers for polyacrylamide gels at various pH. *Anal. Biochem.* **126:** 94–99.

Meyer T.S. and Lamberts B.L. 1965. Use of Coomassie brilliant blue R250 for the electrophoresis of microgram quantities of parotid saliva proteins on acrylamide gel strips. *Biochim. Biophys. Acta* **107:** 144–145.

Mócz G. and Bálint M. 1984. Use of cationic detergents for polyacrylamide gel electrophoresis in multiphasic buffer systems. *Anal. Biochem.* **143:** 283–292.

Morris C.J.O.R. 1966. Gel filtration and gel electrophoresis. In *Protides of the biological fluids* (ed. H. Peeters), vol 14, pp. 543–561. Elsevier, Amsterdam.

Neuhoff V., Arold N., Taube D., and Ehrhardt W. 1988. Improved staining of proteins in polyacrylamide gels including isoelectric focusing gels with clear background at nanogram sensitivity using Coomassie Brilliant Blue G-250 and R-250. *Electrophoresis* **9:** 255–262.

Neuhoff V., Stamm R., Pardowitz I., Arold N., Ehrhardt W., and Taube D. 1990. Essential problems in quantification of proteins following staining with Coomassie Brilliant Blue dyes in polyacrylamide gels, and their solution. *Electrophoresis* **11:** 101–117.

Neville D.M. Jr. 1971. Molecular weight determination of protein-dodecyl sulfate complexes by gel electrophoresis in a discontinuous buffer system. *J. Biol. Chem.* **246:** 6328–6334.

Ornstein L. 1964. Disc electrophoresis. I. Background and theory. *Ann. N.Y. Acad. Sci.* **121:** 321–349.

Panyim S. and Chalkley R. 1969. High resolution acrylamide gel electrophoresis of histones. *Arch. Biochem. Biophys.* **130:** 337–346.

Patton W.F. 2000. A thousand points of light: The application of fluorescence detection technologies to two-dimensional gel electrophoresis and proteomics. *Electrophoresis* **21:** 1123–1144.

——. 2001. Detecting proteins in polyacrylamide gels and on electroblot membranes. In *Proteomics: From protein sequence to function* (ed. S.R. Pennington and M.J. Dunn), pp. 65–86. BIOS Scientific, Oxford, United Kingdom, Springer-Verlag, New York.

Rabilloud T. 2000. Detecting proteins separated by 2-D gel electrophoresis. *Anal. Chem.* **72:** 48A–55A.

Rabilloud T., Girardot V., and Lawrence J.J. 1996. One- and two-dimensional histone separations in acidic gels: Usefulness of methylene blue-driven photopolymerization. *Electrophoresis* **17:** 67–73.

Raymond S. and Weintraub L.S. 1959. Acrylamide gel as a supporting medium for zone electrophoresis. *Science* **130:** 711.

Reynolds J.A. and Tanford C. 1970. Binding of dodecyl sulfate to proteins at high binding ratios. Possible implications for the state of proteins in biological membranes. *Proc. Natl. Acad. Sci.* **66:** 1002–1007.

Righetti P.G. 1995. Macroporous gels: Facts and misfacts. *J. Chromatogr.* **A698:** 3–17.

Schägger H. and von Jagow G.V. 1987. Tricine sodium dodecyl sulfate-polyacrylamide gel electrophoresis for the separation of proteins in the range from 1 to 100 kDa. *Anal. Biochem.* **166:** 368–379.

Segrest J.P. and Jackson R.L. 1972. Molecular weight determination of glycoproteins by polyacrylamide gel electrophoresis in sodium dodecyl sulfate. *Methods Enzymol.* **28B:** 54–63.

Shapiro A.L., Viñuela E., and Maizel J.V. 1967. Molecular weight estimation of polypeptide chains by electrophoresis in SDS-polyacrylamide gels. *Biochem. Biophys. Res. Commun.* **28:** 815–820.

Shi Q. and Jackowski G. 1998. One-dimensional polyacrylamide gel electrophoresis. In *Gel electrophoresis of proteins: A practical approach,* 3rd edition (ed. B.D. Hames), pp. 1–52. IRL Press/Oxford University Press, NewYork.

Sinha P., Poland J., Schnolzer M., and Rabilloud T. 2001. A new silver staining apparatus and procedure for matrix-assisted laser desorption/ionization-time of flight analysis of proteins after two-dimensional electrophoresis. *Proteomics* **1:** 835–840.

Smith B.J. 1994. SDS polyacrylamide gel electrophoresis of proteins. *Methods Mol. Biol.* **32:** 23–34.

Sobieszek A. 1994. Gradient polyacrylamide gel electrophoresis in presence of sodium dodecyl sulfate: A practical approach to muscle contractile and regulatory proteins. *Electrophoresis* **15:** 1014–1020.

Swank R.T. and Munkres K.D. 1971. Molecular weight analysis of oligopeptides by electrophoresis in polyacrylamide gel with sodium dodecyl sulfate. *Anal. Biochem.* **39:** 462–477.

Weber K. and Osborn M. 1969. The reliability of molecular weight determinations by dodecyl sulfate-polyacrylamide gel electrophoresis. *J. Biol. Chem.* **244:** 4406–4412.

Wong C., Sridhara S., Bardwell J.C.A., and Jakob U. 2000. Heating greatly speeds Coomassie blue staining and destaining. *BioTechniques* **28:** 426–432.

FURTHER READING

Righetti P.G., Faupal M., and Wenisch E. 1992. Preparative electrophoresis with and without immobilized pH gradients. In *Advances in electrophoresis* (ed. A. Chrambach et al.), vol. 5, pp. 159–200. VCH, Weinheim, Germany.

WWW RESOURCES

http://www.apbiotech.com Amersham Bioscience home page.
http://www.bio-rad.com Bio-Rad Laboratories home page.
http://invitrogen.com Invitrogen Life Technologies home page.

CHAPTER 3

Preparation of Cellular and Subcellular Extracts*

Extraction of protein from tissues and cells is perhaps the most critical step in any proteomics strategy because this step influences protein yield, biological activity, and the structural integrity of the specific target protein. Thus, care must be taken in selecting the specific extraction conditions employed. The principal aim must be to reproducibly achieve the

*This chapter includes a contribution from Melinda L. Ramsby and Gregory S. Makowski (University of Connecticut Health Center, Framington).

highest degree of cell breakage using minimal disruptive forces while maintaining protein integrity. A summary of commonly used methods for homogenizing cells and tissues is given in Table 3.1.

It is important to avoid altering the native structure of the target protein and, ipso facto, its biological activity. Perturbation of native protein structure during preparation of cell extracts can occur by exposure to extremes of pH, temperature, mechanical stress (shearing forces), pressure, and proteolytic degradation. Although proteolytic degradation may not alter the biological activity per se, it may influence the association of the target protein with other cellular regulatory components. Such nonspecific alterations can result in irreproducibility of behavior of the target protein from one preparation to the next, which makes interpretation of biological studies extremely difficult. For discussions on avoiding proteolytic degradation of proteins in extracts, see Beynon and Oliver (1996) and Beynon and Bond (1989).

Key variables that determine the successful preparation of crude extracts include

- the method of cell lysis,
- the control of pH,
- temperature, and
- avoidance of proteolytic degradation.

Note: A trial-and-error approach in pilot experiments is often required to optimize cell lysis conditions.

It is not always necessary to break open cells to extract recombinant proteins. For example, various high-expression mammalian cell lines (in particular, Chinese hamster ovary [CHO] cells) and strains of yeast (e.g., *Pichia pastoris*) have been readily engineered to secrete recombinant proteins that can be purified directly from cell-conditioned media and culture filtrates. Using these approaches, it is important to concentrate the large volumes of cell-conditioned media into a "protease-free" environment as quickly as possible.

A different problem can occur with *Escherichia coli* expression systems, where the expressed recombinant protein often appears in the crude extract (lysate) as insoluble aggregates, referred to as "inclusion bodies." In this case, the purification of a target protein often involves initial solubilization of the inclusion bodies in a strong denaturant (e.g., guanidine hydrochloride or urea) and subsequent refolding. These procedures are described in Protocols 6 and 7.

When choosing a cell disruption strategy, it is important to consider the intended use of the cell lysate. For example, lysis conditions (choice of buffer, detergents, and so on) will vary markedly, depending on whether the lysate is to be used for

- immunoprecipitation studies,
- western blotting,
- two-dimensional gel electrophoresis,
- native target protein isolation using conventional chromatographic purification procedures, and
- recombinant protein purification procedures that rely on the target protein expressed as a fusion protein including a purification "handle" or "tag."

TABLE 3.1. Methods for homogenizing cells and tissues

Method	Underlying basis of cell disruption	Type of tissue
Gentle		
Osmotic shock	Osmotic disruption of cell membranes	bacteria, erythrocytes
Detergent lysis	Detergent disruption of cell membranes	tissue culture cells
Enzymatic digestion	Digestion of cell wall; contents released by osmotic disruption	bacteria, yeast
Dounce homogenizer and/or Potter-Elvehjem homogenizer[a]	Cells forced through a narrow gap with a clearance of 0.05–0.08 mm (tight fitting) to 0.1–0.3 mm (loose fitting); cell membrane disrupted by liquid shear forces The smaller the clearance, the greater the shearing force; the clearance of a Teflon pestle (Potter-Elvehjem homogenizer) is normally 0.05–0.06 mm[b]	soft animal tissues and cells
Moderately harsh		
Homogenization (Waring Blendor)	Cells broken by rotating blades	most animal, plant tissues
Grinding (with sand, alumina, or glass beads)	Cell walls broken by abrasive action of particles	cell suspensions
Vigorous		
French pressure cell	Cells forced through small orifice at high hydraulic pressure (100–150 Mpa or 15,000–20,000 psi) and disrupted by shear forces	bacteria, yeast, plant cells
Explosive decompression (nitrogen cavitation)	Cells equilibrated with inert gas (e.g., N_2) at high pressure (typically, 5500 kPa or 800 psi); on exposure of cells to 1 atm, disruption occurs	bacteria, yeast, plant cells
Bead mill	Rapid vibration with glass beads disrupts cell wall	cell suspensions
Ultrasonication	High-pressure sound waves cause cell rupture by cavitation and shear forces	cell suspensions

[a]Sometimes called Teflon-and-glass homogenizer; it is power-driven, the pestle typically rotated at 500–1000 rpm.
[b]For soft tissue such as liver, a Potter-Elvehjem homogenizer with a clearance of ~0.09 mm is recommended. Smaller clearances can lead to damage to released organelles (especially nuclei) as well as causing difficulty in moving the glass vessel relative to the pestle in the early stages of homogenization (Graham 1997).

The first part of this introduction describes commonly used methods for cell lysis, including procedures for preparing crude cell extracts for immunoprecipitation and immunoblotting studies. The range of methods discussed is by no means exhaustive and does not include plant tissues and fungi. For more detailed reviews describing the preparation of crude extracts from eukaroytes, prokaryotes, and plants, see Deutscher (1990), Doonan (1996), Graham (1997), and Spector et al. (1998). The second section deals with the problem of solubilizing *E. coli*-derived recombinant proteins from inclusion bodies. The third section discusses the advantages of isolating subcellular fractions and describes the differential detergent fractionation (DDF) of eukaryotic cells for the isolation and analysis of proteins. The protocols described in this chapter are sufficiently general and thus can be applied to a variety of different tissues and cell types with only minor modifications.

TISSUES AND CELLS CAN BE DISRUPTED BY MECHANICAL OR CHEMICAL MEANS

Preparation of Protein Extracts from Mammalian Tissues

The preparation of protein extracts from most animal tissues is relatively simple because the cell membranes are weak and easily disrupted by a combination of osmotic and mechanical forces. The first steps of a typical protein isolation procedure usually consist of washing the tissue, disrupting the tissue in a suitable buffer using a homogenizer, and clarifying the homogenate by centrifugation. The centrifugation step separates the soluble proteins from the membrane fraction and insoluble cell debris. A generalized procedure that may be applied to a number of different tissues and cell types with only minor modifications is given in Protocol 1. Isolation of a particular membrane or subcellular organelle requires more specialized procedures (see Graham and Rickwood 1997; Celis 1998; Spector et al. 1998).

The choice of tissue for extraction of soluble proteins depends on several criteria. Clearly, the tissue distribution of a particular target protein varies markedly. Hence, it is always desirable to perform small-scale pilot experiments to measure the relative content of the target protein in various tissues. However, the final choice of starting tissue will, invariably, be determined by the balance between target protein abundance, ready availability and cost of the tissue, and various technical issues such as minimization of proteolytic activity. It should be borne in mind that certain animal tissues (e.g., liver, spleen, kidney, and macrophages) are rich in lysosomal proteases, particularly cathepsins, and should be avoided unless the goal of the study is the proteolytic enzyme(s) itself. Fresh tissue is preferable, but in some cases, frozen tissue may be acceptable, provided it is frozen rapidly in small pieces and not stored for too long. It is recommended that frozen tissue be stored below –50°C, since some proteolytic degradation of proteins can be expected due to release of proteases from lysosomes as a result of ice-crystal formation (Dignam 1990).

In some cases, it may be desirable to gently disrupt tissues and prepare enriched populations of intact cells, prior to disruption of these cells. For example, the separation of different cell types from normal tissues and tumors may permit comparative information to be obtained (e.g., protein expression profiles and mRNA profiles) that is unavailable if whole tissue (or tumor) is studied. A number of procedures have been well described in various cell biology manuals (e.g., see Celis 1998; Spector et al. 1998) for preparing cell suspensions from tissues and organs using mechanical or enzymatic methods. In general, enzymatic methods are preferred, since there is less damage to the integrity of the cells. In addition, it is usual to add ethylene diaminetetracetic acid (EDTA) to chelate Ca^{2+} ions that are frequently involved in cell-cell adhesion. Purification of cells obtained by these means is usually accomplished by the following:

- ***Differential centrifugation*** (based on cell size and density). This is accomplished using iso-osmotic density gradients generated using nontoxic/nonpermeable media, notably, Percoll, Ficoll, and metrizamide and, more recently, OptiPrep (based on iodixanol) (Graham et al. 1994).
- ***Centrifugal elutriation or counterstream centrifugation.*** This method is based on two opposing forces, namely, media flow and centrifugal force (Bird 1998).
- ***Selective immunoseparation procedures*** employing monoclonal antibody-bound magnetic beads (Dynabeads). This method is based on the efficient selection of specific cells using a simple magnet (Neurauter et al. 1998).

All existing methods have inherent advantages and limitations, the discussion of which is beyond the scope of this manual. For a detailed discussion of these procedures and protocols for enriching cell populations of interest from various tissues, see Celis (1995), Spector et al. (1998), and references therein.

Preparation of Protein Extracts from Mammalian Cultured Cells

Mammalian cultured cells can be lysed by several different methods, the method of choice depending on the final use of the target protein (e.g., immunoprecipitation, immunoblotting, two-dimensional gel electrophoresis, and conventional purification). Mammalian cells lack a cell wall and thus are easily lysed by treatment with mild detergents. If the final preparation of the target protein need not retain its structural integrity (three-dimensional structure) or biological activity, the cells can be lysed under harsh denaturing conditions (e.g., RIPA lysis buffer). If gentler conditions are required, the Nonidet P-40 (NP-40) lysis buffer (or variations thereof) should be used.

Many extraction conditions release proteolytic enzymes in the lysis buffer. If proteolytic degradation of the target protein becomes a problem, two approaches can be employed to lessen its effect:

- Keep the sample cold (temperature has a profound effect on the catalytic activity of most proteases).
- Add protease inhibitors to the lysis buffer (for a list of some of the commonly used protease inhibitors, see Table 3.3).

For immunoprecipitation studies, the conditions used for lysis should be as gentle as possible to maintain the structural integrity of the target protein and to minimize the solubilization of irrelevant proteins. This is best accomplished by using nonionic detergents instead of ionic ones, lower concentrations than higher, and single detergents rather than mixtures. The two most widely used lysis buffers for the extraction of proteins from mammalian culture cells are NP-40 lysis buffer and RIPA lysis buffer. The former buffer releases cytoplasmic and nuclear proteins without releasing chromosomal DNA, which, because of its viscous nature, can cause numerous problems during protein purification and analysis.

The lysis conditions can be easily tailored to suit the target protein. Variables that can affect the release of a target protein from a cell include salt concentration, type of detergent, presence of divalent cations, and pH. To determine the optimum conditions for extracting a target protein, the variables listed below should be monitored in pilot experiments (Harlow and Lane 1999):

- Salt concentration should be varied from 0 to 1 M.
- Nonionic detergent concentrations between 0.1% and 2%.
- Ionic detergent concentrations between 0.01% and 0.5%.
- Divalent cation concentrations between 0 and 10 mM.
- EDTA concentrations between 0 and 5 mM.
- pH values between 6 and 9.

A generalized procedure for lysing tissue culture cells for the purpose of performing immunoprecipitation is given in Protocol 2.

For immunoblotting studies, the conditions used for lysis can be harsher than those used for immunoprecipitation studies. The most widely used buffer system is the Laemmli sample buffer (Laemmli 1970) containing 2% SDS. A procedure for lysing tissue culture cells for the purpose of performing immunoblotting is given in Protocol 3. An alternative procedure for lysing any tissue culture cells (and microorganisms such as bacteria) that relies on gaseous shear to disrupt cells is nitrogen cavitation (see Protocol 4).

Disruption of Bacterial Cells

Due to the advent of recombinant DNA technology, bacteria are now a particularly convenient vehicle for generating large quantities of recombinant protein. Enormous numbers of bacteria can be grown under defined conditions and are relatively easy to break open for extraction purposes. A number of methods, based on mechanical and enzymatic means, are available for lysing bacteria, and to a large extent, the choice will depend on the scale of the process (for a review, see Cull and McHenry 1990).

The mechanical procedures for lysing bacteria (e.g., French pressure cell, nitrogen cavitation, ultrasonication, grinding with abrasive agents, and vortexing with glass beads) rely on shearing forces to disrupt the tough outer cell wall (for a review of these procedures, see Graham 1997). However, mechanical methods are more likely to damage the cellular contents (compared with enzymatic methods) and are not easily scaled-up.

The *French Press* lyses cells by applying hydraulic pressure to the cell suspension (typically, 8,000–20,000 psi, or 550–1400 kg/cm^2), followed by a sudden release to atmospheric pressure. In this device, the cell suspension is propelled by a piston through a narrow orifice, often the annulus around a ball bearing. The rapid change in pressure causes a liquid shear and consequent bursting of cells. In some cases, two to three passes of the cells through the French Press are required to obtain adequate lysis (Cull and McHenry 1990). The French Press method works well for cell suspension volumes of 10–30 ml (ratio of cell wet weight to lysis buffer volume can range from 1:1 to 1:4 g/ml), but it is considered too time-consuming for larger volumes and can be technically difficult for smaller volumes.

In *nitrogen cavitation* devices, the pressure cell consists of a robust stainless steel cylinder with an inlet port for delivery of nitrogen gas from a cylinder and an outlet tube with a needle valve. During pressurization (i.e., 5500 kPa [800 psi] for 10–30 minutes), nitrogen dissolves in the cell suspension buffer as well as in the cytosol of the cells. When the needle valve is opened, the suspension is forced through the outlet tube, and at 1 atm, it experiences a rapid decompression that causes cell disruption due to the sudden formation of bubbles of nitrogen gas. This method can be used with bacteria or any tissue culture cell in volumes ranging from 1 to 1000 ml. Nitrogen cavitation (see Protocol 4) eliminates the heat buildup associated with mechanical and ultrasonic disruption because the cells are actually cooled by the expanding gas.

Sonication disrupts cells by creating vibrations that cause mechanical shearing of the cell wall. This method is suitable for small-scale purifications (up to 1 g of cells or tissue can be lysed at a time). Generation of heat during sonication can be a problem and may result in protein denaturation (as evidenced by foaming); this problem can be overcome by sonicating the cell suspension in short bursts and allowing the sample to cool on ice between treatments.

Like the French Press method, the grinding of cells with *abrasive materials* such as alumina or sand is an efficient method of lysing unicellular organisms, as well as plant cells. Unlike the French Press, this grinding method requires inexpensive materials (e.g., mortar, pestle, and either sand or alumina) and is effective for moderate quantities of cells (up to 30 g wet weight of cells) (Fahnestock 1979; Sebald et al. 1979). Abrasives can also be added to sonication mixtures.

An extension of the grinding method is the *glass bead vortexing procedure*, which has been described for lysing unicellular organisms, particularly yeast (Schatz 1979). Glass bead vortexing is suitable for small samples (~3 g wet weight of cells) that can withstand being repeatedly vortexed with glass beads. A number of instruments are now commercially available for lysing larger quantities of cells (e.g., Manton-Gaulin homogenizer, Braun MSK Glass Bead Mill).

Methods based on *enzymatic breakdown* of the cell wall use the activity of lysozyme, which cleaves the glucosidic linkages in the bacterial cell-wall polysaccharide, to cause disruption (the inner cytoplasmic membrane can be readily disrupted by detergents, osmotic pressure, or mechanical methods). Enzymatic methods are much gentler than mechanical methods and are more easily applied to large-scale processing. Two generic methods for preparing bacterial extracts are given in Protocols 5 and 6.

Disruption of Yeast Cells

A number of procedures exist for preparing yeast extracts, including autolysis (e.g., addition of toluene to yeast suspension), French pressure cell, abrasives (e.g., glass beads), and enzymatic lysis (for an overview, see Jazwinski 1990; Bridge 1996). Of these, the most widely used procedure is the use of abrasives. The abrasive action of well-agitated glass beads (typically, 0.5-mm diameter) on yeast cells yields up to 95% cell breakage (Jazwinski 1990). Although this method is useful for preparing enzymes and some cell organelles, it is considered too harsh to preserve the integrity of nuclei as well as protein complexes (enzymatic methods are recommended for this purpose). Major differences between abrasive procedures reside in the method employed to agitate the glass beads. One of the simplest methods for agitating glass beads involves the use of a vortex mixer (Baker et al. 1988). A generic method for lysing small quantities of yeast cells employing the glass bead vortexing approach is given in Protocol 8. For a discussion on commercially available equipment for lysing yeast, see Jazwinski (1990).

RECOMBINANT PROTEIN CAN BE RECOVERED FROM INCLUSION BODIES BY DENATURATION AND RENATURATION

A significant impediment to overexpressing recombinant proteins in *E. coli* is the tendency for the targeted protein to form inclusion bodies, which are in vivo agglomerates of proteins that appear in the cytoplasm as large, dense bodies in scanning electron micrographs (Williams et al. 1982). Inclusion bodies are relatively rare in nature (most proteins are expressed in soluble form), with sickle cell anemia and other related blood diseases being the notable exception (Carrell et al. 1966). Overexpression of recombinant proteins from strong promoters on multiple-copy plasmids—with expression levels up to 40% of total cell protein—is thought to be the underlying reason for inclusion body formation in *E. coli*. For a review of heterologous protein production in *E. coli* and the purification of recombinant proteins from inclusion bodies, see Hockney (1995) and Marston (1986).

It is thought that inclusion body protein is partially or incorrectly folded, especially a protein containing disulfide bonds. An important difference between the expression of eukaryotic proteins in *E. coli* and their native environment is the inability of *E. coli* to form disulfide linkages, due to the reducing environment of its cytoplasm. Inclusion body formation of overexpressed recombinant protein in *E. coli* is found not only for foreign eukaryotic proteins, but also for overexpressed bacterial proteins that are normally soluble (Gribskov and Burgess 1983). Temperature-sensitive denaturation can be overcome (by reducing the frequency of inclusion body formation, and therefore increasing the soluble fraction of the target protein) by lowering the expression temperature from 37°C to 30°C (Schein and Noteborn 1988).

In addition to growth temperature, a number of other growth parameters have been manipulated to prevent inclusion body formation and increase the soluble fraction of the target protein. Most notable of these are the following:

- ***Varying the media composition and using different host strains*** (Schein and Noteborn 1988).
- ***Coexpression of molecular chaperones.*** For example, coexpression of DNAK increased the percentage of soluble human growth hormone in *E. coli* by ~87% (Blum et al. 1992), and coexpression of GroES and GroEL facilitated the purification of milligram quantities of recombinant $p50^{csk}$ (Amrein et al. 1995).
- ***Fusion of the target protein with a highly soluble protein*** such as glutathione-*S*-transferase (Smith and Johnson 1988) or thioredoxin (La Vallie et al. 1993).

- ***Growing cells in the presence of sorbitol and glycyl betaine.*** Sorbitol facilitates the cellular uptake of the "protein stabilizer" glycyl betaine (Blackwell and Horgan 1991).

It should be emphasized that none of the above-mentioned approaches for increasing the soluble fraction of the target protein are general for all applications (Hockney 1995), and pilot experiments are recommended to ascertain the best approach for a particular protein. The following are two major advantages of inclusion bodies.

- By sequestering recombinant protein, these bodies permit the cell to express the protein at high levels.
- The inclusion bodies can be readily purified away from bacterial cytoplasmic proteins by centrifugation, yielding an effective purification step.

The major disadvantage of inclusion bodies is that extraction of the target protein requires the use of detergents. This problem is exacerbated where natively folded protein is required, particularly if the target protein contains more than three disulfide bonds. The protein contained within inclusion bodies is generally insoluble in nonionic detergents and salts. The most widely used solubilizing agents are the water-soluble chaotropic agents, such as urea (8 M) and guanidine hydrochloride (6 M), that cause complete denaturation and are more compatible with protein refolding. Thus, the target protein can be renatured by simply removing the denaturant under conditions that favor complete refolding of the target protein over the formation of aggregates due to intermolecular protein-protein interactions. Such refolding conditions include low temperature and very low protein concentration (~1 μg protein/ml).

For cysteine-containing target proteins that contain disulfide bonds in their native structure, solubilization of inclusion bodies is usually performed in the presence of a reducing agent, such as dithiothreitol or β-mercaptoethanol, in order to fully reduce the target protein and to prevent the formation of disulfide-bonded aggregates (i.e., to favor intramolecular versus intermolecular disulfide bonds) during the refolding process. The most widely used methods for the correct refolding of recombinant proteins (Marston 1986; Kohno et al. 1990; Marston and Hartley 1990) use dilution and dialysis to reduce the urea or guanidine hydrochloride concentration gradually. In some cases, redox conditions (e.g., reduced/oxidized glutathione) are maintained during the refolding process to accelerate the correct pairing of disulfides and formation of the native structure (Light 1985). This may be critical in some cases (e.g., interleukin-2 [Tsuji et al. 1987; Weir and Sparks 1987]), but not others (e.g., interleukin-6 [Zhang et al. 1992]). In both cases, the target protein is in a fully reduced form in inclusion bodies. Interestingly, the best yields of recombinant prochymosin (which contains three disulfide bonds in the native structure) were obtained by omitting redox reagents at both the solubilization and refolding stages (Marston and Hartley 1990); in fact, little activity was recovered when the molecule was fully reduced and then solubilized in 8 M urea. A generic procedure for the isolation and solubilization of inclusion bodies and renaturation of target protein is given in Protocol 7.

PREPARATION OF SUBCELLULAR EXTRACTS ENRICHES FOR TARGET PROTEINS

Why Isolate Subcellular Fractions?

Subcellular fractionation has been widely practiced by biologists during the past 50 years to gain better insight into the composition and function of cellular organelles and macromolecules. Understanding the biological function of proteins requires knowledge of their subcellular localization and their movement from one compartment in the cell to another in response to external biological stimuli. Proteins must be localized in the same cellular compartment if they are to cooperate toward the execution of common physiological functions, such as metabolism (metabolic pathways), cell signaling (signal transduction pathways), programmed cell death (apoptosis), and maintenance of cell structure (cytoskeleton). For instance, the reversible localization of signal transducing proteins to both the plasma and the internal membranes of cells is critical for the selective activation of downstream functions and depends on both protein/protein and protein/membrane-lipid interactions (Johnson and Cornell 1999; Hurley and Meyer 2001). In this example, in addition to identifying all protein/lipid-mediated subcellular localization mechanisms, it is also important to establish how these targeting mechanisms organize subcellular signaling events in time and space. This latter question represents one of the major challenges of the post-genome era.

Now that the template of the human genome has been fully described, the quest to understand the complete protein complement (the proteome) of cells in various biological states and in response to assorted biological stimuli is gaining greater impetus. However, this is a problem with formidable obstacles, because both the number of proteins in a cell is large and the protein complement within a cell is dynamic. Attempts to analyze an entire cellular proteome without fractionation often fail because the complete protein complement simply cannot be resolved by current techniques, such as two-dimensional gel electrophoresis. Thus, there is a need to tackle the problem of defining a cell's total proteome by breaking it up into "bite size" components and to fully characterize the proteomes of the individual components. A natural choice is to separate the cell into subcellular organelles and compartments and then probe the proteomes of these individual cellular compartments. Already, a systematic identification of many of the various cell organelles has been initiated to understand their function and biogenesis (see Table 3.2) (for a review, see Jung et al. 2000a,b).

2D or Not 2D?

It is apparent that two-dimensional (2D) gel electrophoresis (see Chapter 4), the classical method for studying the global expression of cellular proteins, has major limitations (Gygi et al. 2000):

- 2D electrophoresis gels are unable to resolve all proteins (although improvements in the technology, such as isoelectric focusing [IEF] gels covering only a single pH unit, have increased the resolution of the method) (Corthals et al. 2000; Gorg et al. 2000; Hanash 2000; Hoving et al. 2000; Wildgruber et al. 2000).

TABLE 3.2. Proteomic analyses of subcellular compartments and organelles

Subcellular compartment (organelle)	References
Human placental mitochondria	Rabilloud et al. (1998)
Rat liver mitochondria	Lopez et al. (2000)
Human placental lysosomes	Chataway et al. (1998)
Rat hepatocyte Golgi complex	Taylor et al. (1997; 2000)
Trans-Golgi network-derived carrier vesicles	Fiedler et al. (1997); Morel et al. (2000)
MDCK II cell endocytic vesicles (apical and basolateral)	Fialka et al. (1999)
Burkitt lymphoma BL60 cell line, nuclear fraction	Müller et al. (1999)
Phagosomes	Garin et al. (2001)
Rat liver mitochondrial ribosomes	Cahill and Cunningham (2000)
Human monocytic lysosomes	Journet et al. (2000)
Rat liver caveolae-enriched plasma membranes	Calvo and Enrich (2000)
Epithelial cell plasma membranes	Simpson et al. (2000)
Human myeloid leukemia HL-60 cells microsomal fraction	Han et al. (2001)

- Some protein classes (e.g., the membrane proteins) are underrepresented on 2D electrophoresis because of their poor solubility. Efforts to overcome this problem have been made by using novel detergents (see Santoni et al. 2000) and differential solubilization methods (see Molloy et al. 1998).
- The limited sample capacity and limited detection sensitivity of 2D electrophoresis reduce the ability to detect low-abundance proteins. For example, if a total yeast cell lysate is fractionated by 2D electrophoresis and silver-stained, proteins present at <1000 copies per cell (a large portion of the yeast proteome) are not detected (Gygi et al. 2000).

2D electrophoresis, like most other protein separation methods, is capable of revealing only a "narrow slice" of a cell's total proteome. To address this limitation, some measure of subcellular fractionation prior to 2D electrophoresis is required to bring low-abundance proteins into view. For example, comparative 2D electrophoresis analysis of phagosomes containing living mycobacteria and those containing dead bacilli facilitated the identification of a protein, termed TACO, that allows this intracellular pathogen to evade host defense strategies and survive within macrophages (Ferrari et al. 1999). Such an analysis on intact macrophages would have been nigh on impossible. Another approach to reveal more protein information on 2D electrophoresis involves a combination of subcellular fractionation and the use of narrow pH range gels (Cordwell et al. 2000). Other groups advocate prefractionating complex protein mixtures by solution-based IEF prior to 2D electrophoresis (Zuo and Speicher 2000; Righetti et al. 2001) or free-flow electrophoresis (FFE)/SDS-PAGE (Hoffman et al. 2001). Other non-2D electrophoresis proteomic approaches for identifying low-abundance proteins include a combination of ion-exchange chromatography and reversed-phase high-performance liquid chromatography (RP-HPLC) (see Chapter 8 and Appendix 2).

Two Main Steps in the Subcellular Fractionation Process

Although the aim of subcellular fractionation is to separate cellular compartments with minimal damage to them, it is evident that it is never possible with the use of current fractionation techniques to recover cellular organelles in a completely undamaged state. Indeed, it may never be possible to separate cell organelles in a natural state. At present, there is no single best way to fractionate tissue that applies to all tissues (it cannot be assumed that the subcellular fractionation methods applied to one tissue can be applied to another). To complicate the problem further, methods to determine the subcellular localization of proteins (e.g., fluorescence microscopy, in situ hybridization, and electron microscopy) (Spector et al. 1998) differ from methods used to isolate a particular organelle for proteomic analysis. There are two main steps in the subcellular fractionation process: homogenization of tissues and cells followed by separation of cellular organelles. The principal aims of any homogenization method are to

- achieve maximal cell breakage in a reproducible manner,
- use disruptive forces that minimize damage to the organelles of interest (e.g., protein denaturation and proteolytic degradation), and
- retain the original structure and functional integrity of the organelles of interest.

The principal methods for disrupting cells (osmotic shock, ultrasonic vibration, mechanical grinding or shearing, and nitrogen cavitation) are detailed in earlier parts of this chapter and in the protocols that follow (see also Graham and Rickwood 1997; Celis 1998; Spector et al. 1998). The components of the homogenate are then separated by procedures based on the variations in physical properties of the subcellular components, for example,

- centrifugal methods that separate organelles by size and/or buoyant density in gradient media,
- immunoisolation methods that use antibodies, which bind to specific surface proteins, and
- electrophoresis methods that separate proteins on the basis of surface charge distribution.

For both overviews on subcellular fractionation and isolation of specific cell organelles, see Spector et al. (1998), Graham and Rickwood (1997), and Celis (1998).

Detergents Can Be Used in Combination to Fractionate Eukaryotic Cells

Differential detergent extraction is an established method for cell fractionation, which partitions subcellular constituents into functionally and structurally distinct compartments (Lenk et al. 1977; Lenstra and Bloemendal 1983; Fey et al. 1984; Ramsby et al. 1994; Ramsby and Makowski 1999). The use of differential detergent fractionation (DDF) to obtain cell fractions enriched in cytosolic, membrane, nuclear, and cytoskeletal proteins for direct analysis by 2D electrophoresis has been described recently (Ramsby et al. 1994; Ramsby and Makowski 1999). DDF is applicable for the fractionation of cells grown in suspension or monolayer culture, as well as for fractionation of whole tissue, and can be modified to achieve specific fractionation goals, including further subfractionation.

DDF preserves the structural and functional integrity of cellular proteins, including the cytoskeleton, and permits direct biochemical analysis of detergent extracts by a variety of methods, including enzymatic assays, autoradiography, immunoblotting, immunoprecipitation, 2D electrophoresis, and mass spectrometry (Lenk et al. 1977; Lenstra and Bloemendal 1983; Fey et al. 1984; Ramsby and Kreutzer 1993; Ramsby et al. 1994; Patton 1999; Ramsby and Makowski 1999). Thus, DDF is useful in a variety of proteome research applications, and has been used to

- determine the subcellular localization of biologics,
- semipurify compartment-specific macromolecules,
- enrich for low-abundance proteins,
- investigate dynamic interactions between cytosolic and structural entities (i.e., membranes and the cytoskeleton), and
- monitor treatment-induced compartmental redistributions of macromolecules.

Protocol 9 describes a procedure for the fractionation of proteins from eukaryotic cells, and the preparation of those proteins for analysis by 2D electrophoresis. Additional protocols detail the use of DDF for the concomitant isolation of total cellular RNA and the isolation of tubulins and microtubule-associated proteins present in cytosolic extracts.

PROTOCOL 1

Homogenization of Mammalian Tissue

TO PURIFY OR CHARACTERIZE AN INTRACELLULAR PROTEIN, it is important to choose an efficient method for disrupting the cell or tissue that rapidly releases the protein from its intracellular compartment into a buffer that is not harmful to the biological activity of the protein of interest. One of the most widely used methods for disrupting soft tissues is homogenization. In this protocol, three processes for tissue homogenization using mechanical shear are discussed: chopping the tissue in a Potter-Elvehjem glass-Teflon homogenizer (see Figure 3.1), a Dounce hand homogenizer, or a hand-held Waring Blendor. These methods are rapid and pose little risk to proteins other than the release of proteases from other cellular compartments. Proteolytic degradation can be minimized by the inclusion of protease inhibitors in the homogenization buffers.

MATERIALS

CAUTION: See Appendix 3 for appropriate handling of materials marked with <!>.

Reagents

Dithiothreitol (DTT) (0.5 M) <!>

Prepare a 0.5 M stock solution in cold H_2O and store frozen. Add the reagent to cold buffers at the indicated concentration just prior to use.

Homogenization buffer A

50 mM Tris-Cl (pH 7.5)
2 mM EDTA
150 mM NaCl
0.5 mM DTT

Homogenization buffer B

50 mM Tris-Cl (pH 7.5) <!>
10% (v/v) glycerol (or 0.25 M sucrose)
5 mM magnesium acetate <!>
0.2 mM EDTA
0.5 mM DTT
1.0 mM phenylmethylsulfonyl fluoride (PMSF) <!>

The choice of homogenization buffer will depend on the nature of the extract required. Generally, use a buffer of moderate ionic strength at neutral pH (e.g., 0.05–1.0 M phosphate or Tris, pH 7.0–7.5). The appropriate buffer ionic strength should be chosen by trial and error to optimize the yield of the target protein. For example, the addition of 0.1 M NaCl or KCl will increase the yield of those proteins that have a tendency to attach electrostatically to cell debris/membrane fragments. On the other hand, the association-dissociation behavior of some proteins is influenced markedly by ionic strength. For example, the globular (G) form of actin can be extracted from muscle using low-ionic-strength solutions (0.01 M KCl), but actin aggregates to form fibers (F-actin) if the ionic strength is raised (e.g., to 0.1 M KCl) (Price 1996). If the purpose of the extraction is to isolate organelles, it is important to use low-ionic-strength buffers (e.g., 5–20 mM Tris, HEPES, or TES at

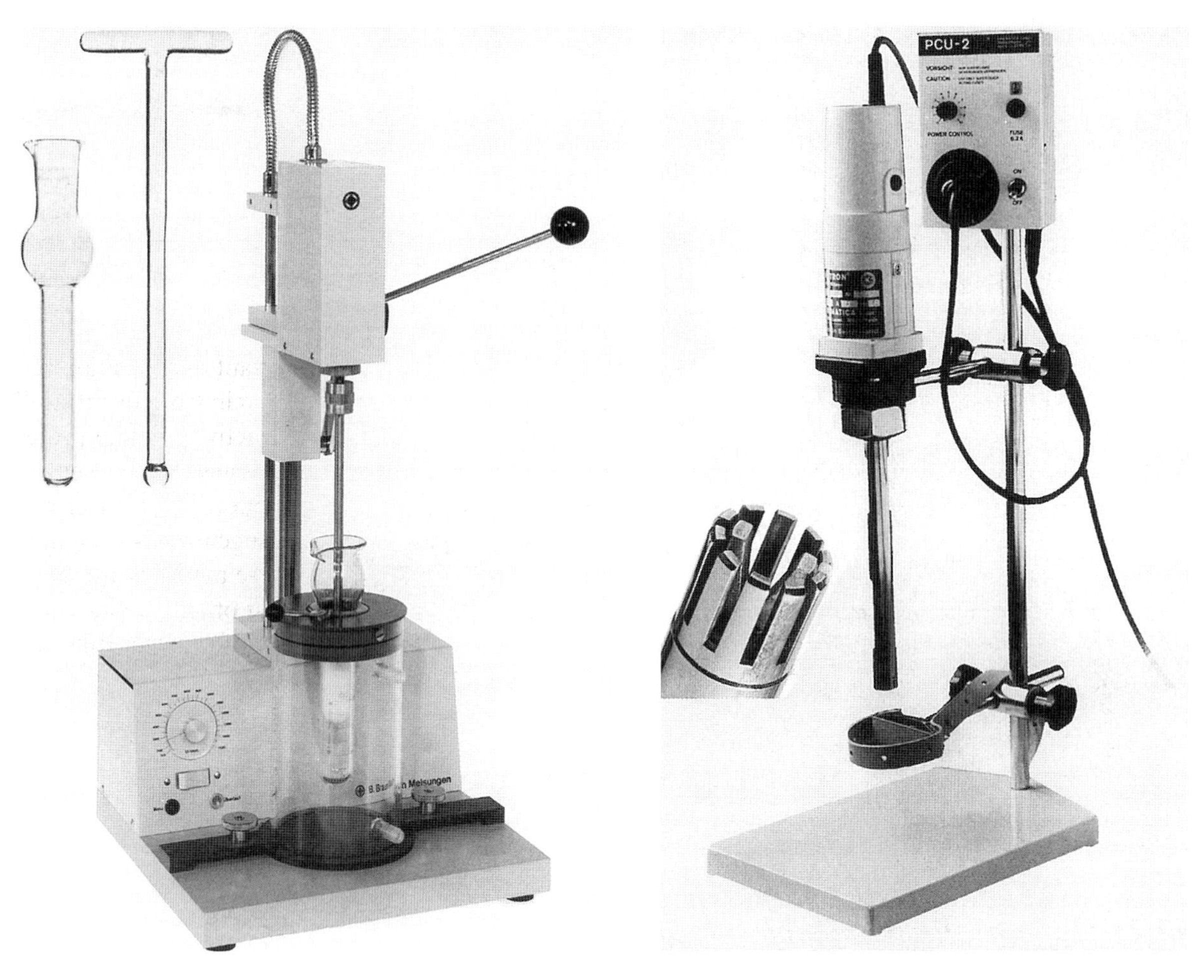

FIGURE 3.1. (*Left*) A Potter-Elvehjem homogenizer (inset Dounce homogenizer). The device holding the Potter-Elvehjem homogenizer allows the glass vessel to be cooled in an ice-water mixture and the pestle can be advanced using a remote handle. (Reprinted, with permission, from Evans 1992, as in Graham and Rickwood 1997.) (*Right*) A Polytron homogenizer. (*Inset*) (inverted) The rotating teeth of the work head. (Reprinted, with permission, from Evans 1992 [©Oxford University Press], as in Graham and Rickwood 1997.)

pH 7.4) containing iso-osmotic sucrose or mannitol (0.25 M). Avoidance of proteolytic degradation of the target protein in a crude extract is a primary concern. In many cases, it may not be essential to add protease inhibitors to the homogenization buffer (due to the protective effect of bulk protein on a target protein), but some proteins are more susceptible to proteolysis than others, and some tissues (e.g., liver and pancreas) have much higher levels of proteases than others (e.g., heart). The use of protease inhibitor cocktails can be expensive if the extract volumes are large. Hence, carry out pilot experiments over a period of a few hours to ascertain whether there are measurable losses of the target protein activity. If proteolytic degradation is a problem, then include protease inhibitors in the homogenization buffers (for a list of protease inhibitors, see Table 3.3 and for the preparation of protease inhibitor cocktails, see Table 3.4). If the target protein is susceptible to oxidation or its activity is inhibited by heavy metals, then add DTT (1 mM) (or 0.1 M β-mercaptoethanol) and EDTA (0.1 M), respectively, to the extraction buffer.

Phenylmethylsulfonyl fluoride (PMSF) (0.2 M)

Prepare a 0.2 M stock solution in 2-propanol, and add the reagent to cold buffers with adequate stirring just prior to use; the reagent will crystallize from 2-propanol when stored at –20°C. Aminoethylbenzenesulfonyl fluoride (AEBSF) is a water-soluble alternative to PMSF that can be used at the same molar concentration (0.1–1.0 mM) for most applications.

TABLE 3.3. Inhibitor cocktails used to control proteolysis during protein isolation

Tissue type	Protease inhibitors (working concentration)	Target protease type[a]	Stock solution[b]
Animal tissues	AEBSF (0.2 mM) (*or* DCI [0.1 mM]or PMSF [0.2 mM])	serine	20 mM in methanol (DCI: 10 mM in DMSO; PMSF: 200 mM in ethanol or isopropanol)
	benzamidine (1 mM)	serine	100 mM
	leupeptin (10 μg/ml)	serine/cysteine	1 mg/ml
	pepstatin (10 μg/ml)	aspartic	5 mg/ml in methanol
	aprotinin (trasylol)(1 μg/ml)	serine	0.1 mg/ml
	EDTA or EGTA[c] (1 mM)	metallo	100 mM
Plant tissues			
	AEBSF (0.2 mM) (*or* DCI [0.1 mM] or PMSF [0.2 mM])	serine	20 mM in methanol (DCI: 10 mM in DMSO; PMSF: 200 mM in ethanol or isopropanol)
	chymostatin (20 μg/ml)	serine/cysteine	1 mg/ml in DMSO
	EDTA or EGTA[c] (1 mM)	metallo	100 mM
Yeasts and fungi			
	AEBSF (0.2 mM) (*or* DCI [0.1 mM]or PMSF [0.2 mM])	serine	20 mM in methanol (DCI: 10 mM in DMSO; PMSF: 200 mM in ethanol or isopropanol)
	pepstatin (15 μg/ml)	aspartic	5 mg/ml in methanol
	1,10-phenanthroline (5 mM)	metallo	1 M in ethanol
Bacteria	AEBSF (0.2 mM) (*or* DCI [0.1 mM]or PMSF [0.2 mM])	serine	20 mM in methanol (DCI: 10 mM in DMSO; PMSF: 200 mM in ethanol or isopropanol)
	EDTA or EGTA[c] (1 mM)	metallo	100 mM

Adapted from North (1989).

Abbreviations: (AEBSF) 4-(2-aminoethyl)-benzenesulfonylfluoride; (DCI) 3,4-dichloroisocoumarin; (DMSO) dimethylsulfoxide; (EDTA) ethylenediamine tetraacetic acid; (EGTA) ethylene glycol bis (β-aminoethyl ether) *N,N,N,'N'*-tetraacetic acid; (PMSF) phenylmethylsulfonyl fluoride.

M_r values of inhibitors: AEBSF (240); PMSF (174); DCI (215); EDTA (disodium salt, dihydrate) (372); benzamidine (hydrochloride) (157); leupeptin (427); pepstatin (686); aprotinin (6500); chymostatin (605); 1,10-phenanthroline (198).

[a]For a review of proteolytic enzymes, see Neurath (1989) and Perlmann and Lorand (1970).

[b]Aqueous solution unless otherwise indicated.

[c]An efficient chelator of divalent metal cations other than Mg^{2+} (for which it has a 10^3-fold lower affinity) (Gegenheimer 1990).

TABLE 3.4. Preparation of a general protease inhibitor cocktail

Inhibitor	Trial working concentration	Stock (100x) concentration	Recipe	Use[a]	Target protease type
AEBSF	0.2 mM	20 mM	239.5 mg/10 ml H_2O (100 mM)	4 ml	serine proteases
EDTA	1 mM	100 mM	19 mg/100 ml H_2O (0.5 M)	4 ml	metallo proteases
Leupeptin <!>	20 μM	2 mM	17 mg/2 ml H_2O (20 mM)	2 ml	cysteine/serine proteases
Pepstatin <!>	1 μM	100 μM	0.86 mg/10 ml methanol (1 mM)	2 ml	aspartic proteases

Adapted from Calbiochem Technical Bulletin CB0578-1294.

[a]Mix the inhibitor solutions and bring to a final volume of 20 ml with H_2O or appropriate aqueous buffer. The resulting solution, 20 ml of a stock 100x protease inhibitor cocktail, can be aliquoted into microfuge tubes and stored at –20°C until required.

Cells and Tissues

Appropriate animal tissue
Working with human tissue presents a unique set of hazards. <!>

Equipment

Centrifuge
Cheesecloth and filter funnel
Knife or meat grinder
Homogenizer (see note to Step 2)
Power-driven Potter-Elvehjem glass-Teflon homogenizer
Clearance range: 0.05–0.6 mm (see Figure 3.1).
Dounce hand homogenizer
Loose fitting: 0.1–0.3-mm clearance. Tight fitting: 0.05–0.08-mm clearance (see Figure 3.1).
Waring Blendor
A mechanical shear homogenizer that uses rotating metal blades or teeth to disrupt the material. There are many variations of the traditional domestic food liquidizer in which the material is placed in a glass reservoir with the blades driven by a motor beneath it. Other models resemble the modern hand-held blenders in which the motor is overhead. The Waring Blendor is typically used to macerate large amounts (100–1000 ml) of hard animal tissue and plant tissue. For smaller volumes (~1–5 ml), the Ultra-Turrex and its successor the Polytron homogenizer are widely used (see Figure 3.1).

METHOD

IMPORTANT: Carry out all procedures at 0–4°C.

1. After the tissue is excised from the animal, trim and discard fat and connective tissue from the tissue. Place the tissue in cold Homogenization buffer A.
2. Dice the washed tissue into small pieces (i.e., 1-cm cubes) with a knife or, alternatively, pass the tissue through a meat grinder twice.

   Tissues such as liver, brain, kidney, and heart are readily homogenized in a Waring Blendor, but tissues such as skeletal muscle and lung are tougher, and it is advisable to grind them in a domestic meat grinder prior to homogenization. Very fibrous tissues such as mammary glands must be frozen prior to homogenization to facilitate disruption (the Ultra-Turrex homogenizer is widely used for this purpose). Cultured mammalian cells and small amounts (1–5 g) of soft tissue such as brain can be homogenized conveniently using a Dounce hand homogenizer (Dignam 1990). In all cases, prechill the homogenizers and glassware to 4°C, and work in a cold room while using the blender.

3. Add 3–4 volumes of Homogenization buffer B per volume of tissue, and transfer the mixture to the homogenizer.
4. Prepare the homogenate using one of the following methods:

   Potter-Elvehjem homogenizer: Homogenize the tissue with the apparatus set at 500–1500 rpm, allowing 5–10 seconds per stroke.

   Dounce hand homogenizer: Homogenize the tissue with 10–20 strokes of the pestle.

   Waring Blendor: Homogenize the tissue three to four times for 20–30 seconds each (no longer), pausing for 10–15 seconds between each homogenization.

5. Pour the homogenate into a glass beaker, place the beaker on ice, and stir the homogenate gently for 30–60 minutes at 4°C to allow further extraction of proteins.

   EXPERIMENTAL TIP: Do not allow the homogenate to foam.

6. Remove cell debris and other particulate matter from the homogenate by centrifugation at 10,000*g* for 10–20 minutes at 4°C.

7. Filter the supernatant through two layers of cheesecloth (or a plug of glass wool) in a filter funnel to remove any fatty material that has floated to the surface. Carefully squeeze the cloth to obtain the maximum amount of filtrate (referred to as the "crude extract").

8. Proceed with the appropriate fractionation or analysis strategy as quickly as possible (see Chapter 1).

ADDITIONAL PROTOCOL: REMOVAL OF MUCIN FROM TISSUE HOMOGENATES

The presence of mucin in bulk biological extracts (e.g., colonic mucosa) complicates subsequent purification because the mucin binds to most chromatographic supports causing blockage of the column. Mucin can be selectively removed from tissue homogenates using this protocol, which was developed for the isolation and characterization of novel growth factors in colonic mucosa (Nice et al. 1991).

Additional Materials

CAUTION See Appendix 3 for appropriate handling of materials marked with <!>.

Ammonium carbonate buffer (7.85 g/liter, pH 9.2, ice-cold) containing the proteolytic inhibitors pepstatin (3 mg/liter) and PMSF (22 mg/liter) <!>

Method

1. Homogenize mucosa scraped from the descending colon, using an Ultra-Turex homogenizer, in 5 volumes of ice-cold carbonate buffer (pH 9.2) containing pepstatin and PMSF.

2. Remove the tissue debris by centrifugation at 10,000*g* for 20 minutes.

3. Slowly adjust the pH of the resulting supernatant to 4.5 (>2 hours) by the dropwise addition of 1 M HCl (~100 ml required) with continuous stirring at 4°C.

   This results in a flocculent precipitate of mucin (Glass 1964).

4. Remove the precipitate by centrifugation at 10,000*g* for 20 minutes at 4°C.

   The supernatant is the crude extract, which can be used as starting material in the desired purification strategy.

PROTOCOL 2

Lysis of Cultured Cells for Immunoprecipitation

Cell lysis with mild detergent is commonly used with cultured animal cells. If low detergent concentrations are sufficient to cause cell lysis (e.g., 1% NP-40 or 1% Triton X-100), this method may be more gentle to the protein of interest than the homogenization methods discussed in Protocol 1. The choice of detergent must be tailored to the nature of the epitope recognized by the immunoprecipitating antibody. If the antibody recognizes a linear peptide epitope (e.g., a synthetic peptide), then use a harsh denaturing lysis buffer (e.g., RIPA buffer). On the other hand, if the antibody is directed toward a conformational epitope, use NP-40 lysis buffer (or 1% Triton X-100) (for lysis buffer details, see Table 3.5). This protocol was contributed by Hong Ji (Joint ProteomicS Laboratory of the Ludwig Institute for Cancer Research, and Walter and Eliza Hall Institute of Medical Research, Melbourne, Australia).

MATERIALS

CAUTION: See Appendix 3 for appropriate handling of materials marked with <!>.

Reagents

Lysis buffer

Good first choices are NP-40 lysis buffer (Triton X-100 may be substituted for NP-40) and RIPA lysis buffer. For additional information on choosing a lysis buffer, see Table 3.5. Chill the lysis buffer to 4°C prior to use.

TABLE 3.5. Commonly used lysis buffers for lysing cultured cells

Buffer	Comments
NP-40 lysis 150 mM NaCl 1.0% NP-40 50 mM Tris-Cl (pH 7.4)	Probably the most widely used lysis buffer. It relies on the nonionic detergent NP-40 as the major solubilizing agent, which can be replaced by Triton X-100 with similar results. Variations include lowering the detergent concentration or using alternate detergents such as digitonin, saponin, or CHAPS.
RIPA lysis 150 mM NaCl 1% NP-40 0.5% sodium deoxycholate 0.1% SDS <!> 50 mM Tris-Cl (pH 7.4)	A much harsher denaturing lysis buffer than NP-40, due to the inclusion of two ionic detergents (SDS and sodium deoxycholate). In addition to releasing most proteins from cultured cells, RIPA lysis buffer disrupts most weak noncovalent protein-protein interactions.

When studying the modification of proteins by phosphorylation, phosphatase inhibitors (e.g., 25 mM NaF, 40 mM β-glycerol phosphate, 100 μM Na_3VO_4, or 1 μM microcystin) should be included. If proteolytic degradation of the target protein is a problem, protease inhibitors should be included in the lysis buffer (some commonly used inhibitors include aprotinin (1 μg/ml), leupeptin (1 μg/ml), pepstatin (1 μg/ml), and PMSF (50 μg/ml). Alternatively, commercially available protease cocktail tablets (e.g., from Boehringer) can be included. Also see Tables 3.3 and 3.4.

Phosphate-buffered saline (PBS)
- 8.0 g of NaCl (final concentration 137 mM)
- 0.2 g of KCl (final concentration 2.7 mM) <!>
- 1.44 g of Na_2HPO_4 (final concentration 10.1 mM) <!>
- 0.24 g of KH_2PO_4 (final concentration 1.8 mM) <!>
- in 800 ml of H_2O

Adjust pH to 7.4, and then adjust volume to 1 liter with H_2O.

◗ Cells

Suspension or monolayer cell cultures

◗ Equipment

Centrifuge

METHODS

Method A: Lysing Cells Grown as Monolayer Cultures

1. Discard the culture medium, and wash the cells twice with ice-cold PBS.
2. Place the culture dishes on ice.
3. Add 1.0 ml of lysis buffer (chilled to 4°C) per 100-mm dish. For culture dishes of other sizes, adjust the volume of lysis buffer accordingly.
4. Incubate the cells for 10–30 minutes (dependent on cell lines being studied) on ice with occasional rocking of the dishes.
5. Tilt a dish on the bed of ice and allow the buffer to drain to one side; remove the lysate with a pipette and transfer it to a microfuge tube or other suitable centrifuge tube. Repeat with all of the remaining dishes.

   Although some researchers prefer to scrape the cells from the tissue-culture dish, this does cause some stress to the cells and is only required in unusual cases.
6. Centrifuge the lysate at 20,000*g* for 10 minutes at 4°C.
7. Carefully remove the supernatant to a fresh tube, making sure not to disturb the pellet. Store the lysate on ice until it is needed for the preclearing and immunoprecipitation (see Harlow and Lane 1999).

   The cell lysate can be snap-frozen using a dry ice/ethanol mixture and then stored at –70°C for long-term storage. However, for the analysis of protein complexes by immunoprecipitation, the use of a freshly prepared cell lysate is recommended.

Method B: Lysing Cells Grown in Suspension

1. Harvest the cells by centrifugation at 480*g* for 10 minutes. Pour off the supernatant and discard.
2. Carefully wash the cell pellet twice with ice-cold PBS, and then place the washed cell pellet on ice.

3. Resuspend the pellet in 1.0 ml of lysis buffer (chilled to 4°C) per 1×10^7 to 5×10^7 cells.

4. Incubate the cells for 15 minutes on ice with occasional vortexing of the tube.

5. Centrifuge the lysate at 20,000*g* for 10 minutes at 4°C.

6. Carefully remove the supernatant to a fresh tube, making sure not to disturb the pellet. Store the lysate on ice until it is needed for the preclearing and immunoprecipitation (see Harlow and Lane 1999).

   The cell lysate can be snap-frozen using a dry ice/ethanol mixture and then stored at –70°C for long-term storage. However, for the analysis of protein complexes by immunoprecipitation, the use of a freshly prepared cell lysate is recommended.

PROTOCOL 3

Lysis of Cultured Animal Cells, Yeast, and Bacteria for Immunoblotting

CELL LYSIS WITH DETERGENTS IS COMMONLY USED with cultured animal cells. Typically, the ionic detergent SDS (e.g., 2% SDS) is sufficient for lysing cells for the purpose of immunoblotting studies. Both cultured animal cells and bacteria such as *E. coli* may be lysed in this manner. If the antigenic determinant recognized by the antibody being studied is dependent on the native spatial conformation and sensitive to reducing conditions, then dithiothreitol should be omitted from the lysis buffer and nonreducing/non-urea gels may need to be employed (Ji et al. 1997; Ji and Simpson 1999). This protocol was contributed by Hong Ji (Joint ProteomicS Laboratory of the Ludwig Institute for Cancer Research, and Walter and Eliza Hall Institute of Medical Research, Melbourne, Australia).

MATERIALS

CAUTION: See Appendix 3 for appropriate handling of materials marked with <!>.

Reagents

Dithiothreitol (DTT) (1 M) (10x stock solution) <!>

Store DTT in aliquots at –20°C.

1x Laemmli sample buffer

2% SDS <!>
10% glycerol
60 mM Tris-Cl (pH 3.8)
0.01% bromophenol blue <!>

It is often convenient to prepare Laemmli sample buffer as a 2x or 5x stock. Just prior to use, add DTT to a final concentration of 100 mM.

Cells

Suspension or monolayer cell cultures, bacteria, or yeast cells

10^9 tissue culture cells is ~1 ml or ~1 g.

Equipment

Centrifuge
Water bath or Heating block preset to 95°C

Additional Reagents

Step 6 of this protocol requires the reagents listed in Chapter 2, Protocol 1.

METHOD

1. Add 1 ml of Laemmli sample buffer containing 100 mM DTT to 1×10^7 to 5×10^7 cells.
2. The cell lysate becomes highly viscous in Laemmli buffer due to the presence of released DNA. Two options are available for surmounting this viscosity problem.
   - Centrifuge the lysate at 100,000*g* for 20 minutes to remove DNA.
   - Lyse the cells by sonicating the mixture using four bursts of 15–30 seconds each. Transfer the samples to ice for 15 seconds between each sonication step.
3. Heat the collected supernatant or sonicated sample for 5 minutes at 95°C.
4. Centrifuge at 20,000*g* for 5 minutes.
5. Transfer the supernatant to a fresh tube.
6. The samples (supernatant) are now ready for electrophoresis (see Chapter 2, Protocol 1) and immunoblotting (see Harlow and Lane 1999). When preparing cell culture extracts for immunoblotting, the protein sample must be
   - in a solution that is compatible with the gel electrophoresis system (e.g., the pH of the solution should be ~7.0 and the salt concentration ~200 mM) and
   - at a protein concentration that does not exceed the loading capacity for a particular gel system (for a discussion of gel electrophoresis variables, see Chapter 2). As a rule of thumb, for a conventional gel, do not load >150 µg of total protein per lane for a minigel.

PROTOCOL 4

Disruption of Cultured Cells by Nitrogen Cavitation

CELL DISRUPTION BY NITROGEN DECOMPRESSION from a pressurized vessel is a rapid and effective way to homogenize cells and tissues, to release intact organelles, and to prepare cell membranes (Hunter and Commerford 1961). The principle of the method is simple: Cells are placed in a pressure vessel, and large quantities of oxygen-free nitrogen are first dissolved in the cells under high pressure (~5500 kPa, which is equivalent to 800 psi). When the gas pressure is suddenly released, the nitrogen comes out of solution as bubbles that expand and stretch the cell membrane, rupturing it, and releasing the contents of the cell. Nitrogen cavitation is well suited for mammalian and plant cells and fragile bacteria (i.e., bacteria treated to weaken the cell wall), but it is less effective at lysing yeast, other fungi, spores, or other cell types with tough cell walls. Features of the nitrogen cavitation method include the following:

- It is a gentle method for homogenizing or fractionating cells because the chemical and physical stresses that it imposes upon enzymes and subcellular compartments are minimized compared to other ultrasonic and mechanical homogenizing methods. For example, functional intact nuclei and mitochondria can be released from most cell types.
- Unlike many cell lysis methods relying on shear stresses and friction, no heat is generated with the nitrogen cavitation because this method is accompanied by an adiabatic expansion that cools the sample instead of heating it. Hence, there is no heat damage to proteins and organelles.
- Any labile cell components are protected from oxidation by the use of an inert gas, nitrogen. Furthermore, nitrogen does not alter the pH of the suspending medium.
- The process is fast and uniform because the same disruptive forces are applied within each cell and throughout the sample, ensuring reproducible cell-free homogenates.
- Variable sample sizes (e.g., from ~1 ml to 1 liter or more) can be accommodated with most commercial systems.

This protocol, adapted from the Kontes Glass Company "instructions for users," is designed for small volumes (1–15 ml) of tissue culture cells using the Kontes Mini-Bomb cell disruption chamber (see Figure 3.2).

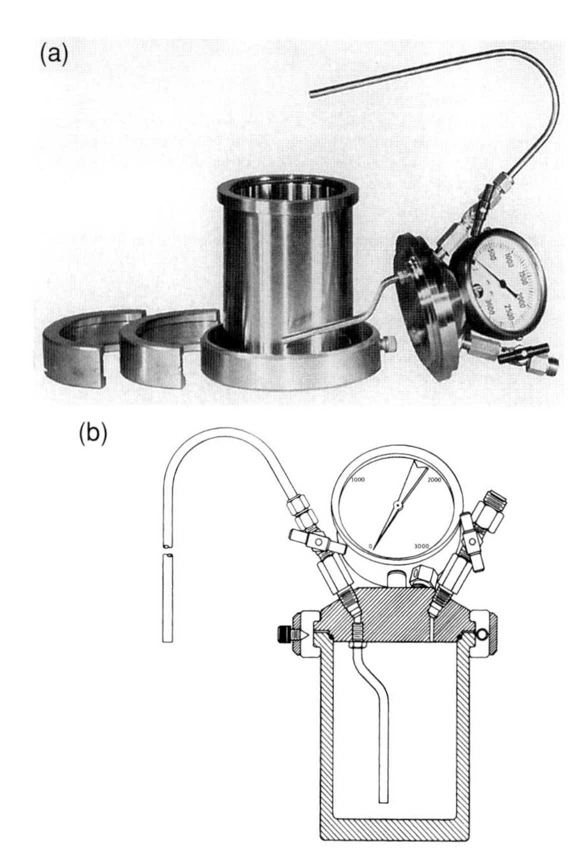

FIGURE 3.2. Nitrogen cavitation vessel. (*a*) Disassembled parts of the homogenizer; (*b*) diagrammatic cross-section. The version shown in *a* is manufactured by Baskervilles (Manchester, U.K.). Artisan Industries (Waltham, Massachusetts) makes a similar version that uses the pressure gauge on the gas cylinder to monitor the pressure. (Reprinted, with permission, from Evans 1992 [©Oxford University Press], as in Graham and Rickwood 1997.)

MATERIALS

CAUTION: See Appendix 3 for appropriate handling of materials marked with <!>.

Reagents

Homogenization medium

0.25 M sucrose
20 mM Tris-Cl (pH 7.4)

For alternative homogenization buffers, see Protocol 1. For nitrogen cavitation, choose any buffer compatible with subsequent purification steps. Isotonic solutions are suitable for most applications. Hypertonic solutions tend to stabilize organelles. Low concentrations of $MgCl_2$, magnesium acetate, or $CaCl_2$ have been reported to stabilize nuclei.

Phosphate-buffered saline

8.0 g of NaCl (final concentration 137 mM)
0.2 g of KCl (final concentration 2.7 mM) <!>
1.44 g of Na_2HPO_4 (final concentration 10.1 mM) <!>
0.24 g of KH_2PO_4 (final concentration 1.8 mM) <!>
in 800 ml of H_2O

Adjust pH to 7.4, and then adjust volume to 1 liter with H_2O.

Cells

Cultured animal cells or animal tissue

Equipment

Cell disruption chamber

Commercially available cell disruption chambers (also called nitrogen pressure vessels) include:

- Mini-Bomb cell disruption chamber (Kontes Glass Company, Vineland, New Jersey; www.kimble-kontes.com).
- Parr Cell Disruption Bomb (Parr Instrument Company, Moline, Illinois; www.parrinst.com).
- Baskervilles (Manchester, U.K.) or Artisan Industries (Waltham, Massachusetts).

Magnetic stirrer

Nitrogen (oxygen-free) <!>

Rubber policeman

METHOD

CAUTION: Because high pressures are generated, carry out this procedure behind a shield.

1. Ensure that the outlet port is closed (do not over-tighten), and cool the cell disruption chamber (if necessary) by immersing the unit in an ice bath.
2. Prepare the cells or tissue for disruption.

   For adherent cultured cells: Wash the cells gently two or three times with phosphate-buffered saline (PBS). Use a rubber cell scraper to scrape the cells from the dish into PBS.

   For nonadherent cells: Collect the cells by centrifugation at 400*g* for 10 minutes. Wash the pellet of cells two or three times with PBS.

   For animal tissue: Mince the tissue using a mechanical homogenizer or by passing through a screen or sieve.
3. Collect the cells by centrifugation at 480*g* for 5 minutes at 4°C.
4. Discard the supernatant and resuspend the cells in 10 volumes (~15 ml) of homogenization medium. Transfer the suspension to a plastic beaker (containing a magnetic stir bar) and gently mix to obtain a uniform suspension.
5. Remove the filter holder from the chilled cell disruption chamber by using a turning and lifting motion.
6. Add up to 15 ml of cell suspension to the cell disruption chamber.

   For larger volumes, an extension is available for the Mini-Bomb. Alternatively, other nitrogen pressure vessels can be used.
7. Replace the filter holder.
8. Attach the unit to the source of oxygen-free nitrogen by screwing on the cap of the pressure vessel, which should already be connected to the source of nitrogen. *HAND TIGHTENING* is all that is required.
9. Replace the pressure vessel in the cooling bath if desired.

10. Pressurize the bomb to the desired pressure by allowing nitrogen to flow from the cylinder into the disruption chamber.

    There must be a gauge on the filling connection that shows the pressure inside the chamber. The amount of pressure required depends on the type of cells being disrupted. Typical pressures used include:

    - 500 psi for KB cells
    - 500 psi for rat liver
    - 1000 psi for chicken red blood cells

    Step 10 must be optimized for each application. In general, lower pressures lead to a greater recovery of intact organelles, whereas higher pressures can result in total homogenization. Cells that are difficult to lyse can be treated more than once for more complete homogenization. For additional information on effective gas pressures to use to disrupt various cells, see Table 3.6.

    CAUTION: Inexperienced workers handling high-pressure gases must always seek expert help before using a gas cylinder.

11. Allow the pressurized apparatus to equilibrate for at least 30 minutes to allow the nitrogen to dissolve and come to equilibrium within the cells. Gentle agitation once or twice during this time period with a magnetic stirrer or periodic gentle shaking keeps the cells in a uniform suspension.

12. Place a collection container (e.g., beaker) in an ice bath by the outlet port.

TABLE 3.6. Suggested working parameters for disruption of cells using nitrogen cavitation

Cell type	Cell suspension	psi	No. cells used/ml	No. times through	No. cells remaining/ml	No. nuclei remaining/ml	% cells totally lysed
Cultured cells	KB	500	3.4×10^6	1	2.8×10^5	n.d.	93
	KB	500		2	2.0×10^3	n.d.	99.95
	KB	250	3.3×10^6	1	5.5×10^5	4.62×10^5	69.4
	KB	250		2	5.4×10^4	2.2×10^4	97.7
	KB	250		3	0	0	100
	KB	0	1.8×10^6	0	1.08×10^6	7.6×10^5	–
	KB	250		1	2.4×10^5	6.4×10^5	52.2
	KB	250		2	3.4×10^4	5.0×10^4	95.4
	KB	250		3	1.6×10^4	4.6×10^4	96.6
Tissue	rat liver	500	9.2×10^5	1	8.5×10^5	n.d.	7.8
		500		2	0		100
Blood	chicken red blood cells	1000	5.6×10^9	1	9.6×10^8	n.d.	83
Bacteria	*E. coli*	1500	1.3×10^{10}	1	1.2×10^{10}	n.d.	8
		1500		2	6.6×10^9	n.d.	50
	E. coli	1500	6.5×10^8	1	3.6×10^8	n.d.	45
		1500		2	3.3×10^8	n.d.	50

These values were determined with the Mini-Bomb cell disruption chamber. (KB cells) Human oral epidermoid carcinoma. n.d. indicates not determined. (Reprinted, with permission, from Kimble/Kontes [www.kimble-kontes.com/html/pg-881455.htm].)

13. With the pressurization unit still activated, *slowly* open the outlet port until the cell suspension begins to flow into the chilled collection container. Make sure that the pressure is maintained until all of the suspension is out.

    A reasonable flow rate is 3–10 drops per second. An indication that the unit is almost empty is the appearance of a "foamier" flow of fluid. The end of the flow is accompanied by a gentle hissing.

    Cell lysis occurs upon release of pressure, but passage through the narrow orifice of the outlet port often aids cell disruption.

    CAUTION: The homogenate is expelled with great force, so this operation must be carried out very carefully, particularly if the sample contains hazardous materials. Because of the potential for creating an aerosol, this procedure must always be performed in a safety hood.

14. Gently stir the "foam" to allow it to subside before any centrifugation or subsequent work up of the homogenate is carried out.

15. Close the outlet port. Do not overtighten it.

16. Turn off the flow of nitrogen at its source.

17. Discharge the gas in the unit by opening the outlet port and venting the unit. Once the pressure inside the bomb has returned to atmospheric pressure, the chamber can be opened for recovery of any remaining sample and thorough cleaning.

    CAUTION: If the lysed cell material is potentially hazardous, vent the gas through a suitable trap.

PROTOCOL 5

Small-scale Extraction of Recombinant Proteins from Bacteria

BACTERIA ARE PARTICULARLY CONVENIENT FOR PRODUCING RECOMBINANT PROTEINS for purification purposes. To monitor induction as well as the levels of recombinant protein expression, it is important to have a rapid, simple method for estimating bacterial protein expression. This protocol describes the preparation of small-scale bacterial extracts using cell lysis with 0.5% Triton X-100.

MATERIALS

CAUTION: See Appendix 3 for appropriate handling of materials marked with <!>.

Reagents

Dithiothreitol (DTT) (1 M) (10x stock solution) <!>
Store DTT in aliquots at –20°C.

1x Laemmli sample buffer
- 2% SDS <!>
- 10% glycerol
- 60 mM Tris-Cl (pH 6.8)
- 0.01% bromophenol blue <!>

It is often convenient to prepare Laemmli sample buffer as a 2x or 5x stock. Just prior to use, add DTT to a final concentration of 100 mM.

Triton X-100 (0.5% v/v) <!>
Cool to 4°C before use.

Cells

Bacterial cells expressing target recombinant protein

Equipment

Boiling-water bath
Sonicator <!>

Additional Reagents

Step 5 of this protocol requires the reagents listed in Chapter 2, Protocol 1.

METHOD

1. Harvest the bacterial cells from 1 ml of culture by centrifugation in a microfuge for 1 minute at room temperature.
2. Pour off the supernatant and resuspend the bacterial cell pellet in 1 ml of chilled aqueous 0.5% (v/v) Triton X-100.
3. Sonicate the suspension for three cycles of 20 seconds each, cooling the cells on ice between treatments.
4. Centrifuge the suspension in a microfuge for 1 minute.
5. Add Laemmli sample buffer containing 100 mM DTT to the supernatant, boil the mixture for 2 minutes, and analyze for the target protein on analytical SDS-PAGE (see Chapter 2, Protocol 1).

PROTOCOL 6

Large-scale Extraction of Recombinant Proteins from Bacteria

BACTERIA ARE PARTICULARLY CONVENIENT FOR PRODUCING RECOMBINANT PROTEINS for purification purposes. Suitable extraction methods for bacterial cells include sonication, glass bead milling, grinding with alumina or sand, high-pressure shearing using the French pressure cell (French Press), and lysozyme treatment. These procedures are applicable for preparing extracts from a variety of gram-negative bacteria such as *E. coli* and gram-positive bacteria such as *Klebsiella pneumoniae* and *Bacillus subtilis*. Disruption of bacterial cells by enzymatic means is commonly used because a relatively uniform treatment is obtained when cells are in suspension. A protocol for enzymatic disruption of *E. coli* follows.

MATERIALS

CAUTION: See Appendix 3 for appropriate handling of materials marked with <!>.

Reagents

Dithiothreitol (DTT) (1 M) (10x stock solution) <!>
  Store DTT in aliquots at –20°C.

DNase I

1x Laemmli sample buffer
  2% SDS <!>
  10% glycerol
  60 mM Tris-Cl (pH 6.8)
  0.01% bromophenol blue <!>

  It is often convenient to prepare Laemmli sample buffer as a 2x or 5x stock. Just prior to use, add DTT to a final concentration of 100 mM.

Lysis buffer
  50 mM Tris-Cl (pH 7.4)
  25% (w/v) sucrose

  EDTA can be included (to a final concentration of 10 mM) to minimize proteolytic degradation due to metalloproteases. Although the inclusion of EDTA and protease inhibitors such as 1 mM PMSF <!> (or AEBSF), 1 μg/ml aprotinin <!>, or 10 μg/ml leupeptin <!> in the lysis buffer may not be necessary when the target protein is packaged in inclusion bodies, they may be important in the solubilization/ refolding steps (unfolded proteins are more susceptible to proteolytic degradation) or when extracting soluble proteins.

Lysozyme (hen egg)

$MgCl_2$ (1 M) <!>

NP-40 (10% v/v)

Cells

Bacterial cells expressing target recombinant protein

Equipment

Centrifuge
Waring Blendor (see Step 4)

Additional Reagents

Step 9 of this protocol requires the reagents listed in Chapter 2, Protocol 1.

METHOD

1. Harvest the bacterial cells by centrifugation at 3000*g* for 15 minutes at 4°C.
2. Wash the cells with lysis buffer to remove the residual culture medium and harvest the washed cells by centrifugation as in Step 1.
3. Pour off the supernatant and weigh the wet pellet.
4. Resuspend the washed *E. coli* cells in ~3 ml of lysis buffer per gram of cell pellet and stir the suspension for 30 minutes at 4°C. If the pellet is not fully resuspended after 30 minutes, mix the suspension in a Waring Blendor at low speed for ~1 minute.
5. Add lysozyme to a concentration of 0.1% (w/v) and incubate for 35 minutes at 4°C, shaking gently.

   A faster rate of lysis may be obtained by increasing the lysozyme concentration to 1.0% (w/v) (10 mg/ml). Under these conditions, satisfactory lysis can be accomplished in as little as 5 minutes at temperatures as low as 4°C (Bollag et al. 1996).

6. Add in the following order:

   NP-40 to a final concentration of 0.5% (v/v)
   $MgCl_2$ to a final concentration of 5 mM
   DNase I to a final concentration of 40 μg/ml

   Stir the suspension for 30 minutes at 4°C to remove the viscous nucleic acid.

   Bacterial extracts are 40–70% protein, 10–30% nucleic acid, 2–10% polysaccharide, and 10–15% lipid (Worrall 1996). The release of DNA upon cell lysis often results in a highly viscous extract that can cause serious problems in subsequent chromatographic purification steps. In addition to DNase I treatment, DNA can be removed from the cell extract (along with other nucleic acids, and in some cases, highly acidic proteins) by the addition of a neutralized solution of positively charged compounds such as protamine sulfate (up to 5 mg/g wet weight of cell pellet) (Scopes 1994) or polyethyleneimine (Burgess and Jendrisak 1975).

   EXPERIMENTAL TIP: Methods for DNA removal involving positively charged compounds should not be used with inclusion body preparations, since the precipitated DNA will cocentrifuge with the inclusion bodies.

7. Centrifuge the suspension at 23,000*g* for 30 minutes at 4°C.
8. Resuspend a small portion of the pellet in Laemmli buffer containing DTT.
9. Analyze aliquots of both the soluble protein fraction (supernatant) and pellet fraction for the presence of target protein using analytical SDS-PAGE (see Chapter 2, Protocol 1).

   If the bulk of the target protein is found in the insoluble pellet fraction, then inclusion bodies have likely formed, and the target protein will need to be solubilized and purified according to Protocol 7. If the target protein is found in the supernatant, this material should be stored at 4°C in readiness for the next purification protocol(s).

PROTOCOL 7

Solubilization of *E. coli* Recombinant Proteins from Inclusion Bodies

BECAUSE MOLECULAR CLONING TECHNIQUES ALLOW HIGH LEVELS of expression in bacteria, this is a particularly convenient system for producing recombinant proteins. Regrettably, these proteins are often difficult to purify due to their tendency to aggregate and precipitate within the bacteria to form insoluble inclusion bodies. The formation of inclusion bodies is especially common for nonbacterial proteins. Although no single method can be applied to every protein, a number of strategies are available to solubilize inclusion body proteins. One of these strategies is described in this protocol. A number of steps must be considered in solubilizing inclusion body proteins:

- cell lysis
- isolation of inclusion bodies
- washing of inclusion bodies
- solubilization of inclusion bodies
- renaturation (if required) of recombinant protein

First, the washing of inclusion bodies prior to solubilization is an important step for removing nontarget proteins, as well as nonproteinaceous material. For example, washing inclusion bodies with 1 M guanidine hydrochloride was used to remove contaminating proteins from recombinant IL-2 (Weir et al. 1987), and 4.0 M urea was used in the case of IL-6 inclusion bodies (Zhang et al. 1992). It is important to remove nonproteinaceous (lipid-like) material in order to prevent the clogging of RP-HPLC columns and therefore prolong column lifetimes. This can be accomplished by extensive washing of inclusion bodies with 4.0 M urea (Zhang et al. 1992) or extraction of the inclusion bodies with butanol-1-ol/10 mM EDTA (Weir et al. 1987). Whereas some target proteins will refold due to air oxidation (e.g., IL-6 [Zhang et al. 1992]), the most common methods to induce proper refolding of proteins use extensive dilution and dialysis in the presence of thiol reagents (e.g., 1 mM DTT) to gradually reduce the concentration of denaturant and promote correct disulfide bond formation.

MATERIALS

CAUTION: See Appendix 3 for appropriate handling of materials marked with <!>.

Reagents

Renaturation buffers
See Step 9 for examples.

Solubilization buffer (8 M guanidine hydrochloride in 50 mM Tris-Cl [pH 7.5]) <!>
The ideal solubilization buffer for a particular protein may have to be determined empirically. See notes to Steps 8 and 9.

Washing buffer
100 mM Tris-Cl (pH 8.0)
2–4.0 M urea <!>
1% (v/v) Triton X-100 <!>

Cells

Bacterial cells expressing the target recombinant protein

Equipment

Centrifuge

Additional Reagents

Step 1 of this protocol requires the reagents listed in Protocol 6 of this chapter.
Step 7 of this protocol requires the reagents listed in Chapter 2, Protocol 1.

METHOD

1. Lyse the cells as described in Protocol 6.
2. Centrifuge the cell lysate at 23,000*g* for 30 minutes at 4°C.
3. Pour off the supernatant and measure the mass of the pellet (wet mass).
4. Resuspend the pellet in ~10 volumes of washing buffer. Stir the suspension for 1 hour at room temperature.
5. Centrifuge the mixture at 23,000*g* for 30 minutes at 4°C.
6. Pour off the supernatant and recover the pellet.
7. Repeat Steps 4–6 three more times (or until the recombinant protein is >80% pure, as judged by analytical SDS-PAGE as described in Chapter 2).
8. Dissolve the pellet from Step 7 in ~8 ml of solubilization buffer per gram wet mass pellet (determined in Step 3), and gently stir the mixture for 16–20 hours at 4°C.

   EXPERIMENTAL TIP: Do not exceed a protein concentration of 2–3 mg/ml.

   Alternative solubilization buffers:

   - 8 M guanidine hydrochloride, 10 mM DTT, 50 mM Tris-Cl (pH 8.5).

     Use 15 ml per 12–15 g wet weight of harvested cells. Incubate the mixture for 1 hour at 37°C (Weir et al. 1987). Renature the protein as described in Step 9a.

   - 8 M urea, 2 mM reduced glutathione/0.2 mM oxidized glutathione.

     Use 9 volumes of buffer per gram wet weight of inclusion body pellet. Incubate the mixture for 1 hour at room temperature. Do not exceed a protein concentration of 2.5 mg/ml (Bollag et al. 1996). Renature the protein as described in Step 9b.

9. Refolding conditions for a specific target protein must be optimized in pilot-scale experiments. Examples of renaturation conditions include the following:

   a. Dilute the solubilized pellet 25-fold with 50 mM Tris-Cl (pH 8.5) containing 1.5 μM copper sulfate to give final concentrations of 0.04 M DTT, 0.24 M guanidine hydrochloride, and ~1.4 μg/ml target protein (Weir et al. 1987). Incubate the mixture for 2 hours at 20°C.

   b. Slowly add 9 volumes of 50 mM phosphate, 50 mM NaCl, and 1 mM EDTA containing 2 mM reduced glutathione/0.2 mM oxidized glutathione to the solubilized pellet. Incubate the mixture for 2–4 hours at 25°C.

10. Recover the folded protein using a concentration step that is appropriate for the target protein (e.g., RP-HPLC, lyophilization, or ultrafiltration).

PROTOCOL 8

Preparation of Extracts from Yeast

BECAUSE YEAST IS EXCEPTIONALLY WELL SUITED TO GENETIC ANALYSIS, both classical and molecular, it is an attractive system for expressing recombinant animal proteins for purification purposes. For a discussion on cell growth and harvesting yeast, especially the genus of the budding yeast *Saccharomyces*, see Jazwinski (1990). A number of methods available for lysing yeast cells include autolysis, pressure cells (e.g., French pressure cell), abrasives (glass bead vortexing), and enzymatic lysis (e.g., zymolase). One of the simplest methods, discussed in this protocol, involves the abrasive action of well-agitated glass beads. This is a very effective method for both low volumes (e.g., <1 ml using a microfuge tube) and many liters using a specialized DynoMill apparatus. Cell breakage is typically >95%, as assessed by phase-contrast microscopy.

MATERIALS

CAUTION: See Appendix 3 for appropriate handling of materials marked with <!>.

Reagents

Lysis buffer (RIPA buffer)
- 150 mM NaCl
- 1% NP-40
- 0.5% sodium deoxycholate
- 0.1% SDS <!>
- 50 mM Tris-Cl (pH 7.2)

Just prior to use, add protease inhibitors to the following final concentrations: 1 μg/ml aprotinin, 1 μg/ml leupeptin, 1 μg/ml pepstatin, and 50 μg/ml PMSF.

Phosphate-buffered saline (PBS)
- 8.0 g of NaCl (final concentration 137 mM)
- 0.2 g of KCl (final concentration 2.7 mM) <!>
- 1.44 g of Na_2HPO_4 (final concentration 10.1 mM) <!>
- 0.24 g of KH_2PO_4 (final concentration 1.8 mM) <!>
- in 800 ml of H_2O

Adjust pH to 7.4, and then adjust volume to 1 liter with H_2O.

Cells

Yeast cells, from a freshly grown culture

Equipment

Glass beads (500 μm) chilled

Prepare glass beads by washing twice in 1 N HCl <!> and twice in lysis buffer. Store the washed beads at 4°C in a small volume of lysis buffer.

METHOD

1. Harvest the cells by centrifugation in a microfuge for 3 minutes.
2. Discard the supernatant, resuspend the cells in PBS, and centrifuge again for 3 minutes.
3. Discard the supernatant. Estimate the volume of the cell pellet. Add 3 volumes of ice-cold lysis buffer per volume of cell pellet. Keep the suspension on ice.
4. Add a volume of chilled glass beads equal to the total volume of the resuspended yeast cells.
5. Vortex the suspension vigorously for 30 seconds.
6. Repeat Step 5 until the bulk of the yeast cells are lysed, as determined by phase-contrast microscopy.
7. Centrifuge the suspension in a microfuge for 5 minutes at 4°C.
8. Carefully pour off the supernatant and transfer to a fresh tube. Keep it on ice until use.

PROTOCOL 9

Differential Detergent Fractionation of Eukaryotic Cells

DIFFERENTIAL DETERGENT FRACTIONATION (DDF) involves the sequential extraction of cells with PIPES buffers containing first digitonin, then Triton, and finally Tween/deoxycholate. The procedure yields four biochemically and electrophoretically distinct fractions (see Figure 3.3 and Table 3.7) composed of the following:

- Cytosolic proteins and extractable cytoskeletal elements: Microtubule components can be semipurified from this fraction by magnesium precipitation (see ADDITIONAL PROTOCOL: PRECIPITATION OF TUBULINS AND MAPS USING MAGNESIUM at the end of this protocol).
- Membrane and organelle proteins: Integral membrane proteins can be enriched by using Triton X-114 (Bordier 1981; Pryde and Phillips 1986).
- Nuclear membrane proteins and extractable nuclear proteins.
- Detergent-resistant cytoskeletal filaments and nuclear matrix proteins: Cytoskeleton-associated proteins and interactions can be investigated further by differential urea extraction.

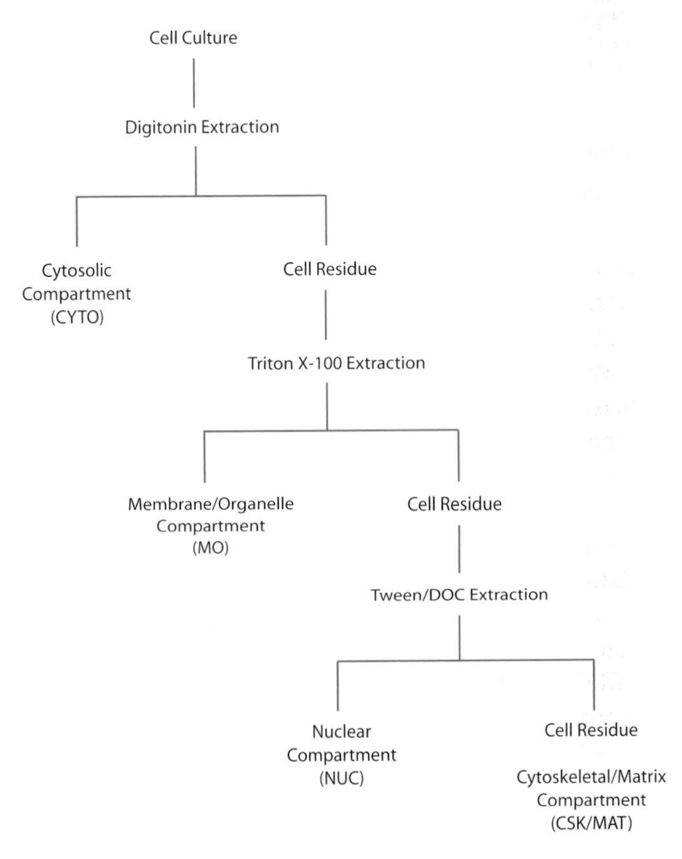

FIGURE 3.3. Schematic for differential detergent fractionation (DDF). (Courtesy of Melinda L. Ramsby and Gregory S. Makowski, University of Connecticut Health Center.)

TABLE 3.7. Protein distribution in detergent fractions of hepatocytes

Cell fraction	Protein distribution in hepatocytes
Digitonin/EDTA extracts (CYTO fraction)	• ~35% total hepatocytic cellular protein • enriched in cytosolic markers (90% LDH activity, 100% carbonic anhydrase immunoreactivity) (Ramsby et al. 1994)
Triton extracts (MO fraction)	• constitutes the bulk of hepatocytic cellular proteins (~50% total protein) • enriched in markers for membrane and organelle proteins (Ramsby et al. 1994)
Tween/DOC (NUC fraction)	• ~5% of total hepatocytic cell protein • contains, exclusively, immunoreactivity for the nuclear protein p38 (Ramsby et al. 1994)
Detergent-resistant fraction (CSK/MAT fraction)	• ~7–10% of hepatocytic cellular protein enriched in intermediate filaments, actin, various cytoskeleton-associating proteins, and nuclear matrix • for monolayer cultures, this fraction also contains extracellular matrix (Ramsby et al. 1994)
Magnesium-precipitable fraction (tubulins and microtubule-associated proteins)	• 1–2% of total hepatocytic cellular protein • represents ~4–5% of the protein in digitonin/EDTA extracts • includes tubulins, actin, and proteins presumed to interact with the cytoskeleton

Hepatocytes have a high metabolic rate, which may affect the percent distributions of protein. Distributions will vary depending on cell type and should be determined empirically. Table courtesy of Melinda L. Ramsby and Gregory S. Makowski (University of Connecticut Health Center).

EXAMPLES OF CELLULAR ANALYSIS USING DIFFERENTIAL DETERGENT FRACTIONATION

DDF has been used to fractionate suspension-cultured hepatocytes for the goal of following intracellular redistribution of mobile cytoskeletal elements in response to nutritional deprivation as a function of autophagy (Ramsby 1989). Fractionation profiles suggest that DDF would also be useful for monitoring compartmental redistribution of heat shock proteins (Ramsby et al. 1994). DDF has been used to fractionate monolayer cultures of bovine corneal endothelial cells toward the goal of monitoring intracellular levels and stores of plasminogen activator (Ramsby and Kreutzer 1993). DDF can be used to selectively extract type IV collagenase (i.e., gelatinase B) from isolated human polymorphonuclear leukocytes and thereby provides a mechanism for differentiating primary and secondary granule content release (Makowski and Ramsby 1997). In rat hepatocytes, DDF has been used to study both regulatory and pathological protein degradation in mitochondrial (Makowski and Ramsby 1999) and cytoskeletal compartments (Makowski and Ramsby 1998), respectively. In addition, DDF can be used to assess RNA distribution in monolayer cultures of osteoblasts for the goal of monitoring induction, turnover, and trafficking of specific mRNA in response to bone cell mitogens (M. Ramsby, unpubl.). RNA isolated from DDF extracts is intact and differentially distributed with respect to ribosomal and transfer RNAs (see Figure 3.4). DDF has also been used extensively for the study of signal transduction pathways leading to protein subcellular translocation, including filamin, nitric oxide synthase, myosin, and NF-κB (Patton 1999 and references therein).

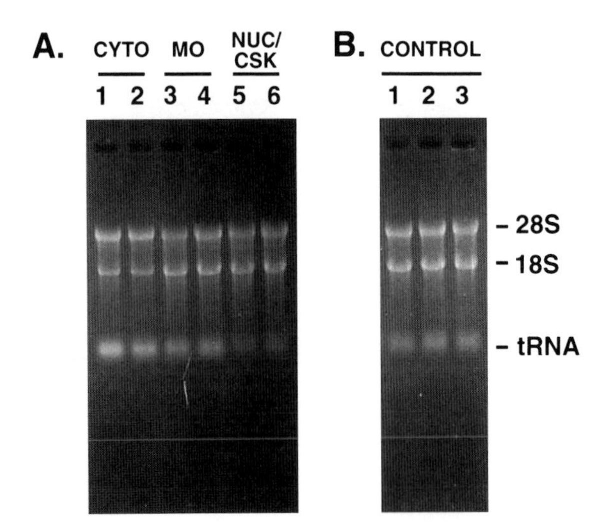

FIGURE 3.4. Agarose gel of osteoblastic RNA isolated from differential detergent cell fractions. RNA was isolated from osteoblastic cell fractions obtained by DDF: digitonin/EDTA cytosolic (CYTO), Triton/EDTA membrane/organelle (MO), and combined nuclear/detergent-resistant cytoskeletal/ matrix (NUC/CSK) (*A*), and compared to total RNA isolated from unfractionated monolayers (*B*). RNA (15 µg) in equal volumes (15 µl) was separated by gel electrophoresis on 1% agarose-formaldehyde gels containing ethidium bromide to visualize ribosomal RNA (28S, 18S) and transfer RNA (tRNA). (Courtesy of Melinda L. Ramsby and Gregory S. Makowski, University of Connecticut Health Center.)

The DDF protocol described here represents a modification of a method used for the fractionation of Madin-Darby canine kidney (MDCK) cells (Fey et al. 1984). Modifications include the addition of a digitonin extraction step, the inclusion of EDTA in digitonin and Triton buffers, and the elimination of a nuclease digestion step (DNA is denatured by shear force in the presence of SDS). Intact RNA can also be isolated from each detergent fraction by routine protocols and subjected to northern blot analysis for investigation of mRNA dynamics (see ADDITIONAL PROTOCOL: ISOLATION OF RNA FROM DETERGENT EXTRACTS at the end of this protocol). The main protocol and additional protocols were generously provided by Melinda Ramsby and Gregory Makowski (University of Connecticut Health Center).

MATERIALS

CAUTION: See Appendix 3 for appropriate handling of materials marked with <!>.

Reagents

Acetone (90%) chilled to –20°C <!>

Detergent extraction buffers

In general, the detergent extraction buffer formulations are for 100 ml of working solution. Adjust the final volumes for specific use, or freeze aliquots for future use. Solutions should be thawed on ice and not refrozen. The PIPES buffer recipes are listed elsewhere in this Materials list. For additional information on the properties of the detergents, see Table 3.8 and the booklet, A guide to the Preparation and Uses of Detergents in Biology and Biochemistry (available at http://www.calbiochem.com).

- Digitonin extraction buffer (0.015% digitonin, pH 6.8 at 4°C) <!>

  Dissolve by heating (boiling) 18.75 mg of digitonin with 10 ml of 4x PIPES buffer in a small flask with a stir bar. Add 1 ml of 100 mM PMSF <!> while the buffer is still warm. Combine with 15 ml of 4x PIPES buffer and 5 ml of 100 mM EDTA. Cool to 4°C and adjust pH to 6.8 with HCl <!>. Add H_2O to a final volume of 100 ml and keep on ice until use. Digitonin buffer without EDTA is stable for 3 hours at 0°C. EDTA prolongs the stability of this buffer.

TABLE 3.8. Detergents used in differential detergent fractionation of cells

Detergent	Cell fraction	Volume of detergent used	Comments
Digitonin	CYTO	*suspension cells:* 5 volumes/g (wet weight) of cell pellet *adherent cells:* 1 ml per T-25 flask-equivalent; 3 ml per 100-mm Petri dish	• Digitonin is a steroidal compound that complexes with plasma membrane cholesterol to cause membrane permeabilization and rapid release of soluble cytosolic components (Mackall et al. 1979). • Low concentrations of digitonin (0.015%) preserve ER, mitochondrial, lysosomal, and organelle membrane integrity, hence, ultrastructure, which are damaged at higher concentrations (~0.1%) (Fiskum et al. 1980; Weigel et al. 1983). • EDTA significantly enhances the effectiveness of low concentrations of digitonin, as evidenced by an increased rate of membrane permeabilization (10 minutes in the presence of EDTA vs. 40 minutes in the absence of EDTA) and inhibits calcium-dependent neutral proteases (calpains), typically enriched in cytosolic extracts (Ramsby et al. 1994). • Digitonin releases proteins larger than 200 kD (Weigel et al. 1983), as evidenced by the presence of myosin (>220 kD), desmoplakin II (>250 kD), and the calpain-inhibitor calpastatin (~300 kD) in cytosolic extracts (Ramsby et al. 1994). • The combination of ice-cold temperatures and EDTA results in microtubule depolymerization, which promotes fractionation of tubulins and microtubule-associated proteins in digitonin/EDTA extracts. Tubulins and microtubule-binding proteins can be harvested by repolymerization at 37°C and magnesium precipitation (see ADDITIONAL PROTOCOL: PRECIPITATION OF TUBULINS AND MAPS USING MAGNESIUM).
Triton X-100	MO	*suspension cells:* 5 volumes/g (wet weight) of cell pellet *adherent cells:* 1 ml per T-25 flask-equivalent; 3 ml per 100-mm Petri dish	• Triton is a nonionic detergent, which solubilizes membrane lipids and releases organelle contents. • It has been used in hyper- or hypotonic buffers to obtain cytoskeletal preparations enriched in intermediate filaments (Franke et al. 1978). • Low concentrations of Triton concomitant with iso-osmolar, isotonic buffer composition, as used here, preserve the integrity of the nuclear membrane as well as the microfilament cytoskeleton (Fey et al. 1984). • Triton extraction is possible using either the X-100 or X-114 detergent series. Fractionation with X-114 allows subfractionation of peripheral versus integral membrane proteins (Bordier 1981; Pryde and Phillips 1986) and lends further flexibility to fractionation goals.
Tween/DOC	NUC	*suspension cells:* 2.5 volumes/g (wet weight) of cell pellet *adherent cells:* 0.5–1 ml per T-25 flask-equivalent	• DOC is a weakly ionic detergent that disrupts nuclear membrane and microfilament integrity (Capco et al. 1982; Fey et al. 1984; Reiter et al. 1985). • Tween/DOC extracts contain proteins common to both the membrane/organelle and detergent-resistant cytoskeletal fractions, perhaps representing a more labile, protein compartment.
–	CSK/MAT (detergent-resistant)	–	• DNA present in the detergent-resistant fraction is viscous and difficult to manage. • Denaturation is achieved by mechanical shear, using either Teflon/glass homogenization (for suspension pellets) or tituration with a pipette (for monolayer residues). • The cytoskeletal filaments in this fraction are intact as evidenced by two-dimensional PAGE (Ramsby et al. 1994). Staircase patterns, indicative of cytoskeletal degradation, however, do arise if PMSF or EDTA is absent from extraction buffers. • Solubilization of detergent-resistant samples in SDS phosphate buffer, in the absence of mercaptoethanol, allows direct assay of protein content by the method of Peterson (1983).

Table courtesy of Melinda L. Ramsby and Gregory S. Makowski (University of Connecticut Health Center).

- Triton X-100 extraction buffer (0.5% Triton X-100, pH 7.4 at 4°C) <!>

  Combine 25 ml of 4x PIPES buffer, 1 ml of 100 mM PMSF, 3 ml of 100 mM EDTA, and 5 ml of freshly prepared 10% (v/v) Triton X-100. Cool to 4°C, adjust pH to 7.4, and add H_2O to 100 ml (final volume). Keep on ice until use.

- Tween-40/deoxycholate extraction buffer (1% Tween-40/0.5% DOC, pH 7.4 at 4°C)

  Dissolve 0.5 g of DOC in 2.5 ml of 10x PIPES buffer, and separately dissolve 1 ml of Tween-40 in 2.5 ml of 10x PIPES buffer (warm if necessary). Combine the Tween and DOC solutions, and then add 5 ml of 10x PIPES buffer and 1 ml of 100 mM PMSF. Cool to 4°C, adjust pH to 7.4, and add H_2O to 100 ml (final volume). Keep on ice until use.

  EXPERIMENTAL TIP: Triton and Tween are both nonionic detergents, which decompose in aqueous solutions to form peroxides that oxidize protein sulfhydryl groups (Lever 1977; Chang and Bock 1980; Lenstra and Bloemendal 1983). Although EDTA prolongs stability, buffers should be used as freshly prepared or freshly thawed and not refrozen. This avoids artifacts and assures reproducibility between different fractionations.

Cytoskeleton solubilization buffer (5% SDS <!>, 10 mM sodium phosphate <!>, pH 7.4)

For nonreducing buffer, dissolve 0.5 g of SDS in 5 ml of 20 mM sodium phosphate buffer, pH 7.4. Add H_2O to 10 ml (final volume). For denaturing buffer, add 1 ml of β-mercaptoethanol <!>. Adjust H_2O appropriately.

1x O'Farrell lysis buffer

Dissolve 5.7 g of ultrapure electrophoretic-grade urea <!>, 0.2 ml of NP-40, 0.2 ml of ampholines (0.16 ml at pH 5–7 and 0.04 ml at pH 3–10), and 0.5 ml of β-mercaptoethanol in H_2O. Adjust final volume to 10 ml with H_2O. Solution may be warmed to facilitate solubilization. Divide into 1-ml aliquots and store at –70°C.

10x O'Farrell lysis buffer

Combine 0.2 ml of NP-40, 0.2 ml of ampholines (0.16 ml at pH 5–7 and 0.04 ml at pH 3–10), and 0.5 ml of β-mercaptoethanol in H_2O. Adjust final volume to 1 ml with H_2O. Solution may be warmed to facilitate solubilization. Divide into 100-μl aliquots and store at –70°C.

Phosphate-buffered saline (PBS, pH 7.4), ice-cold

8.0 g of NaCl (final concentration 137 mM)
0.2 g of KCl (final concentration 2.7 mM) <!>
1.44 g of Na_2HPO_4 (final concentration 10.1 mM) <!>
0.24 g of KH_2PO_4 (final concentration 1.8 mM) <!>
in 800 ml of H_2O

Adjust pH to 7.4, and then adjust volume to 1 liter with H_2O.

PBS (pH 7.4) containing 1.2 mM PMSF, ice-cold

Add the PMSF just before use.

4x PIPES buffer (piperazine-*N,N′*-bis[2-ethanesulfonic acid])

Dissolve 103 g of sucrose and 5.8 g of NaCl in 150 ml of H_2O. Dissolve 3 g of PIPES in a small volume of 1 M NaOH <!>. Mix the PIPES solution with the sucrose/NaCl solution. Add 0.64 g of $MgCl_2 \cdot 6H_2O$ <!> to the mixture. Adjust final volume to 250 ml and filter through a 0.45-μm sterile filter. Store in the dark at 4°C or aliquot and freeze (stable for 2 months at 4°C). Employ aseptic technique when diluting the 4x PIPES buffer for preparation of 1x working solutions or store as single-use aliquots to avoid contamination.

10x PIPES buffer

Dissolve 0.58 g of NaCl in 50 ml of H_2O, and separately dissolve 3 g of PIPES in a small volume of 1 M NaOH. Mix the PIPES solution with the NaCl solution. Add 0.2 g of $MgCl_2 \cdot 6H_2O$ to the solution and adjust final volume to 100 ml with H_2O. Filter and store as described for the 4x PIPES buffer.

Trypan blue

Urea (solid)

Cells

Suspension or monolayer cell cultures

Equipment

Centrifuge (refrigerated)
Homogenizer, Teflon smooth-walled glass
Lyophilizer
Platform mixer

Additional Reagents

Step 4 of Section IV of this protocol requires the reagents listed in Appendix 2, Technique 1. Section V of this protocol requires the reagents listed in Appendix 2.

METHOD

I. Preparation of Cells

1. Cool the cell cultures on ice, and wash them two to three times in ice-cold saline, PBS, or other nondetergent buffer.

   Save an aliquot of the wash buffer (store it at –70°C) to use for future sample normalization or as control material for enzymatic, protein, or RNA analyses.

2. Choose the appropriate method below for working with suspension or monolayer cultures:

   If working with suspension cultures: The volume of DDF solutions for suspension-cultured cells is based on wet weight or cell number.

   a. Transfer an aliquot of the suspension culture to a preweighed plastic tube and centrifuge briefly.

   b. Decant the culture media and determine the wet weight of the cell pellet.

   c. Proceed to Step 1 in Section II.

   If working with monolayer cultures: The volume of DDF buffers for monolayer cell cultures is based on surface area or cell number.

   a. Determine the surface area of the culture flask or dish. Alternatively, count the cells in a representative culture vessel.

   b. Proceed to Step 1 in Section III.

II. Detergent Fractionation of Suspension Cell Cultures

DDF is performed by the sequential addition and removal of DDF buffers. All extractions are performed on ice with gentle agitation. Attention to extraction times is important for reproducibility between different fractionations. Extracts are maintained on ice until used for assay or frozen for storage (–70°C is best). Measure the recovered volume of each detergent extract to calculate net yields or percent distributions. Save an aliquot of each DDF buffer for future sample normalization or control material for assays.

1. Add 5 volumes of ice-cold digitonin extraction buffer per gram wet weight to the washed cell pellets. Gently resuspend the cell pellets by swirling.

2. Incubate the cells on ice with gentle agitation on a platform mixer until 95–100% of the cells are permeabilized (~10 minutes) as assessed by Trypan blue exclusion.

   EXPERIMENTAL TIP: It is important to maintain consistent extraction times between different fractionations.

3. Centrifuge the extraction mixture at 480*g* for 10 minutes.

4. Transfer the supernatant to a clean tube. Record the supernatant (cytosolic fraction) volume, divide the cytosolic fraction into aliquots, and store them at –70°C. This is the CYTO fraction (see Figure 3.3 and Table 3.7).

5. Carefully resuspend the digitonin-insoluble pellets in 5 volumes (relative to starting wet weight of the cell pellet [determined above in Section I, Step 2]) of ice-cold Triton X-100 extraction buffer to obtain a homogeneous suspension.

   Triton X-114 can be substituted for Triton X-100 in DDF protocols without obvious difference on two-dimensional gels. Relative to Triton X-100, Triton X-114 has a lower cloud point (60°C vs. 20°C) and can partition into lipid-rich and lipid-poor phases at nondenaturing temperatures. This enables subfractionation of integral and peripheral membrane proteins as well as soluble organelle contents (Bordier 1981; Pryde and Phillips 1986).

6. Incubate the cells on ice with gentle agitation on a platform mixer for 30 minutes.

7. Centrifuge the extraction mixture at 5000*g* for 10 minutes.

8. Transfer the supernatant to a clean tube. Record the supernatant (membrane and organelle fraction) volume, divide this fraction into aliquots, and store them at –70°C. This is the MO fraction (see Figure 3.3 and Table 3.7).

9. To the Triton-insoluble pellets, add Tween-40/deoxycholate extraction buffer at one-half the volume used for Triton extraction (Step 5). Resuspend the pellets with five strokes at medium speed using a Teflon, smooth-walled glass homogenizer.

10. Centrifuge the extraction mixture at 6780*g* for 10 minutes.

11. Transfer the supernatant to a clean tube. Record the supernatant (nuclear fraction) volume, divide the nuclear fraction into aliquots, and store them at –70°C. This is the NUC fraction (see Figure 3.3 and Table 3.7).

12. Add ice-cold PBS (pH 7.4) containing 1.2 mM PMSF to the detergent-resistant pellets. Resuspend the pellets using three strokes in a Teflon/glass homogenizer. Collect the pellets by centrifugation at 12,000*g* for 10 minutes.

13. Repeat Step 12 two more times to thoroughly wash the pellets.

14. Discard the supernatant, and wash the pellets once with 90% acetone (previously chilled to –20°C).

15. Lyophilize the pellets (cytoskeletal/nuclear matrix fraction) overnight.

16. Determine the pellet weights in tared microfuge tubes. Store the samples at –70°C. This is the CSK/MAT fraction (see Figure 3.3 and Table 3.7).

    Typically, the yield of the CSK/MAT fraction from cell suspensions is large. Thus, it is usually more convenient to store the fraction as dried pellets, and resuspend measured aliquots in nondenaturing cytoskeleton solubilization buffer as needed.

17. Proceed to Section IV.

III. Detergent Fractionation of Monolayer Cell Cultures

DDF is performed by the sequential addition and removal of DDF buffers. All extractions are performed on ice with gentle agitation. Attention to extraction times is important for reproducibility between different fractionations. Extracts are maintained on ice until used for assay or frozen for storage (–70°C is best). Measure the recovered volume of each detergent extract to calculate net yields or percent distributions. Save an aliquot of each DDF buffer for future sample normalization or control material for assays.

1. Add 1 ml of ice-cold digitonin extraction buffer directly to the cell monolayer in a T-25 flask.

   A typical T-25 culture flask contains ~5 x 10^6 cells. Use 3 ml of digitonin extraction buffer for a T-75 flask or 100-mm diameter Petri dish.

2. Incubate the cells on ice with gentle agitation on a platform mixer until 95–100% of the cells are permeabilized (~10 minutes) as assessed by Trypan blue exclusion.

   EXPERIMENTAL TIP: It is important to maintain consistent extraction times between different fractionations.

3. Tilt the culture flask and decant the liquid (the cytosolic fraction) into a culture tube. Use a pipette to transfer any residual liquid from the flask into the culture tube. Record the volume of the cytosolic fraction, divide it into aliquots, and store them at –70°C. This is the CYTO fraction (see Figure 3.3 and Table 3.7).

4. Add 1 ml of Triton extraction buffer per T-25 flask equivalent (~5 x 10^6 cells).

   Triton X-114 can be substituted for Triton X-100 in DDF protocols without obvious difference on two-dimensional gels. Relative to Triton X-100, Triton X-114 has a lower cloud point (60°C vs. 20°C) and can partition into lipid-rich and lipid-poor phases at nondenaturing temperatures. This enables subfractionation of integral and peripheral membrane proteins as well as soluble organelle contents (Bordier 1981; Pryde and Phillips 1986).

5. Incubate the cells on ice with gentle agitation on a platform mixer for 30 minutes.

6. Tilt the culture flask and decant the liquid (the membrane and organelle fraction) into a culture tube. Use a pipette to transfer any residual liquid from the flask into the culture tube. Record the volume of this fraction, divide it into aliquots, and store them at –70°C. This is the MO fraction (see Figure 3.3 and Table 3.7).

7. Extract cell monolayers with 0.5–1.0 ml of Tween-40/deoxycholate extraction buffer per T-25 flask equivalent (~5 x 10^6 cells).

8. Tilt the culture flask and decant the liquid (the nuclear fraction) into a culture tube. Use a pipette to transfer any residual liquid from the flask into the culture tube. Record the volume of the nuclear fraction, divide it into aliquots, and store them at –70°C. This is the NUC fraction (see Figure 3.3 and Table 3.7).

9. Rinse the monolayers in situ with PBS.

10. Add 1–3 ml of nondenaturing cytoskeleton solubilization buffer without β-mercaptoethanol to the detergent-resistant residue in the culture flasks. Suspend the residue by tituration. Record the volume of this material (the cytoskeletal and matrix fraction), divide it into aliquots, and store them at –70°C. This is the CSK/MAT fraction (see Figure 3.3 and Table 3.7).

    The volume of nondenaturing cytoskeleton solubilization buffer required will vary depending on the cell type and length of time in culture (i.e., the amount of extracellular matrix). In general, 1–3 ml is an appropriate range.

11. Proceed to Section IV.

IV. Determination of Protein Concentration

1. Thaw detergent buffers and extracts on ice.

2. Dilute the digitonin/EDTA and Triton/EDTA extracts with H_2O (4 volumes for extracts from suspension cultures, 2–3 volumes for extracts from monolayers). Use the Tween/DOC extracts undiluted.

3. Solubilize the lyophilized CSK/MAT pellets in cytoskeleton solubilization buffer without β-mercaptoethanol. Use 1 ml of buffer per 10 mg (dry weight) of pellet. Dilute CSK/MAT preparations from monolayers as necessary.

4. Assay 20–50 μl of each sample in duplicate, using the BCA method (see Appendix 2).

    An alternative assay is the Folin-phenol method of Peterson (1983), which is not susceptible to interference by detergents. Assay by the standard Lowry method is not recommended because it results in detergent-induced flocculation (M. Ramsby, unpubl.).

V. Analysis of Protein Fractions Using Two-dimensional Gel Electrophoresis

DDF samples obtained from a variety of cell types can be utilized for two-dimensional PAGE under both IEF and nonequilibrium pH gradient electrophoresis (NEpHGE). For examples of two-dimensional gels, see Ramsby et al. (1994) and Ramsby and Makowski (1999). Fresh or thawed DDF extracts (kept on ice) are first normalized to contain protein in equal volumes and then brought to 9.5 M urea by addition of solid urea.

- For a 100-μl sample, add 85 mg of urea and 15 μl of 10x O'Farrell lysis buffer and warm to room temperature.
- For the dried CSK extract, solubilize directly in 1x O'Farrell lysis buffer (O'Farrell 1975; O'Farrell et al. 1977).
- For CSK extracts in nonreducing SDS buffer, adjust to 9.5 M urea by the addition of solid urea and add 10x O'Farrell lysis buffer.

Solid urea and 10x O'Farrell lysis buffer are added to minimize dilutional effects and enable detection of low-abundance proteins. Samples may be warmed slightly to facilitate urea solubilization, but avoid excessive warming (increased temperature and/or prolonged heating), which may result in carbamylation artifacts.

- DDF samples are analyzed by 2D electrophoresis using established methods (O'Farrell 1975; O'Farrell et al. 1977; also see Chapter 4).
- IEF gels (see Chapter 4) contain a total of 3.5% ampholines (2% pH 5–8, 1% pH 3–10, 0.5% pH 2–5), and samples are electrophoresed for a total of 9800 Volt-hours with hyperfocusing at 800 V for the final hour (Duncan and Hershey 1984; Ramsby et al. 1994).
- NEpHGE gels contain 2% pH 3–10 ampholines, and samples are electrophoresed for 2400 Volt-hours (Ramsby et al. 1994).

In general, samples containing 25–100 µg of protein in 15–60 µl are ample for visualization by Coomassie Blue staining or autoradiography; low-abundance proteins are detected by silver staining. In control experiments, the volume of detergent buffer contained in the samples did not adversely affect the linear range of the pH gradient in IEF gels. However, a slight shift to more acidic values may occur with samples containing Tween/DOC.

Degradation Studies

DDF extracts are an excellent source of material for protein degradation studies. Frozen samples are stable (–70°C), and when aliquoted, they provide a reliable and reproducible source of protein. In general, DDF extracts are of sufficient protein concentration (Ramsby et al. 1994) that they may be diluted severalfold in the buffer and pH of choice, thereby limiting potential interference by detergents. DDF detergents are also compatible with a variety of potential protease activators (e.g., metals and sulfhydryl reagents) and inhibitors (e.g., PMSF, pepstatin, leupeptin, and chelators), which may be added from stock solutions (Makowski 1989). DDF extracts have been used successfully to study degradation of specific proteins in various subcellular compartments in rat hepatocytes including cytosolic (Makowski 1989), membrane/organelle (Makowski and Ramsby 1999), and cytoskeletal/matrix fractions (Makowski and Ramsby 1998). Because of the high protein content of DDF extracts, degradation experiments may be simply monitored by standard or modified (Makowski and Ramsby 1993) gel electrophoretic methods (SDS-PAGE), with and without reduction with β-mercaptoethanol, and can be combined with western blotting. In general, the concentration of detergents in the DDF extracts does not interfere with enzymatic and electrophoretic analysis of proteins.

ADDITIONAL PROTOCOL: ISOLATION OF RNA FROM DETERGENT EXTRACTS

Total RNA is isolated from detergent extracts by the acid guanidinium isothiocyanate/phenol/chloroform method of Chomczynski and Sacchi (1987).

IMPORTANT: Strict attention must be paid to the preparation of glassware and reagents to avoid RNA degradation by contaminating RNases. Bake all glassware for >4 hours at 300°C and use for RNA processing only. Formulate RNA buffers/solutions with ultrapure H_2O previously treated with DEPC and autoclaved. Keep samples on ice, and pipette using sterile, cell culture 10-ml plastic pipettes. RNA isolation from 2.0 ml of detergent extract is convenient to work with and affords enough sample for both protein and RNA analysis, including northern blot hydridization. In general, 3 volumes of GIT buffer/volume of detergent extract is the minimal ratio compatible with recovery of intact RNA; lesser ratios are associated with RNA degradation.

Additional Materials

CAUTION: See Appendix 3 for appropriate handling of materials marked with <!>.

Chloroform <!>
Ethanol
Isopropanol
Phenol (H_2O-saturated) <!>
RNA extraction buffer

Prepare RNA extraction buffer according to the method of Chomczynski and Sacchi (1987). To prepare a stock solution, use only molecular-biology-grade reagents. A stock solution contains 4 M guanidinium isothiocyanate <!> (heat at 65°C to dissolve), 0.5 M sodium citrate, and 0.5% (v/v) *N*-lauroylsarcosine (Sarkosyl). The stock solution is stable for 3 months at room temperature. Add β-mercaptoethanol fresh before use (360 µl per 50 ml of RNA extraction buffer) to obtain a working solution.

SDS (0.1%) containing 1 mM EDTA <!>
Sodium acetate (2 M, pH 4.0)
Vacuum dryer (SpeedVac or equivalent)
Water bath preset to 65°C

Method

1. Add 3 volumes of RNA extraction buffer per volume of detergent extract, and titurate four times to denature the proteins.
2. To each sample, add:
   0.1 volume of 2 M sodium acetate (pH 4.0)
   1 volume of H_2O-saturated phenol
   0.2 volume of cholorform

   Mix vigorously after each addition. Allow the samples to sit on ice for 15 minutes.
3. Centrifuge the samples at 10,000*g* for 20 minutes.
4. Carefully remove the upper aqueous phase with a pipette, and transfer it to a clean tube. Add an equal volume of isopropanol, and invert the tube to mix. Store the tubes overnight at –20°C to precipitate RNA.
5. The next day, centrifuge the samples at 10,000*g* for 15 minutes.
6. Carefully discard the supernatant, and invert the tubes to allow them to drain.
7. Wash the RNA precipitates with 1 ml of 75% ethanol, and transfer them to clean 1.5-ml microfuge tubes. Centrifuge them again at 10,000*g* for 10 minutes.
8. Dry the RNA pellet in a centrifugal vacuum dryer for 5 minutes.
9. Resuspend the RNA in 30–50 µl of 0.1% SDS containing 1 mM EDTA. Heat samples for 10 minutes at 65°C.
10. Clarify the solution by centrifugation in a microfuge for 2 minutes.
11. Transfer the clear supernatant to a fresh tube. To quantify and assess purity, dilute an aliquot of the RNA solution 250-fold (e.g., add 4 µl of RNA to 996 µl of H_2O) and measure the absorbance at 260 and 280 nm.

    A_{260}/A_{280} ratios are typically 1.45 ± 0.15.

RESULTS OF RNA ISOLATION USING DDF

- Total RNA yield (milligram of RNA per 100-mm diameter culture dish) is routinely higher for RNA isolated during DDF (459 mg ± 18) than for control RNA (289 mg ± 42.5) prepared in parallel by direct addition of guanidinium isothiocyanate to unfractionated cell cultures.
- In osteoblastic cells, the percent distribution of total cellular RNA is ~17% for digitonin/EDTA extracts, ~29% for Triton extracts, and ~54% for combined nuclear/detergent resistant fraction (M. Ramsby, unpubl.).
- RNA prepared from DDF extracts can be readily fractionated on agarose-formaldehyde gels (see Figure 3.4) and processed for northern blot hybridization by standard protocols. Using this method, single bands of intact mRNA have been visualized and found to vary in a compartmental-specific manner as a function of differentiation stage and treatment (M. Ramsby, unpubl.).
- Enrichment of tRNA in cytosolic extracts and rRNA in nuclear/cytoskeletal fractions is observed, consistent with selective fractionation relative to distinct subcellular compartments (see Figure 3.4).
- Specific mRNAs differentially distribute in DDF fractions relative to differentiation stage and treatment (M. Ramsby, unpubl.), which suggests that RNA isolation using DDF provides a novel approach for investigating mRNA localization and trafficking in parallel with protein analysis.

ADDITIONAL PROTOCOL: PRECIPITATION OF TUBULINS AND MAPS USING MAGNESIUM

Tubulins and microtubule-associated proteins present in cytosolic extracts can be separated from other cytosolic constituents by incubation at 37°C, followed by magnesium-induced precipitation (Sahyoun et al. 1982).

Additional Materials

CAUTION: See Appendix 3 for appropriate handling of materials marked with <!>.

Imidazole buffer (pH 7.5) <!>
$MgCl_2$ (0.1 M)

Method

1. Adjust an aliquot of the digitonin/EDTA extract (the CYTO fraction from the main protocol) to 7–35 mM magnesium by addition of 0.1 M $MgCl_2$. Incubate the samples for 20–30 minutes at 37°C.

   The total volume of starting digitonin/EDTA extract used will depend on experimental goals. Final magnesium concentration will depend on the origin of the starting material. In general, 35 mM magnesium is excessive, but broadly applicable for complete recovery.

2. Harvest tubulins and associated proteins by centrifugation in a microfuge for 8 minutes.
3. Transfer the supernatant to a clean tube. Wash the pellet in a large volume of imidazole buffer (pH 7.5), or other suitable buffer, and resuspend the pellet in cytoskeleton solubilization buffer for determination of protein concentration and electrophoretic analysis (see Sections IV and V of the main protocol and Table 3.7).

REFERENCES

Amrein K.E., Takacs B., Steiger M., Molnos J., Flint N.A., and Burn P. 1995. Purification and characterization of human recombinant p50^{csk} protein-tyrosine kinase from an *Escherichia coli* expression system overproducing the bacterial chaperones GroES and GroEL. *Proc. Natl. Acad. Sci.* **92:** 1048–1052.

Baker D., Hicke L., Rexach M, Schleyer M., and Schekman R. 1988. Reconstitution of SEC gene product-dependent intercompartmental protein transport. *Cell* **54:** 335–344.

Beynon R.J. and Bond J.S. 1989. *Proteolytic enzymes: A practical approach.* IRL Press/Oxford University Press, Oxford, United Kingdom.

Beynon R.J. and Oliver S. 1996. Avoidance of proteolysis in extracts. *Methods Mol. Biol.* **59:** 81–93.

Bird R.C. 1998. Cell separation by centrifugal elutriation. In *Cell biology: A laboratory handbook*, 2nd edition (ed. J.E. Celis), pp. 205–208. Academic Press, San Diego, California.

Blackwell J.R. and Horgan R. 1991. A novel strategy for production of a highly expressed recombinant protein in an active form. *FEBS Lett.* **295:** 10–12.

Blum P., Velligan M., Lin N., and Matin A. 1992. DnaK-mediated alterations in human growth hormone protein inclusion bodies. *Bio/Technology* **10:** 301–304.

Bollag D.M., Edelstein S.J., and Rozycki M.D., eds. 1996. *Protein methods*, 2nd edition. Wiley-Liss, New York.

Bordier C. 1981. Phase separation of integral membrane proteins in Triton X-114 solutions. *J. Biol. Chem.* **256:** 1604–1607.

Bridge P. 1996. Protein extraction from fungi. *Methods Mol. Biol.* **59:** 39–48.

Burgess R. and Jendrisak J. 1975. A procedure for the rapid large-scale purification of *E. coli* DNA dependent RNA polymerase involving Polymin P precipitation and DNA cellular chromatography. *Biochemistry* **14:** 4634–4645.

Cahill A. and Cunningham C.C. 2000. Effects of chronic ethanol feeding on the protein composition of mitochondrial ribosomes. *Electrophoresis* **21:** 3420–3426.

Calvo M. and Enrich C. 2000. Biochemical analysis of a caveolae-enriched plasma membrane fraction from rat liver. *Electrophoresis* **21:** 3386–3395.

Capco D.G., Wan K.M., and Penman S. 1982. The nuclear matrix: Three-dimensional architecture and protein composition. *Cell* **29:** 847–858.

Carrell R.W., Lehmann H., and Hutchinson H.F. 1966. Haemoglobin Koln (beta 98 valine-methionine): An unstable protein causing inclusion body anemia. *Nature* **210:** 915–916.

Celis J.E., ed. 1998. *Cell biology: A laboratory handbook*, 2nd edition, vol. 1. Academic Press, San Diego, California.

Chang H.W. and Bock E. 1980. Pitfalls in the use of commercial nonionic detergents for the solubilization of integral membrane proteins: Sulfhydryl oxidizing contaminants and their elimination. *Anal. Biochem.* **104:** 112–117.

Chataway T.K., Whittle A.M., Lewis M.D., Bindloss C., Davey A., Moritz R.L., Simpson R.J., Hopwood J.J., and Meikle P.J. 1998. Two-dimensional mapping and microsequencing of lysosomal proteins from human placenta. *Placenta* **19:** 643–654.

Chomczynski P. and Sacchi N. 1987. Single-step method of RNA isolation by acid guanidinium thiocyanate-phenol-chloroform extraction. *Anal. Biochem.* **162:** 156–159.

Cordwell S.J., Nouwens A.S., Verrills N.M., Basseal D.J., and Walsh B.J. 2000. Subproteomics based upon protein cellular location and relative solubilities in conjunction with composite two-dimensional electrophoresis gels. *Electrophoresis* **21:** 1094–1103.

Corthals G.L., Wasinger V.C., Hochstrasser D.F., and Sanchez J.C. 2000. The dynamic range of protein expression: A challenge for proteomic research. *Electrophoresis* **21:** 1104–1115.

Cull M. and McHenry C.S. 1990. Preparation of extracts from prokaryotes. *Methods Enzymol.* **182:** 147–153.

Deutscher M.P., ed. 1990. Guide to protein purification. *Methods Enzymol.*, vol. 182.

Dignam J.D. 1990. Preparation of extracts from higher eukaryotes. *Methods Enzymol.* **182:** 194–203.

Doonan S. 1996. Protein purification protocols. *Methods Mol. Biol.* **59:** 1–16.

Duncan R. and Hershey J.W.B. 1984. Evaluation of isoelectric focusing running conditions during two-dimensional isoelectric focusing/sodium dodecyl sulfate-polyacrylamide gel electrophoresis: Variation of gel patterns with changing conditions and optimal isoelectric focusing conditions. *Anal. Biochem.* **138:** 144–145.

Evans W.H. 1992. Isolation and characterization of membranes and cell organelles. In *Preparative centrifugation: A practical approach* (ed. D. Rickwood), pp. 233–270. IRL Press/Oxford University Press, Oxford, United Kingdom.

Fahnestock S.R. 1979. Reconstitution of active 50S ribosomal subunits from *Bacillus lichenformis* and *Bacillus subtilis. Methods Enzymol.* **59:** 437–443.

Ferrari G., Langen H., Naito M., and Pieters J. 1999. A coat protein on phagosomes involved in the intracellular survival of mycobacteria. *Cell* **97:** 435–447.

Fey E.G., Wan K.M., and Penman S. 1984. Epithelial cytoskeletal framework and nuclear matrix-intermediate filament scaffold: Three-dimensional organization and protein composition. *J. Cell Biol.* **98:** 1973–1984.

Fialka I., Steinlein P., Ahorn H., Bock G., Burbelo P.D., Haberfellner M., Lottspeich F., Paiha K., Pasquali C., and Huber L.A. 1999. Identification of syntenin as a protein of the apical early endocytic compartment in Madin-Darby canine kidney cells. *J. Biol. Chem.* **274:** 26233–26239.

Fiedler K., Kellner R., and Simons K. 1997. Mapping the protein composition of trans-Golgi network (TGN)-derived carrier vesicles from polarized MDCK cells. *Electrophoresis* **18:** 2613–2619.

Fiskum G., Craig S.W., Decker G.L., and Lehninger A.L. 1980. The cytoskeleton of digitonin-treated rat hepatocytes. *Proc. Natl. Acad. Sci.* **77:** 3430–3434.

Franke W.W., Schmid E., Osborn M., and Weber K. 1978. The intermediate-sized filaments in rat kangaroo PtK2 cells. II. Structure and composition of isolated filaments. *Cytobiologie* **17:** 392–411.

Garin J., Diez R., Kieffer S., Dermine J.F., Duclos S., Gagnon E., Sadoul R., Rondeau C., and Desjardins M. 2001. The phagosome proteome: Insight into phagosome functions. *J. Cell Biol.* **152:** 165–180.

Gegenheimer P. 1990. Preparation of extracts from plants. *Methods Enzymol.* **182:** 174–194.

Glass G.B.J. 1964. Proteins, mucosubstances, and biologically active components of gastric substances. *Adv. Clin. Chem.* **7:** 235–372.

Gorg A., Obermaier C., Boguth G., Harder A., Scheibe B., Wildgruber R., and Weiss W. 2000. The current state of two-dimensional electrophoresis with immobilized pH gradients. *Electrophoresis* **21:** 1037–1053.

Graham J.M. 1997. Homogenisation of tissue and cells. In *Subcellular fractionation: A practical approach* (ed. J.M. Graham and D. Rickwood), pp. 1–28. IRL Press/Oxford University Press, Oxford, United Kingdom.

Graham J.M. and Rickwood D., eds. 1997. *Subcellular fractionation: A practical approach.* IRL Press/Oxford University Press, Oxford, United Kingdom.

Graham J., Ford T., and Rickwood D. 1994. The preparation of subcellular organelles from mouse liver in self-generated gradients of iodixanol. *Anal. Biochem.* **220:** 367–373.

Gribskov M. and Burgess R.R. 1983. Overexpression and purification of the sigma subunit of *Escherichia coli* RNA polymerase. *Gene* **26:** 109–118.

Gygi S.P., Corthals G.L., Zhang Y., Rochon Y., and Aebersold R. 2000. Evaluation of two-dimensional gel electrophoresis-based proteome analysis technology. *Proc. Natl. Acad. Sci.* **97:** 9390–9395.

Han D.K., Eng J., Zhou H.L., and Aebersold R. 2001. Quantitative profiling of differentiation-induced microsomal proteins using isotope-coded affinity tags and mass spectrometry. *Nat. Biotechnol.* **19:** 946–951.

Hanash S. 2000. Biomedical applications of two-dimensional electrophoresis using immobilized pH gradients: Current status. *Electrophoresis* **21:** 1202–1209.

Harlow E. and Lane D. 1999. *Using antibodies: A laboratory manual.* Cold Spring Harbor Laboratory Press, Cold Spring Harbor, New York.

Hockney R.C. 1995. Recent developments in heterologous protein production in *Escherichia coli. Trends Biotechnol.* **13:** 456–463.

Hoffmann P., Ji H., Connolly L.M., Frecklington D.F., Layton M.J., Moritz R.L., and Simpson R.J. 2001. Continuous free-flow electrophoresis separation of cytosolic proteins from the human colon carcinoma cell line LIM1215: A non two-dimensional gel electrophoresis-based proteome analysis strategy. *Proteomics* **1:** 807–818.

Hoving S., Voshol H., and van Oostrum J. 2000. Towards high performance two-dimensional gel electrophoresis using ultrazoom gels. *Electrophoresis* **21:** 2617–2621.

Hunter M.J. and Commerford S.L. 1961. Pressure homogenization of mammalian tissues. *Biochim. Biophys. Acta* **47:** 580–586.

Hurley J.H. and Meyer T. 2001. Subcellular targeting by membrane lipids. *Curr. Opin. Cell Biol.* **13:** 146–152.

Jazwinski S.M. 1990. Preparation of extracts from yeast. *Methods Enzymol.* **182:** 154–174.

Ji H. and Simpson R.J. 1999. Nonreducing 2-D polyacrylamide gel electrophoresis. *Methods Mol. Biol.* **112:** 255–264.

Ji H., Moriz R.L., Reid G.E., Ritter G., Catimel B., Nice E.C., Heath J.K., White S.J., Welt S., Old L.J., Burgess A.W., and Simpson R.J. 1997. Electrophoretic analysis of the novel antigen for the gastrointestinal-specific monoclonal antibody, A33. *Electrophoresis* **18:** 614–622.

Johnson J.E. and Cornell R.B. 1999. Amphitropic proteins: Regulation by reversible membrane interactions. *Mol. Membr. Biol.* **16:** 217–235.

Journet A., Chapel A., Kieffer S., Louwagie M., Luche S., and Garin J. 2000. Towards a human repetoire of monocytic lysosomal proteins. *Electrophoresis* **21:** 3411–3419.

Jung E., Heller M., Sanchez J.-C., and Hochstrasser D.F. 2000a. Proteomics meets cell biology: The establishment of subcellular proteomes. *Electrophoresis* **21:** 3369–3377.

Jung E., Hoogland C., Chiappe D., Sanchez J.-C., and Hochstrasser D.F. 2000b. The establishment of a human liver nuclei two-dimensional electrophoresis reference map. *Electrophoresis* **21:** 3483–3487.

Kohno T., Carmichael D.F., Sommer A., and Thompson R.C. 1990. Refolding of recombinant proteins. *Methods Enzymol.* **185:** 187–195.

Laemmli U.K. 1970. Cleavage of structural proteins during the assembly of the head of bacteriophage T4. *Nature* **227:** 680–685.

La Vallie E.R., Di Blasio E.A., Kovacic S., Grant K.L., Schendel P.F., and McCoy J.M. 1993. A thioredoxin gene fusion expression system that circumvents inclusion body formation in the *E. coli* cytoplasm. *Bio/Technology* **11:** 187–193.

Lenk R., Ransom L., Kaufman Y., and Penman S. 1977. A cytoskeletal structure with associated polyribosomes obtained from HeLa cells. *Cell* **10:** 67–78.

Lenstra J.A. and Bloemendal H. 1983. Topography of the total protein population from cultured cells upon fractionation by chemical extractions. *Eur. J. Biochem.* **135:** 413–423.

Lever M. 1977. Peroxides in detergents as interfering factors in biochemical analysis. *Anal. Biochem.* **83:** 274–284.

Light A. 1985. Protein solubility, protein modification and protein folding. *BioTechniques* **3:** 298–306.

Lopez M.F., Kristal B.S., Chernokalskaya E., Lazarev A., Shestopalov A.I., Bogdanova A., and Robinson M. 2000. High-throughput profiling of the mitochondrial proteome using affinity fractionation and automation. *Electrophoresis* **21:** 3427–3440.

Mackall J., Meredith M., and Lane L.M. 1979. A mild procedure for the rapid release of cytoplasmic enzymes from cultured animal cells. *Anal. Biochem.* **95:** 270–274.

Makowski G.S. 1989. "Calpains and rat hepatic protein degradation." Ph.D. thesis, University of Connecticut, Storrs.

Makowski G.S. and Ramsby M.L. 1993. pH modification enhances the apparent molecular seiving properties of sodium dodecylsulfate-10% polyacrylamide gels. *Anal. Biochem.* **212:** 283–285.

———. 1997. Binding of type IV collagenase (matrix metalloproteinase-2 and -9) by fibrin *in vitro:* Latent enzyme activation by a plasmin mediated mechanism. *Ann. Clin. Lab. Sci.* **27:** 309–310.

———. 1998. Degradation of cytokeratin intermediate filaments by calcium-activated proteases (calpains) *in vitro:* Implications for formation of Mallory bodies. *Res. Comm. Mol. Pathol. Pharmacol.* **101:** 211–223.

———. 1999. Limited degradation of carbamoyl phosphate synthetase I by calcium-activated protease (calpain): Electrophoretic evidence for removal of the C-terminal N-acetylglutamate regulatory domain. *Res. Commun. Mol. Pathol. Pharmacol.* **101:** 211–223.

Marston F.A.O. 1986. The purification of eukaryotic polypeptides synthesized in *Escherichia coli. Biochem. J.* **240:** 1–12.

Marston F.A.O. and Hartley D.L. 1990. Solubilization of protein aggregates. *Methods Enzymol.* **182:** 264–276.

Molloy M.P., Herbert B.R., Walsh B.J., Tyler M.I., Traini M., Sanchez J.C., Hochstrasser D.F., Williams K.L., and Gooley A.A. 1998. Extraction of membrane proteins by differential solubilisation for separation using two-dimensional gel electrophoresis. *Electrophoresis* **19:** 837–844.

Morel V., Poschet R., Traverso V., and Deretic D. 2000. Towards the proteome of the rhodopsin-bearing post-Golgi compartment of retinal photoreceptor cells. *Electrophoresis* **21:** 3460–3469.

Müller E.C., Schumann M., Rickers A., Bommert K., Wittmann-Liebold B., and Otto A. 1999. Study of Burkitt lymphoma cell line proteins by high resolution two-dimensional gel electrophoresis and nano-electrospray mass spectrometry. *Electrophoresis* **20:** 320–330.

Neurath H. 1989. The diversity of proteolytic enzymes. In *Proteolytic enzymes: A practical approach* (ed. R.J. Beynon and J.S. Bond), pp. 1–13. IRL University Press, New York.

Neurauter A.A., Edward R., Kilaas L., Ugelstad J., and Larsen F. 1998. Immunomagnetic separation of animal cells. In *Cell biology: A laboratory handbook*, 2nd edition (ed. J.E. Celis), pp. 197–204. Academic Press, San Diego, California.

Nice E.C., Fabri L., Whitehead R.H., James R., Simpson R.J., and Burgess A.W. 1991. The major colonic cell mitogenic extractable from colonic mucosa is an N-terminally extended form of basic fibroblast growth factor. *J. Biol. Chem.* **266:** 14425–14430.

North M.J. 1989. Prevention of unwanted proteolysis. In *Proteolytic enzymes: A practical approach* (ed. R.J. Beynon and J.S. Bond), pp. 105–124. IRL Press/Oxford University Press, Oxford, United Kingdom.

O'Farrell P.H. 1975. High resolution two-dimensional electrophoresis of proteins. *J. Biol. Chem.* **250:** 4007–4021.

O'Farrell P.Z., Goodman H.M., and O'Farrell P.H. 1977. High resolution two-dimensional electrophoresis of basic as well as acidic proteins. *Cell* **12:** 1133–1142.

Patton W.F. 1999. Proteome analysis. II. Protein subcellular redistribution: Linking physiology to genomics via the proteome and separation technologies involved. *J. Chromatog. B* **722:** 203–223.

Perlmann G.E. and Lorand L., eds. 1970. Proteolytic enzymes. *Methods Enzymol.*, vol. 19.

Peterson G.L. 1983. Determination of total protein. *Methods Enzymol.* **91:** 95–119.

Price N.C., ed. 1996. Preparation of cell extracts. In *Protein labfax*, pp. 13–20. Academic Press, San Diego, California.

Pryde J.G. and Phillips J.H. 1986. Fractionation of membrane proteins by temperature-induced phase separation in Triton X-114. *Biochem. J.* **233:** 525–533.

Rabilloud T., Kieffer S., Procaccio V., Louwagie M., Courchesne P.L., Patterson S.D., Martinez P., Garin J., and Lunardi J. 1998. Two-dimensional electrophoresis of human placental mitochondria and protein identification by mass spectrometry: Toward a human mitochondrial proteome. *Electrophoresis* **19:** 1006–1014.

Ramsby M.L. 1989. "Autophagy and the intermediate filament cytoskeleton in suspension-cultured rat hepatocytes." Ph.D. thesis, University of Connecticut, Storrs.

Ramsby M.L. and Kreutzer D.L. 1993. Fibrin induction of tissue plasminogen activator in corneal endothelial cells in vitro. *Invest. Ophthalmol. Vis. Sci.* **34:** 3207–3219.

Ramsby M.L. and Makowski G.S. 1999. Differential detergent fractionation of eukaryotic cells. Analysis by two-dimensional electrophoresis. *Methods Mol. Biol.* **112:** 53–66.

Ramsby M.L., Makowski G.S., and Khairallah E.A. 1994. Differential detergent fractionation of isolated hepatocytes: Biochemical, immunochemical and two dimensional gel electrophoretic characterization of cytoskeletal and noncytoskeletal compartments. *Electrophoresis* **15:** 265–277.

Reiter T., Penman S., and Capco D.G. 1985. Shape-dependent regulation of cytoskeletal protein synthesis in anchorage-dependent and anchorage-independent cells. *J. Cell Sci.* **76:** 17–33.

Righetti P.G., Castagna A., and Herbert B. 2001. Prefractionation techniques in proteome analysis. *Anal. Chem.* **73:** 320A–326A.

Sahyoun N., Stenbuck P., LeVine H., Bronson D., Moncarmont B., Henderson C., and Cutrecasas P. 1982. Formation and identification of cytoskeletal components from liver cytosolic precursors. *Proc. Natl. Acad. Sci.* **79:** 7341–7345.

Santoni V., Kieffer S., Desclaux D., Masson F., and Rabilloud T. 2000. Membrane proteomics: Use of additive main effects with multiplicative interaction model to classify plasma membrane proteins according to their solubility and electrophoretic properties. *Electrophoresis* **21:** 3329–3344.

Schatz G. 1979. Biogenesis of yeast mitochondria: Synthesis of cytochrome c oxidase and cytochrome c. *Methods Enzymol.* **56:** 40–50.

Schein C.H. and Noteborn M.H.M. 1988. Formation of soluble recombinant proteins in *Escherichia coli* is favored by lower growth temperature. *Bio/Technology* **6:** 291–294.

Scopes R.K. 1994. *Protein purification: Principles and practice*, 3rd edition. Springer Verlag, New York.

Sebald W., Neupert W., and Weiss H. 1979. Preparation of *Neurospora crassa* mitochondria. *Methods Enzymol.* **59:** 144–148.

Simpson R.J., Connolly L.M., Eddes J.S., Pereira J.J., Moritz R.L., and Reid G.E. 2000. Proteomic analysis of the human colon carcinoma cell line (LIM 1215). *Electrophoresis* **21:** 1701–1732.

Smith D.B. and Johnson K.S. 1988. Single step purification of polypeptides expressed in *E. coli* as fusions

with glutathionine-S-transferase. *Gene* **67:** 31–40.
Spector D.L., Goldman R.D., and Leinwand L.A. 1998. *Cells: A laboratory manual*, vols. 1–3. Cold Spring Harbor Laboratory Press, Cold Spring Harbor, New York.
Taylor R.S., Wu C.C., and Hays L.G. 2000. Proteomics of rat liver golgi complex: Minor proteins are identified through sequential fractionation. *Electrophoresis* **21:** 3441–3459.
Taylor R.S., Fialka I., Jones S.M., Huber L.A., and Howell K.E. 1997. Two-dimensional mapping of the endogenous proteins of the rat hepatocyte Golgi complex cleared of proteins in transit. *Electrophoresis* **18:** 2613–2619.
Tsuji T., Nakagawa R., Sugimoto N., and Fukuhara K.-I. 1987. Characterization of disulfide bonds in recombinant proteins: Reduced human interleukin 2 in inclusion bodies and its oxidative refolding. *Biochemistry* **26:** 3129–3134.
Weigel P.H., Ray D.A., and Oka J.A. 1983. Quantitation of intracellular membrane-bound enzymes and receptors in digitonin-permeabilized cells. *Anal. Biochem.* **133:** 437–449.
Weir M.P. and Sparks J. 1987. Purification and renaturation of recombinant human interleukin-2. *Biochem. J.* **245:** 85–91.
Weir M.P., Sparks J., and Chaplin A.M. 1987. Micropreparative purification of recombinant human interleukin-2. *J. Chromatogr.* **396:** 209–215.
Wildgruber R., Harder A., Obermaier C., Boguth G., Weiss W., Fey S.J., Larsen P.M., and Gorg A. 2000. Towards higher resolution: Two-dimensional electrophoresis of *Saccharomyces cerevisiae* proteins using overlapping narrow immobilized pH gradients. *Electrophoresis* **21:** 2610–2616.
Williams D.C., Van Frank R.M., Muth W.L., and Barnett J.P. 1982. Cytoplasmic inclusion bodies in *Escherichia coli* producing biosynthetic human insulin protein. *Science* **215:** 684–687.
Worrall D.M. 1996. Extraction of recombinant protein from bacteria. *Methods Mol. Biol.* **59:** 31–37.
Zhang J.-G., Moritz R.L., Reid G.E., Ward L.D., and Simpson R.J. 1992. Purification and characterization of a recombinant murine interleukin-6: Isolation of N- and C-terminally truncated forms. *Eur. J. Biochem.* **207:** 903–913.
Zuo X. and Speicher D.W. 2000. A method for global analysis of complex proteomes using sample prefractionation prior to two-dimensional electrophoresis. *Anal. Biochem.* **284:** 266–278.

FURTHER READING

Celis J.E., ed. 1998. *Cell biology: A laboratory handbook*, 2nd edition, vol. 2. Academic Press, San Diego, California.
Deutscher M.P., ed. 1990. Guide to protein purification. *Methods Enzymol.* vol. 182.
Graham J.M. and Rickwood D., eds. 1997. *Subcellular fractionation: A practical approach.* IRL Press/Oxford University Press, Oxford, United Kingdom.
Spector D.L., Goldman R.D., and Leinwand L.A. 1998. *Cells: A laboratory manual*, vol. 1, *Culture and biochemical analysis of cells.* Cold Spring Harbor Laboratory Press, Cold Spring Harbor, New York.

WWW RESOURCES

http://www.kimble-kontes.com Kimble/Kontes home page.
http://www.parrinst.com Parr Instrument company home page.

CHAPTER 4

Preparative Two-dimensional Gel Electrophoresis with Immobilized pH Gradients

Wayne R. Stochaj,* Tom Berkelman,† and Nancy Laird†

*Beyond Genomics, Inc., Waltham, Massachusetts 02451; †Amersham Pharmacia Biotech, San Francisco, California 94107

TWO-DIMENSIONAL (2D) ELECTROPHORESIS IS A POWERFUL PROTEOMICS METHOD for separating complex mixtures of proteins into many more components than is possible with the classical one-dimensional method (see Chapter 2). All proteins in an electric field migrate at a speed that is dependent on their conformation, size, and electric charge. 2D electrophoresis uses the latter two of these characteristics to allow high-resolution separation of proteins. As first described by both P.H. O'Farrell and J. Klose in 1975, the first-dimension step, isoelectric focusing (IEF), separates proteins on the basis of their charge, whereas the second-dimension step, SDS-polyacrylamide gel electrophoresis (SDS-PAGE), separates proteins according to their molecular weights. The result is an array of protein spots (Figure 4.1) that are assigned *x* and *y* coordinates, rather than the protein bands seen in one-dimensional separations. Each spot in the resulting gel represents one to a few proteins, depending on the complexity of the sample. Thus, thousands of proteins can be separated on a single gel, permitting the determination of the isoelectric point, apparent molecular weight, and relative abundance of many of the proteins. In addition, 2D electrophoresis can separate proteins on the basis of their posttranslational modifications. This technique is currently a major field of proteomics where it is applied in conjunction with mass spectroscopy (see Chapter 8) to identify differential expression of proteins (expression proteomics).

The power of 2D electrophoresis has been recognized since its inception; however, it was not widely applied because the original technique was difficult to master and the results were hard to analyze. During the past several years, a number of developments have improved the resolution and reproducibility of the technique and expanded our ability to interpret results (Görg et al. 2000).

- The new 2D method developed by Görg and colleagues (1985, 1988, 2000) uses an improved separation technique in the first dimension that replaces carrier ampholyte-generated pH gradients with immobilized pH gradients within the gel.

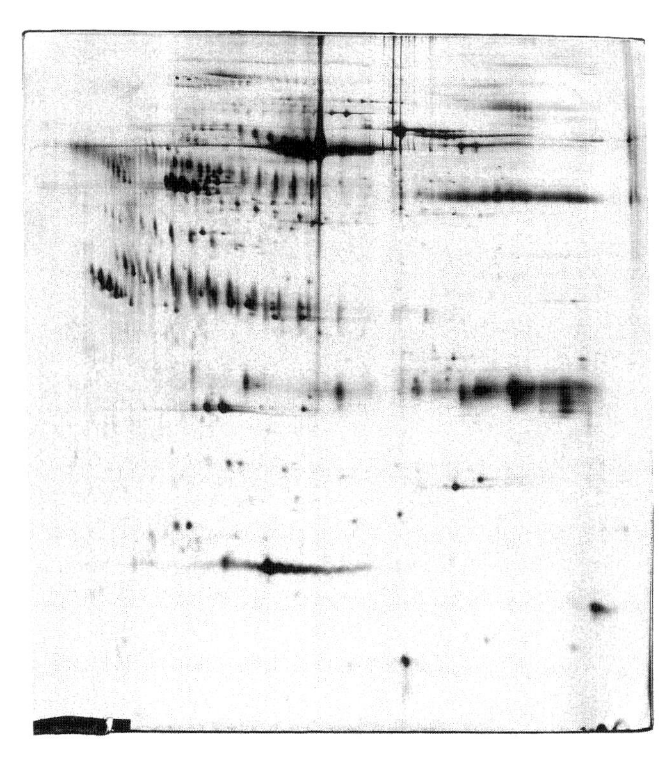

FIGURE 4.1. Separation of human cerebrospinal fluid proteins by 2D gel electrophoresis. pH 3–10 nonlinear first dimension, 14 %T SDS-PAGE second dimension.

- Less expensive and more powerful computers and software have improved our ability to analyze the complex patterns of protein spots within 2D gels.
- Advancements in mass spectrometry techniques have reduced the amount of material needed for identification and characterization of a protein to less than a few hundred nanograms, approximately the amount of material in a single 2D protein spot.
- Ready access, via the World Wide Web, to data on genomes of organisms has allowed for the rapid identification of the genes encoding proteins separated by 2D electrophoresis.

This chapter provides techniques for separating, visualizing, and analyzing complex mixtures of proteins by 2D electrophoresis (for troubleshooting 2D gel electrophoresis, see www.proteinsandproteomics.org). The following topics are covered:

- ***Sample preparation,*** including examples of methods for preparing tissue, cell, and body fluid samples.
- ***Isoelectric focusing*** of proteins using immobilized pH gradients (IPG) for first-dimension separations.
- ***Methods for using mini-, standard-, and large-format SDS-PAGE gels*** for second-dimension separations.
- ***Visualization of proteins on gels,*** including five different staining techniques that can be used on analytical and/or preparative gels.
- ***Methods for analyzing 2D gels by blotting.***
- ***Instrumentation used to obtain images of 2D gels*** and programs for computer analysis of 2D separations.

PREPARATION OF SAMPLES FOR 2D GEL ELECTROPHORESIS

General Strategy

A useful, interpretable, and reproducible 2D electrophoresis separation depends on the appropriate preparation of the sample to be analyzed. Electrophoretic separation requires that the molecules to be separated remain in solution throughout the entire procedure. Various agents are therefore employed in the sample preparation and electrophoresis solutions in order to assure protein solubility. The IEF technique used for the first dimension of 2D electrophoresis imposes a number of constraints on the solubilizing agents that can be used. These agents can neither modify the pI of the proteins to be separated nor result in an excessively conductive solution. Only solubilizing agents that remain substantially uncharged over the pH range of the separation can be used. 2D electrophoresis contrasts in this regard to one-dimensional SDS-PAGE. This technique utilizes the strongly solubilizing detergent SDS both during sample preparation and during the procedure itself. This reagent is exceptionally effective at rendering proteins soluble, but it does so by binding to proteins and imparting a strong charge to the resulting complex. This reagent is therefore unsuitable as a solubilizing agent for IEF, which depends on the unmasked, endogenous charge of the protein. Protein-SDS complexes also remain highly charged, and therefore soluble, throughout electrophoresis. Proteins approach their isoelectric uncharged state during IEF, where they are minimally soluble. For these reasons, there are many proteins that are not amenable to separation by 2D electrophoresis using current methods, as they are not soluble under conditions compatible with IEF. Large proteins and hydrophobic proteins tend to be poorly represented on 2D gels. However, advances continue to be made in identifying reagents and conditions that allow optimal solubility of proteins under IEF conditions.

Sample preparation usually begins with cell lysis or tissue disruption, which should be done rapidly and at as low a temperature as possible in order to minimize proteolysis and other modes of protein degradation. It is often advantageous to lyse cells or disrupt tissue directly in a strongly denaturing solution, which will prevent protein degradation or modification. Protease inhibitors are frequently used as well. Protocols 1–4 describe methods for preparing various mammalian and bacterial samples for 2D electrophoresis.

Complex biological samples generally contain more proteins than can be seen on a single 2D gel. At most, a few thousand of the most abundant proteins are available for analysis. It is therefore advisable to use some kind of prefractionation to lower the complexity of the sample to be analyzed on a single 2D gel. If it is possible to restrict the investigation to a subset of proteins most likely to be of interest, these proteins should be separated from the bulk of irrelevant proteins and analyzed separately. For example, if one is studying a phenomenon known to be restricted to the cell nucleus, then nuclei should be purified and 2D analysis carried out on a nuclear lysate. Alternatively, a complex sample can be fractionated to yield several samples of lower complexity, each of which can then be analyzed separately by 2D electrophoresis (for examples of fractionation schemes utilized in this approach, see Lenstra and Bloemendal 1983; Ramsby et al. 1994; Corthals et al. 1997; Molloy et al. 1998; Taylor et al. 2000) (for preparation of cellular and subcellular extracts, see Chapter 3).

Also present in the sample may be substances that either interfere with the separation, render the proteins less soluble, or interfere with visualization of the result. Steps should be taken in such cases to reduce the amount of these interfering substances, or reduce their effect. The removal of interfering substances requires either additional processing steps (e.g., trichloroacetic acid [TCA] precipitation), which can result in selective losses, or the addition of enzymes (e.g., nucleases), which can result in additional spots in the final result. The investigator must therefore consider the trade-off between improved sample quality and accurate representation of the proteins in the sample.

In general, it is advisable to keep 2D sample preparation as simple as possible. If the sample contains a sufficiently high concentration of protein, and interfering substances are not present in high amounts (as is the case with most animal tissues and cells), it may be sufficient to disrupt the sample directly into a solution containing solubilizing agents, clarify by centrifugation, and apply the supernatant directly to the first-dimension IEF. However, if the sample is dilute, and contains relatively high levels of interfering substances (e.g., plant extracts or urine), proteins in the sample can be precipitated and resuspended in a smaller volume of solubilizing solution. This simultaneously concentrates the sample and removes many of the interfering substances. Flengsrud and Kobro (1989), Guy et al. (1994), Usuda and Shimogawara (1995), Tsugita et al. (1996), and Taylor et al. (2000) provide examples of methods to precipitate proteins for 2D analysis. If the sample source has a high ratio of nucleic acids to protein (as is the case with most bacterial cells), then a nuclease treatment should be employed. Other general guidelines include the following:

- Reagents used in the preparation of samples for 2D electrophoresis should be of the highest quality available. Solutions used in sample preparation should be freshly prepared or stored as frozen aliquots.
- Preparing the sample immediately prior to first-dimension IEF best preserves sample quality. If the sample must be stored, it should be divided into aliquots and kept at –80°C. Samples can be stored this way for up to 1 year.
- All particulate material in the sample should be removed by centrifugation prior to application of the sample to first-dimension IEF.

Protein Solubilization

Proteins in the sample must be fully solubilized, disaggregated, denatured, and reduced to be effectively separated in first-dimension IEF. To achieve this, samples for 2D electrophoresis must always contain:

- ***A neutral chaotrope,*** which is always either urea or a mixture or urea and urea derivatives.
- ***A neutral or zwitterionic detergent.***
- ***A reductant*** capable of reducing protein disulfide bonds to sulfhydryls and maintaining their reduced state.

These reagents are also present during first-dimension IEF. Buffers or ampholytes may be present as well. A number of choices exist for each of these components, which are considered separately below.

Chaotropes

Traditionally, urea is used at a concentration of at least 8 M. More recently, it has been shown that many additional proteins can be solubilized using mixtures of urea and thiourea (Rabilloud et al. 1997). The effect of using these reagents together, as opposed to urea alone, can significantly increase the number of proteins visible in the final 2D gel or membrane. Urea and thiourea are used most commonly at concentrations of 7 M and 2 M, respectively, but the optimal ratio may differ among samples (Musante et al. 1997). The use of thiourea can result in disturbances and loss of resolution in the acidic range (pH 3–5) of IEF separations. This could be an intrinsic effect of thiourea or a result of impurities in commercial preparations.

Full solubilization and denaturation into urea-containing solutions can be slow. The sample should be allowed to remain in this solution for at least 1 hour at room temperature prior to centrifugation and application. Heating urea-containing solutions above 30°C must be avoided because this causes hydrolysis to cyanate, which can result in protein charge modifications.

Detergents

Originally, either of two similar nonionic detergents was used: Triton X-100 or Nonidet P-40 (NP-40). More recently, the use of the zwitterionic cholamidosulfobetaine detergents CHAPS (3-[3-cholamidopropyl)dimethylammonio]-1-propane sulfonate) and CHAPSO has become prominent (Perdew et al. 1983). CHAPS is available in high purity and can be used at concentrations up to 4%. Alkylsulfobetaine detergents such as Zwittergent or SB3-10 have also been used (Adessi et al. 1997; Molloy et al. 1998), but these reagents are relatively insoluble in urea solutions and often require a lower concentration of urea, resulting in potentially less-effective solubilization. Amidoalkylsulfobetaine detergents have been developed that are highly soluble in urea and thiourea solutions and allow solubilization of otherwise undetectable proteins (Rabilloud et al. 1990; Chevallet et al. 1998; Santoni et al. 2000). However, many of these latter detergents are not available commercially.

SDS is often used in sample preparation for 2D electrophoresis, but it must be considered a special case, as it cannot be used as the sole detergent in an IEF separation. SDS solutions are used for their ability to rapidly solubilize and extract proteins while simultaneously inactivating modifying enzymes. The protein sample may also contain SDS because it was originally prepared for SDS-PAGE analysis. In these cases, the interfering effect of SDS must be reduced by diluting the sample into a solution containing an excess of a neutral or zwitterionic detergent prior to application to first-dimension IEF. The final concentration of SDS should be 0.25% (w/v) or lower, and the ratio of the excess detergent to SDS should be at least 8:1.

Reductants

Originally, 2-mercaptoethanol was used to reduce samples for 2D electrophoresis. Its use has been largely superseded by the sulfhydryl reductants dithiothreitol (DTT) and dithioerythritol (DTE), which are effective at lower concentrations and available in higher purity. DTT and DTE are used at concentrations ranging from 20 mM to 100 mM. More recently, it has been reported that tributyl phosphine (TBP) may be more effective with samples that are difficult to solubilize (Herbert et al. 1998). However, the instability and insolubility of this reagent in aqueous solution render it relatively ineffective at maintaining proteins in their reduced state during first-dimension IEF. It is therefore advisable to utilize a sulfhydryl reductant as well during the first-dimension separation.

Ampholytes or buffers

Carrier ampholytes and/or buffers are often used in sample preparation. They act by maintaining the pH at an optimum for protein solubility and by minimizing protein aggregation due to charge-charge interactions. Carrier ampholytes can be used up to 2% (v/v) and buffers can be used at concentrations up to 40 mM during sample preparation. Buffers should be used with care, however, as they can interfere with IEF if present at too high a concentration during the first-dimension separation. Cationic buffers (e.g., Tris) will be drawn toward and concentrate at the cathode, resulting in less-effective separation at the basic end of the pH range. Amphoteric buffers (e.g., HEPES) can actually focus within the pH gradient, resulting in a zone of high conductivity where proteins do not focus. If buffers are used during sample preparation, their final concentration in the first-dimension solution should be less than 5 mM. Samples are often prepared at high pH, since most tissue proteases are inactive above pH 9. Spermine base (used at 20 mM) is reported to be particularly effective for high-pH extractions (Rabilloud et al. 1994a). Different proteins are optimally soluble at different pH values, so the choice of buffering conditions during sample preparation can strongly influence the profile of proteins appearing in the final 2D result.

Interfering Substances

Nonprotein impurities may be present in samples destined for separation by 2D electrophoresis. These impurities can interfere with separation of the proteins or their subsequent visualization, so steps are often taken to rid the sample of these substances. Commonly encountered interfering substances, methods for their removal, and possible consequences if they are not removed are considered below.

Small ionic molecules

Small ionic molecules may be present in a 2D sample either because they are endogenous to the material under investigation (e.g., metabolites in tissue and salts in serum) or because they have been used as additives during sample preparation (e.g., residual buffers and salts). These substances can interfere with first-dimension separations by (1) increasing the conductivity in the first-dimension separation, thereby reducing the overall effectiveness of IEF, and (2) accumulating at the anode or cathode or within the strip, resulting in regions of high conductivity where proteins do not focus.

Such contamination can be removed by precipitating the proteins in the sample and resuspending them in salt-free sample solution. There are many ways of doing this, but the most commonly used precipitants are TCA (Guy et al. 1994) or organic solvents such as acetone (Flegsrud and Kobro 1989), ethanol, or a mixture of chloroform and methanol (Taylor et al. 2000). The combination of TCA (10%) and acetone is particularly effective (Damerval et al. 1986). When using TCA, it is important that this reagent be completely removed prior to resuspension. This is accomplished by extensive washing of the precipitate with acetone or diethyl ether.

Dialysis is also an effective means of desalting a sample, which generally results in minimal loss of protein. Desalting by gel-filtration chromatography may be employed, but it can result in protein losses.

Often, the problem of a highly conductive sample can be solved by modifying upstream sample preparation steps to limit contamination with salts. This is generally preferable to trying to remove these contaminants at a later stage. If high-ionic-strength solutions are used during sample preparation, carryover of salts into subsequent steps can be significant enough to be problematic. Whenever possible, a solution without added salts should be substituted. For example, it is common practice to remove external proteins from cultured cells by washing with phosphate-buffered saline prior to cell lysis. Carryover of salt into the cell lysate can then result in disturbances in the first-dimension IEF separation. This problem can be prevented by substituting a low-ionic-strength washing solution, such as Tris-buffered sorbitol.

Nucleic acids

The presence of DNA and RNA (especially high concentrations of nucleic acid) in a sample can result in poor focusing in the acidic region of the IEF gel. They also generate a significant background if the 2D gel is silver-stained. In addition, the presence of DNA may make the sample too viscous to work with effectively. Large nucleic acids can be removed by ultracentrifugation, but this may also remove larger proteins and proteins bound to nucleic acids. Treating the sample with nucleases often mitigates the negative effects of nucleic acids, which can be accomplished by adding a one-tenth volume of a solution containing 1 mg/ml DNase I, 0.25 mg/ml RNase A, and 50 mM $MgCl_2$ to the protein sample and incubating on ice (Blomberg et al. 1995). High concentrations of urea render these enzymes ineffective, so this treatment must be applied prior to the addition of urea to the sample. Note that the nucleases may appear in the final 2D result as extra spots. These two limitations can be circumvented through the use of Benzonase, a bacterial endonuclease with specificity for both DNA and RNA. Benzonase retains enough activity in the presence of 8 M urea to remain effective, and it can be used at a low enough concentration to not appear on a 2D gel.

Lipids

Lipids can form complexes with proteins, reducing the protein's solubility. They can also form complexes with detergents, which reduces the detergent's effectiveness. Organic solvent precipitation can remove much of the lipid from a sample. Using excess detergent can minimize the effects of lipids.

Phenolic compounds

Phenolic compounds are present in many plant tissues and can modify proteins through an enzyme-catalyzed oxidative reaction. There are a number of ways this can be prevented.

- Employing reductants during sample extraction prevents the oxidative reaction. In addition to the DTT or DTE conventionally used during sample preparation, sulfite or ascorbate can also be used.
- The proteins can be rapidly separated from the phenolic compounds by using precipitation techniques such as those used to remove small ionic molecules.
- The enzyme polyphenol oxidase can be inhibited directly with thiourea.
- Polyphenolic compounds can be sequestered by adsorption to polyvinylpyrrolidone or polyvinylpolypyrrolidone.

Protection against Proteolysis

Cell lysis liberates proteases, which can degrade proteins and complicate 2D analysis. Most proteases are effectively inhibited by the denaturants employed during 2D sample preparation, but some retain activity even in the presence of 8 M urea. In these cases, protease inhibitors may be used. Individual protease inhibitors are active against specific proteases, so it is advisable to use protease inhibitors in combination. Protease inhibitors used during sample preparation for 2D electrophoresis are listed below.

PMSF and AEBSF

PMSF (phenylmethylsulfonyl fluoride) and AEBSF (aminoethylbenzenesulfonyl fluoride) irreversibly inhibit serine and cysteine proteases. They rapidly become inactive in aqueous solutions and must be prepared directly prior to use. Sulfhydryl reagents also inactivate them, so they are best added to an extract to inactivate proteases prior to addition of DTT or DTE. PMSF is used at 1 mM, whereas AEBSF is more soluble and can be used at concentrations of up to 10 mM.

EDTA

EDTA (ethylenediaminetetraacetic acid) inhibits metalloproteases by chelating metal ions required for activity. It should not be present at more than 1 mM during first-dimension IEF.

Peptide protease inhibitors

Reversible protease inhibitors (e.g., leupeptin, pepstatin, aprotinin, and bestatin) are active against various classes of proteases under a variety of conditions. They are relatively expensive, but are cost-effective because they are seldom used at concentrations above 1 μM.

THE FIRST DIMENSION: ISOELECTRIC FOCUSING WITH IMMOBILIZED pH GRADIENTS

Proteins are amphoteric molecules that carry a positive, negative, or zero net charge, depending on the pH of their surroundings. The isoelectric point (pI) of a protein is the pH value of the protein's surroundings at which the protein has a zero net charge. At pH values above its pI, a protein carries a net negative charge, and at pH values below the pI, the protein carries a net positive charge. IEF takes advantage of this phenomenon. By placing proteins in a pH gradient within an electric field, the proteins will migrate to the pH where they have no net charge, their pI. A protein with a positive net charge will migrate toward the cathode through the pH gradient, becoming progressively less positively charged, until it reaches its pI. A negatively charged protein will move toward the anode, progressively becoming less negatively charged, until it reaches its pI. Once there, if a protein drifts from its pI, it gains a net charge and will then migrate back to its pI. As the slope of the pH gradient and the strength of the electric field determine the degree of resolution, IEF is typically carried out under very high voltage (typically in excess of 1000 volts). Additionally, IEF is typically performed under denaturing conditions to provide the highest resolution and the clearest results.

The original IEF method depended on carrier ampholyte-generated pH gradients in polyacrylamide tube gels (Klose 1975; O'Farrell 1975). Carrier ampholytes are small amphoteric molecules, which have high buffering capacities near their pIs. When a voltage is applied across a mixture of these molecules, the carrier ampholytes align themselves according to their pIs. The strong buffering capacity of each carrier ampholyte at its pI results in a continuous pH gradient across the gel. However, the use of carrier ampholytes in 2D electrophoretic studies has several limitations:

- Carrier ampholytes are not a well-defined mixture of molecules and suffer from batch-to-batch variations in manufacturing, creating difficulties with reproducibility in first-dimension separations.
- The pH gradients generated by carrier ampholytes are unstable and have a tendency to drift toward the cathode over time. This drift reduces the reproducibility of the first dimension as it is time-dependent. Additionally, the drift results in a flattening of the pH gradient at each end, especially in the basic region. Thus, it is not possible to resolve proteins at extreme pH values (i.e., above pH 9).
- It is very difficult to work with soft acrylamide tube gels. The gels can stretch or break, resulting in variations between separations. Therefore, the quality of the results of tube-gel-based IEF separations is often dependent on the experience of the operator.

In 1982, Bjellqvist et al. introduced an alternative method for generating immobilized pH gradients for IEF by incorporating acrylamido buffers into a polyacrylamide gel at the time of its casting. During polymerization, the acrylamide portion of the buffers copolymerizes with the acrylamide and bisacrylamide monomers to form a polyacrylamide gel. The pH gradient is formed in a fashion similar to pouring an acrylamide gradient gel, except that rather than using two solutions of different %T, the two solutions contain a relatively acidic mixture and a relatively basic mixture of acrylamido buffers. For improved handling, these gradients are cast onto a plastic backing. After polymerization is complete, the gel is washed to remove catalysts and unpolymerized acrylamide monomers, which can interfere with the separation of proteins and reduce the sample loading capacity. Gianazza (1999) has recently reviewed methods for casting IPG (immobilized pH gradient) gels. Although acrylamido buffers (and carrier ampholytes) are available from several manufacturers, commercial preparations of precast gel strips incorporating immobilized pH gradients are a simpler (but more expensive) alternative. The gel strips can be rehydrated in any desired solution suitable for the first-dimension IEF separation.

IPGs for first-dimension separations provide a number of advantages over carrier ampholyte-based separations.

- The problem of cathodic drift is eliminated, because the pH gradient is covalently fixed within the gel. Thus, broader pH range separations can be preformed, allowing the separation of more highly acidic and highly basic proteins.
- IPG gels have a higher load capacity for proteins than do gels cast with carrier ampholytes (Bjellqvist et al. 1993).
- Acrylamido buffers are a small set of well-characterized molecules, allowing for very reproducible manufacturing, thus eliminating batch-to-batch variation.

Precast IPG gel strips also offer some further advantages.

- Precast IPG gels reduce preparation time and effort and greatly improve the reproducibility of the pH gradient.
- Gel strips are cast onto a plastic backing, preventing stretching or breaking, and generally easing the handling of gels.
- Samples can be applied to the gel strips during gel rehydration (Rabilloud et al. 1994a; Sanchez et al. 1997).

Protocol 5 presents two methods for carrying out the first-dimension IEF step of 2D gel electrophoresis. The first is a flat-bed IEF procedure. The second method uses a high-voltage self-contained IEF system, which requires fewer manipulations of the gel and is capable of faster separations.

THE SECOND DIMENSION: SEPARATION OF PROTEINS BY MOLECULAR WEIGHT

Introduction to SDS-PAGE

SDS-PAGE is a method of separating proteins on the basis of their molecular weight. The technique is performed in a polyacrylamide gel containing the detergent SDS. SDS is an anionic surfactant, with a 12-carbon alkyl tail and an ionic head group that denatures proteins by wrapping around the polypeptide backbone in a ratio of ~1.4 g of SDS per gram of protein. The bound SDS gives the proteins a net negative charge per unit mass, overwhelming the intrinsic electrical charge of the sample protein. In addition, SDS disrupts hydrogen bonds, blocks hydrophobic interactions, and partially unfolds the protein, thus eliminating secondary and tertiary structures. Proteins can be completely unfolded by reacting them with reducing agents, such as DTT or 2-mercaptoethanol, in the presence of SDS. These agents cleave the disulfide bonds within cystine residues. Under these conditions, the unfolded proteins form ellipsoid shapes surrounded by SDS. During electrophoresis in a sieving gel, there is a linear relationship between the logarithm of the molecular mass of the protein and the relative migration distance of the SDS-polypeptide micelle (Weber and Osborne 1969; Laemmli 1970). However, this relationship does not hold for all protein types. For example, glycoproteins with high carbohydrate concentrations migrate at a slower rate than expected based on their molecular mass, because they bind smaller amounts of SDS relative to other proteins of the same molecular weight (Dunbar 1987).

SDS-PAGE is normally performed with a discontinuous buffer system. In this system, the proteins migrate in the order of their mobility in tight zones, called the stack, between the leading and trailing ions of the discontinuous buffer. The most common buffer system uses glycine as the trailing ion as described by Laemmli (1970). Other buffer systems can be used, particularly the Tris-tricine system of Schägger and von Jagow (1987), for resolution of polypeptides below 10 kD in size (for an extensive discussion of SDS-PAGE and alternative buffer systems, see Chapter 2).

Altering the amounts of acrylamide and cross-linker controls the pore size, and thus the sieving properties, of the polyacrylamide gel. Acrylamide and cross-linker concentrations are described in %T and %C.

- %T is the total weight of acrylamide and cross-linker expressed as a percentage of the total volume.
- %C is the weight of cross-linker expressed as the percentage of total weight of both the acrylamide and cross-linker.

For example, a 1-liter stock solution of acrylamide at 30%T and 2.5%C would contain 292.5 g of acrylamide and 7.5 g of cross-linker. When %C remains constant and %T is increased, the pore size of the gel decreases. When %T remains constant and %C is increased, the pore size of the gel follows a parabolic function, where at high and low values of C, the pores are large. The minimum pore size is at C = 4%. Table 4.1 provides recommended acrylamide concentrations for a variety of protein molecular-weight ranges.

Second-dimension gels are made as either homogeneous gels with the same %T and %C throughout or as gradient gels with a varied %T but a constant %C. Homogeneous gels offer the best resolution for a particular molecular-weight window and are the easiest to pour reproducibly. However, a gradient gel is best when proteins over a wide range of molecular weights are to be analyzed. Gradient gels provide a wider linear separation interval, and the resulting protein spots are sharper, because the decreasing pore size functions to minimize diffusion.

TABLE 4.1. Recommended acrylamide concentrations for protein separation

	% Acrylamide in resolving gel	Separation size range (m.w. x 10^{-3})
Single Percentage	5	36–200
	7.5	24–200
	10	14–200
	12.5	14–100[a]
	15	14–60[a]
Gradient	5–15	14–200
	5–12	10–200
	10–20	10–150

[a]The larger proteins fail to migrate significantly into the gel at this percentage of acrylamide.

An Overview of Second-dimension SDS-PAGE

Following the separation of proteins by IEF, the second-dimension separation is carried out by SDS-PAGE. This can be performed on a variety of flatbed or vertical gel electrophoresis systems depending on the throughput and resolution desired.

- Flatbed systems provide excellent resolution and relatively rapid separation and can be used with commercially available precast gels. However, such systems have limited capacity and are only recommended for micro-preparative separations. These systems will not be discussed further.
- Vertical systems have become favored for their relative ease of use and ability to perform multiple separations simultaneously. Vertical 2D is generally run on 1.0- or 1.5-mm thick gels. For rapid results using 7-cm IPG for the first dimension, minigel units are used. Such units are available from a variety of companies including Amersham Biosciences, Bio-Rad, and Invitrogen. In addition, precast vertical SDS-PAGE gels are also available from Bio-Rad, FMC, and Invitrogen. For increased resolution, 11-, 13-, and 17-cm gels can be run on standard size 16 x 16-cm or 20 x 20-cm gels. The multiple gel casters available with these systems improves reproducibility of the second-dimension gels. Precast gels for the Bio-Rad 20 x 20-cm gel system are available.
- For maximal resolution, reproducibility, and capacity, use large-format gels (e.g., the ETTAN DALT II 26 x 20-cm gel from Amersham Biosciences or the 22 x 26-cm Investigator gels from Genomic Solution). These systems are used with 18- and 24-cm IPG gels, and they can run up to 20 second-dimension SDS-PAGE gels simultaneously. They also provide multiple gel-casting systems to help improve the reproducibility of gels. Precast gels for the investigator and the ETTAN DALT II systems are available.

The second dimension consists of four steps:

1. Preparation of the SDS-PAGE gel (Protocols 6 and 7).
2. Equilibration of the IPG strips in SDS buffer (Protocol 8).
3. Placement of the equilibrated IPG gels on the SDS-PAGE gel (Protocol 9).
4. Electrophoresis of the second-dimension gel (Protocol 9).

STAINING 2D GELS

A large number of methods have been developed to detect proteins separated by electrophoresis on 2D gels, with the most common ones involving the binding of a dye or silver ions to the proteins. The choice of method is greatly dependent on the experimental questions being asked. The complexity of the steps, the sensitivity of detection, compatibility with mass spectrometry, cost, and equipment required vary among the staining methods. Five different procedures are presented in Protocols 10 through 14.

- ***The colloidal Coomassie method*** is commonly used with preparative gels, has a detection range of 8–50 ng of protein within a spot, and takes an average of 1–2 days to complete. Although the least sensitive of the methods presented here, it is broadly used because it is simple and compatible with mass spectrometry (see Protocol 10; for conventional Coomassie Blue R250 staining, see Chapter 2).

- ***Classical silver-staining methods*** are commonly used with analytical gels because of their highly sensitive range of detection, typically 2–4 ng of protein per spot. However, classical silver-staining methods are complex and are incompatible with mass spectrometry, because the glutaraldehyde used in the sensitization steps can cross-link proteins, and the silver ions can interfere with mass spectrometric analysis. Protocol 11 provides a very sensitive ammoniacal silver stain based on the method of Merril and Harrington (1989).
- ***Mass spectrometry-compatible silver staining*** is an important variation on conventional silver staining, due to the rising importance of mass spectrometry in proteomics. Protocol 12 provides a silver staining method that is compatible with mass spectrometry, based on the method of Yan et al. (2000b).
- ***Fluorescent stains*** for 2D gel electrophoresis have advanced greatly with the recent development of the ruthenium-based fluorescent dye SYPRO Ruby (Molecular Probes) (see Table 2.1, Chapter 2). This dye provides sensitivity similar to that of ammoniacal silver but without the complex methodology, limitation on linear dynamic range, or the problems with mass spectrometry compatibility. However, these dyes are very expensive and require fluorescent scanners for analysis (see Protocol 13).
- ***Phosphoprotein stains*** are becoming increasingly important as it is clear that the phosphorylation state of proteins is a key regulator of their function (see Chapter 9). Protocol 14 uses the GelCode Phosphoprotein staining kit produced by Pierce.

Although autoradiography has also been used to visualize 2D gels, it is beyond the scope of this chapter. For a recent review, see Link (1999) and Lee and Harrington (1999).

Colloidal Coomassie Staining

Protocol 10 is a modification of the method of Neuhoff et al. (1988) which involves the binding of Coomassie Blue G250 to proteins. This dye complexes with basic amino acids, such as arginine, tyrosine, lysine, and histidine, and produces a very low background due to the colloidal properties of the stain. This property prevents the Coomassie dye from penetrating the gel, eliminating the requirement for the long destaining periods characteristic of Coomassie R250 staining.

Ammoniacal Silver Staining

Silver staining of polyacrylamide gels (Protocol 11) was first introduced by Switzer et al. (1979) and has rapidly become the most popular high-sensitivity staining method for proteins. Problems with high backgrounds, reproducibility, and silver mirrors resulted in many improvements to this initial protocol over the years. In a recent review, Rabilloud et al. (1994b) found more than 100 different protocols in the literature. However, all protocols contain at least the following five basic steps: fixation, sensitization, silver impregnation, image development, and cession of development and image stabilization.

We have found that the ammoniacal silver staining process presented here provides high sensitivity, in the range of 2–4 ng per protein spot, and works well with gels down to a thickness of 1 mm. Due to its high sensitivity, silver staining is very susceptible to interference from a variety of sources. Exceptional cleanliness is necessary, and reagent and water quality is critical. Deionized water with a resistivity of more than 15 mho/cm is necessary for all solutions and washes. Distilled water should not be used, for it tends to give erratic results. All chemicals should be of high grade.

Mass Spectrometry-compatible Silver Staining

Ammoniacal silver staining is one of the most sensitive methods for detecting proteins on an SDS-PAGE gel. However, it and other standard silver-staining methods are incompatible with mass spectrometry, which is fast becoming the best way to identify proteins isolated on 2D gels. This is because the proteins in gels to be analyzed by mass spectroscopy must remain unmodified, so many of the common sensitizing agents used in silver staining procedures (e.g., glutaraldehyde and strong oxidizing agents) cannot be employed. Additionally, many labs find it advantageous to purchase dedicated kits, where reagent quality is assured, to carry out silver staining. Kits for MS-compatible silver staining are available from a variety of sources, including Amersham Biosciences, Bio-Rad, and Pierce. The method provided in Protocol 12 is based on the Amersham Biosciences PlusOne silver stain kit as described by Yan et al. (2000b). These authors found that this method is compatible with MALDI and ESI-MS and works with semipreparative protein loads without resulting in negative staining. However, this process is less sensitive than standard silver-staining methods. Other mass spectrometry-compatible silver-staining methods are described by Shevchenko et al. (1996), Scheler et al. (1998), Gharahdaghi et al. (1999), and Sinha et al. (2001).

Fluorescent Staining of Proteins with SYPRO Ruby

The development of new mass spectrometry instrumentation and its application to the field of proteomics have been a great help in the identification of proteins isolated by 2D electrophoresis (see Chapter 8). Identification of proteins is no longer limited to those that are highly abundant, with protein identification in the femtomole level now being feasible (Jensen et al. 1997; Gatlin et al. 1998). However, to make this sensitivity applicable to 2D gel electrophoresis, sensitive and compatible staining methods are necessary. Silver staining provides very sensitive detection of proteins; however, it is complicated and without modifications that reduce its sensitivity, it is incompatible with mass spectroscopy. A sensitive ruthenium-based fluorescent dye (SYPRO Ruby) has recently been developed by Molecular Probes. Although the structure of this dye is not in the public domain, it has been reported that SYPRO Ruby is a transition metal organic complex that binds directly by electrostatic mechanisms (Patton 2000). Similar to silver stains, SYPRO Ruby interacts with basic amino acids, including lysine, arginine, and histidine (Lopez et al. 2000). Recent studies by Yan et al. (2000a) and Lopez et al. (2000) have shown that this dye is as sensitive as silver staining and completely compatible with MALDI mass spectroscopy. The staining method is also simple, having only a fix, staining, and single wash step. However, SYPRO Ruby is very expensive. Thus, this dye should be limited to micropreparative gels as the cost to run the large number of gels needed for quantitative studies may be prohibitive.

Phosphoprotein Staining with the GelCode Phosphoprotein Staining Kit

The phosphorylation state of a protein has an important role in the regulation of a wide variety of cellular processes. As a result, there has been a great deal of interest in detecting the phosphorylation status of proteins. The method presented in Protocol 14 uses the GelCode phosphoprotein staining kit (Pierce Chemical Company). This method depends on the hydrolysis of the phosphoprotein phosphoester linkage using 0.5 N sodium hydroxide in the presence of calcium ions. The gel containing the newly formed insoluble calcium phosphate is then treated with ammonium molybdate in dilute nitric acid. The resultant insoluble nitrophospho-molybdate complex is finally stained with Methyl Green. After destaining, phosphoproteins are stained green. The detection limits depend on the degree of phosphorylation of the protein.

TRANSFER OF PROTEINS FROM 2D GELS

Proteins can be transferred from 2D gels in the same manner as those transferred from one-dimensional SDS gels (see Chapter 6, Protocol 2). Following the second-dimension run, the gel is placed adjacent to a membrane, and the pair is sandwiched between blotting paper. An electrical field is applied perpendicular to the slab gel, driving the proteins to migrate from the gel onto the membrane. The process is commonly known as western blotting and was first performed by Towbin et al. (1979). Once bound to the membrane, proteins can be stained to examine the total protein pattern (see Protocols 17–20) or specific proteins can be characterized by reactions with antibodies (immunoblotting) or other detection reagents. Processing the membrane rather than the gel with a protein detection reagent provides several advantages.

- The proteins are bound to the membrane's surface, allowing very small volumes of reagents to be used.
- The protein separation pattern on the membrane does not deteriorate due to diffusion.
- Membranes are much more robust than gels, enabling them to survive the rigors of protein detection.
- Membranes can become part of the experimental record.

Membranes containing bound proteins can be used analytically as an end in themselves or they can be a step preceding another analytical method. In the latter case, all of the proteins on the membrane are stained, and selected spots are excised for further analysis by either mass spectrometry or protein sequencing (Matsudaira 1987) (see Chapter 6).

Transfer Equipment

The equipment for transfer of proteins from gels to membranes falls into two categories: tank and semidry. Within the transfer tank, a cassette holds the sandwich of gel, membrane, and blotting paper suspended vertically in buffer. Electrode panels are placed parallel to the cassette. Typically, liters of buffer are required in the tank to maintain the current and pH throughout the transfer. Since these units require high field strength to drive the proteins from the gel onto the membrane and may require extended times to complete efficient transfers, cooling of the transfer unit is typically required to avoid overheating the apparatus. (Overheating will result in poor transfer of proteins to the membrane and, at worst, can cause severe damage to the transfer equipment.) The advantages of tank systems are their efficiency of protein transfer: Transfers can be run for extended periods of time (allowing more complete transfer) and proteins from several gels can be transferred simultaneously.

Semidry units transfer proteins from the gel to the membrane when they are sandwiched between buffer-saturated stacks of blotting paper. The electrical field is applied parallel to the stack. Compared with tank transfer units, semidry transfer units use much less buffer, and because the electrode panels are closer together, less voltage is required for transfer. Typically, protein gels are transferred based on the surface area of the gel exposed to the electrical panels. A recommended current for transferring proteins in a semidry apparatus is 0.8 mA/cm^2. At this current, heating is minimized and the buffering is adequate to transfer gels for up to 2 hours. The disadvantage is that no more than two large gels can be efficiently transferred at one time and that, due to buffer depletion, the units cannot run longer than ~2 hours.

Transfer Membranes

Many types of membranes are available. Nitrocellulose is frequently used for total protein staining methods and immunoblotting, whereas polyvinylidene difluoride (PVDF) is required for protein sequencing. Some nitrocellulose membranes are supplied with material incorporated on or into the membrane to add support. These supported nitrocellulose membranes are preferred as they have more strength than frangible unsupported nitrocellulose. Nylon membranes are also available; however, these are seldom used with proteins due to difficulties with staining. Nylon membranes are more commonly used with nucleic acid transfers.

Important Tips When Transferring Proteins

- Use gloves at all times when handling gels and membranes. Fingerprints can cause staining artifacts and contaminate the membranes. Handle membranes from the edges only.
- To prevent diffusion, transfer gels immediately after the SDS electrophoresis separation is completed.
- Do not equilibrate second-dimension gels when using Tris/glycine/methanol transfer buffer. Although equilibration is sometimes recommended, because the gel may shrink in the methanol buffer, our lab has not found this to be a problem. On the other hand, equilibration may decrease the transfer of high-molecular-mass proteins (>100 kD) and contribute to the loss of lower-molecular-mass proteins (<20 kD). Such losses are much more prevalent in thinner (e.g., 1 mm) gels and lower-percentage gels.
- Equilibrate gels to be transferred in the presence of CAPS buffer with CAPS transfer buffer for 5 minutes before transfer to reduce the amount of Tris and glycine present within the gel. CAPS is the buffer preferred for transfers that will be used for amino acid sequencing, as high concentrations of glycine will interfere with sequencing.
- Take care to remove all air bubbles from the transfer sandwich. Trapped air will block the transfer of proteins in the region of the air bubbles.
- Make sure to wet the nitrocellulose membranes in water and PVDF membranes in methanol prior to equilibrating them in transfer buffer. If the membrane dries out at any time during the process, they must again be wetted in water (nitrocellulose) or methanol (PVDF).
- Wet the membranes by slowly lowering them onto the surface of the liquid. After a moment, gently submerge the membrane. This ensures even wetting without trapping air.
- Nitrocellulose membranes age and will eventually no longer wet properly. An old membrane may show yellowing around the edges. A useable membrane should rapidly absorb the liquid and change color from pure white to a light gray. If the membrane does not wet properly, use a fresh membrane. Fresh membranes always provide the best results.
- Up to 0.1% SDS can be added to the transfer buffer to improve the transfer of proteins out of gels, especially when transferring large proteins. However, the presence of SDS may decrease the efficiency of protein binding to the membrane, resulting in losses during staining. Thus, SDS should be present during protein transfers only when high-molecular-weight proteins are of interest.

IMAGE ACQUISITION AND ANALYSIS OF 2D GELS

Image Acquisition

The varied methods of visualizing proteins on 2D gels require image acquisition techniques that can detect diverse chromogenic, radioactive, and fluorescent signals. A comprehensive review of all of the imaging methods is beyond the scope of this chapter but is available in reviews by Patton et al. (1999), Patton (1995) and Sutherland (1993). Below are described three main types of devices presently used to acquire images of 2D gels.

Document scanner

The least expensive of these devices is a document scanner. For densitometry analysis of gels, a document scanner must have the ability to acquire images in a transmission mode. Additionally, a scanner with high resolution (2400 x 1200 dpi), 14-bit pixel depth, and linearity over a wide optical density range (i.e., above 3.4 OD units) is desirable. The ImageScanner (Amersham Biosciences) and the Model GS-710 Imaging Densitometer (Bio-Rad) meet these requirements.

Charged-coupled device camera

The charged-coupled device (CCD) camera is probably one of the most versatile acquisition devices for electrophoretic applications. CCD cameras have been used to acquire images from a variety of stained gels (including silver, Coomassie, and fluorescent stains), autoradiographs, and stained membranes. CCD-based systems are available from a variety of companies, including Bio-Rad, Amersham Biosciences, BioImage Products, and PerkinElmer Wallac. In the past, these systems have had some limited sensitivity and resolution, especially when dealing with large gels. With the advent of *cooled* CCD cameras, the problem of sensitivity has been greatly reduced. Recent advances in the ARTHUR system (PerkinElmer Wallac) have also decreased the resolution problem with large gels. This system scans the gel using a cooled CCD camera at a fixed focal length, acquiring multiple adjacent overlapping frames that are then stitched together to produce a single image. Because of this scanning mechanism, the resolution is independent of the size of the gel. The addition of transilluminators or top illumination of a specific wavelength has allowed CCD camera-based systems to also be used with fluorescent stains. Increasing the signal to noise can increase the sensitivity of CCD cameras for fluorescent dyes. One way to achieve this is by using camera binning, which is available on some instruments. Binning involves summing data from image pixels either 2 x 2, 3 x 3, or 4 x 4, which has the effect of increasing the signal-to-noise ratio by factors of 4, 9, and 16, respectively. Binning, however, decreases the resolution of the image, and thus may only be useful when the detection of very low-abundance proteins is required. A variety of methods of acquiring 2D gel images using CCD camera systems are available in the recent review by Patton et al. (1999).

Laser-based densitometers

A large number of laser-based densitometers are available for 2D analysis. These systems tend to have a broader linear OD range than scanners and are less susceptible to saturation effects than other detectors. Laser-based systems offer high resolution (50 μm) and sensitivity (many providing >4000 levels of gray scale). Many of these systems have been designed with lasers of multiple wavelengths in order to work with a variety of fluorescent dyes. Additionally, a number of these instruments such as the PhosphorImager from Molecular

Dynamics will also read phosphor storage imaging plates, which can be used with radiolabeled gels to detect [^{32}P]- and [^{35}S]methionine-labeled proteins. The advantage of phosphor imaging plates is that they provide increased sensitivity, good spatial resolution of signal, and a broad linear dynamic range of response as compared to standard X-ray film. Systems with these characteristics are available from Molecular Dynamics, Bio-Rad, and Fuji Medical Systems, as well as other vendors.

Analysis of 2D Gels

Given the inherent complexity of 2D gels, which typically display thousands of proteins on a single gel, they are best analyzed using a computer. Using such systems, the investigator can detect and quantify even faint spots, quantitatively compare 2D images with each other, and identify protein expression changes between gels representing varied conditions. Currently, there are several major commercially available 2D analysis software systems: Z3 (Compugen), Melanie III (Genebio and Bio-Rad), PDQuest (Bio-Rad), and the various versions of Phoretix 2D software marketed as Imagemaster 2D (Amersham Biosciences), HT Analyzer (Genomic Solutions), and Phoretix 2D Advance (PerkinElmer Wallac) (Raman et al. 2002). Each of these software systems differs slightly in various aspects, including the computer platforms on which they run, the graphical user interface, the amount of gel manipulation allowed, and file formats. However, they all follow the same general steps for analysis of the gels. It is beyond the scope of this chapter to contrast these software systems or to give specific instructions of their operations. Detailed information on each software package is available at their respective home pages on the World Wide Web (see Two-dimensional Analysis Software Sites at the end of this chapter). Here, we review the general steps taken in the analysis of 2D gels.

Image visualization and manipulation

First, the image of the gel must be displayed in the program. All three of the programs accept TIFF files, so that is the recommended format for saving the gel images. Once the file is open, the image may need to be edited so that it is in the correct orientation and extraneous areas are removed from the image. This can be achieved using the rotate and crop functions in these programs. In addition, the display of the image may need to be optimized. Modern image acquisition systems are able to scan 2D gels with 12-bit or 16-bit/pixel, with 4,096 to 65,536 gray levels. Common computer screens are only able to display 256 color or gray levels; thus, the programs must map intensity values between the scanned image gray levels and the screen gray levels. To visualize faint spots, increase the contrast of the image by using the transform or adjust contrast feature in these programs. Note that changing the contrast only affects how the image is displayed on the screen and does not alter the underlying data used for spot detection and quantification. Upon completion of the first modification to the gel, the program allows the new image to be saved in a format specific for that program. All three programs also allow manipulation of multiple gel images simultaneously.

Spot detection and quantification

Once the gel images of interest are open and in the desired format, the next step is spot detection. This is the most critical phase of the analysis: All other results are worthless if the spot detection is incorrect. All three programs provide an automated detection system based on unique algorithms. Each program also allows alteration of the system's parameters, such as sensitivity, spot size, background subtraction, and number of smooths, although all of these parameters may not be available with all programs. Once an initial spot detection is complete, all of the programs allow the parameters to be altered again until an optimal detection is achieved. Once the optimal automated detection parameters are established, they can be applied to all of the other gels in the experiment or saved for later application. It is important that the spot detection parameters are consistent across all of the gels analyzed within an experiment. Spot detection may need to be manually edited, adding, deleting, or editing spots that were not detected by the automated spot detection algorithm. The spots may also be fitted to to a Gaussian model, which is done automatically in PDQuest and is an option in Melanie III. Gaussian modeling can help resolve areas of overlapping density and streaking and improve the accuracy of the density measurement of spots. During the spot detection method, all three programs automatically determine a number of parameters for the spot (i.e., spot volume, area, optical density, and coordinates).

Spot matching

Once the spots of interest have been identified on all of the gels in the data set, the next step is to match the spots between gels. The first step is to choose a sample that will serve as the reference gel for the match. The reference gel should have a large number of well-resolved, good-quality spots, be representative of the experiment as a whole, and possibly be a control gel. The purpose of the reference gel is to provide a few starting pairs or landmarks to use for comparison with the other gels. Although these reference points are not required with either the Melanie III or the Phoretix software, using them greatly increases the rate and accuracy of matching. Pairs matching is done in the overlay or stacked mode in the Phoretix and Melanie III software, respectively, and in the match tool mode in PDQuest. It is important, although not always possible, to provide landmarks in all of the regions of the reference gel, especially at the edges where the greatest possible distortion of the gels may have occurred. Once a few starting pairs have been identified in the gel, it can be advantageous to align or warp the gels to the pairs, thus allowing better placement of additional pairs. Once a number of matches have been placed, the automated match mode can then be chosen. This occurs automatically as matches are added in PDQuest.

Editing matches

All three software packages allow visualization of matches using vectors that link the location of the paired spots. The lines in a given region should be reasonably parallel and of the same length. Vector pair lines that cross each other indicate a major problem with matching. These matches can be corrected by using the match editing tools of these programs. It is important to ensure that the spots are matched correctly for accuracy in comparisons between gels.

Molecular weight and pI calibration

In both PDQuest and the Phoretix software, the pI and molecular-weight values of every spot on the gel can be estimated by assigning known values to a few spots. In Phoretix 2D advanced software, this is done by adding values to a protein list of any gel in the matched set. Once these values are added, they are assigned to all other matched spots in the set, and the molecular-weight and pI values are calibrated for all other spots that fall within calculable ranges. This is done in a similar fashion in PDQuest; however, the values are entered using the MrpI data button.

Synthetic and average gels

All three software programs allow the user to create synthetic or average gels. These gels are important when studying variations in protein expression evident among a series of gels. In Melanie III, the synthetic gel is created by selecting set characteristics of all matched gels (i.e., all paired or all nonpaired spots) and creating a gel containing those spots. The spot taken for the group is that which is closest to the average quantification value of the group. PDQuest and Phoretix 2D Advanced offer the same ability; however, they create a true average spot for the average gel. In Melanie III, synthetic gels are commonly used as the reference gel for matching a set of gels, because they contain the most important spots from all gels in the sets. Once a set of gels are matched to the synthetic reference gel, they produce a unique numbering scheme for the spots contained in all considered gels.

Normalization

When comparing 2D gel images, there is often some variation in spot intensity between gels that is not due to differential protein expression. This variation can be caused by differences in sample preparation, loading, staining, and imaging between gels. The process of compensating for this background variation is called normalization. Both PDQuest and Phoretix 2D Advanced allow gels to be normalized to the spot intensity of proteins known not to change in the experimental treatment. In addition, all three of the analysis systems provide data on relative spot quantification, for example:

- % OD = the optical density (OD) of a spot divided by the total OD of the entire image.
- % Volume = the volume of a spot divided by the total volume of the entire image. Volume is defined as (the OD of a spot) x (the area occupied by a spot).

Both of these methods allow for normalization when no stable proteins are known.

Analyzing the data

Once spots have been picked and matched, and the data values have been normalized, all three programs offer a variety of ways of analyzing the data. First, spots may be displayed that have only certain characteristics, e.g., matched spots that show more than a twofold change between gels, or all of the spots found in at least 90% of all members of a gel of a specific group (e.g., treatment group). Once specific groups of spots have been selected for analysis, the changes in spot characteristics (e.g., optical density or spot volume) between individual gels can be visually displayed in histograms. All of the software programs generate reports of statistics (e.g., means, standard deviation, and variation). A variety of simple statistics such as the Student's t-test can also be carried out on the data within these programs. Finally, all programs will allow export of the data to a spreadsheet for formatting, printing, and further analysis.

Annotating the gels

Once information about the protein spots on the gel has been generated, it can be added to the gel image or database using the annotation feature available in each of these programs. As the protein database grows in size and complexity, annotations will be essential for keeping track of the experimental data. Annotations are organized in categories such as accession number, landmark, pI, and molecular weight. In addition, some of the programs allow annotations to be linked to specific data files, 2D databases (including on-line databases), or any other type of file. This is very valuable, because the information from on-line databases, such as SWISS-2DPAGE, can be added to the annotations easily. A fully annotated gel is known as a Master Gel. Master gels can be used to copy data into other gels to identify protein spots by gel comparison. The Phoretix 2D Advanced software also provides a Web page building component that allows production of on-line databases based on the investigator's gels. This can be achieved with PDQuest data using Bio-Rad's PDQWeb software.

PROTOCOL 1

Preparation of Rat Liver Protein Extract for 2D Gel Electrophoresis

LIVER, LIKE MANY ANIMAL TISSUES, CAN BE PREPARED for 2D electrophoresis simply by homogenization in a suitable solubilization solution followed by centrifugation. Additional steps to concentrate protein or reduce the level of interfering substances are not necessary. The extraction solution used here contains urea, thiourea, and the amidosulfobetaine detergent ASB-14, which in combination, maximize the number of proteins solubilized.

MATERIALS

CAUTION: See Appendix 3 for appropriate handling of materials marked with <!>.

Reagents

Dithiothreitol (DTT) 1.2 M <!>

Extraction solution

7 M urea <!>	4.2 g
2 M thiourea <!>	1.52 g
2% (w/v) ASB-14 (Calbiochem)	0.2 g
0.5% (v/v) Pharmalyte 3-10 (AP Biotech)	50 µl
H_2O	to 10 ml

Prepare the solution fresh or store at –80°C. Cool on ice before use. Can be stored at –80°C for up to 6 months.

Phenylmethylsulfonyl fluoride (PMSF) <!> (100 mM) in isopropanol <!>

Store at 4°C or lower for up to 1 month.

Equipment

Liquid nitrogen <!>

Microfuge (refrigerated)

Mortar and pestle

Pestle (plastic or metal) to fit 1.5-ml microfuge tube (e.g., Labscientific)

Cells and Tissues

Rat liver

If the rat is exsanguinated before harvesting the liver, there will be less contamination with serum proteins. Livers should be quick-frozen with liquid nitrogen as soon as possible following harvest, and, if not used immediately, they should be stored at –80°C until needed.

METHOD

1. If not already frozen, place some of the liver in a mortar and fill the mortar with liquid nitrogen to keep the liver frozen. Use a pestle to break the liver into small pieces.
2. Weigh small pieces of frozen liver, and transfer 10–40 mg into a microfuge tube in ice. Work rapidly to minimize thawing of the liver sample. Store the remainder of the liver at –80°C.
3. Pipette 1 ml of cold extraction solution per 40 mg of liver into the tube with the liver sample. Add 10 µl of 100 mM PMSF per 1 ml of extraction solution.
4. Thoroughly grind the liver sample using a small pestle designed to fit in a microfuge tube.
5. Add 50 µl of 1.2 M DTT per 1 ml of extract and vortex.
6. Centrifuge the sample in a microfuge at maximum speed for 10 minutes at 4°C.
7. Measure the protein concentration of the supernatant (see Appendix 2).
8. Divide the supernatant into 100-µl aliquots, and if not used immediately for IEF, store the protein at –80°C.

   Thaw frozen aliquots only once to avoid the adverse effects of multiple freeze-thaw cycles.
9. Proceed to Protocol 5 to perform first-dimension IEF.

   In the case of rat liver proteins, for first-dimension IEF, the protein concentration should be 5–10 mg/ml. The sample may be diluted to the appropriate concentration into rehydration solution containing 7 M urea, 2 M thiourea, and 2% (w/v) ASB-14 for first-dimension IEF. It is, however, possible to load fairly large volumes onto the IEF strips using rehydration loading (see Protocol 5). In addition, the optimal rehydration solution will vary depending on the source of protein and should be determined empirically.

PROTOCOL 2

Preparation of Eukaryotic Lysates for 2D Gel Electrophoresis

CULTURED MAMMALIAN CELL LINES ARE WIDELY USED as study models in biomedical research. The protocol described here has been used successfully for many human colon carcinoma cell lines and fibroblast cell lines (Ji et al. 1994, 1997). Although most cellular proteins can be extracted by this protocol, some DNA-binding proteins may not be recovered. If DNA-binding proteins are to be studied, a nuclease (e.g., Benzonase) should be included, as described in Protocol 3. Expression proteomic studies frequently require the detection of low-abundance proteins. An extremely sensitive and tried-and-true method for detecting these rare proteins is the metabolic labeling of cellular proteins with radioisotopes, which is presented in ADDITIONAL PROTOCOL: RADIOISOTOPIC LABELING OF EUKARYOTIC CELL PROTEINS at the end of this protocol.

This protocol was provided by Hong Ji (Joint ProteomicS Laboratory, Ludwig Institute of Cancer Research/Walter and Eliza Hall Institute for Medical Research, Parkville, Victoria, Australia).

MATERIALS

CAUTION: See Appendix 3 for appropriate handling of materials marked with <!>.

Reagents

Extraction buffer

9 M urea <!>	5.4 g
4% CHAPS <!>	0.4 g
0.5% IPG buffer (pH 3–10) (AP Biotech)	50.0 µl
50 mM DTT <!>	0.077 g
H_2O	to 10 ml

IPG buffer (pH 3–10) (AP Biotech)

Phosphate-buffered saline (PBS), ice-cold

Equipment

Cell scraper

Centrifuge (refrigerated, low-speed)

Filter paper

Ice bath

Ultracentrifuge (refrigerated)

Vortex

Cells

Cultured cells, either adherent or suspension

METHOD

1. Remove the cells from the culture dish.

   If working with adherent cells: Scrape the cells from the culture dish with a cell scraper. Transfer the cells and media into a 15-ml centrifuge tube using a 5-ml pipette.

   If working with suspension cells: Transfer the cells and media into a centrifuge tube.

2. Pellet the cells by centrifugation at 480*g* for 5 minutes at 4°C.
3. Discard the supernatant without disturbing the pellet.

   IMPORTANT: For the remainder of the protocol, keep all cells cold; keep cells on ice when not centrifuging or vortexing them.

4. Wash the cells by adding 10 ml of ice-cold PBS to the centrifuge tube. Resuspend the cell pellet by pipetting the mixture up and down.
5. Centrifuge the cells again at 480*g* for 5 minutes at 4°C.
6. Discard the supernatant without disturbing the pellet.
7. Repeat Steps 4–6 two more times.
8. After the last wash, allow the tube to drain thoroughly. Use a piece of filter paper to draw the remaining PBS away from the pellet.
9. Use a pipette to transfer the extraction buffer into the tube.

   The volume of extraction buffer used is dependent on the cell line being studied. The protein concentration in the lysate should be 10 mg/ml and should not exceed 20 mg/ml.

10. Vortex the tube briefly (10–30 seconds) to resuspend the pellet.
11. Add 50 μl of IPG buffer/10 ml of extraction buffer (to a final IPG buffer concentration of 0.5%).
12. Transfer the lysate to an appropriate ultracentrifuge tube.
13. Centrifuge the lysate at 100,000*g* for 20 minutes.
14. Collect the supernatant, being careful to avoid the clear, viscous blob at the bottom of the tube (which is DNA and RNA).
15. Measure the protein concentration using the BCA assay (see Appendix 2).

    IMPORTANT: If using the BCA assay, omit DTT from the extraction buffer because DTT interferes with the assay. After samples are assayed, add DTT to the sample to a final concentration of 50 mM.

16. Divide the supernatant into 100-μl or 200-μl aliquots.

    The size of the aliquot will depend on the amount of sample to be loaded on the IEF. Store the aliquots at –80°C if they are not used immediately for IEF.

17. Proceed to Protocol 5 to perform first-dimension IEF.

ADDITIONAL PROTOCOL: RADIOSOTOPIC LABELING OF EUKARYOTIC CELL PROTEINS

Radioisotope labeling of proteins is an important technique in expression proteomics studies for detecting low-abundance proteins.

Materials

CAUTION: See Appendix 3 for appropriate handling of materials marked with <!>.

Culture medium for ^{35}S labeling <!>

- methionine-free RPMI 1690 medium (ICN Biomedicals) 90 ml
- dialyzed fetal bovine serum (FBS) (JRH Bioscience) 10 ml

^{35}S-labeled methionine and cystine (ICN Biomedicals) <!>

Culture medium for ^{32}P labeling <!>

- phosphate-free medium (ICN Biomedicals) 90 ml
- dialyzed fetal bovine serum (FBS) 10 ml

^{32}P-labeled orthophosphate (Perkin Elmer Life Science) <!>

Methods

Method A: ^{35}S Labeling of Cells

1. Passage the cells into a 35-mm diameter cell culture dish containing RPMI medium complemented with 10% FBS.
2. Incubate the cell culture for 16–24 hours at 37°C.
3. Discard the RPMI-based medium, and replace with 3 ml of methionine-free medium. Rinse the cells with this medium for a few seconds and discard. Repeat the wash with 3 ml of methionine-free medium.
4. Add 3 ml of ^{35}S-labeling medium to the culture dish and incubate the cells for 30 minutes at 37°C.
5. Add 250–500 μCi/ml of ^{35}S-labeled methionine and cystine.
6. Incubate the cells for 4–16 hours at 37°C.

   The labeling time depends on the metabolic turnover rate of the proteins.
7. Proceed with the preparation of cell lysates as described in the main protocol.

Method B: ^{32}P Labeling of Cells

1. Passage the cells into a 35-mm diameter cell culture dish containing RPMI medium complemented with 10% FBS.
2. Incubate the cell culture for 16–24 hours at 37°C.
3. Discard the RPMI-based medium, and replace with 3 ml of phosphate-free medium. Rinse the cells with this medium for a few seconds and discard. Repeat the wash with 3 ml of phosphate-free medium.
4. Add 3 ml of ^{32}P-labeling medium to the culture dish and incubate for 30 minutes at 37°C.
5. Add 250–500 μCi/ml of ^{32}P-labeled orthophosphate.
6. Incubate the cells for 2–4 hours at 37°C.

   The labeling time depends on the phosphorylation rate of the proteins.
7. Proceed with the preparation of cell lysates as described in the main protocol.

PROTOCOL 3

Preparation of *E. coli* Lysates for 2D Gel Electrophoresis

BACTERIAL LYSATES CONTAIN HIGH QUANTITIES OF NUCLEIC ACIDS and benefit from nuclease treatment prior to analysis by 2D electrophoresis. The lysate is prepared by sonication in an extraction solution containing urea and 3-[3-cholamidopropyl]-dimethylammonio)-1-propanesulfonate (CHAPS). The nuclease used (Benzonase) is active in the presence of urea, but it requires Mg^{2+} for activity. EDTA is added following nuclease treatment to inhibit metalloproteases.

MATERIALS

CAUTION: See Appendix 3 for appropriate handling of materials marked with <!>.

Reagents

Benzonase (*Serratia marcescens* recombinant endonuclease, 250 units/μl; Sigma)
Store at –20°C for up to 1 year.

Dithiothreitol (DTT) (1.2 M) <!>

Extraction solution

8 M urea <!>	4.8 g
4% (w/v) CHAPS <!>	400 mg
20 mM Tris-base	100 μl of 2 M Tris
1 mM $MgCl_2$	100 μl of 100 mM $MgCl_2 \cdot 6H_2O$ <!>
H_2O	to 10 ml

Prepare the solution fresh or store at –80°C. Cool on ice before use.

Na_2EDTA (200 mM)

Phenylmethylsulfonyl fluoride (PMSF) (100 mM) <!>in 10 ml of isopropanol <!>
Store at 4°C or lower for up to 1 month.

Equipment

Centrifuge (refrigerated)

Conical centrifuge tube (15- or 50-ml plastic tube)

Sonicator with microtip (e.g., Branson Sonifier with 3.2-mm [1/8 in] microtip, or equivalent)

Cells

E. coli cells in the growth medium and growth phase of choice

METHOD

1. Pellet *E. coli* by centrifugation of the culture solution in a tared bottle or tube at 12,000*g* for 5 minutes at 4°C.
2. Discard the culture supernatant and allow the bottle or tube to drain thoroughly. Determine the weight of the *E. coli* pellet by weighing the bottle or tube and subtracting the tare weight.
3. Suspend the pellet in cold extraction solution. Use 10 ml of extraction solution per 1 g of cells. Transfer the cell suspension to a plastic conical centrifuge tube. Add 2 µl (500 units) of Benzonase and 100 µl of 100 mM PMSF per 10 ml of extraction solution. Mix by inversion following each addition. Place the tube in a beaker of ice.
4. Sonicate the cells by immersing the microtip in the cell suspension. Set the sonicator to the maximum power output possible for the tip. Sonicate in 15-second bursts but allow the suspension to cool down in the beaker of ice between each burst. Repeat until the cell suspension clarifies.
5. Add 0.5 ml of 1.2 M DTT per 10 ml of extraction solution and mix by inversion. Place the tube on ice for 15 minutes.
6. Add 100 µl of 200 mM Na_2EDTA per 10 ml of extraction solution and mix by inversion.
7. Transfer the lysate to a centrifuge tube and centrifuge at 20,000*g* for 15 minutes at 4°C.
8. Measure the protein concentration of the supernatant (see Appendix 2).
9. Prepare 100-µl aliquots of the supernatant, and if not used immediately for IEF, store the protein at –80°C.

   Aliquots should be used only once to avoid the adverse effects of multiple freeze-thaw cycles.
10. Proceed to Protocol 5 to perform first-dimension IEF.

    For first-dimension IEF, the protein concentration should be 5–10 mg/ml. The sample may be diluted to the appropriate concentration into rehydration solution.

PROTOCOL 4

Preparation of Cerebrospinal Fluid Proteins for 2D Gel Electrophoresis

HUMAN CEREBROSPINAL FLUID IS PRECIPITATED with ethanol to both concentrate the protein and remove salts that would otherwise interfere with first-dimension IEF.

MATERIALS

CAUTION: See Appendix 3 for appropriate handling of materials marked with <!>.

Reagents

Ethanol (absolute, undenatured) <!>

2-Mercaptoethanol <!>

NaOH (1 N) <!>

Solubilization solution

9 M urea <!>	13.5 g
4% (w/v) CHAPS <!>	1 g
0.8% (v/v) Biolyte 3-10 (Bio-Rad)	0.2 ml
0.002% bromophenol blue <!>	~0.5 mg
H_2O	to 25 ml

Prepare the solution fresh or store at –80°C.

Equipment

Centrifuge tubes (15-ml)

Centrifuge (refrigerated)

Sonicator with cup horn (e.g., Branson Sonifier or equivalent)

Cells and Tissues

Human cerebrospinal fluid <!>

Human cerebrospinal fluid is obtained by lumbar puncture, centrifuged to remove cells, and frozen at –80°C in 1-ml aliquots.

METHOD

1. Thaw cerebrospinal fluid samples at room temperature. When the samples have thawed, tighten the tube caps and vortex the tubes.
2. While the samples are thawing, label 15-ml centrifuge tubes, both tube and cap, with the name of the sample to be processed.
3. Combine 1 ml of cerebrospinal fluid with 9 ml of ethanol in each 15-ml centrifuge tube.

   CAUTION: Cerebrospinal fluid is potentially biohazardous. Deposit used pipette tips, tubes, and gloves in an appropriate biohazard receptacle. Carefully clean contaminated surfaces with 1 N NaOH.

4. Place the caps on the centrifuge tubes tightly and vortex. Store the tubes overnight at –20°C.
5. Centrifuge the tubes at 5000*g* for 5 minutes at 4°C.
6. Pour the supernatant into a designated waste bottle. Use a 200-µl pipette to remove as much residual ethanol as possible without disturbing the pellet. Angle the pipette tip away from the pellet.
7. Allow the pellet to air dry in a chemical fume hood for 10–20 minutes. Do not allow the pellet to dry completely.
8. Pipette 100 µl of solubilization solution into each tube. Add 2 µl of 2-mercaptoethanol. Place the caps on the tubes tightly and vortex. Let the tubes stand for 30 minutes at room temperature.
9. Sonicate the tubes for 30 seconds in 5-second bursts using a cooled cup horn sonicator at the maximum setting possible with the cup horn.
10. Centrifuge the tubes at 5000*g* for 5 minutes at 4°C.
11. Transfer each supernatant to a fresh tube, and measure the protein concentration of each supernatant (see Appendix 2).

    The protein concentration will vary widely depending on the cerebrospinal fluid sample. However, 25 µl of this solution is sufficient to detect the proteins on an ammoniacal silver-stained 2D gel (see Protocol 11).

12. If the samples are not used immediately for IEF, store them in 100-µl aliquots at –80°C. Thaw frozen samples *only once* to avoid the adverse effects of multiple freeze-thaw cycles.
13. Proceed to Protocol 5 to perform first-dimension IEF.

PROTOCOL 5

The First Dimension: Isoelectric Focusing of Proteins

TWO METHODS ARE DESCRIBED HERE FOR SEPARATING PROTEINS based on their net charge using the technique of IEF on immobilized pH gradient (IPG) gels. These methods serve as the first dimension of the 2D separation. In addition to the traditional IPG IEF on a flatbed unit (Method A in this protocol), IPG gels can be focused on self-contained instruments for IEF (Method B in this protocol). These high-voltage systems allow fewer manipulations of the IPG gels, resulting in less error, strip mix-up, contamination, air contact, or urea crystallization. Because rehydration and IEF can be performed consecutively within a single unit, these two steps can be performed unattended overnight. Finally, faster separations and sharper focusing are possible due to the higher voltage available in these instruments. IPG gels can be cast in the laboratory (for a recent review, see Gianazza 1999), but it is time-consuming and fraught with problems. Therefore, these protocols assume that the investigator is using dehydrated, precast IPG gels strips available from a variety of suppliers (e.g., Amersham Biosciences, Bio-Rad, and Genomic Solutions).

MATERIALS

CAUTION: See Appendix 3 for appropriate handling of materials marked with <!>.

EXPERIMENTAL TIP: The quality of the reagents used in IEF is critical. Always use electrophoresis grade or better.

Reagents

Carrier ampholytes or IPG buffers

Dithiothreitol (DTT) (1 M) <!>

IPG gel strips

Casting IPG gels in the laboratory is possible, but it is time-consuming and fraught with problems. The more reliable (and more expensive) option is to purchase precast, dehydrated IPG strips (e.g., Bio-Rad or Amersham Biosciences). Rehydrated strips are 3-mm wide and ~0.5 mm thick.

Mineral oil (low viscosity) or paraffin oil

Alternatively, a commercial product such as Immobiline DryStrip Cover Fluid can be used.

Rehydration solution

8 M urea <!>	48 g
1.5% (w/v) CHAPS <!>	1.5 g
H_2O	to 100 ml

Divide the rehydration solution into 2.5–5-ml aliquots, and freeze them at –20°C.

Equipment

Forceps

Pipette (1 ml)

Sample applicators (optional)

Use these applicators if protein samples are loaded onto the IPG strips after the strips have been rehydrated (see Section II, Method A, Step 4). The applicators should be purchased from the manufacturer of the IPG strips.

Screw-cap tubes

The tubes should be of sufficient length to hold the IPG gel strips.

METHOD A: IEF OF PROTEINS IN A MULTIPURPOSE FLATBED ELECTROPHORESIS UNIT

Circulating water bath

Electrophoresis power supply

Choose a power supply capable of running at very low currents (e.g., <0.1 mA or preferably one in which the low current check can be circumvented).

IEF electrode strips

IEF electrophoresis unit with temperature control

IPG strip rehydration tray

METHOD B: IEF OF PROTEINS IN AN IEF-DEDICATED ELECTROPHORESIS UNIT

Dedicated IEF system (Bio-Rad, Genomic Solutions, Amersham Biosciences)

IPG strip holder of appropriate length with cover

Typical lengths are 7, 11, 13, 18, or 24 cm. Wash the strip holders with a suitable cleaner, rinse them thoroughly with H_2O, and either dry the holders with a cotton swab or lint-free tissue or allow them to dry in the air.

Biological Sample

Protein sample, prepared in Protocol 1, 2, 3, or 4

METHODS

SECTION I. REHYDRATION OF IPG STRIPS

1. Add DTT to a final concentration of 0.2% (w/v) to freshly prepared or thawed rehydration solution. Add carrier ampholytes or IPG buffers to a final concentration of 0.5–2.0%. Select the appropriate pH range to match the pH range of the IPG strips to be used in the analysis.

2. If protein samples are to be loaded into gels during rehydration, dilute (or dissolve in) up to 1 mg of sample per strip with the rehydration solution.

   The amount of protein to load will depend on the goals of the experiment. Load smaller amounts of sample when examining abundant proteins; load more sample to detect scarce proteins.

3. Use an indelible marker to label the back (the nongel side) of the IPG gel strip with the name of the protein sample to be loaded.

4. Pipette the appropriate amount of rehydration solution (see Table 4.2) into the slot of the reswelling tray. Deliver the solution slowly to the center of the slot, without producing bubbles. Remove any large bubbles that form. Make sure to record which sample is applied to which IPG gel. Numbers in the rehydration tray can aid in strip identification.

TABLE 4.2. Recommended rehydration solution volumes

IPG strip length	Volume per strip[a] (μl)
7	125
11	200
13	250
18	350
24	480

[a]Includes the volume of the protein sample in cases where the sample is loaded during the IPG gel strip rehydration step.

5. Remove the protective covering from each of the IPG gels and place the strip, gel side down, into the rehydration tray. Make sure that the solution is distributed across the entire gel surface. If necessary, lift the basic end of the strip with a pair of forceps and slide the strip back and forth to ensure complete and even wetting. Make sure that no bubbles are trapped under the IPG gel. If using IEF Method B, make sure that the gel is in contact with the electrodes embedded in the bottom of the strip holder.

6. Pipette ~1.5 ml of mineral oil over each IPG strip to prevent precipitation of the urea. Apply the mineral oil dropwise into one end of the strip holder until it is approximately half full, and then finish filling the strip holder from the other end. Do not overfill the strip holder with oil.

7. Allow the strips to rehydrate for a minimum of 6 hours at room temperature (overnight is preferable). If using IEF Method B, place the plastic cover on top of the strip holder.

   It is also possible with the IEF-dedicated electrophoresis unit to rehydrate the gel strips on the unit platform and to program the rehydration as the first step of the electrophoresis protocol. This permits the focusing steps to automatically follow the rehydration step, without additional need for human intervention.

8. Proceed with either Method A or Method B for isoelectric focusing of the proteins.

SECTION II. ISOELECTRIC FOCUSING OF PROTEINS

Method A: IEF of Proteins in a Multipurpose Flatbed Electrophoresis Unit

1. Set up the 2D electrophoresis unit and temperature controller according to the manufacturer's instructions.

2. Cut two electrode wicks to the appropriate lengths. Soak each strip with 1 ml of H_2O. Blot each strip with filter paper so that they are damp, but not wet.

   Excess water in the electrode strips may cause proteins to streak on the gel strips.

3. Place the IPG and electrode strips into the electrophoresis unit according to the manufacturer's instructions. If necessary, position the two electrodes over the electrode strips.

4. (*Optional*) Apply the protein sample to the gel strips if protein was not loaded into the IPG gel strips during rehydration (Section I, Step 2). Follow the manufacturer's instructions for use of the protein sample applicator.

TABLE 4.3. Recommended electrophoresis conditions for IPG gels on flatbed IEF systems

Step	IPG length (cm)	Voltage	Current (mA)	Power (W)	Mode	Duration (hr:min)	Volt-hours
1	7,11,13,18, 24	200	2	5	step	0:01	1
2	7,11,13,18, 24	3500	2	5	gradient[a]	1:30	2,800
3	7	3500	2	5	step	0:55–1:30	3,200–5,200
3	11	3500	2	5	step	2:20–3:30	8,100–12,100
3	13	3500	2	5	step	3:45–4:20	13,100–18,100
3	18	3500	2	5	step	5:40–7:40	23,000–30,000
3	24	3500	2	5	step	14:30–25:45	50,000–90,000

[a]During Step 2, the voltage should rise from 200 V to 3500 V over the 1:30 time period.

The optimal application point must be empirically determined for each sample type. Sample application at the cathode works well with proteins of acid pI values or when SDS is used in the solubilization of the sample. Application at the anode is required for very basic IPG gels.

Protein samples applied via sample applicators, rather than during rehydration, can precipitate at the application point, reducing the amount of sample loaded onto the gel. To reduce this problem, load dilute protein solutions, which also contain urea, nonionic detergents, and carrier ampholytes. In addition, by applying low current to the strips at the beginning of the electrophoresis run, the sample can enter the strip slowly, which helps to minimize accumulation of sample at the application point.

5. Attach the electrodes to the power supply and program the power supply with the desired run parameters.

   A typical IEF protocol proceeds through a series of voltages that gradually increase to a desired focusing voltage that is held for several hours. The length of time at the focusing voltage is affected by many factors including IPG gel length, pH gradient, sample composition, sample load, and rehydration solution composition, and must be empirically determined for each sample. Table 4.3 provides recommended run conditions for analytical protein loads.

   Overfocusing is seldom a problem below 100,000 total Volt-hours, but on very long runs, it can contribute to horizontal streaking observed in 2D results, especially in the basic end of the gel.

6. Turn on the power supply to begin the electrophoresis. Make sure that the low current check is circumvented if possible.

7. At the completion of the run, turn off the power supply and remove the lid of the electrophoresis unit. Follow the manufacturer's instruction for properly accessing the gel strips.

8. Use a forceps to pick up an IPG gel strip. Allow the oil to drain from the strip and place it in a test tube. Repeat with each gel strip.

9. Proceed to the second-dimension electrophoresis protocols (Protocols 6–9) or store the gel strips in their test tubes at –40°C to –80°C. Gels can be stored up to 1 month.

Method B: IEF of Proteins in an IEF-dedicated Electrophoresis Unit

1. Program the electrophoresis unit for IEF.

   A typical IEF protocol proceeds through a series of voltages that gradually increase to a desired focusing voltage that is held for several hours. The length of time at focusing voltage

TABLE 4.4. Recommended electrophoresis conditions for IPG gels on dedicated IEF systems

Step	IPG length (cm)	Voltage	Duration (hr:min)	Volt-hours	Gradient type
1	7,11,13,18, 24	rehydration	12:00[a]	–	–
2	7	500	0:30	250	step and hold
2	11,13,18, 24	200	1:00	500	step and hold
3	7	8000[b]	1:00	8,000	step and hold
3	11, 13	8000[b]	2:00	16,000	step and hold
3	18	8000[b]	4:00	32,000	step and hold
3	24	8000[b]	6:25–11:25	50,000–90,000	step and hold

[a]The total rehydration time can be varied for convenience, but it must be more than 6 hours.

[b]Depending on the sample, this voltage may not be reached within the suggested step duration. For this reason, it is best to program the steps based on Volt-hours rather than time.

is affected by many factors, including IPG gel length, pH gradient, sample composition, sample load, and rehydration solution composition, and must be empirically determined for each sample. Table 4.4 provides recommended run conditions for analytical protein loads.

Overfocusing is seldom a problem below 100,000 total Volt-hours, but on very long runs, it can contribute to horizontal streaking observed in 2D results, especially in the basic end of the gel. Due to variations in sample types, some samples may not reach maximum voltage during a run. Therefore, it is best to program the run conditions on the basis of total number of Volt-hours, rather than time to ensure that proper focusing occurs.

2. Close the lid to the electrophoresis unit and begin the IEF protocol.
3. At the completion of the run, turn off the power supply and open the lid of the electrophoresis unit.
4. Use forceps to pick up an IPG gel strip. Allow the oil to drain from the strip and place it in a test tube. Repeat with each gel strip.
5. Proceed to the second-dimension electrophoresis protocols (Protocols 6, 7, and 9) or store the gel strips in their test tubes at –40°C to –80°C. Gels can be stored up to 1 month.

PROTOCOL 6

Preparation of Vertical SDS Slab Gels: Casting a Single Homogeneous Gel

FOLLOWING THE SEPARATION OF PROTEINS BY IEF, the second dimension is carried out by SDS-PAGE. This protocol details the method for casting single homogeneous SDS-PAGE gels. Homogeneous gels (with the same %T and %C throughout) offer the best resolution for a particular molecular-weight range and are commonly used because they are the easiest to pour reproducibly.

The second-dimension gels can be conveniently prepared in three different formats (i.e., sizes): minigels, for use with 7-cm IEF first-dimension gels; standard gels, for use with 11-, 13-, and 18-cm IEF first-dimension gels; and large-format gels, for use with 18- and 24-cm IEF first-dimension gels. All of the gels use a common set of reagents, listed below, but differ slightly in the equipment required.

If the gel is subsequently stained using the ammoniacal silver staining method (Protocol 11), consider including thiosulfate in the gel to reduce background levels of stain (Hochstrasser and Merril 1988; see the note to Table 4.5).

MATERIALS

CAUTION: See Appendix 3 for appropriate handling of materials marked with <!>.

EXPERIMENTAL TIP: The quality of the reagents is critical to good separations. Always use electrophoresis grade or better. In addition, always wear gloves when preparing gels, because acrylamide is a neurotoxin, and keratin and other skin and hair proteins can contaminate the gel.

Reagents

Ammonium persulfate (10%) <!>

Fresh ammonium persulfate "crackles" when H_2O is added. If the sample does not crackle, replace it with fresh stock. Prepare 1 ml of ammonium persulfate just prior to use.

n-butanol (H_2O-saturated) <!>

Mix H_2O and butanol in a 1000-ml flask. Once mixed, allow the *n*-butanol and H_2O to separate into two phases. This butanol will be used as a gel overlay (see Step 11).

Ethanol <!>

4x Gel buffer (1.5 M Tris-Cl, pH 8.8)

Dissolve 181.5 g of Tris-base in 750 ml of H_2O. Adjust the pH to 8.8 with HCl <!>, and bring the final volume to 1000 ml with H_2O. Filter the solution through a 0.45-μm filter and store it at room temperature.

Gel storage solution

0.375 M Tris-Cl (pH 8.8)

0.1% SDS <!>

Combine 125 ml of 4x gel buffer, 5 ml of 10% SDS, and 370 ml of H_2O. Store the solution at 4°C.

TABLE 4.5. Recipes for single-percentage gels

	Final gel concentration				
	5%	7.5%	10%	12.5%	15%
Monomer stock solution	16.7 ml	25 ml	33.3 ml	41.7 ml	50 ml
4x Resolving gel buffer	25 ml	25 ml	25 ml	25 ml	25 ml
10% SDS	1 ml	1 ml	1 ml	1 ml	1 ml
H_2O	56.8 ml	48.5 ml	40.2 ml	31.8 ml	23.5 ml
10% Ammonium persulfate	500 µl	500 µl	500 µl	500 µl	500 µl
TEMED	33 µl	33 µl	33 µl	33 µl	33 µl
Total volume	100 ml	100 ml	100 ml	100 ml	100 ml

If the gel will be subsequently stained using the ammoniacal silver staining method (Protocol 11), include sodium thiosulfate in the gel to reduce background levels of stain (Hochstrasser and Merril 1988). Use 180 µl of 10% ammonium persulfate solution, 90 µl of 5% sodium thiosulfate, and 180 µl of TEMED per 100 ml of gel solution. Sodium thiosulfate slightly inhibits polymerization, thus the requirement for higher amounts of TEMED. Ammonium persulfate must be added to the gel solution first, followed by sodium thiosulfate, and TEMED last.

Monomer stock solution (30.8%T, 2.6%C)

acrylamide <!>	150.0 g	(30% final concentration)
N,N′-methylenebisacrylamide <!>	4.0 g	(0.8% final concentration)
H_2O	to 500 ml	

Filter the monomer solution through a 0.45-µm filter. Store it at 4°C away from light.

SDS (10%)

5x SDS electrophoresis buffer

125 mM Tris-base	15.1 g
960 mM glycine <!>	72.1 g
0.5% SDS	5.0 g
H_2O	to 1000 ml

Store the 5x buffer at room temperature. Dilute to a 1x solution just before use.

TEMED <!>

Equipment

Gel-casting apparatus (or clamps and sealing tape)

Glass plates

To clean the glass plates, soak them in a warm solution of laboratory detergent. Scrub the plates with a soft scouring pad, and rinse them in H_2O. Alternatively, plates can be cleaned using a laboratory dishwasher. The plates should be dried and stored in a dust-free environment. Always wear gloves when handling plates to reduce contamination from keratin and other proteins.

Magnetic stir bar

Magnetic stir plate

Pipettes (25 ml, disposable)

Spacers (1.0 or 1.5 mm)

Vacuum flask

Water aspirator or other vacuum system

METHOD

1. Just prior to assembling the glass plates for casting a gel, wipe each plate with an ethanol-soaked lint-free tissue. Make sure that no dust particles are on the surface of the plate.

   EXPERIMENTAL TIP: Wear gloves at all times when handling the plates to reduce the chance of contamination with keratin and other proteins.

2. Align the two glass plates, separating them with a 1.0- or 1.5-mm spacer placed on each side.

3. Clamp the plates together at the spacers using just enough pressure to hold the spacers and plates in place. Stand the entire assembly on its bottom edge on a flat surface, and ensure that the bottom edges of the two glass plates and the spacers are flush with one another. Tighten the clamps securely.

   It is critical that the bottom edge of the plates and spacers *rest flush* on a flat surface. This minimizes the likelihood that the apparatus will leak when casting the gel.

4. Place the clamp assembly into the casting cradle and turn the cams so that the gel form (i.e., the glass plates and spacers that are clamped together) is sealed against the gasket of the casting stand.

5. Use a typewriter or a pencil to prepare a gel label on a small strip of filter paper. Place the filter paper in the bottom left-hand corner of the gel form. Do not use a felt tip pen, as it will bleed into the gel.

6. Mark the outer glass plate ~0.5 cm from its upper edge using an indelible laboratory marker.

7. Fill the gel form with H_2O to determine the volume of gel solution needed before casting the gel. Be sure to remove all H_2O from the gel form before casting the gel, or the acrylamide concentration in the gel will be affected. Prepare the appropriate volume of gel solution (calculate from Table 4.5) in a vacuum flask, omitting the TEMED and ammonium persulfate.

8. Add a small magnetic stir bar to the solution. Use a water aspirator to degas the solution for several minutes while stirring the solution on a magnetic stir plate.

   Although not all investigators degas their acrylamide, it is recommended for more reproducible polymerization.

9. Add the TEMED and ammonium persulfate and gently swirl the flask to mix. Do not generate bubbles.

10. Use a pipette to slowly deliver the solution into one corner of the gel form, filling the form to 0.5 cm from its upper edge.

11. Immediately overlay each gel with a thin layer (100–500 μl) of H_2O-saturated *n*-butanol to minimize exposure of the gel to oxygen and to create a flat gel surface.

    EXPERIMENTAL TIP: It is critical that the surface of the second-dimension gel be flat to allow good contact with the first-dimension gel during transfer. Many investigators overlay with double-distilled H_2O; however, this runs the risk of diluting the concentration of the acrylamide on the top of the gel.

12. Allow the gel to polymerize for a minimum of 1 hour.

13. After polymerization, inspect each gel for complete and even polymerization. Additionally check that the top surface of the gel is straight and flat. Pour off the *n*-butanol and wash the surface of the gel with distilled or deionized H_2O.

14. The gel can now be used for running the second dimension (Protocol 9) or stored for future use for up to 2 weeks at 4°C.

    Gels that are to be stored should be overlaid with gel storage solution and wrapped tightly in plastic wrap.

PROTOCOL 7

Preparation of Vertical SDS Slab Gels: Simultaneous Casting of Multiple Gradient Gels

GRADIENT SDS-PAGE GELS PROVIDE THE BEST RESOLUTION over a wide range of molecular weights, resulting in sharper protein spots, because diffusion is minimized by the decreasing pore size in the gel. However, gradient gels are more difficult to produce reproducibly; thus, they are commonly cast with multiple gel casters, which allows for an identical set of gels to be produced for an experiment. Presented here is a method for casting gradient gels using a multiple gel casting system.

If the gel is subsequently stained using the ammoniacal silver staining method (Protocol 11), consider including thiosulfate in the gel to reduce background levels of stain (Hochstrasser and Merril 1988; see the note to Table 4.6).

TABLE 4.6. Recipes for gradient gels

	Final gel concentration				
Light Solution	5%	7.5%	10%	12.5%	15%
Monomer stock solution	8.4 ml	12.5 ml	16.7 ml	21.0 ml	25 ml
4x Resolving gel buffer	12.5 ml	12.5 ml	12.5 ml	12.5 ml	12.5 ml
10% SDS	0.5 ml	0.5 ml	0.5 ml	0.5 ml	0.5 ml
H_2O	28.5 ml	24.4 ml	20.2 ml	16.0 ml	12.0 ml
10% Ammonium persulfate	165 µl	165 µl	165 µl	165 µl	165 µl
TEMED	16.5 µl	16.5 µl	16.5 µl	16.5 µl	16.5 µl
Total volume	50 ml	50 ml	50 ml	50 ml	50 ml

	Final gel concentration				
Heavy Solution	10%	12.5%	15%	17.5%	20%
Monomer stock solution	16.7 ml	21.0 ml	25.0 ml	29.2 ml	33.3 ml
4x Resolving gel buffer	12.5 ml	12.5 ml	12.5 ml	12.5 ml	12.5 ml
Sucrose	7.5 g	7.5 g	7.5 g	7.5 g	7.5 g
10% SDS	0.5 ml	0.5 ml	0.5 ml	0.5 ml	0.5 ml
H_2O	16.2 ml	11.7 ml	7.7 ml	3.5 ml	0 ml
10% Ammonium persulfate	165 µl	165 µl	165 µl	165 µl	165 µl
TEMED	16.5 µl	16.5 µl	16.5 µl	16.5 µl	16.5 µl
Total volume	50 ml	50 ml	50 ml	50 ml	50 ml

If the gel will be subsequently stained using the ammoniacal silver staining method (Protocol 11), include sodium thiosulfate in the gel to reduce background levels of stain (Hochstrasser and Merril 1988). Use 90 µl of 10% ammonium persulfate solution, 45 µl of 5% sodium thiosulfate, and 90 µl of TEMED per 50 ml of gel solution. Sodium thiosulfate slightly inhibits polymerization, thus the requirement for higher amounts of TEMED. Ammonium persulfate must be added to the gel solution first, followed by sodium thiosulfate, and TEMED last.

MATERIALS

CAUTION: See Appendix 3 for appropriate handling of materials marked with <!>.

EXPERIMENTAL TIP: The quality of the reagents is critical to good separations. Always use electrophoresis grade or better. In addition, always wear gloves when preparing gels, because acrylamide is a neurotoxin, and keratin and other skin and hair proteins can contaminate the gel.

Reagents

Ammonium persulfate (10%) <!>

Fresh ammonium persulfate "crackles" when H_2O is added. If the sample does not crackle, replace it with fresh stock. Prepare 1 ml of ammonium persulfate just prior to use.

n-butanol (H_2O-saturated) <!>

Mix H_2O and *n*-butanol in a 1000-ml flask. Once mixed, allow the *n*-butanol and H_2O to separate into two phases. This butanol will be used as a gel overlay (see Step 13).

Ethanol <!>

4x Gel buffer (1.5 M Tris-Cl, pH 8.8)

Dissolve 181.5 g of Tris-base in 750 ml of H_2O. Adjust the pH to 8.8 with HCl <!>, and bring the final volume to 1000 ml with H_2O. Filter the solution through a 0.45-µm filter and store it at room temperature.

Gel storage solution

0.375 M Tris-Cl (pH 8.8)

0.1% SDS <!>

Combine 125 ml of 4x gel buffer, 5 ml of 10% SDS, and 370 ml of H_2O. Store the solution at 4°C.

Monomer stock solution (30.8%T, 2.6%C)

acrylamide <!>	150.0 g	(30% final concentration)
N,N′-methylenebisacrylamide <!>	4.0 g	(0.8% final concentration)
H_2O	to 500 ml	

Filter the monomer solution through a 0.45-µm filter. Store it at 4°C away from light.

SDS (10%)

5x SDS electrophoresis buffer

125 mM Tris-base	15.1 g
960 mM glycine <!>	72.1 g
0.5% SDS	5.0 g
H_2O	to 1000 ml

Store the 5x buffer at room temperature. Dilute to a 1x solution just before use.

Sucrose

Sucrose is added to the higher %T solution to aid in the formation of the gradient. This solution is denser and is thus frequently called the heavy solution, whereas the lower %T acrylamide solution is referred to as the light solution.

TEMED <!>

Equipment

Glass plates

To clean the glass plates, soak them in a solution of hot laboratory detergent. Scrub the plates with a soft scouring pad, and rinse them in H_2O. Alternatively, plates can be cleaned using a laboratory dishwasher. The plates should be dried and stored in a dust-free environment. Always wear gloves when handling plates to reduce contamination from keratin and other proteins.

Gradient maker with a volume of at least 600 ml

Magnetic stir bars

Magnetic stir plate
Multiple-gel casting chamber
Peristaltic pump capable of 60–75 ml/minute
Pipettes (25-ml, disposable)
Pump tubing and clamp
Spacers (1.0 or 1.5 mm)
Vacuum flask
Water aspirator or other vacuum system

METHOD

1. Just prior to assembling the glass plates for casting a gel, wipe each plate with an ethanol-soaked lint-free tissue. Make sure that no dust particles are on the surface of the plate.

   EXPERIMENTAL TIP: Wear gloves at all times when handling the plates to reduce the chance of contamination with keratin and other proteins.

2. Align the two glass plates, separating them with a 1.0- or 1.5-mm spacer placed on each side.

3. Mark one of the glass plates ~0.5 cm from its upper edge using an indelible laboratory marker.

   In some systems, such as the Bio-Rad Protean II, the glass plates are not of equal length. Make sure that this mark is placed 0.5 cm from the upper edge of the shorter plate.

4. Use a typewriter or a pencil to prepare a gel label on a small strip of filter paper. Place the filter paper in the bottom left-hand corner of the gel form. Do not use a felt tip pen, as it will bleed into the gel.

5. Place the complete gel cassette into the multiple-gel casting chamber. Repeat the assembly of the remaining cassettes, and place each one in the casting chamber separated by a thin plastic sheet. These sheets are usually supplied with the multiple-gel casting chamber, but they can also be cut from a sheet of Mylar. Fill any excess space with acrylic spacers provided with the casting chamber. Clamp the front plate onto the casting chamber to ensure that a good seal is made with the gasket.

6. Fill the casting chamber with H_2O to determine the volume of reagents needed to fill the chamber before casting the gels. Half of the measured volume will be prepared at the low %T (light solution) and the other half will be prepared at the higher %T (heavy solution). Use Table 4.6 to calculate the volumes of the gel components required for casting any number of gels.

7. Prepare the appropriate volume of light and heavy acrylamide solutions, omitting the ammonium persulfate and TEMED in the vacuum flasks, and add a small magnetic stir bar to each flask.

8. Use a water aspirator to degas the solutions (for several minutes) while stirring them on a magnetic stir plate.

   Although not all investigators degas their acrylamide, it is recommended for more reproducible polymerization.

9. While the solutions are degassing, connect a piece of tubing to the peristaltic pump and, using H_2O, ensure that the flow rate of the pump is set to the manufacturer's recommendations. Once the flow rate is set, connect one end of the tubing to the outlet port on the gradient maker and the other end to the inlet port of the multiple-gel caster. Ensure that there is a clamp in place near the inlet port to the multiple-gel caster and that it is open. Make sure that the gradient maker valves are in the closed position.

10. Place a small magnetic stir bar in the mixing chamber and place the gradient maker on a magnetic stir plate.

11. When the gel solutions have been completely degassed, add the ammonium persulfate and TEMED solutions to each solution with a gentle swirl. Pour the light solution into the mixing chamber (the front chamber near the outlet port) and the heavy solution into the reservoir chamber, and turn on the magnetic stirrer.

    The stir bar should form a vortex in the mixing chamber, but it should not spin at such high speeds that bubbles are introduced into the acrylamide solution.

12. Open the outlet valve from the gradient maker and start the peristaltic pump. When the light solution has reached the bottom of the gel, open the valve between the reservoir and the mixing chamber. Check that the stir bar is mixing the solutions adequately. Do not allow air bubbles to enter the tubing.

13. Allow the gels to fill the chambers until there is a gap of ~0.5 cm (the mark previously made on the glass plates) from the top of the gel. This should require nearly all of the solution in the gradient maker. When the casting chamber is full, switch off the pump and close the clamp at the inlet port to the casting chamber. Overlay each gel with an equal volume of H_2O-saturated *n*-butanol.

14. Detach the tubing from the gradient maker and the peristaltic pump. Drain excess acrylamide from the tubing and rinse the gradient maker immediately.

15. Allow the gradient gels to polymerize for a minimum of 2 hours.

16. After polymerization, disassemble the casting chamber. Use a strong plastic spatula to separate the gel cassettes from each other at the plastic sheets. Rinse each cassette in distilled or deionized H_2O to remove any acrylamide adhering to the outer surface. Quickly rinse the top surface of each gel to remove the *n*-butanol.

17. Carefully inspect the gels for air spaces, uneven polymerization, uneven top surfaces, or other defects. Discard any unsatisfactory gels.

18. The gels can now be used for running the second dimension (Protocol 9) or stored for future use for up to 2 weeks at 4°C.

    Gels that are to be stored should be overlaid with gel storage solution and wrapped tightly in plastic wrap.

PROTOCOL 8

IPG Strip Equilibration

THE EQUILIBRATION STEP SERVES TO SATURATE THE IPG STRIP with the SDS buffer system required for the second-dimension separation. The equilibration solution consists of buffer, urea, glycerol, reductant, SDS, and dye. The buffer (50 mM Tris-HCl, pH 8.8) maintains the appropriate pH range for electrophoresis. Urea and glycerol are added to reduce the effects of electroendosmosis (Görg et al. 1985), thus helping improve protein transfer from the IPG strip to the second dimension. The reductant (dithiothreitol) ensures that disulfide bridges are broken. SDS ensures that the proteins are denatured and also provides a net negative charge to all proteins. Iodoacetamide, introduced during a second equilibration step, alkylates thiol groups on the proteins, preventing their reoxidation during electrophoresis, and thus reducing streaking and other artifacts in the second-dimension separation. Iodoacetamide also alkylates residual dithiothreitol, preventing point streaking and other silver staining artifacts (Görg et al. 1988). Finally, a tracing dye (bromophenol blue) is added to allow the electrophoresis to be monitored during the run.

MATERIALS

CAUTION: See Appendix 3 for appropriate handling of materials marked with <!>.

EXPERIMENTAL TIP: The quality of the reagents is critical to good separations. Always use electrophoresis grade or better.

Reagents

SDS equilibration buffer

50 mM Tris-Cl (pH 8.8)	6.7 ml of 1.5 M Tris-Cl
6 M urea <!>	72.07 g
30% (v/v) glycerol	69 ml of 87% (v/v)
2% SDS <!>	4.0 g
bromophenol blue <!>	a few grains

Adjust the buffer to a final volume of 200 ml with H_2O. Store in 40-ml aliquots at –20°C. Immediately prior to use (see Step 1), add dithiothreitol (DTT) <!> to a final concentration of 1% (w/v). If performing the second equilibration (see Step 4), add iodoacetamide <!> to a fresh aliquot of SDS equilibration buffer, to a final concentration of 2.5% (w/v).

Equipment

Blotting paper (3MM)

Culture tubes (screw-cap) of sufficient length to hold the IPG gel strips

Shaker table or rocker

Biological Sample

Focused IPG strips (from Protocol 5)

METHOD

1. Add 100 mg of DTT per 10 ml of SDS equilibration buffer (final DTT concentration is 1% w/v).

2. Place each IPG strip in a separate screw-cap culture tube with the support film against the wall of the tube. Add equilibration buffer (from Step 1) to each tube; 10 ml is recommended for 18- and 24-cm IPG gels, 5 ml for 11- and 13-cm gels, and 2.5–5 ml for 7-cm IPG gels. Place the caps on the tubes tightly, and lay them down on a shaker or rocker.

3. Gently shake or rock the IPG strips for at least 10 minutes. The time to equilibration varies depending on the amount of protein loaded on the strip. Long equilibration times (in excess of 20 minutes) may result in a significant loss of sample due to diffusion from the gel. At the completion of the incubation, proceed to either Step 4, Step 5, or directly to Protocol 9.

   Unfortunately, the only way to know whether the strips are equilibrated is to run the second dimension. Incomplete equilibration will result in vertical streaks in the SDS-PAGE gel.

4. (*Optional*) Perform a second equilibration step, this time with iodoacetamide, to reduce point streaking and other artifacts in the second-dimension separation. This equilibration step is optional for vertical second-dimension separations (but is mandatory for horizontally run SDS gels).

   a. Decant the first equilibration solution into a waste container.

   b. Add ~2 ml of H_2O to each culture tube and quickly rinse the gel.

   c. Discard the H_2O, and add the appropriate volume (see Step 2) of SDS equilibration solution containing 2.5% (w/v) iodoacetamide (e.g., 250 mg per 10 ml of equilibration solution) to each culture tube.

   d. Place the caps on the tubes tightly, and lay them down on a shaker or rocker. Gently shake or rock the IPG strips for 5 minutes.

   e. Proceed to either Step 5 or directly to Protocol 9.

5. (*Optional*) Place the IPG strips on edge on blotting paper moistened with H_2O and allow excess equilibration solution to drain from the surface of the IPG strip for 30–60 seconds.

   Draining excess equilibration solution helps reduce horizontal streaking of highly abundant proteins in the sample. If gels are drained, this should be done immediately before placing them on top of the second-dimension gel. Proceed to Protocol 9.

PROTOCOL 9

The Second Dimension: SDS-PAGE of Proteins

FOLLOWING FIRST-DIMENSION IEF AND EQUILIBRATION OF THE IPG gel strips, the proteins are separated on the basis of their molecular weight in the second dimension on an SDS-PAGE gel. Systems for this separation are available from a variety of suppliers and are commonly found in many protein chemistry laboratories. This protocol describes a method for placement of the IPG strip and gives some recommended electrophoresis conditions for these second-dimension gels.

MATERIALS

CAUTION: See Appendix 3 for appropriate handling of materials marked with <!>.

Reagents

Agarose sealing solution

5x SDS electrophoresis buffer	20 ml
agarose (see note to this recipe)	0.5 g
bromophenol blue <!>	a few grains
H_2O	80 ml

Use a low electroendosmosis form of agarose. A low-melting-temperature form may also be advisable. Add all ingredients to a 500-ml Erlenmeyer flask and swirl to suspend. Heat in a microwave oven on low setting until the agarose is completely dissolved. It is best to do this in 20-second intervals to ensure that the solution does not boil over. Once the solution is warm, keep a close watch, for it will easily boil over. Dispense in 1.5-ml aliquots in screw-cap tubes and store at room temperature or 4°C.

Protein molecular-weight markers

Optional, see Step 3.

5x SDS electrophoresis buffer

125 mM Tris-base	15.1 g
960 mM glycine <!>	72.1 g
0.5% SDS <!>	5.0 g
H_2O	to 1000 ml

Store the 5x buffer at room temperature. Dilute to a 1x solution just before use.

Equipment

Filter paper

Forceps and flexible ruler

The forceps and the ruler must be thin enough to allow the IPG strip to be maneuvered between the glass plates of the slab gel (see Step 2).

Heat blocks preset to 45°C and 100°C
Power supply
SDS-PAGE gel (from Protocol 6 or 7)
Vertical electrophoresis unit

Biological Sample

Focused IPG strips (from Protocol 8)

METHOD

1. Melt agarose sealing solution in a 100°C heat block. Each gel will require ~1 ml of solution. It will take ~10 minutes to completely melt the agarose, so this is best done shortly before equilibration of the IPG strips (Protocol 8). Once melted, allow the agarose to cool to ~40–50°C before using.

2. Use forceps to position the equilibrated IPG strip between the plates on the surface of the second-dimension gel. Ensure that the plastic backing is against one of the glass plates. Use a thin, flexible plastic ruler to slide the IPG strip between the plates until the entire bottom edge of the IPG gel makes a seal with the surface of the slab gel. Work from one end of the IPG strip to the other, progressively sliding the gel down the glass plate. To reduce the chance of tearing the IPG gel, put pressure on the plastic backing of the gel to slide the IPG strip into place.

   Make sure that no air bubbles are trapped between the slab gel and the bottom of the IPG gel, or between the plastic backing and the glass plate. If air bubbles are present, place light pressure on the plastic backing of the IPG strip to force the air bubble out from under the gel. Do not place excessive pressure on the surface of the IPG strip once it is on top of the second-dimension gel; it can potentially collapse the pores on the top surface of the SDS-PAGE gel.

3. If molecular-weight markers are desired, mix an appropriate amount (200–1000 ng of each marker for Coomassie staining and 10–50 ng of each marker for silver staining) with melted agarose sealing solution. Pipette 15–20 µl of this mixture onto a small piece of filter paper.

4. Once the agarose has set, insert the filter paper between the glass plates, so that it contacts the top of the slab gel at one end of the IPG gel.

   Biotinylated standards can also be used if the gels are to be blotted and proteins detected with an avidin-horseradish peroxidase or an avidin-alkaline phosphatase system. See the manufacturer's instructions for suggested amounts of biotinylated standards to use.

5. Slowly pipette the melted agarose sealing solution over the IPG gel strip. Avoid introducing bubbles. Allow the agarose to cool and solidify for at least 1 minute before continuing.

   Embedding the strip in agarose serves to hold the gel strip in place and ensures good contact between the IPG strip and the slab gel.

6. Finish assembling the electrophoresis apparatus as directed by the manufacturer. Connect the cooling apparatus (if available) and set it to maintain a gel temperature of 10–15°C.

   Although cooling is not mandatory, it greatly diminishes gel-to-gel variation and allows the gels to be run at higher currents, thus reducing run times.

TABLE 4.7. Recommended electrophoresis conditions for second-dimension vertical gels

	Step	Current (mA/gel)	Duration (hr:min)
Mini vertical system			
1.5-mm-thick gel	1	15	0:15
	2	30	1:30[a]
1.0-mm-thick gel	1	10	0:15
	2	20	1:30[a]
Standard vertical system			
1.5- mm-thick gel	1	15	0:15
	2	45	5:00–6:00[a]
1.0-mm-thick gel	1	15	0:15
	2	45	5:00–6:00[a]

[a]Time shown here is only an estimate. Electrophoresis should be stopped when the dye is 0.5–1.0 cm from the bottom of the gel.

Cooling is recommended for both standard and large-format gels. Cooling greatly improves gel-to-gel reproducibility and helps reduce artifacts, such as smiling in the gels. Cooling is very important if there are large temperature fluctuations in the laboratory.

7. Use Table 4.7 or 4.8 to set the recommended electrophoresis protocol. Turn on the power supply to begin the electrophoresis.

   Electrophoresis is carried out in two steps. The first step is of a short duration and uses a low current or power setting. During this time, the initial migration and stacking occur. After the initial 5–15 minutes of electrophoresis (see Tables 4.7 and 4.8), check that the bromophenol blue dye front has moved into the gel. The current or power is then increased during the second step to allow for faster separation.

8. Turn off the power supply when the bromophenol blue dye front has reached 0.5–1 cm from the bottom of the gel. Remove the gel assembly from the electrophoresis chamber.

9. Place the assembly on a flat surface and remove the spacers, being careful not to disturb the gel. Use a plastic spatula or spacer to carefully separate the two glass plates of each gel cassette. Do not use a metal spatula, as it will crack or chip the glass plates. The gel should adhere to one of the plates.

10. Proceed immediately with one of the gel-staining procedures (Protocols 10–14) and/or a protein transfer procedure (Protocols 15 and 16).

    EXPERIMENTAL TIP: Do not allow the gel to dry out. Leave in gel cassettes, if necessary, while completing the preparations for the next protocol.

TABLE 4.8. Recommended electrophoresis conditions for large-format second-dimension vertical gels

Temperature	Step	Power (W/gel)	Duration (hr:min)
25°C	1	5	0:15
	2	15	6:00[a]
20°C	1	5	0:15
	2	10	8:00[a]
15°C	1	5	0:45
	2	6.6	10:00–12:00[a]

[a]Time shown here is only an estimate. Electrophoresis should be stopped when the dye is 0.5–1.0 cm from the bottom of the gel. Values given here are based on the limit of the cooling system's capacity at the given temperature and are for 12 gels. Power levels can be increased if running fewer than 12 gels, which will result in reduced run times. These values are optimized for the Ettan DALT II system (Amersham Biosciences).

PROTOCOL 10

Colloidal Coomassie Staining

THIS PROTOCOL IS A MODIFICATION OF THE METHOD of Neuhoff et al. (1988), which involves the binding of Coomassie Blue G250 to proteins. This dye complexes with basic amino acids, such as arginine, tyrosine, lysine, and histidine and produces a very low background due to the colloidal properties of the stain. This property prevents the Coomassie dye from penetrating the gel, eliminating the requirement for the long destaining periods characteristic of Coomassie R250 staining. Colloidal Coomassie is commonly used with preparative gels and has a detection range of 8–50 ng.

MATERIALS

CAUTION: See Appendix 3 for appropriate handling of materials marked with <!>.

Reagents

Coomassie Blue G250 <!> (10 ml, 5% w/v) in H_2O

Stir solution for a few minutes to disperse the dye. The dye will not dissolve completely.

Dye stock solution

ammonium sulfate <!>	50 g
85% phosphoric acid <!>	6 ml
5% Coomassie Blue G250 solution	10 ml
H_2O	to 500 ml

Fixative solution

10% (v/v) acetic acid <!>

40% (v/v) methanol <!>

Prepare 1 liter.

Glycerol (5%)

Methanol

Equipment

Pyrex or plastic dishes

These dishes should be larger than the size of the gel being stained.

Reciprocal or orbital shaking platform

Protein Sample

Gel containing protein samples (from Protocol 9)

METHOD

1. Fix the gel with gentle agitation for at least 1 hour in a Pyrex or plastic dish.

   The gel can be fixed overnight if necessary. If the gels contain labels, multiple gels can be fixed in the same dish; however, the volume of fixative should be sufficient to allow all gels to move freely.

2. Remove the fixative, submerge the gel in H_2O, and wash the gel on an orbital shaker for 10 minutes.

3. Discard the H_2O. Repeat the wash step two more times.

4. During the third wash step, prepare the colloidal Coomassie stain by mixing four parts of Coomassie dye stock with one part methanol.

5. Place the gel in 100–300 ml of colloidal Coomassie stain (depending on the size of the gel, ensuring sufficient volume to just cover the gel). Stain the gel overnight with gentle agitation on an orbital shaker.

   The gel can be left in the stain for a much longer period of time, during which the color of the gel will continue to deepen for up to 7 days.

6. To remove residual stain, transfer the gel to a clean dish of 45–55°C distilled or deionized H_2O. As the H_2O colors, discard it, and wash the gel with gentle agitation again with fresh 45–55°C H_2O. Repeat until the protein bands are at the desired contrast against the background of the gel. Proteins will appear blue on a clear background.

7. Gel can now be scanned for analysis. It is better to scan wet gels, because gels may crack during drying, leading to a loss of data.

8. Prior to drying, soak the gel in 5% glycerol for 30 minutes. Gels can also be stored indefinitely at 4°C wrapped in plastic and sealed in a ZipLock bag.

PROTOCOL 11

Ammoniacal Silver Staining

SILVER STAINING OF POLYACRYLAMIDE GELS was first introduced by Switzer et al. (1979) and has rapidly become the most popular high-sensitivity staining method for proteins. Problems with high backgrounds, reproducibility, and silver mirrors resulted in many improvements to this initial protocol over the years. Recently, Rabilloud et al. (1994b) found more than 100 different protocols in the literature. However, all protocols contain at least the following five basic steps:

- Fixation
- Sensitization
- Silver impregnation
- Image development
- Stopping development and image stabilization

We have found that the ammoniacal silver staining process presented here provides high sensitivity, in the range of 2–4 ng, and it works well with gels down to a thickness of 1 mm. Due to its high sensitivity, silver staining is very susceptible to interference from a variety of sources. Exceptional cleanliness is necessary and reagent and water quality is critical. Deionized H_2O with a resistivity >15 mho/cm is necessary for all solutions and washes. Distilled H_2O must not be used, for it tends to give erratic results. All chemicals should be of the highest grade possible. The background levels obtained with this silver staining method can be reduced greatly by incorporating thiosulfate into the gel during polymerization (Hochstrasser and Merril 1988). This can be achieved by altering the polymerization solution suggested in Tables 4.5 and 4.6. It may be worth comparing this procedure with the silver staining method provided in Protocol 12. Both methods, when properly performed, provide excellent results with low background. The choice on which method to use routinely will lie with the investigator.

MATERIALS

CAUTION: See Appendix 3 for appropriate handling of materials marked with <!>.

IMPORTANT: Prepare all solutions in clean glassware with deionized H_2O. Approximately 250 ml of each solution is required per standard size (160 x 200 x 1.5 mm) gel.

Reagents

Developing solution

citric acid monohydrate <!>	0.1 g
formaldehyde (stock is 36.5–38%) <!>	1 ml
H_2O	to 1000 ml

Mix in a Pyrex bottle, place the cap on the bottle, and shake manually. This solution should be made fresh every time it is needed.

Glutaraldehyde sensitization solution (20% glutaraldehyde solution) <!>

Combine 200 ml of glutaraldehyde (e.g., ACROS Organics, [stock is 50% glutaradehyde in H_2O]) with 800 ml of H_2O. Mix in a Pyrex bottle. A separate waste bottle will be required for glutaraldehyde waste.

Primary fixation solution

10% (v/v) acetic acid <!>

40% (v/v) ethanol <!>

Prepare 1 liter in a graduated cylinder, and mix on a magnetic stirrer.

Secondary fixation solution

5% (v/v) acetic acid

5% (v/v) ethanol

Prepare 1 liter in a graduated cylinder, and mix on a magnetic stirrer.

Silver staining solution

Solution A: In a 250-ml graduated cylinder add:

200 ml of H_2O

800 µl of NaOH <!>

13.3 ml of ammonium hydroxide <!>

Mix by stirring on a magnetic stirrer.

Solution B: In a small beaker (100-ml) add:

8.0 g of silver nitrate <!>

~50 ml of H_2O

Mix by stirring on a magnetic stirrer.

Transfer Solution A to a 1-liter graduated cylinder and let the solution stir in a chemical fume hood. When the contents of Solution B have dissolved, slowly add Solution B to Solution A using a 10-ml pipette. Add Solution B dropwise such that the dark precipitate clears before adding additional drops. If a precipitate forms and does not clear, discard the solution and start over.

When all of Solution B has gone into solution with A, adjust the volume to 1000 ml with H_2O while stirring in the chemical fume hood.

EXPERIMENTAL TIP: This solution stains very easily, and thus must be handled cautiously. Change gloves if they come into contact with silver-staining solution. Wipe area clean after use. This solution must be made fresh each time it is used.

Silver nitrate waste must be stored in a brown bottle used only for silver nitrate.

Stop solution (10% acetic acid)

In a Pyrex bottle, mix 100 ml of glacial acetic acid <!> and 900 ml of H_2O.

Equipment

Plastic wrap or thin polycarbonate plastic sheets

Pyrex or plastic dishes

These dishes should be larger than the size of the gel being stained.

Reciprocal or orbital shaking platform

Protein Sample

Gel containing protein samples (from Protocol 9)

METHOD

Silver staining is very susceptible to interference from a variety of sources. To minimize problems:

- Wear powder-free gloves at all times when preparing solutions or handling gels.
- Do not place pressure on the gels during solution transfers, as this can cause staining artifacts on the gels.
- Do not touch the gel with gloved hands throughout this protocol. Carry out all manipulations of the gel with a piece of plastic wrap or polycarbonate sheet placed between gloved hands and the gel.

1. Place the gel (from Protocol 9) in a dish of H_2O. Agitate the gel on a reciprocal or orbital shaker for 5 minutes. While the gel is on the shaker, prepare fresh primary fixation solution.

   Multiple gels should each be fixed and stained in separate dishes.

2. Drain the H_2O from the dish, holding the gel in the dish with a square of plastic wrap or polycarbonate sheet. Do not allow gloved hands to come in direct contact with the gel. Pour enough primary fixation solution into the dish to adequately cover the gel. Agitate the gel on a shaker for 30–60 minutes.

3. Prepare fresh secondary fixation solution. Drain the primary fixation solution from the dish, holding the gel in the dish with a square of plastic wrap or polycarbonate sheet. Pour enough secondary fixation solution into each dish to cover the gel adequately. Agitate the gel on a shaker for 2 hours if silver staining will be finished on the same day. If not, then cover the dish with a lid or plastic wrap and shake it overnight on an orbital shaker.

   If the gel is shaken overnight, be sure that there is enough solution in each dish to keep the gel from drying out.

4. Drain the secondary fixation solution from the dish, holding the gel in the dish with a square of plastic wrap or polycarbonate sheet. Add sufficient H_2O to cover the gel. Agitate the gel on a shaker for 5 minutes.

5. Prepare ~250 ml of glutaraldehyde sensitization solution per standard size gel (160 x 200 x 1.5 mm). Prepare the solution fresh each time used. Keep volumes to the minimum necessary. Proceed with either Step a or b.

   a. Make sure that the gel is face up in the dish. Drain the H_2O from the dish. Carefully pour ~250 ml of glutaraldehyde solution into the dish containing the gel. Agitate the gel on an orbital shaker for 30 minutes at an agitation speed sufficient to move the gels in the dish, but not sliding out of the solution.

      Knowing the orientation of the gel during the staining process will assist in "reading the gel." Traditionally, the gel is oriented with the acidic end on the left and the basic end on the right. Although this will have no effect on the staining of the gel, it helps to give an idea of where to expect the proteins to appear on the developed gel.

   b. Add ~200 ml of glutaraldehyde solution to a clean dry dish. Carefully transfer the gel from the H_2O to the glutaraldehyde solution. Use caution when lifting the gel; it is fragile. Lift the gel only from its corners to prevent pressure artifacts in critical areas of the gel. Agitate the gel on an orbital shaker for 30 minutes at an agitation speed sufficient to move the gels in the dish, but not sliding out of the solution.

6. Drain the glutaraldehyde solution into a proper waste receptacle. Rinse the gel with H_2O and drain this H_2O into the glutaraldehyde waste receptacle.
7. Generously cover the gel with H_2O and shake it on an orbital shaker for 5 minutes.
8. Drain the H_2O. Repeat the H_2O wash two more times.

   If desired, the number of washes can be increased beyond three, but not decreased.
9. Repeat Steps 7 and 8, but increase each wash time to 30 minutes.

   As with the 5-minute washes, if desired, the number of washes can be increased beyond three, but not decreased.
10. Prepare the silver staining solution 30–60 minutes before it is needed in Step 11.

    One liter of the silver staining solution is sufficient to stain four standard-size gels.
11. Drain the H_2O from the dish, holding the gel in the dish with a square of plastic wrap or polycarbonate sheet. Add ~250 ml of silver staining solution to each dish. Agitate the gel on an orbital shaker for 20 minutes. The agitation time can be shortened to 15 minutes if necessary, but it should be no less than 15 minutes and no more than 20 minutes.
12. Drain the silver staining solution into an appropriate waste bottle, making sure not to touch the gel with gloved hands.
13. Quickly rinse the gel with just enough H_2O to cover the gel. Discard this H_2O in the silver nitrate waste bottle.
14. Add enough H_2O to generously cover the gel. Agitate the gel on an orbital shaker for 5 minutes.
15. Drain the H_2O. Repeat the H_2O wash two more times.
16. While the gel is washing, prepare the developing and stop solutions. Pour the stop solution into a clean dry dish (or dishes, if multiple gels are being stained). Place the dishes in an easily accessible location to avoid any delay between developing and stopping.
17. Drain the H_2O from the dish, holding the gel in the dish with a square of plastic wrap or polycarbonate sheet. Add ~250 ml of developing solution to the dish. Agitate the gel on an orbital shaker that has been lined with white bench paper.

    White bench paper enhances the contrast between stained proteins and the background and will improve the ability to determine the stop point for developing.

    EXPERIMENTAL TIP: Do not develop more than three gels at one time. Keep additional gels in H_2O until ready to develop them.

    The time required for proper development varies with each gel. Watch the gels individually and determine the proper time to stop development, which is, typically, when protein spots appear dark and small protein spots become visible. It may take up to 1 minute before the first protein spots appear. Be careful not to overdevelop. Approximate development time is usually 5–10 minutes.
18. When the gel appears to be sufficiently developed, carefully lift the gel by its bottom corners, and quickly transfer it to a dish containing stop solution. Incubate the gel with gentle agitation in the stop solution for ~10 minutes.
19. Discard the stop solution, and cover the gel with H_2O. Place the gel on an orbital shaker until it can be scanned.

    Gels should be scanned as soon as possible. Residual acid in the gel, from the stop solution, will cause the stained proteins to fade over time. For accurate densitometry, scan the gels within 30 minutes of transfer to H_2O.

PROTOCOL 12

Mass Spectrometry-compatible Silver Staining

THE AMMONIACAL SILVER STAINING METHOD presented in Protocol 11 is one of the most sensitive methods used to detect proteins on an SDS-PAGE gel. However, this and other standard silver staining methods are not compatible with mass spectrometry (MS), which is fast becoming the best way to identify proteins isolated on 2D gels. Because the proteins in gels to be analyzed by mass spectroscopy cannot be modified, many of the common sensitizing agents (e.g., glutaraldehyde and strong oxidizing agents) cannot be used. In addition, many investigators find it advantageous to purchase dedicated kits, where the reagent quality is assured, to carry out the silver staining. Such kits are available from a variety of sources, including Amersham Biosciences, Bio-Rad, Pierce, and other chemical supply companies. The method provided here is based on the Amersham Biosciences PlusOne silver stain kit as described by Yan et al. (2000b). These authors found that this method is compatible with MALDI and ESI-MS, and it shows an increased ability to deal with semipreparative protein loads without negative staining as compared to other silver staining methods (including the method in Protocol 11). However, this process is less sensitive than standard silver staining methods. For further MS-compatible silver staining methods, see Shevchenko et al. (1996), Scheler et al. (1998), Gharahdaghi et al. (1999), and Sinha et al. (2001).

MATERIALS

CAUTION: See Appendix 3 for appropriate handling of materials marked with <!>.

IMPORTANT: Prepare all solutions in clean glassware using deionized H_2O with a resistivity >15 mho/cm.

Reagents

Developing solution

sodium carbonate <!>	6.25 g
formaldehyde (37%) <!>	100 µl
H_2O	250 ml

Fixation solution

10% (v/v) acetic acid <!>

40% (v/v) methanol <!>

Prepare 500 ml of this solution.

Sensitization solution

methanol	75 ml
5% sodium thiosulfate	10 ml
sodium acetate <!>	17 g
H_2O	165 ml

Silver staining solution (0.25% w/v silver nitrate) <!>

Just prior to use, dilute 25 ml of stock silver staining solution (stock concentration is 2.5%) tenfold with 225 ml of H_2O. Alternatively, prepare a fresh 0.25% (w/v) silver nitrate solution from powder just prior to use (see Step 6).

Stop solution (1.46% EDTA)

Dissolve 3.65 g of EDTA in 250 ml of H_2O.

Equipment

Plastic wrap or thin polycarbonate plastic sheets

Pyrex or plastic dishes

These dishes should be larger than the size of the gel being stained.

Reciprocal or orbital shaking platform

Protein Sample

Gel containing protein samples (from Protocol 9)

METHOD

Silver staining is very susceptible to interference from a variety of sources.

- Make sure that all glassware is exceptionally clean.
- Wear powder-free gloves at all times when preparing solutions or handling gels.
- Do not place pressure on the gels during solution transfers, as this can cause artifacts on the gels.
- Prepare all solutions fresh immediately prior to staining. Approximately 250 ml of each solution is required per standard size (160 x 200 x 1.5 mm) gel.

1. Place the gel (from Protocol 9) in a dish containing 250 ml of fixation solution. Agitate on a reciprocal or orbital shaker for 5 minutes.

   Multiple gels should each be fixed and stained in separate dishes.

2. Drain the fixation solution from the dish, holding the gel in the dish with a square of plastic wrap or polycarbonate sheet. Do not allow gloved hands to come in direct contact with the gel. Add another 250 ml of fixation solution and agitate on an orbital shaker for an additional 15 minutes.

3. Discard the fixation solution, and add 250 ml of sensitization solution to the dish. Agitate the gel on an orbital shaker for 30 minutes.

4. Drain the fixation solution from the dish, holding the gel in the dish with a square of plastic wrap or polycarbonate sheet. Add sufficient deionized H_2O to generously cover the gel. Agitate the gel on a shaker for 5 minutes.

5. Drain the H_2O. Repeat the H_2O wash two more times.

6. Prepare the silver staining solution 30–60 minutes before it is needed in Step 7.

   One liter of the silver staining solution is sufficient to stain four standard-size gels.

7. Drain the H_2O from the dish, holding the gel in the dish with a square of plastic wrap or polycarbonate sheet. Add ~250 ml of silver staining solution to each dish. Agitate the gel on an orbital shaker for 20 minutes. The agitation time can be shortened to 15 minutes if necessary, but it should be no less than 15 minutes and no more than 20 minutes.

8. Drain the silver staining solution into an appropriate waste bottle, making sure not to touch the gel with gloved hands.

9. Add enough H_2O to generously cover the gel. Agitate the gel on an orbital shaker for 5 minutes.

10. Drain the H_2O, discarding the first wash in the silver nitrate waste bottle. Repeat the H_2O wash two more times.

11. While the gel is washing, prepare the developing and stop solutions. Place the stop solution in an easily accessible location to avoid any delay between developing and stopping.

12. Drain the H_2O from the dish, holding the gel in the dish with a square of plastic wrap or polycarbonate sheet. Add ~250 ml of developing solution to the dish. Agitate the gel on an orbital shaker that has been lined with white bench paper.

    White bench paper enhances the contrast between stained proteins and the background and will improve the ability to determine the stop point for developing.

    EXPERIMENTAL TIP: Do not develop more than three gels at one time. Keep additional gels in H_2O until ready to develop them.

    The time required for proper development varies with each gel. Watch the gels individually and determine the proper time to stop development, which is, typically, when protein spots appear dark and small protein spots become visible. It may take up to 1 minute before the first protein spots appear. Be careful not to overdevelop. Approximate development time is usually 5–10 minutes.

13. When the gel appears to be developed enough, discard the developer into an appropriate waste container, holding the gel in the dish with a square of plastic wrap or polycarbonate sheet. Quickly pour ~250 ml of stop solution into the dish. Agitate the gel on an orbital shaker for 10 minutes.

14. Discard the stop solution, and generously cover the gel with H_2O. Agitate the gel on an orbital shaker for 5 minutes.

15. Discard the H_2O, and repeat the washes two more times. The gel is now ready for scanning for analysis or spot removal for mass spectrometry.

    Gels should be scanned as soon as possible.

    EXPERIMENTAL TIP: Mass spectrometry results can be improved by incorporating a destaining step to remove the silver prior to in-gel digestions, as described by Gharahdaghi et al. (1999).

PROTOCOL 13

Fluorescent Staining of Proteins with SYPRO Ruby

THE DEVELOPMENT OF NEW MASS SPECTROMETRY INSTRUMENTATION and its application to the field of proteomics has greatly aided in the identification of proteins isolated by 2D gel electrophoresis. Identification of proteins is no longer limited to those that are highly abundant, with protein identification at the femtomole level now being feasible (Jensen et al. 1997; Gatlin et al. 1998). However, to make this level of sensitivity applicable to proteins isolated from 2D gels, sensitive and compatible gel-staining methods are necessary. Silver staining provides very sensitive detection of proteins; however, it is complicated, and without modifications that reduce its sensitivity (see Protocol 12), it is incompatible with mass spectroscopy. A sensitive ruthenium-based fluorescent dye (SYPRO Ruby) has recently been developed by Molecular Probes. Although the structure of this dye is not in the public domain, it has been reported that SYPRO Ruby is a transition metal organic complex that binds directly by electrostatic mechanisms (Patton 2000). Similar to silver stains, SYPRO Ruby interacts with basic amino acids, including lysine, arginine, and histidine (Lopez et al. 2000). Recent studies by Yan et al. (2000a) and Lopez et al. (2000) have shown that this dye is as sensitive as silver staining and completely compatible with MALDI mass spectroscopy. The staining method is also simple, having only three steps: fixation, staining, and destaining. However, SYPRO Ruby is very expensive. Thus, for many laboratories, this dye might be limited to micropreparative gels, as the cost to run the large number of gels done in quantitative studies is prohibitive.

MATERIALS

CAUTION: See Appendix 3 for appropriate handling of materials marked with <!>.

Reagents

Fixative solution
- 7% (v/v) acetic acid <!>
- 10% (v/v) methanol <!>
- Prepare 1000 ml of this solution.

SYPRO Ruby dye (Molecular Probes) <!>

Equipment

Polypropylene tray
- Do not use a glass tray; it will bind the dye.

Reciprocal or orbital shaking platform

UV or blue light transilluminator (e.g., the Dark Reader, Clare Chemical Research, Denver, Colorado) or Laser scanner <!>

Protein Sample

Gel containing protein samples (from Protocol 9)

METHOD

1. Place the gel (from Protocol 9) in a polypropylene dish containing sufficient fix solution to allow the gel to float freely in the dish. Agitate on a reciprocal or orbital shaker for 30 minutes.

   Multiple gels should each be fixed and stained in separate dishes.

2. Drain the fixative solution from the dish, holding the gel in the dish with a square of plastic wrap or polycarbonate sheet. Do not allow gloved hands to come in direct contact with the gel. Incubate the gel in the dark on a reciprocal or orbital shaker in undiluted SYPRO Ruby stain for 90 minutes to overnight.

   The minimum stain volume for standard gel sizes are:

   50 ml for 8 cm x 10 cm x 0.75 mm gel
   330 ml for 16 cm x 20 cm x 1 mm gel
   500 ml for 20 cm x 20 cm x 1 mm gel
   or ~10 times the volume of larger size gels

   Using too little stain will lower the sensitivity. Sensitivity increases with longer staining time, with a minimum of 3 hours for maximum sensitivity. Reusing the stain will greatly reduce sensitivity.

3. (*Optional*) Discard the SYPRO Ruby stain. To remove residual stain that can cause high fluorescent background and speckling, place the gel in a volume of fixative solution equal to the amount of dye that was used in Step 2. Agitate on a reciprocal or orbital shaker table for 30 minutes.

4. Visualize the proteins using a 300-nm transilluminator, a blue light transilluminator, or a laser scanner. SYPRO Ruby has excitation maxima at ~300 and ~420 nm and an emission maximum at ~618 nm.

   The polyester backing of some precast gels is highly fluorescent. For maximum sensitivity using a UV transilluminator, place the gel polyacrylamide side down and use an emission filter to screen out the blue fluorescence of the plastic. The use of a blue-light transilluminator or laser scanner will also reduce the amount of fluorescence from the plastic backing.

   This dye is compatible with imaging systems whose lasers emit at 450, 473, 488, and or 532 nm. Currently, most major suppliers of laser scanners can provide the optimal filter setting for this dye.

PROTOCOL 14

Phosphoprotein Staining with the GelCode Phosphoprotein Staining Kit

THE PHOSPHORYLATION STATE OF A PROTEIN HAS AN IMPORTANT ROLE in the regulation of a wide variety of cellular processes. As a result, there has been a great deal of interest in detecting phosphorylated proteins. The method presented here uses the GelCode phosphoprotein staining kit (Pierce Chemical Company). This method depends on the hydrolysis of the phosphoprotein phosphoester linkage using sodium hydroxide in the presence of calcium ions. The gel containing the newly formed insoluble calcium phosphate is then treated with ammonium molybdate in dilute nitric acid. The resultant insoluble nitrophospho-molybdate complex is stained with Methyl Green. After destaining, the phosphoproteins are colored green to green-blue. The detection limit is in the nanogram range, but depends on the degree of phosphorylation of the protein. This method will detect the phosphoproteins phosvitin and β-casein in the 40–80 ng/band and 80–160 ng/band range, respectively. The method presented here is for staining minigels. Volumes will need to be increased for larger gels.

MATERIALS

CAUTION: See Appendix 3 for appropriate handling of materials marked with <!>.

Reagents

IMPORTANT: These reagents must be purchased as the GelCode phosphoprotein staining kit from the Pierce Chemical Company, as the concentrations of the components are proprietary.

Acetic acid (7%) <!>
Ammonium molybdate solution <!>
Ammonium molybdate solution containing nitric acid <!>
Methyl Green solution <!>
NaOH (0.5 N) <!>
Sulfosalicylic acid solution <!>
Sulfosalicylic acid solution containing calcium chloride <!>

Equipment

Glass or plastic tray with a lid
Oven preheated to 65°C
Reciprocal or orbital shaking platform

Protein Sample

Gel containing protein samples (from Protocol 9)

METHOD

1. Place the gel (from Protocol 9) in a dish containing 50 ml of H_2O. Agitate the gel on a reciprocal or orbital shaker table for 10 minutes.
2. Move the gel to a dish containing 25 ml of sulfosalicylic acid solution, and agitate the gel on a reciprocal or orbital shaker table for 15 minutes.
3. Place the gel in 25 ml of sulfosalicylic acid solution containing calcium chloride, and agitate the gel on a reciprocal or orbital shaker table for 30 minutes.
4. Rinse the gel rapidly with H_2O to remove the calcium chloride adhering to the surface of the gel.
5. Transfer the gel to 25 ml of 0.5 N NaOH, cover the tray with a lid, and incubate it for 20 minutes at 65°C.
6. Place the gel in 25 ml of ammonium molybdate solution, and agitate the gel on a reciprocal or orbital shaker table for 10 minutes.
7. Repeat Step 6.
8. Transfer the gel to 25 ml of ammonium molybdate solution containing nitric acid, and agitate the gel on a reciprocal or orbital shaker table for 20 minutes.
9. Place the gel in 25 ml of Methyl Green solution, and agitate the gel on a reciprocal or orbital shaker table for 20 minutes.
10. To destain, transfer the gel to 25 ml of sulfosalicylic acid solution, and agitate the gel on a reciprocal or orbital shaker table for 15 minutes.
11. Repeat Step 10. At this time, green spots of phosphoproteins should be seen.
12. To completely destain the gel, place it in 25 ml of 7% acetic acid, and agitate the gel on a reciprocal or orbital shaker table overnight. During the destaining, change the acetic acid once, after the solution has become green in color.
13. Following visualization of the phosphoproteins, stain the total proteins on the gel using a Coomassie-Blue-staining method.

PROTOCOL 15

Tank Transfer of 2D Gels

THIS PROTOCOL DESCRIBES A TANK TRANSFER METHOD for blotting proteins, in which a cassette holds the sandwich of gel, membrane, and blotting paper suspended vertically in buffer. Electrode panels are placed parallel to the cassette. Typically, liters of buffer are required in the tank to maintain the current and pH throughout the transfer. The advantages of tank systems are their efficiency of protein transfer; transfers can be run for extended periods of time (allowing more complete transfer) and proteins from several gels can be transferred simultaneously.

MATERIALS

CAUTION: See Appendix 3 for appropriate handling of materials marked with <!>.

Reagents

Transfer buffers (Tris-glycine-methanol [Towbin] or CAPS buffer)

Tris-glycine-methanol buffer should be used for standard transfer and detection procedures. CAPS buffer is used if the proteins will be subjected to amino acid analysis or sequenced from the membrane. If proteins are to be sequenced from a membrane exposed to Tris-glycine-methanol buffer, the membrane must be washed extensively to remove excess glycine and Tris, which will both interfere with the subsequent analysis.

- Towbin transfer buffer

Tris-base	15.1 g
glycine <!>	72.1 g
methanol <!>	500 ml
SDS (optional) <!>	5.0 g
H_2O	to 5 liters

10% methanol is recommended for binding proteins to nitrocellulose membranes. Up to 20% methanol can be included. Methanol helps remove SDS from the proteins and improves the binding of the proteins to the membrane. SDS is optional, but it can be included to help improve the transfer of large proteins.

- CAPS transfer buffer

Add 2.2 g of 3-(cyclohexyl-amino)-1-propanesulfonic acid <!> to ~600 ml of H_2O. Adjust the pH to 11 with concentrated NaOH <!>, and adjust the volume to 1 liter with H_2O.

Equipment

Blotting paper

Membrane, supported nitrocellulose or PVDF

Nitrocellulose is frequently used for total protein stain methods and immunoblotting, whereas polyvinylidene difluoride (PVDF) is required for protein sequencing.

Power supply (400 mA for small tank transfers; 1000 mA for large tank transfer units)
Transfer tank apparatus

Protein Sample

Gel containing protein samples (from Protocol 9)

METHOD

1. Measure the SDS-PAGE gel (from Protocol 9), and cut a transfer membrane to the appropriate size.

   Membranes used for tank transfer can be cut larger or smaller than the actual gel if necessary.

2. Wet the membrane in H_2O (or methanol for PVDF) by slowly lowering the membrane onto the surface of the liquid. Do this slowly to prevent uneven wetting of the membrane and trapped air bubbles beneath the membrane.

3. Transfer the membrane from the wetting solution to a dish of the appropriate transfer buffer. Allow the membrane to equilibrate in the transfer buffer for ~5 minutes.

4. Cut the blotting paper to the appropriate size, according to the transfer tank manufacturer's instructions.

5. Assemble the transfer sandwich. It is critical to avoid trapping air bubbles between any of the layers of the sandwich, because the bubbles will block the transfer of proteins at those sites. It may be best to assemble the sandwich submerged in buffer, as described below in Steps a–h.

   a. Place one side of the transfer cassette in a tray of transfer buffer.

   b. Place a foam sponge onto the cassette and squeeze out any air bubbles trapped in the sponge.

   c. Place one sheet of blotting paper on top of the sponge making sure it wets evenly and completely. Remove any air bubbles by rolling a pipette or glass rod across the surface of the blotting paper.

   d. Place the second-dimension gel onto the surface of the blotting paper aligning the gel so that the sponge, blotting paper, and gel fit within the cassette. Remove any air bubbles by rolling a pipette or glass rod across the surface of the gel. Be careful not to place too much pressure on the gel.

   e. Place the transfer membrane on top of the gel, being careful to place it properly the first time. Moving the membrane across the gel may cause smudged results. Carefully remove any air bubbles from between the membrane and the gel by rolling a pipette or glass rod across the surface of the membrane. It is critical that there be no air pockets between the gel and the membrane or this will prevent transfer of proteins.

   f. Layer another piece of blotting paper on top of the membrane, making sure it wets evenly and completely. Remove any air bubbles by rolling a pipette or glass rod across the surface of the blotting paper.

g. Place a foam sponge onto the cassette and squeeze out any air bubbles trapped in the sponge.

h. Close the cassette and snap it shut.

6. Repeat Step 5 for each gel that is to undergo electrophoretic transfer.

7. Fill the transfer tank approximately two-thirds full with the appropriate transfer buffer.

8. Place the transfer cassette into the tank.

   EXPERIMENTAL TIP: Make sure to orient the assembly so that the electric current will transfer the proteins from the gel to the membrane, and not from the gel to the blotting paper and into the solution!

9. Finish filling the transfer unit to the recommended buffer level, making sure that all the cassettes are submerged in the buffer.

10. Place the lid on the transfer unit so that the red or anode lead is closer to the membrane than the gel.

    During electrophoresis, proteins migrate from the negative cathode (black) toward the positive anode (red).

11. If transfers are carried out for more than 1 hour, attach a recirculating water bath to the heat exchanger and set the temperature to 15°C.

    Tank transfer units are used under very high current conditions, and without adequate cooling, damage to the apparatus can occur. A recirculating water bath will help maintain the temperature during extended transfer times. Quick-fit connectors are recommended to prevent leakage of the coolant when the recirculating water bath is disconnected.

12. Transfer according to the times and currents recommended by the manufacturer of the transfer apparatus.

13. When the transfer is complete, stop the recirculating water bath, turn off the power supply, remove the transfer lid, and remove the cassette.

14. Disassemble the transfer cassette. Rinse the sponges in deionized H_2O and dry in the air for reuse. Dispose of the blotting papers. Use a soft lead pencil to mark the side of the membrane facing the gel.

    Do not use a pen or lab marker, as many inks will solubilize in the buffers used for processing the membrane.

15. Stain or process the blot (see Protocols 17–20) or store the membrane dry in the dark between sheets of blotting paper.

    If stored dry, the membrane must be rewet in water (nitrocellulose) or methanol (PVDF) prior to processing.

PROTOCOL **16**

Semidry Blotting of 2D Gels

SEMIDRY UNITS TRANSFER PROTEINS FROM THE GEL to the membrane when they are sandwiched between buffer-saturated stacks of blotting paper. The electrical field is applied parallel to the stack. Compared with tank transfer units, semidry transfer units use much less buffer, and because the electrode panels are closer together, less voltage is required for transfer.

MATERIALS

CAUTION: See Appendix 3 for appropriate handling of materials marked with <!>.

Reagents

Transfer buffers (Tris-glycine-methanol [Towbin] or CAPS buffer)

Tris-glycine-methanol buffer should be used for standard transfer and detection procedures. CAPS buffer is used if the proteins will be subjected to amino acid analysis or sequenced from the membrane. If proteins are to be sequenced from a membrane exposed to Tris-glycine-methanol buffer, the membrane must be washed extensively to remove excess glycine and Tris, which will both interfere with the subsequent analysis.

- Towbin transfer buffer

Tris-base	15.1 g
glycine <!>	72.1 g
methanol <!>	500 ml
SDS (optional) <!>	5.0 g
H_2O	to 5 liters

10% methanol is recommended for binding proteins to nitrocellulose membranes. Up to 20% methanol can be included. Methanol helps remove SDS from the proteins and improves the binding of the proteins to the membrane. SDS is optional, but it can be included to help improve the transfer of large proteins.

- CAPS transfer buffer

Add 2.2 g of 3-(cyclohexyl-amino)-1-propanesulfonic acid <!> to ~600 ml of H_2O. Adjust the pH to 11 with concentrated NaOH <!>, and adjust the volume to 1 liter with H_2O.

Equipment

Blotting paper
Membrane, supported nitrocellulose or PVDF
Power supply (400 mA)
Semidry transfer apparatus

Protein Sample

Gel containing protein samples (from Protocol 9)

METHOD

1. Measure the SDS-PAGE gel (from Protocol 9), and cut a transfer membrane and six sheets of blotting paper to the same size as the gel.
2. Wet the membrane in H_2O (or methanol for PVDF) by slowly lowering the membrane onto the surface of the liquid. Do this slowly to prevent uneven wetting of the membrane and trapped air bubbles beneath the membrane.
3. Transfer the membrane from the wetting solution to a dish of the appropriate transfer buffer. Allow the membrane to equilibrate in the transfer buffer for ~5 minutes.
4. Assemble the transfer sandwich. It is critical to avoid trapping air bubbles between any of the layers of the sandwich, because the bubbles will block the transfer of proteins at those sites.
   a. Wet three pieces of blotting paper in transfer buffer and stack them onto the anodic electrode panel of the transfer unit. Roll a glass rod across the stack to remove any air bubbles and to make sure the stack is in uniform contact with the electrode panel.
   b. Place the transfer membrane onto the surface of the blotting paper stack.
   c. Lay the gel on top of the membrane, being careful to place it properly the first time. Moving the gel across the membrane may cause smudged results. Roll a glass rod across the surface of the gel to remove any air bubbles.
   d. Wet the other three pieces of blotting paper in transfer buffer, and stack them on top of the gel. Roll a glass rod across the entire stack to remove trapped air.
   e. Place the cathode panel onto the transfer stack and connect the safety interlock to the base.

      During electrophoresis, proteins migrate from the negative cathode (black) toward the positive anode (red).
5. Transfer according to the times and currents recommended by the manufacturer of the transfer apparatus.
6. When the transfer is complete, turn off the power supply and remove the transfer lid.
7. Disassemble the transfer stack. Dispose of the blotting papers. Use a soft lead pencil to mark the side of the membrane facing the gel.

   Do not use a pen or lab marker, as many inks will solubilize in the buffers used for processing the membrane.
8. Stain or process the blot (see Protocols 17–20) or store the membrane dry in the dark between sheets of blotting paper.

   If stored dry, the membrane must be rewet in water (nitrocellulose) or methanol (PVDF) prior to processing.

Detection of Proteins on Membranes

THE TOTAL PROTEIN COMPLEMENT BOUND TO A MEMBRANE can be detected using a variety of staining methods (see Table 4.9). For microsequencing, the membrane can be stained to detect all of the bound protein, and the desired spots are then excised from the membrane for analysis.

For the recognition of specific proteins on a membrane, many different antibodies are available, including radiolabeled antibodies, horseradish-peroxidase-linked or alkaline-phosphatase-linked antibodies, and biotinylated antibodies. In addition, detection can be carried out by several different methods, such as autoradiography, chemiluminescence, or colorimetric procedures. The difficulty of western blotting with 2D electrophoresis is aligning the spots detected by the bound antibody to an identical sample that has been stained with a total protein stain. The following procedures are some of the ways in which the total protein complement on 2D gels can be detected.

TABLE 4.9. Stains for the detection of total protein on a membrane

Stain	Detection limit	Nitrocellulose	PVDF	Comments	Compatible with immunoblotting	Protocol
Ponceau S	1 μg	+	+	reversible	yes	17
Coomassie R250	1 μg	+	+	permanent high background	yes	18
India Ink	100 ng	+	+	permanent	–	19
Colloidal Gold	3 ng	+	+	permanent	no	20

PROTOCOL 17

Staining Membrane-bound Proteins with Ponceau S

BECAUSE PONCEAU S IS RELATIVELY INSENSITIVE (~1 μg of protein), only the most abundant proteins will be visible. However, it is a reversible stain that can be removed completely with H_2O prior to processing the blots (Salinovich and Montelano 1986). After staining, a soft lead pencil can be used to record the presence of visible proteins and molecular-weight markers, which will help when aligning the proteins detected on the membrane by western analysis with those in a total protein-stained gel or membrane. Ponceau S is compatible with both nitrocellulose and PVDF membranes.

MATERIALS

CAUTION: See Appendix 3 for appropriate handling of materials marked with <!>.

Reagent

Ponceau S (0.1% w/v)/acetic acid (5% v/v) <!>

Protein Sample

Membrane containing bound proteins transferred in Protocol 15 or 16

METHOD

1. Submerge the transfer membrane in Ponceau S stain solution with gentle agitation for 5 minutes.
2. Decant the stain and rinse the membrane several times with H_2O until the protein bands are visible.

   Do not reuse the stain; it will result in nonreproducible results due to depletion of the dye after the first use.

3. Use a soft lead pencil to mark the major protein spots and molecular-weight markers.
4. Continue rinsing the membrane with H_2O with gentle agitation until the Ponceau S is removed.
5. Proceed with protein characterization or analysis (see Chapters 6, 7, and 8).

PROTOCOL 18

Staining Membrane-bound Proteins with Coomassie Blue R250

COOMASSIE BLUE R250 PERMANENTLY STAINS MEMBRANE-BOUND PROTEINS and is compatible with PVDF and nitrocellulose membranes, but incompatible with nylon membranes. This technique is relatively insensitive, with a detection limit of ~1.5 µg of protein. One drawback of Coomassie Blue staining is that it produces a high background which can make interpretation of results difficult.

MATERIALS

CAUTION: See Appendix 3 for appropriate handling of materials marked with <!>.

Reagents

Coomassie Blue R250 (0.1% w/v)/methanol (50% v/v) <!>
Methanol (50%) containing 10% acetic acid <!>

Protein Sample

Membrane containing bound proteins transferred in Protocol 15 or 16

METHOD

1. Submerge the membrane in the Coomassie Blue stain solution for 5 minutes with gentle agitation.
2. Decant the stain, and destain the membrane on a rocker table with several changes of 50% methanol containing 10% acetic acid.

   Do not reuse the stain; it will result in nonreproducible results due to depletion of the dye after the first use.

3. Rinse the membrane with H_2O for 5 minutes with gentle agitation.

PROTOCOL 19

Staining Membrane-bound Proteins with India Ink

INDIA INK PERMANENTLY STAINS PROTEINS BLACK against the white background of the membrane. It can be used with both nitrocellulose and PVDF membranes, but it is not recommended for nylon membranes. This technique is sensitive, detecting down to ~100 ng of protein.

MATERIALS

CAUTION: See Appendix 3 for appropriate handling of materials marked with <!>.

Reagents

India Ink stain

Just before use, dilute 1 μl of India Ink (Pelikan 17 black) per 1 ml of Tween-20 solution.

Phosphate-buffered saline (PBS) (pH 7.2)

0.15 M NaCl

0.01 M Na_2HPO_4/NaH_2PO_4 <!>

Tween-20 solution (0.3% v/v) in PBS

Protein Sample

Membrane containing bound proteins transferred in Protocol 15 or 16

METHOD

1. Submerge the membrane in the Tween-20 solution for 10 minutes with gentle agitation.
2. Discard the Tween-20 solution.
3. Repeat Steps 1 and 2 three more times.
4. Submerge the membrane in India Ink stain with gentle agitation. Stain the membrane for 2–18 hours.
5. Discard the stain, and rinse the membrane with Tween-20 solution for 5 minutes with gentle agitation.

PROTOCOL 20

Staining Membrane-bound Proteins with Colloidal Gold

COLLOIDAL GOLD IS THE MOST SENSITIVE STAINING TECHNIQUE, detecting as little as 1–3 ng of protein (Moeremans et al. 1985). Protein spots are permanently stained a dark red after incubation with the colloidal gold solution. Colloidal gold staining can detect proteins on both nitrocellulose and PVDF membranes, but it is not recommended for nylon membranes.

MATERIALS

CAUTION: See Appendix 3 for appropriate handling of materials marked with <!>.

Reagents

Colloidal gold stain solution (Amersham Biosciences, Bio-Rad, and Diversified Biotech)
Phosphate-buffered saline (PBS) (pH 7.2)
  0.15 M NaCl
  0.01 M Na_2HPO_4/NaH_2PO_4 <!>
Tween-20 solution (0.3% v/v) in PBS

Protein Sample

Membrane containing bound proteins transferred in Protocol 15 or 16

METHOD

1. Submerge the membrane in Tween-20 solution for 45 minutes at 37°C with gentle agitation.
2. Wash the membrane with Tween-20 solution for 5 minutes at room temperature with gentle agitation.
3. Discard the wash solution, and repeat Step 2 two more times.
4. Stain the membrane in the colloidal gold stain solution for 2 hours at room temperature with gentle agitation.
5. Rinse the membrane with H_2O for 5 minutes with gentle agitation.

Two-dimensional Electrophoresis Resources on the World Wide Web

A WEALTH OF INFORMATION EXISTS for dealing with 2D electrophoresis and proteomics on the World Wide Web. Below is only a partial list of the Web Sites available with techniques, information on meetings, classes, 2D gel databases, and information on the newest equipment and software for working with 2D electrophoresis.

Commercial Two-dimensional Electrophoresis and Proteomics Sites

Amersham Biosciences
    http://proteomics.apbiotech.com
BioRad Proteomics Workstation
    http://www.proteomeworks.bio-rad.com/
Genomic Solutions
    http://www.genomicsolutions.com/proteomics
Large Scale Biology Home Page
    http://www.lsbc.com

Two-dimensional Analysis Software Sites

Phoretix
    http://www.phoretix.com/
Melanie III
    http://www.genebio.com/Melanie.html
PDQuest:
    http://proteomeworks.bio-rad.com/html/tech5.html
Flicker for 2D gel analysis
    http://www-lecb.ncifcrf.gov/flicker/
NCI/FCRDC LMMB Image Processing Section (GELLAB software)
    http://www.lecb.ncifcrf.gov/lemkin/gellab.html
Compugen (Z3 software)
    http://www.2Dgels.com

Two-dimensional Electrophoresis and Proteomics Databases

Expasy Index to 2D PAGE databases and services
    http://www.expasy.ch/ch2d/2d-index.html
HSC 2-DE Gel Protein Databases list
    http://www.harefield.nthames.nhs.uk/nhli/protein/other_sites.html

Argonne Protein Mapping Group Server:
    http://www.anl.gov/BIO/
Phosophoprotein Database:
    http://www-lecb.ncifcrf.gov/phosphoDB/
Rat Serum Protein Database
    http://linux.farma.unimi.it
Parasite Proteome Maps
    http://www.aber.ac.uk/~mpgwww/Proteome/ProtMaps.html
Cambridge Proteomics Facility
    http://www.bio.cam.ac.uk/proteomics/index.html
Rice 2D Database
    http://semele.anu.edu.au/2d/2d.html
Plant Plasma Membrane 2D Database
    http://sphinx.rug.ac.be:8080/ppmdb/index.html
COMPLUYEAST-2DPAGE Database
    http://babbage.csc.ucm.es/2d/2d.html
Danish Centre for Human Genome Research 2D PAGE Databases (Aarhus)
    http://proteomics.cancer.dk
SIENA-2DPAGE
    http://www.bio-mol.unisi.it/2d/2d.html
PMMA-2D Page—at Purkyne Military Medical Academy, Czech
    http://www.pmma.pmfhk.cz/
Embryonal Stem Cells (Immunobiology, University of Edinburgh)
    http://www.dur.ac.uk/~dbl0nh1/2DPAGE/
Human Colon Carcinoma Protein Database–Joint ProteomicS Laboratory (JPSL), Ludwig Institute for Cancer Research, Melbourne, Australia
    http://www.ludwig.edu.au/jpsl/jpslhome.html
HP-2DPAGE (Max Delbruck Center, Berlin)
    http://www.mdc-berlin.de/~emu/heart/
MitoDat—Mendelian Inheritance and the Mitochondrion
    http://www-lmmb.ncifcrf.gov/mitoDat/
SWISS-2DPAGE at Geneva University Hospital
    http://www.expasy.ch/ch2d/ch2d-top.html
YPD Yeast Protein Database at Proteome, Inc.
    http://www.proteome.com/YPDhome.html
Yeast Proteome Map
    http://www.ibgc.u-bordeaux2.fr/YPM/

Sources of Information and Methods on 2D Electrophoresis and Proteomics

Australian Proteome Analysis Facility
    http://www.proteome.org.au/
The Tubingen Proteome Project
    http://www.uni-tuebingen.de/uni/kxm/Proteome/
University of Aberdeen Protein Lab and Proteomics Facility
    http://www.abdn.ac.uk/~mmb023/proteome/index.htm

The EXPASY Swiss 2D-PAGE site provides methods, reagents, and links to Proteome sites worldwide.

http://www.expasy.ch/

The British Electrophoresis Society site also has excellent links to relevant sites and data bases worldwide.

http://www.proteinworks.com/bes/

The HSC-2DPAGE 2-DE Gel Protein Databases site at Harefield Hospital in London provides links to worldwide databases, as well as information pertaining to upcoming meetings, 2D gel analysis software, and more.

http://www.harefield.nthames.nhs.uk/nhli/protein/

The laboratory of Professor A. Görg in Munich has been instrumental in developing the technology of isoelectric focusing with immobilized pH gradients. The entire Görg laboratory manual is available on-line from this site.

http://www.edv.agrar.tu-muenchen.de/blm/deg/

The laboratory of Dr James R. Jefferies, Parasitology Group, Institute of Biological Sciences, University of Wales at Aberystwyth, Aberystwyth, Ceredigion, SY23 3DA, Wales, UK.

http://www.aber.ac.uk/%7Empgwww/Proteome/Tut_2D.html#Section%201

REFERENCES

Adessi C., Miege C., Albrieux C., and Rabilloud T. 1997. Two-dimensional electrophoresis of membrane proteins: A current challenge for immobilized pH gradients. *Electrophoresis* **18:** 127–135.

Bjellqvist B., Ek K., Righetti P.G., Gianazza E., Görg A., Westermeier R., and Postel W. 1982. Isoelectric focusing in immobilized pH gradients: Principle, methodology, and some applications. *J. Biochem. Biophys. Methods* **6:** 317–339.

Bjellqvist B., Sanchez J.C., Pasquali C., Ravier F., Paquet N., Frutiger S., Hughes G.J., and Hochstrasser D. 1993. Micropreparative two-dimensional electrophoresis allowing the separation of samples containing milligram amounts of proteins. *Electrophoresis* **14:** 1375–1378.

Blomberg A., Blomberg L., Norbeck J., Fey S.J., Larsen P.M., Larsen M., Roepstorff P., Degand H., Boutry M., Posch A., and Görg A. 1995. Interlaboratory reproducibility of yeast protein patterns analyzed by immobilized pH gradient two-dimensional gel electrophoresis. *Electrophoresis* **16:** 1935–1945.

Chevallet M., Santoni V., Poinas A., Rouquie D., Fuchs A., Kieffer S., Rossignol M., Lunardi J., Garin J., and Rabilloud T. 1998. New zwitterionic detergents improve the analysis of membrane proteins by two-dimensional electrophoresis. *Electrophoresis* **19:** 1901–1909.

Corthals G.L., Molloy M.P., Herbert B.R., Williams K.L., and Gooley A.A. 1997. Prefractionation of protein samples prior to two-dimensional electrophoresis. *Electrophoresis* **18:** 317–323.

Damerval C., de Vienne D., Zivy M., and Thiellement H. 1986. Technical improvements in two-dimensional electrophoresis increase the level of genetic variation detected in wheat-seedling proteins. *Electrophoresis* **7:** 52–54.

Dunbar B.S. 1987. *Two-dimensional electrophoresis and immunological techniques.* Plenum Press, New York.

Flengsrud R. and Kobro G. 1989. A method for two-dimensional electrophoresis of proteins from green plant tissues. *Anal. Biochem.* **177:** 33–36.

Görg A., Postel W., and Günther S. 1985. Improved horizontal two-dimensional electrophoresis with immobilized pH gradients. *Electrophoresis* **6:** 599–604.

———. 1988. The current state of two-dimensional electrophoresis with immobilized pH gradients. *Electrophoresis* **9:** 531–546.

Gatlin C.L., Kleemann G.R., Hays L.G., Link A.J., and Yates III J.R. 1998. Protein identification at the low femtomole level from silver-stained gels using a new fritless electrospray interface for liquid chromatography-microspray and nanospray mass spectrometry. *Anal. Biochem.* **263:** 93–101.

Gharahdaghi F., Weinberg C.R., Meagher D.A., Imai B.S., and Mische S.M. 1999. Mass spectrometric identification of proteins from silver-stained polyacrylamide gel: A method for the removal of silver ions to enhance sensitivity. *Electrophoresis* **20:** 601–605.

Gianazza E. 1999. Casting immobilized pH gradients (IPGs). *Methods Mol. Biol.* **112:** 175–188.

Görg A., Obermaier C., Boguth G., Harder A., Scheibe B., Wildgruber R., and Weiss W. 2000. The current state of two-dimensional electrophoresis with immobilized pH gradients. *Electrophoresis* **21:** 1037–1053.

Guy G.R., Philip R., and Tan Y.H. 1994. Analysis of cellular phosphoproteins by two-dimensional gel electrophoresis: Applications for cell signaling in normal and cancer cells. *Electrophoresis* **15:** 417–440.

Hancock K. and Tsang V.C.M. 1983. India ink staining of proteins on nitrocellulose paper. *Anal. Biochem.* **133:** 157–162.

Herbert B.R., Molloy M.P., Gooley A.A., Walsh B.J., Bryson W.G., and Williams K.L. 1998. Improved protein solubility in two-dimensional electrophoresis using tributyl phosphine as reducing agent. *Electrophoresis* **19:** 845–851.

Hochstrasser D.F. and Merril C.R. 1988. Catalysts for polyacrylamide gel polymerization and detection of proteins by silver staining. *Appl. Theor. Electrophoresis* **1:** 35–40.

Jensen O.N., Mortensen P., Vorm O., and Mann M. 1997. Automation of matrix-assisted laser desorption/ionization mass spectrometry using fuzzy logic feedback control. *Anal. Chem.* **69:** 1706–1714.

Ji H., Whitehead R.H., Reid G.E., Moritz R.L., Ward L.D., and Simpson R.J. 1994. Two-dimensional electrophoretic analysis of proteins expressed by normal and cancerous human crypts: Application of mass spectrometry to peptide-mass fingerprinting. *Electrophoresis* **15:** 391-405.

Ji H., Reid G.E., Moritz R.L., Eddes J.S., Burgess A.W., and Simpson R.J. 1997. A two-dimensional gel database of human colon carcinoma proteins. *Electrophoresis* **18:** 605–613.

Klose J. 1975. Protein mapping by combined isoelectric focusing and electrophoresis of mouse tissues. A novel approach to testing for induced point mutations in mammals. *Humangenetik* **26:** 231–243.

Laemmli U.K. 1970. Cleavage of structural proteins during the assembly of the head of bacteriophage T4. *Nature* **227:** 680–685.

Lee K.H. and Harrington M.G. 1999. Double-label analysis. *Methods Mol. Biol.* **112:** 291–298.

Lenstra J.A. and Bloemendal H. 1983. Topography of the total protein population from cultured cells upon fractionation by chemical extractions. *Eur. J. Biochem.* **135:** 413–423.

Link A.J. 1999. Autoradiography of 2-D gels. *Methods Mol. Biol.* **112:** 289–290.

Lopez M.F., Berggren K., Chernokalskaya E., Lazarev A., Robinson M., and Patton W. 2000. A comparison of silver stain and SYPRO ruby protein gel stain with respect to protein detection in two-dimensional gels and identification by peptide mass profiling. *Electrophoresis* **21:** 3673–3683.

Matsudaira P. 1987. Sequence from picomole quantities of proteins electroblotted onto polyvinylidene difluoride membranes. *J. Biol. Chem.* **262:** 10035–10038.

Merrill C.R. and Harrington M.G. 1989. Silver stain protein detection: Mechanisms and applications. In *Two-dimensional electrophoresis: Proceedings of the International Two-dimensional Electrophoresis Conference, Vienna, November 1988* (ed. A.T Endler and S. Hanash), pp. 243– 255. VCH, Weinheim, Germany.

Moeremans M., Daneels G., and De Mey J. 1985. Sensitive colloidal metal (gold or silver) staining of protein blots on nitrocellulose membranes. *Anal. Biochem.* **145:** 315–321.

Molloy M.P., Herbert B.R., Walsh B.J., Tyler M.I., Traini M., Sanchez J.-C., Hochstrasser D.F., Williams K.L., and Gooley A.A. 1998. Extraction of membrane proteins by differential solubilization for separation using two-dimensional gel electrophoresis. *Electrophoresis* **19:** 837–844.

Musante L., Candiano G., and Ghiggeri G.M. 1997. Resolution of fibronectin and other uncharacterized proteins by two-dimensional polyacrylamide electrophoresis with thiourea. *J. Chromatogr.* **705:** 351–356.

Neuhoff V., Arold N., Taube N., and Ehrhardt W. 1988. Improved staining of proteins in polyacrylamide gels including isoelectric focusing gels with clear background at nanogram sensitivity using Coomassie Brilliant Blue G-250 and R-250. *Electrophoresis* **9:** 255–262.

O'Farrell P.H. 1975. High resolution two-dimensional electrophoresis of proteins. *J. Biol. Chem.* **250:** 4007–4021.

Patton W. 1995. Biologist's perspective on analytical imaging systems as applied to protein gel electrophoresis. *J. Chromatogr. A* **698:** 55–87.

——. 2000. A thousand points of light: The application of fluorescence detection technologies to two-dimensional gel electrophoresis and proteomics. *Electrophoresis* **21:** 1123–1144.

Patton W.F., Lim M.J., and Shepro D. 1999. Image acquisition in 2-D electrophoresis. *Methods Mol. Biol.* **112:** 353–362.

Perdew G.H., Schaup H.W., and Selivonchick D.P. 1983. The use of a zwitterionic detergent in two-dimensional gel electrophoresis of trout liver microsomes. *Anal. Biochem.* **135:** 453–455.

Rabilloud T., Valette C., and Lawrence J.J. 1994a. Sample application by in-gel rehydration improves the resolution of two-dimensional electrophoresis with immobilized pH gradients in the first dimension. *Electrophoresis* **15:** 1552–1558.

Rabilloud T., Adessi C., Giraudel A., and Lunardi J. 1997. Improvement of the solubilization of proteins in two-dimensional electrophoresis with immobilized pH gradients. *Electrophoresis* **18:** 307–316.

Rabilloud T., Gianazza E., Cattò N., and Righetti P.G. 1990. Amidosulfobetaines, a family of detergents with improved solubilization properties: Application for isoelectric focusing under denaturing conditions. *Anal. Biochem.* **185:** 94–102.

Rabilloud T., Vuillard L., Gilly C., and Lawrence J.J. 1994b. Silver staining of proteins in polyacrylamide gels: A general overview. *Cell. Mol. Biol.* **40:** 57–75.

Raman B., Cheung A., and Marten M.R. 2002. Quantitative comparison and evaluation of two commercially available, two-dimensional electrophoresis image analysis software packages, Z3 and Melanie. *Electrophoresis* **23:** 2194–2202.

Ramsby M.L., Makowski G.S., and Khairallah E.A. 1994. Differential detergent fractionation of isolated hepatocytes: Biochemical, immunochemical and two-dimensional gel electrophoresis characterization of cytoskeletal and noncytoskeletal compartments. *Electrophoresis* **15:** 265–277.

Salinovich O. and Montelano R.C. 1986. Reversible staining and peptide mapping of proteins transferred to nitrocellulose after separation by sodium dodecyl sulfate-polyacrylamide gel electrophoresis. *Anal. Biochem.* **156:** 341–351.

Sanchez J.C., Rouge V., Pisteur M., Ravier F., Tonella L., Moosmayer M., Wilkins M.R., and Hochstrasser D.F. 1997. Improved and simplified in-gel sample application using reswelling of dry immobilized pH gradients. *Electrophoresis* **18:** 324–327.

Santoni V., Rabilloud T., Doumas P., Rouquié D., Mansion M., Kieffer S., Garin J., and Rossignol M. 2000. Towards the recovery of hydrophobic proteins on two-dimensional electrophoresis gels. *Electrophoresis* **20:** 705–711.

Schägger H. and von Jagow G. 1987 Tricine-sodium dodecyl sulfate-polyacrylamide gel electrophoresis for the separation of proteins in the range from 1 to 100 kDa. *Anal. Biochem.* **166:** 368–379.

Scheler C., Lamer S., Pan Z., Li X., Salnikow J., and Jungblut P. 1998. Peptide mass fingerprint sequence coverage from differently stained proteins on two-dimensional electrophoresis patterns by matrix assisted laser desorption/ionization-mass spectrometry (MALDI-MS). *Electrophoresis* **19:** 918–927.

Shevchenko A., Wilm M., Vorm O., and Mann M. 1996. Mass spectrometric sequencing of protein silver-stained polyacrylamide gels. *Anal. Chem.* **68:** 850–858.

Sinha P., Poland J., Schnölzer M., and Rabilloud T. 2001. A new silver staining apparatus and procedure for matrix/ionization-time of flight analysis of proteins after two-dimensional electrophoresis. *Proteomics* **1:** 835-840.

Sutherland J. 1993. Electronic imaging of electrophoretic gels and blots. *Adv. Electrophor.* **6:** 1–42.

Switzer R.C., Merril C.R., and Shifrin S. 1979. A highly sensitive silver stain for detecting proteins and peptides in polyacrylamide gels. *Anal. Biochem.* **98:** 321–327.

Taylor R.S., Wu C.C., Hays L.G., Eng J.K., Yates J.R.I., and Howell K.E. 2000. Proteomics of rat liver Golgi complex: Minor proteins are identified through sequential fractionation. *Electrophoresis* **21:** 3441–3459.

Towbin H., Staehelin T., and Gordon J. 1979. Electrophoretic transfer of proteins from polyacrylamide gels to nitrocellulose sheets: Procedures and some applications. *Proc. Natl. Acad. Sci.* **76:** 4350–4354.

Tsugita A., Kamo M., Kawakami T., and Ohki Y. 1996. Two-dimensional electrophoresis of plant proteins and standardization of gel patterns. *Electrophoresis* **17:** 855–865.

Usuda H. and Shimogawara K. 1995. Phosphate deficiency in maize. 6. Changes in the two-dimensional electrophoretic patterns of soluble proteins from second leaf blades associated with induced senescence. *Plant Cell Physiol.* **36:** 1149–1155.

Weber K. and Osborne M. 1969. The reliability of molecular weight determination by dodecyl sulfate-polyacrylamide gel electrophoresis. *J. Biol. Chem.* **244:** 4406–4412.

Yan J.X., Harry R.A., Spibey C., and Dunn M.J. 2000a. Postelectrophoresis staining of proteins by two-dimensional gel electrophoresis using SYPRO dyes. *Electrophoresis* **21:** 3657–3665.

Yan J.X., Walt R., Berkelman T., Harry R.A., Westbrook J.A., Wheeler C.H., and Dunn M.J. 2000b. A modified silver staining protocol for visualization of proteins compatible with matix-assisted laser desorption/ionization and electrospray ionization-mass spectrometry. *Electrophoresis* **21:** 3666–3672.

CHAPTER 5

Reversed-phase High-performance Liquid Chromatography

REVERSED-PHASE CHROMATOGRAPHY, WHICH SEPARATES MOLECULES based on their reversible interaction with the hydrophobic surface of a chromatographic medium, is valuable for the separation of proteins and peptides. It has rapidly become the most widely used branch of high-performance liquid chromatography (HPLC) since the mid-1970s, and it has found wide application for the separation, purification, and analysis of small molecules such as peptides. As a result, a great deal of experience has accumulated concerning the physicochemical phenomena underpinning the mechanism of small-molecule separations. Reversed-phase (RP)-HPLC is the method of choice for the final polishing of peptides and is ideal for analytical separations, such as peptide mapping (see Chapter 7).

Since the mid 1980s, RP-HPLC has been used increasingly for the separation of proteins on both analytical and preparative scales. However, RP-HPLC is not recommended for protein purification if recovery of activity and return to a correct tertiary structure are required, because many proteins are irreversibly denatured in the presence of organic solvents. Chromatography of proteins on reversed-phase supports also suffers from problems associated with poor recoveries, broad misshapen peaks, and "ghosting" (i.e., reappearance of a protein peak on subsequent chromatographic runs). Although many of these problems can be overcome by judicious choice of column usage and careful optimization of extra-column variables such as sample pretreatment and mobile-phase and hardware considerations (Hearn 1991b, 1998), the application of this technique to the purification of proteins still requires careful evaluation.

The following issues must be considered before employing RP-HPLC. The biological activities of many proteins are sensitive to extremes of pH, exposure to organic solvents, or high salt concentrations, which lead to unfolding and denaturation. Physical losses of protein, especially at low concentrations, can also occur by adsorption onto glass and hydrophobic matrices. However, for smaller proteins (M_r <30,000), some denaturation effects are minimal (or rapidly reversible), and RP-HPLC can therefore be successfully used to isolate them in a biologically active form (Rivier and McClintock 1989; Simpson and Nice 1989). For a discussion on the biological activity of polypeptides after RP-HPLC and attempts to increase the recovered bioactivity, see Welinder (1988). RP-HPLC is an extremely useful tool for the micropreparative purification of proteins and peptides prior to microsequencing using the Edman degradation procedure (Simpson et al. 1989a; Simpson and Nice 1989) (Chapter 6), mass-spectrometry-based techniques (Chapter 8), and various peptide-mapping procedures (Chapter 7). In this chapter, the basic separation principles of RP-HPLC are discussed, along with protocols for

- the packing of RP-HPLC columns,
- separation of large troublesome polypeptides,
- purification of synthetic peptides,
- desalting of protein samples, and
- computer-assisted method development to optimize gradient conditions.

REVERSED-PHASE CHROMATOGRAPHY SEPARATES MOLECULES BASED ON HYDROPHOBIC INTERACTIONS

RP-HPLC separation of proteins is based predominantly on reversible hydrophobic interactions between amino acid side chains on the protein with the hydrophobic surface of the chromatographic medium, i.e., the stationary phase (Figure 5.1). The nature of the hydrophobic binding interaction is generally considered to be the result of favorable entropy effects. The initial starting conditions of RP-HPLC are primarily aqueous in which a high degree of organized water structure surrounds both the protein molecules and the immobilized ligand (stationary phase). As protein binds to the stationary phase, the amount of hydrophobic area on the surface of the protein(s) exposed to the solvent is minimized. Thus, the degree of organized water is decreased with a concomitant favorable increase in entropy of the system. For this reason, it is advantageous, under these solvent conditions, for proteins to associate with the stationary phase upon loading onto the RP-HPLC column. Mobile-phase composition is then subsequently modified so that bound proteins are differentially eluted (desorbed) back into the mobile phase. The order of protein desorption is based on their relative hydrophobicity (i.e., least hydrophobic proteins elute first, followed by proteins

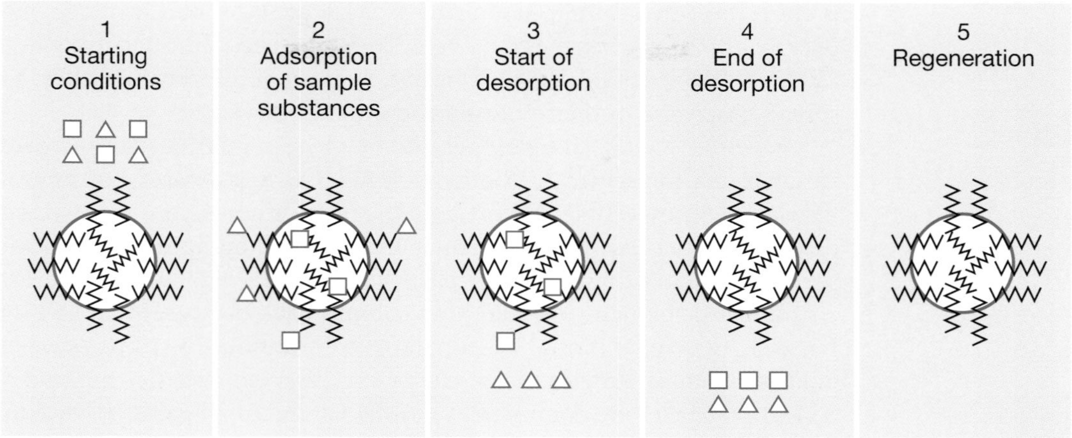

FIGURE 5.1. Principle of RP-HPLC with gradient elution. (Redrawn, with permission, from publ. No. 18-1134-16, Amersham Biosciences 2002.)

in increasing order of their surface hydrophobicity). Because the functional groups on the column support are very hydrophobic in nature, protein binding is usually very strong and requires the use of organic solvents and other additives (ion-pairing agents) in the mobile phase for elution. Column development (elution) is usually performed by increasing the organic solvent concentration (either in a stepwise fashion or gradient manner), typically acetonitrile. By these means, proteins, which are concentrated (or trace enriched; see below) during the binding and separation process, are collected in a purified and concentrated form. For a detailed discussion of chromatographic theory, see the Further Reading section at the end of this chapter.

STANDARD CHROMATOGRAPHIC CONDITIONS FOR RP-HPLC

Before embarking on RP-HPLC of proteins, it is useful to consider the optimal chromatographic support, mobile phase (especially the organic modifier [solvent], mobile-phase pH, and ion-pairing reagent), and column dimensions. These parameters are discussed in turn below. A procedure outlining standard chromatographic conditions for RP-HPLC is given in Protocol 1. A generic column-packing method (slurry packing procedure) for columns of varying internal diameters is given in Protocol 2. Procedures for chromatography of large troublesome polypeptides and also synthetic peptides are given in Protocols 3 and 4, respectively.

Choice of Chromatographic Support

Silica-based reversed-phase materials are the most widely used packings for HPLC of peptides and proteins because they are mechanically strong and provide high separation efficiencies. Of these, large-pore-size, silica-based matrices (>300 Å) or "macroporous" packings with a 5–10-μm range of particle sizes, containing C_4-, C_8 (octyl)-, or C_{18} (octadecyl)-*n*-alkyl chains, provide a good general utility for analytical RP-HPLC. Shorter-chain bonded phases (e.g., C_4 and C_8) are preferred for more hydrophobic samples, and longer-chain phases (e.g., C_{18}) are preferred for hydrophilic samples. For large peptides and proteins of ~10–100K M_r, 300-Å packings seem to offer the best overall performance. Although even larger-pore size packings (1000–4000 Å) have been investigated for very large proteins, they have not gained widespread usage due to their poor mechanical stability.

If working with complex peptide or protein mixtures, and provided all other column variables are the same, it is better to use a 300-Å pore-size packing material to ensure that restricted diffusion or exclusion from the pores is not encountered for very large polypeptides or proteins. Because theory predicts that chromatographic performance should increase as particle size goes down, smaller particles are preferred. However, several limitations may be encountered (namely, cost and high column back pressure) that will influence this decision. For analytical separations (<1 mg), most columns have 5–10-µm particles, and marginal differences in chromatographic efficiency are observed over this range. Although columns with 2–3-µm particles are commercially available, the small gains in performance are often offset by higher column back pressures, greater susceptibility to column blockage (resulting in shorter column life), and cost considerations. For preparative applications involving 10–500 mg of sample, columns containing particle sizes of 10–20 µm offer a reasonable compromise between cost and performance, whereas for large-scale separations (>1 g), columns containing 20–40-µm particles are often the best choice. For a list of some commercially available macroporous columns, suitable for protein separations, see Table 5.1 (see also Table 5.2 for a list of useful URL addresses for seeking HPLC information such as columns, loose packing materials, troubleshooting, and column fittings). For a selection of examples of polypeptides purified by RP-HPLC, along with the column and mobile-phase conditions employed, see Table 5.3.

With the availability of moderately priced column packing equipment (e.g., Shandon pumps) and loose macroporous packings (e.g., Vydac C4; The Separations Group, Hespario, California), it is possible to pack columns that exhibit excellent separation of proteins and peptides (Nice et al. 1979). Indeed, simple separations may be obtained even with relatively inferior column packing technique. For a description of a generic column-packing method (slurry packing procedure) for columns of varying internal diameters, see Protocol 2.

Choice of Mobile Phase

In RP-HPLC, the mobile phase is generally composed of three components: (1) an organic solvent, (2) an aqueous "buffer" component, and (3) an ion-pairing agent to optimize selectivity. Utmost care must be taken in choosing only the highest grade of reagents for RP-HPLC, because any contaminants in the mobile phase will be concentrated (trace-enriched) on the column and will affect the subsequent chromatography by producing unwanted (spurious) extra peaks or ghost peaks, and thus may contaminate the target protein. All organic solvents must be of the highest purity (HPLC-grade), the buffering salts and ion-pairing agents must be of the highest chemical purity (free of metal ions), and highly polished (deionized) water is a necessary requirement for the preparation of mobile phases. It is important that a "blank run" (i.e., no sample injected) be performed *before* running any samples to monitor the baseline for evidence of any contaminants in the mobile phase (see Protocol 1).

Organic solvent

With conventional RP-HPLC, an organic solvent (modifier) is added to the aqueous mobile phase to lower its polarity, thereby causing proteins to elute from the column. The lower the polarity of the mobile phase, the greater its eluting strength in reversed-phase chromatogra-

TABLE 5.1. Representative macroporous reversed-phase packing materials for HPLC

Name	Supplier	Support	Function	Particle size (μm)	Pore diam. (Å)	Comments
RP-300	Brownlee	silica	C8	7	300	
BU-300	Brownlee	silica	C4	7	300	
Phenyl	Brownlee	silica	phenyl	5	300	
Ultrapore-C3	Beckman	silica	C3	5	300	
Ultrapore-C8	Beckman	silica	C8	5	300	
Bakerbond WP_Octadecyl	J.T. Baker	silica	C18	5	300	
Bakerbond WP_Octyl	J.T. Baker	silica	C8	5	300	
Bakerbond WP_Butyl	J.T. Baker	silica	C4	5	300	
Zorbax StableBond 300SB-C18	Agilent	silica	C18	3.5,5,7	300	
Zorbax StableBond 300SB-C8	Agilent	silica	C8	3.5,5,7	300	
Zorbax StableBond 300SB-C3	Agilent	silica	C3	3.5,5,7	300	
Partisil ODS, ODS2, ODS3, C8	Whatman	silica	C18, C8	5,10	350	
Vydac 201TP, 218TP, 238TP	Separations Group	silica	C18	3,5,10	300	Available as loose packing
Vydac 218MS	Separations Group	silica	C18	5,10	300	
Vydac 208TP	Separations Group	silica	C8	5,10	300	Available as loose packing
Vydac 214TP	Separations Group	silica	C4	3,5,10	300	Available as loose packing
Vydac 214MS	Separations Group	silica	C4	5,10	300	
BioBasic C18	Thermo Hypersil-Keystone	silica	C18	5	300	High-purity silica base deactivated
BioBasic C8	Thermo Hypersil-Keystone	silica	C8	5	300	High-purity silica base deactivated
BioBasic C4	Thermo Hypersil-Keystone	silica	C4	5	300	High-purity silica base deactivated
HyperREZ XP RP-300	Thermo Hypersil-Keystone	polystyrene/divinyl-benzene	–	5	300	Rigid bead stable at pH 0–14
Hamilton PRP-3	Hamilton	polystyrene/divinyl-benzene	–	3,5,10	300	
Discovery BIO Wide Pore C18 I	Supelco	silica	C18	3,5,10	300	
Discovery BIO Wide Pore C8 I	Supelco	silica	C8	5,10	300	
Discovery BIO Wide Pore C5 I	Supelco	silica	C5	5,10	300	
PLRP-S	Polymer Labs	polystyrene/divinyl-benzene		5,8	300	
Nucleosil C18, C8, C4	Macherey Nagel	silica	short chain RP	5,7,10	300	
LiChrospher WP 330 RP-18	Merck	silica	C18	5,12,15	300	
HAI*SIL* 300 C4, C8, C18	Higgins Analytical	silica	C4,C8,C18	5,10	300	
TSK Octadecyl-NPR	TSK	silica	C18	2.5	non-porous	
TSK Phenyl-5PW-RP	TSK		phenyl-bonded	10		hydrophilic polymer

Kindly compiled by Robert L. Moritz (Joint ProteomicS Laboratory of the Ludwig Institute for Cancer Research and the Walter and Eliza Hall Institute of Medical Research, Melbourne, Australia).

TABLE 5.2. Useful URL addresses for HPLC information

Chromatographic column suppliers	
ZirChrom Separations, Inc.	http://www.zirchrom.com/
Phenomenex	http://www.phenomenex.com
Waters Corporation	http://www.waters.com
Agilent Technologies	http://we.home.agilent.com
TOSOH-HAAS	http://www.tosohbiosep.com/
Amersham Biosciences	http://www.apbiotech.com
Merck	http://www.merck.de
Macherey-Nagel	http://www.macherey-nagel.com
Thermo-Hypersil	http://www.thermohypersil.com
LC Packings	http://www.lcpackings.com
Purveyors of chromatographic parts and associated equipment	
Western Analytical Products, Inc., (Murrieta, California)	http://www.hplcsupply.com [Old site http://www.netwizards.net/~wap/]
The Nest group	http://www.nestgrp.com/
Ionsource	http://www.ionsource.com/links/ms_links.htm (click on Vendors [LC])
Chromatographic part manufacturers and suppliers	
Upchurch Scientific	http://www.upchurch.com
Alltech	http://www.alltechweb.com

phy. Because there is strong dependence on the relative retention (or capacity factor, k^1; see panel below) of individual proteins and the concentration of organic modifier required for their elution (Figure 5.2), the separation of a mixture of proteins of slightly different relative hydrophobicities requires gradient elution.

The capacity factor (k^1), also referred to as the retention factor, is normalized retention under isocratic elution conditions. It is a unitless term that is a measure of the retention behavior (i.e., the degree of retention) for a particular solute on a particular column, assuming equal flow rates. The capacity factor k can be calculated by the following equation: $k^1 = (t_r\ t_o)/t_o = (v_r v_o)/v_o$, where k^1 is the number of column volumes required to elute a particular solute, and t_o and v_o represent the void time and void volume, respectively. Thus, k^1 is directly related to the distribution coefficient (or partition coefficient) of a solute between the mobile and stationary phases (i.e., moles of solute in stationary phase per moles of solute in mobile phase) and is now well understood in both empirical and thermodynamic terms.

The most common organic solvents used in RP-HPLC, and their order of elutropic strength, are 1-propanol > 2-propanol > tetrahydrofuran ~ dioxane > ethanol > acetonitrile > methanol (Mahoney and Hermodson 1980) (see also Table 5.4). In practice, the most widely used organic modifiers are acetonitrile and methanol, the most popular choice being acetonitrile. Although isopropanol (2-propanol) is often used because of its strong eluting properties, it is limited because of its high viscosity (2.30 cPoise), which results in lower column efficiencies and high back pressures. However, the lower polarity of isopropanol makes it an excellent solvent for cleaning the reversed-phase column. Both acetonitrile (0.37 cPoise) and methanol (0.60 cPoise) are less viscous than isopropanol. Acetonitrile, methanol, and isopropanol are essentially UV-transparent, which is an essential requirement for RP-HPLC because column elution is typically monitored by UV absorbance (Table 5.5). UV transparency is particularly important because most RP-HPLC separations (especially peptides)

TABLE 5.3. Selected examples of polypeptides and proteins purified by RP-HPLC

Column	Hydrophobic chain	Mobile phase	Protein purified	Reference
μBondapak C18	*n*-octadecyl	0.1% TFA, 12–70% CH_3CN	acid phosphatase	Witting et al. (1984)
Nucleosil C18	*n*-octadecyl	0.1% TFA, ethanol/butanol/methoxy-ethanol	aldolase	van der Zee and Welling (1982)
μBondapak C18	*n*-octadecyl	5% formic acid, 40–80% ethanol	bacteriorhodopsin	Gerber et al. (1979)
Nucleosil C18	*n*-octadecyl	50 mM KH_2PO_4, 10–50% 2-methoxy-ethanol	bacteriorhodopsin	Mönch and Dehnen (1978)
LiChrospher C8	*n*-octyl	400 mM Pyr-formate, 0–40% nPrOH	bovine serum albumin	Lewis et al. (1980)
μBondapak alkylphenyl	*n*-propyl-phenyl	1% TEAP, 10–50% CH_3CN	C-apolipoproteins	Hancock et al. (1981)
Ultrasphere C8	*n*-octyl	10 mM TFA, 0–45% nPrOH	carbonic anhydrase	Cooke et al. (1983)
Ultrasphere C3	*n*-propyl	155 mM NaCl (pH 2.1), 0–75% CH_3CN	carbonic anhydrase	Cooke et al. (1983)
μBondapak C18	*n*-octadecyl	0.1% TFA, 0–60% CH_3CN	chorionic gonadotropin	Putterman et al. (1982)
LiChrospher C4	*n*-butyl	10 mM H_3PO_4, 0–45% nPrOH	chymotrypsinogen	Cohen et al. (1984)
Bakerbond diphenyl	diphenyl	0.1% TFA, 0–50% CH_3CN	collagen chains	Skinner et al. (1984)
U-ODS	*n*-octadecyl	0.2% HFBA, 0–50% CH_3CN	epidermal growth factor	Burgess et al. (1982)
Nucleosil C18	*n*-octadecyl	50 mM KH_2PO_4, 10–50% 2-methoxy-ethanol	ferritin	Mönch and Dehnen (1978)
Pharmacia ProRPC	*n*-octyl	0.3% TFA, 39–50% CH_3CN	globin chains	Jeppsson et al. (1984)
LiChrospher C4	*n*-butyl	100 mM NH_4HCO_3, 0–50% CH_3CN	growth hormone	Grego et al. (1984)
μBondapak C18	*n*-octadecyl	0.2% TFA, 0–50% CH_3CN	histone proteins	Gurley et al. (1984)
LiChrospher C8	*n*-octyl	0.8 M Pyr, 1 M formic acid, 0–60% CH_3CN	human fibroblast inter-feron	Stein et al. (1980)
Ultrasphere C3	*n*-propyl	0.1% TFA, 0–50% CH_3CN	inhibin, follicular	Grego and Hearn (1984)
LiChroprep RP8	*n*-octyl	0.9 M Pyr-acetate, 20–60% nPrOH	interleukin-2	Wolfe et al. (1984)
W-DP	*n*-octadecyl	50 mM KH_2PO_4 2-methoxyethanol-PrOH	leukocyte interferon	Herring and Enns (1983)
N-ODS	*n*-octadecyl	33 mM NaOAc, 10–50% ethanol (pH 5.2)	α_2-macroglobulin	Sottrup-Jesen et al. (1984)
Unspecified C8	*n*-octyl	0.1% TFA, 0–60% iPrOH	ovalbumin	Kopaciewicz and Regnier (1983)
LiChrospher C4	*n*-butyl	10 mM H_3PO_4, 0–45% nPrOH	papain	Cohen et al. (1984)
LiChrospher C18	*n*-octadecyl	50 mM Tris-HCl, 0.1 mM $CaCl_2$, 0–70%	parvalbumin	Berchtold et al. (1982)
Ultrasphere C3	*n*-propyl	155 mM NaCl-HCl, 0–50% nPrOH	phosphorylase	O'Hare et al. (1982)
LiChrospher C8	*n*-octyl	0.1% TFA, 0–50% iPrOH	platelet-derived growth factor	Chesterman et al. (1983)
Various	various	0.1% TFA, CH_3CN-PrOH	rhodopsin	Berchtold et al. (1982)
Ultrapore RPSC	*n*-octadecyl	0.1% TFA, 10–80% CH_3CN	ribosomal 50S proteins	Nick et al. (1985)
Synchropak RPP	*n*-octadecyl	0.1% TFA, 15–75% CH_3CN	ribosomal proteins	Nick et al. (1985)
ToyaSoda	trimethyl	0.2% TFA, 25–75% CH_3CN	Sendai viral proteins	van der Zee and Welling (1982)
Bakerbond C18	*n*-octadecyl	10 mM NH_4HCO_3, 0–50% CH_3CN	thyrotropin and subunits	Stanton and Hearn (1987)

Reprinted from Hearn (1998).

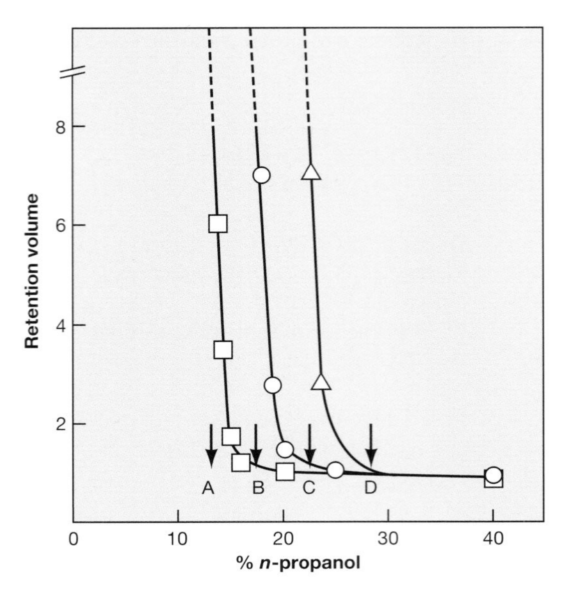

FIGURE 5.2. Effect of varying organic solvent concentrations on protein retention by a reversed-phase support (Ultrapore RPCS C3) (75 x 4.6-mm inner diameter). (*Open box*) Ribonuclease; (*open circle*) cytochrome *c*; (*open triangle*) bovine serum albumin. Protein standards (100 μg) of varying hydrophobicities (ribonuclease, cytochrome *c*, and bovine serum albumin) were applied to the column. Data were obtained under isocratic conditions (0.1% aqueous TFA/*n*-propanol). A decreasing retention volume (*k*) for each protein standard is observed as the concentration of organic modifier is increased. At *n*-propanol concentration: A (12%), all proteins are retained; B (18%), ribonuclease elutes, others retained; C (22%), cytochrome *c* elutes, bovine serum albumin is still retained; D (28%), all proteins elute. Flow rate is 1 ml/min, and temperature is 45°C. (Reproduced, with permission, from Simp-son and Nice 1989.)

are monitored below 220 nm for optimal detection sensitivity of peptides (e.g., proteins and peptides lacking the aromatic amino acids tryptophan, tyrosine, and phenylalanine can only be detected by UV transparency using wavelengths below 225 nm). Because acetonitrile has a lower UV cutoff (190 nm) than other common solvents, it is used almost exclusively when separating peptides.

Minor selective effects can sometimes be obtained with different organic solvents, and in some cases, a combination of different solvents has been used to great effect (Mahoney and Hermodson 1980). Additional optimization of protein and peptide separations can be accomplished by careful manipulation of the flow rate and gradient steepness (Stadalius et al. 1984; Glajch et al. 1986).

Mobile-phase pH

Conventional silica-based reversed-phase packings suffer from two disadvantages: First, they have a limited useable pH range, typically pH 2–8. Below pH 2, the bonded phase is susceptible to hydrolysis. Above pH 8, hydroxide ions can attack and dissolve the silica, which causes the collapse of the packed bed (resulting in decreased column lifetime) and a catastrophic loss of efficiency. Second, basic analytes interact strongly with residual silanols and cause tailing peaks that are detrimental to resolution as well as to the accuracy and precision of quantitation.

Reversed-phase separations are thus most often performed at low pH values, generally between pH 2 and 4. Further advantages of using a low pH include good sample solubility,

TABLE 5.4. Effect of organic modifiers on solute retention by short alkyl chain (C3) packings

	NPA	IPA	THF	CH_3CN	Diox	MeEtOH	EG
Trp	3.8	3.4	4.0	4.3	3.9	3.6	3.6
$ACTH_{1-24}$	13.4	14.5	15.9	20.5	18.1	25.2	52
Ribonuclease	17.0	21.3	22.3	30.0	34.0	n.d.	n.d.
Lysozyme	23.4	29.4	32.8	40.0	50.0	68.0	114.0
Prol (rat prolactin)	32.5	44.5	43.0	59.6	69.5	92.0	n.d.

Reprinted from Simpson and Nice (1989).

Data were obtained with a 75 x 4.6-mm I.D. column packed with Ultrapore RPSC using a linear 1%/min gradient between 0.15 M NaCl (pH 2.1) and various organic modifiers. (NPA) 1-Propanol (*n*-propanol); (IPA) 2-propanol (isopropanol); (THF) tetrahydrofuran; (CH_3CN) acetonitrile; (Diox) dioxane; (MeEtOH) methoxyethanol; (EG) ethyleneglycol. Values are retention times (minutes). n.d. indicates not determined.

ion suppression of both acidic side chains on the sample, and residual silanol groups on the silica support. Commonly used acids in RP-HPLC include trifluoroacetic acid (TFA), heptafluorobutryric acid, and *ortho*-phosphoric acid in the concentration range of 0.0–50.1% (w/v) or 50–100 mM. Unbuffered 1% (w/v) NaCl is a useful mobile phase for high-sensitivity RP-HPLC (Nice and O'Hare 1978) because NaCl is transparent to UV below 220 nm, it is biologically compatible, and it often provides useful alternative selectivities when compared to TFA. For high-sensitivity work, NaCl is considerably cleaner than phosphate in terms of background contaminant peaks. Mobile phases containing ammonium acetate (pH 6–7), ammonium bicarbonate (pH 7), or phosphate salts (e.g., triethylammonium phosphate, pH 6) are suitable for use at pH values closer to neutrality. Note that phosphate buffers and NaCl are not volatile and care must be taken to ensure that the concentration of organic modifier does not exceed 60%, otherwise these salts will precipitate and block the column/flow cell. Although volatile buffer components can be removed from the eluted sample by centrifugal lyophilization along with the organic component, this can be a risky step when dealing with low-microgram amounts of protein/peptide because severe losses are commonly encountered due to irreversible binding of the sample to the sides of glass/polypropylene tubes. Such losses can be avoided by the addition of a nonionic detergent (e.g., Tween-20, Pierce Surfactant-Amps 20 grade) to a final concentration 0.01–0.02% (w/v) prior to the evaporation step (Simpson et al. 1989a). If nonvolatile salts are used in the mobile phase, they must be separated from the recovered sample by an additional desalting step such as RP-HPLC using a volatile salt (e.g., TFA) (see below and Protocol 5) or micro-size exclusion chromatography.

TABLE 5.5. Solvents used in RP-HPLC

Solvent	Polarity index	Boiling point (°C)	UV cutoff (nm)	Viscosity (cPoise at 20°C)	Comments
Acetonitrile	5.8	82	190	0.36	more powerful denaturant than alcohols; toxic
Dioxane	4.8	101	215	1.54	
Ethanol	5.2	78	210	1.20	
Methanol	5.1	65	205	0.60	
1-propanol (*n*-propanol)	4.0	98	210	2.26	viscous
2-propanol (isopropanol)	3.9	82	210	2.30	viscous
Tetrahydrofuran	4.0	65	215	0.55	
Water	9.0	100	<190	1.00	

Why TFA?

TFA is the most commonly used ion-pairing agent in RP-HPLC because

- it is volatile and easily removed by lyophilization,
- it has low absorption within detection wavelengths for peptides and proteins, and
- it has a proven history.

TFA has an absorbancy in water different from that in acetonitrile, and thus the TFA concentration must be made in Buffer B at 85–90% the concentration in Buffer A to avoid a baseline shift during gradient formation. TFA buffer should be fresh, the solvent reservoir should be cleaned occasionally, and the inlet filter replaced periodically.

Because of the problems and limitations associated with conventional silica-based columns (e.g., degradation at high pH values), polymer-based columns have been gaining popularity. Although organic polymer packings are generally stable from pH 1 to pH 14, they yield significantly lower efficiencies than silica-based packings and are not as mechanically strong. Many organic polymer packings also shrink or swell when exposed to different solvents. More recently, a number of silica-based bonded phases (organic-inorganic hybrids) with a long column life have been developed (e.g., Zorbax Extend-C18, Phenomenex Synergi 4μ MAX-RP; for further details, see Tables 5.2 and 5.3) that are stable at high pH (up to pH 11.5, thereby allowing chromatography with ammonium hydroxide and pyrrolidone buffers). An alternative, non-silica-based stationary phase, which can be used in the pH range of 3–14, especially at high pH values, is the Zirconium-based packings manufactured by ZirChrom (see Table 5.2).

Ion-pairing agents

The third component usually added to the mobile phase is an ion-pairing agent (Horváth et al. 1977b; Hancock et al. 1978a; Mant and Hodges 1991a). The addition of millimolar concentrations of ion-pairing agents has the effect of decreasing the retention time of a protein sample and improving peak shapes. This effect can be attributed to the ion-pairing agent binding by ionic interaction to the protein/peptide molecule and changing selectivity. It is thought that ion-pairing agents suppress the interaction between polar groups on the sample and silanol groups on the reversed-phase matrix (Hancock and Sparrow 1983; Bennet 1991). At any given pH, proteins and peptides will exhibit ionized amino acid side chains that are available for ion pairing. For example, over the pH range of 2–8 (the pH range over which most silica-based columns are used), the basic amino acids (arginine and lysine) are fully ionized, and these positively charged side chains (along with the primary amino terminus) associate (or "ion-pair") with anions. Thus, hydrophobic anions such as TFA and alkylsulfonates form hydrophobic ion pairs with the basic amino acid chains and, in so doing, tend to increase the affinity of the protein or peptide for the column (thereby resulting in increased retention times) (for a detailed discussion of ion pairing, see Bennett 1991). The retention behavior of the sample components may be affected by both the type and concentration of the ion-pairing agent used (see Figure 5.3).

Useful hydrophobic ion-pairing agents include pentafluoropropionic acid (PFPA) and heptafluorobutyric acid (HFBA) (for a list of commonly used ion-pairing agents, see Table 5.6). HFBA has been used extensively as an ion-pairing agent under circumstances where the resolving power of TFA to separate a peptide mixture has been inadequate (Bennett 1983). Apart from its effectiveness as an ion-pairing agent, it shares with TFA the advantages of volatility and low UV transparency (see Table 5.5).

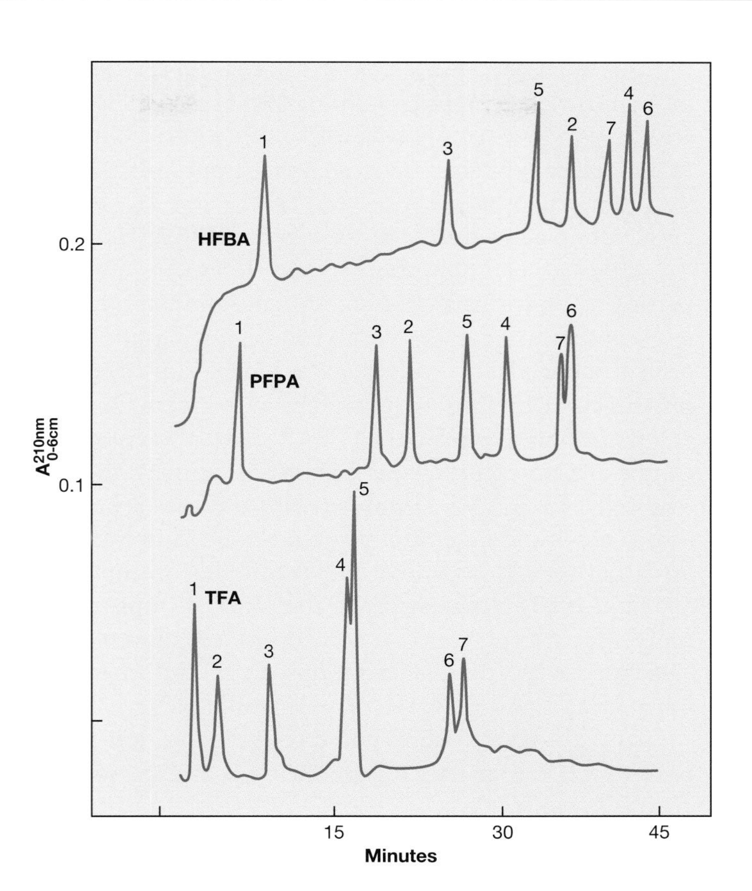

FIGURE 5.3. Comparison of the effectiveness of TFA, pentafluoropropionic acid (PFPA), and heptafluorobutyric acid (HFBA) as hydrophobic ion-pairing reagents in reversed-phase chromatography (RPC) of seven natural and synthetic peptides. Peptides were eluted from the Waters Associates μBondapak C18 column (300 x 3.9-mm I.D.) with a linear AB gradient of 20–40% acetonitrile containing 0.01 M concentrations of each acid during 1 hour (0.33% acetonitrile/min) at a flow rate of 1.5 ml/min. (*Top*) Elution behavior of the peptides with 0.01 M (0.13% v/v) HFBA. (*Middle*) Behavior with 0.01 M (0.10% v/v) PFPA. (*Bottom*) Behavior with 0.01 M (0.07% v/v) TFA. Samples (2 μg) of the following peptides were injected onto the column (number of basic residues at pH 2, including histidine, is shown in parentheses): 1, methionine enkephalin (1); 2, $ACTH_{1\text{-}24}$ (9); 3, α-MSH (3); 4, human $ACTH_{1\text{-}39}$; 5, somatostatin (3); 6, bovine insulin (6); 7, human calcitonin (3). (Reproduced, with permission, from Bennett 1991.)

TABLE 5.6. Ion-pairing agents

Ion-pairing agent	Formula of pairing ion	Comments
Anionic		
Trifluoroacetic acid (TFA)	CF_3COO^-	low UF-absorbance; volatile, low pH,
Pentafluoroproprionic acid (PFPA)	$CF_3CF_2COO^-$	more hydrophobic than TFA; volatile, low pH
Heptafluorobutyric acid (HFBA)	$CF_3CF_2CF_2COO^-$	more hydrophobic than TFA; volatile, low pH
Ammonium acetate	CH_3COO^-	
Phosphoric acid	H_2PO_{4-}, HPO^{4}_{2-}, PO_4^{3-}	less hydrophobic than TFA
Cationic		
Tetramethylammonium chloride	$+N(CH_3)_4$	
Tetrabutylammonium chloride	$+N(C_4H_9)_4$	
Triethylamine	$NH^+(C_2H_5)_3$	

Reprinted from Amersham Biosciences (2002).

It is apparent from Figure 5.3, which shows a comparison of a standard mixture of peptides in solvent systems containing 0.01 M TFA, PFPA, or HFBA, that retention times (also relative retention times or order of peptide elution) for the various peptides are different in each of the solvent systems. In progressing from HFBA to TFA, some peptides increase in retention more than others (e.g., ACTH1-24 and ACTH1-39); this correlates well with the high basic amino acid content of these peptides relative to the others.

Although orthophosphoric acid (H_3PO_4) is nonvolatile, it has proved to be a useful ion-pairing agent for hydrophobic proteins and peptides (Hancock et al. 1978a,b). Because proteins elute from reversed-phase columns at significantly decreased concentrations of organic solvent when H_3PO_4 is used as an ion-pairing agent in the mobile phase, the possibility of protein denaturation or precipitation is greatly diminished. Amine phosphates such as triethylammonium phosphate (TEAP) have also enjoyed great success as ionic modifiers. This can be attributed to a variety of effects, namely, the relatively high salt concentrations used (typically 0.1–0.2 M) combined with the use of both polar anions and cations (which can associate with ionic groups present on the sample); these act to prevent unfavorable interactions between the sample and the silanol groups on the column matrix (Hancock and Sparrow 1983). However, with the marked improvements in bonded-phase chemistry and more effective end-capping procedures for column packings, the chromatographic problems associated with free silanol groups on column matrices (and the usefulness of TEAP as an ion-pairing agent) have decreased significantly.

As with other mobile-phase constituents, ion-pairing agents must be sufficiently pure, have low UV transparency (<220 nm), and be soluble under the low-polarity conditions encountered.

MICROCOLUMN REVERSED-PHASE CHROMATOGRAPHY IS FAVORED FOR STUDIES OF PROTEOMES

Microcolumn reversed-phase chromatography is one of the most commonly used techniques for preparing samples for classical Edman degradation protein microsequencing and the more recent proteomic methods for protein identification and characterization. Depending on the internal diameter (I.D.) of the columns, microcolumn chromatography can be arbitrarily subdivided into microbore (<2.1 mm I.D.), narrowbore (<1.0 mm I.D.), or capillary (<0.5 mm I.D.) column chromatography.

Microbore column chromatography, first proposed in the late 1970s (Ishii et al. 1977; Scott and Kucera 1979a,b), was further developed in the 1980s for preparing samples for gas-phase sequence analysis (Nice et al. 1984, 1985; Simpson and Nice 1987; Simpson et al. 1989a). The success of microbore column chromatography is largely attributable to the following:

- ***Concentrated samples,*** i.e., recovery of samples in low peak volumes of ~20–60 μl compared to 500–1500 μl typically found for their large-diameter (4.6 mm I.D.) conventional column counterparts.
- ***Enhanced sensitivity of protein detection*** (5–10 ng) (Nice et al. 1984). This represents a 5-fold increase in sensitivity (20-fold for 1-mm I.D. columns) when compared with 4.6-mm I.D. columns.
- ***Rapid sample loading capability.*** The low back pressures encountered with short columns (<10 cm in length) permits high flow rates for column reequilibration and sample loading.

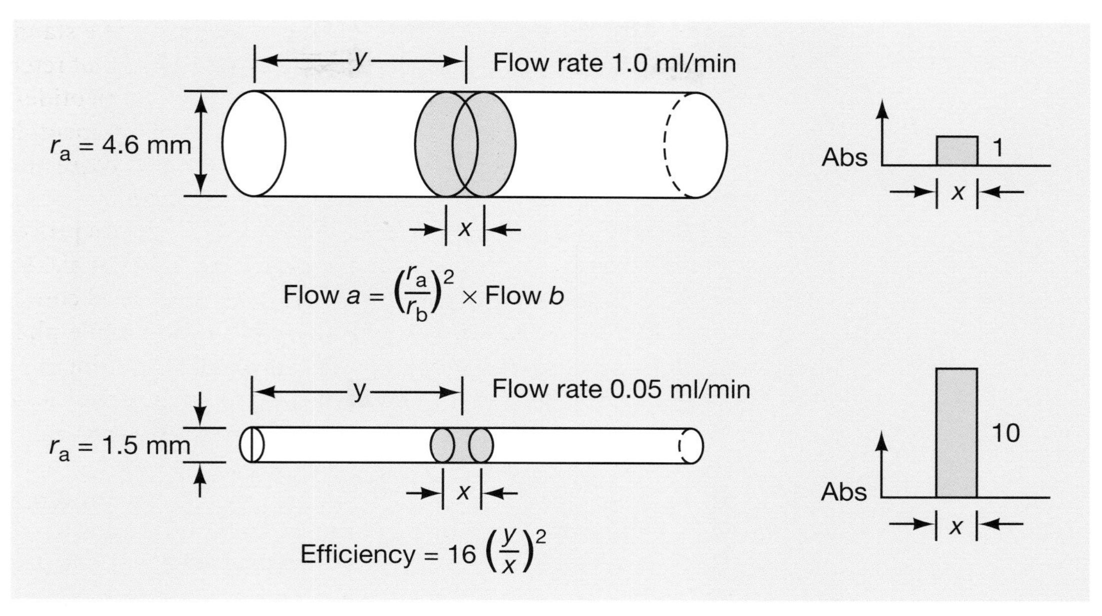

FIGURE 5.4. Principles of microbore column chromatography: The relationship between column diameter and eluent peak concentration. It has been demonstrated that the eluent peak volume from a chromatographic column is inversely proportional to the square of the column diameter and directly proportional to the column length (Scott and Kucera 1979b). In the example shown, if the column diameter is reduced from 4.6 mm to 1.5 mm, the flow rate must be reduced by $(4.6/1.5)^2$, i.e., 9.4-fold (1.0 ml/min to ~0.1 ml/min) to achieve the same linear flow velocity. For concentration-dependent detectors, this will result in an equivalent increase in detector signal, i.e., a 10-fold increase for the example given. (Reproduced, with permission, from Simpson and Nice 1987.)

Principles of Microbore Column Chromatography

To operate microbore columns at linear flow velocities equivalent to those used with wider-bore columns, the flow rate must be decreased in a manner proportional to the square of the reduction in column diameter (Figure 5.4). For example, if the optimal flow rate for a 4.6-mm I.D. column is 1 ml/min, then a 1.5-mm I.D. column would be operated at ~100 μl/min to achieve the same linear flow velocity. If the column efficiencies are maintained (i.e., peak bandwidths remain constant for a given load of protein), which should be true if a column is not overloaded, then proteins will elute from microbore columns in proportionally decreased peak volumes compared with conventional columns (i.e., smaller peak volumes at increased concentrations) (Scott and Kucera 1979a,b; Kucera 1980). The practical relationship between the internal diameter of an HPLC column, protein mass, and the sensitivity of detection is shown in Figure 5.5, which compares the chromatographic performance of α-lactalbumin on a series of columns of the same length (10 cm) and packed with the same support (Brownlee RP-300), but with varying internal diameters (1.0, 2.1, and 4.6 mm I.D.). When compared with a conventional 4.6-mm I.D. column, it is evident that a 20-fold increase in sensitivity of protein detection is achieved by using the 1-mm I.D. column, and a 4-fold increase achieved when using a 2.1-mm I.D. column. All columns were operated at equivalent linear flow velocities as demonstrated by the constant peak bandwidth as a function of time. Peak recovery volumes for the 1-, 2.1-, and 4.6-mm I.D. columns were 25, 100, and 450 μl, respectively.

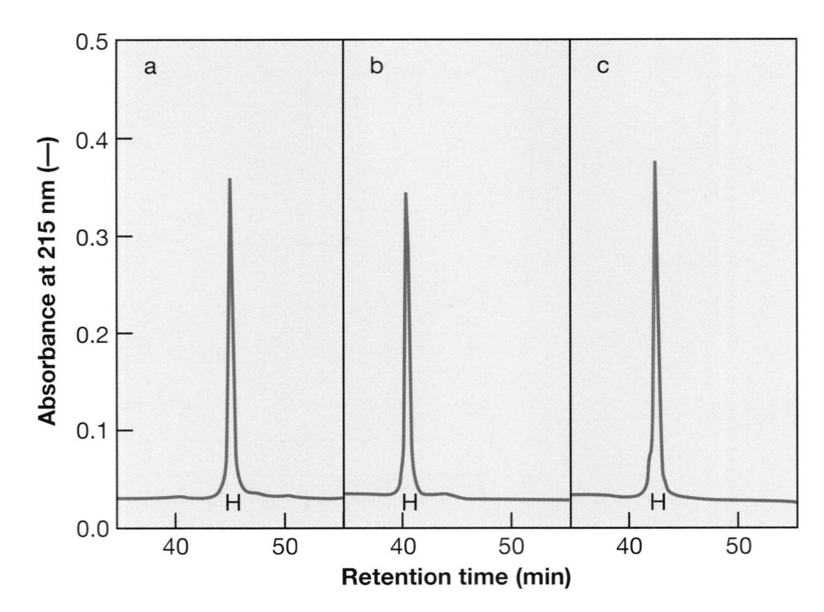

FIGURE 5.5. Effect of column internal diameter and flow rate on detector sensitivity. Chromatographic conditions: column, Brownlee RP-300 (30 nm) 10-μm dimethyloctyl silica; linear 60-minute gradient from 0.15% (v/v) aqueous TFA to 60% acetonitrile/40% H_2O containing 0.12% (v/v) TFA. The column temperature is 45°C. The sample is bovine α-lactalbumin. (*a*) Column, 100 x 1-mm I.D.; flow, 50 μl/min; load, 1 μg; peak volume, 25 μl. (*b*) Column, 100 x 2.1-mm I.D.; flow, 200 μl/min; load, 4 μg; peak volume, 100 μl. (*c*) Column, 100 x 4.6-mm I.D.; flow, 1 ml/min; load 20 μg; peak volume, 450 μl. (Reprinted, with permission, from Simpson and Nice 1987.)

Sample Capacity and Resolution of Microbore Columns

In general, the total amount of protein that can be loaded onto a column (loadability) is a function of the chromatographic support, the dimensions of the column, and the inherent nature of the protein itself. For example, the loadability of microbore column (e.g., 30 x 2.1-mm I.D. column packed with a 30-nm reversed-phase [C_8] support) can be as high as 4 mg using both lysozyme and cytochrome *c* as standard proteins (Nice et al. 1984). However, the recovery volumes under these overload conditions are typically in excess of 1 ml because the optimal loading has been greatly exceeded. The following is the optimal protein loading for columns of varying internal diameters, determined by peak volumes (see Table 5.7):

- 2.5–5 μg (1-mm I.D. column)
- 10–20 μg (2.1-mm I.D. column)
- 30–100 μg (4.6-mm I.D. column)

For 2.1-mm I.D. columns of varying lengths (3 cm and 10 cm), similar peak volumes are obtained under ideal operating conditions regardless of column length. However, band broadening (column overload) occurs more rapidly for the shorter column. To achieve optimal performance, 1.0-mm and 2.1-mm I.D. columns should be operated at flow rates of 50 and 100 μl/min, respectively. Under these conditions, sample recoveries in excess of 95% are routinely achieved. Figure 5.6 illustrates the ability of microbore columns to resolve a low-nanogram mixture of protein standards, the resolution being identical to that obtained on conventional columns (4.6 mm I.D.) packed with the same support. Although steeper gradient rates (e.g., 20% per minute) cause some loss of resolution, samples can be recovered from the column in volumes less than 20 μl.

TABLE 5.7. Effect of protein load on peak bandwidth for columns of varying internal diameters

	Peak volume (μl) for column dimensions			
Protein load (μg)	**1.0 x 10**	**2.1 x 3**	**2.1 x 10**	**4.6 x 10**
0.5	25	–	–	–
1.0	25	100	100	–
2.5	–	–	–	–
5.0	30	100	100	450
10.0	35	120	120	450
20	50	180	140	–
30	–	–	–	–
50	70	390	190	–
100	–	470	240	600
500	–	–	–	1200

Reprinted from Simpson and Nice (1989a).

Support: Brownlee RP-300 (column length 10 cm). Protein standard: bovine α-lactalbumin. Gradient elution (0.15% TFA to 60% CH_3CN/0.12% TFA over 60 minutes) was used at equivalent linear flow velocities: 50 μl/min, 200 μl/min, and 1.0 ml/min for the 1.0-mm, 2.1-mm, and 4.6-mm I.D. columns, respectively.

Trace Enrichment (Concentration) and Buffer Exchange of Samples Using RP-HPLC

During the phase of their development, microbore columns were restricted to sample volume injections of ~1 μl to avoid extra column band broadening. However, this historical practice has not limited the practical use of microbore columns for protein purification involving large sample volumes. This is due to proteins exhibiting very large capacity factors, k^1, that are below the critical secondary solvent composition on reversed-phase supports (see Figure 5.2). Hence, it is possible to trace-enrich large volumes containing nanogram amounts of protein directly onto the microbore column (Nice et al. 1979). Proteins can then be recovered

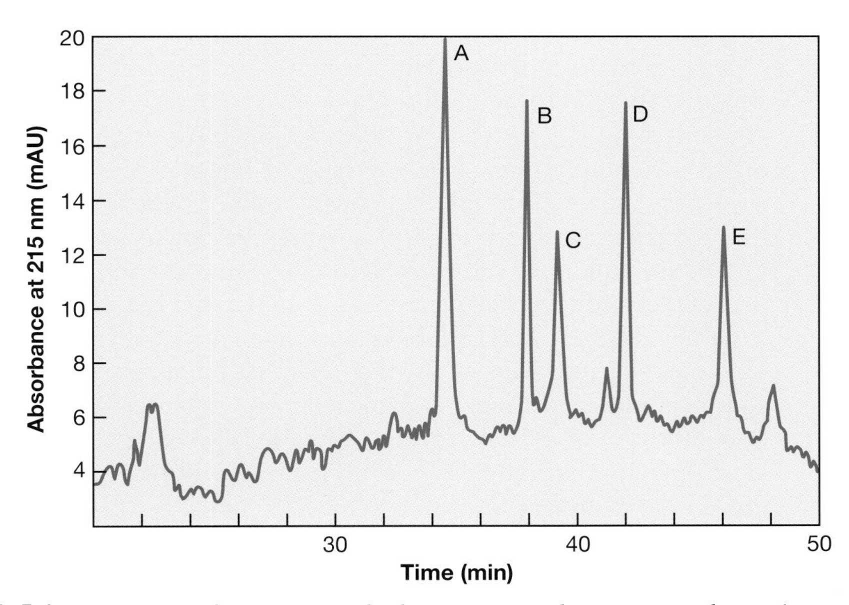

FIGURE 5.6. Separation of protein standards on a Brownlee RP-300 column (50 x 1.0-mm I.D.). Chromatographic conditions: linear 60-minute gradient from 0.15% aqueous TFA to 60% acetonitrile/40% H_2O containing 0.12% (v/v) TFA. The column temperature is 45°C, and the flow rate is 50 μl/min. Protein standards: (*A*) Murine EGF; (*B*) glucagons; (*C*) cytochrome *c*; (*D*) lysozyme; (*E*) bovine α-lactalbumin. The sample load is 25 ng per protein. (Reproduced, with permission, from Simpson and Nice 1987.)

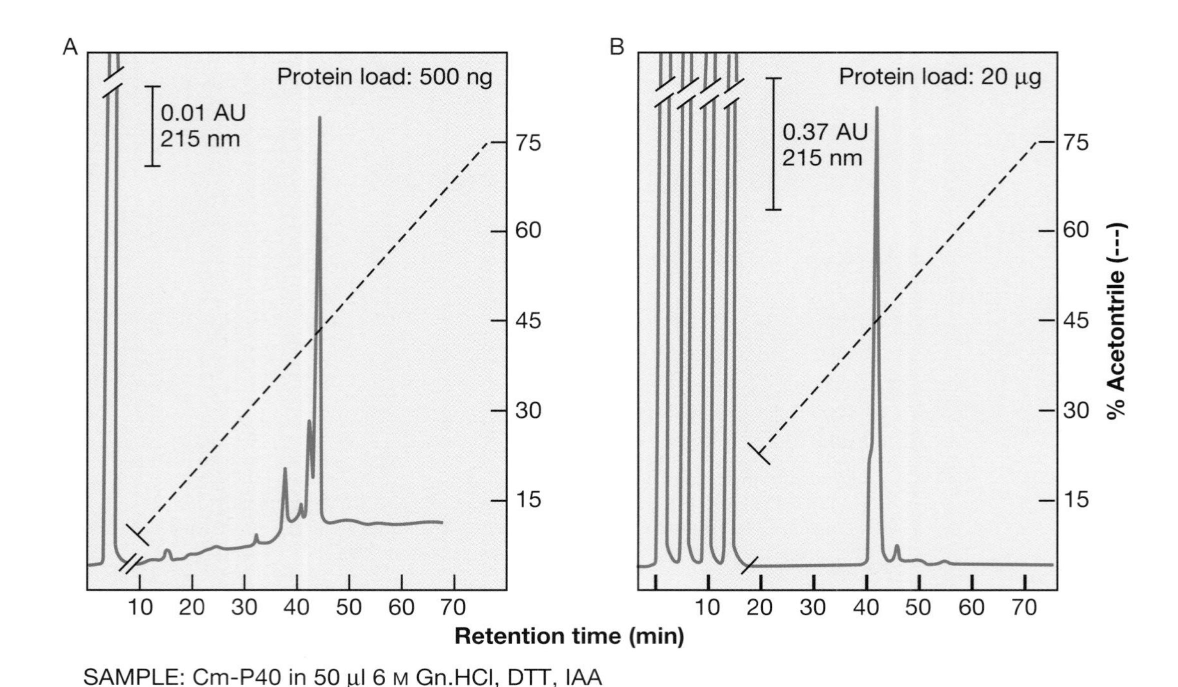

FIGURE 5.7. Desalting of *S*-carboxymethyl P40 (IL-9) using microbore column RP-HPLC. (Reproduced, with permission, from Simpson et al. 1989b.)

from the column by gradient elution (with increasing concentrations of organic solvent) or stepwise elution in small peak volumes (typically, 25–80 μl). Because column back pressure is directly proportional to column length, a practical advantage of short columns (<10 cm length) is that rapid sample loading can be accomplished by using high flow rates (~2 ml/min) without exceeding instrument pressure limits. For example, in Figure 5.6, the protein standards (~1 ml total) were loaded via a 2-ml injection loop at a flow rate of 2 ml/min prior to switching the loop "off-line" and initiating the gradient at a flow rate of 50 μl/min. Because the resolving capability of short columns (<100 cm length) is comparable to that of long (>250 cm) columns, and short columns permit the rapid loading of large sample volumes without exceeding instrument pressure limitations, the use of short-microbore HPLC "guard" columns (30 x 2.1-mm I.D.) has gained popularity for the preparation of samples for protein microsequencing (Nice et al. 1985). A practical application of trace enrichment (and concomitant buffer exchange) of a low-abundance cytokine (20 μg purified from 1 x 10^9 cells) is given for the ~35–40K M_r glycoprotein P40 (see Figure 5.7) (Uyttenhove et al. 1988; Simpson et al. 1989b). A procedure for desalting samples using RP-HPLC is given in Protocol 5.

In the example shown in Figure 5.7, 20 μg of *S*-carboxymethyl (Cm) interleukin-9 (IL-9) in ~300 μl of 7.5 M GdnHCl , 0.2 M Tris-HCl buffer (pH 5.5), 2 mM EDTA, 0.02% Tween-20, 15 mM dithiothreitol (DTT), and 50 mM iodoacetate were concentrated onto a 30 x 2.1-mm I.D. reversed-phase guard column and simultaneously exchanged into a volatile solvent (0.125% v/v TFA) suitable for enzymatic digestion and subsequent sequence analysis of the generated peptides. It is good practice to perform a pilot desalting step (e.g., ~500 ng of material; see Figure 5.7A) before committing the bulk of the sample (~20 μg) to desalting (Figure

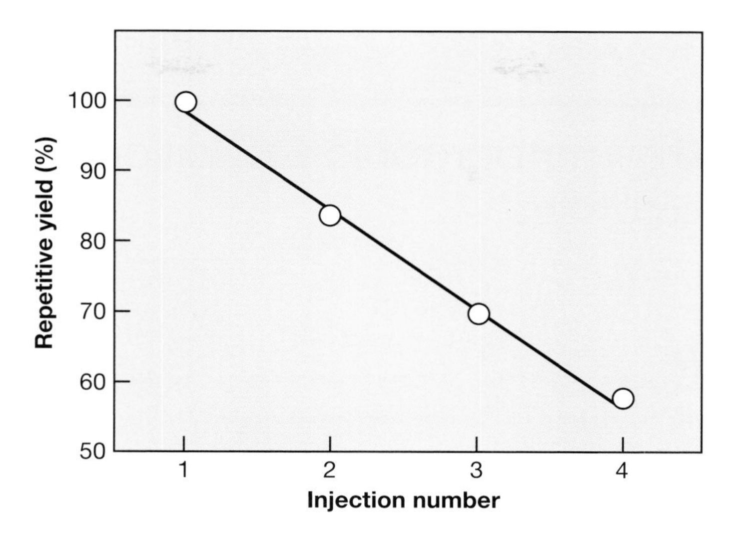

FIGURE 5.8. Repetitive recovery of GM-CSF from a short microbore reversed-phase column. Murine GM-CSF (300 ng) was chromatographed on a Brownlee RP-300 column (30 x 2.1-mm I.D.) at a flow rate of 100 μl/min using the chromatographic conditions described in Figure 5.7. The peak fraction was recovered manually, diluted 1:1 with primary solvent (in the injection syringe), and reinjected. Sample recovery was calculated from peak height measurement. (Reproduced, with permission, from Simpson and Nice 1989.)

5.7B), because some native proteins upon *S*-carboxymethylation are difficult to recover by RP-HPLC operated under the same chromatographic conditions as the native material, for example, granulocyte–colony-stimulating factor (G-CSF).

Sample Recovery

The recovery of proteins from reversed-phase supports is very much dependent on the nature of the support itself, rather than the dimensions of the column. The excellent recoveries (>90%) observed with microbore columns are comparable to those obtained with conventional 4.6-mm I.D. columns packed with the same material. A typical total system recovery (~80%) from repetitive reinjection of a low-M_r growth factor (300 ng of murine granulocyte-macrophage–colony-stimulating factor [GM-CSF], ~14K M_r) from a microbore column is shown in Figure 5.8. This recovery value was based on manual collections, dilution of secondary solvent, and subsequent reinjection (four times).

WWW Resources

Tutorials: RP-HPLC for Proteomics

- Vydac, *The handbook of analysis and purification of peptides and proteins by reversed-phase HPLC* (http://www.vydac.com/vydacpubs/brindex.html)
- Analytical Spectroscopy Research Group, University of Kentucky, *High-performance liquid chromatography (HPLC): A users guide* (http://kerouac.pharm.uky.edu/ASRG/hplc/hplcmytry.html)
- Ionsource, *Reversed phase HPLC basics for LC/MS (Andrew Guzetta)* (http://www.ionsource.com/tutorial/chromatography/rphplc.htm)

PROTOCOL 1

Standard Chromatographic Conditions for RP-HPLC of Proteins*

BEFORE ANALYZING A MIXTURE OF PROTEINS by RP-HPLC, it is extremely important to evaluate the system, both HPLC hardware and column, before it is applied to a biological problem. This is especially true when a new column enters the laboratory and applies to both beginners and more experienced chromatographers, who wish to obtain reproducible peptide maps. In this protocol, only nonspecialized columns, mobile phases, and instrumentation readily available and easily operated are described. For HPLC troubleshooting, see the Further Reading suggestions at the end of this chapter and www.proteinsandproteomics.org.

Before using a new column, the manufacturer's instructions and recommendations for proper conditioning, use, and storage of the column must be read. Often, subtle operating conditions that are crucial for correct column usage are overlooked, resulting not only in irreversible column damage, but also in the possible loss of an important sample. Once the operating procedures and column limitations (e.g., operating pressure and stability with respect to pH) are understood, a test separation should be performed using a set of standards (e.g., peptide or protein mixture) that have been designed to evaluate column performance. For this exercise, it is essential to reproduce the standard set of operating conditions described in this protocol before proceeding with an analysis of the test sample (see Figure 5.9). If an equivalent elution profile cannot be obtained on the new column (see below) using these standards and standard operating conditions, then the new column should be considered suspect and discussed with the manufacturer. (Column separations that have been carried out with a recognized set of standards are more easily returned to a manufacturer than simply telling the manufacturer that their column "cannot separate" an ill-defined complex mixture of interest.) It is important that these standard operating conditions be used on a regular basis (and carefully logged) to monitor column performance and the lifetime of the column. Because stationary phases differ widely from one manufacturer to another (see Table 5.1) in terms of their physical characteristics, the functional ligand used, ligand density, base support (matrix), particle size, and pore size, as well as column dimensions, standard protein/peptide separations are a useful diagnostic of these differences (and can be exploited in a purification strategy). If samples are to be collected for further characterization, careful attention should be given to the "dead volume" of the plumbing between the flow cell and outlet (see below).

TIPS FOR PREPARING AN HPLC

- Read the manufacturer's instructions for correct column usage.
- Ensure that the HPLC is in good working condition by running a blank (i.e., no column installed or sample injected).
- Evaluate the chromatographic system by running a set of "known standards."

*This protocol was contributed by Robert L. Moritz, Joint ProteomicS Laboratory (JPSL), Ludwig Institute of Cancer Research/Walter and Eliza Hall Institute for Medical Research, Melbourne, Australia.

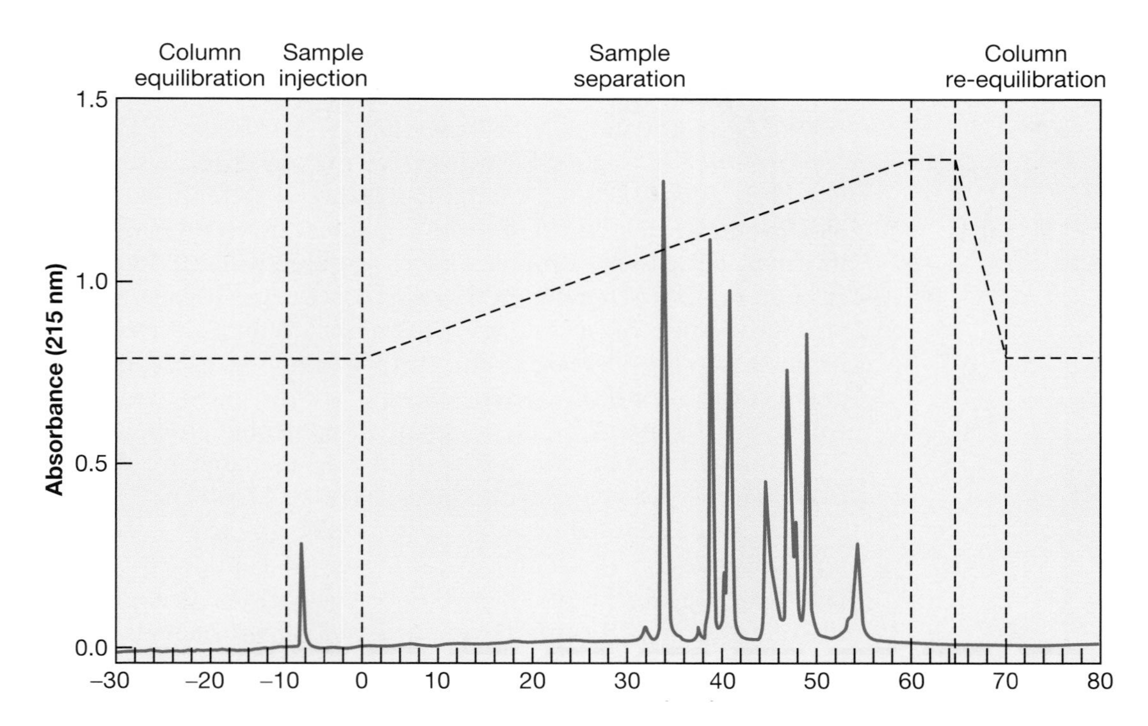

FIGURE 5.9. Standard chromatographic conditions for RP-HPLC separation of protein standards. Chromatographic conditions: Agilent 1100 Binary HPLC with autosampler; Column: Agilent C_{18}, 75 x 2.1-mm I.D.; Buffer A: 0.1% TFA; Buffer B: 0.09% TFA/60% acetonitrile; Temperature: 40°C; Flow rate: 0.1 ml/min; Gradient: 0–100% B in 60 minutes; Sample: 2 µg each of ribonuclease B, cytochrome *c*, lysozyme, BSA, carbonic anhydrase, myoglobin, and ovalbumin. The column volume is ~90 µl. Column equilibration and reequilibration, 20 column volumes.

MATERIALS

CAUTION: See Appendix 3 for appropriate handling of materials marked with <!>.

Reagents

Acetonitrile (HPLC grade) (J.T. Baker, Phillipsburg, New Jersey; Mallinckrodt; or equivalent) <!>

H_2O (deionized, HPLC grade) (18 megaΩ, Milli-Q, Millipore, or equivalent)

Solvent A (14 mM) (aqueous 0.1% v/v TFA) <!>

Add 1.0 ml of neat TFA (using a Gilson pipette, or equivalent) to 1 liter of H_2O in a glass-stoppered measuring cylinder, mix thoroughly, and then pour into a clean liquid chromatograph/HPLC reservoir bottle (extensively washed with Milli-Q H_2O).

EXPERIMENTAL TIP: It is good practice to set aside a 1-liter cylinder to be used only for mixing TFA and H_2O.

Solvent B (60% v/v acetonitrile/40% H_2O containing 0.1% v/v TFA)

Add 600 ml of acetonitrile to a 1-ml glass-stoppered cylinder and adjust to 1 liter with H_2O. Add 1 ml of neat TFA (final concentration 0.1% v/v TFA) and mix the solution thoroughly.

Trifluoroacetic acid (TFA) (HPLC grade) (Pierce, Rockford, Illinois, or equivalent)

SOLVENT PREPARATION

- If both Solvent A and Solvent B are 0.1% (v/v) TFA, then the absorbance at 214 nm of Solvent A will be slightly higher—20–40 milli-absorbance units (mAU)—than that of Solvent B, due to the contributing absorbance of acetonitrile. This will result in a rising baseline during the development of a gradient from 0% to 60% acetonitrile. Although this does not present a problem when working with 50–100-μg amounts of material (using a 4.6-mm I.D. column), the rising baseline can become a serious problem (i.e., the baseline will be off-scale, and late-eluting peaks cannot be detected without adjusting the baseline, which is extremely difficult to accomplish during the course of a chromatographic run), especially when working at low-microgram levels using microbore columns (<2.1 mm I.D.). To overcome this potential problem, it is recommended that the amount of TFA in Solvent B be slightly less than that in Solvent A (e.g., 0.09% or 900 μl of TFA/liter) compared with 1.0% (1000 μl of TFA/liter). By these means, the amount of TFA can be adjusted carefully to accomplish "by trial-and-error" a flat baseline. For high-sensitivity work (submicrogram levels of peptide/protein), where the expected peak heights are in the 10–20-mAU on a 2.1-mm I.D. column (1 μg = 100 mAU and 100 ng = 1 mAU on a 2.1-mm I.D. column), 0.085% or 0.086% (v/v) TFA in Solvent B may be needed.
- If solvents are not degassed, then bubble formation can occur during a chromatographic run as a result of out-gassing. Such bubbles can lodge in the flow cell, resulting in an erratic detector signal (i.e., baseline). To minimize this risk, Solvent B should be a "percentage buffer," i.e., containing H_2O (e.g., 60% TFA, 40% H_2O containing 0.1% TFA). Additionally, solvents should be degassed every 1–2 days with high-purity helium or constant vacuum over a period of ~30 minutes.
- With alcohol-type solvents (e.g., methanol and ethanol), a slow build up of esters occurs at room temperature, resulting in an increased baseline. For this reason, solvents must be prepared every 3 or 4 days.
- Solvents containing salts. Solvent systems such as unbuffered 1% (w/v) NaCl/acetonitrile and 0.1 M phosphate (pH 7.4)/acetonitrile are commonly used in RP-HPLC peptide separations when seeking selectivity other than that achieved with conventional 0.1% (v/v) TFA/acetonitrile (for an example, see peptide mapping of IL-6 in Figure 7.8, Chapter 7). When using salt-containing solvents, the percentage of organic solvent in Solvent B should not exceed 50%; otherwise, there is a high risk that these salts will precipitate and clog the HPLC tubing (typically, the 7/1000-inch and 10/1000-inch capillary tubing used in modern HPLC systems), as well the column. Such damage can be very expensive to repair! To avoid possible precipitation of salts, it is recommended that salt-containing solvents be premixed by adding equal volumes of organic solvent to the buffer salt solution (e.g., 2% w/v aqueous NaCl or 0.2 M phosphate at pH 7.4) to give a final concentration of 1% (w/v) NaCl/50% organic solvent or 0.1 M phosphate/50% organic solvent.

Equipment

HPLC chromatographic system

This system should be equipped with programmed gradient elution, UV detection at 210–220 nm, acquisition of chromatographic data, and integration of peak areas. The system must be capable of accurate and reproducible solvent delivery and gradient formation.

Reversed-phase column (100 Å or 300 Å pore size)

5–10-μm particle size packed into a column of 100 x 4.6-mm I.D. or cartridge of 100 x 2.1-mm I.D. Commonly used columns in this laboratory are Brownlee RP-300 100 x 2.1 7-μm 300A (C8) (Alltech) and Vydac 214TPB 100 x 2.1 3-μm 300A (C_4) (Vydac, Hesparia, California) for samples of 1–100 μg; for samples >100 μg, use the same support, but packed into 4.6-mm I.D. columns.

Biological Sample

Protein standards

hen egg lysozyme (L 6876, Sigma)

ribonuclease B (R 5750, Sigma)

cytochrome *c*

bovine serum albumin (A 3675, Sigma)

carbonic anhydrase (high-purity grade, C 3934, Sigma)

Dissolve the protein standards (1 mg of each protein/ml) in H_2O (can be stored frozen for several weeks). It is recommended that the 1 mg/ml solution of protein standards be stored in small aliquots (e.g., 100 μl) and not freeze-thawed more than five times.

Cautionary Notes

- ***Routine for new column usage.*** Flush the HPLC system completely before attaching the column. Prime both solvent lines with Solvents A and B, respectively, to ensure that there are no bubbles in the system. Connect the solvent delivery tubing from the injector to the top end of the column and the tubing from the column outlet to the flow cell. Pass 30 column volumes of Solvent B though the column followed by 30 volumes of Solvent A. The column is now ready to be evaluated using the procedure described below in Step 1.
- ***Connection of columns to HPLC plumbing.*** Before connecting a column to the HPLC plumbing, make sure that column end-fittings, ferrules, and tubing depths are correctly matched. Incorrect use of column end-fittings can seriously damage a column, resulting in solvent leakage and/or high dead-volumes (and hence poor chromatographic performance). For a technical description and possible problems that ensue from incorrect fittings, see the panel on COLUMN FITTINGS INTERCHANGEABILITY, at the end of this protocol.

METHOD

1. Run a blank or "control chromatogram" (i.e., no sample loaded in Step 1). Typically, this is the first chromatographic run of the day. The following are the recommended standard chromatographic conditions for the RP-HPLC system evaluation:

   - ***Linear gradient:*** From 0% to 100% B, where Solvent A is aqueous 0.1% (v/v), and Solvent B is 0.09% (v/v) TFA in 60% aqueous acetonitrile.
   - ***Gradient rate:*** 1% B/min (i.e., 0–100% Solvent B in 60 minutes).
   - ***Flow rate:*** 1 ml/min, 4.6-mm I.D. column (0.1 ml/min; 2.1-mm I.D. column).
   - ***Temperature:*** 40–45°C.

			Flow rate	
Step no.	Time (minutes)	Mobile-phase composition	4.6-mm I.D. column	2.1-mm I.D. column
1 (sample injection)	0	100% Solvent A	1 ml/min	0.1 ml/min
2	60	100% Solvent B	1 ml/min	0.1 ml/min
3 (re-equilibration)	–*	100% Solvent A	1 ml/min	0.1 ml/min

*The time taken to pass 20 column volumes of Solvent A though the column. Column volumes are calculated by measuring column length and internal diameter and applying the following equation $V = \pi\ r2\ h$, where V is the column volume, r is the internal radius of column, and h is the length of column.

These conditions were used to produce the elution profile of the blank chromatogram shown in Figure 5.9.

Before commencing the blank run, pump 20 column volumes of Solvent B through the column, followed by an equal volume of Solvent A (ensure that baseline is stabilized, i.e., remains flat). Then adjust the flow rate to an operating flow rate of 1 ml/min (or 0.1 ml/min for 2.1-mm I.D. column) and wait until the baseline stabilizes (~z minutes).

2. Run protein standards (control run). In Step 1, inject 100 µl of protein standards (0.1-mg stock standards/ml) for the 4.6-mm I.D. column (i.e., 10 µg of each protein standard; 10 µg = 100 mAU at a flow rate of 1 ml/min). For a 2.1-mm I.D. column, load 10 µl of 0.1-mg protein standards/ml, i.e., 1 µg of each protein standard (1 µg = 100 mAU at a flow rate of 0.1 ml/min).

   These conditions were employed to produce the elution profile of the control protein standard chromatograms shown in Figure 5.9. An alternative to protein standards is a tryptic digest of cytochrome *c* (Figure 7.9c, Chapter 7).

3. Run the test peptide digest mixture, using the same conditions used in Step 2.

4. For sample collection, collect eluting peptide peaks in polypropylene microfuge tubes fitted with tight-fitting caps and store at –20°C for further study. If sample collection is to be performed manually, the dead-volume of the plumbing from the detector flow cell to the outlet tubing must be determined accurately.

   EXPERIMENTAL TIP: A good approximation of the dead volume (d µl) can be made using the following equation: $d = \pi r^2 h$, where h is the length of the tubing from the flow cell to the outlet (in mm), and r is the diameter (in mm) of the tubing. The preferred method for determining the dead volume involves the injection of a dye (aqueous 0.01% w/v Coomassie Blue) and accurately determining the time (using a stopwatch) taken from the first appearance of dye in the flow cell (i.e., off-scale detector signal) and the appearance of blue dye from the outlet tubing (easily seen by spotting the eluent onto white blotting paper). When performing this task, disassemble the column from the HPLC equipment and replace with a comparable length of tubing that is then connected to the flow cell.

5. For long-term column storage, flush the columns and store with an aqueous organic solvent (e.g., 50% v/v propanol/50% H_2O). For overnight storage, pump the columns with Solvent B at a very slow flow rate (e.g., 50 µl/min).

Column Fittings Interchangeability

There are many different fittings in HPLC, and so many that look alike, that interchangeability is a real problem. The following explanation is an attempt to clarify the matter.

1. Take a zero dead-volume union, male nut, and ferrule from one manufacturer.
2. Put the male nut and ferrule on a piece of 1/16-inch O.D. stainless steel tubing.
3. Insert the tubing, ferrule, and nut into the ZDV union. Make sure that the tubing bottoms out in the union.
4. Swage the ferrule onto the tubing by tightening the male nut finger tight, and then make a three-quarters turn with a wrench.
5. Now remove the male nut, ferrule, and tubing from the fitting.
6. Measure the distance from the end of the ferrule to the end of the tubing. This is Dimension X.

Dimension X varies from one manufacturer to another, and it can also vary for fittings from the same manufacturer due to variation in tolerances. Because of the typical variances seen among the array of manufacturers, new fittings, new ferrules, and new connections are recommended each time assemblies are connected, although such care is not necessary in all cases. For example, Valco, Parker, Swagelok, and Upchurch Scientific zero dead-volume fittings are interchangeable at the initial assembly and may be interchangeable in subsequent assemblies. However, Dimension X is long with a Waters fitting. If a ferrule has been swaged in a Waters fitting, and is then inserted into an Upchurch fitting, the ferrule will not seat and it will leak. Dimension X is short with an Upchurch fitting. If the ferrule has been swaged in an Upchurch fitting and it is inserted into a Waters fitting, the tubing will not bottom out and threre will be a dead volume/mixing chamber. (Adapted from the Upchurch Scientific Catalog, 2000 [http://www.upchurch.com].)

PROTOCOL 2

Packing Capillary Columns for RP-HPLC*

CAPILLARY OR MICROCOLUMN (<0.5 mm I.D.) HPLC is an extremely powerful technique for separating small quantities of proteins and peptides. It has distinct advantages over wider-bore column HPLC, such as increased mass sensitivity and lower flow rates, which make it particularly attractive for coupling to detection techniques such as mass spectrometry (see Chapter 8). Although the advantages of column miniaturization have been known for some years now (Ishii et al. 1977; Yang 1982; Novotny and Ishii 1985)—microcolumn (<1 mm I.D.), also known as capillary liquid chromatography—it is only recently that capillary column RP-HPLC has become a widely used technique in mass-spectrometry-based proteomics studies (see Chapter 8). Unlike capillary electrophoresis (for reviews, see Wiktorowicz 1992; Li 1993), which has gained wide acceptance and for which many commercial instruments are available, capillary HPLC still remains a specialist technique. It is only recently that commercial capillary column RP-HPLC instruments and packed capillary columns have become available (e.g., Agilent Capillary LC, WatersCapLC, LC Packings Ultimate; for further information, see Table 5.2).

This protocol describes a procedure for adapting conventional HPLC systems to provide accurate low-flow rates (0.4–4 µl/min) and gradients required to operate slurry-packed capillary columns. A key component of this system is a commercial axial-beam longitudinal flow cell (Chervet et al. 1989) that can be fitted to a number of commercial UV detectors (Moritz et al. 1994). Procedures are described for the fabrication of 0.32-mm I.D. polyimide-coated fused-silica columns, slurry-packed with reversed-phase chromatographic supports. The chromatographic performance of these columns is illustrated using a series of standard proteins and a mixture of peptides derived from a *Staphylococcus aureus* V8 protease digest of recombinant murine IL-6 (see also the alternative "bomb" method in Chapter 8).

MATERIALS

CAUTION: See Appendix 3 for appropriate handling of materials marked with <!>.

Reagents

Methanol (50% aqueous v/v) <!>
n-propanol <!>
Reversed-phase silica <!>
Trifluoroacetic acid (TFA)/acetonitrile solvent system <!>
Solvent A is 0.1% (v/v) aqueous TFA
Solvent B is 60% (v/v) acetonitrile, 40% H_2O containing 0.09% (v/v) aqueous TFA

*This protocol was contributed by Robert L. Moritz, Joint ProteomicS Laboratory (JPSL), Ludwig Institute of Cancer Research/Walter and Eliza Hall Institute for Medical Research, Melbourne, Australia.

Equipment

Bath sonicator

Chromatographic syringe (Microliter 700 series, Hamilton, Reno, Nevada)

Column frits

A variety of porous membranes can be used as column frits to retain the column packing.

Zitex membrane
Glass fiber filter paper
Polyvinylidene difluoride, hydrophilic

End fittings, standard male liquid chromatography (0.5-mm I.D. flowthrough ports, 1/4-inch column end-fittings)

Epoxy resin (24-hour curing-time type)

Ferrules (vespel/graphite) (100/0.4-VG1, Alltech)

Flow cell

Axial-beam longitudinal capillary flow cells with small internal volumes (26.5 nl) can be obtained for various UV detectors such as Spectra-Physics, Applied Biosystems Inc., Carlo Erba, Hitachi, and Waters. The flow cells can be purchased from LC Packings (Amsterdam, The Netherlands; San Francisco, California)

Flow cell holder (LC Packings, Amsterdam, The Netherlands; San Francisco, California)

Forward optics scanning detector (Thermo Galactic/Spectra Physics)

The detector should be quipped with U-shaped longitudinal capillary flow cell (6-mm path length, illuminated volume ~26.5 nl).

Heat gun

Injector

Rheodyne Model 7520 fitted with 0.2–1.0 μl internal sample rotor

or

Model 8125 injector fitted with 0.5–50-μl injection loops

Liquid chromatography column packing pump (Shandon, Cheshire, UK)

Microcolumns

0.8–0.2-mm I.D. microcolumns can be obtained commercially from LC Packings (Amsterdam, The Netherlands, San Francisco, California, Agilent, and Waters, MicroChrom) or constructed as described below.

Fused-silica tubing: Fused-silica tubing is used to provide back pressure to the split tee (see Figure 5.10). Various lengths and internal diameters can be used (e.g., 0.050 mm I.D. x 100 mm or 0.100-mm I.D. x 400 mm). The tubing is held in place at the split tee by sleeving the fused silica with a 10-mm length of snugly fitting Teflon tubing to provide an O.D. of 1/16 inch (see Figure 5.10).

Fused-silica capillary column (Polymicro Technologies Inc., Phoenix, Arizona; Millipore; Selleys Pty. Ltd.,Victoria, Australia; Alltech Associates Inc., Deerfield, Illinois)

- Stationary phase can be purchased loose from many manufacturers or obtained by unpacking prepacked columns.
- One-piece PEEK nut or one-piece fitting (1/16-inch) (32233, Alltech).
- 1/16-inch O.D. x 0.5-mm I.D. Teflon tubing. Other internal-diameter tubing can be used to provide a tight fit around the column body depending on the size of fused silica used (20033, Alltech).
- Polyimide-coated fused-silica tubing (Polymicro Technologies Inc.) of 0.540-mm O.D. x 0.320-mm I.D., 0.340-mm O.D. x 0.200-mm I.D., and 0.200-mm O.D. x 0.100-mm I.D. cut to lengths of 300-mm to provide the column body.
- Frits for these columns are punched from sheets of 0.45-μm porosity hydrophilic PVDF (HVLP 04700, Millipore).
- Epoxy glue is used to join the column body and exit tubing together (04-465, Selleys).
- Exit tubing which fits into these column bodies (i.e., 0.275-mm O.D. x 0.075-mm I.D., 0.190-mm O.D. x 0.050-mm I.D., and 0.090-mm O.D. x 0.050-mm I.D.) cut to lengths of 200 mm.

Nuts, fingertight (1102-1K, Swagelok, Solon, Ohio)

Polypropylene tube (1.5 ml)

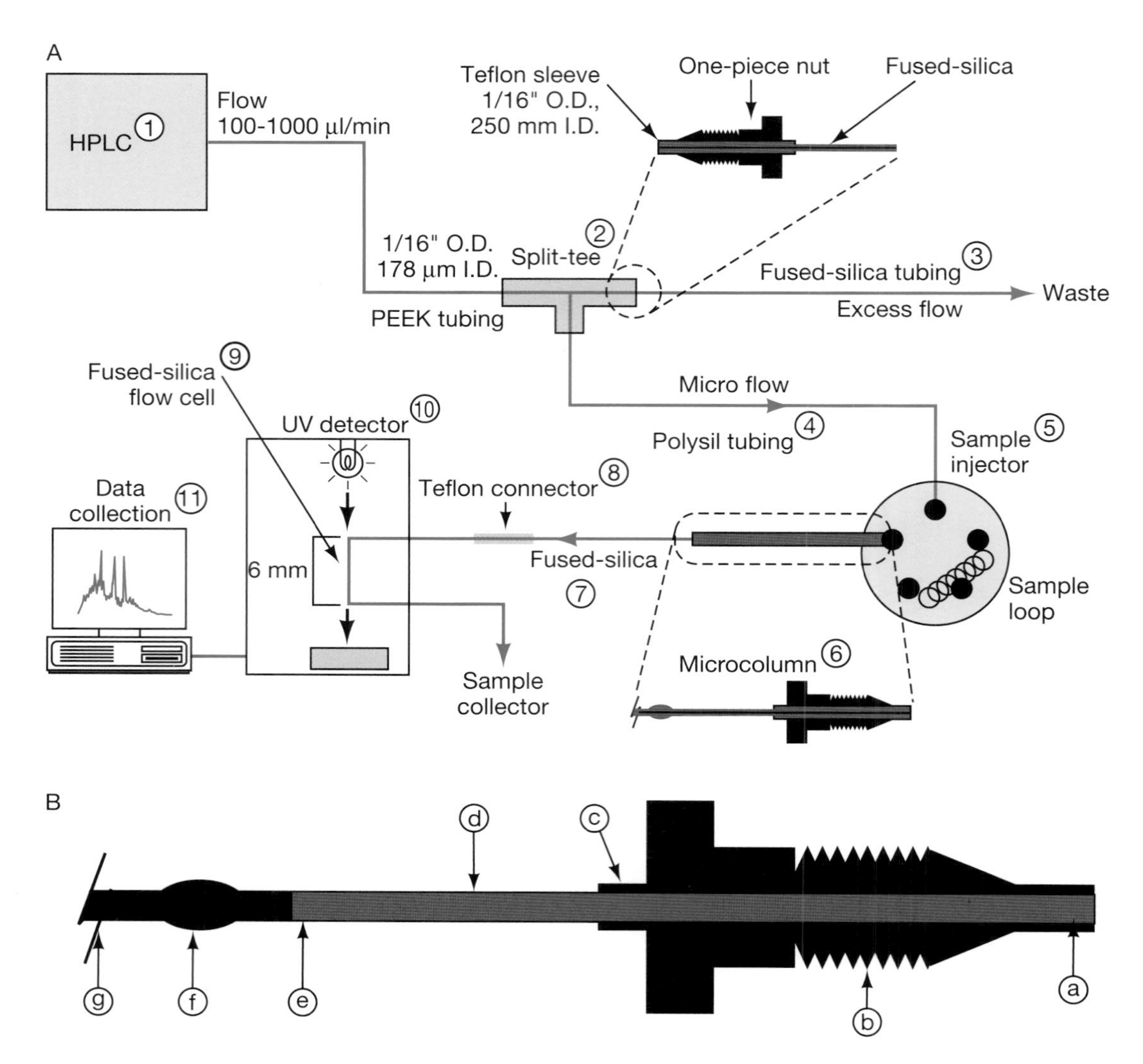

FIGURE 5.10. Gradient capillary HPLC system (*A*). Component parts (indicated by circled numbers): (1) solvent delivery systems; (2) split tee; (3) fused-silica tubing; (4) Polysil tubing; (5) sample injector; (6) microcolumn; (7) fused-silica capillary column; (8) Teflon connector; (9) flow cell; (10) UV detectors; (11) signal data collection. (For more details, see Materials list in this protocol.) (*B*) An expanded view of the polyimide-coated fused-silica capillary column. (*a*) Stationary phase; (*b*) one-piece PEEK nut; (*c*) Teflon tubing 1/16-inch O.D. x 0.5-mm I.D.; (*d*) polyimide-coated fused-silica tubing; (*e*) frits; (*f*) epoxy glue joining; (*g*) exit tubing. (Redrawn, with permission, from Moritz et al. 1994.)

Polysil tubing

Small-bore Teflon-coated, aluminum-clad fused silica (50 x 0.05-mm I.D.) is used to connect the split tee to the injector. This tubing provides a low dead-volume flow path for the column flow (PT50-05, Scientific Glass Engineering Pty. Ltd.,Victoria, Australia)

Preinjection solvent split (1/16-inch) stainless steel tee (see Figure 5.10)

Safety glasses

Sample injector

Small internal port injectors with either internal or external loops (0.06–50 μl) are used to inject samples directly onto the capillary columns. Injectors used in this study are a Rheodyne injector Model 8125 or 7520 or a Valco injector Model C6W or C14W (Rheodyne, Cotati, California; Valco Instruments Co. Inc., Houston, Texas).

Sapphire knife (23740U, Supelco, Bellefonte, Pennsylvania)

Signal data collection

Data collection devices such as strip chart recorders, integrators, or signal processing software can be used for data storage and subsequent manipulation (Kipp & Zonen, Amsterdam, The Netherlands; Hewlett-Packard, Waldbronn, Germany; Spectra Physics, Fremont, California).

Silicon rubber septa

Slurry reservoir

The reservoir is manufactured from an empty 50 x 2-mm I.D. glass-lined stainless steel tube (SGE Inc., Ringwood, Victoria, Australia).

Solvent delivery system

Hewlett-Packard HP1090M (or HP1100 series) liquid chromatograph.

Almost any conventional liquid chromatograph can be used to provide micro flow rates by operating the instrument at standard flow rates (i.e., 1 ml/min) and splitting off the microflow required.

Split tee

Flow split tee junctions with internal ports of 0.5-mm I.D. and 1/16-inch female tubing connections (U-428, Upchurch Scientific Inc., Oak Harbor, Washington).

All tubing used in the micro flow path, originating at the split tee, incorporating the injector and microcolumn, and terminating at the detector flow cell, are kept as short as possible to minimize any flow dispersion.

Teflon connectors

Teflon tubing of 1/16-inch O.D. and various internal diameters are used to provide tight, leak-free tubing unions. If an internal diameter close to size of the outer diameter of the fused-silica tubing cannot be found, slightly larger tubing can be stretched between two pliers to draw out the Teflon tubing and decrease its internal diameter until the required size is obtained (Alltech Associates Inc., Deerfield, Illinois; Upchurch Scientific Inc., Oak Harbor, Washington).

UV detector

Detectors suitable for the capillary HPLC system, described here, can be purchased from Spectra Physics (Model Spectra-100, Spectra Focus) or Applied Biosystems Inc., Model 783 (Spectra Physics, Fremont, California; ABI/Perkin Elmer., Foster City, California).

METHOD

Configuration of Capillary Liquid Chromatography System

The following solvent delivery system was used (see Figure 5.10): A Hewlett-Packard HP1090M liquid chromatograph utilized as a "slave" pump and microgradient source.

Flow Generation

1. For accurate flow rates of 0.05–20 µl/min through the capillary columns and reproducible gradient formation, install a preinjection solvent split (1/16-inch tee) that directs >95% of the solvent flow through a length (~100 cm) of 0.10-mm I.D. x 0.26-mm O.D. fused-silica tubing to waste.

2. Connect capillary columns directly to either a Rheodyne Model 7520 injector fitted with 0.2–1.0-µl internal sample rotor or to a Model 8125 injector fitted with 0.5–50-µl injection loops.

3. Set the flow through the column (0.4–4 μl/min) accurately either by adjusting the length of fused-silica tubing going to waste or by varying the pump flow rate (100–500 μl/min).

   With this split-flow approach, it is important that the actual flow rate through the column be monitored on a regular basis (e.g., every day). This is readily achieved by installing a 10-μl chromatographic syringe at the effluent port and accurately timing the advancing meniscus. Details of the various components used in configuring this system and their commercial sources are listed in Materials.

UV Detection

1. Monitor column eluent from both capillary HPLC systems by absorbance at 214 nm using a Thermo Galactic/Spectra Physics forward optics scanning detector equipped with a U-shaped longitudinal capillary flow cell (6-mm path length, illuminated volume ~26.5 nl) installed in the conventional flow cell holder.
2. Send the data either collected to a strip chart recorder via the single-wavelength mode or collected on a personal computer with Spectra Focus software, or equivalent, installed and operated in the multiple-wavelength mode (195–340 nm, 5-nm intervals).
3. For capillary liquid chromatography/mass spectrometry (LC/MS), connect the eluent from the UV detector directly to the electrospray inlet via a 50-cm length of 0.05-mm I.D. x 0.19-mm O.D. fused-silica tubing.

   This tubing, which replaces the standard stainless steel electrospray needle, extends to the tip of the spray needle assembly.
4. For nanoflow capillary HPLC, connect the outlet directly to the nanoflow tips (New Objective, Inc., Woburn, Massachusetts; http://www.Newobjective.com).

Construction of Capillary Columns

Numerous reports in the literature describe the construction of microcolumns (<0.5 mm I.D.) from such diverse materials as Teflon tubing (Ishii et al. 1977), glass-lined stainless steel tubing (Chervet et al. 1989), and polyimide-coated fused-silica tubing (Yang 1982). We prefer to construct microcolumns from the latter. Compared to the other materials mentioned above, polyimide-coated fused-silica offers a number of distinct advantages (see Table 5.8). Uppermost of these are its flexibility and ease of handling, its transparency, its commercial availability in a wide range of inner and outer diameters, and its relatively low cost.

The following step-by-step procedure for constructing 0.1–0.32-mm I.D. fused-silica microcolumns uses the construction of a 250 x 0.20-mm I.D. column as an example.

1. Cut two lengths of polyimide fused-silica tubing, one 250-mm length (0.200-mm I.D. x 0.340-mm O.D.) to provide the column body and another 200-mm length (0.050-mm I.D. x 0.190-mm O.D.) to provide the column exit tubing.

   CAUTION: Wear safety glasses and adhere to recommended safety precautions during column packing.

   - Thin-walled capillary tubing has a tendency to fracture easily when subjected to high pressures (~400 bar) during the slurry-packing of chromatographic supports. For instance, in our experience, fused-silica tubing of 0.320-mm I.D. x 0.540-mm O.D. and

TABLE 5.8. Fused-silica microcolumn characteristics

Column I.D. (mm)	Slave pump flow rate (ml/min)	Split ratio	Flow rate (µl/min)		Optimal protein load (µg)
			Operating	Loading	
0.3	0.50	1:140	3.6	20	2.0
0.2	0.50	1:310	1.6	5	1.0
0.1	0.50	1:1250	0.4	2	0.2

0.200-mm I.D. x 0.340-mm O.D. can withstand packing pressures of up to 450 bar, whereas tubing of 0.320-mm I.D. x 0.450-mm O.D. fractures readily at these high pressures. In the latter case, slurry-packing pressures should not exceed 50 bar. Finally, it should be noted that superior column efficiencies can be achieved by packing at the higher pressures afforded by using the thick-wall fused-silica tubing.

- To avoid jagged ends and to obtain square-cut capillaries, use a sapphire knife instead of the more common ceramic squares. Tubing with a jagged end will result in imprecise frit positioning.
- Teflon tubing is used to connect lengths of fused-silica tubing (see Figure 5.10). Because there are slight variations in the outer diameters of fused-silica tubing from different manufacturers, it is often necessary to fabricate such connectors to obtain tight-fitting Teflon connectors. This is readily achieved by stretching a slightly oversized inner diameter piece of Teflon tubing by hand with two pairs of pliers.

2. Cut the PVDF frit from a large sheet of PVDF membrane using a short length (50-mm) of fused-silica silica tubing (0.32 [or 0.20]-mm I.D. x 0.45 [or 0.34]-mm O.D.) as a disc cutter. This is accomplished by placing the PVDF sheet on a silicon rubber septum and punching the frit. The column exit tubing is used to insert the fabricated PVDF frit into the microcolumn body to a depth of ~10 mm.

   A variety of porous membranes can be used as column frits to retain the column packing. These include Zitex membrane (Davis and Lee 1992), glass-fiber filter paper (Davis and Lee 1992), and hydrophilic polyvinylidene difluoride (PVDF) (Moritz and Simpson 1992a,b). For the columns described in this study, a 0.45-µm porosity hydrophilic PVDF membrane was used.

3. Position the microcolumn body and the column exit tubing permanently by placing a small drop of precured epoxy resin at the junction of the two pieces of tubing.

   The epoxy resin is partly polymerized, or precured, with a heat gun to increase its viscosity so that it will not seep into the capillary tubing and block the frit.

   Inline MicroFilter assemblies from Upchurch Scientific, comprising a Micro Fingertight Fitting (F-125), a Micro End Fitting (M-120), and a Filter Union Body (M-520-01), can be used as column frits to retain column packing. For instructions on assembling the Inline MicroFilter, see the Upchurch Scientific Web Site: http://www.upchurch.com.

Column Packing

Once the PVDF frit has been positioned and the fused-silica capillary column constructed, a slurry-packing procedure is used to pack the microcolumn.

1. Manufacture a slurry reservoir from an empty 50 x 2-mm I.D. glass-lined stainless steel tube with standard male liquid chromatography (0.5-mm I.D. flowthrough ports, 1/4-inch column end-fittings).

2. Connect the fused-silica microcolumn to the standard 1/4-inch liquid chromatography end-fitting by vespel/graphite and fingertight nuts.

3. Sonicate a slurry of reversed-phase silica in *n*-propanol (40 mg/200 μl) for 15 minutes in a 1.5-ml polypropylene tube.

4. Prior to packing, fill the capillary column with the packing solvent (*n*-propanol) under high pressure (400 bar) using a standard liquid chromatography column-packing pump (Shandon).

   This step prefills the microcolumn with packing solvent and also allows the system to be checked for tight seals.

5. Disconnect the slurry reservoir from both the column packer and the microcolumn, empty it, and reposition it in place. Fill it with 150 μl of the prepared slurry mixture.

6. Pack the column at a constant pressure of 400 bar for 20 minutes, consolidate with 50% (v/v) aqueous methanol for a further 20 minutes at the same pressure, and then allow the column to depressurize.

7. Dismantle the column carefully and seal it by inserting both the top and bottom ends of the column into silicon rubber septa.

COLUMN OPERATION

Standard conditions for operating a capillary column RP-HPLC are given in Protocol 1. All gradient separations were achieved using a standard trifluoroacetic acid/acetonitrile solvent system, i.e., a linear 60-minute gradient from 0% to 100% Solvent B, where Solvent A was 0.1% (v/v) aqueous trifluoroacetic acid and Solvent B was 60% (v/v) acetonitrile, 40% H_2O containing 0.09% (v/v) aqueous trifluoroacetic acid. For mass spectrometric analysis, the trifluoroacetic acid concentration was reduced to 0.05% and 0.045% for Solvents A and B, respectively.

PROTOCOL 3

Purification of Large "Troublesome" Polypeptides

The separation and purification of large synthetic nonpolar polypeptides or polypeptide fragments of proteins derived from cyanogen bromide cleaving procedures can often present formidable challenges due to their low solubility or the propensity of these molecules to self-associate. RP-HPLC provides one avenue to purify such troublesome examples provided certain steps and precautions are taken.

Large polypeptides are often troublesome to handle because of their low solubility or the presence of secondary structural elements that favor supramolecular self-self assembly. Often, these attributes are a direct consequence of the presence in the large polypeptide of a higher content of nonpolar amino acids in the sequence, or alternatively, the order of the amino acids favors the formation of intercalating β-sheets or fibrils, or α-helical coiled-coil dimers or higher oligomers. Hence, different strategies are needed for their purification.

Notwithstanding these challenges, HPLC is still the best technique currently for purifying and analyzing such types of large polypeptides. Due to the vastly different properties of these molecules, four different modes of separation can be used: reversed-phase, ion-exchange, size-exclusion, and hydrophobic interaction. Inevitably, the latter three modes require the use of eluents that contain low concentrations of nonionic or ionic detergents or surfactants, such as 0.1% Brij 25, Tween-20, Triton X-100, or even SDS, or alternatively, a relatively high concentration of a chaotrope such as urea or guanidinium hydrochloride (Hearn 1998). Because of the presence of these additives, ion-exchange, size-exclusion, and hydrophobic interaction HPLC cannot be considered to be the ultimate purification step. Instead, procedures intended to enable other contaminants to be removed selectively from the large polypeptide or protein of interest should be considered. However, as a single technique, RP-HPLC satisfies most of the requirements for such analyses and purifications, and as a consequence, it is the most established and widely used procedure. Thus, RP-HPLC methods not only enable concomitant desalting, removal of additives that aid dissolution of the polypeptide or protein, but also permit resolution as well as maintenance of a reasonably high concentration of the solutes, due to the presence of the (often) low-pH, aquo-organic solvent conditions. This protocol addresses the group of large polypeptides that have proved to be troublesome to handle due to their low solubility or the presence of secondary structural elements that favor supramolecular self-self assembly. As noted above, included in this group are polypeptides with amphipathic α-helical or β-sheet structures with extensive runs of nonpolar (hydrophobic) amino acid side chains or, alternatively, proline-rich sequences. Related procedures are equally germane to polypeptides that have been lipidated or subjected to chemical modifications with nonpolar moieties, as well as to some core cyanogen bromide fragments of large proteins. In the cases described below, the use of the polar "ion-pairing" reagent trifluoroacetic acid with the "dipole clustering" solvent, acetonitrile, is described. Other ion-pairing reagents (Hearn 1980, 1985), high concentrations of formic acid (Heukeshoven and Dernick 1985; Mant and Hodges 1991a), the presence of low concentrations of anionic or cationic detergents (van der Zee et al. 1983; Hancock 1984; Hearn 2000b), or organic solvents of lower polarity, such as 2-propanol or tetrahydrofuran (van der Zee et al. 1983; Mant et al. 1997), can be used to generate alternative mobile-phase systems.

This protocol was contributed by Hooi Hong Keah and Milton T.W. Hearn (Centre for Bioprocess Technology, Department of Biochemistry and Molecular Biology, Monash University, Clayton, Victoria 3800, Australia).

MATERIALS

CAUTION: See Appendix 3 for appropriate handling of materials marked with <!>.

Reagents

Buffer A: Add 1 ml of trifluoroacetic acid (TFA) <!> to 1 liter of Milli-Q H_2O in a 1-liter measuring cylinder

Buffer B: Add separately 600 ml of acetonitrile <!>, 400 ml of Milli-Q H_2O, and 1 ml of TFA in a 1-liter measuring cylinder

Use a sintered glass funnel to filter Buffers A and B separately through a 0.2-μm filter (nylon 47-mm filter membrane, Part No. 2034, Alltech Assoc., Deerfield, Illinois) directly into a graduated flask. Take care at this stage not to over-aspirate or evacuate the flask to prevent evaporation of the organic solvent, and thus reduce the organic solvent content in the Buffer B. Adjust the volume of acetonitrile and H_2O to make a different percentage of acetonitrile content buffer when necessary, but maintain the volume of TFA. Inject ~5 μl of the aliquot for these evaluations.

IMPORTANT: TFA used for protein sequencing may not be suitable for the preparation of RP-HPLC eluents because it can contain anti-oxidants.

Equipment

Columns

- ***For Sample Purification:*** A semipreparative C_{18} or another appropriate *n*-alkylsilica sorbent of suitable particle size (i.e., between 5 and 25 μm), porosity (i.e., between 150 and 1000 Å), and phase ratio packed in a preparative column of appropriate dimensions, e.g., 10 μm, 300 Å, 300 x 21.5-mm I.D. The TSK-ODS-120T (Tosoh Corp.) C_{18} column (300 x 21.5-mm I.D., particle size 10 μm), protected by a guard column, represents one suitable option.
- ***For Sample Analysis:*** An analytical C_{18} or another appropriate *n*-alkylsilica sorbent of suitable particle size (i.e., between 1.5 and 10 μm), porosity (i.e., between 150 and 1000 Å), and phase ratio packed in an analytical column of the appropriate dimensions, e.g., 5 μm, 300 Å, 150 x 4.6-mm I.D. The TSK-ODS-120T (Tosoh Corp.) C_{18} column (150 x 4.6-mm, particle size 5 μm) represents a suitable option or, alternatively, the Agilent SP-C18, Supelco Discovery C_{18}, Hypersil ODS C_{18}, or one of the available types of reversed-phase columns from the 1500 or so vendors.

HPLC system

- Binary/quaternary solvent delivery pumps, such as a Waters Model 601 system or equivalent.
- Gradient controller, such as a Waters 600 controller or equivalent.
- Sample injector (automated injector for analysis, manual injection for separation), such as a Waters Wisp 712 or equivalent.
- Fraction collector, such as the Amersham Biosciences Frac100 or equivalent.
- Variable wavelength or photodiode array UV detector, such as a Waters 486 detector or equivalent.
- Computer system for data collection, and subsequent analysis and presentation, such as the NT-based Millenium software operating on a PC (Pentium system) of adequate Ram (≥64 Mbyte) and hard disk memory (>1 GB) or equivalent.

Lyophilizer

Microfuge

Modular or dedicated HPLC instrumentation (e.g., Beckman System Gold, Agilent 1100 Series HPLC)

This instrumentation can be used in analogous fashion depending on the vendor preference of the investigator.

Nitrogen gas <!>

Biological Sample

Large nonpolar polypeptides or proteins for purification

METHOD

Sample Preparation of Large Nonpolar Polypeptides or Proteins for Purification by RP-HPLC

1. Prepare the polypeptide sample (between 30 and 50 mg if derived from solid-phase chemical synthesis methods, although significantly smaller sample amounts, i.e., down to microgram quantities, can be similarly used if the sample is derived from other sources) by dissolving in the elution buffer (Buffer B, ideally in 1 ml if 30–50 mg of sample is available or in smaller volumes as appropriate to the sample size) in a microfuge tube.

   Depending on the hydrophobicity of the protein and large polypeptide, i.e., the more hydrophobic it is, the higher the acetonitrile content (v/v) will be needed in the buffer. Generally, 60% (v/v) of acetonitrile in H_2O with 0.1% (v/v) TFA is sufficient for this purpose. With polypeptides that are poorly soluble, a solvent of 90% (v/v) acetonitrile-H_2O containing 0.1% (v/v) TFA may be required. It may be necessary to use a detergent-based, urea-based, guanidine-hydrochloride-based solvent or a 60% (v/v) formic-acid-based solvent. Detailed investigations (Hearn 1991a) have shown that the optimal concentration for maximal resolution of many large polypeptide or small proteins in RP-HPLC often occurs near to 15 mM TFA.

2. Centrifuge crude samples in a microfuge at 2000*g* for 5 minutes to ensure that all insoluble material and/or precipitates are removed before injection.

   For additional details on procedures that can be adopted to achieve suitable high-resolution conditions, see Protocols 4, 5, and 6.

Analysis of Large Nonpolar Polypeptides or Proteins

A small quantity of the sample (50 µl) is required for analysis.

1. Prepare ~5–10 aliquots for evaluation of the separation conditions with different types of RP-HPLC sorbents, under various elution conditions, i.e., different flow rates, temperatures, or gradient slope parameters.

2. Inject ~5 µl of the aliquot for these evaluations.

Purification of Large Nonpolar Polypeptides

1. Sparge the buffers by bubbling nitrogen through the buffers for 20 minutes at 100 ml/min per reservior.

   This step is done to exclude any air bubbles before running through the HPLC system. When the system is running, sparging can be slowed down to 20 ml/min per reservoir.

   It is important to ensure that no air bubbles are in the line of the HPLC system. Air bubbles in the system will result in inconsistency of the flow of buffers. The pressure transducer on the pump should indicate stable back-pressure if the system is properly purged and primed; i.e., all of the lines are filled with the new buffer. This is done by pulling some buffer through each pump and checking the consistency of flow at the outlet before the column is connected.

2. Connect the column. Make sure that the column is free of any impurities by running elution buffer through it for 15 minutes at 6 ml/min. Wash the column with buffer for an additional 20 minutes. The system should now be ready for injection.

3. Start the software for data collection and check that the detector is calibrated to the wavelength required.

   For polypeptides containing aromatic side-chain amino acids, 254 nm can be used; otherwise, separation can be carried out at 230 nm. Photodiode array detectors provide a flexible option for obtaining difference, derivative, and higher-order spectra, as well as utilizing peak track algorithmic approaches.

4. Set up solvent gradients for elution on the computer.

   A gradient of 2–98% Buffer B (elution buffer) for 60 minutes is commonly used. Very shallow gradients are normally used to separate peaks that run very close together.

5. Use a manual injector to inject the sample for separation while maintaining the flow rate of the loading buffer (Buffer A) at 6 ml/min.

6. Collect small fractions (~1–2 ml) in test tubes for analysis as the various peaks elute, as shown for the first case study by the high absorbances of the chromatogram as seen in Figure 5.11.

Case Study 1: Purification of a Crude Sample of Synthetic FS-1 Polypeptide

In this case study, the purification of a crude sample of the synthetic FS-1 polypeptide, related to the carboxy-terminal region of the activin-binding protein follistatin-315 (FS315) and implicated as a binding domain, was achieved with a TSK-ODS-120T C_{18} column of dimensions 300 x 21.5 mm with an *n*-octadecyl silica sorbent of average particle size of 10 µm using a linear gradient of 0.1% TFA in H_2O to 100%, 0.1% TFA in H_2O/acetonitrile (10:90 v/v) for 1 hour at a flow rate of 6 ml/min, with detection at 230 nm (Figure 5.11). Despite the fact that it contains several polar amino acid residues, the FS-1 polypeptide has a relatively low solubility in aqueous solutions and tends to precipitate at higher pH values.

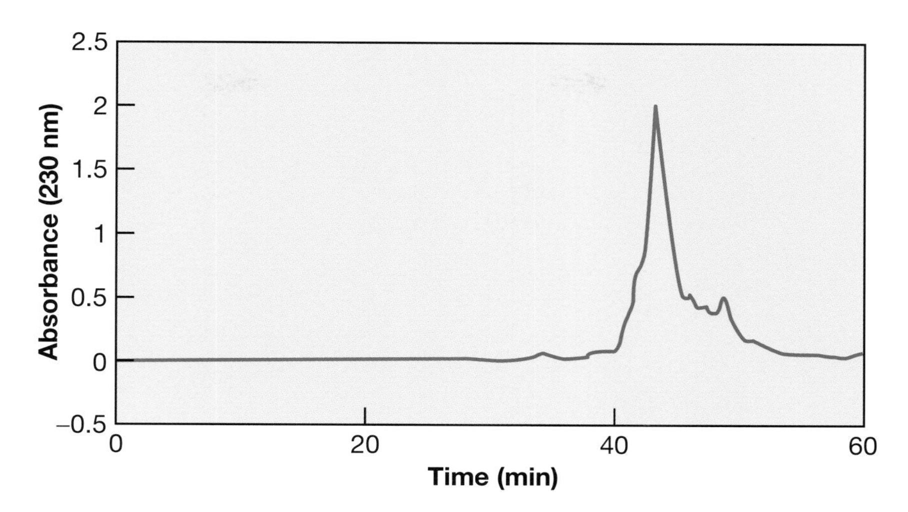

FIGURE 5.11. Purification of crude FS-1 polypeptide was achieved with a TSK-ODS-120T C_{18} columns (300 x 21.5 mm, particle size 10 μm) using a linear gradient of Buffer A (0.1% TFA in H_2O) to 100% Buffer B (0.1% TFA in H_2O/acetonitrile [10:90 v/v]) for 1 hour at a flow rate of 6 ml/min, with detection at 230 nm.

7. Dispense the collected fractions into sample vials for subsequent analysis.

   RP-HPLC analysis, high-performance capillary electrophoresis (HPCZE), capillary electrochromatographic (CEC), or mass spectroscopic (ESI-MS or MALDI-MS) analytical procedures can be routinely employed. Generally, 50 μl per sample is reserved for evaluation of peak purity, etc., and an injection of 30 μl is usually sufficient for all of these analyses.

8. For the analysis by RP-HPLC of the fractions collected, prepare the HPLC system by repeating Steps 1–6, but in this case, reduce the flow rate to 1 ml/min, with the detector wavelength set at 214 nm and a suitable analytical column connected instead of the preparative column.

   For rapid analytical separations, the gradient time can be reduced to 25 minutes for analysis and 10–15 minutes of reequilibration time after that to ensure that the system is ready for the next analysis.

Case Study 2: Chromatogram of Crude Sample Containing the FS-1 Polypeptide and the Purified FS-1 Polypeptide

Figure 5.12, A and B, shows the corresponding analytical chromatograms of the crude sample containing the FS-1 polypeptide and the purified FS-1 polypeptide collected from the preparative run described above (H. Keah and M. Hearn, unpubl.). In this case study, the analytical RP-HPLC separations were carried out using TSK C_{18} columns (150 x 4.6 mm, particle size 5 μm) with detection at 214 nm and with elution achieved using a linear gradient of Buffer A (0.1% TFA in H_2O) to 100% Buffer B (0.1% TFA in H_2O/acetonitrile [10:90 v/v]) for 25 minites at a flow rate of 1 ml/min, followed by 5 minutes of column wash in 100% Buffer B prior to reequilibration with Buffer A.

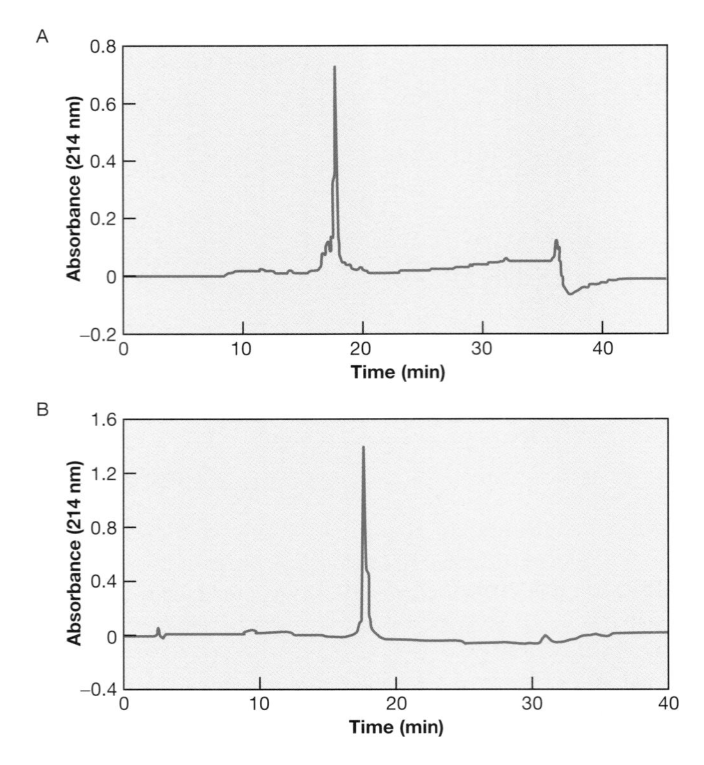

FIGURE 5.12. RP-HPLC profiles of crude (*A*) and purified (*B*) FS-1 peptides. The analytical RP-HPLC separations were carried out on TSK C_{18} columns (150 x 4.6 mm, particle size 5 mm) with samples eluted using linear gradient of Buffer A (0.1% TFA in H_2O) to 100% Buffer B (0.1% TFA in H_2O/acetonitrile [10:90 v/v]) for 25 minutes at a flow rate of 1 ml/min, followed by 5 minutes in 100% Buffer B with detection at 214 nm.

G2 LabExpert

Application of such analytical procedures to the resolution and purification of "difficult" polypeptides and proteins is particularly amenable to the recently developed real-time knowledge-based approach, called G2 LabExpert (I et al. 2002), allowing intelligent automation of HPLC method development. Unlike other computerized method development systems, the G2 LabExpert operates in real time, using an artificial intelligence system and design engine to provide experimental decisions as well as to control the instrumentation, column selection, and mobile-phase choice. The G2 LabExpert has been programmed to control every input parameter to a HPLC data station, such as an Agilent series 1100 HPLC Chemstation system, including operation of the diode array detector and quaternary pump system equipped with degasser, autosampler, and software. It evaluates each output parameter of the HPLC data station as part of its decision process. On the basis of a combination of inherent and user-defined evaluation criteria, the program uses a reasoning process, applying chromatographic principles and acquired experimental observations to provide iteratively a regime for the á priori development of an acceptable HPLC separation method. Because remote monitoring and control are also functions of the G2 LabExpert, the system allows full-time use of analytical chromatographic instrumentation and associated laboratory resources within a peptide synthesis/protein chemistry laboratory. Extensive sample handling is also avoided, a feature important for the purification or analysis of large polypeptides that are troublesome to handle because of their low solubility or propensity to undergo self-self supermolecular assembly due to the presence of secondary structural elements such as coiled-coil motifs.

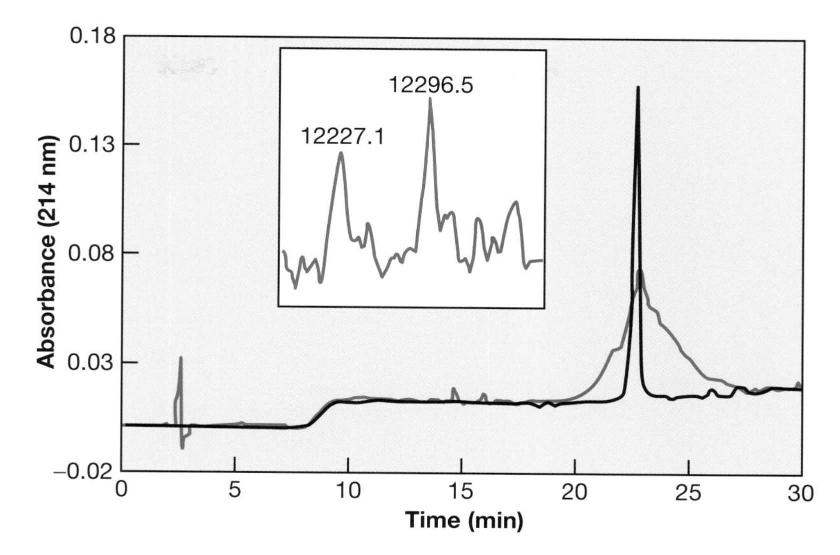

FIGURE 5.13. RP-HPLC profiles of the crude and purified activin β_A subunit polypeptide, Actβ_A (12–115). The analytical RP-HPLC separations were carried out on the TSK C_{18} column (150 x 4.6 mm, particle size 5 µm) with samples eluted using linear gradient of 0.1% TFA in H_2O (Buffer A) to 100% Buffer B (0.1% TFA in H_2O/acetonitrile [20:80 v/v]) for 25 minutes at a flow rate of 1 ml/min, with detection at 214 nm. Purification of the crude Actβ_A (12–115) polypeptide was achieved with a TSK-ODS-120T C_{18} column (300 x 21.5 mm, particle size 10 µm) using an optimized gradient of Buffer A to 100% Buffer B over 2 hours at a flow rate of 6 ml/min, with detection at 230 nm. (*Inset*) ESI-MS transformation spectrum of Actβ_A (12–115) polypeptide with an observed molecular mass of 12296.5 (Calculated mass: 12289.8). The mass peak at 12227.1 corresponds to the des-monoAcm-Actβ_A (12–115).

Case Study 3: Chromatographic Profiles for Crude and Purified Sample of Activin β_A Subunit

Figure 5.13 shows the chromatographic profiles for the crude and purified sample of the activin β_A subunit, Actβ_A (12–115)[13]. This fully synthetic analog of the pleiotropic homodimeric growth factor activin $\beta_A\beta_A$ contains two large antiparallel β-strand loops, built around a catenating cystine knot framework, which creates the potential for self-association under inappropriately selected elution conditions. As evident from the figure, this large polypeptide could be easily purified using the protocols outlined above and summarized in greater detail in the legend to the figure. In particular, the authenticity of this fully synthetic 103 mer could be confirmed from the ESI-MS transformation spectrum, which should be that the purified Actβ_A (12–115) polypeptide had an observed molecular mass of 12296.5 (calculated mass: 12289.8 amu).

Case Study 4: Chromatograms of Crude and Purified MTPX1 Polypeptide

Figure 5.14, A and B, illustrates the corresponding chromatograms of the crude and purified sample of MTPX1 polypeptide, respectively. The MTPX1 polypeptide represents the hydrophobic amino-terminal leader sequence of a CRISP-related mouse sperm tail protein and was required for studies on sperm protein function and localization (Keah et al. 2001a,b). This polypeptide very readily forms aggregates, and following lyophilization, it is extremely difficult to redissolve in low-pH aqueous buffers of low organic solvent content. Generally, an autoinjector instead of a manual injector is used for the analysis. This enables the set of samples to be injected and analyzed automatically with minimal time delays for the operator.

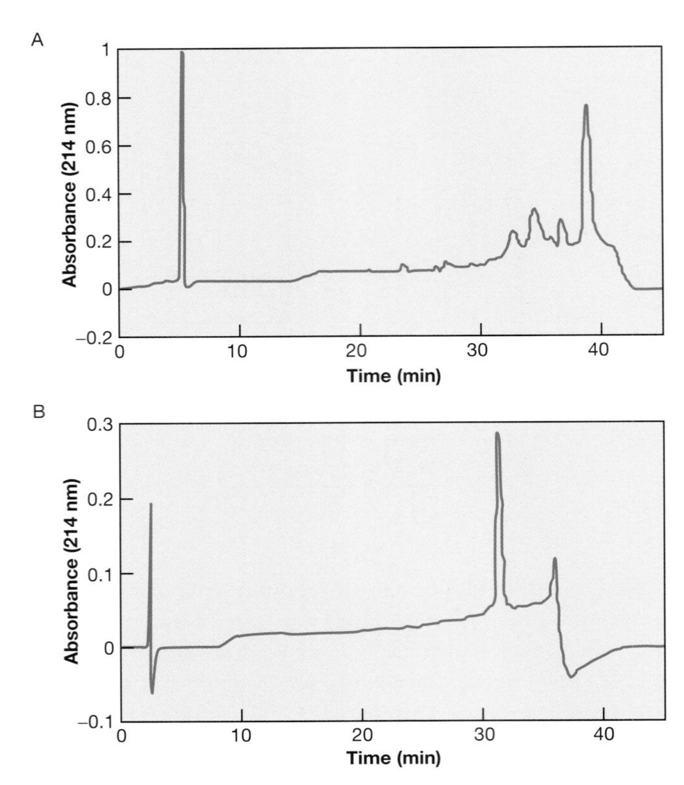

FIGURE 5.14. RP-HPLC profiles of the crude (*A*) and purified (*B*) MTPX1 polypeptide. The analytical RP-HPLC was carried out on TSK C_{18} columns (150 x 4.6 mm, particle size 5 µm) with samples eluted using a linear gradient of Buffer A (0.1% TFA in H_2O) to 100% Buffer B (0.1% TFA in H_2O/acetonitrile [10:90 v/v]) for 30 minutes at a flow rate of 1 ml/min with detection at 214 nm. Also illustrated is an important issue with regard to the control over the mobile-phase composition, temperature, or constancy of flow rate. As evident from this figure, the retention time of the pure MTPX1 polypeptide following gradient elution (*B*) is smaller than that for the separation of the crude MTPX1 sample shown in *A*. This result arose from a deliberate (but small) variation in flow rate to simulate either transient gas bubble formation in the pump head, etc., due to inadequate "degassing" of the mobile phase or pressure flunctations caused by particulate matter lodging in the HPLC tubing/valves, etc., due to inappropriate filtration of the sample/solvents through a 0.2-µm filter, with the appearance of the chromatographic profile and retention time of the desire polypeptide affected. Similar variations can arise when changes in the temperature control of the buffer reservoirs or column occur, or decreases in the organic solvent content of Buffer B used for the gradient elution arise due to evaporation, resulting in a longer retention time for the analytes.

9. When all of the fractions for analysis have been prepared, freeze-dry the samples in the test tubes on a lyophilizer overnight.
10. Carry out the various analytical experiments, and based on the peak profiles and associated mass spectrometric data, pool the relevant peak fraction tubes.
11. If the recovered lyophilized fractions prove to be difficult to resolubilize, use DMSO to bring the dried powder back into solution when conjugation to a carrier protein is planned as part of the preparation of a haptenic immunogen. When the planned studies involve subsequent aliquoting steps, followed by spectroscopic analysis, etc., use DMF, formic acid, or a high content of acetonitrile.
12. Store the polypeptides as dry fractions or as dissolved samples at –20°C.

Case Study 5: Variation of Flow Rate and Retention Time of a Large, Synthetic, Nonpolar Polypeptide

The above general protocols have been applied with various refinements in this laboratory for more than 20 years (Hearn 1984; Forage et al. 1986) with large polypeptides as well as with cyanogen bromide fragments of proteins or with globular proteins themselves. They have been found to be robust in terms of their reliability and reproducibility. However, as illustrated in Figure 5.12, a final "case study" is presented. Here, HPLC conditions were deliberately implemented to demonstrate how the retention time of a large synthetic nonpolar polypeptide, in this case the MTPX1 polypeptide, will be affected if a variation in flow rate occurs, with the appearance of the chromatographic profile thus affected. Similar variations can arise when inadvertent changes in the temperature control of the buffer reservoirs or column occur, or decreases in the organic solvent content of Buffer B used for the gradient elution arise due to evaporation, resulting in a longer retention time for the analytes.

PROTOCOL 4

Purification of Peptides from Solid-phase Peptide Synthesis with RP-HPLC

PURIFICATION OF PEPTIDES DERIVED FROM SOLID-PHASE PEPTIDE SYNTHESIS (SPPS) requires the removal of deletion peptides resulting from incomplete coupling/deprotection steps, from racemization or side-chain rearrangement, and from various chemical substances introduced during the deprotection or cleavage stages of an SPPS procedure. This protocol details the purification and analysis of many synthetic peptides of 2–65 amino acid residues. These peptides contain a reasonable number of ionizable or polar side chains, but do not contain secondary structural elements that favor supramolecular assembly, such a β-sheets, amphipathic structures with extensive runs of nonpolar side chains, proline-rich sequences, or conformationally "fragile" motifs that flip between helical and sheet structures in response to minor changes in mobile phase composition (Hearn 2000b). Protocol 3 described the purification and analysis of very large polypeptides that are troublesome to handle due to the presence of one or more of the above features.

This protocol was contributed by Reinhard I. Boysen and Milton T.W. Hearn (Centre for Bioprocess Technology, Department of Biochemistry and Molecular Biology, Monash University, Clayton, Victoria 3800, Australia).

MATERIALS

CAUTION: See Appendix 3 for appropriate handling of materials marked with <!>.

Reagents

Acetone <!>
Acetonitrile (HPLC grade) <!>
H_2O (deionized, from a Milli-Q water purification system or equivalent)
Phosphoric acid (H_3PO_4) <!>
2-Propanol <!>, methanol <!>, or other suitable organic solvents with excellent transparency in the UV to 210 nm
Sodium nitrate <!>
Sodium phosphate (NaH_2PO_4) <!> or other suitable salts
Thiourea <!>
Trifluoroacetic acid (TFA) <!>

Equipment

Analytical HPLC apparatus

This apparatus consists of a pump module, a mixing chamber, spectrophotometer with an analytical flow cell, an analytical column, a column oven, an autosampler, PC, printer and software, HPLC system with attendant data management systems and system automation controllers. Helium gas is required for buffer degassing and nitrogen for autosampler operation. The column is an analytical C_{18} or another *n*-alkylsilica sorbent of suitable particle size (i.e., between 1.5 and 10 μm), porosity (i.e., between 150 and 1000 Å), and phase ratio packed in an analytical column of the appropriate dimensions (e.g., 5 μm, 300 Å, 150 x 4.6-mm I.D.).

Centrifuge (benchtop, for sample preparation)

Conical vials (for autosampler)

Glassware

- four 1-liter eluent bottles
- two 1-liter measuring cylinders
- two waste bottles

All glassware should be rinsed three times with Milli-Q H_2O.

Hamilton glass syringe (graduated 25 μl, to fit injector port)

Helium gas <!>

Lyophilizer

Mobile-phase filtration facility

- vacuum pump
- 1-liter reservoir
- support base with glass frit and integral vacuum connection
- funnel
- clamp
- 47-mm membrane filter (0.2-μm PTFE)

Nitrogen gas <!>

Preparative HPLC apparatus

This apparatus consists of a preparative pump module, mixing chamber, spectrophotometer with a preparative flow cell, a manual injection valve with preparative (500–1000 μl) sample loop, fraction collector, PC, printer and software with data management system, and system automation controller shared with analytical apparatus. Helium gas is shared with the analytical apparatus. The column is a preparative C_{18} or another *n*-alkylsilica sorbent of suitable particle size (i.e., between 5 and 25 μm), porosity (i.e., between 150 and 1000 Å), and phase ratio packed in a preparative column of appropriate dimensions (e.g., 10 μm, 300 Å, 300 x 21.5-mm I.D.).

Safety glasses

Stopwatch

Syringe (1 ml) with shortened (cutoff) needle

Biological Sample

Peptide or protein for analysis

METHOD

Preparation of the Mobile Phase

1. Prepare Eluent A (weak mobile phase) and Eluent B (strong mobile phase), 1 liter each, for example:

   Eluent A: 0.1% TFA in H_2O
   Eluent B: 0.09% TFA in 60% acetonitrile/40% H_2O (v/v)

   IMPORTANT: TFA employed for protein sequencing may not be suitable for the preparation of RP-HPLC eluents because it can contain anti-oxidants.

   The volumes of organic solvent and H_2O must be measured in two different cylinders and then combined because of volume contraction upon mixing, which may be up to 30 ml/liter prepared solvent (depending on the nature of the organic solvent). Failure to do so can lead to substantial errors in mobile-phase composition.

EXPERIMENTAL TIP: To compensate for the baseline shift in gradient elution (because most organic components absorb more light at low wavelengths than H_2O) when working with H_2O/organic solvent eluents, decrease the amount of ion pair reagent (TFA, H_3PO_4) in Eluent B (Mant and Hodges (1991a) by 10–15% in comparison with Eluent A, yielding a flatter baseline (Dolan 1991).

2. Mix the solvent either by stirring with a Teflon-coated magnetic flea or by shaking in a stoppered cylinder (time depends on volume).

   WARNING: **Do not** use Parafilm (plastic wrap) under any circumstances to cover flasks or cylinders containing the eluents used in RP-HPLC. The organic solvents in combination with the acidic ion-pair reagent components in the eluents will dissolve components in the Parafilm, yielding extraneous peaks in the chromatogram.

3. Filter Eluent A first and then Eluent B through a 0.2-µm PTFE filter.

   Filtering of eluents increases the column life time and contributes to degassing.

4. Seal unused eluent bottles with a stopper to avoid evaporation of the organic solvent.

Preparation of Peptide Sample

1. Dissolve the sample with half the target volume of Eluent A (weak mobile phase) to achieve the desired concentration. If the sample is not soluble, add a small amount (typically <25% of total volume) of Eluent B (strong mobile phase).

   Small amounts of strong mobile phase may cause preelution of the sample depending on the sample loop size, column dimensions, and starting mobile-phase composition of the gradient.

2. Inspect the sample for clearness. Filter the sample through a 0.2-µm PTFE filter if insoluble, opalescent, or solid particles are present. Alternatively, centrifuge the sample using the supernatant for injection.

   IMPORTANT: Do not inject samples that are not fully dissolved because they can block the injector and column.

3. Store the sample if not in use at 4°C or –2°C, depending on the planned storage time and usage. Peptides and protein samples can degrade at room temperature.

   Avoid repeated freezing-thawing of the sample. Instead, prepare small aliquots of the peptide or protein sample, which are kept at –20°C and are used for one task.

Purification of Peptides from Solid-phase Peptide Synthesis with RP-HPLC

1. Test the instrumental HPLC system. The following test allows the evaluation of the column bed integrity (low integrity will be associated with split, fronting, or tailing peaks) and column performance (in terms of plate numbers). This test also allows, if repeated at regular intervals, the monitoring of the performance during the life time of a column, and the assessment of batch-to-batch differences of column fillings (see Protocol 1).

   a. Produce a blank run (inject Eluent A) and run a gradient 100% A → 100% B under the same conditions as intended for the separation of the peptide or protein sample. Repeat if "ghost" peaks occur.

      This procedure cleans the column of peptides and proteins from previous separations, which have not been removed by the flushing process.

b. Measure the dead volume of the column with thiourea or sodium nitrate (or any other noninteractive solute).

c. Test the column performance with a gradient run and an appropriate test mixture: An example of an RP-HPLC test mixture is

0.15% (w/v) dimethylphthalate
0.15% (w/v) diethylphthalate
0.01% (w/v) diphenyl
0.03% (w/v) *O*-terphenyl
0.32% (w/v) dioctylphthalate in methanol

d. Test the column performance with a gradient run and peptide standards for column testing described, for example, in Mant and Hodges (1991b).

2. Perform analytical RP-HPLC of the crude peptide.

a. Prior to sample analysis, conduct at least one blank run under the same conditions as intended for the peptide sample. Repeat if peaks occur.

b. Separate the crude peptide (~100 µg) with an analytical RP-HPLC procedure to assess the sample in terms of purity, peak profile, and elution conditions with the experimental conditions as outlined below in the panel.

c. Identify the component to be purified.

Chromatographic Conditions for the Analytical Separation of Crude Peptide and Protein Mixtures

The following chromatographic conditions are representative of the type of standard analytical protocol that can be employed. Numerous variations in terms of the choice of reversed-phase sorbent, gradient slope and duration, flow rate, mobile-phase composition (including the choice of different organic solvent modifiers, buffer species, or ion-pairing reagents), and pH, temperature, and sample size can be considered. Listed below is a linear gradient procedure that represents a robust protocol around which these variations can be constructed.

- Column: For example, C_4, C_8, C_{18}, etc., reversed-phase sorbent appropriately packed into a suitable column (5 µm, 300 Å, 150 x 4.6-mm I.D.)
- Sample size: <2 mg peptide/protein
- Sample loop size: 20–200 µl
- Linear A → B gradient
- Eluent A: 0.9% aqueous TFA
- Eluent B: 0.1% TFA in acetonitrile/H_2O
- Gradient rate: 1% B/min, for example, 60% acetonitrile/H_2O
- Gradient range and time: 0–100% B in 60 minutes
- Flow rate: 1.0 ml/min
- Detection: 214 nm
- Temperature: Room temperature

3. Prepare RP-HPLC.

   a. Separate the crude peptide mixture (~25–100 mg) with a preparative RP-HPLC procedure using the experimental conditions as outlined in the panel below.

      To avoid detector response overloading effects, wavelengths from 230 to 280 nm are usually chosen. However, small amounts of chemical scavengers (used during the SPPS procedures) present in the crude peptide solution can absorb very strongly in this wavelength range. It may be worthwhile to sacrifice up to a milligram of sample and to perform a preparative separation with detection at 214 nm to determine the retention time of the main peptide product unambiguously.

   b. Collect HPLC fractions (3–7.5 ml).

Chromatographic Conditions for the Preparative Separation of Peptides

The following chromatographic conditions are representative of the type of standard preparative protocol that can be employed. Numerous variations can be contemplated in terms of the choice of reversed-phase sorbent, gradient slope and duration, flow rate, mobile-phase composition (including the choice of organic solvent modifiers, buffer species, or ion-pairing reagents), and pH, temperature, and sample size. Listed below is a linear gradient procedure that represents a robust protocol around which these variations can be constructed.

- Column: For example, C_4, C_8, C_{18}, etc., reversed-phase sorbent appropriately packed into a suitable column (10 μm, 300 Å, 300 x 21.5-mm I.D.)
- Sample size: <150 mg peptide/protein
- Sample loop size: 1 ml, multiple injection
- Linear A → B gradient
- Eluent A: 0.9% aqueous TFA
- Eluent B: 0.1% TFA in acetonitrile/H_2O
- Gradient rate: 0.66% B/min, for example, 60% acetonitrile/H_2O
- Gradient range and time: 0–100% B in 90 minutes
- Flow rate: Activate pumps to deliver 7.5 ml/min
- Detection: 254 nm
- Temperature: Room temperature

4. Perform analytical RP-HPLC.

   a. Analyze aliquots (30–50 μl) of the collected fraction with an analytical RP-HPLC.

      A blank, the crude peptide solution, the fraction of interest, and the two fractions before, and two fractions after are typically analyzed.

   b. Identify the fractions containing the desired protein in sufficient purity (i.e., >95%) by a comparison of the retention time of the purified component with the retention time of the target component in the chromatogram of the crude peptide.

   c. Weigh the peptide into microfuge tubes and store them at –20°C for further investigations.

5. Perform further analysis.

   a. Freeze-dry fractions containing the peptide with a purity >95% as assessed by the analytical RP-HPLC separation, by high-performance capillary zone electrophoresis (HP-CZE) (Sitaram et al. 1999; Hearn et al. 2000), or by capillary electrochromatography (HP-CEC) (Walhagen et al. 2000a,b, 2001) for another appropriate high-resolution analytical technique.

   b. Carry out off-line or on-line electrospray ionization–mass spectroscopy (ESI-MS) of the purest fraction to confirm the correct synthesis and correct identification of the component to be purified.

      If the received mass differs from the expected mass, it is sometimes worthwhile to investigate other fractions with ESI-MS.

CASE STUDY: RP-HPLC AND PURIFICATION OF PRODUCTS FROM SOLID-PHASE PEPTIDE SYNTHESIS

Examples of the use of RP-HPLC in the analysis and purification of products from solid-phase peptide synthesis are shown in Figures 5.15 and 5.16. The preparative isolation of the synthetic 26-mer polypeptide product representing the core of the *Thermus thermophilus* ribosomal L36 protein, NH_2-RIC(Acm)DKC(Acm)KVIRRHGRVYVIC(Acm)ENPKHKQ-COOH (Boysen and Hearn 2000), was achieved on a TSK-ODS-120 T column (300 x 21.5-mm I.D., 120 Å, 10 µm, end-capped) using a flow rate of 7.5 ml/min and a 90-minute gradient from 0% B to 100% B (A: 0.1% TFA in H_2O; B: 60% acetonitrile, 0.09% TFA) at pH 2.1, as depicted in Figure 5.15, with the major peak eluting at 58.77 min. The acetamidomethyl protecting groups were left on the peptide to avoid unwanted disulfide bridge formation.

The resolution of the peptide from the crude mixture, as shown in Figure 5.16A, and the purified product shown in Figure 5.16B with analytical RP-HPLC was achieved with a TSK-ODS-120 T column (150 x 4.6-mm I.D., 120 Å, 5 µm, end-capped) using a flow rate of 1 ml/min and a 60-minute gradient from 100% A (0.1% TFA in H_2O) to 100% B (90% acetonitrile, 0.09% TFA) at pH 2.1. When both the preparative and the analytical run of the crude peptide are compared, it is apparent that the scavenger from the peptide cleavage, which has a relatively strong absorbance at 254 nm, elutes between 80 and 100 minutes in the preparative run (Figure 5.15), while absorbing little in the analytical run at 214 nm (Figure 5.16A). The basis for achieving the obtained high purity of the end product, as seen in Figure 5.16B, was a relatively shallow gradient slope in the preparative RP-HPLC in conjunction with narrow peak fractioning. Adequate equilibration of the analytical column ensures run-to-run reproducibility, facilitating peak identification according to retention times.

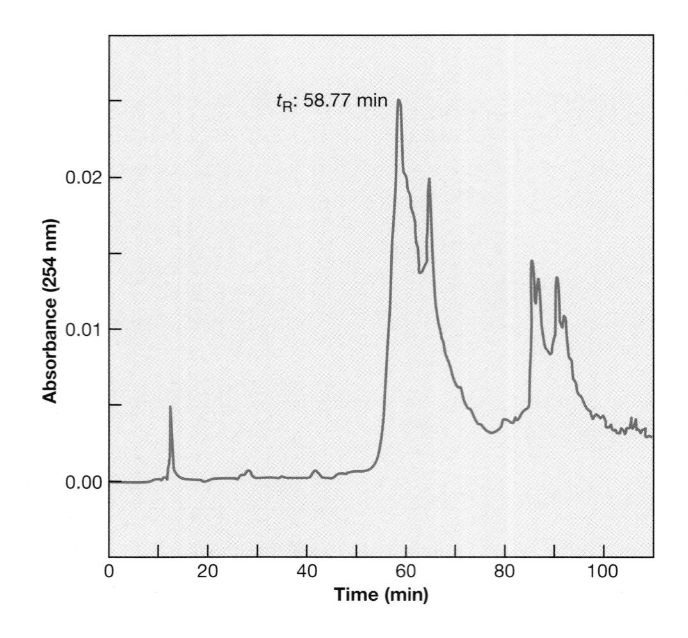

FIGURE 5.15. Use of RP-HPLC in the preparative isolation of synthetic polypeptides. The synthetic 26-mer polypeptide product representing the core of the *T. thermophilus* ribosomal L36 protein, H-RIC(Acm)DKC(Acm)KVIRRHGRVYVIC(Acm) ENPKHKQ-OH (Boysen and Hearn 2000) (70 mg), was purified from the crude mixture obtained from solid-phase peptide synthesis on a TSK-ODS-120 T column (300 x 21.5-mm I.D., 120 Å, 10 µm, end-capped) using a flow rate of 7.5 ml/min and a 90-minute gradient from 0% B to 100% B (A: 0.1% TFA in H_2O; B: 60% acetonitrile, 0.09% TFA) at pH 2.1.

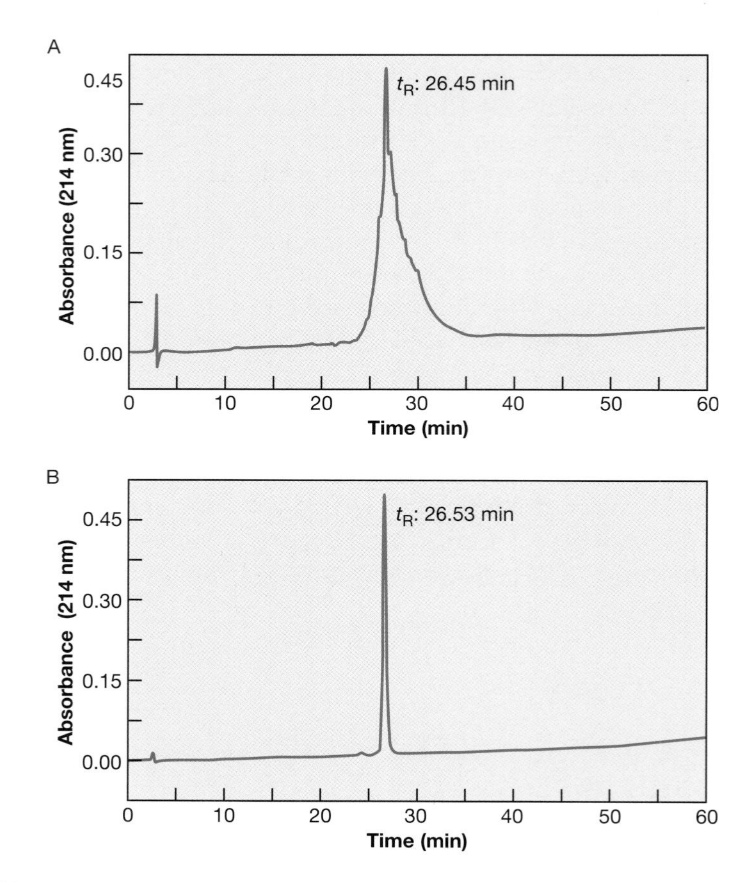

FIGURE 5.16. Use of RP-HPLC in purification of peptides derived from solid-phase peptide synthesis. Resolution of the crude product (*A*), H-RIC(Acm)DKC(Acm)KVIRRHGRVYVI-C(Acm)ENPKHKQ-OH (Boysen and Hearn 2000), and the purified synthetic peptide (*B*) was achieved with a TSK-ODS-120 T column (150 x 4.6-mm I.D., 120 Å, 5 µm, end-capped) using a flow rate of 1 ml/min and a 60-minute gradient from 100% A (0.1% TFA in H_2O) to 100% B (90% acetonitrile, 0.09% TFA) at pH 2.1.

PROTOCOL 5

Desalting of Peptides and Protein Mixtures by RP-HPLC Techniques*

THE RP-HPLC TECHNIQUE CAN BE USED TO "DESALT" PEPTIDE or protein samples derived from extraction procedures, from chemical reactions such as reductive alkylation in the presence of urea or guanidine hydrochloride, citraconylation, iodination, or cyanogen bromide cleavage, or recovered from other chromatographic separation modes (Hearn 1998). The peptide or protein solution is injected onto a small RP-HPLC column. An aqueous buffer is used to elute the salts while the peptides or proteins are concentrated at the top of the column. After elution of the salts, monitored by UV detection, the peptides or proteins are eluted with H_2O–acetonitrile or H_2O–2-propanol mobile phases. The loading capacity of an analytical column (100–300 x 4 mm I.D.) is typically ~8 mg, whereas the loading capacity for a semipreparative column (30 x 16 mm I.D.) is ~34 mg (Pohl and Kamp 1987).

This protocol was contributed by Reinhard I. Boysen and Milton T.W. Hearn (Centre for Bioprocess Technology, Department of Biochemistry and Molecular Biology, Monash University, Clayton, Victoria 3800, Australia).

MATERIALS

CAUTION: See Appendix 3 for appropriate handling of materials marked with <!>.

Reagents

Acetonitrile (HPLC grade) <!>
H_2O (deionized, from a Milli-Q water purification system or equivalent)
2-Propanol <!>
Sodium nitrate <!>
Thiourea <!>
Trifluoroacetic acid (TFA) <!>

Equipment

Analytical HPLC apparatus

This apparatus consists of a pump module, mixing chamber, spectrophotometer with an analytical flow cell, an analytical column, PC, printer and software, HPLC system with attendant data management systems, and system automation controllers (see Figure 5.10).

Column

Analytical C_4, C_8, C_{18}, or another appropriate alkylsilica sorbent appropriately packed into a suitable column (e.g., 5 μm, 300 Å, 40 x 4-mm I.D.).

Eluent bottles (two, 1-liter)

Hamilton glass syringe (graduated 25 µl)
Helium for buffer degassing <!>
Measuring cylinder (two, 1-liter)
All glassware should be rinsed three times with Milli-Q H_2O.
Microfuge (benchtop) (capable of ~20,000*g*, for sample preparation)
Mobile-phase filtration facility
vacuum pump
1-liter reservoir
support base with glass frit and integral vacuum connection
funnel
clamp
47-mm membrane filter (0.2-µm PTFE)
Safety glasses
Stopwatch
Syringe (1 ml, with truncated needle)
Vials (conical for autosampler)
Waste bottle

Biological Sample

Protein or peptide sample for analysis

METHOD

Preparation of the Mobile Phase

1. Prepare Eluent A (weak mobile phase) and Eluent B (strong mobile phase), 1 liter each, for example,

   Eluent A: 0.1% TFA in H_2O
   Eluent B: 0.09% TFA in 2-propanol

   IMPORTANT: TFA employed for protein sequencing may not be suitable because it can contain anti-oxidants.

2. Mix the solvent either by stirring with a Teflon-coated magnetic flea or by shaking a stoppered cylinder (time depends on volume).

3. Filter Eluent A and then Eluent B through a 0.2-µm PTFE filter.

4. Seal unused eluent bottles with a stopper to avoid evaporation of the organic solvent.

Preparation of Peptide and Protein Samples

1. Inspect the sample for clearness.

2. Filter the sample through a 0.2-µm PTFE filter if insoluble, opalescent, or solid particles are present. Alternatively, centrifuge the sample using the supernatant for injection.

Desalting of Proteins with RP-HPLC

1. Test the instrumental HPLC system.

   a. Produce a blank run (inject Eluent A) and run the step elution procedure first with isocratic 100% Eluent A and then with isocratic 100% Eluent B under the same conditions as intended for the protein sample.

   b. Measure the dead volume of the column with thiourea or sodium nitrate (or any other noninteractive solute, e.g., salt solution) to identify the elution time of the salt.

2. "Desalt" the proteins.

   a. Inject the salt-containing sample onto the reversed-phase column using 100% Eluent A. The salt elutes at or near to the column dead (void) time.

      Analogous methods can be used to elute urea or guanidine hydrochloride following reductive alkylation or solubilization/refolding studies, and many low-molecular-weight chemical compounds present in the sample as a consequence of chemical derivatization, backbone cleavage, or partial reduction reactions, subunit-subunit dissociation or following ion exchange, biospecific affinity, or hydrophobic interaction chromatographic steps.

   b. Elute the protein either with a step of up to 100% Eluent B or, alternatively, by gradient elution procedures.

   c. Collect the fraction containing the protein for further analysis.

Chromatographic Conditions for the Desalting of Peptides and Proteins

The following chromatographic procedures are representative of the type of standard analytical protocol that can be employed.

- Chemicals: Acetonitrile, 2-propanol, TFA
- Column: For example, C_4, C_8, C_{18}, or another appropriate alkylsilica sorbent appropriately packed into a suitable column (10 µm, 300 Å, 300 x 4-mm I.D.)
- Sample size: 8-mg peptide or protein sample
- Sample loop size: 1 ml
- Step elution
- Eluent A: 0.1% aqueous TFA
- Eluent B: 0.1% TFA in acetonitrile or 2-propanol
- Elution conditions: 100% Eluent A for 3 minutes, then 100% Eluent B for 3 minutes
- Flow rate: 2.5 ml/min
- Detection: 230 nm
- Temperature: Room temperature

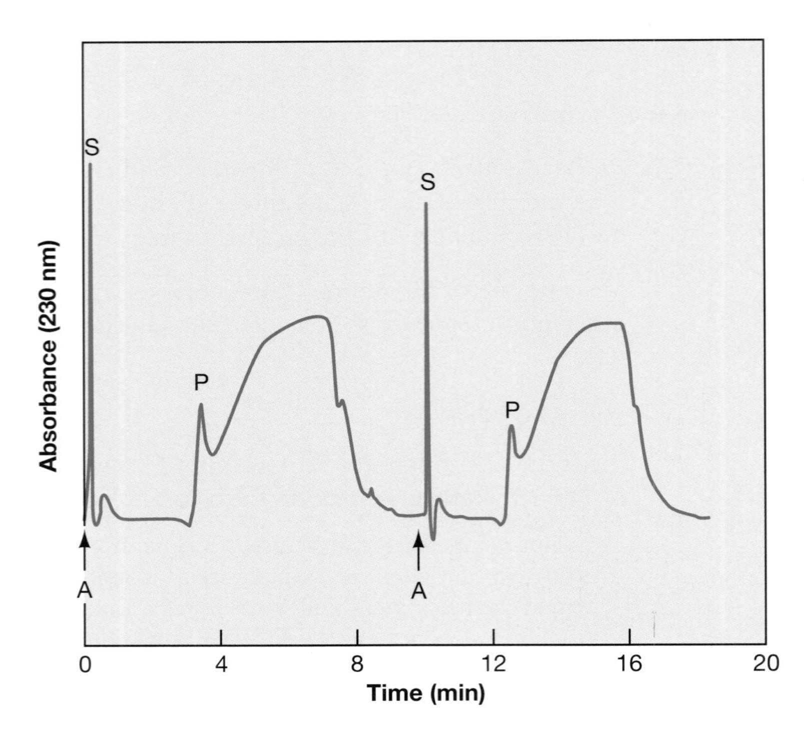

FIGURE 5.17. Use of RP-HPLC in the desalting of peptide and protein samples isolated by other techniques. In this example, the 50S ribosomal protein from *T. aquaticus* on a C_4 column (40 x 4-mm I.D., 300 Å, 5 μm) was desalted at a flow rate of 2.5 ml/min using a step elution protocol with 3 minutes of elution with Eluent A (0.1% TFA in H_2O) and 3 minutes of elution with Eluent B (0.1% TFA in 2-propanol), pH 2.1. (S) Salt; (P) protein; (A) time of injection.

Case Study: RP-HPLC in the Desalting of Peptide and Protein Samples

The example shown in Figure 5.17 is illustrative of the use of RP-HPLC in the desalting of peptide and protein samples isolated by other techniques for the step elution of a 50S ribosomal protein sample derived from *Thermus aquaticus* preparation on a C_4 column (40 x 4-mm I.D., 300 Å, 10 μm) using a flow rate of 2.5 ml/min.

PROTOCOL 6

Separation of Proteins with RP-HPLC Using a Computer-assisted Method to Optimize Gradient Conditions

FOR THE SEPARATION OF DIVERSE COMPONENTS OF A SAMPLE containing peptides or proteins of unknown composition, with a given solvent system and a given column, separation conditions that result in different retention times of the various components must be optimized. Empirical concepts describe the retention behavior of peptides and proteins with a ligand in the presence of different solvent combinations. The most commonly adapted concepts are based on the Solvophobic Theory (Horváth et al. 1976, 1977a; Hearn and Zhao 1999; Hearn 2002) and the Linear Solvent Strength Theory (Snyder 1980; Purcell et al. 1999). These concepts allow the development of fast, robust, and cost-effective separation methods (Boysen et al. 1998).

This protocol was contributed by Reinhard I. Boysen and Milton T.W. Hearn (Centre for Bioprocess Technology, Department of Biochemistry and Molecular Biology, Monash University, Clayton, Victoria 3800, Australia).

MATERIALS

CAUTION: See Appendix 3 for appropriate handling of materials marked with <!>.

Reagents

Acetone <!>

Acetonitrile (HPLC grade) <!>

H_2O (deionized, from a Milli-Q water purification system or equivalent)

Phosphoric acid (H_3PO_4) <!>

Sodium dihydrogen phosphate (NaH_2PO_4) <!> or other suitable salts

Thiourea <!> or sodium nitrate <!>

Trifluoroacetic acid (TFA) <!>

Equipment

Analytical HPLC apparatus

This apparatus consists of a pump module, a mixing chamber, spectrophotometer with an analytical flow cell, an analytical column, a column oven, a manual injector with sample loop, PC, printer and software, e.g., Beckman, System Gold; Agilent HP-1090A (or 1100 series) liquid chromatograph, or Waters 600/486 HPLC system with attendant data management systems and system automation controllers.

Column

Analytical C_{18} or another appropriate *n*-alkylsilica sorbent appropriately packed into a suitable column (e.g., 5 μm, 300 Å, 250 x 4-mm I.D.).

Conical vials (for autosampler)
Eluent bottles (two, 1-liter)
Hamilton glass syringe (graduated, 25 µl)
Helium gas (for buffer degassing) and nitrogen (for autosampler operation) <!>
Measuring cylinder (two, 1-liter)
All glassware should be rinsed three times with Milli-Q H_2O.
Microfuge (benchtop)
Mobile-phase filtration facility
vacuum pump
1-liter reservoir
support base with glass frit and integral vacuum connection
funnel
clamp
47-mm membrane filter (0.2-µm PTFE)
Safety glasses
Stopwatch
Syringe (1 ml, with truncated needle)
Waste bottle

Biological Molecules

HPLC peptide standards
See respective sections below.
Protein sample

METHOD

Preparation of the Mobile Phase

The following considerations are relevant to the preparation of mobile phases to be used with samples containing peptides and proteins.

1. Prepare Eluent A (weak mobile phase) and Eluent B (strong mobile phase), 1 liter each, for example,

   Eluent A: 0.1% TFA in H_2O
   Eluent B: 0.09% TFA in 60% acetonitrile/40% H_2O (v/v)

   The volumes of organic solvent and H_2O must be measured in two different cylinders and then combined because of volume contraction upon mixing, which may be up to 30 ml/liter prepared solvent (depending on the nature of the solvent). Failure to do so can lead to substantial errors in mobile-phase composition.

   To compensate for the baseline shift in gradient elution (because organic components absorb more light at low wavelengths) when working with H_2O/organic eluents, the amount of ion-pair reagent (TFA, H_3PO_4) in Eluent B is usually decreased by 10–15% in comparison with Eluent A, yielding a flat baseline (Dolan 1991).

   IMPORTANT: TFA employed for protein sequencing may not be suitable because it can contain anti-oxidants.

2. Mix the solvent either by stirring with a Teflon-coated magnetic flea or by shaking a stoppered cylinder (time depends on volume).

   WARNING: *Do not* use Parafilm under any circumstances to cover the eluents, because the organic solvents in combination with acidic ion-pair reagent components in the eluents will dissolve components in the Parafilm, yielding extra peaks in the chromatogram.

3. Filter Eluent A and then Eluent B through a 0.20-µm PTFE filter.

   Filtering of eluents increases the column lifetime and contributes to degassing.

4. Seal unused eluent bottles with a stopper to avoid evaporation of the organic solvent.

Preparation of the Protein Sample

1. Dissolve the sample with half the target volume of Eluent A (weak mobile phase). If the sample is not soluble, add a small amount (typically <25% of total volume) of Eluent B (strong mobile phase).

   Small amounts of strong mobile phase may cause pre-elution of the sample depending on the sample loop size, column dimensions, and starting mobile-phase composition of the gradient.

2. Inspect the sample for clearness. Filter the sample through a 0.2-µm PTFE filter if insoluble, opalescent, or solid particles are present. Alternatively, centrifuge the sample using the supernatant for injection.

   Do not use samples that are not fully dissolved for injections, because they can block the injector and column.

3. Store sample if not in use at 4°C or –20°C, depending on the planned storage time and usage. Peptides and protein samples can degrade at room temperature.

   Avoid repeated freezing-thawing of the sample. Instead, prepare small aliquots of the peptide or protein sample, which are kept at –20°C and are used for one task.

Test the Instrumental HPLC System

The following considerations are relevant to the testing of the HPLC system for use with samples containing peptides and proteins.

1. Produce a blank run (inject Eluent A) and run a gradient 100% A → 100% B under the same conditions as intended for the peptide or protein sample. Repeat if peaks occur.

   This procedure cleans the column of peptides and proteins from previous separations, which have not been removed by the flushing process.

2. Measure the dead volume of the column with thiourea or sodium nitrate (or any other noninteractive solute).

3. Test the column performance with a gradient run and an appropriate test mixture.

   An example of the RP-HPLC test mixture is 0.15% (w/v) dimethylphthalate, 0.15% (w/v) diethylphthalate, 0.01% (w/v) diphenyl, 0.03% (w/v) *O*-terphenyl, 0.32% (w/v) dioctylphthalate in methanol.

This test allows the evaluation of the column bed integrity (low integrity will be associated with split, fronting, or tailing peaks) and column performance (in terms of plate numbers). It also allows (if repeated at regular intervals) monitoring of performance during the lifetime of a column, and the assessment of batch-to-batch differences of column fillings.

Peptide standards for column testing are described, for example, in Mant and Hodges (1991b).

4. Measure the gradient delay (dwell volume) of the HPLC system (for another way of going about this task, see Chapter 7). The following steps must be done without a column connected to the HPLC instrumentation:

   a. Connect the injector directly to the detector with union piece (zero-length column).

   b. Prepare a special Eluent A and Eluent B, 200 ml each:

      Eluent A: acetonitrile
      Eluent B: acetonitrile, 0.2% acetone

   c. Run a gradient of 10–90% Eluent B for 10 minutes at a flow rate of 2.0 ml/min. Detection is carried out at 254 nm.

      The measured value of the dwell volume can be influenced by the injection technique. If after the injection, the valve remains in the inject position, the dwell volume will include the volume of the sample loop; if the valve is put back in load position, the dwell volume will not. This effect can produce errors with sample loops >100 µl. The same consideration is valid if the sample loop is exchanged from an analytical separation (e.g., a sample loop of 50 µl) to a semipreparative separation (e.g., a sample loop of ≥500 µl) on the same column.

   d. Determine the gradient delay and present the results graphically in a format similar to that shown as Figure 5.18.

      The dwell volume, V_D, is the volume of eluent from the pump heads to the column inlet (including the mixing chamber volume). The dwell volume values range from 2 ml to 7 ml; autosamplers in particular make a large contribution to the delay volume. It should be determined with an accuracy of ±0.5 ml. The obtained profile can be used for diagnostic purposes, because the volume accuracy of the pump delivery is also monitored.

      Knowledge of the gradient delay is essential for method development, because it allows the accurate calculation of the S and k_0 values (loc. cit). Its determination is particularly important when establishing segmented gradients (since various errors can accumulate here) and when an established HPLC method is transferred from one instrument to another instrument.

Perform Analytical RP-HPLC of the Protein Sample

1. Prior to sample analysis, conduct at least one blank run under the same conditions as intended for the protein sample. Repeat if peaks occur.

2. Separate the peptide or protein sample (~100 µg) with two analytical RP-HPLC runs differing by a factor of 3.

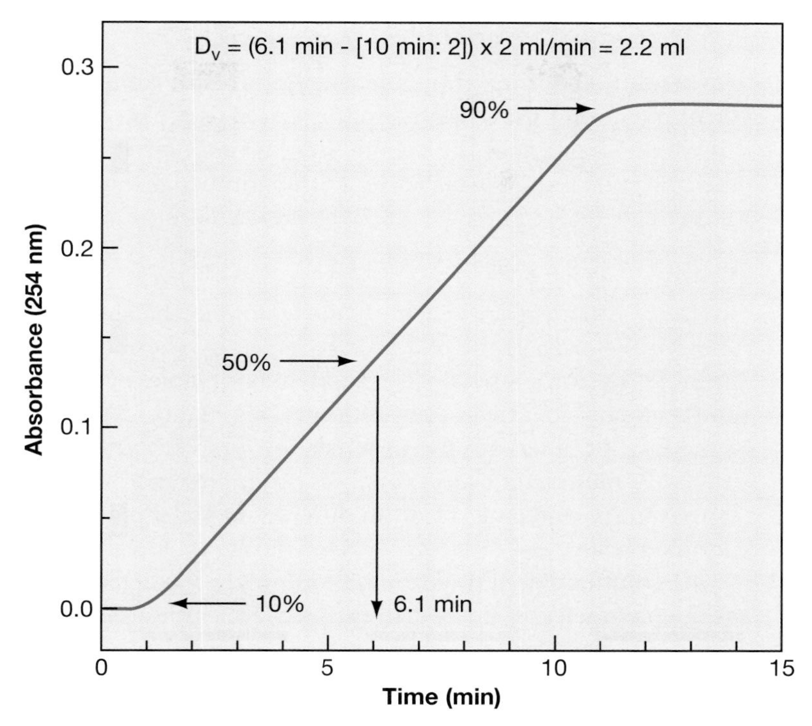

FIGURE 5.18. Graphical illustration of the approach employed to determine the gradient delay. The pregradient (10%) and postgradient baselines (90%) are used to determine the time when the 50% value is reached. This is then used to calculate the gradient delay by subtracting the theoretical half-way time under consideration of the flow rate as depicted in the formula.

Chromatographic Conditions for the Analytical Separation of Crude Peptides and Proteins

- Column: For example, C_4, C_8, C_{18}, or another appropriate *n*-alkylsilica sorbent appropriately packed into a suitable column etc (5 µm, 300 Å, 250 x 4.0-mm I.D.)
- Sample size: <2 mg peptide/protein
- Sample loop size: 100–500 µl
- Linear A → B gradient
- Eluent A: 0.9% aqueous TFA <!>
- Eluent B: 0.1% TFA in acetonitrile <!>/H_2O
- Gradient rate: 1%B/min, for example, 60% acetonitrile/H_2O
- Gradient range and time: 0–100% B in 60 minutes
- Flow rate: 1.0 ml/min
- Detection: 214 nm
- Temperature: Room temperature

Optimization of the Gradient Conditions

For resolution optimization, the strategy takes advantage of the relationship between the gradient retention time of a protein (expressed as the median capacity factor, $\bar{k}$) and the median volume fraction of the organic solvent modifier $\bar{\varphi}$ in regular RP-HPLC systems based on the concepts of the Linear Solvent Strength theory (Horváth et al. 1976; Snyder 1980; Hearn and Grego 1983c; Purcell et al. 1999; Hearn 2002), such that

$$\ln \bar{k} = \ln k_O - S\bar{\varphi}$$

where k_O is the capacity factor of the solute in the absence of the organic solvent modifier, and S is the slope of the plot of $\ln \bar{k}$ versus $\bar{\varphi}$. The values of $\ln k_O$ and S can be calculated by linear regression analysis. The underlying principles of an intuitively performed optimization and manually achieved optimization (using Excel spreadsheets to calculate, e.g., the $\ln k_O$ and S values), or, alternatively, optimization via computer simulation software (e.g., Simplex methods, multivariate factor analysis programs, and DryLab G/*plus*) are essentially the same. However, the outcomes result in different levels of precision. The approaches are collectively outlined below.

Resolution, R_S, of peak zones is optimized through adjustment of $\bar{k}$ by successive change of the parameters t_G and $\Delta\varphi$ in the gradient elution mode according to the steps:

1. Initial experiments.
2. Choice of starting conditions.
3. Calculation of $\ln k_O$ and S values from initial chromatograms.
4. Peak tracking and assignment of the peaks.
5. Optimization of gradient run time t_G over the whole gradient range.
6. Determination of new gradient range.
7. Calculation of new gradient retention times t_g.
8. Change of gradient shape (*optional*).
9. Verification of the results.

1. Initial experiments

The peptide or protein sample in initial experiments is separated under two linear gradient conditions differing by a factor of 3 in their gradient run times t_G (all other chromatographic parameters being held unchanged) (Dolan et al. 1989; Boysen et al. 1998) to obtain the RP-HPLC retention times of each of the peptides or proteins. Irrespective of what optimization strategy will then be used, it is advisable to separate any sample with at least two different gradient run times to identify overlapping peaks. For the optimization of the gradient shape and to achieve maximum resolution between adjacent peak zones, the ability to determine the retention times of the peptides or proteins and to classify the parameters that reflect the contributions from the mobile-phase composition and column dimensions (see Table 5.9) is essential (Ghrist and Snyder 1988a,b; Ghrist et al. 1988). Although the determination of the volume, V_{mix}, is useful (Stadalius et al. 1984), determination of dead volume and gradient delay is essential (Snyder 1980; Boysen et al. 1998).

With these compiled input parameters, various algorithms, including DryLab G/*plus*, can generate the relative resolution map (RRM), based on the calculation of the corresponding S values and k_o values for each component. If DryLab G/*plus* is not available, the resolution information can be extracted directly from the vertical distances of the individual R_S versus

TABLE 5.9. Parameters required for the calculation of $\ln k_O$ and S values, for manual optimization, and optimization with DryLab G/*plus*

Parameter	Optimization utilizing basic equations (Excel)	Computer-assisted optimization (DryLab)
Gradient delay volume V_D (ml)	√	√
Dead volume t_O (ml)	√	–
Gradient range (% B)	√	√
Gradient run time t_{G1},t_{G2} (min)	√	√
Column length L (mm)	–	√
Column diameter d_C (mm)	–	√
Flow rate F (ml/min)	–	√
Column plate number N	–	√
Retention times t_{g1} for all peaks	√	√
Retention times t_{g2} for all peaks	√	√

t_G plots, whereby a crossover of lines reflects a peak overlap as shown in Figure 5.19a,b,c. Over a small range of $\overline{\varphi}$ values, the S values and k_o values for an individual peptide or protein will remain essentially constant. It is assumed that no significant conformational change occurs for the individual peptides or proteins as a consequence of the differences in column residency time and gradient time (Purcell et al. 1993, 1999). The typical retention behavior of the peptides and proteins is depicted in Figure 5.19 as plots of $\ln \overline{k}$ versus $\overline{\varphi}$. These plots become nonparallel, and frequently intersect, as the mobile-phase composition is varied.

2. Choice of starting conditions

The S values of peptides and proteins can be empirically correlated with their molecular weights (Snyder et al. 1983), based on a correlation of

$$S = 0.48(MW)^{0.44}$$

Thus, S values, estimated with this equation or, alternatively, values of similar peptides or proteins taken from the literature, can be used to calculate a reasonable starting point for the initial experiments according to the following equation:

$$t_G = \frac{V_m * \Delta\varphi * S * \overline{k}}{0.87 * F}$$

The required gradient run time t_G for a separation of peptides and proteins with expected S values of ~20 can be calculated for a gradient of 0–100% (60% acetonitrile) at a flow rate of 1 ml/min, and an ideal k value of 5 as follows:

$$t_G = \frac{1.9 \text{ ml} * 0.6 * 20 * 5}{0.87 * 1 \text{ ml/min}} = 131 \text{ min}$$

On the basis of this calculation, the initial gradients from 0% to 100% B for gradient times of 1- and 3-hour duration can be selected. On the other hand, availability of the S values, derived from two gradient elution RP-HPLC experiments, can be used as an analytical criterion early in the separation of the target peptide or protein from other components in soluble extracts of biological sources to distinguish low- from high-molecular-weight mole-

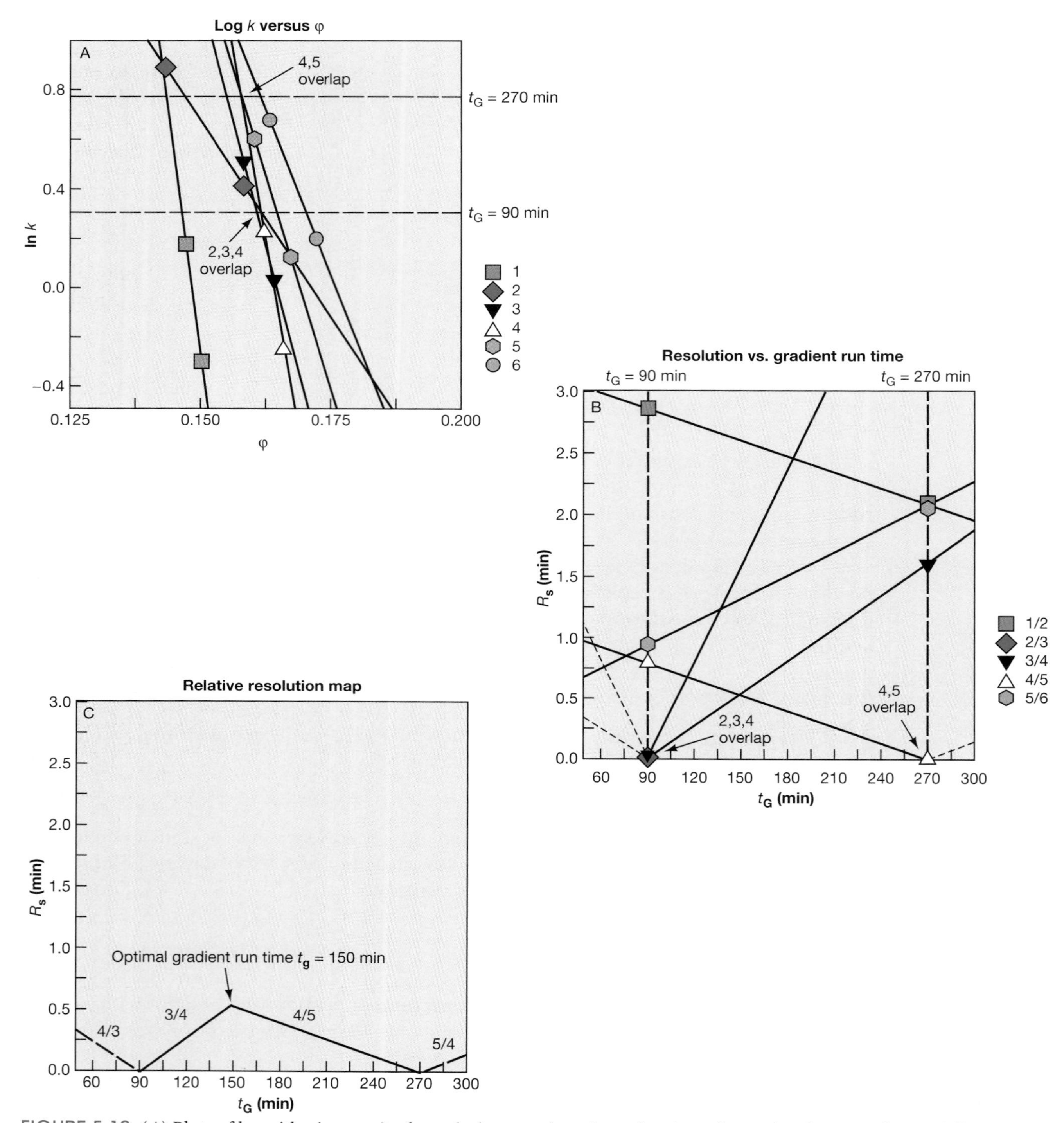

FIGURE 5.19. (*A*) Plots of logarithmic capacity factor ln *k* versus the volume fraction of organic solvent, φ, for six different ribosomal proteins, denoted 1–6, obtained from *T. aquaticus* on a C_{18} column (Molnar et al. 1989a). As evident from this plot, the intersection of two plots represents complete overlap of the peak zones at the specific gradient run times t_G = 90 and 270 minutes. (*B*) Plots of the resolution R_S (resolution taken as the difference of two adjacent peaks) versus the gradient run time t_G of these six different proteins from *T. aquaticus*, which is the blueprint for generation of the relative resolution map (RRM). Because the plotted values of the resolution are absolute values, negative resolution values of peaks that change their elution order are depicted as broken lines. (*C*) RRM of the six different proteins from *T. aquaticus*, depicting only the critical peak pairs, including their change in retention order. The resolution optimum for this specific set of chromatographic conditions occurs at t_G = 150 minutes.

cules and, for example, to exclude peptide fragments participating in the optimization process, without having to resort to SDS-PAGE experiments. The RP-HPLC behavior of small peptides with molecular weights from 300 to 1000 are approximately correlated (Hearn and Grego 1981; Snyder 1990) with S values between 3 and 10. For medium-molecular-weight globular proteins, S values above 20 are expected (Aguilar et al. 1985). As an example, cytochrome c, with a molecular weight of ~12,000, has an S value of 28.8 on a Nucleosil C_{18} column (Stadalius et al. 1984).

3. Calculation of ln k_O and S values from initial chromatograms

The retention times t_{g1} and t_{g2} for a peptide or protein solute separated under conditions of two different gradient run times (t_{G1} and t_{G2}, whereby $t_{G1} < t_{G2}$) can be given by the equations (Quarry et al. 1986; Ghrist et al. 1988; Purcell et al. 1999):

$$t_{g1} = \left(\frac{t_0}{b_1}\right)\log(2.3k_0b_1) + t_0 + t_D$$

$$t_{g2} = \left(\frac{t_0}{b_2}\right)\log(2.3k_0b_2) + t_0 + t_D$$

with

$$\frac{b_1}{b_2} = \frac{t_{G2}}{t_{G1}} = \beta$$

Here, t_{G1}, t_{G2} are the gradient run time values of t_G for two different gradient runs, resulting in different values of b (b_1, b_2) and t_g (t_{g1}, t_{g2}) for a single solute; t_{g1}, t_{g2} are the gradient retention times for a single solute in two different gradient runs; b_1, b_2 are the gradient steepness parameters for a single solute and two gradient runs differing only in their gradient times. Steep gradients correspond to large b values and small $\bar{k}$ values; k_0 is the solute capacity factor at the initial mobile-phase composition; β is the ratio of t_{G2} and t_{G1} which is equivalent to the ratio of b_1 and b_2; t_0 is the column dead time; and t_D is the gradient delay time.

For peptides and proteins, there is an explicit solution (Stadalius et al. 1984; Quarry et al. 1986) for b and k_0, namely:

$$b_1 = \frac{t_0 \log \beta}{\left[t_{g1} - \left(\frac{t_{g2}}{\beta}\right) + (t_0 + t_D)\left(\frac{1-\beta}{\beta}\right)\right]}$$

$$\log k_0 = \left(\frac{b_1}{t_0}\right)(t_{g1} - t_0 - t_D) - \log(2.3b_1)$$

From the knowledge of b and k_0, the values of $\bar{k}$ and $\bar{\varphi}$ can be calculated (Snyder 1980; Hearn 1991c):

$$\bar{k} = \frac{1}{1.15b_1}$$

$$\bar{\varphi} = \frac{\left[t_{g1} - t_0 - t_D - \left(\frac{t_0}{b_1}\right)\log 2\right]}{t^{0}_{G1}}$$

Here, $\bar{k}$ is the value of k' (capacity factor) for a solute when it reaches the column midpoint during elution; φ is the volume fraction of solvent in the mobile phase; $\Delta\varphi$ is the change in φ for the mobile phase during the gradient elution ($\Delta\varphi = 1$ for a 0–100% gradient); $\bar{\varphi}$ is the effective value of φ during gradient elution and the value of φ at band center when the band is at the midpoint of column, and t_G^0 is the normalized gradient time with $t_{G1}^0 = t_{G1}/\Delta\varphi$.

By linear regression analysis, using $\bar{k}$ and $\bar{\varphi}$, the S value (empirically related to the hydrophobic contact area between solute and ligand) can be derived from the slope of the log $\bar{k}$ versus $\bar{\varphi}$ plots, and $\ln k_O$ (empirically related to the affinity of the solute toward the ligand) as the y intercept (Horváth et al. 1976; Hearn 2000a, 2002).

$$S = \frac{\ln k_O - \ln \bar{k}}{\bar{\varphi}}$$

4. Peak tracking and assignment of the peaks

Complex chromatograms that result from reversed-phase gradient elution can often exhibit changes in peak order when the gradient steepness is changed. Before $\ln k_O$ and S values are calculated, or computer simulation is used, the peaks from the two initial runs must be correctly assigned. Several approaches to peak tracking have been described, using algorithms based on relative retention and peak areas (Glajch et al. 1986; Lankmayr et al. 1989; Molnar et al. 1989b) or based on diode-array detection (DAD) (Berridge 1986; Strasters et al. 1989; Round et al. 1994). The assignment of peaks can be done in the following steps:

- ***Integrate the chromatograms*** (using the integration software of the HPLC system or an integrator) of the initial runs and correct integration due to baseline drift or other instrumental causes where necessary.
- ***Print out both chromatograms*** including the peak area percent reports.
- ***Number all relevant peaks*** in the chromatogram with the better resolution. Relevant peaks have, for example, an area greater than 0.5% of the overall peak area.
- ***Assign the peaks*** according to their peak areas allowing a reasonable elution window.

The total peak areas, A_{T1} and A_{T2} for initial runs 1 and 2 (whereby $t_{G1} < t_{G2}$) are determined as

$$A_T = \sum_{i=1}^{n} A_i$$

and their ratio is calculated:

$$R_T = \frac{A_{T2}}{A_{T1}}$$

Consequently, the ratio of the peak areas, A_{i1} and A_{i2} of a baseline-separated single component in initial runs 1 and 2, respectively, is

$$R_i = \frac{A_{i2}}{A_{i1}}$$

If the peaks are correctly assigned, $R_T = R_i$ difficulties arise when peaks partially or completely overlap. If $R_T << R_i$ and the peak in run 2 is composed of a single component, the peaks are wrongly matched. If $R_T >> R_i$, then the peaks might be correctly matched, but the peak area in run 1 could be enlarged by a hidden additional peak. To resolve these difficulties, the deviation ($\%R_T$) of R_i from R_T can be calculated and taken as a measure of the likelihood that a peak assignment hypothesis is correct:

$$\%R_T = \frac{100R_i}{R_T} - 100$$

5. Optimization of the gradient time, t_G, over the whole gradient range

The capacity factor $\bar{k}$ is a linear function of the gradient run time t_G if $\Delta\varphi$ is kept constant. Hence,

$$\frac{\bar{k}}{t_G} = \frac{0.87F}{V_m * \Delta\varphi * S} = \text{const.} = C$$

The optimized gradient run time t_{GRRM} can be obtained from the RRM or, alternatively, from the plot of R_S versus t_G (see Figure 5.19), and yields for each peptide or protein the new values of $\bar{k}_{new}$ by t_{GRRM} being multiplied with C:

$$C\, t_{GRRM} = \bar{k}_{new}$$

6. Determination of the new gradient range

If the gradient run time t_{GRRM} is changed in relation to $\Delta\varphi$ with $t^0_{G1} = \text{const.}$, the $\bar{k}$ values do not change, as can be seen from the following equation:

$$t^0_{G1} = \frac{t_{GRRM}}{\Delta\varphi} = \frac{V_m * S * \bar{k}}{0.87 * F}$$

where

$$\Delta\varphi_{opt} = \frac{t_{Gopt}}{t^0_{G1}}$$

and the retention time t_g of the first peak is greater than $(t_0 + t_D)$ and the retention time t_g of the last peak is less than Δt_{Gopt}.

7. Calculation of the new gradient retention times, t_g

On the basis of knowledge of the S and the $\ln k_0$ values, new gradient retention times can then be calculated.

8. Change of gradient shape (optional)

The multisegmented gradient should only be performed when the gradient delay has been measured. With multisegmented gradients, an error in the gradient delay will reoccur at the beginning and at the end of each gradient step. In addition, the effect of V_{mix} (determined according to the procedures described in Ghrist et al. 1988), which modifies the composition of the gradient at the start and end (rounding of the gradient shape), can lead to deviation of the experimentally determined retention times from the predicted "ideal" retention times in DryLab G/*plus* simulations.

9. Verification of the results

After completion of the optimization process, the simulated chromatographic separation can now be verified experimentally using the predicted chromatographic conditions.

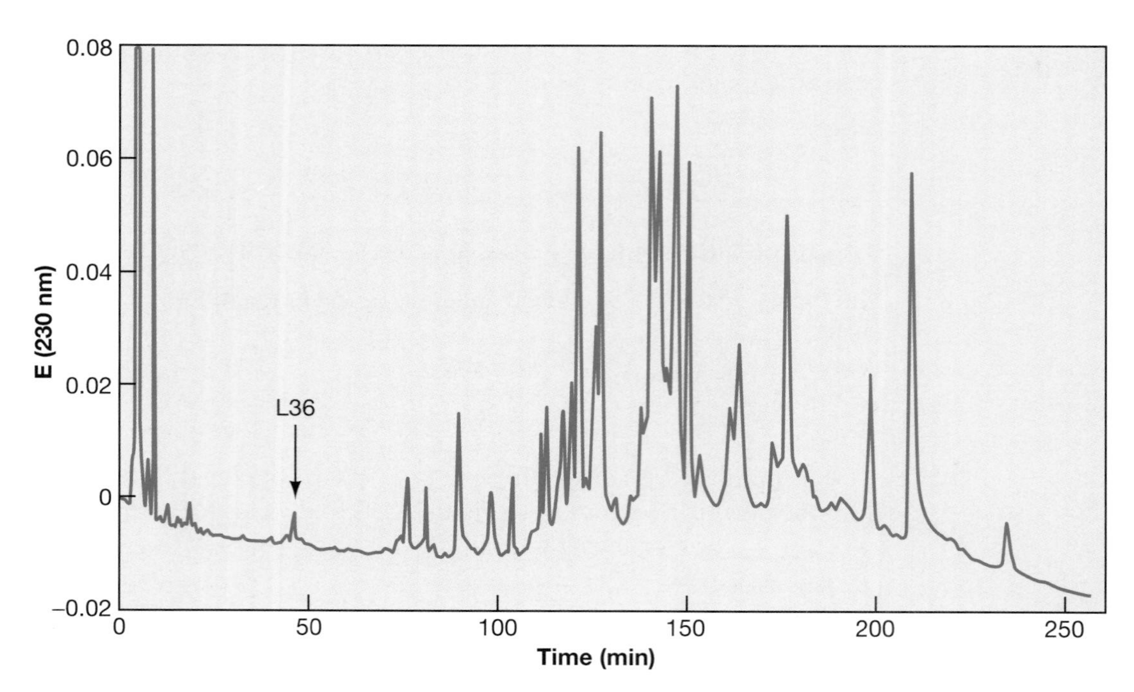

FIGURE 5.20. 50S ribosomal subunit proteins of *T. thermophilus* were separated by gradient elution RP-HPLC, using Eluent A (500 mM NaH_2PO_4, 500 mM H_3PO_4 at pH 2.1) and Eluent B (200 mM H_3PO_4 at pH 2.1):(acetonitrile) (25:75) (v/v) (Boysen et al. 1998). The initial linear gradients for optimization were from 10% to 80% B of either 3 hours or 9 hours duration with a flow rate of 1.0 ml/min. All experiments were carried out at 20°C. The location of the zinc finger ribosomal protein L36 of *T. thermophilus* is indicated by the arrow. The sequences of all other *T. thermophilus* 50S ribosomal subunit proteins have been similarly determined by automated Edman microsequencing following optimization of this RP-HPLC separation.

Case Study: RP-HPLC of *Thermus thermophilus* 50S Ribosomal Proteins

The method of de novo optimization of the resolution of complex protein mixtures using the computer-assisted method development was successfully applied to the RP-HPLC of *T. thermophilus* 50S ribosomal proteins (Figure 5.20) (Boysen et al. 1998). The separation approach for these basic ribosomal proteins used a mobile phase of high ionic strength to suppress silanophilic interactions with a non-end-capped LiChrospher RP-18 sorbent. These conditions were found to be a key requirement for achieving good resolution with minimal peak-tailing. The retention times of the 50S ribosomal proteins were observed to be in very close agreement with values predicted by computer simulation procedures based on linear solvent strength concepts, with an average error of only 0.5%.

REFERENCES

Aguilar M.I., Hodder A.N., and Hearn M.T.W. 1985. High-performance liquid chromatography of amino acids, peptides and proteins. LXV. Studies on the optimization of the reversed-phase gradient elution of polypeptides: Evaluation of retention relationships with β-endorphin related polypeptides. *J. Chromatogr.* **327:** 115–138.

Alvarez V.L., Roitsch C.A., and Henriksen O. 1981. Purification of H-2a heavy chain and β-microglobulin by reversed-phase high-performance liquid chromatography. *Anal. Biochem.* **115:** 353–358.

Amersham Pharmacia Biotech. 2002. *Reversed phase chromatography: Principles and methods*, publication no. 18-1134-16. Amersham Biosciences, Buckinghamshire, United Kingdom.

Bennett H.P. 1983. Isolation of pituitary peptides by reversed-phase high performance liquid chromatography. Expansion of the resolving power of reversed-phase columns by manipulating pH and the nature of the ion-pairing reagent. *J. Chromatogr.* **266:** 501–510.

——. 1991. Manipulation of pH and ion-pairing reagents to maximize the performance of reversed-phase columns. In *High-performance liquid chromatography of peptides and proteins: Separation, analysis, and conformation* (ed. C.T. Mant and R.S. Hodges), pp. 319–326. CRC Press, Boca Raton, Florida.

Berchtold M.W., Wilson K.J., and Heizmann C.W. 1982. Isolation of neuronal parvalbumin by high-performance liquid chromatography. Characterization and comparison with muscle parvalbumin. *Biochemistry* **21:** 6552–6557.

Berridge J.C. 1986. *Techniques for the automated optimization of HPLC separations.* Wiley Interscience, Chichester, United Kingdom.

Boysen R.I. and Hearn M.T.W. 2000. Direct characterisation by electrospray ionisation mass spectroscopy of mercuro-polypeptide complexes after deprotection of acetamidomethyl groups from protected cysteine residues of synthetic polypeptides. *J. Biochem. Biophys. Methods* **45:** 157–168.

Boysen R.I., Erdmann V.A., and Hearn M.T.W. 1998. Systematic, computer-assisted optimisation of the isolation of *Thermus Thermophilus* 50S ribosomal proteins by reversed-phase high-performance liquid chromatography. *J. Biochem. Biophys. Methods* **37:** 69–89.

Burgess A.W., Knesel J., Sparrow L.G., Nicola N.A., and Nice E.C. 1982. Two forms of murine epidermal growth factor: Rapid separation by using reversed-phase HPLC. *Proc. Natl. Acad. Sci.* **79:** 5753–5757.

Chervet J.P., Ursem M., Salzmann J.P., and Vannort R.W. 1989. Ultra-sensitive UV detection in micro separation. *HRC J. High Resolut. Chromatogr.* **12:** 278–281.

Chesterman C.N., Walker T., Grego B., Chamberlain K., Hearn M.T., and Morgan F.J. 1983. Comparison of platelet-derived growth factor prepared from release products of fresh platelets and from outdated platelet concentrates. *Biochem. Biophys. Res. Commun.* **116:** 809–816.

Cohen K.A., Schellenberg K., Benedek K., Karger B.L., Grego B., and Hearn M.T. 1984. Mobile-phase and temperature effects in the reversed phase chromatographic separation of proteins. *Anal. Biochem.* **140:** 223–235.

Cooke N.H.C., Archer B.G., O'Hare M.J., Nice E.C., and Capp M. 1983. Effects of chain length and carbon load on the performance of alkyl-bonded silicas for protein separations. *J. Chromatogr.* **255:** 115–123.

Davis M.T. and Lee T.D. 1992. Analysis of peptide mixtures by capillary high performance liquid chromatography: A practical guide to small-scale separations. *Protein Sci.* **1:** 935–944.

Dolan J.W. 1991. Preventive maintenance and troubleshooting LC instrumentation. In *High-performance liquid chromatography of peptides and proteins: Separation, analysis, and conformation* (ed. C.T. Mant and R.S. Hodges), pp. 23–29. CRC Press, Boca Raton, Florida.

Dolan J.W., Lommen D.C., and Snyder L.R. 1989. DryLab computer simulation for high-performance liquid chromatographic method development. II. Gradient elution. *J. Chromatogr.* **485:** 91–112.

Forage R.G., Ring J.M., Brown R.W., McInerney B.V., Cobon G.S., Gregson P., Robertson D.M., Morgan F.J., Hearn M.T.W., Findlay J.K., Wettenhall R.E.H., Burger H.G., and de Kretser D.M. 1986. Cloning and sequence analysis of cDNA species coding for the two subunits of inhibin from bovine follicular fluid. *Proc. Natl. Acad. Sci.* **83:** 3091–3095.

Gerber G.E., Anderegg R.J., Herlihy W.C., Gray C.P., Biemann K., and Khorana H.G. 1979. Partial primary structure of bacteriorhodopsin: Sequencing methods for membrane proteins. *Proc. Natl. Acad. Sci.* **76:** 227–231.

Ghrist B.F.D. and Snyder L.R. 1988a. Design of optimized high-performance liquid chromatographic gradients for the separation of either small or large molecules. II. Background and theory. *J. Chromatogr.* **459:** 25–41.

——. 1988b. Design of optimized high-performance liquid chromatographic gradients for the separation of either small or large molecules. III. An overall strategy and its application to several examples. *J. Chromatogr.* **459:** 43–63.

Ghrist B.F.D., Coopermann B.S., and Snyder L.R. 1988. Design of optimized high-performance liquid chromatographic gradients for the separation of either small or large molecules. Minimizing errors in computer simulations. *J. Chromatogr.* **459:** 1–23.

Glajch J.L., Quarry M.A., Vasta J.F., and Snyder L.R. 1986. Separation of peptide mixtures by reversed-phase gradient elution. Use of flow rate changes for controlling band spacing and improving resolution. *Anal. Chem.* **58:** 280–285.

Grego B. and Hearn M.T. 1984. High-performance liquid chromatography of amino acids, peptides and proteins. LXIII. Reversed-phase high-performance liquid chromatographic characterisation of several polypeptide and protein hormones. *J. Chromatogr.* **336:** 25–40.

Grego B., Baldwin G.S., Knessel J.A., Simpson R.J., Morgan F.J., and Hearn M.T. 1984. High-performance liquid chromatography of amino acids, peptides and proteins. LVIII. Application of reversed-phase high-performance liquid chromatography to the separation of tyrosine-specific phosphorylated polypeptides related to human growth hormone. *J. Chromatogr.* **297:** 21–29.

Gurley L.R., D'Anna J.A., Blumenfeld M., Valdez J.G., Sebring R.J., Donahue P.R., Prentice D.A., and Spall W.D. 1984. Preparation of histone variants and high-mobility group proteins by reversed-phase high-performance liquid chromatography. *J. Chromatogr.* **297:** 147–165.

Hancock W.S., ed. 1984. *CRC handbook of HPLC for the separation of amino acids, peptides and proteins.* CRC Press, Boca Raton, Florida.

Hancock W.S. and Sparrow J.T. 1983. The separation of proteins by reversed-phase high-performance liquid chromatography. In *High-performance liquid chromatography. Advances and perspectives* (ed. C. Horváth), vol. 3, pp. 50–87. Academic Press, New York.

Hancock W.S., Bishop C.A., Prestige R.L., Harding D.R.K., and Hearn M.T.W. 1978a. Reversed-phase high-pressure liquid chromatography of peptides and proteins with ion-pairing ragents. *Science* **200:** 1168–1170.

Hancock W.S., Bishop C.A., Prestige R.L., Harding D.R.K., and Hearn M.T.W. 1978b. High-pressure liquid chromatography of peptides and proteins. II. The use of phosphoric acid in the analysis of underivatized peptides by reversed-phase high-pressure liquid chromatography. *J. Chromatogr.* **153:** 391–398.

Hancock W.S., Bishop C.A., Gotto A.M., Harding D.R., Lamplugh S.M., and Sparrow J.T. 1981. Separation of the apoprotein components of human very low density lipoproteins by ion-paired, reversed-phase high performance liquid chromatography. *Lipids* **16:** 250–259.

Hearn M.T.W. 1980. Ion-pair chromatography on normal and reversed-phase systems. *Adv. Chromatogr.* **18:** 59–100.

——. 1984. Reversed-phase high performance liquid chromatography of proteins. *Methods Enzymol.* **104:** 190–212.

——. 1985. Ion-pair chromatography of amino acids, peptides, and proteins. In *Ion-pair chromatography: Theory and biological and pharmaceutical applications* (ed. M.T.W. Hearn), pp. 207–257. Marcel Dekker, New York.

——. 1991a. Current status and future challenges of high performance liquid chromatographic techniques for biopolymer analysis and purification. In *HPLC of peptides, proteins and polynucleotides* (ed. M.T.W. Hearn), pp. 1–35. VCH, New York.

——. 1991b. High-performance liquid chromatography of peptides and proteins: General principles and basic theory. In *High-performance liquid chromatography of peptides and proteins: Separation, analysis, and conformation* (ed. C.T. Mant and R.S. Hodges), pp. 95–104. CRC Press, Boca Raton, Florida.

——. 1991c. High-performance liquid chromatography of peptides and proteins: Quantitative relationships for isocratic and gradient elution of peptides and proteins. In *High-performance liquid chromatography of peptides and proteins: Separation, analysis, and conformation* (ed. C.T. Mant and R.S. Hodges), pp. 105–122. CRC Press, Boca Raton, Florida.

——. 1998. High-resolution reversed-phase chromatography of polypeptides and proteins. In *Protein purification: Principles, high-resolution methods, and applications*, 2nd edition. (ed. J.-C. Janson and L.

Rydén), pp. 239–282. Wiley-Liss, New York.

——. 2000a. Conformational behaviour of polypeptides and proteins in reversed-phase environments. In *Theory and practice of biochromatography* (ed. M.A. Vijayalakshmi), pp. 72–235. Harwood Academic Publishers, Switzerland.

——. 2000b. Physicochemical factors in polypeptide and protein purification and analysis by high-performance liquid chromatographic techniques: Current status and future challenges. In *Handbook of bioseparation* (ed. S. Ahyja), pp. 72–235. Academic Press, San Diego, California.

——. 2002. Reversed-phase and hydrophobic interaction chromatography of proteins and peptides. In *HPLC of biological macromolecules* (ed. K.M. Gooding and F.E. Regnier), pp. 99–245. Marcel Dekker Inc., New York.

Hearn M.T.W. and Grego B. 1981. Organic solvent modifier effects in the separation of unprotected peptides by reversed-phase liquid chromatography. *J. Chromatogr.* **218:** 497–507.

Hearn M.T.W. and Grego B. 1983a. High performance liquid chromatography of amino acids, peptides and proteins. XLVI. Selectivity effects of peptidic positional isomers and oligomers separated by reversed-phase high-performance liquid chromatography. *J. Chromatogr.* **266:** 75–87.

Hearn M.T.W. and Grego B. 1983b. High performance liquid chromatography of amino acids, peptides and proteins. LIII. Evaluation of the effect of several stationary phase parameters on the chromatographic separation of polypeptides on alkylsilicas. *J. Chromatogr.* **282:** 541–560.

Hearn M.T.W. and Grego B. 1983c. High-performance liquid chromatography of amino acids, peptides and proteins. XL. Further studies on the role of the organic modifier in reversed-phase high-performance liquid chromatography of polypeptides. Implications for gradient optimisation. *J. Chromatogr.* **255:** 125–136.

Hearn M.T.W. and Zhao G.L. 1999. Investigations into the thermodynamics of polypeptide interaction with nonpolar ligands. *Anal. Chem.* **71:** 4874–4885.

Hearn M.T.W., Keah H.H., Boysen R.I., Messana I., Misiti F., Rossetti D.V., Giardina B., and Castagnola M. 2000. Determination of biophysical parameters of polypeptide retro-inverso isomers and their analogues by capillary electrophoresis. *Anal. Chem.* **72:** 1964–1972.

Herring S.W. and Enns R.K. 1983. Rapid purification of leukocyte interferons by high-performance liquid chromatography. *J. Chromatogr.* **266:** 249–256.

Heukeshoven J. and Dernick R. 1985. Characterization of a solvent system for separation of water-insoluble poliovirus proteins by reversed-phase high-performance liquid chromatography. *J. Chromatogr.* **326:** 91–95.

Horváth C., Melander W., and Molnár I. 1976. Solvophobic interactions in liquid chromatography with nonpolar stationary phases. *J. Chromatogr.* **125:** 129–156.

——. 1977a. Liquid chromatography of ionogenic substances with nonpolar stationary phases. *Anal. Chem.* **49:** 142–154.

Horváth C., Melander W., Molnár I., and Molnár P. 1977b. Enhancement of retention by ion-pair formation in liquid chromatography with nonpolar stationary phases. *Anal. Chem.* **49:** 2295–2305.

I T.O., Guhan S., Taksen K., Vavra M., Myers D., and Hearn M.T.W. 2002. Intelligent automation of HPLC method development using a real-time knowledge-based approach. *J. Chromatogr. A* (in press).

Ishii D., Asai K., Hibi K., Jonakuchi T., and Nagaya M. 1977. A study of micro-high-performance liquid chromatography. I. Development of technique for miniaturization high-performance liquid chromatography. *J. Chromatogr.* **144:** 157–168.

Jeppsson J.O., Källman I., Lindgren G., and Fägerstam L.G. 1984. Hb-linköping (β36 Pro→Thr): A new hemoglobin mutant characterized by reversed-phase high-performance liquid chromatography. *J. Chromatogr.* **297:** 31–36.

Keah H.H., O'Bryan M., de Kretser D.M., and Hearn M.T.W. 2001a. Synthesis and application of peptide immunogens related to the sperm tail protein tpx-1, a member of the CRISP superfamily of proteins. *J. Peptide Res* **57:** 1–10.

Keah H.H., Allen N., Clay R., Boysen R.I., Warner T., and Hearn M.T.W. 2001b. Total chemical synthesis of activin $\beta A_{[12\text{-}115]}$ and related large loop polypeptides. *Biopolym. Pept. Sci.* **60:** 279–289.

Kopaciewicz W. and Regnier F.E. 1983. A system for coupled multiple-column separation of proteins. *Anal. Biochem.* **129:** 472–482.

Kucera P. 1980. Design and use of short microbore columns in liquid chromatography. *J. Chromatogr.* **198:** 93–109.

Lankmayr E.P., Wegscheider W., and Budna K.W. 1989. Global optimization of HPLC separations. *J. Liq. Chromatogr.* **12:** 35–58.

Lewis R.V., Fallon A., Stein S. Gibson K.D., and Udenfriend S. 1980. Supports for reverse-phase high-performance liquid chromatography of large proteins. *Anal Biochem.* **104:** 153–159.

Mahoney W.C. and Hermodson M.A. 1980. Separation of large denatured peptides by reverse phase high performance liquid chromatography. Trifluoroacetic acid as a peptide solvent. *J. Biol. Chem.* **255:** 11199–11203.

Mant C.T. and Hodges R.S. 1991a. The effects of anionic ion-pairing reagents on the peptide retention in reversed-phase chromatography. In *High-performance liquid chromatography of peptides and proteins: Separation, analysis, and conformation* (ed. C.T. Mant and R.S. Hodges), pp. 327–341. CRC Press, Boca Raton, Florida.

———. 1991b. Requirements for peptide standards to monitor column performance and the effect of column dimensions, organic modifiers, and temperature in reversed-phase chromatography. In *High-performance liquid chromatography of peptides and proteins: Separation, analysis, and conformation* (ed. C.T. Mant and R.S. Hodges), pp. 289–295. CRC Press, Boca Raton, Florida.

Mant C.T., Kondejewski L.H., Cachia P.J., Monera O.D., and Hodges R.S. 1997. Analysis of synthetic peptides by high-performance liquid chromatography. *Methods Enzymol.* **289:** 426–469.

Molnar I., Boysen R.I., and Erdmann V.A. 1989a. Reversed-phase high-performance liquid chromatography of *Thermus aquaticus* 50S and 30S ribosomal proteins. *Chromatographia* **28:** 39–44.

Molnar I., Boysen R.I., and Jekow P. 1989b. Peak tracking in high-performance liquid chromatography based on normalized band areas. A ribosomal protein sample as an example. *J. Chromatogr.* **485:** 569–579.

Mönch W. and Dehnen W. 1978. High-performance liquid chromatography of polypeptides and proteins on a reversed-phase support. *J. Chromatogr.* **147:** 415–418.

Moritz R.L. and Simpson R.J. 1992a. Application of capillary reversed-phase high-performance liquid chromatography to high-sensitivity protein sequence analysis. *J. Chromatogr.* **599:** 119–130.

———. 1992b. Purification of proteins and peptides for sequence-analysis using microcolumn liquid-chromatography. *J. Microcolumn. Sep.* **4:** 485–489.

Moritz R.L., Reid G.E., Ward L.D., and Simpson R.J. 1994. Capillary HPLC: A method for protein isolation and peptide mapping. *Methods* **6:** 213–226.

Nice E.C. and O'Hare M.J. 1978. Selective effects of reversed-phase column packings in high-performance liquid chromatography of steroids. *J. Chromatogr.* **166:** 263–267.

Nice E.C., Capp M., and O'Hare M.J. 1979. Use of hydrophobic interaction methods in the isolation of proteins from endocrine and paraendocrine tissues and cells by high performance liquid chromatography. *J. Chromatogr.* **185:** 413–427.

Nice E.C., Grego B., and Simpson R.J. 1985. Application of short microbore HPLC guard columns for the preparation of samples for protein microsequencing. *Biochem. Int.* **11:** 187–195.

Nice E.C., Lloyd C.J., and Burgess A.W. 1984. Role of short microbore high-performance liquid chromatography columns for protein separation and trace enrichment. *J. Chromatogr.* **296:** 153–170.

Nick H.P, Wettenhall R.E., and Hearn M.T. 1985. Isolation of protein S6 from rat liver ribosomes by reversed-phase high-performance liquid chromatography. *Anal Biochem.* **148:** 93–100.

Novotny M.V. and Ishii D., eds. 1985. Microcolumn separations: Columns, instrumentation and ancillary techniques. *J. Chromatogr.*, vol. 30. Elsevier, The Netherlands.

O'Hare M.J., Capp M.W., Nice E.C., Cooke N.H., and Archer B.G. 1982. Factors influencing chromatography of proteins on short alkylsilane-bonded large pore-size silicas. *Anal Biochem.* **126:** 17–28.

Pohl T. and Kamp R.M. 1987. Desalting and concentration of proteins in dilute solution using reversed-phase high-performance-liquid-chromatography. *Anal. Biochem.* **16:** 388–391.

Purcell A.W., Aguilar M.I., and Hearn M.T.W. 1993. High-performance liquid chromatography of amino acids, peptides, and proteins. 123. Dynamics of peptides in reversed-phase high-performance liquid chromatography. *Anal. Chem.* **65:** 3038–3047.

Purcell A.W., Zhao G.L., Aguilar M.I., and Hearn M.T.W. 1999. Comparison between the isocratic and gradient retention behaviour of polypeptides in reversed-phase liquid chromatographic environments. *J. Chromatogr.* **852:** 43–57.

Putterman G.J., Spear M.B., Meade Cobun K.S., Widra M., and Hixson C.V. 1982. A rapid isolation of human chorionic gonadotropin and its subunits by reversed-phase high performance liquid chromatography. *J. Liq. Chromatogr.* **5:** 715–730.

Quarry M.A., Grob R.L., and Snyder L.R. 1986. Prediction of precise isocratic retention data from two or more gradient elution runs. Analysis of some associated errors. *Anal. Chem.* **58:** 907–917.

Rivier J. and McClintock R. 1989. Isolation and characterization of biologically active peptides and proteins using reversed-phase HPLC. In *The use of HPLC in receptor biochemistry* (ed. A.R. Kerlavage), pp. 77–103. A.R. Liss, New York.

Round A.J., Aguilar M.I., and Hearn M.T. 1994. High-performance liquid chromatography of amino acids, peptides and proteins. CXXXIII. Peak tracking of peptides in reversed-phase high-performance liquid chromatography. *J. Chromatogr. A* **661:** 61–75.

Scott R.P.W. and Kucera P. 1979a. Mode of operation and performance characteristics of microbore columns for use in liquid chromatography. *J. Chromatogr.* **169:** 51–72.

———. 1979b. Use of microbore columns for the separation of substances of biological origin. *J. Chromatogr.* **185:** 27–41.

Simpson R.J. and Nice E.C. 1987. The role of microbore HPLC in the purification of subnanomole amounts of polypeptides and proteins for gas-phase sequence analysis. In *Methods in protein sequence analysis–1986* (ed. K. Walsh), pp. 213–228. Humana Press, Clifton, New Jersey.

———. 1989. Strategies for the purification of subnanomole amounts of protein and polypeptides for microsequence analysis. In *The use of HPLC in receptor biochemistry* (ed. A.R. Kerlavage), pp. 201–244. A.R. Liss, New York.

Simpson R.J., Moritz R.L., Reid G.E., and Ward L.D. 1991. Current strategies for microscale purification of protein and peptides for sequence analysis. In *Methods in protein sequence analysis* (ed. H. Jornvall et al.), pp. 67–77. Birkhauser Verlag, Basel, Switzerland.

Simpson R.J., Moritz R.L., Begg G.S., Rubira M.R., and Nice E.C. 1989a. Micropreparative procedures for high sensitivity sequencing of peptides and proteins. *Anal. Biochem.* **177:** 221–236.

Simpson R.J., Moritz R.L., Rubira M.R., Gorman J.J., and Van Snick J. 1989b. Complete amino acid sequence of a new murine T-cell growth factor P40. *Eur. J. Biochem.* **183:** 715–722.

Sitaram B.R., Keah H.H., and Hearn M.T.W. 1999. Studies on the relationship between structure and electrophoretic mobility of alpha-helical and beta-sheet peptides using capillary zone electrophoresis. *J. Chromatogr.* **857:** 263–273.

Skinner S.J.M., Grego B., Hearn M.T.W., and Liggins G.C. 1984. The separation of collagen α-chains by reversed-phase high-performance liquid chromatography. Comparison of column alkyl stationary phases and temperature effects. *J. Chromatogr.* **308:** 111–119.

Snyder L.R. 1980. Gradient elution. In *HPLC: Advances and perspectives* (ed. C. Horvath), vol. 1, pp. 208–316. Academic Press, New York.

———. 1990. Gradient elution separation of large biomolecules. In *HPLC of biological macromolecules: Methods and applications* (ed. K.M. Gooding and F.E. Regnier), pp. 231–259. Marcel Dekker, New York.

Snyder L.R., Stadalius M.A., and Quarry M.A. 1983. Gradient elution in reversed-phase HPLC separation of macromolecules. *Anal. Chem.* **55:** 1413–1430.

Sottrup-Jensen L., Stepanik T.M., Jones C.M., Lonblad P.B., Kristensen T., and Wierzbicki D.M. 1984. Primary structure of human alpha 2-macroglobulin. I. Isolation of the 26 CNBr fragments, amino acid sequence of 13 small CNBr fragments, amino acid sequence of methionine-containing peptides, and alignment of all CNBr fragments. *J. Biol. Chem.* **259:** 8293–8303.

Stadalius M.A., Gold H.S., and Snyder L.R. 1984. Optimization model of for the gradient elution separation of peptide mixtures by reversed-phase high performance liquid chromatography. Verification of retention relationships. *J. Chromatogr.* **296:** 31–59.

Stanton P.G. and Hearn M.T. 1987. The iodination sites of bovine throtropin. *J. Biol. Chem.* **262:** 1623–1632.

Stein S., Kenny C., Friesen H.J., Shively J., Del Valle U., and Pestka S. 1980. NH2-terminal amino acid sequence of human fibroblast interferon. *Proc. Natl. Acad. Sci.* **77:** 5716–5719.

Strasters J.K., Billiet H.A.H., de Galan L., Vandeginste B.G.M., and Kateman G. 1989. Automated peak recognition from photodiode array spectra in liquid chromatography. *J. Liq. Chromatogr.* **12:** 3–22.

Uyttenhove C., Simpson R.J., and Van Snick J. 1988. Functional and structural characterization of P40, a mouse glycoprotein with T-cell growth factor activity. *Proc. Natl. Acad. Sci.* **85:** 6934–6938.

van der Zee R. and Welling G.W. 1982. Molecular sieving during reversed-phase high-performance liquid chromatography of proteins. *J. Chromatogr.* **244:** 134–136.

van der Zee R., Welling-Wester S., and Welling G.W. 1983. Purification of detergent-extracted Sendai virus proteins by reversed-phase high-performance liquid chromatography. *J. Chromatogr.* **266:** 577–584.

Walhagen K., Unger K.K., and Hearn M.T.W. 2000a. Capillary electroendoosmotic chromatography of peptides. *J. Chromatogr.* **887:** 165–185.

——. 2000b. Influence of temperature on the behaviour of small linear peptides in capillary electrochromatography. *J. Chromatogr.* **893:** 401–409.

——. 2001. Capillary electrochromatography analysis of hormonal cyclic and linear peptides. *Anal. Chem.* **73:** 4924–4936.

Welinder K.G. 1988. Generation of peptides suitable for sequence analysis by proteolytic cleavage in reversed-phase high-performance liquid chromatography solvents. *Anal. Biochem.* **1:** 54–64.

Wiktorowicz J.E., ed. 1992. Capillary electrophoresis. *Methods*, vol. 4, issue 3. Academic Press, New York.

Witting L.A., Gisch D.J., Ludwig R., and Eksteen R. 1984. Bonded-phase selection in the high-performance liquid chromatography of proteins. *J. Chromatogr.* **296:** 97–105.

Wolfe R.A., Casey J., Familletti P.C., and Stein S. 1984. Isolation of proteins from crude mixtures with silica and silica-based adsorbents. *J. Chromatogr.* **296:** 277–284.

Yang F.J. 1982. Fused-silica narrow-bore micro-particle-packed-column high-performance liquid chromatography. *J. Chromatogr.* **236:** 265–277.

FURTHER READING

Frenz J., Hancock W.S., and Henzel W.J. 1990. Reversed phase chromatography in analytical biotechnology of proteins. In *HPLC of biological macromolecules: Methods and applications* (ed. K.M. Gooding and F.E. Regnier), pp. 145–177. Marcel Dekker, New York.

Hearn M.T.W. 1998. High-resolution reversed-phase chromatography. In *Protein purification: Principles, high-resolution methods, and applications*, 2nd edition (ed. J.-C. Janson and L. Rydén), pp. 239–282. Wiley-Liss, New York.

Mant C.T. and Hodges R.S., eds. 1991. *High-performance liquid chromatography of peptides and proteins: Separation, analysis, and conformation.* CRC Press, Boca Raton, Florida.

WWW RESOURCE

For additional Web addresses, see Table 5.2.

http://www.upchurch.com Upchurch Scientific, Inc. home page.

CHAPTER 6

Amino- and Carboxy-terminal Sequence Analysis*

*This chapter includes contributions by Alastair Aitken (University of Edinburgh) and Thomas Bergman, Ella Cederlund, and Hans Jörvall (Karolinska Institutet).

IN 1954, FREDERICK SANGER AND HIS COLLEAGUES at the University of Cambridge provided a landmark in protein chemistry when, after 10 years of intensive work on insulin, they succeeded in achieving the first complete description of the chemical structure of insulin, a low-molecular-weight protein molecule (for review, see Thompson 1955). Insulin, the pancreatic hormone that governs sugar metabolism in the body, is a small protein of just 51 amino acids. It comprises two chains that are held together by interchain disulfide bonds. "For his work on the structure of proteins, especially that of insulin," Sanger was awarded the Nobel prize for Chemistry in 1958. (Sanger also received a share of the 1980 Nobel prize in Chemistry for his method of determining the base sequences in nucleic acids.) The ability of the protein chemist to determine the covalent structure of protein molecules has been instrumental both to the development of three-dimensional structures of proteins and to furthering our understanding of protein structure-function relationships.

One of the key technologies that enabled Sanger and his team to solve the structure of insulin was the DNP (for dinitrophenyl group)-labeling method that they developed for covalently modifying the end amino acid in a peptide (Sanger 1945). The DNP group, which gives the peptide a distinctive yellow color, acts as a chemical marker that remains attached to the amino group after the peptide is hydrolyzed with either dilute acid or enzymes. For example, Sanger first broke each chain of the insulin molecule into small peptides and then determined the sequence of each of the peptides in turn. A critical factor in this strategy was to identify overlaps in sequences that would enable the different pieces of the jigsaw puzzle to be assembled correctly, i.e., build up a contiguous sequence by aligning all of the peptides. Analysis of the peptide sequences proceeded as follows.

First, the peptides were treated with the DNP-labeling reagent 2,4-dinitro-1-fluorobenzene (DNFB) and then broken down into their constituent amino acids by acid hydrolysis. α-DNP amino acid derivatives are resistant to acid hydrolysis and can be recovered with yields ranging from 25% to 90% after boiling in azeotropic hydrochloric acid (5.7 N HCl) or heating with 12 N HCl for 12 hours at 105°C. The amino-terminal amino acid, say *A*, was identified by its yellow color using two-dimensional paper chromatography (Figure 6.1). The process is repeated with a second sample of the peptide, but this time, the peptide is only partially hydrolyzed so that two amino acids remain as a dipeptide derivative colored yellow. If amino acid *B* is partnered with, for example, amino acid *A* in this peptide fragment, then one knows that the sequence must be *AB*, and the order in the original sample, say tripeptide, was therefore *ABC*.

A second key technology that had an indispensable role in solving the insulin sequence was the partition chromatography method (also called filter paper chromatography) for separating amino acids and peptides, invented by A.J.P. Martin and R.L.M Synge. "For their invention of partition chromatography"—the foundation of modern chromatography—Martin and Synge were awarded the 1952 Nobel prize in Chemistry (for a review, see Moore and Stein 1951).

During the intervening years since Sanger's Herculean effort in determining the primary structure of insulin, the methodology for polypeptide sequencing has developed at a fantastic rate. This has been largely due to the development of the Edman degradation procedure whereby amino acids in a polypeptide can be removed in a stepwise fashion from the amino-terminal end. This subject, along with carboxy-terminal sequencing, is discussed in this chapter. The strategy of the Edman degradation relies upon the polypeptide having a free α-amino group at the amino terminus. In many situations, the α-amino group is derivatized or blocked (e.g., by acetylation and formylation), making the polypeptide refractory to Edman

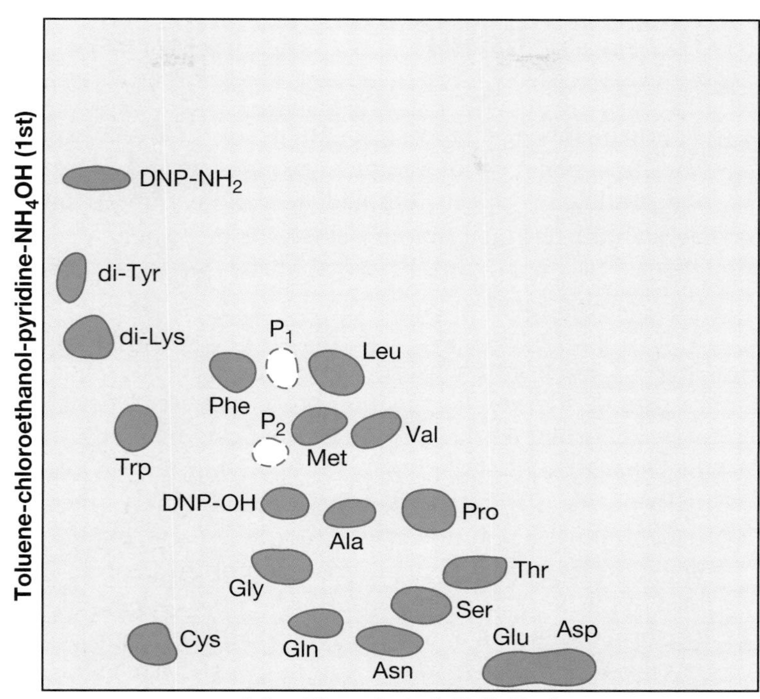

FIGURE 6.1. Two-dimensional paper chromatogram of a synthetic mixture of DNP amino acids according to a method described by Levy (1954). Development was first carried out with the "toluene" method (first dimension) and then with 1.5 M phosphate buffer (second dimension). P1 and P2, respectively, denote the position of α-DNP-amino–δ-chlorovaleric acid and δ-DNP-amino–α-chlorovaleric acid. (Redrawn, with permission, from Narita et al. 1975 [©Springer-Verlag].)

degradation. Methods for deblocking are also discussed. The actual day-to-day operation and maintenance of automated sequencing instruments require much technical prowess, and it is thus recommended that these applications be left in the hands of the aficionado. For a detailed discussion of protein sequencer maintenance and troubleshooting, see Dunbar (1997) and Geisow and Aitken (1989). In this chapter, emphasis is placed on instrument and chemical limitations, as well as sample-loading constraints.

Core Facilities for Protein Sequencing

To find a laboratory that offers biomolecular analytical services on an commercial basis, check the searchable yellow pages of the Association of Biomolecular Resources Facilities (ABRF) Web Site (http://www.abrf.org).

With the advent of mass spectrometric procedures for sequencing small peptides (as described in Chapter 8), the use of the automated Edman degradation procedure has declined markedly. However, it must be stressed that many biological problems require knowledge of the amino or of the carboxyl terminus of a polypeptide and that mass spectrometric methods for assigning the amino or the carboxyl terminus of a protein or peptide are limited.

AMINO-TERMINAL SEQUENCING OF POLYPEPTIDES USING EDMAN DEGRADATION

The stepwise degradation of peptides was first successfully applied in 1930 by Abderhalden and Brockmann, who used phenylisocyanate (PIC) as an amino group coupling reagent for the production of an intermediate that rearranged under acidic conditions, cleaving the derivatized terminal amino acid from the parent peptide (Abderhalden and Brockmann 1930). This method was extended by the Swedish scientist Pehr Victor Edman, who changed the coupling agent to phenylisothiocyanate (PITC), a modification that yields a more readily cyclized intermediate (and hence more easily cleaved amino-terminal amino acid) than that derived from phenylisocyanate (Edman 1949). For an anecdotal account of the history of peptide stepwise degradation procedures, see Doolittle (1982).

All current amino-terminal sequencing methods are based on the procedure developed by Pehr Edman, referred to by the eponym, "Edman degradation" procedure. The characteristics of this method are that the reaction removes the derivatized amino acid (phenylthiohydantoin [PTH] amino acid) from the protein, but does not destroy the remaining peptide chain, so that a sequential degradation of the peptide can be performed. Thus, each amino acid of the peptide chain is identified by one cycle of Edman chemistry and one cycle of high-performance liquid chromatography (HPLC) to analyze the PTH amino acid. This method has been developed in a variety of modes (Edman and Henschen 1975), and eventually in the form of an automated sequencing instrument in 1967 (Edman and Begg 1967). Not long afterward, the solid-phase sequencer was constructed for the degradation of peptides that were covalently attached to a solid support (Laursen 1971).

The Edman degradation procedure is divided into three steps: coupling, cleavage, and conversion (Figure 6.2). In the coupling reaction, the Edman PITC chemically modifies the free amino-terminal α-amino group of a polypeptide to form a phenylthiocarbamyl (PTC) polypeptide. At pH 9, coupling is favored at α-amino groups and takes place within 15–30 minutes at a temperature of 40–55°C in a very high yield. Coupling is inhibited by amino-terminal modifications such as formylation, acetylation, fatty acid acylation, and cyclization of glutamine residues to form pyroglutamate (Brown and Roberts 1976; Brown 1979; Hirano et al. 1992).

In the cleavage reaction, the PTC amino-terminal residue is rapidly cleaved from the polypeptide chain with anhydrous acid. This process is facilitated by the proximity of the nucleophilic sulfur atom of the derivatized amino terminus to the carbonyl carbon of the first peptide bond to yield a five-membered heterocyclic derivative, an anilinothiazolinone (ATZ) amino acid, and the $n-1$ polypeptide. The shortened $n-1$ polypeptide has a reactive amino-terminal α-amino group, which can undergo another cycle of coupling and cleavage. This procedure is repeated in an iterative manner.

The solubility of the small, hydrophobic ATZ amino acid is significantly different from that of the hydrophilic polypeptide and can be extracted selectively by a nonpolar solvent such as chlorobutane or ethylacetate. In the third, conversion, step of the cycle, the unstable ATZ derivative amino acid is converted to a more stable PTH derivative. Conversion is a two-step reaction and occurs by treatment with aqueous acid. Cleavage and conversion can be accomplished with just one aqueous acid reaction, but the anhydrous acid optimizes the specific cleavage of the peptide bond of the amino-terminal amino acid. First, the unstable ring structure of the ATZ amino acid is opened by aqueous acid and increased temperature (~60°C) to form a PTC amino acid. Rearrangement then takes place, yielding the more stable PTH amino acid. It has been suggested (Farnsworth and Steinberg 1993) that the thiohydantoin is actually formed preferentially when cleavage is carried out in the presence of a thiol such as dithiothreitol (DTT). The Edman degradation cycle is schematically summarized in Figure 6.3.

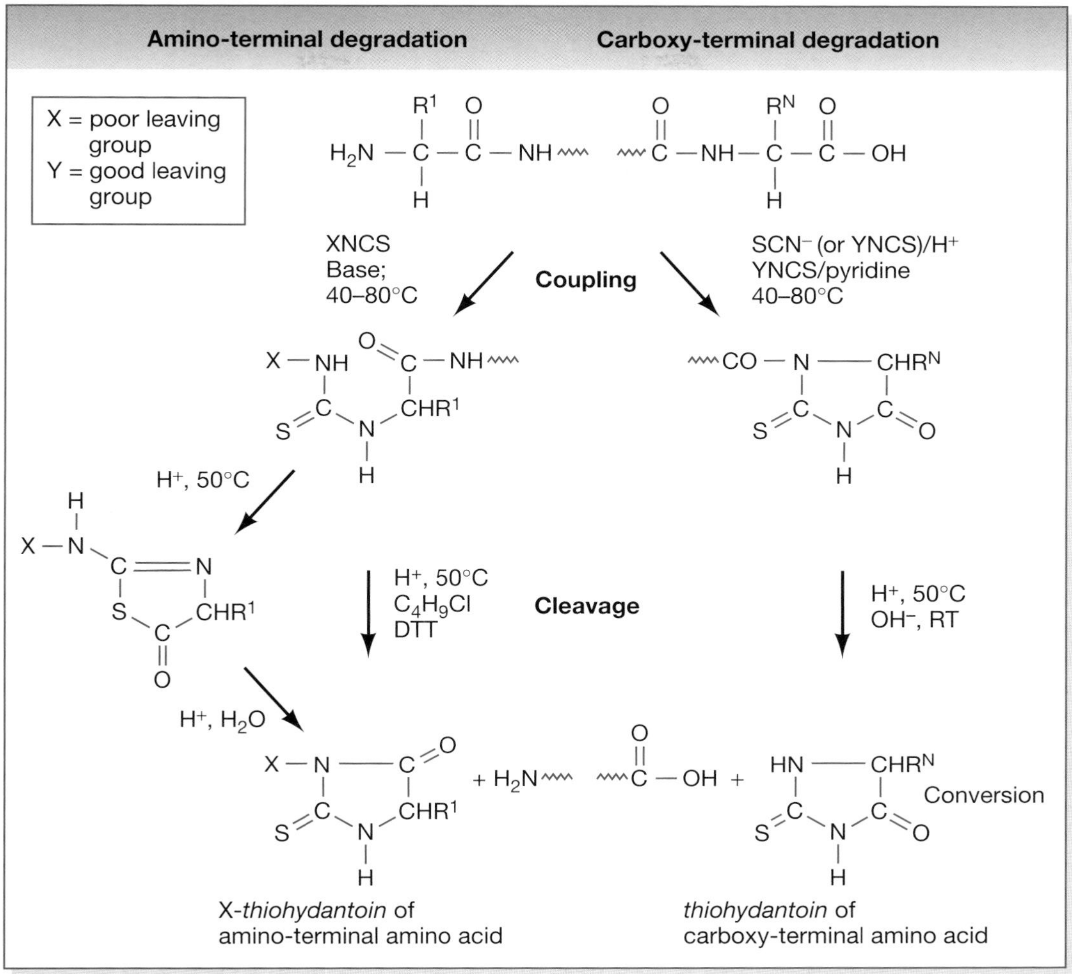

FIGURE 6.2. Procedures for removal of amino- and carboxy-terminal amino acids as thiohydantoins. Both derivatives may be detected in the UV because of their high absorbance at 268 nm (E_{max} ~17,500). Classical Edman chemistry used phenylisothiocyanate (*X* = phenyl) for amino-terminal degradations. (Redrawn, with permission, from Inglis et al. 1995.)

Automation of Edman Degradation Using a Spinning-cup Sequencer

The repetitive nature of Edman degradation suggested to Edman's technical assistant Geoffrey Begg that the procedure could be automated. Edman and Begg produced a prototype protein sequenator in 1961. Automation of the coupling and cleavage steps (Edman and Begg 1967; Laursen 1971) enabled researchers to determine the sequence of the first 30–40 amino residues in a protein (0.3 μmole) routinely. Figure 6.4 illustrates the instrumental setup devised by Edman and Begg—the spinning-cup sequencer—which consisted of a solvent delivery system that delivered solvents and reagents by nitrogen pressure to the reaction cartridge (the spinning cup) via an electronically operated valve block. In the spinning-cup reaction vessel (Figure 6.5), the protein was held against the inner wall by centrifugal force (Edman and Henschen 1975). Reagents and solvents delivered to the spinning cup (by nitrogen pressure) were precisely measured to wet only the protein film. Excess reagents and solvents were removed by evaporation in vacuo. A pick-up line was installed in an upper groove of the cup to remove the waste. One of the early limitations of the Edman sequencer was the

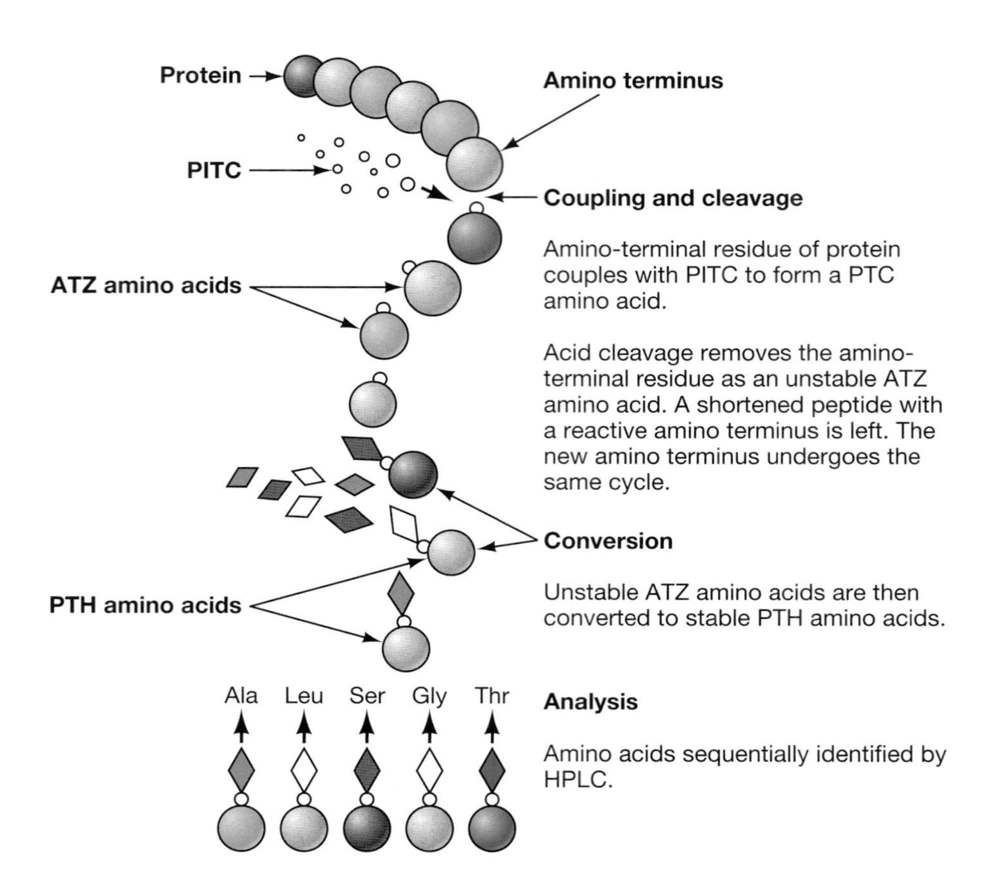

FIGURE 6.3. How amino acids are sequentially cleaved in Edman degradation: A conceptual view of the Edman degradation chemistry cycle. (Redrawn, with permission, from the handbook, *Preparing samples for protein sequencing: A newcomer's guide* [©The Perkin-Elmer Corporation].)

poor performance of short peptides due to sample "wash out" from the reaction vessel. This problem was overcome by the introduction of the polymeric quarternary ammonium salt Polybrene (Klapper et al. 1978; Tarr et al. 1978), which anchored both proteins and peptides, thereby allowing the sequencing of even short hydrophobic peptides to completion. The Edman spinning-cup sequencer was adapted for a commercial instrument by Beckman Inc. in 1969 and was in use until 1980. Edman did not patent his sequenator.

After coupling and cleavage, the released ATZ amino acids were delivered to a cooled fraction collector. ATZ amino acids were manually converted off-line to the PTH derivatives in batches. Initially, PTH amino acids were separated and identified by chromatography. For this purpose, paper chromatography, thin-layer chromatography (TLC), and partition chromatography, in either a liquid-liquid system or a gas-liquid system, were proposed (Edman and Henschen 1975). For a representative TLC separation and identification of PTH amino acids, see Figure 6.6. The inherent limitations of the TLC and hydrolytic methods for PTH analysis were eventually overcome by the development of gas chromatographic and HPLC PTH amino acid separation methods (Zimmerman et al. 1973; Pisano 1975). Later workers added devices for automatic conversion of the separated thiazolinones to the PTH amino acids (Wittmann-Liebold et al. 1976). These devices were followed by on-line detection of the PTHs using HPLC. The performance of the original spinning-cup sequencer (sensitivity limit, 0.3 μmole; repetitive yield, 92–95%) is listed in Table 6.1.

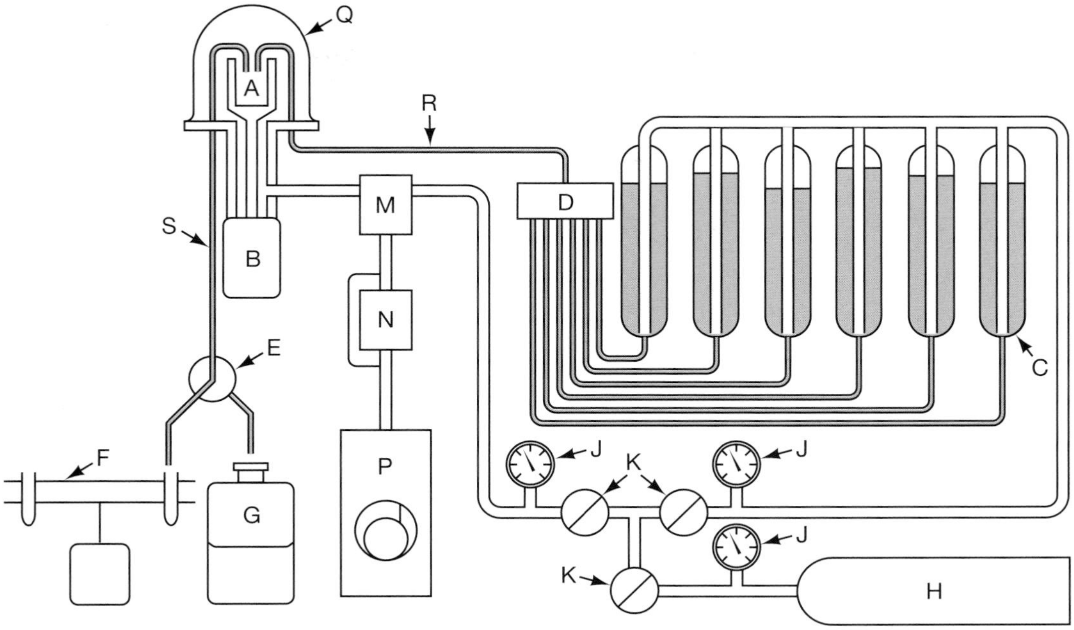

FIGURE 6.4. Diagram of spinning-cup sequencer. (*A*) Spinning cup; (*B*) electric motor; (*C*) reagent (solvent) reservoir; (*D*) valve asssembly; (*E*) outlet stopcock assembly; (*F*) fraction collector; (*G*) waste container; (*H*) nitrogen cylinder; (*J*) pressure gauges; (*K*) pressure regulators; (*M*) 3-way valve; (*N*) 2-way valve with bypass; (*P*) rotary vacuum pump; (*Q*) bell jar; (*R*) feed line; (*S*) effluent line. Gas lines are open, and liquid lines are filled. (Redrawn, with permission, from Edman and Begg 1967 [©Blackwell Science Ltd.].)

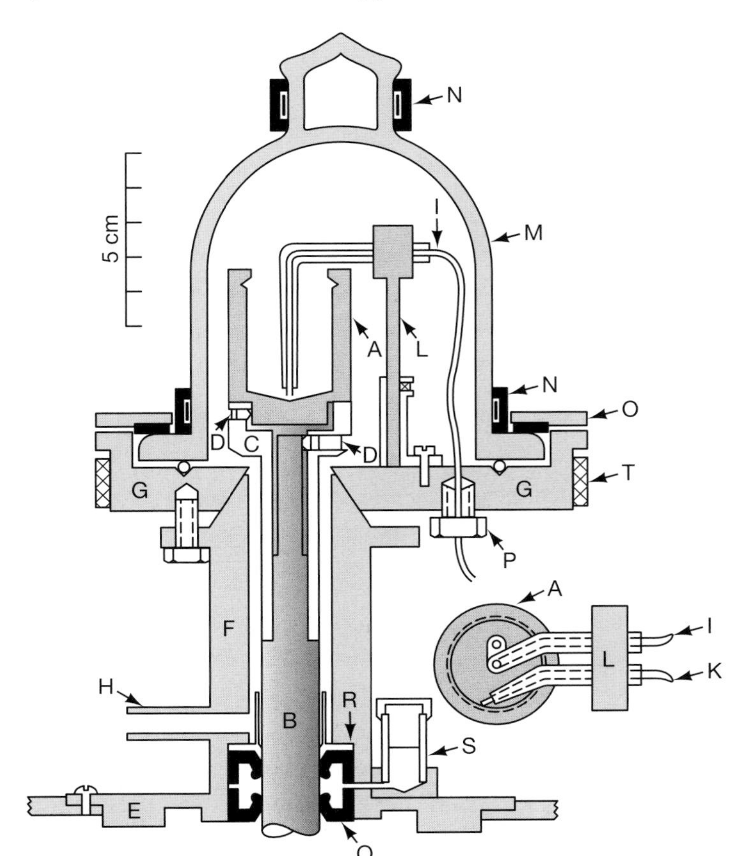

FIGURE 6.5. Spinning-cup reaction chamber. (*A*) Glass cup; (*B*) motor shaft extension; (*C*) cup support; (*D*) grub screws; (*E*) motor support; (*F*) column; (*G*) base plate; (*H*) side tube; (*I*) feed line; (*K*) effluent line; (*L*) adjustable stand; (*M*) bell jar; (*N*) electrodes; (*O*) rubber padded ring; (*P*) sealing bolt; (*Q*) oil seal; (*R*) PTFE sleeve; (*S*) oil reservoir; (*T*) band heater. A top view of the cup, the feed line, and the effluent line is shown in the lower right hand corner. (Redrawn, with permission, from Edman and Begg 1967 [©Blackwell Science Ltd.].)

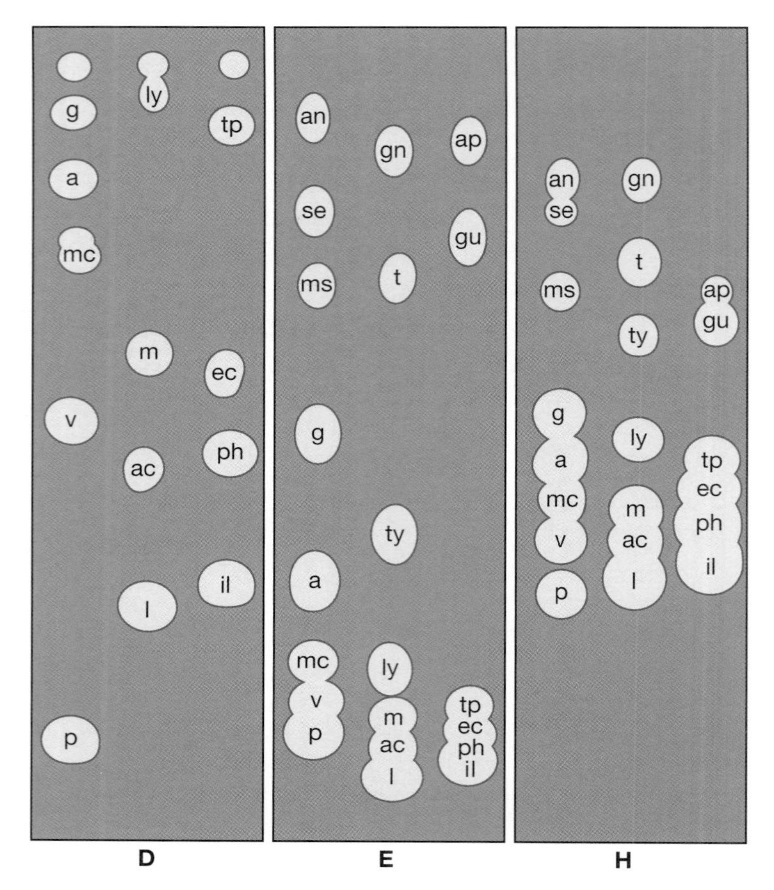

FIGURE 6.6. Chart showing the TLC of PTH amino acids in systems D (xylene/formamide), E (*n*-butyl acetate/ propionic acid/ formamide), and H (ethylene chloride/acetic acid). PTH derivatives of asparagine (an), serine (se), methionine sulfone (ms), glycine (g), alanine (a), *S*-methyl cysteine (mc), valine (v), proline (p), glutamine (gn), threonine (t), tyrosine (ty), lysine (ly), methionine (m), *S*-allyl cysteine (ac), leucine (l), aspartic acid (ap), glutamic acid (gu), tryptophan (tp), *S*-ethyl cysteine (ec), phenylalanine (ph), and isoleucine (il). (Redrawn, with permission, from Edman and Henschen 1975 [©Springer-Verlag].)

Microsequencing with a Gas-phase Sequencer

Automated microsequencing came of age with the release of a commercial instrument, the gas-phase sequencer, which used miniaturized components and gaseous coupling and cleavage reactions of the protein immobilized on a glass fiber disc (see Table 6.1) (Hewick et al. 1981). This instrument revolutionized protein microsequencing by affording an almost 1000-fold increase in sensitivity compared with the spinning-cup instrument and, in so doing, eliminated most other protein sequencing techniques based on modified Edman reagents (see below Alternative Amino-terminal Reaction Methods). Sample wash out was minimized by addition of a polyamide (Polybrene) to the glass support. The two key reagents, a coupling base (triethylamine) and a cleavage acid (trifluoroacetic acid, TFA), which are both capable of dissolving proteins/peptides, are delivered to the reaction vessel (e.g., the glass cartridge block in Figure 6.7A and the reaction cartridge in Figure 6.7B) in the gaseous phase via a stream of argon or nitrogen. By these means, the appropriate pH conditions for coupling and cleavage could be achieved without risk of sample wash out from the reaction vessel. Selective extraction of reaction by-products and ATZ amino acids (but not protein/peptide) was accomplished by judicious choice of organic solvents that were transferred to the reaction

TABLE 6.1. Advances in amino acid sequence automation

Instrument		Sample required	Repetitive yield (RY)	Comments and drawbacks
Spinning cup[a]	1970	0.3 μmole	92–95%	Manual conversion only; peptide wash out
	1973			Auto conversion; improved valves
	1978			Polybrene peptide anchor
	1978			HPLC detection of PTH amino acids
Gas Phase Applied Biosystems[b] Model 470A	1981	100 pmoles	92–95%	Miniaturized components; glass fiber support; low-volume reaction vessel; on-line conversion of ATZ amino acids
Applied Biosystems Model 477A	1985			On-line PTH amino acid analysis; 0.5 pmole detection
Applied Biosystems Model 49xProcise		2 pmoles	>94%	x = 1, 2, or 4 depending on the amount of cartridges; RY based on L, I, and V with 10-pmole load of β-lactoglobulin; threshold of detection ~250 fmoles; signal/noise (S/N) ratio of 2:1 for PTH alanine; cycle time <35 min; 9-mm reaction cartridge for volumes up to 15 μl
Applied Biosystems Model 49xProcise cLC		500 fmoles	>92%	x = 1, 2, or 4 depending on the amount of cartridges; RY based on L, I, and V with 10-pmole load of β-lactoglobulin; threshold of detection ~50 fmoles; S/N ratio of 2:1 for PTH alanine; cycle time <50 min; 6-mm reaction cartridge for volumes <7.5 μl
Hewlett-Packard Model HPG1001A	1992	3–5 pmoles	97%	Biphasic column reaction vessel; on-line PTH amino acid analysis; 0.5 pmole detection

[a]Original prototype reported 98% repetitive yield for myoglobin.
[b]Applied Biosystems, Foster City, California. Protocols and other electronic information on protein sequencing from Applied Biosystems can be found at: http://www.appliedbiosystems.com/apps/techniques.cfm.

vessel in the liquid phase. Extracted ATZ amino acids were automatically converted to their stable PTH derivatives following their transfer from the reaction vessel to a conversion flask. A combination of the miniaturized glass reaction vessel (cartridge), zero-dead volume and inert valves, and especially purified chemicals contributed largely to a user-friendly and highly reliable instrument that enabled sequencing of less than 100-pmole samples. With the subsequent addition of an on-line HPLC system and optimized PTH amino acid separation, sequence analyses at the 10-pmole level or even lower were readily achievable. With the development of capillary columns (<0.32 mm I.D.), the sensitivity of PTH amino acid analysis has been extended to the low femtomole level (see Figure 6.8) (Moritz and Simpson 1992; Erdjumen-Bromage et al. 1993).

A More-refined Pulsed-liquid-phase Sequencer

Further refinements of the gas-phase sequencer (simplification of the valve blocks, gas regulation, software analysis, and reagent/solvent delivery protocols) led to the development of the pulsed-liquid-phase sequencer. The basic principles of these two instruments were essentially the same, the main difference being that the cleavage acid (TFA) was delivered as a liquid pulse. The amount of acid delivered was precisely controlled, sufficient to wet the sample but not enough to wash the sample from the reaction vessel. Liquid-phase cleavage resulted in faster cleavage times (i.e., shorter cycle times and accelerated sequencing). Following cleav-

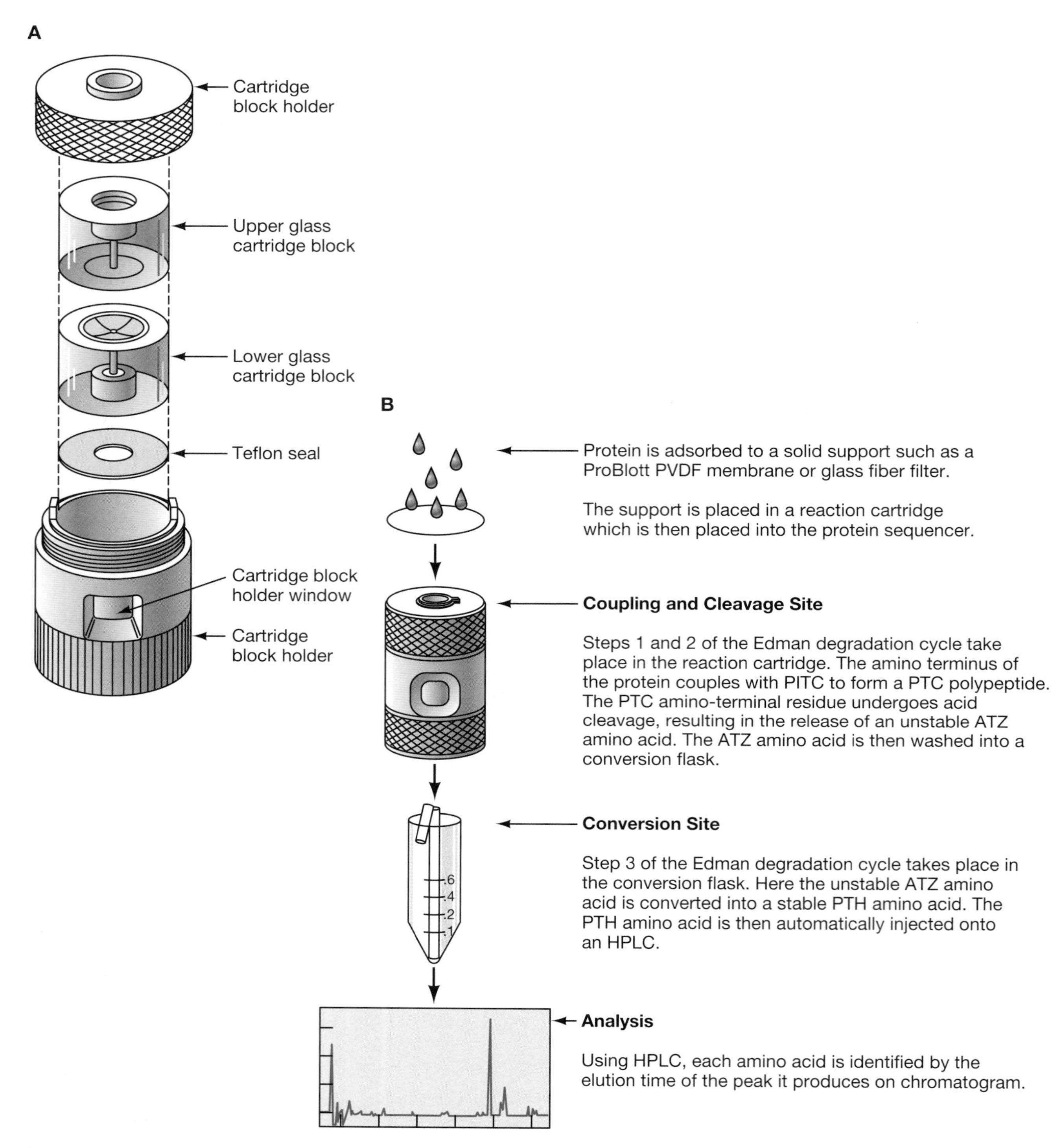

A
Cartridge
block holder
Upper glass
cartridge block
Lower glass
cartridge block
Teflon seal
Cartridge block
holder window
Cartridge
block holder
B
Protein is adsorbed to a solid support such as a
ProBlott PVDF membrane or glass fiber filter.
The support is placed in a reaction cartridge
which is then placed into the protein sequencer.
Coupling and Cleavage Site
Steps 1 and 2 of the Edman degradation cycle take
place in the reaction cartridge. The amino terminus of
the protein couples with PITC to form a PTC polypeptide.
The PTC amino-terminal residue undergoes acid
cleavage, resulting in the release of an unstable ATZ
amino acid. The ATZ amino acid is then washed into a
conversion flask.
Conversion Site
Step 3 of the Edman degradation cycle takes place in
the conversion flask. Here the unstable ATZ amino
acid is converted into a stable PTH amino acid. The
PTH amino acid is then automatically injected onto
an HPLC.
.6
.4
.2
.1
Analysis
Using HPLC, each amino acid is identified by the
elution time of the peak it produces on chromatogram.

FIGURE 6.7. Automated Edman sequencing using a gas-phase sequencer. (*A*) Exploded view of the reaction cartridge system. The protein sample is introduced into the reaction cartridge either on a PVDF membrane or glass fiber filter. The membrane/filter is sandwiched between the upper and lower glass cartridge blocks, and the entire assembly is placed into the protein sequencer. (Redrawn, with permission, from Applied Biosystems Procise Protein Sequencer user's manual.) (*B*) Coupling, cleavage, and conversion reactions (see Figure 6.2) all occur automatically within the protein sequencer. The identity and sequence of the cleaved amino acids are determined by HPLC. (Reproduced, with permission, from the handbook, *Preparing samples for protein sequencing: A newcomer's guide* [©The Perkin-Elmer Corporation].) (Courtesy of Lynne Zieske, Applied Biosystems.)

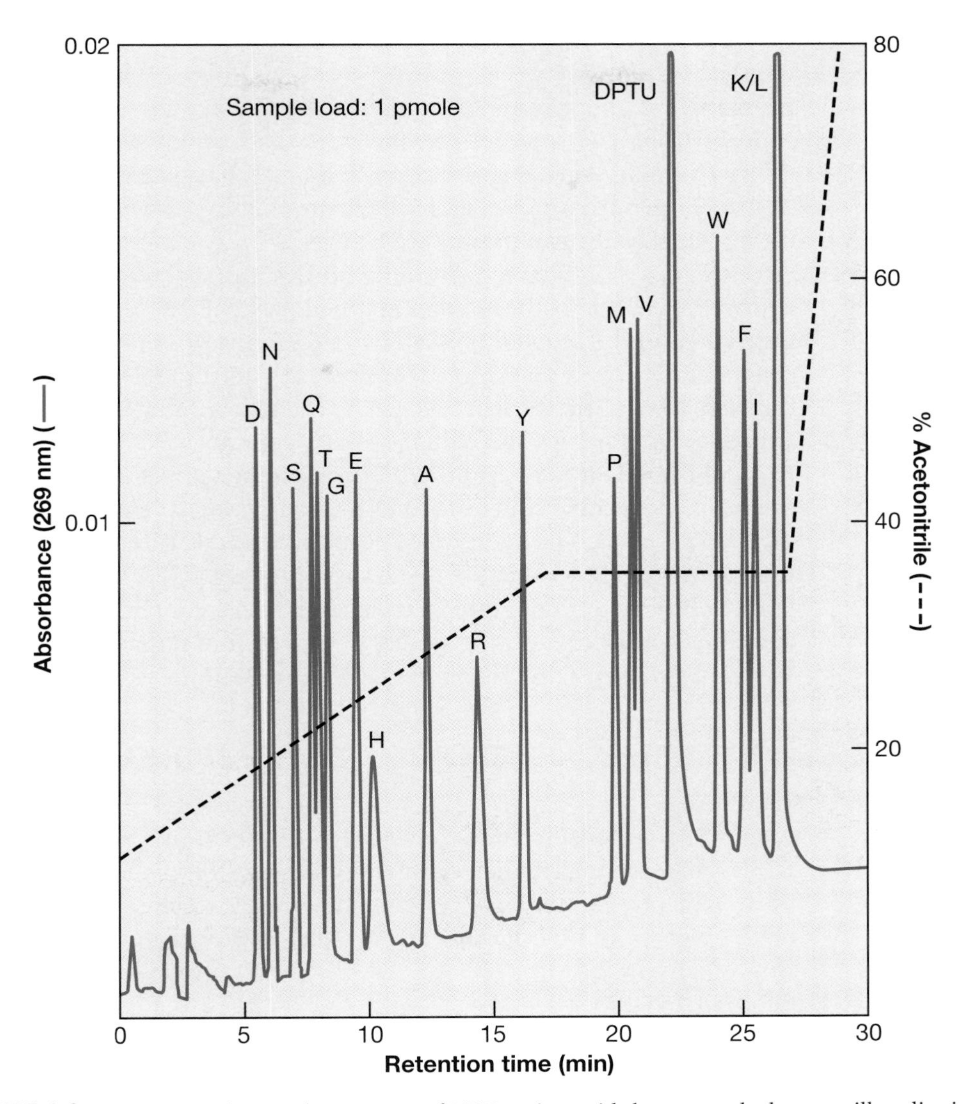

FIGURE 6.8. Separation of 1-pmole amounts of PTH amino acids by reversed-phase capillary liquid chromatography. Column: 150 x 0.32-mm I.D. Applied Biosystems PTH-C_{18} (packed by LC packings). (Solvent A) 8.3 mM sodium acetate/5% (v/v) tetrahydrofuran (pH 4.1); (Solvent B) acetonitrile. Column temperature is 55°C and flow rate is 5 µl/min. PTH amino acid notation is shown using the one-letter code for amino acids. DPTU denotes diphenylthiourea. (Reprinted, with permission, from Moritz and Simpson 1992 [©Elsevier Science].)

age, the volatile TFA was removed (without loss of sample). With careful optimization of reaction cycle protocols and reagent/solvent deliveries, cycle times of less than 30 minutes are achievable (Totty et al. 1992). With the implementation of microcolumn (1.0 mm I.D.) PTH amino acid analysis technology, these instruments (Applied Biosystems, model Procise sequencer) can operate in the range of 0.5 pmole and even lower with a repetitive yield of ~95% (see Table 6.1).

Improvements Provided by the Biphasic Column Sequencer

Although analyses on the gas-phase sequencer showed improved repetitive yields when compared with the spinning-cup instruments, they did not provide the high repetitive yields (>95%) essential for long sequencer runs. To overcome this problem, in the early 1990s,

Hewlett-Packard (now Agilent) introduced an automated sequencer that provided a different approach for immobilizing protein/peptides for Edman chemistry. The reaction vessel in this instrument comprises an adsorptive biphasic column—one half of the column contains a solid hydrophobic support and the other half contains a hydrophilic support (see Figure 6.9A). In a manner analogous to loading a reversed-phase HPLC column, very large volumes of protein/peptide are applied to the hydrophobic portion of the biphasic sequencer column under dilute acidic conditions (see Figure 6.9B). The sample is retained on the top of the column, and any inorganic salts and buffers can be washed away. The hydrophobic and hydrophilic halves of the column are then reassembled, and the resulting biphasic column is positioned in the sequencer (see Figure 6.9A, inset). Allowing the flow direction of the solvents to be reversed minimizes sample wash out from the column. Aqueous solvent flows are directed toward the hydrophobic half of the column, which immobilizes the sample via hydrophobic interactions. Alternatively, organic solvent flows are directed toward the hydrophilic half of the column, which retains the sample when organic solvents are employed to elute hydrophobic contaminants/reaction products (e.g., diphenylthiourea [DPTU] or the ATZ amino acid). The addition of nonvolatile alkylamines to the biphasic reactor column in this sequencer allows state-of-the-art repetitive yields (>98%), presumably by both ensuring that the protein amino group is uncharged for the coupling reaction and competing with the amino terminus for blocking species in the reaction system (see Figure 6.2). For a description of sample loading onto the biphasic column reaction vessel, see Protocol 1.

SEVERAL RATE-LIMITING STEPS ARE ASSOCIATED WITH SEQUENCE ANALYSIS

A number of considerations must be borne in mind before engaging in protein/peptide sequencing, especially when dealing with low-picomole levels of material, including sample purity, chemical limitations (i.e., contaminants that either block the amino terminus of a protein/peptide or inhibit the Edman degradation procedure), and sample loading conditions.

Sample Preparation (Purity)

Sample purity is one of the rate-limiting steps for successful protein sequencing (for a review, see Simpson and Nice 1989). Samples should be homogeneous with respect to the polypeptide chain, i.e., contain one protein component only; sample microheterogeneity due to differential glycosylation (such samples often electrophorese as a broad band on SDS-PAGE or a broad peak on reversed-phase HPLC) or phosphorylation will not interfere with amino-terminal sequence analysis. However, under normal operating conditions, the PTH derivatives of glycosylated or phosphorylated amino acids are not extracted from the reaction vessel, and these cycles appear as "blank cycles." Analysis of sites of protein phosphorylation using Edman degradation are best performed using a solid-phase sequencing approach (see Aitken and Learmonth 1997 and below for protocols). Likewise, a blank cycle is observed with cysteine residues unless the sample is alkylated (e.g., using iodoacetic acid, iodoacetamide, or vinyl pyridine) prior to sequencing; in this situation, cysteine is detected as the *S*-alkylated PTH cysteine. In some cases, seemingly homogeneous samples prepared by HPLC or SDS-PAGE yield ambiguous sequencing results. For example, more than one sequencing signal will be produced if samples contain more than one protein component or if a single protein has been proteolytically cleaved during purification but is held together by disulfide bonds. For this reason, samples should be reduced with DTT and the cysteines alkylated with a suitable alkylating agent prior to Edman degradation. If the multiple components in the sample are present in nearly equivalent amounts, the sequence of any one component is almost

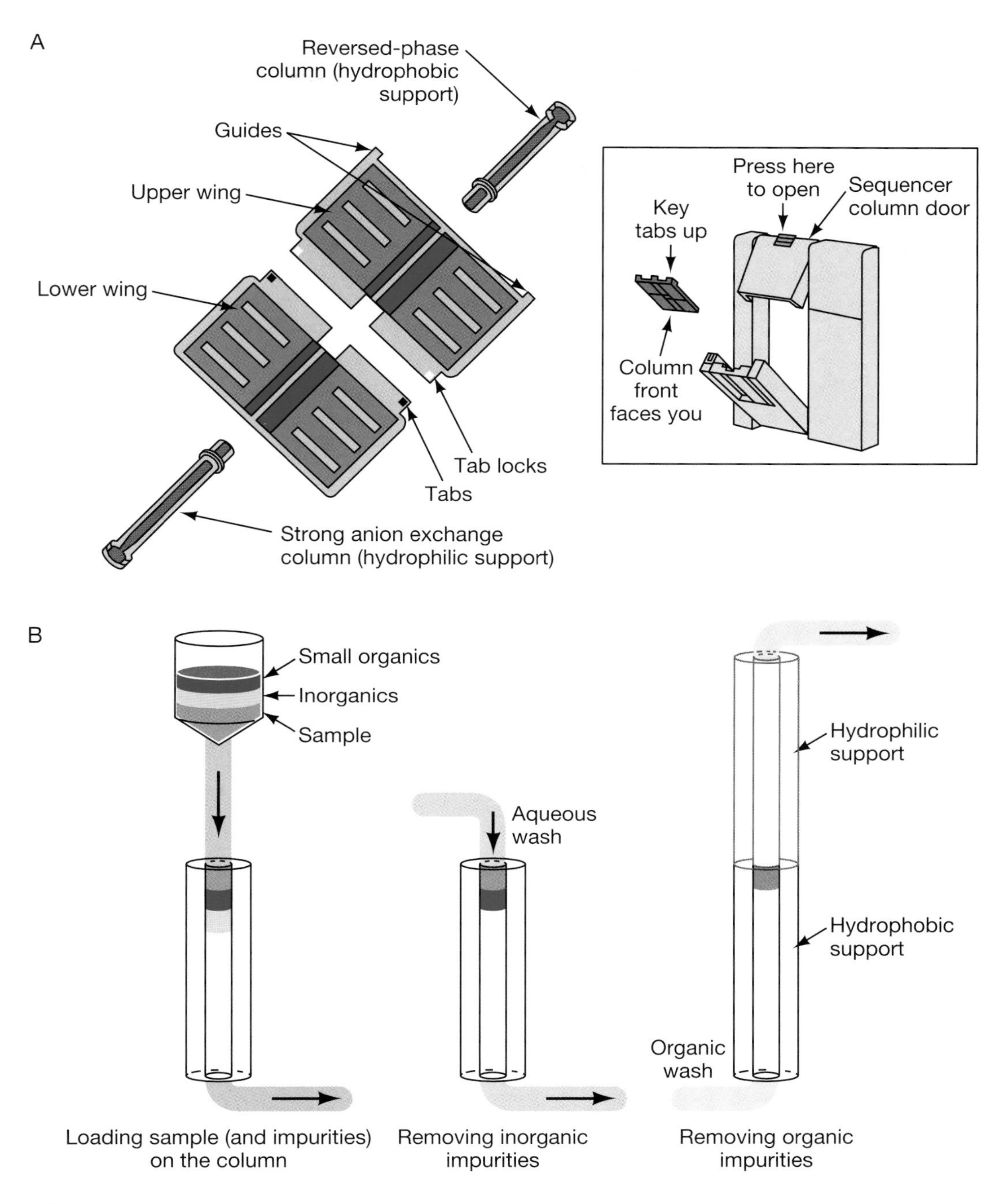

FIGURE 6.9. Biphasic sequencer column. (*A*) Exploded view of the biphasic column. (*Inset*) Insertion of the biphasic column into the sequencer. (Redrawn, with permission, from The Agilent model G100A Protein Sequencer manual.) (*B*) Protein sample preparation using the biphasic column prior to sequencing. The protein sample is applied to the top of hydrophobic component of the reaction cartridge, which resembles a reversed-phase column. Large volumes can be loaded; the sample is retained on the top of the column and excipients such as buffer salts, which otherwise interfere with the Edman chemistry, are washed away. After sample loading, the hydrophobic and hydrophilic components of the reaction cartridge are reassembled and inserted into the sequencer. For a full description of sample loading onto the biphasic column sequencer, see Protocol 1. (Redrawn, with permission, from Hewlett-Packard HP G1005A advertising literature.)

impossible to decipher. However, if the molar ratio of major to minor components is significantly different, then it is possible to distinguish among the primary, secondary, and even tertiary components in the mixture (Simpson and Nice 1984).

Amino-Terminal Blocking

Various blocking reagents might render the amino group of a peptide/protein unreactive to the Edman degradation procedure. Partial amino-terminal blockage of a sample during purification can reduce sequencing yields by reducing the initial sequencing yield. Important blocking reagents are aldehydes that tend to form Schiff bases. Needless to say, precautions must be taken to remove traces of aldehydes, which occur universally, from solvents and reagents used during purification (and, especially, the sequencer procedure). Cyanate is another common contaminant that may react with amino termini of peptides/proteins. (*Note:* Ammonium cyanate is generated in significant quantities in concentrated alkaline urea solutions that are often used during protein purification.) For methods by which amino-terminal modifications can occur, see http://www.urmc.rochester.edu/research/propep/microseq.html.

Chemical Contaminants

Chemical contaminants that interfere with the Edman degradation must be removed from the sample prior to loading onto the sequencer. This is particularly important when using the pulsed-liquid instruments; in the case of biphasic column instruments, many contaminants can be readily eliminated during the sample loading process (see below). Sample contaminants that interfere with the conditions required for efficient Edman degradation include the following:

- ***Reagents that cause destruction of or side reaction with the Edman reagent PITC.*** For example, large amounts of primary amines will compete with the peptide/protein for PITC and will inhibit or prevent the coupling reaction. PITC is also destroyed by oxidizing reagents.
- ***Contaminants that affect the coupling pH*** (e.g., buffer salts with pKa values of <8.0). Such reagents disturb the coupling step by lowering the pH. This results in inefficient coupling and hence poor repetitive yields. Because buffer salts are not very soluble in the organic solvents used in sequencing instruments, they remain on the sample filter (especially in pulsed-liquid instruments) throughout the entire sequencing run.
- ***Reagents that affect the cleavage reaction by neutralizing the cleavage acid*** (TFA). For example, large quantities of buffer or other salts in the sample that remain on the sample support may inhibit the cleavage reaction by neutralizing the cleavage acid TFA. This results in incomplete cleavage and hence poor repetitive yields.

In addition to chemical contaminants that interfere with the Edman degradation, reagents that interfere with the reversed-phase HPLC analysis of PTH amino acids must be avoided, including any UV-absorbing (A_{269nm}) compounds that can be recovered from the sequencer along with the PTH amino acids. Such compounds (e.g., primary amines that react with PITC) can complicate the PTH amino acid chromatograms by obscuring the PTH amino acid peaks.

A list of common reagents that can interfere with the Edman chemistry and which should be removed from samples prior to sequencing is shown in Table 6.2 (this list is by no means exhaustive); sample attributes for successful sequencing are given in Table 6.3.

It should be remembered that microsequencing, in particular, is particularly sensitive to any factor that causes destruction of PITC, alteration of the pH of either the coupling or cleavage reaction, or blockage of the α-amino group of the amino-terminal amino acid.

TABLE 6.2. Reagents that interfere with sequencing

Glycerol or sucrose	These reagents are often added to purified samples for storage and handling purposes.
SDS	SDS may cause instrument blockage (and hence malfunction) and also loss of sample from the sample support (especially in the pulsed-liquid instrument).
Nonionic detergents	Common nonionic detergents such as Triton X-100, Tween, and Brij often contain aldehydes, oxidants, and other contaminants (e.g., UV-absorbing substances) that can interfere with the Edman chemistry and reversed-phase HPLC analysis of PTH amino acids.
Ammonium sulfate, other ammonium salts, and guanidine salts	These salts are often used for concentrating and fractionating proteins (see Chapter 4).
Buffers and primary amines	Glycine, Tris, and ethanolamine. Tris and glycine buffers are frequently used for protein purification and are common in samples recovered from SDS-PAGE.

Reagents that interfere with protein sequencing must be avoided or removed from samples prior to sequencing (see Protocol 1 for the biphasic column sequencing instrument, and Protocol 6 for pulsed-liquid instruments). Where possible, avoid dialysis as a last step in a sample preparation since dialysis tubing is often a source of contaminants that interfere with the Edman chemistry (also, sample recoveries from dialysis are often unacceptably low). If dialysis cannot be avoided, use thoroughly cleaned (follow manufacturer's instructions), high-quality tubing (e.g., Spectropor) and perform the dialysis in the presence of a low-concentration counterion salt or acid (otherwise the contaminants will remain adsorbed to the protein and dialysis membrane). When a sample is electroblotted onto a PVDF-type membrane for sequencing (see Protocol 2), contaminating reagents can be easily washed away from the immobilized sample with H_2O and 50% aqueous methanol. Similarly, samples that are applied to a biphasic sequencer reaction column or to the sample membrane prior to analysis using a gas-liquid-phase sequencer are immobilized, which permits removal (by washing) of excipient compounds that otherwise interfere with the sequence analysis.

TABLE 6.3. Sample attributes for successful sequencing

Sample purity	Peptide/protein must be amino-terminally homogeneous. The interpretation of PTH amino acid profiles obtained from homogeneous samples is usually straightforward, with a unique PTH amino acid being detected at each cycle. For impure (or partially digested) samples, several PTH amino acid derivatives can be displayed at each cycle, thereby making interpretation of the original (or major) polypeptide chain very difficult, or impossible. Even a sample ratio of 2:1 makes correct amino acid assignment uncertain (this is essentially due to variable yields of many of the PTH amino acids).
Sample volume	Ideally, the sample volume should be <150 µl. This is particularly important for the pulsed-liquid sequencers where the sample is loaded onto a glass fiber disk that can only absorb 30 µl of liquid (typically, multiple 30-µl aliquots are applied with drying of the disk between each aliquot). Sample volume is not a limiting factor when using the Agilent biphasic column sequencer (see Protocol 1) or the Applied Biosystems Problott apparatus in conjunction with the pulsed-liquid sequencer.
Sample buffer	The sample should be in a volatile solvent (e.g., H_2O, acetic acid, acetonitrile, propanol, formic acid, trifluoroacetic acid, or triethylamine) or buffer (e.g., ammonium bicarbonate if lyophilized repeatedly). The sample can contain a small quantity of detergent (e.g., 0.01–0.02% Tween-20 or <0.1% SDS). Care should be taken with SDS since larger amounts can wash the sample out of the reaction vessel and cause disruption of reagent/solvent delivery the instrument. All solvents and reagents used in protein sequencing should be of the highest purity available (e.g., HPLC-grade solvents, "sequencing-grade" and "electrophoresis-grade" reagents).
Sample amount	Although most modern sequencing instruments can sequence 0.5–1.0 pmole of sample, it is recommended that a minimum of 10–50 pmoles of sample be analyzed to ensure that the sequence interpretation is correct or to determine whether the sample is amino-terminally blocked if no sequence is obtained. Invariably, the amount of sequenceable material is underestimated due to sample loss (e.g., irreversibly "sticking" to plasticware during manipulation and storage*), inaccurate quantitation, or amino-terminal blockage during isolation/purification. It is always preferable to err on the side of too much sample (e.g., 50% more than required for analysis) than too little sample. Sample quantitation can be estimated from silver- or Coomassie Blue-stained gels, micro BCA or absorbance. For samples prepared by microbore reversed-phase HPLC, sample amounts can be estimated by their absorbance at 214 nm (e.g., 100 mAU on a 2.1-mm I.D. column is equivalent to ~1 µg; and 100 mAU on a 4.1-mm I.D. column is equivalent to ~10 µg).

*It is recommended that liquid samples be stored in the presence of 0.01–0.02% Tween-20 in tightly capped polypropylene tubes (e.g., microfuge tubes) and samples isolated by reversed-phase HPLC not be dried down, but stored in their collected state (e.g., aqueous TFA/acetonitrile) at –20°C until analyzed.

HOW SEQUENCER PERFORMANCE HAS IMPROVED

In the 50 years since Edman's first report of a manual procedure for the amino-terminal sequencing of peptides, there has been an almost 10,000-fold increase in sensitivity (compare the ~10 nmoles starting material required in 1950 to the ~1 pmole required today). This has been largely due to four factors:

1. Automation drastically reduced sample loss, an inherent problem with any manual procedure.
2. Increased sensitivity of PTH amino acid analysis (e.g., compare modern HPLC, especially the use of microcolumns [see Figure 6.8] with the original TLC method for PTH amino acid detection that was sensitive to ~1 nmole of PTH derivative in a spot).
3. Miniaturization of instrumentation.
4. Reagent/solvent quality.

Collectively, these factors influence the overall quality of sequence analysis. The initial yield and the repetitive yield, together with the sensitivity limit of the instrument, determine the amount of sequence that can be obtained.

- ***Initial yield*** is a measure of the percentage of the total sample loaded onto the instrument that can be sequenced (typically 50–80% unless the amino terminus is blocked). The reasons for this are poorly understood, but factors such as sample preparation (e.g., proteins can be amino-terminally blocked by exposure to poor-quality reagents/solvents that contain aldehydes) during purification, the sequencer itself, and the reaction parameters of the instrument clearly influence initial yield.
- ***Repetitive yield*** is the percentage of sequence detected after each cycle of the Edman reaction. This value is measured by a linear regression fit to the amount of PTH amino acid at each cycle. Extrapolation to cycle 0 gives the initial amount of sequenceable peptide. Whereas highly skilled technicians using the manual Edman degradation procedure achieved repetitive yields of up to 90%, >95% repetitive yields could be readily achieved after automation. It can be calculated that an average repetitive yield (i.e., from one cycle to the next) of 90% and 95% would allow 20 and 40 residues, respectively, to be identified from a 10-pmole peptide before reaching a sensitivity limit of 1 pmole (lowest amount of readable sequence). However, with a 100-pmole sample load, 40 and 90 cycles are allowed with instruments performing at 90% and 95% repetitive yield, respectively (Figure 6.10). Thus, the higher the repetitive yield, the longer the sequences that can be determined before background peaks generated by side reactions, spurious cleavage of the peptide chain during sequencing (especially acid-labile peptide bonds such as Asp-Pro), and out-of-frame sequences (generated by incomplete cleavage or coupling reactions, referred to as the "lag") obscure the PTH amino acid peak derived from the amino-terminal residue.

  In practice, 20–25 cycles of sequence can be expected in most cases from 1 to 5 μg of sample, but occasionally 60 residues or more can be assigned for large peptides using instruments performing at >96% repetitive yield (Figure 6.11).
- ***Repetitive yield*** (RY), expressed as a percentage, is calculated as follows: RY = 10(M) x 100.

  For a multiple point calculation:

  M is the slope of the linear regression line calculated from a set of points [log (pmole) vs. cycle].

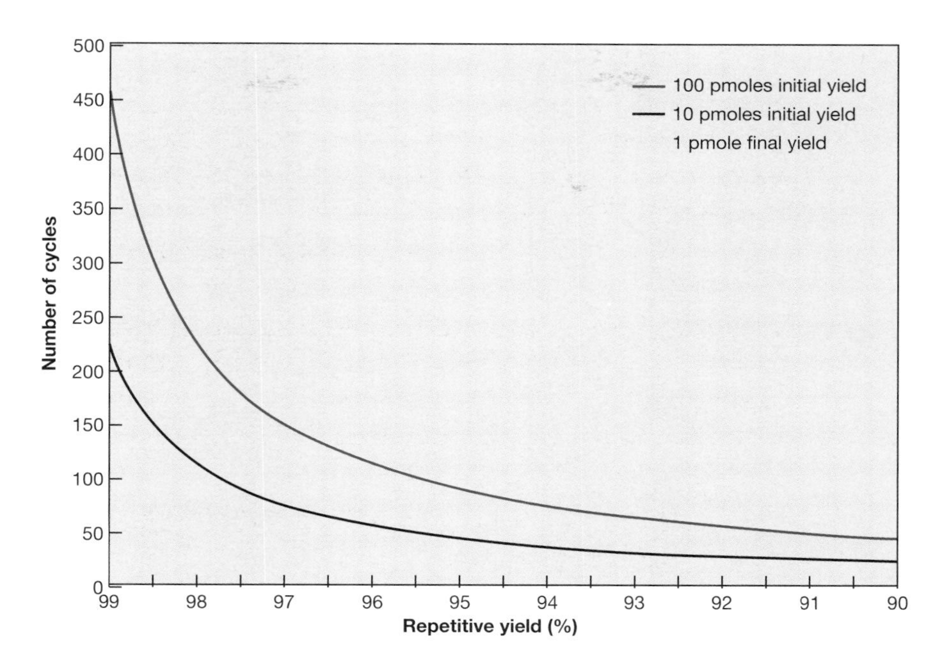

FIGURE 6.10. Maximum length of determinable sequence with respect to repetitive yield. The amount of sequence that can be obtained is determined by a combination of the initial and repetitive yields. The top line shows the theoretical expected number of residues (*y* axis) that can be established from an initial yield of 100 pmoles with respect to the different achievable repetitive yields (*x* axis). Similarly, the bottom line shows the number of residues that can theoretically be determined given an initial yield of 10 pmoles with respect to varying achievable repetitive yield. The level of PTH amino acid detection is 1 pmole final yield. The calculation for determination of repetitive yield is shown below. In practice, initial yields are typically 30–50%; i.e., only 30–50% of molecules contain a free amino-terminal amino group that is amenable to Edman chemistry.

Calculation of Repetitive Yield (RY)

RY = 10(M) x 100

For a point-to-point calculation without background correction:

$$M = \frac{\log_{10}(Y_2) - \log_{10}(Y_1)}{X_2 - X_1}$$

For a point-to-point calculation with background correction:

$$M = \frac{\mathrm{Log}_{10}(Z_2) - \log_{10}(Z_1)}{X_2 - X_1}$$

where

$Y_2 =$ pmole value (uncorrected) of 2nd point
$Y_1 =$ pmole value (uncorrected) of 1st point
$X_2 =$ cycle number of 2nd point
$X_1 =$ cycle number of 1st point
$Z_2 =$ pmole value (background corrected) of 2nd point
$Z_1 =$ pmole value (background corrected) of 1st point

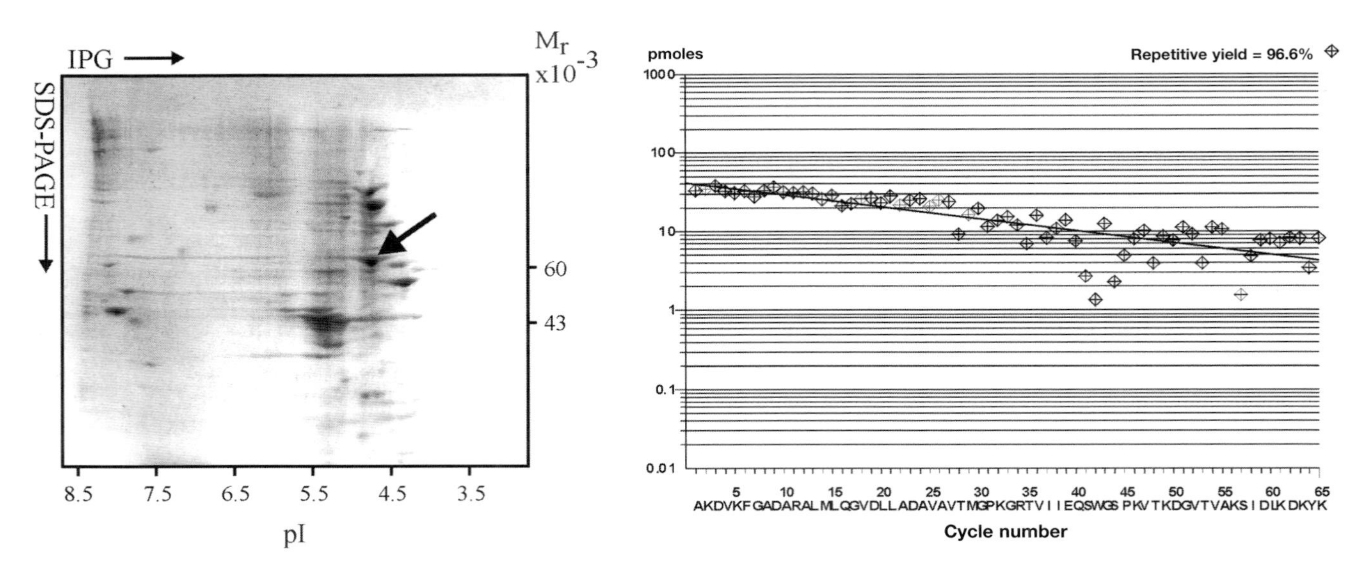

FIGURE 6.11. Amino acid sequence analysis of heat shock protein 60, from a reducing two-dimensional (2D) gel electrophoresis of LIM 1215 cells. Protein spots (such as the one indicated by the arrow on the gel) from three identical 2D gel/PVDF electroblots were combined for sequence analysis using the biphasic-column sequencer. The PTH amino acids are indicated by the one-letter notation used for amino acids. (Reproduced, with permission, from Ji et al. 1997 [©Wiley-VCH].)

HPLC AND PAGE ARE USED TO PREPARE SAMPLES FOR MICROSEQUENCING

The two high-resolution techniques that are commonly used to prepare samples for microsequencing, especially trace-abundant proteins obtainable at low-nanogram amounts only, are high-performance liquid chromatography (HPLC) and polyacrylamide gel electrophoresis (PAGE) (for details of these methodologies, see Chapter 5 and Chapter 4, respectively). Whereas reversed-phase (RP)-HPLC is usually considered applicable at any stage in a purification protocol, electrophoresis is usually reserved for the last purification step.

RP-HPLC

With the steadily growing variety of reversed-phase, ion-exchange, and affinity chromatography supports available for HPLC, the task of the protein chemist in purifying trace-abundant proteins is becoming much easier. Of these, RP-HPLC remains the method of choice for most separation strategies, particularly for peptides and low-M_r (<40,000) hydrophilic proteins. The separation of proteins and peptides by reversed-phase supports is based predominantly on the differential hydrophobic interactions between amino acid side chains on the protein or peptide and the functional groups on the column support. For details on typical column packings used in peptide/protein separations, see Chapter 5. The following are typical chromatographic conditions that could be applied, as an initial approach, for the separation of proteins and peptides:

- Brownlee RP-300 column, made of dimethyloctyl silica at 10-μm particle size and 300 Å pore size.
- Linear 60-minute gradient from 0.1% aqueous TFA to 60% acetonitrile/40% H_2O containing 0.089% TFA.

Several options are available for improving RP-HPLC separations, including changing the mobile-phase solvent and pH, and, especially in the case of peptide separation, switching to a different column packing. Selected examples of these options are given in Chapter 5 and Chapter 7.

Recently, there has been an increasing awareness of the need for faster chromatography, especially for large-scale, high-volume sample throughput. This has led to the development of polystyrene divinylbenzene-based polymeric supports such as the perfusive support (see Chapter 5). These supports have very large pore structures (>8000 Å) that transect the particle and confer two distinct advantages: (1) increased surface area, and hence higher mass loading, and (2) the minimization of slow diffusive flow that permits rapid partitioning, and hence resolution, of proteins and peptides within the supports over a wide range of high flow velocities (1000–9000 cm/hour). It should be noted that for micropreparative-scale purifications, fast chromatographic analysis of proteins and peptides could be performed using conventional silica-based supports and standard liquid chromatographs (Moritz et al. 1994, 1995)

Microbore column chromatography enhances recovery and detection of picomole amounts of protein

The ability to purify and subsequently manipulate (e.g., buffer-exchange, concentrate, reduce and alkylate, and fragment) material at low-microgram (i.e., picomole) levels, and in sufficiently small volumes for direct loading onto Edman chemistry sequencing instruments and mass spectrometers, presents a considerable technical challenge. Moreover, samples purified by conventional high-resolution procedures, although homogeneous with respect to proteinaceous material, frequently contain excipients that impede the performance of these instruments (e.g., buffer salts and detergents). One chromatographic approach for concentrating (or trace-enriching) samples derived from the most widely used HPLC columns (i.e., 4.6 mm I.D.) employs short microbore columns (e.g., 30 x 2.1 mm I.D.) (for a review, see Simpson et al. 1989).

The principles and operation of microbore column chromatography are covered in detail in Chapter 5. Microbore HPLC provides two key advantages for preparing samples for microsequence analysis:

- The ability to concentrate and subsequently recover material in small peak volumes suitable for direct loading onto the gas-phase/pulsed-liquid sequencing instruments.
- An increased level of sensitivity that enhances protein detection, thereby permitting low-level analytical studies to be undertaken.

Microbore column trace enrichment of samples derived from conventional columns avoids the poor recoveries associated with classical concentration procedures such as lyophilization and precipitation (e.g., TFA or acetone). RP-HPLC-derived samples usually contain sufficient quantities of secondary solvent (e.g., acetonitrile) to prevent their retention on a similar support; thus, samples require severalfold dilution prior to injection. This is readily accomplished either by diluting the sample two- to threefold with primary solvent in the sample syringe immediately prior to injection or by repetitive injection of small aliquots (one tenth of loop volume) where dilution occurs in the precolumn hydraulics of the instrument.

One of the attractive innovations of the sample-loading procedure for the biphasic column sequencer is that large-volume samples can be concentrated, and desalted, in a manner akin to loading a sample onto an RP-HPLC column (see Figure 6.9). Thus, samples containing excipients such as buffer salts (e.g., 1 ml of 20 mM Tris-HCl buffer containing 1.5 M NaCl), guanidine and amine-containing buffers (e.g.,1 ml of buffer consisting of 6 M guanidine-HCl, 1 M Tris-HCl, and 10 mM EDTA) can be applied directly onto the biphasic reaction vessel without the need for any prior trace enrichment or desalting; after the sample is loaded on the sequencer column, these excipients are easily washed away prior to sequence analysis.

PAGE

Once an acceptable polyacrylamide-gel-based protocol has been established for separating a mixture of proteins (see Chapters 2 and 4), a number of critical decisions must be made with respect to which strategy should be adopted for obtaining sequence information: amino-terminal sequencing or internal sequencing of peptides (see Figure 6.12). A major consideration is the losses incurred when electroblotting a protein from a gel onto an immobilizing matrix such as polyvinylidene difluoride (PVDF). Parameters influencing losses during electroblotting are the physical characteristics of the membrane (e.g., specific surface area, pore size distribution, pore volumes, and sequencing solvent/reagent permeability considerations) (Eckerskorn and Lottspeich 1993). For reviews of commercially available blotting membranes, the blotting process, and gel electrophoresis-induced protein modifications, see Simpson et al. (1989), Aebersold (1993), Eckerskorn (1994), Patterson (1994), and Patterson and Aebersold (1995).

An example of a sequence analysis of a relatively abundant gel-separated protein (housekeeping protein), heat shock protein 60, following electrotransfer onto a PVDF membrane

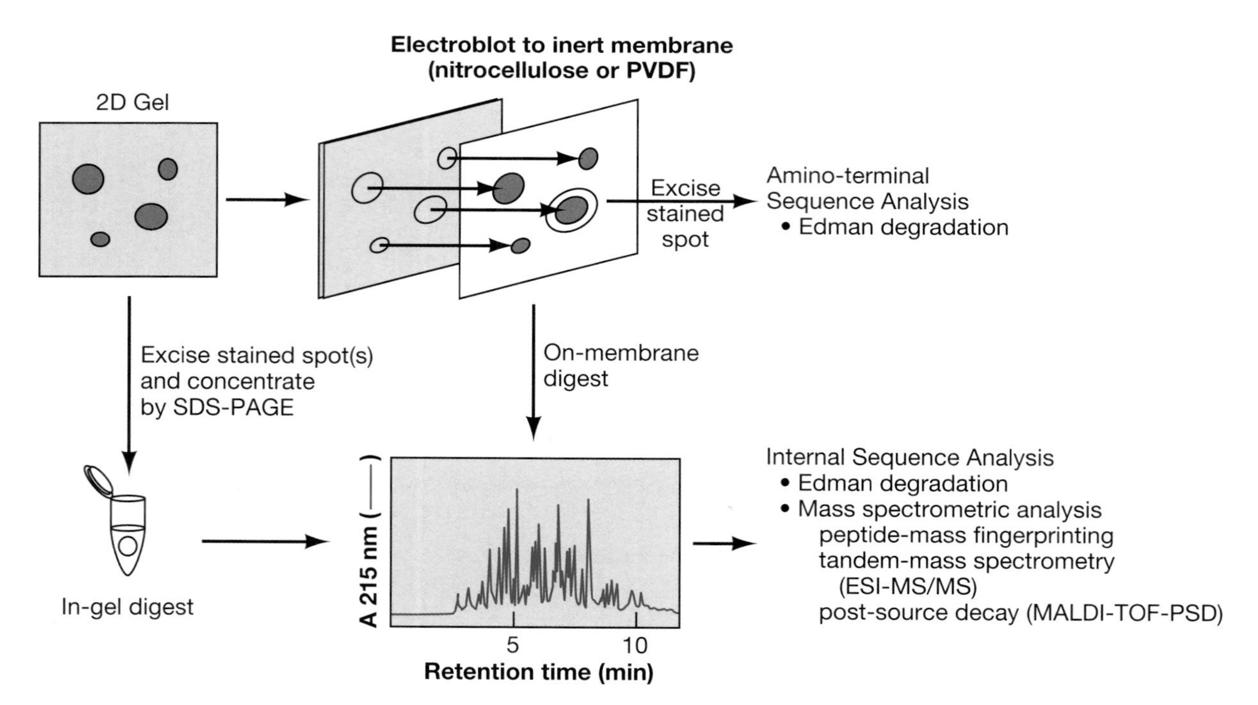

FIGURE 6.12. Experimental approaches for obtaining sequence information from gel-separated proteins. For in-gel proteolytic digestion strategies, see Chapter 7; mass-spectroscopic-based protein and peptide identification techniques are described in Chapter 8. (Redrawn, with permission, from Simpson and Reid 1997.)

(see Protocols 2 and 3) is shown in Figure 6.11. For the above-mentioned reasons, and taking into consideration the effort and cost involved in obtaining a highly purified protein, researchers must decide whether to take the risk in attempting direct amino-terminal sequence of an electroblotted protein. The alternative strategy outlined in Figure 6.12 is internal sequence analysis (see Chapters 8 and 9). Although this procedure is more time-consuming, it is a safer option for a precious sample, especially if it is not known whether the protein is blocked at the amino terminus. Moreover, an internal sequence strategy yields higher sequence coverage of the polypeptide chain (cf. amino-terminal sequence analysis) and increases the likelihood of providing posttranslational information.

Concentration of gel-separated proteins prior to amino-terminal sequence analysis

Empirical observations from a number of proteomics laboratories indicate that concentration of stained excised gel spots (or bands) prior to electroblotting onto PVDF membranes is required to achieve high initial sequencing yields. Likewise, concentration prior to in-gel proteolytic (or chemical) fragmentation (see Chapters 7 and 9) is necessary to obtain high peptide yields. Preconcentration of protein spots/bands can be accomplished by reelectrophoresis of the stained gel pieces in a second conventional SDS-polyacrylamide gel or, alternatively, in a secondary gel matrix in the tip of a conventional Pasteur pipette (Gavaert et al. 1996) as described in Protocol 4. For this procedure, it is important to ensure that the total volume of combined gel pieces and equilibration buffer is smaller than 0.5 ml. When handling larger volumes (e.g., 1.2 ml), this procedure can be performed in 400–500-μl batches, and these batches are then combined for further concentration in a single Pasteur pipette. Alternatively, larger volumes can be handled by similar means using a simple apparatus described by Gevaert et al. (1995) and Dainese et al. (1997).

An alternative approach to obtaining sufficient material is to enrich low-abundance proteins prior to running a 2D gel. For example, this can be achieved by applying a total cell lysate onto a slab SDS-PAGE, excising a region of the gel comprising a narrow-M_r range containing the protein of interest, passively eluting the proteins, and then loading them onto a high-resolving 2D gel for subsequent separation and microsequence analysis (Ji et al. 1993). Sequence data can thus be obtained from proteins in the submicrogram range.

PROTEINS THAT WILL NOT SEQUENCE MUST BE DEBLOCKED

Some proteins naturally undergo posttranslational modification, resulting in amino-terminal blockage with formyl, acetyl, or pyroglutamyl groups (formation of a pyroglutamyl group may occur by cyclization of amino-terminal glutamine residues under acidic conditions) (Brown and Roberts 1976; Brown 1979). This blockage prevents the amino terminus from coupling with PITC. Empirical estimates from a number of laboratories suggest that 40–50% of naturally occurring proteins resolved by 2D electrophoresis are amino-terminally blocked. For example, 50% of the proteins from human myocardial tissue prepared by electroblotting 2D electrophoresis gels were found to be refractory to Edman degradation (Baker et al. 1992), and ~80% of the soluble proteins from Ehrlich ascites cells are *N*-acetylated (Brown and Roberts 1976). If the protein of interest is not naturally blocked, care must be taken to ensure that chemical blockage of the amino terminus does not occur during sample preparation, especially during electrophoresis.

Tips to Avoid Amino-terminal Blockage in Gels

The following precautions can be adopted to minimize the chemical modification of proteins in polyacrylamide gels (Moos et al. 1988):

- Use HPLC-grade glycerol in the sample buffer. Heat sample only to 37°C for 10 minutes.
- Add 0.1–1 mM sodium thioglycolate or 10 mM reduced glutathione in the Laemmli buffer.
- Allow running gel to polymerize overnight and/or preelectrophorese it before adding the stacking gel (acrylamide monomer can react with free amino termini of proteins).
- Use Towbin buffer (Towbin et al. 1979) if the protein of interest is reduced and alkylated; do not use a pH 11 electroblotting buffer.

For these reasons, erring on the side of caution is warranted when estimating the quantities of protein required for sequence analysis by Edman degradation. Although a number of elegant methods exist for deblocking electroblotted proteins (Wellner et al. 1990; Hirano et al. 1992, 1993), they have not proven to be very efficient (<40%) or generally applicable. Hence, these methods are not recommended for low-abundance proteins.

Deblocking Methods

The following deblocking methods are listed by the group(s) they remove. The conditions apply to proteins in solution or stained protein bands on a PVDF-type membrane (for further details, see Hirano et al. 1997).

- ***Formyl group removal.*** Incubate with 0.6 N HCl for 24 hours at 25°C.
- ***Amino-terminal acetylserine or acetylthreonine removal.*** Incubate with TFA for 30 minutes at 60°C in a sealed microfuge tube under nitrogen. Some cleavage of internal serines and threonines may also occur.
- ***Amino-terminal pyroglutamic acid removal.*** Pyroglutamate aminopeptidase (Boehringer Mannheim or TaKaRa) is reported to remove amino-terminal pyroglutamic acid from proteins bound to PVDF membranes or absorbed to glass fiber filters.
- ***Acetylated amino-terminal amino acid removal.*** Acylamino-acid-releasing enzyme (TaKaRa) is reported to remove acetylated amino-terminal amino acids from peptide fragments recovered in solution from in situ trypsin digestion.

Enzymatic removal of amino-terminal blocking groups is more efficient from some samples than for others. The efficiency of the deblocking method can sometimes be improved if the sample is first reduced and alkylated (for reduction/alkylation conditions, see Chapter 9).

MICROSEQUENCE ANALYSIS AROUND PHOSPHORYLATION SITES CAN HELP DELINEATE SIGNALING PATHWAYS

Reversible protein phosphorylation is a fundamental intracellular mechanism for transducing extracellular stimuli to the nucleus. The class of enzymes that phosphorylate proteins and (i.e., protein kinases) exhibits widely preferred substrate specificities (for a review, see Aitken 1999) and can be broadly classified into groups depending on which amino acid they phosphorylate (e.g., tyrosine protein kinases and serine/threonine protein kinases). In general, protein kinases are subject to a variety of regulatory mechanisms such as phosphorylation/dephosphorylation, second messengers, calcium ions, and the presence of pseudosequences in their regulatory domain, to name a few (for a review of protein kinases, see Hunter 1987). To delineate the circuitry of a signaling pathway, it is important to establish the class of protein kinase(s) responsible for regulating the various protein components of that pathway. This can be accomplished from knowledge of the amino acid sequence surrounding a site of phosphorylation.

Analysis of sites of protein phosphorylation (serine, threonine, or tyrosine) can be accomplished by various procedures such as thin-layer electrophoresis (see Appendix 2), derivitization of phosphoamino acids to facilitate PTH amino acid identification (e.g., phosphoserine to *S*-ethylcysteine; see Appendix 2), manual Edman degradation of ^{32}P-labeled phosphopeptides, mass-spectrometry-based techniques (see Chapter 8), and microsequencing of phosphopeptides. For a description of these procedures, see Aitken and Learmonth (1997). Of these methods, phosphopeptide mapping by thin-layer electrophoresis and solid-phase sequencing of ^{32}P-labeled phosphopeptides (see Protocol 5) are considered to be the most sensitive.

THE BEST RECOVERY OF PHOSPHOAMINO ACIDS IS ACHIEVED WITH SOLID-PHASE SEQUENCERS

The characteristics of solid-phase sequencing are the removal of the thiazolinone derivatives of the residues at each cycle of Edman degradation, along with a high percentage of the ^{32}P-labeled phosphate, by washing with anhydrous TFA. The covalent linkage of the peptide or protein to the derivatized PVDF membrane supports allows such a strong wash. These mem-

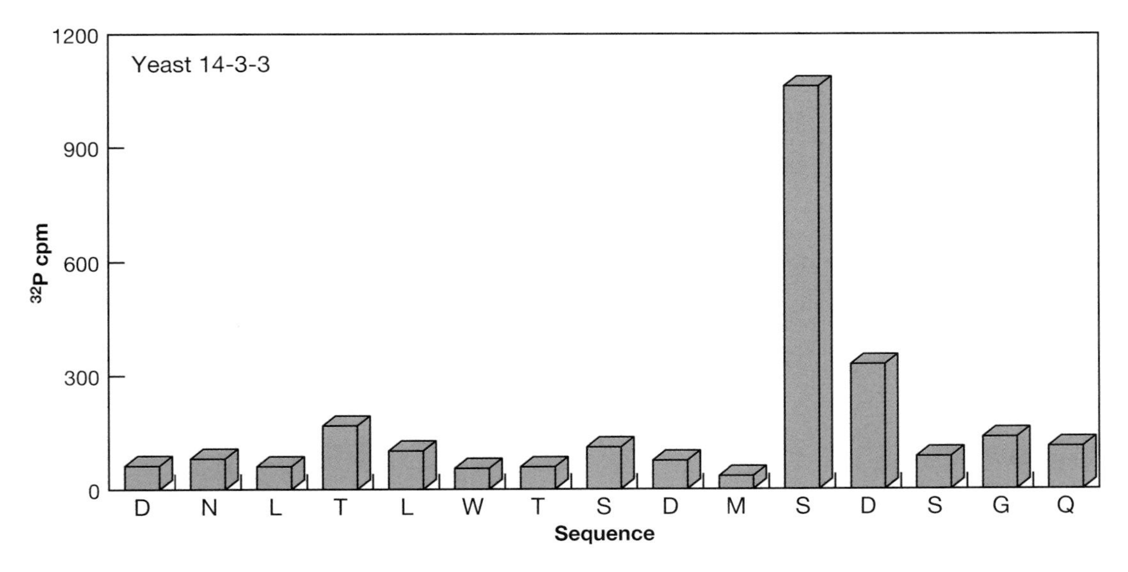

FIGURE 6.13. Automated sequencer analysis on PE Biosystems 477 of a phosphopeptide labeled with ^{32}P orthophosphate. The radioactive protein band from an immunoprecipitate was digested in-gel with trypsin, and the ^{32}P-labeled phosphopeptide was recovered by microbore HPLC on a 1-mm Vydac C_{18} column (Dubois et al. 1997). The material extracted from the sequencer by the aqueous methanol after each cycle of Edman degradation was measured by Cerenkov counting. The amount of ^{32}P radioactivity loaded onto the instrument was ~1000 cpm by Cerenkov counting, and the recovery of 1100 cpm in the peak by scintillation counting (which gives more than double the efficiency) indicated a typical recovery of 40–50% at cycle 11. The analysis shows that the threonine residues at cycles 4 and 7 and the other serines at cycles 8 and 13 are not phosphorylated. (Courtesy of Alastair Aitken, University of Edinburgh.)

branes enable the attachment of proteins via side-chain lysines and amino termini (Sequelon-DITC) or via side-chain carboxyl groups (Sequelon-AA). Sequelon-AA membranes can be used in sequencing phosphorylated peptides in conventional gas- or pulsed-liquid-phase protein sequencers. With the use of the Applied Biosystems sequencer, butyl chloride is replaced with 9:1 methanol:H_2O (containing 2 mM phosphate), resulting in almost quantitative recovery of very low levels of radiolabeled phosphate (see Figure 6.13). The procedure for solid-phase microsequencing of a ^{32}P-labeled peptide on an Applied Biosystems sequencer is described in Protocol 5.

ALTERNATIVE AMINO-TERMINAL REACTION METHODS

A number of variations on the Edman reagent, PITC, have been proposed, primarily to increase the analytical sensitivity of the method. Both colored (e.g., dabsyl-isothiocyanate reagent) (Chang and Creaser 1976) and fluorescent isothiocyanates have been used for specific purposes, but none have superseded the Edman reagent for routine sequencing purposes, primarily because they have not provided quantitative coupling, presumably due to either the increased size of the reagent or deactivation of the isothiocyanate group.

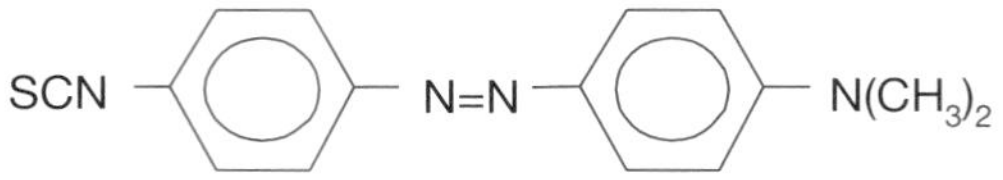

SCN–C_6H_4–N=N–C_6H_4–$N(CH_3)_2$

CARBOXY-TERMINAL SEQUENCING METHODS ARE MORE PROBLEMATIC

Although there were active efforts to develop chemistry for sequential removal of carboxy-terminal amino acids from a protein some 25 years prior to Edman's first report on amino-terminal sequencing, advances have lagged badly behind amino-terminal methodology. Despite a plethora of potentially useful chemical modifications of the carboxy-terminal amino acid, which include conversions of the carboxyl group to thiohydantoins, carboxylic acid esters, alcohols, acylureas, isothioureas, azides, and hydrazides, investigators have found the published procedures difficult (for review, see Inglis 1991). Consequently, enzymatic methods using carboxypeptidases have provided most of the carboxy-terminal sequence data obtained over the years. An alternative solution has been to isolate the carboxy-terminal peptide from a digest of the protein and analyze it using either the amino-terminal degradation procedure or mass spectrometry.

AUTOMATED CARBOXY-TERMINAL SEQUENCING

Carboxy-terminal sequence analysis complements mass spectrometry and Edman degradation for protein characterization. The carboxy-terminal portion of a protein sequence can be directly analyzed without previous proteolytic cleavage and without identification of the carboxy-terminal peptide for analysis by mass spectrometry.

The most widely studied method for carboxy-terminal sequencing has been one in which the thiohydantoin of the carboxy-terminal amino acid is first formed and the adjacent peptide bond is then cleaved in either acidic or basic medium. This yields an amino acid thiohydantoin and a shortened protein that may be subjected to a further degradation cycle. The carboxy-terminal thiohydantoin procedure can be used repetitively, and there are no major interfering side reactions. This approach has similarities to the Edman reactions for amino-terminal sequencing, as shown in Figure 6.2. Although the latter employ initial coupling with "stable" isothiocyanates such as PITC, the carboxy-terminal procedure can use either thiocyanate salts or organic isothiocyanates.

The reaction mechanisms may differ—the amino group is best coupled at pH 9 (to remove H^+) in a polar environment, whereas the carboxyl group has been found to couple under anhydrous conditions over a wider range of pH values (from slightly basic to acidic conditions). Both coupling reactions cause a weakening of the adjacent peptide bond, which is subsequently cleaved to release relatively stable thiohydantoin derivatives of the terminal amino acids. The common PTH amino acids (x = phenyl in Figure 6.2) are less polar than the thiohydantoin amino acids, but they have similar properties, the UV spectra especially being important because of the high extinction coefficient (E_{max} ~17,500) at ~269 nm that may be used for identification purposes. This chemistry has been moderately successful in the commercial carboxy-terminal sequencers developed by Hewlett Packard (now Agilent) and Perkin Elmer (now Applied Biosystems).

CARBOXY-TERMINAL ANALYSIS CAN BE PERFORMED WITH INCREASED SENSITIVITY AND IMPROVED PERFORMANCE

Bergman et al. (2001) have modified parameters in the protocol for a carboxy-terminal protein sequencer operated according to the principles of Boyd et al. (1992, 1995). The methodology has been applied to carboxy-terminal analysis of a broad range of polypeptides with improved results with respect to standard protocols, particularly regarding sensitivity and length of degradation, and the ability to sequence through prolines and to perform both amino- and carboxy-terminal sequence analyses on the same sample.

Dilution of the chemicals and reagents used in the sequencer and during chromatography (Table 6.4) lowers the background and produces a flatter baseline in the chromatograms of the alkylated thiohydantoin (ATH) amino acids generated during the degradation cycle. The sensitivity is thereby increased, allowing the use of 10 pmoles instead of the manufacturer´s recommended 100 pmoles of ATH amino acid standard (Figure 6.14). During pretreatment with phenylisocyanate (PIC), which blocks the amino groups, 1% PIC in diisopropylethylamine (DIEA) is used, instead of the recommended 2% PIC. In addition, a single PIC treatment, instead of the three applications of PIC recommended by the manufacturer, is sufficient to lower the levels of artifact peaks in the chromatograms and improve the sensitivity.

For the reversed-phase separation of the ATH amino acids, ethanol is used in both solvents A and B instead of tetrahydrofuran (THF) (Table 6.4). Ethanol changes the elution time for most ATH amino acids, and the ethanol concentration is thus a parameter that can be used to optimize separation and improve resolution. The separation of ATH-Asn and ATH-Gln is significantly improved, and baseline resolution is achieved for the pairs ATH-Asp/ATH-Glu and ATH-Trp/ATH-Thr (Figure 6.14) if ethanol replaces THF. Ethanol is also less toxic and less hazardous than THF (THF generates peroxides and poses a risk of explosion).

Only acetone is used in solvent A to balance the baseline (DIEA is excluded), improving chromatographic behavior. Instead of a U-shaped baseline at high sensitivity with DIEA present, a flat baseline is obtained allowing sensitive detection of all ATH amino acids. In addition, the column temperature is decreased from 45°C to 40°C, which improves resolution and stability of the ATH amino acids, in particular of ATH-Ser. The recovery of ATH-Ser and ATH-Thr is further improved by also lowering the temperature of the transfer flask, from 45°C to 40°C (Table 6.4).

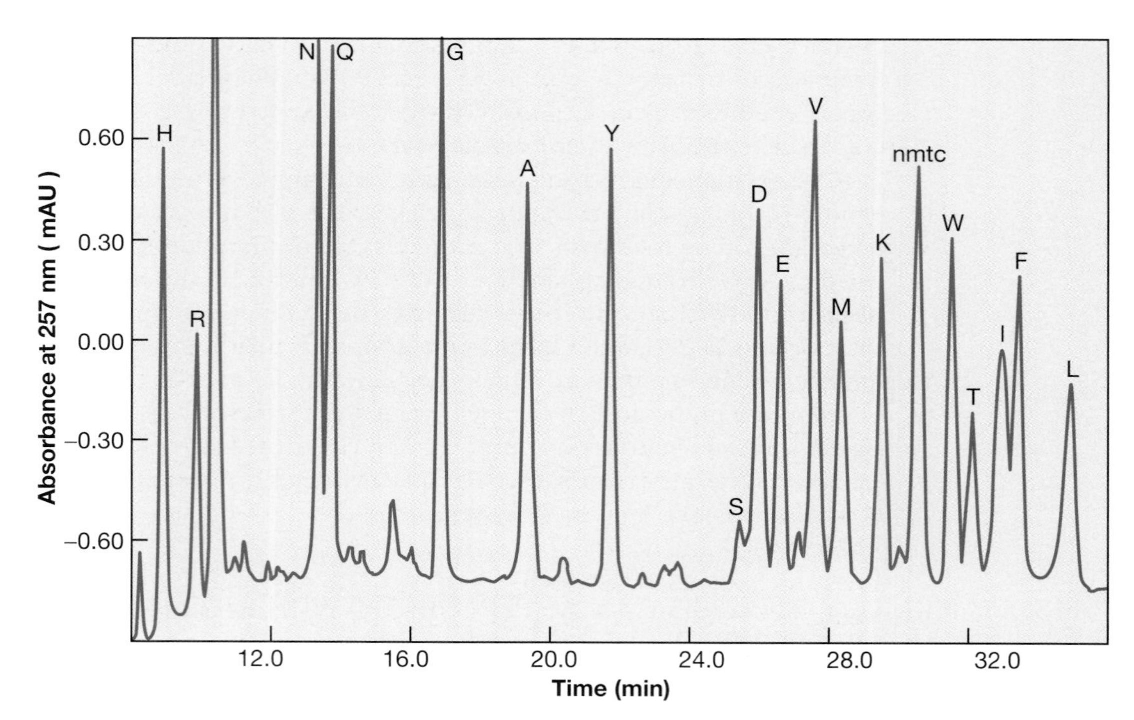

FIGURE 6.14. Chromatography of ATH standard amino acids (10 pmoles) using the modified protocol specified in Table 6.4. nmtc denotes naphthylmethylthiocyanate, the major by-product formed during sequence analysis. Note the resolution for the pairs N/Q, D/E, and W/T that are normally difficult to separate. (Courtesy of Tomas Bergman, Ella Cederlund, and Hans Jörnvall, Karolinska Institutet.)

TABLE 6.4. Parameters altered with respect to the manufacturer´s protocol for improved carboxy-terminal sequencing

Parameter changed	Before	After
Dilution		
(C1[a]) ATH amino acid standard	100 pmoles/injection	10 pmoles/injection
(C3) *N*-Methylimidazole/Acetonitrile	as delivered	diluted 1:1 with acetonitrile
(C4) Piperidine thiocyanate/Acetonitrile	as delivered	diluted 1:1 with acetonitrile
(C6) Acetic anhydride/Lutidine/Acetonitrile	as delivered	diluted 1:1 with acetonitrile
(C8) Bromomethyl-naphthalene/Acetonitrile	as delivered	diluted 1:1 with acetonitrile
(C10) Tetrabutylammonium thiocyanate/Acetonitrile	as delivered	diluted 1:1 with acetonitrile
(C11) DIEA/Heptane	as delivered	diluted 1:1 with heptane
2% Phenylisocyanate/DIEA	according to Applied Biosystems recommendation	diluted 1:1 with acetonitrile
Chromatography		
Solvent A	3.5% THF in 82.5 mM sodium acetate (pH 3.8), DIEA/acetone	7.1% ethanol in 30 mM sodium acetate (pH 3.8), acetone
Solvent B	18% THF in acetonitrile	7.1% ethanol in acetonitrile
Temperature		
Column	45°C	40°C
Transfer flask	45°C	40°C

[a]Label according to Applied Biosystems nomenclature.

To minimize the yield of by-products from degradation and alkylation, the sequencer program is altered to accommodate two additional washing steps with ethyl acetate in each cycle. With three washes instead of one wash, chromatography is clean with a low background.

A reaction cartridge with a vertical slit for the sequencer membrane, instead of a horizontal slit, is recommended. The liquid thus flows along the entire length of the membrane instead of perpendicular to it, resulting in more efficient solvent extraction and reagent penetration. The increased liquid hold-up time improves the recovery of ATH amino acids and gives better sensitivity. The results show that this type of cartridge, designed for amino-terminal sequence analysis of electroblotted samples, further increases the overall yield (Figure 6.15).

More than 200 proteins and fragments ranging from 10 to 600 residues have been analyzed using the modified protocol (Bergman et al. 2001). The average length of degradation is five residues, but extended degradation lengths have also been achieved (Figure 6.16). Samples are applied to PVDF membranes, either by electroblotting the proteins that have been separated on polyacrylamide gels or by direct application of polypeptides in solution. The amounts of protein vary from a few hundered pmoles down to 10 pmoles (Figure 6.17). The average initial yield is 15% and the average repetitive yield is less than 70%. The recoveries and lengths of degradation are very much dependent on the protein sequence. In particular, proline and acidic and hydroxylated residues can stop degradation or lower the yields. However, using the procedures in Protocol 6, sequencing past proline and identification of the next residue in the carboxy-terminal sequence are possible and have been shown with several proline-containing sequences.

COMBINED AMINO- AND CARBOXY-TERMINAL SEQUENCE ANALYSIS CAN MAXIMIZE SEQUENCE INFORMATION

Chemical sequence analysis of proteins from both ends of each molecule using the same sample application is an attractive way to maximize the sequence information recovered from limited amounts of sample, particularly when the sensitivity in carboxy-terminal analysis is down to the 10-pmole level. This has been tested quite extensively and the following procedure is recommended. The sample is applied to a PVDF membrane and analyzed in the amino-terminal sequencer for 5–25 cycles. The membrane is removed from the sequencer reaction cartridge and treated with PIC to block the α-amino group, followed by transfer to the reaction cartridge of the carboxy-terminal sequencer. The program begins with an ethyl acetate wash in the cartridge to remove by-products from the Edman chemistry (mainly diphenylthiourea). Lysine residues will not be derivatized with PIC since their ε-amino group is already derivatized with PITC in the Edman degradation, resulting in the formation of phenylthiocarbamyl (PTC) lysine derivatives. The corresponding ATH-PTC-Lys derivative, or a breakdown product thereof, elutes close to ATH-Asp and is clearly detected using the modified protocol. Carboxy-terminal sequence information can regularly be recovered for four to eight residues (Figure 6.18).

Amino-terminally blocked proteins are common and often acetylated. Sequence analysis of blocked proteins can be performed using tandem mass spectrometry applied to the corresponding amino-terminal proteolytic peptide (Jonsson et al. 2000) and by sequencer degradation after application of deblocking protocols to the intact protein (Gheorghe et al. 1997). In this respect, carboxy-terminal sequence analysis is an attractive alternative, particularly because carboxy-terminal blocking of proteins is not commonly encountered.

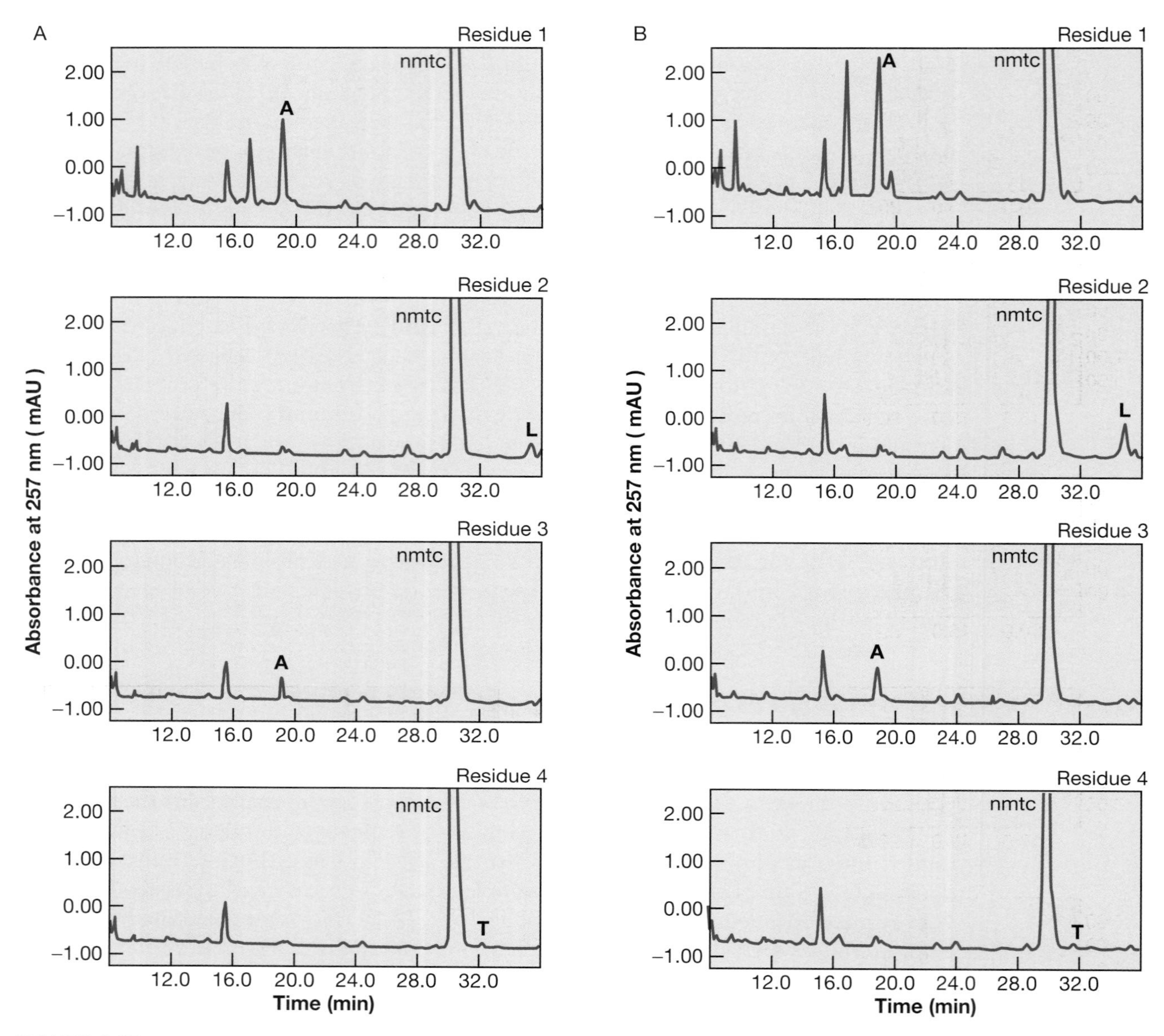

FIGURE 6.15. Carboxy-terminal sequence analysis using a standard cartridge with liquid flow perpendicular to the membrane (*A*) and a reaction cartridge (intended for amino-terminal analysis of electroblotted samples) with liquid flow along the membrane (*B*). The same amount of protein (49 pmoles of bovine serum albumin) was applied from solution in both runs. The carboxy-terminal sequence is -Thr-Ala-Leu-Ala-OH. (Courtesy of Tomas Bergman, Ella Cederlund, and Hans Jörnvall, Karolinska Institutet.)

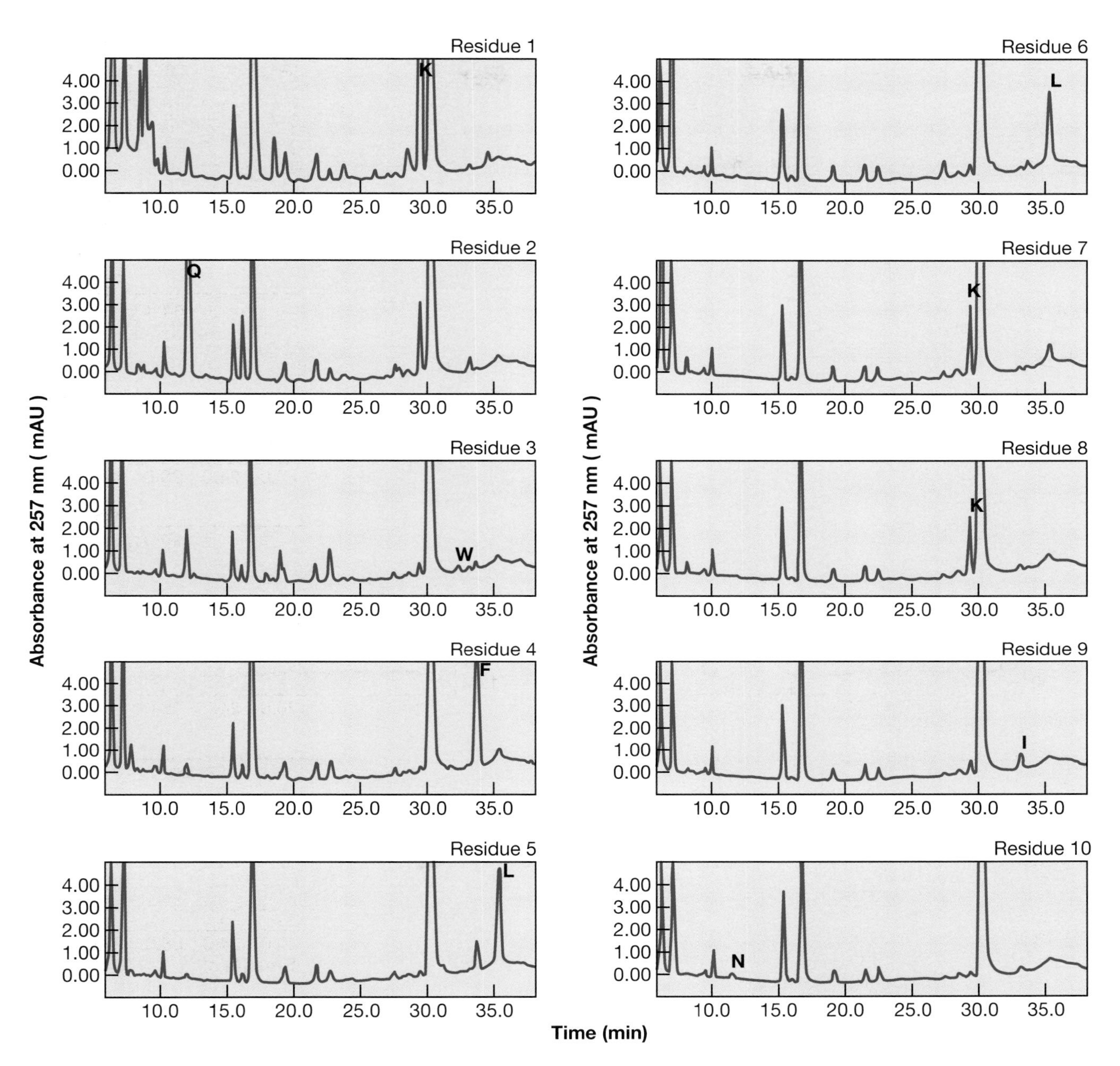

FIGURE 6.16. First ten cycles of carboxy-terminal sequence analysis of a 37-kD protein (400 pmoles) illustrating the improved chromatography with a flat baseline after application of the modified protocol (Table 6.4). The carboxy-terminal sequence was analyzed for 15 residues and the part shown is -Asn-Ile-Lys-Lys-Leu-Leu-Phe-Trp-Gln-Lys-OH. (Courtesy of Tomas Bergman, Ella Cederlund, and Hans Jörnvall, Karolinska Institutet.)

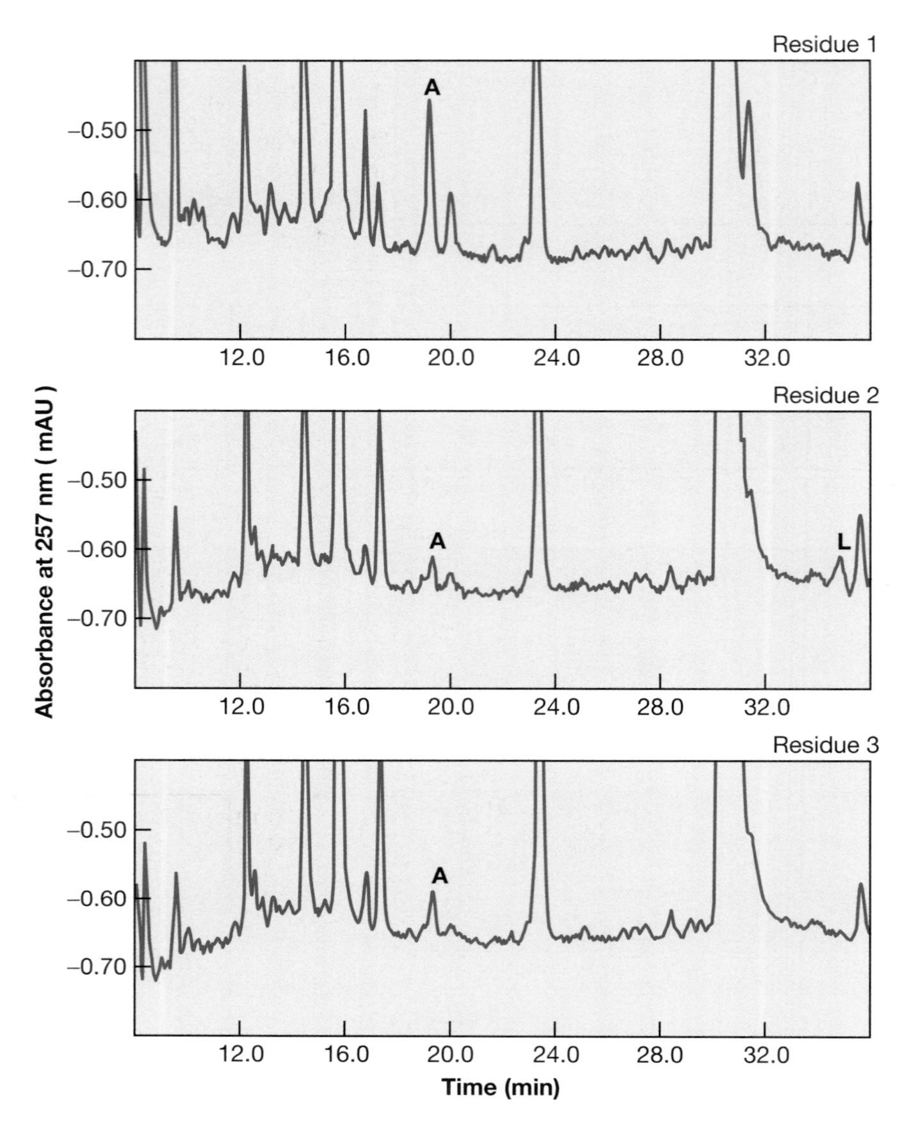

FIGURE 6.17. Carboxy-terminal sequence analysis using 11 pmoles of a 67-kD protein (bovine serum albumin) illustrating the achievable sensitivity. The sequence shown is -Ala-Leu-Ala-OH. (Courtesy of Tomas Bergman, Ella Cederlund, and Hans Jörnvall, Karolinska Institutet.)

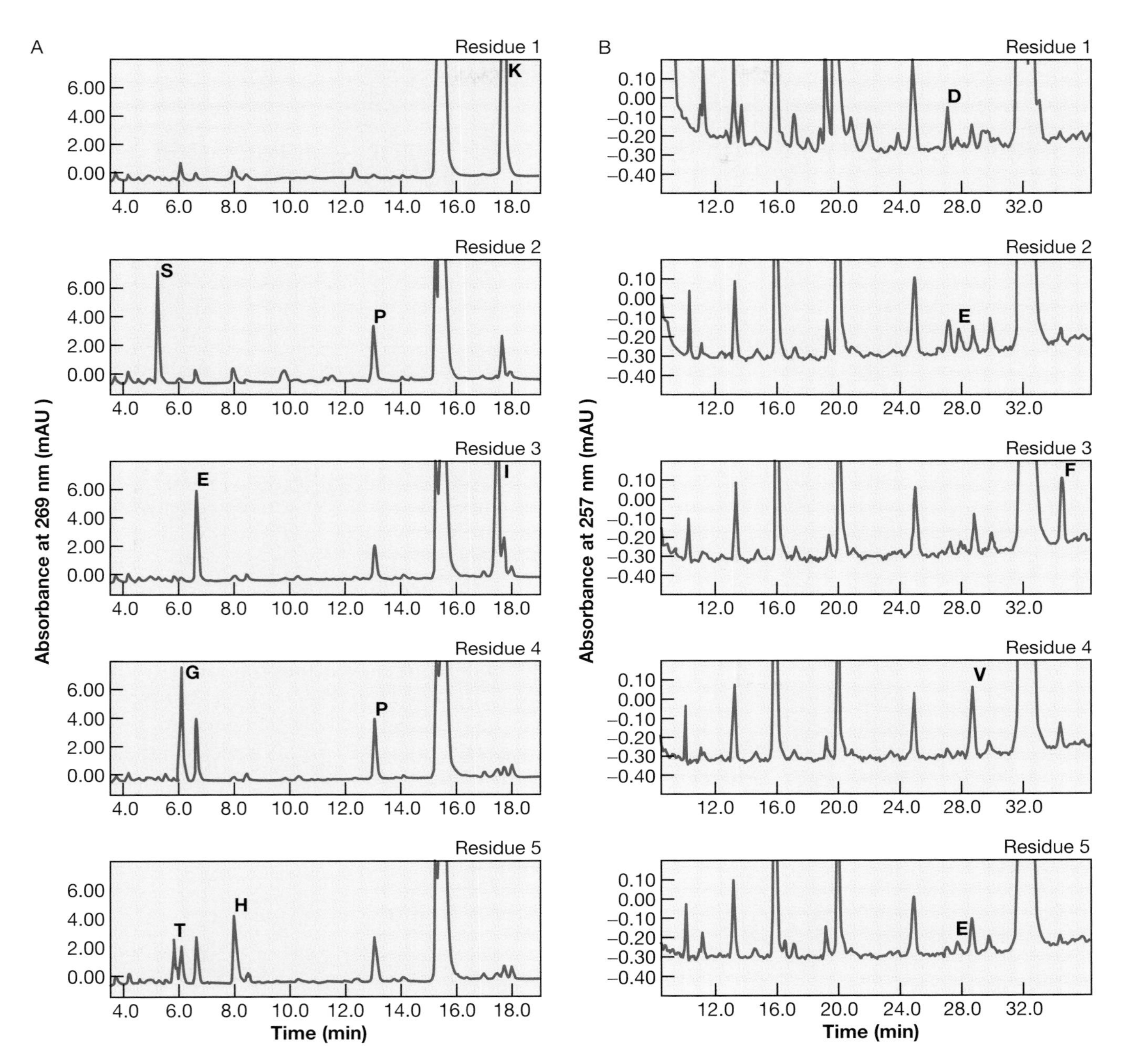

FIGURE 6.18. Combined amino- and carboxy-terminal sequence analysis of a preparation of an amino-terminally truncated protein (corresponding to a mixture of 28- and 27-kD polypeptides in amounts of 140 and 60 pmoles, respectively). The sample first underwent amino-terminal degradation for five cycles (*A*), followed by carboxy-terminal analysis for an additional five cycles (*B*). The combined sequence result is *N*-Lys-Ser/Pro-Ile/Glu-Gly/Pro-His/Thr- - - - -Glu-Val-Phe-Glu-Asp-OH. The amino-terminal sequences both had lysine in position 1, whereas positions 2–5 revealed two residues each. For the latter, the first residue given belongs to the major sequence (28-kD component) and the second residue given belongs to the minor sequence (27-kD component). (Courtesy of Tomas Bergman, Ella Cederlund, and Hans Jörnvall, Karolinska Institutet.)

PROTOCOL 1

Sample Loading onto a Biphasic Column Sequencer

THE BIPHASIC COLUMN APPROACH FOR IMMOBILIZING PROTEINS and peptides for the Edman degradation procedure was introduced by Hewlett Packard (now Agilent). The absorptive biphasic column consists of separate halves: One component is hydrophobic and the other is hydrophilic. These two components are assembled together to form the reaction cartridge and are mounted in the sequencer (see Figure 6.9A). The sample is applied to the top of the disassembled hydrophobic component, which resembles a reversed-phase column (see Figure 6.9B). Thus, like reversed-phase chromatography (see Chapter 5), very large sample volumes can be applied; the sample is retained on top of the column and inorganic salts and buffers, which otherwise interfere with the Edman chemistry, can be readily washed away. After sample loading, the hydrophobic and hydrophilic column halves are reassembled and inserted into the sequencer. The biphasic column acts as a reaction cartridge with all Edman reagents and solvents flowing through. A novel feature of the instrument is that the flow direction of the solvents can be reversed, helping to prevent sample washout. All aqueous solvent flows are directed against the hydrophobic half of the column, which retains sample by hydrophobic interactions. With organic solvent flows, the flow direction is against the hydrophilic half of the column, which retains the sample, whereas hydrophobic contaminants, by-products, and products of the Edman procedure (e.g., DPTU and ATZ amino acid derivatives) are eluted to waste or analysis.

MATERIALS

CAUTION: See Appendix 3 for appropriate handling of materials marked with <!>.

Reagents

Methanol (HPLC grade) <!>
Trifluoroacetic acid (TFA) (2%) in H_2O <!>

Equipment

Biphasic sequencing column (Agilent)
Nitrogen gas supply (pressure regulator to allow 80 psi) <!>
Polypropylene tube (50 ml) or small beaker
Sample applicator
Sample loading funnel

Biological Sample

Sample solution
This solution can be any aqueous solution that will bind to a reversed-phase column, i.e., can tolerate high salt concentrations but not high concentrations (>10%) of organic solvents.

METHOD

Wetting the Column

1. Remove the hydrophilic column half (male connector) from the prepared biphasic column and set it aside.
2. Mate the hydrophobic column half (female connector) to the sample loading funnel.
3. Install the sample loading funnel/column half into the sample loading chamber by pushing down on the sample loading funnel and turning it clockwise to lock it into the chamber.
4. Fill the sample loading funnel with 1 ml of HPLC-grade methanol.
5. Tighten the sample loader screw top.
6. Turn on the nitrogen (80 psi) long enough to just empty the sample loading funnel, but not long enough to dry the column packing.
7. Remove the screw top and repeat Steps 4–6, replacing the methanol with 1 ml of 2% TFA in H_2O. If the column packing is allowed to dry, repeat this step.

Preparing and Loading the Sample Solution

8. Adjust the sample solution to a total volume of 1–5 ml by diluting it with 2% TFA in H_2O. The dilution should be at least 1:1. If the sample solution is 1 ml before dilution, make sure to save the loading effluent.
9. Place a clean small beaker or 50-ml polypropylene tube under the column to collect the flowthrough.
10. Remove the screw top from the sample applicator.
11. Fill the sample loading funnel with sample solution. One milliliter of sample solution should pass through the column within 2 minutes.

    Up to 5 ml of sample solution may be loaded at one time, but will require a longer loading time.

12. Tighten the screw top and switch on the nitrogen pressure to 80 psi.

    EXPERIMENTAL TIP: Save the effluent from Steps 11 and 12 until the sample has been sequenced.

13. Repeat Steps 10–12 as necessary to load larger sample solution volumes.
14. Once the entire sample has been applied to the column, dry the column with nitrogen at 80 psi for 30 seconds.
15. Remove the screw top. Twist the sample loading funnel counterclockwise to unlock it.
16. Lift out the sample loading funnel and attached column half. Disconnect the funnel from the hydrophobic column half.

PROTOCOL 2

Electrotransfer of Proteins from Gels to PVDF Membranes

ELECTROTRANSFER (ELECTROBLOTTING) OF ELECTROPHORETICALLY RESOLVED PROTEINS to immobilizing matrices, first described in 1979 by Renart et al. (1979) and Towbin et al. (1979), is a routine method for immunodetection (western blotting) of proteins (see immunoblotting section in Spector et al. [1998] and Sambrook and Russell [2001]). For a detailed discussion of transfer and immobilization optimization methods and protein visualization, see reviews by Beisiegel (1986) and Tovey et al. (1987). In the mid 1980s, electroblotting was adapted for the direct amino acid sequence analysis of proteins resolved by one- and two-dimensional (1D and 2D) PAGE (Vandekerckhove et al. 1985; Aebersold et al. 1986; Bauw et al. 1987; Bergman and Jörnval 1987; Matsudaira 1987; Eckerskorn and Lottspeich 1989). The following are key features of this electroblotting method for direct sequencing using Edman chemistry:

- Sequencing of proteins in complex protein mixtures is simplified, especially that of membrane proteins, by being coupled to high-resolving protein separation methods such as 1D- and 2D-PAGE.
- Sample handling procedures are minimized due to the direct electrotransfer of acrylamide-gel-resolved proteins onto an inert membrane.
- Samples can be prepared in a relatively short time, free of incipients such as buffer salts.
- Low-picomole amounts of protein can be handled and sequenced with high yields (Matsudaira 1987).

In addition to direct sequencing, immobilized samples can be subjected to alternate characterization methods, such as amino acid composition analysis (Eckerskorn et al. 1988; Nakagawa and Fukuda 1989; Ploug et al. 1989; Tous et al. 1989), internal amino acid sequence analysis following proteolytic cleavage (Aebersold et al. 1987), chemical fragmentation (Scott et al. 1988; Jahnen et al. 1990; Patterson et al. 1992), and mass spectrometry (Eckerskorn et al. 1992; Strupat et al. 1994).

The critical step of this method is the quantitative transfer and immobilization of the proteins onto a suitable immobilizing matrix (for reviews, see Xu and Shively 1988; Simpson et al. 1989; Jungblut et al. 1990; Eckerskorn 1994). This electrotransfer protocol has been adapted from Ji et al. (1997).

MATERIALS

CAUTION: See Appendix 3 for appropriate handling of materials marked with <!>.

Reagents

CAPS (3-[cyclohexylamino]-1-propane sulfonic acid) (2.2% w/v) in H_2O
Adjust pH to 11.0 with HCl. <!>

Methanol <!>

Transfer buffer

Mix 1 part 2.2% CAPS, 1 part methanol, and 8 parts H_2O. For alternative transfer buffers and tips on improving electrotransfer of proteins, refer to the panel below.

Equipment

Electroblotting apparatus

The Trans-Blot (Bio-Rad), Xcell II Blot Module (NOVEX, Invitrogen), TE 22 Mighty Small Transphor Tank Transfer Unit (Hoefer), and Transphor Tank Transfer Unit (Hoefer) are a few examples of suitable electroblotters.

Plastic containers

Polyvinylidene difluoride (PVDF) membrane

Almost quantitative retention of proteins during electrotransfer from acrylamide gels can be obtained with PVDF membranes with a high specific surface area (<1500 m^2) and narrow pores (<0.350 μm) such as Immobilon PSQ (Millipore), Trans-Blot (Bio-Rad), and Fluorotrans (Pall, Dreieich, Germany), whereas PVDF membranes with a relatively low specific area (e.g., Immobilon P from Millipore, 380 m^2) exhibit significantly reduced yields (Eckerskorn and Lottspeich 1993).

Biological Sample

Polyacrylamide gel (unstained) containing the desired protein samples

Procedures for preparing 1D and 2D polyacrylamide gels, sample loading, and electrophoresis are described in Chapters 2 and 4, respectively.

Electrotransfer Buffers

The choice of electrotransfer buffer is not critical provided ionic strengths similar to those of the CAPS buffer in the Materials list are employed. Alternative electroblotting buffers include:

- 50 mM sodium borate (pH 8.0), 0.02% β-mercaptoethanol, 20% methanol (Vandekerckhove et al. 1985)
- 25 mM *N*-ethylmorpholine (pH 8.3), 0.5 mM dithiothreitol (Aebersold et al. 1988)
- 25 mM Tris-HCl, 10 mM glycine (pH 8.3), 0.5 mM dithiothreitol (Aebersold et al. 1986)

For the sodium borate buffer system, significantly decreased electrotransfer yields occur if the ionic strength is reduced below 10 mM or increased above 100 mM (Jungblut et al. 1990).

In SDS-free transfer buffers, high-M_r proteins and hydrophobic proteins such as membrane proteins are poorly soluble and elute from acrylamide gels in poor yield. The bulk of these proteins remain precipitated in the gel. Although increased solubilization of this class of protein can be improved by the addition of SDS (up to 0.01% final concentration) to the transfer buffer, it must be balanced against the fact that SDS generally prevents protein absorption to the immobilizing matrix (especially hydrophilic and low-M_r proteins).

Because the stability of SDS-protein complexes is influenced by methanol (e.g., such complexes dissociate more readily at increasing concentrations of methanol, leading to increased interaction of protein with the immobilizing matrix), varying the amount of methanol used for different proteins is a useful way to increase the recovery of a target protein onto a PVDF-type membrane. Typically, PVDF membranes require the presence of 10–20% methanol for optimal protein absorption. For proteins >80,000 daltons, use 5% methanol; for proteins between 20,000 and 80,000 daltons, use 10% methanol; and for low-M_r proteins, use 20% methanol.

METHOD

IMPORTANT: Wear gloves throughout this procedure to avoid contaminating the gel or membrane with proteins.

1. Immediately following gel electrophoresis, transfer the gel to a plastic container.

   Gels should not be stained before electroblotting because exposure to methanol/acetic acid will fix the proteins in the acrylamide gel, thereby significantly reducing the yield of protein transferred out of the gel during electroblotting.

2. Add sufficient transfer buffer to cover the gel. Allow the submerged gel to stand for 5–10 minutes at room temperature.

3. Cut a piece of PVDF membrane to the size of the gel. Wet the membrane with 100% methanol and then transfer it quickly to another plastic container containing transfer buffer. Soak the membrane in the transfer buffer for 5 minutes at room temperature.

4. Remove the gel and the PVDF membrane from the transfer buffer and assemble them in the electroblotting apparatus according to the manufacturer's instructions.

5. Electrophoretically transfer the proteins from the gel to the membrane for 3 hours at 4°C and 500 mA. Consult manufacturers' instructions for specific conditions. Suggested transfer conditions for specific equipment are as follows:

   If using the Mini Trans-Blot Electrophoretic Transfer Cell (Bio-Rad): Set the voltage at 90–100 V for 1 hour or 30 V overnight.

   If using the Trans-Blot Electrophoretic Transfer Cell (Bio-Rad): Transfer time is dependent on the percentage of the gels and the size of a particular protein and can last from 30 minutes to 4 hours. Perform the transfer at 4°C, and using Tris-glycine buffer, set the voltage: up to 100 V produces ~360 mA, 150 V produces ~550 mA, or 200 V produces ~850 mA.

6. Proceed with Protocol 3 to detect the proteins on the membrane.

TROUBLESHOOTING GUIDE FOR ELECTROBLOTTING

The PVDF membrane does not wet properly.

- PVDF-type membranes are very hydrophobic.
- Immerse the membranes in 100% methanol until completely wet.
- Quickly place them into electroblotting buffer for equilibration.

Coomassie staining the gel after electroblotting shows that the target protein has not completely left the gel.

- Reduce the concentration of methanol in the transfer buffer.
- Reduce the amount of SDS removed from the gel by incubating the gel in electrotransfer buffer no longer than 5 minutes prior to transfer.
- Increase the current while keeping the transfer sandwich as close to room temperature as possible. Electroblotting for a longer time period is not likely to help because the target protein may be unable to leave the gel after the SDS has been driven away.
- If the separation of the target protein from other proteins in the mixture is sufficient, use a lower acrylamide concentration in the separating gel; e.g., proteins of ~100 kD usually require 8% acrylamide and proteins of 10–70 kD often need 12% acrylamide.
- Try an electroblotting buffer with a higher (or lower) pH in case the pI of the target protein happens to be near the pH of the transfer buffer first tried.

Although the gel contains little or no protein after electroblotting, the stained membrane also shows little or no protein.

- Protein, especially those with low M_r, may transfer through some PVDF-type membranes (e.g., wide pore). When this occurs, a second sheet of membrane placed behind the first sheet in the transfer sandwich may bind some of the protein that has "blown through." To avoid the problem, use PVDF-type membranes with a high binding capacity (e.g., ProBlott membranes from Perkin Elmer, Immobilon PSQ from Millipore, Trans Blott from Bio-Rad, Fluorotrans from Pall).
- Reduce the SDS concentration, equilibrate the gel in electroblotting buffer for a longer time prior to transfer, and/or increase the methanol concentration of the electrotransfer buffer.
- Because PVDF membranes are very hydrophobic, they must be immersed in 100% methanol before they are equilibrated with buffer. Proteins will not bind well to dry PVDF.

Blank patchy areas on the membrane remain after staining.

- Remove any air bubbles that are trapped between the layers of the transfer sandwich because they can prevent transfer from occurring in small regions.
- Use only high-quality PVDF-type membranes. Poor-quality membranes may contain irregularities.
- After complete immersion in methanol, the entire membrane must be thoroughly wetted with buffer. Proteins will not bind well to areas of PVDF that are not properly wetted.

No signal results from direct sequence analysis.

- Run a sample evaluation gel (e.g., SDS-PAGE; see Chapter 2) to reassess the amount of protein in the gel prior to electroblotting, or if the amount of material allows, measure the amount of protein in an aliquot of the sample by amino acid analysis or other methods (e.g., BCA method, Appendix 2). If an insufficient amount of sample was delivered to the sequencer, prepare a new sample.
- Fragment the unsuccessful sample on its sequencing matrix with cyanogen bromide and attempt sequencing again. If the sample was amino-terminally blocked, several signals will be apparent. See reference(s) to "deblocking" of proteins mentioned in the chapter introduction.
- Prepare fragments for internal sequencing (see Chapter 7).

More than one signal results from direct sequence analysis.

- The primary signal from the major protein can sometimes be distinguished from the secondary signals of contaminating proteins by careful analysis of the data.
- Improve the resolution of the target protein on a gel of different acrylamide concentrations, a gradient gel, or a 2D gel (see Chapters 2 and 4).
- Protect the sample from excessive heat and acid, which can cause nonspecific cleavage of internal peptide bonds.

PROTOCOL 3

Detecting Proteins on PVDF Membranes

A NUMBER OF DIFFERENT STAINS CAN BE USED to visualize proteins on PVDF membranes, including Coomassie Brilliant Blue R250 (CBR-250), Ponceau S, Amido Black, colloidal metals, and transillumination. Apart from colloidal metal stains such as gold and silver, all of these stains are compatible with Edman chemistry sequencing. CBB R250 is versatile and sensitive enough for detecting sequenceable amounts of protein. Ponceau S and Amido Black are less sensitive than CBR-250 (Sanchez et al. 1992), but they can be removed from the gel with H_2O. A drawback of Amido Black is that it contributes a UV-absorbing contaminant that appears in the hydrophobic region of reversed-phase chromatograms, thereby limiting its usefulness for subsequent peptide mapping and PTH amino acid analysis. Transillumination is a simple method, but it is not very sensitive for detecting proteins. (Transillumination works as follows: dried PVDF membranes are rewetted with 20% methanol and viewed on a light table. The membrane remains opaque except where protein is bound; the protein bands appear translucent, wetter, and shinier than the surrounding membrane.)

Gels should not be stained before electroblotting because exposure to methanol/acetic acid will fix the proteins in the acrylamide gel, thereby significantly reducing the yield of protein transferred out of the gel during electroblotting. This protocol describes the CBR-250 method for staining blotted proteins, and was adapted from Ward et al. (1990).

MATERIALS

CAUTION: See Appendix 3 for appropriate handling of materials marked with <!>.

Reagents

Coomassie Brilliant Blue staining solution

Dissolve 1 g of CBR-250 <!> into 1 liter of aqueous 50% methanol <!> (0.1% w/v) final concentration. Stir the solution for 3–4 hours and then filter through Whatman filter paper.

Destaining solution

Mix 50% methanol, 10% acetic acid <!>, and 40% H_2O. To avoid amino-terminal blockage of proteins, use only high-quality acetic acid (free of peroxides, aldehydes, etc.) during the staining and destaining procedures.

Equipment

Plastic container

Biological Sample

PVDF membrane from Protocol 2 containing the desired proteins

METHOD

1. Immediately following electrotransfer in Protocol 2, place the PVDF membrane in a plastic container and add enough Coomassie Brilliant Blue staining solution to cover the membrane. Allow it to stand for 5–10 minutes at room temperature to stain the membrane.

2. Discard the staining solution. Cover the membrane with destaining solution. Change the destaining solution until the PVDF membrane is fully destained.

   Approximately three to four changes are usually required for a total time of 30 minutes to 1 hour.

3. Wash the PVDF membrane with H_2O for 5–10 minutes.

4. Repeat the wash five more times.

PROTOCOL 4

Concentrating Acrylamide Gel Spots

A NUMBER OF REPLICATE PROTEIN BANDS FROM 1D acrylamide gels (or 2D gel spots) are often required to provide sufficient material for sequencing. In these situations, it is necessary to excise several stained protein bands or spots from the gels and preconcentrate them prior to electroblotting for amino-terminal sequencing or internal sequence analysis following in-gel proteolytic (or chemical) fragmentation. Empirical observations indicate that to obtain efficient cleavage of a protein in a gel, it is necessary to keep the ratio of protein to acrylamide as high as possible (Eckerskorn and Lottspeich 1989; Tempst et al. 1990). This can be accomplished by reelectrophoresis of several stained gel pieces (the number dictated by the volume of the sample well) either in an analytical SDS-PAGE gel or in a simple SDS-PAGE gel poured in a single conventional glass Pasteur pipette (Gevaert et al. 1996). The Pasteur pipette concentrating procedure is described in this protocol, which is adapted, with permission, from Gevaert et al. (1996).

MATERIALS

CAUTION: See Appendix 3 for appropriate handling of materials marked with <!>.

Reagents

Acrylamide mix (30%) <!>
- To prepare the 30% acrylamide mix, combine 30% (w/v) acrylamide and 1.0% *N,N′*-methylene-bisacrylamide <!> in H_2O. The final mixture is 30%T and 3.3%C.

Agarose (5% w/v)
- Melt the agarose and hold at 55–65°C.

Coomassie Brilliant Blue staining solution
- Dissolve 1 g of CBB-R250 <!> into 1 liter of aqueous 50% methanol <!> (0.1% w/v) final concentration. Stir the solution for 3–4 hours and then filter through Whatman filter paper.

Destaining solution
- Mix 50% methanol, 10% acetic acid <!>, and 40% H_2O. To avoid amino-terminal blockage of proteins, use only high-quality acetic acid (free of peroxides, aldehydes, etc.) during the staining and destaining procedures.

Equilibration buffer
- Mix 12.5 ml of 0.5 M Tris-HCl (pH 6.8), 20 ml of 10% SDS, 5 ml of 2-mercaptoethanol <!>, and 10 ml of glycerol, and then add a few grains of bromophenol blue.

10x Laemmli electrophoresis running buffer
- Dissolve 30.0 g of Tris base, 144.0 g of glycine <!>, and 10.0 g of SDS in 1000 ml of H_2O. The pH of the buffer should be 8.3 and no pH adjustment is required. Store the running buffer at room temperature and dilute to 1x before use.

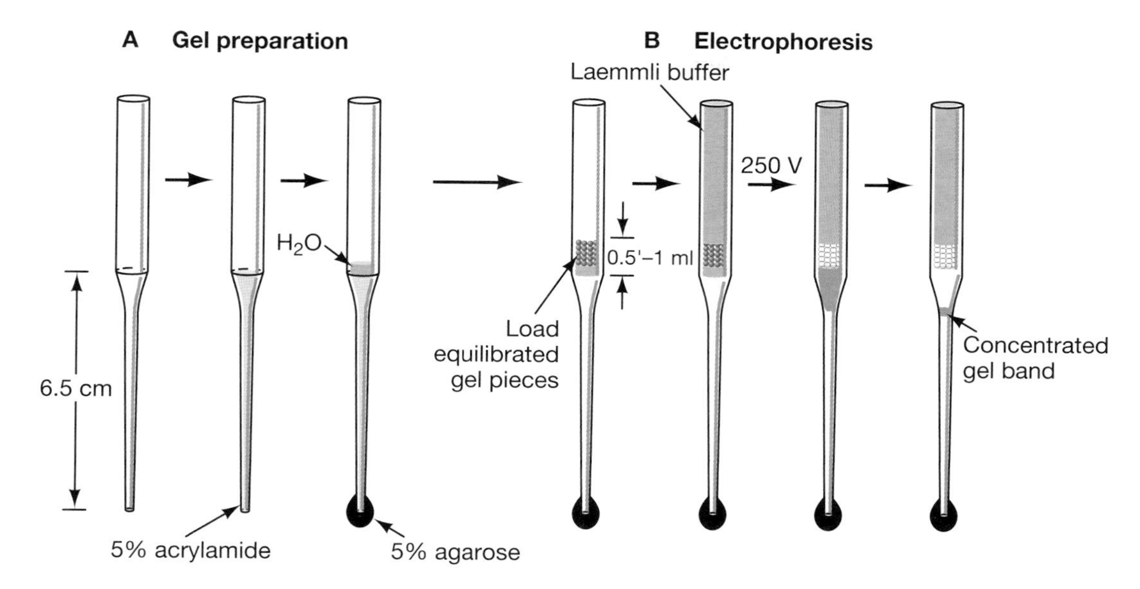

FIGURE 6.19. Design and operation of the Pasteur pipette concentration gel system. (Redrawn, with permission, from Simpson and Reid 1997; originally adapted from Gaevert et al. 1996.)

Stacking gel (5%)
To prepare 5% stacking gel mixture, combine in the following order:
2 ml of 30% acrylamide mix
3 ml of 0.5 M Tris-HCl (pH 6.8)
0.12 ml of 10% (w/v) SDS <!>
6.76 ml of H_2O
0.12 ml of 10% ammonium persulfate <!>
0.006 ml of *N,N,N′,N′*-tetramethylelthylenediamine (TEMED) <!>

Equipment

Pasteur pipettes
Polypropylene tubes (10 ml)
Protean II xi 2-D cell (Bio-Rad) or equivalent
Syringe (2 ml)
Tubes (50 ml)
Tube gel adaptor

Biological Sample

2D electrophoresis gel spots containing the protein of interest

METHOD

1. Excise the desired 2D electrophoresis gel spots and place each spot in its own 10-ml polypropylene tube. Rehydrate the spots in 9 ml of H_2O.

   Take care not to mix up the spots. Multiples of identical spots from replicate gels can be combined and treated in one tube.

2. Wash the rehydrated gel spots with 9 ml of H_2O five times for ~5 minutes per wash.
3. Discard the H_2O from the final wash, and replace it with enough equilibration buffer to cover the gel pieces. Equilibrate the gel spots for at least 1 hour at room temperature.
4. Dip the tip of a Pasteur pipette into a solution of melted 5% agarose and hold the pipette for a short while until the agarose solidifies. Pipette the stacking gel mixture into the Pasteur pipette to cast the concentrating gel.
5. Insert the pipettes into the Protean II xi 2-D cell with a tube gel adaptor. Layer 0.5 ml of H_2O on top of each gel.
6. Allow the gels to polymerize (for gel details, see Figure 6.19A). Remove the H_2O from the top of the pipette and load the equilibrated gel spots and equilibration buffer (from Step 3) into the pipette (see Figure 6.19B). The maximum total volume of combined gel pieces and equilibration buffer is 0.5 ml. Fill the remaining volume of the pipette with Laemmli electrophoresis running buffer.
7. Perform electrophoresis in the Protean II xi 2-D cell at 250 V at room temperature until the dye reaches the end of the pipette.
8. Stop the electrophoresis. Use a 2-ml syringe to remove the concentrating gel from the pipette by applying gentle pressure from the bottom of the pipette.
9. Transfer the concentrating gel into a 50-ml tube filled with 30 ml of Coomassie Brilliant Blue staining solution. Stain the gel for 10–20 minutes at room temperature.
10. Destain the gel with destaining solution.
11. Excise the stained protein band for further characterization, for example, internal sequence analysis (see Chapter 7) or direct Edman degradation following electroblotting onto PVDF (see Protocol 2).

PROTOCOL 5

Solid-phase Microsequencing of Phosphopeptides

ANALYSIS OF PHOSPHORYLATED PEPTIDES isolated from endogenously phosphorylated tissue samples where no ^{32}P or other radiolabel is present may be particularly difficult if the amount is below detection levels for mass spectrometry. The identification of phosphorylation sites where ^{32}P is present may also be problematical. However, excellent recovery of radioactivity from phosphorylated residues can be achieved with solid-phase sequencing, which allows the attachment of proteins via side-chain ε-lysine and amino-terminal NH_2 groups (to Sequelon-DITC) or via side-chain and carboxy-terminal carboxyl groups (to Sequelon-AA). The latter PVDF arylamine supports are particularly applicable to the analysis of phosphorylated peptides in conventional gas- or pulsed-liquid-phase protein sequencers. This method results in almost quantitative recovery of very low levels of radiolabeled phosphate using 9:1 methanol:H_2O (containing 2 mM phosphate as a "carrier") to replace the hydrophobic solvent, *n*-butanol. The hydrophilic solvent does not wash away the peptide due to the covalent linkage of the peptide or protein to the derivatized PVDF membrane supports. If sufficient quantities of sample *and* radioactivity are available, the thiazolinone derivatives can be converted at each cycle of Edman degradation to the PTH amino acids and the sample can be split—part to the conventional HPLC and part to a fraction collector to measure the ^{32}P-labeled phosphate released at each cycle. This protocol was provided by Alastair Aitken (University of Edinburgh, UK).

MATERIALS

CAUTION: See Appendix 3 for appropriate handling of materials marked with <!>.

Reagents

Carbodiimide reagent (1-[3-dimethylaminopropyl]-3-ethylcarbodiimide hydrochloride, EDC) <!>

Coupling buffer (pH 5) supplied in Sequelon kit (nonreactive with carbodiimide, e.g., pyridinium-HCl or MES <!>)

Methanol:H_2O (9:1) containing 2 mM sodium phosphate <!>

Peptide solvent (e.g., 20% acetonitrile/0.1% TFA) <!>

S3 (*n*-butyl chloride/*n*-butanol) <!>

Scintillation fluid (e.g., Universol-ES, ICN) <!>
This fluid is added to the samples recovered at each step for counting after collection, but note the exceptions within the protocol.

Equipment

ATZ-collect program
Heating block preset to 55°C
Micropipette
Mylar sheet (supplied with kit)
Protein sequencer (Applied Biosystems or equivalent)
Scintillation counter
Sequelon-AA kit (GEN 920033, Applied Biosystems)
Test tubes

Biological Sample

Peptide, radiolabeled <!>

The quantity is not important. Approximately 200 cpm (Cerenkov counts of ^{32}P radioactivity) is the minimum to observe a reasonable "burst of counts," depending on the length of the sequencer run.

METHOD

1. Dissolve the radiolabeled peptide of interest in a small tube in 20 µl of appropriate solvent (e.g., aqueous acetonitrile up to 30% with or without 0.1% TFA, which is a typical solution in which a peptide might elute from a RP-HPLC column). Heat the mixture at 55°C and mix to dissolve.
2. Count the tube in a scintillation counter with the Cerenkov program. *Do not add* any scintillation fluid for Cerenkov counting.
3. Place the Sequelon-AA disk from the kit onto a Mylar sheet placed on a heating block at 55°C.
4. Pipette the radiolabeled peptide sample onto the Sequelon-AA disk. Allow the disk to dry thoroughly. Count the disk and the sample tube in a scintillation counter. Again *do not use* any scintillation fluid.
5. Rinse the tube three times with 10–20 µl of peptide solvent. Pipette the washings onto the Sequelon-AA disk and dry between each application. Repeat the wash if significant counts are left in the tube.
6. Count the disk in the scintillation counter without using any scintillation fluid.
7. Pipette 100 µl of coupling buffer into a test tube and add 1 mg of the H_2O-soluble carbodiimide reagent EDC.
8. Pipette 5 µl of the EDC solution onto the sample disk. Allow the reaction to proceed for 20 minutes at room temperature.
9. After the sample has been coupled, rinse the disk a few times with a peptide solvent similar to that in Step 5 to remove noncovalently linked peptide, which could give rise to background during the sequencing run. Count the disc in a scintillation counter.

10. Replace S3 (*n*-butyl chloride) with methanol:H_2O (9:1) containing 2 mM sodium phosphate.

11. Place the disc in the reaction cartridge of the protein sequencer using methanol:H_2O (9:1) containing 2 mM sodium phosphate. Carry out standard Edman sequencing using an ATZ-collect program.

    This delivers the resulting ATZ products and the ^{32}P-labeled phosphate directly to the fraction collector. It is advisable to check through and modify the program to optimize the delivery times of the reagents and solvents. The aqueous methanol has a higher viscosity than butyl chloride and will require a longer delivery time. The "Procise" model has sensors that should ensure correct delivery, but it is strongly advisable to verify that sufficient solvent delivery occurs.

12. Count the samples from the fraction collector in scintillation fluid in the counter.

    Scintillation fluid may be used here and it will approximately double the radioactivity counted. The background will increase similarly; therefore, a significant improvement in the ratio of signal to noise is not always achieved.

13. To check the efficiency of Edman degradation, remove the disc from the sequencer cartridge, allow it to dry, and count it in a scintillation counter to check the level of radioactivity remaining.

    Depending on the residues encountered and the length of the sequencer run, typically ~10–20% of the initial quantity might be left, due to the normal repetitive yield and accumulative blockage during each cycle of degradation.

PROTOCOL 6

Carboxy-terminal Sequence Analysis

CARBOXY-TERMINAL SEQUENCING IS A LARGELY AUTOMATED PROCESS, typically capable of determining the identity of five residues from 0.5 nmole of protein. This carboxy-terminal sequence analysis protocol is a modification of the protocol from Applied Biosystems and is capable of longer runs with increased sensitivity. It has been tested with more than 600 proteins and protein fragments in sizes ranging from 10 to 600 residues (Bergman et al. 2001). The technique provides quantitative characterization of carboxy-terminal sequences and truncation patterns and is efficient for analysis of polypeptides blocked at the amino terminus. The following are its performance characteristics:

- Sensitivity down to the 10-pmole level.
- Extended runs: up to 10 residues using 100–500 pmoles.
- Detection of proline and PTC-Lys.
- Efficient combination of amino- and carboxy-terminal degradation.

This technique provides quantitative characterization of carboxy-terminal sequences and truncation patterns and is efficient for analysis of amino-terminally blocked polypeptides. This protocol was provided by Tomas Bergman, Ella Cederlund, and Hans Jörnvall (Department of Medical Biochemistry and Biophysics, Karolinska Institutet).

MATERIALS

CAUTION: See Appendix 3 for appropriate handling of materials marked with <!>.

Reagents

Reagents supplied by Applied Biosystems are listed with their bottle position (C) on the sequencer and part number (PN). Table 6.4 contains additional information about the reagents required to perform carboxy-terminal sequencing.

Acetic anhydride/lutidine in acetonitrile (C6, PN 402142) <!>
Dilute 1:1 with acetonitrile.

Alkylated thiohydantoin (ATH) amino acid standard (C1, PN 403088)
Store at –20°C.

Bromomethylnaphthalene in acetonitrile (C8, PN 401703) <!>
Dilute 1:1 with acetonitrile. Store at 4°C.

Coomassie Brilliant Blue (R250) <!>

Diisopropylethylamine (DIEA) in heptane (C11, PN 401702) <!>
Dilute 1:1 with heptane.

Ethyl acetate <!>

Methanol <!>

N-methylimidazole in acetonitrile (C3, PN 402141) <!>
Dilute 1:1 with acetonitrile.

Phenyl isocyanate (PIC) (1% v/v) in acetonitrile <!>

Piperidine thiocyanate in acetonitrile (C4, PN 401700) <!>
Dilute 1:1 with acetonitrile.

Tetrabutylammonium thiocyanate in acetonitrile (C10, PN 401854) <!>
Dilute 1:1 with acetonitrile.

Trifluoroacetic acid (TFA) (C12, PN 401701) <!>

Equipment

Amino-terminal sequencing instrument
During the development of this protocol, when carboxy-terminal degradation was combined with amino-terminal analysis, a Procise HT sequencer (Applied Biosystems) was used for amino-terminal degradations.

Argon <!>

Beaker (25–50 ml)
See Step 1f, if adsorbing the protein sample to the PVDF membrane.

Carboxy-terminal sequencing instrument
During the development of this protocol, carboxy-terminal sequencing was carried out using a Procise C instrument (Model 494, Applied Biosystems), basically operated according to the manufacturer's recommendations but with important modifications introduced as detailed in this protocol. Analysis of ATH amino acids from the degradations was performed using a 2 x 220-mm reversed-phase C_{18} column (Brownlee Spheri-5 PTC, 5 μm) with detection at 257 nm.

Heating block preset to 60°C

Microfuge tubes (1.5 ml)

Petri dish (glass)

PVDF membrane (e.g., Immobilon-P, Millipore) or Sample preparation cartridge (ProSorb, Applied Biosystems)
The ProSorb sample preparation cartridge is essentially a microfuge tube with its rounded end removed and a PVDF membrane affixed to the bottom of the tube, which is held against a filter by a plastic cylinder that serves as a coupling device. Solutions applied to the upper surface of the membrane are pulled through the membrane by the wicking action of the filter below.

Reaction cartridge (Applied Biosystems, PN 603993)
This cartridge should be of the type with a vertical slit intended for electroblotted samples.

Reversed-phase C_{18} column (2 x 220 mm) (PN 07110204, Brownlee Spheri-5 PTC, 5 μm)

Biological Sample

Proteins separated on a polyacrylamide gel using either 1D- or 2D-PAGE, or proteins supplied in solution

Additional Reagents

Step 1 of this protocol may require the reagents listed in Protocols 2 and 3 of this chapter.

METHOD

Transfer of Proteins to PVDF Membrane

1. Apply 50–100 pmoles of the protein sample to a PVDF membrane either by electroblotting or by adsorption directly from solution (using the ProSorb sample preparation cartridge).

For electroblotting protein to a PVDF membrane:

a. Electrotransfer the gel-separated proteins to a PVDF membrane (see Protocol 2 and Mozdzanowski et al. 1992).

b. Stain the PVDF membrane containing the proteins of interest with Coomassie Brilliant Blue R250 (see Protocol 3). Allow the membrane to dry in the air.

EXPERIMENTAL TIP: Use a paper clip to hang the sample in a chemical fume hood, or dry it on the bench, but cover the membrane with filter paper to protect it from contamination.

c. Excise the desired protein band(s) or spot(s).

If not immediately analyzed, the bands/spots can be stored up to several months at –20°C.

d. If a combination of amino- and carboxy-terminal sequence analysis is required, proceed to Step 2 to perform the amino-terminal sequencing on the PVDF-immobilized sample. Alternatively, if only carboxy-terminal sequencing is required, place the excised membrane-bound protein spots in a 1.5-ml microfuge tube, and proceed to Step 3.

For adsorption of protein in solution to a PVDF membrane (ProSorb):

a. Wet the PVDF membrane with 5 µl of methanol.

b. Add 100 µl of 0.5% TFA to the sample reservoir.

c. Add 5–800 µl of the protein sample solution to the sample reservoir.

d. Wash the membrane with 100 µl of 0.1% TFA if the sample solution contains salt or other low-molecular-weight contaminants. Repeat the wash two more times each with 100 µl of 0.1% TFA.

e. Remove the sample reservoir housing with the PVDF membrane from its holder and invert the reservoir onto a Petri dish. Make sure that the PVDF membrane with the immobilized sample faces up.

f. Cover the inverted membrane (in its reservoir) with a glass beaker and place the Petri dish on a heating block set at 60°C. Dry the PVDF membrane completely (~20 minutes).

The intent is to keep contaminants from the air away from the PVDF membrane while it is drying. Do not allow the beaker to touch the membrane either.

g. Carefully free the PVDF membrane from its housing.

h. If a combination of amino- and carboxy-terminal sequence analysis is required, proceed to Step 2 to perform the amino-terminal sequencing on the PVDF-immobilized sample. Alternatively, if only carboxy-terminal sequencing is required, place the membrane-bound protein in a 1.5-ml microfuge tube, and proceed to Step 3.

Amino Acid Sequencing

2. To sequentially perform amino- and carboxy-terminal sequencing:

   a. Perform amino-terminal sequence analysis for 5–25 cycles.

   b. Transfer the PVDF membrane to a 1.5-ml microfuge tube, and proceed to Step 3.

3. To prepare the sample for carboxy-terminal sequencing:

   a. Add 5 μl of 5% DIEA in heptane to the PVDF membrane. Allow the heptane to evaporate for ~1 minute at 60°C.

   b. Add 5 μl of 1% PIC in acetonitrile and incubate for 5 minutes at 60°C.

   c. Transfer the PVDF membrane with the PIC-derivatized sample to a reaction cartridge for a carboxy-terminal sequencing instrument.

MODIFICATIONS TO THE CARBOXY-TERMINAL SEQUENCER PROGRAM

1. Start with an ethyl acetate wash in the reaction cartridge.

   30-second wash and 30-second argon dry
   60-second wash and 30-second argon dry

2. Insert additional washing steps immediately before the C10/C12 combined cleavage/cyclization step. Instead of a single ethyl acetate wash, include four washing steps with intervening drying steps.

   90-second wash and 30-second argon dry
   90-second wash and 30-second argon dry
   45-second wash and 30-second argon dry
   45-second wash and 30-second argon dry

3. The chemistry of the degradation cycle follows the thiohydantoin method (Stark 1968). However, follow the procedures listed below:

   - Activate the carboxyl terminus only once with acetic anhydride/lutidine in acetonitrile.
   - Add bromomethylnaphthalene in each cycle to *S*-alkylate the thiohydantoin to make it a better leaving group and allow a mild combination of peptide cleavage and activation for the next cycle by using tetrabutylammonium thiocyanate in TFA (Boyd et al. 1992, 1995).
   - Use a reaction cartridge with a vertical slit intended for amino-terminal degradation of electroblotted samples (e.g., the reaction cartridge for the Procise HT, Applied Biosystems) to provide improved sensitivity in carboxy-terminal degradation.
   - Use a gradient with linear segments, typically: 14–27% Solvent B (see Table 6.4) for 0–4 minutes, 27–44% Solvent B from 4–28 minutes, 44–47% Solvent B for 28–34 minutes, and 47–90% Solvent B for 34–35 minutes. Follow this with an isocratic washing of the column at 90% Solvent B for 3 minutes. The optimal elution program may vary slightly and is dependent on the particular reversed-phase column used.

For additional changes to the manufacturer's recommended parameters (involving dilutions of solvents, choice of solvents used in the chromatography, and reaction temperatures), see Table 6.4.

Microgram/Picomole Conversion Guide

For microsequencing purposes, proteins are measured in units of micrograms (μg) or in terms of picomoles (pmoles). Note that half as many micrograms of a 50,000-dalton protein are required than are required of a 100,000-dalton protein. This is because a given number of micrograms of a 50,000-dalton protein contains twice as many picomoles (i.e., twice as many amino termini) as the same number of micrograms of a 100,000-dalton protein.

General Formula

The following general formula can be used for picomole/microgram conversions.

1000/molecular weight of protein in kD = number of pmoles/μg

Given the molecular weight of the protein, quick estimates can be made using this mnemonic device.

100 kD:	100 μg = 1 nmole		10 μg = 100 pmoles
50 kD:	50 μg = 1 nmole		5 μg = 100 pmoles
25 kD:	25 μg = 1 nmole		2.5 μg = 100 pmoles
10 kD:	10 μg = 1 nmole		1 μg = 100 pmoles

From Micrograms to Picomoles

The number of picomoles in a particular sample is equal to the number of micrograms divided by the molecular weight of the protein times 10^{-6}. To convert micrograms of protein to picomoles:

1. Calculate the number of micrograms per picomole of a particular protein by multiplying the protein's molecular weight in daltons by 10^{-6}.
2. Divide the measured or estimated number of micrograms by the number of micrograms per picomole. The following formula can be used:

   $M/(W \times 10^{-6}) = n$

   where M is the number of micrograms of sample available, W is the molecular weight of the protein, and n is the number of picomoles of sample available. For example, a 45,000-dalton protein weighs 0.045 μg per picomole (45,000 x 10^{-6}). Therefore, 2.5 μg is equal to 55.6 pmoles (2.5/0.045). Similarly, a band in a gel estimated to contain 4.0 μg of a 45,000-dalton protein provides ~90 pmoles for sequencing (4.0/0.045 = 88.9).

From Picomoles to Micrograms

The number of micrograms of protein needed for a sample consisting of a specified number of picomoles is equal to the number of picomoles multiplied by the molecular weight times 10^{-6}. To convert picomoles of protein to micrograms:

1. Multiply the molecular weight of the protein by 10^{-6}.
2. Multiply the sum from Step 1 by the number of picomoles desired.

In other words, to determine the amount of protein required to prepare a 50-pmole sample for sequencing, multiply the number of micrograms per picomoles by 50. The following formula can be used:

$n(W \times 10^{-6}) = M$

where M is the number of micrograms of sample required, W is the molecular weight of the protein, and n is the number of picomoles desired. For example, a 60,000-dalton protein weighs 0.06 μg per picomole (60,000 x 10^{-6}). Therefore, the preparation of a 50-pmole sample for sequencing requires 3 μg (0.06 x 50). Similarly, preparing a 50-pmole sample of a 30,000-dalton protein for sequencing requires 1.5 μg (0.03 x 50).

(Reprinted, with permission, from the handbook, *Preparing samples for protein sequencing: A newcomer's guide* [©The Perkin-Elmer Corporation].)

REFERENCES

Abderhalden E. and Brockmann H. 1930. The contribution determining the composition of proteins especially polypeptides (German). *Biochem. Z.* **225:** 386–408.

Aebersold R. 1993. Internal amino acid sequence analysis of proteins after *in situ* protease digestion on nitrocellulose. In *A practical guide to protein and peptide purification for microsequencing*, 2nd edition (ed. P. Matsudaira), pp. 103–124. Academic Press, San Diego, California.

Aebersold R., Leavitt J., Hood L.E., and Kent S.B. 1987. Internal amino acid sequence analysis of proteins separated by one- or two-dimensional gel electrophoresis after in situ protease digestion on nitrocellulose. *Proc. Natl. Acad. Sci.* **84:** 6970–6974.

Aebersold R., Teplow D.B., Hood L.E., and Kent S.B. 1986. Electroblotting onto activated glass: High efficiency preparation of proteins from analytical sodium dodecyl sulfate-polyacrylamide gels for direct sequence analysis. *J. Biol. Chem.* **261:** 4229–4238.

Aitken A. 1999. Protein consensus sequence motifs. *Mol. Biotechnol.* **12:** 241–253.

Aitken A. and Learmonth M. 1997. Analysis of sites of protein phosphorylation. *Methods Mol. Biol.* **64:** 293–306.

Baker C.S., Crobett J.M., May A.J., Yacoub M.H., and Dunn M.J. 1992. A human myocardial two-dimensional electrophoresis database: Protein characterization by microsequencing and immunoblotting. *Electrophoresis* **13:** 723–726.

Bauw G., De Loose M., Inze D., Van Montagu M., and Vandekerckhove J. 1987. Alterations in the phenotype of plant cells studied by NH_2-terminal amino acid—Sequence analysis of proteins electroblotted from two-dimensional gel-separated total extracts. *Proc. Natl. Acad. Sci.* **84:** 4806–4810.

Beisiegel U. 1986. Protein blotting. *Electrophoresis* **7:** 1–18.

Bergman T. and Jörnval H. 1987. Electroblotting of individual polypeptides from SDS/polyacrylamide gels for direct sequence analysis. *Eur. J. Biochem.* **169:** 9–12.

Bergman T., Cederlund E., and Jörnvall H. 2001. Chemical C-terminal protein sequence analysis: Improved sensitivity, length of degradation, proline passage, and combination with Edman degradation. *Anal. Biochem.* **290**: 74–82.

Boyd V.L., Bozzini M., Guga P.J., DeFranco R.J., and Yuan P.-M. 1995. Activation of the carboxy terminus of a peptide for carboxy-terminal sequencing. *J. Org. Chem.* **60:** 2581–2587.

Boyd V.L., Bozzini M., Zon G., Noble R.L., and Mattaliano R.J. 1992. Sequencing of peptides and proteins from the carboxy terminus. *Anal. Biochem.* **206:** 344–352.

Brown J.L. 1979. A comparison of the turnover of α-*N*-acetylated and nonacetylated mouse L-cell proteins. *J. Biol. Chem.* **254:** 1447–1449.

Brown J. and Roberts W. 1976. Evidence that approximately eighty percent of the soluble proteins from Ehrlich ascites cells are N-acetylated. *J. Biol. Chem.* **251:** 1009–1014.

Chang J.Y. and Creaser E.H. 1976. A novel method for protein sequence analysis. *Biochem. J.* **157:** 77–85.

Dainese P., Staudenmann W., Quadroni M., Korostensky C., Gonnet G., Kertesz M., and James P. 1997. Probing protein function using a combination of gene knockout and proteome analysis by mass spectrometry. *Electrophoresis* **18:** 432–442.

Doolittle R.F. 1982. An anecdotal account of the history of peptide stepwise degradtion procedures. In *Methods in protein sequence analysis* (ed. M. Elzinga), pp. 1–24. Humana Press, Clifton, New Jersey.

Dubois T., Rommel C., Howell S., Steinhussen U., Soneji Y., Morrice N., Moelling K., and Aitken A. 1997. 14-3-3 is phosphorylated by casein kinase I on residue 233. Phosphorylation at this site in vivo regulates Raf/14-3-3 interaction. *J. Biol. Chem.* **272:** 28882–28888.

Dunbar B. 1997. Protein sequencer maintenance and troubleshooting. *Methods Mol. Biol.* **64:** 217–233.

Eckerskorn C. 1994. Blotting membranes as the interface between electrophoresis and protein chemistry. In *Microcharacterization of proteins* (ed. R. Kellner et al.), pp. 75–89. VCH, Weinheim, Germany.

Eckerskorn C. and Lottspeich F. 1989. Internal amino acid sequence analysis of proteins separated by gel electrophoresis after tryptic digestion in polyacrylamide matrix. *Chromatographia* **28:** 92–94.

———. 1993. Structural characterization of blotting membranes and the influence of membrane parameters for electroblotting and subsequent amino acid sequence analysis of proteins. *Electrophoresis* **14:** 831–838.

Eckerskorn C., Jungblut P., Mewes W., Klose J., and Lottspeich F. 1988. Identification of mouse brain proteins after two-dimensional electrophoresis and electroblotting by microsequence analysis and amino acid composition. *Electrophoresis* **9:** 830–838.

Eckerskorn C., Strupat K., Karas M., Hillenkamp F., and Lottspeich F. 1992. Mass spectrometric analysis of blotted proteins after gel electrophoretic separation by matrix-assisted laser desorption/ionization. *Electrophoresis* **13:** 664–665.

Edman P. 1949. A method for the determination of the amino acid sequence in peptides. *Arch. Biochem.* **22:** 475–476.

Edman P. and Begg G.S. 1967. A protein sequenator. *Eur. J. Biochem.* **1:** 80–91.

Edman P. and Henschen A. 1975. Sequence determination. *Mol. Biol. Biochem. Biophys.* (2nd edition) **8:** 232–279.

Erdjumen-Bromage H., Geromanos S., Chodera A., and Tempst P. 1993. Successful peptide sequencing with femtomole level PTH-analysis: A commentary. In *Techniques in protein chemistry IV* (ed. R.H. Angeletti), pp. 419–426. Academic Press, San Diego, California.

Farnsworth V. and Steinberg K. 1993. A generation of phenylthiocarbamyl or anilinothiazolinone amino acids from the postcleavage products of the Edman degradation. *Anal. Biochem.* **215:** 200–210.

Geisow M.J. and Aitken A. 1989. Gas- or pulsed liquid-phase sequence analysis. In *Protein sequencing: A practical approach* (eds. J.B.C. Findlay and M.J. Geisow), pp. 85–98. IRL Press, Oxford, United Kingdom.

Gevaert K., Rider M., Puype M., Van Damme J., De Boeek K., and Vandekerekhove J. 1995. New Strategies in high sensitivity chacterization of proteins separated from 1-D or 2-D gels. In *Methods in protein structure analysis* (ed. M.Z. Atassi and E. Appella), pp. 15–25. Plenum Press, New York.

Gevaert K., Verschelde J.-L., Puype M., Van Damme J., Goethals M., De Boeck S., and Vandekerckhove J. 1996. Structural analysis and identification of gel-purified proteins, available in the femtomole range, using a novel computer program for peptide sequence assignment, by matrix-assisted laser desorption ionization-reflectron time-of-flight-mass spectrometry. *Electrophoresis* **17:** 918–924.

Gheorghe M.T., Jörnvall H., and Bergman T. 1997. Optimized alcoholytic deacetylation of *N*-acetyl-blocked polypeptides for subsequent Edman degradation. *Anal. Biochem.* **254:** 119–125.

Hewick R.M., Hunkapiller M.W., Hood L.E., and Dreyer W.J. 1981. A gas-liquid solid phase peptide and protein sequenator. *J. Biol. Chem.* **256:** 7990–7997.

Hirano H., Komatsu S., and Tsunasawa S. 1997. On-membrane deblocking of proteins. *Methods Mol. Biol.* **64:** 285–292.

Hirano H., Komatsu S., Kajiwara H., Takagi Y., and Tsunasawa S. 1993. Microsequence analysis of the amino-terminally blocked proteins immobilized on polyvinylidene difluoride membrane by Western blotting. *Electrophoresis* **4:** 839–846.

Hirano H., Komatsu S., Takakura H., Sakiyama F., and Tsunasawa S. 1992. Deblocking and subsequent microsequence analysis of N^{α}-blocked proteins electroblotted onto PVDF membrane. *J. Biochem.* **111:** 754–757.

Hunter T. 1987. A thousand and one protein kinases. *Cell* **50:** 823–829.

Inglis A.S. 1991. Chemical procedures for C-terminal sequencing of peptides and proteins. *Anal. Biochem.* **195:** 183–196.

Inglis A.S., Reid G.E., and Simpson R.J. 1995. Chemical techniques employed for the primary structural analysis of proteins and peptides. In *Interface between chemistry and biochemistry* (ed. P. Jollés and H. Jörnvall), pp. 141–171. Birkhäuser Verlag, Basel, Switzerland.

Jahnen W., Ward L.D., Reid G.E., Moritz R.L., and Simpson R.J. 1990. Internal amino acid sequencing of proteins by in situ cyanogen bromide cleavage in polyacrylamide gels. *Biochem. Biophys. Res. Commun.* **166:** 139–145.

Jahnen W., Ward L.D., Reid G.E., Moritz R.L., and Simpson R.J. 1990. Internal amino acid sequencing of protein by *in situ* cyanogen bromide cleavage in polyacrylamide gels. *Biochem. Biophys. Res. Commun.* **166:** 139–145.

Ji H., Moritz R.L., Reid G.E., Ritter G., Catimel B., Nice E., Heath J.K., White S.J., Welt S., Old L.J., Burgess A.W., and Simpson R.J. 1997. Electrophoretic analysis of the novel antigen for the gastrointestinal-specific monoclonal antibody, A33. *Electrophoresis* **18:** 614–621.

Ji H., Baldwin G.S., Burgess A.W., Moritz R.L., Ward L.D., and Simpson R.J. 1993. Epidermal growth factor induces serine phosphorylation of stathmin in a human colon carcinoma cell line (LIM 1215). *J. Biol. Chem.* **268:** 13396–13405.

Jonsson A.P., Griffiths W.J., Bratt P., Johansson I., Strömberg N., Jörnvall H., and Bergman T. 2000. A novel

Ser *O*-glucuronidation in acidic proline-rich proteins identified by tandem mass spectrometry. *FEBS Lett.* **475:** 131–134.

Jungblut P., Eckerskorn C., Lottspeich F., and Klose J. 1990. Blotting efficiency investigated by using two-dimensional electrophoresis, hydrophobic membranes and proteins from different sources. *Electrophoresis* **11:** 581–588.

Klapper D.G., Wilde C.E., and Capra J.D. 1978. Automated amino acid sequence of small peptides utilizing polybrene. *Anal. Biochem.* **85:** 126–131.

Laursen R.A. 1971. Solid-phase Edman degradation—An automatic peptide sequencer. *Eur. J. Biochem.* **20:** 89–102.

Levy A.L. 1954. A paper chromatographic method for the quantitative estimation of amino-acids. *Nature* **174:** 126–127.

Matsudaira P. 1987. Sequence from picomole quantities of proteins electroblotted onto polyvinylidene fluoride membranes. *J. Biol. Chem.* **262:** 10035–10038.

Matsudaira P., ed. 1993. Introduction. In *A practical guide to protein and peptide purification for microsequencing,* 2nd edition, pp. 1–13. Academic Press, San Diego, California.

Moore S. and Stein W.H. 1951. Chromatography. *Sci. Am.* **521:** 546.

Moos M., Nguyen N.Y., and Liu T.-Y. 1988. Reproducible high yield sequencing of proteins electrophoretically separated and transferred to an inert support. *J. Biol. Chem.* **263:** 6005–6008.

Moritz R.L. and Simpson R.J. 1992. Application of capillary reversed-phase high-performance liquid chromatography to high-sensitivity protein sequence analysis. *J. Chromatogr.* **599:** 119–130.

Moritz R.L., Eddes J., Ji H., Reid G.E., and Simpson R.J. 1994. High-speed chromatographic separation of proteins and peptides: Application to rapid peptide mapping of in-gel digested proteins. *J. Protein Chem.* **13:** 486–487.

________. 1995. Rapid separation of proteins and peptides using conventional silica-based supports: Identification of 2-D gel proteins following in-gel proteolysis. In *Techniques in protein chemistry VI* (ed. J.W. Crabb), pp. 417–425. Academic Press, Orlando, Florida.

Mozdzanowski J., Hembach P., and Speicher D.W. 1992. High yield electroblotting onto polyvinylidene difluoride membranes from polyacrylamide gels. *Electrophoresis* **13:** 59–64.

Nakagawa S. and Fukuda T. 1989. Direct amino acid analysis of proteins electroblotted onto polyvinylidene fluoride membranes from sodium dodecyl sulfate-polyacrylamide gel. *Anal. Biochem.* **181:** 75–78.

Narita K., Matsuo H., and Nakajima T. 1975. End group determination. *Mol. Biol. Biochem. Biophys.* (2nd edition) **8:** 30–103.

Patterson S. 1994. From electrophoretically separated protein to identification: Strategies for sequence and mass analysis. *Anal. Biochem.* **221:** 1–15.

Patterson S.C. and Aebersold R. 1995. Mass spectrometric approaches for the identification of gel-separated proteins. *Electrophoresis* **16:** 1791–1814.

Patterson S.D., Hess D., Yungwirth T., and Aebersold R. 1992. High-yield recovery of electroblotted proteins and cleavage fragments from a cationic polyvinylidene fluoride-based membrane. *Anal. Biochem.* **202:** 193–203.

Pisano J.J. 1975. Analysis of amino acid phenylthiohydantoins by gas chromatography and high performance liquid chromatography. *Mol. Biol. Biochem. Biophys.* (2nd edition) **8:** 280–297.

Ploug M., Jensen A.L., and Barkholt V. 1989. Determination of amino acid compositions and NH_2-terminal sequences of peptides electroblotted onto PVDF membranes from tricine-sodium dodecyl sulfate-polyacrylamide gel electrophoresis: Application to peptide mapping of human complement component C3. *Anal. Biochem.* **181:** 33–39.

Renart J., Reiser J., and Stark G.R. 1979. Transfer of proteins from gels to diazobenzyloxymethyl paper and detection with antisera: A method for studying antibody specificity and antigen structure. *Proc. Natl. Acad. Sci.* **76:** 3116–3120.

Sambrook J. and Russell D.W. 2001. *Molecular cloning: A laboratory manual,* 3rd edition. Cold Spring Harbor Laboratory Press, Cold Spring Harbor, New York.

Sanchez J.C., Ravier F., Pasquali C., Frutiger S., Paquet N., Bjellquist B., Hochstrasser D.F., and Hughes G.J. 1992. Improving the detection of protein after transfer to polyvinylidene fluoride membranes. *Electrophoresis* **13:** 715–717.

Sanger F. 1945. The free amino groups of insulin. *Biochem. J.* **39:** 507–515.

Scott M.G., Crimmins D.L., McCourt D.W., Tarrand J.J., Eyerman M.C., and Nahm M.H. 1988. A simple in situ cyanogen bromide cleavage method to obtain internal amino acid sequence of proteins electroblotted to polyvinylidene fluoride membranes. *Biochem. Biophys. Res. Commun.* **155:** 1353–1359.

Simpson R.J. and Nice E.C. 1984. *In situ* cyanogen bromide cleavage of amino-terminally blocked proteins in a gas-phase sequencer. *Biochem. Int.* **8:** 787–791.

_______. 1989. Strategies for the purification of subnanomole amounts of protein and polypeptides for microsequence analysis. In *The use of HPLC in receptor biochemistry* (ed A.R. Kerlavage), pp. 201–244. A.R. Liss, New York.

Simpson R.J. and Reid G.E. 1997. Sequence analysis of gel-resolved protein. In *Gel electrophoresis of proteins—A practical approach*, 3rd edition (ed. B.D. Hames), pp. 255–263. Humana Press, Totawa, New Jersey.

Simpson R.J., Moritz R.L., Begg G.S., Rubira M.R., and Nice E.C. 1989. Micropreparative procedures for high sensitivity sequencing of peptides and proteins. *Anal. Biochem.* **177:** 221–236.

Spector D.L., Goldman R.D., and Leinwand L.A. 1998. *Cells: A laboratory manual*, vol. 1: *Culture and biochemical analysis of cells.* Cold Spring Harbor Laboratory Press, Cold Spring Harbor, New York.

Stark G.R. 1968. Sequential degradation of peptides from their carboxyl termini with ammonium thiocyanate and acetic anhydride. *Biochemistry* **7:** 1796–1807.

Strupat K., Karas M., Hillenkamp F., Eckerskorn C., and Lottspeich F. 1994. Matrix-assisted laser desorption/ionisation mass spectrometry of proteins electroblotted after polyacrylamide-gel electrophoresis. *Anal. Chem.* **66:** 464–470.

Tarr G.E., Beecher J.F., Bell M., and McKean D.J. 1978. Polyquarternary amines prevent peptide loss from sequenators. *Anal. Biochem.* **84:** 622–627.

Tempst P., Link A.J., Riviere L.R., Fleming M., and Elicone C. 1990. Internal sequence analysis of proteins separated on polyacrylamide gels at the submicrogram level: Improved methods, applications and gene cloning strategies. *Electrophoresis* **11:** 537–553.

Thompson E.O.P. 1955. The insulin molecule. *Sci. Am.* **192:** 36–41.

Totty N.F., Waterfield M.D., and Hsuan J.J. 1992. Accelerated high-sensitivity microsequencing of proteins and peptides using a miniature reaction cartridge. *Protein Sci.* **1:** 1215–1224.

Tous G.I., Fausnaugh J.L., Akinyosoye O., Lackland H., Winter-Cash P., Vitorica F.J., and Stein S. 1989. Amino acid analysis on polyvinylidene difluoride membranes. *Anal. Biochem.* **179:** 50–55.

Tovey E.R. and Baldo B.A. 1987. Comparison of semi-dry and conventional tank-buffer electrotransfer from polyacrylamide gels to nitrocellulose membranes. *Electrophoresis* **8:** 384–387.

Tovey E.R., Ford S.A., and Baldo B.A. 1987. Protein blotting on nitrocellulose: Some important aspects of the resolution and detection of antigens in complex extracts. *J. Biochem. Biophys. Methods* **14:** 1–17.

Towbin H., Staehelin T., and Gordon J. 1979. Electrophoretic transfer of proteins from polyacrylamide gels to nitrocellulose sheets: Procedure and some applications. *Proc. Natl. Acad. Sci.* **76:** 4350–4354.

Vandekerckhove J., Bauw G., Puype M., Van Damme J., and Van Montagu M. 1985. Protein-blotting on Polybrene-coated glass-fiber sheets. A basis for acid hydrolysis and gas-phase sequencing of picomole quantities of protein previously separated on sodium dodecyl sulfate/polyacrylamide gel. *Eur. J. Biochem.* **152:** 9–19.

Ward L.D., Reid G.E., Moritz R.L., and Simpson R.J. 1990. Strategies for internal amino acid sequence analysis of proteins separated by polyacrylamide gel electrophoresis. *J. Chromatogr.* **519:** 199–216.

Wellner D., Panneerselvam C., and Horecker B.L. 1990. Sequencing of peptides and proteins with blocked N-terminal amino acids: *N*-acetylserine or *N*-acetylthreonine. *Proc. Natl. Acad. Sci.* **87:** 1947–1949.

Wittmann-Liebold B., Graffunder H., and Kohls H. 1976. A device coupled to a modified sequenator for the automated conversion of anilinothiazolinones into PTH amino acids. *Anal. Biochem.* **75:** 621–633.

Xu Q.-Y. and Shively J. 1988. Improved electroblotting of proteins onto membranes and derivatized glass-fiber sheets. *Anal. Biochem.* **170:** 19–30.

Zimmerman C.L., Pisano J.J., and Appella E. 1973. Analysis of amino acid phenylthiohydantoins by high speed liquid chromatography. *Biochem. Biophys. Res. Commun.* **55:** 1220–1224.

FURTHER READING

Kellner R., Lottspeich F., and Meyer H.E., eds. 1994. *Microcharacterization of proteins.* VCH, Weinheim, Germany.

Matsudaira P., ed. 1993. *A practical guide to protein and peptide purification for microsequencing*, 2nd edition, Academic Press, San Diego, California.

WWW RESOURCES

http://www.abrf.org The Association of Biomolecular Research Facilities homepage.

http://www.appliedbiosystems.com/apps/techniques.cfm Applied Biosystems protein sequencing applications.

http://www.urmc.rochester.edu/research/propep/microseq.html University of Rochester Medical Center MicroChemical Protein/Peptide Facility.

CHAPTER 7

Peptide Mapping and Sequence Analysis of Gel-resolved Proteins

Peptide mapping (or protein fingerprinting) refers to the generation of peptides from a protein by enzymatic or chemical means, followed by separation and analysis of the fragments. This technique provides useful information about the primary structure of a protein (analytical peptide mapping), as well as the conformation and the relationships between proteins (comparative peptide mapping). Analytical peptide mapping can be used for the following:

- ***Identifying specific internal peptide sequences*** within a protein such as posttranslational modification sites, antigenic determinants, active sites, and ligand binding regions.
- ***Obtaining sufficient internal sequence information*** to allow identification of a protein, which is especially important for amino-terminally blocked proteins that are refractory to the Edman degradation procedure.
- ***Determining cysteine connectivities*** (disulfide bonds).

Comparative peptide mapping is useful for comparing the sequences or conformations of related proteins. A major advantage of this peptide mapping approach is that it can yield much useful information without much investment of time and equipment.

NOT ALL PROTEINS YIELD MEANINGFUL AMINO-TERMINAL sequence information using the Edman degradation procedure (outlined in Chapter 6). For instance, many polypeptide chains have blocked amino termini (due to, e.g., *N*-acetyl and *N*-formyl derivitization of α-amino groups or pyroglutamate formation of an amino-terminal glutamine residue under acidic conditions) and are refractory to the Edman degradation procedure. A free α-amino group is a necessary prerequisite for the Edman degradation method. It has been estimated that greater than 75% of intracellular proteins isolated from eukaryotic cells have blocked amino termini, mainly as a result of posttranslational modification, especially acetylation (Brown and Roberts 1976; Brown 1979). Furthermore, for many high-molecular-mass proteins (>50 kD), amino-terminal sequence data may be uninterpretable due to a rapid build up of phenylthiohydantoin (PTH) amino acid background levels during the Edman degradation procedure.

When working with subpicomole amounts of material (i.e., proteins in the low-microgram range), an amino-terminal sequence strategy based on Edman degradation of the intact material is particularly risky because it may provide an "all-or-none" opportunity to obtain useful sequence data. If a protein is amino-terminally "blocked," and no sequence information is obtained, the sample is lost. (However, it is possible to glean some sequence information by removing the sample from the sequencer, treating it in situ with cyanogen bromide on the sample disk, and resequencing the cleaved material [Simpson and Nice 1984].) Therefore, for many reasons, the preferred strategy is to obtain *internal sequence information* on low-M_r peptides following enzymatic or chemical fragmentation of the mature protein. This strategy offers the likelihood of obtaining multiple peptide sequences from various regions of the polypeptide chain.

Although limited amino-terminal sequence information can be extremely useful for protein identification purposes, extensive sequence coverage of a polypeptide chain is often required to

- confirm the integrity of a recombinant protein,
- identify posttranslational modification sites (e.g., phosphorylation or glycosylation),
- identify artificial modifications (e.g., chemical cross-linking studies),

- map antibody recognition sites (epitopes), and
- determine cysteine connectivities (disulfide bonds) within a polypeptide chain.

Treatment with a specific protease such as trypsin can generate a set of peptides covering the entire polypeptide chain. Visualization of peptide mixtures separated by chromatography (typically, reversed-phase chromatography, see Protocol 15 and Chapter 5) or electrophoresis (see Cleveland "limited digests" procedure in Protocols 16 and 17) is referred to as peptide mapping or "fingerprinting." Although mapping techniques are commonly used for purifying peptides from a complex digest, they are also used for comparing one protein with another by the distinctive pattern that the peptides form on the map, hence the name "fingerprinting."

MACROSCALE AND MICROSCALE METHODS ARE USED TO GENERATE PEPTIDE MAPS

The usual approach for generating a peptide map of a protein for the purpose of obtaining peptide sequence information is described below:

- ***Reduction and alkylation of the purified protein.*** This can be performed either "in solution" or in situ in a one- or two-dimensional acrylamide gel.
- ***Fragmentation of the reduced and alkylated protein*** by chemical (e.g., cyanogen bromide or limited acid hydrolysis) or enzymatic (e.g., trypsin) means.
- ***Separation of the peptides generated.***
- ***Sequence analysis of the peptides*** using classical Edman degradation (see Chapter 6) or mass-spectrometry-based methods (see Chapter 8).

The strategy for generating peptides for sequence analysis very much depends on the nature of the source material (see Figure 7.1). There are three key methods for purifying proteins:

- Chromatographic procedures such as reversed-phase high-pressure liquid chromatography (RP-HPLC; see Chapter 5) and ion exchange chromatography, which yield proteins in solution.
- One- and two-dimensional acrylamide gel electrophoresis, from which proteins can be isolated as bands or spots, respectively (see Chapter 2 or 4).
- Gel electrophoresis in combination with electroblotting of proteins to an inert immobilizing membrane such as polyvinylidene difluoride (PVDF; see Chapter 4).

In this chapter, attention is given to both macroscale (i.e., milligram amounts) and microscale (i.e., low-microgram amounts) peptide mapping, which require different strategies. For instance, any desalting, reduction, alkylation, and concentration steps carry the likely risk of sample loss due to adsorption onto surfaces, especially after a drying step, such as lyophilization. These losses are particularly significant when working with microgram amounts of protein (2 μg of a 20-kD protein is equivalent to 100 pmoles), presenting the investigator with technical challenges in both sample handling and manipulation.

Although extremely high purity is essential for Edman sequencing, mass-spectrometry-based sequencing strategies are more forgiving. In fact, mass spectrometry provides an additional separation step based on the mass:charge ratio of the proteins or peptides. Even so, when attempting microscale purification and sequencing, the investigator must plan ahead to ensure the compatibility of buffers and detergents with downstream reduction and alkylation, fragmentation, and separation steps. Special consideration for sample preparation for microscale sequencing is given in this chapter.

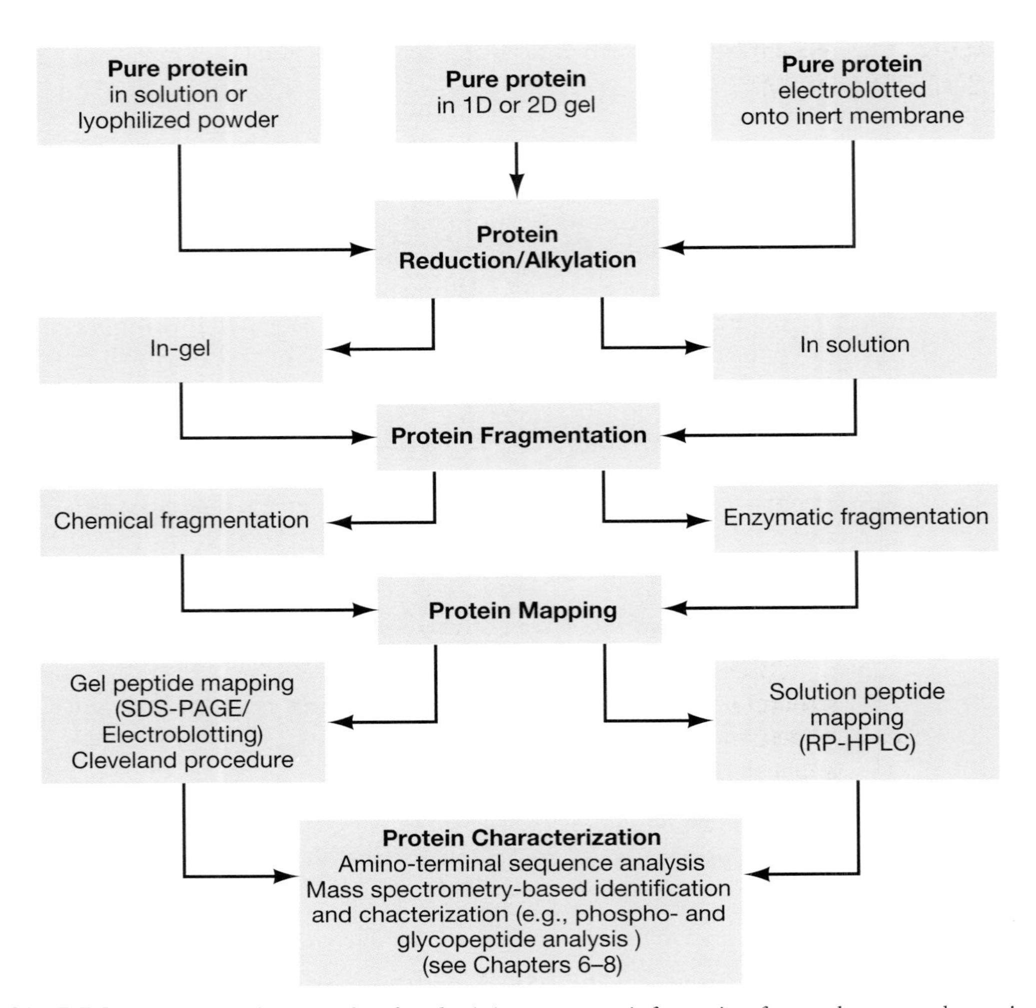

FIGURE 7.1. Experimental approaches for obtaining sequence information from gel-separated proteins.

Some Considerations When Handling Samples for Microsequencing

- Avoid drying samples prepared by RP-HPLC. If the volume must be reduced, concentrate the sample by trace enrichment using a narrow-bore reversed-phase column (i.e., <2.1-mm internal volume) (for experimental procedures, see Protocol 15). Using this procedure, a large volume (several milliliters) can be concentrated to <50 µl with >90% yield.
- Minimize the number of handling steps.
- Store RP-HPLC samples at –20°C in tightly capped polypropylene tubes (e.g., 1.5-ml microfuge tubes) to avoid losses that occur by the sample sticking irreversibly to the sides of the tube upon evaporation of the organic phase. To avoid this potential problem, add a small amount of detergent to the sample (e.g., 0.01% Tween-20).
- Proteolysis of RP-HPLC-purified samples can be conducted in the same collection tube by simply adjusting the detergent concentration to ~0.02% Tween-20, reducing the volume to ~50 µl by centrifugal lyophilization, and then adding the appropriate digest buffer and protease (see Protocols 3 and 5).

COVALENT CROSS-LINKS WITHIN PROTEINS MUST BE BROKEN TO IMPROVE PEPTIDE MAPPING

Once a protein has been purified to homogeneity using one of the approaches outlined in Figure 7.1, it must be readied for fragmentation (either chemical or enzymatic), the first stage of peptide mapping, and subsequent sequence analysis. However, one major problem frequently encountered during fragmentation of a protein sample is the obstruction of protease action by covalent inter- and intrapolypeptide cross-links (e.g., disulfide bonds). For peptide mapping studies, such cross-links must be broken to render the polypeptide chain more uniformly accessible to the protease and to increase the likelihood of obtaining cleavage of all susceptible peptide bonds. In addition, proteolytic cleavage fragments comprising two or more cross-linked peptides are often difficult to interpret during peptide mapping and subsequent structural analysis. However, maintaining the integrity of cross-links is often useful in obtaining limited cleavage of a protein in its native state. Limited digests are a useful first step in strategies aimed at determining either domain structures of a protein (Polverino de Laureto et al. 1999) or conformational epitopes (Jemmerson and Paterson 1986). In this chapter, disulfide bonds are the only type of covalent polypeptide cross-links that are discussed.

Other types of less frequently occurring cross-links (those involved in maintaining the structural integrity of collagen, elastin, and other fibrous proteins, e.g., desmosine and isodesmosine) (Traub and Piez 1971) are not described here because there are no generally applicable methods for their selective cleavage.

Performic Acid Is Used to Cleave Disulfide Bonds and Modify Sulfhydryl Groups

Procedures for the cleavage of the covalent disulfide linkages of cystine include oxidation, reduction, and other nucleophilic displacement reactions (for a review, see Liu 1977; Fontana and Gross 1987). Oxidative cleavage of disulfide bonds with performic acid was a frequently used procedure in the early days of protein sequence analysis (Sanger 1949). This method (see Protocol 1) has many attractions, especially for amino acid analysis. Performic acid oxidation is a simple single-step chemical modification procedure that quantitatively converts cysteine to cysteic acid and methionine to methionine sulfone in yields of 90–95%. Amino acid analysis of performic-acid-oxidized proteins is often used to determine the total half-cystine (as cysteic acid) and methionine content of a protein.

One of the major attractions of performic acid cleavage of disulfide bonds is the formation of cysteic acid, a strongly acidic sulfonate moiety with a highly polar character that increases the solubility of the polypeptide chain. A similar functional group (*S*-sulfo) can be introduced onto the β-carbon of cysteine without the oxidation of tryptophan indole and methionine thioether side chains, which accompanies performic acid oxidation. This is accomplished by sulfitolysis (i.e., treating disulfides with sulfite [SO_3^{2-}] at alkaline pH) whereby *S*-sulfonated derivatives and thiols are formed

$$RSSR + RSO_3^{2-} \longleftrightarrow RS^- + RSSO_3^{2-}$$

One half of the original disulfide is converted to *S*-sulfocysteine and the other half is converted to the cysteine mercaptide. With a large excess of sulfite, the reaction continues until disulfides are completely converted into *S*-sulfonate. Although sulfitolysis has been used infrequently as a mild procedure for the cleavage of disulfide linkages in proteins, the *S*-sulfonate derivatives are stable at neutral and mildly acidic conditions and are water-soluble. Because the *S*-sulfo group is readily removed by reduction with an excess of reducing agent, this group has been used for the reversible protection of the thiol function of cysteine (Liu 1977). Treatment of protein hydrolysates with dithiothreitol (DTT) followed by an excess of sodium tetrathionate ($Na_2S_4O_6$) converts cystine and cysteine to *S*-sulfocysteine, a derivative that is suitable for quantitative analysis (Inglis and Liu 1970; Simpson et al. 1976).

The disadvantages of performic acid oxidation are that it can lead to the modification of amino acids besides cystine and methionine (e.g., the destruction of the indole side chain of tryptophan) and cleavage of acid-labile peptide bonds (e.g., Asp-Pro bonds). Furthermore, PTH–cysteic acid derivatives are difficult to identify during sequencing by the Edman degradation method. Carboxymethylation (see Protocols 2 and 3) is not ideal for quantitative amino acid analysis involving cystine or cysteine residues because it is difficult to assess the completeness of the reaction. Pyridylethylation is the preferred method of cystine/cysteine modification both for sequence analysis and for those amino acid analyzers that use phenylisothiocyanate as an amino-terminal derivatizing agent (see Protocol 12).

Performic acid oxidation cannot be applied prior to cyanogen bromide cleavage of the polypeptide chain (see Protocol 6) because methionine residues are converted to their sulfone derivatives, which, unlike the methionine thioether, is not alkylated by cyanogen bromide.

Many Reagents Are Available for Reducing and Alkylating Proteins

By far the most selective and useful method for splitting disulfide bonds in proteins is reductive cleavage (Cecil 1963; Liu 1977). Reduction is customarily performed in the absence of oxygen, at a slightly alkaline pH (~pH 8) and under denaturing conditions where the protein is unfolded, for example, in the presence of 8 M urea or 6 M guanidine hydrochloride. The classical approach for reducing a protein involves treating the protein with a high concentration of a low-molecular-weight thiol reagent. It is clear from the reaction represented below that a large molar excess (typically, 100–1000-fold) of thiol reagent must be used to drive the reaction to completion.

$$\text{Protein}(\text{S—S}) + 2\text{RSH} \rightleftharpoons \text{Protein}(\text{SH}\ \ \text{SH}) + \text{RSSR}$$

Thiol reagents used for this purpose include cysteine, glutathione, 2-mercaptoethylamine, thioglycollic acid, 2-mercaptoethanol, dithiothreitol (and its isomer dithioerythritol), tributylphosphine, and tris(2-carboxyethyl)phosphine (see Figure 7.2) (Liu 1977).

Although 2-mercaptoethanol was the traditional reducing agent for protein structural analysis, DTT, also referred to as Cleland's reagent after its discoverer (Cleland 1964), is now the reagent of choice (Konigsberg 1972). DTT (and its isomer, dithioerythritol) is a superior alternative to the mercaptans, requiring much lower concentrations (20-fold molar excess) for maintaining monothiols completely in the reduced state and of reducing disulfides quantitatively (Cleland 1964). DTT favors the reaction to go to the right because its oxidized product forms a thermodynamically stable six-membered ring (containing a disulfide bridge) that

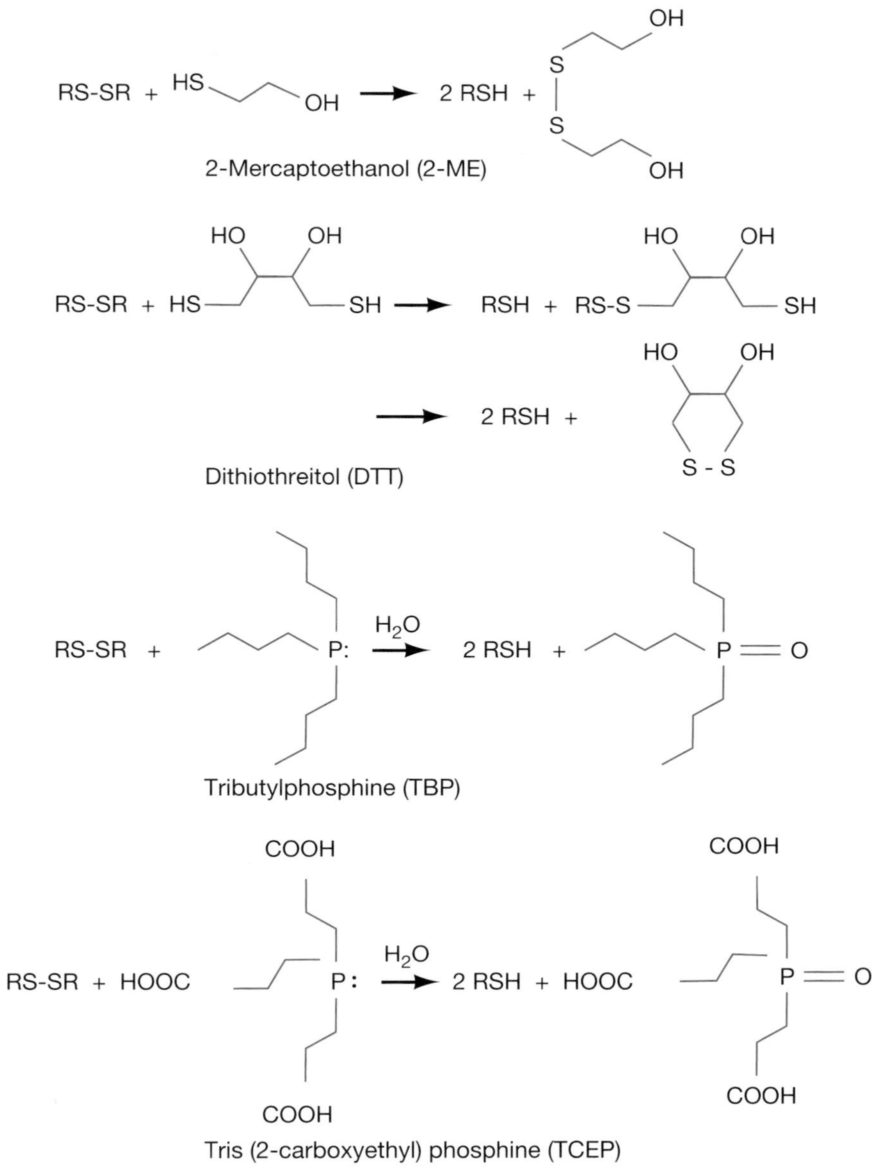

2-Mercaptoethanol (2-ME)

Dithiothreitol (DTT)

Tributylphosphine (TBP)

Tris (2-carboxyethyl) phosphine (TCEP)

FIGURE 7.2. Reducing agents commonly used in protein chemistry. (Portion reprinted, with permission, from Kellner et al. 1994.)

is energetically favored over the mixed disulfide (Figure 7.2). Due to its low redox potential (–0.33 V at pH 7), DTT has gained wide acceptance because it can fully reduce cystine at pH 8 in a few minutes. Additionally, it is a highly water-soluble solid that is resistant to air oxidation, and it has a relatively low level of unpleasant odor compared to 2-mercaptoethanol.

Trialkylphosphines such as tributylphosphine have been used extensively for the reductive cleavage of protein disulfides, especially wool cystine (see Maclaren and Sweetman 1966; Sweetman and Maclaren 1966; for review, see Rüegg and Rudinger 1977). Trialkylphosphines reduce disulfides rapidly and stoichiometrically in aqueous solution (see Figure 7.2); they are highly specific for disulfides and are essentially unreactive toward other amino acid functional groups found in proteins (Rüegg and Rudinger 1977). Tributylphosphine is a powerful reducing agent, and it can readily reduce very stable disulfides such as oxidized DTT. Tributylphosphine has been used to reduce proteins prior to two-dimensional gel electrophoresis (Herbert et al. 1998; Herbert 1999), but it has not gained universal acceptance by protein chemists because of its toxicity, unpleasant odor, and poor solubility in water. The newly introduced sulhydryl reducing agent tris(2-carboxyethyl)phosphine (TCEP) (Burns et al. 1991), which is now commercially available, is a more attractive alternative because it is water soluble and does not have an unpleasant odor, and it is a more effective reducing agent than DTT or 2-mercaptoethanol (Getz et al. 1999). For a comparison of reducing agents available for protein chemistry, see Table 7.1.

Procedures designed for the complete reduction of a protein usually require strong denaturants (see below) and elevated temperatures (e.g., 45–50°C) (Crestfield et al. 1963; Hirs 1967) (see Protocols 2 and 3). Even if a protein does not contain disulfide linkages, a reduction step should be performed, because any free thiols in the protein may have been oxidized during the purification procedure and thus would not be available for alkylation. By these means, *S*-sulfocysteine (and methionine sulfone) is regenerated to form cysteine and methionine. The extent of reduction of disulfide bonds in a protein may be estimated with Ellman's reagent, 5,5′-dithiobis(2-nitrobenzoic acid) (DTNB) (Habeeb 1972; Riddles et al. 1983) (see Protocol 4).

Once the reducing reagent is removed, protein thiol groups readily reform disulfide bonds due to air oxidation. This can be prevented by modification of the thiol groups to form stable derivatives (e.g., by alkylation). Several sulfhydryl alkylating groups may be used for

TABLE 7.1. Reagents for the reductive cleavage of disulfide bonds in proteins

Reagent	Comments	References
2-Mercaptoethanol	Weak reducing agent; original reagent of choice for reducing disulfides; large molar excess required; unpleasant odor (must be used in a chemical fume hood), water soluble, inexpensive.	Liu (1977)
Dithiothreitol (DTT)	Stronger reducing agent than 2-mercapto-ethanol; water soluble, little odor; also known as Cleland's reagent after its discoverer.	Cleland (1964); Konigsberg (1972)
Tributylphosphine (TBP)[a]	Powerful reducing agent (even reduces oxidized DTT); reduces disulfides stoichiometrically (1 mole of TBP reduces 1 mole of disulfide); rapid, low water solubility, unpleasant odor, toxic; works at low pH; does not react with some alkylating reagents.	Rüegg and Rudinger (1977)
Tris(2-carboxyethyl)phosphine (TCEP)[a,b]	Powerful reducing agent (even reduces oxidized DTT); reduces disulfides rapidly; highly water soluble, odor-free, low-toxicity; works at low pH; does not react with some alkylating reagents.	Burns et al. (1991)

The extent of reduction of disulfide bonds in a protein may be estimated with Ellman's reagent, 5,5′-dithiobis(2-nitrobenzoic acid) (DTNB) (Habeeb 1972; Riddles et al. 1983; see Protocol 4).

[a]TBP and TCEP are compatible with many alkylating agents (e.g., 4-vinylpyridine, but not iodoacetic acid or its amide). Therefore, TBP (or TCEP) and 4-vinylpyridine can be added together. When using iodoacetic acid or iodoacetamide in conjunction with TCEP or tributylphosphine, it is advisable to carry out reduction and alkylation in two steps (Rüegg and Rudinger 1977).

[b]For a comparison of the use of TCEP and DTT in protein chemistry, see Getz et al. (1999).

this purpose following disulfide reduction. For a list of the most commonly used alkylating reagents, see Table 7.2. The reduced protein is usually not isolated, and alkylation is generally accomplished in the same reaction mixture containing reduced protein, denaturant, and excess reducing agent. The two most frequently used alkylating agents are 4-vinylpyridine (Raftery and Cole 1966), which yields *S*-pyridylethyl cysteine, and iodoacetic acid (or iodoacetamide), yielding *S*-carboxymethyl cysteine (Crestfield et al. 1963). The frequently observed modification of protein thiols by acrylamide monomers during gel electrophoresis has also been adapted for use in *S*-alklyation (Brune 1992).

To cleave disulfide bonds efficiently in a protein, it is often necessary to use a denaturing agent that renders the protein in a conformational state that exposes these linkages sufficiently to the reducing agent. Commonly used denaturing agents utilized during the reduction/alkylation step include urea and guanidine-HCl at concentrations in the range of 6–8 M, as well as detergents. Care should be exercised in selecting the appropriate grade of denaturant, especially when working with microscale amounts of sample. For example, the grade of

TABLE 7.2. Reagents for the alkylation of cysteine residues in proteins

Reagent	Comment	References
Iodoacetic acid	Classical alkylating reagent for cysteines. Modified cysteines (*S*-carboxymethylcysteine) are negatively charged. *S*-carboxymethylcysteine may be quantitated in the intact protein by amino acid analysis; stable in acid. PTH derivative is subject to β-elimination during sequencing; can react with other amino acids (mainly methionine, histidine, and lysine).	Crestfield et al. (1963); Hirs (1967)
Iodoacetamide	Classical alkylating reagent for cysteines. *S*-carboximadomethylcysteine has a neutral charge; easily quantitated in the intact protein by amino acid analysis; stable in acid. PTH derivative is subject to β-elimination during sequencing; can react with other amino acids (mainly methionine, histidine, and lysine).	
4-Vinylpyridine	*S*-(2-pyridylethyl) cysteine is stable in acid. Well resolved in amino acid analysis; amenable to spectral quantitation in the intact protein (absorbance at 254 nm) (Renlund et al. 1990). PTH derivative is stable during sequencing and readily detected during mass spectrometry by the release of the labile 106-dalton *S*-pyridylethyl moiety (Moritz et al. 1996). *S*-pyridylethyl modification of cysteine introduces a polar nature to the side chain that increases the solubility of the protein. *S*-pyridylethylation must be followed immediately by separation of the reaction mixture because prolonged incubation may cause side reactions with histidine, tryptophan, and methionine.	Raftery and Cole (1966); Cavins and Friedman (1970)
Acrylamide	PTH derivative is stable during Edman sequencing.	Brune (1992)

urea is important because poor-quality material (especially aged urea) is usually contaminated by cyanate ions (Stark et al. 1960; Cole 1961). Ammonium cyanate formation in freshly prepared urea is an inherent property of the reagent that occurs upon long-term storage. Cyanate present in 8 M urea is sufficient to react with amino groups of proteins, rendering them refractory to the Edman chemistry (Stark et al. 1960) and altering the electrophoretic behavior of proteins in gels containing urea (Cole and Mecham 1966). This phenomenon is particularly troublesome for two-dimensional gel analysis in which proteins are separated in the first dimension on the basis of their charge (see Chapter 4). Cyanate-free urea, which is commercially available, can also be produced from aged material by recrystallization from hexane or by simple treatment of aqueous urea solutions with a mixed bed resin (see Appendix 2).

For chromatographic purposes, it is important to use ultrapure guanidine-HCl free of metal ions and materials that absorb in the 225–300-nm range. Both high-grade urea (e.g., Sequanal grade from Pierce) and guanidine-HCl suitable for microscale protein manipulation and sequencing are commercially available either as the free salt or, in the case of guanidine-HCl, as an aqueous 8.0 M solution that can be diluted with an appropriate buffer to the desired molarity.

Procedures for the reduction and alkylation of proteins at large-scale levels (>1 mg) and microscale levels (1–20 μl) are given in Protocols 2 and 3, respectively.

THE POLYPEPTIDE CHAIN MUST BE FRAGMENTED BY ENZYMATIC OR CHEMICAL MEANS

A fundamental step in peptide mapping is the fragmentation of the polypeptide chain to yield low-molecular-weight peptides that are conducive to sequence analysis by Edman chemistry and mass spectrometric methods. Either enzymatic or chemical means can accomplish this. Different proteases of high and reproducible specificity, like trypsin, are commonly used to fragment proteins. In general, proteases like trypsin can be used to produce small peptides, which are an advantage for proteomic analysis methods such as mass spectrometry (see Chapter 8). Chemical cleavage is usually a compromise between efficient cleavage and specificity. The most common chemical cleavage method employs cyanogen bromide, which is specific for Met-X bonds. Due to the low abundance of Met residues in a protein, cyanogen bromide cleavage usually yields large peptide fragments. Although large peptides are preferred for Edman degradation, such peptides are often hydrophobic and difficult to purify.

Fragmenting Proteins by Enzymatic Means

Ideally, for peptide mapping studies, proteolytic cleavage should be selective as well as specific. The selectivity of a protease reflects the number of sites in the substrate at which peptide bond hydrolysis may occur. The complexity of the peptide map can thus be anticipated from the selectivity of the protease and the amino acid composition of the protein. Proteases can be selected to cleave at either basic, acidic, or hydrophobic side chains. Commonly used proteases that are commercially available are described in Tables 7.3 and 7.4. The nomenclature used for protease selectivity was introduced in 1967 by Schechter and Berger and is now widely used in the literature (Schechter and Berger 1967) (see Figure 7.3). For an estimate of the number and length of peptide fragments for a hypothetical 300-residue protein, see Table 7.5.

The specificity of a protease reflects the efficiency of cleavage at susceptible sites (see Tables 7.3 and 7.4). Highly specific proteases effect complete scission of certain peptide bonds in the polypeptide substrate without partially hydrolyzing any other bonds. Thus, specific

TABLE 7.3. Selected chemical and enzymic cleavage reagents for protein chains

Reagent	Site	Comments
Enzymatic cleavage		
Trypsin	Arg, Lys	Specific, often used first; bonds to proline are not cleaved; amino-ethylcysteine provides additional site.
Endoproteinase Lys-C	Lys	Very high specific activity; very effective in gel digests.
Clostripain	Arg	Lys bonds remain intact; yields may be lower than with trypsin; cleaves Arg Pro.
A. mellea protein	Lys	Splits on amino side; sometimes after Arg.
S. aureus protein	Glu	Sometimes cleaves after Asp; also cleaves cysteic acid.
Endoproteinase Asp-N	Asp	Cleaves amino-terminal to Asp; also cleaves cysteic acid.
Asparaginylendopeptidase	Asn-X	pH 5.5–6.5; stable to denaturants.
Chymotrypsin	Trp, Tyr, Phe, Leu, Met	Largely specific for hydrophobic residues; also useful for limited digestion applications.
Thermolysin	hydrophobic	Splits on amino side; may be used at 60°C.
Pepsin	Phe-Phe	Relatively nonspecific; prefers aromatics or Leu; low pH optimum; no disulfide interchange occurs.
Subtilisin, Elastase	various	Nonspecific; for providing small peptides.
Thrombin, Bromelain	Arg-Gly, Gly-Val	Limited proteolytic activity; give large fragments.
Chemical-Enzymic		
Succinylation-trypsin	Arg	Lysine blocked; changes its polarity; citraconylation provides reversible modification.
Chemical cleavage		
Cyanogen bromide	Met	Converted to homoserine lactone; yields >70%, Ser, Thr bonds difficult; Trp-X can cleave.
Dilute acid (pH 2)	Asp	Use formic acid for insoluble proteins and dilute; some amide cleavage; yields >60%.
Formic acid (80%)	Asp-Pro	Yields of 40% with high specificity.
HCl (6 M)	multiple	For random generation of small peptides.
Hydroxylamine	Asn-Gly	Also breaks imide link to Gly; yields ~50%.
DMSO/HCl/HBr	Trp	Several alternative reagents available; oxidation and modification of amino acids likely.
NTCB	Cys	Splits on the amino side; Cys amino blocked.

Cleavage occurs at the carboxyl end of the amino acid unless indicated otherwise.

TABLE 7.4. List of proteases commonly used for fragmenting proteins

Protease[a]	EC no.[b]	Class[c]	Peptide bond selectivity	pH optimum	Molecular mass (kD)[d]	Accession no.[e]	Commercial availability[f]
Endoproteinase							
Trypsin (bovine)	3.4.21.4	serine	P_1-P_1^1- (P_1 = Lys,Arg)	8.0–9.0	23.5	P00760^{S}	C* R* S W
Chymotrypsin (bovine)	3.4.21.1	serine	P_1-P_1^1- (P_1 = aromatic, P_1^1= nonspecific)	7.5–8.5	25	P00766^{S}	C* R* S W
Endoproteinase Asp-N (*Pseudomonas fragi*)	3.4.24.33	metallo	P_1-Asp- (and -P_1-cysteic acid)	6.0–8.0	27	φ	C* T R* S*
Endoproteinase Arg-C (mouse submaxillary gland)	φ	serine	-Arg-P_1-	8.0–8.5	30	n.a.	C R S
Endoproteinase Glu-C (V8 protease) (*Staphylococcus aureus*)	3.4.21.19	serine	-Glu-P_1^1- (and -Asp-P_1^1-) (2)	8.0	27	P04188^{S}	C* R* S* W
Endoproteinase Lys-C (*Lysobacter enzymogenes*)	3.4.21.50	serine	-Lys-P_1^1-	8.0	30^{NR} 33^{R}	S77957^{P}	C* S*
Pepsin (porcine)	3.4.23.1	aspartic	P_1-P_1^1- (P_1 = hydrophobic preferred)	2.0–4.0	34.5	P00791^{S}	C R S W
Thermolysin (*Bacillus thermo-proteolyticus*)	3.4.24.27	metallo	P_1-P_1^1- (P1 = Leu,Phe,Ile,Val,Met,Ala)	7.0–9.0	37.5	P00800^{S}	C R S
Elastase (porcine)	3.4.21.36	serine	P_1-P_1^1- (P_1 = uncharged, nonaromatic)	7.8–8.5	25.9	P00772^{S}	C*R S W
Papain (*Carica papaya*)	3.4.22.2	cysteine	P_1-P_1^1- (P_1 = Arg,Lys preferred)	6.0-7.0	23	P00784^{S}	R S W
Proteinase K (*Tritirachium album*)	3.4.21.64	serine	P_1-P_1^1- (P_1 = aromatic, hydrophobic preferred)	7.5–12.0	18.5	P06873^{S}	C T R* S
Subtilisin (*Bacillus subtilis*)	3.4.21.62	serine	P_1-P_1^1- (P_1 = neutral/acidic preferred)	7–11	30^{S} 27.3^{L}	P04189^{S}	C^{L} R^{S} S^{L,S}
Clostripain (endoproteinase-Arg-C) (*Clostridium histolyticum*)	3.4.22.8	cysteine	-Arg-P_1- (P_1 = Pro preferred)	7.1–7.6	59	P09870^{S}	C* R* S* W
Exopeptidase							
Carboxypeptidase A (bovine)	3.4.17.1	metallo	P_1-P_1^1- (P_1 cannot be Arg,Lys,Pro)	7.0–8.0	34.5	P00730^{S}	R S W
Carboxypeptidase B (porcine)	3.4.17.2	metallo	P_1-P_1^1- (P_1 = Lys,Arg)	7.0–9.0	34.6	P00732^{S}	C R S W
Carboxypeptidase P (*Penicillium janthinellum*)	φ	serine	P_1-P_1^1- (nonspecific)	4.0–5.0	51	n.a.	C* T R* S
Carboxypeptidase Y (yeast)	3.4.16.5	serine	P_1-P_1^1- (nonspecific)	5.5–6.5	61	P00729^{S}	C* R* S*
Cathepsin C	3.4.14.1	cysteine	X-P_1-P_1^1- (removes amino-terminal dipeptide)	5.5	210	n.a.	C^{T} R^{*B} S^{B}
Acylamino-acid-releasing enzyme (porcine)	3.4.19.1	serine	Ac-P_1-P_1^1- (P_1 = Ser,Ala,Met preferred)	7.5	80^{H} 360^{P}	P19205^{S} +	C^{P} T^{P} R^{H}
Pyroglutamate aminopeptidase (bovine)	3.4.19.3	cysteine	P_1-P_1^1- (P_1 = 5-oxoproline or pyroglutamate)	7.0–9.0	70–80^{B}	n.a.	C^{P} T^{E} R^{B} S^{B}

For additional information on proteases, see Harlow and Lane (1999).

[a]Source shown in parentheses.

[b]Enzyme nomenclature information can be accessed at the following Web Sites : http://delphi.phy.univ-tours.fr/Prolysis; http://www.expasy.ch/enzyme/; http://www.chem.qmw.ac.uk/iubmb/enzyme/. The MEROPS database (http://www.merops.ac.uk) provides a catalog and a structure-based classification of peptidases (i.e., all known proteolytic enzymes) (Rawlings et al. 2002). φ indicates EC not yet included in IUBMB recommendations.

[c]For Schecter/Berger nomenclature, see Figure 7.3.

[d] NR = Nonreduced; R = reduced; H = Source Equine; P = Source Porcine; B = Source Bovine; S = Source *Bacillus subtilis*; L = Source *Bacillus licheniformis*.

[e]S = SwissProt; P = PIR; + = porcine sequence; φ = partial sequences of Asp-N are described Hagmann M.-L. et al. (1995, 1998). Accession numbers: AAB35279, AAB35280, AAB35281, AAB35282. n.a. indicates primary structure not available.

[f]C = Calbiochem (http://www.calbiochem.com); R = Roche (Roche Molecular Biochemicals; http://biochem.roche.com); S = Sigma-Aldrich (http://www.sigma-aldrich.com); T = TaKaRa (http://bio.takara.co.jp/BIO_EN/); W = Worthington (http://www.worthington-biochem.com); * = Supplier specified sequencing grade enzyme available; P = porcine; E = *Escherichia coli*; B = bovine; H = equine; T = turkey; S = *Bacillus subtilis*; L = *B. licheniformis*.

TABLE 7.5. Calculated number and length of peptide fragments for a hypothetical 300-residue protein

Amino acid	Specific cleavage	No. fragments	Average length
Phe, Tyr, Trp, Leu	chymotrypsin	54	6
Lys, Arg	trypsin	35	9
Glu	Glu-C	20	15
Lys	Lys-C	19	16
Arg	Arg-C	17	18
Asp	Asp-N	17	18
Met	cyanogen bromide (CNBr)	8	38
Trp	BNPS-skatole	5	60

The hypothetical 300-residue protein was calculated from the NBRF-PIR protein database by the total number of each amino acid versus the number of database entries. (Reprinted, with permission, from Kellner 1994.)

cleavage results in each region of the polypeptide chain being represented by a single peptide with no overlapping peptides (i.e., two or more peptides representing the same region of the amino acid sequence) or redundant peptides.

Proteolytic cleavage is facilitated by first denaturing the polypeptide chain of interest and, if a protein contains disulfide bonds, reducing the –S–S– bridges and alkylating the cysteines (Crestfield et al. 1963). Susceptible protease cleavage sites are not equally accessible in native proteins, and denaturation is required to disrupt the secondary and tertiary structures of the protein so that its conformation assumes that of a random coil. However, peptide mapping strategies that involve proteolytic cleavage of a protein under nonreducing conditions are extremely useful for determining the cysteine connectivities (disulfide bonds) in a protein. (Knowledge of disulfide linkages is a useful prerequisite for determining the three-dimensional structure of proteins [see below, Peptide Mapping Can Be Used to Determine Disulfide Linkages in Proteins; also Simpson et al. 1988; Cole et al. 1999; Moritz et al. 2001].)

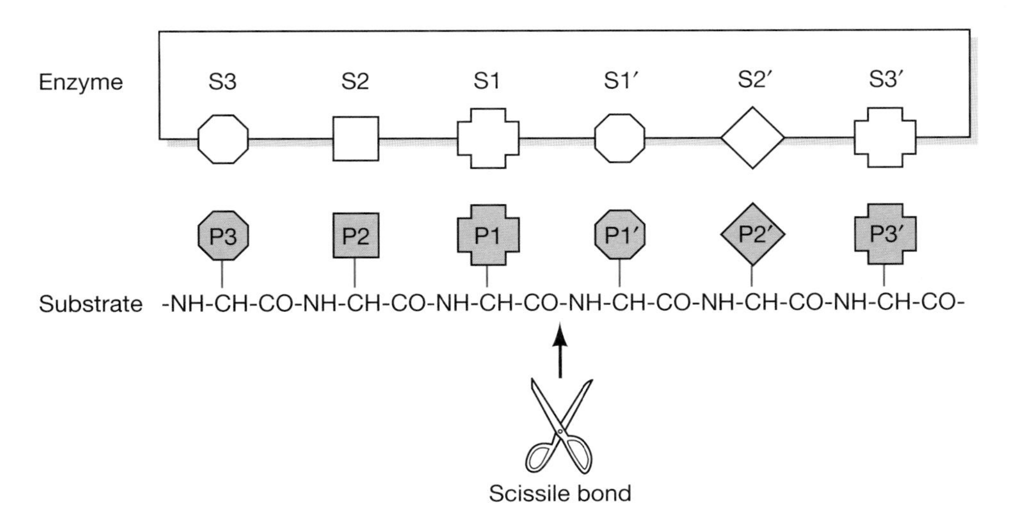

FIGURE 7.3. The Schechter and Berger nomenclature for the description of protease subsites. In this system, it is considered that the amino acid residues of the polypeptide substrate bind in enzyme subsites of the active site. By convention, these subsites on the protease are called S (for subsites) and the substrate amino acid residues are called P (for peptide). The amino acid residues of the amino-terminal side of the scissile bond are numbered P3, P2, and P1, and those residues of the carboxy-terminal side are numbered P1′, P2′, P3′... . P1 or P1′ are those residues located adjacent to the scissile bond. The substrate residues around the cleavage site can then be numbered up to P8. The subsites on the protease that complement the substrate-binding residues are numbered S3, S2, S1, S1′, S2′, S3′... .

TABLE 7.6. Optimized proteolysis conditions in the presence of chaotropes, detergents, and organic solvents

Conditions	Enzyme				
	Trypsin	Chymotrypsin	Lys-C (*Achromobacter* protease)	Glu-C (V8)	Subtilisin
Native					
buffer	0.1 M AB[a]	0.1 M AB	0.1 M AB	0.05 M AB	0.1 M AB
time	2 hours	2 hours	2 hours	5 hours	1 hour
Urea	4 M	2 M	8 M	2 M (R)[b]	8 M (R)
buffer	+5 mM Ca^{++}		0.2 M TC[a]		0.1 M TC
time	15 hours	5 hours	5 hours	5 hours	1 hour
Guanidine-HCl	(1–2 M)	2 M	2 M	1 M (R)	2 M (R)
buffer	0.2 M TC	0.2 M AB	0.2 M TC		
time	24 hours	24 hours	5 hours	18 hours	2 hours
SDS	<0.1%	<0.1% (R)	1%	0.1% (R)	1%
CHAPS	2%	1%	2%	2% (R)	2% (R)
Acetonitrile[c]	40%	30%	40%	20% (R)	40%

Adapted from Riviere et al. (1991).

Buffers and incubation times are the same as those under native conditions, except where indicated. The highest concentrations of excipients permitting adequate proteolysis are listed; concentrations higher than 2% CHAPS and 40% acetonitrile are not listed.

[a]AB = Ammonium bicarbonate; TC = Tris-Cl.

[b]R = Restricted digest.

[c]For a comprehensive study on the effects of organic cosolvents, see Welinder (1988).

Optimized proteolysis conditions for a number of commonly used proteases, including their activity in the presence of widely used denaturants, are given in Table 7.6. Additional details about proteolytic enzymes are contained in the information panel on NOMENCLATURE ON PROTEASES, PROTEINASES, AND PEPTIDASES at the end of this chapter.

USEFUL WEB SITES FOR OBTAINING INFORMATION ABOUT PROTEASES

Enzyme: http://www.expasy.ch/enzyme/ or http://www.chem.qmw.ac.uk/iubmb/enzyme/

Enzyme nomenclature. Primarily based on the recommendations of the nomenclature committee of IUBMB, and it describes each type of characterized enzyme for which an EC (Enzyme Commission) number has been provided.

Prolysis: http://delphi.phys.univ-tours.fr/Prolysis/

Protease/peptidase nomenclature, classification, and chemistry (T. Moreau).

Prowl: http://prowl.rockefeller.edu/recipes/contents.htm

Protocols for using proteases for mass spectrometric analysis (R.C. Beavis).

MEROPS: http://www.merops.ac.uk

Catalog and structure-based classifications of peptidases (i.e., all proteolytic enzymes).

For general reviews of proteolytic enzymes and their uses, see Lorand (1981) and Bond and Beynon (1989).

How to choose a proteolytic enzyme

One of the most commonly used and best-characterized proteinases in protein structural analysis is trypsin (EC 3.4.21.4); the use of this enzyme is described in Protocol 5. Other useful proteinases for fragmenting proteins are listed in Tables 7.3 and 7.4, and optimized conditions for their use are given in Table 7.6. For very basic proteins that are rich in lysine and/or arginine residues, trypsin can generate a large number of peptides of small average size that may not be amenable to sequence analysis using the Edman procedure (or mass spectrometry sequencing procedures). The action of trypsin can be modified to restrict cleavage to arginine residues alone by selectively modifying all lysine residues in the protein substrate. Two common methods for modifying lysyl side chains—succinylation and citraconylation (this modification has the added advantage in that it is fully reversible under mildly acidic conditions)—are described in the additional protocols in Protocol 5. The need for lysyl side-chain modification strategies has been ameliorated in recent times due to the advent of commercially available proteinases that selectively cleave only lysyl bonds (Lys-C, EC 3.4.21.50; also called lysyl endopeptidase) or argininyl bonds (Arg-C from mouse submaxillary gland, EC 3.4.22.8, and Clostripain, EC 3.4.22.8). Note that Clostripain, unlike trypsin, cleaves Arg-Pro bonds. For properties and optimal cleavage conditions, see Table 7.4.

A large variety of proteinases can be used in the same manner as trypsin; attention must be paid to the buffer (optimal pH) conditions (see Tables 7.3, 7.4, and 7.6). Described below are other commonly used proteinases.

- ***Glu-C*** (EC 3.4.21.9) (or "V8" protease) isolated from *Staphylococcus aureus* strain V8 is a very useful proteinase because, like trypsin (and Lys-C and Arg-C), it shows good specificity (Houmard and Drapeau 1972; Drapeau 1977). It cleaves on the carboxy-terminal side of glutamyl residues (using 50 mM ammonium bicarbonate, pH 7.8, or 50 mM ammonium acetate, pH 4.0); however, this proteinase will cleave Asp-P′ bonds (albeit at low frequency) as well as Glu-P′ bonds in phosphate buffer (50 mM) at pH 7.8. Glu-C will not cleave Glu-Pro bonds, and potentially labile bonds in a multiple sequence (e.g., Glu-Glu-X) may not each be cleaved in a good yield. Endoproteinase Glu-C is active in 0.1–0.2% SDS, 2 M urea, 1 M guanidine-HCl, and 10% acetonitrile. Its amino acid sequence is known (Drapeau 1978; SwissProt accession number P04188).
- ***Asp-N*** endoproteinase (EC 3.4.24.33) from a *Pseudomonas fragi* mutant is a metalloproteinase with excellent specificity. In phosphate, acetate, or Tris buffers at pH 6.0–8.5, it cleaves peptide bonds amino-terminally at aspartic acid or cysteic acid (Drapeau 1980). Asp-N will cleave P_1-Glu bonds (at very low frequency) as well as P_1-Asp- bonds if high enzyme:substrate ratios and long incubation times are employed. Additional cleavage at glutamyl residues can be prevented by reducing the enzyme concentration (enzyme:substrate ratio of 1:1000) at an incubation time of 2–6 hours. Endoproteinase Asp-N is active in 0.1% (w/v) SDS, 1 M urea, 1 M guanidine-HCl, and 10% acetonitrile. Partial amino acid sequences of Asp-N have been described previously (Hagmann et al. 1995; Hagmann 1998). In addition to proteinases with high selectivity, it is worth noting two enzymes of broad specificity, namely, pepsin and thermolysin.
- ***Pepsin*** (EC 3.4.23.1) works at very low pH (pH 2–3), which makes it suitable for disulfide bond determination studies because, at low pH, disulfide bonds rearrange less frequently than under alkaline conditions. Moreover, low-pH conditions may favor the solubility of some proteins.

- ***Thermolysin*** (EC 3.4.24.4) is extremely useful because it is active at high temperatures (up to 80°C) and in 8 M urea. Thus, thermolysin is active under conditions highly favorable to protein unfolding. Thermolysin has broad specificity for the amino-terminal side of hydrophobic residues, particularly, leucine, isoleucine, phenylalanine, and valine in descending order of preference (Heinrickson 1977). Thermolysin has been described by many reports in the literature to be successful in fragmenting proteins that were refractory to commonly used endoproteinases (see, e.g., Du et al. 1996).

Other broad-specificity endoproteinases include papain (useful for the preparation of $F(ab)_2$ fragments), proteinase K, and subtilisin (for properties, see Tables 7.3 and 7.4). For example, papain activated prior to use with cysteine, at pH 5.5 (0.1 M acetate, 3 mM EDTA), is added to immunoglobulin G (IgG) (concentration of ~10 mg/ml) at 37°C at time zero, and again later (after 9 hours) at an enzyme:substrate ratio of 1:20 (w/w) (Parham et al. 1982).

In-gel proteolytic digestion

Most "sequencing-grade" endoproteinases used in conventional solution digests (e.g., trypsin, *Staphylococcus* protease V8, endoproteinase Lys-C, and endoproteinase Asp-N) give comparable peptide maps for in-gel digests. Of these, endoproteinase Lys-C (*Achromobacter* protease I) gives considerably higher in-gel cleavage efficiencies. Lys-C also retains full activity in the presence of 0.1% SDS, whereas other proteases are inhibited to varying extents by this detergent, trypsin being the most sensitive. A serious disadvantage of Lys-C is that several of the generated peptides will contain internal arginine residues that complicate the interpretation of mass spectrometric collision-induced dissociation fragmentation patterns (see Chapter 8). This, in turn, limits the usefulness of some mass-spectrometric-based peptide identification procedures. For these reasons, trypsin is recommended for methods relying on mass-spectrometric-based identification procedures. Prior to in-gel digestion, it is customary to first visualize the protein and, if necessary, concentrate it to optimize yields of peptide fragments.

Visualization of Gel-separated Proteins

The most widely used method for the detection of proteins in primary gels is the conventional Coomassie Blue staining procedure (Wilson 1983), which can detect proteins in the low-microgram range (~0.5–1.0 μg) (see Chapters 2 and 4). Care must be exercised with this staining method (see Protocol 9), because overstaining can interfere with subsequent in-gel proteolysis and peptide mapping steps. To minimize these potential problems, the gel should be stained for only 20 minutes; the gel background should be destained overnight to visualize the proteins and remove excess Coomassie Blue. After the protein spot has been excised, additional extended washing steps must be performed with deionized H_2O to remove Coomassie Blue associated with the protein and excess SDS.

Another category of protein visualization methods gaining wide acceptance is negative staining, which relies on selective precipitation of metal ions in gels, leaving protein bands unstained and readily amenable for structural analysis. Common negative staining methods include potassium acetate/chloride (Nelles and Bamburg 1976; Bergmann and Jörnvall 1987), sodium acetate (Higgins and Darmus 1979), copper chloride (Lee et al. 1987), aurodye (Casero et al. 1985), zinc chloride (Lonnie and Weaver 1990), and one of the most widely used methods, zinc-imadazole (Fernandez-Patron et al. 1995a,b) (see Protocol 10), which is approximately ten times more sensitive than Coomassie Blue. Like Coomassie Blue, this negative staining procedure does not result in detectable protein modification. A sensitive (~0.5 ng) procedure for the negative staining of proteins, based on the precipitation of methyltrichloroacetate in gels, has been reported by Candiano et al. (1996). In this procedure, after separation, gels are incubated with 8% methyltrichloroacetate ester in 38% isopropanol and then washed in H_2O to produce a negative image of colorless proteins against an opaque background (Candiano et al. 1996). Because this process is reversible, gels can be restained (e.g., silver staining, see Protocol 11) after rapid visualization.

Silver staining is the most widely used high-sensitivity staining method for the detection of proteins in polyacrylamide gels (Switzer et al. 1979), with a sensitivity reported to be ~100 times better than Coomassie Blue (for detailed reviews, see Rabilloud 1990; Rabilloud et al. 1994). Silver staining has not been used for protein microanalysis until recently because the detection limits (~1–10 ng) were much too low for conventional sequence analysis by automated Edman degradation. However, with the improved sensitivity of protein structural analysis by mass spectrometric methods (low femtomole levels) (McCormack et al. 1997), silver staining is gaining broader acceptance (Muzio et al. 1996; Shevchenko et al. 1996).

The silver staining procedure designed for visualizing proteins for mass spectrometric analysis purposes is significantly less sensitive than the classical staining method, due to the elimination of the glutaraldehyde fixing step (see Rabilloud 1990, 1999, 2000, 2001; Rabilloud et al. 1994; Shevchenko et al. 1996).

When using any silver staining protocol for primary structural analysis, a critical parameter is chemical modification of the protein. Such modifications are likely to occur during the presensitizing step, which involves pretreatment of the gel with sensitizers (e.g., sulfosalicylic acid, sodium thiosulfate, DTT, glutaraldehyde, and chelators) (for a review on various silver staining protocols, see Rabilloud 1990; Rabilloud et al. 1994) to improve the contrast (and hence sensitivity) between the stained protein and the gel background. The "acidic" silver staining method (Rabilloud 1990, 2000), adapted for primary structural analysis by omission of the fixation/sensitization treatment with glutaraldehyde, is described in Protocol 11 (for additional silver staining methods compatible with mass spectrometry, see Chapter 2, Protocol 7 and Chapter 4, Protocol 12). This silver staining method has been used recently for a number of biological applications (Muzio et al. 1996; McCormack et al. 1997) and results in avoidance of protein modification.

Empirical observations from a number of proteomics laboratories indicate that preconcentration of stained excised gel spots (or bands) prior to electroblotting onto PVDF is necessary to achieve high initial sequencing yields. Similarly, preconcentration prior to in-gel proteolytic (or chemical) fragmentation is necessary to obtain high peptide yields. For a discussion of methods suitable for concentrating gel-resolved proteins, see Chapter 6, and for a description of a method, see Protocol 4 in Chapter 6.

Internal Sequencing of Gel-separated Proteins

Several methods have been reported for obtaining peptides from proteins separated by polyacrylamide gel electrophoresis. Earlier approaches involved the electrophoretic or passive extraction of proteins from the gel, followed by proteolytic digestion in solution (Hunkapiller et al. 1983; Simpson et al. 1987; Ward et al. 1990). These approaches were often technically difficult due to the problem of removing detergents that interfere with both proteolytic digestion and chromatographic fractionation of peptide mixtures (Simpson et al. 1987), and they have now been superseded by two general techniques:

- In-gel digestion and extraction of the resultant peptides (Eckerskorn and Lottspeich 1989; Kawasaki et al. 1990; Ward et al. 1990; Rosenfeld et al. 1992; Hellman et al. 1995; Moritz et al. 1995).
- In situ cleavage of proteins electroblotted onto inert membranes such as nitrocellulose (Aebersold et al. 1987) or PVDF (Fernandez et al. 1992).

Methods for chemical cleavage of proteins (e.g., cyanogen bromide) have been reported for the following approaches: in situ on the sample disk of a gas-phase sequencer (Simpson and Nice 1984), in-gel (Jahnen et al. 1990), and on-membrane (Stone et al. 1992). Although each of these procedures has some disadvantages, the recent impetus has been toward the in-gel approach, due mainly to protein losses incurred by overblotting as well as inefficient transfer to the membrane (particularly of high-M_r proteins), i.e., huge losses incurred by electroblotting onto inert membranes. A comparison of peptide maps of proteins digested in-gel and on-membrane usually reveals fewer peptides for the on-membrane-digested proteins. Anecdotal observations suggest that this is due to the protease having limited access to the substrate embedded in the nitrocellulose or PVDF membrane. Several attempts have been made to improve digests of membrane-bound proteins, including the use of a cationic PVDF membrane, which allows high recovery of peptides (Patterson et al. 1992), and the use of the detergent Zwittergent 3-16 (Lui et al. 1996). For reviews of in-gel and on-membrane digestion procedures, see Patterson and Aebersold (1995) and Williams et al. (1993).

Reduction/Alkylation and Proteolytic Digestion of Protein Gel Spots

Crucial to many of the peptide-based methods for identifying proteins separated by two-dimensional electrophoresis is the ability to achieve efficient digestion of proteins immobilized in the acrylamide gel matrix or upon a blotting membrane. It is well recognized that attempts to digest mature proteins composed of disulfide-bonded structures (in contrast to nondisulfide-bonded proteins) are often unsuccessful, regardless of whether the digest was performed on-membrane or in-gel. Moreover, for disulfide-bonded proteins that do digest, the peptide maps are often extremely complex. For these reasons, it is useful to reduce and alkylate gel-resolved proteins prior to proteolytic digestion. This additional manipulation step prior to performing in-gel digestion has the additional advantage of lowering the level of background artifacts (Moritz et al. 1996), an important consideration for any subsequent mass spectrometric analysis. A number of procedures for in situ reduction and alkylation of proteins on PVDF blots (Iwamatsu 1992; Henzel et al. 1993; Jenö et al. 1995) and in-gel digestion protocols (Moritz et al. 1996) have been described. A procedure for the *S*-pyridylethylation of proteins in intact polyacrylamide gels that is compatible with subsequent mass spectrometric analysis techniques is given in Protocol 12 (see Figure 7.4). Procedures for performing in-gel and on-membrane digestion of gel-separated proteins are described in Protocols 13 and 14, respectively.

Fragmenting Proteins by Chemical Means

There are a variety of chemical reactions known to cleave polypeptide chains. Some of these methods are nonspecific (e.g., 6 M HCl for 24 hours at 110°C hydrolyzes a protein into its constituent amino acids), whereas others cleave at specific amino acid residues (see Tables 7.3 and 7.7). A variety of chemical methods for cleavage have been reviewed (Fontana and Gross 1986). Unfortunately, the high specificity required for a fragmentation technique to be useful for peptide mapping (i.e., specificity in peptide bond cleavage) is rarely achievable, with

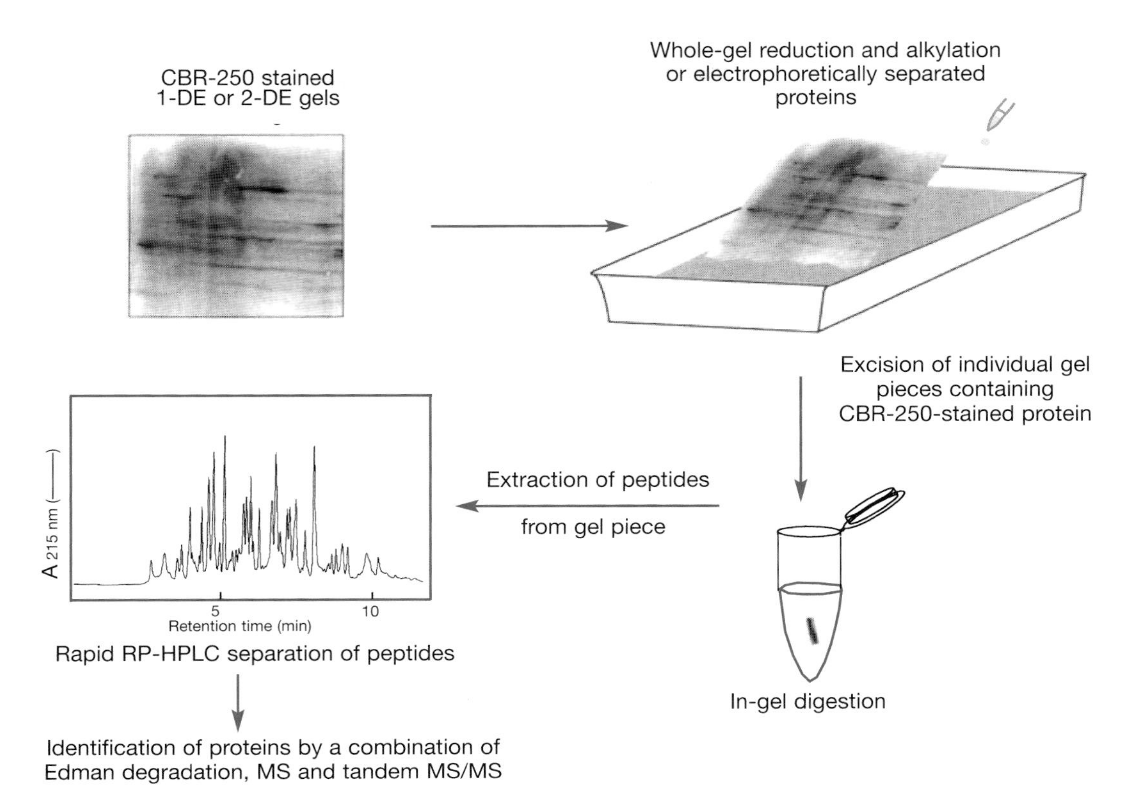

FIGURE 7.4. Flow diagram for whole-gel reduction, *S*-pyridylethylation, in situ digestion, and identification of acrylamide gel-resolved proteins. (Reprinted, with permission, from Moritz et al. 1996.)

the exception of cyanogen bromide cleavage of methionyl bonds. Most chemical fragmentation methods suffer from one or more of the following disadvantages:

- low cleavage yield
- lack of specificity
- undesirable side reactions
- wide variability in the reactivity of the sensitive bond (Kasper 1975)

TABLE 7.7. Chemical methods for selective peptide bond cleavage in proteins

Reagent	Peptide bond cleaved	Abbreviated method	Comments
Cyanogen bromide (CNBr)	-Met-X-	Incubate in CNBr (20–100-fold molar excess over Met) in acid (e.g., 70% formic acid or 85% TFA) for 18–24 hours at room temperature in the dark.	Excellent specificity and yield. Met-Ser and Met-Thr may give <100% yield. Cys may be oxidized to cysteic acid, and acid labile bonds (eg., Asp-Pro) cleaved. Excess CNBr can cause cleavage (and degradation) of Trp and Tyr side chains. (*Note: CNBr is toxic and must be used in a chemical fume hood.*)
Hydroxylamine	-Asn-Gly	Incubate in 2 M hydroxylamine, 6 M guanidine-HCl (adjusted to pH 9 with 4 M LiOH) for 4 hours at 45°C.	Good specificity. Some Asn-Gly bonds are resistant to cleavage, and some Asn-X bonds may cleave.
BNPS-skatole	-Trp-X	Incubate in BNPS-skatole (100-fold molar excess over Trp) in 50% (v/v) acetic acid for 48 hours at 25°C in the dark.	BNPS-skatole reagent is unstable under acidic conditions and must be freshly prepared prior to use to minimize side reactions (e.g., with Tyr, Met, and Cys oxidation).

Cyanogen bromide cleavage

The most specific and generally applicable chemical fragmentation method available is cyanogen bromide cleavage of the polypeptide chain at methionyl residues (Gross and Witkop 1961) (see Protocol 6). Cyanogen bromide reacts with the sulfur of the thioether side chain of methionine to yield a mixture of homoserine and homoserine lactone plus methylthiocyanate (Gross and Witkop 1961). When the reaction occurs with methionine in peptide linkage, the bond involving the carboxyl group of methionine is cleaved (see the reaction mechanism in Figure 7.5). All peptide fragments produced by cyanogen bromide treatment of a protein contain a carboxy-terminal homoserine or its lactone, except the carboxy-terminal peptide of the polypeptide chain (unless methionine is the carboxy-terminal amino acid). It is difficult to find chromatographic conditions for the separation of homoserine lactone and homoserine during amino acid analysis using ion-exchange chromatography, because homoserine lactone elutes with the basic amino acids between NH_3 and arginine, and homoserine coelutes with glutamic acid. Thus, it is useful to convert homoserine lactone to the free acid, by treating homoserine lactone with 0.1 M pyridine acetate (pH 6.5) for 1 hour at 105°C (Ambler 1965) or by dissolving it in 0.1 M NaOH for 1 hour at room temperature (Schechter et al. 1976).

Because methionine sulfoxide does not react with cyanogen bromide, the sample may need to be reduced (e.g., with β-mercaptoethanol) prior to treatment with cyanogen bromide to improve cleavage yields (methionine sulfoxide is readily converted to methionine by reduction). The cleavage yields at Met-Ser and Met-Thr bonds may be low because of the involvement of the β-hydroxyl groups of seryl and threonyl residues in alternate reactions that do not lead to cleavage (see the reaction mechanism in Figure 7.5).

Hydroxylamine cleavage of Asn-Gly bonds

At alkaline pH, hydroxylamine can be used to specifically cleave Asn-Gly bonds in proteins to generate peptide fragments (Blumenfeld et al. 1965; for a review, see Bornstien and Balian

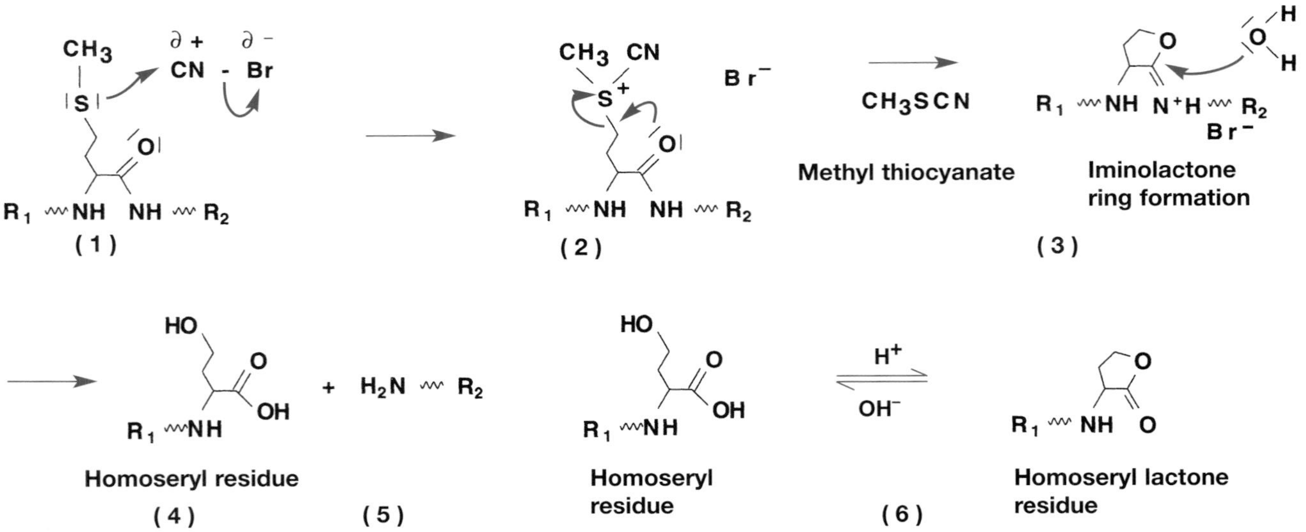

FIGURE 7.5. Reaction mechanism for cyanogen bromide cleavage. The nucleophilic thioether sulfur from the methionine side chain reacts with cyanogen bromide (*1*) to form a sulfonium ion (*2*). Methyl thiocyanate is released while an intermediate imino ring (iminolactone) is formed involving the carbonyl group from methionine (*3*). The iminolactone is hydrolyzed and the Met-X bond is cleaved (*4*), resulting in a homoserine at the carboxyl terminus and the release of a new peptide fragment (with an α-amino amino terminus) (*5*). Homoserine and homoserine lactone are interconvertible and form a mixture (*6*). (Adapted from Inglis and Edman 1970; modified, with permission, from Kellner et al. 1994.)

1977) (see Protocol 7). The Asn-Gly sequence in proteins occurs with low frequency (about every 350 amino acids) and thus usually results in the generation of very large peptide fragments. The reaction mechanism (Figure 7.6) involves the attack of the side-chain carbonyl group of asparagine by the carboxy-terminally adjacent backbone amine group to form a cyclic imide (succinimide) intermediate. Under the nucleophilic attack of hydroxylamine at

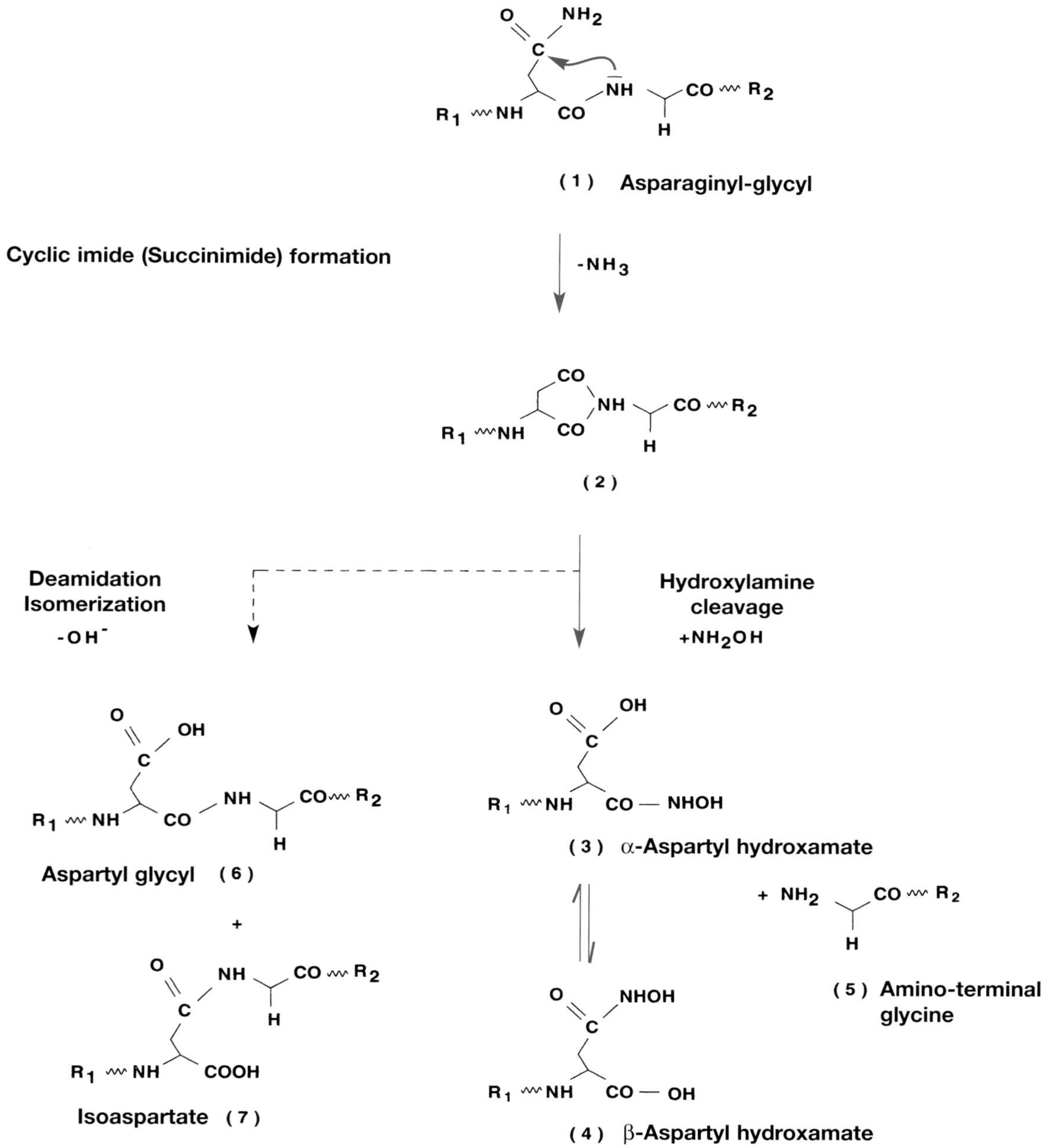

Cyclic imide (Succinimide) formation

(1) Asparaginyl-glycyl

$-NH_3$

(2)

Deamidation Isomerization $-OH^-$

Hydroxylamine cleavage $+NH_2OH$

Aspartyl glycyl (6)

(3) α-Aspartyl hydroxamate

Isoaspartate (7)

(4) β-Aspartyl hydroxamate

(5) Amino-terminal glycine

FIGURE 7.6. Cleavage of Asn-Gly bonds using hydroxlyamine. In this reaction, the side-chain carbonyl group of asparagine is subjected to nucleophilic attack by the α-amino group of the carboxy-terminally neighboring residue (glycine) (*1*) to form the cyclic imide, anhydroaspartylglycine (*2*). Cleavage with hydroxylamine yields (through the intermediary of the hydroxamate) a mixture of α-aspartyl (*3*) and β-aspartyl (*4*) hydroxamates and a new amino-terminal glycine (*5*). Alternatively, nucleophilic attack of the cyclic imide (succinimide) intermediate (*2*) by hydroxide ion leads to a mixture of aspartate (*6*) or an isoaspartate residue (*7*). Isoaspartate residues are refractory to the Edman degradation procedure and will block ongoing amino-terminal sequence analysis. (Modified, with permission, from Kellner et al. 1994.)

alkaline pH, the Asn-Gly peptide bond can be hydrolyzed to form a mixture of α-3-aspartyl and β-aspartyl hydroxamate, as well as a new amino-terminal glycine. The cyclic imide ring may also open and form either an aspartate residue (i.e., a deamidation reaction) or an isoaspartate residue (i.e., isomerization reaction). These latter phenomena are known to occur spontaneously in protein aging (Stephenson and Clarke 1989), especially in eye lens proteins.

Cleavage at tryptophan

The high reactivity of the indole side chain of tryptophan, and the relative rarity of this residue in most proteins, has evoked much interest in finding methods for its selective cleavage. A large number of cleavage reactions for tryptophan have been described. These include BNPS-skatole (2-[2-nitrophenylsulfonyl]-3-methyl-3-bromoindolenine) (Omenn et al. 1970), dimethylsulfoxide and halogen acid (Savige and Fontana 1977), treatment with a high concentration of cyanogen bromide in heptofluorobutyric acid (HFBA) (Ozols and Gerard 1977), and exposure to tribromocresol (Burnstein and Patchornik 1972) and o-iodosobenzoic acid (Mohoney et al. 1981). For a comprehensive review of tryptophan cleavage procedures, see Fontana and Gross (1986). Of these methods, the o-iodosobenzoic acid procedure has become accepted as the most reliable (see Protocol 8).

PEPTIDE MAPPING BY SDS-PAGE (CLEVELAND METHOD) IS A RELATIVELY EASY APPROACH

Peptide separation by one-dimensional SDS-PAGE is a relatively easy approach for comparing the peptide fragments of proteins generated by proteolytic or chemical cleavage (i.e., peptide mapping). The similarity/dissimilarity of the resultant peptide pattern reflects the similarity/dissimilarity of the parent proteins. This peptide mapping approach, first described by Cleveland (1977), requires no special equipment, and it can also be combined with western blotting to locate epitopes (i.e., epitope mapping). Protein fragmentation can be performed prior to loading the gel (the preferred approach) (see Protocol 16) or, alternatively, in the stacking gel with the resultant peptides separated directly in the running gel (Protocol 17). The conventional Laemmli SDS-PAGE gel system is used to resolve peptide fragments <3000 M_r, and the tricine gel system (Schägger and von Jagow 1987) is used for smaller peptides (for basic PAGE protocols, see Chapters 2 and 4). Typically, several lanes with increasing incubation times or enzyme concentrations are employed to optimize the proteolytic conditions. In contrast to traditional peptide mapping procedures (e.g., RP-HPLC methods), which require low-microgram amounts of protein, the Cleveland method requires very small amounts of material (e.g., sufficient for visualization by silver staining; sensitivity can be further enhanced by radiolabeling). For commonly used protein fragmentation procedures, see Tables 7.6 and 7.7.

PEPTIDE MAPPING CAN BE USED TO DETERMINE DISULFIDE LINKAGES IN PROTEINS

Analyzing the cysteine connectivities in proteins is an important facet of the structural characterization of a protein. Disulfide linkages between cysteine residues are an important structural element of the extracellular domains of membrane-associated receptors, as well as secreted proteins and peptides such as plasma proteins, hormones, protease inhibitors, and venom proteins. Disulfide linkages can have an important role in establishing and maintaining the structural fold of a protein.

Before embarking on such a quest, it is important first to establish the number of free thiols and disulfide bonds in the protein of interest. Traditionally, Ellman's reagent (DTNB) has been used to assay the number of free thiols in a protein (Ellman 1959). The method is based on the reaction of a free thiol in a protein with DTNB to give the mixed disulfide and 2-nitro-5-thiobenzoate (TNB), which is quantified by the absorbance of the anion (TNB^{2-}) at 420 nm. The reagent has also been used to quantitate the number of disulfides by first blocking any free thiols with an alkylating agent (e.g., iodoacetic acid, iodoacetamide, and 4-vinylpyridine) and then reducing the protein (e.g., with DTT) prior to reaction with DTNB (Anderson and Wetlaufer 1975). For the analysis of free thiols/disulfide thiols, see Protocol 4.

Traditional methodology for determining disulfide linkages in proteins involves cleavage of the polypeptide chain with proteinases between half-cysteinyl residues to obtain peptides that are tethered by one disulfide bond. The resulting mixture of peptides is separated by RP-HPLC. Cysteine-linked peptides are identified in peptide maps (e.g., by comparison of nonreduced and reduced maps) by the difference in their RP-HPLC retention times. The retention times of nondisulfide-containing peptides will be the same in both nonreducing and reducing peptide maps, whereas under reducing conditions, disulfide bridges will be broken with the resultant loss of bridged peptides in the reduced peptide map and concomitant appearance of their reduced peptide components. Typically, fragmentation is performed under low pH conditions (e.g., pepsin at pH 2–3) to minimize disulfide exchange. However, the native protein is first alkylated (e.g., with 4-vinylpyridine or iodoactetate) (for a list of commonly used alkylating agents, see Table 7.2) to prevent possible disulfide exchange. Fragmentation can then be performed using a specific proteinase(s) that requires basic conditions (e.g., trypsin and Glu-C). Disulfide-containing peptides can be identified by their sequences using classical Edman chemistry or tandem mass spectrometry, or molecular masses. For an example of the classical approach, see Figure 7.7, which outlines the strategy involved in the determination of the disulfide connectivities of interleukin-6 (IL-6), a 20-kD glycoprotein with two disulfides (Simpson et al. 1988). A more recent strategy, employing both Edman chemistry and mass spectrometry, for the characterization (free thiol, disulfide linkages, and *N*-glycosylation sites) of the membrane-associated IL-6 receptor components, IL-6 receptor, and gp130 is outlined in Cole et al. (1999) and Moritz et al. (2001).

When cysteines are well dispersed in the polypeptide chain (as was the case with the example of IL-6 given in Figure 7.7), a single proteinase or multistep enzymatic or chemical digestions may be used to obtain a full set of diagnostic peptide fragments that will allow the determination of their connectivities. However, when cysteines are adjacent or tightly clustered, it is almost impossible to obtain peptides containing a single disulfide bond. In 1993, Gray devised a strategy for analyzing disulfide patterns in highly bridged small peptides (Gray 1993). In this strategy, peptides are partially reduced with the water-soluble reducing agent TCEP at pH 3.0, the nascent free thiols are alkylated with iodoacetamide, and their positions are identified from sequence analysis. A strategy to locate free thiols in proteins using 2-nitro-5-thiocyanobenzoic acid (NTCB) (Jacobson et al. 1973) has been described recently (Wu and Watson 1997). NTCB selectively cyanylates cysteine thiols. The amino-terminal peptide bond of cysteines modified with NTCB can be readily cleaved under alkaline conditions to form an amino-terminal peptide and a series of 2-iminothiazolidone-4-carboxyl peptides (Jacobson et al. 1973) that can be mass mapped using matrix-assisted laser desorption/ionization time-of-flight mass spectrometry (MALDI-TOF) to the sequence of the original molecule (Wu and Watson 1997). A variation on this approach using the cyanylating agent, 1-cyano-4-dimethylamino-pyridium tetrafluoroborate (CDAP), has been described for assigning disulfide pairings in proteins (Wu and Watson 1997).

PEPTIDE MAPPING USING RP-HPLC

The fractionation of peptide mixtures to yield homogeneous peptides suitable for sequence analysis (especially for the Edman degradation procedure) usually requires a number of chromatographic steps employing various columns as well as solvent-mediated selective steps. For a review of reversed-phase peptide separation procedures, see Simpson et al. (1989), and references therein. A recommended initial separation is

- performed on a microbore Brownlee RP-300 (2.1 mm I.D. x 30 mm, Perkin-Elmer) or Vydac C4 (Vydac, Hesperio, California) reversed-phase column, or

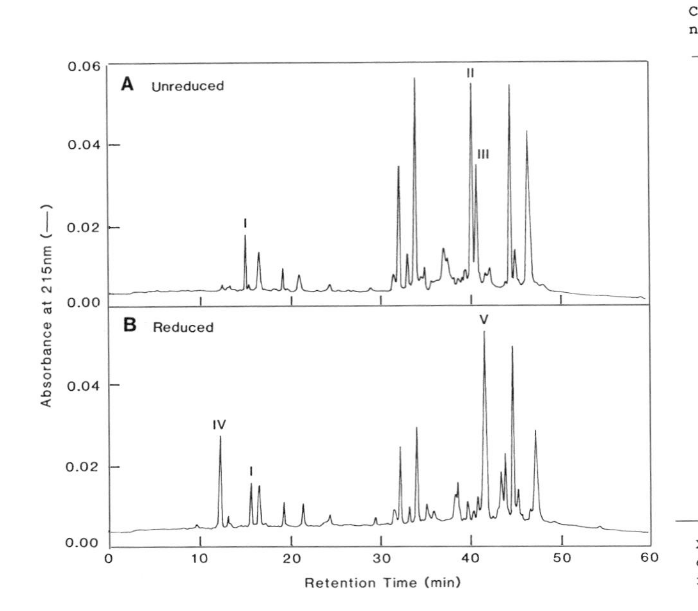

Cycle number	Pth-Xaa (pmol) Fraction I	Fraction II	Fraction III
1	Leu (280)	Ile (133)	Ile (46)
2	X	Gln (36)	Gln (26)
3	Asn (187)	Arg (36), Leu (30)	Arg (16)
4	Gly (115)	Asn (61), Leu (54)	Asn (31)
5	Asn (104)	Asp (80), Lys (39)	Asp (27)
6	Ser (10)	Gly (50), Ile (26)	Gly (41)
7	Asp (42)	Ser (10), X[a]	X
8	X[a]	Ser (10), Tyr (30)	Tyr (20)
9	Met (76)	Gln (24), Gly (14)	Gln (19)
10	Asn (63)	Leu (13), Thr (2)	Thr (7)
11	Asn (105)	Leu (20), Gly (10)	Gly (28)
12	Asp (38)	Tyr (15), Glu (8)	Tyr (18)
13	Asp (53)	Asn (15), Tyr (7)	Asn (16)
14	Ala (32)	Gln (7), His (5)	Gln (12)
15	Leu (38)	Glu (4), Ser (1)	Glu (11)
16	Ala (24)	Tyr (3)	Ile (11)
17	Glu (5)	Leu (5)	X
18			Leu (11)
19			Leu (13)
20			Lys (5)
21			Ile (6)
22			Ser (1)
23			Ser (1)
24			Gly (2)
25			Leu (2)

X, unidentified Pth-amino acid derivative. [a]The di-Pth derivitive of cystine (eluted in tyrosine position) was identified in cycle 8 of fraction I and cycle 7 of fraction II.

E L C N G N S D C M N N D D A L A E N N L K L P E I Q R N D G C Y Q T S Y N Q E I C L L K

45 50 55 60 65 70 75 80 85

Peptide fraction	Structural data	Disulfide bond Cys-Cys
I	S4 LXNGNSDCMNDDALAE	46-52
II	S1 IQRNDGCYQTGYNQE	75-85
	S2 IXLLKISSGLLEY	
III	S3 IQRNDGXYQTSYNQEIXLLKISSGLL	75-85

Sequences indicate the results of Edman degradation of unreduced peptide fractions. X denotes a Cys residue expected from the known sequence but not observed; C denotes the di-Pth derivative of cystine.

FIGURE 7.7. Disulfide assignment of IL-6 using a combination of peptide mapping and Edman chemistry. (*Top left panel*) Recombinant mIL-6 protein was digested with *Staphylococcus aureus* V8 protease, and 10 µg of peptides was loaded onto a Brownlee RP-300 column (100 x 4.6 mm I.D., 7 µm dp). (*A*) Unreduced *S. aureus* digest of rmIL-6 (10 µg of peptide) loaded directly. (*B*) Prior to loading, *S. aureus* digest of rmIL-6 (10 µg of peptide) was treated with 100 mM DTT. Peaks (400–600 µl) were collected manually in microfuge tubes and stored at –20°C. For sequence analysis, peptides were concentrated to 40–60 µl by microbore RP-HPLC. (*Top right panel*) Sequence analysis of peptide fractions I, II, and III from the peptide map shown in the top left panel. (*Bottom panel*) Summary of sequence data and disulfide bridge identifications. (Reprinted, with permission, from Simpson et al. 1988b.)

- using a trifluoroacetic acid (TFA)/acetonitrile gradient system composed of Solvent A (0.1% v/v aqueous TFA) and Solvent B (60% acetonitrile containing 0.085% v/v aqueous TFA), which is usually developed with a linear 60-minute gradient from 0% to 100% Solvent B at a flow rate of 100 µl/min.

For additional information on choosing the appropriate column and flow rate, see Chapter 5, Figures 5.4 and 5.5 (these two figures cover the principles of varying column internal diameters/flow rates and a practical example). Peptides are usually detected in the column eluate by their absorbance at UV_{214nm}. Aromatic amino-acid-containing peptides (particularly peptides containing tryptophan and tyrosine), as well as *S*-pyridylethylated-cysteine-containing peptides, are readily identified by their characteristic UV spectra using photodiode-array detection (Simpson et al. 1989; Moritz et al. 1996). Peptide fractions should be collected in 1.5-ml tubes according to UV absorbance at 214 nm and immediately capped to prevent loss of organic solvent. To prevent sample loss during long-term storage (e.g., irreversible sticking of peptide to the tube walls), the addition of Tween-20 (final concentration, 0.01–0.02% w/v) to peptide fractions is recommended. This is especially important when working with low-picomole amounts of sample. To reduce contaminant levels that interfere with RP-HPLC and mass spectrometry, it is recommended that extremely clean tubes be used (e.g., microfuge tubes). For high-sensitivity work, the tubes should be rinsed with Solvent B prior to usage. A standard protocol for the RP-HPLC of peptides is given in Protocol 15.

In situations where peptide fractions are clearly impure (as judged from peak shape and/or UV spectrum across the peak), samples are rechromatographed using a separation system with a different selectivity (Simpson et al. 1989). Solvent systems that can be used for the second chromatographic system include

- aqueous 1% (w/v) NaCl/acetonitrile,
- aqueous 0.1% (v/v) TFA/methanol, and
- 50 mM (w/v) sodium phosphate/acetonitrile.

These solvent systems are compatible with most Edman chemistry sequencers. To prevent sample loss, peptide fractions should not be dried between chromatographic steps. Because peptide fractions usually contain sufficient quantities of secondary solvent to prevent their retention on a similar interactive support, it is necessary to dilute samples (one- to twofold) with the primary solvent to facilitate their retention (trace enrichment) onto the reversed-phase column used in the subsequent chromatographic step. Dilution can be readily accomplished in a large-volume (1.5 ml) sample-loading syringe immediately prior to injection. An illustration of the selective effects achieved with various mobile-phase conditions, using IL-6 peptide mapping as an example, is given in Figure 7.8.

In the example shown in Figure 7.8, IL-6 (15 µg) was *S*-carboxymethylated before enzymatic digestion. *S*-carboxymethyl (Cm) IL-6 was recovered from the reaction mixture by RP-HPLC using a microbore column (50 x 1-mm I.D.). Before committing the bulk of the sample to this desalting procedure, the chromatographic recovery of Cm-IL-6 was evaluated using an analytical microbore-column step (Figure 7.8A,B). This pilot desalting step is extremely important because some proteins, upon *S*-carboxymethylation, exhibit different chromatographic behaviors (e.g., are more hydrophobic and hence retained more strongly) than their native form. Most of the IL-6 sequence was established by analysis of peptides by treatment of Cm-IL-6 with trypsin, chymotrypsin, V8 protease, and cyanogen bromide. A peptide map, using the conventional TFA/acetonitrile solvent system, of a tryptic digest of Cm-IL-6 is shown in Figure 7.8C. Peptides were fractionated in a second chromatographic dimension, using an unbuffered saline/acetonitrile solvent system, to achieve homogeneity (Figure 7.8D). If the peptides were still not homogeneous (as judged from peak shape or spectral analysis

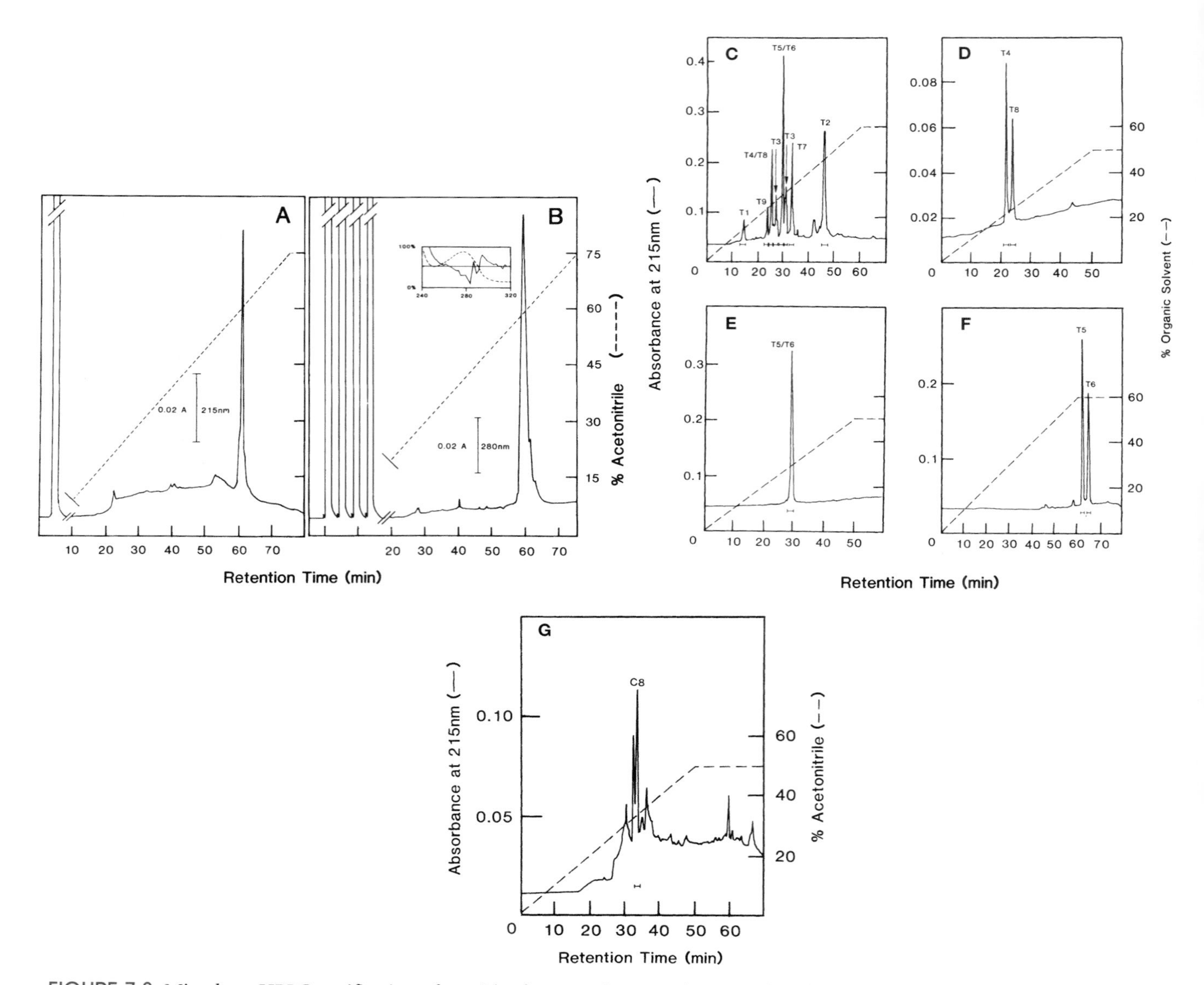

FIGURE 7.8. Microbore HPLC purification of peptides from murine IL-6. (Reprinted, with permission, from Simpson et al. 1988a.)
(*A*) Chromatographic conditions: column, Brownlee RP-300 (50 x 1-mm I.D.); 5 µl of the reduction/alkylation reaction mixture (containing 300 ng of Cm-IL-6, see Protocol 3) was loaded directly at a flow rate of 0.5 ml/min onto the column that had been equilibrated in aqueous 0.1% (by volume) TFA. The flow rate was decreased to 40 µl/min, and after the baseline had stabilized (~10 minutes) the column was developed with a linear 75-minute gradient of 0–100% B where Solvent A was aqueous 0.1% TFA and Solvent B was 75% (by volume), acetonitrile/25% H_2O containing 0.085% TFA. The column temperature was 45°C. Peaks were collected manually.
(*B*) The remainder of the reaction mixture (245 µl containing 14 µg of Cm-IL-6) was loaded in 60-µl aliquots onto the same column using identical chromatographic conditions to those described in *A*. Real-time spectral data of eluting peaks were obtained using a diode array detector (Hewlett-Packard Model 1040). (*Inset*) Absorption spectrum of Cm-IL-6: vertical axis, absorbance normalized to relative absorbance on a scale 0–100%; (- - - -) zero-order derivative spectrum; (——) second-order derivative spectrum. The extremum at 290 nm is characteristic of a tryptophan residue.
(*C*) Separation of tryptic peptides of Cm-IL-6 by microbore RP-HPLC. Chromatographic conditions: column, Brownlee RP-300 (30-nm pore size, 7-µm dimethyloctylsilica packed into a column (30 x 2.1-mm I.D.): linear 60-minute gradient of 0–100% B, where Solvent A was 0.1% (by volume) aqueous TFA and Solvent B was 60% actonitrile/40% H_2O containing 0.09% (by volume) TFA. Flow rate was 100 µl/minute; column temperature was 45°C. Sample, 8 µg (370 pmoles) in 1050 µl, was loaded at 1 ml/min. All peaks (~60–100 µl) were collected manually and stored at –20°C. Pooling of peptide fractions is indicated by horizontal bars (–). Peptides found in each pool (T4, T5, T6, etc.) have been labeled in accordance with their position in the sequence as proposed by Simpson et al. (1988a).
(*Continued on facing page.*)

using diode array detection; Grego et al. 1985), then a third chromatographic dimension using a different packing (e.g., ODS-Hypersil) and organic mobile-phase component (e.g., methanol) was used (Figure 7.8E,F). To isolate short, highly charged peptides that are not usually retained on conventional reversed-phase chromatographic systems and elute with the column breakthrough, the ion-pairing agent sodium hexylsulfonate was used (Figure 7.8G).

Ion pair reagents are strong hydrophobic ions that form neutral ion pairs with oppositely charged sample molecules (see Chapter 5). By these means, the simultaneous separation of charged and noncharged molecules is possible. For further information, see the Merck Web Site at http:www.merck.de/english/services/chromatographie/reagents/lichropu.htm.

In the example shown, unretained peptides from a first chromatographic separation of a chymotryptic digest of Cm-IL-6 (Brownlee RP-300 column, 30-nm pore size/TFA-acetonitrile system; for details, see Figure 7.8B) were chromatographed on an ODS-Hypersil column employing 0.02 M sodium phosphate (pH 2.5) and 0.03 M sodium hexylsulfonate in the mobile phase. Peptides recovered by these means (e.g., chymotryptic peptide C8 Arg-Ser-Thr-Arg-Gln-Thr) were directly sequenced without the need for desalting.

For peptide mapping of trace-abundant proteins, we recommend the use of microbore columns (≤2.1 mm I.D.) or capillary columns (<0.5-mm I.D.) (see Chapter 8). Microbore columns can be operated on conventional HPLC systems that can deliver accurate low flow rates (~50 μl/min) (Simpson et al. 1989). In addition, commercial HPLC systems designed for their usage (e.g., the Ultimate LC pump, LC Packings Inc., and the Agilent Cap1100LC system) are now available. However, conventional HPLC systems can be readily (and economically) modified to provide the accurate low flow rates (0.4–4 μl/min) and gradients necessary to operate capillary columns (Moritz and Simpson 1992a,b, 1993; Moritz et al. 1994; Tong et al. 1997). Such columns are essential for the preparation of samples for a number of mass spectrometric methods for peptide identification. Detailed protocols for the facile fabrication of <0.32-mm I.D. polyimide-coated fused-silica columns, with a detection limit of ~500 pg amounts of protein, are described by Moritz et al. (1994) and Tong et al. (1997). A method for packing capillary columns using a conventional column-packing device is given in Chapter 5, Protocol 2.

FIGURE 7.8. (*Continued from facing page.*)

(*D*) Rechromatography of peak fraction T4/T8 from *A* by RP-HPLC. Peptides T4 and T8 were resolved by chromatography using the same chromatographic conditions described in *A*, but using unbuffered sodium chloride as the mobile phase. Peak fraction T4/T8 (~100 μl), was diluted twofold with H_2O and then applied to the column that had been previously equilibrated with 1% (mass/volume) aqueous sodium chloride. The column was developed with a linear 60-minute gradient of 0–50% B, where Solvent A was 1% aqueous sodium chloride and Solvent B was acetonitrile. Flow rate was 100 μl/min; column temperature was 45°C. Pooling of fractions is indicated by horizontal bars (–).

(*E*) Rechromatography of peak fraction T5/T6 from *A* by RP-HPLC. Chromatographic conditions were the same as those described in *B*. Pooling of fractions is indicated by a horizontal bar (–).

(*F*) Rechromatography of peak fraction T5/T6 from *C* by RP-HPLC. Chromatographic conditions: column, ODS-Hypersil 12-nm pore size, 5-μm particle diameter octadecylsilica packed into a stainless steel column (50 x 1-mm I.D.); the peptide fraction T5/T6 from *C* was diluted twofold with H_2O and applied at a flow rate of 200 μl/min to the column that had been previously equilibrated with 0.1% (by volume) TFA. The column was developed at a flow rate of 40 μl/min, with a linear 60-minute gradient of 0–100% *B*, where Solvent A was aqueous 0.1% TFA and Solvent B was 60% methanol/40% H_2O containing 0.09% TFA. Pooling of fractions is indicated by horizontal bars (–).

(*G*) Ion-pair chromatography of nonretained chymotryptic peptides of Cm-IL-6. Chromatographic conditions: column, ODS-Hypersil 12-nm pore size, 5-μm particle size octadecylsilica packed into a stainless steel column (50 x 1-mm I.D.); linear 50-minute gradient from 0% to 100% B, where Solvent A was aqueous 0.02 M sodium phosphate, 0.1% (by volume) phosphoric acid (pH 2.5), and 0.03 M sodium hexylsulfonate and Solvent B was 50% acetonitrile in Solvent A. Flow rate was 40 μl/min; column temperature was 45°C. Sample load: eluent breakthrough (~2 ml) was diluted twofold with primary Solvent A and applied to the column in 400-μl aliquots at a flow rate of 1 ml/min.

PROTOCOL 1

Performic Acid Oxidation of Proteins

It is customary to cleave disulfide bonds as a prerequisite to many peptide-mapping strategies. This will facilitate protein unfolding and thus optimize proteolytic digestion. In addition, cleavage of disulfide bonds simplifies the interpretation of peptide maps by removing possible peptide fragments(s) in the mixture that are held together by disulfide bonds. Described in this protocol is the performic acid oxidation method for cleaving disulfide bonds, originally described by Sanger (1949) in his pioneering work on the structure of insulin.

Performic acid is an extremely powerful oxidizing reagent that oxidizes tryptophan, the sulfur-containing amino acids (cysteine, cystine, and methionine), the phenolic group of tyrosine, and the hydroxyl groups of serine and threonine. Of these reactions, the most rapid is the conversion of cysteine to cysteic acid and of methionine to the sulfoxides, followed by the oxidation of cystine to cysteic acid and methionine sulfoxide to methionine sulfone. Somewhat slower is the oxidation of tryptophan to several derivatives including *N*-formylkynurenine. Conversion of cystine (disulfide bonds) and cysteines to cysteic acid is ~92%, and oxidation of methionine to the sulfone is essentially quantitative. Because cysteic acid and methionine sulfone can be analyzed quantitatively using the ninhydrin detection method of amino acid analysis (CHP quantitation), perfomic acid oxidation provides a convenient determination of the combined cysteine and cystine content and the total methionine content of a protein.

Performic acid oxidation, however, is not a recommended procedure for cleaving disulfide bonds for general peptide-mapping strategies due to its significant limitations, especially the loss of tryptophan and modifications to tyrosine, serine, and threonine. The preferred method for cleaving disulfide bonds for general peptide mapping is by reduction (see Protocols 2 and 3). For further details of the performic acid oxidation procedure, see Hirs (1967), from which this protocol was adapted.

MATERIALS

CAUTION: See Appendix 3 for appropriate handling of materials marked with <!>.

Reagents

Formic acid (88% w/v) <!>

Hydrobromic acid (48% w/v) <!>

WARNING: Because it gives off a caustic, irritating vapor, use this acid in a chemical fume hood.

Hydrogen peroxide (30% w/v) <!>

NaOH pellets (see Step 6) <!>

Equipment

Evaporator, rotatory (SpeedVac or equivalent) or Lyophilizer, centrifugal
Tubes (glass or polypropylene, capped)

Biological Sample

Purified protein of interest, lyophilized

METHOD

1. Add 100 µl of hydrogen peroxide to 900 µl of formic acid. Allow the mixture to stand for 1 hour at room temperature. This reaction produces performic acid (HCOOOH).

2. Chill the performic acid on ice to ~0°C.

   During performic acid oxidation, the temperature of all the reactants should be at 0°C to minimize side reactions such as oxidation of phenolic groups (tyrosine and phenylalanine) and the hydroxyl groups of serine and threonine.

3. Dissolve the protein in 50 µl of performic acid (~500 µg/ml) in a precooled tube.

4. Place the cap on the tube and incubate the mixture for ~4 hours at 0°C. If the protein is insoluble, allow the reaction to proceed for up to 20 hours.

5. Add 30 µl of cold hydrobromic acid to destroy the excess performic acid.

   Care must be taken since a small quantity of bromine will be liberated.

6. Dry the sample thoroughly using a rotatory evaporator or centrifugal lyophilizer. Removal of residual bromine and formic acid is facilitated by the addition of a few pellets of sodium hydroxide in the condenser. Evaporation to dryness takes ~10 minutes.

   Because many amino acids are modified, it is recommended that performic-acid-oxidized proteins be used for the estimation of S-S/S-H (as cysteic acid) and methionine (as methionine sulfone) groups by amino acid analysis (but not sequence analysis).

PROTOCOL 2

Reduction and *S*-carboxymethylation of Proteins: Large-scale Method

INTACT INTERCHAIN AND/OR INTRACHAIN DISULFIDE LINKAGE in a protein can present problems during proteolytic or chemical fragmentation procedures.

- Disulfide-linked peptides elute as single peaks during RP-HPLC-based peptide mapping, making the interpretation of such maps difficult.
- It is desirable to separate peptide chains linked through -S–S- bridges prior to sequence work, otherwise the task of interpreting the sequence data may be too complicated.
- Proteins whose disulfide bonds are not split are less susceptible to proteolytic fragmentation.

The most commonly used method for cleaving disulfide bonds involves the reduction of cystine to yield cysteine residues. However, cysteine residues are highly reactive, which can cause complications (e.g., random disulfide bond formation and susceptibility to oxidation upon removal of the reducing agent) during sequence work. For this reason, it is customary to stabilize them as suitable derivatives (e.g., by alkylation).

In this protocol, the intact protein (>1 mg) is reduced and then *S*-alkylated with iodoacetic acid (or iodoacetamide). The resulting cysteine derivative *S*-carboxymethylcysteine (or *S*-carboximadomethylcysteine) is easily detectable during chemical sequencing. Another frequently used alkylating agent is 4-vinylpyridine, which yields 4-pyridylethylcysteine (Raftery and Cole 1966). 4-Pyridylethylcysteine is readily detected in the Edman degradation process as its PTH derivative, as well as by mass-spectrometry-based sequencing methods. For microscale (<100 µg of protein) reduction and *S*-alkylation procedures, see Protocol 3.

MATERIALS

CAUTION: See Appendix 3 for appropriate handling of materials marked with <!>.

Reagents

Acetic acid (1 M) (ultrapure grade) <!>

Alkylation buffer

Alkylation buffer is 1.5 M Tris-HCl <!> (pH 8.5) containing 2.5 mM EDTA and 6 M guanidine-HCl <!>. Urea <!> (8 M) can be substituted for guanidine-HCl. Stock solutions of reduction buffer made with urea may be stored frozen for up to 2 weeks prior to use.

NH_4HCO_3 (0.1 M or 1% w/v, Analar grade) <!>

Ammonium bicarbonate (1% w/v) gives a pH of 7.8, and pH adjustment is not necessary.

Dithiothreitol (DTT) <!>

Guanidine hydrochloride, solid (Mallinckrodt) or aqueous 8 M solution (Pierce)

HCl, concentrated <!>

Iodoacetic acid <!> or Iodoacetamide <!>

The iodoacetic acid used must be colorless. Any free iodine (revealed by yellow coloration of the material) causes oxidation of cysteine residues (thereby, preventing alkylation) and possibly tyrosine residues. High-quality iodoacetic acid can be obtained from Fluka (Buchs, Switzerland). If the iodoacetic acid is not available or if it is not colorless, recrystallize it from hexane before use. 4-Vinylpyridine may be substituted for the iodoacetic acid or iodoacetamide (see the panel at the end of the protocol).

2-Mercaptoethanol <!>

N′,*N*-Ethylmorpholine acetate buffer (0.2 M, pH 8.0) <!>

Add 230 µl of *N*-ethylmorpholine to ~8.0 ml of H_2O. Add acetic acid to give a pH of 8.0. Adjust the volume to a total of 10 ml.

Reduction buffer

Reduction buffer is 0.2 M Tris-HCl (pH 8.5) containing 2.5 mM EDTA and 6 M guanidine-HCl. Urea (8 M) can be substituted for guanidine-HCl. Stock solutions of reduction buffer made with urea may be stored frozen for up to 2 weeks prior to use.

Equipment

Chromatography system and size-exclusion column

For example, Sephadex G-10, G-25 (medium), G-50 SF, or Bio-Gel P-10 size-exclusion resin, using 1 M acetic acid to develop the column.

Nitrogen <!> (purified grade) or Argon <!>

pH indicator paper

Polypropylene tube (tightly capped)

Tygon tubing (see Step 3)

Biological Sample

Protein sample, purified (lyophilized)

METHOD

Sample Reduction

1. Dissolve the protein in reduction buffer (~10 mg protein/ml buffer) in a polypropylene tube.

2. Add an accurately weighed amount of solid DTT to give a 60-fold molar excess (i.e., approximately half the weight of the protein) or to give a final concentration of 0.1 M.

3. Gently blow a stream of nitrogen (or argon) over the top of the solution for 30 seconds.

   This can be accomplished by attaching a Pasteur pipette to a length of Tygon tubing connected to the nitrogen cylinder regulator, and adjusting the nitrogen flow such that it can be barely detected on the back of the hand.

4. Incubate the protein solution for 3 hours at 37°C.

5. If desired, the number of –SH groups can be quantitated using the Ellman procedure (see Protocol 4).

Sample Alkylation

In some cases, reduced and alkylated proteins can be rapidly desalted and recovered in a small volume of volatile solvent, by RP-HPLC. This alternative procedure is particularly attractive for low-M_r protein (15–25K) such as growth factors and cytokines. Desalting (and sample concentration) by RP-HPLC (TFA/acetonitrile solvent system) has been successfully employed for low-microgram amounts of material (see Protocol 3).

6. Cool the reduced protein substrate solution on ice.

7. Add iodoacetic acid (approximately the same weight as the protein substrate) freshly dissolved in a small volume (~0.2 ml) of alkylation buffer. Alternatively, use a small aliquot from a stock solution of sodium iodoacetic acid (15 mg of iodoacetic acid per 100 µl of 0.1 M NaOH) to give a slight molar excess over the total SH groups in the mixture (i.e., protein SH groups plus DTT).

8. Mix the contents of the tube by gently tapping the side of the tube.

9. Flush the reaction tube with nitrogen for 30 seconds.

10. Incubate the reaction for 15 minutes at room temperature in the absence of light.

    The reaction is carried out in the dark to prevent the formation of free iodine from the breakdown of iodoacetic acid.

11. Add a molar excess of 2-mercaptoethanol (~50 µl per 5 mg of DTT) to "mop up" the excess alkylating reagent and thus stop the reaction.

12. Adjust the pH of the solution to 2–3 by the careful addition of concentrated HCl. Check the pH by spotting ~0.5 µl of the reaction mixture onto a strip of pH indicator paper.

    A pilot study is recommended to check the solubility of the alkylated protein (which may differ significantly from that of the native protein). Alternative solutions and buffer salts that are volatile include 1 M acetic acid (pH 2.1), 1% (w/v) ammonium bicarbonate (pH 7.8), and 0.2 M *N*-ethylmorpholine acetate buffer (pH 8.0). Protein can be readily recovered from these solvents and buffer salt solutions by lyophilization.

Recovery of *S*-carboxymethyl Protein (Desalting)

Alternatively, reduced and alkylated protein can be desalted by extensive dialysis against a volatile solution (e.g., 1 M acetic acid) or buffer salt solution (e.g., 1% ammonium bicarbonate). However, this approach is time-consuming (typically, 24 hours with several changes of dialysis solution are essential) and high losses can occur. Following dialysis, protein can be recovered by lyophilization.

13. To remove the excess reagents, rapidly desalt the protein mixture by size-exclusion chromatography.

14. Recover the reduced and alkylated protein from the 1 M acetic acid eluent as a dried powder using lyophilization.

Alternatively, reduced protein can be alkylated with 4-vinylpyridine. This is accomplished by treating the reduced protein substrate for 90 minutes with 3.0 moles of 4-vinylpyridine per 1 mole of DTT at room temperature. Desalting of the *S*-pyridylethyl protein is performed as described above for *S*-carboxymethylated protein. For details, see Step 12.

PROTOCOL 3

Reduction and *S*-carboxymethylation of Proteins: Microscale Method

REDUCTION AND ALKYLATION OF MICROSCALE AMOUNTS OF PROTEIN (10–50 μg) require special attention to avoid losses (especially those due to the sample sticking irreversibly to the sides of the sample vessel). Adding a small quantity of detergent (e.g., 0.02% Tween-20) to the protein solution during all manipulations can prevent such losses. Classical methods for desalting (e.g., size-exclusion chromatography) and dialysis should be avoided due to large sample losses.

In this protocol, a procedure for reduction and *S*-carboxymethylation of low-microgram amounts of protein is described. Samples are reduced and alkylated in small volumes (<200 μl) in the presence of 0.02% Tween-20 and desalted (and concentrated) by RP-HPLC in yields of >90%. Avoid using Tween-20 if the subsequent step involves mass spectrometry (MS) because Tween-20 tends to stick to the RP-HPLC supports and leach off the column resulting in artifact ion peaks in the MS spectra. This procedure has been successfully applied to the complete amino acid sequence determination of the glycoproteins IL-6 and IL-9 (Simpson et al. 1988, 1989).

MATERIALS

CAUTION: See Appendix 3 for appropriate handling of materials marked with <!>.

Reagents

Acetonitrile <!>

Iodoacetic acid <!>

Prepare a 0.5 M solution of iodoacetic acid (Puriss grade from Fluka [Buchs, Switzerland] recrystallized prior to use) in 1 M NaOH <!> immediately before use.

Store iodoacetic acid dry by placing the bottle in a plastic container containing silica gel in the dark at –20°C. Allow the bottle to come to room temperature before opening. With time, solid iodoacetic acid will develop a yellow coloration (free iodine) and should be recrystallized.

n-Propanol <!>

Reduction buffer

Stock buffer adjusted to 0.15 M DTT <!> (prepared fresh immediately before use).

Stock buffer

0.2 M Tris-HCl (pH 8.3) containing 2 mM EDTA, 6 M guanidine-HCl <!>, and 0.02% (w/v) Tween-20. Mix 1 volume of 1 M Tris-HCl (pH 8.3) and 5 volumes of aqueous 8 M guanidine-HCl (solution provided by Pierce); adjust to 2 mM EDTA.

Trifluoroacetic acid (TFA) <!>

Equipment

Centrifuge, bench-top

HPLC system microbore reversed-phase column

For example, Brownlee RP-300 30-nm pore size, 7-μm particle diameter, octyl silica packed into a stainless steel column, either a 50 x 1-mm I.D. or 30 x 2.1-mm I.D. (PE Brownlee). For basic RP-HPLC protocols, see Chapter 5.

Lyophilizer, centrifugal (Thermo Savant)

Polypropylene tubes (1.5-ml, capped)

Water bath or incubator preset to 40°C

Biological Sample

Pure protein samples (~10-15 μg)

Two cytokines are mentioned in this protocol (IL-6 and IL-9) because different strategies were used for the last stage of their purification, prior to reduction/alkylation. The last stage of purification for IL-6 (32 μg) involved immunoaffinity chromatography and the protein was recovered from the column in ~130 μl using 1 M acetic acid <!> containing 0.01% (w/v) Tween-20 (Simpson et al. 1988). For IL-9, a reversed-phase column was used and pure IL-9 (15 μg) was recovered in 120 μl of 35% aqueous acetonitrile <!> containing 0.1% (by volume) TFA <!> (the active fraction was immediately adjusted to a final concentration of 0.02% Tween-20) (Simpson et al. 1989).

METHOD

Sample Reduction

1. Evaporate solutions of purified proteins (containing 0.01–0.02% Tween-20) to near dryness (~10 μl) by centrifugal lyophilization.
2. Add 150 μl of stock buffer to the microfuge tube containing the "near" dried protein sample. Mix thoroughly by gently tapping the microfuge tube or by vortexing it and then briefly centrifuging the sample in a microfuge.
3. Add 16 μl of reduction buffer to give a final concentration of 15 mM DTT.
4. Incubate the sample for 4.5–5 hours at 40°C.

Sample Alkylation

5. Allow the reduced protein solution to come to room temperature.
6. Add 20 μl of stock 0.5 M iodoacetic acid solution to give a final concentration of ~0.05 M iodoacetic acid.
7. Incubate the reaction for 30 minutes at 25°C in the dark.

S-carboxymethyl Protein Recovery (Desalting/Concentration)

8. Inject 2–3% (~300 ng of protein) of the total mixture onto a reversed-phase microbore column (50 x 1-mm I.D. HPLC system (see Protocols 1 and 2 in Chapter 5).

   Before committing the bulk of the *S*-carboxymethylated proteins to RP-HPLC, conduct a pilot experiment using 2–3% of the total sample (~300 ng) to ascertain whether the material can be recovered from the reversed-phase column. This pilot desalting step is extremely important because some proteins display different chromatographic behaviors upon *S*-carboxymethylation (e.g., they are more hydrophobic than the native protein). If the relative hydrophobicity of an *S*-carboxymethylated protein is significantly higher than that of the native protein, it may be difficult to recover from the reversed-phase column using a TFA/acetonitrile solvent system. In this case, try substituting the acetonitrile with *n*-propanol (see Chapter 5, Protocol 5). If this does not work, try desalting the *S*-carboxymethylated protein using micro-size exclusion chromatography or fast micro-desalting.

9. Once it is established that the *S*-carboxymethylated protein can be recovered from the reversed-phase column, inject the remainder of the material.

10. Adjust the *S*-carboxymethyl protein-containing fraction (typically, in 300–500 µl of aqueous 0.1% TFA/acetonitrile) to 0.02% (w/v) with respect to Tween-20. For proteolytic digestion of *S*-carboxymethylated proteins collected in this manner, proceed to Protocol 5.

PROTOCOL 4

Estimation of Free Thiols and Disulfide Bonds Using Ellman's Reagent

IN THIS PROTOCOL, A METHOD FOR QUANTITATING THE AMOUNT OF FREE THIOLS and disulfide bond linkages in proteins is described. The method uses the sulfhydryl reagent 5,5′-dithiobis (2-nitrobenzoic acid), which is also referred to as the Ellman reagent after its originator (Ellman 1959). For a review outlining the use of Ellman's reagent for determination of total protein sulfhydryl, and the amount of sulfhydryl groups in tissues, see Habeeb (1972).

MATERIALS

CAUTION: See Appendix 3 for appropriate handling of materials marked with <!>.

Reagents

Denaturing buffer
- 6 M guanidine hydrochloride <!>
- 0.1 M Na_2HPO_4 (pH 8.0) <!>

Dithiothreitol (DTT) (200 mm) in deionized H_2O (Calbiochem or Roche) <!>

Ellman's reagent (DTNB)
- Prepare 10 mM DTNB (4 mg/ml) in 0.1 M sodium phosphate buffer (pH 8.0). <!>

Iodoacetic acid <!>

Nitrogen <!>

Reaction buffer (0.1 M sodium phosphate buffer, pH 8.0)

4-Vinylpyridine <!>

Equipment

Dialysis equipment
- For further information concerning dialysis membranes and equipment, see http://www.spectra-por.com.

Size-exclusion chromatography equipment

Spectrophotometer and cuvettes

Biological Sample

Protein sample
- The sample should contain an accurate amount of protein (for protein estimation protocols, see Appendix 2)—at least 2 nmoles in 100 µl of reaction buffer or denaturing buffer (e.g., buried thiols may need to be exposed by denaturing the protein).

METHODS

Method 1: Free Thiols

1. Add 3 ml of reaction buffer to the sample and reference cuvettes.
2. Measure the absorbance at 412 nm (the absorbance should be adjusted to zero, A_{buffer}).
3. Add 100 µl of reaction buffer to the reference cuvette.
4. Add 100 µl of Ellman's reagent to the sample cuvette. Measure the absorbance at 412 nm (A_{DTNB}).
5. Add 100 µl of protein solution to the reference cuvette.
6. Add 100 µl of protein solution to the sample cuvette. Mix thoroughly. Measure the absorbance when there is no further increase (after 3–5 minutes). This is the A_{final}.
7. Calculate the concentration of thiols as follows: $\Delta A_{412\text{nm}} = E_{412}$ TNB^{2-} [RSH], where $\Delta A_{412} = A_{\text{final}} - (3.1/3.2)\ (A_{\text{DTNB}} - A_{\text{buffer}})$ and $E_{412\text{nm}}$ $TNB^{2-} = 1.415 \times 10^4$ cm^{-1} M^{-1} (for denaturing buffer, $E_{412\text{nm}}$ $TNB^{2-} = 1.37 \times 10^4$ cm^{-1} M^{-1}).

   The $E_{412\text{nm}}$ TNB^{2-} values given are from Riddles et al. (1983).

Method 2: Disulfide Thiols

This method is essentially the same as that described above in Method 1, except that the protein sample is first alkylated (without prior reduction) to derivatize any free thiols, but leaving the disulfide links intact. Alkylation can be performed using iodoacetic acid or 4-vinylpyridine (see Protocol 2).

1. Dissolve the sample in reaction buffer or denaturing buffer (at least 2 nmoles in 100 µl).
2. Add freshly prepared DTT (10–100 mM final concentration), and displace the oxygen with a stream of nitrogen. Incubate the reactions for 1–2 hours at 25°C.
3. Desalt the reduced sample by dialysis against the reaction buffer or by size-exclusion chromatography.
4. Measure the protein concentration (for basic protocols, see Appendix 2), and analyze the newly exposed thiols using the procedure described in Method 1 above.

PROTOCOL 5

Fragmentation of Protein Using Trypsin

IN THIS PROTOCOL, THE HIGHLY SPECIFIC PROTEASE TRYPSIN is used to hydrolyze a protein completely. Proteolysis is carried out with high levels of trypsin to ensure total proteolysis. (Trypsin is a robust enzyme; hence, proteolysis can also be carried out under denaturing conditions such as 2 M guanidine-HCl, 0.1% SDS, and >10% acetonitrile to ensure complete digestion.) Trypsin cleaves the peptide bond between the carboxyl group of arginine or the COOH group of lysine and the amino group of the adjacent amino acid. The rate of cleavage occurs more slowly when the lysine and arginine residues are adjacent to acidic amino acids in the sequence or cystine. Cleavage does not occur when lysine or arginine is followed by proline.

MATERIALS

CAUTION: See Appendix 3 for appropriate handling of materials marked with <!>.

Reagents

$CaCl_2$ <!>

NH_4HCO_3 (Analar grade) <!> (1% w/v) in deionized H_2O <!>

The pH of this solution is 7.8 and does not require further adjustment. The solution may be stored refrigerated or frozen for long periods of time.

Phenylmethylsulfonyl fluoride (PMSF) <!>

Prepare this reagent just prior to use by dissolving it in 1-propanol <!> to a final concentration of 1 M PMSF. Dilute PMSF 1000-fold to 1 mM in the reaction mixture. PMSF is an effective inhibitor of serine proteinases due to its high reactivity with serine residues at the active site of this class of enzymes.

Trypsin stock solution (1 mg/ml trypsin in 1 mM HCl, 20 mM $CaCl_2$)

Trypsin is available from various commercial suppliers. It is stable for years as a dry solid at –20°C. Most grades of commercially available trypsin (EC 3.4.21.4) are from bovine pancreas. It is recommended that TPCK-treated trypsin be used for peptide-mapping studies because this treatment specifically inhibits chymotrypsin, a common contaminant of trypsin preparations. Trypsin activity can vary from batch to batch. For reproducible peptide maps, the same batch of trypsin should be used throughout the studies.

Use fresh or divide the solution into small aliquots (~50 μl) and store frozen at –20°C. A stock solution may be thawed and refrozen a few times, but for consistent results, thaw only once.

Tween-20

Equipment

Polypropylene tube (tightly capped)
Water bath or Incubator capable of maintaining temperature to ±1°C

Biological Sample

Lyophilized protein of interest

METHOD

1. Dissolve the lyophilized protein substrate in 1% ammonium bicarbonate using the minimal volume necessary to achieve a high substrate concentration.

   When working with micro amounts of protein substrate (i.e., <1 mg), add Tween-20 to the ammonium bicarbonate to a final concentration of 0.1% (w/v), otherwise serious losses will occur due to the substrate (and enzyme) sticking irreversibly to the walls of the tube. Ammonium bicarbonate is a simple volatile buffer that can be readily removed by lyophilization. Nonvolatile buffers at pH 8 such as 0.05 M Tris-HCl (pH 8.0–8.3) may be readily substituted for ammonium bicarbonate.

2. Add trypsin to the substrate at a substrate:enzyme ratio of 200:1 to 50:1 for denatured (e.g., reduced and alkylated proteins), or as high as 1:1 for native protein substrates. For in-gel proteolysis, use 0.5–1.0 µg, regardless of the amount of substrate (typically, <0.5 µg for Coomassie-Blue-stained proteins).

3. Set up a parallel reaction mix lacking only the protein sample (and if there is sufficient material, a second reaction lacking only trypsin) as a control for RP-HPLC peptide mapping.

4. Incubate the reactions for 16 hours (overnight is convenient) at 37°C.

5. Stop the reaction by placing the sample onto an RP-HPLC column to begin peptide mapping, or chill the incubation mixture on ice to slow the reaction.

   Alternatively, stop the reaction by adding a specific inhibitor (e.g., *N*α-tosyl-L-lysyl chloromethyl ketone, or *N*-tosyl-L-phenylalanine chloromethyl ketone [TPCK], or PMSF) in molar excess to the trypsin used. Because trypsin is inactive at low pH values, the reaction can also be stopped by adjusting the pH of the reaction mixture to pH 2–3 (e.g., by the addition of volatile acids such as acetic acid or trifluoroacetic acid).

Troubleshooting Trypsin Digests

- Autodigestion of trypsin during incubation at neutral pH produces ψ-trypsin, among other products. The intrinsic chymotrypsin-like activity of ψ-trypsin (Keil-Dlouhá et al. 1971) explains the occasional "nontryptic"-type cleavage (i.e., at some tyrosyl, phenylalanyl, or tryptophanyl bonds) sometimes observed with tryptic digests. Thus, the rare nonspecific cleavages that can occur with trypsin, adjacent to aromatic or hydrophobic residues, are most likely due to the presence of small amounts of ψ-trypsin, and not to chymotrypsin contamination. The addition of low levels of Ca^{2+} (0.1 mM $CaCl_2$) to the incubation buffer is recommended to reduce trypsin autodigestion (and ψ-trypsin production).
- For optimal digestion, it is important that the protein substrate be as soluble as possible, or as evenly suspended as possible, in the digestion buffer. This can be accomplished by adjusting the protein substrate solution to 8 M urea and then diluting to 2 M urea for digestion with urea. Trypsin is a robust proteinase, being active in the presence of 2 M guanidine-HCl, 0.1% (w/v) SDS or in the organic solvent, acetonitrile (up to ~40% v/v) (see Table 7.6 and Welinder 1988). In the case of acetonitrile, low concentrations (e.g., <10% v/v) may facilitate more rapid proteolysis compared to that of buffer alone, whereas high concentrations (e.g., 50%) usually slow the reaction.
- The activity of trypsin in acetonitrile is an important attribute, because fractions from RP-HPLC (aqueous acetonitrile/0.1% v/v TFA system) may be readily digested after adding Tween-20 (to a final concentration of 0.02% w/v), reducing the volume to ~50 μl in a centrifugal vacuum concentrator (to lower the acetonitrile concentration), and simply diluting to 1 ml with 0.1% ammonium bicarbonate containing 1 mM $CaCl_2$ and 0.02% Tween-20 (Simpson et al. 1988, 1989; see also Protocols 2 and 3).
- Native protein substrates may be tightly folded, and thus they may need to be unfolded to facilitate proteolytic attack. This can be accomplished by reduction/alkylation, which unravels the protein, making it readily accessible to trypsin. Alternatively, denaturation of the protein substrate can be accomplished by boiling or with the use of urea, guanidine-HCl, SDS, or an organic solvent (see text above about optimal digestion).

Additional Protocol: Modification of Lysyl Side Chains Using Succinic Anhydride

The sites of trypsin cleavage can be restricted when the ε-amino group of lysine is blocked, for example, by succinic anhydride. This can be useful for obtaining overlapping peptides during classical protein sequencing strategies. For example, in selective peptide cleavage with trypsin, it may be desirable to block the ε-amino groups of lysine so that the enzyme attacks only at arginine peptide bonds. Another advantage of succinylation is that the reagent places substantial negative charge on the protein because the positively charged lysine side chain is replaced by a negatively charged carboxyl group of the succinate half-amide.

$$\text{Protein}-(NH_3^+)_n + \text{O}{<}\begin{matrix}\text{C(=O)}-\text{CH(R)} \\ | \\ \text{C(=O)}-\text{CH}_2\end{matrix}$$

$$\downarrow$$

$$\underset{(NH_3^+)_{n-y}}{\text{Protein}}-\left(\text{NH}-\overset{\text{O}}{\overset{\|}{\text{C}}}-\overset{\text{R}}{\overset{|}{\text{CH}}}-\text{CH}_2-\text{C}{\begin{matrix}\text{//O}\\ \text{\\O}^-\end{matrix}}\right)_y + 2y\,\text{H}^+$$

Hence, succinylation may be used to solubilize a protein while blocking its lysine groups (in general, succinylated proteins are soluble at pH >7). The attached succinyl moiety forms a stable linkage to most protein treatments, but may be cleaved off with 6 N HCl at 100°C. For a review of succinylation, see Klotz (1967).

Materials

CAUTION: See Appendix 3 for appropriate handling of materials marked with <!>.

HPLC system, microdialysis, or microsize-exclusion chromatography for desalting
Lyophilized protein of interest
Micro pH probe or pH indicator paper
NaOH (1 M) <!>
NH_4HCO_3 buffer (0.1%) containing 6.4 M guanidine-HCl
Mix 2 volumes of 0.5% ammonium bicarbonate <!> with 8 volumes of 8 M guanidine-HCl <!>.
Succinic anhydride <!>

Method

1. Dissolve ~10 mg /ml protein in 200 µl of 0.1% NH_4HCO_3 buffer containing 6.4 M guanidine-HCl, and adjust the pH to 9.0 with 1 M NaOH.

2. Crush one or two crystals of solid succinic anhydride (~2 mg) and slowly add it to the mixture with gentle tapping (or using a micromagnetic stir bar) over a period of ~15–60 minutes. Monitor the pH of the solution using pH paper or a micro pH probe, and maintain the pH at 9.0 by adding 1 M NaOH as needed.

   If the succinic anhydride is added too quickly, it will be difficult to maintain the pH at 9.0.

3. Remove excess reagents by RP-HPLC, microsize-exclusion chromatography, or microdialysis.

ADDITIONAL PROTOCOL: MODIFICATION OF LYSYL SIDE CHAINS USING CITRACONIC ANHYDRIDE

Reversible masking of ε-amino groups of lysine is a valuable procedure for limiting trypsin hydrolysis of proteins to arginine peptide bonds. Dixon and Perham (1968) introduced the use of citraconic anhydride as a reversible blocking agent of lysyl side chains. A major advantage of citraconylation is the ease with which the blocking group can be removed under conditions that will not lead to protein denaturation. For a review of citraconylation, and a comparison with other reversible lysyl blocking reagents (e.g., maleic anhydride and trifluorosuccinic anhydride), see Atassi and Habeeb (1972).

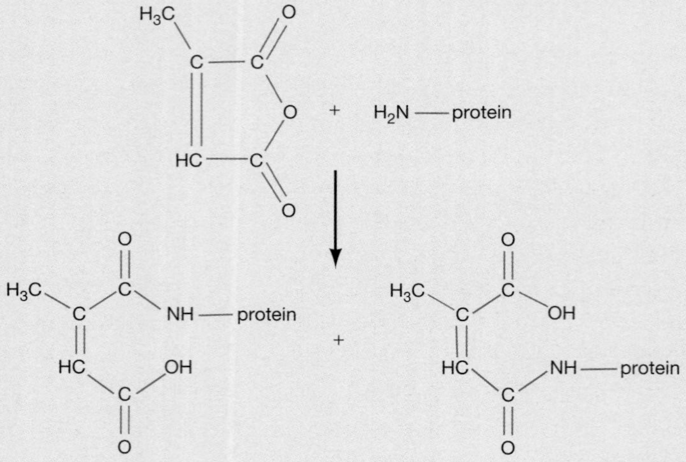

Materials

CAUTION: See Appendix 3 for appropriate handling of materials marked with <!>.

Citraconic anhydride <!>

HPLC system, microdialysis, or microsize-exclusion chromatography for desalting

Lyophilized protein of interest

Micro pH probe or pH indicator paper

NaOH <!>

NH_4HCO_3 buffer (0.1%) containing 6.4 M guanidine-HCl

Mix 2 volumes of 0.5% ammonium bicarbonate <!> with 8 volumes of 8 M guanidine-HCl <!>.

Method

1. Dissolve ~10 mg/ml protein in ~200 µl of 0.1% NH_4HCO_3 buffer containing 6.4 M guanidine-HCl, and adjust the pH to 8.0, if necessary, with 1 M NaOH.
2. Add citraconic anhydride (~1 µg/mg protein) to the solution in small aliquots, while maintaining the pH at 8 by the addition of 1 M NaOH.

   If the citraconic anhydride is added too quickly, it will be difficult to maintain the pH at 8.0.
3. Allow the reaction to proceed for 2 hours at room temperature at pH 8 with gentle mixing.

   The reaction is usually complete in ~30 minutes, but it is preferable to incubate for the more conservative 2 hours.

   Although most proteins become solubilized during modification, the reaction can be conducted in the presence of urea or guanidine-HCl.
4. Remove excess reagents by RP-HPLC, microsize-exclusion chromatography, or microdialysis.

   Removal of citraconyl groups can be accomplished by incubation at pH 3.5 (e.g., 10% acetic acid) at room temperature (100% deblocking occurs within 3 hours, although it may be convenient to incubate overnight). For example, Gibbons and Perham (1970) recommend deblocking by dialysis against 10 mM HCl (pH 2.0) for 6 hours at room temperature.

PROTOCOL 6

Cleavage at Met-X Bonds by Cyanogen Bromide

SELECTIVE CLEAVAGE OF A PROTEIN BY CYANOGEN BROMIDE, first introduced by Gross and Witkop in 1961, generates a distinctive set of peptide fragments. Cyanogen bromide cuts peptide bonds on the carboxy-terminal side of methionine residues (see Figure 7.5 for mechanism of cyanogen bromide cleavage), and because this amino acid is relatively infrequent in proteins, this cleavage tends to produce relatively large and relatively few peptides. Met-X amino acid bonds are cleaved specifically and almost quantitatively, with the exception of Met-Ser and Met-Thr bonds where the hydroxyl group of neighboring side chain may interfere with ring opening of the iminolactone. (The use of 70% trifluoroacetic acid instead of formic acid has been reported to improve cleavage yields [Titani et al. 1972].)

Typical procedures use ~100-fold molar excess of cyanogen bromide over the methionine, and the reaction is performed under mild acidic conditions (e.g., 70% formic acid), under nitrogen, and in the dark for 4–24 hours. All reagents are volatile and can be readily removed by lyophilization.

MATERIALS

CAUTION: See Appendix 3 for appropriate handling of materials marked with <!>.

Reagents

Cyanogen bromide <!> (Aldrich)

Store in a dry refrigerated container (e.g., place reagent bottle in a wide-necked plastic bottle containing silica gel) in the dark. Use only white crystals.

WARNING: Cyanogen bromide is very toxic. Wear protective gloves and safety glasses and work in a chemical fume hood.

Formic acid, minimum assay 98% (Aristar grade)<!>

β-Mercaptoethanol <!>

Store refrigerated.

NH_4HCO_3 (0.2 M) <!>

Sodium hypochlorite solution (domestic bleach) <!>

Equipment

Chemical fume hood

Concentrator (SpeedVac)

Microfuge tubes (with caps)

Nitrogen (pressurized) <!>

Biological Sample

Lyophilized protein

METHOD

Reduction

1. Dissolve the protein in 0.2 M ammonium bicarbonate to a final concentration of 10–20 mg/ml.
2. Add β-mercaptoethanol to between 1% and 5% (v/v).
3. Blow nitrogen over the solution to displace oxygen, cap the tube, and incubate overnight at room temperature.
4. Dry the sample using a SpeedVac concentrator.

   Warming the sample will help evaporate all of the ammonium bicarbonate.

Cyanogen Bromide Cleavage

5. Redissolve the dried sample (10–20 mg/ml) in 70% formic acid.
6. Add solid cyanogen bromide (2 mg of reagent per milligram of protein) directly to the protein solution. For very small amounts of protein, add one small crystal of cyanogen bromide.

   Cyanogen bromide is usually present at 20–100-fold excess with respect to methionyl residues, which is equivalent to approximately equal amounts of protein and cyanogen bromide.
7. Cap the tube and incubate the reaction for 18–20 hours at room temperature in the dark.
8. Terminate the reaction by diluting the reaction mixture with ~5 volumes of H_2O and removing the excess reagents by lyophilization with a SpeedVac concentrator.

   The products can then be separated by liquid chromatography or gel electrophoresis (or stored at –20°C).
9. To remove all traces of cyanogen bromide, decontaminate equipment (e.g., spatulas tubes) by immersion in sodium hypochlorite (bleach) solution for a few minutes until effervescence stops.

TROUBLESHOOTING

- ***Cyanogen bromide:*** Some side reactions may occur if impure cyanogen bromide (colored yellow or orange) is used. This can be avoided by using the white crystals that have sublimed on the inner surface of the reagent bottle. Excess cyanogen bromide (e.g., 1000-fold or higher) will oxidize methionine to methionine sulfoxide, especially in strong acid such as heptafluorobutyric acid (HFBA). This phenomenon has been used to effect tryptophan cleavage (albeit with poor yields), but not Met-X bonds. Under these conditions, tyrosyl residues may become brominated.
- ***Formic acid:*** Alternative acids to formic acid include 0.1 M HCl, 75–85% TFA, or HFBA. The stronger acids may cause Asp-Pro cleavage and increased acid hydrolysis.
- ***Met-X cleavage and sequencing:*** It is possible to cleave Met-X bonds in samples that have been applied to the glass fiber disks of the gas phase/pulsed liquid sequencers (Simpson and Nice 1984). This cleavage has proven to be particularly useful for obtaining sequence information from amino-terminally blocked proteins. It is also possible to chemically fragment proteins in situ in acrylamide gels (Jahnen et al. 1990) for obtaining sequence information.
- ***Lactone derivative coupling and sequencing:*** Because the lactone derivative can be coupled selectively, and in good yield, to solid supports of the amine type (e.g., 3-amino propyl glass) (Horn and Laursen 1973), this is a useful method for the isolation of methionine-containing peptides for sequence analysis.

PROTOCOL 7

Cleavage of Asn-Gly Bonds by Hydroxylamine

SELECTIVE CLEAVAGE OF ASN-GLY BONDS IN PROTEINS by hydroxylamine generates a distinctive set of peptide fragments (for the mechanism of hydroxylamine cleavage of Asn-Gly bonds, see Figure 7.6). Because the Asn-Gly sequence occurs statistically about every 350 amino acids in a protein, cleavage of this bond will yield relatively large and relatively few peptide fragments (Kellner 1994). In this protocol, a procedure for performing Asn-Gly bonds cleavage in solution is described. The hydroxylamine cleavage method can be adapted to fragment proteins within acrylamide gels (Saris et al. 1983). A cleavage buffer comprising 2 M hydroxylamine and 6 M guanidine-HCl in 15 mM Tris titrated to pH 9.3 with 4.5 M lithium hydroxide will yield ~25% cleavage.

MATERIALS

CAUTION: See Appendix 3 for appropriate handling of materials marked with <!>.

Reagents

Cleavage buffer

2 M hydroxylamine HCl <!>

4.3 M guanidine-HCl <!>

0.2 M potassium carbonate (pH 9.0)<!>

Dissolve 3.5 g of hydroxylamine (Analar grade, BDH) in 12 ml of prechilled 6 M aqueous guanidine-HCl (use an ice bath), and then add 2 ml of 50% (w/v) NaOH <!> slowly with vigorous stirring (magnetic stir bar) followed by 5 ml of 1 M potassium carbonate. Adjust the pH to 9.0 with 50% NaOH, and then adjust the volume to 25 ml with 6 M guanidine-HCl.

The inclusion of guanidine-HCl is reported to improve the efficiency of Asn-Gly cleavage (Kwong and Harris 1994).

Formic acid (Puriss grade, Merck) <!>

Trifluoroacetic acid (TFA) <!>

Optional, see Step 3.

Equipment

Column (0.8 cm I.D. x 30 cm; Sephadex G-25 medium) (Amersham Biosciences)

Alternatively, a disposable desalting column from Bio-Rad can be used.

Liquid chromatography equipment (RP-HPLC or size-exclusion)

Magnetic stir plate

Polypropylene tubes (tightly capped)

Water bath or incubator preset to 45°C

Biological Sample

Lyophilized protein

METHOD

1. Dissolve the protein directly in the cleavage buffer to give a final concentration of ~5 mg/ml.
2. Incubate the mixture for 4 hours at 45°C.
3. Stop the reaction by cooling the sample and adjusting the pH to 2.5 by the addition of concentrated formic acid (or 3 volumes of 2% v/v TFA in H_2O).
4. Desalt the sample by chromatography on a column. Analyze the peptide mixture immediately by liquid chromatography, or, alternatively, store the samples at –20°C for later analysis.
5. If analyzing by gel electrophoresis, the reaction itself may be stopped (Step 3), not by acidification, but by the addition of SDS-PAGE sample buffer.

PROTOCOL 8

Cleavage at Tryptophan by o-iodosobenzoic Acid

CHEMICAL REAGENTS THAT SELECTIVELY CLEAVE AT TRYPTOPHAN RESIDUES have contributed greatly to the elucidation of protein structure and function (Fontana and Gross 1986). Selective cleavage of a protein at tryptophan, a relatively rare amino acid, generates a distinctive set of large peptide fragments. The use of o-iodosobenzoic acid for specific cleavage at tryptophan residues was first proposed by Mahoney and Hermodson in 1979. This protocol has been adapted from the method of Mahoney et al. (1981). The yield of cleavage is moderate to high (up to 80%) and, besides some cleavage at tyrosine residues, there are very few side reactions (Fontana et al. 1983). For a detailed review of other, albeit less specific, procedures for cleaving proteins at tryptophan, see Fontana and Gross (1986).

MATERIALS

CAUTION: See Appendix 3 for appropriate handling of materials marked with <!>.

Reagents

ρ-Cresol <!>
Glacial acetic acid (Puriss grade, Merck)<!>
Guanidine-HCl <!>
o-Iodosobenzoic acid <!>

Equipment

Column, size-exclusion (Sephadex G-25, Amersham Biosciences or equivalent)
Concentrator (SpeedVac, Thermo Savant)
Nitrogen supply <!>

Biological Sample

Lyophilized protein

METHOD

1. Dissolve the o-iodosobenzoic acid (10 mg) in 1.0 ml of 80% (v/v) acetic acid containing 4 M guanidine-HCl and 20 μl of p-cresol.
2. Incubate the mixture for 2 hours at room temperature.
3. Add the protein to a final concentration 5–10 mg/ml. Flush the tube with a stream of nitrogen and incubate for 24 hours at room temperature in the dark.
4. Terminate the reaction by adding H_2O (~10 volumes) and dry using a SpeedVac concentrator (or equivalent). Alternatively, obtain the peptides by applying the digest directly to a size-exclusion column.

PROTOCOL 9

Staining Proteins in Gels with Coomassie Blue

THE MOST COMMONLY USED DYE FOR VISUALIZING PROTEINS in SDS-PAGE is Coomassie Brilliant Blue R250 (CBR-250) because of its relatively high sensitivity (Meyer and Lambert 1965; Syrovy and Hodny 1991). This protocol describes the standard CBR-250 staining method, along with a simple method for preparing stained gels for long-term storage. The limit of detection of standard CBR-250 is generally quoted as 50–100 ng, but if sufficient destaining can be achieved, then as little as 10 ng of protein can be detected. By increasing the temperature of both the staining and destaining steps, the length of the procedure can be shortened to ~25 minutes with a concomitant increase in sensitivity to 2.5–5 ng of protein (for other Coomassie Blue staining procedures, see Chapter 2, Protocols 2, 3, and 4, and Chapter 4, Protocol 10). This protocol was adapted from Simpson and Reid (1998).

MATERIALS

CAUTION: See Appendix 3 for appropriate handling of materials marked with <!>.

Reagents

Coomassie Brilliant Blue staining solution <!>

Mix 1 g of Coomassie Brilliant Blue (Bio-Rad) in 1 liter of 50% (v/v) methanol <!>, 10% (v/v) glacial acetic acid <!>, and 40% H_2O. Stir the solution for 3–4 hours and then filter through Whatman filter paper.

Destaining solution

12% (v/v) methanol

7% (v/v) glacial acetic acid

81% H_2O

Equipment

Cellophane

Gel dryer frame, plastic (Amersham Biosciences)

Mechanical shaker

Plastic containers

Plastic wrap

Tissue paper, fine grade (Kimwipes)

Biological Sample

Proteins within a gel that have been separated by electrophoresis

METHOD

Staining of Gels

Perform the staining steps at room temperature unless otherwise indicated.

1. Place the gel containing the proteins of interest in a plastic container that has sufficient Coomassie Brilliant Blue staining solution to cover the gel. Place the container on a mechanical shaker and allow the gel to stain for 20 minutes at room temperature.

2. Remove the staining solution and add destaining solution and three sheets of fine-grade tissue paper. Destain with shaking. Replenish the destain solution several times until the gel is fully destained.

   The tissue paper binds released stain.

Storage of Coomassie-Blue-stained Gels

3. If the gel will be processed in less than 1 month, store it in 200–250 ml of destaining solution at room temperature in a sealed plastic container.

4. For long-term storage, place the gel between two sheets of cellophane in a plastic gel frame. Air dry the gel in a chemical fume hood for ~16 hours.

5. Wrap the dried gel in plastic wrap and store it at room temperature.

PROTOCOL 10

Zinc/Imidazole Procedure for Visualization of Proteins in Gels by Negative Staining

THE ZINC/IMIDAZOLE STAINING PROCEDURE FOR VISUALIZING PROTEINS in acrylamide gels is based on differential salt binding. In this protocol, the negative stain uses the heavy divalent cation zinc for making a precipitate with dodecyl sulfate. Because protein-bound salts (e.g., dodecyl sulfate or the heavy cation zinc) are chemically less active than the free zinc ions in the gel, precipitation of an insoluble salt is much slower in those regions of the gel occupied by proteins than in the gel background where zinc dodecyl sulfate precipitates. The result is a "negative stain," with translucent proteins and an opaque gel background, due to zinc dodecyl sulfate precipitation. The sensitivity of the method was markedly improved by altering the composition of the precipitated salt, from zinc dodecyl sulfate to a complex of zinc and imidazole (Ortiz et al. 1992; Fernandez-Patron et al. 1998). Typical protein detection sensitivities with the zinc/imidazole staining procedure approach the low-nanogram range. The stain is reversible when divalent chelators, such as EDTA, are included. The differential salt-binding method for protein visualization has also been adapted for proteins blotted onto membranes (Patton et al. 1994) and for general proteome analysis of gel-resolved proteins (Castellanos-Serra L et al. 1999).

This protocol, which was adapted from a procedure described by Fernandez-Patron et al. (1995a,b) and from Simpson and Reid (1998), provides two methods: reverse stain using imidazole, SDS, and zinc, and a double-staining procedure using Coomassie Blue stain followed by zinc/imidazole.

MATERIALS

CAUTION: See Appendix 3 for appropriate handling of materials marked with <!>.

Reagents

Fixing solution

- 50% (v/v) methanol <!>
- 5% (v/v) glacial acetic acid <!>
- 45% H_2O

Imidazole (0.2 M) containing 0.1% SDS

Prepare 6.8 g of imidazole <!>, 0.5 g of SDS <!>, and 500 ml of H_2O.

Zinc sulfate (0.2 M)

Dissolve 28.7 g of $ZnSO_4 \cdot 7H_2O$ in 500 ml of H_2O.

Equipment

Shaker

Biological Sample

Proteins within a gel that have been separated by electrophoresis

These proteins should be either unstained (Method 1) or stained with Coomassie Brilliant Blue (Method 2; see Protocol 9 for staining procedure).

METHODS

Method 1: Direct Reverse Staining with Imidazole, SDS, and Zinc

1. Submerge the gel in fixing solution for 20 minutes with gentle shaking.
2. Discard the fixing solution. Wash the gel twice with deionized H_2O with gentle shaking for 15 minutes.
3. Incubate the gel in 0.2 M imidazole containing 0.1% SDS for 15 minutes with gentle shaking.
4. Remove the imidazole-SDS solution. Add 0.2 M zinc sulfate to the gel and agitate it for 30–60 seconds.
5. When the gel has stained satisfactorily, remove the zinc sulfate solution and submerge the gel in deionized H_2O.

Method 2: Double Staining of Coomassie-Blue-stained Polyacrylamide Gels with Imidazole, SDS, and Zinc

1. After Coomassie Blue staining (see Protocol 9), wash the destained gel twice in deionized H_2O with gentle shaking for 15 minutes.
2. Incubate the gel in 0.2 M imidazole, 0.1% SDS solution for 15 minutes with gentle shaking.
3. Remove the imidazole-SDS solution. Add 0.2 M zinc sulfate to the gel, and agitate it for 30–60 seconds.
4. When the gel has stained satisfactorily, remove the zinc sulfate solution and submerge the gel in deionized H_2O.

PROTOCOL **11**

Staining Proteins in Gels with Silver Nitrate

SILVER STAINING IS ONE OF THE COMMONLY USED PROCEDURES for visualizing proteins in acrylamide gels (for two excellent reviews, see Merril and Washart 1998; Rabilloud 2000). All silver staining methods rely on the reduction of ionic to metallic silver to provide metallic silver images, the selective reduction at gel sites occupied by proteins compared to nonprotein sites being dependent on differences in the oxidation-reduction potentials at these sites (Merril 1987). There are two broad methodologies for silver staining. One approach (nondiamine silver nitrate stains) employs silver nitrate as the silvering agent and formaldehyde in alkaline carbonate solution as the developing agent, whereas the other approach (diamine or ammoniacal stains) uses ammoniacal silver as the silvering agent and formaldehyde in dilute citric acid as the developing agent. Although protocols using ammoniacal silver are arguably more sensitive and give darker hues than those based on silver nitrate, they are more prone to negative staining, resulting in hollow or "doughnut" spots, give unacceptable backgrounds with tricine-based gel systems, and are not very robust due to their reliance on the ammonia-silver ratio (Rabilloud 2000). Additionally, ammoniacal silver staining is more sensitive for basic proteins, but less so for very acidic proteins. This protocol describes an adaptation of the silver nitrate procedure developed by Rabilloud (1990, 2001), Shevchenko et al. (1996), and Simpson and Reid (1998). The sensitivity of the method is in the low-nanogram range, which is 50–100 times more sensitive than classical Coomassie Blue staining (see Protocol 9), ~10 times better than colloidal Coomassie Blue staining (Neuhoff et al. 1988; Chapter 4, Protocol 10), and at least twice as sensitive as the negative staining zinc/imidazole method (see Protocol 10).

MATERIALS

CAUTION: See Appendix 3 for appropriate handling of materials marked with <!>.

Reagents

Acetic acid (1%) in H_2O <!>
Developing solution
  2% (w/v) sodium carbonate <!>
  0.04% (v/v) formaldehyde <!> in H_2O
Fixing solution (50% v/v methanol <!> and 5% v/v glacial acetic acid <!> in H_2O)
Methanol (50%)
Silver nitrate (0.1% w/v) (chilled on ice) in H_2O <!>
Sodium thiosulfate (0.02% w/v) in H_2O
Stopping solution (5% v/v glacial acetic acid in H_2O)

Equipment

Plastic containers
Shaker

Biological Sample

Proteins within a gel that have been separated by electrophoresis

METHOD

1. Place the gel in a plastic container (use a disposable plastic container or a thoroughly washed container) and add sufficient fixing solution to cover the gel. Fix the gel for 20 minutes with gentle shaking.
2. Remove the fixing solution and add enough 50% methanol to cover the gel. Shake the gel gently for 10 minutes.
3. Remove the 50% methanol and add deionized H_2O. Shake the gel gently for a further 10 minutes.
4. Remove the deionized H_2O and soak the gel in 0.02% sodium thiosulfate for 1 minute.
5. Rinse the gel twice with deionized H_2O for 1 minute.
6. Submerge the gel in chilled 0.1% silver nitrate solution and incubate it for 20 minutes at 4°C.
7. Rinse the gel twice with deionized H_2O for 1 minute.
8. Submerge the gel in developing solution and shake the container vigorously until the desired intensity of staining is reached (see Step 9). Discard the developing solution if it turns a yellow color and replenish it with fresh developing solution.

   EXPERIMENTAL TIP: It is critical that the development solution remains transparent during the development step.
9. When the desired intensity of staining is reached, stop the development by replacing the developing solution with stopping solution.
10. Store the silver-stained gel in 1% acetic acid until further use.

PROTOCOL 12

In-gel *S*-pyridylethylation of Gel-resolved Proteins

IN-GEL PROTEOLYTIC DIGESTION OF ONE-DIMENSIONAL or two-dimensional gel-separated proteins (see Protocol 13) is a commonly used method for generating peptide fragments for identifying and characterizing proteins by classical Edman chemistry (see Chapter 6) or mass-spectrometry-based methods (see Chapter 8). Although most gel methods for separating proteins (e.g., SDS-PAGE and conventional two-dimensional gel electrophoresis) are conducted under reducing conditions, there is a strong propensity for disulfide bond formation to occur during protein digestion and/or peptide isolation. This can result in cysteine-containing peptides being cross-linked by disulfide bridges and increased complexity of peptide maps. In this protocol, adapted from Simpson and Reid (1998), a method for performing in-gel reduction and *S*-alkylation of Coomassie-Blue-stained proteins is described. The method is applicable for either the whole gel or individually excised, stained protein spots/bands. Reduction is performed with dithiothreitol, and alkylation with 4-vinylpyridine. (Treatment of free cysteines with 4-vinylpyridine yields the *S*-β-(4-pyridylethyl) cysteinyl derivative.) *S*-β-(4-pyridylethyl) cysteine-containing peptides can be readily identified during RP-HPLC by their characteristic absorbance at 254 nm and during electrospray ionization tandem mass spectrometry by the appearance of a characteristic pyridylethyl fragment ion of 106 daltons (Moritz et al. 1996). The position of cysteine residues in a polypeptide sequence can be determined either by mass spectrometry (see Chapter 8) or as phenylthiohydantoin *S*-β-(4-pyridylethyl) cysteine during Edman degradation (see Chapter 6).

MATERIALS

CAUTION: See Appendix 3 for appropriate handling of materials marked with <!>.

Reagents

2-Mercaptoethanol <!>

Reduction buffer

- 0.2 M Trizma (base) (pH 8.4)
- 10 mM dithiothreitol (DTT) <!>
- 2 mM EDTA
- Add DTT just before using this buffer.

4-Vinylpyridine (Aldrich) <!>

Equipment

Centrifugal lyophilizer (e.g., SpeedVac, Thermo Savant) (optional)

H_2O (deionized) (Milli-Q)

Incubator or oven preset to 40°C

Lyophilizer (Thermo Savant)

Microfuge tube (1.5-ml)

Pipette (micropipettor or Pasteur)

Plastic container

Biological Sample

Gel containing the proteins of interest separated by electrophoresis

METHODS

Method 1: Whole Gel Reduction and *S*-pyridylethylation

1. Ensure that the gel has been appropriately stained and destained (see Protocol 9).
2. Wash the intact gel extensively in three changes of 400–500 ml of H_2O for ~1.5 hours.

   This step removes any acetic acid resulting from the destain protocol that would otherwise adversely affect the pH of the reduction.

3. Transfer the intact gel to a clean container.
4. Immerse the gel in 50 ml of reduction buffer. (The volume depends on the size of both the gel and the container.) Incubate the gel for 2 hours at 40°C.
5. Add 4-vinylpyridine to the reduction buffer to a final concentration of 2% (v/v). Incubate the gel in the dark for 1 hour at room temperature.
6. Add excess 2-mercaptoethanol (2% v/v final concentration) to stop the alkylation.
7. Wash the gel extensively in three changes of 400–500 ml of H_2O for ~1.5 hours.

   This step is included to remove β-mercaptoethanol, which would otherwise interfere with proteolysis.

8. If the protein spots are no longer visible, repeat the staining and destaining procedures (see Protocol 9).

Method 2: Individual Gel Spot Reduction and S-pyridylethylation

1. Inspect the gel to make sure it has been appropriately stained and destained (see Protocol 9).
2. Excise the spot of interest. Place the excised gel piece in a clean 1.5-ml polypropylene microfuge tube.
3. Wash the excised gel piece extensively using three changes of 1 ml each of H_2O for ~45 minutes. Alternatively, completely dehydrate the gel by centrifugal lyophilization.
4. Remove the deionized H_2O from the microfuge tube.
5. Add enough reduction buffer to cover the gel piece(s) completely (typically 100–200 μl).

   Dehydrated gel pieces will swell as they absorb the reduction buffer. Make sure that there is enough buffer to accommodate this.

6. Incubate the gel for 1 hour at 40°C.
7. Add 4-vinylpyridine to the reduction buffer to a final concentration of 2% (v/v). Incubate the gel in the dark for 1 hour at room temperature.
8. Add excess β-mercaptoethanol (2% v/v final concentration) to stop the alkylation.
9. Carefully remove the reduction buffer using a micropipettor or a Pasteur pipette and discard.
10. Wash the gel pieces extensively in three changes of 1 ml each of H_2O for ~45 minutes.

PROTOCOL 13

In-gel Proteolytic Digestion and Extraction of Peptides

POLYACRYLAMIDE GEL ELECTROPHORESIS IS ONE OF THE MOST COMMONLY USED TOOLS for proteome analysis, especially when the method is combined with peptide mapping and internal sequence analysis procedures for identifying and characterizing proteins. Two main approaches for digesting gel-separated proteins to generate peptide maps are (1) protein blotting onto immobilizing membranes followed by on-membrane digestion (Simpson et al. 1989; Gevaert and Vandekerckhove 2000; see Protocol 14) and (2) digesting proteins directly in the gel matrix and extracting the peptides (Ward et al. 1990). It is now generally recognized that the overall recoveries of peptides from in-gel digestion methods are significantly greater than that from on-membrane digest strategies. In this protocol, adapted from Simpson and Reid (1998), a general method for in situ digestion of gel-separated proteins is provided. Because in-gel digestions might fail when the visualized protein is below ~10 μg of protein/cm^2, it is customary to combine several weakly stained Coomassie Blue gel spots and reconcentrate them electrophoretically in a new gel, either a one-dimensional SDS-PAGE gel (see Chapter 2) or by using the Pasteur pipette gel procedure for concentrating proteins described in Chapter 6, Protocol 4.

MATERIALS

CAUTION: See Appendix 3 for appropriate handling of materials marked with <!>.

Reagents

Buffers for in situ trypsin cleavage
- Digestion buffer (0.05 M NH_4HCO_3 <!>)
- Trypsin (0.1 μg/ml) in digestion buffer
  - Prepare a stock solution of trypsin (~0.1 μg/ml) in 1 mM HCl <!> and store at –20°C. Prior to use, dilute the trypsin stock solution 1:1 with digestion buffer to ensure correct concentration and that the pH of the enzyme solution is 7.8.
- Wash buffer
  - 0.1 M NH_4HCO_3
  - 50% acetonitrile <!>

Buffers for in situ *Achromobacter lyticus* protease I (Lys-C cleavage)
- *A. lyticus* protease I (0.1 μg/ml) in digestion buffer
- Digestion buffer
  - 0.1 M Tris-HCl (pH 9.3)
- Wash buffer
  - 0.05 M Tris-HCl (pH 9.3)
  - 50% acetonitrile

Trifluoroacetic acid (TFA) <!> (0.1% v/v) aqueous containing 60% acetonitrile
Prepare this reagent by adding 600 ml of acetonitrile to a l-liter glass measuring cylinder and then adjusting the volume to 1 liter with H_2O. Add 1.0 ml of TFA to achieve a final concentration of 0.1% (v/v).

Equipment

Lyophilizer, centrifugal (e.g., SpeedVac, Thermo Savant)
Microfuge tubes
Water bath preset to 30°C

Biological Sample

Proteins separated on a two-dimensional gel

METHOD

1. Excise the protein gel spots of interest and place in microfuge tubes.
2. Remove excess Coomassie Blue stain by washing twice with 1 ml of either 0.1 M NH_4HCO_3, 50% acetonitrile (for trypsin) or 0.05 M Tris-HCl (pH 9.3), 50% acetonitrile (for Lys-C) each for 30 minutes at 30°C.
3. Dry each gel piece *completely* by centrifugal lyophilization. The gel piece *should not* stick to the walls of the microfuge tube when completely dry.
4. Rehydrate the gel piece by adding 10 µl of digestion buffer, containing 0.5 µg of the appropriate protease, directly onto the dried gel piece.

   For other enzymes, see Table 7.4.
5. Wait ~10–20 minutes, until the solution has been absorbed by the gel piece.
6. If necessary, repeat Steps 4 and 5 to allow the gel piece to fully swell.
7. Add 200 µl of digestion buffer to fully immerse the gel piece.
8. Incubate for ~16 hours at 35°C.
9. Carefully remove the digestion buffer (now called the extract), and place it into a clean microfuge tube.

   The digestion buffer contains >80% of the extractable peptides.
10. Add 200 µl of 0.1% TFA, 60% acetonitrile to the gel piece.
11. Incubate the tube with the gel piece for 30 minutes at room temperature.
12. Carefully remove the extract away from the gel piece and combine it with the extract from Step 9.
13. Reduce the volume of the pooled extracts by centrifugal lyophilization. Do not dry the pooled extracts completely otherwise sample loss may result.

    The objective of this step is to remove acetonitrile from the pooled extracts and reduce the volume for subsequent dilution and injection onto a capillary RP-HPLC system.

PROTOCOL 14

On-membrane Proteolytic Digestion of Electroblotted Proteins

THERE ARE TWO MAIN APPROACHES FOR DIGESTING PROTEINS to generate peptide maps for gel-separated proteins: (1) protein blotting onto immobilizing membranes (see Chapter 6, Protocol 2 and Chapter 4, Protocols 15 and 16) followed by on-membrane digestion (Simpson et al. 1989; Gevaert and Vandekerckhove 2000), and (2) digesting proteins directly in the gel matrix and extracting the peptides (Ward et al. 1990) (see Protocol 13). In this protocol, an adaptation of the Aebersold (1993) method for performing on-membrane protein digestion is provided.

MATERIALS

CAUTION: See Appendix 3 for appropriate handling of materials marked with <!>.

Reagents

Acetic acid (1% aqueous) <!>

Amido Black 10B dye (0.1%) in H_2O/acetic acid/methanol <!> (45:10:45, v/v/v)

Digestion buffer

For trypsin (bovine or porcine), chymotrypsin, LysC, and AspN, use 100 mM NH_4HCO_3 <!> (pH 7.9), 10% (v/v) acetonitrile <!>, at 37°C overnight. Use an enzyme concentration of ~100 ng of enzyme/μl of digestion buffer, irrespective of the amount of substrate.

NaOH (200 mM) <!>

Ponceau S dye (0.1%) in 1% aqueous acetic acid

PVP-40 (0.5% w/v) in 100 mM acetic acid

PVP-40 is used to prevent absorption of the protease to the nitrocellulose during digestion. This detergent can be replaced by 1% Zwittergent 3-16 (Calbiochem) (Lui et al. 1996), which is fully compatible with downstream processing methods such as RP-HPLC (Zwittergent 3-16 is UV_{214} transparent) and MALDI-TOF mass spectrometry.

WARNING: Zwittergent 3-16 is not compatible with nanoseparation techniques such as capillary RP-HPLC because the detergent competes with the peptides for the binding sites on the reversed-phase column.

Equipment

Electroblotting apparatus

Microfuge tubes (1.5, 0.5, 0.2 ml)

Nitrocellulose or PVDF membrane

RP-HPLC column

Biological Sample

Gel containing the proteins of interest separated by electrophoresis

METHOD

Electroblotting and Staining the Proteins

1. Electroblot the proteins from the gel onto a nitrocellulose or PVDF membrane.

   Typically, this takes 2 hours for 0.5-mm-thick gels. For proteins that are difficult to transfer, add up to 0.005% SDS <!> to the transfer buffer. Nitrocellulose membranes are preferred to PVDF membranes due to the hydrophobic surface of PVDF, which limits the recovery of peptide fragments, and the higher yields of recovered peptides from nitrocellulose.

   It is not possible to perform on-membrane digestion (PVDF) after amino-terminal sequence analysis of the protein. Because nitrocellulose is not inert to Edman chemistry reagents and solvents, this membrane cannot be used for direct amino-terminal sequence analysis. Both Amido Black and Ponceau S staining procedures do not interfere with proteolytic digestion, the extraction of peptides from the membrane, or subsequent RP-HPLC analysis of peptides.

2. Stain the membrane with either Amido Black or Ponceau S.

   For staining with Amido Black

   a. Immerse the nitrocellulose membrane or PVDF membrane in 0.1% Amido Black 10B for 1–3 minutes.

   b. Rapidly destain with several washes of H_2O/acetic acid/methanol.

   c. Rinse the destained blots thoroughly with deionized H_2O to remove any excess acetic acid.

   d. Cut out the stained protein bands (or for two-dimensional gel spots, up to 40 spots from identical gels may be required) and transfer these bands to 1.5-ml microfuge tubes for immediate processing (begins with Step 3) or for storage at –20°C.

   For staining with Ponceau S

   a. Immerse the nitrocellulose membrane in 0.1% Ponceau S for 1 minute.

   b. Gently agitate the blot for 1–3 minutes in 1% aqueous acetic acid to remove excess stain.

   c. Cut out the protein bands of interest and transfer them to microfuge tubes.

   d. Destain the protein bands by washing the membrane pieces with 200 mM NaOH for 1–2 minutes.

   e. Wash the membrane pieces with deionized H_2O and process them immediately (begins with Step 3) or store them wet at –20°C (avoid excessive drying).

Digestion of the Membrane-bound Protein

3. Add 1.2 ml of 0.5% (w/v) PVP-40 in 100 mM acetic acid to each tube.

4. Incubate the tube for 30 minutes at 37°C.

5. Centrifuge the tube at ~1000*g* for 5 minutes.

6. Remove the supernatant solution and discard.
7. Add ~1 ml of H_2O to the tube.

   It is essential to remove excess PVP-40 before RP-HPLC peptide mapping because of the strong UV absorbance of this detergent. Moreover, breakdown products of PVP-40 also produce major contaminant peaks in electrospray ionization mass spectrometry (ESI-MS).

8. Vortex the tube for 5 seconds.
9. Repeat Steps 5 and 6.
10. Repeat Steps 7–9 five more times.
11. Cut the nitrocellulose strips into small pieces (~1 x 1 mm) and place them in a fresh tube (0.5- or a 0.2-ml tube).
12. Add the minimal quantity of digestion buffer (10–20 μl) to submerge the nitrocellulose pieces.
13. After digestion, typically 16 hours or overnight at 37°C, load the total reaction mixture onto an appropriate RP-HPLC column for peptide fractionation (or store the peptide mixture at –20°C until use) (see Protocol 15 and Chapter 5).

PROTOCOL 15

Reversed-phase HPLC of Peptides: Standard Conditions

IN THIS PROTOCOL, ONLY NONSPECIALIZED COLUMNS, MOBILE PHASES, and instrumentation that are readily available and easily operated are described. For more details concerning chromatographic theory, see Mant and Hodges (1991) and Snyder and Kirkland (1979), and for HPLC troubleshooting, see www.proteinsandproteomic.org.

EVALUATING THE RP-HPLC SYSTEM

Before analyzing a mixture of peptides by RP-HPLC, it is extremely important to evaluate the system—both HPLC hardware and column—*before* it is applied to a biological problem. This is especially true when a new column enters the laboratory, and applies to both beginners and more experienced chromatographers who wish to obtain reproducible peptide maps.

When using a new column, make sure first to read the manufacturer's instructions and recommendations for proper conditioning, use, and storage of the column. Often, subtle operating conditions that are crucial for correct column usage are overlooked, resulting in irreversible column damage and the possible loss of an important sample. Once the operating procedures and column limitations (e.g., operating pressure and stability with respect to pH) are understood, perform a test separation using a set of standards (e.g., peptide or protein mixture) that have been designed to evaluate column performance. For this exercise, it is essential to reproduce the standard set of operating conditions described in this protocol before proceeding with an analysis of the test sample. If an equivalent elution profile cannot be obtained on the new column (see below) using these standards and standard operating conditions, consider the new column suspect and discuss the problem with the manufacturer. (Columns that do not perform with a recognized set of standards are more easily returned to a manufacturer than simply reporting to the manufacturer that the column "cannot separate" an ill-defined complex mixture of interest.) These standard operating conditions should be used on a regular basis (and carefully logged) to monitor column performance and the lifetime of the column.

Because stationary phases differ widely from one manufacturer to another (see Table 5.1) in terms of their physical characteristics—the functional ligand used, ligand density, base support (matrix), particle size, and pore size, as well as column dimensions—standard protein/peptide separations are a useful diagnostic of these differences (and can be exploited in a purification strategy). For a review of current columns for RP-HPLC separations in highly aqueous mobile phases, see Majors and Przybyciel (2002). If samples are to be collected for further characterization, careful attention should be given to the "dead volume" of the plumbing between the flow cell and outlet (see below).

For useful RP-HPLC Web Site tutorials, see Chapter 7, and for HPLC troubleshooting and accessories, see http://www.ionsource.com.

Tips for Preparing HPLC for Peptide Mapping

- Read the manufacturer's instructions for correct column usage.
- Ensure that the HPLC system is in good working condition by running a blank (i.e., no column installed or sample injected).
- Evaluate the chromatographic system by running a set of "known standards."

MATERIALS

CAUTION: See Appendix 3 for appropriate handling of materials marked with <!>.

Reagents

Acetonitrile (HPLC grade) (J.T. Baker, Mallinckrodt, or equivalent) <!>
Deionized H_2O (HPLC grade, 18 mega Ω) (Milli-Q, Millipore, or equivalent)
Protein standards (in order of their elution from the suggested reversed-phase columns)
  Ribonuclease B (R 7884, Sigma)
  Hen egg lysozme (L 6876, Sigma)
  Bovine serum albumin (A 3675, Sigma)
  Carbonic anhydrase (C 3934, Sigma)
  Myoglobin (M 0630, Sigma)
  Ovalbumin (albumin chicken egg) (high-purity grade from Sigma, A 7641)
    Dissolve the protein standards (1 mg of each protein/ml) in H_2O, and divide them into 100-µl aliquots. These standards can be stored frozen for several weeks, but do not freeze-thaw them more than five times. Upon thawing aliquots to use as HPLC controls, dilute them 1:10 with Solvent A just prior to use (see Step 2).
Trifluoroacetic acid (TFA, HPLC grade) (Pierce or equivalent) <!>

Equipment

HPLC chromatographic system
  The system should be equipped with programmed gradient elution, UV detection at 210–220 nm, components for acquisition of chromatographic data, and integration of peak areas.
  The system must be capable of accurate and reproducible solvent delivery and gradient formation. For a schematic of a generic HPLC system, see Figure 7.10A (at the end of the protocol).
Polypropylene tubes (1.5 ml) with tight-fitting caps
Reversed-phase column (100 Å or 300 Å pore size)
  For a description, see Chapter 5. Use a 5–10-µm particle size packed into a column of 100 x 4.6-mm I.D. or cartridge of 100 x 2.1-mm I.D.
Commonly used columns for samples of >50 µg used in the author's laboratory:
  Brownlee RP-300 (100 x 2.1 7 µm, 300 Å) (C_8) (from Alltech, http://www.alltechweb.com)
  Vydac 218TP3215 (150 x 2.1 3 µm, 300 Å) (C_{18}) (Vydac)
    For samples >100 µg, use the same support, but packed into 4.6-mm I.D. columns.

Biological Sample

Peptide mixture generated by digestion of protein (as in Protocol 5, 6, 7, 8, or 13)

Solvent Preparation

- If both Solvents A and B are 0.1% (v/v) TFA, then the absorbance at 214 nm of Solvent B will be slightly higher (20–40 milli-absorption units [mAU]) than that of Solvent A, due to the contributing absorbance of acetonitrile. This will result in a rising baseline during the development of a gradient from 0% to 60% acetonitrile. This does not present a problem when working with 50–100-μg amounts of material (using a 4.6-mm I.D. column). The rising baseline, however, can become a serious problem, especially when working at low-microgram levels using microbore columns (<2.1 mm I.D.). The baseline will be off-scale, and late-eluting peaks cannot be detected without adjusting the baseline, which is extremely difficult to accomplish during the course of a chromatographic run (10 ng of protein/peptide has an absorbance at 214 nm of ~1 mAU).

  To overcome this potential problem, it is recommended that the amount of TFA in Solvent B be slightly less than that in Solvent A (e.g., 0.09%) (900 μl of TFA/1 liter) compared with 0.1% (1000 μl of TFA/1 liter). By these means, the amount of TFA can be adjusted carefully to accomplish, "by trial-and-error," a flat baseline. For high-sensitivity work (submicrogram levels of peptide/protein), where the expected peak heights are in 10–20 mAU on a 2.1-mm I.D. column (1 μg = 100 mAU and 100 ng = 1 mAU on a 2.1-mm I.D. column), 0.085% or 0.086% (v/v) TFA in Solvent B may be needed. (The concentration of TFA used in Solvent B will depend on the quality, i.e., UV transmittance of the acetonitrile used.)

- If solvents are not degassed, then bubble formation can occur during a chromatographic run as result of out-gassing. Such bubbles can lodge in the flow cell, resulting in an erratic detector signal (i.e., baseline). To minimize this risk, it is recommended that Solvent B be a "percentage buffer," i.e., containing H_2O (e.g., 60% TFA, 40% H_2O containing 0.1% TFA). Additionally, solvents should be degassed by purging with high-purity helium or applying constant vacuum over a period of ~30 minutes.

- With alcohol-type solvents (e.g., methanol and ethanol), a slow buildup of esters occurs at room temperature, resulting in an increased baseline. For this reason, solvents must be prepared every 1–2 days.

Solvents

Solvent A (aqueous 0.1% v/v trifluoroacetic acid [TFA]; 14 mM)

Use a micropipettor to add 1.0 ml of undiluted TFA to 1 liter of H_2O in a stoppered glass measuring cylinder. Mix thoroughly and then pour the diluted TFA into a clean liquid chromatograph/HPLC reservoir bottle that has been washed extensively with Milli-Q H_2O.

EXPERIMENTAL TIP: It is good practice to set aside a 1-liter glass cylinder to be used for this purpose only.

Solvent B (60% v/v acetonitrile and 40% H_2O containing 0.1% v/v TFA)

Add 600 ml of acetonitrile to a 1-liter glass-stoppered cylinder and adjust to 1 liter with H_2O. Add 1 ml of undiluted TFA (final concentration 0.1% v/v TFA) and mix the solution thoroughly.

EXPERIMENTAL TIP: Due to increasing accumulation of UV-absorbing contaminants, use acetonitrile-based solvents for no longer than 5 days, and alcohol/acid-based solvents for no more than 24 hours.

METHOD

1. Run a blank or "control chromatogram" (in which no sample is loaded in Step 1). Typically, this is the first chromatographic run of the day.

   The recommended standard chromatographic conditions for RP-HPLC system evaluation are as follows:

   Linear gradient from 0% to 100% B, where Solvent A is aqueous 0.1% (v/v) TFA, and Solvent B is 0.09% (v/v) TFA in 60% aqueous acetonitrile.

   Gradient rate: 1% B/minute (i.e., 0–100% solvent B in 60 minutes).
   Flow rate: 1 ml/min for a 4.6-mm I.D. or 0.1 ml/min for a 2.1-mm I.D. column
   Temperature: 45°C.

   Timetable for linear gradient chromatography

Step number	Time	Mobile-phase composition	Flow rate 4.6-mm I.D. col	Flow rate 2.1-mm I.D. col
1. Sample injection	0 min	100% Solvent A	1 ml/min	0.1 ml/min
2.	60 min	100% Solvent B	1 ml/min	0.1 ml/min
3. Reequilibration	*a*	100% Solvent A	1 ml/min	0.1 ml/min

   a is the time taken to pass 20 column volumes of Solvent A through the column. As an approximate guide, column volumes can be calculated by dividing the internal volume of the column by 2 to take into account the volume of the chromatographic packing. The internal volume of the column can be determined by measuring the column length and internal diameter and applying the equation $V = \pi r^2 h$, where V = column volume, r = internal radius of column, and h = length of the column.

   These conditions were used to produce the elution profile of the blank chromatogram shown in Figure 7.9A.

   Before commencing the blank run, pump 20 column volumes of Solvent B through the column and then follow this with an equal volume of Solvent A (ensure that the baseline is stabilized, i.e., remains flat). Adjust the flow rate to an operating flow rate of 1 ml/min (or 0.1 ml/min for 2.1-mm I.D. column) and wait until the baseline stabilizes (~5 minutes).

2. Run the protein standards (control run) as described in Step 1, but inject 100 µl of 0.1 mg/ml protein stock standards for the 2.1-mm I.D. column (i.e., 10 µg of each protein standard; 10 µg = ~1000–2000 mAU at a flow rate of 0.1 ml/min). For a 4.6-mm I.D. column, load 100 µl of 1.0 mg/ml protein stock standards, i.e., 100 µg of each protein standard (100 µg = ~1000–2000 mAU at a flow rate of 1.0 ml/min).

   These conditions were used to produce the elution profile of the control protein standards chromatogram shown (only 2.1-mm I.D. column profile shown) in Figure 7.9B. An alternative to protein standards is a tryptic digest of cytochrome *c* (Figure 7.9C).

3. Run the test peptide digest mixture, using the same conditions as in Step 2 above (see Figure 7.8 for peptide maps of IL-6).

4. Collect the eluting peptide peaks in polypropylene tubes with tight-fitting caps and store them at –20°C for further analysis.

   If sample collection is to be performed manually, the dead volume of the plumbing from the detector flow cell to the outlet tubing must be determined accurately.

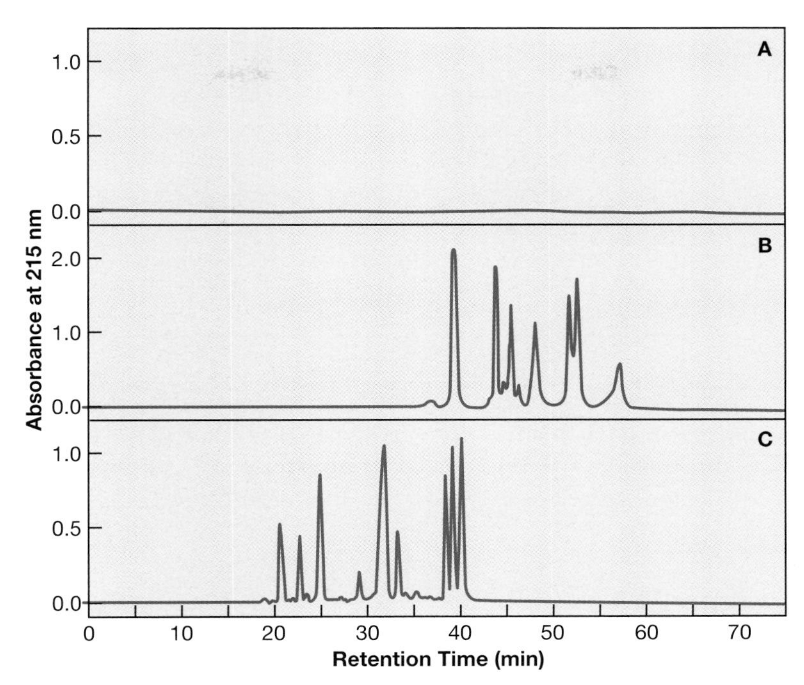

FIGURE 7.9. Standard conditions for RP-HPLC of peptides. (*A*) Blank run; (*B*) 10 µg each protein standard (in order of elution): ribonuclease B, cytochrome *c*, lysozyme, bovine serum albumin, carbonic anhydrase, myoglobin, and ovalbumin; (*C*) tryptic digest of cytochrome *c* (10 µg) in aqueous 0.1% TFA. Chromatographic conditions: linear 60-minute gradient of 0–100% Solvent B, where Solvent A is aqueous 0.1% TFA, and Solvent B is aqueous 0.09% TFA containing 60% acetonitrile. Flow rate was 0.1 ml/min and column temperature was 45°C. Chromatographic support: Brownlee RP-300 2.1-mm I.D. x 100-mm cartridge. HPLC system: Agilent 1100 HPLC equipped with diode-array detector. (Figure provided courtesy of Robert L. Moritz.)

Determining Dead Volume of the Plumbing from the Detector Flow Cell to the Tubing Outlet

A good approximation of the dead volume (d µl) can be made using the following equation: $d = \pi r^2 h$, where h is the length of the tubing from the flow cell to the outlet (in millimeters), and r is the radius (in millimeters) of the tubing.

The preferred method for determining the dead volume involves the injection of a visible dye (aqueous 0.01% w/v Coomassie Blue) and accurately determining the time (using a stopwatch) taken from the first appearance of dye in the flow cell (i.e., off-scale detector signal) and the appearance of blue dye from the outlet tubing (easily visualized by spotting the eluent onto white blotting paper). When performing this task, the column should be disassembled from the HPLC and replaced by a comparable length of small-bore tubing that is then connected to the flow cell.

Column Storage

For long-term storage, columns should be flushed and stored with an aqueous organic solvent (e.g., 30% v/v methanol/70% H_2O). For overnight storage, pump the columns with Solvent B at a very slow flow rate (e.g., 50 µl/min).

Cautionary Notes!

- ***Routine for new column usage:*** Flush the HPLC system completely before attaching the column. Prime both solvent lines with Solvents A and B, respectively, to ensure that there are no bubbles in the system. Connect the solvent delivery tubing from the injector to the top end of the column and the tubing from the column outlet to the flow cell. Pass 30 column volumes of Solvent B through the column and then follow this with 30 volumes of Solvent A. The column is now ready to be evaluated using the procedure described in Steps 1–4.
- ***Connection of columns to HPLC plumbing:*** Before connecting a column to the HPLC plumbing, take care to ensure that column end-fittings, ferrules, and tubing depths are correctly matched. If column end-fittings are used incorrectly, the column can be seriously damaged, resulting in solvent leakage and/or high dead volumes (and hence poor chromatographic performance). For a technical description of the interchangeability of column fittings and possible problems that ensue from incorrect fittings, see Chapter 5 and the Upchurch catalog or proceed to their Web Site at www.upchurch.com/Techinfo/interchange.asp.

Solvents Containing Salts

Neutral pH solvent systems such as unbuffered 1% (w/v) NaCl/acetonitrile and 0.1 M sodium phosphate (pH 7.4)/acetonitrile are commonly used in RP-HPLC peptide separations when seeking selectivity other than that achieved with conventional acidic solvents such as 0.1% TFA/acetonitrile (for an example, see the peptide mapping of IL-6 in Figure 7.8). Always choose salts of the highest quality, such as Aristar-grade reagents from BDH. When using salt-containing solvents, the percentage of organic solvent in Solvent B should not exceed 50%; otherwise, there is a high risk that these salts will precipitate and clog the HPLC tubing (typically, 0.007- and 0.010-inch capillary tubing is used in modern HPLC systems), as well the column. Such damage can be very expensive to repair! To avoid possible precipitation of salts, salt-containing solvents should be premixed by adding equal volumes of organic solvent to the buffer salt solution (e.g., 2% w/v aqueous NaCl or 0.2 M sodium phosphate at pH 7.4) to give a final concentration 1% (w/v) NaCl/50% organic solvent or 0.1 M phosphate/50% organic solvent.

Equipment for Gradient Elution

There are two major devices for generating HPLC gradients. The binary pump-type shown in Figure 7.10B (left-hand panel) involves the high-pressure mixing of two solvents to generate a mobile-phase gradient by computer programming the delivery from the high-pressure system. The output from two high-pressure pumps (reciprocating and/or displacement-type pumps are used for this purpose) is programmed into a low-volume mixing chamber before flowing into the column. In the low-pressure gradient former shown in Figure 7.10B (right-hand panel), solvent gradients are first formed by mixing two or more solvents at atmospheric pressure and then pumped to the column via a single high-pressure pump. The gradient is produced by sequentially opening valves leading to the various reservoirs. The concentration of the various solvents can be selected by proportioning the time that the valves to the various reservoirs are opened or closed, using an electronic gradient programmer and controller. After the desired solvent mixture is generated, it is mixed and fed to a single reciprocating high-pressure pump for pressurization before flowing into the column. Like the binary pump-type device, useful gradients can be generated with two reservoirs containing solvents of different strengths. In general, the binary pump devices have a lower flow-path volume than the low-pressure gradient-forming devices and are considered to be technically superior.

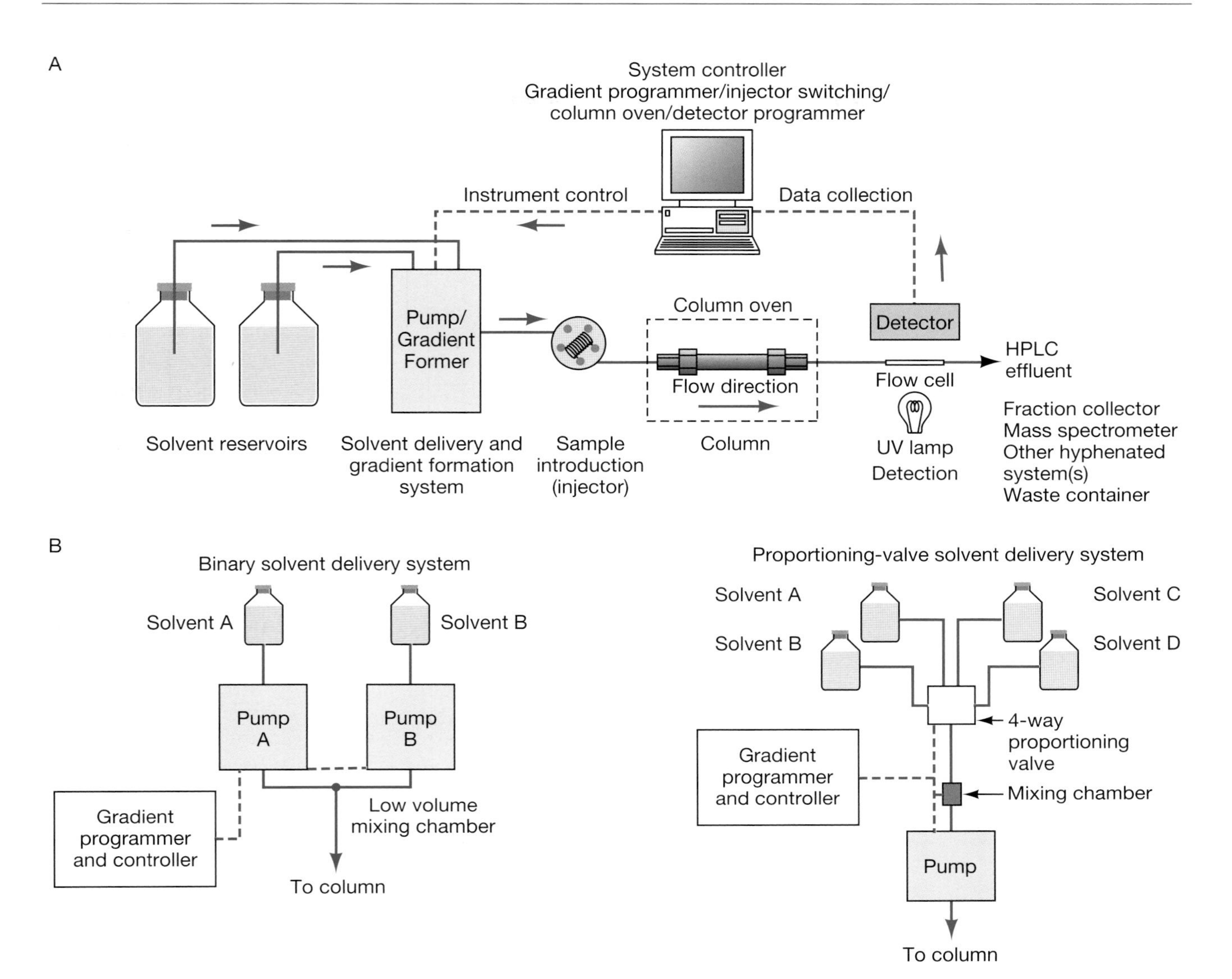

FIGURE 7.10. Generic HPLC systems. (*A*) Schematic diagram of an HPLC system. An HPLC system principally consists of four components: a solvent delivery system, a separation device, a detection unit, and computer-controlled systems controller. (*B*) Solvent delivery systems. In the *binary-solvent delivery system*, each solvent is fed into a mixing chamber by a dedicated pump in a manner that allows formation of an accurate gradient with differing compositions of Solvent A and Solvent B (accomplished by differentially adjusting the flow rates of Solvent A and Solvent B delivery to achieve the desired solvent gradient). Because this type of gradient-forming system is generated at pressures greater than atmospheric pressure, it is referred to as a "high-pressure gradient former." The basic principle of a *proportioning-valve solvent delivery system* is to "suck" the solvents from the solvent reservoirs through a proportioning valve. The desired solvent composition of the gradient is accomplished by accurately timing the delivery of each solvent, i.e., volume per unit time, through the proportioning valve. Because the solvents are at atmospheric pressure, this solvent delivery system is referred to as a "low-pressure gradient former." Since the solvents in this system are drawn at atmospheric pressure (or, in some cases, slightly negative pressure), these systems are prone to cavitation (i.e., air bubble formation), requiring that the solvent reservoirs be constantly purged with an inert gas such as helium or, alternatively, on-line vacuum-sparged.

PROTOCOL 16

SDS-PAGE Peptide-mapping Procedure (Cleveland Method)

ONE OF THE MOST QUOTED PROCEDURES FOR PEPTIDE MAPPING of proteins is the Cleveland method, named after its originator (Cleveland et al. 1977). This method relies on partial hydrolysis of the protein yielding a considerable number of large peptides that can be readily identified by SDS-PAGE (see Chapter 2, Protocol 1) or tricine-SDS-PAGE (see Chapter 2, Protocol 11). (Small peptides obtained with a "total digest" are typically poorly fixed in the gel and may be washed out and lost, thereby contributing nothing to the peptide map.) There are two main approaches for performing Cleveland peptide maps. One approach, adapted from Cleveland et al. (1977), and described in this protocol, relies on performing a partial digest of the protein in solution and then separating the peptide fragments by SDS-PAGE. In the other approach (see Protocol 17), sample proteins are first separated by gel electrophoresis (either one- or two-dimensional gel electrophoresis), and then in situ hydrolysis is performed, which is followed by a second gel step to separate the peptides.

MATERIALS

CAUTION: See Appendix 3 for appropriate handling of materials marked with <!>.

Reagents

Buffers and acrylamide <!> solutions necessary for running SDS-PAGE
 Prepare buffers and an SDS-PAGE gel as described in Chapter 2, Protocol 1.

Fixing/staining/destaining solutions for visualizing peptide fragments
 See Step 7.

2-Mercaptoethanol <!>

Protein standards of known molecular weights in SDS sample buffer
 For example, a low-molecular-weight kit from Bio-Rad or equivalent or peptide molecular-weight markers (Amersham Biosciences or equivalent).

Proteinase of choice (see Table 7.4)
 The activity of many proteinases is hindered by SDS (e.g., trypsin, chymotrypsin, and thermolysin; see Table 7.6). Endoproteinase Glu-C (or V8 protease), which can tolerate high concentrations of SDS, is recommended for the Cleveland method.

SDS (10%) <!>

SDS-PAGE sample buffer
 0.125 M Tris-Cl (pH 6.8)
 0.5% SDS
 10% (v/v) glycerol
 0.001% bromophenol blue <!>

Equipment

Polypropylene tubes (tightly capped)

Slab gel electrophoresis equipment (see Chapter 2)

Either use a uniform concentration 15%T 2.6%C gel, or a concentration gradient 5–20%T (or 10–25%T) gel in the Laemmli buffer system.

Water bath (boiling)

Biological Sample

Protein, purified sample (lyophilized or in solution)

METHOD

1. Prepare the protein samples (0.5 mg/ml) in SDS sample buffer and boil them for 2–5 minutes.
2. Add proteinase to each sample, typically to 1–100 µg/ml final concentration.
3. Incubate for a set period of time, typically 10–60 minutes at 37°C.

   Alternatively, load 10–30 µl (10–20 µg) of each protein to be compared into three separate wells of an SDS-PAGE gel. Overlay samples with 10–20 µl of digestion buffer (1% aqueous SDS, 1 mM EDTA, 1% glycerol, 0.125 M Tris-Cl buffer, pH 6.8) containing 0.005, 0.05, and 0.5 µg of proteinase, respectively. Carefully fill sample wells with running buffer and electrophorese until the dye reaches the bottom of the stacking gel. Turn off the power and incubate for 2 hours at 37°C (partial digestion). Then continue electrophoresis until the dye reaches the bottom of the gel. Fix, stain, and destain gel (or electroblot onto PVDF or nitrocellulose for western analysis).

4. Stop proteolysis by adding 2-mercaptoethanol to 10% final concentration and SDS to 2% final concentration, and boiling for 2 minutes.
5. Load 20–30 µl of each sample (10–25 µg of peptide) onto the SDS-PAGE slab gel. Also load standard proteins of known molecular weights in parallel tracks of the gel.
6. Carry out the electrophoresis until the bromophenol blue nears the end of the gel.
7. Detect the separated peptides by Coomassie Blue staining or silver staining (see Protocol 9 or 11 and Chapter 2, Protocols 2 and 7).

   Radiolabeled peptides can be detected by autoradiography or fluorography.

PROTOCOL 17

In Situ SDS-PAGE Peptide-mapping Procedure (Cleveland Method)

THERE ARE TWO MAIN APPROACHES FOR PERFORMING CLEVELAND peptide maps. One approach, described in Protocol 16, relies on performing a partial digest of the protein in solution and then separating the peptide fragments by SDS-PAGE. In the other approach (this protocol), sample proteins are first separated by either one- or two-dimensional gel electrophoresis. The gel is briefly stained and destained, and the proteins are excised for peptide mapping. Peptide fragmentation (either enzymically or chemically) is performed in situ in the gel pieces, and the partial digest is separated in a second gel.

MATERIALS

CAUTION: See Appendix 3 for appropriate handling of materials marked with <!>.

Reagents

Chemical cleavage reagent (for Method 2)

See Tables 7.3 and 7.7. For peptide mapping after chemical cleavage at aspartyl-prolyl peptide bonds, see Rittenhouse and Marcus (1984).

Coomassie Brilliant Blue (0.1% w/v)

Dissolve the Coomassie Brilliant Blue in 50% (v/v) methanol <!> and 10% (v/v) acetic acid <!>.

Ethanol containing 1 M Tris-Cl (pH 6.8) <!>

Polyacrylamide peptide-mapping slab gel <!>

Either a uniform concentration gel (e.g., 15%T or 2.6%C) or a concentration gradient gel (e.g., 5–10%T or 10–25%T) prepared in Laemmli buffer with a conventional stacking gel (see Chapter 2, Protocol 1).

Proteinase digestion buffer (for Method 1)

0.125 M Tris-CI (pH 6.8)
0.1% SDS <!>
20% glycerol
0.005% bromophenol blue <!>

Proteinase solution (for Method 1)

0.1% SDS
10% glycerol
0.005% bromophenol blue
1–100 μg/ml proteinase in 0.125 M Tris-HCI (pH 6.8)

Endoproteinase Glu-C (or V8 protease), which can tolerate high concentrations of SDS, is recommended for the Cleveland method.

Equipment

Light box and scalpel (or razor blade)
Polypropylene tubes
Slab gel electrophoresis equipment (see Chapter 2)
SpeedVac concentrator

Biological Sample

Polyacrylamide slab gel containing the separated proteins

Additional Reagents

Electrophoresis buffers (Laemmli system)
These buffers are necessary for running SDS-PAGE (see Chapter 2, Protocol 7).

Fixing/staining and destaining solutions
These solutions are needed for visualizing peptide fragments (see Protocol 9 or 11).

METHODS

Method 1: Enzymatic Fragmentation

1. Fix the proteins that have been separated in the polyacrylamide gel. Stain with Coomassie Brilliant Blue (5–10 minutes) and destain.

   If the protein bands are too faint, recycle the gel through the staining and destaining procedure again or lengthen the period of staining.

2. Place the gel on the light box. Excise the protein bands with a scalpel, and place the bands in separate polypropylene tubes.

3. Soak the excised gel bands, containing the proteins to be compared, in 10 ml of ethanol containing 1 M Tris-Cl (pH 6.8) for 30 minutes at room temperature to shrink the gel.

   This faciltates loading the gel into the second slab gel. The washing time should be reduced if the resulting peptide maps are too faint.

4. Cut the gel bands into fragments and place them in separate wells of the second slab gel formed in the long (5 cm) stacking gel. Use a clean spatula to guide the fragments carefully to the bottom of the wells.

5. Fill the regions around the gel fragments with proteinase digestion buffer delivered from a syringe.

6. Overlay each sample with 10 μl of proteinase solution.

   Use sufficient proteinase to complete the digestion. Try a 1:50 ratio of enzyme to protein substrate initially.

7. Carry out electrophoresis until the bromophenol blue nears the bottom of the stacking gel.

8. Switch off the electrical current for 20–30 minutes to allow proteinase digestion of the protein samples.

9. Resume electrophoresis and continue until the bromophenol blue nears the bottom of the slab gel.

10. Visualize the separated peptides by Coomassie Blue or silver staining (see Protocol 9 or 11). Alternatively, detect radiolabeled peptides using autoradiography or fluorography.

Method 2: Chemical Fragmentation

1. Fix the gel, stain with Coomassie Brilliant Blue (5–10 minutes), and destain.

   If the protein bands are too faint, recycle the gel through the staining and destaining procedure again or lengthen the period of staining.

2. Place the gel on the light box. Excise the protein bands with a scalpel, and place the bands in separate 1.5-ml polypropylene tubes.

3. Dry the protein-containing gel pieces using a SpeedVac concentrator (or equivalent device; e.g., air-dry).

4. Add ~10 μl of the appropriate chemical cleavage reagent (1 mg/ml) directly onto the dried gel pieces, followed by 90 μl of cleavage reagent.

5. Incubate the mixture for the required period of time (for incubation conditions, see Protocol 6, 7, or 8).

6. Carefully aspirate the peptide-containing supernatant liquor.

7. Dry the sample in a SpeedVac, add 50 μl of H_2O, and dry the sample again.

8. Repeat the H_2O wash and drying two more times.

9. Add 10–30 μl of SDS-PAGE sample buffer, and boil the samples for 5 minutes.

10. Load the samples onto the SDS-PAGE gel and proceed with the electrophoresis until the bromophenol blue nears the bottom of the slab gel.

11. Visualize the separated peptides by Coomassie Blue or silver staining (see Protocol 9 or 11). Alternatively, detect radiolabeled peptides using autoradiography or fluorography.

NOMENCLATURE ON PROTEASES, PROTEINASES, AND PEPTIDASES

In 1984, the International Union of Biochemistry and Molecular Biology (IUBMB) recommended the use of the term peptidase for the subset of peptide bond hydrolases (Subclass EC 3.4.). The widely used term protease is synonymous with the term peptidase. Proteases comprise two groups of enzymes: the endoproteases and the exoproteases.

- ***Endoproteases*** cleave peptide bonds at points within the polypeptide chain, generating peptide fragments. They are used mainly for peptide-mapping studies. The endoproteases include enzymes that have a high selectively for cleaving peptide bonds exclusively adjacent to charged amino acids (the basic amino acids arginine and lysine, e.g., trypsin) and the acidic residues glutamic acid (e.g., Glu-C) and aspartic acid (e.g., Asp-N). For a more extensive list of proteases, see Tables 7.3 and 7.4. Proteases like chymotrypsin, thermolysin, or pepsin cleave adjacent to several amino acid residues and hence yield a more extended fragmentation. For a list of endoproteases and their selectivities, see Table 7.3.
- ***Exoproteases*** remove amino acids sequentially from either the amino or carboxyl terminus, respectively. They are of special importance for the removal of amino-terminal blocking groups such as pyroglutamate (e.g., pyroglutamate aminopeptidase, 5-oxoprolylo-peptidase, EC 3.4.19.3; Armentrout and Doolittle 1969) or an *N*-acetyl group (acylamino-acid-releasing enzyme, acylaminoacyl peptidase, EC 3.4.19.1; Kobayashi and Smith 1987). For the acylpeptide hydrolases, the neighboring amino acid residues affect the efficiency of fragmentation considerably, and applications are generally restricted to peptides rather than proteins. Similarly, the acyl amino-acid-releasing enzyme (AARE) (Mitta et al. 1989), used to cleave *N*-acetylated amino-terminal residues, is restricted to peptides of ~40 residues in length; hence, the polypeptide chain must be fragmented first and the amino-terminal peptide isolated prior to AARE digestion. Cathepsin C cleaves repetitively dipeptides from the amino terminus of a polypeptide chain. The carboxypeptidases are of particular importance for the stepwise removal of amino acid residues from the carboxyl terminus.

Proteinase is also a term used as a synonym for endopeptidase. There are four classes of proteinases recognized by the IUBMB, and they have been grouped according to their catalytic mechanism: serine proteinases, cysteine proteases, aspartic proteinases, and metalloproteinases. It has been suggested that this classification by catalytic types be extended by a classification by families based on the evolutionary relationships of proteases (Rawlings and Barrett 1993). Additionally, there is a section of the *Enzyme Nomenclature* that is allocated for proteases of unidentified catalytic mechanism. Although this indicates that the catalytic mechanism for this group of proteases has not been identified, the possibility remains that novel types of proteases may exist.

Proteases are involved in a great variety of physiological processes. Their action can be divided into two different categories:

- ***Limited proteolysis.*** This group of proteases cleaves only one or a limited number of peptide bonds of a target protein, leading to the activation or maturation of the formerly inactive protein (e.g., conversion of prohormones to hormones).
- ***Unlimited proteolysis.*** Proteases belonging to this group degrade proteins totally into their amino acid constituents. Proteins to be degraded are usually first conjugated with ubiquitin, a modification that marks them for rapid hydrolysis by the proteasome in the presence of ATP (for a review of ubiquitinylation, see Hershko and Ciechanover 1998; for a review of degradation, see Baumeister et al. 1998). Another pathway consists of the compartmentation of proteases, e.g., in lysosomes. Proteins transferred into this compartment undergo rapid proteolytic degradation (Baumeister et al. 1998).

REFERENCES

Aebersold R. 1993. Internal amino acid sequence of proteins after in situ protease digestion on nitrocellulose. In *A practical guide for protein and peptide purification for microsequencing*, 2nd edition (ed. P. Matsudaira), pp. 73–81. Academic Press, New York.

Aebersold R.H., Leavitt J., Saavedra R.A., Hood L.E., and Kent S.B. 1987. Internal amino acid sequence analysis of proteins separated by one- or two-dimensional gel electrophoresis after in situ protease digestion on nitrocellulose. *Proc. Natl. Acad. Sci.* **84:** 6970–6974.

Ambler R.P. 1965. The behaviour of peptides formed by cyanogen bromide cleavage of proteins. *Biochem. J.* **96:** 32.

Anderson W.L. and Wetlaufer D.B. 1975 A new method for disulfide analysis of peptides *Anal. Biochem.* **67:** 493–502.

Armentrout R.W. and Doolittle R.F. 1969. Pyrrolidonecarboxylyl peptidase: Stabilization and purification. *Arch. Biochem. Biophys.* **132:** 80–90.

Atassi M.Z. and Habeeb A.F.S.A. 1972. Reaction of proteins with citraconic anhydride. *Methods Enzymol.* **25:** 546–553.

Baumeister W., Walz J., Zühl F., and Seemüller E. 1998. The proteasome: Paradigm of a self-compartmentalizing protease. *Cell* **92:** 367–380.

Bergman T. and Jörnvall H. 1987. Electroblotting of individual polypeptides from SDS/polyacrylamide gels for direct sequence analysis. *Eur. J. Biochem.* **169:** 9–12.

Blumenfeld O.O., Rojkind M., and Gallop P.M. 1965. Subunits of hydroxylamine-treated tropocollagen. *Biochemistry* **4:** 1780–1788.

Bond J.S. and Beynon R.J., eds. 1989. *Proteolytic enzymes: A practical approach.* IRL Press/Oxford University Press, Oxford, United Kingdom.

Bornstein P. and Balian G. 1977. Cleavage at Asn-Gly bonds with hydroxlyamine. *Methods Enzymol.* **47:** 132–145.

Brown J.L. 1979. A comparison of the turnover of alpha-N-acetylated and nonacetylated mouse L-cell proteins. *J. Biol. Chem.* **254:** 1447–1449.

Brown J.L. and Roberts W.K. 1976. Evidence that approximately 80 per cent of the soluble proteins from Ehrlich ascites cells are N^{α}-acetylated. *J. Biol. Chem.* **251:** 1009–1014.

Brune D.C. 1992 . Alkylation of cystein with acrylamide for protein sequence analysis. *Anal. Biochem.* **207:** 285–290.

Burns J.A., Butler J.C., Moran J., and Whitesides G.M. 1991. Selective reduction of disulfides by tris(2-carboxyethyl)phosphine. *J. Org. Chem.* **56:** 2648–2650.

Burnstein Y. and Patchornik A. 1972. Selective chemical cleavage of tryptophanyl peptide bonds in peptides and proteins. *Biochemistry* **11:** 4641–4650.

Candiano G., Porotto M., Lanciotto M., and Ghiggeri G.M. 1996. Negative staining of proteins in polyacrylamide gels with methyl trichloroacetate. *Anal. Biochem.* **243:** 245–248.

Casero P., Del Campo G.B., and Righetti P.G. 1985. Negative aurodye for polyacrylamide gels: The impossible stain. *Electrophoresis* **6:** 362–372.

Castellanos-Serra L., Proenza W., Huerta V., Moritz R.L, and Simpson R.J. 1999. Proteome analysis of polyacrylamide gel-separated proteins visualized by reversible negative staining using imidazole-zinc salts. *Electrophoresis* **20:** 732–737.

Cavins J.F. and Friedman M.A. 1970. An internal standard for amino acid analyses: S-β-(4-pyridylethyl)-L-cysteine. *Anal. Biochem.* **35:** 489–493.

Cecil R. 1963. Intramolecular bonds in proteins. The role of sulfur in proteins. In *The proteins*, 2nd edition (ed. H. Neurath), vol. 1, pp. 379–476. Academic Press, New York.

Cleland W.W. 1964. Dithiothreitol, a new protective reagent for SH groups. *Biochemistry* **3:** 480–482.

Cleveland D.W., Fischer S.G., Kirschner M.W., and Laemmli U.K. 1977. Peptide mapping by limited proteolysis in sodium dodecyl sulfate and analysis by gel electrophoresis. *J. Biol. Chem.* **252:** 1102–1106.

Cole A.R., Hall N.E., Eddes J.S., Reid G.E., Moritz R.L., Treutlein H., and Simpson R.J. 1999. Disulfide bond structure and *N*-glycosylation sites of the extracellular domain of the human interleukin-6 receptor. *J.*

Biol. Chem. **274:** 7207–7215.

Cole E.G. and Mecham D.K. 1966. Cyanate formation and electrophoretic behaviour of proteins in gels containing urea. *Anal. Biochem.* **14:** 215–222.

Cole R.D. 1961. On the transformation of insulin in concentrated solutions of urea. *J. Biol. Chem.* **236:** 2670–2671.

Crestfield A.M., Moore S., and Stein W.H. 1963. The preparation and enzymatic hydrolysis of reduced and *S*-carboxymethylated proteins. *J. Biol. Chem.* **238:** 622–627.

Dixon H.B.F. and Perham R.N. 1968. Reversible blocking of amino groups with citraconic anhydride. *Biochem. J.* **109:** 312–314.

Drapeau G.R. 1977. Cleavage at glutanaic acid with staphylococcal protease. *Methods Enzymol.* **47:** 189-191.

———. 1978. The primary structure of staphylococcal protease. *Can. J. Biochem.* **56:** 534–544.

———. 1980. Substrate specificity of a proteolytic enzyme isolated from a mutant of *Pseudomonas fragi*. *J. Biol. Chem.* **255:** 839–840.

Du H., Simpson R.J., Clarke A.E., and Bacic A. 1996. Molecular characterization of a stigma-specific gene encoding and arabinogalactan-protein (AGP) from *Nicotiana alata*. *Plant J.* **9:** 313–323.

Eckerskorn C. and Lottspeich F. 1989. Internal amino acid sequence analysis of proteins separated by gel electrophoresis after tryptic digestion in polyacrylamide matrix. *Chromatographia* **28:** 92–94.

Ellman G.L. 1959. Tissue sulfhydryl groups. *Arch. Biochem. Biophys.* **82:** 70–77.

Fernandez J., DeMott M., Atherton D., and Mische S.M. 1992. Internal protein sequence analysis: Enzymatic digestion for less than 10 micrograms of protein bound to polyvinylidene fluoride or nitrocellulose membranes. *Anal. Biochem.* **201:** 255–264.

Fernandez-Patron C., Hardy E., Sosa A., Seoane J., and Castellanon L. 1995a. Double staining of Coomassie blue-stained polyacrylamide gels by imidazole-sodium dodecyl sulfate-zinc reverse staining: Sensitive detection of Coomassie blue-undetected proteins. *Anal. Biochem.* **224:** 263–269.

Fernandez-Patron C., Castellanos-Serra L., Hardy E., Guerra M., Estevez E., Mehl E. and Frank R.W. 1998. Understanding the mechanism of the zinc-ion stains of biomacromolecules in electrophoresis gels: Generalization of the reverse-staining technique. *Electrophoresis* **19:** 2398–2406.

Fernandez-Patron C., Calero M., Collazo P.R., Garcia J.R., Madrazo J., Musacchio A., Soriano F., Estrada R., Frank R., Castellanos-Serra L.R., and Mendez E. 1995b. Protein reverse staining: High-efficiency microanalysis of unmodified proteins detected on electrophoresis gels. *Anal. Biochem.* **224:** 203–211.

Fontana A. and Gross E. 1986. Fragmentation of polypeptides by chemical methods. In *Practical protein chemistry* (ed. A. Darbyre), pp. 67–120. Wiley, Chichester, United Kingdom.

———. 1987. Fragmentation of polypeptides by chemical methods. In *Practical protein chemistry: A handbook* (ed. A. Darbre), pp. 67–120. Wiley, New York.

Fontana A., Dalzoppo D., Grandi C., and Zambonin M. 1983. Cleavage at tryptophan with o-iodosobenzoic acid. *Methods Enzymol.* **91:** 311–318.

Getz E.B., Xiao M., Chakrabarty T., Cooke R., and Selvin P.R. 1999. A comparison between sulfhydryl reductants tris(2-carboxyethyl)phosphine and dithiothreitol for use in protein biochemistry. *Anal. Biochem.* **273:** 73–80.

Gevaert K. and Vandekerckhove J. 2000. Protein identification methods in proteomics. *Electrophoresis* **21:** 1145–1154.

Gibbons I. and Perham R.N. 1970. The reaction of aldolase with 2-methylmaleic anhydride. *Biochem. J.* **116:** 843–849.

Gray W.R. 1993. Disulfide structures of highly bridged peptides: A new strategy for analysis. *Protein Sci.* **2:** 1732–1748.

Grego B., Van Driel I.R., Stearne P.A., Goding J.W., Nice E.C., and Simpson R.J. 1985 A microbore high performance liquid chromatography strategy for the purification of polypeptides for gas-phase sequence analysis: Structural studies on the murine trasferrine receptor. *Eur. J. Biochem.* **148:** 485–491.

Gross E. and Witkop B. 1961. Selective cleavage of the methionyl peptide bonds in ribonuclease with cyanogen bromide. *J. Am. Chem. Soc.* **83:** 1510–1511.

Habeeb A.F.S.A. 1972. Reaction of protein sulfhydryl groups with Ellman's reagent. *Methods Enzymol.* **25:** 457–464.

Hagel P., Gerding J.J.T., Fieggen W., and Bloemendal H. 1971. Cyanate formation in solutions of urea. I. Calculation of cyanate concentrations at different temperature and pH. *Biochem. Biophys. Acta* **243:** 366–373.

Hagmann M.-L. 1988. *Handbook of proteolytic enzymes* (ed. A.J. Barrett et al.), pp. 1542–1543. Academic Press, New York.

Hagmann M.-L., Geuss U., Fischer S., and Kresse G.-B. 1995. Peptidyl-Asp metalloendopeptidase. *Methods Enzymol.* **248:** 782–787.

Harlow E. and Lane D., eds. 1999. *Using antibodies: A laboratory manual.* Cold Spring Harbor Laboratory Press, Cold Spring Harbor, New York.

Heinrickson R.L. 1977. Applications of thermolysin in protein structural analysis. *Methods Enzymol.* **47:** 175–189.

Hellman U., Wernstedt C., Gonez J., and Heldin C.-H. 1995. Improvement of an "in-gel" digestion procedure for the micropreparation of internal protein fragments for amino acid sequencing. *Anal. Biochem.* **224:** 451–455.

Henzel W.J., Billeci T.M., Stults J.T., Wong S.C., Grimley C., and Watanabe C. 1993. Identifying proteins from two-dimensional gels by molecular mass searching of peptide fragments in protein sequence databases. *Proc. Natl. Acad. Sci.* **90:** 5011–5015.

Herbert B. 1999. Advances in protein solubilisation for two-dimensional gel electrophoresis. *Electrophoresis* **20:** 650–663.

Herbert B., Molloy M.P., Gooley A.A., Walsh B.J., Bryson W.G., and Williams K.L. 1998. Improved protein solubility in two-dimensional-gel electrophoresis using tributyl phosphine as a reducing agent. *Electrophoresis* **19:** 845–851.

Hershko A. and Ciechanover A. 1998. The ubiquitin system. *Annu. Rev. Biochem.* **67:** 425–480.

Higgins R.C. and Dahmus M.E. 1979. Rapid visualization of protein bands in preparative SDS-polyacrylamide gels. *Anal. Biochem.* **93:** 257–260.

Hirs C.H.W. 1967. Determination of cystine as cysteic acid. *Methods Enzymol.* **11:** 59–62.

Horn M. and Laursen R.A. 1973. Solid-phase Edman degradation. Attachment of carboxy-terminal homoserine peptides to an insoluble resin. *FEBS Lett.* **36:** 285–288.

Houmard J. and Drapeau G.P. 1972. Staphylococcal protease: A proteolytic enzyme specific for glutamoyl bonds. *Proc. Natl. Acad. Sci.* **69:** 3506–3509.

Hunkapiller M.W., Luhan E., Ostrander F., and Hood L.E. 1983. Isolation of microgram quantities of proteins from polyacrylamide gels for amino acid sequence analysis. *Methods Enzymol.* **91:** 227–236.

Inglis A.S. and Edman P. 1970. Mechanism of cyanogen bromide reaction with methionine in peptides and proteins. *Anal. Biochem.* **37:** 73–80.

Inglis A.S. and Liu T.-Y. 1970. The stability of cysteine and cystine during acid hydrolysis of proteins and peptides. *J. Biol. Chem.* **245:** 112–116.

Iwamatsu A. 1992. S-carboxymethylation of proteins transferred onto polyvinylidene difluoride membranes followed by in situ protease digestion and amino acid microsequencing. *Electrophoresis* **13:** 142–147.

Jacobson G.R., Schaffer M.H., Stark G.R., and Vanaman T.C. 1973. Specific chemical cleavage in high yield at the amino peptide bonds of cysteine and cystine residues. *J. Biol. Chem.* **248:** 6583–6591.

Jahnen W., Ward L.D., Reid G.E., Moritz R.L., and Simpson R.J. 1990. Internal amino acid sequencing of proteins by *in situ* cyanogen bromide cleavage in polyacrylamide gels. *Biochem. Biophys. Res. Commun.* **166:** 139–145.

Jemmerson R. and Paterson Y. 1986. Mapping epitopes on a protein antigen by the proteolysis of antigen-antibody complexes. *Science* **232:** 1001–1004.

Jenö P., Mini T., Moes S., Hintermann E., and Horst M. 1995. Internal sequences from proteins digested in polyacrylamide gels. *Anal. Biochem.* **224:** 75–82.

Kasper C.B. 1975. Fragmentation of proteins for sequence studies and separation of peptide mixtures. In *Protein sequence determination*, 2nd edition. *Mol. Biol. Biochem. Biophys.* **8:** 114–161.

Kawasaki H., Emori Y., and Suzuki K. 1990. Production and separation of peptides from proteins stained with Coomassie brilliant blue R-250 after separation by sodium dodecyl sulfate-polyacrylamide gel electrophoresis. *Anal. Biochem.* **191:** 332–336.

Keil-Dlouhá V., Zylba N., Imhoff J.-M., Tong N.-T., and Keil B. 1971. Proteolytic activity of pseudotrypsin. *FEBS Lett.* **16:** 291–295.

Kellner R. 1994. Chemical and enzymatic fragmentation of proteins. In *Microcharacterization of proteins* (ed. R. Kellner et al.), pp. 11–27. VCH, New York.

Kellner, R., Lottspeich F., and Meyer H.E., eds. 1994. *Microcharacterization of proteins.* VCH, New York.

Klotz I.M. 1967. Succinylation. *Methods Enzymol.* **11:** 576–580.

Kobayashi K. and Smith J.A. 1987. Acyl-peptide hydrolase from rat liver. Characterization of enzyme reaction. *J. Biol. Chem.* **262:** 11435–11445.

Konigsberg W. 1972. Reduction of disulfide bonds in proteins with dithiothreitol. *Methods Enzymol.* **25:** 185–188.

Kwong M.Y. and Harris R.J. 1994. Identification of succinimide sites in proteins by N-terminal sequence analysis after alkaline hydroxylamine cleavage. *Protein Sci.* **3:** 147–149.

Laemmli U.K. 1970. Cleavage of structural proteins during the assembly of the head of bacteriophage T4. *Nature* **227:** 680–696.

Lee C., Levin A., and Branton D. 1987. Copper staining: A five-minute protein stain for sodium dodecyl sulfate-polyacrylamide gels. *Anal. Biochem.* **166:** 308–312.

Liu T.-Y. 1977. The role of sulfur in proteins. In *The proteins*, 3rd edition (ed. H. Neurath and R.L. Hill.), pp. 240–403. Academic Press, New York.

Lonnie D.A. and Weaver K.M. 1990. Detection and recovery of proteins from gels following zinc chloride staining. *Appl. Theor. Electrophoresis* **1:** 279–282.

Lorand L., ed. 1981. Proteolytic enzymes. *Methods Enzymol.*, vol. 80.

Lui M., Tempst P., and Erdjument-Bromage H. 1996. Methodical analysis of protein-nitrocellulose interactions to design a refined digestion protocol. *Anal. Biochem.* **241:** 156–166.

Maclaren J.A. and Sweetman B.J. 1966. The preparation of reduced and *S*-alkylated wool keratins using TRI-n-butylphosphine. *Aust. J. Chem.* **19:** 2355–2360.

Mahoney W.C. and Hermodson M.A. 1979. High yield cleavage of tryptophanyl bonds with o-iodosobenzoic acid. *Biochemistry* **18:** 3810–3814.

Mahoney W.C., Smith P.K., and Hermodson M.A. 1981 Fragmentation of proteins with o-iodosobenzoic acid: Chemical mechanisms and identification of o-iodosobenzoic acid as a reactive contaminant that modifies tyrosyl residues. *Biochemistry* **20:** 443–448.

Mant C.T. and Hodges R.S., eds. 1991. *High-performance liquid chromatography of peptides and proteins: Separation, analysis, and conformation.* CRC Press, Boca Raton, Florida.

Majors R.E. and Przybyciel M. 2002. Columns for reversed-phase LC separations in highly aqueous mobile phases. *LCGC* **20:** 584–593 (www.chromatographyonline.com).

McCormack A.L., Schieltz D.M., Goode B., Yang S., Barnes G., Drubin D., and Yates J.R. 1997. Direct analysis and identification of proteins in mixtures by LC/MS/MS and database searching at the low-femtomole level. *Anal. Biochem.* **69:** 767–776.

Merril C.R. 1987. Detection of proteins separated by electrophoresis. *Adv. Electrophoresis* **1:** 111–139.

Merril C.R. and Washart K.M. 1998. Protein detection methods. In *Gel electrophoresis of proteins. A practical approach,* 3rd edition (ed. B.D. Hames), pp. 53–92. Oxford University Press, Oxford, United Kingdom.

Meyer T.S. and Lambert B.L. 1965. Use of Coomassie brilliant blue R250 for the electrophoresis of microgram quantities of parotid saliva proteins on acrylamide-gel strips. *Biochem. Biophys. Acta* **107:** 144–145.

Mitta M., Asada K., Uchimura Y., Kimizuka F., Kato I., Sakiyama F., and Tsunasawa S. 1989. The primary structure of porcine liver acylamino acid-releasing enzyme deduced from cDNA sequences. *J. Biochem.* **106:** 548–551.

Moritz R.L. and Simpson R.J. 1992a. Application of capillary reversed-phase high-performance liquid chromatography to high-sensitivity protein sequence analysis. *J. Chromatogr.* **599:** 119–130.

———. 1992b. Purification of proteins and peptides for sequence analysis using microcolumn liquid chromatography. *J. Microcol. Sep.* **4:** 485–489.

———. 1993. Capillary liquid chromatography: A tool for protein structural analysis. In *Methods in protein sequence analysis* (ed. K. Imahori and F. Sakiyama), pp. 3–10. Plenum Press, New York.

Moritz R.L., Eddes J.S., Reid G.E., and Simpson R.J. 1996. S-pyridylethylation of intact polyacrylamide gels and *in situ* digestion of electrophoretically-separated proteins: A rapid mass spectrometric method for identifying cysteine-containing peptides. *Electrophoresis* **17:** 907–917.

Moritz R.L., Hall N.E., Connolly L.M., and Simpson R.J. 2001. Determination of the disulfide structure and N-glycosylation sites of the extracellular domain of the human signal transducer gp130. *J. Biol. Chem.* **276:** 8244–8253.

Moritz R.L., Reid G.E., Ward L.D., and Simpson R.J. 1994. Capillary HPLC: A method for protein isolation and peptide mapping. *Methods* **6:** 213–226.

Moritz R.L., Eddes J.S., Ji H., Reid G.E., and Simpson R.J. 1995. Rapid separation of proteins and peptides using conventional silica-based supports: Identification of 2-D gel proteins following in-gel proteolysis. In *Techniques in protein chemistry VI* (ed. J.W. Crabb), pp. 311–319. Academic Press, San Diego, California.

Muzio M., Chinnaiyan A.M., Kischkel F.C., O'Rourke K., Shevchenko A., Ni J., Scaffidi C., Bretz J.D., Zhang M., Gentz R., Mann M., Krammer P.H., Peter M.E., and Dixit V.M. 1996. FLICE, a novel FADD-homologous ICE/CED-3-like protease, is recruited to the CD95 (Fas/APO-1) death-inducing signaling complex. *Cell* **85:** 817–827.

Nelles L.P. and Bamburg J.R. 1976. Rapid visualization of protein-dodecyl-sulfate complexes in polyacrylamide gels. *Anal. Biochem.* **73:** 522–531.

Neuhof V., Arold N., Taube D., and Ehrhardt W. 1988. Improved staining of proteins in polyacrylamide gels including isoelectric focusing gels with clear background at nanogram sensitivity using Coomassie Brilliant Blue G-250 and R-250. *Electrophoresis* **9:** 255–262.

Omenn G.S., Fontana A., and Anfinsen C.B. 1970. Modification of the single tryptophan residue of staphylococcal nuclease by a new mild oxidizing agent. *J. Biol. Chem.* **245:** 1895–1902.

Ortiz M.L., Calero M., Fernandez-Patron C., Patron C.F., Castellanos L., and Mendez E. 1992. Imidazole-SDS-Zn reverse staining of proteins in gels containing or not SDS and microsequence of individual unmodified electroblotted proteins. *FEBS Lett.* **296:** 300–304.

Ozols J. and Gerard C. 1977. Covalent structure of the membranous segment of horse cytochrome b_5: Chemical cleavage of the native hemoprotein. *J. Biol. Chem.* **252:** 8549–8553.

Parham P., Matthew J., Androlewicz M.J., Brodsky F.M., Holmes N.J., and Ways J.P. 1982 Monoclonal antibodies: Purification, fragmentation, and application to structural and functional studies of class I MHC antigen. *J. Immunol. Methods* **53:** 133–173.

Patterson S.D. and Aebersold R. 1995. Mass spectrometric approaches for the identification of gel-separated proteins. *Electrophoresis* **16:** 1791–1814.

Patterson S.D., Hess D., Yungwirth T., and Aebersold R. 1992. High-yield recovery of electroblotted proteins and cleavage fragments from a cationic polyvinylidene fluoride-based membrane. *Anal. Biochem.* **202:** 193–203.

Patton W.F., Lam L., Su Q., Lui M., Erdjument-Bromage H., and Tempst P. 1994. Metal chelates as reversible stains for detection of electroblotted proteins: Application to protein microsequencing and immunoblotting. *Anal. Biochem.* **220:** 324–335.

Polverino de Laureto P., Scarmella E., Frigo M., Wondrich F.G., De Filippis V., Zambonin M., and Fontana A. 1999. Limited proteolysis of bovine α-lactalbumin: Isolation and characterization of protein domains. *Protein Sci.* **8:** 2290–2303.

Rabilloud T. 1990. Mechanisms of protein silver staining in polyacrylamide gels: A 10-year synthesis. *Electrophoresis* **11:** 785–794.

———. 1999. Silver staining of 2-D electrophoresis gels. *Methods Mol. Biol.* **112:** 297–305.

———. 2000. Detecting proteins separated by 2-D gel electrophoresis. *Anal. Chem.* **72:** 48A–55A.

———. 2001. A new silver staining apparatus and procedure for matrix/ionization-time of flight analysis of proteins after two-dimensional electrophoresis. *Proteomics* **1:** 835–840.

Rabilloud T., Vuillard L., Gilly C., and Lawrence J.-J. 1994. Silver-staining of proteins in polyacrylamide gels: A general overview. *Cell. Mol. Biol.* **40:** 57–75.

Raftery M.A. and Cole R.D. 1966. On the aminoethylation of proteins. *J. Biol Chem.* **241:** 3457–3461.

Rawlings N.D. and Barrett A.J. 1993. Evolutionary families of peptidases. *Biochem. J.* **290:** 205–218.

Rawlings N.D., O'Brien E., and Barrett A.J. 2002. MEROPS: The protease database. *Nucleic Acids Res.* **30:** 343–346.

Renlund S., Klintrot I.-M., Nunn M., Schrimsher J.L., Wernstedt C., and Hellmann U. 1990. Peptide mapping of HIV polypeptides expressed in *Escherichia coli*—Quality control of different batches and identification of tryptic fragments containing residues of aromatic amino acids or cysteine. *J. Chromatogr.* **512:** 325–335.

Riddles P.W., Blakeley R.L., and Zerner B. 1983. Reassessment of Ellman's reagent. *Methods Enzymol.* **91:** 49–60.

Rittenhouse J. and Marcus F. 1984. Peptide mapping by polyacrylamide gel electrophoresis after cleavage at aspartyl-prolyl peptide bonds in sodium dodecyl sulfate-containing buffers. *Anal. Biochem.* **138:** 442–448.

Riviere L.R., Fleming M., Elicone C., and Tempst P. 1991. Study and applications of the effects of detergents and chaotropes on enzymatic proteolysis. In *Techniques in protein chemistry II* (ed. J.J. Villafranca), pp. 171–179. Academic Press, San Diego, California.

Rosenfeld J., Capdevielle J., Guillemot J., and Ferrara P. 1992. In-gel digestion of proteins for internal sequence analysis after one- or two-dimensional gel electrophoresis. *Anal. Biochem.* **203:** 173–179.

Rüegg U.T. and Rudinger J. 1977. Reductive cleavage of cystine disulfides with tributylphosphine. *Methods Enzymol.* **47:** 111–116.

Sanger F. 1949. Fractionation of oxidized insulin. *Biochem. J.* **44:** 126–128.

Saris C.J.M., van Eenbergen J., Jenks B.G., and Bloemers H.P.J. 1983. Hydroxylamine cleavage of proteins in polacrylamide gels. *Anal. Biochem.* **132:** 54–67.

Savige W.E. and Fontana A. 1977. Cleavage of the tryptophanyl peptide bond by dimethyl sulfoxide-hydrobromic acid. *Methods Enzymol.* **47:** 459–469.

Schaffner W. and Weissman C. 1973. A rapid, sensitive, and specific method for the determination of protein in dilute solution. *Anal. Biochem.* **56:** 502–514.

Schägger H. and von Jagow G. 1987. Tricine-sodium dodecyl sulfate-polyacrylamide gel electrophoresis for the separation of proteins in the range 1 to100 kDa. *Anal. Biochem.* **166:** 368–397.

Schechter I. and Berger A. 1967. On the size of the active site in proteases. *Biochem. Biophys. Res. Commun.* **27:** 157–162.

Shechter Y., Patchornik A., and Burstein Y. 1976. Selective chemical cleavage of tryptophanyl peptide bonds by oxidative chlorination with N-chlorosuccinimide. *Biochemistry* **15:** 5071–5075.

Shevchenko A., Wilm M., Vorm O., and Mann M. 1996. Mass spectrometric sequencing of proteins silver-stained polyacrylamide gels. *Anal. Chem.* **68:** 850–858.

Simpson R.J. and Nice E.C. 1984. *In situ* cyanogen bromide cleavage of N-terminally blocked proteins in a gas-phase sequencer. *Biochem. Int.* **8:** 787–791.

Simpson R. and Reid G.E. 1998. Sequence analysis of gel-resolved proteins. In *Gel electrophoresis of proteins: A practical approach*, 3rd edition (ed. B.D. Hames), pp. 237–267. Oxford University Press, Oxford, United Kingdom.

Simpson R.J., Neuberger M.R., and Liu T.-Y. 1976. Complete amino acid analysis of proteins from a single protein hydrolysate. *J. Biol. Chem.* **251:** 1936–1940.

Simpson R.J., Moritz R.L., Nice E.C., and Grego B. 1987. A high-performance liquid chromatography procedure for recovering subnanomole amounts of protein from SDS-gel electroeluates for gas-phase sequence analysis. *Eur. J. Biochem.* **165:** 21–29.

Simpson R.J., Moritz R.L., Rubira M.R., and Van Snick J. 1988a. Murine hybridoma/plasmacytoma growth factor. *Eur. J. Biochem.* **176:** 187–197.

Simpson R.J., Moritz R.L., Van Roost E., and Van Snick J. 1988b. Characterization of a recombinant murine interleukin-6: Assignment of disulphide bonds. *Biochem. Biophys. Res. Commun.* **157:** 364–372.

Simpson R.J., Moritz R.L., Begg G.S., Rubira M.R., and Nice E.C. 1989. Micropreparative procedures for high sensitivity sequencing of peptides and proteins. *Anal. Biochem.* **177:** 221–236.

Snyder L.R. and Kirkland J.J. 1979. *Introduction to modern liquid chromatography*, 2nd edition. Wiley, New York.

Stark G.R., Stein W.H., and Moore S. 1960. Reactions of the cyanate present in aqueous urea with amino acids and proteins. *J. Biol. Chem.* **235:** 3177–3181.

Stephenson R.C. and Clarke S. 1989. Succinimide formation from aspartyl and asparaginyl peptides as a model for the spontaneous degradation of proteins. *J. Biol. Chem.* **264:** 6164–6170.

Stone K.L., McNulty D.E., LoPresti M.L., Crawford J.M., DeAngelis R., and Williams K.R. 1992. Elution and internal amino acid sequencing of PVDF-blotted proteins. In *Techniques in protein chemistry III* (ed. R. Angeletti), pp. 23–34. Academic Press, San Diego, California.

Sweetman B.J. and Maclaren J.A. 1966. The reduction of wool keratin by tertiary phosphines. *Aust. J. Chem.* **19:** 2347–2354.

Switzer III R.C., Merril C.R., and Shifrin S. 1979. A highly sensitive silver stain for detecting proteins and peptides in polyacrylamide gels. *Anal. Biochem.* **98:** 231–237.

Syrovy I. and Hodny Z. 1991. Staining and quantification of proteins separated by polyacrylamide gel electrophoresis. *J. Chromatogr.* **569:** 175–196.

Titani K., Hermodson M.A., Ericsson L.H., Walsh K.A., and Neurath H. 1972. Amino acid sequence of thermolysin. Isolation and characterization of the fragments obtained by cleavage with cyanogen bromide.

Biochemistry **11:** 2427–2435.

Tong D., Moritz R.L., Eddes J.S., Reid G.E., Rasmussen R.K., Dorow D.S., and Simpson R.J. 1997. Fabrication of stable packed capillary reversed-phase columns for protein structural analysis. *J. Prot. Chem.* **16:** 425-431.

Traub W. and Piez K.A. 1971. The chemistry and structure of collagen. *Adv. Protein Chem.* **25:** 243–352.

Vorm O., Roepstorff P., and Mann M. 1994. Improved resolution and very high sensitivity in MALDI TOF of matrix surface made by fast evaporation. *Anal. Chem.* **66:** 3281–3287.

Ward L.D., Reid G.E., Moritz R.L., and Simpson R.J. 1990. Strategies for internal amino acid sequence analysis of proteins separated by polyacrylamide gel electrophoresis. *J. Chromatogr.* **519:** 199–216.

Welinder K.G. 1988. Generation of peptides suitable for sequence analysis by proteolytic cleavage in reversed-phase high-performance liquid chromatography solvents. *Anal. Biochem.* **174:** 54–64.

Williams K., Kobayashi R., Lane W., and Tempst P. 1993. Internal amino acid sequencing: Observations from four different laboratories. *Assoc. Biomol. Res. Fac. News* **4:** 7.

Wilson C.M. 1983. Staining of proteins on gels: Comparisons of dyes and procedures. *Methods Enzymol.* **91:** 236–247.

Wu J. and Watson J.T. 1997. A novel methodology for assignment of disulfide bond pairings in proteins. *Protein Sci.* **6:** 391–398.

FURTHER READING

Bond J.S. and Beynon R.J., eds. 1989. *Proteolytic enzymes: A practical approach.* IRL Press/Oxford University Press, Oxford, United Kingdom.

Fontana A. and Gross E. 1986. Fragmentation of polypeptides by chemical methods. In *Practical protein chemistry* (ed. A. Darbyre), pp. 67–120. Wiley, Chichester, United Kingdom.

Liu T.-Y. 1977. The role of sulfur in proteins. In *The proteins*, 3rd edition (ed. H. Neurath and R.L. Hill.), pp. 240–403. Academic Press, New York.

Lorand L., ed. 1981. Proteolytic enzymes. *Methods Enzymol.,* vol. 80.

Snyder L.R. and Kirkland J.J. 1979. *Introduction to modern liquid chromatography*, 2nd edition. Wiley, New York.

WWW RESOURCES

http://biochem.roche.com/ Roche Diagnostics Corporation homepage.

http://biotakara.co.jp/BIO TaKaRa Biotechnology products catalog online.

http://prowl.rockefeller.edu/recipes/contents.htm A collection of protocols for protein mass spectrometry.

http://www.alltechweb.com Alltech homepage.

http://www.calbiotech.com Calbiotech home page.

http://www.chem.qmw.ac.uk/iubmb/enzyme/ Recommendations of the Nomenclature Committee of the International Union of Biochemistry and Molecular Biology (NC-IUBMB) on the nomenclature and classification of enzyme-catalysed reactions.

http://www.delphi.phys.univ-tours.fr/Prolysis/ University of Tours, France, Prolysis homepage.

http://www.expasy.ch/enzyme/ Enzyme nomenclature database.

http://www.ionsource.com Mass spectrometry and biotechnology resource.

http://www.merck.de/services/chromatographie/reagents/lichropur.htm LiChropur® reagents for analytical HPLC.

http://www.merops.ac.uk Catalog and structure-based classification of proteolytic enzymes.

http://www.sigma-aldrich.com Sigma-Aldrich homepage.

http://www.spectrapor.com Spectrum Laboratories homepage.

http://www.upchurch.com/Techinfo/interchange.asp Interchangeability.

http://www.worthington-biochem.com Worthington Biochemical Corporation homepage.

CHAPTER 8

The Use of Mass Spectrometry in Proteomics*

*The chapter text includes contributions by Martin R. Larsen (Macquarie University, Sydney, Australia); David M. Schieltz and Michael P. Washburn (The Torrey Mesa Research Institute, San Diego, California); Catherine Déon, Willy Bienvenut, Jean-Charles Sanchez, and Denis F. Hochstrasser (Geneva University Hospital, Geneva, Switzerland); Markus Müeller, Robin Gras, and Ron D. Appel (Swiss Institute of Bioinformatics, Geneva, Switzerland); Andrew A. Gooley, Femia G. Hopwood, Janice L. Duff, Parag S. Ghandi, Cameron J. Hill, Wendy L. Holstein, Paul E. Smith, Andrew J. Sloane, and Keith L. Williams (Proteome Systems, Sydney, Australia); Patrick W. Cooley and David B. Wallace (MicroFab Technologies Inc, Plano, Texas); Gavin E. Reid, David Frecklington, Eugene A. Kapp, and James S. Eddes (Ludwig Institute for Cancer Research and the Walter and Eliza Hall Institute of Medical Research, Parkville, Australia); Richard O'Hair (School of Chemistry, University of Melbourne, Australia); Melanie P. Gygi, Larry J. Licklider, Junmin Peng, and Steven P. Gygi (Harvard Medical School, Boston, Massachusetts); Ted Bures (Amgen Inc., Thousand Oakes, California); Gavin Reid and Scott McLuckey (Department of Chemistry, Purdue University, Indiana).

PROTOCOLS

INFORMATION PANELS

MASS SPECTROMETRY (MS)-BASED METHODS FOR THE IDENTIFICATION of gel-separated proteins are fundamental platform technologies for proteomics. Indeed, MS has essentially replaced the Edman degradation procedure, the classical technique for protein identification (see Chapter 6), because it is much more sensitive, can cope with protein mixtures, and offers much higher sample throughput. Current MS-based proteomic strategies rely primarily on digestion of either one-dimensional (1D) (see Chapter 2) or two-dimensional (2D) (see Chapter 4) gel-separated proteins into peptides using a sequence-specific protease such as trypsin. The reasons for analyzing proteolytically derived peptides rather than intact proteins by MS are that proteins are inherently more difficult to elute from polyacrylamide gels and to analyze by MS. (Generally, the molecular weight of the protein alone is not sufficient for unambiguous protein identification. This problem is exacerbated when posttranslational modifications must be taken into account. Methods and instrumentation for the fragmentation of whole-protein ions have not been readily available until recently; see information panel on "TOP-DOWN" PROTEIN SEQUENCE ANALYSIS USING MS/MS.) In contrast, peptides are readily eluted from gels following in situ proteolysis (see Chapter 7, Protocol 13), and the masses of even a small subset of tryptic peptides from a protein, or sequence information obtained from the fragmentation of selected peptides, usually provide sufficient information to allow unambiguous protein identification.

The steps typically taken during an MS-based proteomics analysis are illustrated in Figure 8.1. Although the proteomic strategy outlined in Figure 8.1 is based on the characterization of proteolytically derived peptides for subsequent protein identification (referred to as a "bottom-up" MS approach), direct MS/MS of intact protein molecular ions, which avoids the need to digest proteins prior to MS analysis, can also provide sufficient fragment masses for database searching (Mortz et al. 1996; Meng et al 2001; Reid and McLuckey 2002; also see the information panel on "TOP-DOWN" PROTEIN SEQUENCE ANALYSIS USING MS/MS). This latter approach to protein characterization is referred to as "*top-down*" protein characterization. More recently, a comprehensive method for proteome analysis that integrates both "bottom-up" and "top-down" approaches has been described (see VerBerkmoes et al. 2002).

GENERAL PRINCIPLES

Mass spectrometry, also called mass spectroscopy, is an instrumental approach that allows for the mass measurement of ions derived from molecules. Mass spectrometers are capable of forming, separating, and detecting molecular ions based on their mass-to-charge ratio (*m/z*). (Note that, formerly, *e* and not *z* was used to denote the charge on an ion. Accordingly, many earlier publications in the literature represent the mass-to-charge ratio of an analyte as *m/e*. Here, the term *m/z* is used to conform to IUPAC recommendations.) Mass spectrometers consist of a set of essential modular components (Figure 8.2):

- ***Ionization source.*** The ionization source converts and transfers molecules (analytes) into gas-phase ions.
- ***Mass analyzer.*** This device is used for separating gas-phase ions. Ions of a particular *m/z* value (for a definition of *m/z*, see the panel on GLOSSARY OF MASS SPECTROMETRY TERMS below) are separated one from another using physical properties such as electric or magnetic fields. Typically, the physical field in the mass analyzer is changed as a function of time. The major types of analyzers are quadrupole analyzers, quadrupole ion-trap analyzers, Fourier transform-ion cyclotron resonance (FT-ICR) mass analyzers, and time-of-flight (TOF) analyzers.
- ***Detector.*** Ions of a particular *m/z* value pass from the mass analyzer to strike the detector. The magnitude of current produced at the detector as a function of time is used to determine the intensity (abundance) and *m/z* value of the ion. In general, the relative abundances of ions generated in an ionization source are measured in a detector, and the results are displayed with *m/z* values as abscissae and *ion abundances* as ordinates. Mass spectra are usually depicted in either a *normalized* or *percentage relative abundance* (%RA) format or as a *percentage of total ion current* (%TIC). In a *normalized* MS spectrum, the largest peak (i.e., the most abundant ion, which is not necessarily the analyte being studied) in a spectrum is called the *base peak* and its height is adjusted to 100 units. The relative heights of all other peaks in the spectrum are *normalized* to this base peak and thus lie between 0 and 100 units.
- ***Data recorder/processor.***

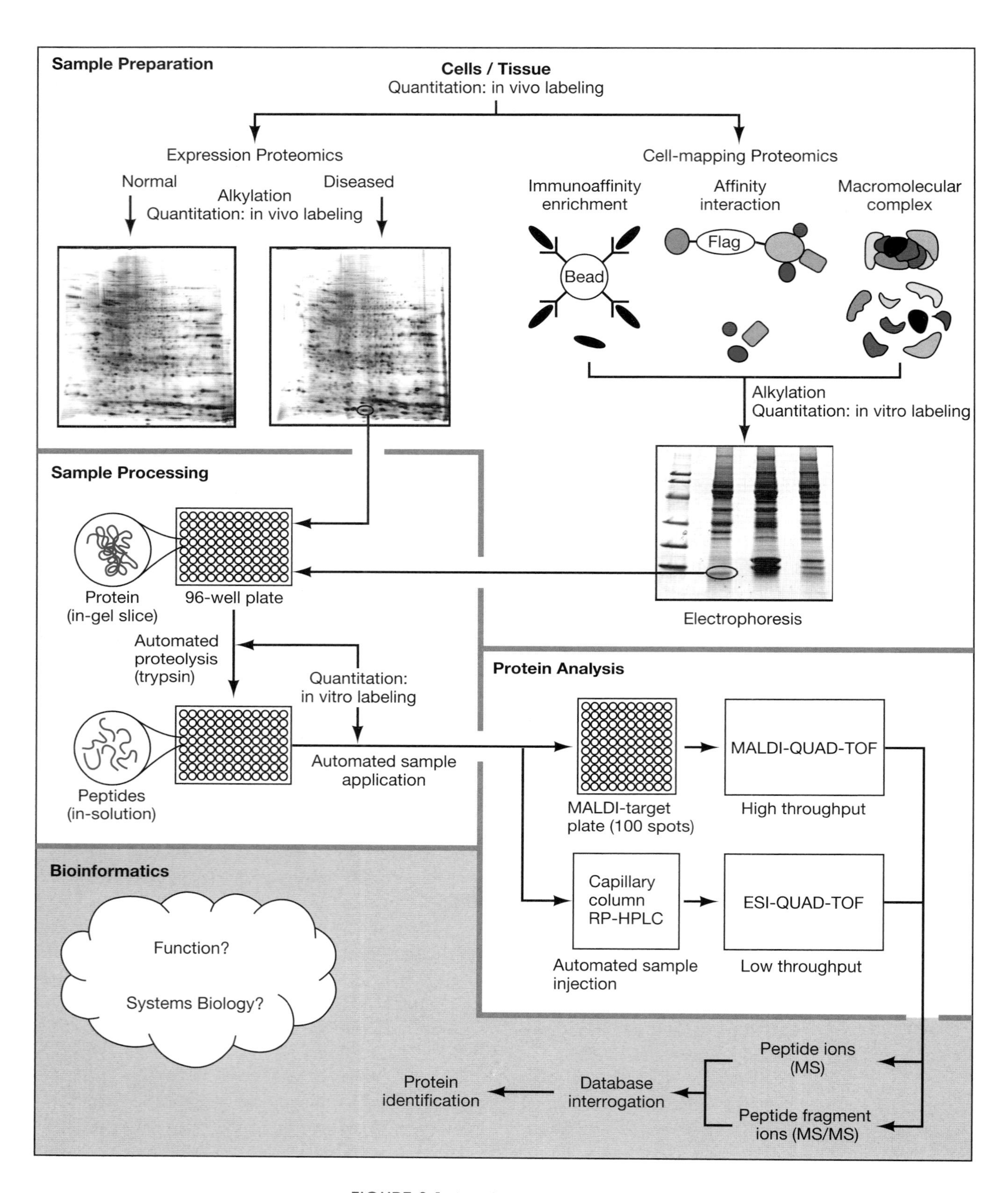

FIGURE 8.1. (*See facing page for legend.*)

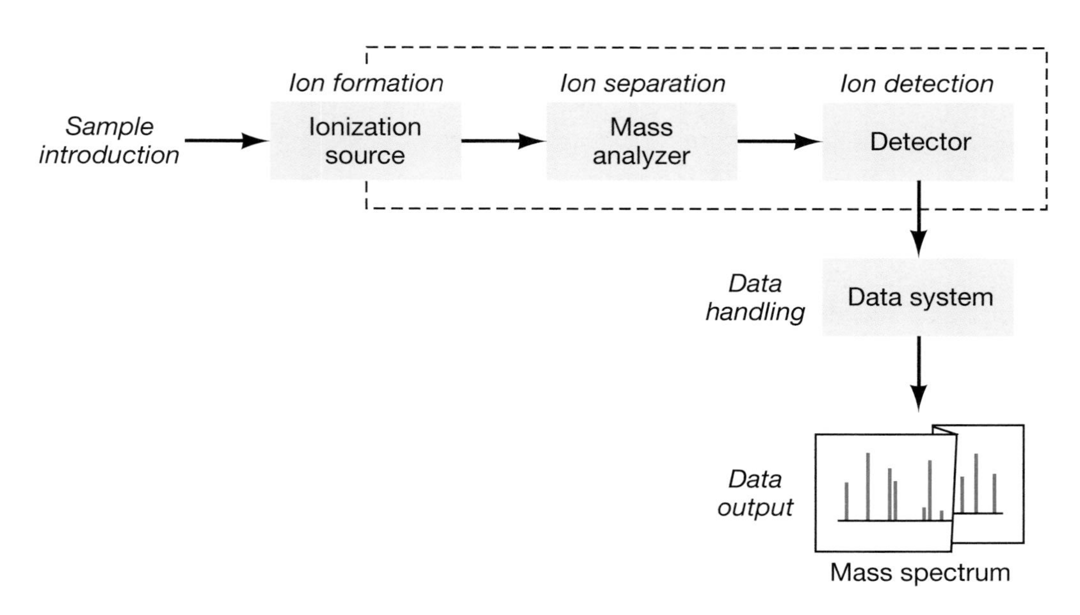

FIGURE 8.2. Modular components of a mass spectrometer. (Adapted, with permission, from www.asms.org [Fig 1] the American Society for Mass Spectrometry [©1998 ASMS; What is Mass Spectrometry, 3rd edition].)

Mass spectrometers are able to produce two types of mass information: They can determine both the mass of a molecule and the masses of product ions derived from the parent molecule following fragmentation, data that are especially valuable for determining the amino acid (or carbohydrate) sequence of biomolecules (see the information panel on FRAGMENTATION MECHANISMS OF PROTONATED PEPTIDES IN THE GAS PHASE).

FIGURE 8.1. Protein identification and characterization by mass spectrometry. Using global protein expression approaches, proteins are separated by 2D gel electrophoresis. Following visualization by Coomassie Brilliant Blue, fluorescence staining, or silver staining (for staining protocols, see Chapter 4), proteins of interest are then excised, alkylated, and subjected to in-gel proteolysis using trypsin (see Chapter 7, Protocol 13). For differential expression proteomics, quantitation can be accomplished by first isotopically labeling proteins, either in vivo or in vitro (see also Chapter 1). For cell-mapping proteomics or protein-interaction studies, proteins of interest are obtained by (1) classical "pull-down" experiments using an antibody directed toward the protein of interest; (2) tagging a gene of interest with a small-peptide epitope that can be used to affinity purify the protein following transfection of the tagged gene into cells; or (3) isolation of macromolecular complexes (see Chapter 10). Separation of protein complexes is carried out by either 1D electrophoresis (Chapter 2) or, in the case of immunoaffinity approaches, by 2D electrophoresis (see Chapter 4). Visualized proteins of interest are processed using the same method as described for global profiling. A hierarchical MS approach is outlined, where tryptic digests are first analyzed in high-throughput using matrix-assisted laser desorption/ionization (MALDI)–quadrupole time-of-flight (qTOF) instrumentation. This approach should yield many protein identifications by interrogation of the databases using peptide ion masses and peptide fragment ion masses. For proteins that remain unidentified by this approach, tryptic digests are subjected to low-throughput ESI methods. (Reproduced, with permission, from Simpson and Dorow 2001 [©Elsevier Science].)

An Historical Overview of Mass Spectrometry

The fundamental physics behind modern mass spectrometry originated more than 100 years ago with the seminal work of J.J. Thomson (Cavendish Laboratory, University of Cambridge), which led to the discovery of the electron in 1897 during investigations of cathode rays. While Thomson was measuring the effects of electric and magnetic fields on ions generated by residual gases in cathode ray tubes, he noticed that the ions moved through parabolic trajectories that were proportional to their "mass-to-charge" values. This observation led to the construction of the first mass spectrometer (then called a parabola spectrograph). In 1906, Thomson received the Nobel prize in Physics, "in recognition of the great merits of his theoretical and experimental investigations on the conduction of electricity by gases."

More sophisticated mass spectrometers were subsequently designed and constructed by Arthur J. Dempster (1918) and Francis W. Aston (1919) (Grayson 2002). Later, Alfred Nier incorporated these developments along with advances in vacuum technology and electronics to greatly decrease the instrument's size. During the period from the late 1930s to the early 1970s, further accomplishments were made in the field of mass spectrometry. For example, in 1946, W.F. Stephens proposed the concept of time-of-flight (TOF) analyzers that separated molecular ions by measuring their velocities as they moved in a straight path toward a detector. In the mid 1950s, Wolfgang Paul developed the quadrupole mass analyzer, which separates molecular ions by use of an oscillating electrical field. Another Paul innovation was the quadrupole ion trap, which is a device specifically designed to trap and measure ions. The first ion trap became available commercially in 1983, and now both the quadrupole and quadrupole ion trap (along with TOF) are the most widely used mass analyzers today. For his innovative work with the quadrupole mass analyzer and the ion trap device, Paul was awarded the 1989 Nobel prize in Physics.

Fast atom bombardment ionization (FAB), developed in the early 1980s (Barber et al. 1981; Hunt et al. 1981) (along with liquid secondary ion mass spectrometry [LSIMS]) were the first *soft-ionization* MS techniques that could be used to ionize large, nonvolatile, polar compounds such as proteins and peptides (Dell and Morris 1982; Barber et al. 1983; Desiderio and Katakuse 1983; Cottrell and Frank 1985; Hyver et al. 1985; Biemann et al. 1986). In FAB, the analyte is dissolved in a nonvolatile liquid matrix, such as glycerol, and placed under vacuum in the instrument. The sample is bombarded with a stream of fast neutral atoms (usually formed by ionization/neutralization of argon or xenon) to induce a shock wave within the solution that ejects analyte ions already present in the solution into the gas-phase. In LSIMS, the sample is also dissolved in a liquid matrix. However, the sample is then bombarded with heavy ions, such as those formed from cesium, instead of neutral atoms, as in FAB. Ions of 10,000 daltons and above can be observed with these techniques.

In 1988–1989, two other soft-ionization techniques were introduced that allowed the transfer of large biomolecules, such as proteins and peptides, into the gas phase: matrix-assisted laser desorption/ionization (MALDI) (Karas and Hillenkamp 1988) and electrospray ionization (ESI) (Fenn et al. 1989). MALDI uses a pulsed laser to desorb analytes from a solid matrix containing a highly UV-absorbing substance, producing primarily singly charged ions. With ESI, a high-voltage potential is applied to a liquid as it passes through a small capillary. Ions are desorbed into the gas phase following evaporation of the droplet as it exits the capillary. A unique feature of electrospray, compared to other ionization techniques, is that both singly and multiply charged ions can be formed from a single analyte molecule population. The development of MALDI and ESI, along with the development of mass spectrometry instrumentation with improved performance characteristics (e.g., improved mass range and resolution) during the last 15 years, has been fundamental to the advancement of mass spectrometry as a core technology in proteomics. For an excellent historical perspective of mass spectrometry, see Grayson (2002).

Glossary of Mass Spectrometry Terms

- ***Mass-to-charge ratio* (m/z):** Mass spectrometers measure the mass-to-charge values of molecular ions. For peptide ions, these are typically formed by ionization of the molecule via the addition of one (in the case of MALDI) or more protons (in the case of ESI). Thus, a peptide with a molecular mass of 2000 daltons or atomic mass units (amu) will have an *m/z* value of 2001 ($[M+H]^{1+}$) after ionization by the addition of one proton and an *m/z* value of 1001 with the addition of two protons ($[M+2H]^{2+}$).
- ***Resolution:*** Resolution can be defined as the ability to separate and measure the masses of ions of similar, but not identical, molecular mass. Mass resolution is generally expressed as the ratio of $m/\Delta m$ and has historically been defined in two ways (Figure 8.3):
  1. The "10% valley" definition, in which a pair of adjacent ion peaks at mass m_1 and m_2 (in atomic mass units or daltons) are resolved so that a valley of 10% is observed between them, relative to the least most abundant ion of the pair (Δm is defined as the main difference between m_1 and m_2).
  2. The "full-width, half-maximum" (FWHM) definition, where the width of an ion at mass m, at 50% relative abundance, is used to define Δm.

  Note that the resolution values determined using the FWHM definition are about twice that of those calculated using the 10% valley definition. For example, a resolution of 1000 obtained using the 10% valley definition will correspond to a resolution of ~2000 using the FWHM definition. Hence, it is essential to know what definition is being applied when comparing resolution values for different mass spectrometers. The influence of resolution on the ability to separate isotopic peaks in molecular ion clusters is illustrated for the $[M+H]^+$ ions of angiotensin (~1283 daltons), the reduced A and B chains of bovine insulin (~2340 and ~3400 daltons, respectively), bovine insulin (~5734 daltons), and horse heart myoglobin (~16,952 daltons) in Figures 8.4 and 8.5. An example of the utility of high resolution performance for the analysis of peptide mixtures is given in Figure 8.6, where the separation of two peptides of ~1846-dalton molecular mass, which differ in mass by only 0.26 dalton, is shown.
- ***Monoisotopic mass versus average mass:*** It is important when expressing the molecular weight of a compound to appreciate how it was calculated. A common way of defining

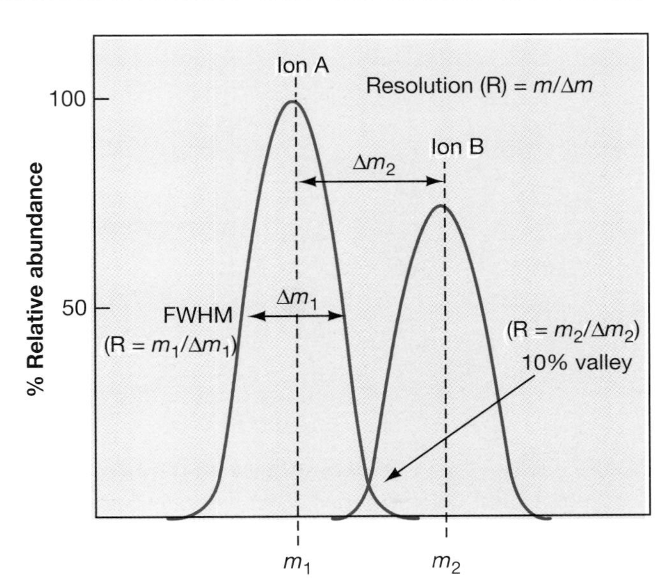

FIGURE 8.3. Definition of resolution. Simple schematic diagram of procedures for calculating resolution ($R = m/\Delta m$). (1) 10% Valley definition, where Δm is the mass difference between two adjacent peaks at m_1 and m_2, resolved to a 10% valley, and (2) FWHM, full-width, half-maximum definition, i.e., where Δm is the peak width of ion m_1 at 50% relative abundance.

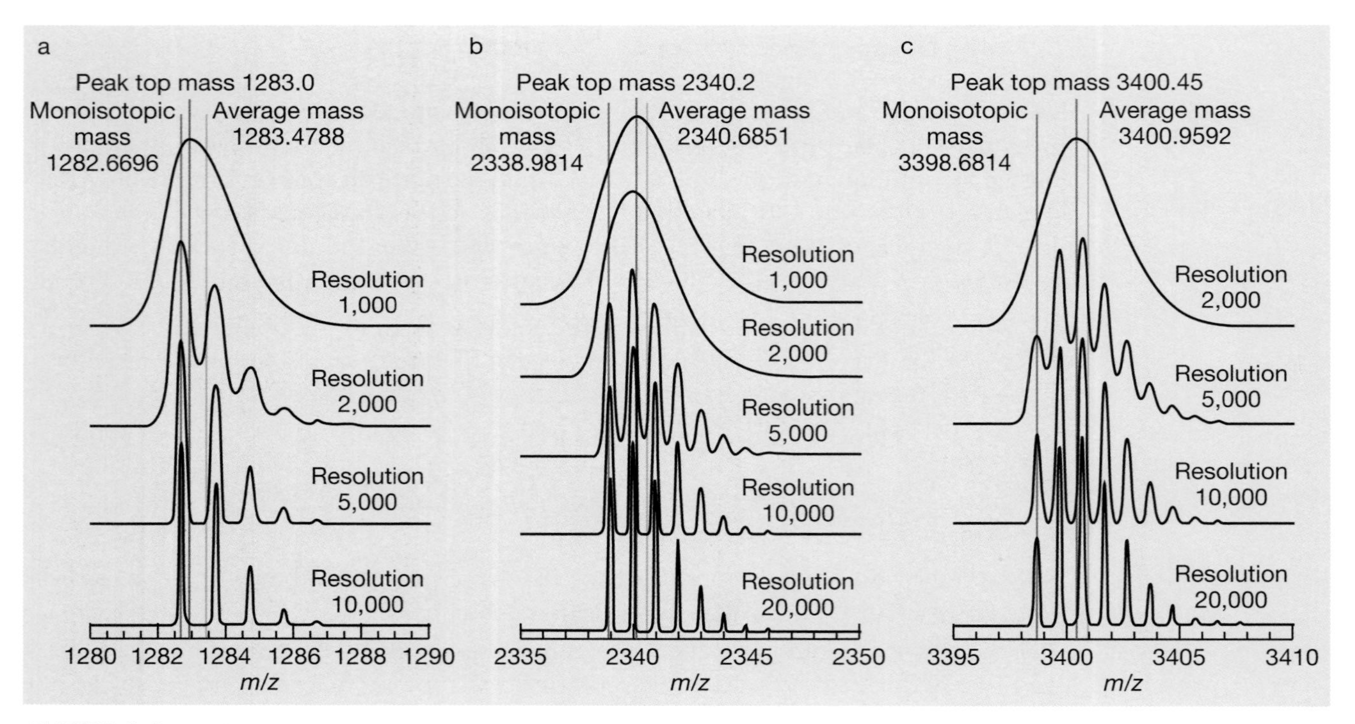

FIGURE 8.4. Resolution: The hypothetical $[M+H]^+$ molecular ion clusters for angiotensin (molecular mass, 1283) (*a*), reduced insulin A chain (molecular mass, 2340) (*b*), and reduced insulin B chain (molecular mass, 3400) (*c*) are shown at various resolutions. Monoisotopic mass, peak top mass, and average mass are indicated by the dark blue, gray, and light blue lines, respectively.

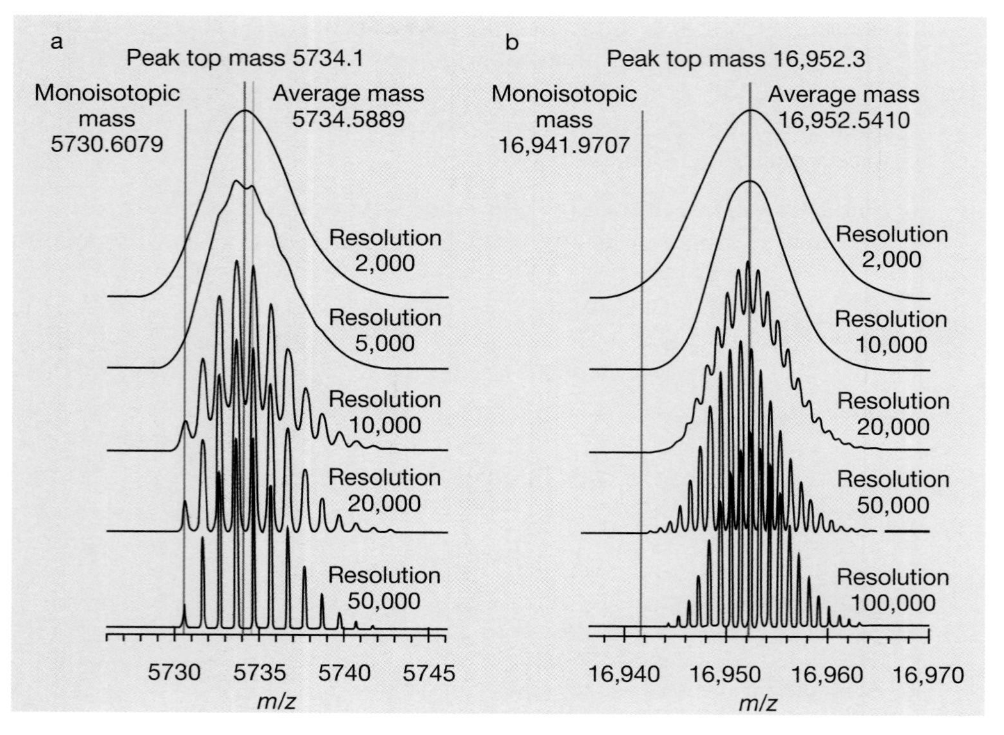

FIGURE 8.5. Resolution: The hypothetical $[M+H]^+$ molecular ion clusters for bovine insulin (molecular mass 5734) (*a*), and horse heart myoglobin (molecular mass 16,952) (*b*) are shown at various resolutions. Monoisotopic mass, peak top mass, and average mass are indicated by the dark blue, gray, and light blue lines, respectively.

molecular weight is *monoisotopic mass*. The monoisotopic mass of an element refers to the lightest stable isotope of that element. For example, the monoisotopic mass of carbon is 12.00000. (There are two major isotopes of carbon, namely, ^{12}C and ^{13}C, with masses of 12.00000 and 13.003355, with natural abundances of 98.92% and 1.08%, respectively.) Similarly, there are two major isotopes of nitrogen: ^{14}N (14.003074; natural abundance, 99.63%) and ^{15}N (15.000109; natural abundance, 0.37%). A list of common elements and their naturally occurring isotopic abundances is given in Table 8.1. The *monoisotopic mass* of an analyte is thus determined by summing the monoisotopic masses of each element present in the analyte. In practice, an ion comprises not only molecular species having the lightest isotopes of the elements present, but also a small percentage containing the heavier isotopes. The contribution of these heavier isotopes in a molecular ion cluster will clearly depend on the abundance-weighted sum of each element present in the analyte. Hence, in a mass spectrometer capable of resolving the constituent isotopes of an analyte ion, a molecular ion cluster will be observed (see Figures 8.4 and 8.5). Another way of defining the molecular mass of an element is the *chemical average mass*, which is the sum of the abundance-weighted masses of all of its stable isotopes (e.g., 98.9% ^{12}C and 1.1% ^{13}C, to give the isotope-weighted average mass of 12.011 for carbon). The *average mass* of an analyte is therefore determined by summing the *chemical average masses* of its component elements. Yet another mass term, the *peak top mass*, often referred to as the maximum mass, is simply the mass of the peak maximum of the unresolved molecular ion isotope cluster. An important consequence of the contribution of heavier isotope peaks in a molecular ion cluster relates to whether the investigator should use monoisotopic mass or the average mass when measuring and reporting molecular masses of analytes. Clearly, this decision will depend on both the mass of the analyte and the resolving capability of the mass spectrometer used to conduct the measurement. For instance, it can be seen in Figures 8.4 and 8.5 for peptides of mass >1750–1800 that peaks corresponding to the monoisotopic mass no longer represent the most abundant in the isotopic cluster, and that for analytes with masses greater than ~8000, the monoisotopic mass has an insignificant contribution to the isotopic ion cluster envelope.

- ***Mass accuracy:*** The accuracy of a mass measurement can be presented as a percentage of the measured mass (e.g., molecular mass = 1000 ± 0.01%) or as parts-per-million (e.g., molecular mass = 1000 ± 100 ppm) (Carr and Burlingame 1996). Clearly, as the mass of an analyte increases, the absolute mass error corresponding to the ppm or percent error will also increase proportionally. For example, a mass accuracy of 0.01% (or 100 ppm) will correspond to a mass error of 0.1 dalton at *m/z* 1000, 0.5 dalton at *m/z* 5000, and 5 daltons at *m/z* 50,000. Mass accuracy can be significantly affected by how data are treated. For example, in the mass range 1000 to ~5000, pronounced errors can be introduced if unresolved isotope clusters must be measured. As illustrated in Figures 8.4 and 8.5, for the mass range 1000–15,000, an unresolved isotopic cluster will be asymmetric in shape due to the intensities of the underlying isotopic masses in the molecular ion clusters. In this situation, if the mass to be reported is the average mass, the centroid of the distribution should be determined and used instead of the measured peak top mass since these will not be the same in this mass range. Note that as the mass of the analyte increases, the peak profile becomes more symmetrical and the average mass and peak top mass become almost identical (see Figure 8.5b,c). In general, a resolution of at least several thousand (FWHM) is required for good mass measurement when working in the mass range of 1000 to ~5000 daltons.

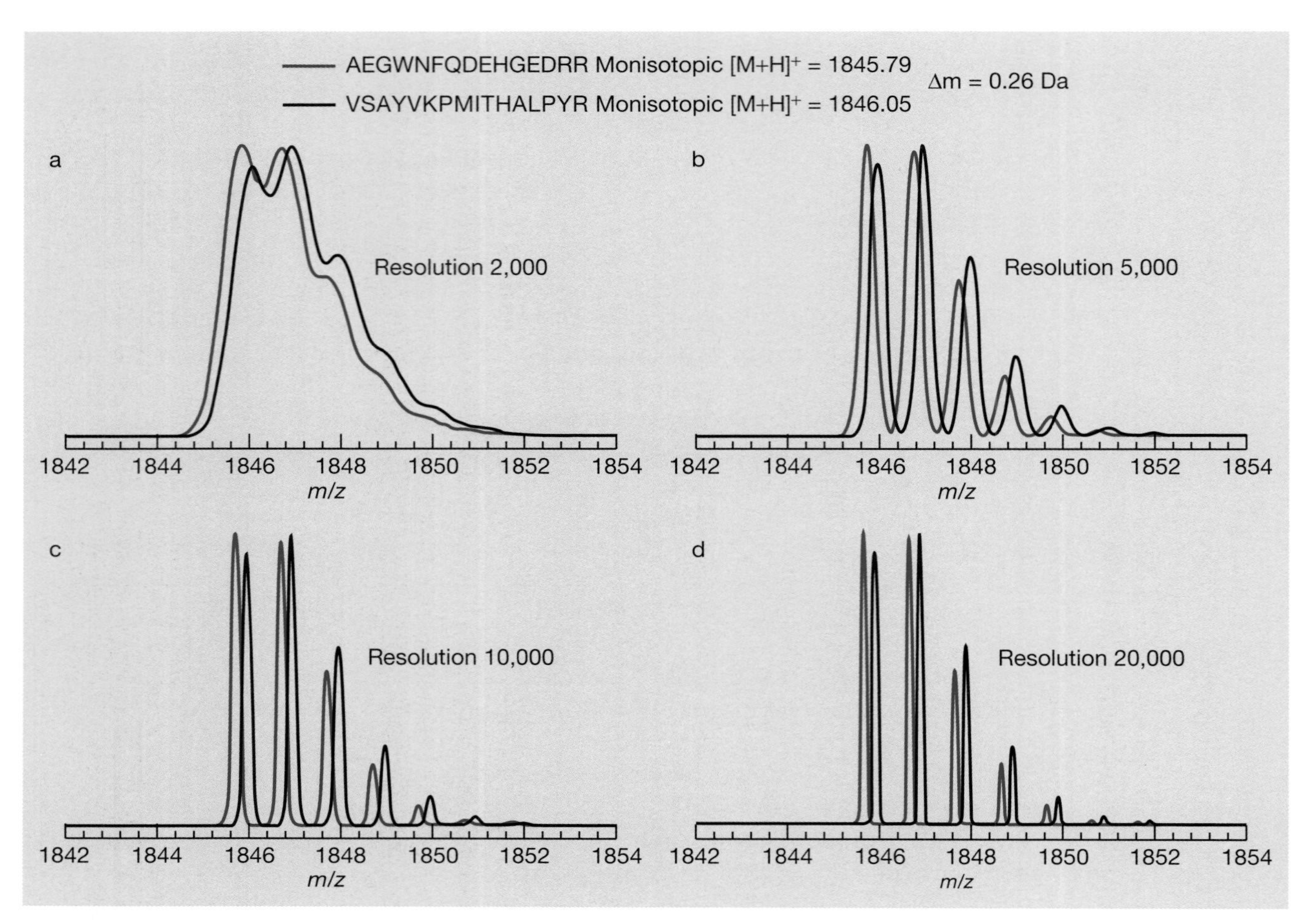

FIGURE 8.6. Resolution: Issues associated with the analysis of peptides in mixtures. The hypothetical mass spectra of two peptides (AEGWNFQDEHGEDRR, monoisotopic $[M+H]^+$ ion *m/z* of 1845.79, and VSAYVKPMITHALPYR, monoisotopic $[M+H]^+$ ion *m/z* of 1846.05), whose masses differ by 0.26 dalton are shown at a resolution of 2,000 (*a*), 5,000 (*b*), 10,000 (*c*), and 20,000 (*d*).

MODERN IONIZATION METHODS: HOW IONS ARE FORMED

The ability to create gas-phase ions from nonvolatile, polar, or charged molecules is quintessential to the mass spectrometric analysis of all biomolecules. Ionization (the act of placing a charge on a neutral molecule), in general, can occur by either of two fundamental processes: the loss/gain of an electron or the loss/gain of a charged particle (e.g., a proton), generating odd or even electron ions, respectively. For a general review, see Johnstone and Rose (1996), and for ESI and MALDI details, see Gaskell (1997) and Busch (1995), respectively.

Traditionally, the transfer of small organic molecules into the gas phase has been accomplished by thermal vaporization followed by electron impact (EI) or chemical ionization (CI) (Harrison and Cotter 1990; Harrison 1992). The major obstacle in applying these approaches to biological mass spectrometric analysis, however, was the difficulty of vaporizing such molecules into the mass spectrometer without extensive thermal decomposition (Daves 1979). Throughout 1950–1970, several workers demonstrated that the polarity of di-, tri-, and tetrapeptides could be reduced by acylation, esterification, and reduction, or *N*- and *O*-permethylation (Biemann 1962, 1978; Morris 1972; Morris et al. 1973). By these means, small peptides could be introduced into the gas phase for sequence analysis. An excellent personal recollection from Klaus Biemann discusses these developments (Biemann 1995).

TABLE 8.1. Common elements and their naturally occurring isotopic relative abundance values

Element	Isotope	Mass[a]	Natural abundance (%)[b]
Hydrogen, H	^{1}H	1.007 825 035	99.985
	^{2}H	2.014 101 779	0.015
Carbon, C	^{12}C	12.000 000 000	98.90
	^{13}C	13.003 354 826	1.10
Nitrogen, N	^{14}N	14.003 074 002	99.63
	^{15}N	15.000 108 97	0.37
Oxygen, O	^{16}O	15.994 914 63	99.76
	^{17}O	16.999 131 2	0.04
	^{18}O	17.999 160 3	0.200
Fluorine, F	^{19}F	18.998 403 22	100
Sodium, Na	^{23}Na	22.989 767 7	100
Silicon, Si	^{28}Si	27.976 927 1	92.23
	^{29}Si	28.976 494 9	4.67
	^{30}Si	29.973 770 1	3.10
Phosphorus, P	^{31}P	30.973 762 0	100
Sulfur, S	^{32}S	31.972 070 698	95.02
	^{33}S	32.971 458 428	0.75
	^{34}S	33.967 866 650	4.21
	^{36}S	35.967 080 620	0.02
Chlorine, Cl	^{35}Cl	34.968 852 728	75.77
	^{37}Cl	36.965 902 619	24.23
Potassium, K	^{39}K	38.963 707 4	93.2581
	^{40}K	39.963 999 2	0.012
	^{41}K	40.961 825 4	6.7302
Bromine, Br	^{79}Br	78.918 336 1	50.69
	^{81}Br	80.916 289	49.31
Iodine, I	^{127}I	126.904 473	100

Adapted from Burlingame and Carr (1996).
[a] Wapstra et al. (1985). Standard errors are omitted from the present table.
[b] Weast et al. (1989).

In the early 1980s, fast atom bombardment (FAB) (Morris et al. 1981; Barber et al. 1982; Seifert and Caprioli 1996) provided a major step forward in the development of MS-based procedures for peptide and protein characterization. However, it was the commercial introduction of two new ionization methods after 1988 that made MS routinely available to researchers in the biological arena: MALDI, discovered by Karas and Hillenkamp (1988), and ESI, developed by Fenn and co-workers (1989).

Matrix-assisted Laser Desorption/Ionization

MALDI (Karas and Hillenkamp 1988; Beavis and Chait 1996) is performed by embedding analyte molecules in an excess of a specific wavelength-absorbing matrix (usually UV_{337nm}; typically, a 1:1000 molar ratio of analytes to matrix is employed), which is dried to produce a cocrystallized mixture. Typical matrices that are amenable to peptide and protein analysis include α-cyano-4-hydroxy-cinnamic acid and 3,2-dimethoxy-4-hydoxy-cinnamic (sinapinic) acid (see Table 8.8 in Protocol 1). The chemical structure of some commonly used matrices are shown in Figure 8.7. Ions are produced by bombarding the sample with short-duration (1–10 ns) pulses of UV light from a nitrogen laser. The interaction of the laser pulse with the sample results in ionization (usually protonation) of both matrix and analyte molecules via an energy transfer mechanism from the matrix to the embedded analyte, rather than by direct laser ionization. A high potential electric field (typically ±30 kV) applied between the sample probe and an orifice is used to accelerate the ions into the mass analyzer (see Figure 8.8) (Beavis and Chait 1989a,b,c; Cohen and Chait 1996; Kussmann and Roepstorff 2000). MALDI primarily produces singly charged ions ($[M+H]^+$ or $[M-H]^-$). An interesting model to explain why only singly charged molecular ions survive the MALDI ionization process, which can produce multiply charged molecular ions under certain circumstances, has recently been proposed by Karas et al. (2000).

Electrospray Ionization

ESI (Fenn et al. 1989; Banks and Whitehouse 1996) is unique compared to other MS ionization techniques in that both singly and multiply charged ions may be formed. To form ions by electrospray, a fine spray of charged droplets is created from a solution introduced through a small-diameter needle in the presence of a strong electric field (typically ±3–5 kV) (see Figure 8.9). The charged droplets are desolvated by the application of a countercurrent flow of gas and/or heat causing the droplet to evaporate. When the surface charge density of the droplet exceeds the Rayleigh stability limit, electrostatic repulsion greater than the surface tension of the droplet results in the explosion of the droplet and expulsion of gas-phase ions. The most striking feature of electrospray is that a family of multiple charge states of the ion can arise from a single precursor. The extent of multiple charging of an analyte ion is influenced by factors such as the composition and pH of the electrospray solvent, as well as the chemical nature of the analyte. For peptides (<2000 daltons), ESI typically generates singly or doubly charged ions, whereas for proteins (>2000 daltons), this ionization process usually gives rise to a series of multiply charged species ("charge envelope"). These multiple charge states can be used to provide independent verification of the mass or to provide access to different reaction channels in tandem MS experiments through the selection of different charge states for fragmentation (see the information panel on FRAGMENTATION MECHANISMS OF PROTONATED PEPTIDES IN THE GAS PHASE). Importantly, this charging phenomenon has the effect of lowering the *m/z* values of the analyte to a range that can be readily measured by many different types of mass analyzers. Thus, quadrupole and ion-trap mass analyzers, which generally have an upper limit of 2,000–10,000 *m/z*, may be used to determine the mass of proteins of much larger molecular mass. For instance, Figure 8.10 shows a hypothetical ESI mass spectrum of horse heart apomyoglobin (average molecular mass, 16,951.5 daltons). The molecular ions observed in the spectrum correspond to that of apomyoglobin with a varying number of protons attached. (Note that the charge envelope of a protein profoundly changes when the protein becomes unfolded [or denatured].)

Using an instrument that is capable of high resolution (>10,000), it is often possible to determine the charge state of a particular ion in a mass spectrum as well as the molecular mass of the parent molecule (provided the molecular mass is somewhat less than the resolving power of the instrument), by analyzing the isotopic cluster of an individual ion (see Figure 8.10b). Figure 8.11 shows the mass spectrum of a hypothetical charge state distribution for a typical tryptic peptide (2484.95 daltons), whose predicted *m/z* values of the charge states can be calculated using the simple formula shown in the legend to Figure 8.11.

All commercial mass spectrometers that are equipped with ESI capability provide a deconvolution algorithm for mathematically processing the charge state and/or isotopic envelope in order to obtain a statistically averaged (and more accurate) molecular mass. Additional mathematical methods such as smoothing, background subtraction, and noise filtering options are also incorporated in most commercial algorithms (Bonner and Shushan 1995; Horn et al. 2000b).

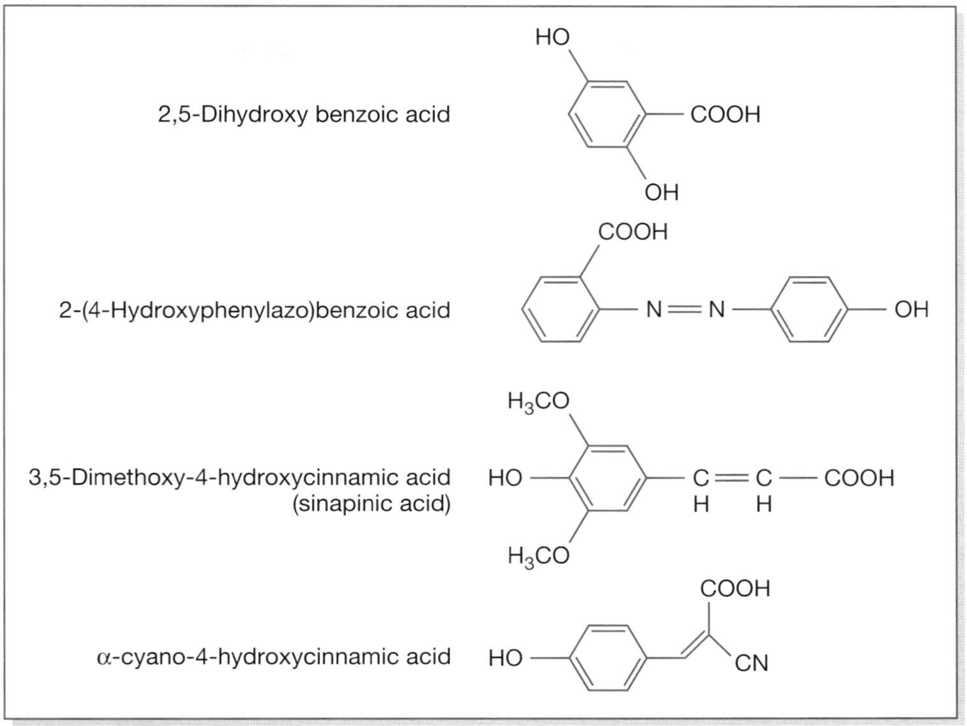

2,5-Dihydroxy benzoic acid
HO
COOH
OH
2-(4-Hydroxyphenylazo)benzoic acid
COOH
N=N
OH
3,5-Dimethoxy-4-hydroxycinnamic acid (sinapinic acid)
H_3CO
HO
C=C—COOH
H H
H_3CO
α-cyano-4-hydroxycinnamic acid
HO
COOH
CN

FIGURE 8.7. Chemical structures of some commonly used matrices.

UV laser λ 337 nm
MALDI plate
Cocrystals
Mass analyzer
Analyte
Matrix

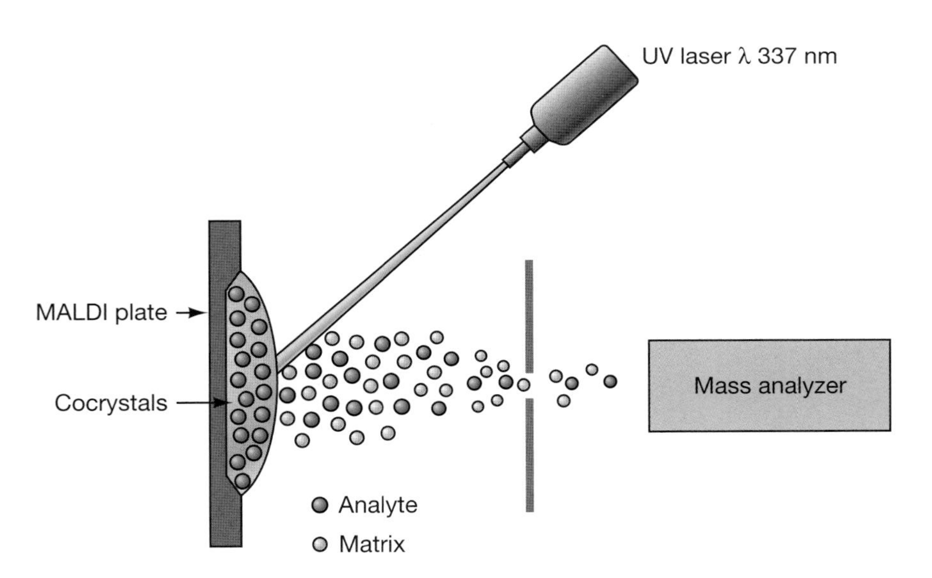

FIGURE 8.8. Schematic representation of the MALDI process. To enable MALDI, an analyte is first dissolved (or mixed) with an excess amount of matrix (which has a strong absorption at the laser wavelength) and then dried. Upon laser irradiation of the cocrystallized analyte/matrix sample, a plume of neutral molecules and ions is desorbed. By applying an electrostatic potential between the sample slide and a sampling orifice, the desorbed ions are then guided into the mass analyzer by a series of lenses. Because MALDI generates mainly singly charged ions (unlike ESI, which yields multiply charged ions), this MS approach is particularly suitable for high-throughput and for analysis of complex mixtures.

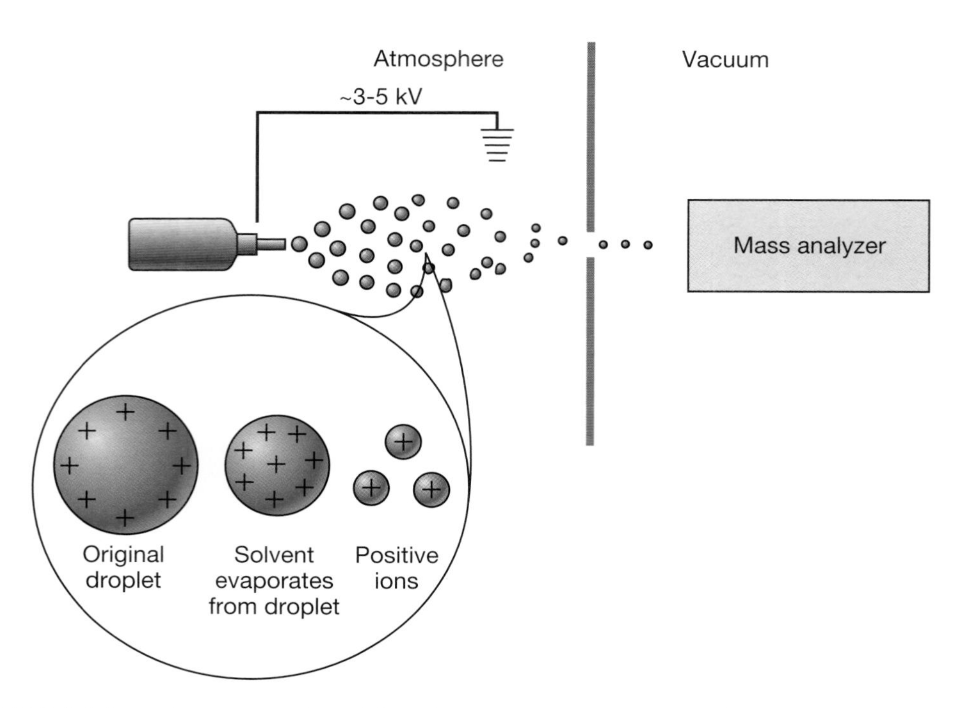

FIGURE 8.9. Schematic representation of the ESI process. An electrospray is formed by the application of a high-voltage potential (3–5 kV) between a solution containing an analyte passing through a fine capillary, and a counter electrode situated ~0.5–1.0 cm from the needle. A spray of fine charged droplets that contain the analyte and solvent molecules are generated, which can be desolvated to yield gas-phase ions. These ions are then transferred into the mass analyzer. A unique feature of ESI is that highly multiply charged ions can be produced from these droplets, thereby allowing analytes such as proteins or large peptides to be readily analyzed by mass spectrometers with limited *m/z* range (e.g., 2000).

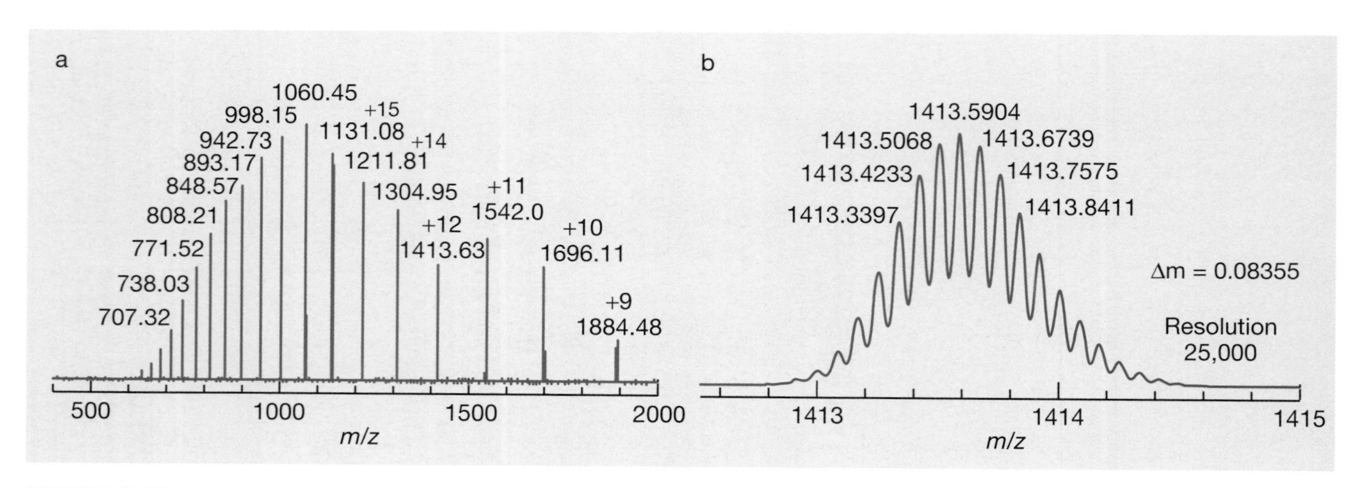

FIGURE 8.10. Hypothetical ESI mass spectra of horse heart myoglobin (mass 16,951.5 daltons). (*Panel a*) Hypothetical multiple charge state distribution of myoglobin. A multiply charged ion series ranging from +9 (*m/z* 1884.48) to +27 (*m/z* 628.83) is shown. A simple method for calculation of the charge states of individual ions in the distribution, and determination of the mass of the protein, is described in the text. (*Panel b*) Hypothetical isotopic distribution of the $[M+12H]^{12+}$ charge state of myoglobin at a resolution of 25,000. The mass difference of 0.08355 dalton between adjacent isotopic peaks readily allows determination of the charge state.

CALCULATION OF THE CHARGE STATE OF A MULTIPLY CHARGED ION AND THE MASS OF ITS PARENT PROTEIN

In general, the charge state of a multiply charged ion and the molecular mass of the protein from which it has been derived can be calculated as follows: $P = m/z$, where P is a peak in a mass spectrum with a protonation (charge) state m/z. The charge state of an ion is defined as the number of charges a molecular ion carries such that a singly charged ion has a charge state of 1, a doubly charged ion has a charge state of 2, a triply charged ion has a charge state of 3, and so on. Two adjacent peaks in the ESI spectrum separated by a single charge (z) can be denoted P_1 and P_2, corresponding to $(m/z)_1$ and $(m/z)_2$, respectively.

$$P_1 = (M + z_1)/\ z_1$$
$$P_2 = [M + (z_1 - 1)]/(z_1 - 1)$$

where M is the molecular mass of the molecule, and the unknown charges are z. Thus, two successive ion peaks yield two equations and two unknowns, which can be solved to reveal the molecular mass of the analyte. For the example shown in Figure 8.10a for the multiple charge state distribution of horse heart apomyoglobin, if $P_1 = 1304.95$ and $P_2 = 1413.63$, then

$$1304.95\ z_1 = M + z_1$$
$$1413.63\ (z_1 - 1) = M + (z_1 - 1)$$

Solving for M and z_1 for the ion at m/z 1304.95 yields $M = 16{,}951.35$ daltons and the charge state, $z_1 = 13$. The ion at m/z 1413.63 therefore corresponds to a charge state of $z = 12$ and mass $M = 16{,}951.56$ daltons. Typically, the experimental molecular masses observed for each pair of molecular ions in an ESI spectrum are averaged in order to obtain a more accurate value. In this example, this yields an average mass of 16,951.455.

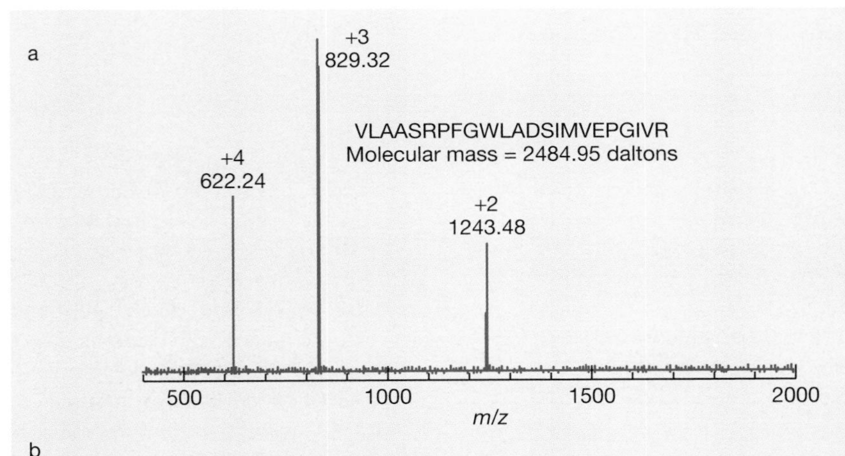

b

To calculate the m/z values of predicted charge states in a mass spectrum for any given peptide when the molecular weight is known:

$$m/z = \frac{[M + (nX)]}{n}$$

where M is the molecular mass of the peptide, X is the mass of the cation added to the molecule (usually a proton, thus X has a mass of 1.008 daltons), and n is an integer representing the number of charges.

FIGURE 8.11. (*Panel a*) Hypothetical ESI mass spectrum of a typical tryptic peptide with a molecular mass 2484.95 daltons, acquired on a quadrupole ion-trap mass spectrometer. The m/z of the singly protonated ion ($[M+H]^+$) of 2485.95 is beyond the mass range of the mass spectrometer (up to m/z 2000). However, the +2, +3, and +4 charge states of the peptide can be observed. (*Panel b*) Simple method for calculation of the m/z ratios of various charge states of a peptide of known molecular weight. For example, the m/z ratio of the +3 charge state of the peptide sequence shown in *panel a* can be readily calculated: $m/z = 2484.95 + (3 \times 1.008)/3 = 829.32$.

Often times, the question arises: Which instrument should I buy? Although the ionization techniques are complementary, often a choice must be made. To assist in this decision-making process, Table 8.2 provides a summary of various advantages and disadvantages of MALDI and ESI.

TABLE 8.2. Advantages and disadvantages of ESI-MS and MALDI-MS

Advantages	Disadvantages
ESI	
Practical mass limit ~70,000 daltons.	Low salt tolerance. Typically, samples are desalted prior to analysis using RP-HPLC, either "on-line" (see Protocol 6) or "off-line" (see Protocol 7) (see also Chapter 5).
Good sensitivity. Femtomole to low picomole sensitivity is typical.	Difficulty in cleaning overly contaminated instrument due to high sensitivity for certain compounds.
Softest ionization. Capable of observing biologically native noncovalent interactions.	Low tolerance for mixtures. Simultaneous mixture analysis can be poor. The purity of the sample is important. To overcome this problem, samples are generally introduced into the instrument via on-line RP-HPLC (see Protocol 5).
Easily adaptable to microbore liquid chromatography. Capable of directly analyzing LC effluent at a flow rate of 50 nl/min.	Multiple charging, which can be confusing, especially with mixture analysis.
No matrix interference.	Quantitation requires internal standard.
Easy adaptability to triple quadrupole analysis, conducive to structural analysis (for various scan modes, see Figure 8.12)	
Multiple charging, allowing for the analysis of high-mass ions with a relatively low-*m/z* range instrument.	
Multiple charging, giving better mass accuracy through averaging.	
Excellent for determining peptide modifications.	
Sample preparation: Samples can be analyzed directly from salt-free solution or via RP-HPLC (coupled on line).	
MS/MS capabilities (e.g., peptide sequencing) are excellent.	
MALDI	
Practical mass limit ~300,000 daltons. Species of much greater mass have been reported.	Generally low resolution. Some MALDI instruments are capable of high resolution; however, this is only in a relatively low-mass range and is accomplished at the expense of sensitivity.
Typical sensitivity on the order of low femtomole to low picomole. Reports have indicated that attomole sensitivity is possible.	Matrix background, which can be a problem for compounds below a mass of 1000 daltons. This background interference is highly dependent on the matrix material. For a list of commonly used matrices, see Table 8.8.
Soft ionization with little to no fragmentation observed.	MS/MS capability is minimal unless the instrument is equipped with a reflectron device (post-source decay).
Suitable for the analysis of complex mixtures.	Possibility of photodegradation by laser desorption/ionization.
Sample preparation: Samples are added directly to appropriate matrix.	Not possible to study noncovalent interactions.
	Quantitation requires internal standard.
	Poor for determining peptide modifications (especially linear mode MALDI instruments; however, MALDI instruments with reflectron capability are good for determining some peptide modifications, e.g., phosphorylation).
	Low tolerance for various excipients (e.g., phosphate buffers, salts >150 mM); see Table 8.4.

TANDEM MASS SPECTROMETRY

The revolution in biological mass spectrometry has its origins with the development in the 1980s of ionization techniques suited to biomolecular analysis. As these "soft ionization methods" produce pseudomolecular ions (e.g., protonated $[M+H]^+$ ions) with little or no fragmentation, renewed interest has occurred in the use of tandem mass spectrometry (MS/MS) methods to induce fragmentation and thereby obtain structural information. The mass analyzer types that are commonly configured for MS/MS experiments are shown in Figures 8.12 through Figure 8.17.

In broad terms, MS/MS can be described as a series of events consisting of mass selection of a precursor ion in a first stage of analysis, an *intermediate reaction event*, followed by analysis of the product ions in a second stage of analysis (Busch et al. 1988; de Hoffmann 1996). (A reaction event is defined as an event that converts a precursor ion to a product ion having a different *m/z* value.) With the development of multistage tandem mass spectrometric (MS^n) experiments (Busch et al. 1988; McLuckey et al. 1995), the structure of product ions can be further interrogated using additional stages of MS. Note that although the term "tandem mass spectrometry" has traditionally been used to describe reaction events involving energetic dissociation of the mass selected ion, many alternative reaction types that also fit the definition of a tandem mass spectrometric reaction sequence are commonly used to obtain information about the structure and reactivity of ions in the gas phase. Thus, an MS/MS (or MS^n) experiment may also include, for example, ion-molecule, ion-ion, and ion-mobility reaction events.

In the context of proteomics, the major type of intermediate reaction event employed in an MS/MS experiment involves energetic dissociation of the ion to cause fragmentation, and thereby derives structural information relating to the sequence of the peptide. (Note that from herein, the term "product ion" is used interchangeably with the more technically correct "fragment ion.") Many methods have been used to deposit energy into an otherwise stable mass-selected ion to induce fragmentation (Busch et al. 1988; McLuckey and Goeringer 1997). These include:

- Collisions with an inert gas (known as collision-induced dissociation [CID] or collisionally activated dissociation [CAD]) (Hayes and Gross 1990; McLuckey 1992).
- Collisions with a surface (known as surface-induced dissociation or SID) (Dongre et al. 1996).
- Interaction with photons (e.g., via a laser) resulting in photodissociation (Bowers et al. 1984; Little et al. 1994).
- Thermal/black body infrared radiative dissociation (BIRD) (Price and Williams 1997).
- Electron-capture dissociation (ECD), whereby multiply charged cations are allowed to interact with low-energy electrons in an FT-ICR mass spectrometer (Zubarev et al. 1998).

By far the most widely used technique for performing energetic dissociation is CID. The process of CID involves collisional activation of the ion, by energetic collisions with a neutral target gas, to convert translational energy into internal energy, thus placing the ion into an activated (excited) state, followed by unimolecular dissociation of the activated ion to yield products (McLuckey 1992).

Both low-energy (<100 eV) (Hunt et al. 1986) and high-energy (keV) (Biemann 1990) collisional activation processes can be observed, depending on the instrumentation used. A well-defined nomenclature scheme (which is summarized in Table 8.3) has been proposed for the product ions arising from the fragmentation of protonated peptides. Product ions resulting from backbone cleavage of the αC-C, the C-N amide linkage, or the N-αC bond are called a-, b-, and c-type ions, respectively, if the charge is retained on the amino-terminal

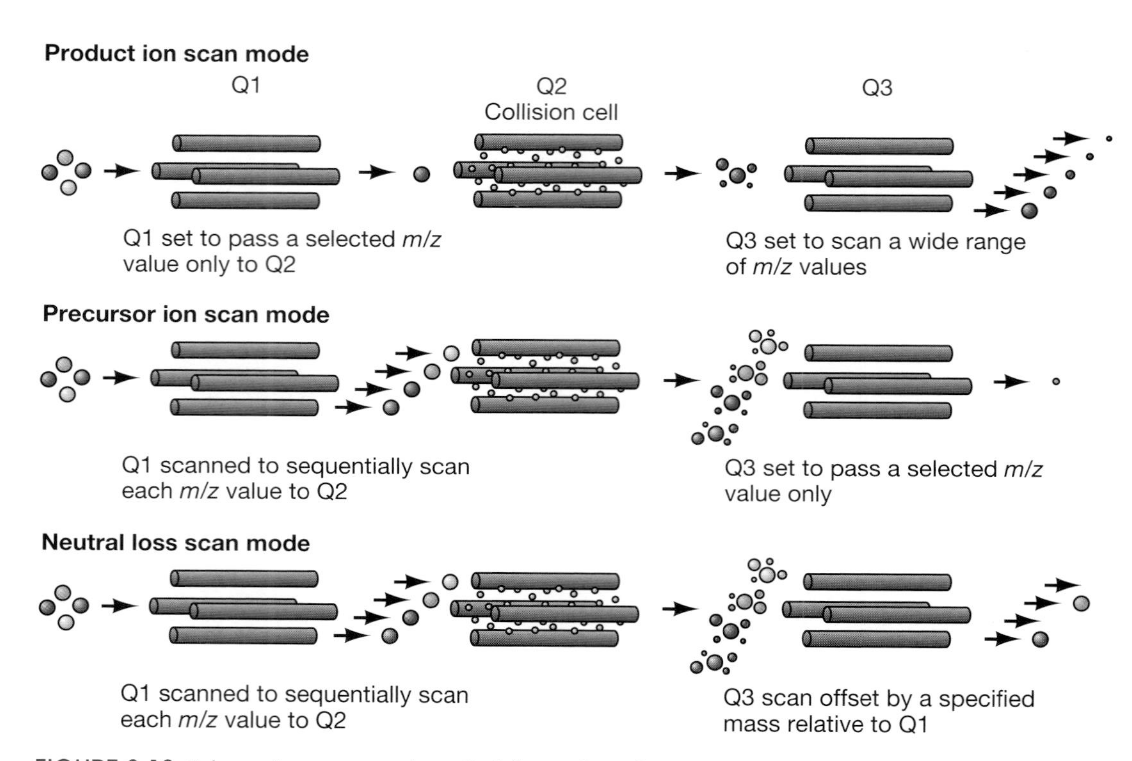

FIGURE 8.12. Schematic representation of triple quadrupole MS/MS scan modes. *The product ion scan:* An ion of interest is mass-selected in the first mass analyzer (Q1) and is transferred to the collision cell (Q2) where collision-induced dissociation (CID) takes place. The resultant product ions are then separated by scanning the second mass analyzer (Q3). The observed signal is therefore the result of the mass-analyzed product ions derived from a mass-selected precursor ion. *The precursor ion scan:* The first mass analyzer is scanned to sequentially transmit all ions into the collision cell for fragmentation. The second mass analyzer is set to only transmit a single *m/z* product ion of interest to the detector. Q3 and Q1 are linked so that the precursor mass in Q1 that gives rise to a product ion signal in Q3 may be determined upon detection of any product ion signal. The mass spectrum obtained therefore reveals all precursor ions that fragment to a yield a common product ion. *The neutral loss scan:* Mass analysis in both Q1 and Q3 mass analyzers is performed where the scan range of Q3 is linked to Q1 by a constant mass difference. As for the other scan modes, fragmentation of precursor ions occurs in the collision cell region. In this scan mode, the mass spectrum obtained reveals those precursor ions that fragment via a constant neutral loss (or gain).

fragment, or x-, y-, and z-type ions, respectively, if the charge is retained on the carboxy-terminal fragment. The product ions are numbered according to their position from their respective terminal end. Under high-energy CID conditions, d-, v-, and w-type ions corresponding to side-chain cleavages may also be formed. In general, the most commonly observed product ions resulting from low-energy CID belong to the b- and y-type ion series. These ions can be regarded as being related to each other in that they both arise via cleavage of the amide bond (for a review of peptide fragmentation, see the information panel on FRAGMENTATION MECHANISMS OF PROTONATED PEPTIDES IN THE GAS PHASE).

TABLE 8.3. Nomenclature for product ions arising from the fragmentation of protonated peptides

	Product ion nomenclature	
Site of backbone cleavage	if charge is retained on amino-terminal fragment	if charge is retained on carboxy-terminal fragment
αC—C bond	a	x
C—N amide bond	b	y
N—αC bond	c	z

MASS ANALYZER CONFIGURATIONS USED IN TANDEM MASS SPECTROMETRY

Critical to the development of techniques for probing the gas-phase structure of biomolecules have been advances in the development of mass analyzers that not only are more versatile, but offer ever-increasing levels of sensitivity (i.e., decreasing detection limits) (Brunnee 1987; Jennings and Dolnikowksi 1990). Mass analyzers that are applicable to MS/MS fall into two basic categories: tandem-in-space and tandem-in-time (Busch et al. 1988).

- ***Tandem-in-space mass spectrometers*** have discrete mass analyzers for each stage of mass spectrometry. Examples include double-focusing-sector analyzers, hybrid combinations of sector and quadrupole analyzer instruments, time-of-flight, and triple quadrupole instruments.
- ***Tandem-in-time mass spectrometers*** have only one mass analyzer. Each stage of mass spectrometry takes place in the same region, but is separated in time by a sequence of events. Examples of tandem-in-time mass spectrometers include FT-ICR and quadrupole ion-trap mass spectrometers.

Tandem-in-Space Mass Spectrometers

Triple quadrupole

The triple-quadrupole mass spectrometer (Yost and Boyd 1990; Wysocki 1992) consists of three quadrupoles connected in series (see Figure 8.12). Both the first (Q1) and third (Q3) quadrupoles have a combination of DC (variable voltage) and RF (constant frequency, variable amplitude) applied to the four rods, but adjacent rods have opposite polarity. If the magnitudes of the DC and RF voltages are increased while maintaining a constant RF-DC ratio, stable trajectories are created for ions of sequentially increasing *m/z*. Thus, a mass range may be "scanned" in order to obtain a spectrum. For MS-only scans, a fixed RF-only voltage is applied to the second (Q2) and third (Q3) quadrupoles, resulting in the transmission of a wide *m/z* range of ions that occurs simultaneously through these sections of the instrument.

For simple MS/MS experiments, the RF applied to Q1 is maintained at a frequency corresponding to the stable trajectory of only one ion, enabling it to be selectively isolated. All other ions experience unstable trajectories and are lost. CID occurs in the second quadrupole region, where an offset DC voltage is applied to ions entering the collision cell (typically an enclosed octapole in commercial instruments), so that the ions experience a net potential, and hence increase in kinetic energy, resulting in energetic collisions with a heavy background gas of argon, and subsequent fragmentation. Ions exiting Q2 then pass through quadrupole Q3, which is scanned (in the same manner as described in the preceding paragraph) in order to analyze the product ion population and obtain the mass spectrum. To date, only ESI sources are available for commercial triple-quadrupole instruments.

In addition to the *product ion scan* event described above, two other scan modes are commonly performed on triple-quadrupole instruments. In a *precursor ion scan*, Q1 can be scanned and, following a reaction event in Q2, detection of a single product ion can be performed in Q3. As long as Q3 and Q1 are linked, the precursor mass in Q1 that gives rise to a product ion signal in Q3 may be determined. Alternatively, both Q1 and Q3 may be scanned at the same time, where Q3 is offset by a constant value below the mass of Q1. If Q3 and Q1 are linked, after a reaction event in Q2, a *neutral loss* scan may be obtained. Similarly, if Q3 is offset by a constant value above the mass of Q1, and the other conditions hold, then a *neutral gain* scan may be obtained. For further information concerning the use of scan modes for detecting specific diagnostic fragment ions, see the panel below on USE OF SCAN MODES FOR DETECTING PEPTIDE IONS THAT PRODUCE DIAGNOSTIC FRAGMENT IONS UPON CID and Figure 8.12.

Use of Scan Modes for Detecting Peptide Ions That Produce Diagnostic Fragment Ions upon CID

A salient feature of triple-quadrupole mass spectrometers is the highly sensitive scan mode feature (see Figure 8.12). For example, it is possible to use the precursor ion scan (also known as a parent ion scan) to detect in a mixture of peptides only those precursors that efficiently produce a specific product ion upon dissociation of the parent ion commonly performed in the collision cell. Using this approach, the first quadrupole (Q1) is used for scanning and the third quadrupole (Q3) is fixed at the *m/z* of a specific diagnostic product ion (e.g., for phosphorylation, at *m/z* 79, the mass of the PO_3^- ion; for detection of cysteine residues that have been *S*-alkylated using the alkylating reagent 4-vinylpyridine [Moritz et al. 1996], at *m/z* 106, the mass of the *S*-pyridylethyl ion). If these fragment ions are produced in the collision cell (Q2), they are transmitted to the detector. Thus, when the detector registers a signal, it is not using the mass of Q3 (which is fixed at the *m/z* of the specific diagnostic fragment ion, and always the same) but the mass of Q1, which just transmitted the intact *S*-pyridylethylated peptide ion to Q2 and Q3. This task is accomplished by coupling Q1 data to Q3. Hence, only when an *S*-alkylated peptide ion is transmitted through Q1 is the *m/z* 106 product ion produced in Q2 and a signal recorded at the detector. This scan mode can be used for any structure that efficiently produces a specific diagnostic product ion.

Time of flight

This mass analyzer measures the *m/z* values of analytes by pulsing molecular ions from the ionization source into a flight tube. The *m/z* values are calculated by the time it takes ions to travel a set distance and strike a detector (i.e., their time of flight). In 1991, a method of peptide sequencing by MALDI-TOF-MS was described by Spengler et al. (1991, 1992). The method is based on *metastable* decay of laser-desorbed ions, which occurs in the first field-free drift region of a reflectron time-of-flight (TOF) mass spectrometer. Also referred to as post-source decay (PSD), this method can be used to obtain structural information in peptides (see Figure 8.13).

Quadrupole time of flight

Historically, TOF mass analyzers had been exclusively coupled with MALDI, due to the complementary pulsed nature of both the ion source and mass analyzer. However, the develop-

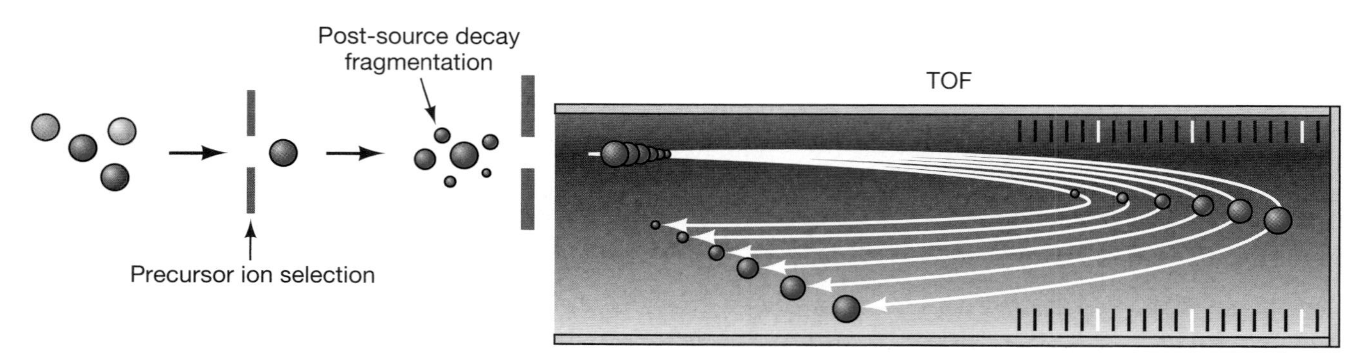

FIGURE 8.13. Schematic representation of post-source decay (PSD) MS/MS via reflectron time-of-flight mass spectrometry.

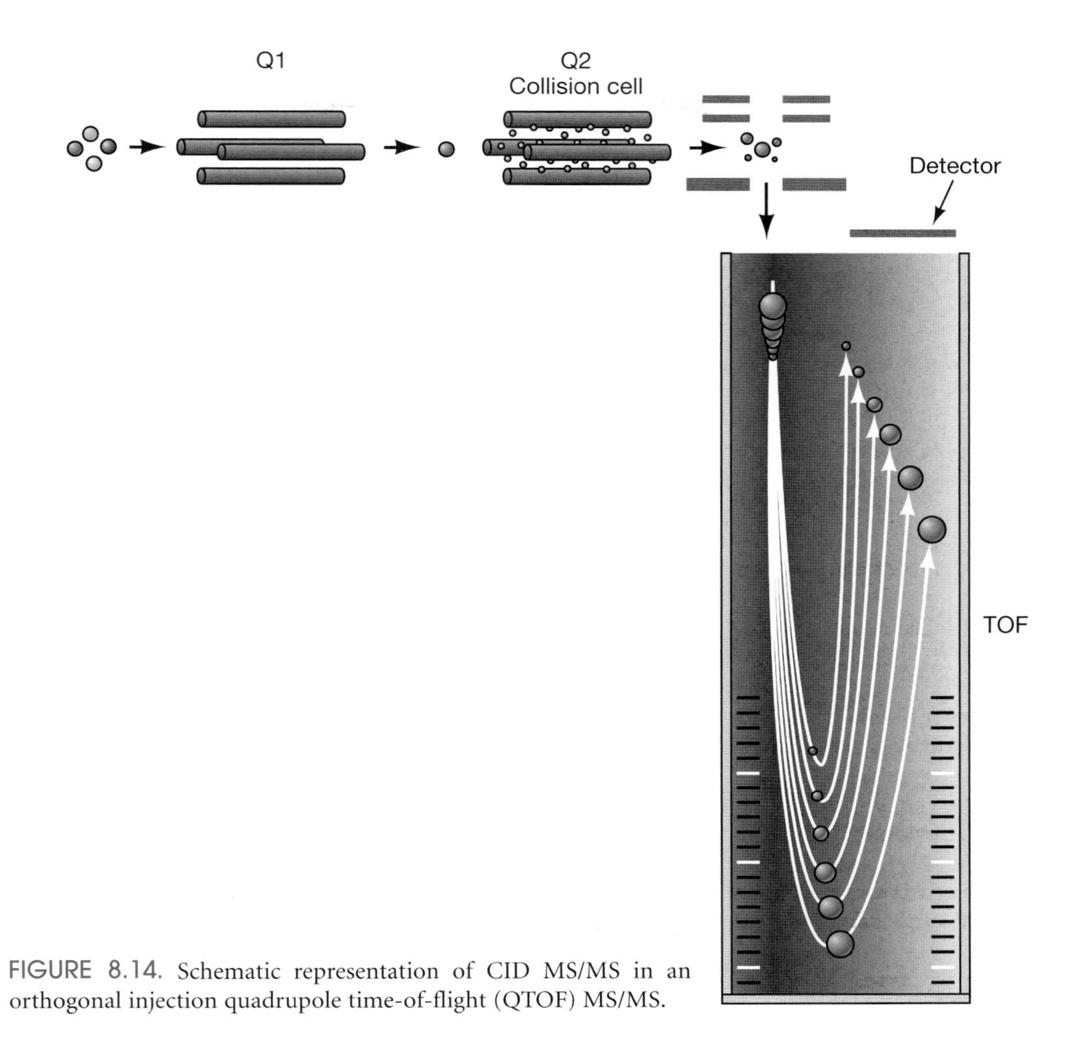

FIGURE 8.14. Schematic representation of CID MS/MS in an orthogonal injection quadrupole time-of-flight (QTOF) MS/MS.

ment of pulsed orthogonal ion injection into TOF mass analyzers has now allowed ESI sources to be coupled with TOF for high-resolution MS and MS/MS experiments (Morris et al. 1996). The operation of this type of tandem mass spectrometer is essentially the same as that for a triple-quadrupole mass spectrometer, except that product ions exiting Q2 following CID are orthogonally pulsed into a TOF mass analyzer to acquire the mass spectrum (see Figure 8.14).

Both ESI and MALDI sources have been successfully coupled to quadrupole TOF (qTOF) instruments (Loboda et al. 2000; Shevchenko et al. 2000, 2001; Wattenberg et al. 2002; Baldwin et al. 2001). The qTOF mass spectrometer was developed in order to take advantage of the high-performance resolution capabilities of the TOF mass analyzer for product ion assignment, together with the ability to obtain efficient precursor ion selection and dissociation using the quadrupole mass filter. This instrument therefore overcomes many of the limitations of triple-quadrupole (poor resolution and low sensitivity due to poor duty cycles) and TOF-based (poor product ion formation via post source decay) MS/MS methods. Recent applications of this type of instrumentation have included studies in phosphopeptide mapping (Bateman et al. 2002) and quantitative proteomics (Griffin et al. 2001b).

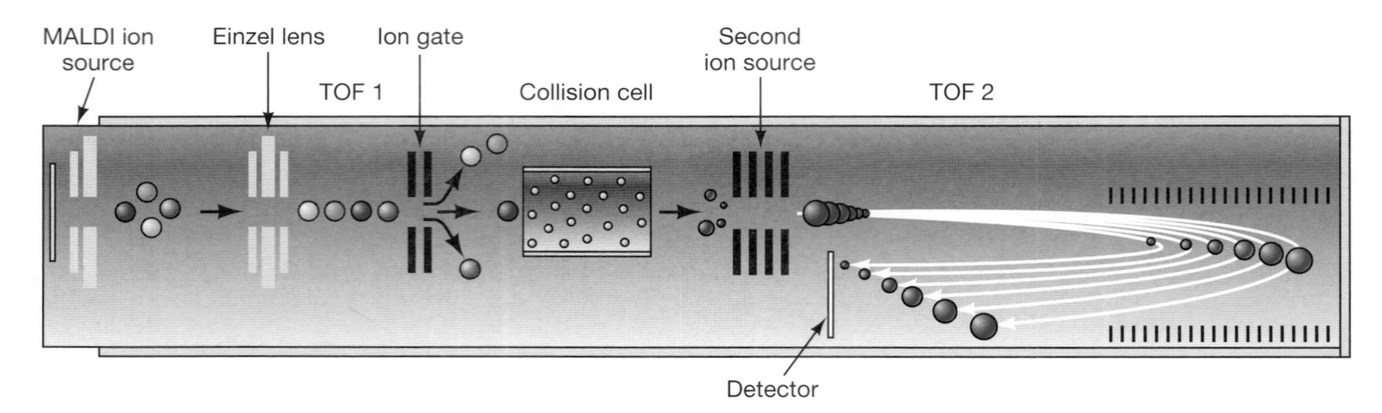

FIGURE 8.15. Schematic representation of CID-MS/MS by TOF/TOF-MS. The MALDI ion source comprises a sample plate, laser source, and a set of focusing lenses that direct the ions to the TOF mass analyzer.

Time-of-flight/time-of-flight

A MALDI-TOF/TOF high-resolution tandem mass spectrometer has recently been introduced for high-throughput, high-sensitivity peptide sequencing applications (Medzihradszky et al. 2000; Bienvenut et al. 2002). Ions desorbed from the MALDI target are accelerated to 3 keV kinetic energies through the first TOF region. Precursor ion selection is then performed by a timed ion gate prior to CID in a gas-filled collision cell. The high kinetic energy of the ions entering the collision cell allows high-energy CID processes to be observed. Thus, a wider range of product ion types can be observed compared to those formed under low-energy CID conditions, including d-, v-, and w-type ions resulting from side-chain cleavages, thereby allowing the isobaric amino acids isoleucine and leucine to be differentiated. Mass analysis is performed in the second TOF mass analyzer as indicated in Figure 8.15.

Tandem-in-Time Mass Spectrometers

Quadrupole ion trap

The quadrupole ion trap, originally developed by Paul (March 1997, 1998; Schartz and Jardine 2000), is essentially a three-dimensional quadrupole, consisting of a hyberbolic ring electrode, and two hyperbolic end-cap electrodes. Small holes in the end-cap electrodes allow the passage of ions into and out of the trap. A high-voltage RF potential is applied to the ring electrode while the end caps are held at ground. The oscillating potential difference established between the ring and end-cap electrodes forms a substantially quadrupolar field (Figure 8.16). Ions injected into the trap above a certain *m/z* value, determined by the amplitude of the ring electrode RF, have stable trajectories. Collisions of the injected ions with helium (present at $\sim 10^{-3}$ torr) cause "damping" of their trajectories so that they can be trapped by the RF field into the center of the trap. To obtain a mass spectrum, the ring electrode RF voltage amplitude is increased while applying a supplementary RF voltage to the end caps, sequentially bringing ions of increasing *m/z* into resonance with the applied frequency, whereupon they become unstable and their velocity and trajectory increase in the axial direction until they are ejected from the trap and detected.

For MS/MS analysis, precursor ions of a selected *m/z* may be isolated by increasing the amplitude of the ring electrode RF to the point that all ions below the mass of the selected ion are ejected. During this "RF ramp process," ions above the precursor mass are also eject-

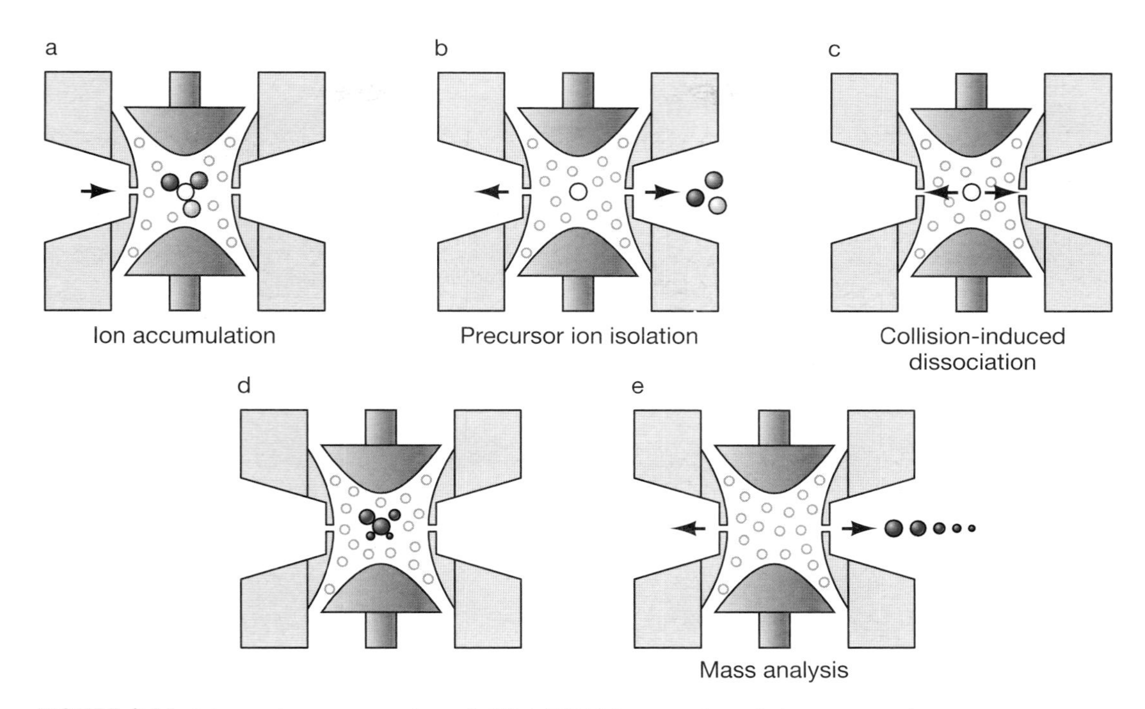

FIGURE 8.16. Schematic representation of CID MS/MS in a quadrupole ion-trap tandem mass spectrometer. (*a*) *Precursor ion accumulation:* Ions injected from either ESI or MALDI external ion sources collide with a background gas of helium and are trapped in the center of the ion trap. (*b*) *Precursor ion isolation:* Ions at the *m/z* of interest may be isolated by sequentially ejecting all other ions higher and lower in *m/z* by increasing the amplitude of the ring electrode RF while applying a supplementary RF voltage to the end caps. (*c*) *Collision-induced dissociation (CID):* CID is achieved by increasing the motion of the isolated ion in the axial direction via the application of an RF resonance excitation voltage to the end-cap electrodes to enable energetic collisions with the background helium. (*d*) *Collisional cooling:* Product ions are cooled and trapped by collisions with the background helium gas. (*e*) *Mass analysis:* The resultant product ions are sequentially ejected from the trap in order of increasing *m/z* to acquire a product ion spectrum. Note that by cycling back to Step b from Step d, subsequent product ion isolation and further CID reaction events can be performed to enable further interrogation of first-generation product ions (i.e., an MS^n experiment).

ed by applying a supplementary RF voltage to the end caps, whose frequency and amplitude are set just above the mass of the precursor. Thus, as the ring electrode RF amplitude is increased, these ions of higher mass are successively brought into resonance with the supplementary voltage and ejected until only the precursor ion remains. CID of the selected precursor ion is accomplished generally by increasing the motion of the ion in the axial direction by applying an RF resonance excitation voltage to the end-cap electrodes, at a frequency comensurate with the ion of interest, to enable energetic collisions with the background helium. The resultant product ion population is sequentially scanned out of the trap in order of increasing *m/z* to acquire a product ion spectrum. MS^n experiments can easily be performed by adding subsequent ion isolation and reaction event steps to the experimental sequence following the initial MS/MS reaction. Note that whereas only product ion scans can be performed in an ion-trap mass spectrometer, precursor and neutral loss scan spectra can be constructed by post-acquisition data processing following the acquisition of full-product ion scan mode spectra. Both ESI and MALDI ion sources are now available on commercial ion-trap instruments.

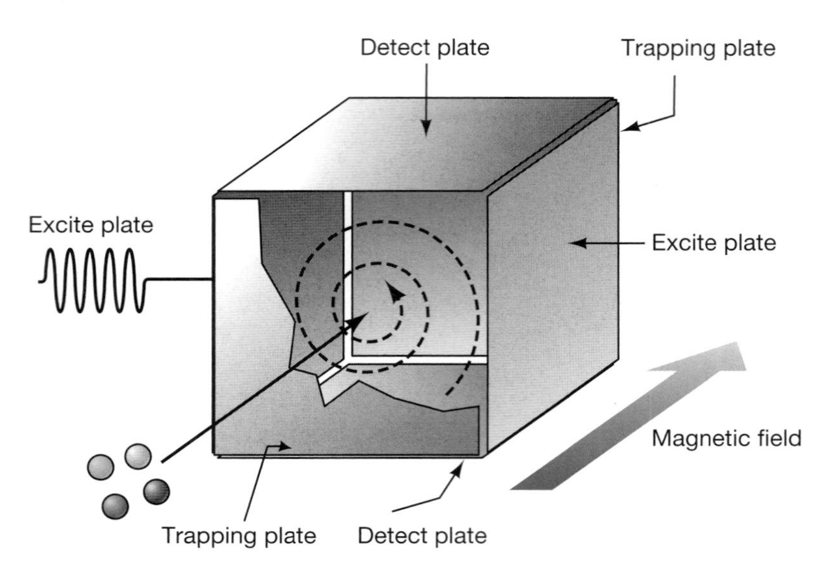

FIGURE 8.17. Schematic representation of a Fourier transform-ion cyclotron resonance (FT-ICR) mass spectrometer.

Fourier transform-ion cyclotron resonance

MS/MS using FT-ICR (Marshall and Verdun 1990; Amster 1996) is similar to quadrupole ion-trap MS/MS because in both cases, ions are trapped in three-dimensional space. However, reactions in the FT-ICR are carried out in a cell bound by electrodes (known as trapping, excite, and detect plates) located in a magnetic field, where the *m/z* value of an ion is directly related to its cyclotron frequency (see Figure 8.17). For MS/MS experiments, a "tailored" stored waveform inverse Fourier transform excitation pulse (or SWIFT, for short) is applied to the excite plates of the cell (Marshall et al. 1985). This causes ions other than the desired precursor ion to be excited and ejected. Following isolation of the precursor ion, CID is effected by accelerating the precursor ion in the presence of a pulsed collision gas to higher kinetic energies (but not ejecting it from the cell) by the application of a sustained off resonance irradiation (SORI) frequency shifted away from the cyclotron resonance frequency of the trapped ion (Gauthier et al. 1991). To acquire a product ion mass spectrum, ions are excited by a broadband RF excitation pulse applied to the excite plates, and an image current (the current induced in the walls of the cell due to the proximity of charged particles) is detected by the detect plates as a function of time. This time domain signal is transformed into a frequency domain signal and the masses are determined. Note that MS^n experiments may be performed by simply adding additional isolation and reaction events steps to the sequence in a manner similar to that for the quadrupole ion trap. For further description of FT-ICR, see the information panel on "TOP-DOWN" PROTEIN SEQUENCE ANALYSIS USING MS/MS.

COMPLEX MIXTURES OF PROTEINS CAN BE SEPARATED AND ANALYZED BY COUPLING MS WITH PROTEIN SEPARATION METHODS

A number of technologies have been successfully coupled with MS for the purpose of separating complex mixtures of macromolecules and analyzing the components of the mixture. New technologies continue to be developed as the demand for efficient, high-throughput methods increases. The following sections explain some of these strategies for separating and analyzing protein mixtures. For details for carrying out the methods, see the protocols in this chapter and Appendix 2, and for related information and protocols, see Chapters 1, 2, 4, 5, and 9–11.

- MS can be preceded by 1D or 2D liquid chromatography (LC) to separate complex mixtures and/or remove contaminants that interfere with MS analysis.
- Proteins can be separated using 1D or 2D PAGE, followed by MS analysis of selected proteins taken from the gel.
- Gel-separated proteins can be transferred to membranes prior to analysis by MS.
- Affinity chromatography can enrich for specific classes of proteins prior to MS analysis.

Coupling MS/MS to 1D Liquid Chromatography

Separation and concentration of peptides from mixtures prior to sequence analysis by MS/MS are almost universally accomplished by on-line reversed-phase high-performance liquid chromatography (RP-HPLC). A protein digest can be loaded onto an HPLC system, separated on-line by RP-HPLC, and the eluate analyzed by MS/MS. With the mass spectrometer's ability to select peptide ions based on their *m/z* values, it is well suited as a detector for analysis of mixtures and provides more specific information (e.g., amino acid sequence and carbohydrate structure) than that of a peptide mass map. As ion signals increase above a chosen threshold, MS/MS is performed on the ion until it drops back below the threshold value. It is possible to collect MS/MS data on several of the most intense ions per unit time and to incorporate an exclusion list of ions that have already been selected for CID. (In practice, the mass spectrometer can be programmed to perform one scan to determine the peptide masses and then to sequence the three to eight most abundant peptides [a method known as data-dependent acquisition].) With an LC system coupled to the tandem mass spectrometer through an ESI source and the ability to perform data-dependent scanning, it is now possible to distinguish proteins in complex mixtures containing more than 50 components without first purifying each protein to homogeneity (McCormack et al. 1994, 1997).

In MS/MS, a peptide ion is selected and fragmented, making it theoretically irrelevant as to how complex the original protein digest is unless there are isobaric peptide masses. Therefore, the coupling of LC to data-dependent MS/MS and automated database searching is a very powerful technique for "shotgun" identification of proteins in very complex mixtures. This approach now allows biologists to explore the cell with unprecedented speed, determining protein identities, sites of posttranslational modifications, and protein-protein interactions, along with measuring the relative quantities of proteins in a sample (Gygi et al. 1999; Griffin et al. 2001a; Natsume et al. 2002).

The use of capillary columns for LC increases the sensitivity of the MS/MS analysis. LC-MS/MS can be made exquisitely sensitive by using capillary columns and capillary needle columns instead of traditional large-bore HPLC columns (Griffin et al. 1991; Emmett and Caprioli 1994; Davis et al. 1995; Gatlin et al. 1998; Moritz and Simpson 1992a,b, 1993; Natsume et al. 2002) (see also Chapter 5). Referred to as nano-LC or nLC, these narrow-bore columns provide two advantages:

- Flow rates through these very narrow columns are 50–400 nl/min.
- The capillary column serves as the ionization source.

These features improve the sensitivity by reducing the volume of solvent per unit time that is sprayed into the mass spectrometer, which in effect concentrates the sample as it enters the ion source of the instrument. Protocols 4 and 5 detail methods for preparing and using microcapillary columns for proteomic studies.

MS Sensitivity: What Is the Typical Sensitivity Expected?

For most nano-LC (50–400 nl/min) systems coupled to triple-quadrupole mass spectrometers, 1–10-fmole levels of sensitivity are achievable (Cox et al. 1994). Further improvements in sensitivity (and increased resolution) can be gained using the newer quadrupole TOF instruments. Although 1–5-fmole sensitivity is achievable using ion-trap mass spectrometers (full-scan MS mode), this mass analyzer may be limited by the introduction of chemical noise accumulating in the trap. With the introduction of the variable-flow LC (5–200 nl/min), also referred to as *peak parking* (Davis and Lee 1997), which is possible with the advent of fully automated capillary LC systems, 5–10-attomole levels of sensitivity have been reported for MS/MS analysis in ion-trap instruments (Martin et al. 2000). These detection levels are as sensitive as those obtained with silver staining of proteins on SDS-PAGE gels (Link et al. 1999). However, it should be stressed that the inherent sensitivities of the MS configuration mentioned above were determined on material directly applied to nano-LC column and not necessarily the material visualized in a gel band. (Sample losses are frequently incurred during electrophoretic separation and subsequent sample handling steps prior to MS analysis.)

Coupling MS/MS to 2D Liquid Chromatography

Although tandem mass spectrometers inherently possess the ability to sequence peptides directly from complex mixtures, highly complex peptide mixtures (e.g., proteome-scale) are best analyzed first by 2D chromatography techniques. This is most commonly achieved by the combination of strong cation exchange (SCX) (Mant and Hodges 1985; Alpert and Andrews 1988) and RP-HPLC (Link et al. 1999; Gygi et al. 2000a,b; Washburn and Yates 2000; Washburn et al. 2001). By exploiting two of the physical properties of peptides (i.e., charge and hydrophobicity), complex peptide mixtures can be effectively resolved and concentrated prior to sequence analysis by MS. SCX chromatography, with its greater loading capacity, is generally used first. The eluent from the SCX chromatography is loaded onto an RP-HPLC column, which removes salts from the solution while performing a second separation of the peptides based on their hydrophobicity. The RP-HPLC column eluent is applied directly to the mass spectrometer using ESI.

2D chromatography can be accomplished by either an on-line or an off-line approach. For the on-line approach, an acidified complex peptide mixture is applied to an SCX column, and discrete fractions of the adsorbed peptides are sequentially displaced directly onto the RP-HPLC column using a salt step gradient. Peptides are then analyzed by LC-MS/MS. This approach can utilize as many as 15 (or more) stepwise increases in salt concentration "bumps" to fractionate a peptide mixture (Link et al. 1999; Washburn et al. 2001). The on-line approach is amenable to being almost completely automated. Essentially, the sample is loaded onto the SCX column, and 24 hours later, the analysis is finished (see Protocol 6).

The off-line approach is performed by applying the acidified complex peptide mixture to the SCX chromatography column followed by a binary gradient to high salt to elute the peptides. Fractions are typically collected every minute into a 96-well plate and reduced in volume, and then each fraction is loaded onto a reversed-phase chromatographic column automatically via an autosampler and finally analyzed by LC-MS/MS. The off-line technique has several advantages over the on-line approach:

- Peptide separation is superior using true chromatography instead of salt steps.
- SCX chromatography is ideally performed with >20% acetonitrile in the buffers to linearize peptides (Burke et al. 1989), yet the on-line approach can tolerate only 5–10% acetonitrile.

- A greater number of peptide fractions can be collected with the off-line approach (e.g., 96 vs. 15).
- The investigator has the discretion to choose which fractions to analyze, and interesting fractions can be analyzed more than once.

Protocol 7 provides the details of performing the off-line approach of 2D-LC. The on-line technique is discussed in great detail in Protocol 6 (see also Appendix 2).

Microscale LC Cleanup of Protein Samples Improves the Value of Mass Spectra

A major problem in MS analysis is the presence of nonvolatile low-molecular-mass contaminants that increase ion suppression effects and chemical noise (see Figure 8.18). In MALDI-MS, the traditional dried-droplet sample preparation method (Hillenkamp et al. 1991) is per-

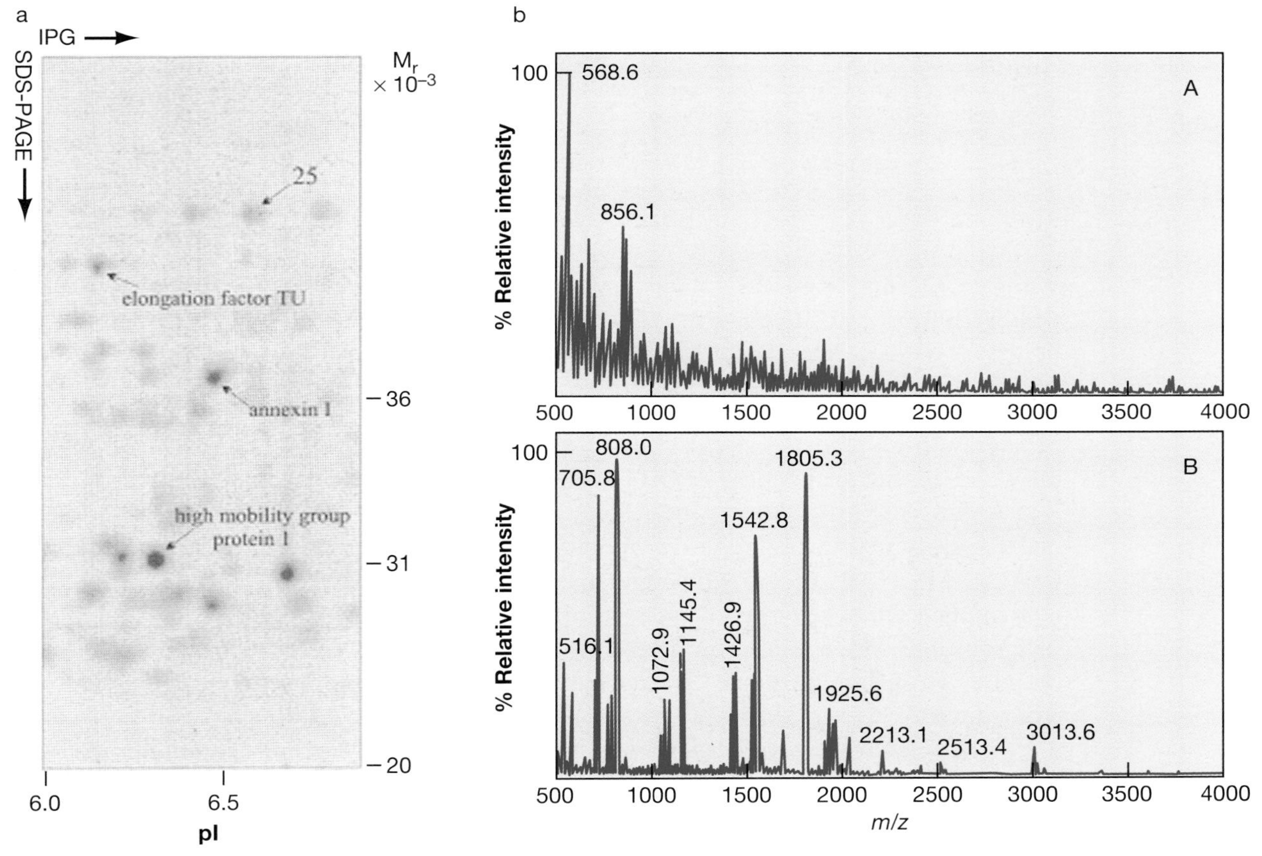

FIGURE 8.18. (*a*) 2D gel electrophoresis pattern of enriched M_r 25,000–45,000 MDA-MB231 proteins. Shown is a MALDI-TOF-MS analysis of a total tryptic digest of protein #25 from MDA-MB231 breast carcinoma cells. Proteins from a total cellular lysate of ~6 × 10^7 MDA-MB231 cells were separated on a 10% SDS-PAGE slab gel. The M_r 25,000–45,000 region was excised and proteins were passively eluted from the gel slice (H. Ji et al. 1993), concentrated, and further resolved by 2D electrophoresis using an 18-cm linear pH 3–10 IPG (Amersham Biosciences) in the first electrophoretic dimension and 10% SDS-PAGE in the second dimension. The pH gradient profile was calibrated using a carbonic anhydrase carbamylate calibration kit (Amersham Biosciences), according to the manufacturer's instructions. Proteins were visualized by staining with CBR-250, and spot #25 was selected for in-gel proteolysis (see Chapter 7, Protocol 13) and subsequent MS analysis. (*b*) The total peptide mixture was analyzed by MALDI-TOF-MS either directly (*A*) (i.e., no concentration/desalting step) by loading 0.5 µl of the peptide mixture (total volume ~100 µl, ~500 pmoles) or indirectly (*B*) following a simple capillary (0.32 mm I.D.) column RP-HPLC concentration/desalting step (ON/OFF), where the total peptide mixture (total volume ~100 µl, ~500 pmoles) was applied to the column, the column washed extensively with Solvent A (aqueous 0.1% TFA), and the total peptide mixture was then recovered in ~25 µl by developing the column with 100% Solvent B (60% acetonitrile in aqueous 0.1% TFA) in a batchwise manner. MALDI-TOF-MS sample load: 0.5 µl. (Reproduced, with permission, from Reid et al. 1998.)

TABLE 8.4. Excipients compatible with MALDI-MS

Excipient	Maximum recommended concentration
Urea	0.5 M
Guanidine-HCl	0.5 M
Glycerol	1%
Alkali metal salts	0.1 M
Tris buffer	0.05 M[a]
NH_4HCO_3	0.05 M
Phosphate buffer	0.01 M[a]
Detergents	n.c.[b]
SDS	0.005%
Sodium chloride	10 mM[c]
Hexafluoroisopropanol	up to 40%

Useful Web Sites for sample preparation: University of Illinois: www.biotech.uiuc.edu/mass_guide.htm; www.srsmaldi.com/Maldi/Step/Desalt.htmlStanford Research; www.srsmaldi.com; www.astbury.leeds.ac.uk/Facil/MStut/mstutorial.htm.

[a]Phosphate buffers should be avoided, because they discriminate salts that are hygroscopic (e.g., sodium acetate) because of the resultant wet surface.

[b]n.c. indicates not compatible. Some samples require the addition of a detergent for solubilization (especially if the sample has a hydrophobic character. In these situations, Triton X-150 or Tween-80 can be added at a concentration of 0.1–0.6% (Börnsen 2000).

[c]Addition of 2-propanol improves signal intensities of high-salt-containing samples.

formed by mixing the analyte and matrix solutions directly on the MALDI target (see Protocol 1). Alternatively, the matrix and analyte solutions can be mixed in a tube prior to application onto the target. The tolerance of this method toward nonvolatile contaminants is limited by the efficiency of the crystallization process and by sample washing following matrix crystallization (see Table 8.4) For reviews on sample preparation techniques for peptides and proteins analyzed by MALDI-MS, see Börnsen (2000) and Kussmann and Roepstorff (2000). It has also been reported that matrix solution conditions (especially the solvent) have a large influence on the quality of the MALDI-MS analysis (Cohen and Chait 1996).

An increase in sensitivity and resolution over the dried-droplet technique was obtained by the introduction of the fast evaporation or thin-layer method (Vorm et al. 1994) for MALDI-MS matrix preparations. The matrix is dissolved in a highly volatile organic solvent and applied onto the target to create a thin homogeneous layer of matrix crystals. The analyte solution is then placed on top of the film and dried at ambient temperature. A higher number of analyte molecules are presumably desorbed upon laser irradiation, as they are embedded exclusively in the outermost layer of crystals. In addition, increased signal intensities and sensitivity, together with a higher tolerance toward contaminants, have been observed. The fast evaporation method was further improved with the inclusion of nitrocellulose in the matrix (Jensen et al. 1996a,b; Kussmann et al. 1997). The presence of nitrocellulose reduced the intensity of the alkali metal ion adducts observed in the mass spectra, presumably because it prevented the alkali metal ions from being desorbed into the gas phase.

Nonvolatile contaminants interfere even more readily with ESI-MS than with MALDI-MS. The exact mechanism by which the nonvolatile contaminants suppress the ionization of the analyte has not yet been determined. Recently, it has been postulated that the suppression may be partly the result of attractive forces holding drops tightly together, thus inhibiting smaller droplets from forming as they enter the mass spectrometer. In addition, contaminating salts cause the analyte to precipitate, further hampering the analysis of the sample (King et al. 2000). Consequently, a chromatographic step must be included to remove nonvolatile contaminants from biological samples prior to their analysis by ESI-MS.

A significant improvement in sample preparation for both MALDI-MS and ESI-MS came with the introduction of custom-made disposable microcolumns as a fast cleanup step prior to MS. This approach was first proposed by Wilm and Mann (1996), who used a thin capillary needle packed with a small volume of reversed-phase chromatographic material and eluted the analyte molecules directly into a nano-ESI capillary needle by centrifugation. In this way, small amounts of sample could be rapidly concentrated and desalted prior to MS analysis. Low-molecular-mass contaminants are easily removed from the column, resulting in an increased signal-to-noise ratio in the mass spectrum and thus improved sensitivity. In addition, significantly higher amino acid sequence coverage from peptide mass maps is observed, which is important for the complete characterization of a protein. More recently, this technique has been further simplified by using either GELoader tips (Eppendorf) packed with chromatographic material or prepacked microcolumns (e.g., ZipTips, Millipore) (Gobom et al. 1997; Kussmann et al. 1997; Erdjument-Bromage et al. 1998; Gobom et al. 1999; Shaw et al. 1999; John et al. 2000; P.M. Larsen et al. 2001; M.R. Larsen et al. 2001b,c).

Customized Microcolumns and Strategies for MALDI-MS Sample Cleanup

Reversed-phase microcolumns are an efficient and fast way to clean up and concentrate protein samples prior to MALDI-MS analysis (Kussmann et al. 1997; Gobom et al. 1999). In addition to reversed-phase resins, a variety of other chromatographic resins can be used in microcolumns for sample cleanup prior to MS analysis. Protocols 2 and 3 detail the preparation of microcolumns containing either reversed-phase or immobilized enzyme resin, respectively. Some popular resins include:

- ***Poros R1 and R2 and Oligo R3*** (Applied Biosystems) are reversed-phase packings used to desalt and concentrate proteins and, especially, peptides. Poros R1 is designed for very hydrophobic proteins and peptides. The binding strength is similar to low carbon-loading C_4 supports. Poros R2 is designed for general separation of proteins, peptides, and nucleic acids. The binding strength is similar to low carbon-loading C_8 or C_{18} supports. The Oligo R3 medium is designed for hydrophilic peptides and nucleic acids and is similar to high carbon-loading C_{18} supports. For other reversed-phase packings and capillary column fabrication, see Chapter 5.
- ***Immobilized metal affinity chromatography*** (IMAC) material is used for purification of His-tagged proteins and characterization of phosphorylated proteins and peptides (see Chapters 9 and 10 and Protocol 1 in Chapter 9).
- ***Graphite powder*** can be used as a peptide cleanup medium, especially for small hydrophilic peptides or phosphopeptides (Chin and Papac 1999). However, it also works very well for all other peptides if they are eluted with the matrix, α-cyano-4 hydroxy-cinnamic acid (HCCA) (Larsen et al. 2002a). Graphite powder can also be used to desalt and concentrate carbohydrates prior to MS analysis (Larsen et al. 2002b).
- ***Immobilized proteolytic enzymes*** can be used for rapid and efficient protein/peptide digestion (Gobom et al. 1997; M.R. Larsen et al. 2001b).
- ***Immobilized RNA or DNA*** can be used to affinity-purify very small amounts of nucleic-acid-binding proteins prior to SDS-PAGE. This has been used to isolate proteins binding to the human insulin-like growth factor II leader 2 mRNA (Pedersen et al. 2002). Only a limited amount of cell material is necessary with this approach (M.R. Larsen et al. 2001a).

In a significant number of studies, the use of microcolumns for sample cleanup prior to MS has improved sensitivity and enhanced the identification of low-abundance proteins separated by gel electrophoresis (e.g., see Shaw et al. 1999; John et al. 2000; P.M. Larsen et al. 2001; M.R. Larsen et al. 2001b,c; Natsume et al. 2002). The next six sections provide examples of ways in which microcolumns can be used to facilitate sample cleanup prior to analysis by MS.

Desalting and concentration of proteins

Microcolumns have been used to desalt and concentrate trypsinogen prior to MALDI-MS analysis. Trypsinogen was diluted from a stock solution in phosphate-buffered saline (PBS) buffer to a final concentration of 1 pmole/μl; 1 μl of this solution was mixed with the matrix solution, 2,5-dihydroxybenzoic acid (DHB) in 50% acetonitrile/1% trifluoroacetic acid (TFA) on the MALDI target. The resulting MALDI-MS spectrum from this preparation is shown in Figure 8.19a. The signal-to-noise ratio is low, probably due to the high-salt content in the PBS buffer, resulting in very low signals from the protein. In addition, the salt molecules contribute to a decreased resolution of signal due to overlapping peaks from sodium and potassium adducts. If the same amount of sample is desalted and concentrated on a Poros R1 microcolumn and eluted with DHB, a pronounced increase in the signal-to-noise ratio is evident (Figure 8.19b) leading to improved signal intensity and resolution. Additionally, some fragments originating from the protein could also be detected.

Desalting and concentration of peptides

Another reversed-phase resin, Poros R2, can be used in microcolumns for desalting and concentrating samples prior to MALDI-MS analysis. The advantages of using this resin are illustrated by the analysis of a tryptic digest of a yeast protein separated by 2D gel electrophoresis. Figure 8.20 compares the performance of the traditional MALDI-MS sample preparation

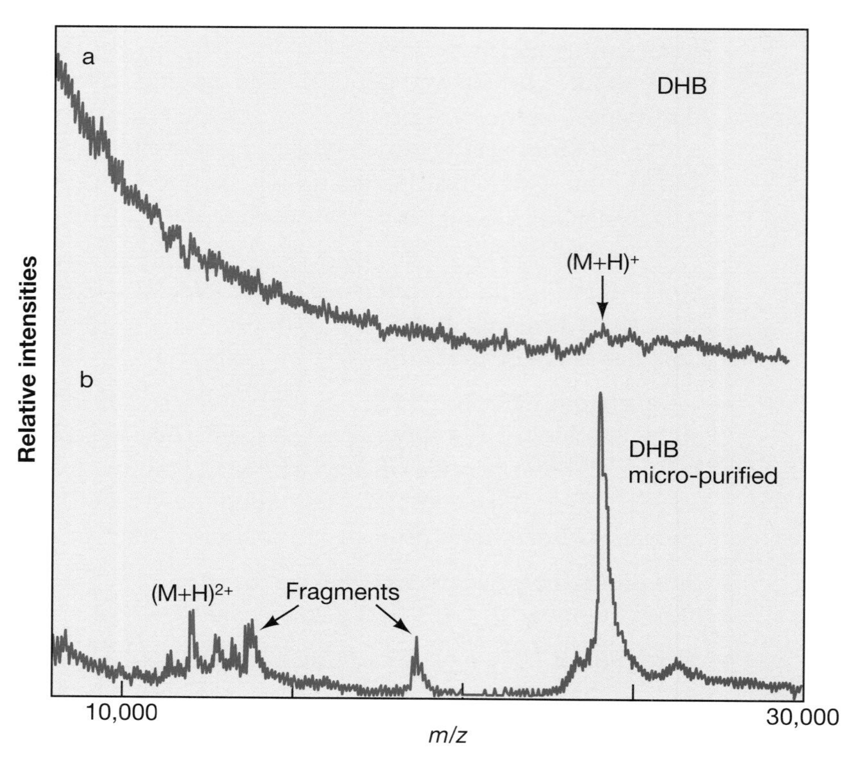

FIGURE 8.19. Comparison of the dried-droplet and micropurification sample preparation using trypsinogen dissolved in PBS buffer. (*a*) MALDI spectrum of 1 μl of a stock solution of trypsinogen (1 pmole/μl) using DHB dissolved in 50% acetonitrile/1% TFA as matrix. (*b*) MALDI spectrum of the same amount of trypsinogen desalted and concentrated on a Poros R1 microcolumn prior to MS. The protein was eluted from the microcolumn with DHB dissolved in 50% acetonitrile/0.1% TFA. (Courtesy of Martin R. Larsen, Macquarie University.)

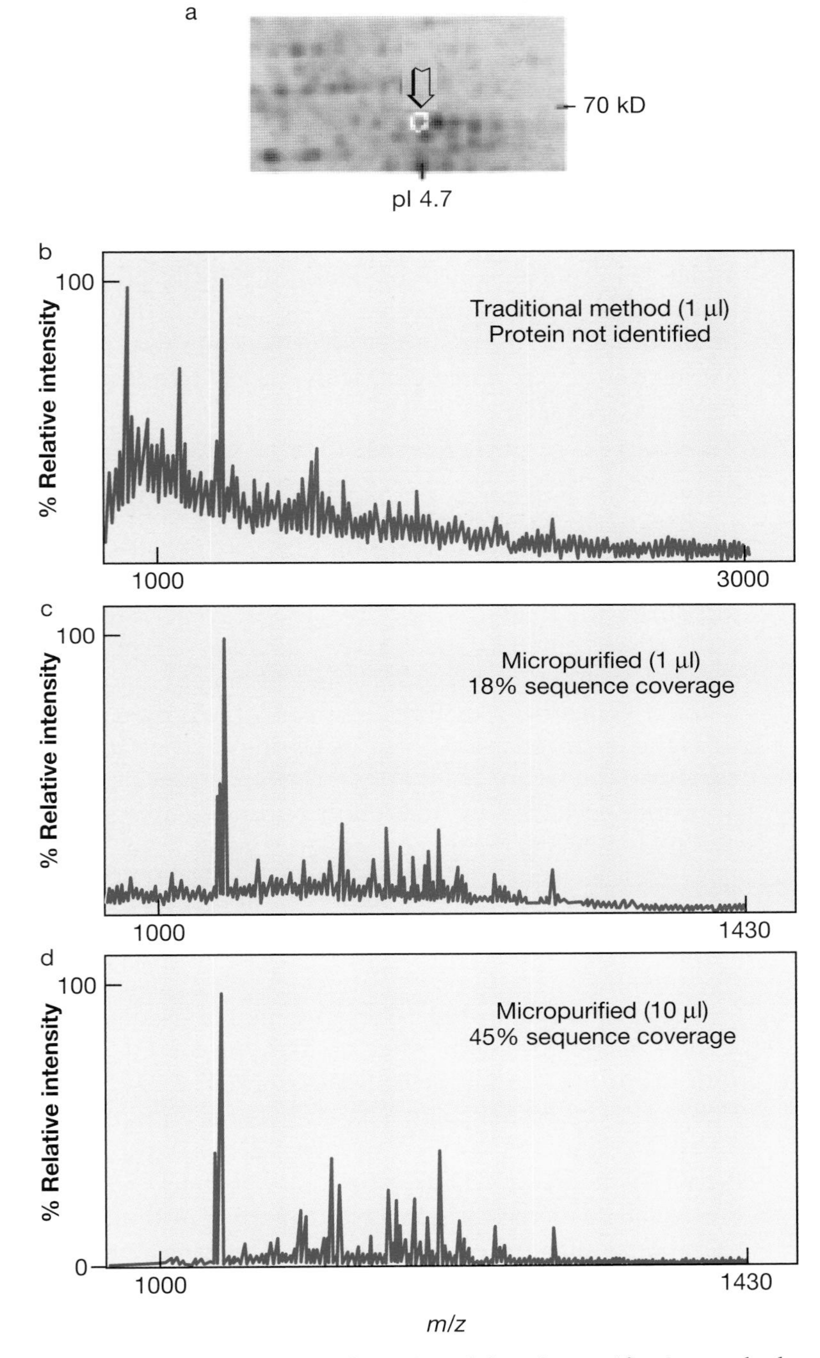

FIGURE 8.20. Comparison of the traditional matrix and the micropurification methods using the peptide mixture from an in-gel tryptic digest of a weak [^{35}S]methionine-labeled yeast protein. (*a*) Autoradiogram illustrating the "intensity" of the excised protein spot. The total protein load on the gel equaled 200 µg. (*b*) MALDI-MS of an aliquot (1 µl) of the tryptic digest supernatant using the traditional matrix preparation method. (*c*) MALDI-MS of 1 µl of the tryptic digest supernatant desalted and concentrated on a Poros R2 microcolumn. (*d*) MALDI-MS as in *c*, but with an aliquot of 10 µl. (Courtesy of Martin R. Larsen, Macquarie University.)

method (see Protocol 1) with the micropurification method described in Protocol 2. The excised spot is shown on the magnified image of a 2D gel containing 200 μg of [^{35}S]methionine-labeled yeast proteins (Figure 8.20a) (Nawrocki et al. 1998). The protein spot was excised and subjected to in-gel digestion with trypsin (see Protocol 13 in Chapter 7). The MALDI-MS peptide mass map obtained using the traditional matrix preparation method with an aliquot of the digest supernatant (1 μl of 20 μl) is shown in Figure 8.20b. In this example, a large proportion of low-molecular-mass contaminants (e.g., salts) yielded a poor signal-to-noise ratio in the low-mass area, and consequently, only a few peptide signals were detected. The protein could not be identified from this peptide mass map. When the same amount of sample was desalted and concentrated on a Poros R2 microcolumn, the signal-to-noise ratio and the overall signal intensities increased, resulting in the detection of a sufficient number of peptides to unambiguously identify the protein (Figure 8.20c). The peptides in the latter peptide mass map covered 18% of the amino acid sequence. If a larger aliquot (10 μl) of the sample was micropurified, a further improvement in the signal-to-noise ratio and signal intensities were observed (Figure 8.20d), resulting in detection of more peptides (45% sequence coverage).

Sequential elution of peptides

Even though the use of microcolumns prior to MS increases the signal-to-noise ratio and the overall signal intensities, resulting in higher sequence coverage, some peptides may still be lost during the analysis. One of the major reasons is the pronounced suppression effect generally observed in mass spectrometric analysis (Figure 8.22). The initial competition for protons in the ionization processes results in some components giving a much lower signal than if single-component analysis were performed. By using sequential extraction of proteins from the microcolumn with an increasing concentration of acetonitrile, a complex peptide mass map can be divided into several simpler peptide mass maps with an increased number of peptides detected. In addition to a greater sequence coverage, this approach is especially useful for reducing the number of components in the sample prior to de novo tandem mass spectrometric sequencing and post-source decay (PSD) analysis.

The sequential extraction concept is illustrated by the analysis of a previously unknown protein originating from a pig sperm membrane protein preparation separated by 2D gel electrophoresis. The protein spot was excised and subjected to in-gel digestion with trypsin (see Chapter 7, Protocol 13). A small aliquot of the digest supernatant (1%) was applied onto a Poros R2 microcolumn. After washing the column, the peptides were sequentially eluted with increasing acetonitrile concentrations. The resulting peptide mass maps are shown in Figure 8.22. When compared with the "normal" elution using HCCA in 70% acetonitrile and 0.1% TFA, illustrated by the lowest peptide mass map, significantly more peptide signals are detected with the sequential elution (nine peptides). In addition, the signal-to-noise ratio and the overall quality of the spectra are improved. Although it is very useful for protein characterization, sequential elution is rather time-consuming. Recently, it has been shown that a two-step elution using 16% and 30% acetonitrile is sufficient for a significant increase in the number of detected peptide signals (Erdjument-Bromage et al. 1998).

Use of increasingly hydrophobic microcolumns improves sequence coverage

Small peptides that are very hydrophilic or have been posttranslationally modified are difficult to detect by MS, which may be due to either peptide loss during purification or suppression by matrix signals. However, by trying different chromatographic materials in the

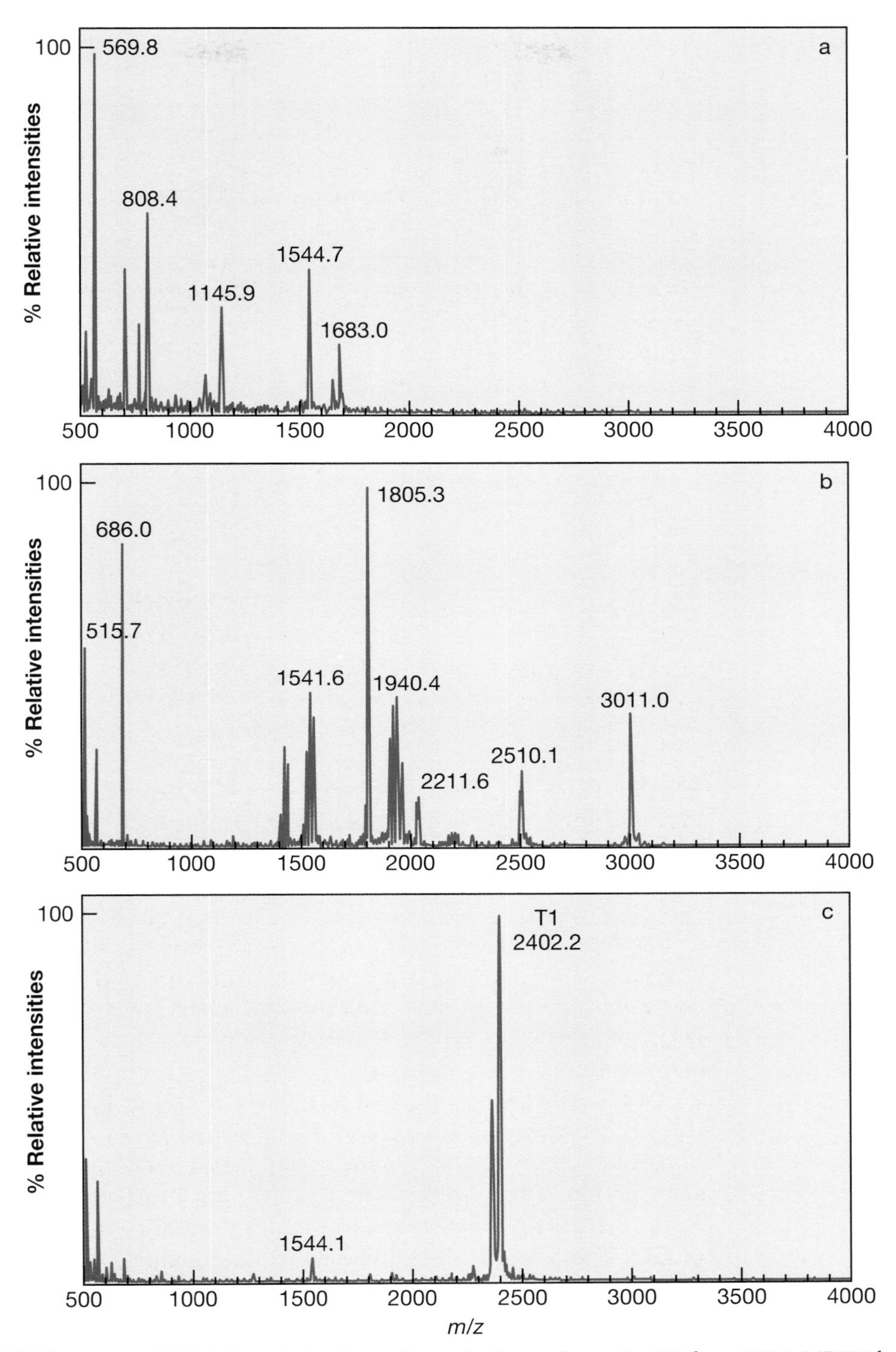

FIGURE 8.21. MALDI-TOF-MS analysis of a total tryptic digest of protein #25 from MDA-MB231 breast cancer carcinoma cells following stepped elution capillary column (0.32 mm I.D.) RP-HPLC. The total peptide mixture (total volume ~100 µl, ~500 pmoles) was applied to the column at 16 µl/min, and the column was developed in a stepwise manner with 40% Solvent B (*a*), 70% Solvent B (*b*), and 100% Solvent B (*c*), where Solvent A was aqueous 0.1% TFA, and Solvent B was 60% acetonitrile in aqueous 0.1% TFA. The volume of collected fractions was ~25 µl. MALDI-TOF-MS sample load: 0.5 µl. Note in panel *c* that the ionization of tryptic peptide T1 (*m/z* 2402.2) is completely suppressed in the presence of other peptides in the mixture (cf. panels *a* and *b*). (Reproduced, with permission, from Reid et al. 1998.)

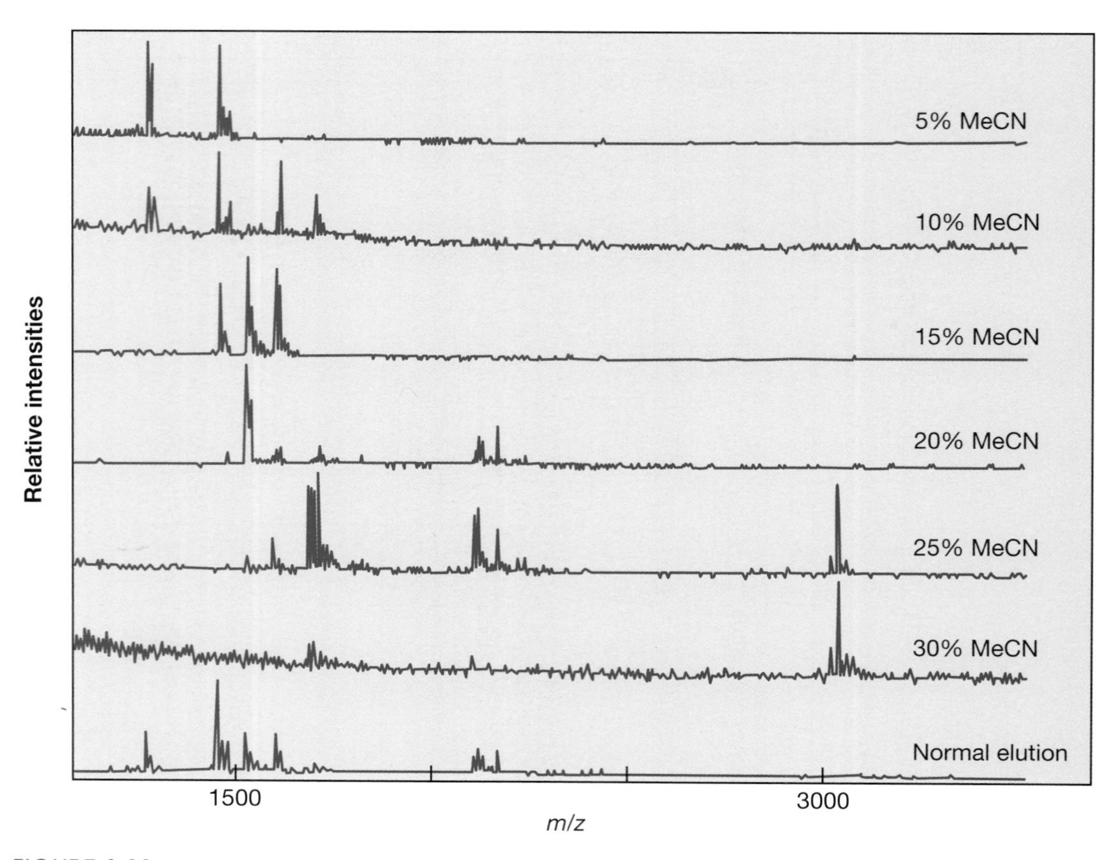

FIGURE 8.22. MALDI-MS peptide mass maps of peptides eluted from a Poros R2 column with increasing concentrations of acetonitrile; 1% of a peptide mixture originating from digestion of a gel-separated protein was applied to a Poros R2 microcolumn. After washing, the peptides were eluted stepwise using increasing concentrations of acetonitrile. The MALDI-MS peptide mass map in the bottom of the figure illustrates results from elution using 4HCCA in 70% acetonitrile/0.1% TFA. (Courtesy of Martin R. Larsen, Macquarie University.)

microcolumn, some of these peptides can be recovered and analyzed. Two resins, Oligo R3 and graphite powder, which are more hydrophobic than Poros R2, can be used efficiently for this purpose.

The peptide mass maps of in-gel-generated tryptic peptides from in-vitro-phosphorylated F-STOP protein illustrate the advantages of using Poros R2 together with Oligo R3 (Figure 8.23). After digestion, an aliquot of the digest supernatant was applied onto a Poros R2 microcolumn. The peptides were desalted and concentrated and analyzed by MALDI-MS (Figure 8.23a). The flowthrough from the R2 column was passed through an Oligo R3 microcolumn. The MALDI-MS peptide mass map obtained from the R3 column eluant (Figure 8.23b) contained five peptide signals that were not detected in the R2 map. The amino acid sequence of these signals revealed a high number of hydrophilic amino acids. In addition to an increase in sequence coverage of almost 10%, a phosphorylated peptide could also be easily detected in the R3 peptide map (M.R. Larsen et al. 2001c). The usefulness of using Oligo R3 for enrichment of phosphorylated peptides prior to MS has also been shown by Neubauer and Mann (1999) and Larsen and Roepstorff (2000).

Graphite columns have been used in HPLC for desalting hydrophilic peptides prior to MALDI-MS (Chin and Papac 1999) and can also be packed into microcolumns and used in a sample preparation step prior to MS (Larsen et al. 2002a,b). The advantages of graphite powder are its low cost and compatibility with both Poros R2 and Oligo R3, permitting recovery of

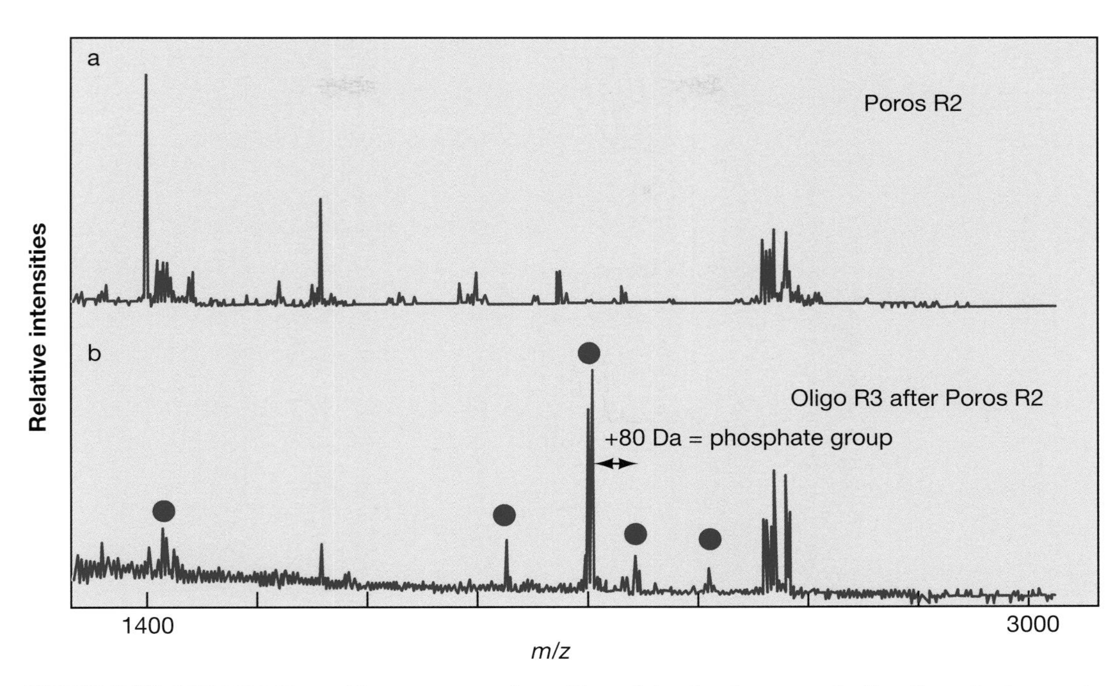

FIGURE 8.23. MALDI-MS peptide mass maps of peptides originating from tryptic digestion of gel-separated F-STOP protein. (*a*) MALDI-MS peptide mass map of a 5% aliquot of peptide mixture from the digestion of F-STOP protein, after desalting and concentration on a Poros R2 microcolumn. (*b*) MALDI-MS peptide mass map of the flowthrough after desalting and concentration on an Oligo R3 microcolumn. The peptide signals marked with a circle illustrate those not retained by the Poros R2 chromatographic material. (Courtesy of Martin R. Larsen, Macquarie University.)

smaller peptides. Note, however, that the choice of matrix in MALDI-MS has a significant role in distinguishing the smaller peptides from the matrix background (Chin and Papac 1999).

An example of the use of graphite microcolumns is given in Figure 8.24. Only the low-mass region from 500 to 1200 daltons is shown. A protein spot derived from a 2D gel of enriched pig sperm membrane proteins was submitted to in-gel tryptic digestion. A small aliquot of the peptides (2%) was desalted and concentrated on a Poros R2 column and analyzed by MALDI-MS (Figure 8.24a). The protein was identified as voltage-dependent anion channel 2 protein with 53% sequence coverage. The flowthrough from the Poros R2 column was collected and concentrated on an Oligo R3 column. The MALDI-MS peptide mass map of the peptides bound to the Oligo R3 (Figure 8.24b) revealed only one peptide not detected in the Poros R2 map, increasing the sequence coverage by a further 3%. The flowthrough from the Oligo R3 column was collected and concentrated on a microcolumn packed with graphite powder. The MALDI-MS peptide mass map of the peptide eluted with HCCA in 70% acetonitrile and 0.1% TFA from the graphite column is shown in Figure 8.24c. Five peptides not detected in either the Poros R2 or the Oligo R3 maps were found in this map. The five peptides are relatively hydrophilic and are probably not bound to either of the previous columns or they are suppressed in the peptide mass maps. The five peptides cover 10% of the protein sequence and thereby increase the overall sequence coverage to 66%.

The use of a multi-tiered approach with increasingly more hydrophobic chromatographic material packed in microcolumns for characterizing proteins is advantageous for detection of a larger number of components in the sample and thus improved sequence coverage. The analysis is fast and uses only a limited amount of the total sample. Graphite powder can also be used in a one-tiered approach for desalting and concentration of peptides prior to MALDI-MS if the peptides are eluted off of the column with HCCA in 70% acetonitrile and 0.1% TFA (Larsen et al. 2002a).

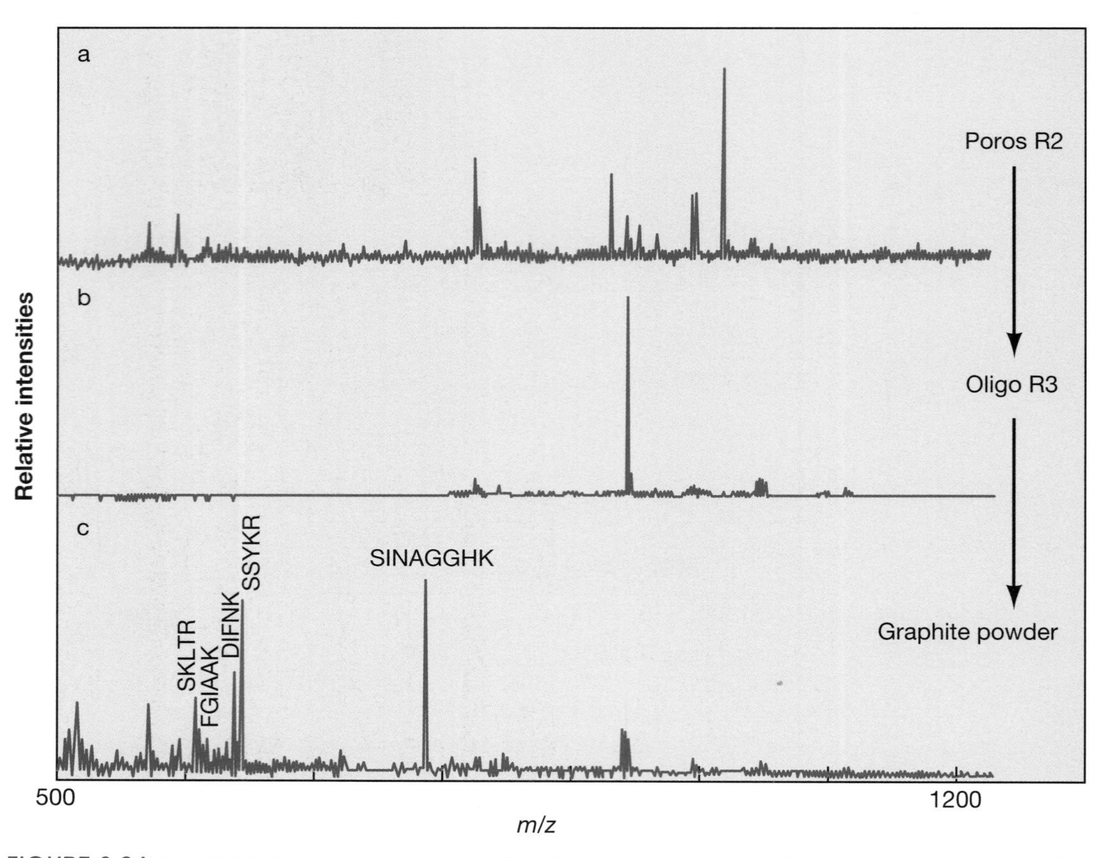

FIGURE 8.24. MALDI-MS peptide mass maps of a micropurified peptide mixture originating from an in-gel digestion of voltage-dependent anion channel 2 protein. Only the low-mass region from 500 to 1200 daltons is shown. (*a*) Peptide mass map of the peptides eluted off the Poros R2 microcolumn. (*b*) Peptide mass map of the flowthrough from the Poros R2 desalted and concentrated on an Oligo R3 microcolumn. (*c*) Peptide mass map of the flowthrough from the Oligo R3 column desalted and concentrated on a graphite microcolumn. The sequences of the peptides not retained on either the Poros R2 or Oligo R3 are shown. (Courtesy of Martin R. Larsen, Macquarie University.)

Comparison of Poros R2 microcolumns with prepacked μC_{18} ZipTips

The effectiveness of homemade microcolumns (packed in GELoader tips) for desalting and concentration of samples prior to MS was compared with commercially available prepacked μC_{18} ZipTips (Millipore). The results are summarized in Figure 8.25. Peptides originating from an in-gel tryptic digest of a previously unknown protein were prepared with either a Poros R2 microcolumn or μC_{18} ZipTips. The MALDI-MS peptide mass maps of increasing amount of digestion supernatant (1–4 µl) using the GELoader tip approach are shown in Figure 8.25a. The number of peptides that could be detected with increasing amount of digestion supernatant was 10, 17, 23, and 25, respectively. For comparison, the same amount of peptides was desalted and concentrated using the μC_{18} ZipTips, according to the manufacturer's protocol (Figure 8.25b), and the number of peptides that could be detected with increasing amounts of digestion supernatant was 0, 0, 3, and 9, respectively.

The reasons for the different performances of the two columns are only speculative. The GELoader tip columns have a much lower bed volume, permitting the peptides to be efficiently eluted with a very low volume (typically 0.5 µl) of HCCA in 70% acetonitrile containing 0.1% TFA, whereas the ZipTip requires a higher elution volume, which dilutes the sample. In addition, the elution volume can in the GELoader tip approach be divided into

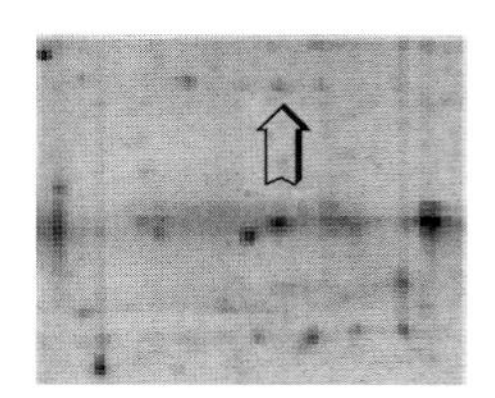

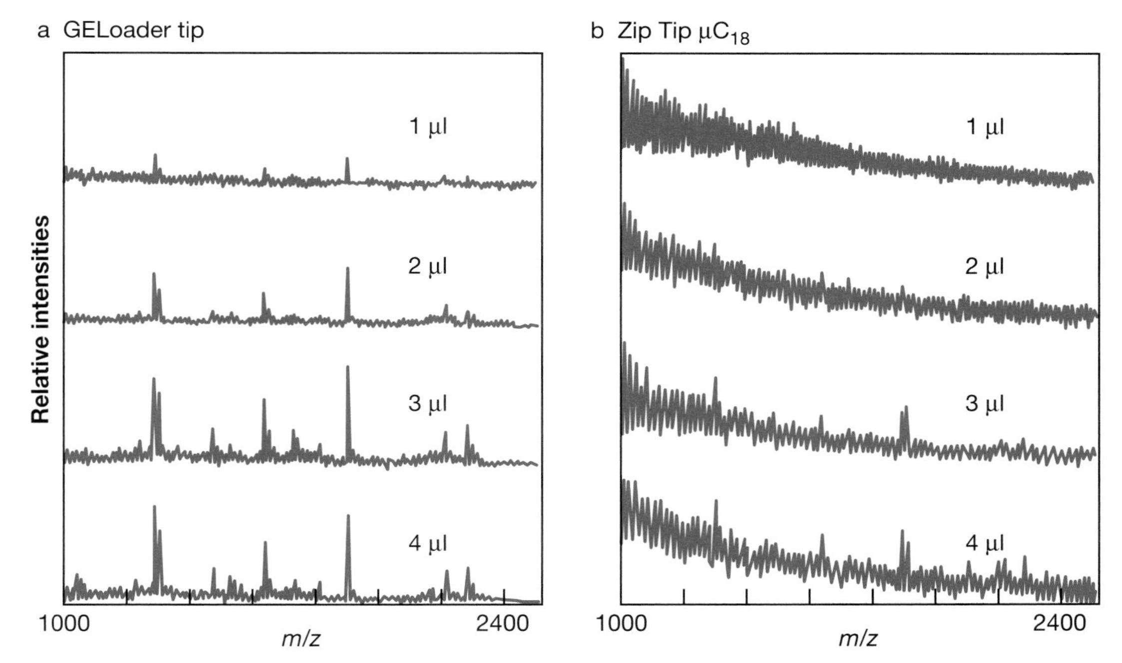

FIGURE 8.25. Comparison of the performance of GELoader tip microcolumns and the commercially available μC_{18} ZipTip. The peptide mixture was generated from the protein marked on the gel picture. (*a*) GELoader tip microcolumns: MALDI-MS peptide mass maps of increasing volumes of a peptide mixture originating from a pig sperm membrane protein, 1–4 μl, respectively. (*b*) μC_{18} ZipTip: MALDI-MS peptide mass maps of increasing volumes of a peptide mixture originating from a pig sperm membrane protein, 1–4 μl. (Courtesy of Martin R. Larsen, Macquarie University.)

several spots on the MALDI target (up to ten), where the peptides are only found in the first and second spots. This cannot be done using the ZipTips as the peptides seem to be eluted randomly in all of the small fractions (data not shown). The low bed volume also increases the ratio of surface area to volume, resulting in increased loading capacity in the GELoader tip. The effect described above is most noticeable when characterizing low-abundance proteins. The efficiency of the ZipTips can be significantly improved if the buffers and protocol described for the GELoader tips above are used, i.e., loading the peptide solution from the top of the ZipTip (data not shown).

A ZipTip pipette tip (from Millipore) is a microcolumn with the resin prepacked into the narrow end of a 10-μl pipette tip. The resins currently available are reversed-phase (C_{18} and C_4), immobilized metal ion, and hydrophilic media. ZipTips can be used where microcolumns are called for in any of the protocols in this chapter. Millipore provides complete protocols at their Web Site (http://www.millipore.com/ziptip) on sample preparation of peptides or proteins prior to MALDI-TOF-MS, sample preparation of peptides or proteins prior to ESI-MS, enrichment of phosphopeptides prior to MS, and removal of detergents and Coomassie Blue stain from peptide mixtures, among others.

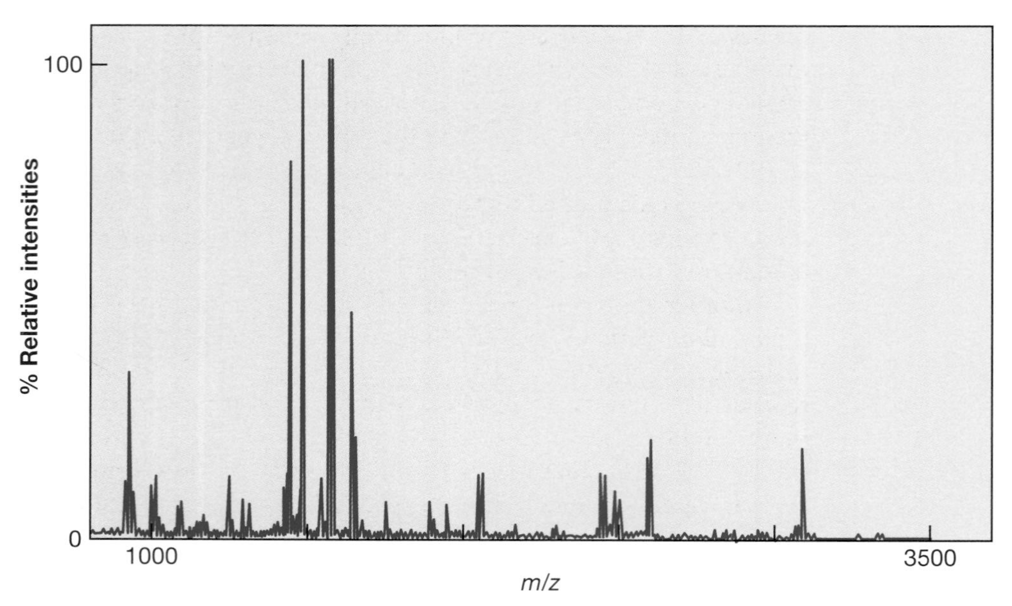

FIGURE 8.26. MALDI-MS peptide mass map of peptides originating from an on-column tryptic digestion of 2 pmoles of BSA. An aliquot (4 µl) of a solution of BSA (0.5 pmole/µl) was diluted in 20 µl of 50 mM NH_4HCO_3 (pH 7.6) and applied onto a microcolumn containing immobilized trypsin packed as described in Protocol 3. The peptides were desalted and concentrated on a Poros R2 microcolumn prior to MALDI-MS analysis. (Courtesy of Martin R. Larsen, Macquarie University.)

On-column enzymatic digestion

GELoader tips packed with immobilized enzyme material can also be used to rapidly digest proteins in solution with high efficiency (Gobom et al. 1997; M.R. Larsen et al. 2001b). For a procedure for preparing immobilized enzyme microcolumns, see Protocol 3.

As an example, 2 pmoles of bovine serum albumin (BSA, M_r ~67 kD) was digested on a microcolumn containing immobilized trypsin. The flowthrough was concentrated and desalted on a Poros R2 microcolumn, and the peptides were eluted directly onto the MALDI target using HCCA in 70% acetonitrile and 0.1% TFA. The resulting MALDI-MS peptide mass map is shown in Figure 8.26 (only the mass range of 800–3500 daltons is shown). From the peptide mass map, peptides covering ~40% of the BSA sequence could be assigned readily, even though the protein had not previously been reduced and alkylated, thus leaving most of the disulfide bonds intact. In a similar experiment using carbamidomethylated BSA, the sequence coverage from the on-column digestion was 82.3% compared to only 72.7% from the in-solution digest (Gobom et al. 1997). This approach is fast and in many cases more efficient than traditional solution digestion.

Global Protein Arrays: Identifying Proteins Immobilized on Membranes

One consequence of the increased sensitivity of modern mass spectrometers is that only small quantities of protein are needed for identification. Hence, clinical samples such as tissue biopsies can be prepared for electrophoresis and many of the separated proteins can be identified. Another advantage of separating proteins by electrophoresis is that the proteins can be electrophoretically transferred to a membrane for archiving valuable clinical samples.

Various groups have identified proteins directly from electroblotted proteins with MALDI-TOF-MS. Infrared MALDI-TOF-MS has been used to determine intact protein masses of 1D and 2D electrophoretically separated proteins electroblotted onto a variety of membranes (Eckerskorn et al. 1992). The sensitivity of analysis was comparable to silver-stained gels, and some posttranslational modifications were identified by comparison of the theoretical mass (predicted from the DNA sequence) and the apparent mass (Eckerskorn et al. 1997a,b). UV MALDI-TOF analysis has been used to measure intact protein masses of 1D-separated bands blotted onto polyvinylidene difluoride (PVDF) membranes (Strupat et al. 1994; Vestling and Fenselau 1994; Patterson 1995; Blais et al. 1996) and derivatized PVDF membranes (Immobilon CD) (Patterson 1995).

Another approach for the identification of proteins directly from the membrane surface is an endoproteinase digest of the protein followed by MS analysis of the resulting peptides (Schleuder et al. 1999). Immobilon CD, a cationic durapore membrane, showed the most reproducible results for infrared MALDI. It was thought that the hydrophilic surface was more conducive to an efficient tryptic digestion than the hydrophobic surface of underivatized PVDF (Schleuder et al. 1999). The results for underivatized PVDF demonstrated higher signal intensities and better reproducibility, and larger numbers of tryptic fragment signals were observed with an increase of the pore size diameter (Schleuder et al. 1999). Enzyme-compatible matrices have also been investigated to acquire an intact protein mass by MALDI and then perform an enzymatic digestion on the protein (Schleuder et al. 1999).

The Molecular Scanner

One automated technology that has emerged is the molecular scanner (Bienvenut et al. 1999; Binz et al. 1999), a process where all proteins separated on the 2D polyacrylamide gel are digested simultaneously during the electroblotting process. Endoproteinase digestion occurs via a membrane that contains immobilized trypsin, which is intercalated between the gel and a PVDF capture membrane. The membrane is then scanned directly by MALDI-TOF-MS. Minimal sample manipulation occurs compared to "in-gel" digestion, and the gel image is created by direct matching of the MS scanning results with the corresponding robotic coordinate. The molecular scanner is only suitable for proteins below 60,000 daltons with a pI below 8.5 because electroblotting is performed at a pH that is suitable for trypsin activity. The technique has been demonstrated on 1D bands of standard proteins, small sections of 2D blots of human plasma, and *Escherichia coli* (Bienvenut et al. 1999; Binz et al. 1999). Protocol 12 describes the use of the molecular scanner technology. A disadvantage of the molecular scanner is that it is a "shotgun" technique whereby all proteins are fully digested by the treatment, and it requires global scanning to find the proteins present.

The Chemical Printer

An electroblotted archive of 2D-PAGE-separated proteins (bound to a membrane) is a type of protein chip, albeit a macroarray of proteins. The membrane-derived macroarray is identical to a chip-based protein microarray except that rather than the coordinates of each protein being predetermined in robotic pixels as in the microarray, they are determined by the physical attributes of the protein's apparent molecular mass and isoelectric point. Once the coordinates of each protein within the protein macroarray have been identified by an image-capture device, each protein spot now has a defined *X,Y* position and can be manipulated by any state-of-the-art robotic platform.

Protocol 13 describes the use of a technology that combines the advantages of both protein chips and 2D electrophoresis, termed a "chemical printer" (Figure 8.27). The chemical printer uses a microdispensing device, such as drop-on-demand type ink-jet technology

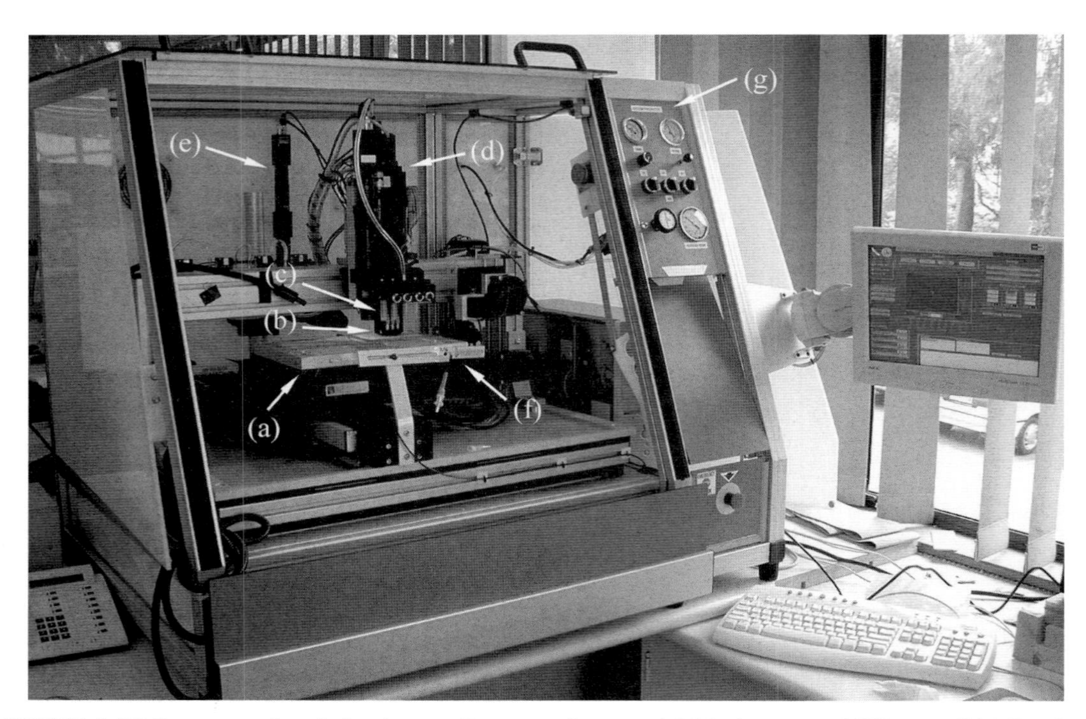

FIGURE 8.27. Prototype chemical printer at Proteome Systems. (*a*) Motion control XY stage, (*b*) MicroJet piezoelectric devices, (*c*) sample reservoirs, (*d*) membrane viewing camera, (*e*) elbow drop formation camera, (*f*) pneumatics control panel; (*g*) delay strobe for droplet viewing. The fluid dispensing, XY motion of the target platform, and Z motion of the piezoelectric devices are controlled by integrated software and controller boxes. As many as 4 piezoelectric devices can be mounted on the instrument and up to 4 MALDI targets with membranes attached. (Courtesy of Andrew A. Gooley, Proteome Systems.)

FIGURE 8.28. Enlarged view of the MicroJet piezoelectric device assembly. The assembly has three of four possible devices installed (*a*) connected to the reagent delivery system (*b*) with vacuum connection (*c*). In this view, device 4 is currently active with the vacuum toggle in the "on" position (*d*). A MALDI target with attached electroblot is held in position via a vacuum on the XY motion stage (*e*). (Courtesy of Andrew A. Gooley, Proteome Systems.)

Details of the Chemical Printer

A prototype chemical printer was developed by Gooley and colleagues at Proteome Systems in collaboration with the Shimadzu Corporation (Kyoto, Japan) (see Figure 8.27). The instrument incorporates a motion control stage and ink-jet printing technologies (Adams and Roy 1984; Bogy and Talke 1984; Dijksman 1984) for microdispensing reagents. The prototype chemical printer has an embedded 450-MHz computer with a LINUX operating system. The prototype software controls the precision motion control system: a Galil card controller (Galil, California) with Tol-O-Matic actuators (Tol-O-Matic, Minnesota) and Oriental Motor motors (Oriental Motor, Japan), and the piezoelectric MicroJet III controller (MicroFab Technologies). The instrument can hold four Axima CFR MALDI targets (80 x 120 mm) where the PVDF electroblots adhere to the surface of the target.

Solutions are delivered to the proteins on the membrane with drop-on-demand type ink-jet devices (Adams and Roy 1984; Bogy and Talke 1984; Dijksman 1984). A single droplet of ~100 pl is delivered to the protein spot with a MicroJet device (see Figure 8.28) that consists of a glass capillary (55-μm orifice) surrounded with PZT material. A droplet is dispensed when a voltage pulse is applied to the PZT material, which imparts an acoustic wave to the liquid sample; the pressure/voltage changes in the fluid cause a droplet to be dispensed from the orifice of the glass capillary. Two different trigger modes can be used with the MicroJet device: single or continuous. In continuous mode, droplets are produced continuously at a specified frequency, typically 250 Hz, until a stop command is given. This mode is usually used for optimization of the voltage pulse shape; parameters can be altered dynamically while the device is dispensing. Single mode delivers a finite number of drops (1–999) per trigger at a specified frequency (1 Hz to 20 kHz). In an automated procedure, the device is triggered at each protein spot position the XY stage is moved to.

The droplet generated from the MicroJet device is viewed with a digital camera that has a 3 x 3-mm field of view. A strobe LED, pulsing at the same frequency the MicroJet device is operating, is used to backlight the droplet. A variable time delay can be applied to the strobe, with a strobe-delay electronics box, and thus the droplet can be observed at any time from formation at the device orifice to delivery onto the membrane. An overhead camera is on a motorized W-axis to align it with the active device. Four MicroJet devices are mounted on a multichannel liquid delivery unit that is connected to a pneumatics control console. Pressure is necessary to purge the devices for reagent exchange as well as loading, whereas a minor vacuum is necessary to control the fluid meniscus at the orifice. Pneumatics control corrects for imbalances in surface tension of the fluid, capillary action, or hydrostatic pressure that prevents solution from being microdispensed.

(Adams and Roy 1984; Bogy and Talke 1984; Dijksman 1984). Small volumes of liquid can be dispensed by application of a voltage pulse to a piezoelectric material (PZT, lead, zirconium, and titanium) that imparts an acoustic wave to a liquid sample. A motion-control stage enables the drop to be accurately dispensed on a predefined target location. Microdispensing of solutions onto a membrane of electroblotted proteins at defined coordinates permits localized endoproteinase digestion and subsequent MALDI-TOF-MS peptide mass fingerprinting directly from the membrane surface. This approach bypasses the multiple liquid-handling steps associated with "in-gel" digestion procedures. Importantly, the ability to analyze immobilized proteins allows both for archiving of samples pre/post analysis and for multiple chemical reactions, such as several endoproteinase reactions or reactions which uncover protein modifications (e.g., glycosylation) that can be performed on the same protein spot.

IN VIVO AND IN VITRO LABELING METHODS PERMIT MS TO BE USED FOR QUANTITATION AND EXPRESSION PROTEOMICS

It is well acknowledged that the traditional 2D gel electrophoresis approach for differential protein profiling has significant, inherent limitations, especially when analyzing whole-cell lysates (Gygi et al. 2000b) (see also Chapter 1). The following are the major drawbacks:

- Limited sample capacity.
- Limited detection sensitivity (see Figure 1.8).
- Difficult to resolve hydrophobic proteins (e.g., membrane-associated proteins).
- A significant portion of 2D electrophoresis-separated gel spots contain more than one protein due to coelectrophoresis and/or differentially modified (or processed) forms of the same protein.
- 2D electrophoresis is a very labor-intensive, slow, and technically demanding method that is not readily amenable to high-throughput or automation.

These limitations severely restrict the detection of low-abundance proteins, which make up a large proportion of a given proteome (Gygi et al. 2000b). Thus, 2D electrophoresis does not give a true representation of all expressed proteins. Rather, it reflects only a *small slice* of the proteome.

An alternative proteomic approach that has the potential to identify trace-abundant proteins involves the affinity capture of cysteine-containing peptides (Spahr et al. 2000). Described in Protocol 8, this in vitro "labeling" method, which relies on an affinity tag for cysteines, simplifies the mixture of peptides (generated by proteolytic digestion of a whole lysate), such that the mixture of peptides contains approximately tenfold fewer peptides than the original mixture. (This simplification can facilitate the MS analysis of the sample.) By extending this approach to include a coded affinity tag (ICAT)—with three functional moieties (a cysteine reactive moiety, a linker that contains an isotope or mass "tag," and a biotin moiety, i.e., the affinity tag)—quantitation information can be obtained (Gygi and Aebersold 1999, 2000; Gygi et al. 1999, 2000a) (see Protocol 9).

It has been shown that global isotope labeling can also be achieved by intrinsically labeling proteins in vivo while they are being expressed (Oda et al. 1999; Chen et al. 2000; Veenstra et al. 2000; Peng and Gygi 2001). For example, isotopic labeling can be performed by growing identical strains of bacterium, such as *E. coli*, in normal and ^{15}N-enriched media where proteins isolated from the two cultures are differentially labeled (Oda et al. 1999; Conrads et al. 2001). Protocol 10 outlines a simple protocol for intrinsic ^{15}N-labeling of protein in *E. coli*.

In vivo isotopic labeling can also be performed by growing bacteria in media depleted of the naturally abundant isotopes of carbon (^{13}C) and nitrogen (^{15}N), and using FT-ICR-MS (Marshall et al. 1997; Pasa-Tolic et al. 1999). Bacteria grown in isotope-depleted versus normal media yield proteins whose peaks are shifted toward a monoisotopic mass. For example, a protein with 1000 carbon atoms (~22 kD) yields a protein with an average molecular mass that is 11 daltons lighter than at natural abundance. Only FT-ICR-MS is capable of resolving such small differences in molecular mass (Miranker 2000); these differences in mass are used to measure differential protein expression. For further discussion on quantification as a tool for measuring global protein changes, see Chapter 1.

In many proteomic problems, it is necessary to define the carboxyl terminus of a protein (e.g., to determine the carboxy-terminal processing site). Although this can be accomplished by automated carboxy-terminal sequencing (see Chapter 6), the anhydrotrypsin method outlined in Protocol 11 has the potential for high-throughput and enhanced sensitivity.

SEARCH ENGINES FOR IDENTIFYING PROTEINS USING MS DATA

This section was kindly contributed by Eugene A. Kapp and James E. Eddes from the Joint ProteomicS Laboratory (JPSL) of the Ludwig Institute for Cancer Research and the Walter and Eliza Hall Institute of Medical Research, Melbourne, Victoria, Australia.

History

In 1984, Biemann et al. (Gibson and Biemann 1984) proposed a strategy for mass spectrometric verification of three protein structures based on their DNA sequence. This strategy included digesting the proteins using the enzyme trypsin, separating the pool of peptides by HPLC, and then determining the masses of the peptides by fast atom bombardment (FAB-MS) (Morris et al. 1981). These values were then compared with molecular masses of tryptic peptides predicted from the DNA-deduced amino acid sequence using all three reading frames.

The computer program (FRAGFIT) described by Henzel et al. (1993) identified proteins by matching two or more molecular masses of peptide fragments obtained from chemical or enzymatic cleavages with all the fragment masses in a protein sequence database. Similar strategies were presented by C.J. March and T. Farrah (pers. comm.), using plasma desorption MS, and J.R. Yates et al. (pers. comm.). In 1993, five groups outlined their strategies for protein identification using MS by relying on the fact that multiple observed peptide fragment masses could be correlated with the theoretical peptide fragment masses obtained from a protein sequence database. The groups of Pappin et al. (1993) and James et al. (1993) outlined superior scoring schemes, taking into account the size of protein entries in the database. Searching of expressed sequence tag (EST) libraries and their six-frame translation was described by James et al. (1994). Probability scoring was introduced to take into account the large number of possible matches, as well as the concept of adding extra information to the search by means of dual digests or hydrogen-deuterium exchange experiments. Masses alone no longer afforded the discrimination required for a positive identification due to the size of the database being searched. During the last 10 years, the main focus on improving these strategies has been in providing reliable input to these search engines and developing improved scoring schemes to improve confidence in the search result.

A software algorithm for interpreting MS/MS data and hence deducing the amino acid sequence of a protein was first published by Biemann et al. (1966). The use of amino acid sequence information, derived by Edman sequencing or MS/MS, as well as the mass of a peptide in a database search, was first described by Mann (1994). Rather than searching multiple peptide masses, it was suggested that one peptide mass plus a sequence tag was sufficiently discriminating in searching a sequence database. The characteristic fragment ion patterns of peptides, previously described by Biemann (1988) and modified by Roepstorff and Fohlman (1984), were used to pinpoint the tag (usually two to four residues in length). In 1994, Eng et al. (1994) proposed searching protein sequence databases using uninterpreted MS/MS data, which required searching the mass of the peptide and correlating the observed MS/MS fragment ions against the theoretical fragment ion patterns for a peptide (SEQUEST program). This strategy has stood the test of time and has proven popular due to the relative ease and nonexpert nature of operation.

Sequence Databases

The most sensitive comparisons between sequences are made at the protein level. Detection of distantly related sequences is easier in protein translation, because the redundancy of the genetic code of 64 codons is reduced to 20 distinct amino acids. However, the loss of degeneracy at this level is accompanied by a loss of information that relates more directly to the

TABLE 8.5. Internet addresses for nonredundant protein and DNA sequence databases

NCBI	http://www.ncbi.nlm.nih.gov
EBI (EMBL)	http://www.ebi.ac.uk
Japan	http://www.ddbj.nig.ac.jp
SwissProt	http://www.expasy.ch
NCBInr from NCBI	ftp://ftp.ncbi.nih.gov/blast/db (click nr.Z)
dbEST from NCBI	ftp://ftp.ncbi.nih.gov/blast/db (click est_human.Z, or est_mouse.Z, or est_others.Z)
GenPept (protein translation of GenBank)	ftp://ftp.ncbi.nih.gov/genbank (click genpept.fsa.qz)
OWL	ftp://ftp.ncbi.nih.gov/repository/OWL (click owl.fasta.Z)
PIR	http://pir.georgetown.edu/pirwww/search/pirnref.shtml

evolutionary process, because proteins are a functional abstraction of genetic events that occur in DNA (Attwood and Parry-Smith 1999). DNA sequence databases contain genomic sequence data, including information at the level of the untranslated sequence, introns and exons, mRNA, cDNA, and translations. ESTs are generally 200–400 bases in length, obtained from cDNA libraries, and represent the expressed genome of a specific cell type. A search of a DNA or EST database therefore requires conceptual translation in six reading frames because it is not known whether the first base marks the start of the coding sequence. There are three forward frames, which are achieved by beginning to translate at the first, second, and third bases, respectively; the three reverse frames are determined by reversing the DNA sequence and again beginning on the first, second, and third bases. ESTs by their very nature are incomplete and to a certain degree inaccurate, with an error frequency of 2–5%. Error rates of 0.01–0.05% are reported for genomic sequence data. Currently, with more than 6 million EST entries in the public dbEST database, it is prudent to first search a nonredundant protein sequence database before a six-frame DNA translation is attempted.

The major available public domain sequence databases are maintained at NCBI, EBI (EMBL outstation), and Japan. Although extensive interchange of information occurs between these groups, there is no rigorous system to ensure that any one database is truly nonredundant and contains a complete set of known protein sequences. Table 8.5 lists Internet addresses of commonly used protein and DNA sequence databases.

Peptide Mass Fingerprinting: Protein Identification by Peptide Mass Mapping

MALDI-TOF-MS is sensitive, capable of high-throughput, and currently the first choice for protein identification from 2D electrophoresis gels. A sample protein is digested with a specific protease, such as trypsin, and the peptide fragment masses are determined by MS. The peptide fragment masses provide a fingerprint of the protein of interest and are not unilaterally affected by the presence of posttranslational modifications. The masses of the measured proteolytic peptides are compared to predicted proteolytic peptides from sequence databases (Figure 8.29). In detail, a protein sequence entry from the database is theoretically digested according to the enzyme cleavage rules, taking into account user-specified parameters such as number of missed cleavage sites, monoisotopic or average masses, and known modifications (e.g., gel-induced). A theoretical mass spectrum is therefore constructed for each protein entry from the sequence database and compared to the mass-measured spectrum. The comparison or scoring function can be simple (i.e., number of matching ions) or

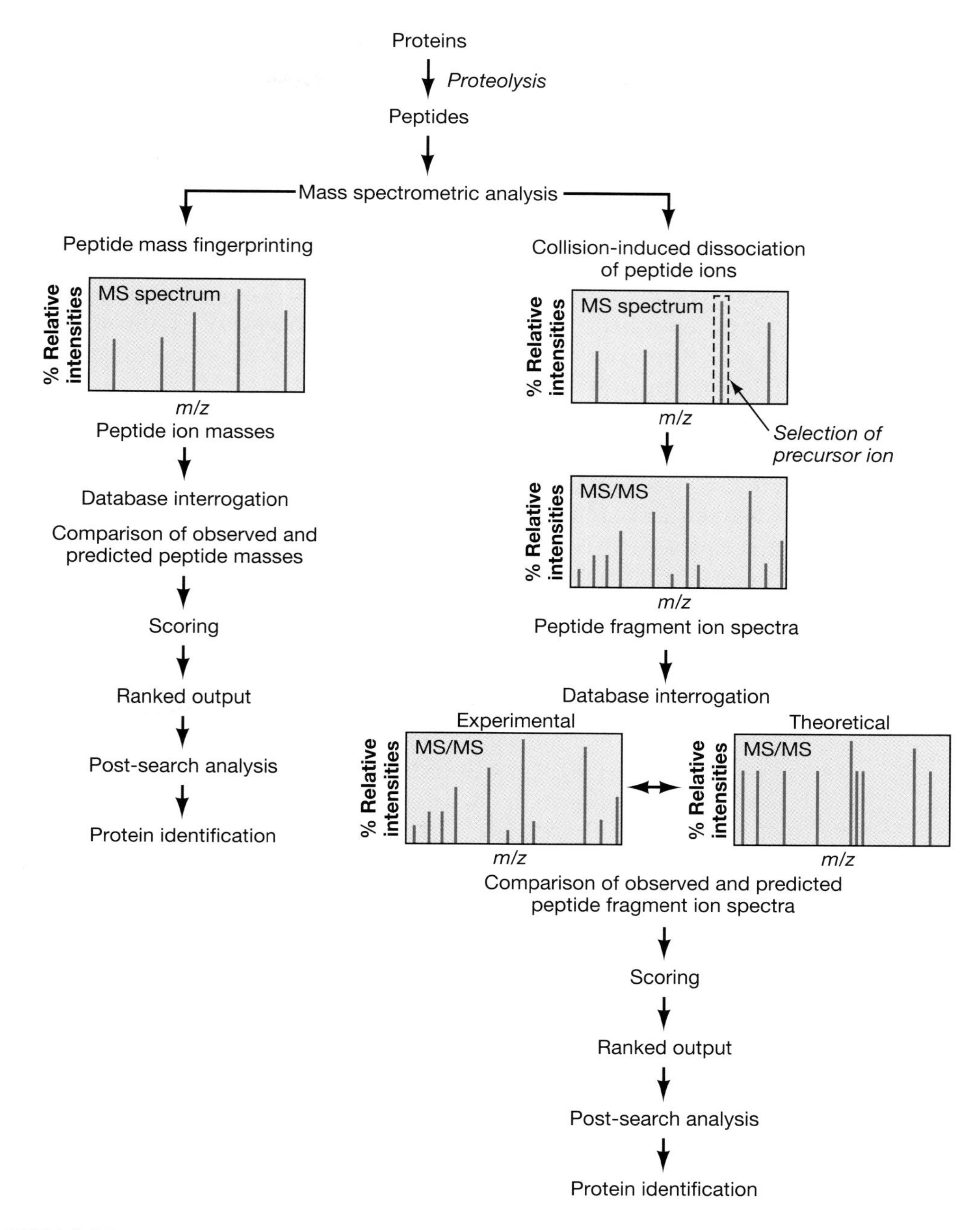

FIGURE 8.29. Schematic of peptide mass fingerprinting (PMF) versus MS/MS-based protein identification strategies. PMF algorithms compare the measured masses of intact peptide ions with the calculated masses of peptides derived from *in silico* proteolysis of proteins from a database. Each protein is scored primarily by counting the number of matching peptide masses within a user-prescribed mass tolerance. PMF protein identification strategies are well suited to analysis of relatively pure samples containing few proteins. MS/MS-based algorithms attempt to correlate uninterpreted CID spectra to theoretical spectra of peptides derived from *in silico* proteolysis of proteins from a database. As with PMF, CID spectra are initially matched to peptide sequences with an intact peptide ion mass similar to that of the observed precursor ion. The calculated peptide fragment ion spectrum of the peptide sequence is then compared to the observed CID spectrum. As CID spectra yield sequence information, unambiguous protein identifications can be made with significantly fewer matching peptides than required by PMF approaches. MS/MS-based protein identification strategies therefore allow confident analysis of samples comprising complex mixtures of proteins.

Peptide mass fingerprinting is generally restricted to identifying simple protein mixtures (one or two proteins) from microorganisms with fully sequenced genomes. The success rate for mammalian systems is typically lower than that for microorganisms due to incomplete genomes.

more complex, taking into account a multitude of variables such as the molecular mass of the protein, pI, digestion conditions, and posttranslational modifications. The difference in the score between the top two ranked proteins often suggests protein identity. The success of peptide mass fingerprinting (or mass mapping) depends on the detection of a representative set of peptide masses derived from a protein and that the protein in question is known (i.e., it exists in a protein sequence database).

Figure 8.30a shows an example MALDI spectrum from a tryptic digest of a 2D gel spot. Two proteins were positively identified using a hierarchical-based search, in which the second protein was identified by searching unmatched masses from the top scoring protein. Figure 8.30b shows a MALDI spectrum that resulted in a false-positive hit (i.e., an incorrect identification). MS/MS was used to identify the protein successfully. Table 8.6 displays a comprehensive list of PMF search engines.

Peptide Fragment Ion Searches: Protein Identification by MS/MS of Peptides

The set of fragment ions generated by MS/MS act as a fingerprint for an individual peptide. Two or more peptides identified in this way are usually sufficient to unambiguously identify a protein. This approach allows searching of databases that contain incomplete gene information (e.g., ESTs) and also allows the identification of proteins from complex mixtures. In the CID process, peptides fragment in a semipredictable manner; thus, sequences from the database can be used to predict an expected fragmentation pattern and match the expected pattern to that observed in the acquired data. The matching of observed against theoretical (expected) fragment ion masses (as depicted in Figure 8.29) involves counting and summing the matching ions. A thorough understanding of the various fragmentation pathways or ions resulting from an MS/MS experiment under particular ionization conditions is required for successful correlation (see the information panel on FRAGMENTATION MECHANISMS OF PROTONATED PEPTIDES IN THE GAS PHASE). At present, the mass spectrometric idiosyncrasies of individual amino acids, present in a peptide, as well as the rather subjective assessment of the importance (abundance) of an ion, remain unresolved and require further work. Notwithstanding this issue, peptide fragment ion searches are generally more reliable and successful at identifying peptides and therefore proteins, especially if the proteins are of mammalian origin. Example ESI spectra are shown in Figure 8.31 with the rankings of the correct peptide identifications based on two MS/MS search algorithms. Both search engines correctly identified the peptide shown in Figure 8.31a, whereas the identifications from Figure 8.31b,c were ambiguous, with both search engines failing due to selective enhanced cleavages at specific amino acid residues. Table 8.7 shows a comprehensive list of MS/MS search engines.

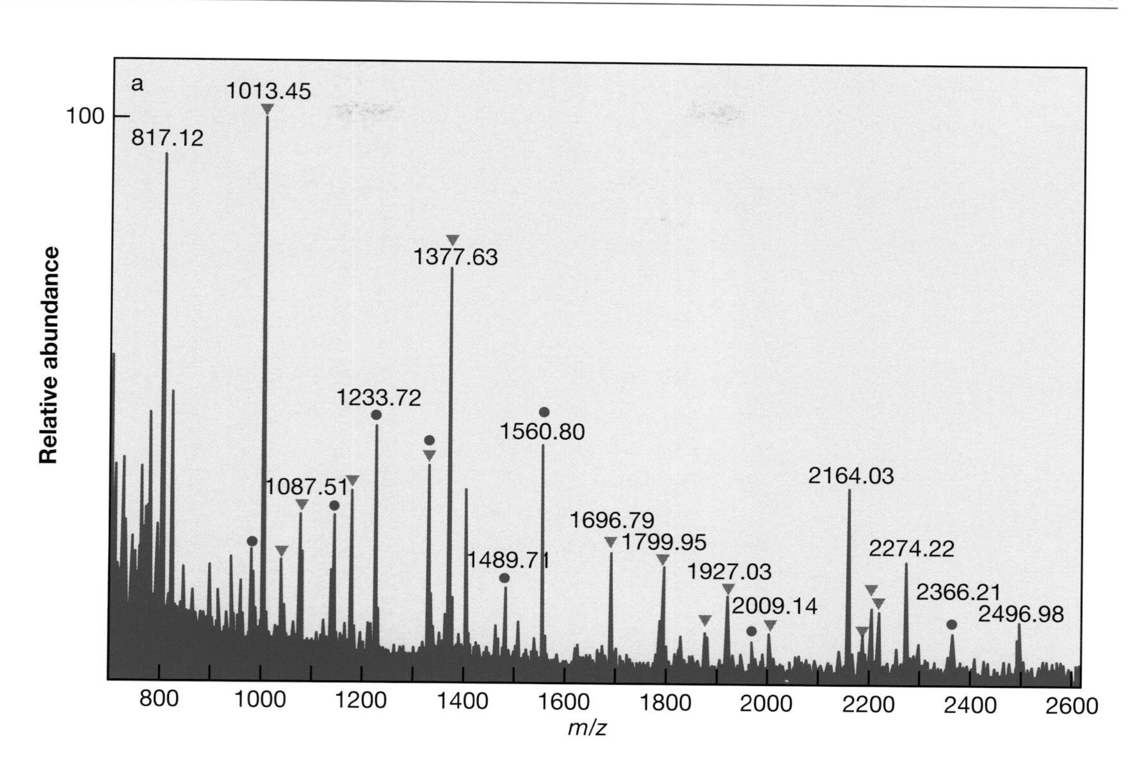

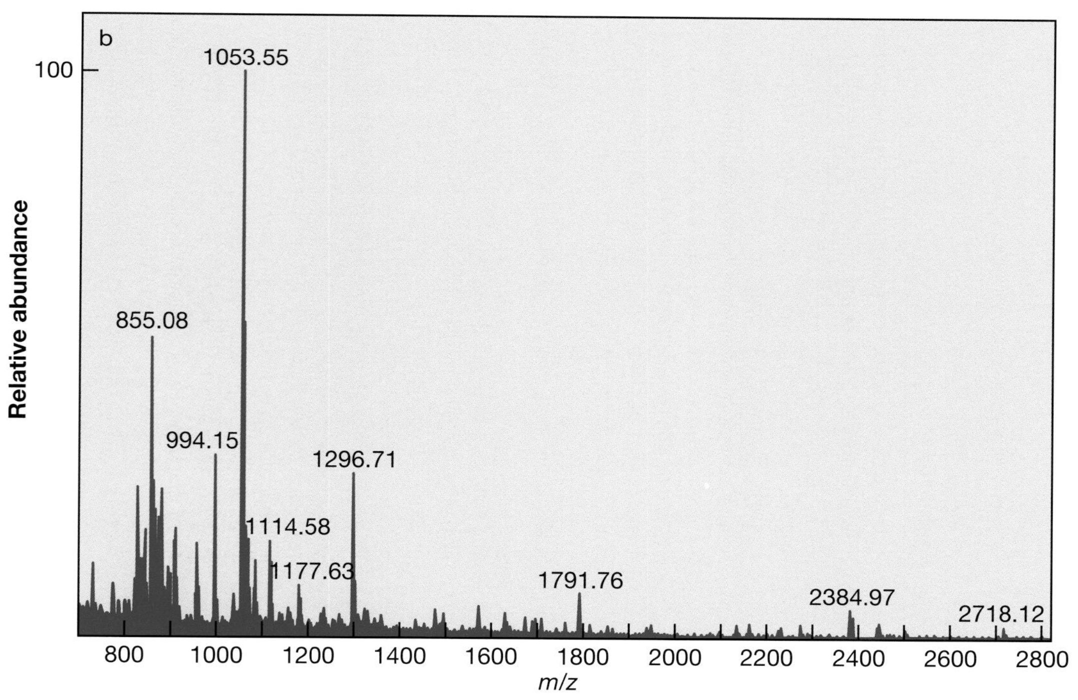

FIGURE 8.30. (*a*) MALDI-MS spectrum of an in situ digest of a 2D electrophoresis gel spot. A PMF algorithm was used to compare the observed peptide ion masses with those in a nonredundant protein database. Peaks identified with a gray triangle belong to the top scoring protein match, heterogeneous nuclear ribonucleoproteins A2/B1 (accession number: P22626). Peaks labeled with a blue circle are from the second place protein match, malate dehydrogenase (accession number: P40926). (*b*) MALDI-MS spectrum of an in situ digest of a 2D electrophoresis gel spot. Interrogation of the database using the observed peptide ion masses yields a false-positive identification.

TABLE 8.6. Peptide mass fingerprinting: Protein identification search engines

Program name/ Web Site/e-mail	Comments	Availability	Reference
FRAGFIT ckw@gene.com	First "on-the-fly" database searches for 2D electrophoresis. Output ranking is based on number of matches.	C-language source code available via e-mail.	Henzel et al. (1993)
MassSearch http://cbrg.inf.ethz.ch	Coined the term mass profile fingerprinting. A complete description of their algorithm is given in Chapter 20 of "A Tutorial on Computational Biochemistry using Darwin" available from the MassSearch URL address. A score was derived for matching proteins with emphasis placed on the number of masses used in the search, the resultant score, and the number of matches found. This search was extended in 1994 to searching DNA sequence databases and the scoring improved to reflect the probability of each match happening at random.	Automated server at ETH, Zurich.	James et al. (1993)
PeptideSearch http://195.41.108.38	Allowed one missed-cleavage site, which effectively doubled the size of the database but suggested that the mass tolerance should be better than 0.2% and that masses larger than 900 daltons should be used in the search. The algorithm, written in the THINK C environment using object-oriented programming, was used to transform the PIR database for rapid search times.	Free Web searches. Also part of ABI suite of software tools	Mann et al. (1993); Mortz et al. (1994)
MOWSE (molecular-weight search) http://srs.hgmp.mrc.ac.uk/cgi-bin/mowse	Proposed that only four masses were required to uniquely identify a protein with a mass accuracy of 3–4 daltons. The molecular weight of the protein was taken into account based on a normalized distribution frequency value calculated for different enzymes. A theoretical digestion of a large protein produces many more tryptic fragments than a small protein and hence the number of random matches increases for larger proteins. The MOWSE scoring scheme was developed that compensated for the nonrandom distribution of fragment molecular weights in proteins of different sizes and that larger peptides carried more scoring weight. The number of matching proteins decreases rapidly as a function of increasing peptide mass, showing that the higher the mass of the peptide, the better its value as a constraint for protein identification (Fenyo et al. 1998). However, the likelihood of encountering unknown modifications or extended partials increases as the peptides increase in size.	Free Web searches.	Pappin et al. (1993)
MS-Fit (part of the ProteinProspector suite of tools) http://prospector.ucsf.edu	Ranks output based on the MOWSE scoring scheme as well as number of matches. Output is comprehensive with easy navigation to other tools for further analysis. A unique feature of this search engine is in the scoring of methionine-oxidized peptides, which requires that the unmodified and modified peptides be simultaneously matched for the modified peptide to be counted as a match.	Free Web searches. Intranet license available.	

Mascot http://www.matrix-science.com	An extension of the original MOWSE search engine but with "on-the-fly" searching using a multithreaded algorithm. Data from various MS instruments can be used as input, as well as MS vendor-specific data formats. A probability-based MOWSE score and level of significance are indicated for each search. Four differential modifications as well as any number of constant modifications are simultaneously allowed in a search. A protein summary report is comprehensive with further links to other search engines.	Free Web searches. Per processor license.	Perkins et al. (1999)
PepIdent2 http://expasy.ch/tools/peptident.html	Uses an optimized peak detection algorithm and genetic algorithm for "on-the-fly" searches allowing one missed cleavage site and two differential modifications. All input parameters, including hydrophobicity coefficient, contribute to the score, however, with different weights. The weights are computed using a genetic algorithm and a training set of 91 spectra.	Free Web searches.	Gras et al. (1999)
ProteinLynx Global SERVER http://www.micromass.co.uk	PMF searches for a specific MALDI-TOF mass spectrometer. Mass spectra are processed such that isotope clusters are reduced to a single peak (isotope-depleted), and all ions above a specified mass are considered in the scoring scheme. Probability-based protein scoring is output as a Web-based interface using XML tags (the XML tag formats can be downloaded from the Web Site). The algorithm is multithreaded for fast execution on a multiprocessor machine, and modifications are well supported.	Commercial product.	
ProFound (part of the Prowl software suite) http://prowl.rockefeller.edu	An expert system using a Bayesian algorithm for protein identification. A Bayesian algorithm can take into account properties of proteins, sample preparation, estimated molecular weight, pI, etc., and be used in the scoring scheme. A significance level is indicated (Eriksson et al. 2000) and "Z" score calculated (Tang et al., pers. comm.) to indicate the quality and significance of the database search result. The search engine, which is fast, can also be guided to consider mixtures of proteins.	Free Web searches. Also part of the Amersham mass spectrometry suite.	Zhang and Chait (2000)
PeptideMass (part of the ExPASy server tools) http://www.expasy.org/tools/	Designed to take into account the annotation information available in the SwissProt database. This tool is useful for accounting for unassigned peaks in a mass mapping experiment that could be due to modifications. Signal peptides are removed and known modifications applied to the list of theoretical digest fragment masses.	Free Web searches.	Wilkins et al. (1997)

TABLE 8.7. Peptide ion fragment spectra (MS/MS): Protein identification search engines

Program name/ Web Site/e-mail	Comments	Availability	Reference
SEQUEST http://www.thermoquest.com	Finds all peptides in the database that match the input mass, and then theoretically calculates and matches the expected fragment ion masses against the observed MS/MS data. Two scoring schemes are used for ranking the best matching peptides: a preliminary scoring (Sp) scheme which selects the top 500 peptide sequences and an independent fast Fourier transform (FFT) cross-correlation analysis to provide a final ranking of best matches. TurboSEQUEST is much faster, using indices created for speed or size. SEQUEST has been modified over the years, first to allow the inclusion of modifications (Yates et al. 1995a,b), searching of DNA databases (Yates et al. 1995), searching of MALDI-PSD data (Griffin et al. 1995), and searching of high-energy CID data (Yates et al. 1996a,b). Lee et al. (Moore et al. 2000) have developed several Perl scripts for categorizing spectra based on their quality and found that database search speeds and overall quality of the search results can be significantly improved by screening MS/MS spectra before submitting to a search.	Commercial product. Packaged as part of BioWorks 3.	Eng et al. (1994)
MS-Tag (part of the ProteinProspector suite of tools) http://prospector.ucsf.edu	Originally developed for MALDI-PSD by Clauser et al., but the MS-Tag can be used for any MS/MS data. MS-Tag uses the fragment ion masses and the mass of the peptide, input by the user, to identify a protein. Fragment ion masses need not be of the same ion type, and an advanced MS-Tag allows a homology or error-tolerant mode. MS-Tag Simple or the advanced "identity" mode should always be used first. There are no statistical significance values associated with this search engine, since no scoring function is used other than a simple rank ordering based on the number of matching fragment ions. MS-Tag allows the user to choose an instrument type for those not familiar with the different ion types observed under different ionization conditions. Selecting several intense peaks at the high end of the mass scale as well as a few low-mass immonium ions (between 50 and 200) is a reasonably specific search. The number of unmatched fragment masses can be specified, allowing some control over the number of random matches.	Free Web searches. Intranet license available.	Clauser et al. (1999)
PepFrag (Sonar MSMS) (part of the Prowl software suite) http://prowl.rockefeller.edu	Designed based on unique properties of tandem MALDI ion-trap mass spectrometry. PepFrag allows searching of both protein or nucleotide sequence databases (SwissProt, PIR, GENPEPT, OWL, or dbEST). It had previously been found that fragmentation of peptide ions was highly selective and that Arg-containing peptides undergo facile, preferential cleavage adjacent to amino acid residues with acidic side chains, producing exclusively b- and/or y-type ions. Two dominant and informative fragment ions (i.e., adjacent to Asp/Glu), plus the mass of the peptide and that the peptide is tryptic are all that is needed for positive identification. Selecting the correct peaks is sometimes difficult due to the complicating factor of neutral loss peaks (difference of 17 or 18 daltons). A further enhancement to this program is the ability to enter fragment ion masses generated by cleavage at	Free Web searches. Also part of the Amersham mass spectrometry suite.	Qin et al. (1997)

	unspecified amino acid residues. Generally, four or more masses must be entered, of which at least three ions (b-type and/or y-type) must match. Searches are fast, but there is no error-tolerant mode of operation and no statistically significant scoring.		
Mascot http://www.matrix-science.com	Allows uninterpreted MS/MS ion searches. Data from various MS instruments can be used as input, as well as MS vendor-specific data formats. Input of fragment ion masses and intensities is via a file of specified format through a Web interface. Mascot iteratively selects subsets of the most intense peaks, looking for the subset that most clearly differentiates the top score of the matched protein. A probability-based MOWSE score and level of significance are indicated for each search. Four differential (variable) modifications as well as any number of constant (fixed) modifications are allowed in each search. Searches are fast and designed for batch-mode analysis with a protein and peptide summary report.	Free Web searches. Per processor licenses.	Perkins et al. (1999)
ProteinLynx™ Global SERVER http://www.micromass.co.uk	MS/MS searches for the hybrid quadrupole-TOF mass spectrometer. MS/MS spectra are processed (deconvoluted and de-isotoped) prior to searching, and all ions are used in the scoring system. Immonium ions, which are indicative of particular amino acids, add weight to the peptide score and improve the significance of a search result. The deconvolution (preprocessing) step improves searches for multiply charged peptides (>2). Output is either Web-based (using XML tags which can be downloaded from the ProteinLynx Web pages within the Micromass site) or application-driven. The algorithm is multithreaded and thus searches are fast and modifications well supported.	Commercial product.	
MassFrag Jeanluc.Verschelde@rug.ac.be	Matches experimental MALDI-PSD spectra with calculated fragmentation patterns of peptide sequences. The program uses an "all-ion pattern" and/or "a, b, and y ion pattern" for fragment ion scoring during the database search. A combination of the two scoring schemes is used to improve confidence in the identification.		Gevaert and Vandekerckhove (2000)
FRAGFIT	Originally developed for peptide mass fingerprinting, FRAGFIT has recently been modified and enhanced for MS/MS searches. Modifications, partial cleavages, and a six-frame translation of an EST database have all been incorporated into the new search engine. The more abundant high-mass ions, above the precursor mass, are used as "b," "y," or neutral loss ions in the algorithm.		Arnott et al. (1998)
Find Protein	Allows searching of multiplexed MS/MS data. In a single experiment. several (seven) peptides could be simultaneously identified using FT-ICR-MS combined with the use of highly accurate mass measurements for both the precursor and fragment ion masses. A peptide was considered as "identified" when three or more predicted fragment ion masses matched masses measured in the multiplexed MS/MS spectrum.		Masselon et al. (2000)

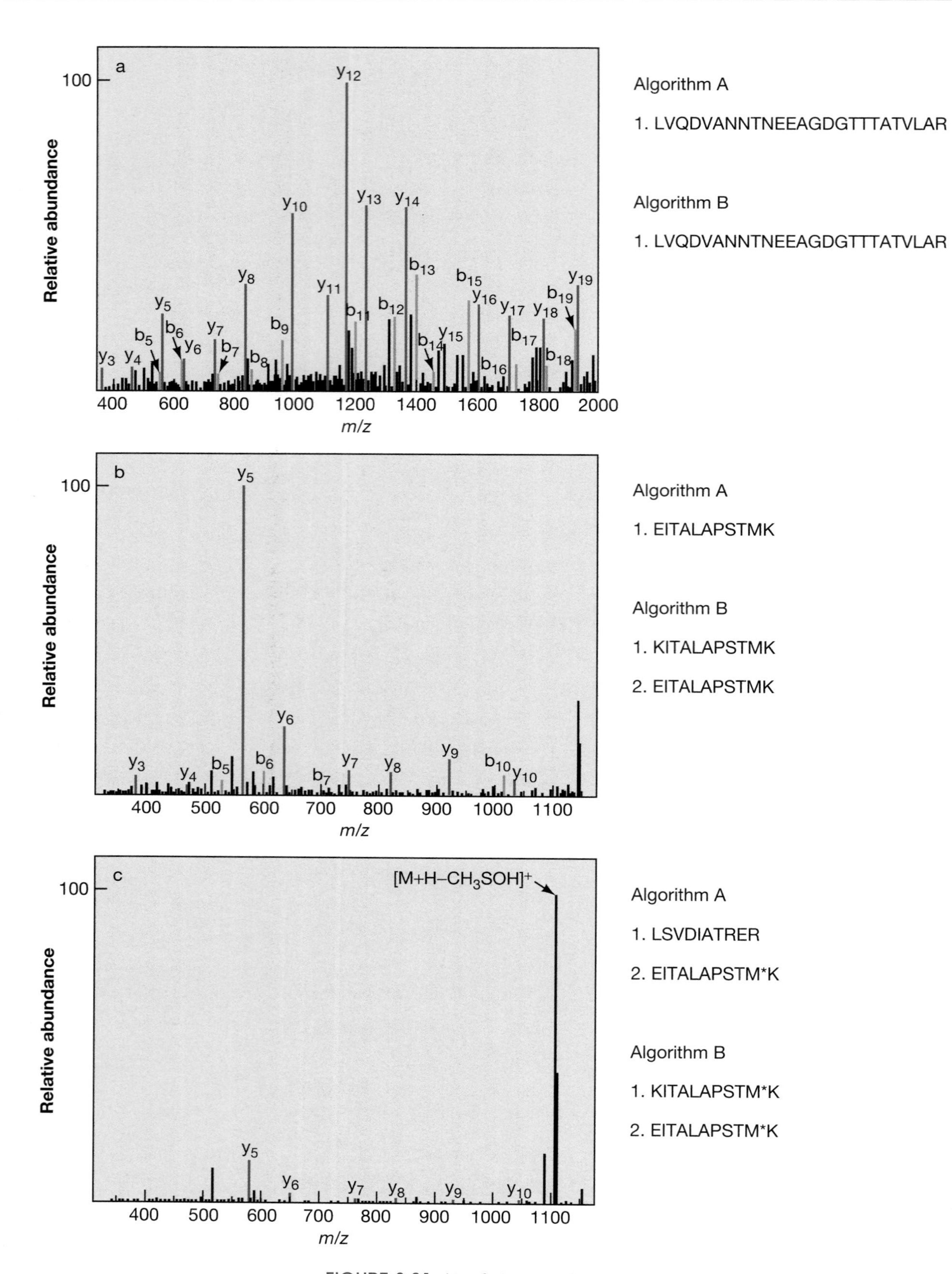

FIGURE 8.31. (*See facing page for legend.*)

SUMMARY

Highly reliable and fully automated identification procedures are required in the modern proteomics laboratory. The balance between the effort to identify as many proteins as possible with the effort to avoid misidentifying a protein requires integration among sample handling (chemistry), experimental design and goals, and bioinformatics workflow. A technique or methodology, which was previously used in isolation and which had little or no impact on downstream processes, might now require a radical re-think when integrated into the modern proteomics workflow. For example, particular chemistries that result in a nonspecific modification will certainly hinder fast database searches. The impact of each technique or methodology on downstream techniques therefore requires careful and strategic planning on the basis of the overall objective of the study. Currently, valuable time and resources are being spent on verifying database search results and on manually inferring MS/MS data that do not give reliable search results, due in many cases to unknown modifications. Our lack of understanding of the fundamental gas-phase ion chemistry of biomolecules and the mechanisms that yield observed fragment ions also hinders the development of robust software tools for the required high-throughput proteomics arena.

Many improvements to database search engines have been made in the last few years, particularly with respect to scoring schemes for peptide mass fingerprint searches but also the development of faster algorithms, particularly multithreaded versions. The success of these algorithms relies on the completeness of databases and the availability of a good scoring mechanism. Correct usage and a better understanding of these search engines will also lead to higher-quality search results and fewer false-positive identifications. With increased automation and less interactive assessment of results come the need for quality control, especially in developing methods to assess the reliability of protein identification. The impact of hybrid search strategies, particularly PMF combined with MS/MS of selected peptides, is expected to have a major role in proteomics applications. Labeling experiments, whereby fragmentation mechanisms of peptides could be better controlled or directed, and the simultaneous measurement of the amount of protein in question are expected to be further refined for the high-throughput environment. Protein/peptide information obtained by these MS-based methodologies can be further analyzed for its biological significance (such as structure, functional context, and interacting partners) using the bioinformatics tools described in Chapter 11.

FIGURE 8.31. (*a*) The ESI-IT CID spectrum shown was searched against a nonredundant protein database using two separate algorithms. Both algorithms return the peptide sequence LVQDVANNTNEEAGDGTTTATVLAR as the best match. Fragment ions corresponding to b- and y-type ions are labeled and shown in pale blue and dark blue, respectively. (*b*) The ESI-IT CID spectrum shown was searched against a nonredundant protein database using two separate algorithms. Algorithm A returns the correct peptide sequence, EITALAPSTMK, as the top match. Algorithm B returns the correct sequence as its second match. Although both sequences returned by Algorithm B are confirmed by the dominance of the y_5 ion, corresponding to preferential cleavage of the peptide backbone amino-terminal to the proline residue at position seven, the identification of EITALAPSTMK as the correct peptide is inferred from the frequency of identified peptides from the same protein in analysis. (*c*) The ESI-IT CID spectrum shown was searched against a nonredundant protein database using two separate algorithms. Both Algorithm A and Algorithm B return the correct peptide sequence, EITALAPSTMK, as the second match (asterisk denotes oxidized methionine). The identification can be confirmed by the characteristic loss of CH_3SOH from the precursor ion associated with the oxidized methionine residue at position ten.

PROTOCOL 1

General Method for MALDI-MS Analysis of Proteins and Peptides

FOR MALDI-TOF-MS ANALYSIS OF PROTEINS AND PEPTIDES, samples are cocrystallized with an excess of organic matrix that absorbs at a specific wavelength (usually, UV_{337nm}). Typically, sinapinic acid (SA) is the matrix of choice for large proteins, whereas α-cyano-4-hydroxycinnamic acid (HCCA) is the preferred matrix for peptide mapping (for matrix selection, see Table 8.8). Following a short laser pulse, analytes are protonated and desorbed into the gas phase, and their *m/z* values are determined in a TOF mass analyzer. Mass accuracy determinations vary from ±0.01% to 0.1% depending on the sample preparation technique and the method used for mass calibration. This protocol was contributed by David Frecklington (Joint ProteomicS Laboratory [JPSL] of the Ludwig Institute for Cancer Research and the Walter and Eliza Hall Institute of Medical Research, Melbourne, Australia).

MATERIALS

CAUTION: See Appendix 3 for appropriate handling of materials marked with <!>.

Reagents

Calibration standards (1–10 pmoles/μl in 60% acetonitrile, <!> aqueous 0.1% v/v TFA <!>)

To obtain the best interpretation of MALDI-TOF-MS data, standards of known molecular mass close to the molecular mass of the unknown sample are required to ensure a linear calibration curve. These must be prepared and chosen carefully. Typically, angiotensin I or angiotensin II, whose accurate molecular masses are known, can be used for peptide mapping studies. Tables 8.9 and 8.10 contain a range of peptides and proteins that can be used routinely as standards in MALDI-MS analysis. All of these standards are commercially available, and therefore, stock solutions can be accurately prepared. Stock solutions of peptide calibrants are prepared at a concentration of 100 pmoles/μl aliquots (frozen stocks can be kept for several years). For working solutions, these aliquots are diluted to 1–10 pmoles/μl with aqueous 0.1% (v/v) TFA. Working solutions of calibrant can be used for 1 week when kept at 4°C.

The concentration of the calibration standards must be similar to that of the unknown sample (range 1–10 pmoles/μl) to ensure accurate results.

Matrix solution (20 mg/ml matrix in 60% acetonitrile, 0.1% TFA)

Under these conditions, the matrices are saturated solutions and must be centrifuged prior to use. For uniform sample/matrix formation, avoid using undissolved matrix crystals. HCCA <!>, DHB <!>, and SA <!> can be obtained from Sigma.

Methanol <!>

TABLE 8.8. Choice of matrix for MALDI-MS analysis of different analytes

Matrix[a]	Peptide[b] Mapping	Small proteins	Large proteins	Glyco peptides	Glyco proteins	References
HCCA	1	3	–	2	–	Beavis et al. (1992)
SA	–	1	1	3	2	Beavis and Chait (1989b)
DHB	2	2	1	1	1	Strupat et al. (1991); Karas et al. (1993)
THAP	2	–	–	–	–	Kussman and Roepstorff (2000)
DHAP/DAHC	2	–	–	–		Gorman et al. (1996)

Portion adapted, with permission, from Kussmann and Roepstorff (2000).

The choice of matrix must be adapted to the physical properties of the analyte. First choice is denoted by 1, second choice by 2, and third choice by 3.

[a](HCCA) α-cyano-4-hydroxy-cinnamic acid; (SA) sinapinic acid; (DHB) 2,5-dihyroxybenzoic acid; (THAP) 2,4,6-trihydroxyacetophenone; (DHAP) 2,6-dihydroxyacetophonone; (DAHC) diammonium hydrogen citrate.

[b]The choice of matrix can considerably influence the sequence coverage in peptide mapping. For example, peptide mapping data obtained with the matrix THAP can complement those obtained with DHB and HCCA (Kussman and Roepstorff 2000).

TABLE 8.9. Peptide mass calibration standards

Name	Source	Monoisotopic mass $[M+H]^+$	Average mass $[M+H]^+$	Sigma product number	SwissProt Reference
Leu enkephalin	Human	556.277	556.6	L 9133	PENK_HUMAN
Met enkephalin	Human	574.234	574.7	M 6638	PENK_HUMAN
[D-Ala-2]-Deltorphin II		782.396	782.9		DAD2_PHYBI
Angiotensin II	Human	1046.542	1047.2	A 9535	ANGT_HUMAN
Bradykinin	Human	1060.5692	1061.23	B 3259	KNH_HUMAN
Lutienizing hormone-releasing hormone	Human	1182.581	1183.3	L 7134	GONL_HUMAN
Ala-Pro-Gly-[Ile-3,Val-5] Angiotensin II	Human	1271.654	1272.5	A 0289	ANGT_HUMAN
Angiotensin I	Human	1296.6853	1297.50	A 9650	ANGT_HUMAN
Substance P		1347.7360	1348.7	S 6883	TKNB_HUMAN
α-endorphin	Human	1745.842	1747.0	E 6136	COLI_HUMAN
Peptide sequencing standard		1657.841	1658.8	P 2046	
Neurotensin		1690.928	1692.0	N 6383	NEUT_HUMAN
Renin substrate	Porcine	1758.933	1760.1	R 8129	ANGT_PIG
Renin substrate	Human	1759.940	1761.0	R 5880	ANGT_HUMAN
Dynorphin		2147.1988	2148.52		
Adrenocorticotropic hormone fragment 18-39	Human	2465.199	2466.7	A 0673	COLI_HUMAN
Insulin A-chain, oxidized	Bovine	2530.921	2532.7	I 1633	INS_BOVIN
Melittin	Bee Venom	2845.762	2847.5	M 1407	MEL1_APIME
β-endorphin		3463.8296	3466.04		
Glucagon	Bovine	3481.624	3483.8	G 7774	GLUC_BOVIN
Insulin B-chain, oxidized	Bovine	3494.651	3496.9	I 6383	INS_BOVIN
Adrenocorticotropic hormone fragment 7-38		3657.929	3660.2	A 1527	COLI_HUMAN
Insulin	Sheep	5700.598	5704.5	I 9254	INS_SHEEP
Insulin	Bovine	5730.609	5734.6	I 5500	INS_BOVIN
Insulin	Equine	5744.624	5748.6		INS_HORSE
Insulin	Porcine	5774.635	5778.6		INS_PIG
Insulin	Human (Recombinant in *E. coli*)	5804.646	5808.6	I 0259	INS_HUMAN
Ubiquitin	Bovine	8560.625	8565.8	U 6253	UBIQ_BOVIN

Reproduced from: http://www.srsmaldi.com/Maldi/Res/Pep_Stds.html.

TABLE 8.10. Protein mass calibration standards

Name	Source	Average Mass $[M+H]^+$	Sigma product number	SwissProt Reference
Cytochrome *c*	Tuna	12043.6	C 2011	CYC_KATPE
Cytochrome *c*	Bovine	12232.0	C 3131	CYC_BOVIN
Cytochrome *c*	Equine	12361.2	C 7752	CYC_HORSE
Ribonuclease A	Bovine	13683.4	R 5500	RNP_BOVIN
α-lactalbumin	Human	14071.1	L 7269	LCA_HUMAN
α-lactalbumin	Bovine	14179.1	L 6010	LCA_BOVIN
Lysozyme c	Chicken egg	14306.1	L 6876	LYC_CHICK
Hemoglobin α-chain	Human	15127.3	G 5890	HBA_HUMAN
Hemoglobin β-chain	Human	15868.7	G 5890	HBB_HUMAN
Apomyoglobin	Equine	16952.5	A 8673	MYG_HORSE
β-lactoglobulin B	Bovine	18278.2	L 8005	LACB_BOVIN
β-lactoglobulin A	Bovine	18364.3	L 7880	LACB_BOVIN
Trypsin Inhibitor, α-chain	Soybean	20037.6	T 9003	ITRB_SOYBN
Trypsin Inhibitor, β-chain	Soybean	20091.7	T 9003	ITRA_SOYBN
Trypsin Inhibitor, χ-chain	Soybean	20163.8	T 9003	ITRA_SOYBN
Growth hormone	Human	22126.2	S 4776	SOMA_HUMAN
Trypsinogen	Bovine	23982.0	T 1143	TRYP_BOVIN
Chymotrypsinogen A	Bovine	25657.1	C 4879	CTRA_BOVIN
Aldolase	Rabbit	39204.7	A 7145	ALFA_RABIT
Pepsinogen	Porcine	39607.9	P 4656	PEPA_PIG
Alcohol dehydrogenase, S	Equine	39712.4	A 6128	ADHS_HORSE
Alcohol dehydrogenase, E	Equine	39833.4	A 6128	ADHE_HORSE
Ovalbumin	Chicken	44401.0	A 2512	OVAL_CHICK
Glucose-6-phosphate dehydrogenase	Yeast	57433.5	G 4134	G6PD_YEAST
Albumin	Bovine	66431.0	A 0281	ALBU_BOVIN
Albumin	Human	66439.2	A 3782	ALBU_HUMAN
Urease	Jack Bean	90762.9	U 0376	UREA_CANEN
Phosphorylase B	Rabbit	97219.5	P 6635	PHS2_RABIT
Albumin dimer	Bovine	132859.0	A 9039	ALBU_BOVIN

Reproduced from http:// www.srsmaldi.com/Maldi/Res/Prot_Stds.html.

Equipment

MALDI mass spectrometer

See the note to Step 6.

MALDI plate

Biological Sample

Lyophilized samples (dissolved in 60% acetonitrile, 0.1% v/v TFA at a concentration of 1–10 pmoles/μl) or RP-HPLC-purified fractions, which can be used directly

Samples containing excipients such as buffers, salts, detergents, and denaturants must be desalted prior to analysis (Rabilloud 1990; Burlet et al. 1992; Papayannopoulos 1995; Patterson 1997) (see Protocol 2). Even minor quantities of sodium (*m/z* 23) and potassium (*m/z* 39) ions, readily generated via laser ionization, can cause significant ion suppression.

The detergent *n*-octyl glucoside (0.1%) can be added at different stages of sample preparation (e.g., digestion or solubilization of digested peptides) to prevent adsorption of peptides on the sample tube wall and/or pipette tip, thereby yielding increased peptide peak number and improved sequence coverage (Katayama et al. 2001).

METHOD

1. Pipette 0.5 μl of the matrix solution (using a 2-μl pipette for accuracy) onto the sample wells of the metal sample plates used for MALDI-MS analysis.

2. Immediately add 0.5 μl of the standard or sample to the matrix before it dries.

3. Allow the solvent to evaporate and the samples to dry for ~5 minutes either at room temperature or in a 40°C oven.

4. Transfer the metal sample plate to the vacuum chamber of the mass spectrometer.

5. Acquire an initial mass spectrum at a laser power well above the ionization threshold to "warm up" the calibration standard.

6. Decrease the laser power until a good spectrum is obtained.

   Steps 5 and 6 apply to MALDI-TOF instruments that are equipped with a high-vacuum ionization source (e.g., Kompact MALDI IVTM™, Kratos Analytical/Shimadzu, or equivalent). For instruments equipped with a CCD camera for viewing the cocrystallized sample (e.g., o-MALDI QSTAR™ Pulsar I, Applied Biosystems, or equivalent), select a large crystal for analysis (e.g., large crystals form when using the matrix DHB; this is not so important when using HCCA or SA, which usually yield uniformly small crystals). HCCA is preferred as a MALDI matrix for peptide mapping because it yields the highest sensitivity and forms a uniform matrix layer on a MALDI sample plate, which makes it amenable for automated analysis (Beavis et al. 1992). For optimal results, run the unknown samples at approximately the same laser power as the calibration standard. Although a high laser power may give a good signal, resolution may be compromised due to peak broadening. The use of a low laser power at the ionization threshold may give high resolution, but result in poor signal-to-noise ratios. Typically, the thresholds of ionization for HCCA and DHB are 20 and 30 μJoules, respectively. Thus, HCCA is considered to be a "hotter" matrix than DHB, which gives rise to increased metastable ion formation and concomitant PSD. A consequence of the latter is broader peak formation and reduced resolution. Although a sample may appear to be uniform, it is recommended that different regions of the spot be examined to find "sweet spots," i.e., regions of the sample spot that give superior signal-to-noise ratios.

7. Obtain a linear external two-point calibration using an appropriate calibrant (refer to Tables 8.9 and 8.10).

   When using HCCA, the matrix-derived ion (e.g., $[M+H]^+ - OH$, *m/z* 173.2) and the singly charged ion of the appropriate calibration standard can also be used as calibrants.

8. Repeat Steps 5 and 6 for unknown samples and compare with the calibration values to obtain accurate masses.

9. After use, clean the MALDI plate by rinsing thoroughly with H_2O to remove any visible crystals and then with methanol; wipe with lint-free tissues (e.g., Kimwipes).

PROTOCOL 2

Preparation of Reversed-phase Microcolumns

A POWERFUL PROTEOMIC TOOL FOR IDENTIFYING AND CHARACTERIZING PROTEINS and peptides is MALDI-TOF-MS. It is well recognized that sample preparation is a crucial step in MALDI-MS analysis of proteins and peptides and that a single component or mixture must be free of excipients such as buffers, salts, detergents, or denaturants. One versatile strategy for sample cleanup prior to MALDI-MS analysis uses microscale columns designed for direct sample elution onto the MALDI target plate. This protocol describes the fabrication of a reversed-phase microcolumn designed for this purpose.

The microcolumns are prepared from GELoader tips essentially as described previously (Gobom et al. 1997; Kussmann et al. 1997). This protocol has been optimized for sample cleanup prior to MALDI-MS. However, with slight modifications, it works equally well with samples destined for ESI-MS. This protocol was provided by Martin R. Larsen (Macquarie University, Sydney, Australia).

MATERIALS

CAUTION: See Appendix 3 for appropriate handling of materials marked with <!>.

IMPORTANT: All reagents used in this protocol must be sequence-grade.

Reagents

Acetonitrile <!>

Matrix solutions (see note to Step 8)

- Protein matrix solutions: SA <!> or DHB <!> dissolved in 50% acetonitrile containing 0.1% TFA <!>
- Peptide matrix solution: HCCA <!> in 70% acetonitrile containing 0.1% TFA

Methanol <!>

Trifluoroacetic acid (TFA) (0.1%) or 1% Formic acid <!>

See note to Step 4.

Equipment

Forceps, blunt tip

GELoader pipette tips (Eppendorf)

MALDI-MS target

Pipette tip (disposable, 20–200 μl size)

Poros R1, R2, or Oligo R3 chromatography resins (Applied Biosystems) <!>

Syringe (1 ml)

Biological Sample

Protein or peptide sample to be analyzed by MS

METHOD

1. Partially constrict a GELoader pipette tip by squeezing the narrow end. The two most common ways to do this are illustrated in Figure 8.32 and listed below:

   - Place the narrow end of a GELoader tip flat on a hard surface. Roll a 1.5-ml microfuge tube over the final 1 mm of the tip.
   - Squeeze the narrow end of a GELoader tip using blunt forceps. Turn the tip once while holding it with the forceps, to close the end.

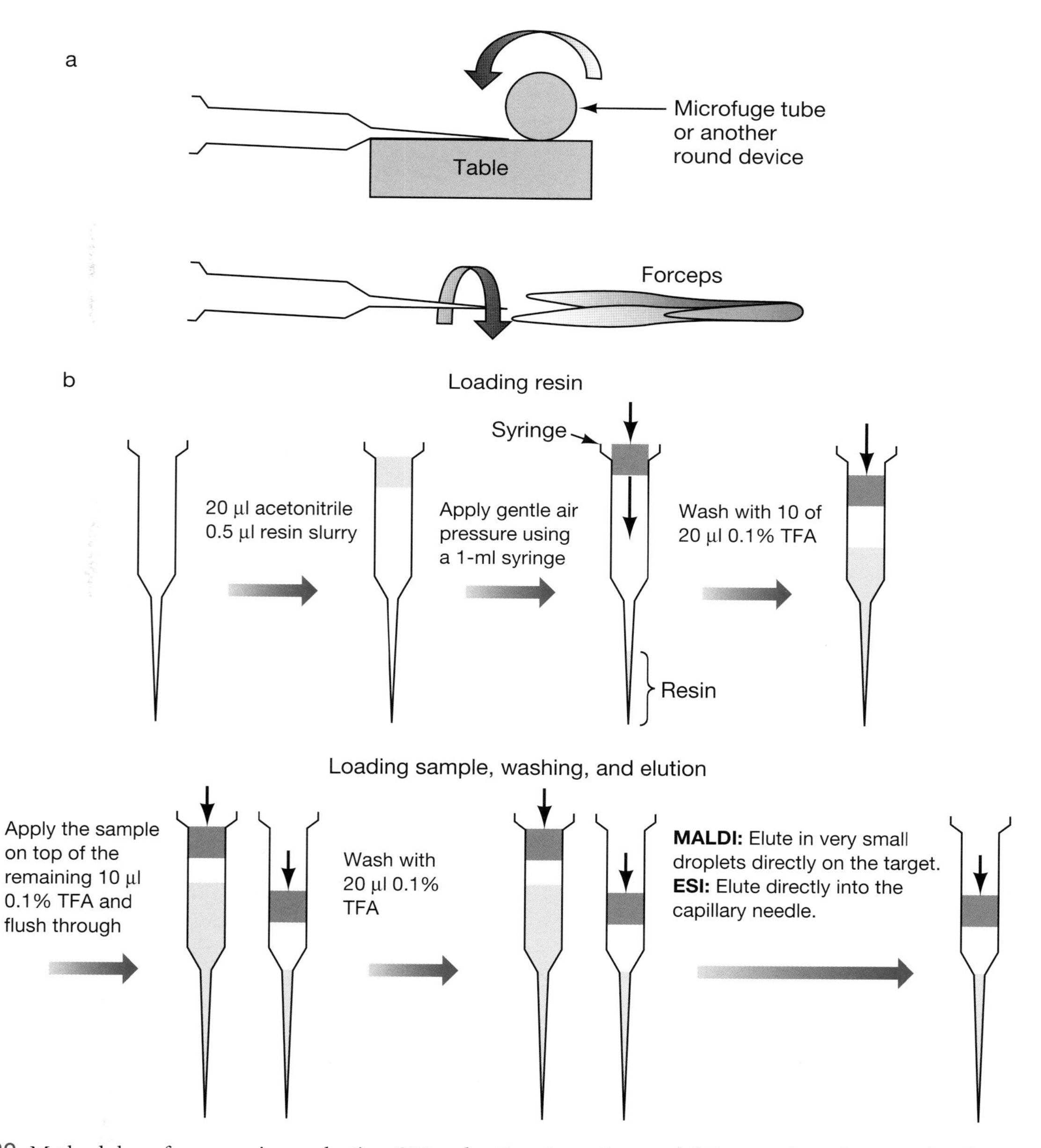

FIGURE 8.32. Methodology for preparing and using GELoader tip microcolumns. (*a*) Preparation of a constricted GELoader tip. (*b*) Generation of the column, application of the analyte sample, and elution of the analyte molecules. (Courtesy of Martin R. Larsen, Macquarie University.)

2. Prepare a slurry of 100–200 μl of chromatographic resin (Poros R1, R2, or Oligo R3) in 70% acetonitrile (use ~1.5 mg of resin/100 μl of acetonitrile).

   Steps 3–8 are illustrated in Figure 8.32b.

3. Load 20 μl of 70% acetonitrile in the top of the constricted GELoader tip, and add 0.5 μl of the resin slurry on top of the acetonitrile. Use a 1-ml syringe fitted to the GELoader tip with a disposable pipette tip to gently press the liquid down to create a small column at the end of the constricted microcolumn. Dry the column by letting all of the liquid escape from the bottom of the column before performing the next step.

   The yellow pipette tip must be cut twice to fit both the syringe and the GELoader tip. The amount of resin-slurry used to create the column should be varied with the approximate concentration of the sample. In general, the column height should be 1–6 mm (approximate bed volume is 10–60 nl) when working with peptides generated from poorly abundant, gel-separated proteins.

4. Apply 20 μl of 0.1% TFA to the top of the column. Equilibrate the column by gently pushing 10 μl of 0.1% TFA through it, using gentle air pressure generated by the syringe. The remaining 10 μl of 0.1% TFA should remain on top of the column bed.

   When the microcolumn is used as a cleanup step prior to nano-ESI-MS, 1% formic acid should be used instead of TFA, which is incompatible with ESI-MS.

5. Apply the protein/peptide sample on top of the remaining 10 μl of 0.1% TFA.

6. Press the liquid gently through the column by applying air pressure with the syringe. Do not allow the column to dry out; leave ~2 μl of solution on top of the column bed.

7. Wash the column with 20 μl of 0.1% TFA, and allow the column to run dry.

8. Elute the analytes, using 0.5 μl of matrix solution, directly onto the MALDI-MS target. The 0.5 μl of matrix solution should be spotted as several droplets (5–10) on the target. Alternatively, if the analytes are to be analyzed by ESI-MS, elute the peptides from the column using methanol/formic acid/H_2O (50:1:49) either directly into the capillary needle or into a microfuge tube.

   The preferred MALDI matrices used for eluting protein from the column are either SA or DHB dissolved in 50% acetonitrile/0.1% TFA. For eluting peptides from the column, HCCA in 70% acetonitrile containing 0.1% TFA is preferentially used. Alternatively, the analytes can be eluted directly into a microfuge tube for storage or further analysis, using any percentage of organic solvent. When eluted in several small spots, only the first two or three spots will contain the analytes, resulting in a further concentration of the sample.

   The column can be reused after washing it extensively with 100% acetonitrile. Depending on the size of the column and the abundance/concentration of the analyte molecules that have been loaded onto it, the column can be reused two to ten times without observing any memory effects.

PROTOCOL 3

Preparation of Immobilized Enzyme Microcolumns

PEPTIDE MAPPING BY MALDI-TOF-MS has emerged as a powerful proteomics tool for identifying and characterizing proteins. One of the key steps in this method involves the proteolytic cleavage of proteins, followed by MS analysis of the generated peptides. A limitation of the method, especially for high-throughput proteomics, is the speed and efficiency of proteolytic cleavage.

In this protocol, a method for rapid tryptic digestion using immobilized enzyme microcolumns is described. This protocol was provided by Martin R. Larsen (Macquarie University, Sydney, Australia).

MATERIALS

CAUTION: See Appendix 3 for appropriate handling of materials marked with <!>.

Reagents

NH_4HCO_3 (50 mM, pH 7.8, sequence grade) <!>

NH_4HCO_3 (50 mM, pH 7.8, sequence grade) containing 10% (v/v) acetonitrile <!>

Equipment

Forceps (blunt tip)

GELoader pipette tips (Eppendorf)

Immobilized trypsin chromatography resin

Immobilized enzymes are available from several different manufacturers. However, they are also easily prepared using, for example, cyanogen-bromide-activated Sepharose 4B fast-flow media (Amersham Biosciences).

Pipette tip (yellow disposable)

Syringe (1 ml)

Biological Sample

Protein or peptide sample to be analyzed by MS

Additional Reagents

Step 6 of this protocol may require the reagents listed in Protocol 2 of this chapter.

METHOD

1. Partially constrict a GELoader pipette tip by squeezing the narrow end. The two most common ways to do this are illustrated in Figure 8.32a and listed below:

   - Place the narrow end of a GELoader tip flat on a hard surface. Roll a 1.5-ml microfuge tube over the final 1 mm of the tip.

   - Squeeze the narrow end of a GELoader tip using blunt forceps. Turn the tip once and squeeze it with the forceps again to close the end.

2. Prepare a slurry of the immobilized trypsin resin using ~3 mg of resin per 100 µl of 50 mM NH_4HCO_3 (pH 7.8).

   The concentration of the immobilized enzyme resin should be varied with the amount of enzyme immobilized onto the resin.

3. Apply the slurry to the empty GELoader tip (from Step 1). Use a 1-ml syringe fitted to the GELoader tip with a yellow pipette tip to gently push the slurry down to create a column at the end of the constricted tip. Do not allow the column to dry out.

   The yellow pipette tip must be cut twice to fit both the syringe and the GELoader tip. NH_4HCO_3 (50 mM, pH 7.8) can be used with the majority of enzymes having a pH optimum of ~7.5. For enzymes active at other pH optima or buffer conditions, alternative buffers should be considered.

4. Wash the column with several 30-µl aliquots of 50 mM NH_4HCO_3 (pH 7.8). Do not allow the column to dry out during the wash steps.

   Alternatively, the column can be washed with 50 mM NH_4HCO_3 (pH 7.8) containing 10% (v/v) acetonitrile.

5. Apply the protein/peptide sample to be digested onto the column in 50 mM NH_4HCO_3 (pH 7.8).

6. Use a syringe to slowly and gently press the liquid through the column bed. Collect the flowthrough in a microfuge tube for further analysis. Optionally, for very small amounts of protein or peptides or very diluted samples, collect the liquid in a micro-reversed-phase column for desalting and concentration (see Protocol 2).

   EXPERIMENTAL TIP: The flowthrough from the enzyme column can be re-applied to the column several times to ensure a more complete digestion.

7. If the amount of protein or peptides in the flowthrough is low, or for the sake of completeness, wash the column with 5 µl of 50 mM NH_4HCO_3 (pH 7.8) containing 10% (v/v) acetonitrile. The wash can be collected together with the flowthrough.

8. If picomole amounts of protein (0.02–0.05 µg) or peptides (0.001–0.005 µg) have been recovered, mix a small amount of the flowthrough with 2% formic acid and a suitable matrix on the MALDI target. If the amount of material is too low, first desalt and concentrate the material recovered in Steps 6 and 7 using a reversed-phase microcolumn, as described in Protocol 2.

PROTOCOL 4

Preparation and Use of an Integrated Microcapillary HPLC Column and ESI Device for Proteomic Analysis

Mass spectrometry, and in particular the application of ESI coupled on-line with high-performance separation techniques such as capillary electrophoresis (CE) and microcapillary HPLC (μLC), has had a dramatic effect on the sensitivity and the speed with which the primary structure of proteins and peptides can be determined (Aebersold and Goodlett 2001). Advances in separation techniques, particularly their implementation in miniaturized formats on-line with high-performance mass spectrometers (Hunt et al. 1991; Kennedy and Jorgenson 1991; Moseley et al. 1991; Wahl et al. 1992, 1993; Emmett and Caprioli 1994) and the development of miniaturized sprayers as ESI ion sources (Wahl et al. 1994; Wilm and Mann 1994; Emmett and Caprioli 1994; Wilm and Mann 1996) have reduced the amount of peptide required for complete and routine sequence characterization from several picomoles of peptide (Hunt et al. 1986; Matsudaira et al. 1987) in the mid 1980s to a few femtomoles and below by the mid 1990s (Davis and Lee 1997; McCormack et al. 1997; Figeys et al. 1998; Lazar et al. 1999; Martin et al. 2000).

Arguably, much of this gain in sensitivity of the mass spectrometer is due to the combination of concentration-dependent (Goodlett et al. 1993) type ionization devices such as ESI and on-line capillary separation devices of very small internal diameter (I.D.). The increase in sensitivity with small capillary internal diameter can be primarily attributed to a reduced mass flow rate of solvents and other background constituents into the ESI source, which allows for greater sample ionization efficiency (Wahl et al. 1992, 1993). For reasons of robustness, most μLC work is done with 75- or 100-μm I.D. capillary columns that clog less frequently than 50-μm I.D. capillary columns. The following protocol describes the preparation of an integrated C_{18}-packed capillary column–ESI microspray device. This type of integrated device was first reported in 1994 (Emmett and Caprioli 1994) and has recently gained popularity from work by Yates and co-workers (Gatlin et al. 1998) most notably as a clever biphasic capillary column (Link et al. 1999). One of the key steps in preparing packed μLC columns is devising a means to hold the C_{18}-derivatized particles in place. In this protocol, a polyimide capillary tip tapered to ~5-μm is used to hold the C_{18}-derivatized particles in place (Emmett and Caprioli 1994; Gatlin et al. 1998; Link et al. 1999). An alternative method uses underivatized silica particles tamped into place and then sintered in a flame to form a porous frit (Hunt et al. 1991; Kennedy and Jorgenson 1991; Moseley et al. 1991).

This protocol describes the preparation of an integrated microcapillary column–microelectrospray ionization device, system suitability testing, and some typical results that investigators could achieve with practice. The protocol was provided by David R. Goodlett and Eugene C. Yi (Institute for Systems Biology, Seattle, Washington) and Philippe Mottay (Brechbuhler, Inc., Spring, Texas).

MATERIALS

CAUTION: See Appendix 3 for appropriate handling of materials marked with <!>.

Reagents

Acetic acid <!>
Acetonitrile <!>
Ethanol (70%) <!>
Methanol (50%) <!>

Equipment

Autosampler (e.g., DIONEX FAMOS, Sunnyvale, California)
C_{18} trap cartridge (Michrome Bioresources, Auburn, California)
Capillary cutter (Scientific Instrument Services, Ringoes, New Jersey)
Derivatized silica resin (5 μm, C_{18}) <!>
Derivatized silica resin is available from numerous suppliers, including Michrome Bioresources (Auburn, California), Western Analytical (Murrieta, California), and Appelera (Framingham, Massachusetts).
Divert valve (six-port)
Forceps
Helium gas (pressurized) <!>
Hex driver
Used to seal the pressure cell.
HPLC system (e.g., the Agilent 1100 HPLC)
Mass spectrometer (e.g., ThermoFinnigan ITMS, San Jose, California)
MicroCross with 0.006-inch through-holes (Upchurch Scientific, Oak Harbor, Washington)
A Micro-electrospray ionization device available from Brechbuhler, Inc. (Spring, Texas) is used to hold the MicroCross in place.
Microscope (low-resolution)
pH paper
Polyimide-coated capillary (75 μm I.D. x 360 μm O.D.) (PolymicroTechnologies, Phoenix, Arizona)
Pressure cell (Brechbuhler, Inc., Spring, Texas)
The pressure cell is used for capillary packing (see Figure 8.33).
Propane torch
Regulator (two-stage, high-pressure)
Sonicator
Teflon ferrules (Chromatography Research Supplies, Louisville, Kentucky)

Biological Sample

Protein sample of interest

METHOD

Microcapillary Column Construction

A microcapillary column inside a polyimide capillary has one end manually tapered to 5 μm to serve as a frit. The tapered end also serves as the point from which ESI emanates. A low-resolution microscope will be needed to monitor capillary packing and the tapering process.

1. Cut a 30-cm long piece of 75 μm I.D. x 360 μm O.D. polyimide-coated capillary. Hold one end of the capillary with forceps and remove 2–3 cm of polyimide by passing it through a flame from a propane torch. Pull the capillary slowly to make a straight tapered tip.

2. Place the tapered end on a clean index finger (rinse it with 70% ethanol) and carefully trim the end with a capillary cutter. Rinse the tapered tip with 70% ethanol. To ensure good ESI performance, inspect the tip under a low-resolution (5–10x magnification) microscope to ensure that the tapered tip is blunt (i.e., square and not jagged) and ~5 μm across.

   There is an art to manually tapering the tip so be patient. Alternatives to manually tapering the tip are to purchase the tips from New Objectives Inc. (Woburn, Massachusetts) or to purchase a laser puller from Sutter Instruments Inc. (Novato, California).

3. Use a spatula to place ~50–100 μl (dry volume) of 5 μm of C_{18}-derivatized silica resin in a 1.7-ml microfuge tube.

   EXPERIMENTAL TIP: Due to the low back pressure of POROS (Appelera) beads, it may be preferable for the novice to begin with these.

4. Add 500 μl of 70% ethanol to the C_{18} resin, vortex to mix, and sonicate it briefly (i.e., no more than 1 minute).

   EXPERIMENTAL TIP: Due to the occurrence of apparently irreversible changes in the C_{18} beads that adversely affect separation, prepare the slurry daily.

5. Place the tube containing the slurry in the pressure cell (Figure 8.33).

   Forceps are useful for handling the slurry-filled tube. The chamber is drilled to accept tubes of various lengths so it may be necessary to place some Kimwipes in the bottom to bring the tube containing the slurry to a height inside the chamber that is easy to use.

   The viscosity of the solvent can be modified (in Step 5) to affect the packing rate.

FIGURE 8.33. Pressure cell. The cell is shown with a polyimide-coated capillary coming out the top and the high-pressure valve for turning gas in the cell on/off in the front.

6. Thread the blunt untapered end of the capillary into the top of the pressure cell; i.e., through the nut, the Teflon ferrule, and the top of the pressure cell (Figure 8.33). The end of the capillary inside the pressure cell should sit about halfway down into the 500 µl of slurry. Looking across the top of the pressure cell chamber, it is possible to observe when the capillary begins to bend and thus determine that it is at the bottom of the vial. Pull it up ~0.5 cm. Close the top of the cell making sure that the O-ring is in place and then tighten all four bolts with the hex driver. Now secure the capillary in place by tightening the nut on top of the cell.

   WARNING: Always secure all four bolts before turning on the helium gas. If a leak occurs, turn off the gas and re-seat the O-ring or replace it with a new one and try again.

7. Set the two-stage, high-pressure regulator to 1000 psi. The capillary column is now ready to be packed. Slowly turn on the pressure in the cell by turning the valve 180°. Pack the capillary to 10 cm and stop, or pack the entire length of the capillary. Trim the length of the capillary column as desired.

If the beads do not appear to flow, try one or all of the following:

- Turn the pressure inside the cell on/off several times to dislodge an air pocket in the slurry.
- Vortex the slurry again; the slurry may have settled to the bottom of the tube. Alternatively, add a micro-stir bar and perform the packing on a magnetic stir plate. This keeps the slurry suspended, but will not be necessary with experience. Use of the stir bar can fragment the C_{18} beads so it is best to prepare the slurry fresh daily.
- Cut off a 0.5-cm length of capillary inside the cell and reposition; it may have become clogged.
- Hold the capillary taut between the top of the cell and the microscope stage and vigorously flick the capillary so that it vibrates like a violin string between a finger and the top of the pressure cell.

8. When the packing has come to within 0.5–1.0 cm of the desired length, turn off the pressure inside the cell. Remove the slurry and replace it with a vial of 0.1 M acetic acid. Seal the cell, turn the pressure on, and let the capillary rinse at least until the effluent turns acidic (check with pH paper). This wash will finish the packing because there remains some unpacked silica in the back of the capillary.

9. The column can be stored indefinitely by placing the end inside a 15-ml polypropylene tube filled with 50% methanol and sealing the top.

The following are some general comments on the selection of derivatized silica and capillary dimensions. In general, the larger the I.D. of the capillary, the longer the column can be packed without exceeding the HPLC operating back pressure. The complete novice should start out with Appelera's POROS IIR packing, which is very easy to work with because it generates lower back pressure than other particles, due to a large pore size. Investigators who branch out to more specific applications will find that there are many derivatized beads from which to choose. In addition, be aware that particle size directly affects back pressure and chromatographic fidelity. In general, 5-µm particles provide good results. Many of these final caveats can be readily explained by seeking out an expert on chromatography.

Microsprayer Setup

- The capillary column is ready for testing and use if it passes the system suitability test. Figure 8.34 shows a μESI source with an integrated capillary column–ESI sprayer such as the one just packed with the blunt end of the packed capillary column inside the Upchurch all PEEK four-way union at *Port 1*. The tapered tip should be placed to within 1–2 cm of the MS orifice and will be the point from which ESI emanates due to the high voltage applied via a platinum lead at *Port 2* in Figure 8.34, which for this particular setup will be ~2 kV.
- The sample is loaded and washed while the six-port divert valve is open so that flow enters the four-way union at *Port 3* and exits at *Port 4*. After suitable washing, the divert valve closes and flow is redirected to the C_{18} capillary column in *Port 1*. This type of setup using a trap cartridge provides extended life for the separation column. The device shown in Figure 8.34 can be used without the trap cartridge in *Port 3* if desired, but remember to keep the six-port divert value closed if not using the trap.
- Begin by running fast linear gradients, i.e., 0–60% acetonitrile in 15 minutes. Most investigators, after becoming comfortable with the loading and elution, find that the μcapillary columns work well with all sorts of gradients.
- Flow rate through the column will vary depending on internal diameter and packed length. For a 75-μm capillary column packed to 7 cm, the flow rate can be as low as 150 nl/min when optimized. However, investigators should begin with higher flow rates (e.g, 500 nl/min) until they are comfortable with the system and how flow rate affects ESI stability/sensitivity. To compensate for the decrease in flow that occurs with a decrease in viscosity (i.e., higher-percentage acetonitrile has lower viscosity relative to H_2O) when using a restrictive flow splitter of constant length, the investigator may program increasing flow rate at the HPLC pump over the duration of the linear acetonitrile gradient.

EXPERIMENTAL TIP: The same setup can be used with an empty tapered tip and a syringe pump for tuning the mass spectrometer. For optimal MS performance, the MS should be tuned under conditions as nearly identical to operation conditions as possible.

System Suitability Test

Once the LC-MS system is set up, it must be tested routinely to determine if all of the components are functioning properly before any precious sample is analyzed. If the system is routinely tested using the same conditions and it fails, then logical systematic procedures can be followed to correct the problem. Thus, before using the LC-MS setup, it is recommended that a system suitability check be carried out. This will mean analysis of a sample that is well-characterized. Often, the easiest sample is a purified peptide, but it could be any sample that is well-characterized. For LC-MS systems, there are four simple questions to ask to assess the system performance. These follow along with some idea of what system parameters to monitor:

1. ***Is the mass spectrometer calibrated?*** Routinely compare observed mass to theoretical mass (Figure 8.35a).
2. ***Is the HPLC and μLC column functioning properly?*** Compare measured to expected retention times, monitor back pressure at the HPLC, and calculate peak width of a single analyte by plotting the single ion current trace (Figure 8.35c).

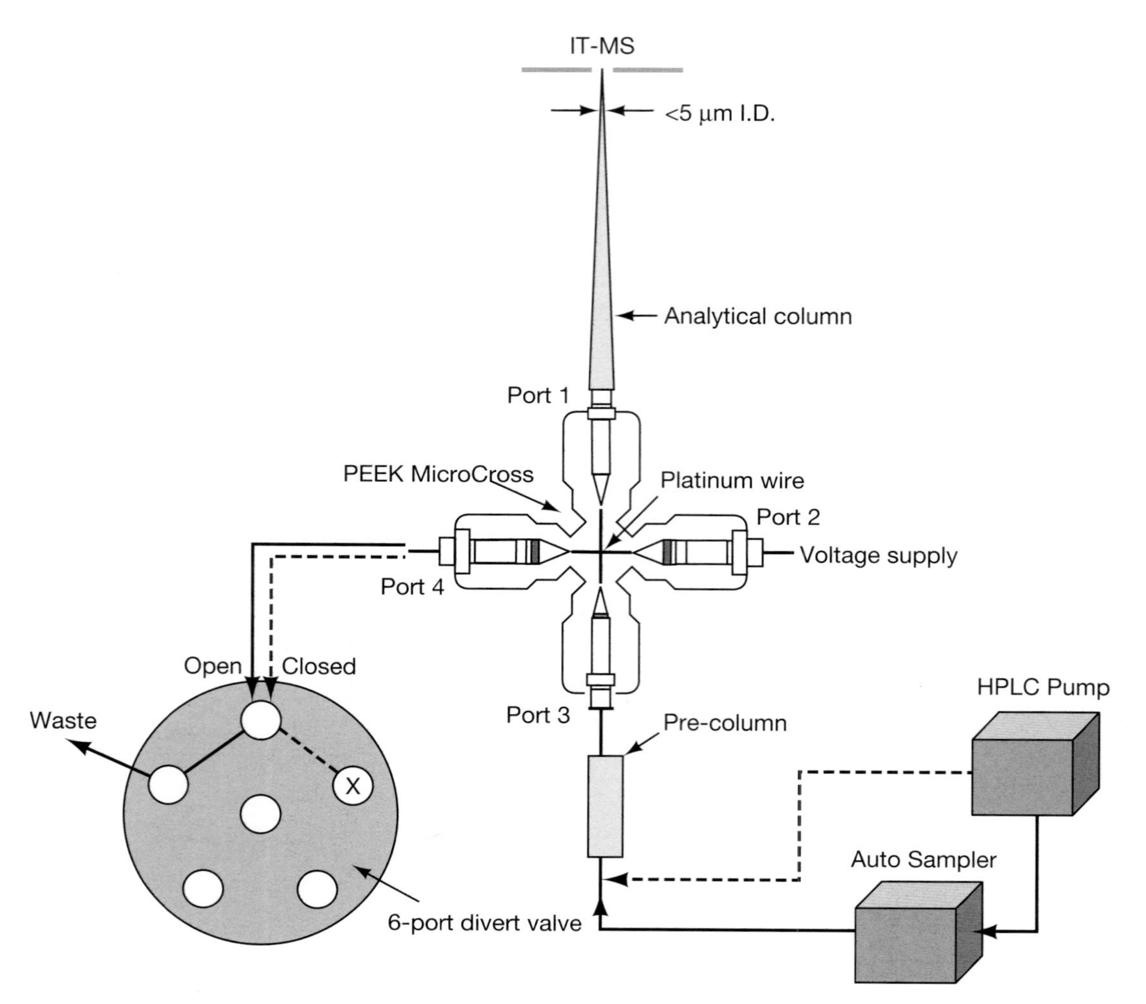

FIGURE 8.34. Diagram of an automated μTrap-μLC-ESI-MS/MS system. A tapered ESI emitter packed with C_{18} (*Port 1*), a high-voltage lead (*Port 2*), a miniature C_{18}-packed peptide μtrap cartridge (*Port 3*), and a six-port divert valve (*Port 4*) were connected to a PEEK four-way union. Peptides were loaded for desalting on the C_{18} μTrap cartridge with the six-port divert valve open. When closed, the flow is redirected to the ESI emitter/C_{18} μLC column.

3. ***Are the solvents (and sample) clean?*** Measure signal to noise (Figure 8.35b) at a point where an analyte elutes by summing mass spectra across the point of elution. An analyte may "drag" along uncharacterized components, by electrostatic and hydrophobic interactions, that decrease the ratio of signal to noise; aging solvents in general increase the noise floor.

4. ***Is the mass spectrometer tuned properly?*** Measure the ratio of signal to noise at the point when an analyte elutes. Because mass spectrometric data are not absolute, monitoring the signal-to-noise ratio will be more meaningful than signal response alone.

All of these questions can be answered using data from a single LC-MS analysis performed prior to analyzing real samples. To provide a basis for comparing operating performance of multiple instruments within and between laboratories, it is best to designate a single system suitability test. It should be an analyte in the same class as that being analyzed and be well-characterized. For example, the data shown in Figure 8.35 are from injection of 500 fmoles of neurotensin onto a μLC column as described above and analyzed using an ion-trap mass spectrometer. The results obtained are analyzed in light of the above four questions.

- If the HPLC system has functioned properly to deliver solvents as programmed, then the elution time for the peptide will be reproducible.
- If elution time is outside of the expected range, then either the HPLC system is not functioning as programmed or perhaps the column is beginning to fail.
- If the peak shape is not symmetrical or too broad (it should be ~10–30 seconds for a 0.5–1.0-pmole injection of peptide standard) (Figure 8.35), then there may be a problem with the packing material, or dead volume has been introduced somewhere in the μLC column.
- If the HPLC-MS system is working properly, then the signal-to-noise ratio of the analyte should be above an empirically determined level. An example of how to calculate this is shown in Figure 8.35c, where the ordinate axis is expanded for demonstration purposes to be 200x of normal. The ratio of signal to noise is estimated by expanding the ordinate axis to the point at which individual charge states are obscured by the noise. If the signal-to-noise

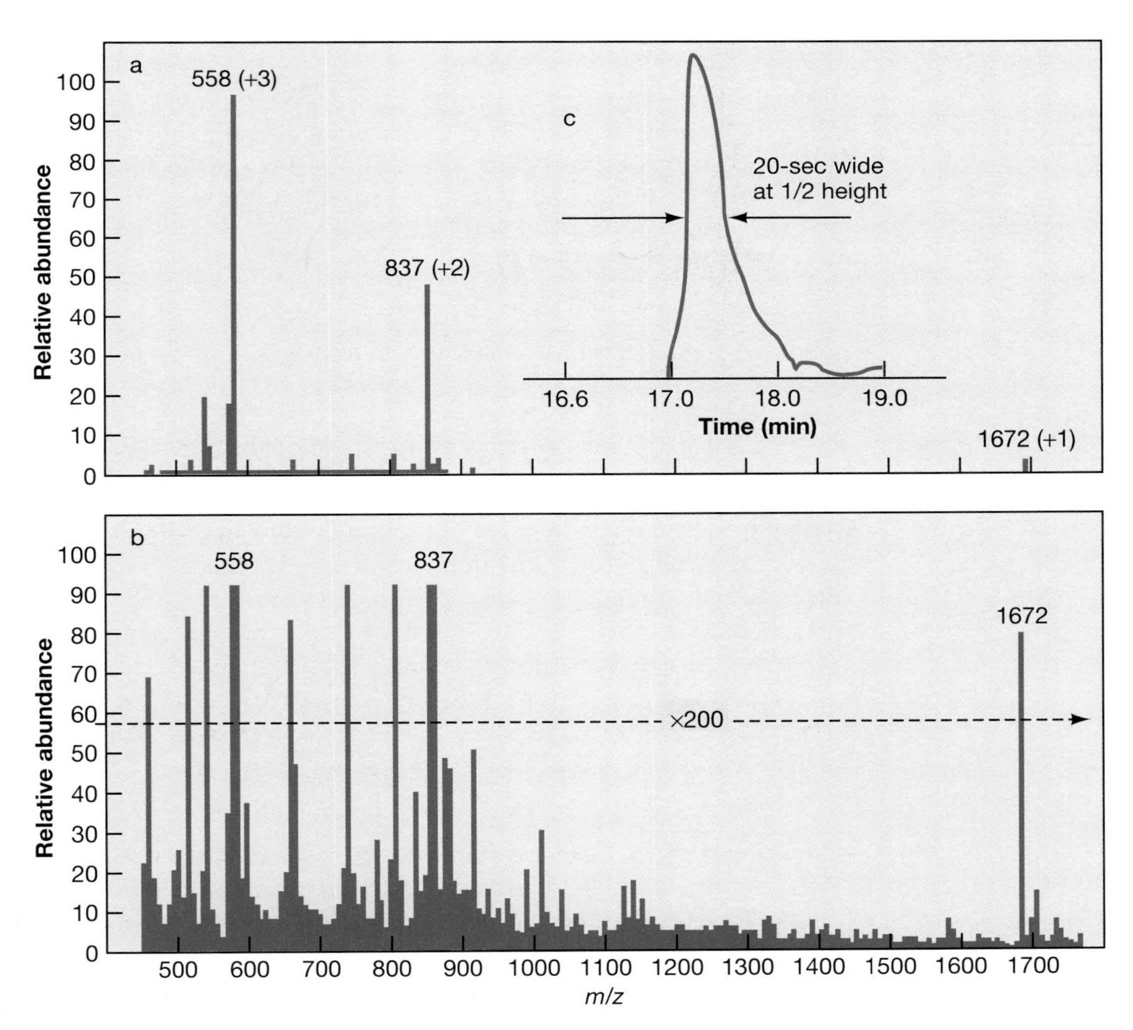

FIGURE 8.35. Example of multiple-charged peptide mass spectrum. The three most abundant charge states of neurotensin (*a*), the chromatographic total ion current trace at the point where neurotensin eluted showing an average peak width at half-height (*b*), and the same mass spectrum as in *a* but with the ordinate axis expanded 200x normal to demonstrate how to check for signal/noise (*c*).

ratio of the analyte is below an acceptable value, then perhaps the solvents should be changed, or the mass spectrometer requires re-tuning, or the lenses involved in focusing ions must be cleaned. When a lens becomes coated with even a thin invisible film (e.g., from repeated analysis of samples), then the applied voltage is no longer equal to the field voltage experienced by the ions during transmission, and a decrease in performance (e.g., signal to noise) is observed.

- If the calculated and observed molecular weight for the analyte does not match within the tolerance of the mass spectrometer, the instrument may simply need to be cleaned and tuned, or it may need to be tuned and calibrated after cleaning.

Following these simple steps will help the novice operator troubleshoot the complex LC-MS setup. As a challenge, the best sensitivity for a peptide standard will be ~5 attomoles loaded on a column.

Analysis of a Complex Yeast Lysate

Proteins from a yeast whole-cell lysate were subjected to proteolysis with trypsin and prepared for MS analysis as follows:

- Protein-containing pellets from total yeast cell lysates were resuspended in 0.5 ml of 50 mM NH_4HCO_3 (pH 8.3) and 0.5% SDS and solubilized by incubating them for 30 minutes at 60°C with occasional vortexing. The resulting mixture was diluted to have 0.05% SDS with 50 mM NH_4HCO_3 (pH 8.3). Modified trypsin was added at an enzyme:substrate ratio of 100:1, and the samples were incubated overnight at 37°C.
- Prior to LC-MS analysis, the resulting peptides were purified by an OASIS MCX cartridge according to the manufacturer's protocol.
- Acidified peptide mixtures were injected onto a C_{18} trap cartridge connected to *Port 3* (see Figure 8.34) from a DIONEX FAMOS autosampler, and the flow was directed through the six-port divert valve (*Port 4*) to waste.
- After a 5-minute washing period at 100% A, the divert valve closed, redirecting flow through the 10 cm x 75 μm I.D. μLC column (*Port 1*) that also served as the ESI emitter. Simultaneous to the redirection of flow to the capillary column, the linear gradient flow of acetonitrile (Solvent B) from an Agilent 1100 HPLC began. To decrease chemical noise in the mass spectrometer, no acid was included in Solvent B. The linear gradient was formed over 60 minutes from 2% B to 45% B. The flow rate at the pump was initially 100 μl/min.
- After restrictive flow splitting using an empty piece of polyimide-coated capillary (the dimensions of which must be set empirically) placed between the autosampler and the spray source, the flow at the device shown in Figure 8.34 was 300 nl/min. The Upchurch Scientific MicroCross with 0.006-inch through-holes shown in Figure 8.34 was held in place using a device (from Brechbuhler, Inc., Spring, Texas) that mounted on a ThermoFinnigan ITMS.
- ESI was initiated by delivering high voltage to *Port 2*. Ion selection for CID was via an automated routine that used top-down data-dependent ion selection of the three most intense ions in a survey scan, including a 3-minute dynamic exclusion period to prevent reselection of previously selected ions and repeated continuously throughout the μLC-ESI-MS/MS analysis.
- A total of 1 μg of the subsequent complex peptide mixture, inferred from measured protein concentration, was separated using the μLC column described above, and CID of the eluting peptides was conducted in series using an ion-trap mass spectrometer. The analytical setup was identical to that shown in Figure 8.34.

- Peptide ions for CID were selected from a broad *m/z* range of 400–2000 by a data-dependent routine that included dynamic exclusion for 3 minutes to prevent immediate reselection of the most intense ions already subjected to CID.
- Yeast proteins were identified from the tandem mass spectra of peptides by a database search using SEQUEST (Eng et al. 1994). Figure 8.36 shows an example of a typical base peak trace from such an experiment.

Triplicate μLC-ESI-MS/MS analysis of this sample identified 401 proteins by requiring all correctly matched peptide sequences to have tryptic amino and carboxyl termini and using a standard set of scores known to give correct identifications (Keller 2002). The peptide tandem mass spectra used to identify two different proteins are shown in Figure 8.36 along with their SEQUEST Xcorr and dcorr scores. Currently, data from a SEQUEST database search must be manually confirmed by an expert analyst even after applying the above specified scoring criteria because the top-scoring SEQUEST score is not always the correct peptide sequence match. However, in the future, analysts using SEQUEST and other "expert" routines will have access to programs that screen tandem mass spectral quality prior to a database search (Moore et al. 2000) and provide probabilities of the correctness of a database match such as QSCORE (Moore et al. 2002). These types of tools will greatly advance the expert's ability to process large amounts of data.

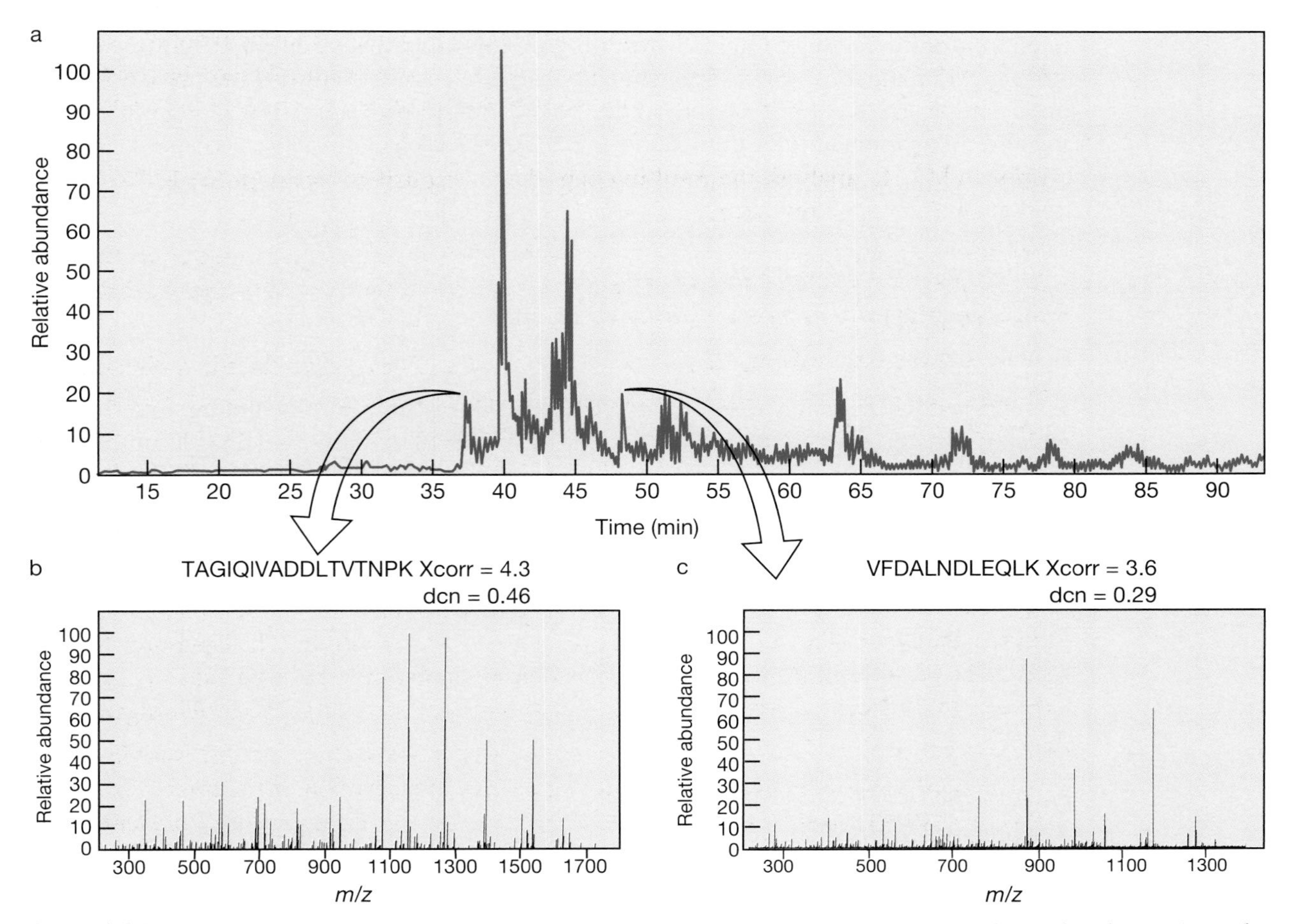

FIGURE 8.36. Yeast peptides analyzed by μLC-ESI-IT-MS. All soluble proteins from yeast were digested with trypsin, and 1 μg total of the peptide mixture was analyzed by the μTrap-μLC-ESI-IT-MS/MS setup shown in Figure 8.34 (*a*). (*b,c*) Examples of peptide tandem mass spectra and the corresponding peptide sequences identified by a sequence database search using SEQUEST, along with the Xcorr and dcn scores used by experts to judge the correctness of the sequence match.

PROTOCOL 5

Analysis of Complex Protein Mixtures Using Nano-LC Coupled to MS/MS

NANOLITER-LC COUPLED TO TANDEM MASS SPECTROMETRY (nano-LC-MS/MS) permits the rapid and sensitive determination of protein-protein interactions (see Chapter 10). By using a specific purification technique such as coimmunoprecipitation or affinity purification in conjunction with nano-LC-MS/MS not only are proteins identified, but specific protein-protein interactions are elucidated as well. This was first demonstrated in *Saccharomyces cerevisiae* by using a macromolecular complex as "bait" to enrich for proteins that interacted with the complex (McCormack et al. 1997). The macromolecular complex was generated by polymerizing bovine tubulin and then incubating it with a yeast cell lysate. The lysate was removed, and the microtubule complex was treated with two different concentrations of salt to elute different sets of interacting proteins. These mixtures of proteins were enzymatically digested and subjected to nano-LC-MS/MS and database searching, which identified the proteins. In a second example, the yeast mitochondrial nucleoid complex, which is responsible for the packaging and the maintenance of mitochondrial DNA (Meeusen et al. 1999), was analyzed by

- fractionating a mitochondrial lysate using differential sucrose centrifugation,
- treating the enriched nucleoids to free the proteins from the DNA,
- subjecting them to a sucrose cushion,
- digesting the protein mixture with trypsin, and
- analyzing the fractions by nano-LC-MS/MS.

Thirty-four proteins were identified (J. Nunnari, unpubl.) including one, mgm101, which was determined to be responsible for the repair of oxidatively damaged mitochondrial DNA (Meeusen et al. 1999). This protocol was provided by David M. Schieltz, Michael P. Washburn, and Lara G. Hays (The Torrey Mesa Research Institute, San Diego, California).

MATERIALS

CAUTION: See Appendix 3 for appropriate handling of materials marked with <!>.

Reagents

Acetonitrile (HPLC grade) <!>
Acetonitrile (5%)/5% formic acid <!>
HPLC solvents
  Solvent A: 5% acetonitrile and 0.02% heptafluorobutyric acid (HFBA) <!>
  Solvent B: 80% acetonitrile and 0.02% HFBA
Methanol (HPLC grade) <!>

Equipment

Alcohol burner

Binary or quaternary pump (e.g., series 1100, Agilent Technologies)

C_{18} reversed-phase packing material (5 µm) (e.g., Zorbax XDB, Agilent Technologies)

CAUTION: Do not inhale; use in a chemical fume hood.

C_{18} solid-phase extraction pipette tips (e.g., SPEC Plus PT C18, ANSYS Technologies, Lake Forest, California)

These C_{18} solid-phase disk pipette tips have a 0.4 µg sorbent capacity and a loading volume of up to 800 µl.

Ceramic scribe

Fused-silica capillary (100 µm I.D. x 365 µm O.D.)

The capillary can be obtained from Agilent Technologies or Polymicro Technologies.

Gold wire (0.025 diameter) (Scientific Instrument Services, Inc., Ringoes, New Jersey)

Graduated glass capillaries

Helium tank with regulator (at least 1000 psi pressure) <!>

High-performance liquid chromatography equipment

Laser puller (e.g., P-2000, Sutter Instruments)

Nano-LC ion sources (ThermoFinnigan, The Scripps Research Institute, and Cytopea, Inc.)

PEEK MicroCross, Microtight tubing sleeves (Upchurch Scientific)

Stainless steel pressurization bomb (The Scripps Research Institute and Cytopea, Inc.)

Tandem mass spectrometer

A variety of tandem mass spectrometers are suitable from a number of manufacturers (e.g., ThermoFinnigan and Micromass, Inc.)

Biological Sample

Protein fractions

METHOD

Preparation of a Nano-LC Column

1. Make a window in the center of an ~12–15-inch, 100 x 365-µm fused-silica capillary by holding it over an alcohol flame until the polyimide coating has been charred (see Figure 8.37). Remove the charred material by wiping the capillary with a tissue soaked in methanol (see Figure 8.38).

   Unlike an alcohol burner, a Bunsen burner flame is too hot and will seal the inside of the capillary.

2. To pull a needle, place the window portion of the capillary into the P-2000 laser puller (Figure 8.39). Position the window in the mirrored chamber of the puller, where the laser will concavely focus and melt the fused silica.

   Arms on each side of the mirror have grooves and small vises, which properly align the fused silica and hold it in place. Table 8.11 includes typical parameters for pulling ~3-µm tips from a 100 x 365-µm capillary.

TABLE 8.11. Typical parameters for pulling ~3-µm needle tips from 100 x 365-µm fused-silica capillary

	Heat	Filament	Velocity	Delay	Pull
1	320	0	40	200	0
2	310	0	30	200	0
3	300	0	25	200	0
4	290	0	20	200	0

These values will differ slightly from instrument to instrument. For a very useful section on the construction of capillary needles, see the Sutter Instrument P-2000 manual.

FIGURE 8.37. Preparation of a window in the fused-silica capillary. The capillary is held over an alcohol flame to char the polyimide coating. The length of the charred portion is ~1–3 cm.

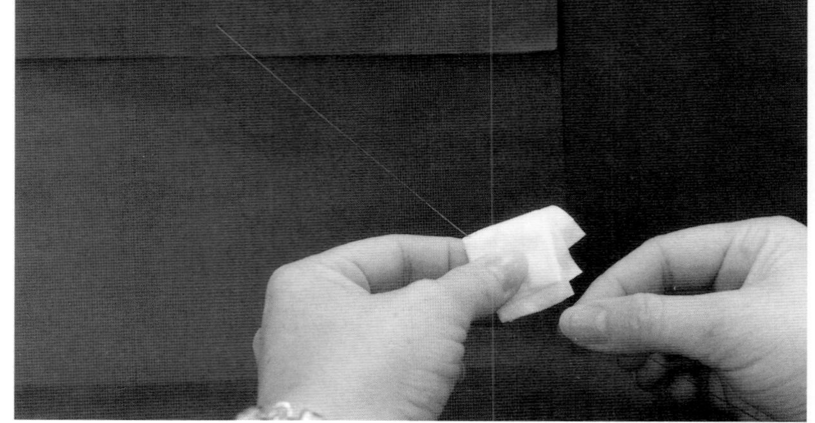

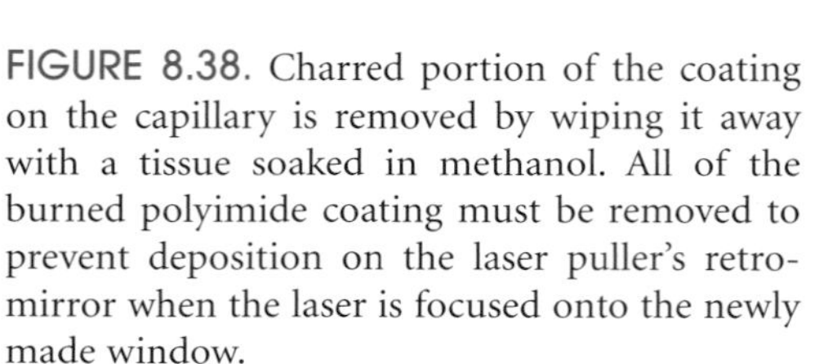

FIGURE 8.38. Charred portion of the coating on the capillary is removed by wiping it away with a tissue soaked in methanol. All of the burned polyimide coating must be removed to prevent deposition on the laser puller's retromirror when the laser is focused onto the newly made window.

FIGURE 8.39. The fused-silica capillary is placed into the laser puller to produce two pulled needle capillaries. The windowed area of the capillary is placed within the "shroud" containing the retro-mirror. The ends of the capillary are fastened within the vises to hold the capillary in position.

3. Pack the pulled needle capillary with C_{18} reversed-phase packing material using a stainless steel pressurization bomb.

   WARNING: Always wear safety glasses during this step. In the event that the capillary has not been seated properly, it will be forced out of the bomb at great velocity due to the pressure.

   This packing step is similar to the method described by McCormack et al. (1994).

   a. Place ~5 mg of C_{18} reversed-phase packing material into a 1.7-ml microfuge tube and add ~1 ml of methanol. Shake the tube to suspend the particles and place it into the bomb. Secure the bomb lid by tightening the five bolts.

      It is very important to secure all five bolts. Both the lid of the bomb and the bomb itself have grooves that fit an O-ring; this provides an air-tight seal when the lid is seated against the bomb.

      WARNING: High-pressure gas will escape violently if the lid is not secured tightly.

   b. The lid has a Swagelok fitting containing a Teflon ferrule. Feed the fused-silica, pulled needle capillary down through the ferrule until the end of the capillary reaches the bottom of the microfuge tube. Tighten the ferrule to secure the pulled needle capillary. Figure 8.40 shows the pressurization bomb and the microcapillary to be packed with reversed-phase material.

   c. Apply pressure to the bomb by first setting the regulator on the helium gas cylinder to ~400–800 psi and then opening a valve on the bomb to pressurize it.

      The packing material will begin filling the pulled needle capillary. This now becomes the capillary microcolumn. To achieve good chromatographic separation, pack the capillary with 10–15 cm of material.

4. Attach the needle to the HPLC system through an Upchurch PEEK MicroCross. For a layout of the connections for the Upchurch MicroCross, see Figure 8.41.

   - ***The first connection point*** of the cross contains the transfer line from the HPLC pump. This consists of a 50 x 365-µm fused-silica capillary, where the length is sufficient to reach from the HPLC pump to the mass spectrometer.

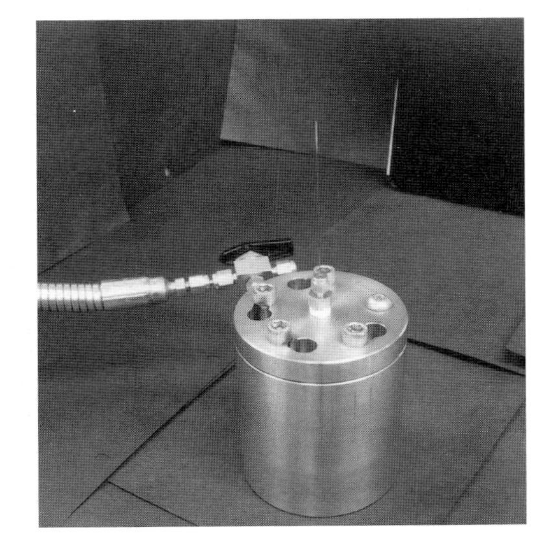

FIGURE 8.40. Pressurization device or "bomb" with high-pressure line and valve. The inner portion of the bomb contains an open area to allow a microfuge tube to stand upright with the cap open. The bomb and the lid have a groove, which holds a viton O-ring to ensure a high-pressure seal when the lid is tightened down. The lid contains five holes for bolts and in the center, a Swagelok fitting. Within the fitting sits a Teflon ferrule that allows the capillary to be inserted down into the bomb and into the microfuge tube. The ferrule is tightened to hold the capillary in place and provide a high-pressure seal.

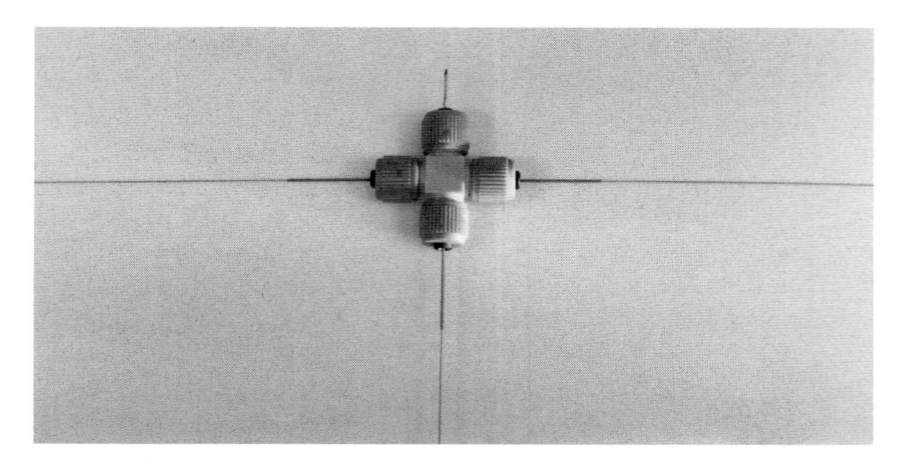

FIGURE 8.41. Layout of the Upchurch MicroCross. The first connection at the bottom is for a transfer line to bring the solvent flow from the HPLC pump to the MicroCross. Moving clockwise to the second connection is the split line, which is used to control the final flow rate of the solvent through the microcolumn. The next connection is to hold a small section of gold wire, which makes electrical contact with the solvent. The final connection is where the microcolumn is attached.

- ***The second connection point*** contains a length of fused-silica capillary that is used as a split line. This split line allows a majority of the flow to exit through the split; therefore, very low flow rates can be achieved through the packed capillary microcolumn. The size and length of this section of capillary depend on the flow rate from the pump and the length of the microcolumn. A good starting point is to use a 2-foot section of 100 x 365 μm for the split line (but see Step 6).
- ***The third connection point*** contains a section of gold wire, which will raise the voltage of the solvent entering the needle from zero to ~1800 V, thus allowing electrospray to occur.
- ***The fourth connection point*** is for the packed capillary microcolumn.

5. Before loading the sample, equilibrate the column by pushing the methanol out with 100% Solvent A for 5 minutes at a flow rate of 150 μl/min at the pump.
6. After 5 minutes, measure the flow from the tip of the capillary microcolumn, using graduated glass capillaries. The target flow rate at the tip should be ~100–300 nl/min. If the flow rate is above this value, use a ceramic scribe to cut off a portion of the split line capillary. This will allow more of the flow to exit out of the split and less flow through the microcolumn. If the flow is less than 100 nl/min, use a longer piece of capillary or a section with a smaller inner diameter to force more flow through the microcolumn.

   Measuring the flow rate and adjusting the split line may have to be repeated a number of times until the target flow rate is reached.

Concentration of Sample

Peptide samples can be solubilized in any number of reagents, including Tris, ammonium bicarbonate, acetic acid, formic acid, and urea. However, peptide samples are typically the product of a digested protein or protein mixture, in which a variety of reagents may be present including some that will interfere with the performance of the reversed-phase column and the mass spectrometer (see the Experimental Tip below Step 11).

Peptide sample volumes range from several microliters to 1 ml or more. The bed volume of the microcolumn is ~1.5 μl, which allows samples up to 50 μl to be loaded directly onto the column. For sample volumes greater than 50 μl, concentration of the sample (as detailed in Steps 7–11) is necessary.

7. Wet the SPEC Plus PT C_{18} solid-phase extraction pipette tip with 1 ml of methanol as follows. Push approximately half of the methanol through the disk, and then wait for 15–30 seconds to allow the disk to activate. Push the remainder of the methanol through, but do not push air through the disk.

8. Equilibrate the disk with 1 ml of Solvent A by pushing it through the disk.

9. Pull the peptide solution into the pipette tip and then push the solution back out.

   This can be repeated two to three times. Peptides remain in the tip and are concentrated onto the disk. The flowthrough can be discarded.

10. Elute the peptides with ~100 μl of 90% acetonitrile/0.5% acetic acid. Push the eluting solvent through the disk, pull the solution back up through the disk, and then push it through one final time, to give three passes across the disk.

11. Use a vacuum concentrator to remove the acetonitrile until the peptides are nearly dry. Resuspend the peptides in ~10–15 μl of 5% acetonitrile/5% formic acid. The sample is ready to load onto the reversed-phase capillary microcolumn.

    EXPERIMENTAL TIP: Avoid detergents whenever possible or remove them prior to loading the sample onto the reversed-phase column. Once a detergent enters the column, it will bind to the reversed-phase and "leak off" in the elution gradients, contaminating subsequent analysis. Detergents ionize more readily than peptides and therefore will mask any peptide ions. If the sample contains solubilization or fractionation agents such as urea, glycerol, or sucrose, loading may be somewhat difficult due to the viscosity of the solution. In such cases, once the sample is loaded, wash the column extensively with 100% Solvent A to remove these chemicals before starting the gradient.

Loading the Sample onto the C_{18} Column

12. Centrifuge the samples in a microfuge at 14,000 rpm for 10 minutes to pellet any solid material.

13. If any pellet is visible, transfer the supernatant (peptide sample) to a fresh microfuge tube. Even minute amounts of solid material will plug the microcolumn. Place the tube containing the peptide sample into the pressurization bomb, and tighten the lid to the bomb (for instructions on tightening the bomb lid, see Step 3a).

14. Feed the reversed-phase capillary microcolumn down through a Teflon ferrule until the end of the capillary reaches the bottom of the microfuge tube. Tighten the ferrule to secure the capillary microcolumn.

15. Set the regulator on the gas cylinder to 400–800 psi, and then pressurize the bomb by opening the valve on the bomb that connects to the gas cylinder. The peptide sample will begin to flow into the column.

16. Measure the loaded volume at the tip of the needle using a graduated glass capillary.

17. When 8–10 μl are loaded, release the pressure, remove the column from the bomb, and place it back into the Upchurch MicroCross.

Ion-source Setup

18. Place the Upchurch MicroCross with the connections into a stage, which is designed in this case for the ThermoFinnigan LCQ series mass spectrometer. This stage performs a threefold purpose:

    - Supports the MicroCross and holds it in place along with the connections.
    - Insulates the MicroCross from electrical contact with its surroundings when it is held at high-voltage potential.
    - Allows for fine position adjustment of the microcolumn with respect to the entrance of the mass spectrometer (heated capillary) by using an XYZ manipulator.

    Figure 8.42 shows the stage comprising the Plexiglas support, MicroCross, high-voltage connection, microcolumn, and entrance to the mass spectrometer. Plastic tabs with small Teflon screws are used to hold the connections in place. The potential needed to raise the solvent in the cross and the microcolumn to 1800 V is delivered through an insulated cable that attaches to an aluminum block seated in the Plexiglas support. Electrical contact is made when the plastic tab is tightened down onto the gold wire, which presses down onto the aluminum. This allows only the MicroCross to be energized and not the XYZ manipulator or metal support plate.

19. Position the microcolumn using the XYZ manipulator so that the needle tip is ~2–5 mm from the orifice of the mass spectrometer's heated capillary, and set the voltage at 1.5–1.8 kV.

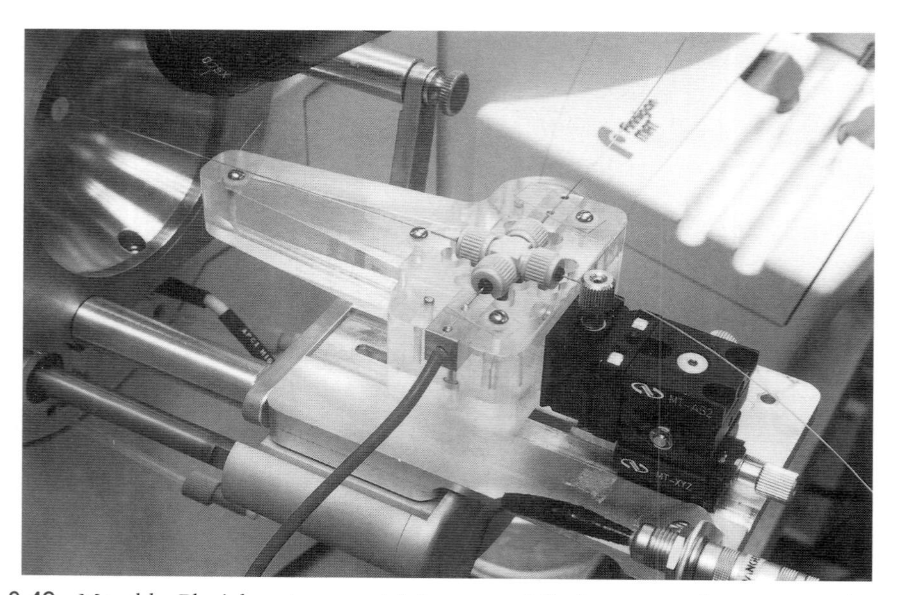

FIGURE 8.42. Movable Plexiglas stage containing nano-LC electrospray ion source. The Upchurch MicroCross with HPLC connections is held in position in the Plexiglas stage with plastic tabs. The solvent enters the MicroCross from the transfer line of the HPLC pump. A majority of the flow leaves the cross through the split line, but a small fraction moves through the microcolumn toward the opening of the mass spectrometer. An insulated cable supplies the high voltage that is connected to the aluminum portion of the stage. The aluminum makes contact with the gold wire, energizing the solvent flowing through the cross. This provides a large voltage potential between the tip of the microcolumn and the opening of the mass spectrometer, allowing electrospray ionization to occur. An XYZ manipulator is used to provide fine positioning of the microcolumn with respect to the entrance of the mass spectrometer.

HPLC Programming

20. Program the HPLC system as follows:

    - Start the gradient with 100% Solvent A (0% Solvent B) and a flow rate of 150 μl/min.
    - Over a period of time (call it *X* minutes), ramp up the concentration of Solvent B to 60% along with an increase in the flow rate to 250 μl/min.
    - Set a ramp-down period of 5 minutes, during which the concentration of Solvent B returns to 0% and the flow rate slows to 150 μl/min.

    As a guide, if the sample contains ~15–20 proteins, (*X*) should be a 30-minute gradient; if the sample contains ~40–50 proteins, (*X*) would be either a 60- or 90-minute gradient. The gradient profile can be programmed through the ThermoFinnigan Xcalibur software or the Micromass Inc. Masslynx software, depending on the mass spectrometer being used.

Tandem Mass Spectrometry Analysis

Data-dependent acquisition of tandem mass spectra during the HPLC gradient is also programmed through the LCQ Xcalibur or the QTOF2 Masslynx software. Guidance is provided here for setting the parameters for data-dependent acquisition using the ThermoFinnigan LCQ system. For experimental design of data-dependent tandem mass spectra acquisition for the QTOF2, refer to the Masslynx manual (version 3.5).

The settings in Step 21 are for a typical data-dependent MS/MS acquisition analysis. The method consists of a continual cycle beginning with one scan of MS (scan one), which records all of the *m/z* values of the ions present at that moment in the gradient, followed by

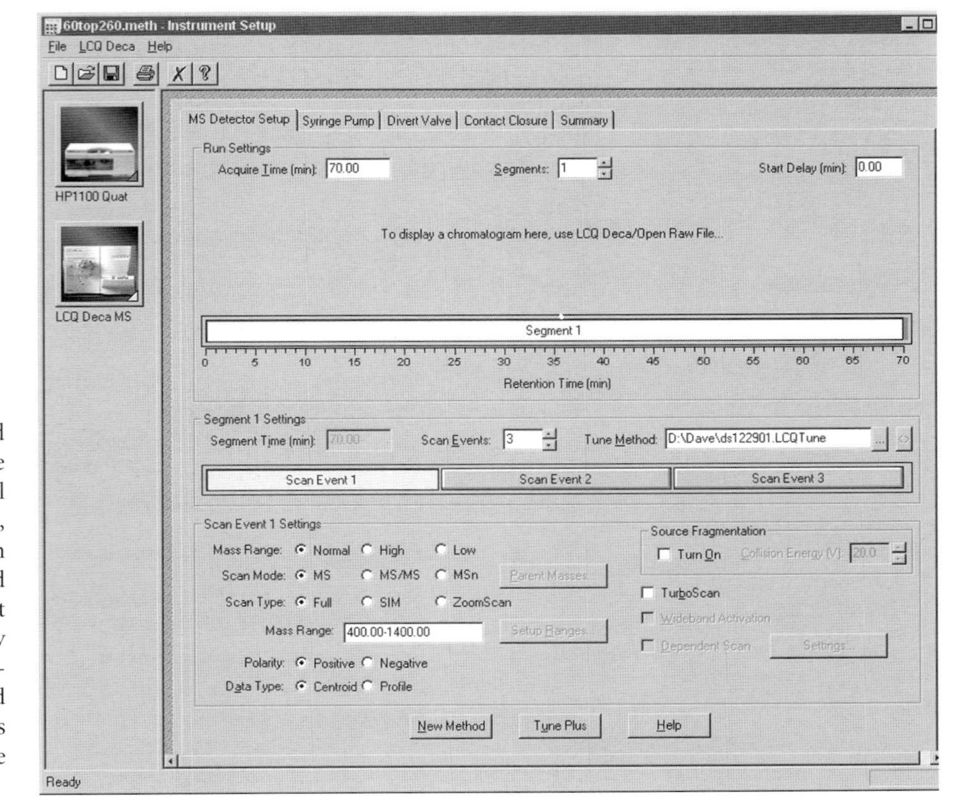

FIGURE 8.43. Main LCQ method page editor. This window allows the user to design a method for several types of mass spectrometric analyses, including data-dependent tandem mass spectra acquisition. The method consists of one 70-minute segment where three scan events continuously cycle as the analysis runs. The acquisition consists of an MS scan followed by two data-dependant MS/MS scans of the first and second most-intense ions.

two rounds of MS/MS. The initial MS/MS scan is of the first most-intense ion recorded from the MS scan. The second MS/MS scan is of the second most-intense ion recorded from scan one. Dynamic exclusion is activated to improve the protein identification capacity during the analysis (described in Step 21i).

21. Use the settings below to create a typical data-dependent MS/MS method.

    a. In the main Xcalibur software page, select Instrumental Setup.

    b. In the next window, select the button labeled Data Dependent MS/MS.

    c. The next window will contain the design parameters for the acquisition setup (see Figure 8.43). At the top, input the time of the acquisition.

       Typically, the time of the acquisition is the same as the duration of the HPLC gradient plus time to allow for dead volume in the lines.

    d. The Segments setting is generally set to 1 and the Start Delay setting for the acquisition is set to 0.

    e. The Scan Event is set to 3. The first scan event is a full scan of MS, and the next two scan events are MS/MS, to allow the mass spectrometer to perform one round of MS followed by two rounds of MS/MS, of the first and second most-intense peaks. These parameters will be set in the following sections.

    f. Highlight the bar showing Scan Event 1 Settings. Below this, check Normal Mass Range as Normal, Scan Mode as MS, and Scan Type as full. Set the Mass Range to 400–1400, the Polarity to Positive, and the Data Type to Centroid. This range generally gives good coverage for peptides from a tryptic digest.

    g. The Tune Method box specifies the path for a file containing the parameters for the electrostatic lenses and ion trap. These parameters are established through the "LCQ Tune" as described in the ThermoFinnigan LCQ Getting Started Manual.

    h. Highlight the bar labeled as Scan Event 2, check the box next to Dependent Scan, and then click the Settings button. A window will come up where the parameters can be set for all of the MS/MS scan events (Global and/or Segment) and for individual MS/MS scan events.

    i. At the side of the window, press the Global button. These are parameters that are common among all segments and scan events. Several tabs will be visible.

       - The Global tab sets the values for both the MS and the MS^n data-dependent masses and are left at the default settings.

       - In the next tab (Mass Widths) for high-throughput protein identification, these values are all left at the default settings.

       - The Dynamic Exclusion tab contains several parameters that are important for effective data-dependent tandem mass analysis. Dynamic exclusion prevents an abundant peptide ion, with a broad elution profile, from being continually selected for MS/MS. The intense ions prevent other lower-abundance peptides from being selected. Ions that are selected for MS/MS are placed onto a list. While on this list, they will be excluded for a period of time so that other ions may be selected.

The Enable box is checked and typical settings include a Repeat Count of 2, a Repeat Duration of 0.5 minute, and an Exclusion Duration of 3 minutes.

The Exclusion Mass Width is set for "by mass" and a width of 3 daltons is set around the peak, 0.8 dalton on the low-mass side, and 2.2 daltons on the high-mass side. This ensures that the entire isotopic distribution of the peak is put into the exclusion list.

These parameters would be appropriate for a 30-minute gradient. If a longer gradient were used such as a 90-minute gradient, then the Exclusion Duration could be increased to 10 minutes.

j. The Analog tab and the Isotopic Data Dependence tab are not enabled and are not used in the analysis.

k. Again, on the left side, press the Segment button to bring up the parameters for the current segment (in this case, there is only one segment containing three scan events).

- Under the Current Segment tab, leave blank the Parent Masses and the Reject Masses and therefore do not check the box next to Most Intense If No Parent Mass Found.
- The Normalized Collision Energy is set at 35%. This is the amount of energy delivered to the peptide to cause fragmentation.
- The Activation Q and Activation Time are both left at their default setting of 0.250 and 30, respectively.
- The Default Charge State is set to 2, which makes the assumption that all peptides entering the mass spectrometer are doubly charged.
- The Min MS Signal (10^4 counts) is set to 10, which is the signal intensity needed for a peak recorded in an MS scan to trigger the mass spectrometer to select this peak for MS/MS. The Min MSn Signal (10^4 counts) is set to 0.5; this is the threshold to trigger the mass spectrometer to perform higher-order MS/MS. In this type of analysis, no higher-order MS/MS is programmed into the segments so this value does not get utilized.
- The last value is the Isolation Width, which is the window around the ion in the MS scan that is selected for MS/MS. This is typically set to 2.5 daltons. The Add/Sub tab is not enabled.

l. Press the Scan Event button to access the last set of values, which defines parameters for this specific scan event: Scan Event 2. The user can then select the ranking of ion abundances per MS scan (Scan Event 1) to be selected for MS/MS. In this case, check the Nth Most Intense Ion and set to 1 (this is the first most-intense ion). After pressing the okay button, repeat the process by going back to the "method page" and pressing the Scan Event 3 bar, checking Dependent Scan and pressing the Settings button. The Scan Event button is then selected. Use Previous List of Ions is checked so the selection for MS/MS is from Scan 1 and is set to select the second most-intense ion. Press okay to accept.

During the gradient, the mass spectrometer will continually cycle through these three scans collecting tandem mass spectra in a data-dependent fashion.

Database Searching

Searches are not constrained for only tryptic peptides, and indexed databases (databases only containing tryptic peptides) are not used. In cases where there are very complex mixtures, such as cell lysates, nonspecific cleavages can occur. Therefore, nontryptic peptides would be missed in the database search. If tryptic peptides are matched using Lys-C and trypsin, this adds a degree of confidence in the result since tryptic peptides should occur. Search results that lead to positive identification of a protein consisted of the following:

- Multiple unique peptides from the same protein.
- High-value SEQUEST $<X_{corr}>$ scores.
- Peptides that are tryptic.

In general, for +1 peptides, good $<X_{corr}>$ scores start at 1.5 and go up and for +2 peptides, good scores start at 2.4 and go up; however, for +3 peptides, their resulting SEQUEST outputs are more difficult to interpret and should be accepted only if they are fully or partially tryptic and have an $<X_{corr}>$ above 3.7. The SEQUEST delta $<C_n>$ values should be larger than 0.08 (Eng et al. 1994).

It is important to recognize that the unique aspects of peptides and their respective SEQUEST scores are dependent on their charge state. In general, when more than four peptides for a protein were identified by these criteria, the presence of protein was assured. In these cases, spectra were manually confirmed to determine if at least one of the SEQUEST results fit criteria described by Link et al. (1999). Two of these criteria are that (1) any given MS/MS spectrum had to clearly be above the baseline noise and (2) there must be continuity to a b or y ion series (Link et al. 1999). When fewer than four peptides are identified from a particular protein by SEQUEST, each SEQUEST result must be manually confirmed via these same criteria (Link et al. 1999). Following these parameters should assist in the rapid correct identification of a large number of proteins.

IDENTIFICATION OF PROTEINS INTERACTING WITH THE *SACCHAROMYCES CEREVISIAE* PROTEASOME BY NANO-LC-MS/MS AND DATABASE SEARCHING

The yeast 26S proteasome is a complex of ~32 proteins contained in a 19S regulatory cap and a 20S core. This complex is responsible for ubiquitin-dependent proteolysis, but the assembly of the components in the complex and the mechanism of function are not well understood. A one-step affinity purification protocol of the 26S proteasome was developed to look at the proteins that interact with the 26S proteasome and the 19S cap in both the absence of ATP and the presence of the ATP analog ATP-γ-S, which is much more difficult to hydrolyze. Direct analysis of large protein complexes (DALPC) (Link et al. 1999) also referred to as multidimensional protein identification technology (MuDPIT) was used to identify these proteins without prior separation by gel electrophoresis(Verma et al. 2000). A method for identifying interacting proteins using 1D nano-LC coupled to MS/MS is given in Protocol 5.

1. The proteasome samples (20 μg from 375 μg/ml) were buffer-exchanged into 50 mM Tris-Cl (pH 8.5) containing 8 M urea.
2. Digestion was initiated by adding 0.3 μg of endoproteinase Lys-C and incubating in the dark for 4 hours at 37°C.
3. The samples were diluted fourfold by the addition of 50 mM Tris-Cl (pH 8.5) to reduce the urea concentration to 2 M, where trypsin still maintains activity. In some cases, when proteins are solublized in 8 M urea (without prior digestion with Lys-C), the proteins precipitate upon dilution to 2 M urea. Lys-C therefore provides a useful method for altering the tertiary structure of some protein enough that there should not be any aggregation when the solution is diluted to 2 M urea.
4. $CaCl_2$ (10 mM) was added to a final concentration of 1 mM along with 0.2 μg of trypsin. The samples were incubated in the dark overnight at 37°C.
5. The samples each were separately loaded onto a nano-LC column after a 5-minute equilibration in a manner similar to that described in Protocol 5, Steps 12–17.
6. Once the sample was loaded, the column was washed for ~5 minutes. The HPLC pump was set to ramp the gradient from 2% to 60% Buffer B over a 90-minute period. During this time, the ThermoFinnigan LCQ Classic was programmed to perform four rounds of MS. The first was to take an MS scan from *m/z* 400 to 1400 to measure masses and intensities of ion signals. This scan was followed by three MS/MS scans: MS/MS of the first most-intense ion recorded in the MS scan, then MS/MS of the second most-intense ion, and then MS/MS of the third most-intense ion. This cycle repeated itself for the duration of the gradient. The exclusion option was set to exclude an ion for 5 minutes after it had been selected.
7. Tandem mass spectra were extracted from the LCQ's raw file and searched using the SEQUEST program. The search was done against the yeast ORF (open reading frame) database that is publicly available from Stanford University. Common contaminant sequences such as keratin, IgG, trypsin, and Lys-C were added to this database. Criteria for a positive identification followed the method described in the panel above on DATABASE SEARCHING.

The analysis determined that 71 proteins copurified with the cap in the absence of ATP. Of these proteins, 19 were known 19S subunits, and the remaining 52 proteins were identified as possible proteasome-interacting proteins. Upon subsequent analysis, 25 were discounted because they were considered to be nonspecific interactors. This left 27 potential proteasome-interacting proteins, which were validated by generating fusion proteins with either *myc* or hemagglutinin epitopes for coimmunoprecipitation; 24 of these were categorized as novel proteasome subunits, chaperones, regulatory proteins (including transcriptional regulators), and abundant proteins (Verma et al. 2000).

PROTOCOL 6

Analysis of Complex Protein Mixtures Using Multidimensional Protein Identification Technology

THE ANALYSIS OF HIGHLY COMPLEX MIXTURES of proteins, like the ribosomes of yeast (Link et al. 1999) or the yeast proteome (Washburn et al. 2001), requires resolution capabilities beyond that of LC-MS/MS. 2D-PAGE followed by MS is the most commonly used method for resolving and identifying proteins from such highly complex mixtures (Hanash 2000; Pandey and Mann 2000). 2D-PAGE (see Chapter 4) separates proteins in one dimension by their pI values and in the other dimension by their molecular masses. However, proteins with extremes in pI and molecular mass (Corthals et al. 2000; Oh-Ishi et al. 2000), low-abundance proteins (Fountoulakis et al. 1999a,b; Gygi et al. 2000b), and membrane-associated or bound proteins (Molloy 2000; Santoni et al. 2000) are rarely identified in a 2D-PAGE study.

Like 2D-PAGE, an alternative 2D separation system must subject proteins or peptides to two independent separation modalities and must during the second dimension maintain the separation of any two components resolved during the first dimension (Giddings 1987). A variety of efforts are under way to utilize multidimensional chromatography coupled with MS to characterize proteomes (Nilsson and Davidsson 2000; Washburn and Yates 2000). Link et al. (1999) developed an on-line method called multidimensional protein identification technology (MuDPIT), which couples 2D-LC to MS/MS, to resolve and identify peptides from complex mixtures. In this method, a pulled capillary microcolumn is packed with two independent chromatography phases, a strong cation exchanger, and reversed-phase matrix material. Once a complex peptide mixture is loaded onto the system, no additional sample handling is required, because as the peptides elute from the column, they are directed into the ESI ion-trap mass spectrometer, where they are ionized, mass selected, and fragmented. Finally, advanced search algorithms match the fragmented peptides to their respective proteins in a database. Using MuDPIT, 1484 proteins and 5540 peptides from the *S. cerevisiae* proteome were resolved and identified, including low-abundance proteins such as protein kinases and transcription factors, proteins with extremes in pI and molecular mass, and integral membrane proteins (Washburn et al. 2001). For an alternative method for performing 2D-LC-MS, see Appendix 2.

This protocol, which was kindly provided by David M. Schieltz and Michael P. Washburn (The Torrey Mesa Research Institute, San Diego, California), describes the analysis of a proteome using MuDPIT.

MATERIALS

CAUTION: See Appendix 3 for appropriate handling of materials marked with <!>.

Reagents

Acetonitrile (5%)/0.5% acetic acid <!>

Ammonium bicarbonate (1 M) <!>

This solution should be freshly made each time.

$CaCl_2$ <!>
Chromatography solvents
- Solvent A: 5% acetonitrile/0.02% heptafluorobutyric acid (HFBA, Pierce) <!>
- Solvent B: 80% acetonitrile/0.02% HFBA
- Solvent C: 250 mM ammonium acetate<!>/5% acetonitrile/0.02% HFBA
- Solvent D: 500 mM ammonium acetate/5% acetonitrile/0.02% HFBA

Dithiothreitol (DTT) <!>
Endoproteinase Lys-C (sequencing grade)
Iodoacetamide (IAA) <!>
Modified trypsin (sequencing grade) or Poroszyme-immobilized trypsin (Applied Biosystems)
  See Step 7.
Urea <!>

Equipment

C_{18} solid-phase extraction pipette tips (e.g., SPEC Plus PT C18, ANSYS Technologies, Lake Forest, California)
  These C_{18} solid-phase disk pipette tips have a 0.4-μg sorbent capacity and a loading volume of up to 800 μl.
Gold wire (0.025 diameter) (Scientific Instrument Services, Inc., Ringoes, New Jersey)
High-performance liquid chromatography equipment
- Quaternary HPLC pump (e.g., a quaternary Hewlett-Packard 1100 series)
- Fused-silica capillary (100 μm I.D. x 365 μm O.D.)
  The capillary can be obtained from Agilent Technologies or Polymicro Technologies.
- Laser puller (e.g., P-2000, Sutter Instruments)
- Stainless steel pressurization bomb (The Scripps Research Institute and Cytopea, Inc.)
- Helium tank with regulator (at least 1000 psi pressure) <!>
- C_{18} reversed-phase packing material (5 μm) (e.g., Zorbax XDB, Agilent Technologies)
  CAUTION: Do not inhale; use in a chemical fume hood.
- Strong cation exchange resin (e.g., PartiSphere SCX, Whatman)

Nano-LC ion sources (ThermoFinnigan, Scripps Research Institute, and Cytopea, Inc.)
PEEK MicroCross, Microtight tubing sleeves (Upchurch Scientific)
Tandem mass spectrometer
  A variety of mass spectrometers are suitable, including the LCQ Classic, Deca, Duo, or TSQ series (ThermoFinnigan), or the QTOF1 or QTOF2 (Micromass, Inc.)
Water bath preset to 50°C

Biological Sample

Soluble protein fractions
  If proteins found in insoluble fractions are to be analyzed using MuDPIT, begin with the ADDITIONAL PROTOCOL: DIGESTION OF INSOLUBLE PROTEIN FRACTIONS FOR MUDPIT ANALYSIS.

Additional Reagents

Step 1 of this protocol requires the reagents listed in Appendix 2, Techniques 1 or 2.

METHOD

Digestion of Soluble Protein Extracts for MuDPIT

1. Adjust the pH of the protein extract to 8.5 with 1 M ammonium bicarbonate. Determine the protein concentration.

   The concentration of the protein mixture is needed for future reference and to determine the amount of protease to add in Steps 5 and 7.

2. Add solid urea to the protein solution to a final concentration of 8 M.

3. Add DTT to the protein solution to a final concentration of 1 mM. Incubate the solution for 20 minutes at 50°C.

4. Allow the solution to cool to room temperature. Add iodoacetamide to the denatured protein to a concentration of 10 mM. Incubate the solution for 20 minutes at room temperature in the dark.

5. Add endoproteinase Lys-C to a final ratio of substrate to enzyme of 100:1 and incubate the reaction overnight at 37°C.

6. Dilute the sample to 2 M urea with 100 mM ammonium bicarbonate (pH 8.5). Add $CaCl_2$ to a final concentration of 1 mM.

7. Digest the protein mixture with trypsin using either of the following two methods:

   For Method A:

   a. Add modified trypsin to the protein solution to a final ratio of substrate to enzyme of 50:1. Incubate the reaction overnight at 37°C.

   b. Centrifuge mixture in a microfuge at maximum speed to remove insoluble material.

   c. Transfer the supernatant to a fresh microfuge tube.

   For Method B:

   a. Add ~1 µl of Porosyzme immobilized trypsin slurry to every 50 µg of protein starting material. Incubate the reaction overnight at 37°C on a rotating wheel.

   b. Centrifuge the reaction in a microfuge at maximum speed to remove insoluble material and the trypsin beads.

   c. Transfer the supernatant to a fresh microfuge tube.

      Method B has the advantage that there is less trypsin contaminating the protein sample during the subsequent analysis.

8. Load the supernatant onto a SPEC Plus PT C_{18} solid-phase extraction cartridge according to the manufacturer's instructions. Exchange the supernatant into 5% acetonitrile/0.5% acetic acid.

   This essential step removes buffer components from the complex peptide mixture (i.e., ammonium bicarbonate, urea, etc.) that are incompatible with ESI-MS/MS. Removal of salt from the sample is critical, because salt prevents the peptides from binding to the strong cation-exchange portion of the 2D column. The peptides will bypass the ion-exchange resin and bind to the reversed-phase material, effectively reducing the 2D-LC to a single-dimension reversed-phase column. Salts are used to bump populations of peptides from the cation exchanger to the reversed-phase resin in a controlled manner. A sample containing 1 M salt will fail in a MuDPIT analysis. In addition, the extraction cartridge concentrates the sample, which is important because it can take ~1.5 hours to load 15 μl onto the 2D microcolumn.

MuDPIT

To carry out MuDPIT, a system must be assembled with a tandem mass spectrometer and a quaternary HPLC pump. In our case, a quaternary Agilent HP1100 series HPLC is directly coupled to a ThermoFinnigan LCQ Deca ion-trap mass spectrometer equipped with a nano-LC-ESI source (Gatlin et al. 1998).

9. Prepare the column as described in Protocol 5. In summary:

   a. Pull a fused-silica capillary microcolumn (100 μm I.D. x 365 μm O.D.) with a P-2000 laser puller.

   b. Pack the microcolumn with 10 cm of C_{18} reversed-phase material in methanol.

   c. Pack the column with 4 cm of 5 μm strong cation-exchange material in methanol.

10. Connect the column to an Upchurch PEEK MicroCross (for details, see Protocol 5, Step 4, and Figure 8.41) to which is also attached

    - a gold wire to supply a spray voltage of ~1800 V,
    - a microcapillary split line to provide an effective flow rate of 100–300 nl/min, and
    - a solvent line from the HPLC.

    Measure the flow from the tip of the microcolumn by following Protocol 5, Step 6.

11. Equilibrate the column with 100% Solvent A for 5 minutes at a flow rate of 150 μl/min at the pump.

12. Follow Protocol 5, Steps 12–17 to load the sample onto the column.

    Because MuDPIT columns are longer than single-phase LC microcolumns, a pressure of ~1000 psi is required to load the sample.

13. Place the MicroCross with the connections into a stage, as per Protocol 5, Steps 18 and 19, and Figure 8.42.

14. Begin the MuDPIT analysis of the peptide sample.

    The key feature of MuDPIT is the 2D capillary microcolumn chromatography, which performs better than when the resins are in separate columns and the buffers shunted through the columns. By having the beds integrated into the same column, the sensitivity is much higher and the sample losses are much fewer. The chromatography is controlled by the mass spectrometer itself, through the use of Xcalibur software. A typical analysis is a fully automated, 15-step chromatography run on highly complex mixtures, but any number of steps can be applied to a column. The gradient of each individual step is built within the Instrument Setup portion of Xcalibur (refer to Protocol 5, Steps 20 and 21). The following is a typical 15-step setup:

Step 1

70-minute gradient from 0% to 80% Solvent B
10 minutes held at 80% Solvent B

Steps 2–13

Each of these steps follows the sequence:
5 minutes of 100% Solvent A
2 minutes of x% Solvent C*
3 minutes of 100% Solvent A
10-minute gradient from 0% to 10% Solvent B
90-minute gradient from 10% to 45% Solvent B

*The 2-minute Solvent C percentages (x) in Steps 2–13 are as follows: 10%, 20%, 30%, 40%, 50%, 60%, 70%, 80%, 90%, 90%, 100%, and 100%, respectively.

Step 14

5-minute 100% Solvent A wash
20-minute 100% Solvent C wash
5-minute 100% Solvent A wash
10-minute gradient from 0% to 10% Solvent B
90-minute gradient from 10% to 45% Solvent B

Step 15

5-minute 100% Solvent A wash
20-minute 100% Solvent D wash
5-minute 100% Solvent A wash
10-minute gradient from 0% to 10% Solvent B
90-minute gradient from 10% to 45% Solvent B

Also within the Instrument Setup of Xcalibur and along with the building of an individual step are built the MS and MS/MS acquisition rules (see Steps 20 and 21 in Protocol 5). Typically, the instrument is operated here in a manner very similar to that in Steps 20 and 21, Protocol 5; however, instead of the top two ions being selected for MS/MS, the top three ions are selected during MuDPIT analysis.

SEQUEST Analysis

The SEQUEST (Eng et al. 1994) output from a MuDPIT run is very complex, and substantial computational resources are required to generate and analyze this output. Due to the large number of spectra typically generated from a MuDPIT run, MS/MS data must be analyzed in a conservative fashion. The SEQUEST algorithm is run on any data set generated against the pertinent protein, DNA, or EST database, depending on the organism analyzed. When cyanogen bromide is used, the MS/MS data resulting from the two samples treated with cyanogen bromide/formic acid must be independently analyzed twice with SEQUEST because cyanogen bromide cleaves at methionine residues and leaves either homoserine or homoserine lactone (Aitken et al. 1989). For each of the two runs, engage the differential search modification and set to either –30 for homoserine or –48 for homoserine lactone. SEQUEST results are evaluated using the criteria described in the panel on DATABASE SEARCHING in Protocol 5.

Additional Protocol: Digestion of Insoluble Protein Fractions for MuDPIT Analysis

Proteins associated with insoluble particulate matter from a whole-cell lysate or carefully prepared membrane samples can also be analyzed using MuDPIT. However, preliminary steps are required to prepare a usable complex protein mixture from these types of samples.

Additional Materials

CAUTION: See Appendix 3 for appropriate handling of materials marked with <!>.

Ammonium bicarbonate, saturated <!>
Cyanogen bromide <!>
Formic acid (90%) <!>

WARNING: These steps MUST be carried out in a chemical fume hood exercising proper safety precautions.

Method

1. Lyophilize the insoluble fraction.
2. Add 100 μl of 90% formic acid to the lyophilized pellet, and incubate the tube for 5 minutes at room temperature.
3. Add 100 μg of cyanogen bromide to the protein, mix the solution well, and incubate it overnight at room temperature in the dark (wrap the tube in aluminum foil).
4. Add saturated ammonium bicarbonate to the tube until the pH reaches 8.5. Add the ammonium bicarbonate solution very slowly, because a large amount of bubbling occurs during this step. Use pH paper to check the pH of the sample.
5. Concentrate the fraction to ~200 μl by lyophilization in a vacuum concentrator.
6. Proceed with Step 2 in the main protocol.

Analysis of the *S. cerevisiae* Proteome Using MuDPIT

The genome of *S. cerevisiae* is arguably the most highly characterized of any organism. Ongoing efforts in the *S. cerevisiae* community include developing extensive on-line databases that provide functional information concerning the proteins of *S. cerevisiae* (Mewes et al. 2000; Ball et al. 2001; Costanzo et al. 2001b). Because of these resources, the results of a MuDPIT analysis of the *S. cerevisiae* proteome could be characterized (Washburn et al. 2001).

1. Strain BJ5460 (Jones 1991) was grown to mid-log phase (O.D. 0.6) in YPD at 30°C and the pellets were washed three times with 1x phosphate-buffered saline (1.4 mM NaCl, 0.27 mM KCl, 1 mM Na_2HPO_4, 0.18 mM KH_2PO_4 at pH 7.4).
2. Three unique fractions were generated by lysing two separate groups of cells in a Mini-BeadBeater (BioSpec Products, Bartlesville, Oklahoma) followed by centrifugation. Three cycles of lysis and supernatant removal were carried out. After concentration by lyophilization to ~200 μl, the supernatant was digested and prepared for MuDPIT analysis as described in Protocol 6, Steps 1–7.
3. The remaining two pellets were washed by the addition of 1x phosphate-buffered saline to the tube, vortexed for 2 minutes, and pelleted by centrifugation at 14,000 rpm in a microfuge for 10 minutes. One pellet (named the lightly washed insoluble pellet) was washed once, and a second pellet (named the heavily washed insoluble pellet) was washed three times in this fashion. Each sample was lyophilized to dryness and prepared for MuDPIT analysis as described in the ADDITIONAL PROTOCOL: DIGESTION OF INSOLUBLE PROTEIN FRACTIONS FOR MUDPIT ANALYSIS.
4. Each of the three fractions was then analyzed separately by MuDPIT as described in Protocol 6.

The analysis of the yeast proteome by MuDPIT is the most comprehensive proteome analysis to date. After combining the MS/MS data generated from all three samples, 5540 peptides were assigned to MS spectra leading to the identification of 1484 proteins from the *S. cerevisiae* proteome. Each of the three prepared fractions provided unique hits to the final data set. After analysis of the data set using on-line *S. cerevisiae* databases (Mewes et al. 2000; Ball et al. 2001; Costanzo et al. 2001b), MuDPIT proved to be largely unbiased, because proteins predicted to be of low abundance and proteins with extremes in pI, molecular mass, and hydrophobicity were detected and identified. Of particular note was the detection and identification of 131 proteins with three or more predicted transmembrane domains, many of which were known integral membrane proteins (Washburn et al. 2001). These results demonstrated the potential for MuDPIT to rapidly detect and identify thousands of proteins from an organism in a largely unbiased fashion.

PROTOCOL 7

Combining 2D Chromatography and MS for the Separation of Complex Peptide Mixtures: An Off-line Approach

THE COMBINATION OF MULTIDIMENSIONAL CHROMATOGRAPHY for peptide separation and MS/MS for further separation and peptide sequence analysis represents an important tool for proteome analysis. The following protocol is based on the separation and analysis of 1 mg of soluble yeast protein, but the method can be optimized for the separation of any complex protein mixture. This protocol was kindly provided by Steven P. Gygi and his colleagues, Melanie P. Gygi, Larry J. Licklider, and Junmin Peng (Harvard Medical School, Boston, Massachusetts).

MATERIALS

CAUTION: See Appendix 3 for appropriate handling of materials marked with <!>.

Reagents

$CaCl_2$ <!>

Denaturing solution

- 8 M urea <!>
- 0.2% SDS <!>
- 10 mM dithiothreitol (DTT) <!>
- 10 mM Tris-Cl (pH 8.5)

  For resolubilization, the SDS is optional. It is removed in the void volume of the SCX chromatography column. Its final concentration in the protein solution must be below 0.1% for effective trypsin digestion (see Steps 5–7).

Dialysis solution

- 2 M urea
- 10 mM Tris-Cl (pH 8.5)

Iodoacetamide (IAA) <!>

RP chromatography solvents

- Solvent A
  - 0.4% acetic acid <!>
  - 0.005% heptafluorobutyric acid (HFBA) <!>
  - 5% acetonitrile <!>
- Solvent B
  - 0.4% acetic acid
  - 0.005% HFBA

SCX chromatography buffers
- Buffer A
  - 5 mM KH_2PO_4 <!>
  - 25% acetonitrile (pH 2.7)
- Buffer B
  - 5 mM KH_2PO_4
  - 25% acetonitrile (pH 2.7)
  - 350 mM KCl <!>
  - 95% acetonitrile

Trifluoroacetic acid (TFA) <!>

Equipment

Dialysis tubing

Gold wire

LCQ DECA ion-trap mass spectrometer (ThermoFinnigan, San Jose, California) or equivalent

MicroCross (Upchurch Scientific)

Restrictor capillary for flow splitting (50 µm I.D. x 50 cm)

RP capillary chromatography column

This column is an in-house-packed 75 µM x 12-cm fused-silica capillary column packed with Magic C_{18} (5 µM, 200 Å, Michrom BioResources, Auburn, California).
See Steps 15 and 16.

SCX chromatography column

This column is a commercially packed 2.1 x 150-mm polysulfoethyl aspartamide (PolyLC, Columbia, Maryland) column and an in-line guard column of the same material.

SEQUEST software (ThermoFinnigan, San Jose, California)

This algorithm can be utilized to match acquired tandem mass spectra with peptide sequences from databases.

Biological Molecules

Peptide standards (Sigma)

Peptide standards are used to evaluate the performance of the SCX column. The following six peptides were individually dissolved in H_2O, mixed together, and diluted to a concentration of 1 pmole peptide/µl in SCX chromatography Buffer A. The predicted net charge state of each peptide at pH 2.7 is shown in parentheses.

- angiotensin I (4+)
- bradykinin (3+)
- neurotensin (3+)
- bradykinin fragment 1–5 (2+)
- leucine enkephalinamide (2+)
- leucine enkephalin (1+)

Soluble yeast protein or other complex protein mixture

Soluble yeast protein can be obtained as described by Garrels et al. (1994).

Trypsin, TPCK-modified sequencing grade (Promega)

METHOD

Proteolysis of Soluble Yeast Protein

1. Solubilize 1 mg of lyophilized soluble yeast protein for 15 minutes at 37°C in 0.5 ml of denaturing solution.
2. Add IAA to a final concentration of 50 mM and place the sample in the dark for 30 minutes at room temperature to alkylate the cysteine residues.
3. Dialyze the solution against 100 ml of dialysis solution for 1 hour at room temperature to remove salts and other low-molecular-weight material.
4. Repeat Step 3 twice with fresh changes of dialysis solution.
5. Add 20 µg of trypsin to the sample, and incubate it for 10 minutes at 37°C.
6. Dilute the sample twofold (to 1.0 ml) with 10 mM Tris-Cl (pH 8.5). Add $CaCl_2$ to a final concentration of 1 mM. Use a strip of pH paper to verify that the pH of the sample is between 8 and 9.

   If SDS was included in the denaturation solution (Step 1), its final concentration in the protein solution must be below 0.1% for effective trypsin digestion.

7. Allow the trypsin to digest the protein sample overnight at 37°C.
8. Stop the digestion by acidifying the sample with 2 µl of TFA. Use a strip of pH paper to verify that the pH of the solution is <3.

   Alternative methods for solubilizing (Step 1) and proteolyzing proteins may be used (e.g., omitting SDS, using a different reducing reagent, and omitting dialysis if salt concentration is low). However, it is critical that the final salt concentration be <20 mM (this includes Tris-Cl and any salt from protein harvesting) and the solution pH be ~3 immediately prior to peptide separation by SCX chromatography. If the salt concentration is greater than 20 mM, use either membrane dialysis (see Steps 3 and 4) or a C_{18} solid-phase extraction (SPE) column (Vydac) according to the manufacturer's instructions for removal of salt.

Preparation of SCX Chromatography Column

9. Set up the HPLC with the SCX buffers and SCX column.
10. Set the flow rate of the column to 200 µl/min, and acquire a blank gradient.

TABLE 8.12. Shallow gradient for SCX chromatography of tryptic peptides

Time (min)	%B	Flow (µl/min)
0	0	200
1	5	200
80	35	200
90	100	200
100	100	300
110	0	300
120	0	200

Conditions: 2.1 x 200-mm SCX column; Buffer A was 5 mM KH_2PO_4, 25% acetonitrile (pH 2.7); Buffer B was 5 mM KH_2PO_4, 25% acetonitrile, 350 mM KCl (pH 2.7).

The gradient for peptide elution should be shallow in the area from ~5% to 35% Buffer B (see Table 8.12). This permits superior separation of the peptides with a net charge of 2^+ (which is most tryptic peptides) than does a linear gradient.

11. Test the performance of the SCX column.

    a. Load 200 pmoles of each standard peptide in Buffer A onto the SCX column.

    b. Analyze the peptides using a UV spectrophotometer with the absorbance set at 214 nm.

       Six completely resolved chromatographic peaks should be observed with peak widths of <1 minute.

SCX Chromatography of Yeast Peptides

12. Load the acidified yeast peptide sample onto the column. Wash the column for 5–10 minutes with Buffer A until the UV trace returns to baseline.

13. Start the gradient and collect 200-µl fractions per minute.

    The peptide elution profile for 1 mg of yeast separated by this technique is shown in Figure 8.44. It is possible to load and separate as much as 5 mg of total peptides by this technique. Other column sizes either larger or smaller can be used depending on the amount of starting material.

14. Reduce the fraction volumes to ~50–100 µl by centrifugal evaporation.

    This step also removes most of the acetonitrile, permitting peptides to adsorb to the reversed-phase chromatography column.

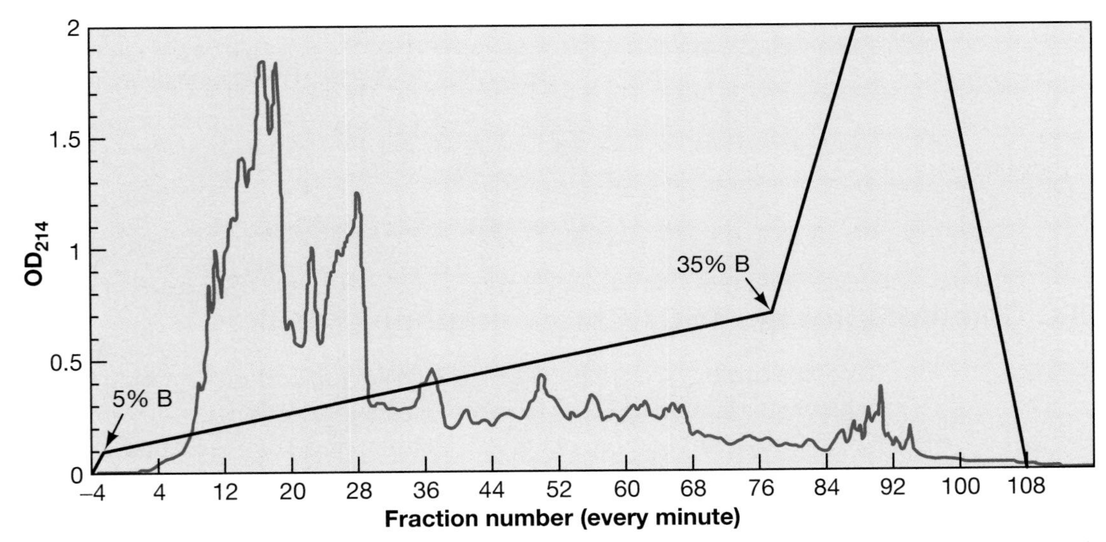

FIGURE 8.44. Separation of 1 mg of yeast peptides by strong cation exchange (SCX) chromatography. Soluble yeast protein (1 mg) was harvested, reduced, denatured, and proteolytically cleaved with trypsin as described in this protocol. Fractions (200 µl) were collected every minute into a 96-well plate. The elution salt gradient has been superimposed and is from Table 8.12. Specific fractions (16, 32, 48, and 64) were later examined by reversed-phase nanoscale microcapillary LC-MS/MS for peptide identification (Table 8.13). Conditions: 2.1 x 200-mm SCX chromatography column, 200 µl/min flow rate, monitored UV absorbance at 214 nm.

TABLE 8.13. Peptide sequence analysis of four SCX fractions by nanoscale reversed-phase capillary LC-MS/MS and database searching

Fraction number[a]	Five representative peptides identified from each fraction[b]	Peptide charge state (mean ± S.D.)[c]	% Peptides identified with charges (2:3:4:5)
16	(2) R. .WAGNANELNAAYAADGYA**R**. .I (2) K. .NDLESYLNQAVEVS**R**. .D (2) R. .SNYDTDVFTPLFE**R**. .I (2) R. .SSVLADALNAINNAE**K**. .T (2) R. .NPSDITQEEYNAFY**K**. .S	2.0±0.10	99:1:0:0
32	(3) K. .LPLVGG**H**EGAGVVVGMGENV**K**. .G (3) R. .FPSNGAFAEYSAISSETAY**K**PA**R**. .E (3) K. .IQEIQDAV**H**NQALQLP**R**. .V (3) K. .ATDGGA**H**GVINVSVSEAAIEAST**R**. .Y (3) K. .ALENPT**R**PFLAILGGA**K**. .V	3.0±0.00	0:100:0:0
48	(3) K. .**K**PQVTVGAQNAYL**K**. .A (3) R. .**H**LNDQPNADIVTIGD**K**. .I (3) K. .S**H**VVDYDDSSITENT**R**. .C (3) R. .N**K**DDVIAQNLIES**K**. .Y (3) R. .**H**ALALSPLGAEGE**R**. .K	3.02±0.15	0:98:2:0
64	(3) K. .**H**SNGDAWVEA**R**. .G (4) R. .YGANP**H**Q**K**PAQAYVSQQDSLPF**K**. .V (4) R. .IEEELGDNAVFAGENF**HH**GD**K**L. .– (4) K. .STSGNT**H**LGGQDFDTNLLE**H**F**K**. .A (4) K. .LNS**K**PVEL**K**PVSLDN**K**. .T	3.92±0.31	0:9:90:1

[a]Fraction numbers correspond to those in Figure 8.44.

[b]Peptides were identified by database matching using the SEQUEST algorithm against the yeast protein database with no enzyme specified. Matches were considered significant if the Xcorr score was >3.0, and the peptide was fully tryptic. Hundreds of peptides were automatically identified from each fraction. Only five representative peptides are shown here for method instruction. The predicted solution charge state at pH 2.7 is shown in parentheses and was calculated using the amino acids underlined and the amino terminus. For thoroughness, the amino acid both immediately before and after the identified peptide is also shown.

[c]Mean (positive) solution charge state was calculated for each peptide at pH 2.7, with each occurrence of histidine, arginine, and lysine providing one charge and the amino terminus always providing an additional charge.

RPC Chromatography of SCX Chromatography Fractions

For maximum sensitivity and peptide identification, combine nanoscale reversed-phase chromatography capillary LC with on-line detection via an ion-trap mass spectrometer. Using this combination, peptide detection levels are ~1–5 fmoles, with tandem mass spectra (sequencing attempts) being acquired at a rate of more than 2000 spectra/hour.

15. Pull a 20-cm-long piece of 75-μM I.D. fused-silica capillary to form a needle tip.

    There are many ways to pull an electrospray needle. The simplest is to attach a weight (paper-binder) to the needle and then use a microtorch flame to stretch the capillary to a point. Alternatively, the procedure using a laser puller described in Protocol 5 is effective. In addition, the pulled tips or the pre-poured columns may be purchased (e.g., New Objective, Cambridge, Massachusetts).

16. Pack a piece of 75-μM fused-silica capillary that has a flame-drawn needle tip with a bed of 5 μM, 200 Å C_{18} particles to a length of 12 cm. (For details on packing the capillary column, see Protocol 4 or 5.)

    The pulled tip serves to contain the packed bed in the capillary and functions as an electrospray needle for MS.

17. Connect the four arms of a microcross as follows:

    a. A flow from the HPLC (100 μl/min).

    b. A restrictor capillary for flow splitting (50 μm I.D. x 50 cm) 99.7 μl/min.

    c. A gold wire connected to a high-voltage supply (1.8 kV).

    d. A nano-scale reversed-phase capillary column with pulled tip (from Step 16) receiving 300 nl/min.

18. Operate the HPLC at 100 μl/min and split down the flow to only 300 nl/min using the microcross before the reversed-phase column.

19. Pressure-load off-line 10 μl (or more) from a single 50-μl SCX chromatography fraction onto the column, followed by gradient elution.

20. Subject eluting peptide peaks to ESI and place them in the ion-trap mass spectrometer.

    The MS operates in two modes, namely, MS mode (measuring the *m/z* ratios of peptide ions) and MS/MS mode (sequencing peptides).

    Thousands of sequencing attempts can be made in a single 1-hour reversed-phase LC-MS/MS analysis by acquiring the MS/MS scans in a data-dependent fashion, where the peptide ions to be sequenced from each MS scan are chosen "on the fly" and five peptide ions are selected for sequencing (fragmentation) from each MS spectrum.

The setup described above can be further automated by using an autosampler to inject the SCX chromatography fractions directly from the 96-well collection plate. A peptide trap (LC Packings or Michrom BioResources) can also be incorporated into the strategy before the resolving reversed-phase chromatography capillary column to allow for increased sample volumes and improved concentration and desalting of samples. In addition, we have designed a high-sensitivity method for the automated analysis by nanoscale capillary LC-MS/MS directly from 96-well plates utilizing a device termed a "V-column," or "vented column," which is derived from the direct fusion of a peptide trap and a nanoscale capillary column (Licklider et al. 2002).

The separation power of the technique can be appreciated by the total number of sequencing attempts that can be obtained from the original peptide sample. When 20 fractions are analyzed, more than 40,000 peptides could potentially be sequenced. Furthermore, the length of the analyses in both dimensions are highly scalable, which can result in an increase in total sequencing attempts of perhaps an order of magnitude.

PROTOCOL 8

Cysteinyl Peptide Capture of Proteins

ANALYZING AN UNKNOWN PROTEOLYTIC MIXTURE by LC-MS/MS can be difficult because of the potential complexity of the mixture as well as the dynamic range of proteins that are represented. Peptides from the most-abundant proteins may be repeatedly selected by the mass spectrometer for fragmentation, thereby preventing the identification of peptides from less-abundant proteins. One strategy to aid in the identification of less-abundant proteins from a complex proteolytic mixture is to fractionate the mixture prior to its analysis in a mass spectrometer.

Cysteine represents a mere 1.7% of all amino acids in proteins (Coligan et al. 1999). By affinity tagging cysteine residues prior to digesting the protein mixture, the resulting cysteinyl peptides can be selectively captured by affinity chromatography. The goal is to capture one cysteinyl peptide from each protein in the mixture, due to the low abundance of cysteine within proteins. The fraction containing cysteinyl peptides can then be analyzed by LC-MS/MS and should result in the identification of low-abundance proteins that would ordinarily not be selected for fragmentation within the mass spectrometer.

Recently, sulfhydryl-specific reagents have been used for simplification of complex peptide mixtures (Spahr et al. 2000) as well as for quantitative analysis (Gygi et al. 1999). The sulfhydryl-specific reagent used in this protocol, Biotin-HPDP (*N*-[6-(biotinamido)hexyl]-3´-[2´-pyridyldithio] propionamide; Pierce Chemical Company), contains a sulfhydryl-reactive moiety separated from a biotin moiety by a thiol-cleavable spacer arm. Reduction of the biotinylated cysteinyl peptides bound to immobilized avidin permits isolation of the cysteinyl peptides in their native state.

This protocol works well with complex protein mixtures across a broad dynamic range even in the presence of undesirable buffer components. However, to optimize the results, the protocol must be adapted for specific applications. For the analysis of relatively simple proteolytic mixtures by MS (especially where identifying components is the goal), it may be more suitable to rely on the data-dependent functions of the mass spectrometer for fragmenting as many peptides as possible, rather than using the cysteinyl peptide capture strategy. For simple mixtures, some of the mass spectrometer software utilities such as "dynamic exclusion" should be sufficient for looking deeper into a sample. Dynamic exclusion puts an ion that has been previously fragmented into a "don't fragment" list for a user-defined period of time, thereby allowing the selection of less-intense precursor ions. This protocol was provided by Chris Spahr (Amgen Inc., Thousand Oaks, California).

MATERIALS

CAUTION: See Appendix 3 for appropriate handling of materials marked with <!>.

Reagents

Biotinylation reagent (EZ-Link Biotin-HPDP from Pierce)

Prepare a 4 mM Biotin-HPDP stock solution by adding 2.2 mg of Biotin-HPDP to 1 ml of dimethyl sulfoxide (DMSO) <!>. This stock can be stored frozen.

Biotinylation/tryptic digestion buffer (PBS [pH 7.4] containing 1 mM EDTA and 2 M urea)

Prepare a 100 mM EDTA stock solution by dissolving 0.186 g of EDTA in 5 ml of H_2O. Combine 2.4 g of urea <!> with 200 µl of the EDTA stock solution, and add 1x PBS to a final volume of 20 ml.

Denaturation buffer (40 mM Tris-Cl [pH 7.0] containing 8 M urea)

Combine 2.4 g of urea with 200 µl of 1 M Tris-Cl (pH 7.0) stock. Add H_2O to give a final volume of 5 ml.

Denaturation/reducing buffer

Just prior to use, add 10 µl of 1 M dithiothreitol <!> to 1 ml of denaturation buffer.

Dithiothreitol (DTT) <!>

Formic acid <!>

Iodoacetamide (IAA, an alkylating reagent) <!>

Prepare a 1 mM stock solution by adding 0.184 g of IAA to 1 ml of H_2O.

PBS containing 1 mM EDTA

Phosphate-buffered saline (PBS)

Trypsin (sequencing-grade, from Roche Molecular Biochemicals)

Prepare a 1 µg/µl stock solution by diluting the dry enzyme in H_2O. The stock should be stored frozen at 80°C.

Equipment

Centricon YM-3 (3000-m.w. cutoff; Millipore)

Centrifuge (refrigerated) (Beckman GS-6R centrifuge or equivalent)

ImmunoPure immobilized avidin AffinityPak columns (1-ml prepacked columns, Pierce Endogen)

Oasis HLB cartridge

Use 1-cc (30 mg) extraction cartridge (Waters, Milford, Massachusetts) or equivalent desalting device.

Vacuum concentrator (e.g., SpeedVac, Thermo Savant)

Biological Sample

Complex protein mixture

This protocol is based on the analysis of protein released from mouse liver mitochondria following treatment with atractyloside (Patterson et al. 2000). Protein (120 µg) was used for the cysteinyl peptide capture. Although varying amounts of a complex protein mixture can be used in this protocol, the limiting factor may be the binding capacity of the immobilized avidin column. The amounts of denaturant/reductant, biotin, trypsin, and avidin should be optimized for each protein sample.

METHOD

Denaturation, Reduction, and Buffer Exchange

1. Concentrate 120 µg of the complex protein mixture under vacuum in a vacuum concentrator until the sample is completely dry.

2. Resuspend the dried sample in ~1 ml of denaturation/reducing buffer. Denature and reduce the sample by incubating it for 30 minutes at 37°C.

   The precise volume is unimportant, because the sample buffer will be exchanged in the next step. The smaller the sample volume, however, the less time it will take to exchange the buffer.

3. To remove the reductant, decrease the urea concentration, and remove any undesirable buffer components. Exchange the sample buffer with ~4 ml of biotinylation/tryptic digestion buffer using a Centricon YM-3 concentrator at 4°C in a centrifuge.

Labeling with Biotin-HPDP and Buffer Exchange

4. To label all of the free sulfhydryls, add 50 µl of 4 mM Biotin-HPDP in DMSO to the sample (the volume of the sample from Step 3 was ~1 ml) and incubate the labeling mixture for 90 minutes at room temperature.

   An excess of Biotin-HPDP was used because of the uncertainty as to whether or not the labeling reaction would work in the presence of 2 M urea. This amount of Biotin-HPDP greatly exceeds the binding capacity of the immobilized avidin column used (theoretical binding capacity of 1 mg of avidin is 59 nmoles of biotin).

5. Remove all of the unbound Biotin-HPDP by exchanging the buffer with fresh biotinylation/tryptic digestion buffer at 4°C in a Centricon YM-3 concentrator.

   The concentrated sample volume should be ~100 µl.

   Unbound Biotin-HPDP will compete with the biotinylated peptides for binding to avidin. The Centricon YM-3 used in Step 3 can be used again in this step.

Digestion of the Labeled Complex Mixture

6. Add 1.5 µg of trypsin to the biotinylated sample and incubate the digest overnight at 37°C.

7. Dilute the biotinylated digest in PBS containing 1 mM EDTA to reduce the urea concentration to <0.5 M.

   If the digest is to be applied to the avidin affinity column within a few hours, it can be held at room temperature, 4°C, or frozen. For extended storage, freeze the peptides at –80°C.

Immobilized Avidin Affinity Capture and Elution

8. Equilibrate a 1-ml immobilized avidin column with 5 ml of PBS containing 1 mM EDTA.

9. Apply the biotinylated digest (from Step 7) to the column, close the column, and incubate the material for 30 minutes at room temperature. Do not allow the column to run dry.

10. To collect noncysteinyl peptides, wash the column with 10 ml of PBS containing 1 mM EDTA and collect the eluate. This yields the flowthrough fraction.

11. Elute the cysteinyl peptides from the immobilized avidin:

    a. Apply 2 ml of PBS containing 50 mM DTT (added just prior to use) to the column and allow it to enter the column.

    b. Close the column and incubate it for 30 minutes at room temperature.

    c. Open the column, and collect the eluate in 2-ml fractions. This yields the bound fraction.

Alkylation and Desalting

12. If LC-MS/MS analysis is desired, alkylate the flowthrough and bound fractions with IAA (10 mM and 100 mM, respectively) for 30 minutes at room temperature in the dark.

    The flowthrough fraction is alkylated in case any cysteinyl peptides made it into this fraction.

13. Acidify both fractions with 90% formic acid (add 20 µl to the bound fraction and 100 µl to the flowthrough fraction) to give a final concentration of 1% formic acid. Freeze the fractions at –80°C until they are to be desalted.

14. Desalt each fraction by passing the material through an Oasis HLB cartridge or equivalent device prior to LC-MS/MS according to the manufacturer's instructions.

PROTOCOL 9

Isotope-coded Affinity Tagging of Proteins

QUANTITATIVE PROTEOMICS HAS TRADITIONALLY BEEN PERFORMED using 2D gel electrophoresis where quantitation is accomplished by recreating differences in the staining patterns of proteins derived from two states of cell populations or tissues from a similar biological system. More recently, MS methods based on stable isotope quantitation have been developed that show significant potential for differential expression proteomic studies. One such in vitro method, described in this protocol, involves the use of isotope-coded affinity tags (ICATs). This approach, illustrated schematically in Figure 8.45, uses a protein tag with three functional moieties: a cysteine reactive moiety, a linker with either eight hydrogens (the *light* form of the reagent) or eight deuteriums (the *heavy* form of the reagent, having an isotope code or mass tag of 8 daltons), and a biotin moiety (the affinity tag). Using this technique, the cysteine side chains in complex mixtures of proteins from two different states of a cell population (e.g., normal vs. disease) are reduced and alkylated using the light form of the reagent (d0-labeled tag) in one cell state and the heavy form of the reagent (d8-labeled tag) for proteins in the second cell state. The two mixtures (d0- and d8-labeled) are then combined and subjected to proteolytic digestion (typically, with trypsin and/or Lys-C). Generated cysteine-containing peptides are affinity-purified using an avidin column resulting in a "simplified" mixture of peptides that contains about tenfold fewer peptides than the original mixture. These peptides are analyzed by MS, and quantitation information based on the relative abundance of the d0 and d8 isotopes is obtained. The identification of proteins is obtained from the peptide molecular mass and MS/MS-derived amino acid sequence. This protocol was kindly provided by Ruedi Aebersold, Timothy J. Griffin, and Sam Donohoe (The Institute for Systems Biology). Four stages comprise this protocol:

- ***Stage 1:*** Labeling Proteins with ICAT
- ***Stage 2:*** Cation Exchange Cleanup of ICAT Samples
- ***Stage 3:*** Selection of Tagged Proteins Using an Avidin Column
- ***Stage 4:*** Analysis of ICAT-labeled Peptides by MS

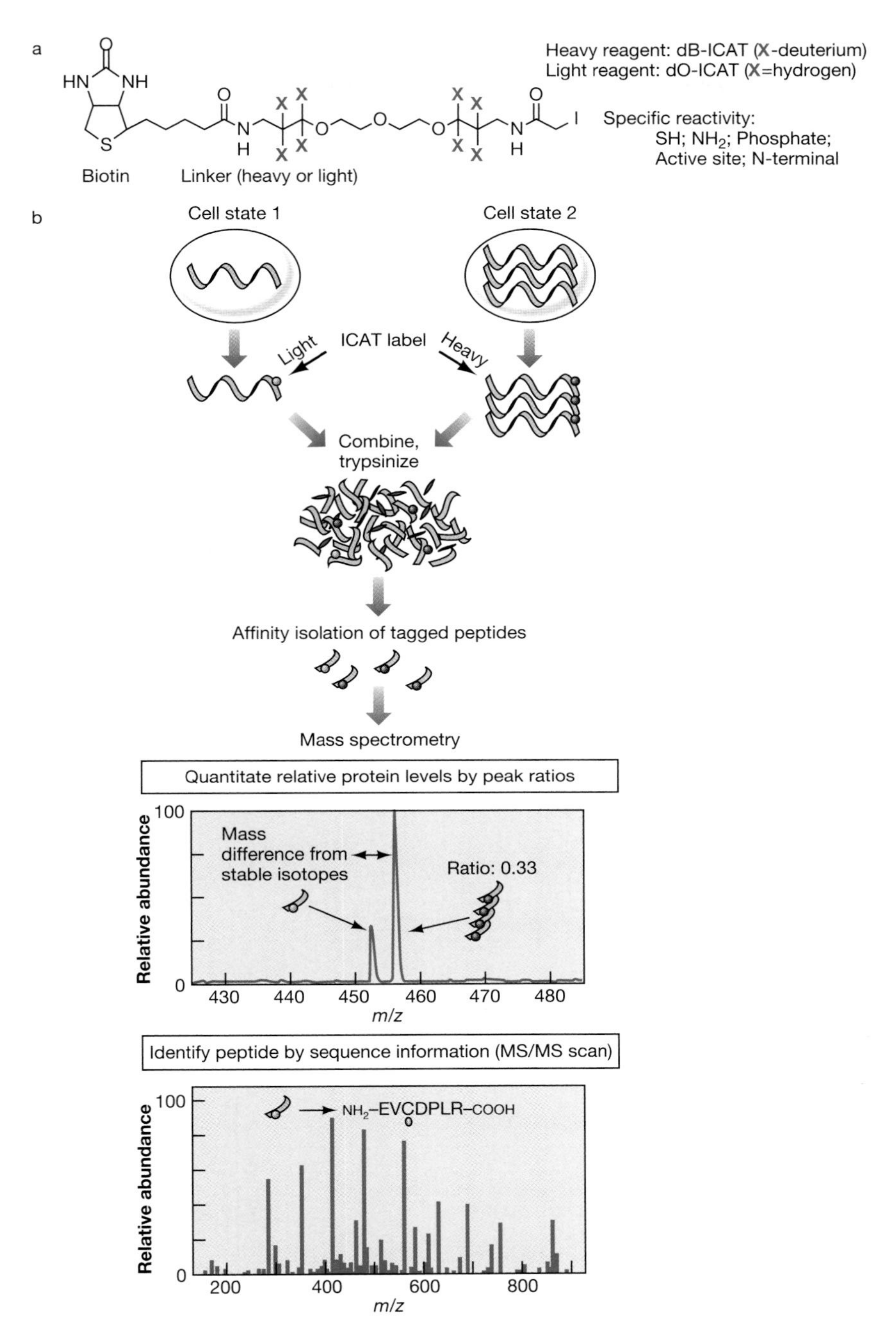

FIGURE 8.45. ICAT strategy for quantifying differential protein expression. (*a*) Structure of the ICAT reagent. The reagent consists of three elements: an affinity tag (biotin), which is used to isolate ICAT-labeled peptides; a linker that can incorporate stable isotopes; and a reactive group with specificity toward thiol groups (cysteines). The reagent exists in two forms: heavy (contains eight deuteriums) and light (contains no deuteriums). (*b*) Two protein mixtures representing two different cell states are treated with the isotopically light (*gray*) and heavy (*black circle*) ICAT reagents, respectively; an ICAT reagent is covalently attached to each cysteinyl residue in every protein. The protein mixtures are combined and digested to peptides, and ICAT-labeled peptides are isolated utilizing the biotin tag. These peptides are separated by microcapillary HPLC. A pair of ICAT-labeled peptides are chemically identical and are easily visualized because they essentially coelute, and there is an 8-dalton mass difference measured in a scanning mass spectrometer (four *m/z* units difference for a doubly charged ion). The ratios of the original amounts of proteins from the two cell states are strictly maintained in the peptide fragments. The relative quantification is determined by the ratio of the peptide pairs. Every other scan is devoted to fragmenting and then recording sequence information about an eluting peptide (tandem mass spectrum). The protein is identified, with the aid of a computer, by searching with the recorded sequence information against large protein databases. (Adapted, with permission, from Gygi et al. 1999.)

STAGE 1: Labeling Proteins with ICAT

MATERIALS

CAUTION: See Appendix 3 for appropriate handling of materials marked with <!>.

Reagents

Dithiothreitol (DTT) (1 M) <!>

EDTA (0.5 M)

ICAT reagents

The ICAT reagents are either deuterated ("heavy") or native ("light") and can either be purchased from Applied Biosystems or synthesized (see Gygi et al. 1999). Dissolve each ICAT reagent in methanol <!> prior to adding it to proteins in solution. The volume of methanol used to dissolve the reagent should be no more than 10% of the final volume of the proteins in solution.

Tributylphosphine (TBP) (Aldrich Chemical Company) <!>

If using small amounts of TBP, prepare a 1:10 dilution of TBP in 1-propanol (HPLC grade) <!>.

WARNING: TBP is extremely toxic. Always work in a chemical fume hood.

Tris(2-carboxyethyl) phosphine hydrochloride (TCEP) <!> can substitute for TBP as the reducing agent.

Tris (50 mM, pH 8.3) containing 0.5% SDS <!>, 5 mM EDTA, and 6 M urea <!>

Trypsin (Promega sequencing-grade modified trypsin or equivalent)

Prepare the trypsin at 1 µg/µl.

Biological Samples

Lyophilized or precipitated protein samples

Additional Reagents

Step 8, if performed, requires the reagents from Chapter 7, Protocol 11.

METHOD

- Be extremely careful with the protein concentrations throughout the experiment.
- Handle samples on ice unless otherwise noted.
- The pH should remain at ~8.3 during the labeling steps.
- Work with equivalent amounts of sample throughout this stage of the protocol.

1. Solubilize each protein pellet (300 µg) in 50 µl of 50 mM Tris (pH 8.3) containing 0.5% (w/v) SDS, 6 M urea, and 5 mM EDTA.

   The SDS is used to solubilize hydrophobic membrane proteins. If the sample does not include such proteins, omit the SDS.

   (*Optional*) Set aside ~1–2 µg of native/unlabeled protein to load on a gel when checking the efficiency of the labeling reaction (see Step 8).

2. When the proteins are in solution, dilute each sample 1:10 with H_2O, adding EDTA to keep it at 5 mM. In the end, 0.05% SDS, 5 mM Tris, and 5 mM EDTA are needed.

   The urea must be <1.5 M for the tryptic digest (Step 7) to work.

3. Add TBP to the protein samples to a final TBP concentration of 5 mM to reduce the proteins. Incubate the solutions for 10 minutes at room temperature.

   TCEP can substitute for TBP to reduce proteins.

4. Add either light or heavy ICAT reagent to its respective protein sample to a final ICAT concentration of 1–1.5 mM. Incubate the reaction with stirring for 1–5.2 hours at room temperature in the dark.

   For example, place mini-stir bars in the sample tubes, wrap the tubes in aluminum foil, and place the tubes on a magnetic stir plate.

5. Add DTT to the reaction to a final concentration of 10 mM to inactivate residual ICAT reagent and limit the formation of adducts between the ICAT reagent and TBP.

6. Combine equivalent amounts of heavy and light peptide samples.

   Methods for determining equivalent amounts vary (e.g., cell counts and protein assay, performed after sample preparation and before labeling).

7. Digest the pooled samples with trypsin at a mass ratio of 1:25 (trypsin:protein) for 3 hours at 37°C.

8. (*Optional*) Check the efficiency of the labeling reaction by running the proteins on a polyacrylamide gel and visualizing the separated proteins with a silver stain (see Chapter 2, Protocol 7 or Chapter 7, Protocol 11). Use a 4–15% or 4–20% gradient minigel. Include unlabeled protein (from Step 1), labeled protein (from Step 5), and post-digest peptides on the gel.

   If labeling works, some bands should increase in molecular weight (and hence migrate a shorter distance from the origin) when comparing the unlabeled and labeled protein lanes. If and how much they increase depends on the number of accessible cysteines within each protein.

9. Proceed to the cation exchange procedure (Stage 2).

STAGE 2: Cation Exchange Cleanup of ICAT Samples

MATERIALS

CAUTION: See Appendix 3 for appropriate handling of materials marked with <!>.

Reagents

Buffer A (10 mM KH_2PO_4, 25% acetonitrile [pH 3.0]) <!>
Buffer B (10 mM KH_2PO_4, 25% acetonitrile, 350 mM KCl [pH 3.0])

Equipment

HPLC column (e.g., 2.1 mm x 20 cm, 5 μm particles, 300 Å pore size Polysulfoethyl A, a strong cation-exchange material; PolyLC Inc.)
HPLC system

Biological Sample

ICAT-labeled peptides (from Stage 1, Step 7)

METHOD

1. Set up the cation exchange column.
2. Acidify the peptide samples (from Stage 1, Step 7) to pH ≤3 with Buffer A to a final volume of 2 ml.

   Acidification of the samples is important, because the peptides will not be fully charged at higher pH values and may not stick to the column!

3. Load from 200 μg to ~5 mg of digested ICAT-labeled total protein onto the column, using a 2-ml sample loop.
4. Set flow rate to 200 μl/min and gradient to the times and percentages listed below:

Time (min)	% Buffer B
0	0
30	25
50	100

5. Collect fractions at 1–2-minute intervals.

   The gradient shown is designed to spread out the elution of doubly charged peptides, with these peptides usually eluting starting at ~8–9 minutes into the run until ~15–16 minutes, after which triply charged peptides begin to elute.

6. Proceed to the avidin column procedure (Stage 3) when all of the samples have eluted from the column.

   Varying numbers of cation-exchange fractions have been pooled prior to running them over the avidin column. If and how much is pooled depends on the complexity of the sample and the sensitivity of the mass spectrometer. Consider running a single cation exchange fraction over the avidin column and analyzing the resulting eluent in the mass spectrometer. The MS data should indicate whether or not to subsequently pool the remaining cation exchange fractions.

   The fractions selected for avidin purification should be those that show UV absorbance in the cation-exchange HPLC fractionation, indicating a significant amount of peptide in those fractions.

STAGE 3: Selection of Tagged Proteins Using an Avidin Column

MATERIALS

CAUTION: See Appendix 3 for appropriate handling of materials marked with <!>.

Reagents

Biotin blocking reagent (2 mM d-biotin <!> in PBS)

Elution reagent (30% acetonitrile <!> containing 0.4% trifluoroacetic acid [TFA]) <!>

Use glass to handle the elution reagent.

NH_4HCO_3 (1 M and 50 mM, pH 8.3) <!>

Prepare 50 mM NH_4HCO_3 in 20% methanol <!>.

2x Phosphate-buffered saline (PBS)

Regeneration buffer (100 mM glycine <!> at pH 2.8)

The regeneration buffer must be sterile, and it does not last long. Autoclave before use, and store at 4°C, where it will last 1 week.

Equipment

Avidin beads (e.g., Ultralink Monomeric Avidin, Pierce Chemical)

Pierce provides helpful information for packing and usage of their avidin columns (see their technical literature at www.piercenet.com).

At any time during the column preparation, it is acceptable to wash the column with 2x PBS containing 0.01% NaN_3 <!> until the pH of the flowthrough is ~7. The column can then be stored for an extended period at 4°C (for overnight storage, the NaN_3 is unnecessary). Make sure that the column packings are submerged in buffer during storage.

Centrifugal concentrator (e.g., SpeedVac, Thermo Savant)

Glass pipette

Glass wool

Peptide collection tubes (glass, four needed per column)

See Steps 17–20.

pH meter or paper

Biological Sample

Peptide fractions from the cation-exchange column

From Stage 2, Step 5. See the note at Stage 2, Step 6.

METHOD

Preparation of the Avidin Column

1. Pack a small piece of glass wool into the neck of a glass pipette. Pack 400 μl of avidin into the tube.

   The avidin slurry comes as a 50% dilution, so 800 μl of 50% slurry must be added in order to get 400 μl of avidin.

   The actual amount of avidin used can vary, although 400 μl of avidin beads will bind ~500 μg of peptides.

2. Allow the avidin beads to settle, and then wash the column with 2x PBS to dispel the beads from the side of the tube.

3. Wash the column with 30% acetonitrile containing 0.4% TFA until the flowthrough pH changes to ~1. Continue to wash the column with one additional column volume (400 μl) of the acetonitrile/TFA.

   This acidic wash step removes polymers associated with the beads.

   EXPERIMENTAL TIP: Use a glass pipette when pipetting acetonitrile or TFA.

4. Wash the column with 2x PBS (pH 7.2) until the flowthrough pH is ~7.2. Continue to wash the column with 3 more column volumes (1200 μl) of PBS.

5. Wash the column with 3–4 column volumes of d-biotin blocking reagent.

   The d-biotin blocks the more retentive avidin sites on the column, ensuring recovery of the sample from the remaining binding sites in Steps 17–19.

6. Wash off loosely bound biotin with ~6 column volumes (2400 μl) of regeneration buffer until the flowthrough pH is ~2.8.

7. Wash the column with 6 column volumes of 2x PBS to return the column to an approximate pH of 7.2 (as measured by the pH of the flowthrough).

Preparation of the Biotinylated Protein

8. Place the sample (from Stage 2, Step 5) in a centrifugal concentrator to remove the acetonitrile. Reduce the original sample volume by at least 25%.

9. Adjust the pH of the sample to 7.2 with 1 M NH_4HCO_3 (<10 μl should be sufficient to adjust the pH).

10. Add 2x PBS to increase the sample volume to approximately three quarters of the total column volume (sample volume = 250–300 μl).

Loading and Incubation of the Sample on the Column

11. Load the sample onto the column, and allow it to flow through until the solution is level with the top of the column.

12. Add 100 μl of 2x PBS to the loaded sample, and cover the top of the column to stop the flow.

13. Allow the sample to incubate in the column for ~20 minutes.

Washing Unbound Material from the Column

14. Wash the column with 5 column volumes of 2x PBS (pH 7.2) and start collecting fractions.

   The unbound material washed from the column in this step and the two subsequent steps can be pooled into a single fraction.

15. Wash the column with 5 column volumes of 1x PBS to reduce the salt concentration.

16. Wash the column with 6 column volumes of 50 mM NH_4HCO_3 (pH 8.3) in 20% methanol while continuing to collect fractions.

    The NH_4HCO_3 further reduces the salt concentration and the methanol removes hydrophobic peptides.

Collection of the Biotinylated Peptides

17. Place the first peptide collection tube underneath the outlet of the column. Add 1 column volume of elution reagent onto the top of the column, and allow it to flow into the column until the reagent is flush with the top of the column.

18. Add another column volume of elution reagent while placing the second peptide collection tube beneath the column.

19. Repeat Step 18 two more times so that a total of four fractions have been collected.

20. The fractions that include the peptides will be the first collection tube to have an acidic pH and the collection tube that follows it. Pool these two samples.

21. Proceed to analyze the peptides by MS (Stage 4).

STAGE 4: Analysis of ICAT-labeled Peptides by MS

MATERIALS

CAUTION: See Appendix 3 for appropriate handling of materials marked with <!>.

Reagents

Reversed-phase Solvent A
- 0.5% acetic acid <!>
- 0.005% heptafluorobutyric acid <!>
- 5% acetonitrile <!>

Reversed-phase Solvent B
- 0.4% acetic acid
- 0.005% heptafluorobutyric acid
- 100% acetonitrile

Equipment

Centrifugal concentrator (e.g., SpeedVac, Thermo Savant)

Mass spectrometer

Reversed-phase column packing material (100 Å spherical silica, C_{18} [5 µm mean particle size])

RP-HPLC column

Use a column with one of the following set of dimensions:
- 50 µm I.D./200 µm O.D.
- 75 µm I.D./360 µm O.D.
- 100 µm I.D./350 µm O.D.

Biological Sample

Pooled fractions from avidin column

From Stage 3, Step 20.

METHOD

1. Concentrate the pooled avidin fractions in a centrifugal concentrator. *Do not* allow the sample to dry out.
2. Centrifuge the sample in a microfuge at 14,000*g* for 15 minutes to pellet any particulates (there should be no visible or a barely visible pellet).
3. Transfer the supernatant to a fresh tube.
4. Pressure-load varying percentages of the sample onto a reversed-phase column that is in-line with the mass spectrometer (for additional details on using a LC-MS setup, see Protocols 4 and 5).

   The percentages of sample to load will depend on the complexity of the sample and should be determined empirically.

5. Wash the column with Reversed-phase Solvent A for at least 10 minutes to remove salts.

6. Elute the sample from the column directly into the mass spectrometer using the following elution and MS conditions:

   Flow rate: At the column tip, the flow rate is 200–300 nl/min (i.e., the HPLC pump is run at ~0.1 ml/min and the flow is split prior to the column).

   Gradient protocol (example):

% Solvent B	Minutes
0	0
10	5
40	45
80	50
80	55
0	56
0	66

   Mass spectrometer operating conditions: The total mass spectrum (*m/z* 400–1800) is scanned in 1 second. If the most-intense peak in the full scan is above a dictated threshold, the following scan will be an MS/MS scan of that most-intense peak. If no single peak surpasses the MS/MS threshold value, then a full scan per second is repeatedly made until a large enough peak is found. An example of a threshold value is 500,000 counts minimum signal (required for MS/MS). Masses of analyzed peaks are put into an exclusion table for several minutes to prevent recurrent MS/MS analysis of the same peak.

   Heated capillary temperature: 170°C

   Collision energy to form MS/MS fragments: 27

   These operating parameters were established on an LCQ ion trap mass spectrometer (ThermoFinnigan MAT [San Jose, California]). Other instruments may require slightly different parameters that will have to be determined empirically.

PROTOCOL 10

In Vivo Isotopic Labeling of Proteins for Quantitative Proteomics

STABLE ISOTOPE CODING STRATEGIES ARE OF IMMENSE VALUE in determining protein concentration changes in cells and tissues triggered by regulatory stimuli (e.g., drugs and toxins) and disease (e.g., caused by mutational changes). Recognizing and identifying the small number of proteins whose expression levels differ as a consequence of disease or external stimuli are complicated by the complexity of biological extracts and the fact that protein posttranslational modifications are numerous and can occur at many sites on a protein.

One way of quantifying global protein expression patterns involves in vivo labeling. Using the following protocol, stable isotopes can be incorporated into metabolic products, and the relative difference between these products from cells grown in normal or isotope-enriched media can be readily quantified by MS analysis. This protocol was provided by Yoshiya Oda (Eisai Co., Ltd., Ibaraki, Japan).

MATERIALS

CAUTION: See Appendix 3 for appropriate handling of materials marked with <!>.

Reagents

Acetone, ice-cold <!>

Acetonitrile <!>

Acetonitrile/H_2O/TFA (66/33/0.1 [v/v/v] and 5/95/0.1 [v/v/v])

Ammonium bicarbonate (50 mM) <!>

Ammonium bicarbonate (25 mM) containing 0.05 μg of trypsin and 0.1% *n*-octylglucoside

Ammonium bicarbonate (25 mM) containing 0.1% *n*-octylglucoside

Cell culture growth media

The experiment requires growth medium capable of supporting the cells under study (bacterial, yeast, other fungi, insect, or mammalian). The medium must be available in two varieties, which are identical except for the inclusion of ^{15}N-labeled metabolites in one. Two sources of such media are Spectra Stable Isotopes (http://www.spectrastableisotopes.com), which produces the Celtone and Yeastone products; and Cambridge Isotope Laboratories, Inc. (Andover, Massachusetts, http://www.isotope.com/cil/index.html), which produces the Bio-Express line of growth media.

Experience with Bio-Express 1000 medium indicates that it is preferable to add tryptophan to the medium to a final concentration of 40 mg/ml. Bio-Express 1000 media are provided as sterile, 10x concentrates in H_2O. Dilution experiments indicate a linear growth response over the range of 0.25x to 2x concentrations. Because growth rates of *S. cerevisiae* are slightly slower in Bio-Express 1000 medium, prepare a 2x solution as the working stock. However, growth conditions must be optimized empirically for each cell type and medium.

Dry ice or liquid nitrogen <!>

H_2O, ice-cold

Ice

2-Mercaptoethanol (10 mM) <!>
Methanol (50%) <!>/40% H_2O/10% acetic acid
Potassium ferricyanide (15 mM) <!>
Protein extraction reagent (e.g., Y-PER reagent, Pierce)
Protease inhibitor cocktail
Sodium thiosulfate (50 mM)

MATERIALS FOR PROTEIN SEPARATION USING 2D GEL ELECTROPHORESIS

See Step 22. For additional details on running 2D gel electrophoresis, see Chapter 4.

Rehydration buffer A
- 0.5 M Tris-Cl (pH 8.3)
- 8 M urea <!>
- 0.5% Triton X-100 <!>
- 2 mM dithiothreitol (DTT) <!>
- 0.2% Ampholine (Amersham Biosciences)*

*Choose the Ampholine set with the appropriate pH range for the proteins to be separated.

Immobiline dry strip gel (Amersham Biosciences)
Polyacrylamide gel (12.5%T, 2.7%C)
Electrophoresis apparatus (Hoefer DALT system, Amersham Biosciences, or equivalent)

MATERIALS FOR PROTEIN SEPARATION USING HPLC AND SDS-PAGE

See Step 22. For additional details on running ion-exchange HPLC and SDS-PAGE, see Chapters 5 and 2, respectively.

Rehydration buffer B
- 20 mM Tris-Cl (pH 8.3)
- 8 M urea <!>
- 0.5% Triton X-100 <!>

Mono-Q column (HR 5/5, Amersham Bioscience, or equivalent)
Mobile phase A: 20 mM Tris-Cl (pH 8.3) containing 1 M urea
Mobile phase B: 20 mM Tris-Cl (pH 8.3) containing 1 M urea and 1 M NaCl
HPLC system
Trichloroacetic acid (TCA) (10%) <!>
SDS-PAGE sample buffer (2x sample buffer for SDS-PAGE: 125 mM Tris-Cl [pH 6.8], 20% glycerol, 4% SDS <!>, 100 mM DTT <!>, 0.04% bromophenol blue)
Triglycine 8–16% gradient gel (Invitrogen-NOVEX)

MATERIALS FOR PROTEIN SEPARATION USING 2D HPLC

See Step 22. For additional details on running HPLC, see Chapter 5.

Rehydration buffer B (see recipe above)
Mono-Q column (HR 5/5, Amersham Bioscience, or equivalent)
Mobile phase A: 20 mM Tris-Cl (pH 8.3) containing 1 M urea <!>
Mobile phase B: 20 mM Tris-Cl (pH 8.3) containing 1 M urea and 1 M NaCl
HPLC system
RP-HPLC column (YMC-Pack CN-AP 4.6 x 150 mm, Ap-502 [YMC Inc, Kyoto, Japan] or equivalent)
Solvent A: 5% acetonitrile containing 0.1% TFA
Solvent B: 95% acetonitrile containing 0.1% TFA
Trypsin

Equipment

Centrifugal evaporator (SpeedVac, Thermo Savant or equivalent)

Centrifuge tubes (100-ml capacity)

Hemocytometer

Optional, see Step 8.

Incubator preset to 30°C

MALDI-TOF mass spectrometer and an LC-MS/MS system

See Step 34 and Protocols 4–7.

Sonicator

Spectrophotometer

Biological Sample

Saccharomyces cerevisiae, freshly grown colonies on a Petri plate

Other types of cells can be labeled using this protocol, provided the growth conditions are adjusted to optimize growth of the cells under study.

Additional Reagents

If a gel is used for the separation of proteins, Step 22 will require the silver stain reagents from Chapter 2, Protocol 7 or Chapter 4, Protocol 12.

METHOD

Culturing of Cells

1. Beginning with a freshly grown plate, choose a single yeast colony, and inoculate 2 ml of ^{15}N-labeled culture medium with the colony. Choose a second colony as a control sample, and inoculate 2 ml of unlabeled culture medium with that colony.
2. Vigorously vortex the tubes of medium for ~1 minute to disperse the cells thoroughly.
3. Incubate the two cultures with shaking at 230–270 rpm overnight at 30°C.
4. Following overnight growth, measure the culture growth with a spectrophotometer. The cultures should be dense ($OD_{600} > 1.5$).
5. Vortex the overnight cultures for ~1 minute to disperse any cell clumps.
6. Inoculate 50 ml of ^{15}N-labeled medium with the entire overnight culture grown in ^{15}N-labeled medium. Likewise, inoculate 50 ml of the unlabeled medium with the entire control culture.
7. Incubate with shaking at 220–250 rpm at 30°C until the cultures have reached mid-log phase ($OD_{600} < 1.0$).

Preparation of Cultures for Protein Extraction

8. Determine the cell density by measuring the culture at 600 nm in a spectrophotometer or by counting cells in a hemocytometer.

9. Using as much of each culture as possible, combine equal numbers of the ^{15}N-labeled culture and the unlabeled control culture together.

10. Quickly chill the culture by pouring it into two chilled 100-ml centrifuge tubes filled halfway with ice.

11. Immediately place the tubes in a chilled rotor and centrifuge them at 1000*g* for 5 minutes at 4°C.

12. Pour off the supernatant and resuspend the cell pellet in 50 ml of ice-cold H_2O.

    Any unmelted ice pours off with the supernatant.

13. Recover the pellet by centrifugation at 1000*g* for 5 minutes at 4°C.

14. Again, discard the supernatant, and immediately freeze the cell pellet by placing the tubes on dry ice or in liquid nitrogen. Store the cells at –80°C until they are needed.

Preparation of Protein Extracts

15. Add an appropriate volume of protein extraction reagent containing a protease inhibitor cocktail and 10 mM 2-mercaptoethanol to the frozen cell pellet. Usually 2.5–5 ml of the extraction reagent, Y-PER reagent, is used for a 1-ml (1 g) cell pellet.

16. Incubate the cells gently for 30 minutes at room temperature.

17. Spin down the debris and collect the supernatant.

18. Add 5 volumes of ice-cold acetone to the supernatant and vortex the tube.

19. Incubate the protein extract for 10 minutes at –80°C.

20. Centrifuge the proteins in a microfuge at 20,000*g* for 5 minutes at 4°C.

21. Discard the supernatant and immediately carry out Step 22 or store the precipitated proteins at –80°C until they are needed.

Separation of Proteins

22. Three different ways to separate the protein mixture into individual proteins are 2D gel electrophoresis, HPLC combined with SDS-PAGE, and 2D HPLC. Perform one of these procedures for separating the proteins.

For 2D gel electrophoresis: The following steps are a summary of the 2D gel electrophoresis procedure. For step-by-step instructions, see Chapter 4.

a. Dissolve 500 µg of precipitated protein (Step 21) in 400 µl of Rehydration buffer A.

b. Rehydrate on an Immobiline dry strip gel.

c. Perform IEF in the first dimension.

d. Apply the strip to a 12.5%T, 2.7%C polyacrylamide gel, and separate the proteins in the second dimension.

e. Following the completion of 2D electrophoresis, stain the gel with silver using a procedure that is compatible with subsequent analysis of the proteins by MS (see Chapter 4, Protocol 12).

f. Proceed to Step 23.

For HPLC-SDS/PAGE: The following steps are a summary of the ion-exchange HPLC and 1D gel electrophoresis procedures. For step-by-step instructions, see Chapters 5 and 2, respectively.

a. Dissolve 10 mg of precipitated protein (Step 21) in 2 ml of Rehydration buffer B.

b. Establish a column flow rate of 1 ml/min. Set the concentration gradient for Mobile phase B at:

   0–5 minutes, 0% B
   5–45 minutes, 0–50% B
   45–60 minutes, 50–100% B

c. Collect 30 2-ml fractions.

d. After the HPLC is complete, precipitate the proteins from each 2-ml fraction with 1 ml of 10% (w/v) TCA. Vortex the tubes and incubate for 30 minutes on ice.

e. Centrifuge the tubes at 3000*g* for 30 minutes at room temperature.

f. Discard the supernatant. Add SDS-PAGE sample buffer, and run each fraction in a separate lane of an 8–16% gradient triglycine gel.

g. Stain the gels with silver using a procedure that is compatible with subsequent analysis of the proteins by MS (see Chapter 2, Protocol 7).

h. Proceed to Step 23.

For 2D-HPLC: The following steps are a summary of the ion-exchange and RP-HPLC procedures. For step-by-step instructions, see Chapter 5.

a. Dissolve 10 mg of precipitated protein (Step 21) in 2 ml of Rehydration buffer B.

b. Establish a column flow rate of 1 ml/min. Set the concentration gradient for mobile phase B at:

   0–5 minutes, 0% B
   5–45 minutes, 0–50% B
   45–60 minutes, 50–100% B

c. Collect 30 2-ml fractions.

d. Inject each 2-ml fraction into the second HPLC column, a reversed-phase column.

   Solvent A is 5% acetonitrile containing 0.1% TFA, and Solvent B is 95% acetonitrile containing 0.1% TFA.

e. Establish a column flow-rate of 1 ml/min. Set the concentration gradient for Mobile phase B at:

   0–5 minutes, 0% B
   5–60 minutes, 0–55% B

f. Collect 30 2-ml fractions.

g. After the second HPLC is complete, dry the fractions in a centrifugal evaporator.

h. Dissolve the fractions in 20 µl of 20 mM NH_4HCO_3 and 5 mM *n*-octylglucoside. Vortex and sonicate the samples. Check that the pH is ~8.5.

i. Add 1 µl of 0.05–1 µg of trypsin to the fractions. Incubate overnight at 37°C.

   The amount of trypsin to use will depend on the protein concentration. The final substrate-to-trypsin ratio should be 50:1.

j. Proceed to Step 34.

In-gel Digestion

23. Destain the silver-stained gels for a few minutes with a freshly prepared mixture of 15 mM potassium ferricyanide and 50 mM sodium thiosulfate.
24. Wash the gels vigorously with 500 µl of 50% methanol/40% H_2O/10% acetic acid for 30 minutes.
25. Repeat Step 24 four more times, using fresh washing solution each time.
26. Incubate the gels with 500 µl of 50 mM ammonium bicarbonate solution for 5 minutes.
27. Incubate the gels with 500 µl of acetonitrile for 5 minutes.
28. Dry the gel pieces completely in a centrifugal evaporator.
29. Swell the dried gel pieces in 2 µl of 25 mM ammonium bicarbonate containing 0.05 µg of trypsin and 0.1% *n*-octylglucoside.
30. After all solvents have penetrated into the gels (usually 5–10 minutes), add 10 µl of 25 mM ammonium bicarbonate containing 0.1% *n*-octylglucoside, and allow the mixture to stand for 2 hours at 37°C.
31. Extract the tryptic peptides twice with 40 µl of acetonitrile/H_2O/TFA (66/33/0.1 [v/v/v]) solution in a 350 W sonicator for 5–10 minutes.
32. Dry the combined extracts (~90 µl) in a centrifugal evaporator.

Mass Spectrometric Analysis of the Proteins

33. Dissolve the dried tryptic peptides in 5 µl of acetonitrile/H_2O/TFA (5/95/0.1 [v/v/v]) solution.
34. Subject 10% of the digest to MALDI-TOF-MS, and use the remainder for LC-MS/MS analysis.

    Proteins separated by 2D-HPLC and digested in solution must be passed through a ZipTip prior to MALDI or loaded onto an RP-HPLC column prior to MS/MS.

PROTOCOL **11**

Anhydrotrypsin Affinity Capture of Proteins

MUTATION OF SER-195, LOCATED WITHIN THE ACTIVE SITE of trypsin, to a dehydroalanine residue renders the enzyme catalytically inert. However, the resulting anhydrotrypsin derivative retains a strong affinity for peptides that contain a carboxy-terminal arginine or lysine residue. This interaction is such that a column prepared of the immobilized form of anhydrotrypsin can be effective in capturing peptides with basic carboxyl termini from a typical protein digest. This anhydrotrypsin affinity capture technique was developed more than two decades ago (Yokosawa and Ishii 1976; Ishii et al. 1979) and remains useful today, especially when used in conjunction with MS for the characterization of the carboxyl termini of proteins (Sechi and Chait 2000; Bures et al. 2001). Anhydrotrypsin affinity capture is a reliable technique for isolating the carboxy-terminal peptide of a tryptically digested protein (Kumazaki et al. 1986, 1987). Because the carboxy-terminal tryptic peptides of proteins typically do not contain a carboxy-terminal arginine or lysine, these peptides can be selectively isolated from complex tryptic digests of proteins by their inability to bind to an anhydrotrypsin affinity capture column (all other carboxy-terminal arginine or lysine peptides will be retained). The following protocol, which was kindly provided by Ted Bures (Amgen Inc., Thousand Oaks, California), involves five stages:

- Preparation of an affinity resin (salmine-trypsin [ST]-Sepharose) containing a tryptic digest of salmine, an arginine-rich polypeptide, coupled to Sepharose.
- Preparation of anhydrotrypsin from the alkaline treatment of phenylmethylsulfonyltrypsin.
- Purification of anhydrotrypsin on ST-Sepharose.
- Preparation of an anhydrotrypsin affinity column.
- Use of the anhydrotrypsin affinity column for capturing basic carboxy-terminal peptides.

MATERIALS

CAUTION: See Appendix 3 for appropriate handling of materials marked with <!>.

Reagents

Acetic acid (1 M) <!>
Bovine trypsin (Roche Molecular Biochemicals)
$CaCl_2$ (1 M) <!>
Cyanogen-bromide-activated Sepharose 4B (Amersham Biosciences) <!>
HCl (1 M) <!>
KOH (1 M) <!>
$NaHCO_3$ (1 M, pH 8.3)
$NaHCO_3$ (1 M, pH 8.3) containing 5 M NaCl
NaOH (20 mM) <!>
Phenylmethylsulfonyl fluoride (PMSF) (87.5 mg) <!> in 2.5 ml of acetone <!>

Phosphate-buffered saline (PBS) containing 0.02% NaN_3 <!>
Salmine sulfate (protamine sulfate, Grade X, isolated from salmon, Sigma)
Sodium acetate (0.1 M, pH 4.0) containing 0.5 M NaCl
Sodium acetate (50 mM, pH 5.0) containing 0.02 M $CaCl_2$
Trifluoroacetic acid (TFA) (0.05% v/v) <!>
Tris-Cl (0.05 M, pH 7.0) containing 20 mM $CaCl_2$
Tris-Cl (0.1 M, pH 8.0) containing 0.5 M NaCl

Equipment

Dialysis membrane (6000–8000 m.w. cutoff)
Sintered glass funnel
Water bath (boiling)

METHOD

Preparation of ST-Sepharose

1. Dissolve 10 mg of bovine trypsin and 2.5 g of salmine sulfate in 75 ml of 0.1 M $NaHCO_3$ (pH 8.3) and incubate for 100 minutes at 37°C.
2. Heat the solution for 3 minutes to 100°C to arrest the digestion. Allow the salmine-trypsin solution to cool to room temperature.
3. Weigh out 14.3 g of cyanogen-bromide-activated Sepharose 4B resin.
4. Swell and wash the resin in 3 liters of 1 mM HCl on a sintered glass filter for 15 minutes.
5. Mix the salmine-trypsin solution (from Step 2) with the washed cyanogen-bromide-activated Sepharose 4B on a rotating wheel for 1 hour at room temperature. This allows the salmine-trypsin to bind to the activated Sepharose.
6. Add 10 gel volumes of 0.1 M $NaHCO_3$ (pH 8.3) to the Sepharose solution to dilute the excess ligand (i.e., salmine-trypsin) and aid in its subsequent removal.
7. Block the remaining active groups by allowing the resin to stand in 50 ml of 0.1 M Tris (pH 8.0) containing 0.5 M NaCl for 2 hours at room temperature.
8. Aspirate all of the liquid from atop of the settled resin.
9. Carry out three cycles of alternating pH washes using:

   0.1 M sodium acetate (pH 4.0) containing 0.5 M NaCl
   0.1 M Tris-Cl (pH 8.0) containing 0.5 M NaCl

   Perform each wash with 50 ml of buffer. Between each washing, allow the resin to settle, and aspirate off the liquid.
10. Store the completed ST-Sepharose in PBS containing 0.02% NaN_3.

    The ST-Sepharose may be stored for up to 1 year at 4°C with no loss of binding affinity.

    As an alternative to batch processing of the ST-Sepharose, Steps 6–10 can be carried out in a column.

Preparation of Anhydrotrypsin

11. Inactivate 500 mg of bovine trypsin (in 250 ml of 0.05 M Tris [pH 7.0] containing 20 mM $CaCl_2$) with 2.5 ml of PMSF (87.5 mg in acetone) for 30 minutes at room temperature with gentle stirring. Using a pH probe kept continuously in the solution, maintain the pH at 7.0 with 0.02 M NaOH throughout this step (Ishii et al. 1983).

12. To terminate the reaction, add 1 M HCl until the pH of the trypsin solution is 3.0.

13. Dialyze the trypsin overnight at 4°C against 15 liters of 1 mM HCl. Use a dialysis membrane rated with a 6000–8000 molecular-weight cutoff.

14. Adjust the resulting 290 ml of phenylmethylsulfonyl trypsin to an alkaline pH with the addition of a 1/19th volume (or 15.3 ml) of 1 M KOH and stir for 20 minutes at 0°C.

15. Adjust the pH to 5.0 with the addition of ~15.6 ml of 1 M acetic acid. Add 6.4 ml of 1 M $CaCl_2$ to the phenylmethylsulfonyl trypsin.

    The final mixture should contain 50 mM acetate and 20 mM $CaCl_2$.

16. Filter the 327 ml of phenylmethylsulfonyl trypsin solution through a 0.22-μm filter.

17. Purify the anhydrotrypsin on an ST-Sepharose column (Kasai and Ishii 1975):

    a. Prepare a column of ST-Sepharose (using the resin from Step 10). Wash the resin thoroughly to remove the NaN_3 and equilibrate the column in 50 mM sodium acetate buffer (pH 5.0) containing 0.02 M $CaCl_2$.

    b. Load the phenylmethylsulfonyl trypsin solution onto the column at a flow rate of 1 ml/min.

    c. Wash the column with 15 column volumes of 50 mM sodium acetate buffer (pH 5.0) containing 0.02 M $CaCl_2$. Wash with a column flow rate of 3 ml/min.

    d. Elute the anhydrotrypsin from the column with 3 column volumes of 5 mM HCl, operating the column at a flow rate of 1 ml/min. Monitor the eluent with a spectrophotometer set at A_{280}.

Preparation of Anhydrotrypsin Affinity Column

18. Dilute 10 ml of the concentrated anhydrotrypsin solution with 10 ml of 0.1 M $NaHCO_3$ (pH 8.3) containing 0.5 M NaCl.

    The following coupling method is based on the technique of Axen and Ernback (1971).

19. Add 1 ml of 1 M $NaHCO_3$ (pH 8.3) containing 5 M NaCl to the diluted anhydrotrypsin, and filter the solution through a 0.22-μm filter.

    This adjusts the solution to 0.1 M $NaHCO_3$ (pH 8.3) containing 0.5 M NaCl. Filtering the anhydrotrypsin removes any precipitates and prevents the possibility of clogging the column.

20. Wash 0.314 g of cyanogen-bromide-activated Sepharose 4B with 70 ml of 1 mM HCl.

21. Mix the diluted anhydrotrypsin (from Step 19) with the cyanogen-bromide-activated Sepharose 4B for 3 hours at room temperature on a rotating wheel.

22. Pack a column with the affinity resin according to the manufacturer's instructions.

23. Wash excess ligand from the affinity resin with 5 column volumes of 0.1 M $NaHCO_3$ (pH 8.3) containing 0.5 M NaCl.

24. Block any free active groups on the activated Sepharose with 10 column volumes of 0.1 M Tris-Cl (pH 8.0) containing 0.5 M NaCl.

25. Carry out three cycles of alternating pH washes for ~5 minutes each at room temperature using:

    20 ml of 0.1 M sodium acetate (pH 4.0) containing 0.5 M NaCl
    20 ml of 0.1 M Tris-Cl (pH 8.0) containing 0.5 M NaCl

26. Store the completed anhydrotrypsin-Sepharose in PBS containing 0.02% NaN_3.

    The anhydrotrypsin-Sepharose can be stored for up to 1 year at 4°C.

Sample Preparation, Loading, and Elution

27. Before loading the digested protein sample onto the anhydrotrypsin column, acidify the protein sample to pH 5.0 with 0.1 N acetic acid.

28. Divide the sample into 2-ml aliquots (2 ml is the maximum load capacity of the column).

    A sample containing a calibrated amount of enkephalin-Arg was passed through the column to determine the column capacity. The amount of sample added is such that the column capacity is not exceeded. The column is equilibrated and washed with 50 mM sodium acetate (pH 5.0)/20 mM $CaCl_2$/0.02% NaN_3.

29. Elute bound peptides from the anhydrotrypsin affinity column using 0.05% (v/v) TFA.

    Alternative elution methods include eluting with 5 mM HCl or with 0.1 M formic acid (Kumazaki et al. 1987), or generating a pH gradient from pH 5.0 to pH 2.5 (Kumazaki et al. 1986).

PROTOCOL 12

Analysis of Proteomes Using the Molecular Scanner

THE MOLECULAR SCANNER IS A HIGHLY AUTOMATED PROTEOMICS method, which simultaneously processes thousands of proteins under identical conditions and offers a flexible and powerful visualization tool that can create a fully annotated 2D gel electrophoresis map (Bienvenut et al. 1999). Proteins separated by 2D gel electrophoresis are then simultaneously digested while undergoing electrotransfer from the gel to a membrane. The peptides are subjected to peptide mass fingerprint (PMF) analysis to identify proteins directly from the PVDF membranes by MALDI-TOF-MS scanning. An ensemble of dedicated tools is then used to create, analyze, and visualize a proteome as a multidimensional image (Binz et al. 1999). The molecular scanner method reduces to a minimum the sample handling prior to mass analysis and decreases the sample size to a few tens of micrometers, i.e., the size of the MALDI-TOF-MS laser beam impact. This protocol has been kindly provided by Catherine Déon, Willy Bienvenut, Jean-Charles Sanchez, and Denis F. Hochstrasser (Geneva University Hospital, Geneva, Switzerland) and Markus Müeller, Robin Gras, and Ron D. Appel (Swiss Institute of Bioinformatics, Geneva, Switzerland). The process can be divided into four parts (see Figure 8.46): separation and digestion of proteins, acquisition of PMF data, processing of the MS data and protein identification, and creation of multidimensional proteome maps (see below).

- ***Separation and digestion of proteins.*** Digestion of gel-electrophoresis-separated proteins occurs while they are being transferred onto a PVDF membrane. Following either 1D or 2D separation of proteins, electrotransfer of the proteins is performed through a field of trypsin covalently bound to a membrane interposed between the gel and the PVDF membrane (Figure 8.46a). Consequently, the proteins separated on the gel are endoproteolytically cleaved during their migration and the corresponding peptides are collected on the PVDF membrane. The protein digestion is improved by performing an in-gel predigestion step prior to the electrotransfer. The combination of digestion steps (called the double-parallel digestion process or the DPD process) improves the proteolysis of high-molecular-weight and basic proteins. Furthermore, an alternating square-shape voltage is applied during the electrotransfer to increase the migration time of the proteins through the enzymatic membrane, which improves tryptic hydrolysis efficiency.

  This technique allows additional analyses to be carried out. For example, the collecting membrane can be reused for MS scans under different conditions or additional chemistry to improve protein identification (e.g., derivatization of peptides). The DPD process can also be used with other enzymes such as phosphatase or glycosidase followed by any endoproteinase. Finally, with the development of new mass spectrometric technologies such as the MALDI-TOF/TOF (Medzihradszky et al. 2000) and the MALDI-QqTOF (Loboda et al. 2000), future promising developments of the molecular scanner process could combine the MALDI scanning technique with MS/MS acquisition and generation of sequence data for protein identification.

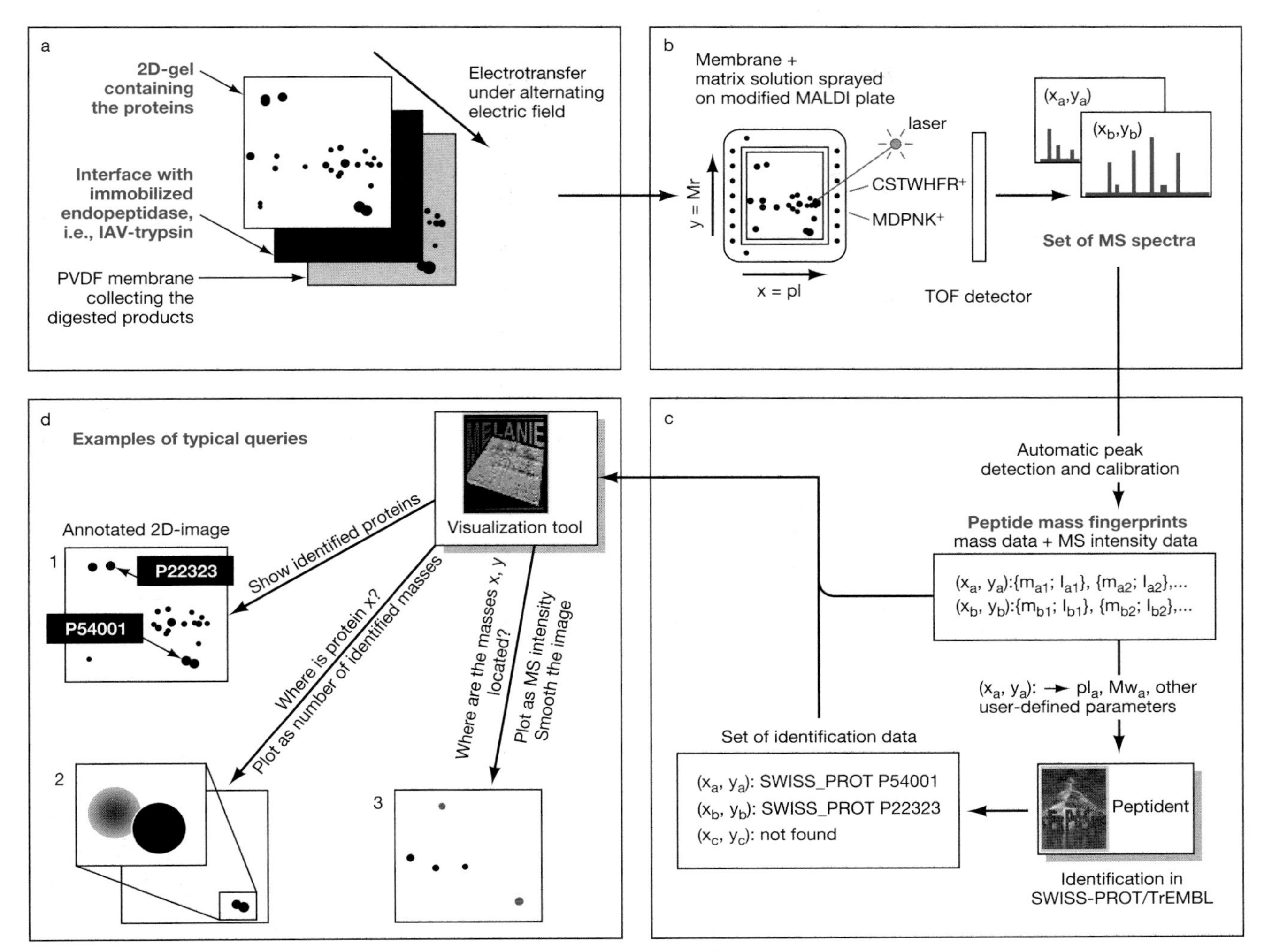

FIGURE 8.46. Scheme of the molecular scanner. (*a*) Parallel digestion and simultaneous electrotransfer of proteins from a 2D electrophoresis gel using the DPD process. (IAV) Immobilon AV membrane. (*b*) MALDI-TOF-MS scanning of PVDF collecting membrane after spraying with matrix solution. (x_i, y_i) refers to the position where MS spectra were measured on the PVDF membrane. (*c*) Identification procedure. The peak detection and mass calibration yields sets of PMF. The MS signal measured at each (x_i, y_i) coordinate is represented by its *m/z* value m_{ix} and its MS intensity I_{ix}. The x_i and y_i values are interpreted as pI and M_r values. The PMF data are submitted to SmartIdent. Identification results are collected together with the PMF data. (*d*) A visualization tool allows representation of the analyzed data in different forms. Three examples of typical queries and representations are described here. (*1*) An MS intensity image can be created that contains the identification data as database labels. It is generated in a Melanie readable format. (*2*) Other options allow the user to search for a particular protein and to visualize it as an intensity plot. In this plot, the gray levels represent the number of masses identified that belong to the protein at each position. (*3*) The program further allows the user to search for a set of predefined masses and to generate an intensity image where the gray levels represent the total intensity of the mass peaks found at each (*x*, *y*) position. This image can be smoothed if needed. (Reprinted, with permission, from Binz et al. 1999 [©American Chemical Society].)

- ***Acquisition of PMF data.*** The PVDF membrane, containing the digestion products of all the proteins at discrete positions on the membrane surface, is sprayed with an MS matrix solution and then scanned directly by MALDI-TOF-MS to obtain PMF data over the entire surface of the membrane (Figure 8.46b).
- ***Processing of the MS data and protein identification.*** The identification process is managed by an interactive tool, which automatically treats all of the generated MS data and performs the various steps of the analysis, starting with peak detection and calibration (Figure 8.46c). PMF data combined with pI and M_r information (extracted from the position on the 2D gel) are sent for identification to SmartIdent-adapted software, an identification tool developed by the Swiss Institute of Bioinformatics (Gras et al. 1999; http://ch.expasy.org/tools/).
- ***Creation of virtual maps.*** A multidimensional image is generated corresponding to a virtual annotated "2D map" (Figure 8.46d). *x* and *y* coordinates are related, respectively, to pI and M_r values. A *z* axis is added that represents different data or information obtained during the acquisition and interrogation processes. The resulting images containing identification results are readable by the Melanie 2D electrophoresis analysis software. The user can choose to filter and visualize only particular features from all of the data contained in the multidimensional image (Figure 8.46d).

MATERIALS

CAUTION: See Appendix 3 for appropriate handling of materials marked with <!>.

Reagents

Acetonitrile (preparative HPLC grade, 50% and 80%) <!>

α-cyano-4-hydroxy-cinnamic acid (HCCA) (10 mg/ml in 70% methanol and 0.1% TFA) <!>

α-tosyl-L-arginine methyl ester (TAME) (10 mM)

Amido Black (0.5% w/v)

Dissolve the dye in 25% (v/v) 2-propanol <!> and 10% (v/v) acetic acid <!>.

Capping reagent

1 M ethanolamine (pH 10.5) <!>

1 M sodium bicarbonate buffer (pH 9.5)

Coomassie Brilliant Blue R250 (0.1% w/v)

Prepare the dye in a 60/30/10 (v/v/v) mixture of H_2O, methanol, and acetic acid.

Destain the solution (50/40/10 [v/v/v] mixture of H_2O, methanol, and acetic acid).

HCl (1 mM) <!>

Paraffin oil

Phosphate-buffered saline (PBS) containing Tween (PBS-Tween, pH 7.4)

20 mM NaH_2PO_4 <!>

140 mM NaCl

0.5% Tween 20

0.5x Transblotting buffer (Towbin buffer containing 0.01% w/v of SDS) <!>

Tris-Cl (46 mM, pH 8.1) containing 1 mM $CaCl_2$ <!> and 0.1% sodium azide <!>

Tris-Cl (460 mM, pH 8.1) containing 11.5 mM $CaCl_2$

Trypsin (2 mg/ml) in 20 mM NaH_2PO_4 (pH 7.8)

Trypsin (0.05 mg/ml) in 10 mM Tris-Cl (pH 8.2)

All of the trypsin used in this experiment should be TPCK-treated trypsin (Type IX from porcine pancreas, Sigma Chemical).

Gel electrophoresis reagents:

- One-dimensional minigel
    - 60 mM Tris-Cl (pH 6.8) containing 10% (v/v) glycerol, 2% (w/v) SDS, and 3% (v/v) mercaptoethanol
    - 12%T, 2.6%C, 1-mm thick linear polyacrylamide gels
- Two-dimensional minigel
    - 2D Equilibration solution
        - 50 mM Tris-Cl (pH 6.8)
        - 6 M urea <!>
        - 30% (v/v) glycerol
        - 2% (w/v) SDS
        - 2% (w/v) dithioerythritol (DTE)
    - Sulfhydryl blocking solution
        - 50 mM Tris-Cl (pH 6.8)
        - 6 M urea
        - 30% (v/v) glycerol
        - 2% (w/v) SDS
        - 2.5% (w/v) iodoacetamide <!>
        - a trace of bromophenol blue
    - Overlay solution
        - 0.5% (w/v) agarose
        - 25 mM Tris (pH 8.3)
        - 198 mM glycine <!>
        - 0.1% (w/v) SDS

        Shortly before use, boil the entire overlay solution to melt the agarose. Hold the solution at ~70°C.
    - 2D gel rehydration solution
        - 8 M urea
        - 4% (w/v) CHAPS
        - 2% (v/v) Resolytes 4–8
        - 65 mM DTE
        - a trace of bromophenol blue
- Two-dimensional full-size
    - First-dimension sample loading buffer
        - 8 M urea
        - 4% (w/v) CHAPS
        - 40 mM Tris
        - 65 mM DTE
        - 0.05% (w/v) SDS
        - a trace of bromophenol blue
    - Sodium thiosulfate (5 mM)
    - 2D Equilibration solution
        - 50 mM Tris-HCl (pH 6.8)
        - 6 M urea
        - 30% (v/v) glycerol
        - 2% (w/v) SDS
        - 2% (w/v) DTE
    - Sulfhydryl blocking solution
        - 50 mM Tris-HCl (pH 6.8)
        - 6 M urea

30% (v/v) glycerol
2% (w/v) SDS
2.5% (w/v) iodoacetamide
a trace of bromophenol blue
Overlay solution (for recipe, see Two-dimensional minigel above)

Equipment

Electroblotting apparatus
The apparatus can be purchased (e.g., Bio-Rad) or laboratory made (Bienvenut et al. 1999).
Gel electrophoresis system:

- One-dimensional minigel
  1D minigel system (e.g., Mini-Protean II Cell, Bio-Rad)
  Water bath preset to 95°C
- Two-dimensional minigel
  2D minigel system (e.g., Multiphor II, Amersham Biosciences)
  Ready-made IPG strips (7-cm) (Amersham Biosciences)
  Water bath preset to 70°C
  Precast or homemade vertical or horizontal SDS slab gels for the second dimension
  Dimension: 80 x 60 x 1.0 mm or 100 x 80 x 1.0 mm
  Resolving gel: acrylamide/piperazinediacrylamide (12%T/2.6%C)
  Stacking gel: no stacking
  Leading buffer: 0.375 M Tris-Cl (pH 8.8)
  Trailing buffer: 25 mM Tris, 198 mM glycine, 0.1% (w/v) SDS (pH 8.3)
  Additive: 5 mM sodium thiosulfate
  Polymerization agents: 0.05% TEMED, 0.1% ammonium persulfate
- Two-dimensional full size
  Ready-made IPG strips (18-cm) (Amersham Biosciences)
  Sample cups (Amersham Biosciences)
  Large electrophoresis system (e.g., Protean II Cell, Bio-Rad)
  Water bath preset to 70°C

Immobilon AV membrane (Millipore)
The IAV membrane is a modified PVDF matrix presenting activated carboxylic groups able to react with nucleophiles such as amine groups of proteins.
MALDI sample plate (modified)
See Step 24.
Mass spectrometer
See Step 26.
PVDF membrane
Rotating hybridizer (e.g., model HB-2D, Techne, Cambridge, England)
UV-visible spectrophotometer (Ultrospec III, Amersham Biosciences)
Water bath preset to 35°C

Additional Reagents and Equipment

If using high-resolution 2D gel electrophoresis to separate proteins:

- Step 9a of this protocol requires the reagents and equipment in Chapter 4, Protocol 5 (Method A)
- Step 9b, those in Chapter 4, Protocol 7
- Step 9c, those in Chapter 4, Protocol 8
- Step 9f, those in Chapter 4, Protocol 9

METHOD

Preparation of IAV-Trypsin Membranes

1. Submerge a 10 x 12-cm² IAV membrane in a solution of 2 mg/ml trypsin in 20 mM NaH_2PO_4 (pH 7.8); incubate in a rotating hybridizer for 3 hours at room temperature.
2. Wash the membrane three times briefly but with agitation in 10 ml of PBS-Tween to remove unbound trypsin.
3. Incubate the membrane in 10 ml of capping reagent for 3 hours at 4°C to block the remaining free carboxylic groups.
4. Wash the membrane three times briefly but with agitation in 10 ml of PBS-Tween to remove the capping reagent.
5. Gently wash the membrane twice for 30 minutes each in 10 ml of PBS-Tween.

   Membranes can be stored for 2–3 years at 4°C in a 46 mM Tris-HCl, 1 mM calcium chloride, 0.01% sodium azide buffer (pH 8.1).

Testing the Activity of the Immobilized Trypsin

6. Add 1-cm² of IAV-trypsin membrane to a mixture of:

   2.6 ml of 460 mM Tris-HCl (pH 8.1) containing 11.5 mM $CaCl_2$
   0.3 ml of 10 mM TAME
   0.1 ml of 1 mM HCl

7. Stir the mixture for 40 seconds. Measure the absorbance of the solution at 247 nm with a UV-visible spectrophotometer.
8. After 3 minutes of constantly stirring the solution, take a second absorbance reading. The value of ΔA_{247}/minute is used to calculate the equivalent amount of active trypsin expressed per unit of surface area as described by Hummel (1959).

Protein Separation by Gel Electrophoresis

9. Separate the proteins by either 1D or 2D gel electrophoresis. The choice of method will depend on the complexity of the sample and the abundance of the protein(s) of interest.

For 1D gel electrophoresis:

a. Dilute proteins to the appropriate amount in 60 mM Tris-HCl (pH 6.8) containing 10% glycerol (v/v), 2% SDS (w/v), and 3% mercaptoethanol (v/v).

b. Reduce the proteins by heating the sample for 5 minutes at 95°C just before loading the samples onto the gel.

   12%T, 2.6%C, 1-mm-thick linear polyacrylamide gels are typically used.

c. Separate the proteins at 200 V for 45 minutes.

For 2D minigel electrophoresis:

a. Combine 1 mg of the protein sample with 150 µl of rehydration solution.

b. Apply the entire diluted protein sample to 7-cm ready-made IPG strips.

c. Peel off the protective cover sheets from the IPG strips, and position them in the rehydration chamber such that the gel of the strip is in contact with the sample.

d. Cover the IPG strips with low-viscosity paraffin oil, and allow them to rehydrate for at least 6 hours (overnight is fine) at room temperature.

e. Use tweezers to remove the rehydrated IPG gel strips from the chamber, rinse them with H_2O, and place the strips gel side up in the electrophoresis running tray according to the manufacturer's instructions.

f. Increase the voltage linearly from 300 to 3500 V over a 10-minute period, followed by 1 additional hour at 3500 V.

g. Equilibrate the strips for 12 minutes in the rehydration tray using 3 ml per groove of 2D equilibration solution. Discard the equilibration solution and replace it with 3 ml per groove of sulfhydryl blocking solution. Incubate the strips for 5 minutes.

h. After the equilibration, cut the IPG gel strips to size.

i. Overlay the second-dimension gels with ~70°C overlay solution.

j. Immediately load the IPG gel strips through the overlay solution.

k. Run the gels at 200 V (constant) for 30 minutes at 12°C using a minigel electrophoresis system.

For high-resolution 2D gel electrophoresis:

a. Follow the steps in Chapter 4, Protocol 5 (Method A) to separate the proteins by IEF (the first dimension). Incorporate the following details:

- Use IPG strips that are 3 mm x 18 cm.
- Rehydrate the strips in the absence of the protein sample, which should be applied using sample cups.
- Mix 4 mg of protein sample with 60 µl of first-dimension sample loading buffer.
- Separate the proteins using a linear voltage gradient that increases from 300 to 3500 V over 3 hours, followed by 3 more hours at 3500 V. Then increase the voltage to 5000 V. Separate the proteins by electrophoresis overnight for a total of 100,000 volt-hours.

b. Follow the steps in Chapter 4, Protocol 7 to prepare the second-dimension SDS-PAGE gels. Incorporate the following details:

- Gel dimensions are 160 x 200 x 1.5 mm.
- Each gel is 9–16%T/2.6%C acrylamide and bisacrylamide.
- Sodium thiosulfate (5 mM) is added to acrylamide/bisacrylamide to initiate/catalyze polymerization.
- Pour the gels to within 0.7 cm from the top of the plates and overlay them with 1-butanol.
- Allow the gels to polymerize for ~2 hours.
- Remove the butanol, overlay the gels with H_2O, and incubate them overnight.

c. Follow the steps in Chapter 4, Protocol 8 to prepare the proteins for the second-dimension separation by SDS-PAGE. Incorporate the following details:

- Equilibrate the strips for 12 minutes with 100 ml of 2D equilibration solution.
- Block all sulfhydryl groups within the proteins for 5 minutes with 100 ml of sulfhydryl blocking solution.

d. Cut the IPG gel strips to size so that they fit atop the SDS-PAGE gels. Remove 6 mm from the anodic end and 14 mm from the cathodic end.

e. Layer each SDS-PAGE gel with overlay solution heated at ~70°C.

f. Immediately apply the IPG gel strips to the top of the SDS-PAGE gels, according to Chapter 4, Protocol 9.

g. Run the gels at 40 mA/gel under constant current for 5 hours at 8–12°C using a large electrophoresis system.

Staining the Gel-bound Proteins

10. Stain the proteins with 0.1% (w/v) of Coomassie Brilliant Blue R250 solution by submerging the gel in the stain solution for 30 minutes.

11. Destain the gel in repeated washes of destain solution.

Double-parallel Digestion of Proteins

12. Soak the gel in H_2O for 5 minutes.

13. Repeat Step 12 two more times.

14. Soak the gel in 50% acetonitrile for 20 minutes, followed by 10 minutes in 80% acetonitrile.

15. Air-dry either the entire wet gel or a selected piece at room temperature.

16. Incubate the gel with 0.05 mg/ml of trypsin in 10 mM Tris-Cl (pH 8.2) for 30 minutes at 35°C.

    The volume of trypsin solution to use equals 3–5 times the initial volume of the gel. This step both rehydrates the gel and begins digesting the proteins.

17. After 30 minutes of incubation, discard the trypsin solution.

18. Just prior to electroblotting the proteins, equilibrate the IAV trypsin membrane and the PVDF membrane (in separate containers) in transblotting buffer for 5 minutes.

19. Carry out the electrotransfer overnight (12–18 hours) at room temperature in a semi-dry apparatus by inserting a double layer of IAV-trypsin membrane between the gel and the PVDF membrane.

    The migration time of the proteins through the enzymatic membranes can be increased by applying an alternating square-shape voltage during the transfer: +12.5 V for 125 msec followed by –5 V for 125 msec.

20. Wash the PVDF membrane in H_2O for 5 minutes.

Staining the Membrane-bound Proteins

21. (*Optional*) Stain the PVDF membranes with 0.5% Amido Black for 1 minute, and destain them with repeated washes of H_2O.

Acquisition of PMF Data

22. Wash the PVDF membrane containing the bound proteins with H_2O and air-dry.

23. Spray the membrane surface with a solution of 10 mg/ml of HCCA until it is completely wetted. Allow the membrane to air dry.

24. Use a very small amount of high-vacuum grease to attach a 4 x 4-cm^2 piece of PVDF membrane onto a modified MALDI sample plate. Make sure that the grease does not penetrate the membrane.

25. Define absolute coordinates on the membrane, and establish a grid that represents positions where mass spectra will be acquired.

    For a given experiment, the distance between distinct mass spectrum acquisitions should be defined as a constant ranging between 0.2 and 0.5 mm, the exact distance determined empirically.

26. Insert the MALDI plate into the mass spectrometer and set the parameters for mass spectra acquisition, according to the manufacturer's instructions.

    Our mass spectra were acquired on a Voyager DE-STR MALDI-TOF mass spectrometer (Applied Biosystems) equipped with a 337-nm UV nitrogen laser, a delayed extraction device, and an acquisition rate of 20 Hz. The acquisition was performed with an acceleration voltage of 20 kV and a delay extraction time settled to 150 nsec. The mass range was generally defined from 850 to 4000 daltons with a low-mass gate fixed at 750 daltons. The laser power was set ~20% above threshold. Spectra (50–100) were acquired at each position depending on the amount of analyzed material.

Protein Identification

27. Using the mass spectrometer's acquisition software, set the peak detection threshold to the value optimized for a set of calibrated spectra (for standards, see Tables 8.9 and 8.10). Then, convert the positions on the MALDI sample plate to apparent molecular mass (M_r) and pI values.

    An interactive tool that enables the steps of the identification process to be performed automatically has been developed (Gras et al. 1999; Perkins et al. 1999).

28. Combine the PMF data from all of the spectra with calculated pI and M_r, and also with user-defined interrogation criteria, such as peptide mass tolerance and chemical modifications. The combined data are sent automatically to SmartIdent (http://ch.expasy.org/tools/.

    The process of identification takes into account weak expression of some proteins but also overlapping protein spots, so the minimal number of matching peptides for the PMF search should be set as low as possible (i.e., 3 peptides). Set the number of missed cleavages to 1. In any case, peptides with one missed cleavage, or other peptides carrying artifactual modifications such as propionamide cysteine, will have a less important weight in the final protein identification score than the nonmissed cleavage and nonmodified peptides. It is well known that the position of mass spectrum acquisition on the MALDI sample plate influences the calibration error (Egelhofer et al. 2000). In addition, the membrane pasted on the sample plate could be also slightly warped, resulting in a variation in the calibration of the spectra. Therefore, expect mass differences up to 0.6 dalton, although usually, this will not hinder protein identification because the SmartIdent tool is able to compensate for calibration errors.

29. Write peak-list and identification results into a text file.

Creation of Virtual Maps

30. Analyze and cluster the identified proteins according to the procedure described by Bienvenut et al. (2001). Several criteria are defined to trace and exclude false identifications from the final list of identified proteins.
    - First, identifications of the same protein should be clustered in regions that have the shape of a spot on a 2D electrophoresis gel. Therefore, identifications that are isolated or spread out over a large portion of the scanned membrane should be discarded.
    - Second, identifications that require matches with a matrix cluster or impurity masses should be excluded, as well as weakly expressed proteins that required peptide masses of abundant proteins.
    - Third, identifications that have a low average score are not considered.

Results

- The identification results are represented as a virtual, annotated 2D electrophoresis map. The program generates a multidimensional image where x and y coordinates are related to pI and M_r, respectively, and the z axis can be related to numerous data or information associated with the identified protein such as PMF spectra intensity, SmartIdent identification score, number of matching peptides, etc.
- The image files are stored in a graphical format that can be read by the Melanie 2D image analysis software. Those images contain numerous distinct attributes such as pI, M_r, identification labels (SwissProt or TrEMBL AC numbers, I.D. labels), peptide masses, and MS intensities. The user can choose to filter and visualize only particular aspects, as desired.

APPLICATION: ANALYSIS OF A 2D GEL FROM *E. COLI*

The molecular scanner approach was applied to the analysis of *E. coli* proteins. Three 1-mg samples of *E. coli* were separated on mini-2D gels.

- The first gel was stained with Coomassie Brilliant Blue (Figure 8.47a).
- The proteins on the second gel were electrotransferred to a PVDF membrane without any exposure to trypsin. The PVDF membrane containing the bound proteins was stained with Amido Black (Figure 8.47b).
- The proteins on the third gel were treated as in the protocol. The results are:
  1. No proteins were detectable after staining the membrane with Amido Black (Figure 8.47c).
  2. A 9 x 13-mm^2 piece of the collecting PVDF membrane corresponding to the 5.1–5.2 pI range and the 35,000–45,000-dalton M_r range was cut out and pasted on the modified sampling plate of a MALDI-TOF mass spectrometer. The membrane was scanned on a 48 x 32 mm grid with a sampling distance of 0.25 mm in both horizontal and vertical directions.
  3. Sixty-four laser shots were fired at a frequency of 3 Hz, leading to an acquisition time of ~9 hours. Almost 1540 spectra were acquired. The disc space needed to store all the spectra was 350 MB.
  4. All of the acquired PMF data (Figure 8.47e) were used for protein identification in the SwissProt protein database with the SmartIdent program, and seven distinct proteins were unambiguously identified including proteins from overlapping 2D electrophoresis spots:

     Isocitrate dehydrogenase (IDH_ECOLI, P08200)
     Isocitrate lyase (ACEA_ECOLI, P05313)
     Aldehyde dehydrogenase A (ALDA_ECOLI, P25553)
     6-Phosphogluconate dehydrogenase (6PGD_ECOLI, P00350)
     S-Adenosylmethionine synthetase (METK_ECOLI, P04384)
     Phosphoglycerate kinase (PGK_ECOLI, P11665)
     Hypothetical protein (YBHE_ECOLI, P52697)
  5. Position on the sample plate (pI, M_r coordinates), PMF data, and result scores of identification process were used to generate a multidimensional image (Figure 8.47d). Figure 8.47 shows that a close correlation exists between the Coomassie Brillant Blue-stained gel image, the Amido Black-stained membrane image, and the reconstructed image obtained from MS scanning of the membrane. These scanning results were confirmed by in-gel digestion and MS analysis of the corresponding spots excised from the Coomassie Brilliant Blue-stained mini-2D electrophoresis gel.

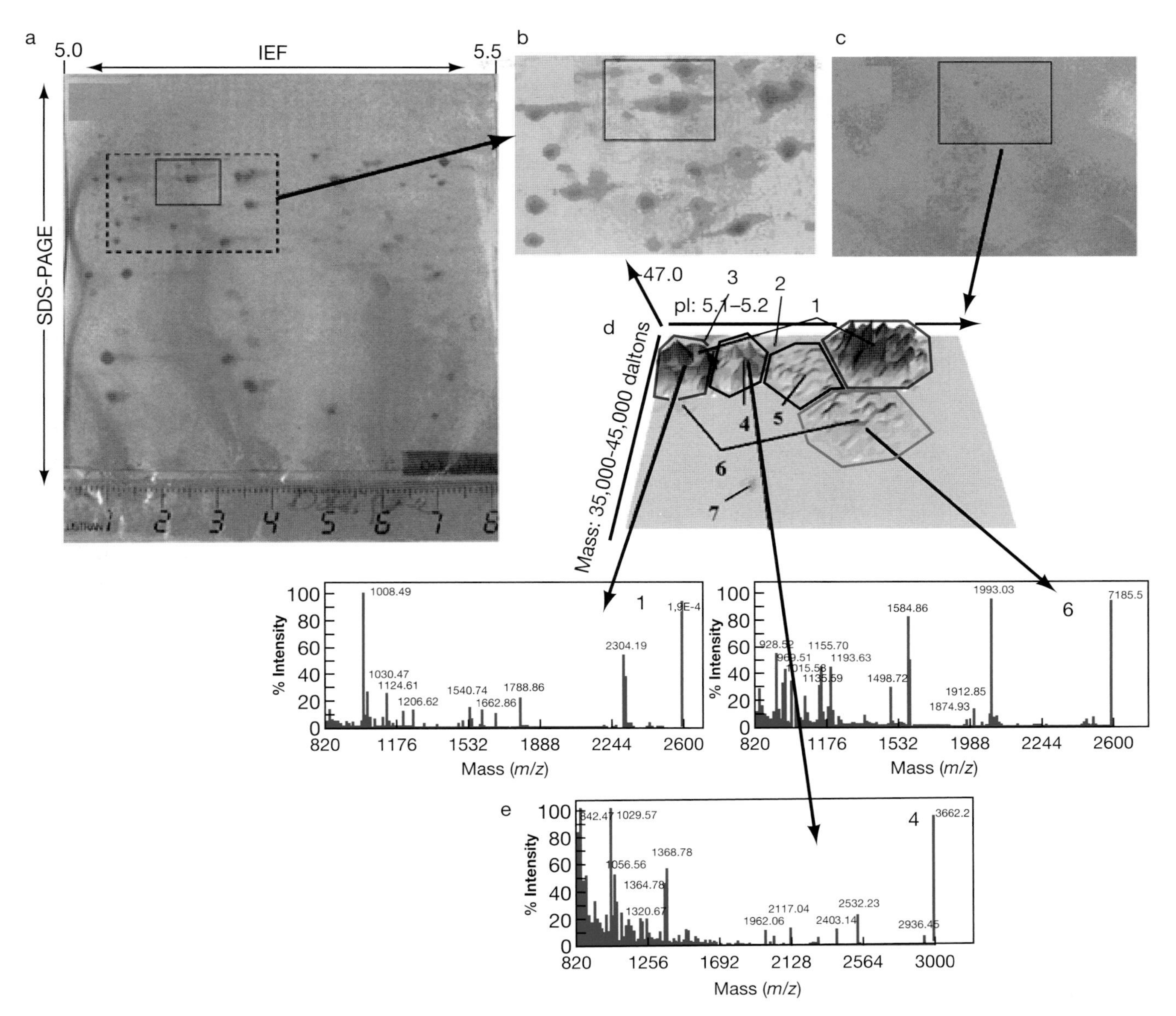

FIGURE 8.47. DPD result from a mini-2D electrophoresis of *E. coli* extract. (*a*) Coomassie Blue-stained mini-2D electrophoresis gel. (*b*) Amido Black-stained PVDF membrane after electrotransfer. The image corresponds to the dashed box in *a*. (*c*) Amido-Black-stained PVDF membrane after DPD transfer. The image corresponds to the dashed box in *a*. (*d*) 3D image obtained after MS scanning of the straight line delimited area of *c* (pI range: 5.1–5.2, m.w. range: 35–45 kD, real size: ~9 x 13 mm, 1536 MS spectra). The vertical axis corresponds to the SmartIdent identification score. Proteins identified in the *E. coli* scanning experiment are (1) isocitrate dehydrogenase (IDH_ECOLI, P08200); (2) isocitrate lyase (ACEA_ECOLI, P05313); (3) aldehyde dehydrogenase A (ALDA_ECOLI, P25553); (4) 6-phosphogluconate dehydrogenase (6PGD_ECOLI, P00350); (5) *S*-adenosylmethionine synthetase (METK_ECOLI, P04384); (6) phosphoglycerate kinase (PGK_ECOLI, P11665); (7) hypothetical protein (YBHE_ECOLI, P52697). Note that proteins 1 and 6 appear as two different spots. (*e*) Peptide mass fingerprint spectra obtained during the scanning process for isocitrate dehydrogenase, 6-phosphogluconate dehydrogenase, and phosphoglycerate kinase.

PROTOCOL 13

Peptide Mass Fingerprinting Using the Chemical Printer

THE MOST POPULAR METHOD FOR THE ANALYSIS OF PROTEINS separated by 2D-PAGE is "in-gel" digestion (Pandey and Mann 2000). This method typically involves the excision of the protein spot from the gel and deposit into the well of a microtiter plate. A number of liquid-handling steps are performed before and after endoproteinase digestion. The most sensitive results are achieved if the sample is cleaned and concentrated with microtips packed with reversed-phase resin, such as the Millipore ZipTip (see also Protocol 2). The chemical printer approach for peptide mass fingerprinting differs significantly from traditional "in-gel" digestion methods in that the endoproteinase chemistry is brought to the membrane-bound proteins, rather than excising the gel-entrapped protein spots and delivering them to a microtiter plate. Peptide mass fingerprinting is achieved by inserting the manipulated membrane into the MALDI-TOF mass spectrometer and desorbing the peptides directly from the membrane surface.

Electroblotting gel-separated proteins onto membranes is an ideal method for generating high-resolution protein arrays that are suitable for long-term storage. When coupled with the chemical printer technology, membrane-based protein arrays represent a robust medium for chip-based proteomics. Archiving protein arrays on membranes is particularly important as proteomic studies move into clinical studies where patient samples are unique and rarely available for subsequent collections. Microdispensing reagents with piezoelectric devices onto protein spots at precise coordinates ensures minimal cross-contamination of the samples, with the additional advantage that unused protein within the spot can be archived for future analyses.

This protocol was provided by Andrew A. Gooley and his colleagues, Femia G. Hopwood, Janice L. Duff, Parag S. Ghandi, Cameron J. Hill, Wendy L. Holstein, Paul E. Smith, Andrew J. Sloane, and Keith L. Williams (Proteome Systems, Sydney, Australia), and Patrick W. Cooley and David B. Wallace (MicroFab Technologies Inc., Plano, Texas).

MATERIALS

CAUTION: See Appendix 3 for appropriate handling of materials marked with <!>.

Reagents

ACTH peptide

Denaturing sample buffer

The composition of the denaturing sample buffer will vary depending on the source of the proteins. Example buffers:

For *E. coli:* 7 M urea, 2 M thiourea, 1% (v/v) ASB-14, 2 mM TBP (tributylphosphine), 40 mM Tris

For human plasma proteins: 7 M urea, 2 M thiourea, 2% (w/v) CHAPS, 5 mM Tris-Cl

Direct Blue-71 (Sigma)

Direct Blue-71 is more sensitive than Coomassie Blue and has a very low background. A fluorescent stain is not suitable because the blot cannot be registered on the chemical printer.

IEF-strip equilibration buffer
6 M urea <!>
2% SDS <!>
375 mM Tris-Cl (pH 8.8)
Iodoacetamide <!>
Matrix solution (α-cyano-4-hydroxy-cinnamic acid) <!>
Before using, recrystallize the matrix solution and dissolve it to 10 mg/ml in an organic solvent mixture of 30% (v/v) acetonitrile <!>, 20% (v/v) isopropanol <!>, and 20% (v/v) butanol<!>/0.5% (v/v) formic acid <!>.
Octyl-glucopyranoside (OGP) (1% v/v)
Porcine modified trypsin, sequencing grade (Promega)
TBP (tributylphosphine)

Equipment

Centrifuge
Chemical printer
Conductive adhesive tape (3M, product 9073)
Humidified container (see Step 21)
Immobilized pH gradient strips (7 cm, pI 4–7) (Amersham Biosciences)
MALDI target
Mass spectrometer (e.g., Axima-CFR MALDI-TOF, Kratos, Manchester)
NuPAGE (4–12% minigels) (Invitrogen)
Off-line scanner
PVDF membrane (e.g., Immobilon P^{sq}, Millipore, and other protein-sequencing grade PVDF membranes)

Biological Sample

Cells or proteins of interest

Additional Equipment and Reagents

Steps 5 and 6 require the materials used in Chapter 4, Protocol 5 (Method A); Step 8, those in Chapter 4, Protocol 7; Step 9, those in Chapter 4, Protocol 8; Step 10, those in Chapter 4, Protocol 9; and Step 11, those in Chapter 4, Protocol 15 or 16.

METHOD

2D PAGE and Electroblotting of Samples

1. Resuspend the cells or proteins in denaturing sample buffer to a concentration of ~3 mg/ml. Incubate the sample for 2 hours at room temperature.

2. If necessary, reduce the proteins with 3 mM TBP (final concentration from 20 mM stock solution) for 2 hours at room temperature.

3. Alkylate the proteins by adding iodoacetamide (1 mM stock solution) to a final concentration of 15 mM, and incubating the sample in the dark for 1 hour at room temperature.

4. Centrifuge the sample at 21,000*g* for 5 minutes at 4°C.

5. Apply 115 µl of the supernatant to a 7-cm IPG (pH 4–7) strip, and rehydrate the strip for 4 hours at room temperature.
6. Follow the steps in Chapter 4, Protocol 5 (Method A) to separate the proteins by IEF (the first dimension). Incorporate the following detail: Separate the proteins by electrophoresis for a total of 36,000 volt-hours.
7. Following IEF, equilibrate the strip for 20 minutes in IEF-strip equilibration buffer.
8. Follow the steps in Chapter 4, Protocol 7 to prepare the second-dimension SDS-PAGE gels.
9. Follow the steps in Chapter 4, Protocol 8 to ready the proteins for the second-dimension separation by SDS-PAGE.
10. Apply the IPG gel strip to the top of the 4–12% minigel SDS-PAGE gel, and separate the proteins in the second dimension according to Chapter 4, Protocol 9.
11. Following electrophoresis, electroblot the gel onto an Immobilon-P^{sq} membrane at 400 mA for 1.3 hours, according to Chapter 4, Protocol 15 or 16.
12. Stain the membrane with Direct Blue-71 (see Hong et al. 2000).
13. Dry the membrane under vacuum, and store it in a sealed plastic bag to prevent keratin contamination.

Preparation of the Membrane for Chemical Printing

14. Attach the dried membrane to a MALDI target with conductive adhesive tape (3M).
15. Acquire an image of the electroblotted proteins adhered to the MALDI target using an off-line scanner.
16. Detect the protein spots either automatically using a proprietary segmentation analysis algorithm or manually with a simple "point and click" on the scanned electroblot image.

    The commercial version of this instrument (Shimadzu Biotech) includes this software.

17. To achieve registration of the MALDI-membrane target on the chemical printer, select three registration points of the MALDI target (see Figure 8.48). View these points with an overhead digital camera and use the XY position of each registration point to calculate the exact positions of the proteins that have been selected for digestion.

    The overhead camera can also be used to view the protein spots after microdispensing solutions onto the membrane.

Automated On-membrane Tryptic Digestion of Protein Spots

18. Program the chemical printer to dispense 1.5 nl (15 drops) of 1% (v/v) OGP onto each protein spot for three iterations (total volume equals 4.5 nl).

    The use of the nonionic detergent OGP is necessary to wet the hydrophobic PVDF membrane surface, as well as to serve as a blocking reagent to minimize adsorption of the endoproteinase to the PVDF membrane.

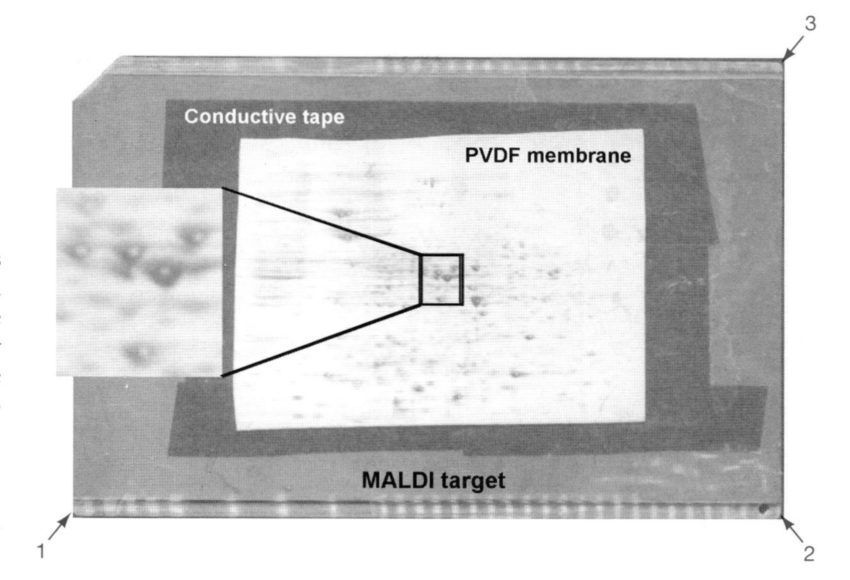

FIGURE 8.48. An electroblot of *E. coli* proteins separated on a 2D polyacrylamide gel and attached to a MALDI target with conductive tape. Proteins are identified with PMF directly from the intact membrane. The corners of the MALDI target, labeled 1, 2, and 3 are used to register the protein spots in the chemical printer and in the Axima-CFR mass spectrometer. (*Inset*) Enlarged view of a section of the membrane where endoproteinase solutions have been dispensed onto the electroblotted proteins.

19. Program the chemical printer to deliver 1 nl (10 drops) of 200 ng/μl trypsin in 25 mM ammonium hydrogen carbonate on top of the detergent for 50 iterations (a final volume of 50 nl). This is equivalent to 10 ng of trypsin delivered to each protein spot.

    For experiments acquiring PMF data from two different endoproteinases dispensed onto the same protein spot, deliver 0.1% (w/v) polyvinylpyrrolidone/50% (v/v) methanol using one iteration of 60 drops.

20. Place the printed membrane into a humidified container and incubate it for 3 hours at 37°C.
21. Following digestion, return the membrane to the chemical printer.
22. Add 50 fmoles/μl ACTH peptide to the matrix solution.
23. Program the printer to dispense 2 nl (20 drops) of matrix solution onto the digested proteins spots with 50 iterations.

Using the dispensing iterations described above, the digested protein area on the protein spots is ~0.2 mm^2 (400–500-μm diameter). If the solutions were dispensed continuously, the digested area would be significantly larger and the resulting peptides diffused over the area. Figure 8.48 shows the PVDF electroblot of the *E. coli* proteins adhered to the MALDI target after the matix has been applied to the digested proteins. The expanded region clearly shows the positions where the protein has been digested.

MALDI-TOF-MS

24. Introduce the entire PVDF membrane (adhered to the MALDI target) into the mass spectrometer. A coordinate file corresponding to the jetting positions is converted into an Axima-CFR stage file (see Figure 8.49 a,b), with the Axima-CFR software Ascii2plate.
25. Acquire all of the mass spectra in positive-ion mode. Ionize the peptides with a 337-nm nitrogen laser and accelerate them with a 25-kV time-delayed extraction pulse.

    Despite the membrane having a surface that is not flat, quality spectra are obtained by acquiring all spectra from a single position (see Figure 8.49c).

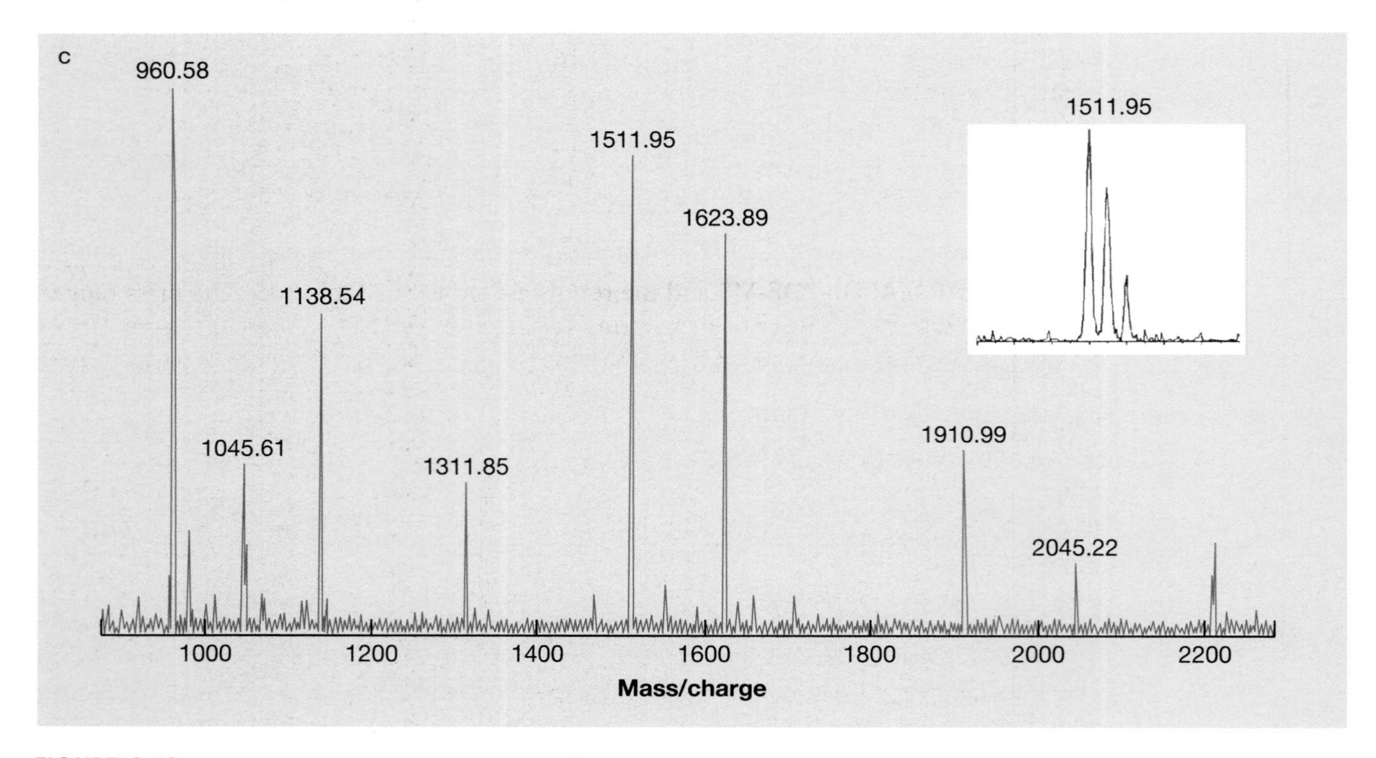

FIGURE 8.49. Automated peptide mass collection from proteins digested on-membrane. (*a*) Axima CFR control window; (*b, inset*) a MALDI target coordinate file (defining the positions of each of the piezoelectric dispensing locations) created using the chemical printer and converted into an Axima CFR stage file; (*c*) resultant PMF spectrum of albumin fragment (see Table 8.15) acquired directly from the membrane.

26. To obtain mass spectra, average 50–100 individual laser shots on on-membrane-digested samples. Use an internal 2-point calibration for on-membrane-digested samples on one trypsin autolysis peptide (monoisotopic mass of 842.51) and ACTH peptide (monoisotopic mass of 2465.19).

27. Perform monoisotopic peak picking using a custom peak harvesting program (Breen et al. 2000) and identify proteins using either a PMF search engine (which searches GenBank, www.ncbi.nlm.nih.gov/Genbank/index.html) or ExPASy PeptIdent www.expasy.ch.

PEPTIDE MASS FINGERPRINT ANALYSIS OF *E. COLI* AND HUMAN PLASMA PROTEOMES

MALDI-TOF mass spectrometric analysis of the proteins digested on an *E. coli* electroblot (Figure 8.50) resulted in peptide mass fingerprint (PMF) identification of 56 protein spots from a search of the SwissProt database (see Table 8.14). The mass range covered was evenly distributed from DNA-binding protein H-NS (predicted mass 15,530 daltons) to HSP-70 protein (predicted mass 68,983 daltons). The average coverage was 31.5% with a sample high of 65.4% (adenylate kinase, predicted mass 23,571) and a sample low of 8.3% (Hsp70 protein, predicted mass of 689,383 daltons). Several protein spots appeared to be in low abundance based on the staining of Direct Blue-71: spot 46 (stringent starvation protein, 53.3% amino acid coverage), spot 7 (prolyl-tRNA synthetase, 11.4% amino acid coverage), and spot 23 (coding sequence of gene *lpd*, 21.1% amino acid coverage).

As part of an ongoing study to understand the influence of glycosylation on the separation properties of glycoproteins, proteins were identified from a 2D electroblot of a human plasma sample (Figure 8.51). Whereas *E. coli* proteins migrate essentially as predicted from their gene sequence (the correlation coefficient for the apparent mass of the proteins in Figure 8.50 was $R^2 = 0.966$), the same is not apparent for plasma-derived glycoproteins. The correlation coefficient for the apparent mass of the plasma proteins in Figure 8.51 is $R^2 = 0.74$. Whereas the pI is often difficult to measure, the observed correlation coefficient for the *E. coli* proteins was quite good at $R^2 = 0.8$. For the plasma proteins, there is essentially no correlation with an $R^2 = 0.38$.

Despite the poor predictability of glycoprotein migration, 32 protein spots were identified with PMF MALDI-TOF-MS and the results are shown in Table 8.15. The mass range covered was distributed from transthyretin (predicted mass 13,761 daltons) to serotransferrin (predicted mass 75,181 daltons). The average coverage was 37%, with a sample high of 66.3% (apolipoprotein A-I, predicted mass 28,078 daltons) and a sample low of 15.2% (α-2 HS glycoprotein, predicted mass of 30,221 daltons).

Overall, the sequence coverage for proteins identified with on-membrane digestion is typically slightly lower than that of the standard "in-gel" digestion procedures practiced in our laboratories. This is not unexpected for two reasons. First, the area digested for "in-gel" digestion is typically three times larger than the "on-membrane" digestion: ~0.6 mm^2 compared to ~0.2 mm^2. Second, the in-gel digests are normally purified and concentrated with C_{18} ZipTips. Both of these factors would contribute to a larger coverage obtained for in-gel-digested samples. However, the advantage of the on-membrane approach is that most of the protein remains available for subsequent experiments, such as a second endoproteinsase digestion (see panel on MULTIPLE ENDOPROTEINASE REACTIONS WITH THE CHEMICAL PRINTER following Table 8.15) or characterization of the sites of N-linked glycosylation using sequential PNGase F and trypsin digestion (Sloane et al. 2002).

TABLE 8.14. PMF results from the on-membrane digestion of *E. coli*

Spot No.	Protein ID	Accession number[a]	pI	MW (Da)	AA coverage (%)	Number of peptides found
1	Hsp70 protein	216440	4.83	68983	13.3	9
2	Hsp70 protein	216440	4.83	68983	8.3	6
3	GroEL protein	536987	4.85	57293	21.2	16
4	GroEL protein	536987	4.85	57293	27.6	16
5	GroEL protein	536987	4.85	57293	33.6	16
6	analog of ATP-dependent protease regulatory subunit	41114	5.42	65858	13.5	10
7	prolyl-tRNA synthetase	147362	5.12	63662	11.4	7
8	ATP synthase beta chain (EC 3.6.1.34)	P00824	4.9	50294	19.4	6
9	aldehyde dehydrogenase, NAD-linked	1787684	5.07	52240	28	12
10	tol-8	882565	5.46	53967	30.7	16
11	ketol-acid reductoisomerase	148181	5.2	54034	23	13
12	ketol-acid reductoisomerase	148181	5.2	54034	24.9	14
13	6-phosphogluconate dehydrogenase	146942	5.1	51447	20.9	12
14	glutamate decarboxylase isozyme	1787769	5.29	52634	16.7	9
15	glutamate decarboxylase isozyme	1787769	5.29	52634	27.7	15
16	glutamate decarboxylase isozyme	1787769	5.29	52634	35.6	19
17	isocitrate dehydrogenase	9664346	5.26	43178	26.5	10
18	enolase	563868	5.66	46417	17.9	8
19	enolase	1789141	5.33	45626	9.5	7
20	elongation factor Tu	147969	5.3	43286	34.3	15
21	alkaline phosphatase	147225	5.75	49408	41.6	13
22	alkaline phosphatase	147225	5.75	49408	34.6	15
23	coding sequence of gene lpd	434012	5.79	50788	21.1	11
24	periplasmic oligopeptide-binding protein.	1742032	5.85	58359	36.6	12
25	ATP synthase F1 α-subunit	290583	5.8	55188	32.4	16
26	sn-glycerol 3-phosphate transport system; periplasmic-binding protein	1789862	6.28	48418	23.7	10
27	sn-glycerol 3-phosphate transport system; periplasmic-binding protein	1789862	6.28	48418	38.4	17

28	serine hydroxymethyltransferase	1788902	6.04	45288	25.4	15
29	phosphate-binding protein	147256	6.92	34134	52.3	15
30	transaldolase A	1788807	5.9	35636	35.1	9
31	OmpA	42161	6	37178	29.8	10
32	OmpA	42161	6	37178	21.7	8
33	OmpA	42161	6	37178	32.7	16
34	aspartate aminotransferase	1787159	5.55	43546	31.1	10
35	outer-membrane protein II	146981	6	37178	21.4	12
36	"elongation factor EF-Ts"	473825	5.22	30404	19.4	6
37	RNA polymerase alpha subunit	147715	4.97	36489	52.3	16
38	phoE protein	42391	4.93	38898	17.7	6
39	putative lipoprotein	1786396	4.93	27236	32.1	6
40	peripheral membrane protein U	290572	5.14	27329	41.9	13
41	FliY protein	687652	5.29	26068	29.7	11
42	alkyl hydroperoxide reductase small subunit	216543	5.03	20748	64.7	12
43	PTS enzyme III glc	145623	4.73	18240	52.1	11
44	orf, hypothetical protein	1789043	5.18	23086	36.8	10
45	orf, hypothetical protein	1787124	5.2	23086	36.5	9
46	stringent starvation protein	606168	5.22	24289	53.3	10
47	putative actin	1789289	6.12	26619	26.8	10
48	orf, hypothetical protein	1788278	5.63	31171	24.7	7
49	adenylate kinase	216516	5.56	23571	65.4	21
50	triosephosphate isomerase	305022	5.64	26955	25.1	10
51	superoxide dismutase, iron	1787946	5.59	21253	31.1	4
52	peptidyl-prolyl cis-trans isomerase b	145290	5.32	18172	36.6	8
53	DPS	41295	5.71	18684	63.5	13
54	orf, hypothetical protein	1789019	5.69	16053	55	11
55	DNA-binding protein H-NS	1651637	5.44	15530	45.3	14
56	outer membrane protein X	1787034	5.3	16382	30.4	6
				Average	**31.5**	**11.4**

[a]Accession number from GenBank.

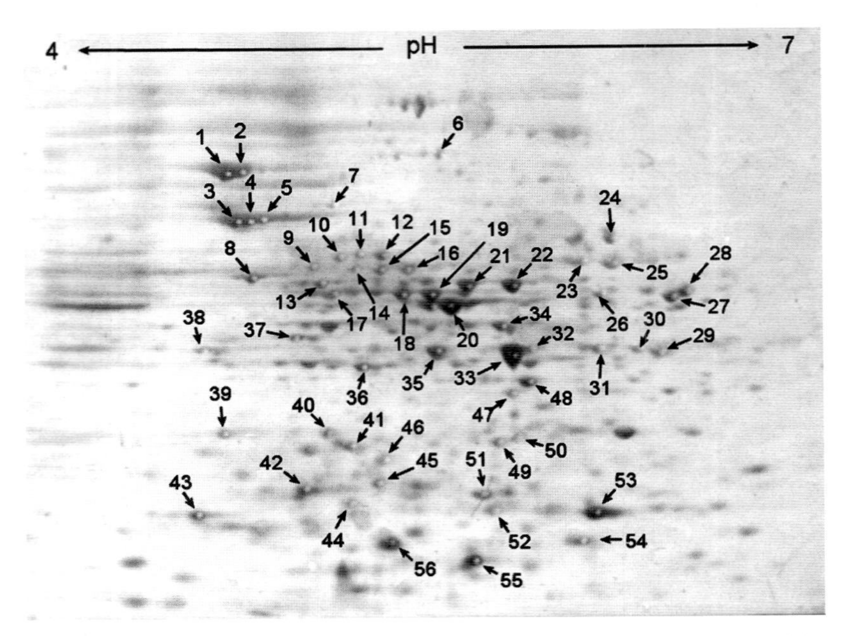

FIGURE 8.50. A 2D polyacrylamide gel of *E. coli* proteins electroblotted onto PVDF (Immobilon-P^{sq}). Labeled protein spots were digested on-membrane using the chemical printer, and peptide masses were collected using MALDI-TOF-MS.

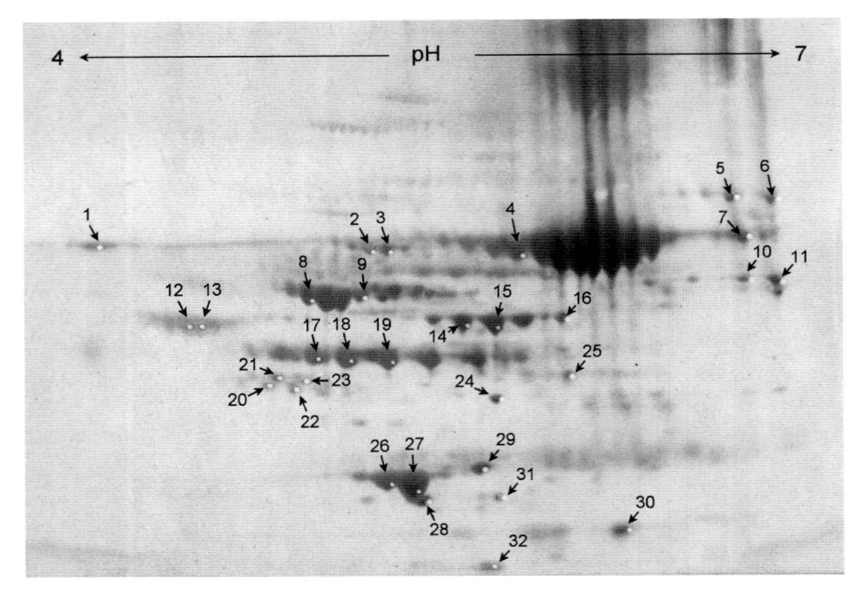

FIGURE 8.51. Protein spots identified on a 2D polyacrylamide gel of human plasma proteins electroblotted onto Immobilon-P^{sq}. Labeled protein spots were digested on-membrane using the chemical printer and peptide masses were collected using MALDI-TOF-MS. Protein spots 1 and 7 were identified as human serum albumin that had not completely focused due to the high concentration of albumin in plasma.

TABLE 8.15. PMF analysis of 2D electrophoresis human plasma proteins electroblotted onto PVDF

Spot No.	Protein ID	Accession number	pI	MW	Amino acid coverage (%)	Number of peptides found	MC[a]
2	α-1B-glycoprotein	P04217	5.7	51941	39.7	18	5
3	α-1B-glycoprotein	P04217	5.7	51941	25.3	11	2
4	hemopexin	P02790	6.4	49245	23.5	12	3
5	serotransferrin	P02787	6.7	75181	29.5	18	5
6	serotransferrin	P02787	6.7	75181	16.9	12	0
8	α-1-antitrypsin	412141	5.4	44324	34.8	15	0
9	α-1-antitrypsin	412141	5.4	44324	31	10	4
10	fibrinogen β chain	P02675	7.95	50762	28.6	12	2
11	fibrinogen β chain	P02675	8	50762	32.9	14	2
12	α-2-HS-glycoprotein (Fetuin-A)	P02765	4.5	30221	18.4	6	1
13	α-2-HS-glycoprotein (Fetuin-A)	P02765	4.5	30221	15.2	7	1
14	fibrinogen γ A chain	P02679	5.2	48468	31.4	17	0
15	fibrinogen γ A chain	P02679	5.2	48468	24.4	11	0
16	fibrinogen γ A chain	P02679	5.2	48468	26	12	4
17	haptoglobin-1/2- β chain	P00738	6.32	27265	40	10	3
18	haptoglobin-1/2- β chain	P00738	6.32	27265	46.1	12	1
19	haptoglobin-1/2- β chain	P00738	6.32	27265	40	9	2
20	clusterin (apolipoprotein J) α chain	P10909	5.66	25883	45	9	6
21	clusterin (apolipoprotein J) β chain	P10909	6.27	24197	37.6	8	5
22	clusterin (apolipoprotein J) α chain	P10909	5.66	25883	46.4	9	2
23	clusterin (apolipoprotein J) β chain	P10909	6.27	24197	37.6	7	3
24	apolipoprotein E	P02649	5.5	34236	50.2	17	0
25	fragment of serum albumin	P02768	6.16	27265	48.6	5	2
26	apolipoprotein A-I	P02647	5.3	28078	66.3	23	1
27	apolipoprotein A-I	P02647	5.3	28078	60.9	20	3
28	plasma retinol-binding protein	P02753	5.2	21071	29.8	8	9
29	apolipoprotein A-I	P02647	5.3	28078	51	15	7
30	haptoglobin-1/2- α chain	P00738	5.57	15946	29.8	4	0
31	plasma retinol-binding protein	P02753	5.3	21072	29	5	2
32	transthyretin (prealbumin)	P02766	5.5	13761	74	9	3
				Average	**37.0**	**11.5**	

[a]MC indicates Number of single-miss cleavages.

Multiple Endoproteinase Reactions with the Chemical Printer

Control of the ink-jet devices allows exquisite precision of the volumes dispensed. Hence, multiple endoproteinase digestions can be performed on many of the protein spots. Proteins from a human plasma sample were separated on an 11-cm IPG with a pI range of 5–6.5. Under these conditions, apolipoprotein A-IV is clearly resolved from the β-chain of haptoglobin (Figure 8.52) and was analyzed on the chemical printer with both endoproteinase Glu-C and trypsin. The two printed positions are shown in Figure 8.52.

Despite the successful dispensing of the acetonitrile-based matrix solution, this reagent diffuses into the membrane easily due to its nonpolar properties. Combined with a small error in the registration between the chemical printer point file and the Axima-CFR MALDI stage file, as well as the added variable of the size of the MALDI laser beam (~110-µm diameter), we discovered that better quality results were obtained by aspirating the multiple digests from the membrane surface onto a micro-C_{18} ZipTip (Millipore) rather than desorbing them directly in the Axima-CFR MALDI-TOF-MS. Peptides were desorbed from the micro-C_{18} ZipTip directly onto an Axima CFR MALDI target, and the resulting PMF data obtained a sequence coverage of 29.8% for trypsin and 21.5% for endoproteinase Glu-C, a combined amino acid coverage of 41.2% (see Figure 8.52). This increase, as well as the matching being derived from two endoproteinases, increases the confidence of protein identification.

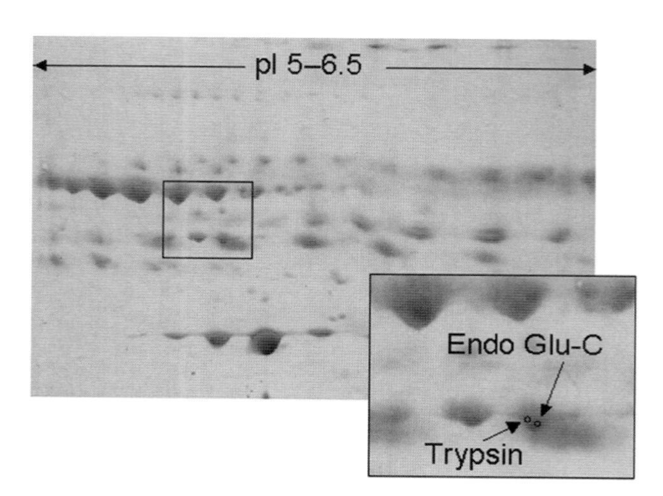

Apolipoprotein A-IV P06727
M_r 43374.5 pI 5.18
Trypsin 29.8%
Endo Glu-C 21.5%
Combined 41.2%

		EVSADQVATV	MWDYFSQLSN	NAKEAVEHLQ
KSELTQQLNA	LFQDKLGEVN	TYAGDLQKKL	VPFATELHER	LAKDSEKLKE
EIGKELEELR	ARLLPHANEV	SQKIGDNLRE	LQQRLEPYAD	QLRTQVNTQA
EQLRRQLTPY	AQRMERVLRE	NADSLQASLR	PHADELKAKI	DQNVEELKGR
LTPYADEFKV	KIDQTVEELR	RSLAPYAQDT	QEKLNHQLEG	LTFQMKKNAE
ELKARISASA	EELRQRLAPL	AEDVRGNLKG	NTEGLQKSLA	ELGGHLDQQV
EEFRRRVEPY	GENFNKALVQ	QMEQLRQKLG	PHAGDVEGHL	SFLEKDLRDK
VNSFFSTFKE	KESQDKTLSL	PELEQQQEQQ	QEQQQEQVQM	LAPLES

FIGURE 8.52. Apolipoprotein A-IV was digested with the endoproteinases trypsin and Endo Glu-C on the marked positions of the protein spot. Peptide mass fingerprinting analysis generated a combined amino acid coverage of 41.2%. The gray-shaded peptides were identified with trypsin, and the blue-shaded peptides were identified with Endo Glu-C.

"TOP-DOWN" PROTEIN SEQUENCE ANALYSIS USING MS/MS*

Gavin E. Reid and Scot A. McLuckey
Department of Chemistry, Purdue University, Indiana

Recently, approaches toward protein identification and characterization have been developed that enable primary sequence information to be obtained directly from the gas-phase dissociation of intact protein ions without prior recourse to extensive separation or proteolytic digestion steps [1]. In this scenario, intact multiply charged protein ions derived from ESI are isolated and then fragmented in the mass spectrometer; if necessary, the resulting product ions are subjected to further fragmentation, until sufficient product ions are generated to allow for the identification of the protein. Protein identification can be carried out by derivation of a "sequence tag" [2–4] via database searching of the uninterpreted product ion spectrum [5,6] or through "de novo" assignment of the complete amino acid sequence [7,8]. This approach has been termed "top-down" protein characterization (Figure 8.53) [1]. The major

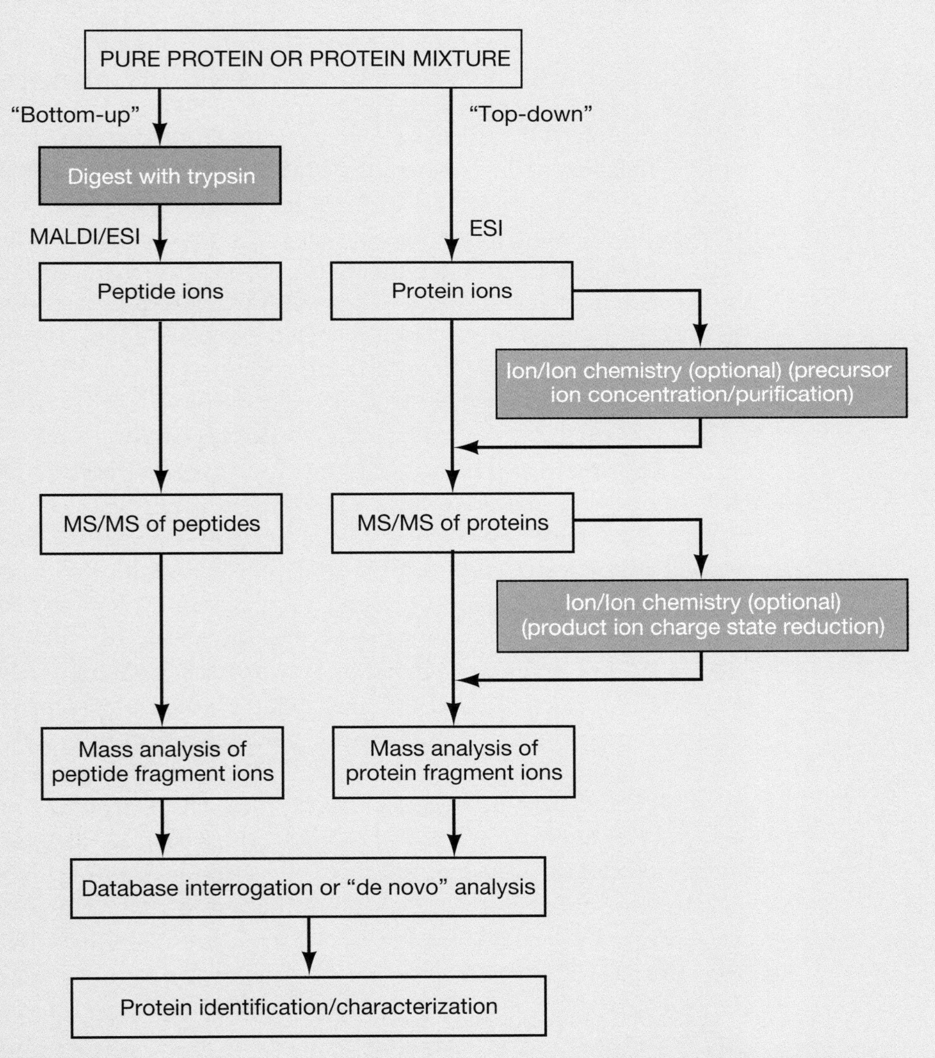

FIGURE 8.53. Schematic overview comparing "bottom-up" and "top-down" protein characterization approaches.

*The numbers in brackets indicate reference citations.

advantage of the top-down approach is that performing an MS/MS experiment on intact protein ions makes the entire sequence available for examination, better enabling complete characterization of the protein and any associated posttranslational modifications [5,9-12]. Additionally, the large number of redundant identifications that are usually obtained with conventional "bottom-up" peptide sequencing strategies can be avoided. Furthermore, the masses of intact proteins are spread over a wider range than those of proteolytically derived peptides, potentially simplifying the task of analyzing proteins in complex mixtures.

In the first whole-protein ion-dissociation studies, triple-quadrupole instrumentation was used to collect MS/MS data from multiply charged protein ions derived from ESI [13–19]. Product ion spectra derived from these ions are typically composed of ions with charges ranging from unity up to the charge of the parent ion. However, the limited resolving power associated with mass measurements in triple-quadrupole instruments was insufficient to resolve the isotope spacings in product ion signals, thereby precluding charge state determination of the individual product ions. The fact that product ion charge is not known a priori can complicate interpretation of the spectrum. Therefore, the application of MS/MS to whole-protein ions using triple-quadrupole MS has been largely restricted to protein fingerprinting measurements (i.e., the comparison of uninterpreted product ion spectra).

Fourier Transform–Ion Cyclotron Resonance Mass Spectrometry

The issue of product ion charge state determination can be overcome by the use of high-magnetic-field-strength Fourier transform-ion cyclotron resonance (FT-ICR) MS [1,2,4,5,7,8, 10–12,20–50]. This form of instrumentation enjoys sufficient resolving power (typically $>10^5$) to measure the isotope spacings in high-mass product ions and provides high-mass accuracy (<10 ppm) for facile product ion identification. This technology therefore allows for the interpretation of product ion spectra such that reliable sequence information can be obtained from the fragmentation of highly multiply charged protein ions. A variety of activation techniques for protein ions have been examined in conjunction with FT-ICR, including collision-activated dissociation (CAD) [20–31], photodissociation [32–36], surface-induced dissociation (SID) [37], and blackbody infrared radiative dissociation (BIRD) [38].

An alternate dissociation technique for implementation on FT-ICR instruments, termed electron-capture dissociation (ECD), was introduced by McLafferty et al. [39–50]. ECD involves capture of low-energy (<1 eV) electrons by multiply charged $[M+nH]^{n+}$ protein ions to yield odd electron-reduced $[M+nH]^{(n-1)+\cdot}$ products which rapidly dissociate by cleavage between the N-αC linkage of the peptide backbone. ECD has been shown to result in more random fragmentation along the peptide backbone for small proteins compared to CAD [44,45]. Additionally, the loss of labile posttranslational modifications, such as carboxylation, glycosylation, phosphorylation, and sulfation, is less evident in ECD-derived product ion spectra, thereby better enabling their characterization [42,43,48]. Recently, an improvement of the ECD method has been described [47], whereby ions are collided with a background gas while subjecting them to electron capture. This activated ion ECD (AI-ECD) method has the effect of increasing the number of observed fragment ions, presumably by breaking intramolecular noncovalent bonds of the secondary and tertiary structures of the ions that otherwise would prevent separation of the fragmentation products arising from ECD cleavage of the protein backbone. For example, Figure 8.54 shows the AI-ECD spectrum obtained from the $[M+20H]^{20+}$ to $[M+40H]^{40+}$ ions of the 29-kD protein carbonic anhydrase. This activation method produces 116 inter-residue cleavages of the possible 258 inter-residue amino acid bonds, whereas CAD of this protein produced only 66 of 258 cleavages [1], and

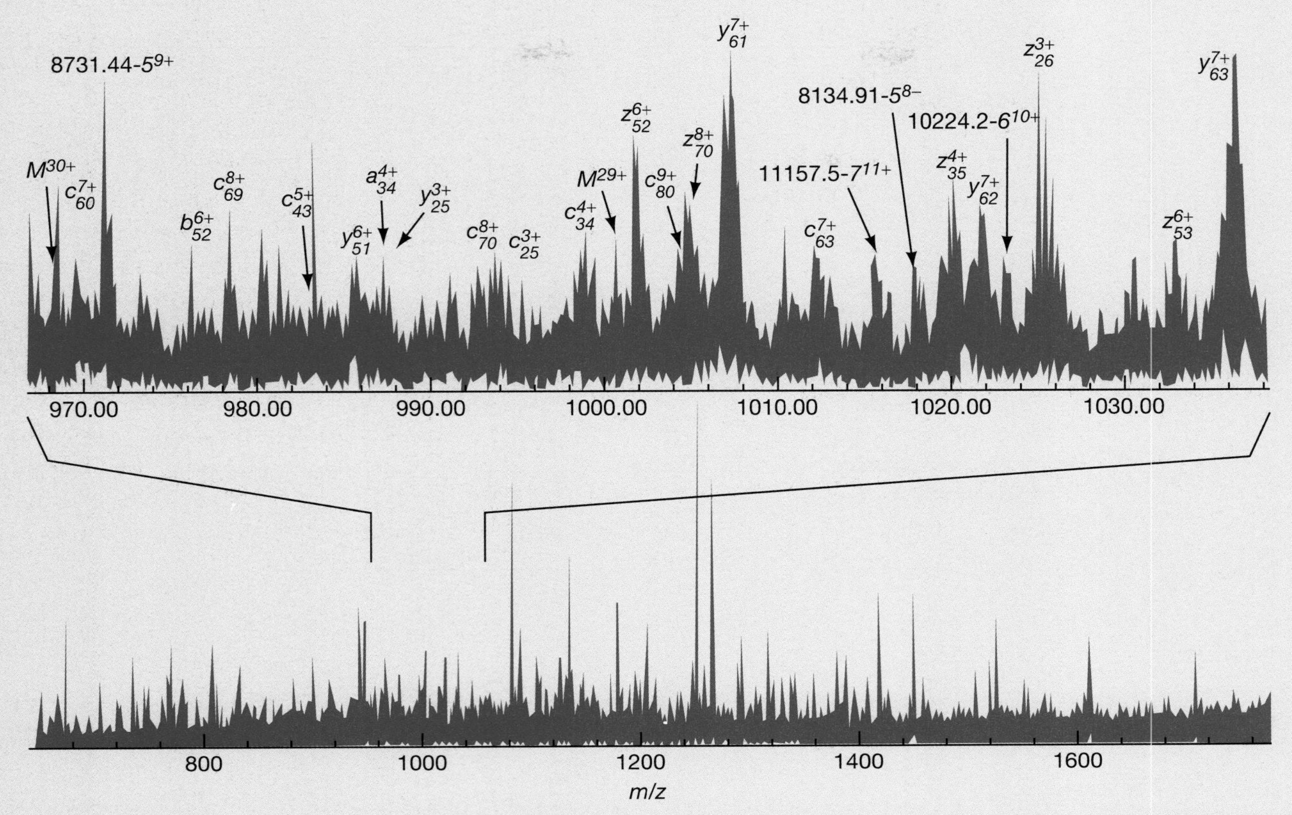

FIGURE 8.54. Activated ion-electron capture dissociation (AI-ECD) spectrum of ions from ESI of carbonic anhydrase B subjected to in-beam collisions during exposure to low energy electrons. (Reproduced, with permission, from [47] Horn et al. 2000a [©American Chemical Society].)

ECD of the isolated $[M + 34H]^{34+}$ ion produced no cleavages [46]. To date, extensive sequence information on proteins as large as 45 kD has been obtained using the AI-ECD technique [50].

Two recent papers have demonstrated the utility of FT-ICR MS for the identification of unknown proteins from simple mixtures. Kelleher and co-authors have identified a number of archaeal or bacterial proteins ranging from 7 kD to 36 kD, present in a mixture of modest complexity. Interrogation of the uninterpreted product ion spectra, using only three or four nonadjacent fragment ions, a mass accuracy of ±0.1 dalton, and no intact mass bias, resulted in correct protein identification from a database containing 5000 proteins [5]. Fenselau et al. have also used intact protein ion dissociation in an FT-ICR, coupled with database searches using a sequence tag, to identify the major ~7-kD biomarker derived from an extract of *Bacillus cereus* T spores, as well as its methionine-oxidized derivative [4].

Ion/Ion Reactions and Quadrupole Ion-trap Mass Spectrometry

An alternative approach to resolving product ion charge state ambiguity relies on the use of gas-phase ion chemistry, rather than high resolving power, as a tool for measuring product ion masses. Such an approach enables the use of relatively low-resolving-power mass analyzers for performing whole-protein MS/MS measurements. The most fully developed example

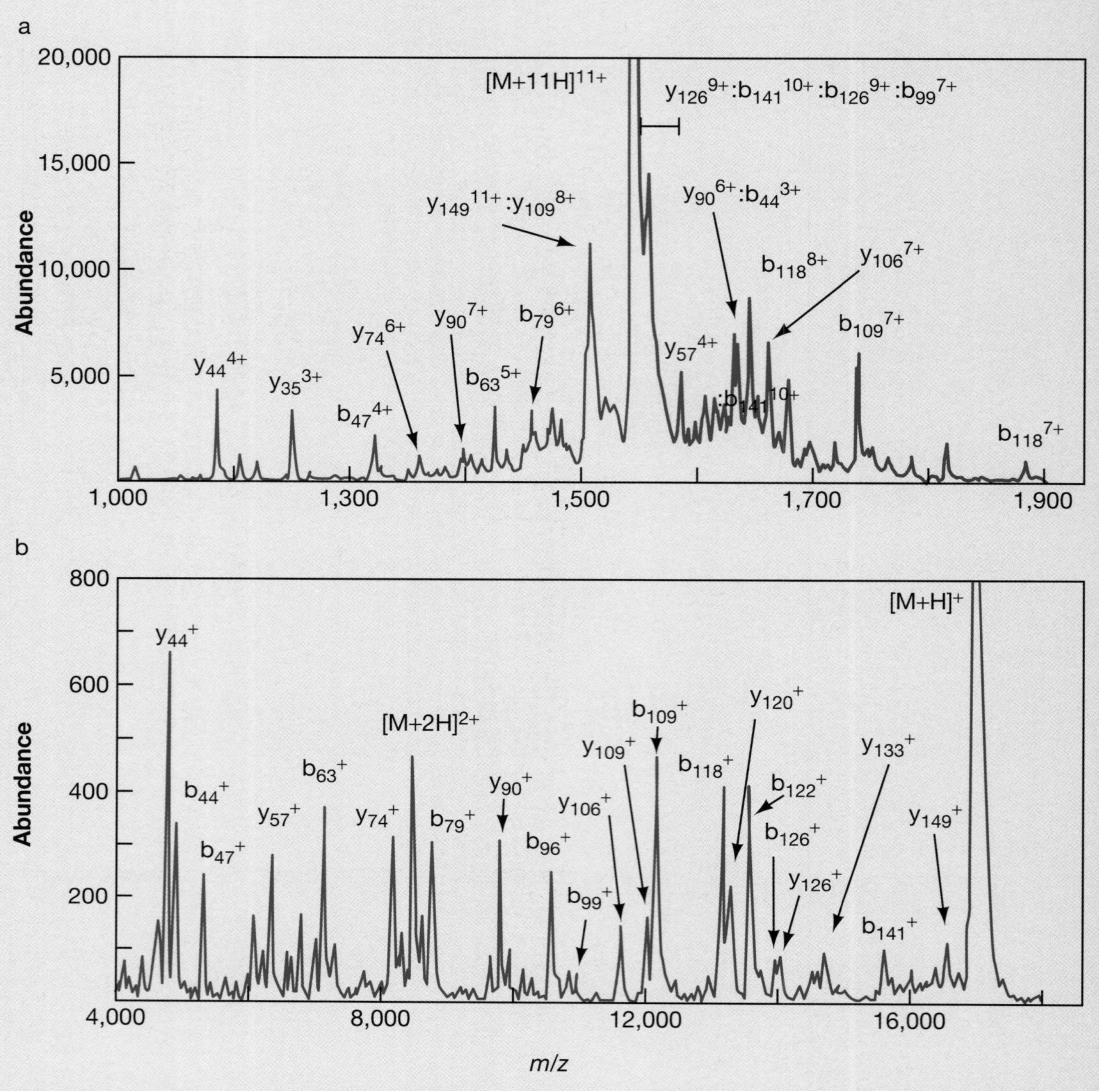

FIGURE 8.55. Quadrupole ion-trap CAD MS/MS spectra of the $[M+11H]^{11+}$ ions of apomyoglobin. (*a*) MS/MS product ion spectra obtained using on-resonance collisional activation. (*b*) Post-ion/ion reaction MS/MS product ion spectra using the same collisional activation conditions employed in *a*, and obtained after ion/ion reactions with PDCH anions. (Adapted, with permission, from [69] Newton et al. 2002.)

in which chemical reactions are used to overcome product ion charge state ambiguities involves the use of ion/ion proton transfer reactions to reduce the multiply charged product ions largely to the +1 charge state [51–72]. Conversion of product ions to the +1 charge state after parent ion dissociation greatly relaxes the mass resolution requirements for product ion identification, thereby allowing interpretable MS/MS spectra of proteins of at least 20 kD to be obtained in instruments with modest resolving power (e.g., $M/\Delta M = 1000–2000$) [63–72]. To date, this approach has been used exclusively in quadrupole ion-trap instrumentation. Figure 8.55a provides an illustrative case whereby the multiply charged ion of a protein of modest size, i.e., the $[M+11H]^{11+}$ ion of the 17-kD protein apomyoglobin, can be readily dissociated in the ion trap, but the spectrum cannot be interpreted a priori due to charge state ambiguity associated with the limited mass resolving power of the ion-trap mass analyzer. Note that it is only with the benefit of the post-ion/ion MS/MS data (discussed below) that the identity of many of the multiply charged product ions may be assigned. Additionally, a number of the potential product ions labeled in the spectrum cannot be unambiguously assigned. For example, a resolving power of greater than 16,000 would be required to unambiguously assign the y_{90}^{6+} (m/z 1635.39) and b_{44}^{3+}(m/z 1635.49) ions, whose multiply charged pre-ion/ion product ion masses differ by only 0.1 dalton. This problem is readily overcome by the use of ion/ion proton transfer reactions to reduce the multiply charged products largely to +1 charge states. The post-ion/ion proton transfer reaction product ion spectrum of the $[M+11H]^{11+}$ ion of apomyoglobin, acquired after subjecting the multiply charged product ions shown in Figure 8.55a to ion/ion proton transfer reactions with singly charged anions derived from atmospheric sampling glow discharge ionization (ASGDI) of perfluoro-1,3-dimethylcyclohexane (PDCH), is shown in Figure 8.55b. These product ions may be readily assigned as either *b*- or *y*-type product ions by comparison with the expected product ion masses of apomyoglobin [69].

To date, the fragmentation of a wide range of proteins has been studied, including ribonuclease A [9], mellitin [59], lysozyme [63], hemoglobin β-chain [64], insulin [66], ubiquitin [59,68], holo- and apomyoglobin [69,70], ferri-, ferro-, and apocytochrome *c* [67,71], and bacteriophage MS2 coat protein [65,72]. In several of these cases, collisional activation of individual precursor ion charge states in the quadrupole ion trap has resulted in dissociation of more than 50% of the amide bonds along the protein backbone [69], with more than 80% coverage obtained from several charge states [71]. The degree of sequence coverage obtained for a protein of interest can be further extended by the use of multistage MS/MS (MS^n) of selected first-generation product ions [68]. It has also been demonstrated that dissociation of whole-protein ions under ion-trap collisional activation conditions can be used to characterize posttranslational modifications, such as the N-linked glycosylation site of ribonuclease B [9].

Importantly, ion/ion reactions may also be used to manipulate the charge states of multiply charged precursor ion populations, to form a range of precursor ion charge states lower than those produced by ESI alone [9,68,69,71,72], as well as to facilitate the gas-phase purification of individual components of complex protein mixtures. In the latter case, a "double-isolation" experiment is performed, whereby an initial m/z region containing a charge state of the protein of interest is first isolated and then subjected to a short ion-ion reaction period. Following the ion/ion reaction, isolation of a second m/z region corresponding to a lower-charge state of the protein of interest can be used to "charge state" resolve the protein from all other proteins of different mass initially present in the first isolated m/z region [6,65,72]. One limitation of this "double-isolation" technique is that ions from a given protein charge state are distributed over several lower-charge states during the ion/ion reaction, thereby diluting the protein ion signal and decreasing the sensitivity for subsequent dissociation. Recently, however, it has been demonstrated that the rates of ion/ion reactions in a quadrupole ion trap may be selectively inhibited in a mass-to-charge selective fashion during the ion/ion reaction period, allowing essentially all of the ion current of a protein of interest to be concentrated into a single-charge state. This technique is referred to as "ion parking" [73]. When combined with the double-isolation technique, the ion-parking approach can be used to facilitate the gas-phase concentration and purification of selected precursor ions from complex protein mixtures for subsequent dissociation. In an example of this approach, multiply charged precursor ions of five of the abundant protein components present in a relatively complex mixture derived from a whole-cell lysate HPLC fraction of *E. coli* containing ~30 components, ranging in mass from 7 kD to 10 kD, were subjected to this concentration/purification procedure and then dissociated by collisional activation. Interrogation of the uninterpreted product ion spectra against a partially annotated protein sequence database, coupled with a scoring scheme based on the relative abundances of the experimentally observed product ions and the frequency of fragmentations occurring at preferential cleavage sites, resulted in the unambiguous identification of four out of five of these proteins [6].

Future Developments in Top-Down Protein Characterization

Understanding the fragmentation behavior of whole-protein ions

Protein ion dissociation behavior is currently less well understood than peptide ion dissociation in part because the body of observations associated with protein ion dissociations is still relatively small. Therefore, if intact protein sequencing approaches are to find more general use, the factors governing the fragmentation of multiply charged protein ions, such as the influence of precursor ion charge state on fragmentation, the role of primary, secondary, and tertiary structures, as well as the presence of posttranslational modifications, must be explored. A number of studies on the factors affecting the dissociation of multiply charged protein ions have been performed in a quadrupole ion-trap mass spectrometer. It has been shown that the fragmentation of whole-protein ions under ion-trap collisional activation conditions is strongly influenced by the precursor ion charge state [63–72]. Generally, a limited number of intermediate charge states give rise to nonspecific cleavages of the protein backbone, thereby allowing derivation of sequence tags for subsequent database searching. At other charge states, the facile loss of NH_3 or H_2O (very low-charge states) and preferential cleavage at the carboxyl terminus of aspartic acid and lysine residues (low-charge states) and at the amino terminus of proline residues (high-charge states) are often the dominant fragmentation products observed. Although many of these cleavages are also observed in the fragmentation of peptide ions, cleavage at the carboxyl terminus of lysine residues, particularly when present at the amino terminus of an adjacent histidine residue or in a sequence of residues containing a number of adjacent basic sites, has not been noted previously [69,71]. Thus, further examination of the fragmentation of whole proteins is warranted in order to identify the general charge state dependence of whole-protein ion dissociations and the mechanisms behind these fragmentation reactions, and to maximize the amount of structural information that can be obtained from selected charge states.

Novel instrumentation

The application of MS/MS for peptide-sequencing applications, coupled with the development of modern biomolecular ionization methods, has been the major driving force in commercial instrument development throughout much of the past decade. Although new ionization methods have enabled the study of large ionic species in the gas phase, with the increasing recognition of the value of "top-down" MS/MS approaches for protein identification/characterization, it is likely that future instrument developments will be directed toward improving the instrument capabilities of MS/MS specifically for high-mass ions. Mass analyzer characteristics that are most important for high-mass ions include mass resolution, mass accuracy, mass range, sensitivity, dynamic range, and speed (i.e., mass spectral acquisition rate) [61]. To date, most protein ion dissociation studies have been conducted using high-field FT-ICR, triple-quadrupole tandem mass spectrometers, and specialized quadrupole ion-trap instrumentation. The use of high-magnetic-field-strength FT-ICR for tandem mass spectrometry of high-mass ions will continue to grow due to an increasing number of practitioners making use of the remarkable flexibility and mass analysis capabilities of this tool. However, it seems likely that new instruments of lower cost and, perhaps, higher speed than FT-ICR will be developed for MS/MS of high-mass ions. For example, state-of-the-art commercial tandem mass spectrometers employing high-performance time-of-flight mass analyzers are beginning to be applied to whole-protein ion analysis. The significantly superior resolving power and mass accuracy afforded to product ion measurement in modern quadrupole time-of-flight instruments relative to triple-quadrupole instruments allow for improved performance in "top-down" applications using beam-type tandem mass spectrometers [74]. Electrodynamic ion-trapping tools, in both two- and three-dimensional versions, are likely to be used prior to the time-of-flight instrument to facilitate precursor ion accumulation, purification, and dissociation.

Novel chemistries

Realization of the potential of mass spectrometry for whole-protein identification and characterization will require further advances in the understanding of the behavior of gas-phase macro-ions derived from proteins. Given the high dimensionality presented by gaseous protein ions (i.e., primary, secondary, tertiary, and quaternary structures, charge state, post-translational modifications, etc.), it is expected that a rich array of ion chemistries might be utilized. By far, the most widely studied aspect of protein ion chemistry has been the unimolecular dissociation behavior of the ions. Although a variety of ion activation methods intended to induce dissociation have been applied to protein ions, only a few have been characterized with a number of different proteins. Therefore, the relative merits of the vast array of activation methods thus far developed for simple polyatomic ions have not yet been clearly delineated for macro-ions.

It has already been demonstrated that ion/electron and ion/ion reactions can have important roles in "top-down" protein identification approaches. To date, ion/ion reactions have been used largely as a means for manipulation of ion charge. However, it is likely that novel ion/ion chemistries will be developed that can provide structural information. One such example might be the selective covalent modification of a protein ion in the gas phase followed by ion activation/dissociation. Such reactions might prove to be an attractive alternative to the modification of protein ions in solution for subsequent tandem mass spectrometry.

The use of ion/molecule reactions also has a potential role in protein structural analysis. To date, ion/molecule reactions involving protein ions have largely been restricted to proton transfer reactions [75–77] and hydrogen/deuterium exchange [78–82]. However, other types of ion/molecule reactions may also prove to be useful in the study of high-mass ions. For example, it has been shown that the sum of the protein ion charge and the maximum number of molecules of hydroiodic acid (HI) that attach to an ion in the gas phase correlates with the number of basic sites in the protein, therefore providing information about the composition of a protein [83–87]. Other ion/molecule reactions of potential use to protein sequence analysis involve gas-phase derivatization [88–90]. Although these types of reactions have thus far been applied only to peptides, this class of ion/molecule reaction could be of use in directing the dissociation of whole protein ions to desired fragmentation channels, thereby facilitating sequence determination. Given the relatively limited attention thus far paid to the ion/ion and ion/molecule chemistries of protein ions, it seems likely that other types of reactions with relatively high specificity might also be discovered that could have useful roles in "top-down" sequencing strategies.

REFERENCES

1. Kelleher N.L., Lin H.Y., Valaskovic G.A., Aaserud D.J., Fridriksson E.K., and McLafferty F.W. 1999. *J. Am. Chem. Soc.* **121:** 806–812.
2. Mortz E., O'Connor P.B., Roepstorff P., Kelleher N.L., Wood T.D., McLafferty F.W., and Mann M. 1996. *Proc. Natl. Acad. Sci.* **93:** 8264–8267.
3. Cargile B.T., McLuckey S.A., and Stephenson J.L., Jr. 2001. *Anal. Chem.* **73:** 1277–1285.
4. Demirev P.A., Lin J.S., Pineda F.J., and Fenselau C. 2001. *Anal. Chem.* **73:** 4566–5573.
5. Meng F., Cargile B.J., Miller L.M., Forbes A.J., Johnson J.R., and Kelleher N.L. 2001. *Nat. Biotechnol.* **19:** 952–957.

6. Reid G.E., Shang H., Hogan J., Lee G.U., and McLuckey S.A. 2002. *J. Am. Chem. Soc.* **124:** 7353–7362.
7. Horn D.M., Zubarev R.A., and McLafferty F.W. 2000. *Proc. Natl. Acad. Sci.* **97:** 10313–10317.
8. Horn D.M., Zubarev R.A., and McLafferty F.W. 2000. *J. Am. Soc. Mass Spectrom.* **11:** 320–332.
9. Reid G.E., Stephenson J.L., Jr., and McLuckey S.A. 2001. *Anal. Chem.* **74:** 577–583.
10. Kelleher N.L., Zubarev R.A., Bush K., Furie B., Furie B.C., McLafferty F.W., and Walsh C.T. 1999. *Anal. Chem.* **71:** 4250–4253.
11. Fridriksson E.K., Beavil A., Holowka D., Gould H.J., Baird B., and McLafferty F.W. 2000. *Biochemistry* **39:** 3369–3376.
12. Shi S.D.-H., Hemling M.E., Carr S.A., Horn D.M., Lindh I., and McLafferty F.W. 2001. *Anal. Chem.* **73:** 19–22.
13. Loo J.A., Edmonds C.G., Udseth H.R., and Smith R.D. 1990. *Anal. Chim. Acta* **241:** 167–173.
14. Loo J.A., Edmonds C.G., and Smith R.D. 1990. *Science* **248:** 201–204.
15. Smith R.D., Loo J.A., Barinaga C.J., Edmonds C.G., and Udseth H.R. 1990. *J. Am. Soc. Mass Spectrom.* **1:** 53–65.
16. Loo J.A., Edmonds C.G., and Smith R.D. 1991. *Anal. Chem.* **63:** 2488–2499.
17. Loo J.A., Edmonds C.G., and Smith R.D. 1993. *Anal. Chem.* **65:** 425–438.
18. Light-Wahl K.J., Loo J.A., Edmonds C.G., Smith R.D., Witkowska H.E., Shackleton C.H.L., and Wu C.S. 1993. *Biol. Mass Spectrom.* **22:** 112–120.
19. Feng R. and Konishi Y. 1993. *Anal. Chem.* **65:** 645–649.
20. Loo J.A., Quinn J.P., Ryu S.I., Henry K.D., Senko M.W., and McLafferty F.W. 1992. *Proc. Natl. Acad. Sci.* **89:** 286–289.
21. Beu S.C., Senko M.W., Quinn J.P., Wampler F.M., and McLaffery F.W. 1993. *J. Am. Soc. Mass Spectrom.* **4:** 557–565.
22. Senko M.W., Speir J.P., and McLafferty F.W. 1994. *Anal. Chem.* **66:** 2801–2808.
23. Speir J.P., Senko M.W., Little D.P., Loo J.A., and McLafferty F.W. 1995. *J. Mass Spectrom.* **30:** 39–42.
24. O'Connor P.B., Speir J.P., Senko M.W., Little D.P., and McLafferty F.W. 1995. *J. Mass Spectrom.* **30:** 88–93.
25. Wood T.D., Chen L.H., Kelleher N.L., Little D.P., Kenyon G.L., and McLafferty F.W. 1995. *Biochemistry* **34:** 16251–16254.
26. Wood T.D., Chen L.H., White C.B., Babbitt P.C., Kenyon G.L., and McLafferty F.W. 1995. *Proc. Natl. Acad. Sci.* **92:** 11451–11455.
27. Kelleher N.L., Costello C.A., Begley T.P., and McLafferty F.W. 1995. *J. Am. Soc. Mass Spectrom.* **6:** 981–984.
28. Kelleher N.L., Taylor S.V., Grannis D., Kinsland C., Chiu H.-J., Begley T.P., and McLafferty F.W. 1998. *Protein Sci.* **7:** 1796–1801.
29. Jensen P.K., Pasa-Tolic L., Anderson G.A., Horner J.A., Lipton M.S., Bruce J.E., and Smith R.D. 1999. *Anal. Chem.* **71:** 2076–2084.
30. Fridriksson E.K., Beavil A., Holowka D., Gould H.J., Baird B., and McLafferty F.W. 2000. *Biochemistry* **39:** 3369–3376.
31. Maier C.S., Yan X., Harder M.E., Schimerlik M.I., Deinzer M.L., Pasa-Tolic L., and Smith R.D. 2000. *J. Am. Soc. Mass Spectrom.* **11:** 237–243.
32. Little D.P., Speir J.P., Senko M.W., O'Connor P.B., and McLafferty F.W. 1994. *Anal. Chem.* **66:** 2809–2815.
33. Guan G., Kelleher N.L., O'Connor R.B., Aaserud D.J., Little D.P., and McLafferty F.W. 1996. *Int. J. Mass Spectrom. Ion Processes* **157/158:** 357–364.
34. Price W.D., Schnier P.D., and Williams E.R. 1996. *Anal. Chem.* **68:** 859–866.
35. Li W., Hendrickson C.L., Emmett M.R., and Marshall A.G. 1999. *Anal. Chem.* **71:** 4397–4402.
36. Freitas M.A., Hendrickson C.L., and Marshall A.G. 2000. *J. Am. Chem. Soc.* **122:** 7768–7775.
37. Chorush R.A., Little D.P., Beu S.C., Wood T.D., and McLafferty F.W. 1995. *Anal. Chem.* **67:** 1042–1046.
38. Ge Y., Horn D.M., and McLafferty F.W. 2001. *Int. J. Mass Spectrom.* **210/211:** 203–214.
39. Zubarev R.A., Kelleher N.L., and McLafferty F.W. 1998. *J. Am. Chem. Soc.* **120:** 3265–3266.
40. Zubarev R.A., Kruger N.A., Fridriksson E.K., Lewis M.A., Horn D.M., Carpenter B.K., McLafferty F.W. 1999. *J. Am. Chem. Soc.* **121:** 2857–2862.
41. Cerda B.A., Horn D.M., Breuker K., Carpenter B.K., and McLafferty F.W. 1999. *Eur. Mass Spectrom.* **5:** 335–338.
42. Kelleher N.L., Zubarev R.A., Bush K., Furie B., Furie B.C., McLafferty F.W., and Walsh C.T. 1999. *Anal. Chem.* **71:** 4250–4253.

43. Mirgorodskaya E., Roepstorff P., and Zubarev R.A. 1999. *Anal. Chem.* **71:** 4431–4436.
44. Kruger N.A., Zubarev R.A., Carpenter B.K., Kelleher N.L., Horn D.M., and McLafferty F.W. 1999. *Int. J. Mass Spectrom.* **182/183:** 1–5.
45. Zubarev R.A., Fridriksson E.K., Horn D.M., Kelleher N.L., Kruger N.A., Lewis M.A., Carpenter B.K., and McLafferty F.W. 2000. *Mass spectrometry in biology and medicine* (eds. A.L. Burlingame et al.), pp. 111–120. Humana Press, Totowa, New Jersey.
46. Zubarev R.A., Horn D.M., Fridriksson E.K., Kelleher N.L., Kruger N.A., Lewis M.A., Carpenter B.K., and McLafferty F.W. 2000. *Anal. Chem.* **72:** 563–573.
47. Horn D.M., Ge Y., and McLafferty F.W. 2000. *Anal. Chem.* **72:** 4778–4784.
48. Shi S.D.-H., Hemling M.E., Carr S.A., Horn D.M., Lindh I., and McLafferty F. W. 2001. *Anal. Chem.* **73:** 19–22.
49. McLafferty F.W., Horn D.M., Breuker K., Ge Y., Lewis M.A., Cerda B., Zubarev R.A., and Carpenter B.K. 2001. *J. Am. Soc. Mass Spectrom.* **12:** 245–249.
50. Ge Y., Lawhorn B.G., ElNagger M., Strauss E., Park J.-H., Begley T.P., and McLafferty, F.W. 2002. *J. Am. Chem. Soc.* **124:** 672–678.
51. Stephenson J.L. and McLuckey S.A. 1996. *Anal. Chem.* **68:** 4026–4032.
52. Stephenson J.L. and McLuckey S.A. 1996. *J. Am. Chem. Soc.* **118:** 7390–7397.
53. Herron W.J., Goeringer D.E., and McLuckey S.A. 1996. *Anal. Chem.* **68:** 257–262.
54. Stephenson J.L. and McLuckey S.A.. 1997. *Int. J. Mass Spectrom. Ion Proc.* **162:** 89–106.
55. Stephenson J.L. and McLuckey S.A. 1997. *Anal. Chem.* **69:** 3760–3766.
56. Stephenson J.L., Jr. and McLuckey S.A. 1998. *Anal. Chem.* **70:** 3533–3544.
57. Stephenson J.L. and McLuckey S.A. 1998. *J. Mass Spectrom.* **33:** 664–672.
58. Stephenson J.L. and McLuckey S.A. 1998. *J. Am. Soc. Mass Spectrom.* **9:** 585–596.
59. McLuckey S.A., Stephenson J.L, and Asano K.G. 1998. *Anal. Chem.* **70:** 1198–1202.
60. McLuckey S.A. and Stephenson J.L. 1998. *Mass Spectrom. Rev.* **17:** 369–407.
61. McLuckey S.A. and Wells J.M. 2001. *Chem. Rev.* **101:** 571–606.
62. Stephenson J.L., McLuckey S.A., Reid G.E., Wells J.M., and Bundy J.L. 2002. *Curr. Opin. Biotechnol.* **13:** 57–64.
63. Stephenson J.L., Jr., Cargile B.J., and McLuckey S.A. 1999. *Rapid Commun. Mass Spectrom.* **13:** 2040–2048.
64. Schaaff T.G., Cargile B.J., Stephenson J.L., Jr., and McLuckey S.A. 2000. *Anal. Chem.* **72:** 899–907.
65. Cargile B.T., McLuckey S.A., and Stephenson J.L. 2001. Jr. *Anal. Chem.* **73:** 1277–1285.
66. Wells J.M. and McLuckey S.A. 2000. *Int. J. Mass Spectrom.* **203:** *A1–A9*.
67. Wells J.M., Reid G.E., Engel B.J., Pan P., and McLuckey S.A. 2001. *J. Am. Soc. Mass Spectrom.* **12:** 873–876.
68. Reid G.E., Wu J., Chrisman P.A., Wells J.M., and McLuckey S.A. 2001. *Anal. Chem.* **73:** 3274–3281.
69. Newton K.A., Chrisman P.A., Reid G.E., Wells J.M., and McLuckey S.A. 2001. *Int. J. Mass Spectrom.* **212:** 359–376.
70. Chrisman P.A., Newton K.A., Reid G.E., Wells J.M., and McLuckey S.A. 2001. *Rapid. Commun. Mass Spectrom.* **15:** 2334–2340.
71. Engel B.J. Pan P., Reid G.E., Wells J.M., and McLuckey S.A. 2002. *Int. J. Mass Spectrom.* **219:** 171–187.
72. He M, Reid G.E., Shang H.-S., Lee G.U., and McLuckey S.A. 2002. *Anal. Chem.* **74:** 4653–4661.
73. McLuckey S.A., Reid G.E., and Wells J.M. 2002. *Anal. Chem.* **74:** 336–346.
74. Nemeth-Cawley J.F. and Rouse J.C. 2002. *J. Mass Spectrom.* **37:** 270–282.
75. McLuckey S.A., Glish G.L., and Van Berkel G.J. 1991. *Anal.Chem.* **63:** 1971–1978.
76. McLuckey S.A. and Goeringer D.E. 1995. *Anal. Chem.* **67:** 2493–2497.
77. Williams E.R. 1996. *J.Mass Spectrom.* **31:** 831–842.
78. Suckau D., Shi Y., Beu S.C., Senko M.W., Quinn J.P., Wampler F.M., and McLafferty F.W. 1993. *Proc. Natl. Acad. Sci.* **90:** 790–793.
79. Freitas M.A., Hendrickson C.L., Emmett M.R., and Marshall A.G. 1998. *J. Am. Soc. Mass Spectrom.* **9:** 1012–1019.
80. Freitas M.A., Hendrickson C.L., Emmett M.R., and Marshall A.G. 1999. *Int. J. Mass Spectrom.* **185/186/187:** 565–575.
81. Wyttenbach T. and Bowers M.T. 1999. *J. Am. Soc. Mass Spectrom.* **10:** 9–14.
82. Schaaff T.G., Stephenson J.L., and McLuckey S.A. 2000. *J. Am. Soc. Mass Spectrom.* **11:** 167–171.
83. Stephenson J.L. and McLuckey S.A.. 1997. *Anal. Chem.* **69:** 281–285.
84. Stephenson J.L. and McLuckey S.A.. 1997. *J. Am. Chem. Soc.* **119:** 1688–1696.

85. Schaaff T.G., Stephenson J.L., and McLuckey S.A. 1999. *J. Am. Chem. Soc.* **121:** 8907–8919.
86. Stephenson J.L., Schaaff T.G., and McLuckey S.A. 1999. *J. Am. Soc. Mass Spectrom.* **10:** 552–556.
87. Schaaff T.G., Stephenson J.L., and McLuckey S.A. 2000. *Int. J. Mass Spectrom.* **202:** 299–313.
88. O'Hair R.A.J. and Reid G.E. 2000. *J. Am. Soc. Mass Spectrom.* **11:** 244–256.
89. Reid G.E., Tichy S.E., Pérez J., O'Hair R.A.J., Simpson R.J., and Kenttämaa H.I. 2001. *J. Am. Chem. Soc.* **123:** 184–1192.
90. Freitas M.A., O'Hair R.A.J., Dua S., and Bowie J.H. 1997. *Chem. Commun.* 1409–1410.

FRAGMENTATION MECHANISMS OF PROTONATED PEPTIDES IN THE GAS PHASE*

Richard A.J. O'Hair
School of Chemistry, University of Melbourne, Australia

The successful sequencing of peptides using MS/MS hinges upon the formation of a complete set of sequence ions via random cleavage of each of the peptide bonds. This is typically achieved via collision-induced dissociation (CID) of protonated peptides, which yields the complementary *b* and *y* sequence ion series (see the section on CID in the chapter introduction). The universally accepted nomenclature for the sequence ions formed in these experiments is shown in Figure 8.56 [1,2]. In an ideal world, each of the sequence ions would be formed in equal abundance, allowing the peptide sequences to be "read out" from the MS/MS spectrum. In reality, this is not the case, because some sequence ions are not observed and some side reactions or rearrangements complicate the mass spectrum, which in the worst case prevents the investigator from being able to assign the sequence of a peptide. To understand the causes of these complications, it is necessary to examine the mechanisms of protonated peptide fragmentation and how they control the types of fragment ions observed in an MS/MS experiment. One of the most important considerations is the location of the site of protonation (governed by the local proton affinities of the various sites within a peptide [3]) and how this relates to the fragmentation reactions observed. To illustrate key concepts for understanding fragmentation processes of protonated peptides, the structures and mechanisms associated with the formation of five key types of ions are briefly discussed: (1) a_1 ions, (2) b_1 ions, (3) b_n ions ($n \geq 2$), (4) y_n ions, and (5) nonsequence ions. For more detailed discussions of fragmentation mechanisms of simple protonated molecules and peptides, the reader is directed to several recent reviews [3–7].

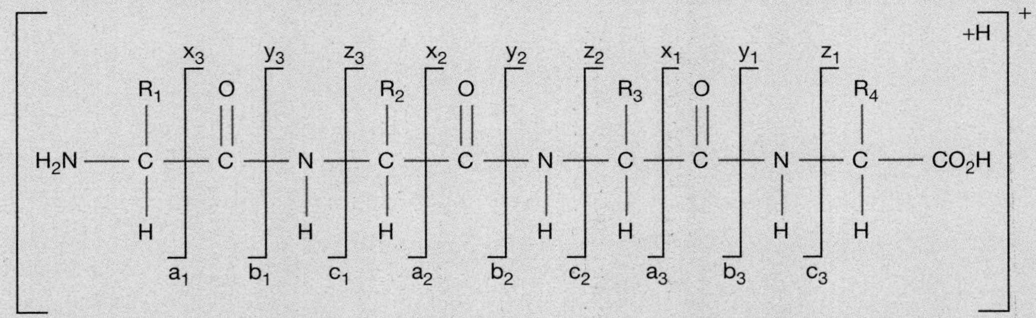

FIGURE 8.56. Sequence ion nomenclature for a protonated tetrapeptide.

a_1 Ions

A common fragment ion that identifies the amino acid residue at the amino terminus is the a_1 ion (see Equation 1). Several experimental and theoretical studies have been carried out on simple systems ranging from glycine to simple dipeptides and tripeptides to determine the mechanisms of these reactions, and key findings are discussed in this section.

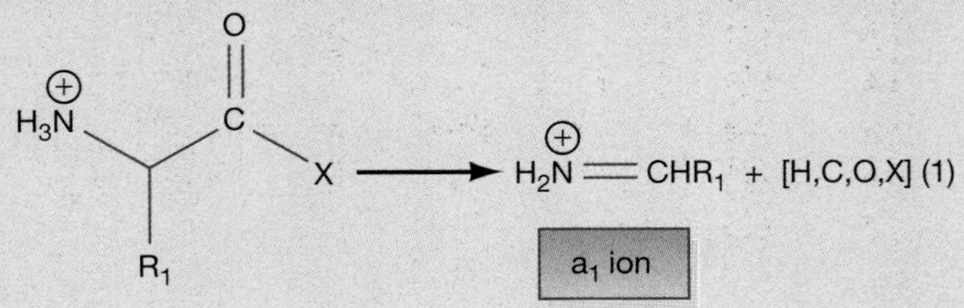

$$\longrightarrow H_2N^{+}{=}CHR_1 + [H,C,O,X] \quad (1)$$

a_1 ion

(1)

*The numbers in brackets indicate reference citations.

Glycine (X = OH)

Three possible mechanisms for the formation of the a_1 ion of glycine are shown in Scheme 1, and these can be distinguished by their neutral losses [8]. Each of these losses occurs via different sites of protonation within glycine, which are governed by their relative "local proton affinities." Mechanism A involves the 1,2 elimination of formic acid from *N*-protonated glycine, 1, which is the thermodynamically favored site of protonation. In contrast, the other two mechanisms involve intramolecular proton transfer to form CO-protonated glycine, 2 (the second most favored site of protonation), or HO-protonated glycine, 3 (the least favored site of protonation), which then fragment via dihydroxycarbene loss (Mechanism B) or the combined losses of H_2O and CO (Mechanism C). Using CID ionization (CIDI), Wesdemiotis [8] discovered neutral products due to H_2O and CO loss, consistent with Mechanism C, but could find no neutral products which would support the other mechanisms (i.e., Mechanism A or B).

Mechanism A (Scheme 1)

1 → [transition state]‡ → H_2N^+= + HCO_2H

Mechanism B

1 —Proton transfer→ 2 —Fragment→ H_2N^+= + $(HO)_2C$:

Mechanism C

1 —Proton transfer→ 3 —Fragment→ H_2N^+= + H_2O + CO

Support for Mechanism C comes from experimental [9] and theoretical [10] measurements of the activation energy required to induce formation of the a_1 ion. Using collision-induced threshold energy measurements, Klassen and Kebarle have shown that 43.9 kcal mol^{-1} is required to induce this reaction [9], which nicely agrees with a recent theoretical estimate from *ab initio* calculations of 39.1 kcal mol^{-1} [10]. In contrast, *ab initio* calculations predict that Mechanism A has an activation barrier of >90 kcal mol^{-1} and Mechanism B also has a higher activation barrier (67.0 kcal mol^{-1}) [10].

Even a simple system such as protonated glycine thus demonstrates that several experimental techniques further bolstered by theoretical calculations are required to prove mechanistic details of fragmentation reactions. Furthermore, it demonstrates that the initial site of protonation may not be that at which fragmentation occurs. Instead, "mobilization" of the proton may be required to induce fragmentation at another site (i.e., Mechanism C) [6]. Further support for proton mobility under CID conditions comes from experiments on $[M+D]^+$ ions from simple amino acids such as alanine and leucine, which lose the combined elements of (H_2O + CO) and (HOD + CO) in a statistical ratio. This finding indicates that proton transfer from 1 to 3 is reversible [11].

Glycinamide (X = NH_2)

Using a combination of experiment and theory, Kinser et al. [12] have shown that a similar mechanism (compare Mechanism C in Scheme 1) operates for the formation of the a_1 ion from glycinamide. Reaction of neutral glycinamide with deuteron (D^+) donors again demonstrates the mobilization of the proton (both $CH_2{=}NH_2^+$ and $CH_2{=}NHD^+$ are observed). By changing the energy of protonation, these authors were able to demonstrate that >35.8 kcal mol^{-1} is required to induce fragmentation, which is consistent with their own *ab initio* prediction of 41.1 kcal mol^{-1} as well as the activation energy of 44.2 kcal mol^{-1} measured by Klassen and Kebarle [9] using collision-induced threshold energy measurements.

(Scheme 2)

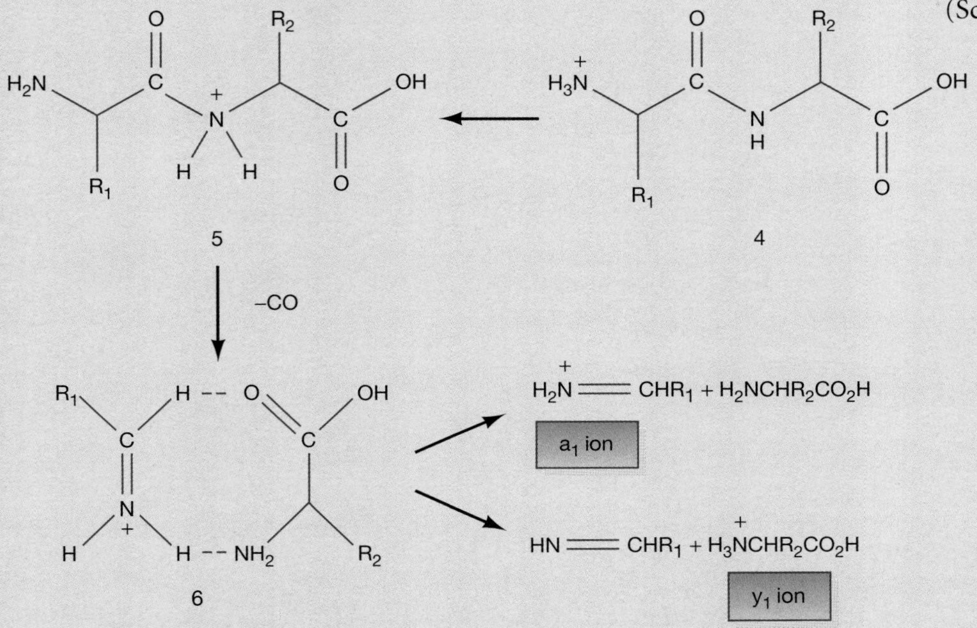

Glycylglycine (X = $NHCH_2CO_2H$) and other dipeptides

Simple di- and tripeptides can also fragment to form a_1 ions directly, which are often in competition with y_n ions [9,12–15]. Recently, Paizs and Suhai have suggested the mechanism shown in Scheme 2, which accounts for the combined formation of a_1 and y_1 ions in protonated glycylglycine using *ab initio* calculations [12]. Thus, intramolecular proton transfer from the themodynamically favored amino-terminal protonated species 4 to the N-amide protonated form 5 precedes expulsion of CO. Once CO departs, an ion-molecule complex between the a_1 ion and the neutral amino acid is formed (6 in Scheme 2). This ion-molecule complex, 6, can then fragment to give either the a_1 ion or the y_1 ion.

Their model can also be used to rationalize some interesting experimental results [13,14]. For example, Isa et al. have examined the competition between a_1 and y_1 ion formation in the CID spectra of a series of protonated dipeptides: Xxx-Gly, Gly-Xxx, Xxx-Leu, Leu-Xxx (where Xxx represents various amino acid residues) [13]. They found that for Gly-Xxx, both the y_1 and a_1 ions were always observed, whereas the y_1 ion was rarely seen for Xxx-Gly, where the a_1 ion usually dominated. When Leu was substituted for Gly, the abundances of the a_1 ion increased for Leu-Xxx, as did the abundances of the y_1 ion for Xxx-Leu, suggesting that the proton affinity differences between the conjugate bases within the ion-molecule complex 6 in Scheme 2 has a role in determining their observed relative abundances. A study of the protonated dipeptides Val-Xxx and Xxx-Phe found similar influences of the proton affinity differences between the amino- and carboxy-terminal amino acids on the relative abundances of y_1 and a_1 ions [14]. Once again, this is consistent with a competition for the proton within the ion-molecule complex 6 in Scheme 2.

b_1 Ions

Why, when ion 5 fragments, are two bonds broken (at *a* and *b* in Scheme 3), forming an a_1 ion, as opposed to only one bond (at *a*) being broken to form a b_1 ion with an acylium structure as shown? The answer comes from several experimental and theoretical studies, which have shown that for simple aliphatic amino acids, the acylium ion structures of b_1 ions are unstable with respect to CO loss [16].

(Scheme 3)

5

a_1 ion

b_1 ion

Competition between b_n Ions ($n \geq 2$) and y_n Ion Formation: Peptide Bonds as Neighboring Groups

Given that the simple acylium ion structures of b_1 ions are unstable, the commonly encountered b_n ions ($n \geq 2$) are unlikely to adopt a simple acylium ion structure. One of the first studies to examine the structure of b_2 ions experimentally and theoretically is that of Harrison et al. [17]. These authors found that a five-membered ring oxazolone structure is favored. The role that this structure has in the fragmentation reactions of protonated GGG has recently been examined via *ab initio* calculations [18]. Key features of this study are shown in Scheme 4. Once again, intramolecular proton transfer is required to mobilize the proton from the first amide bond of species 7 (it is interesting to note that this O-protonated form is now favored over protonation at the amino terminus, indicating that internal hydrogen bonding can influence the relative local proton affinities of various groups in simple peptides) to the second amide bond to yield the N-protonated form, 8. The first amide bond can then act as an internal nucleophile to induce cleavage of the amide bond to give an ion-molecule complex between the oxazolone b_2 ion and the neutral amino acid, 9. This intramolecular process is called a neighboring group process. Once again, this ion-molecule complex 9 can fragment to form either a b_2 ion (path a of Scheme 4) or a y_1 ion (path b of Scheme 4).

In Schemes 2 and 4, we have seen that y_1 ions can be formed in competition with other a_n and b_n sequence ions. These y_1 ions are amino acids. Several studies have shown that higher analogs (y_n ions, where $n \geq 2$) correspond to truncated peptides [19]. Thus, the mechanism shown in Scheme 4 can be regarded as a general mechanism for the competition between y and b ion formation in protonated peptides. More importantly, these neighboring group processes have direct solution phase analogies for cleavage of peptide bonds under acidic conditions [20].

(Scheme 4)

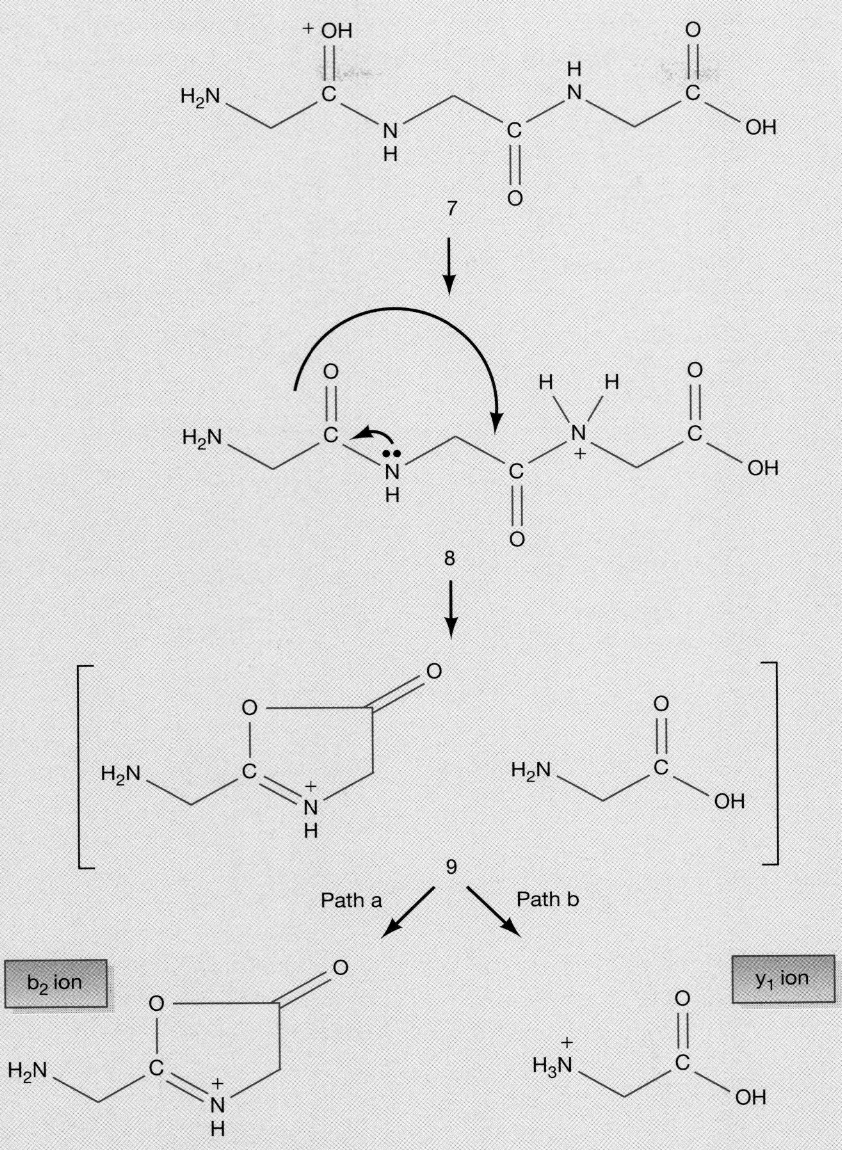

Path a

Path b

b_2 ion

y_1 ion

Other Types of Processes Leading to b_n and y_n Ion Formation

The section above has shown how the adjacent peptide bond can facilitate peptide bond cleavage via the formation of an oxazolone structure. The presence of certain residues within peptides can enhance the effect of cleavage of adjacent peptide bonds. These types of reactions can be further classified as follows:

- ***b_1 ion formation:*** Stabilized b_1 ions are observed for peptides that contain amino-terminal lysine, histidine, and methionine peptides [21]. These residues facilitate cleavage of the adjacent peptide via a neighboring group pathway involving their nucleophilic side chains resulting in ring structures. Thus, these represent channels competing with those discussed for a_1 ion formation.
- ***The proline effect:*** Enhanced amide bond cleavage occurs at the amino-terminal side of a proline residue. This is known as the "proline effect" [22].
- ***The aspartic acid effect:*** The "aspartic acid effect" occurs when the amide bond carboxy-terminal to an aspartic acid residue undergoes facile cleavage [23]. A possible mechanism is

shown in Scheme 5 and involves proton transfer from the adjacent acidic side chain to the amide bond in 10 to form a salt bridge structure 11, which then undergoes cleavage to yield the products shown. Indirect evidence for the salt bridge structure comes from experiments in which the aspartic acid residues are converted to esters. This leads to a substantial decrease in cleavage at the esterified aspartic acid residue [23]. Enhanced cleavage has also been reported recently for the amide bond between Gln-Gly residues within peptides [24].

(Scheme 5)

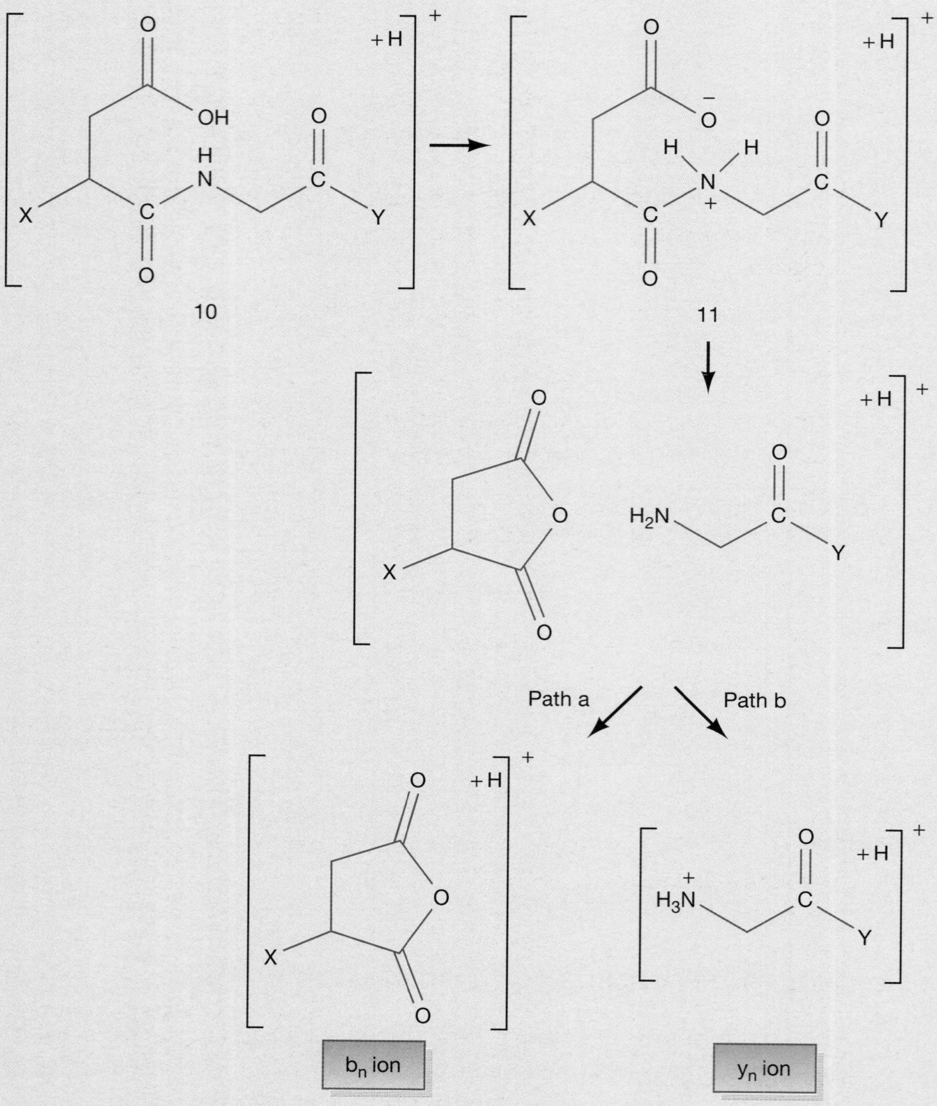

10

11

Path a

Path b

b_n ion

y_n ion

Nonsequence Ions

Neighboring group processes similar to those described above for sequence ion formation can also operate for the formation of nonsequence ions. For example, (1) a side-chain nucleophile can attack a protonated backbone site of a peptide to liberate either H_2O or ammonia or (2) the peptide backbone can attack a protonated side chain to induce fragmentation of the side chain. The latter class of reaction can prove to be useful in identifying a posttranslationally modified amino acid residue. For example, oxidized methionine loses CH_3SOH, whereas phosphoserine loses H_3PO_4. Two recent publications discuss neighboring group mechanisms associated with these losses [25,26].

Summary of Important Concepts in Peptide Fragmentation Reactions

The preceding sections have illustrated how various tools can be used to reveal the key concepts associated with the fragmentation mechanisms of protonated peptides. One of the most useful tools—theroretical calculations—not only provides energetics, but also allows the visualization of reactants, intermediates, products, and the transition states connecting them. From these studies, the following key concepts emerge:

- Under low-energy collision conditions, most fragmentations are directed by the site of the charge.
- If cleavage occurs at multiple sites throughout the peptide, the charge must be allowed to migrate from the initial site of protonation to the site of cleavage. This has been termed the mobile proton model and is one of the central tenants in peptide fragmentation mechanisms [7]. The energy required to mobilize the proton from the thermodynamically favored site of protonation to less favored sites of protonation is provided by the collision energy imparted to the ion.
- Once appropriate protonated precursors have been formed, cleavage of the amide bond can be facilitated by a neighboring group pathway in which an adjacent nucleophile (amide bond or nucleophilic side chain) facilitates cleavage.
- Initial cleavage yields an ion-neutral complex and proton transfer within this complex can account for the competition between formation of two types of ions (e.g., b and y ions).
- Arginine has the highest proton affinity of all the amino acids, and thus the proton of a singly protonated peptide containing an arginine is sequestered to the arginine side chain. As a consequence, the $[M+H]^+$ ion of an arginine-containing peptide often fragments poorly (i.e., the proton is not readily mobilized).
- Other side chains can introduce new fragmentation pathways that differ from those shown in Schemes 1 and 2, facilitating peptide bond cleavage or yielding nonsequence ions due to the loss of small molecules (such as NH_3 or H_2O) [4].

Future Developments

With recent advances in mass spectrometry instrumentation, it is now becoming a realistic proposition to sequence small proteins using a top-down approach. Some of the key concepts introduced here are likely to have an important role in fragmentation pathways of protein ions (see the information panel on "TOP-DOWN" PROTEIN SEQUENCE ANALYSIS USING MS/MS). Other approaches for gaining structural information on peptides use CID on $[M-H]^-$ ions [27] or $[M+Cat]^+$ (where Cat means a metal ion). The recently introduced electron capture dissociation (ECD) method opens up new fragmentation pathways involving radical chemistry [28].

REFERENCES

1. Roepstorff P. and Fohlman J. 1994. *Biol. Mass. Spectrom.* **11:** 601.
2. Papayannopoulos I.A. and Biemann K. 1994. *Accts. Chem. Res.* **27:** 370.
3. O'Hair R.A.J. 2001. Ion chemistry and fragmentation. In *Mass spectrometry in drug discovery* (ed. D.T. Rossi et al.), p. 85. Marcel Dekker, New York.
4. O'Hair R.A.J. 2000. *J. Mass Spectrom.* **35**: 1377.
5. Schlosser A. and Lehmann W.D. 2000. *J. Mass Spectrom.* **35:** 1382–1390.
6. Polce M.J., Ren D., and Wesdemiotis C. 2000. *J. Mass Spectrom.* **35**: 1391–1398.

7. Wysocki V.H., Tsaprailis G., Smith L.L., and Breci L.A. 2000. *J. Mass Spectrom.* **35:** 1399–1406.
8. Beranova S., Cai J., and Wesdemiotis C. 1995. *J. Am. Chem. Soc.* **117:** 9492.
9. Klassen J.S. and Kebarle P. 1997. *J. Am. Chem. Soc.* **119:** 6552.
10. O'Hair R.A.J., Broughton P.S., Styles M.L., Frink B.T., and Hadad C.M. 2000. *J. Am. Soc. Mass Spectrom.* **11:** 687–696.
11. Harrison A.G. and Yalcin T. 1997. *Int. J. Mass Spectrom. Ion Processes* **165/166:** 339.
12. Paizs B. and Suhai S. 2001. *Rapid Commun. Mass Spectrom.* **15:** 651.
13. Isa K., Omote T., and Amaya M. 1990. *Org. Mass Spectrom.* **25:** 620–628.
14. Ambipathy K., Yalcin T., Leung H.-W., and Harrison A.G. 1997. *J. Mass Spectrom.* **32:** 209–215.
15. Harrison A.G., Csizmadia I.G., Tang T.-H., and Tu Y.-P. 2000. *J. Mass. Spectrom.* **35:** 683–688.
16. O'Hair R.A.J. and Reid G.E. 2000. *Rapid Commun. Mass Spectrom.* **14**: 1220 and references cited therein.
17. Yalcin T., Khouw C., Csizamadia I.G., Peterson M.R., and Harrison A.G. 1995. *J. Am. Soc. Mass Spectrom.* **6:** 1164.
18. Rodriquez C.F., Cunje A., Shoeib T., Chu I.K., Hopkinson A.C., and Siu K.W.M. 2001. *J. Am. Chem. Soc.* **123:** 3006.
19. Reid G.E., Simpson R.J., and O'Hair R.A.J. 1999. *Int. J. Mass Spectrom.* **190/191:** 209 and references cited therein.
20. Urban J., Vaisar T., Shen R., and Lee M.S. 1996. *Int. J. Pept. Protein Res.* **47:** 182 and references cited therein.
21. For a summary of the relevant literature see: Farrugia J.M., O'Hair R.A.J., and Reid G.E. 2001. *Int. J. Mass Spectrom.* **210–211:** 71.
22. Schwartz B.L. and Bursey M.M. 1992. *Biol. Mass Spectrom.* **21:** 92.
23. Yu W., Vath J.E., Huberty M.C., and Martin S.A. 1993. *Anal. Chem.* **65:** 3015–3023.
24. Jonsson A.P., Bergman T., Jornvall H., and Griffiths W.J. 2001. *Rapid Commun. Mass Spectrom.* **15:** 713–720.
25. O'Hair R.A.J. and Reid G.E. 1999. *Eur. Mass Spectrom.* **5:** 325–334.
26. Reid G.E., Simpson R.J., and O'Hair R.A.J. 2000. *J. Am. Soc. Mass Spectrom.* **11:** 1047–1060.
27. Waugh R.J. and Bowie J.H. 1994. *Rapid Commun. Mass Spectrom.* **8:** 169.
28. Zubarev R.A., Horn D.M., Fridriksson E.K., Kelleher N.L., Kruger N.A., Lewis M.A., Carpenter B.K., and McLafferty F.W. 2000. *Anal. Chem.* **72:** 563–573.

REFERENCES

Adams L.R. and Roy J.A. 1984. A one dimensional numerical model of a drop-on-demand in jet. *J. Appl. Mech.* **53:** 193–197.

Aebersold R. and Goodlett D.R. 2001. Proteomics and mass spectrometry. *Chem. Rev.* **101:** 269–295.

Aitken A., Geisow M.J., Findlay J.B.C., Holmes C., and Yarwood A. 1989. Peptide preparation and characterization. In *Protein sequencing: A practical approach* (ed. J.B.C. Findlay and M.J. Geisow), pp. 43–68. IRL Press, New York.

Alpert A.J. and Andrews P.C. 1988. Cation-exchange chromatography of peptides on poly(2-sulfoethyl aspartamide)-silica. *J. Chromatogr.* **443:** 85–96.

Amster I.J. 1996. Fourier transform mass spectrometry. *J. Mass. Spectrom.* **31:** 1325–1337.

Arnott D., Henzel W.J., and Stults J.T. 1998. Rapid identification of comigrating gel-isolated proteins by ion trap-mass spectrometry. *Electrophoresis* **19:** 968–980.

Attwood T.K. and Parry-Smith D.J. 1999. DNA sequence analysis In *Introduction to bioinformatics*, pp. 81–107. Addison-Wesley Longman, Harlow, Essex, United Kingdom.

Axen R. and Ernback S. 1971. Chemical fixation of enzymes to cyanogen halide activated polysaccharide carriers. *Eur. J. Biochem.* **18:** 351–360.

Baldwin M.A., Medzihradszky K.F., Lock C.M., Fisher B., Settineri T.A., and Burlingame A.L. 2001. Matrix-assisted laser desorption/ionization coupled with quadrupole/orthogonal acceleration time-of-flight mass spectrometry for protein discovery, identification, and structural analysis. *Anal. Chem.* **73:** 1707–1720.

Ball C.A., Jin H., Sherlock G., Weng S., Matese J.C., Andrada R., Binkley G., Dolinski K., Dwight S.S., Harris M.A., Issel-Tarver L., Schroeder M., Botstein D., and Cherry J.M. 2001. Saccharomyces Genome Database provides tools to survey gene expression and functional analysis data. *Nucleic Acids Res.* **29:** 80–81.

Banks J.F. and Whitehouse C.M. 1996. Electrospray ionization mass spectrometry. *Methods Enzymol.* **270:** 486–518.

Barber M., Bordoli R.S., Sedgwick R.D., and Tyler A.N. 1981. Fast atom bombardment of solids as an ion source in mass spectroscopy. *Nature* **293:** 270–275.

Barber M., Bordoli R.S., Elliott G.J., Horoch N.J., and Green B.N. 1983. Fast atom bombardment mass spectrometry of human proinsulin. *Biochem Biophys. Res Commun.* **110:** 753–757.

Barber M., Bordoli R.S., Elliot G.J., Sedgewick R.D., and Tyler A. 1982. Fast atom bombardment: Mass spectrometry. *Anal. Chem.* **54:** 645–657.

Bateman R.H., Carruthers R., Hoyes J.B., Jones C., Langridge J.I., Millar A., and Vissers J.P.C. 2002. A novel precursor ion discovery method on a hybrid quadrupole orthogonal acceleration time-of-flight (Q-TOF) mass spectrometer for studying protein phosphorylation. *J. Am. Soc. Mass Spectrom.* **13:** 792–803.

Beavis R.C. and Chait B.T. 1989a. Matrix-assisted laser-desorption mass spectrometry using 355 nm radiation. *Rapid Commun. Mass Spectrom.* **3:** 436–439.

——. 1989b. Cinnamic acid derivatives as matrices for ultraviolet laser desorption mass spectrometry of proteins. *Rapid Commun. Mass Spectrom.* **3:** 432–435.

——. 1989c. Factors affecting the ultraviolet laser desorption of proteins. *Rapid Commun. Mass Spectrom.* **3:** 233–237.

——. 1996. Matrix-assisted laser desorption ionization mass spectrometry. *Methods Enzymol.* **270:** 519–551.

Beavis R.C., Chaudhary T., and Chait B.T. 1992. Cyano-4-hydroxycinnamic acid as a matrix for matrix-assisted laser desorption mass spectrometry. *Org. Mass. Spectrom.* **27:** 156–158.

Biemann K. 1962. *Mass spectrometry: Organic chemical applications.* McGraw-Hill, New York.

——. 1978. Mass spectrometric sequencing of peptides and proteins. *Pure Appl. Chem.* **50:** 149–158.

——. 1988. Contributions of mass spectrometry to peptide and protein structure. *Biomed. Environ. Mass Spectrom.* **16:** 99–111.

——. 1990. Sequencing of peptides by tandem mass spectrometry and high energy collision-induced dissociation. *Methods Enzymol.* **193:** 455–479.

——. 1995. The coming of age of mass spectrometry in peptide and protein chemistry. *Protein Sci.* **4:**

1920–1927.

Biemann K., Cone C., Webster B.R., and Arsenault G.P. 1966. Determination of the amino acid sequence in oligopeptides by computer interpretation of their high-resolution mass spectra. *J. Am. Chem. Soc.* **88:** 5598–5606.

Biemann K., Martin S., Scoble H., Johnson R., Papayannopoulos I., Biller J., and Costello C. 1986. How to obtain and how to use mass spectral data at high mass. In *Mass spectrometry in the analysis of large molecules* (ed. C.J. McNeal), pp. 131–149. Wiley, Chichester, United Kingdom.

Bienvenut W.V., Deon C., Pasquarello C., Campbell J.M., Sanchez J.C., Vestal M.L., and Hochstrasser D.F. 2002. Matrix-assisted laser desorption/ionization-tandem mass spectrometry with high resolution and sensitivity for identification and characterization of proteins. *Proteomics* **2:** 868–876.

Bienvenut W.V., Sanchez J.C., Karmime A., Rouge V., Rose K., Binz P.A., and Hochstrasser D.F. 1999. Toward a clinical molecular scanner for proteome research: Parallel protein chemical processing before and during western blot. *Anal. Chem.* **71:** 4800–4807.

Bienvenut W.V., Müeller M., Palagi P.M., Heller M., Gasteiger E., Binz P.A., Giron M., Gay S., Jung E., Gras R., Hugues G.J., Sanchez J.C., Appel R.D., and Hochstrasser D.F. 2001. Proteomics and mass spectrometry: Some aspects and recent developments. In *Mass spectrometry and genomic analysis* (ed. N. Housby), pp. 1–53. Kluwer, The Netherlands.

Binz P.A., Muller M., Walther D., Bienvenut W.V., Gras R., Hoogland C., Bouchet G., Gasteiger E., Fabbretti R., Gay S., Palagi P., Wilkins M.R., Rouge V., Tonella L., Paesano S., Rossellat G., Karmime A., Bairoch A., Sanchez J.C., Appel R.D., and Hochstrasser D.F. 1999. A molecular scanner to automate proteomic research and to display proteome images. *Anal. Chem.* **71:** 4981–4988.

Blais J.C., Nagnanlemeillour P., Bolbach G., and Tabet J.C. 1996. MALDI-TOFMS identification of odorant binding proteins (OBPs) electroblotted onto poly(vinylidene difluoride) membranes. *Rapid Comm. Mass Spectrom.* **10:** 1–4.

Bogy D.B. and Talke F.E. 1984 Experimental and theoretical study of wave propagation phenomena in drop-on-demand in jet devices. *IBM J. Res. Dev.* **29:** 314–321.

Bonner R. and Shushan B. 1995. The characterization of proteins and peptides by automated methods. *Rapid Commun. Mass Spectrom.* **9:** 1067–1076.

Börnsen K.O. 2000. Influence of salts, buffers, detergents, solvents, and matrices on MALDI-MS protein analysis in complex mixtures. *Methods Mol. Biol.* **146:** 387–404.

Bowers W.D., Delbert S.-S., Hunter R.L., and McIver R.T.J. 1984. Fragmentation of oligopeptide ions using ultraviolet laser radiation and Fourier transform mass spectrometry. *J. Am. Chem. Soc.* **106:** 7288–7289.

Breen E.J., Hopwood F.G., Williams K.L., and Wilkins M.R. 2000. Automatic Poisson peak harvesting for high throughput protein identification. *Electrophoresis* **21:** 2243–2251.

Brunnee C. 1987. The ideal mass analyzer: Fact or fiction? *Int. J. Mass Spectrom. Ion Proc.* **76:** 121–237.

Bures E.J., Courchesne P.L., Douglass J., Chen K., Davis M.T., Jones M.D., McGinley M.D., Robinson J.H., Spahr C.S., Wahl R.C., and Patterson S.D. 2001. Identification of incompletely processed potential carboxypeptidase E substrates from CpE^{fat}/CpE^{fat}mice. *Proteomics* **1:** 79–92.

Burke T.W., Mant C.T., Black J.A., and Hodges R.S. 1989. Strong cation-exchange high-performance liquid chromatography of peptides. Effect of non-specific hydrophobic interactions and linearization of peptide retention behaviour. *J. Chromatogr.* **476:** 377–389.

Burlet O., Yang C.Y., and Gaskell S.J. 1992. Influence of cysteine to cysteic acid oxidation on the collision-activated decomposition of protonated peptides—Evidence for intraionic interactions. *J. Am. Soc. Mass Spectrom.* **3:** 337–344.

Burlingame A.L. and Carr S.A., eds. 1996. *Mass spectrometry in the biological sciences*, p. 535. Humana Press, Totowa, New Jersey.

Busch K.L. 1995. Mass spectrometric detectors for samples separated by planar electrophoresis. *J. Chromatogr. A* **692:** 275–290.

Busch K.L., Glish G.L., and McLuckey S.A. 1988. *Mass spectrometry/mass spectrometry. Techniques and applications of tandem mass spectrometry*. Wiley, New York.

Carr S.A. and Burlingame A.L. 1996. The meaning and usage of the terms monoisotopic mass, average mass, mass resolution, and mass accuracy for measurements of biomolecules. In *Mass spectrometry in the biological sciences* (ed. A.L. Burlingame and S.A. Carr), pp. 546–553. Humana Press, Totowa, New Jersey.

Chen X., Smith L.M., and Bradbury E.M. 2000. Site-specific mass tagging with stable isotopes in proteins for accurate and efficient protein identification. *Anal. Chem.* **72**: 1134–1143.

Chin E.T. and Papac D.I. 1999. The use of a porous graphitic carbon column for desalting hydrophilic peptides prior to matrix-assisted laser desorption/ionization time-of-flight mass spectrometry. *Anal. Biochem.* **273:** 179–185.

Clauser K.R., Baker P., and Burlingame A.L. 1999. Role of accurate mass measurement (+/– 10 ppm) in protein identification strategies employing MS or MS/MS and database searching. *Anal. Chem.* **71:** 2871–2882.

Cohen S.L. and Chait B.T. 1996. Influence of matrix solution conditions on the MALDI-MS analysis of peptides and proteins. *Anal. Chem.* **68:** 31–37.

Coligan J.E., Dunn B., Ploegh H., Speicher D., and Wingfield P., eds. 1999. *Current protocols in protein science*, p. A.1A. Wiley, New York.

Conrads T.P., Alving K., Veenstra T.D., Belov M.E., Anderson G.A., Anderson D.J., Lipton M.S., Pasa-Tolic L., Udseth H.R., Chrisler W.B., Thrall B.D., and Smith R.D. 2001. Quantitative analysis of bacterial and mammalian proteomes using a combination of cysteine affinity tags and 15N-metabolic labeling. *Anal. Chem.* **73:** 2132–2139.

Corthals G.L., Wasinger V.C., Hochstrasser D.F., and Sanchez J.C. 2000. The dynamic range of protein expression: A challenge for proteomic research. *Electrophoresis* **21:** 1104–1115.

Costanzo G., Camier S., Carlucci P., Burderi L., and Negri R. 2001a. RNA polymerase III transcription complexes on chromosomal 5S rRNA genes in vivo: TFIIIB occupancy and promoter opening. *Mol. Cell. Biol.* **21:** 3166–3178.

Costanzo M.C., Crawford M.E., Hirschman J.E., Kranz J.E., Olsen P., Robertson L.S., Skrzypek M.S., Braun B.R., Hopkins K.L., Kondu P., Lengieza C., Lew-Smith J.E., Tillberg M., and Garrels J.I. 2001b. YPD, PombePD and WormPD: Model organism volumes of the BioKnowledge library, an integrated resource for protein information. *Nucleic Acids Res.* **29:** 75–79.

Cottrell J.S. and Frank B.H. 1985. Fast atom bombardment mass spectrometry of bovine proinsulin. *Biochem. Biophys. Res. Commun.* **127:** 1032–1038.

Cox A.L., Skipper J., Chen Y., Henderson R.A., Darrow T.L., Shabanowitz J., Engelhard V.H., Hunt D.F., and Slingluff C.L., Jr. 1994. Identification of a peptide recognized by five melanoma-specific human cytotoxic T cell lines. *Science* **264:** 716–719.

Daves G.D. 1979. Mass spectrometry of involatile and thermally unstable molecules. *Accts. Chem. Res.* **12:** 359–365.

Davis M.T. and Lee T.D. 1997. Variable flow liquid chromatography-tandem mass spectrometry and the comprehensive analysis of complex protein digest mixtures. *J. Am. Soc. Mass Spectrom.* **8:** 1059–1069.

Davis M., Stahl D., Hefta S., and Lee T. 1995. A microscale electrospray interface for on-line capillary liquid chromatography/tandem mass spectrometry of complex peptide mixtures. *Anal. Chem.* **67:** 4549–4556.

de Hoffmann E. 1996. Tandem mass spectrometry: A primer. *J. Mass. Spectrom.* **31:** 129–137.

Dell A. and Morris H.R. 1982. Fast atom bombardment—High field magnet mass spectrometry of 6000 dalton polypeptides. *Biochem. Biophys. Res. Commun.* **106:** 1456–1462.

Desiderio D.M. and Katakuse I. 1983. Fast atom bombardment-collision activated dissociation-linked field scanning mass spectrometry of the neuropeptide substance P. *Anal. Biochem.* **129:** 425–429.

Dijksman J.F. 1984. Hydrodynamics of small tubular pumps. *J. Fluid Mech.* **139:** 173–191.

Dongre A.K., Somogyi A., and Wysocki V.H. 1996. Surface induced dissociation: An effective tool to probe structure, energetics and fragmentation mechanisms of protonated petpides. *J. Mass Spectrom.* **31:** 339–350.

Eckerskorn C., Strupat K., Karas M., Hillenkamp F., and Lottspeich F. 1992. Mass spectrometric analysis of blotted proteins after gel electrophoretic separation by matrix-assisted laser desorption/ionization. *Electrophoresis* **13:** 664–665.

Eckerskorn C., Strupat K., Kellermann J., Lottspeich F., and Hillenkamp F. 1997a. High-sensitivity peptide mapping by micro-LC with on-line membrane blotting and subsequent detection by scanning-IR-MALDI mass spectrometry. *J. Protein Chem.* **16:** 349–362.

Eckerskorn C., Strupat K., Schleuder D., Hochstrasser D., Sanchez J.C., Lottspeich F., and Hillenkamp F. 1997b. Analysis of proteins by direct-scanning infrared-MALDI mass spectrometry after 2D-PAGE separation and electroblotting. *Anal. Chem.* **69:** 2888–2892.

Egelhofer V., Bussow K., Luebbert C., Lehrach H., and Nordhoff E. 2000. Improvements in protein identification by MALDI-TOF-MS peptide mapping. *Anal. Chem.* **72:** 2741–2750.

Emmett M.R. and Caprioli R.M.J. 1994. Micro-electrospray mass spectrometry: Ultra-high-sensitivity analysis of peptides and proteins. *J. Am. Soc. Mass Spectrom.* **5:** 605–613.

Eng J.K., McCormack A.L., and Yates J.R., III. 1994. An approach to correlate tandem mass-spectral data of peptides with amino-acid-sequences in a protein database. *J. Am. Soc. Mass Spectrom.* **5:** 976–989.

Erdjument-Bromage H., Lui M., Lacomis L., Grewal A., Annan R.S., McNulty D.E., Carr S.A., and Tempst P. 1998. Examination of micro-tip reversed-phase liquid chromatographic extraction of peptide pools for mass spectrometric analysis. *J Chromatogr. A* **826:** 167–181.

Eriksson J., Chait B.T., and Fenyo D. 2000. A statistical basis for testing the significance of mass spectrometric protein identification results. *Anal. Chem.* **72:** 999–1005

Fenn J.B., Mann M., Meng C.K., Wong S.F., and Whitehouse C.M. 1989. Electrospray ionization for mass spectrometry of large biomolecules. *Science* **246:** 64–71.

Fenyo D., Qin J., and Chait B.T. 1998. Protein identification using mass spectrometric information. *Electrophoresis* **19:** 998–1005.

Figeys D., Gygi S.P., McKinnon G., and Aebersold R. 1998. An integrated microfluidics-tandem mass spectrometry system for automated protein analysis. *Anal. Chem.* **70:** 3728–3734.

Fountoulakis M., Takacs M.F., and Takacs B. 1999a. Enrichment of low-copy-number gene products by hydrophobic interaction chromatography. *J. Chromatogr. A* **833:** 157–168.

Fountoulakis M., Takacs M.F., Berndt P., Langen H., and Takacs B. 1999b. Enrichment of low abundance proteins of *Escherichia coli* by hydroxyapatite chromatography. *Electrophoresis* **20:** 2181–2195.

Garrels J.I., Futcher B., Kobayashi R., Latter G.I., Schwender B., Volpe T., Warner J.R., and McLaughlin C.S. 1994. Protein identifications for a *Saccharomyces cerevisiae* protein database. *Electrophoresis* **15:** 1466–1486.

Gaskell S.J. 1997. Electrospray: Principles and practice. *J. Mass Spectrom.* **32:** 677–688.

Gatlin C.L., Kleemann G.R., Hays L.G., Link A.J., and Yates J.R., III. 1998. Protein identification at the low femtomole level from silver-stained gels using a new fritless electrospray interface for liquid chromatography-microspray and nanospray mass spectrometry. *Anal. Biochem.* **263:** 93–101.

Gauthier J.W., Trautman T.R., and Jacobson D.B. 1991. Sustained off-resonance irradiation for collision-activated dissoication involving Fourier transform mass spectrometry. Collision-activated dissociation technique that emulates infrared multiphoton dissociation. *Anal. Chim. Acta* **246:** 211–225.

Gevaert K. and Vandekerckhove J. 2000. Protein identification methods in proteomics. *Electrophoresis* **21:** 1145–1154.

Gibson B.W. and Biemann K. 1984. Strategy for the mass spectrometric verification and correction of the primary structures of proteins deduced from their DNA sequences. *Proc. Natl. Acad. Sci.* **81:** 1956–1960.

Giddings J.C. 1987. Concepts and comparisons in multidimensional chromatography. *J. High Res. Chromatogr.* **10:** 319–323.

Gobom J., Nordhoff E., Ekman R. and Roepstorff P. 1997. Rapid micro-scale proteolysis of proteins for MALDI-MS peptide mapping using immobilized trypsin. *Int. J. Mass Spectrom. Ion Proc.* **169/170:** 153–163.

Gobom J., Nordhoff E., Mirgorodskaya E., Ekman R., and Roepstorff P. 1999. Sample purification and preparation technique based on nano-scale reversed-phase columns for the sensitive analysis of complex peptide mixtures by matrix-assisted laser desorption/ionization mass spectrometry. *J. Mass Spectrom.* **34:** 105–116.

Goffeau A., Barrell B.G., Bussey H., Davis R.W., Dujon B., Feldmann H., Galibert F., Hoheisel J.D., Jacq C., Johnston M., Louis E.J., Mewes H.W., Murakami Y., Philippsen P., Tettelin H., and Oliver S.G. 1996. Life with 6000 genes. *Science* **274:** 546, 563–546, 567.

Goodlett D.R., Wahl J.H., Udseth H.R., and Smith R.D. 1993. Reduced elution speed detection for capillary electrophoresis-mass spectrometry. *J. Microcol. Sep.* **5:** 57–62.

Goodlett D.R., Keller A., Watts J.D., Newitt R., Yi E.C., Purvine S., Eng J.K., von Haller P., Aebersold R., and Kolker E. 2001. Differential stable isotope labeling of peptides for quantitation and de novo sequence derivation. *Rapid Commun. Mass Spectrom.* **15:** 1214–1221.

Gorman J.J., Ferguson B.L., and Nguyen T.B. 1996. Use of 2,6-dihydroxyacetophenone for analysis of fragile peptides, disulphide bonding and small proteins by matrix-assisted laser desorption/ionization. *Rapid Commun. Mass Spectrom.* **10:** 529–536.

Gras R., Müeller M., Gasteiger E., Gay S., Binz P.A., Bienvenut W., Hoogland C., Sanchez J.C., Bairoch A., Hochstrasser D.F., and Appel R.D. 1999. Improving protein identification from peptide mass fingerprinting through a parameterized multi-level scoring algorithm and an optimized peak detection.

Electrophoresis 20: 3535–3550.

Grayson M.A., ed. 2002. *Measuring mass: From positive rays to proteins.* Chemical Heritage Press, Philadelphia, Pennsylvania.

Griffin P.R., MacCoss M.J., Eng J.K., Blevins R.A., Aaronson J.S., and Yates J.R. III. 1995. Direct database searching with MALDI-PSD spectra of peptides. *Rapid Commun. Mass Spectrom.* **9:** 1546–1551.

Griffin P.R., Coffman J.A., Hood L.E., and Yates J.R., III. 1991. Structural analysis of proteins by capillary HPLC electrospray tandem mass spectrometry. *Int. J. Mass Spectrom. Ion Processes* **111:** 131–149.

Griffin T.J., Han D.K., Gygi S.P., Rist B., Lee H., Aebersold R., and Parker K.C. 2001a. Toward a high-throughput approach to quantitative proteomic analysis: Expression-dependent protein identification by mass spectrometry. *J. Am. Soc. Mass Spectrom.* **12:** 1238–1246.

Griffin T.J., Gygi S.P., Rist B., Aebersold R., Loboda A., Jilkine A., Ens W., and Standing K.G. 2001b. Quantitative proteomic analysis using a MALDI quadrupole time-of-flight mass spectrometer. *Anal. Chem.* **73:** 978–986.

Gygi S.P. and Aebersold R. 1999. Absolute quantitation of 2-D protein spots. *Methods Mol. Biol.* **112:** 417–421.

———. 2000. Mass spectrometry and proteomics. *Curr. Opin. Chem. Biol.* **4:** 489–494.

Gygi S.P., Rist B., and Aebersold R. 2000a. Measuring gene expression by quantitative proteome analysis. *Curr. Opin. Biotechnol.* **11:** 396–401.

Gygi S.P., Corthals G.L., Zhang Y., Rochon Y., and Aebersold R. 2000b. Evaluation of two-dimensional gel electrophoresis-based proteome analysis technology. *Proc. Natl. Acad. Sci.* **97:** 9390–9395.

Gygi S.P., Rist B., Gerber S.A., Turecek F., Gelb M.H., and Aebersold R. 1999. Quantitative analysis of complex protein mixtures using isotope-coded affinity tags. *Nat. Biotechnol.* **17:** 994–999.

Hanash S.M. 2000. Biomedical applications of two-dimensional electrophoresis using immobilized pH gradients: Current status. *Electrophoresis* **21:** 1202–1209.

Harrison A.G. 1992. *Chemical ionization mass spectrometry.* CRC Press, Boca Raton, Florida.

Harrison A.G. and Cotter R.J. 1990. Methods of ionization. *Methods Enzymol.* **193:** 3–36.

Hayes R.N. and Gross M.L. 1990. Collision-induced dissociation. *Methods Enzymol.* **193:** 237–263.

Henzel W.J., Billeci T.M., Stults J.T., Wong S.C., Grimley C., and Watanabe C. 1993. Identifying proteins from two-dimensional gels by molecular mass searching of peptide fragments in protein sequence databases. *Proc. Natl. Acad. Sci.* **90:** 5011–5015.

Hillenkamp F., Karas M., Beavis R.C., and Chait B.T. 1991. Matrix-assisted laser desorption/ionization mass spectrometry of biopolymers. *Anal. Chem.* **63:** 1193A–1203A.

Hong H.Y., Yoo G.S., and Choi J.K. 2000. Direct Blue 71 staining of proteins bound to blotting membranes. *Electrophoresis* **21:** 841–845.

Horn D.M., Ge Y., and McLafferty F.W. 2000a. Activated ion electron capture dissociation for mass spectral sequencing of larger (42 kDa) proteins. *Anal. Chem.* **72:** 4778–4784.

Horn D.M., Zubarev R.A., and McLafferty F.W. 2000b. Automated reduction and interpretation of high resolution electrospray mass spectra of large molecules. *J. Am. Soc. Mass Spectrom.* **11:** 320–332.

Hummel B.C.W. 1959. A modified spectrophotometric determination of chymotrypsin, trypsin, and thrombin. *Can. J. Biochem. Physiol.* **37:** 1393–1398.

Hunt D.F., Yates J.R., III, Shabanowitz J., Winston S., and Hauer C.R. 1986. Protein sequencing by tandem mass spectrometry. *Proc. Natl. Acad. Sci.* **83:** 6233–6237.

Hunt D.F., Bone W. M., Shabanowitz J., Rhodes J., and Ballard J.M. 1981. Sequence analysis of oligopeptides by secondary ion/collision activated dissociation mass spectrometry. *Anal. Chem.* **53:** 1704–1706.

Hunt D.F., Alexander J.E., McCormack A.L., Martino P.A., Michel H., Shabanowitz J., Sherman N., Moseley M.A., Jorgenson J.W., and Tomer K.B. 1991. *Techniques in protein chemistry II*, p. 441. Academic Press, New York.

Hyver K.J., Campana J.E., Cotter R.J., and Fenselau C. 1985. Mass spectral analysis of murine epidermal growth factor. *Biochem. Biophys. Res. Commun.* **130:** 1287–1293.

Ishii S., Yokosawa H., Kumazaki T., and Nakamura I. 1983. Immobilized anhydrotrypsin as a biospecific affinity adsorbent for tryptic peptides. *Methods Enzymol.* **91:** 378–383.

Ishii S., Yokosawa H., Shiba S., and Kasai K. 1979. Specific isolation of biologically-active peptides by means of immobilized anhydrotrypsin and anhydrochymotrypsin. *Adv. Exp. Med. Biol.* **120A:** 15–27.

James P., Quadroni M., Carafoli E., and Gonnet G. 1993. Protein identification by mass profile fingerprinting. *Biochem. Biophys. Res. Commun.* **195:** 58–64.

——. 1994. Protein identification in DNA databases by peptide mass fingerprinting. *Protein Sci.* **3:** 1347–1350.

Jennings K.R. and Dolnikowski G.G. 1990. Mass analyzers. *Methods Enzymol.* **193:** 37–60.

Jensen O.N., Podtelejnikov A., and Mann M. 1996a. Delayed extraction improves specificity in database searches by matrix-assisted laser desorption/ionization peptide maps. *Rapid Commun. Mass Spectrom.* **10:** 1371–1378.

Jensen O.N., Kulkarni S., Aldrich J.V., and Barofsky D.F. 1996b. Characterization of peptide-oligonucleotide heteroconjugates by mass spectrometry. *Nucleic Acids Res.* **24:** 3866–3872.

Ji H., Baldwin G.S., Burgess A.W., Moritz R.L., Ward L.D., and Simpson R.J. 1993. Epidermal growth factor induces serine phosphorylation of stathmin in a human colon carcinoma cell line (LIM 1215). *J. Biol. Chem.* **268:** 13396–13405.

John N.E., Andersen H.U., Fey S.J., Larsen P.M., Roepstorff P., Larsen M.R., Pociot F., Karlsen A.E., Nerup J., Green I.C., and Mandrup-Poulsen T. 2000. Cytokine- or chemically derived nitric oxide alters the expression of proteins detected by two-dimensional gel electrophoresis in neonatal rat islets of Langerhans. *Diabetes* **49:** 1819–1829.

Johnstone R.A.W. and Rose M.E. 1996. *Mass spectrometry for chemists and biochemists*, 2nd edition. Cambridge University Press.

Jones E.W. 1991. Tackling the protease problem in *Saccharomyces cerevisiae. Methods Enzymol.* **194:** 428–453.

Karas M. and Hillenkamp F. 1988. Laser desorption ionization of proteins with molecular masses exceeding 10,000 daltons. *Anal. Chem.* **60:** 2299–2301.

Karas M., Gluckmann M., and Schafer J. 2000. Ionization in matrix-assisted laser desorption/ionization: Singly charged molecular ions are the lucky survivors. *J. Mass Spectrom.* **35:** 1–12.

Karas M., Ehring H., Nordhoff E., Stahl B., Stupat K., Grehl M., et al. 1993. Matrix-assisted laser desorption mass spectrometry with additives to 2.5-dihydroxy benzoic acid. *Org. Mass Spectrom.* **28:** 1476–1481.

Kasai K. and Ishii S. 1975. Affinity chromatography of trypsin and related enzymes. I. Preparation and characteristics of an affinity adsorbent containing tryptic peptides from protamine as ligands. *J. Biochem.* **78:** 653–662.

Katayama H., Nagasu T., and Oda Y. 2001. Improvement of in-gel digestion protocol for peptide mass fingerprinting by matrix-assisted laser desorption/ionization time-of-flight mass spectrometry. *Rapid Commun. Mass Spectrom.* **15:** 1416–1421.

Kennedy R. and Jorgenson J.W. 1989. Preparation and evaluation of packed capillary liquid chromatography columns with inner diameter from 20 to 50 μm. *Anal. Chem.* **61:** 1128–1135.

King R., Bonfiglio R., Fernandez-Metzler C., Miller-Stein C., and Olah T. 2000. Mechanistic investigation of ionization suppression in electrospray ionization. *J. Am. Soc. Mass Spectrom.* **11:** 942–950.

Kumazaki T., Terasawa K., and Ishii S. 1987. Affinity chromatography on immobilized anhydrotrypsin: General utility for selective isolation of C-terminal peptides from protease digests of proteins. *J. Biochem.* **102:** 1539–1546.

Kumazaki T., Nakako T., Arisaka F., and Ishii S. 1986. A novel method for selective isolation of C-terminal peptides from tryptic digests of proteins by immobilized anhydrotrypsin: Application to structural analysis of the tail sheath and the tube proteins from bacteriophage T4. *Proteins Struct. Funct. Genet.* **1:** 100–107.

Kussmann M., Lassing U., Sturmer C.A., Przybylski M., and Roepstorff P. 1997. Matrix-assisted laser desorption/ionization mass spectrometric peptide mapping of the neural cell adhesion protein neurolin purified by sodium dodecyl sulfate polyacrylamide gel electrophoresis or acidic precipitation. *J. Mass Spectrom.* **32:** 483–493.

Kussmann M. and Roepstorff P. 2000. Sample preparation techniques for peptides and proteins analyzed by MALDI-MS. *Methods Mol. Biol.* **146:** 405–424.

Larsen M.R. and Roepstorff P. 2000. Mass spectrometric identification of proteins and characterization of their post-translational modifications in proteome analysis. Fresenius. *J. Anal. Chem.* **366:** 677–690.

Larsen M.R., Cordwell S.J., and Roepstorff P. 2002a. Graphite powder as an alternative to reversed-phase material for desalting and concentration of peptide mixtures prior to mass spectrometric analysis. *Proteomics* **2:** 1277–1287.

——. 2002b. Characterization of modified proteins using graphite powder microcolumns in combination with gel electrophoresis and advanced mass spectrometry. In *Proceedings of the 5th Siena Meeting: From Genome to Proteome: Functional Proteomics*, Siena, Italy, p. 72.

Larsen M.R., Pedersen S.K., and Walsh B. 2001a. The use of GELoader tip microcolumn technology in combination with mass spectrometry. In *Proceedings of the 49th ASMS Conference on Mass Spectrometry and Allied Topics*, Chicago, Illinois, poster no. MPF146.

Larsen M.R., Larsen P.M., Fey S.J., and Roepstorff P. 2001b. Characterization of differently processed forms of enolase 2 from *Saccharomyces cerevisiae* by two-dimensional gel electrophoresis and mass spectrometry. *Electrophoresis* **22:** 566–575.

Larsen M.R., Sorensen G.L., Fey S.J., Larsen P.M., and Roepstorff P. 2001c. Phospho-proteomics: Evaluation of the use of enzymatic de-phosphorylation and differential mass spectrometric peptide mass mapping for site specific phosphorylation assignment in proteins separated by gel electrophoresis. *Proteomics* **1:** 223–238.

Larsen P.M., Fey S.J., Larsen M.R., Nawrocki A., Andersen H.U., Kahler H., Heilmann C., Voss M.C., Roepstorff P., Pociot F., Karlsen A.E., and Nerup J. 2001. Proteome analysis of interleukin-1beta-induced changes in protein expression in rat islets of Langerhans. *Diabetes* **50:** 1056–1063.

Lazar J.M., Ramsey R.S., Sundberg S., and Ramsey J.M. 1999. Subattomole-sensitivity microchip nanoelectrospray source with time-of-flight mass spectrometry detection. *Anal. Chem.* **71:** 3627–3631.

Licklider L.J., Thoreen C.C., Peng J., and Gygi S.P. 2002. Automation of nanoscale microcapillary liquid chromatograpy-tandem mass spectrometry with a vented column. *Anal. Chem.* **74:** 3076–3083.

Link A.J., Eng J., Schieltz D.M., Carmack E., Mize G.J., Morris D.R., Garvik B.M., and Yates J.R., III. 1999. Direct analysis of protein complexes using mass spectrometry. *Nat. Biotechnol.* **17:** 676–682.

Little D.P., Speir J.P., Senko M.W., O'Connor P.B., and McLafferty F.W. 1994. Infrared multiphoton dissociation of large multiply charged ions for biomolicule sequencing. *Anal. Chem.* **66:** 2809–2815.

Loboda A.V., Krutchinsky A.N., Bromirski M., Ens W., and Standing K.G. 2000. A tandem quadrupole/time-of-flight mass spectrometer with a matrix-assisted laser desorption/ionization source: Design and performance. *Rapid Commun. Mass Spectrom.* **14:** 1047–1057.

Mann M. 1994. Sequence database searching by mass spectrometric data. In *Microcharacterisation of proteins* (ed. R. Kellner et al.), pp. 223–245. VCH, Weinheim, Germany.

Mann M., Hojrup P., and Roepstorff P. 1993. Use of mass spectrometric molecular weight information to identify proteins in sequence databases. *Biol. Mass Spectrom.* **22:** 338–345.

Mant C.T. and Hodges R.S. 1985. Separation of peptides by strong cation-exchange high-performance liquid chromatography. *J. Chromatogr.* **327:** 147–155.

March R.E. 1997. An introduction to quadrupole ion trap mass spectrometry. *J. Mass Spectrom.* **32:** 351–369.

———. 1998. Quadrupole ion trap mass spectrometry: Theory, simulation, recent developments and applications. *Rapid. Commun. Mass Spectrom.* **12:** 1543–1554.

Marshall A.G. and Verdun F.R. 1990. *Fourier transforms in NMR, optical and mass spectrometry: A user's handbook.* Elsevier, New York.

Marshall A.G., Wang T.-C.L., and Ricca T.L. 1985. Tailored excitation for fourier transform ion cyclotron mass spectrometry. *J. Am. Chem. Soc.* **107:** 7893–7897.

Marshall A.G., Senko M.W., Li W., Li M., Dillon S., Guan S., and Logan T.M. 1997. Protein molecular mass to 1 Da by ^{13}C, ^{15}N double-depletion and FT-ICR mass spectrometry. *J. Am. Chem. Soc.* **119:** 433–434.

Martin S.E., Shabanowitz J., Hunt D.F., and Marto J.A. 2000. Subfemtomole MS and MS/MS peptide sequence analysis using nano-HPLC micro-ESI fourier transform ion cyclotron resonance mass spectrometry. *Anal. Chem.* **72:** 4266–4274.

Masselon C., Anderson G.A., Harkewicz R., Bruce J.E., Pasa-Tolic L., and Smith R.D. 2000. Accurate mass multiplexed tandem mass spectrometry for high-throughput polypeptide identification from mixtures. *Anal. Chem.* **72:** 1918–1924.

Matsudaira P. 1987. Sequence from picomole quantities of proteins electroblotted onto polyvinylidene difluoride membranes. *J. Biol. Chem.* **262:** 10035–10038.

McCormack A., Eng J., and Yates J.R. 1994. Peptide sequence analysis on quadrupole mass spectrometers. *Methods* **1994:** 274–83.

McCormack A.L., Schieltz D.M., Goode B., Yang S., Barnes G., Drubin D., and Yates J.R., III. 1997. Direct analysis and identification of proteins in mixtures by LC/MS/MS and database searching at the low-femtomole level. *Anal. Chem.* **69:** 767–776.

McLafferty F.W., Fridriksson E.K., Horn D.M., Lewis M.A., and Zubarev R.A. 1999. Techview: Biochemistry. Biomolecule mass spectrometry. *Science* **284:** 1289–1290.

McLuckey S.A. 1992. Principles of collisional activation in analytical mass spectrometry. *J. Am. Soc. Mass Spectrom.* **3:** 599–614.

McLuckey S.A. and Goeringer D.E. 1997. Slow heating methods in tandem mass spectrometry. *J. Mass Spectrom.* **32:** 461–474.

McLuckey S.A., Van Berkel G.J., Glish G.L., and Schwartz J.C. 1995. Electrospray and the quadrupole ion trap. In *Practical aspects of ion trap mass spectrometry. II: Ion trap instrumentation* (ed. R.E. March and J.F.J. Todd), pp. 89–141. CRC Press, Boca Raton, Florida.

Medzihradszky K.F., Campbell J.M., Baldwin M.A., Falick A.M., Juhasz P., Vestal M.L., and Burlingame A.L. 2000. The characteristics of peptide collision-induced dissociation using a high-performance MALDI-TOF/TOF tandem mass spectrometer. *Anal. Chem.* **72:** 552–558.

Meeusen S., Tieu Q., Wong E., Weiss E., Schieltz D., Yates J.R., and Nunnari J. 1999. Mgm101p is a novel component of the mitochondrial nucleoid that binds DNA and is required for the repair of oxidatively damaged mitochondrial DNA. *J. Cell Biol.* **145:** 291–304.

Meng F., Cargile B.J., Miller L.M., Forbes A.J., Johnson J.R., and Kelleher N.L. 2001. Informatics and multiplexing of intact protein identification in bacteria and the archaea. *Nat. Biotechnol.* **19:** 952–957.

Mewes H.W., Frishman D., Gruber C., Geier B., Haase D., Kaps A., Lemcke K., Mannhaupt G., Pfeiffer F., Schuller C., Stocker S., and Weil B. 2000. MIPS: A database for genomes and protein sequences. *Nucleic Acids Res.* **28:** 37–40.

Miranker A.D. 2000. Protein complexes and analysis of their assembly by mass spectrometry. *Curr. Opin. Struct. Biol.* **10:** 601–606.

Molloy M.P. 2000. Two-dimensional electrophoresis of membrane proteins using immobilized pH gradients. *Anal. Biochem.* **280:** 1–10.

Moore R.E., Young M.K., and Lee T.D. 2000. Method for screening peptide fragment ion mass spectra prior to database searching. *J. Am. Soc. Mass Spectrom.* **11:** 422–426.

———. 2002. Qscore: An algorithm for evaluating SEQUEST database search results. *J. Am. Soc. Mass Spectrom.* **13:** 378–386.

Moritz R.L. and Simpson R.J. 1992a. Application of capillary reversed-phase high-performance liquid chromatography to high-sensitivity protein sequence analysis. *J. Chromatogr.* **599:** 119–130.

———. 1992b. Purification of proteins and peptides for sequence analysis using microcolumn liquid chromatography. *J. Microcol. Sep.* **4:** 485–489.

———. 1993. Capillary liquid chromatography: A tool for protein structural analysis. In *Methods in protein sequence analysis* (ed. K. Imahori and F. Sakiyama), pp. 3–10. Plenum Press, New York.

Moritz R.L., Eddes J.S., Reid G.E., and Simpson R.J. 1996. S-pyridylethylation of intact polyacrylamide gels and in situ digestion of electrophoretically separated proteins: A rapid mass spectrometric method for identifying cysteine-containing peptides. *Electrophoresis* **17:** 907–917.

Morris H.R. 1972. Complete sequence determination of proteins by mass spectrometry. Rapid procedure for the successful permethylation of histidine-containing peptides. *FEBS Lett.* **22:** 257–260.

Morris H.R., Dickinson R.J., and Williams D.H. 1973. Studies toward the complete sequence determination of proteins by mass spectrometry. Derivatization of methionine, cysteine, and arginine containing peptides. *Biochem. Biophys. Res. Commum.* **51:** 247–255.

Morris H.R., Panico M., Barber M., Bordoli R.S., Sedgwick R.D., and Tyler A. 1981. Fast atom bombardment: A new mass spectrometric method for peptide sequence analysis. *Biochem. Biophys. Res. Commun.* **101:** 623–631.

Morris H.R., Paxton T., Dell A., Langhorne J., Berg M., Bordoli R.S., Hoyes J., and Bateman R.H. 1996. High sensitivity collisionally-activated decomposition tandem mass spectrometry on a novel quadrupole/orthogonal-acceleration time-of-flight mass spectrometer. *Rapid Commun. Mass Spectrom.* **10:** 889–896.

Mortz E., Vorm O., Mann M., and Roepstorff P. 1994. Identification of proteins in polyacrylamide gels by mass spectrometric peptide mapping combined with database search. *Biol. Mass. Spectrom.* **23:** 249–261.

Mortz E., O'Connor P.B., Roepstorff P., Kelleher N.L., Wood T.D., McLafferty F.W., and Mann M. 1996. Sequence tag identification of intact proteins by matching tanden mass spectral data against sequence data bases. *Proc. Natl. Acad. Sci.* **93:** 8264–8267.

Moseley M.A., Deterding L.J., Tomer K.B., and Jorgenson J.W. 1991. Nanoscale packed-capillary liquid-chromatography coupled with mass-spectrometry using a coaxial continuous-flow fast-atom-bombardment interface. *Anal. Chem.* **63:** 1467–1473.

Natsume T., Yamauchi Y., Nakayama H., Shinkawa T., Yanagida M., Takahashi N., and Isobe T. 2002. A direct nanoflow liquid chromatography-tandem mass spectrometry system for interaction proteomics. *Anal. Chem.* **74:** 4725–4733.

Nawrocki A., Larsen M.R., Podtelejnikov A.V., Jensen O.N., Mann M., Roepstorff P., Gorg A., Fey S.J., and Larsen P.M. 1998. Correlation of acidic and basic carrier ampholyte and immobilized pH gradient two-dimensional gel electrophoresis patterns based on mass spectrometric protein identification. *Electrophoresis* **19:** 1024–1035.

Neubauer G. and Mann M. 1999. Mapping of phosphorylation sites of gel-isolated proteins by nanoelectrospray tandem mass spectrometry: Potentials and limitations. *Anal. Chem.* **71:** 235–242.

Newton K.A., Chrisman P.A., Reid G.E., Wells J.M. and McLuckey S.A. 2001. Loss of charged versus neutral heme from gaseous holomyoglobin ions. *Rapid Commun. Mass Spectrom.* **15:** 2334–2340.

Nilsson C.L. and Davidsson P. 2000. New separation tools for comprehensive studies of protein expression by mass spectrometry. *Mass Spec. Rev.* **19:** 390–397.

Oda Y., Huang K., Cross F.R., Cowburn D., and Chait B.T. 1999. Accurate quantitation of protein expression and site-specific phosphorylation. *Proc. Natl. Acad. Sci.* **96:** 6591–6596.

Oh-Ishi M., Satoh M., and Maeda T. 2000. Preparative two-dimensional gel electrophoresis with agarose gels in the first dimension for high molecular mass proteins. *Electrophoresis* **21:** 1653–1669.

Pandey A. and Mann M. 2000. Proteomics to study genes and genomes. *Nature* **405:** 837–846.

Pappin D.J., Hojrup P., and Bleasby A. 1993. Rapid identification of proteins by peptide-mass fingerprinting. *Curr. Biol.* **3:** 327–332.

Papayannopoulos I.A. 1995. The interpretation of collision-induced dissociation tandem mass-spectra of peptides. *Mass Spectrom. Rev.* **14:** 49–73.

Pasa-Tolic L., Jensen P.K., Anderson G.A., Lipton M.S., Peden K.K., Martinovic S., Tolic N., Bruce J.E., and Smith R.D. 1999. High throughput proteome-wide precision measurements of protein expression using mass spectrometry. *J. Am. Chem. Soc.* **12:** 7949–7950.

Patterson S.D. 1995. Matrix-assisted laser-desorption/ionization mass spectrometric approaches for the identification of gel-separated proteins in the 5–50 pmol range. *Electrophoresis* **16:** 1104–1114.

———. 1997. Identification of low to subpicomolar quantities of electrophoretically separated proteins: Towards protein chemistry in the post-genome era. *Biochem. Soc. Trans.* **25:** 255–262.

Pedersen S.K, Christiansen J., Hansen T.O., Larsen M.R., and Nielsen F.C. 2002. Human insulin-like growth factor II leader 2 mediates internal initiation of translation. *Biochem. J.* **363:** 37–44.

Peng J. and Gygi S.P. 2001. Proteomics: The move to mixtures. *J. Mass Spectrom.* **36:** 1083–1091.

Perkins D.N., Pappin D.J., Creasy D.M., and Cottrell J.S. 1999. Probability-based protein identification by searching sequence databases using mass spectrometry data. *Electrophoresis* **20:** 3551–3567.

Price W.D. and Williams E.R. 1997. Activation of peptide ions by blackbody radiation: Factors that lead to dissociation kinetics in the rapid energy exchange limit. *J. Phys. Chem. A.* **101:** 8844–8852.

Qin J., Fenyo D., Zhao Y., Hall W.W., Chao D.M., Wilson C.J., Young R.A, and Chait B.T. 1997. A strategy for rapid, high-confidence protein identification. *Anal. Chem.* **69:** 3995–4001.

Rabilloud T. 1990. Mechanisms of protein silver staining in polyacrylamide gels: A 10-year synthesis. *Electrophoresis* **11:** 785–794.

Reid G.E., Rasmussen R.K., Dorow D.S., and Simpson R.J. 1998. Capillary column chromatography improves sample preparation for mass spectrometric analysis: Complete characterization of human alpha-enolase from two-dimensional gels following in situ proteolytic digestion. *Electrophoresis* **19:** 946–955.

Reid G.E. and McLuckey S.A. 2002. "Top down" protein characterization via tandem mass spectrometry. *J. Mass Spectrom.* **37:** 663–675.

Roepstorff P. and Fohlman J. 1984. Proposal for a common nomenclature for sequence ions in mass spectra of peptides. *Biomed. Mass Spectrom.* **11:** 601.

Santoni V., Molloy M., and Rabilloud T. 2000. Membrane proteins and proteomics: Un amour impossible? *Electrophoresis* **21:** 1054–1070.

Schartz J.C. and Jardine I. 2000. Quadrupole ion trap mass spectrometry. *Methods Enzymol.* **270:** 552–586.

Schleuder D., Hillenkamp F., and Strupat K. 1999. IR-MALDI-mass analysis of electroblotted proteins directly from the membrane: Comparison of different membranes, application to on-membrane digestion and protein identification by database searching. *Anal. Chem.* **71:** 3238–3247.

Sechi S. and Chait B.T. 2000. A method to define the carboxyl terminal of proteins. *Anal. Chem.* **72:** 3374–3378.

Seifert W.E. and Caprioli R.M. 1996. Fast atom bombardment mass spectrometry. *Methods Enzymol.* **270:** 453–485.

Shaw A.C., Rossel L.M., Roepstorff P., Holm A., Christiansen G., and Birkelund S. 1999. Mapping and identification of HeLa cell proteins separated by immobilized pH-gradient two-dimensional gel electrophoresis and construction of a two-dimensional polyacrylamide gel electrophoresis database. *Electrophoresis* **20:** 977–983.

Shevchenko A., Loboda A., Shevchenko A., Ens W., and Standing K.G. 2000. MALDI quadrupole time-of-flight mass spectrometry: A powerful tool for proteomic research. *Anal. Chem.* **72:** 2132–2141.

Shevchenko A., Sunyaev S., Loboda A., Shevchenko A., Bork P., Ens W., and Standing K.G. 2001. Charting the proteomes of organisms with unsequenced genomes by MALDI-quadrupole time-of-flight mass spectrometry and BLAST homology searching. *Anal. Chem.* **73:** 1917–1926.

Simpson R.J. and Dorow D.S. 2001. Cancer proteomics: From signaling networks to tumor markers. *Trends Biotechnol.* **19:** S40–S48.

Sloane A.J., Duff J.L., Wilson N.L., Gardhi P.S., Hill C.J., Hopwood F.G., Smith P.E., Thomas M.L., Cole R.A., Packer N.H., Breen E.J., Cooley P.W., Wallace D.B., Williams K.L., and Gooley A.A. 2002. High throughput peptide mass fingerprinting and protein macroarray analysis using chemical printing strategies. *Mol. Cell. Proteomics* **1:** 490–499.

Spahr C.S., Susin S.A., Bures E.J., Robinson J.H., Davis M.T., McGinley M.D., Kroemer G., and Patterson S.D. 2000. Simplification of complex peptide mixtures for proteomic analysis: Reversible biotinylation of cysteinyl peptides. *Electrophoresis* **21:** 1635–1650.

Spengler B., Kirsch D., and Kaufmann R. 1991. Metastable decay of peptides and proteins in matrix-assisted laser-desorption mass spectrometry. *Rapid Commun. Mass Spectrom.* **5:** 198–202.

Spengler B., Kirsch D., Kaufmann R., and Jaeger E. 1992. Peptide sequencing by matrix-assisted laser-desorption mass spectrometry. *Rapid Commun. Mass Spectrom* **6:** 105–108.

Strupat K., Karas M., and Hillenkamp F. 1991. 2,5-Dihydroxybenzoic acid: A new matrix for laser desorption-ionization mass spectrometry. *Int. J. Mass Spectrom. Ion Processes* **111:** 89–102.

Strupat K., Karas M., Hillenkamp F., Eckerskorn C., and Lottspeich F. 1994. Matrix-assisted laser desorption ionization mass spectrometry of proteins electroblotted after polyacrylamide gel electrophoresis. *Anal. Chem.* **66:** 464–470.

Tang C., Zhang W., Chait B.T., and Fenyö D. 2000. A method to evaluate the quality of database search results. ABRF Conference Proceedings. Bellevue, Washington.

Veenstra T.D., Martinovic S., Anderson G.A., Pasa-Tolic L., and Smith R.D. 2000. Proteome analysis using selective incorporation of isotopically labeled amino acids. *J. Am. Soc. Mass Spectrom.* **11:** 78–82.

VerBerkmoes N.C., Bundy J.L., Hauser L., Asano K.G., Razumovskaya J., Larimer F., Hettich R.L., and Stephenson J.L. 2002. Integrating"“top-down” and “bottom-up” mass spectrometric approaches for proteomic analysis of *Shewanella oneidensis*. *J. Proteome Res.* **1:** 239–252.

Verma R., Chen S., Feldman R., Schieltz D., Yates J., Dohmen J., and Deshaies R.J. 2000. Proteasomal proteomics: Identification of nucleotide-sensitive proteasome-interacting proteins by mass spectrometric analysis of affinity-purified proteasomes. *Mol. Biol. Cell* **11:** 3425–3439.

Vestling M.M. and Fenselau C. 1994. Polyvinylidene difluoride (PVDF): An interface for gel electrophoresis and matrix-assisted laser desorption/ionization mass spectrometry. *Biochem. Soc. Trans.* **22:** 547–551.

Vorm O., Roepstorff P., and Mann M. 1994. Matrix surfaces made by fast evaporation yield improved resolution and very high sensitivity in MALDI TOF. *Anal. Chem.* **66:** 3281–3287.

Wahl J.H., Gale D.C., and Smith R.D. 1994. Sheathless capillary electrophoresis-electrospray ionization mass spectrometry using 10 μm I.D. capillaries: Analyses of tryptic digests of cytochrome *c*. *J. Chromatogr. A* **659:** 217–222.

Wahl J.H., Goodlett D.R., Udseth H.R., and Smith R.D. 1992. Attomole level capillary electrophoresis mass spectrometric protein analysis using 5 μm id capillaries. *Anal. Chem.* **64:** 3194–3196.

———. 1993. Use of small-diameter capillaries for increasing peptide and protein detection sensitivity in capillary electrophoresis-mass spectrometry *Electrophoresis* **14:** 448–457.

Wapstra A.H., Audi G., and Hoekstra R. 1985. The 1983 atomic mass evaluation. IV. Evaluation of input values, adjustment procedures. *Nucl. Phys. A* **A432:** 185–362.

Washburn M.P. and Yates J.R., III. 2000. Analysis of the microbial proteome. *Curr. Opin. Microbiol.* **3:** 292–297.

Washburn M.P., Wolters D., and Yates J.R., III. 2001. Large-scale analysis of the yeast proteome by multidimensional protein identification technology. *Nat. Biotechnol.* **19:** 242–247.

Wattenberg A., Organ A.J., Schneider K., Tyldesley R., Bordoli R., and Bateman R.H. 2002. Sequence dependent fragmentation of peptides generated by MALDI quadrupole time-of-flight (MALDI Q-TOF) mass spectrometry and its implications for protein identification. *J. Am. Soc. Mass Spectrom.* **13:** 772–783.

Weast R.C., Lide D.R., Astle M.J., and Beyer W.H., eds. 1989. *CRC handbook of chemistry and physics: A ready Reference book of chemical and physical data*, 70th edition, p. B-227. CRC Press, Boca Raton, Florida.

Wilkins M.R., Lindskog I., Gasteiger E., Bairoch A., Sanchez J.C., Hochstrasser D.F., and Appel R.D. 1997. Detailed peptide characterization using PEPTIDEMASS—A World-Wide-Web-accessible tool. *Electrophoresis* **18:** 403–408.

Wilm M.S. and Mann M. 1994. Electrospray and Taylor-Cone theory, Dole's beam of macromolecules at last? *Int. J. Mass Spectrom. Ion Processes* **136:** 167–180.

——. 1996. Analytical properties of the nanoelectrospray ion source. *Anal. Chem.* **68:** 1–8.

Wysocki V.H. 1992. Triple quadrupole mass spectrometry. In *Mass spectrometry in the biological sciences: A tutorial* (ed. M.L.Gross), pp. 59–78. Kluwer, Dortrecht, The Netherlands.

Yates J.R., III, Eng J.K., and McCormack A.L. 1995a. Mining genomes: Correlating tandem mass spectra of modified and unmodified peptides to sequences in nucleotide databases. *Anal. Chem.* **67:** 3202–3210.

Yates J.R., III, Eng J.K., Clauser K.R., and Burlingame A.L. 1996. Search of sequence databases with uninterpreted high-energy collision-induced dissociation spectra of peptides. *J. Am. Soc. Mass Spectrom.* **7:** 1089–1098.

Yates J.R., III, Eng J.K., McCormack A.L., and Schieltz D. 1995b. Method to correlate tandem mass spectra of modified peptides to amino acid sequences in the protein database. *Anal. Chem.* **67:** 1426–1436.

Yokosawa H. and Ishii S. 1976. The effective use of immobilized anhydrotrypsin for the isolation of biologically active peptides containing L-arginine residues in C-termini. *Biochem. Biophys. Res. Commun.* **72:** 1443–1449.

Yost R.A. and Boyd R.B. 1990. Tandem mass spectrometry: Quadrupole and hybrid instruments. *Methods Enzymol.* **193:** 154–200.

Zhang W. and Chait B.T. 2000. ProFound: An expert system for protein identification using mass spectrometric peptide mapping information. *Anal. Chem.* **72:** 2482–2489.

Zubarev R.A., Kelleher N.L., and McLafferty F.W. 1998. Electron capture dissociation of multiply charged protein cations. A nonergodic process. *J. Am. Chem. Soc.* **120:** 3265–3266.

WWW RESOURCES

For additional Web resources, see Tables 8.5 and 8.6 in this chapter.

http://www.asms.org American Society for Mass Spectrometry. What is mass spectrometry?

http://www.astbury.leeds.ac.uk/Facil/Mstut/mstutorial.htm An introduction to mass spectrometry, A.E. Ashcroft, University of Leeds.

http://www.biotech.uiuc.edu/mass_guide.htm Protein Analysis by MALDI-TOF Mass Spectrometry, Guidelines for Sample Preparation, Protein Sciences Facility, University of Illinois at Urbana–Champaign Protein

http://www.isotope.com/cil/index.html Cambridge Isotope Laboratories, Inc. homepage

http://www.srsmaldi.com Stanford Research Systems, Inc., MALDI

http://www.millipore.com/ziptip Millipore Corporation ZipTip pipette tips

http://www.srsmaldi.com/Maldi/Step/StepbyStep.html MALDI: Desalting and Purification

http://www.spectrastableisotopes.com Spectra Stable Isotopes homepage

www.piercenet.com Pierce Biotechnology homepage

www.ncbi.nlm.nih.gov/Genbank/index.html Submit to GenBank. National Center for Biotechnology Information (NCBI).

www.expasy.ch ExPASy (Expert Protein Analysis System) Molecular Biology Server

CHAPTER 9

Proteomic Methods for Phosphorylation Site Mapping

The chapter text contains contributions by Hanno Steen, Allan Stensballe, and Ole Nørregaard Jensen (Department of Biochemistry & Molecular Biology, University of Southern Denmark, Odense University, DK-5230 Odense M); Leesa J. Deterding, Jenny M. Cutalo, and Kenneth B. Tomer (Laboratory of Structural Biology, National Institute of Environmental Health Sciences, National Institutes of Health, P.O. Box 12233, Research Triangle Park, North Carolina 27709); Garry L. Corthals (Geneva Proteomics Centre, 24 rue Micheli-du-Crest 24, LCCC/Geneva University Hospital, Geneva, Switzerland), Valerie C. Wasinger (Bioanalytical MS Facility, University of New South Wales, Kensington, NWS, Australia 2052), and David R. Goodlett (The Institute for Systems Biology, 1441 North 34th Street, Seattle, Washington 98103-8904).

The systematic analysis of large sets of proteins isolated from tissues, cells, cellular organelles, and protein complexes often includes studies of posttranslationally modified proteins (Krishna and Wold 1993) (see Chapter 1). Posttranslational modification (PTM) may encompass enzyme-catalyzed attachment of molecular entities onto the protein (e.g., phosphorylation, acylation, and glycosylation), formation of intramolecular cross-links (e.g., Cys-Cys disulfide bridges), as well as enzymatic processing of the protein (e.g., removal of amino-terminal signal sequences). Such modifications maintain protein integrity, modulate protein activity and interactions, or serve as targeting signals for delivery of proteins to specific organelles inside the cell or for export across the cell membrane. The field of "modification-specific proteomics," for example, phosphoproteomics and glycoproteomics, will have a major role in elucidating the highly complex molecular mechanisms in cells, tissues, and whole organisms, because PTMs are not readily predicted from DNA or amino acid sequences (Jensen 2000). Since covalent modifications of proteins lead in most cases to an altered protein molecular mass, direct evidence for such modifications can be determined accurately by mass spectrometry (MS) (Shou et al. 2002) (see Chapter 8).

Phosphorylation of proteins at specific amino acid residues is a key regulatory mechanism in cells. Phosphorylation of particular proteins by specific protein kinases controls basic cellular processes, such as cell cycle regulation, cell growth, and cell differentiation (Hunter 2000; Schlessinger 2000; Blume-Jensen and Hunter 2001b). For a historical perspective of protein phosphorylation, see Krebs (1994) and Cohen (2002); for a discussion on the kinetic and catalytic mechanisms of protein kinases, see Francis and Corbin (1993), Adams (2001), and Johnson and Lewis (2001). Protein phosphatases catalyze the removal of phosphate groups from proteins, completing the regulatory cycle (Hunter 1995). Temporal and spatial regulation of cellular events and processes is achieved via reversible phosphorylation of proteins involved in signal transduction, gene regulation, and metabolism.

Proteins are usually phosphorylated at specific serine, threonine, or tyrosine residues (Posada and Cooper 1992), but other phosphoamino acids exist (e.g., phosphoarginine, phosphohistidine, and phospholysine) (see Figure 9.1) (Shrecker et al. 1975; Pas and Robillard 1988; Yan et al. 1998; Sickmann and Meyer 2001). The chemical stability of known phosphorylated amino acids is shown in Table 9.1. Of these, phosphoserine is by far the most

TABLE 9.1. Chemical stability of phosphorylated amino acids

Nature of	Stability in			
phosphoamino acid	Acid	Alkali	Hydroxylamine	Pyridine
O-Phosphates				
Phosphoserine	+	–	+	+
Phosphothreonine	+	±	+	+
Phosphotyrosine	+	+	+	+
N-Phosphates				
Phosphoarginine	–	–	–	–
Phosphohistidine	–	+	–	–
Phospholysine	–	+	–	–
Acylphosphates				
Phosphoaspartate	–	–	–	–
Phosphoglutamate	–	–	–	–
S-Phosphates				
Phosphocysteine	(+)	+	+	+

Reprinted, with permission, from Sickmann and Meyer (2001), originally from Martensen (1971).

O-phosphomonoesters

Phosphoserine

Phosphothreonine

Phosphotyrosine

Phosphoamidates

Phosphoarginine

Phosphohistidine

Phospholysine

Phosphate anhydrides

Phosphoaspartate

Phosphoglutamate

S-phosphothioester

Phosphocysteine

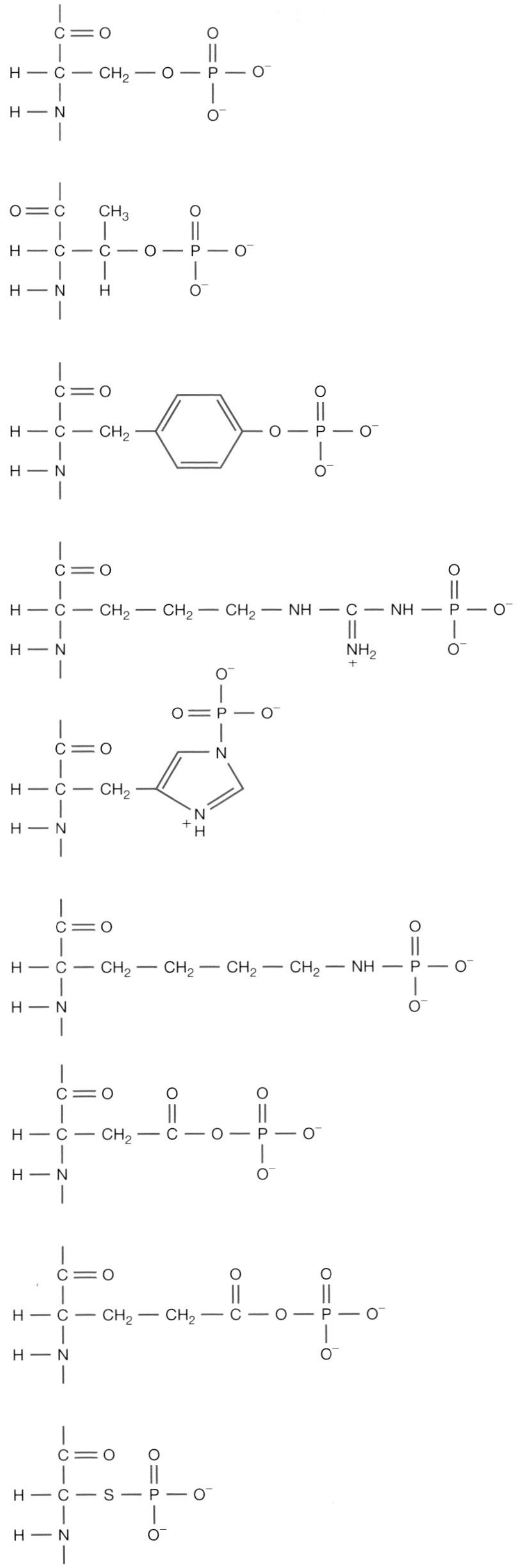

FIGURE 9.1. Chemical structures of phosphoamino acids.

common phosphoamino acid, followed by phosphothreonine and phosphotyrosine (Hunter 1998). Phosphorylation leads to a mass increase of 80 daltons (or more accurately, 79.9663 daltons) as a result of the addition of HPO_3, usually donated by ATP or GTP in a process catalyzed by protein kinases. Protein kinases exhibit sequence specificity, i.e., they recognize serine, threonine, or tyrosine residues in the context of a particular sequence motif. For example, protein kinase C (PKC) recognizes [ST]-x-[RK] (where S or T is the phosphorylation site). Prediction of candidate phosphorylation sites is therefore possible to a certain extent using bioinformatic tools, but this approach is insufficient to establish whether or not a putative phosphorylation site is occupied in a given situation. However, phosphorylation site prediction tools can be helpful in designing experiments aimed at direct determination of phosphorylation sites in proteins.

Hundreds of protein kinases and phosphatases exist that differ in their substrate specificities, kinetic properties, tissue distribution, and association with regulatory pathways. For details see the following Web Sites:

http://www.cbs.dtu.dk/databases/PhosphoBase/
http://www.lecb.ncifcrf.gov/phosphoDB/
ProSite at http://ca.expasy.org/prosite/
NetPhos at http://www.cbs.dtu.dk/services/NetPhos/
ScanSite at http://scansite.mit.edu

For a compendium of information on protein kinases, computational and structural resources, and links to other sites, see the Protein Kinase Resource (http://pkr.sdsc.edu/html/index.shtml).

Labeling proteins with radioactive isotopes such as ^{32}P or ^{33}P is the most common procedure for the analysis of phosphorylation sites. However, ineffective incorporation of label due to the presence of endogenous ATP or GTP and the fact that radioactive labeling itself does not necessarily lead to the exact localization of the phosphorylation site warrant the need for alternative approaches. MS is increasingly becoming the method of choice for nonradioactive phosphorylation analysis (Aebersold and Goodlett 2001; Ahn and Resing 2001; Chen et al. 2002; Mann et al. 2002; Shou et al. 2002).

Most MS-based strategies for identifying phosphorylation sites in proteins include the following stages:

- Isolation and characterization of intact phosphoproteins.
- Generation of peptides by sequence-specific proteolytic cleavage of phosphoproteins.
- Characterization of the peptide mixtures and identification of the phosphopeptides by MS and/or tandem MS (MS/MS).
- Phosphopeptide sequencing by MS/MS.

The remainder of this introduction describes these four stages in greater detail. Techniques used in the first two stages are outlined next, followed by a more detailed discussion of the latter two stages, emphasizing optimized protein chemistry techniques combined with matrix-assisted laser desorption/ionization (MALDI) and electrospray ionization (ESI) MS (for an overview of methods for enrichment, identification, and sequencing of phosphopeptides, see Figure 9.2). Methods for purification and functional analysis of phosphoproteins and protein kinases are beyond the scope of this chapter (see, e.g., Aebersold and Goodlett 2001). The introduction closes with several examples of multidimensional phosphoprotein characterization and a look at several emerging MS techniques for sequencing phosphopeptides.

SAMPLE PREPARATION

Complex Mixture of Phosphorylated and Nonphosphorylated Proteins

Proteins may be metabolically labeled with radioisotopes in vivo by incubating cells or tissue with [^{32}P]phosphate

ENRICHMENT/ISOLATION OF PHOSPHOPROTEINS

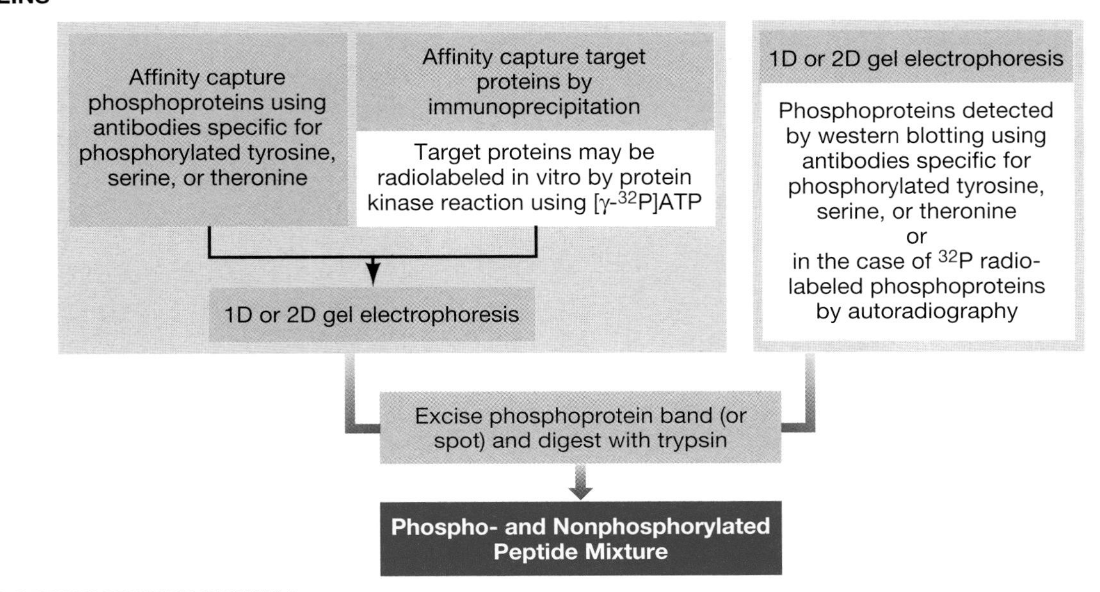

IDENTIFICATION OF PHOSPHOPROTEINS/ PEPTIDES AND SITES OF PHOSPHORYLATION

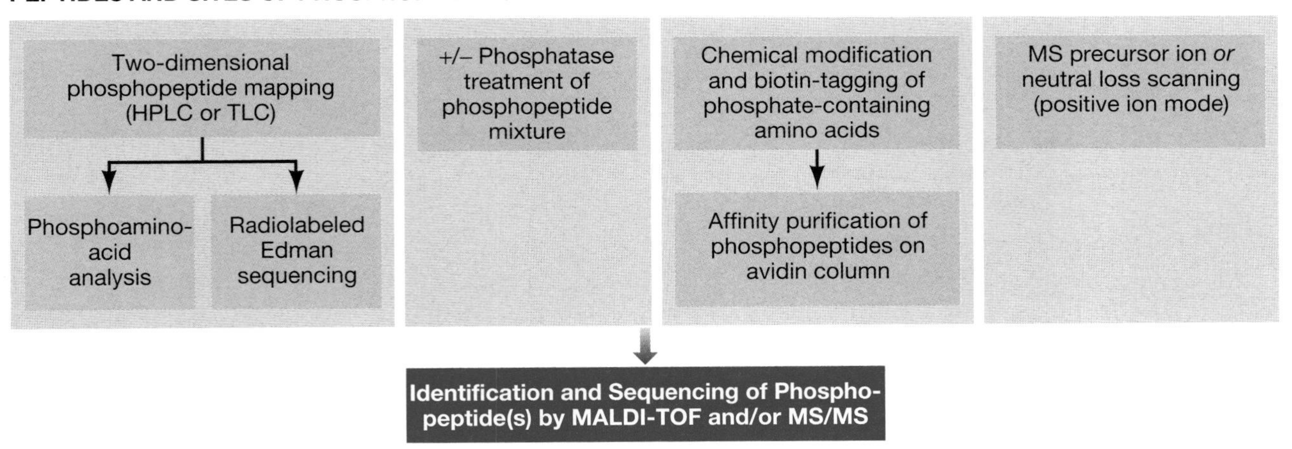

FIGURE 9.2. Summary of techniques for enrichment, identification, and sequencing of phosphopeptides. For additional information on mass spectrometry methods and tandem mass spectrometry, see Chapter 8.

DETECTION AND ANALYSIS OF INTACT PHOSPHOPROTEINS

Phosphoproteins can be isolated from complex protein preparations by a variety of techniques, including classical biochemical methods, immunoprecipitation, affinity chromatography, and detergent extraction/solubilization techniques. Typically, SDS-polyacrylamide gel electrophoresis (SDS-PAGE) or two-dimensional (2D) PAGE (Ji et al. 1993; Gatti and Traugh 1999) (see Chapters 2 and 4) is used as a final purification step when the goal is to identify and characterize the primary structure and natural modifications of small amounts of phosphoprotein (low microgram or nanogram levels) (Powell et al. 2000; Aebersold and Goodlett 2001; Godovac-Zimmermann and Brown 2001; Larsen et al. 2001; Mann et al. 2001). A variety of methods for detecting phosphorylated proteins will be discussed in turn.

Phosphoprotein Detection by ^{32}P Radiolabeling

Proteins can be labeled in vivo using [^{32}P]orthophosphate (Ji et al. 1993) or in vitro using purified protein kinase and [γ-^{32}P]ATP (Zhao et al. 1989; Aebersold et al. 1991; Boyle et al. 1991; Wettenhall et al. 1991; Yan et al. 1998). Proteins metabolically labeled in vivo or in vitro with ^{32}P can be detected on 2D gels or one-dimensional (1D) SDS-PAGE by autoradiography or phosphorimaging. After incorporation, the protein is chemically or enzymatically cleaved, and the phosphopeptides are separated by high-performance liquid chromatography (HPLC) (Craig 2001; see also Chapter 5) or 2D thin-layer chromatography (TLC) and detected by scintillation counting or autoradiography. Phosphopeptide-containing HPLC fractions or excised TLC spots are subjected to phosphoamino acid analysis (Sickmann and Meyer 2001; Blume-Jensen and Hunter 2001a) (see also Appendix 2) and analyzed by Edman sequencing, monitoring for a loss of radioactivity at each cycle of the Edman degradation (MacDonald et al. 2002) (see also Chapter 6, Protocol 5). Because there is typically insufficient sample to detect the unlabeled amino acids (even by ESI-MS or MALDI-MS technologies), the phosphorylation site must be deduced by comparing the Edman cycle number at which the radioactivity is released with a list of predicted peptides to identify those that contain serine, threonine, or tyrosine at the same position in the peptide sequence. (Note that traditional methods such as TLC and Edman sequencing are limited in that they typically require radiolabeled samples and either prior knowledge of the protein sequence or the ability to make assumptions about the peptide sequence. Incorporation of radioactive phosphate into the target protein may alter the physiological state of the cells, for example, by inducing expression of the tumor suppressor protein p53 [Bond et al. 1999]. If the protein is already fully phosphorylated in vivo, further addition of phosphate will not occur, and the results can therefore be misleading.)

Phosphoprotein Detection by Western Blotting

Phosphorylated proteins can be detected, following their electrophoretic separation, by western blotting using antibodies against the phosphoamino acids. This is one of the most sensitive and widely used techniques for detecting specific phosphorylation sites (Arad-Dann et al. 1993; Abu-Lawi and Sultzer 1995). However, even when the phosphoamino acid is part of the recognition sequence, in some cases, the antibody may not recognize the epitope due to steric hindrance within the recognition site.

If sufficient amounts of a phosphoprotein are present in an acrylamide gel for MS analysis, then the protein can be proteolytically digested in situ in the gel (see Chapter 7, Protocol 17) and the phosphopeptide (including the phosphorylation site) identified. (Typically, phosphorylated peptides are of low abundance in a peptide mixture and need to be enriched prior to MS analysis.) It should be noted that detailed characterization of a phosphoprotein currently requires significantly more material than is normally used for protein identification alone (see below Characterization of Phosphopeptides by MS or MS/MS).

Intact Phosphoprotein Analysis

MS is very useful for accurate molecular-mass determination of solubilized, intact phosphoproteins. Such an analysis indicates whether a protein is actually modified and to what extent, as well as establishing the homogeneity of the sample. For mass analysis of intact proteins, the preparation preferentially should contain only low millimolar amounts of salts and other low-molecular weight contaminants, as they will interfere with sample preparation and ionization in MS.

Detailed analysis of protein modifications is facilitated by having a homogeneous phosphoprotein sample; i.e., little or no contaminating protein species should be present. If in doubt, a fraction of the intact phosphoprotein sample should be checked for homogeneity by SDS-PAGE or MS. When relatively high amounts of sample are available (tens of micrograms), a solubilized sample of phosphoprotein is preferred, but buffers and detergents that are incompatible with MS must be avoided (Jensen et al. 1997) (for methods to remove MS-interfering contaminants, see Chapter 8). Recovery of intact protein by elution from the polyacrylamide gel matrix after SDS-PAGE for subsequent MS analysis by MALDI or ESI is sometimes possible, but usually requires microgram amounts of protein to be successful (Haebel et al. 1995).

Both ESI (Fenn et al. 1989) and MALDI (Karas and Hillenkamp 1988) MS can be used for intact molecular-mass determination of phosphoproteins (see, e.g., Weijland et al. 1997; Garnier et al. 2001; Merrick et al. 2001). Analysis of phosphoproteins is preferably performed in a "differential" experiment where two identical phosphoprotein aliquots are investigated in parallel: One sample is treated with a phosphatase enzyme (e.g., alkaline phosphatase) and the other left intact (see Protocol 2) (Annan and Carr 1997). Both samples are analyzed by MS, and on the basis of the mass difference (if any) between the two samples, it can be established whether the protein is phosphorylated and the number of phosphate groups on the protein can be quantitated. In a similar experiment, a protein sample can be incubated in vitro with a kinase and ATP and another identical but untreated sample serves as the control.

GENERATION OF PEPTIDES FROM PHOSPHOPROTEINS BY PROTEOLYTIC CLEAVAGE

Peptides are usually produced by cleavage of the protein with a sequence-specific protease, but a range of chemical agents are also available. Trypsin is a favorite protease in most laboratories due its high activity and specificity, and because it generates peptides of a size appropriate for mass analysis, i.e., in the range of 500–4000 daltons. Additionally, it is advantageous for MS/MS analysis to have a highly basic amino acid residue at one of the termini, because this greatly simplifies the fragmentation pattern in MS/MS experiments (see the information panel on FRAGMENTATION MECHANISMS OF PROTONATED PEPTIDES IN THE GAS PHASE in Chapter 8). Endoproteinases Lys-C, Asp-N, and Glu-C are also very useful proteases for the generation of phosphopeptides. A number of protocols for enzymatically digesting solubilized proteins or proteins isolated by gel electrophoresis have been established (see, e.g., Shevchenko et al. 1996; Jensen et al. 1999).

CHARACTERIZATION OF PHOSPHOPEPTIDES BY MS OR MS/MS

Analysis of protein phosphorylation typically involves:

- Detection and identification of the phosphorylated peptides in a complex mixture.
- Unambiguous localization of phosphorylation site(s) by peptide sequencing.

Whether only the first step or both steps must be performed will depend on the level of information that is desired. For certain biological questions, it is sufficient to detect and identify the phosphopeptides, which can be accomplished using the peptide mass fingerprinting (PMF) strategy outlined in Chapter 8. However, a comprehensive phosphorylation analysis requires detection and exact localization of the phosphorylation sites, and MS/MS experiments must be conducted to accomplish this (for details on MS/MS, see Chapter 8). Knowledge of the amino acid(s) phosphorylated in vivo in a protein and the surrounding amino acid sequence (i.e., the sequence motif) can often be used to identify the kinase(s) and phosphatase(s) responsible for catalyzing the specific reaction. This information, in turn, is central to understanding the functional significance of any phosphorylation event in a cell.

The most significant problem encountered in phosphorylation peptide mapping is having sufficient amounts of sample. More sample than is required for protein identification by MS is necessary for two reasons: (1) Many times, knowing the sequence of the protein is a prerequisite for modification mapping. (2) For modification mapping, the modified peptides must be present in the sample and observable in the mass spectrum to enable the phosphorylation site analysis, whereas for identification purposes, any set of peptides can be sufficient as long as the criteria for an unambiguous database search are fulfilled. Phosphorylated proteins can be present in cells at very low concentrations. In such cases, the amount of phosphoprotein available for analysis may challenge the limits of even the most sensitive MS systems. This problem can be further compounded by a low stoichiometry of phosphorylation at a given site (Katze et al. 2000); i.e., only a small fraction of the population of molecules of the protein of interest is phosphorylated at a particular site. Often, multiple differentially phosphorylated forms of a protein exist, which also complicates site-specific analysis (Storm and Khawaja 1999). Thus, the analysis of the in vivo phosphorylation state of a protein often presents a formidable technical challenge.

For MS analysis of phosphorylated samples, MALDI-time of flight (TOF)-MS has proven to be very useful because it is one of the most sensitive methods for the detection of phosphopeptides (after appropriate enrichment or treatment of the protein sample), and it is fast and relatively easy to perform. However, as MALDI-TOF-MS analysis of phosphopeptides does not readily provide sequence information, it relies completely on relating expected and observed m/z values of ion signals. Therefore, this approach has serious limitations if additional modifications or nonspecific enzyme cleavages occurred, because both give rise to unexpected peptide masses. Under these circumstances, additional experiments, such as phosphatase treatment, are required to confirm the presence of phosphopeptides (Protocol 2).

Problems Associated with MS Analysis of Phosphopeptides

Mass spectrometric analysis of phosphopeptides is complicated by the presence of phosphogroups in three ways.

- ***The ionization efficiency of phosphopeptides in the positive-ion mode is strongly reduced*** when compared with their nonphosphorylated counterparts, which is thought to be the result of increased hydrophilicity and acidity because of the phospho group (Liao et al. 1994). Even though MS analysis of phosphopeptides can suffer from suppression effects, this method has still proven to be very useful in mapping posttranslational phosphorylation of proteins (Annan and Carr 1997; Qin and Chait 1997). The advantages of the MS-based approach include generally rapid analysis, and the fact that it does not require radiolabeling. Both MALDI-MS and ESI-MS have been used to determine the phosphorylation sites on proteins (Liao et al. 1994; Annan and Carr 1996, 1997; Carr et al. 1996; Qin and

Chait 1997; Immler et al. 1998; Zhang et al. 1998; Asara and Allison 1999; Neubauer and Mann 1999). Two recent studies have quantified the differences in ionization efficiencies for MALDI and ESI in positive-ion mode. For MALDI, the response factor for a particular nonphosphorylated peptide was ten times higher than that for the tyrosine-phosphorylated counterpart, a result that is valid for the majority of phosphopeptides (Chen et al. 2001). To compensate for this discrimination in ionization during MALDI, ammonium salts such as ammonium citrate can be added (Asara and Allison 1999). The difference in ionization efficiencies is not as great for ESI. Miliotis et al. (2001) determined an intensity ratio of 1:1.5:5 for equimolar amounts of the doubly, singly, and nontyrosine-phosphorylated peptide ALGADDSYYTAR. Therefore, the ion signals of phosphopeptides are normally only of low abundance when present in a crude mixture with nonphosphorylated peptides. Recently, two groups independently showed that phosphopeptides are sometimes more efficiently ionized by MALDI in negative-ion mode as compared with nonphosphorylated peptides. This difference in ionization efficiencies in positive- and negative-ion modes can be used to identify phosphopeptides in simple mixtures of peptides (Janek et al. 2001; Ma et al. 2001).

- ***Phosphorylated amino acid residues can interfere with enzymatic cleavage behavior.*** Studies have shown that phosphoserine or phosphothreonine residues in the +2 position downstream from the cleavage side (R/K-X-pS/pT) reduce significantly the cleavage rate of trypsin, the most widely used enzyme for protein characterization by MS (Benore-Parsons et al. 1989; Schlosser et al. 2000). The reduced cleavage efficiency results in larger peptides, which make MS analysis and MS/MS sequencing more complicated.
- ***Phosphopeptides easily lose the phospho moiety during mass spectrometric experiments,*** which is especially true for serine- and threonine-phosphorylated peptides, which readily lose phosphoric acid (H_3PO_4) upon low-energy collision-induced dissociation (CID). Even laser irradiation for MALDI experiments can be sufficient to initiate the gas-phase loss of phosphoric acid. Although this characteristic feature can sometimes be used for the identification of phosphopeptides in mixtures (Annan and Carr 1996), it generally causes problems for the mass spectrometric analysis, due to the additional decrease in sensitivity in MALDI-MS and the complicated fragmentation pattern in MS/MS studies.

Circumventing the Problem of Ionization Suppression of Phosphopeptides

There are several ways to circumvent the difficulties associated with ionization suppression of phosphopeptides and the lability of the phosphoester group.

Derivatization of phosphopeptides

Some of the derivatization strategies are based on methods devised in the early 1970s to avoid problems caused by alkylphosphates during peptide/protein sequencing by Edman degradation (for an early example, see Kolesnikova et al. 1974). The basic idea is to replace the labile, highly acidic phosphate group by a more stable, less acidic moiety. In the case of phosphoserine and phosphothreonine residues, this can be accomplished by performing a β-elimination reaction under strong alkaline conditions. The resulting unsaturated amino acid residue (dehydroalanine or 2-aminodehydrobutyric acid) can be analyzed either directly by MS (Resing et al. 1995) or following a Michael addition with a nucleophile such as ethanethiol performed prior to mass spectrometric analysis of the phosphopeptides (Meyer et al. 1986; Lapko et al. 1997; Jaffe et al. 1998; Weckwerth et al. 2000; see also Appendix 2). The derivatization strategies, including their advantages and disadvantages, are summarized in Table 9.2.

TABLE 9.2. Strategies for the mass spectrometric analysis of phosphoproteins and phosphopeptides

Strategy	Approach	Advantages	Disadvantages	References
Derivatization	β-elimination	• Removal of labile, acidic phospho-moiety results in improved mass spectrometric behavior since losses of phosphopeptide signal due to ion suppression are reduced. • Only one derivatization step.	• Reaction conditions very harsh, resulting in partial protein hydrolysis. • Reaction not specific to phosphorylation but also to *O*-glycosylation and fatty acid adducts. • Side reactions with methionine and cysteine residues (oxidation required to present this). • Only phosphoserine and phosphothreonine amenable to this approach. • Reaction with phosphothreonine suboptimal.	Resing et al. (1995)
	β-elimination and Michael addition	• See above. • Prevention of partial hydrolysis of the Michael system. • Appropriate thiols allow isotope labeling for relative quantification.	• See above (same disadvantages as β-elimination); additional side reaction conceivable.	Meyer et al. (1986); Lapko et al. (1997); Jaffe et al. (1998); Weckwerth et al. (2000); Goshe (2001)
Enrichment	Antiphospho-amino acid antibodies	• Very specific for intact tyrosine-phosphorylated proteins; little background. • Works very well even in the presence of a vast excess of unphosphorylated protein species.	• Yields for phosphopeptide immunoprecipitation very low. • Possible IgG contamination due to antibodies. • Specificity of anti-phosphoserine and anti-phosphothreonine antibodies is limiting.	Gold et al. (1994); Godovac-Zimmermann et al. (1999); Marcus et al. (2000); Pandey et al. (2000a,b)
	IMAC	• Batch enrichment of phosphopeptides and on-line coupling with LC possible. • Very good sensitivity.	• Problems with selectivity; acidic peptides may be retained. • In combination with MALDI-MS, other phosphopeptide specific reactions, such as phosphatase treatment, necessary for unambiguous confirmation.	Nuwaysir and Stults (1993); Posewitz and Tempst (1999); Ahn and Resing (2001); Larsen et al. (2001); Stensballe et al. (2001)

	β-elimination/ Michael addition of biotin linker	• Proteome-wide application. • Very specific. • Purification from vast excess of unphosphorylated peptide/protein species.	• See above (β-elimination).	Oda et al. (2001)
	Phosphoramidate derivatization	• Proteome-wide application. • Very specific. • All phosphoamino acids amenable to this approach.	• Many derivatization steps. • Repeated HPLC purification required.	Zhou et al. (2001)
MS/MS methods	Constant neutral loss (80 and 98 daltons)	• Working in positive-ion mode, i.e., immediate sequencing possible. • Applicable to all kinds of phosphorylation. • Can easily be combined with LC; also post-acquisition. • No sample splitting.	• One experiment only good for one charge state, i.e., several experiments required for comprehensive phosphopeptide mapping. • Collision energy adjustment important for good sensitivity. • Lower sensitivity for tyrosine-phosphorylated peptides.	Covey et al. (1991b); Schlosser et al. (2001)
	Precursors of *m/z* -79	• Applicable to all kinds of phosphorylation. • Very sensitive. • Easy adjustment of collision energy. • Off-line very sensitive.	• Working in negative-ion mode, i.e., immediate sequencing suboptimal or not possible. • Spraying in negative-ion mode from acidic solutions compromises sensitivity (when on-line coupling with LC). • Sample splitting required.	Huddleston et al. (1994); Carr et al. (1996); Wilm et al. (1996); Neubauer and and Mann (1999); Annan et al. (2001)
	Precursors of *m/z* 216	• Good sensitivity. • Working in positive-ion mode, enabling immediate sequencing. • No sample splitting.	• Requires expensive qTOF technology. • Only tyrosine-phosphorylated peptides amenable to this approach.	Steen et al. (2001a,b)

Immunological methods for enriching phosphopeptides

Another way to circumvent the problem of ionization suppression is to enrich the sample for phosphopeptides. This approach reduces the excess (or bulk) of unmodified peptides that can suppress ionization. One enrichment strategy utilizes immunological methods, such as antibodies against phosphoamino acids, in immunoprecipitation experiments. However, at the moment, only anti-phosphotyrosine antibodies are of sufficient quality to allow the immunoprecipitation of phosphotyrosine-containing peptides only. Several studies have been published in which phosphoproteins have been immunoprecipitated prior to separation of the proteins by PAGE, protein analysis, and phosphorylation mapping (Gold et al. 1994; Godovac-Zimmermann et al. 1999; Pandey et al. 2000a,b). Although this approach in theory should work with immunoprecipitation of phosphopeptides, a recent study showed that the yield of phosphopeptides was ~2% only (Marcus et al. 2000).

Enrichment of phosphopeptides by immobilized metal ion affinity chromatography

Phosphopeptides can also be enriched by employing immobilized metal ion affinity chromatography (IMAC). Suppression effects can be greatly reduced by selectively enriching for phosphopeptides on metal ion affinity media prior to MS analysis. This approach makes use of the high affinity of phosphate groups toward specific trivalent metal ions, in particular Fe^{3+} and Ga^{3+} (Andersson and Porath 1986; Nuwaysir and Stults 1993; Olcott et al. 1994; Posewitz and Tempst 1999; Zhou et al. 2000; Stensballe et al. 2001). Two types of metal-chelating stationary phases, iminodiacetic acid (IDA) and nitrilotriacetic acid (NTA), are coupled to a support, such as Sepharose, agarose, or macroporous silica, permitting affinity chromatography based on the noncovalent binding of phosphopeptides (Neville et al. 1997). IMAC can be highly efficient, selective, and sensitive, but it also shows limitations, including poor adsorption of certain phosphopeptide species and nonspecific adsorption of nonphosphorylated peptides (Ahn and Resing 2001). This often necessitates efficient means of verification of phosphopeptide candidates such as enzymatic dephosphorylation or MS/MS (Larsen et al. 2001; Stensballe et al. 2001). An example of IMAC enrichment in combination with alkaline phosphatase treatment, MS, and MS/MS analysis is shown in Figure 9.3. Several methods for using IMAC to enrich for phosphoproteins prior to protein analysis by MS are given in Protocols 1, 3, and 4. IMAC can be utilized both on-line and off-line followed by MS analysis directly or coupled to either capillary electrophoresis or μLC systems.

Off-line micro-IMAC. IMAC-μLC-ESI-MS/MS has been used successfully even though optimal low-level analysis cannot be easily achieved due to the relative disparity between IMAC and μLC flow rates. This disparity can be linked in part to the necessity of using IMAC columns of much higher binding capacity than the subsequent μLC step to avoid overloading and subsequent loss of phosphopeptides. IMAC performed off-line from μLC-ESI-MS permits a wide range of sample volumes to be loaded (1–100 μl) using a sample pressure vessel (Goodlett et al. 2000b) or an autosampling system equipped with a sample-trapping cartridge (Hayashi et al. 2001) (see Protocols 4 and 5). The use of a pressure vessel (Figure 9.4) is preferred because (1) the sample moves directly from the vial containing the sample onto the IMAC column, avoiding nonspecific sample loss that occurs by pipetting a sample, and (2) the sample does not come in contact with stainless steel fluidics tubing that will selectively remove acidic peptides such as phosphopeptides.

Whether using a pressure cell or gel-loader tips (see Chapter 8, Protocol 2), the IMAC-purified peptides can then be injected into a μLC separation system and analyzed by μLC-ESI-MS/MS as described in Protocol 5 and Chapter 8 (see also Becker et al. 1998; Katze et al. 2000). Note that phosphopeptides behave and elute similarly to nonphosphorylated peptides during RP-HPLC, but because of a decrease in hydrophobicity due to the addition of the

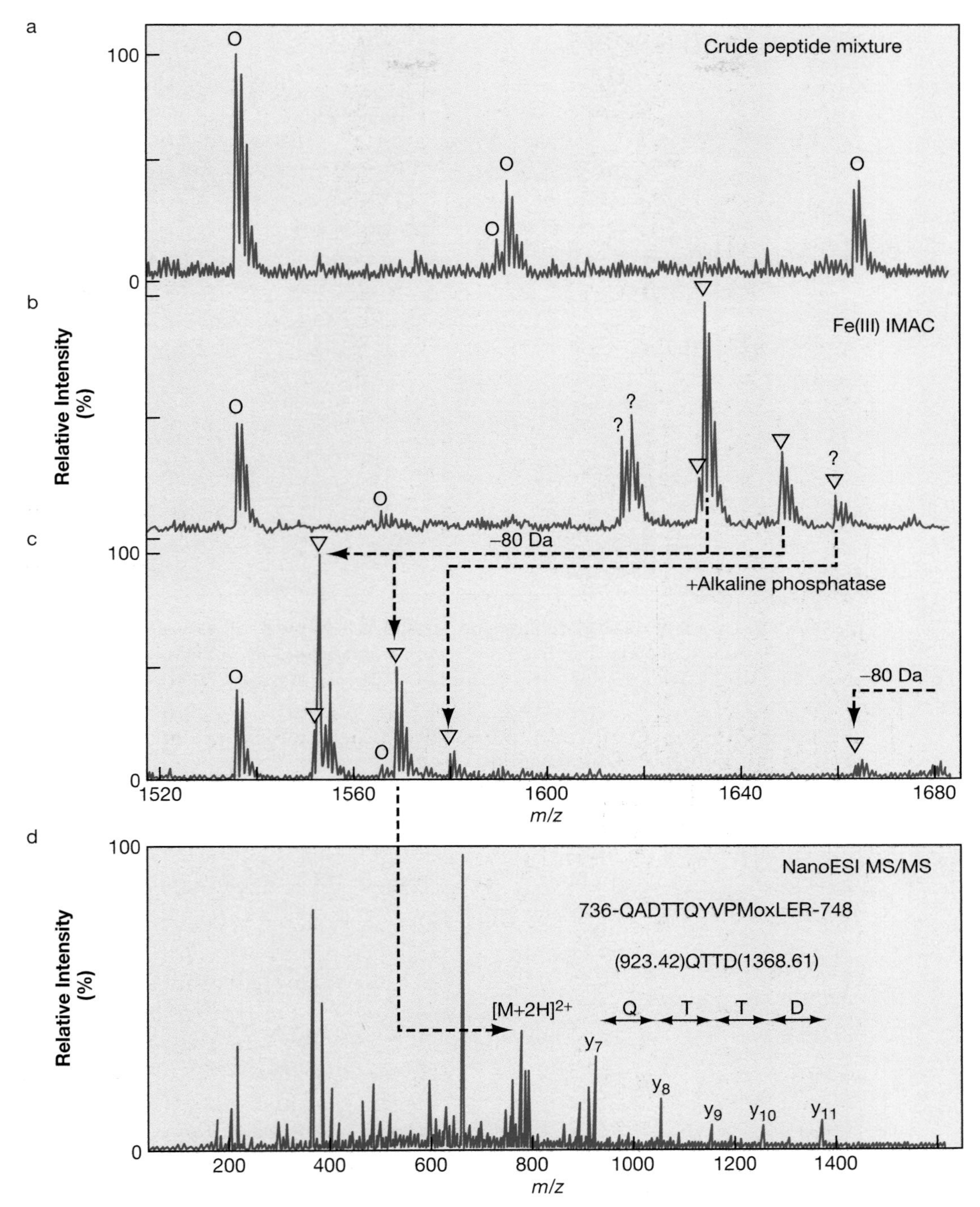

FIGURE 9.3. Phosphopeptide recovery by Fe(III)-IMAC from phosphoprotein isolated by gel electrophoresis. Mouse platelet-derived growth factor receptor a/b (PDGF-R) was isolated by SDS-PAGE and in-gel-digested with trypsin. (*a*) MALDI peptide mass map recorded from the crude tryptic peptide mixture using 2,5-DHB as the matrix. Signals from PDGF-R peptides (*open circles*) were evident. (*b*) Purification of phosphopeptides by nanoscale Fe(III)-IMAC followed by off-line MALDI mass analysis of the eluate using 2,5-DHB as the matrix. A number of peptides can be assigned as nonphosphopeptides (*open circles*), several putative phosphopeptides (*triangles*) based on mass accuracy alone as well as unassignable peptides (*question mark*). (*c*) Alkaline phosphatase treatment of the analyte/matrix deposit from which the spectrum in panel *b* was obtained. The phosphatase assay was employed, and the resulting peptide mixture was analyzed by MALDI-MS using 2,5-DHB matrix. Mass shifts of –80 daltons corresponding to removal of one phosphate group confirmed the modified peptides. (*d*) Recovery of the analyte/matrix deposit from which the spectra in panels *b* and *c* were obtained followed by nanoscale desalting and nanoelectrospray MS/MS confirmed the identity of the dephosphorylated peptide PDGF-R α (residues 736–748). (Courtesy of Hanno Steen, Allan Stensballe, and Ole N. Jensen, University of Southern Denmark.)

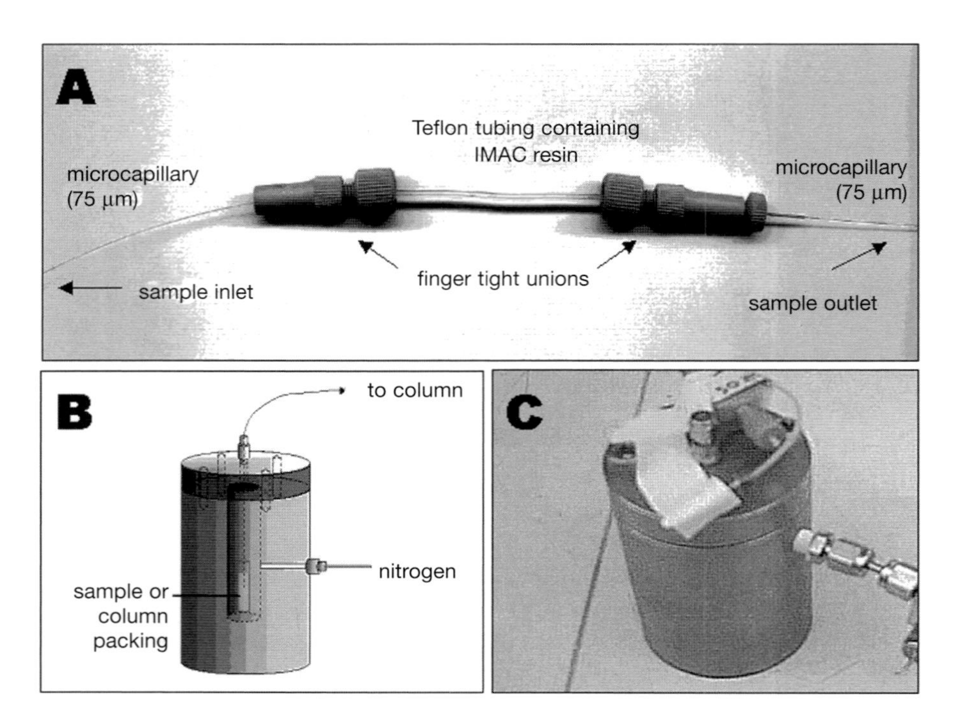

FIGURE 9.4. Diagram of IMAC column and pressure cell. Photograph of an IMAC column (*A*) setup for use with a pressure cell and diagram of pressure cell cross-section (*B*) with photograph of top view of pressure cell in use (*C*). (Courtesy of Garry L. Corthals, Valerie C. Wasinger, and David R. Goodlett.)

phosphate moiety, they will generally elute before the nonphosphopeptide with the identical amino acid sequence. This difference in elution time between phosphopeptide and nonphosphopeptide of identical amino acid sequence is most pronounced for short peptides. Thus, when analyzing phosphopeptides on-line with ESI-MS/MS analysis, it is a good idea to acquire data during the wash step (i.e., prior to gradient elution) because some short phosphopeptides may not even bind to the C_{18} resin in 100% aqueous solutions. Otherwise, these very hydrophilic peptides will be missed.

The following are typical μLC conditions for the separation of IMAC-enriched phosphopeptides on-line with ESI-MS/MS:

- ***Column:*** Use a 50-μm (I.D.) microcapillary column for optimum sensitivity or a 100-μm (I.D.) column for optimum ruggedness, pressure packed with, for instance, ODS-AQ (Waters Corp.).
- ***Solvents:*** Solvent A is aqueous with 0.4% acetic acid/0.005% heptafluorobutyric acid (HFBA)/H_2O. Solvent B is acetonitrile only because addition of acid here is not necessary for good chromatography (it is used with UV detection to avoid a rise in the base line with increasing acetonitrile) and adds substantially to chemical noise in the mass spectrometer.
- ***Gradient:*** Use a linear gradient from 0% to 60% Solvent B over 50 minutes.
- ***Flow rate:*** The flow rate depends on the column used, but, in general, ~200 nl/min and 500 nl/min across 50- and 100-μm (I.D.) columns, respectively, which is achieved by restrictive flow splitting using a length (empirically determined) of capillary tubing set in a tee before the capillary column. This setup allows the use of normal flow HPLC pumps capable of operating between 0.5 and 4.0 ml/min to be used for microcapillary column operation.

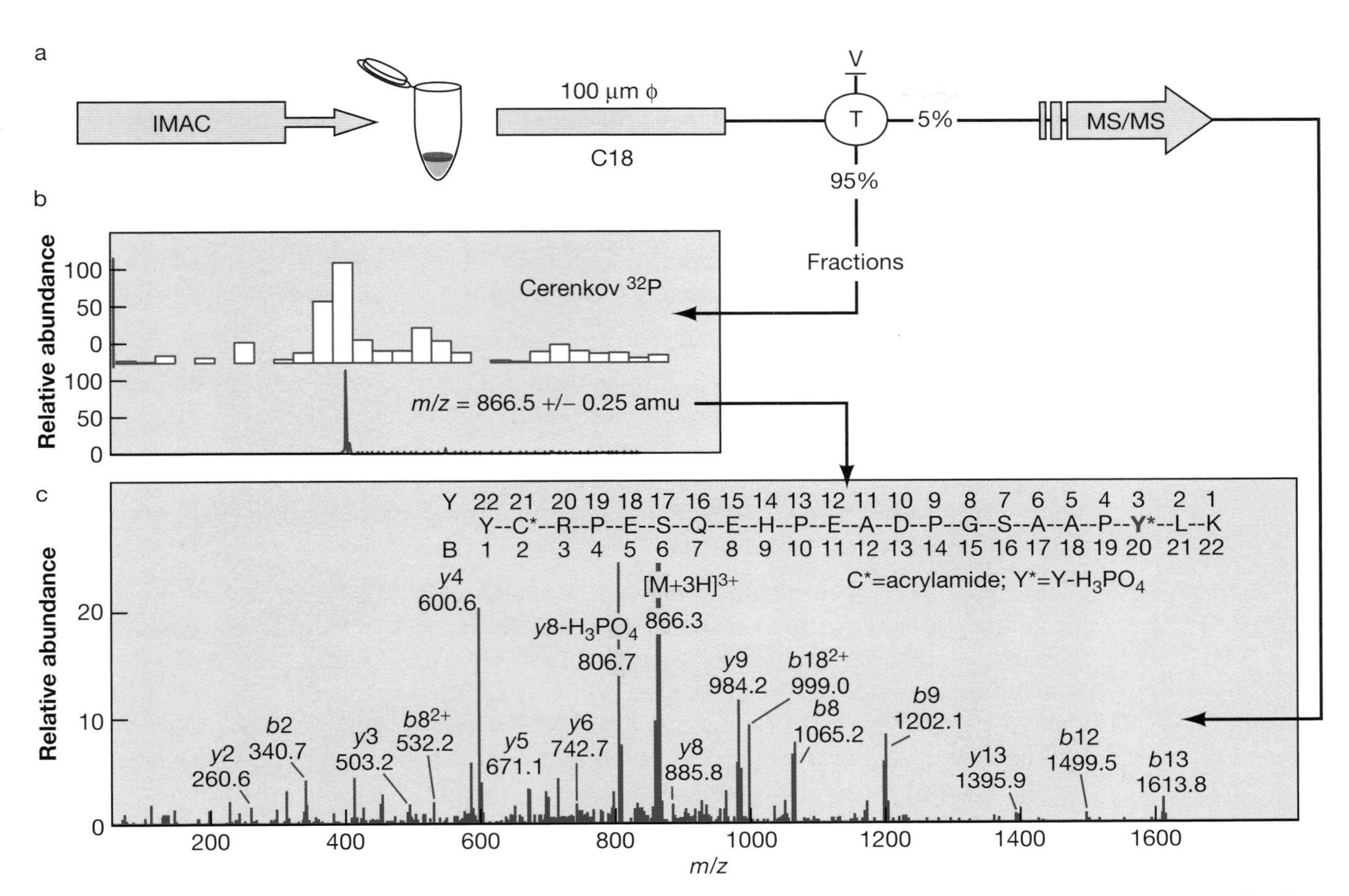

FIGURE 9.5. Off-line IMAC and MS/MS. Flow diagram (*a*) of off-line IMAC followed by µLC-ESI-MS/MS sequencing of a phosphopeptde. Tryptic digest of complex phosphoprotein sample was separated by IMAC using a pressure cell as in Figure 9.4. Pooled phosphopeptides underwent further fractionation by µLC on-line to ESI-MS. The post-column split (*a*) sent 95% of the sample to microfuge tubes for fractionation by time to produce a radioactive profile (*b*) and the remainder went to the mass spectrometer for sequence identification by database search using the tandem mass spectrum (*c*). (Courtesy of Garry L. Corthals, Valerie C. Wasinger, and David R. Goodlett.)

Peptides eluting from the column are introduced into the MS via a micro-ESI interface that can be built in the laboratory (Goodlett et al. 2000b) or purchased (contact garry.corthals@dim.hcuge.ch). Peptides for CID are subsequently selected for fragmentation using automated data-dependent MS/MS protocols available from most MS manufacturers. Where phosphopeptides are radiolabeled with ^{32}P, it is recommended that a post-column flow-splitter be integrated into the micro-ESI device so that a portion of each ^{32}P-labeled peptide is captured into a microtiter plate. The recovered ^{32}P-labeled peptides are used to create a radioactive profile versus chromatographic time by counting each fraction without scintillant (Gallis et al. 1999; Katze et al. 2000) as shown in Figure 9.5. The radioactivity profile generated this way is useful for confirming the elution time of the phosphopeptides and, under certain circumstances, to quantify the amount of ^{32}P-labeled peptide present in specific fractions (Goodlett et al. 2000a). As ESI-MS is essentially a concentration-dependent detection method (Goodlett et al. 1993; Wahl et al. 1993), post-column flow-splitting only minimally reduces the detection sensitivity. ^{32}P-labeled peptide samples collected off-line have also been successfully further separated and analyzed by solid-phase extraction capillary electrophoresis (SPE-CE) connected on-line with ESI-MS/MS. "Peak parking" has been found during capillary electrophoresis (i.e., data-dependent modulation of electrophoretic voltage) to be extremely advantageous for the detection of very low-abundance phosphopeptides by data-dependent MS/MS (Figeys et al. 1999). This advantage is achieved because a reduction of the voltage reduces the electrophoretic mobility and therefore enhances the time available for the mass spectrometer to detect and select for CID peptides without adversely affecting sensitivity (Goodlett et al. 1993). This technique has been applied to the analysis of in vivo phosphorylation sites of endothelial nitric oxide synthase (Gallis et al. 1999). The utility of this method is highlighted in Figure 9.6. The system consists of the following:

- Off-line IMAC/HPLC pre-enrichment setup for phosphopeptides as described above.
- SPE-CE device for peptide concentration and separation.
- ESI-MS/MS instrument (e.g., ThermoFinnigan TSQ) that can be controlled via software protocols (e.g., ICL from ThermoFinnigan) such that capillary electrophoretic voltage is modulated in a data-dependent fashion in response to ion selection for CID; i.e., when a peptide is selected for CID, the ICL protocol decreases the capillary electrophoretic voltage (for ICL protocols, see Geneva Proteomics Web Site www.expasy.org/people/gpc.html or the companion Web Site to this book).

In laboratories where a tandem mass spectrometer is unavailable for sequence analysis of phosphopeptides by CID, phosphatase treatment (Zhang et al. 1998) of IMAC-enriched peptide samples has been used in conjunction with MALDI-TOF analysis to identify the phosphopeptides in the sample (see Protocols 2 and 3). In these experiments, the collective masses of all peptides enriched by micro-IMAC are first recorded in the single-stage MALDI-TOF-MS. Next, the same sample is treated with a broad specificity phosphatase (e.g., calf intestinal phosphatase) in solution, or the identical sample still on the MALDI probe is treated and masses are recorded after an incubation time appropriate to the phosphatase being used. The two MALDI-TOF mass spectra for the original untreated and phosphatase-treated samples are then compared to identify any mass that shows a decrease of 80 daltons (or multiples thereof), implying loss of phosphate due to phosphatase treatment. This experiment is technically simple, and the interpretation of the spectra is straightforward due to the fact that MALDI-TOF-MS produces almost exclusively singly protonated peptide ions.

On-line micro-IMAC. The analysis of phosphopeptides using IMAC on-line with an MS or MS/MS system is desirable because the number of sample handling steps is decreased and thus peptide loss is minimized. Although the advantages of using IMAC to selectively enrich for phosphopeptides is evident, direct coupling of IMAC columns to an ESI-MS instrument is undesirable, because the solvent conditions used to elute phosphopeptides from the IMAC resin are not directly compatible with high-sensitivity ESI-MS. Successful use of on-line IMAC with ESI-MS/MS has been nevertheless achieved with both capillary electrophoresis (Cao and Stults 1999) and μLC (Affolter et al. 1994). In general, conditions for on-line IMAC are very similar to those used for off-line except that the eluted peptides pass directly onto the μLC column (Affolter et al. 1994). In a clever integration of the enrichment/separation system by IMAC-CE-ESI-MS/MS, the selective enrichment of casein phosphopeptides and their analysis was performed on-line using an ion trap that allowed low-picomole levels of sensitivity. To accomplish this, the IMAC column (5 cm x 150 μm I.D., 360 μm O.D.) was fitted over a smaller (75 cm x 75 μm I.D., 150 μm O.D.) capillary electrophoresis column containing a low-binding PVDF frit, and a 1-cm bed of activated Fe(III) Poros beads was pulled into the IMAC capillary by applying a vacuum.

Chemical derivatization and selective enrichment

Three approaches that combine both derivatization and selective enrichment of phosphorylated species were recently introduced by Oda et al. (2001), Zhou et al. (2001), Gosche et al. (2001), and Steen and Mann (2002). The methods of Oda et al. (2001) and Gosche et al.

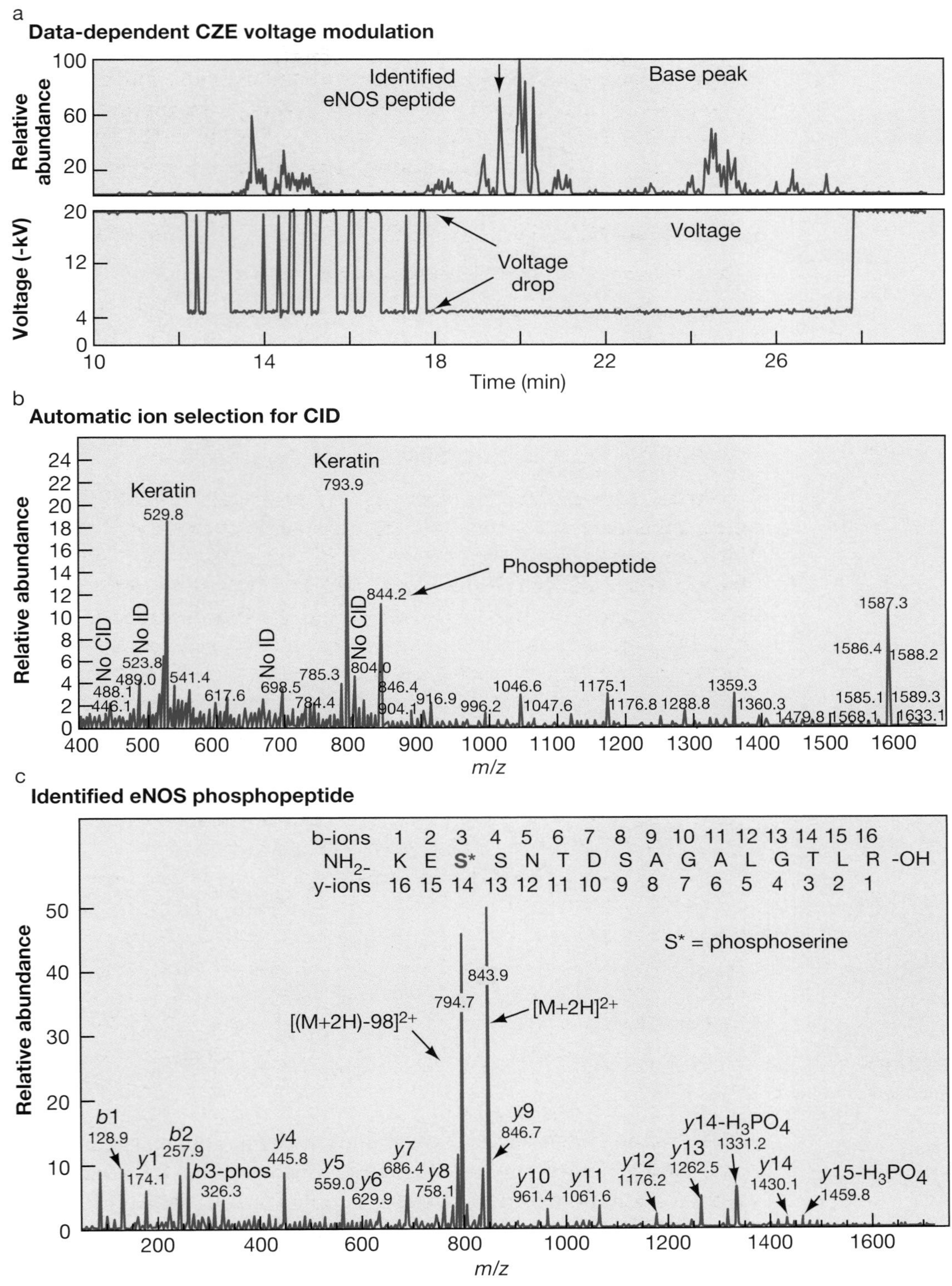

FIGURE 9.6. Modulation of capillary electrophoretic voltage in a data-dependent fashion. The sequence of events involved in peak parking with solid-phase extraction capillary electrophoresis (SPE-CE) on-line with a mass spectrometer capable of data-dependent ion selection for collision-induced dissociation (CID). The base peak ion current trace (*a*) is shown above a trace of the electrophoretic voltage (V_{CE}) that was modulated based on data-dependent ion selection for CID; i.e., a decrease in V_{CE} results in data-dependent ion selection to provide more time for CID. A full-scan mass spectrum (*b*) acquired prior to selection of the ion at 844.2 *m/z* for CID and (*c*) the tandem mass spectrum of the phosphorylated peptide selected in *b*. (Courtesy of Garry L. Corthals, Valerie C. Wasinger, and David R. Goodlett.)

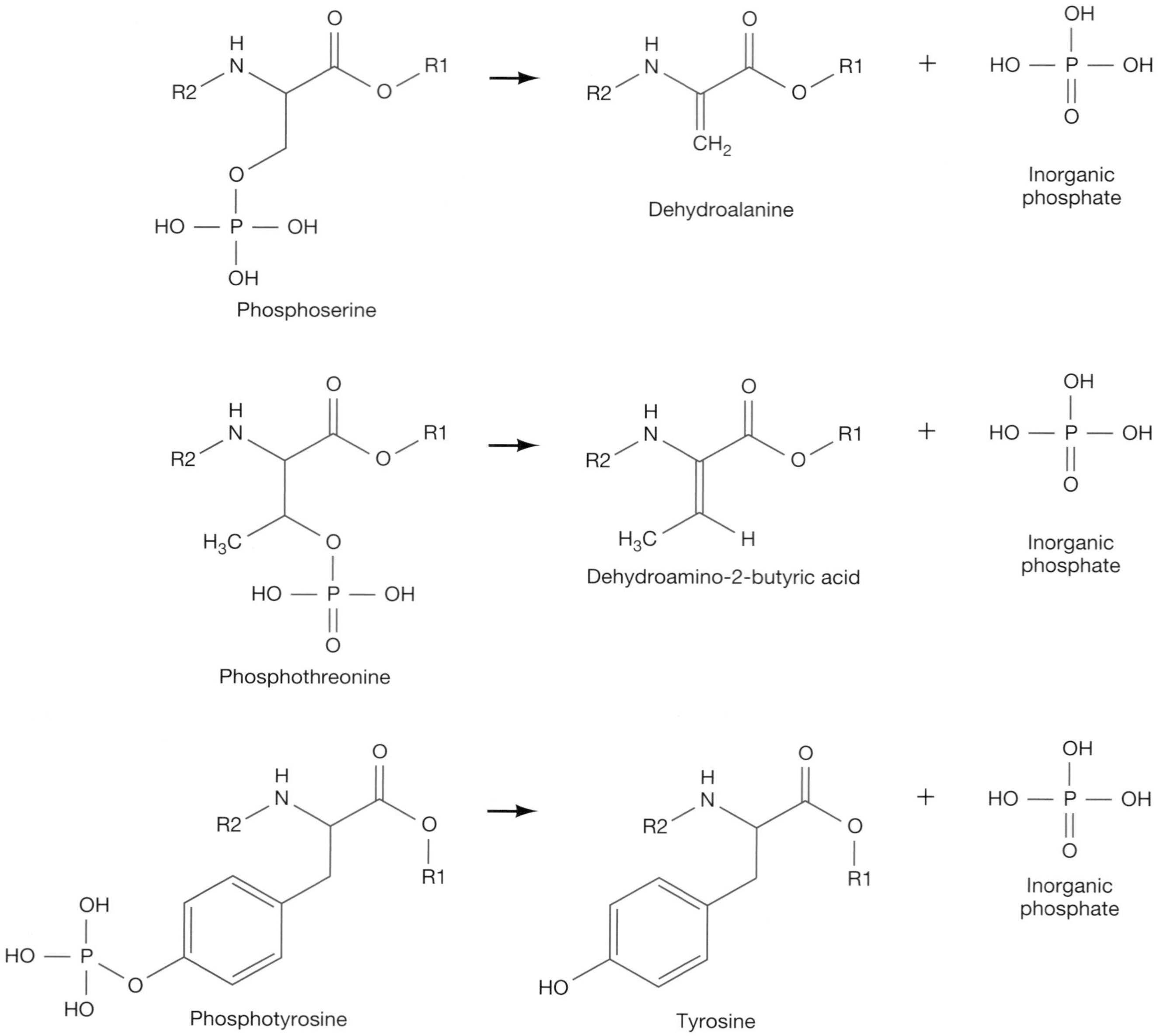

FIGURE 9.7. Phosphoamino acid structures before and after phosphate elimination. (Courtesy of Garry L. Corthals, Valerie C. Wasinger, and David R. Goodlett.)

(2001) are limited to the isolation of phosphoserine and phosphothreonine, because they rely on β-elimination of phosphate (see Figure 9.7) and the introduction of a biotin moiety by Michael addition. The derivatized peptides can be selectively purified based on the high affinity of avidin to biotin. The method of Steen and Mann (2002) also makes use of β-elimination/Michael addition reactions to introduce a functional group at the original site of phosphorylation, which gives rise to a dimethylamine-containing sulfenic acid derivative with a characteristic fragment ion at *m/z* 122.06 upon low-energy CID. This enables detection of the phosphorylated species within complex peptide mixtures by sensitive and specific precursor ion-scanning mode (see Chapter 8 and Figure 9.8) (Steen and Mann 2002). The method of Gosche et al. (2001) allows quantification of phosphopeptides between two different mixtures of proteins using differential stable isotope labeling. The method and reagent name,

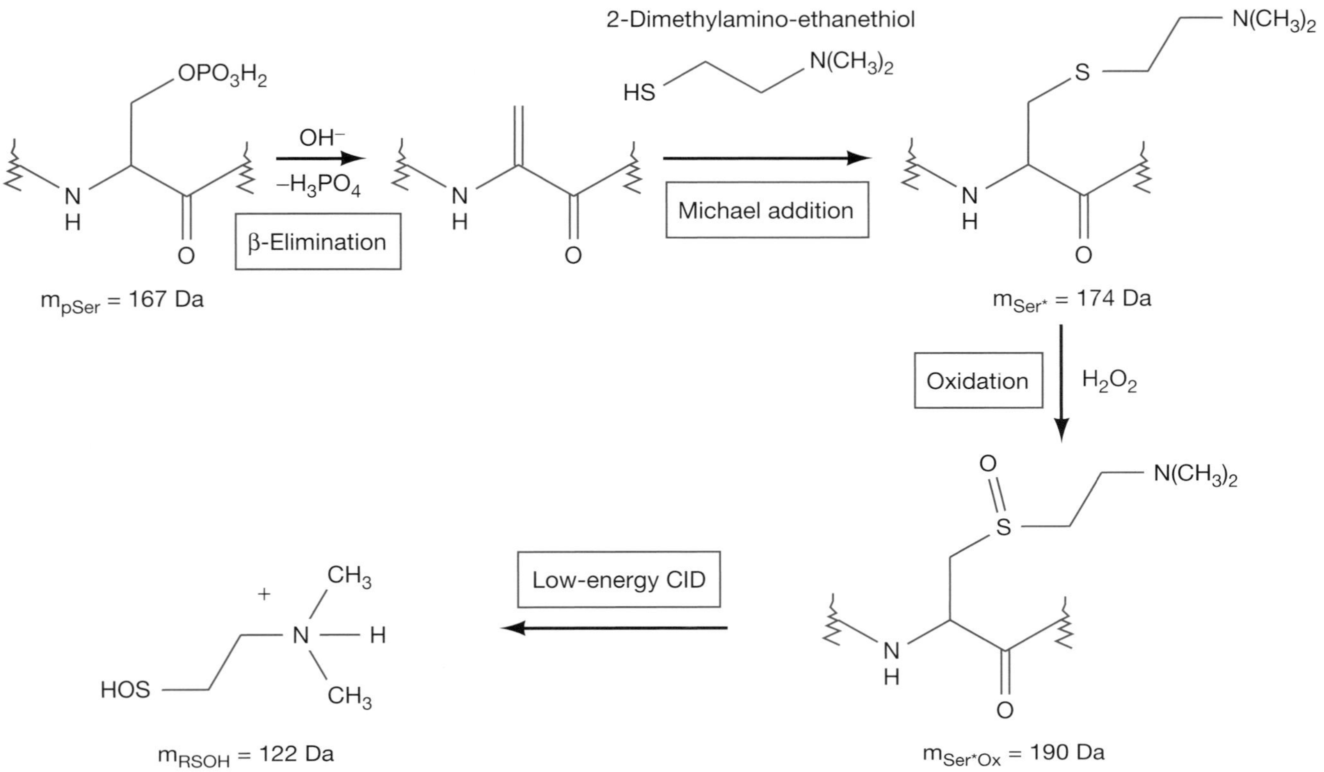

FIGURE 9.8. Derivatization strategy for serine-phosphorylated peptides. Serine-phosphorylated peptides undergo β-elimination upon treatment with strong bases. The resulting Michael system reacts with the nucleophilic thiol group of 2-dimethylaminoethanethiol, which introduces an additional highly basic functional group at the former phosphorylation site. Controlled oxidation of the thioether to the sulfoxide is accomplished by short incubation with 3% H_2O_2. The generated 2-dimethylaminoethanesulfoxide derivative gives rise to a characteristic fragment ion (sulfenic acid derivative) at *m/z* 122.06 upon low-energy CID. (Reprinted, with permission, from Steen and Mann 2002 [©Elsevier Science].)

phosphoprotein isotope-coded affinity tags (PhIAT), are derived from a new class of compound, isotope-coded affinity tags (ICAT), that allow the determination of relative abundance of proteins from two different biological states (Gygi et al. 1999). Zhou et al. (2001) made use of phosphoramidate chemistry to introduce free thiol groups, which facilitated immobilization of the phosphopeptides on iodoacetyl-activated beads. For more details on these methods, see Table 9.2. Of these three new methods, only the method of Zhou et al. (2001) allows for the chemical isolation of phosphoserine and phosphothreonine, as well as phosphotyrosine. Quantitation without isolation had been demonstrated previously using this approach (Weckwerth et al. 2000). It is expected that these new phosphopeptide selective methods will eventually gain widespread use and perhaps eliminate the use of IMAC resins for phosphopeptide isolation.

Mass Spectrometry of Phosphopeptides

There are a number of different mass spectrometric methods for detecting phosphopeptides in complex peptide mixtures and for determining which amino acid residue(s) in a peptide is phosphorylated. These fall into two general categories. The first relies on the propensity of the phosphoester bonds of phosphoserine, phosphothreonine, and, to a lesser extent, phosphotyrosine to undergo fragmentation in a mass spectrometer. This type of fragmentation can be induced in the collision cell (Figure 9.9a) or in the ion source region (Figure 9.9b) of

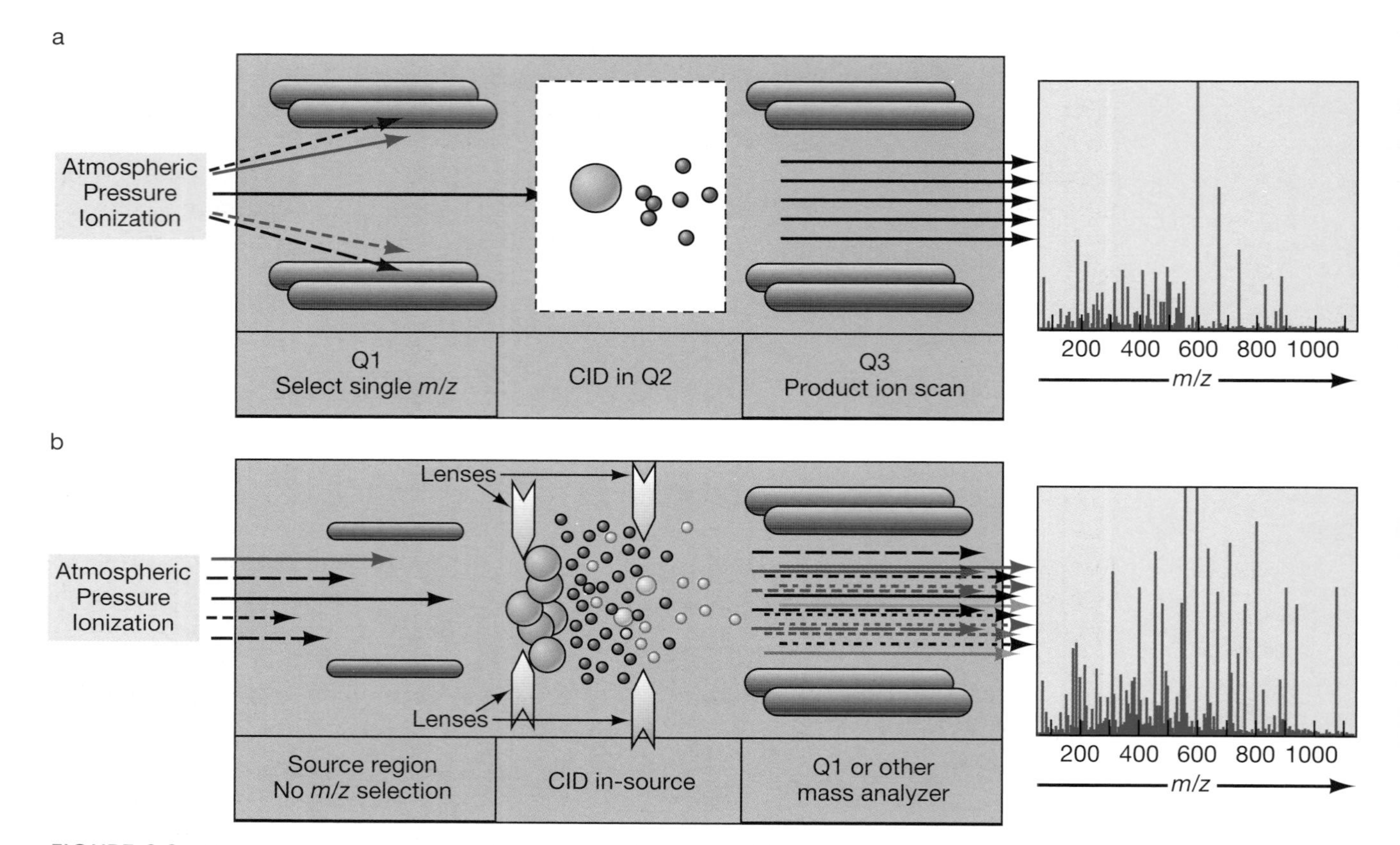

FIGURE 9.9. Diagramatic comparison of CID in a collision cell and in-source. (*a*) MS/MS where a mass selection device (Q1) purifies a single *m/z* that is fragmented in a collision cell (Q2) by the standard CID process to produce a product ion mass spectrum containing fragment ions specific to the parent ion selected by Q1. (*b*) In-source CID at atmospheric pressure without mass selection to produce a mass spectrum in Q1 that contains fragments from all ions that entered the source at a given moment in time. Note that arrows with different lines represent ions at a unique *m/z* and that the process in panel *b* is often used in diagnostic ion scanning procedures described in the text. (Courtesy of Garry L. Corthals, Valerie C. Wasinger, and David R. Goodlett.)

ESI instruments or during post-source decay (PSD) in a MALDI-MS (for further details, see Chapter 8). Phosphopeptides that lose phosphate can then be identified by implementing one of several possible phosphate-specific diagnostic ion scans, which include precursor ion scans, neutral loss scans, and in-source dissociation. The phosphate-specific diagnostic ions (i.e., $H_2PO_4^-$, 97 daltons; PO_3^-, 79 daltons; and PO_2^-, 63 daltons) generated by ESI in negative-ion mode during in-source CID (Katta et al. 1991) can be monitored to identify phosphopeptides (Huddleston et al. 1993; Hunter and Games 1994). In general, methods that produce some sort of phosphate-specific ion (i.e., a diagnostic ion) are useful in cases where incorporation of ^{32}P is not possible or in which the radiolabel has decayed past the point of detection (Meyer et al. 1993).

The second type of detection method is based on the additional mass of a peptide that has been phosphorylated. Typically, in protein phosphorylation studies, the amino acid sequence of the protein investigated is known. Therefore, phosphopeptides derived from the protein can, in principle, be detected by a net mass differential of 80 amu that occurs when phosphate is added to the side chains of serine, threonine, or tyrosine. Thus, a peptide mass map of the proteolysed phosphoprotein can potentially identify the phosphorylated peptide by comparison to the theoretical peptide map. Neither method, however, identifies the phosphorylated amino acid residue(s) within the peptide directly, except in cases where the peptide sequence contains only a single possible phosphorylation site; in which case, the phosphorylated residue is effectively located by default. Generally, the precise amino acid that is phosphorylated is most readily identified by CID in an ESI-MS/MS instrument. Often, the information obtained by the specific scanning methods is designed to differentiate phosphopeptides from nonphosphopeptides, rather than to provide sequence information. The remainder of this section describes several types of MS-based approaches for phosphopeptide detection and sequence analysis following IMAC enrichment. For optimum results, it is recommended that the mass spectrometer be specifically tuned with phosphopeptide standards before using any of the methods described below.

Direct MS analysis of phosphoproteins and peptides bound to solid supports

Elution of phosphopeptides from IMAC columns before MS analysis can result in significant sample loss. However, affinity-bound analytes, including phosphopeptides, can be directly analyzed by MALDI-MS without a separate elution step from the affinity media (Papac et al. 1994; Raska et al. 2002). For example, affinity-bound phosphoproteins and phosphopeptides, like human apotransferrin and phosphokemptide, could be analyzed directly from Fe^{3+}-loaded Sepharose using MALDI-MS. (It has recently been reported that MALDI-TOF-MS analysis of IMAC-retained phosphopeptides is in fact due to ligand displacement of the monophosphopeptide-NTA chelate, rather than laser-induced direct desorption from the chelate [Hart et al. 2002], and that IMAC-retained phosphopeptides can be solubilized by solutions of CHCA and DHB, with 2,5-DHB being more effective in obtaining a larger subset of monophosphopeptides.) The utility of this method has been extended by the observation that consecutive enzymatic reactions, such as phosphatase or carboxypeptidase Y digestion, can be carried out on affinity-bound peptides (Papac et al. 1994; Qian et al. 1999; Li and Dass 1999; Zhou et al. 2000; Merrick et al. 2001) (see Protocol 3). When the affinity-bound phosphopeptides are treated with phosphatase, the number of phosphorylation sites can be determined based on the observation of 80-dalton (or multiples of 80-dalton) mass shifts in the MALDI-MS of the reaction mixture. Carboxypeptidase Y treatment of the affinity-bound phosphopeptides can also be used to cleave the amino acids from the carboxyl terminus, with subsequent direct analysis of the enzymatic products by MALDI-MS to locate the phosphorylation sites on the bound phosphopeptides. Cleavage of amino acids from the amino terminus with aminopeptidase does not work with IMAC, because metal ions are required for this enzyme's activity, and the IMAC media removes the metals from the solution. High sensitivity, elimination of the need for radiolabeling or HPLC separation, ease of use, and the ability to analyze extremely complicated phosphopeptide mixtures make this method attractive.

Lability of phospho-moiety in phosphopeptides: A blessing or a curse?

Removal of the phospho-moiety and the selective enrichment of the phosphorylated species are strategies devised to circumvent the problems caused by the labile nature of the phosphate esters, which is often considered a curse. However, the lability of the phospho-moiety can also be considered a blessing, because the loss of HPO_3 or H_3PO_4 provides a characteristic "signature" for phosphopeptides, which can be utilized in MS/MS-based approaches for selective detection of phosphopeptides present in crude peptide mixtures. The neutral loss of H_3PO_4 (98 daltons) or HPO_3 (80 daltons) can be used for phosphopeptide-specific constant *neutral loss* experiments (Schlosser et al. 2001). This approach can also be applied to on-line LC-MS/MS experiments, where the data are screened "on the fly" for ion pairs with mass differences corresponding to the loss of the phospho-moiety. All peptide ion spectra can be screened after their acquisition for this characteristic peak pair discrepancy, to identify those product ion spectra that are derived from phosphopeptides. A distinct disadvantage of this method, however, is the bias against tyrosine-phosphorylated species resulting from the relatively high stability of arylphosphates with respect to the loss of phosphoric acid (–98 daltons) as compared to alkylphosphoesters.

Less discriminating is the use of the characteristic anion at –79 daltons (PO_3^-) observed in negative-ion mode, which is generated by all phosphopeptides. Hence, this characteristic "reporter" ion can be used in precursor ion experiments to identify phosphopeptides in crude peptide mixtures. This approach is reliable, highly sensitive, and very specific. It must be kept in mind, however, that peptide sequencing is not efficient in negative-ion mode; therefore, the instrument must be switched to the positive-ion mode and the pH must be adjusted before sequencing the phosphopeptide by MS/MS (Carr et al. 1996; Wilm et al. 1996).

In the special case of tyrosine-phosphorylated peptides, a characteristic "reporter" cation at 216.043 daltons can be used for precursor ion experiments in positive-ion mode. The highest selectivity is achieved on high-accuracy, high-resolution tandem mass spectrometers, such as the quadrupole TOF-type instrument (Steen et al. 2001a,b). An example of a phosphotyrosine-specific precursor ion experiment is shown in Figure 9.10.

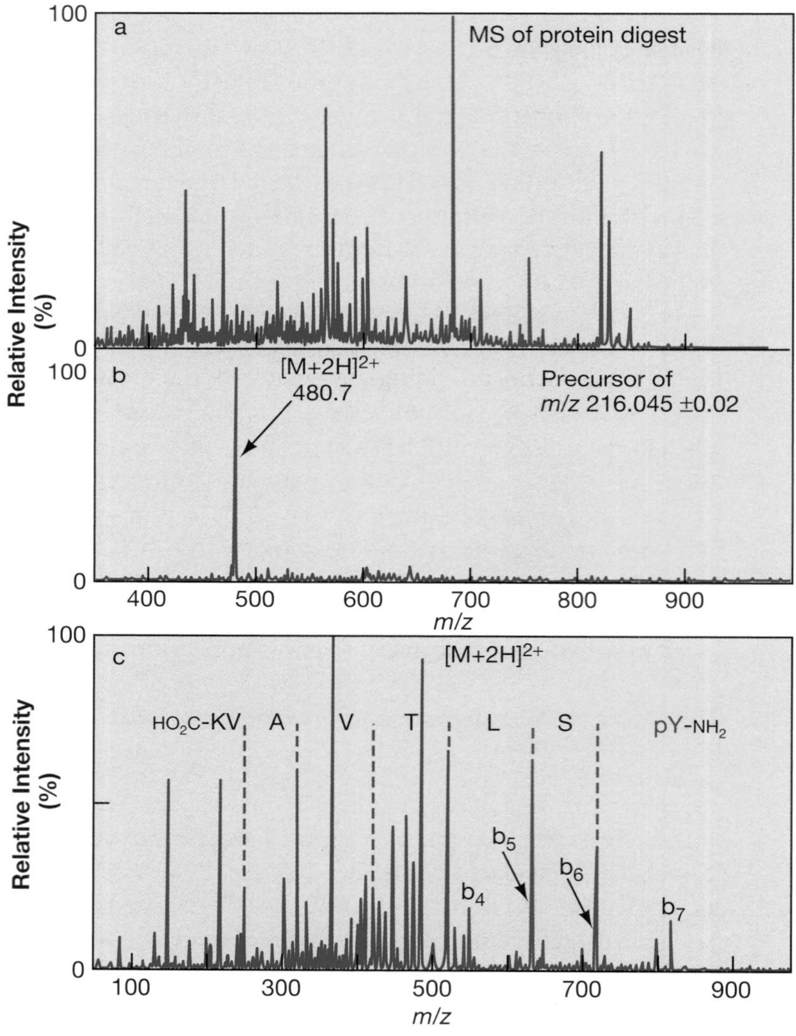

FIGURE 9.10. Mass spectrometric analysis of a protein digest for the detection of tyrosine-phosphorylated peptides. (*a*) Mass spectrum of the tryptic in-gel digest of Bcr/Abl protein, eluted from an Oligo R3 microcolumn with 40% methanol/5% formic acid. (*b*) The specific precursor of *m/z* 216.045 ± 0.020 experiment. (*c*) Product ion spectrum of the doubly charged tryptic peptide $T_{664\text{-}671}$ (**pY**)SLTVAVK) at *m/z* 480.7 from which the "reporter" cation (*m/z* 216.045) was derived. (Courtesy of Hanna Steen, Allan Stensballe, and Ole N. Jensen, University of Southern Denmark.)

In-source CID

Several MS methods for detecting phosphopeptides in complex mixtures rely on MS/MS (see Protocols 7 and 8; for further details on MS/MS and instrument configuration, see also Chapter 8). One of the early successful methods used fragmentation in the ion source region of ESI instruments. These methods used a higher than normal difference in voltage across the lenses (see Figure 9.9b) or other devices used for desolvation/focusing of ions, together with the higher than normal gas pressure present in the ion source region of an atmospheric pressure ionization device. By these means, peptide fragmentation is effected similar to that produced in a collision cell. However, as depicted in Figure 9.9b, no ion selection is available, and thus all ions that enter simultaneously are fragmented together. It was thus important to correlate loss of phosphate in the source region with a chromatographic trace. If carried out as done by Carr and co-workers (1996) in a triple-quadrupole (TQ) mass spectrometer with a nozzle-skimmer-type interface design in the ion source region (e.g., SCIEX TQ), then both the chromatographic marker and the phosphopeptide molecular weight can be determined in the same scan. The generation of fragment ions was accomplished by use of a high orifice potential across the nozzle-skimmer prior to Q1, whereas the low m/z range was scanned for the low-mass diagnostic ions. When the TQ began to scan across higher m/z, then the nozzle-skimmer potential was returned to a normal voltage that did not induce fragmentation. This allowed both the diagnostic ion scan and the scan to determine peptide molecular weight to be done in a continuously alternating fashion through the chromatographic analysis. A similar experiment can also be done on instruments in which a heated capillary-skimmer design (e.g., ThermoFinnigan TQ) is used in place of the nozzle-skimmer (Aebersold et al. 1998). Here, an alternating scan approach was used, where selected ion monitoring of appropriate diagnostic ions at a high octapole offset voltage was followed by two full scans. The first full scan was conducted at the same high offset voltage as the single-ion monitoring (SIM) experiment, providing signals for the deprotonated phosphopeptide molecular ion and the phosphopeptide molecular ion minus phosphate. The second full scan was conducted at a normal octapole offset voltage to provide a reference to the full scan at high octapole offset. This series of MS scans was repeated continuously throughout the LC separation. Such an experiment provided the same information as the method of Carr and co-workers (1996), but because SIM was used for the diagnostic ion detection, rather than simply scanning across the low m/z range, the protocol had a lower duty cycle and was potentially more sensitive. The first full scan at high octapole offset was compared to the full scan at low octapole offset to provide an indication as to which peptide ion was phosphorylated. This was particularly useful if several peptides coelute, a situation that is common if complex peptide mixtures are separated by μLC. Such techniques generally achieved a sensitivity limit of a few fmoles for phosphopeptide standards loaded on the column. However, the sensitivity of detection of "real" peptide samples generated by the digestion of in vitro or in vivo phosphorylated proteins usually was in the picomole range.

It would be advantageous to detect phosphopeptides by the release of phosphate-specific diagnostic ions in negative-ion mode and then switch to positive-ion mode for a CID of the detected phosphopeptide. This switch between negative- and positive-ion mode is necessary because the phosphate-specific fragment ions are difficult to detect in positive-ion mode and because negative-ion CID spectra generally produce insufficient fragment ions for sequence elucidation. With current instruments, rapid switching between negative- and positive-ion mode is technically difficult to achieve. It can be anticipated that this problem will be overcome in the future as better electronics are developed.

Neutral loss scanning

Neutral loss scanning for phosphopeptide detection and analysis was first described by Covey et al. (1991a,b) and further developed by Carr and co-workers (Huddleston et al. 1993). It was carried out in positive-ion mode with ESI in a triple-quadrupole MS. Instead of using Q1 to select specific ions for fragmentation in Q2, Q1 and Q3 were scanned such that Q1 and Q3 *m/z* scan ranges were offset by the neutral mass of interest (e.g., 49 amu for loss of phosphate from a $[M+2H]^{2+}$ parent ion) (see also Chapter 8). The two quadrupoles were then continuously scanned to monitor for loss of the neutral ion. As shown in Figure 9.7, only phosphoserine and phosphothreonine may undergo neutral loss of 98 amu by β-elimination (Gibson and Cohen 1990; Bateman et al. 2002). Phosphotyrosine remains intact under these conditions because the α-carbon proton that must be abstracted by a lone pair of electrons from the phosphate moiety is now too far removed. The method has not been as popular as the aforementioned in-source CID methods because of false positives and the need to know the charge state of the phosphate-containing peptide ion. An advantage of the method is that it is carried out in positive-ion mode and can be used with data-dependent scanning to acquire CID spectra for phosphopeptide sequencing and determination of the site of phosphorylation in the same experiment as a neutral loss scan for the detection of phosphorylated peptides.

Precursor ion scanning

In this method, negative-ion ESI is carried out with continuous scanning of Q1. All ions are fragmented in Q2, and Q3 passes only one ion, which for phosphopeptides is usually *m/z* 79 (i.e., loss of PO_3^-) (see also Chapter 8). Consequently, the resultant full-scan mass spectrum shows only ions that lost *m/z* 79 (Wilm et al. 1996; Neubauer and Mann 1999). This greatly simplifies mixture analysis and is best done during direct infusion with a nanospray source. As described for the phosphate diagnostic ion scans, there is a problem associated with sequencing in positive-ion mode immediately after detecting the loss of phosphate in negative-ion mode. This method requires a mass spectrometer with a collision cell positioned between two mass separation devices such as a TQ or a quadrupole TOF (qTOF) arrangement (see Chapter 8).

A novel precursor ion discovery method using an ESI hybrid quadrupole orthogonal acceleration TOF (QTOF) mass spectrometer has been described recently (Bateman et al. 2002). In this study, the observed phosphotyrosine immonium ion at *m/z* 216.043 was used to discover phosphopeptides containing phosphotyrosine. (A similar approach for identifying phosphotyrosine-containing phosphopeptides has been described using a MALDI-qTOF mass spectrometer [Bennett et al. 2002].)

PHOSPHOPEPTIDE SEQUENCING BY MS/MS

During low-energy CID, the amide bonds of peptides fragment to yield *y*- and *b*-type ion series (see the information panel on FRAGMENTATION MECHANISMS OF PROTONATED PEPTIDES IN THE GAS PHASE in Chapter 8). The *b* fragments contain the original amino terminus of the peptide, whereas *y*-fragment ions include the original carboxyl terminus (for the nomenclature of peptide fragment ions, see Chapter 8) (Roepstorff and Fohlmann 1984; Biemann 1988). The mass difference between two consecutive ions of the same peptide fragment ion series corresponds to the mass of an amino acid residue, which allows the sequence of the parent peptide to be determined. Phosphopeptides are sequenced in the same manner, taking into account the mass alteration of serine, threonine, and tyrosine residues due to the phosphorylation (+80 daltons). However, as serine- and threonine-phosphorylated peptides normally show a predominant loss of phosphoric acid (H_3PO_4, 98 daltons) under low-energy CID conditions, resulting from gas-phase β-elimination of the labile alkylphosphoester, the unsaturated amino acid residues must be considered as well. Therefore, the position of phosphoserine residues can be identified based on the mass difference of two successive fragments (of the same fragment ion series) of 167 daltons (phosphoserine) or 69 daltons (dehydroalanine) if a β-elimination reaction occurred.

Phosphothreonine residues can be localized in the same manner, i.e., based on a fragment mass difference of 181 daltons (phosphothreonine) or 83 daltons (2-aminodehydrobutyric acid) after β-elimination. Because phosphotyrosine cannot undergo β-elimination reactions, tyrosine-phosphorylated peptides are stable under low-energy CID conditions. The location of phosphotyrosine residues can be determined by the mass difference of two successive peptide fragment ions of 243 daltons. Precursor ion scans, in conjunction with a nanoelectrospray source, provide a sensitive MS/MS tool for the detection of phosphorylated peptides and the determination of phosphorylation sites (Neubauer and Mann 1999; Craig 2001; Bateman et al. 2002; Bennett et al. 2002) (see Protocols 7 and 8).

MULTIDIMENSIONAL STRATEGIES FOR PHOSPHOPROTEIN CHARACTERIZATION

For comprehensive phosphorylation analysis, one method alone is often insufficient. This was demonstrated by Vihinen and Saarinen in their study of the phosphorylation of Semliki Forest virus nonstructural protein 3 (Vihinen and Saarinen 2000). In addition to traditional biochemical methods such as ^{32}P-labeling, TLC separation, and Edman degradation for sequencing, they used IMAC purification and several mass spectrometric methods including precursor-ion scanning (*m/z* 79) and MALDI analysis before and after alkaline phosphatase treatment to confirm 8 phosphorylation sites and to localize up to 12 additional phosphorylation sites, depending on the degree of phosphorylation. Other studies in which a combination of techniques was used for phosphorylation analysis include the following:

- Zhang et al. (1998) used MALDI-TOF analysis before and after alkaline phosphatase treatment of an in-gel protein digest to identify the phosphorylated peptides. Subsequently, LC-MS and LC-MS/MS with preset *m/z* values corresponding to the expected *m/z* values of the identified phosphopeptides were performed for an unambiguous localization of the phosphorylation site.
- Cao and Stults (1999, 2000) used a plug of IMAC material to retain the phosphopeptides generated by *in-gel* digestion of phosphoproteins. All peptides bound to the IMAC plug were eluted into a capillary electrophoresis column for separation. This setup was combined on-line with a mass spectrometer, allowing for the unambiguous identification of the phosphopeptides including the site of phosphorylation.
- Chen et al. (2002) used a variety of methods including IMAC, alkaline phosphatase treatment, MALDI PSD, and multidimensional electrospray-based approaches for the phosphorylation site mapping of hyperphosphorylated proteins associated with Net1, a regulator of exit from mitosis in yeast. These authors concluded that with the use of existing technologies, ***no single method*** was able to identify all sites in highly phosphorylated Net1.

For additional examples of different multidimensional protein phosphorylation studies, see Amankwa et al. (1995), Annan et al. (2001), Steen et al. (2001b), and Stensballe et al. (2001).

EMERGING MS AND MS/MS TECHNIQUES FOR PHOSPHOPEPTIDE SEQUENCING

Several new types of MS/MS methods for peptide sequencing have been developed during the past few years. McLafferty and co-workers observed that multiply charged peptide and protein ions generated by ESI can dissociate upon electron capture in the cell of a Fourier transform-ion cyclotron resonance (FT-ICR) mass spectrometer (see Chapter 8) (Zubarev et al. 1998). Electron capture dissociation (ECD) leads to extensive fragmentation of the polypeptide backbone generating *c*- and *z*-type ions, providing good amino acid sequence coverage for the fragmented peptides (Kelleher et al. 1999). ECD is a very gentle dissociation method that enables detailed analysis of posttranslational modifications, including identification of phosphorylated residues (Stensballe et al. 2000; Shi et al. 2001). In contrast to other MS dissociation methods, ECD does not induce the neutral loss (β-elimination) of phosphoric acid from phosphoserine and phosphothreonine even when multiple phosphorylation sites are present in a single peptide (Stensballe et al. 2000).

Although ECD seems to be the method of choice for the fragmentation of peptides containing labile modifications such as phosphopeptides, problems arise when the phosphopeptide is present in a mixture such that the phosphorylated species must first be identified before further analysis can be performed. Flora and Muddiman (2001) recently introduced a new method for the selective detection of phosphopeptides in FT-ICR MS based on the loss of the phospho-moiety. Performing ESI FT-ICR experiments in negative-ion mode, mass spectra of the peptide mixture were acquired before and after low-intensity infrared irradiation. Although this level of irradiation did not induce cleavage of the peptide backbone, it was sufficient to induce the loss of the phospho-moiety. Comparing the two spectra and searching for peak pairs differing by 98 daltons permitted the identification of phosphorylated species in the mixture.

External MALDI sources mounted onto qTOF and ion-trap tandem mass spectrometers provide the ability to generate amino acid sequence information from individual peptide species present in complex mixtures while still providing the advantages of MALDI (Qin and Chait 1997; Loboda et al. 2000; Shevchenko et al. 2000). MALDI tandem MS/MS analysis of singly protonated peptide ions produces somewhat complex mass spectra that nevertheless allow automatic protein identification via database searching (Qin et al. 1997; Shevchenko et al. 2002). MALDI-MS/MS is a very attractive method for characterization of posttranslationally modified proteins. Several researchers have demonstrated that MALDI-qTOF-MS/MS enables localization of phosphorylation sites in peptides up to 3.2 kD (Bennett et al. 2000; Baldwin et al. 2001; Lee et al. 2001), and Chait and co-workers have demonstrated the utility of MALDI ion-trap MS/MS for the characterization of several types of modified peptides (Qin and Chait 1997).

To avoid the problem of ionization suppression of phosphorylated peptides observed for MALDI as well as for ESI methods, Wind et al. (2001) utilized inductively coupled plasma mass spectrometry (ICPMS) in combination with liquid chromatography for the selective detection of phosphorylated species in a protein digest. Although no information about the phosphopeptides apart from the retention time is obtained, ICPMS has the significant advantage that the signal intensity is proportional to the molar amount of incorporated phosphorus, i.e., the degree of phosphorylation can reliably be measured and quantified.

As phosphorylation is a reversible process, there is significant interest in comparing the degree of phosphorylation between two different sample sets, such as different cell states, before and after drug treatment, or with and without exposure to certain external conditions. To address this question, preliminary studies providing proof of principle have been published recently (Oda et al. 1999; Weckwerth et al. 2000; Goshe et al. 2001). The first method is based on metabolic labeling with stable isotopes, whereas the latter two approaches are based on β-elimination followed by the incorporation of nucleophilic groups that are either unlabeled or labeled with stable isotopes. For a discussion of new tools for quantitative phosphoproteome analysis, see Conrads et al. (2002) and Mann et al. (2002).

SUMMARY

A wide range of MS-based techniques for detailed analysis of phosphoproteins are now available, but there is presently no "*best method*" for this purpose. The main challenge is the development of robust and efficient sample preparation methods for phosphopeptide analysis by MS. The most successful approaches utilize a combination of complementary methods for the analysis of phosphopeptides. Analytical strategies for global investigation and quantitation of phosphoproteins, i.e., *phosphoproteome analysis*, are in high demand. The research activity in the protein mass spectrometry community aimed at phosphoprotein characterization holds promise for development of much improved methods in the near future.

PROTOCOL 1

Phosphopeptide Purification by IMAC with Fe(III) and Ga(III)

IMMOBILIZED METAL ION AFFINITY CHROMATOGRAPHY (IMAC), first introduced by Porath (1975), makes use of matrix-bound metals to affinity-purify phosphoproteins and phosphopeptides. Commonly used metals in early studies such as Ni^{2+}, Co^{2+}, Zn^{2+}, and Mn^{2+} were shown to bind strongly to proteins with a high density of histidines (Porath et al. 1975). More recently, immobilized Fe^{3+}, Ga^{3+}, and Al^{3+} metal ions have been used for the selective enrichment of phosphopeptides from complex proteolytic digest mixtures containing both phosphorylated and nonphosphorylated components (Stensballe et al. 2001; Posewitz and Tempst 1999). The use of a nitrilotriacetic acid (NTA) matrix over iminodiacetic-acid-modified matrices has been reported to provide an advantage in selectivity (Neville et al. 1997). The development of elution conditions that are directly compatible with MS analysis of the enriched phosphopeptide samples provides the option to interface IMAC and MS on-line. In general, the strength and the selectivity of the interaction between the immobilized metal ion and the phosphopeptide depend on numerous factors, including the degree of phosphorylation, pH, salt concentration and composition of the sample solution, peptide concentration, type of chelated metal ions, temperature, and degree of exposure of chelated ions interacting with the peptide side chains. This protocol, kindly provided by Hanno Steen, Allan Stensballe, and Ole N. Jensen (University of Southern Denmark), describes the enrichment of phosphopeptides by IMAC using Fe^{3+}- and Ga^{3+}-NTA resin.

MATERIALS

CAUTION: See Appendix 3 for appropriate handling of materials marked with <!>.

IMPORTANT: Use the purest chemicals and H_2O available.

Reagents

Acetic acid (0.2 M)

Add 0.12 ml of glacial acetic acid <!> to 9.88 ml of H_2O to prepare a stock solution. The working solution throughout the protocol is 0.1 M acetic acid.

Acetic acid (0.1 M) and acetonitrile (3:1 v/v) <!>

EDTA (0.05 M) in 1 M NaCl

Add 0.19 g of EDTA and 0.58 g of NaCl to 10 ml of H_2O.

Ferric chloride ($FeCl_3$) <!> (0.2 M) or 0.2 M gallium chloride ($GaCl_3$) <!>

Add either 0.54 g of $FeCl_3$ or 0.35 g of $GaCl_3$ to 10 ml of H_2O.

WARNING: $GaCl_3$ reacts violently with H_2O. Before use (Step 6) combine equal parts 0.2 M $FeCl_3$ or 0.2 M $GaCl_3$ with 0.2 M acetic acid.

Formic acid (5%) <!>

Add 0.5 ml of formic acid to 9.5 ml of H_2O.

MALDI matrix (semisaturated DHB <!> in 50% acetonitrile and 2.5% formic acid, or CHCA in 70% acetonitrile and 0.1% trifluoroacetic acid [TFA] <!>)
If using ESI-MS instead of MALDI-MS, see Step 20 for alternative elution reagents.
Methanol (50%) <!>
Proteolytic enzyme (e.g., trypsin, Lys-C, or Asp-N)
Solvent (pH 10.5) (~0.1% ammonia solution at pH 10.5 <!>)

Equipment

Centrifuge (benchtop)
Desalting resin, Poros 10R2; Poros Oligo R3; Poros 20MC (Applied Biosystems)
IMAC resin, QIAGEN nitrilotriacetic acid (NTA)-silica <!> (16–24-µm particle size, QIAGEN) or Poros 20MC (Applied Biosystems)
MALDI probe (sample plate), polished stainless steel or AnchorChip type
Mass spectrometer, preferably a MALDI-TOF with delayed extraction and reflectron, or an ESI-qTOF (e.g., Bruker, Micromass, and Applied Biosystems)
Pipette tips (long narrow tips) (e.g., GELoader tips, Eppendorf)
Software: GPMAW (http://welcome.to/gpmaw)
Syringe (plastic, 1 ml)

Biological Sample

Proteins of interest separated either in-gel or in-solution
The protocol is intended for low- or subpicomole sample amounts per experiment; however, the method can be scaled up by increasing volumes of resins and reagents appropriately.

Additional Reagents

Step 11 of this protocol requires the reagents listed in Chapter 7, Protocol 5.

METHOD

Preparation of the IMAC Resin

In Steps 1–10, the slurry is gently mixed with 1 ml of each reagent, in turn, in a microfuge tube for 30 seconds at ambient temperature. At the end of each step, centrifuge the slurry and discard the supernatant before proceeding with the next step.

1. In a microfuge tube, resuspend the IMAC resin in H_2O to a concentration of ~1 mg/ml.
2. Wash the resin with H_2O.
3. Incubate the resin twice with 50 mM EDTA in 1 M NaCl at room temperature.
4. Wash the resin with H_2O.
5. Wash the resin with 0.1 M acetic acid.
6. Incubate the resin twice with 0.1 M $FeCl_3$ or $GaCl_3$ in 0.1 M acetic acid at room temperature.
7. Wash the resin with 0.1 M acetic acid.

8. Wash the resin with 0.1 M acetic acid and acetonitrile (3:1 v/v).
9. Wash the resin with 0.1 M acetic acid.
10. Resuspend the resin in 0.1 M acetic acid.

    The ready-to-use slurry of IMAC resin can be stored for at least 40 days at 4°C without reducing the performance of the resin.

Preparation of the Peptide Sample

11. Perform in-gel or in-solution digestion of the sample protein with the desired enzyme (trypsin, Lys-C, Asp-N, etc.) as detailed in Chapter 7, Protocol 5.
12. Dilute an aliquot of the digest corresponding to 0.1–10 pmoles of the sample into 30–50 µl of 0.1 M acetic acid.

Phosphopeptide Purification by Nanoscale IMAC

13. Prepare an IMAC column by loading enough IMAC resin (from Step 10) into a long, narrow pipette tip to pack a 15–20-mm column of IMAC resin (for the preparation and use of this column, see the panel on ADDITIONAL PROTOCOL: PREPARATION AND USE OF MICRO-COLUMNS FOR SAMPLE DESALTING OR NANOSCALE IMAC).
14. Load the protein sample (30–50 µl) *slowly* onto the IMAC column (1–3 µl/min) using air pressure from a syringe or a micropipettor.
15. Collect the flowthrough in the upper end of a nanoscale desalting column (Poros R2 or Oligo R3 prepared according to the additional protocol at the end of this protocol) prior to desalting and MALDI-TOF-MS peptide mass mapping.

    This fraction contains all of the peptides that were not retained on the IMAC column.

16. Wash the IMAC column with 20 µl of 0.1 M acetic acid.

    Here, and in the following wash steps, use gentle air pressure from an attached syringe to push the solutions through the column. Discard the eluents from each wash step.

17. Wash the IMAC column with 20 µl of 0.1 M acetic acid and acetonitrile (3:1 v/v).
18. Wash the IMAC column with 20 µl of 0.1 M acetic acid.
19. Elute the retained phosphopeptides with two times 5-µl volumes of pH 10.5 solvent directly into 20 µl of 5% formic acid placed in the upper end of a fresh nanoscale desalting column (Poros R2 or Oligo R3).

    Alternative eluting reagents are MALDI matrices such as DHB in 50% acetonitrile/2.5% formic acid or CHCA in 70% acetonitrile/0.1% TFA. If these are used, the eluent should be spotted directly onto the MALDI probe.

20. Elute the desalted IMAC eluate from the desalting column with 1 µl of MALDI matrix solution (see Step 19).

    For the preparation of samples for MALDI-MS analysis, see Chapter 8, Protocol 1. The eluate is spotted as a series of 5–8 droplets onto the MALDI probe. The major part of the analyte will be present in droplet 1 and 2. Alternatively, the peptides can be eluted for nano-ESI analysis using either 1–2 µl of 50% methanol/1% formic acid *or* 1–2 µl of 5% NH_3/50% methanol plus acidification for negative-ion-mode analysis (acidification by 50% formic acid/50% methanol).

ADDITIONAL PROTOCOL: PREPARATION AND USE OF MICROCOLUMNS FOR SAMPLE DESALTING OR NANOSCALE IMAC

This method, which is very similar to the preparation of microcolumns detailed in Chapter 8, Protocols 2 and 3, is based on the principles of Gobom et al. (1999) and Stensballe et al. (2001). Microcolumns can be prepared using any kind of chromatographic resin. The resin is held in place by making a constriction at the end of a GELoader pipette tip. Sample loading, washing, and elution is performed by loading liquid on top of the resin and applying air pressure to generate a low flow through the column. No frits are necessary.

Additional Materials

CAUTION: See Appendix 3 for appropriate handling of materials marked with <!>.

Chromatographic resin (e.g., Poros R1, R2, Oligo R3, or IMAC)
GELoader tips (Eppendorf)
Methanol <!>
Plastic syringe (1 ml)
Syringe adaptor

The syringe is attached to the GELoader tips by using the top part of a 200-µl pipette tip modified to connect the syringe and the GELoader tip.

Method

1. Prepare a slurry of 100-200 µl of chromatographic resin in 1 ml of methanol.
2. Partially constrict a GELoader pipette tip by gently squeezing or twisting the end of the tip. This allows liquid to flow through the tip while retaining the chromatographic resin.
3. Load 5 µl of slurry into the GELoader tip from the top and pack it by applying air pressure with the 1-ml syringe adapted to fit the GELoader tip microcolumn. The column height should be 2–4 mm.

   In the case of the IMAC column prepared for the main protocol, use enough resin to prepare a column of 15–20 mm in height.
4. Equilibrate the resin by flushing the column with 10–20 µl of 5% formic acid through the GELoader tip using air pressure from the syringe.

   TFA or acetic acid can be used instead of formic acid in the mobile phases, whereas acetonitrile can substitute for methanol. Formic acid and methanol are recommended for nano-electrospray mass spectrometry (see Chapter 8).
5. Dissolve the peptide or protein sample in 20–40 µl of 5% formic acid.
6. Load 5–20 µl of sample onto the microcolumn and gently press it through the column by air pressure using the syringe.
7. Wash the resin by flushing 10–20 µl of 5% formic acid through the packed GELoader tip by air pressure (syringe).
8. Elute the sample using a small volume of 5% formic acid/50% methanol.

   The eluate can be collected in a microfuge tube, deposited directly onto the MALDI probe tip, or eluted directly into a nano-electrospray needle. In the latter two cases, the elution volume should be 1–2 µl only.

   Peptide separation may be improved by eluting the sample from the microcolumn by a step gradient using a series of mobile phases containing 5% formic acid in 15% methanol, 30% methanol, and 50% methanol, respectively. For MALDI-MS analysis, elute the sample directly onto the MALDI probe using matrix solution, e.g., HCCA, SA, or DHB dissolved in 30–50% methanol or acetonitrile. Deposit the eluate in a series of tiny droplets rather than one large drop (see Chapter 8, Protocol 11).

PROTOCOL 2

Alkaline Phosphatase Treatment of Phosphopeptides Either Before or After MALDI Analysis

THE USE OF THE ENZYME ALKALINE PHOSPHATASE allows identification of phosphopeptides in a mixture of predominantly nonphosphopeptides. Using a MALDI-MS instrument, the masses of peptides are acquired both pre- and postalkaline phosphatase treatment, which removes phospho-moieties from serine, threonine, and/or tyrosine. (Any peptides whose mass decreases by 80 daltons, or multiples thereof, is a phosphopeptide.) An advantage of using MALDI-MS for these experiments is that the peptide ions produced tend to be singly charged rather than multiply charged (as with ESI), thus making the interpretation easier (see Chapter 8). In this protocol, provided by Hanno Steen, Allan Stensballe, and Ole N. Jensen (University of Southern Denmark), three methods are given:

- ***Method A:*** The in-solution dephosphorylation method prior to MALDI-MS analysis.
- ***Method B:*** In-solution dephosphorylation after MALDI-MS analysis based on the protocol of Stensballe et al. (2001).
- ***Method C:*** On-probe dephosphorylation after MALDI-MS analysis based on the protocol of Larsen et al. (2001).

MATERIALS

CAUTION: See Appendix 3 for appropriate handling of materials marked with <!>.

IMPORTANT: Use the purest chemicals and H_2O available.

Reagents

Acetonitrile (70%) <!> in 50 mM NH_4HCO_3

Alkaline phosphatase (AP) (0.05 unit AP/µl in 50 mM NH_4HCO_3)

Stock concentrations of calf intestinal alkaline phosphatase (Roche) of 20 units/µl rather than 1 unit/µl is recommended due to a lower concentration of storage buffer in the usage concentration.

Formic acid (5%) <!>

Add 0.5 ml of formic acid to 9.5 ml of H_2O.

MALDI matrix (semisaturated DHB <!> in 50% acetonitrile and 2.5% formic acid or CHCA in 70% acetonitrile and 0.1% trifluoroacetic acid <!>)

For choice of MALDI matrix, see Table 8.8 in Chapter 8.

NH_4HCO_3 (50 mM, pH 7.8) <!>

Add 0.040 g of NH_4HCO_3 to 10 ml of H_2O.

Trifluoroacetic acid (5%) (TFA)

Equipment

Centrifuge (benchtop)

Desalting resin, Poros 10R2; Poros Oligo R3; Poros 20MC (Applied Biosystems)

MALDI probe

Mass spectrometer

The spectrometer should be capable of generating MALDI reflector-TOF and MALDI linear-TOF spectra.

Plastic box

The box should have a lid able to keep the MALDI metal probe in a humid environment. See Method C, Step 2.

Pipette tips (long narrow tips) (e.g., GELoader tips, Eppendorf)

Syringe (disposable plastic, 10 ml)

Biological Sample

Phosphopeptide sample either in solution (Method A) or bound to a MALDI probe (Methods B and C)

The protocol is intended for low- or subpicomole sample amounts per experiment.

METHODS

Method A: In-solution Dephosphorylation Prior to MALDI-MS Analysis

1. Prepare two nanoscale desalting columns (Poros R2 or R3) according to the additional protocol in Protocol 1. Make sure that the resin remains wet with 5% formic acid to keep it humid during the incubation (Step 3). The upper end of the GELoader tip serves as a reaction chamber.
2. Place 20 µl of alkaline phosphatase in 50 mM NH_4HCO_3 in one of the reaction chambers. Dilute the peptide sample (typically 1–2 µl) into the alkaline phosphatase solution.

   The final pH of the reaction must be ~8, but if necessary adjust it with 50 mM NH_4HCO_3 (may be checked with pH paper).

   IMPORTANT: Prepare a control reaction within a GELoader tip containing 50 mM NH_4HCO_3. Add an equivalent amount of peptide, but no alkaline phosphatase.
3. Incubate the GELoader tips for 45 minutes at 37°C to dephosphorylate the peptides.
4. Acidify the samples by addition of 20 µl of 5% formic acid and load each sample onto a Poros resin by application of air pressure from a syringe.
5. Wash the columns with 20 µl of 5% formic acid and then elute with 1 µl of saturated MALDI matrix. Spot the eluates as a series of 5–8 droplets onto the MALDI probe for optimal sensitivity. The major part of the analytes will be present in droplets 1 and 2.
6. Record MALDI reflector-TOF and MALDI linear-TOF spectra on a portion of the phosphopeptides, on the material from Step 5, and the control sample.

Method B: In-solution Dephosphorylation After MALDI-MS Analysis

This method is applicable for both DHB and CHCA MALDI-MS matrices.

1. Prepare two nanoscale desalting columns (Poros R2 or R3) according to the additional protocol in Protocol 1. Make sure that the resin remains wet with 5% formic acid to keep it humid during the incubation (Step 4).

2. Place 20 μl of alkaline phosphatase in 50 mM NH_4HCO_3 in the upper end of the GELoader tips, which serves as a reaction chamber.

   IMPORTANT: Prepare a control reaction within a GELoader tip containing 50 mM NH_4HCO_3.

3. Record MALDI reflector-TOF and MALDI linear-TOF spectra on the peptide mixture.

4. Use a micropipettor to carefully dissolve the analyte/matrix deposits using a maximum volume of 0.5–1.5 μl of 70% acetonitrile in 0.05 M NH_4HCO_3. Transfer equal volumes of dissolved analyte/matrix deposits directly to each reaction chamber.

   The stainless steel surface of the MALDI probe must be of a "polished" type or AnchorChip for the recovery of analyte to be successful.

5. Incubate the GELoader tips for 45 minutes at 37°C to dephosphorylate the peptides.

6. Acidify the samples by addition of 2 volumes of 5% formic acid and load each sample onto a Poros resin by application of air pressure from a syringe.

7. Wash the columns with 20 μl of 5% formic acid and then elute with 1 μl of saturated MALDI matrix. Spot the eluates as a series of 5–8 droplets onto the MALDI probe for optimal sensitivity. The major part of the analytes will be present in droplets 1 and 2.

8. Record MALDI reflector-TOF and MALDI linear-TOF spectra on the alkaline phosphatase-treated peptides.

Method C: On-probe Dephosphorylation After MALDI-MS Analysis

This method is applicable only for CHCA MALDI-MS matrix.

1. After obtaining MALDI reflector-TOF and MALDI linear-TOF spectra from an analyte/matrix deposit, dissolve the matrix on the target using 1.5 μl of 0.05 M NH_4HCO_3 containing alkaline phosphatase (0.05 unit/μl).

2. Place the target in a closed plastic box containing a wet tissue to prevent the sample from drying. Incubate the box for 30 minutes at 37°C.

3. Acidify the sample with 0.5 μl of 5% TFA, thus allowing the matrix to recrystallize.

4. Prior to MALDI analysis of the dephosphorylated peptides, wash the surface of the matrix microcrystals gently with 0.1% TFA to wash away salts and glycerol originating from the dephosphorylation buffer. Apply the 10 μl of 0.1% TFA as a droplet onto the matrix; deposit and remove immediately using the edge of a tissue.

PROTOCOL 3

Characterization of Phosphopeptides Using a Combination of Immobilized Metal Ion Affinity Media and Direct Analysis by MALDI-TOF-MS

PHOSPHOPROTEINS AND PEPTIDES CAN BE BOUND with high specificity to immobilized metal ions, such as Fe^{3+} and Ni^{2+} (Andersson and Porath 1986; Muszynska et al. 1992; Olcott et al. 1994). Recently, Tempst reported that Ga^{3+} has better selectivity than Fe^{3+} for phosphopeptides (Posewitz and Tempst 1999). This technique can be used with either on-line or off-line MS analysis (Michel et al. 1988; Nuwaysir and Stults 1993; Neville et al. 1997; Cleverley et al. 1998; Hanger et al. 1998; Li and Dass 1999; Posewitz and Tempst 1999). Elution of the phosphopeptides from the metal ion column prior to MS analysis can, however, result in sample loss. Affinity-bound analytes, including phosphopeptides, can be directly analyzed by MALDI-MS without prior elution from the affinity media (Papac et al. 1994). The utility of this method has been extended by the observation that consecutive enzymatic reactions, such as phosphatase or carboxypeptidase Y digestion, can be carried out on affinity-bound peptides (Papac et al. 1994; Qian et al. 1999; Zhou et al. 2000; Merrick et al. 2001). When the affinity-bound phosphopeptides are treated with phosphatase, the number of phosphorylation sites can be determined based on the observation of 80-dalton (or multiples of 80 daltons) mass shifts in the MALDI-MS of the reaction mixture. Carboxypeptidase Y treatment of the affinity-bound phosphopeptides can also be used to cleave the amino acids from the carboxyl terminus, with subsequent direct analysis of the enzymatic products by MALDI-MS to locate the phosphorylation sites on the bound phosphopeptides. This protocol details the preparation and use of Fe^{3+} or Ga^{3+} metal IMAC with the on-bead analysis of phosphopeptides by MALDI-MS (see Figure 9.11). Enzymatic digestion of affinity-bound peptides is also described. This protocol was provided by Leesa J. Deterding, Jenny M. Cutalo, and Kenneth B. Tomer (National Institute of Environmental Health Sciences, Research Triangle Park, North Carolina).

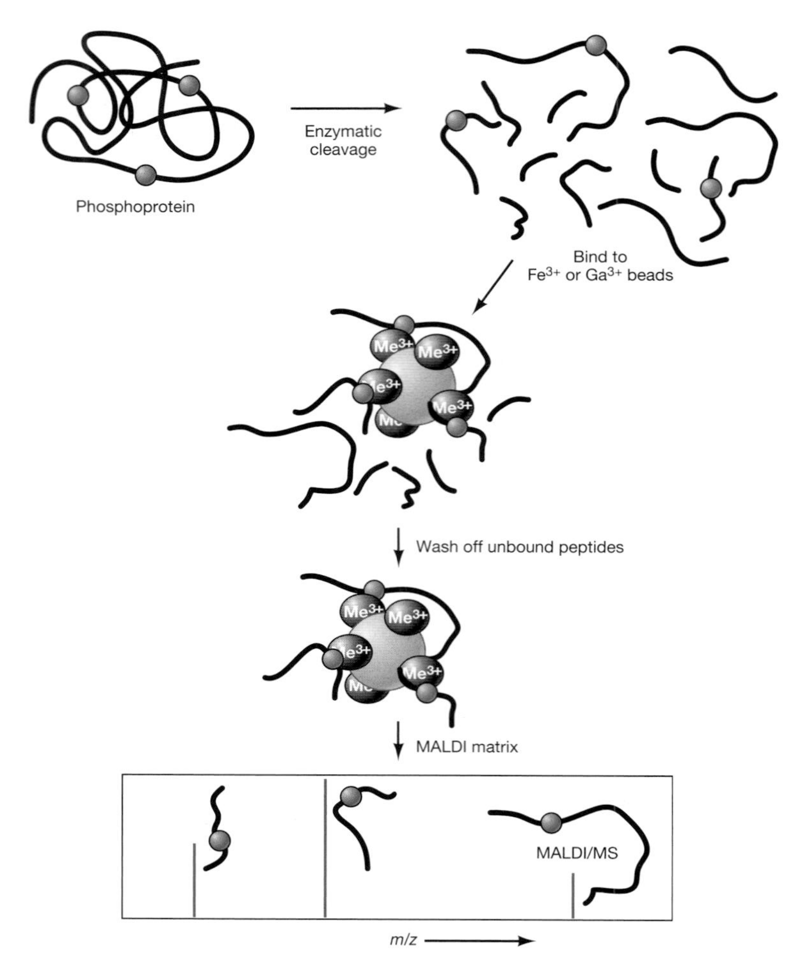

FIGURE 9.11. Schematic of affinity binding of phosphopeptides to immobilized metal ion affinity columns.

MATERIALS

CAUTION: See Appendix 3 for appropriate handling of materials marked with <!>.

Reagents

Acetic acid (100 mM) <!>

Ammonium bicarbonate (50 mM, pH 8.0) <!>

EDTA (100 mM in H_2O, pH 8.0)

Gallium chloride ($GaCl_3$) (60 mM) <!> or 100 mM ferric chloride ($FeCl_3$) <!>

MALDI matrix

The MALDI matrix is a saturated solution of α-cyano-4-hydroxy-cinnamic acid <!> in 45:45:10 ethanol:H_2O:formic acid <!> (v/v/v). Store it in the dark at room temperature.

Nickel-nitrilotriacetic acid (Ni-NTA) resin (QIAGEN)

Sodium citrate (50 mM, pH 6.0) <!>

Equipment

Compact reaction columns (CRC) and filters (35-μm pore size) (USB Corporation)

Incubator (slow rotation, i.e., rotating wheel)

MALDI mass spectrometer (e.g., the Voyager DE-STR) or equivalent

The Voyager DE-STR mass spectrometer (Applied Biosystems) is equipped with a nitrogen laser (337 nm) to desorb and ionize the samples. A close external calibration, using two points to bracket the mass range of interest, should be performed prior to analyzing the protein sample (see Chapter 8, Protocol 1).

MALDI target

Microfuge tubes (0.7 ml)

Biological Molecules

Calf intestinal alkaline phosphatase (Roche)

Carboxypeptidase Y (1 μg/μl in H_2O)

EXPERIMENTAL TIP: Because MALDI-MS sensitivity is significantly reduced in the presence of glycerol, it is highly recommended that enzymes not be in solutions containing glycerol.

Phosphoprotein of interest

TPCK-modified trypsin (sequence-grade, e.g., Promega)

METHOD

Preparation of the Immobilized Metal Ion Affinity Column

1. Insert the CRC filter (35 μm) into the CRC, and place the CRC in a standard microfuge tube.
2. Add ~30 μl of the Ni-NTA resin slurry to the bottom of the CRC tube and drain the column by centrifugation at 120*g* for ~1 minute at room temperature.

   These centrifugation conditions are the same for all washes unless otherwise stated.
3. Wash the Ni-NTA resin three times each with 30 μl of 100 mM EDTA (pH 8.0) to remove the bound Ni^{2+} metal.
4. Wash the column three times each with 30 μl of H_2O.
5. Wash the column three times each with 30 μl of 100 mM acetic acid.
6. To regenerate the columns with either Fe^{3+} or Ga^{3+} metal, wash the column three times each with 30 μl of either 60 mM $GaCl_3$ or 100 mM $FeCl_3$.
7. To remove any unbound metal ions, wash the column three times each with 30 μl of H_2O followed by three times each with 30 μl of 100 mM acetic acid.

Affinity Binding of Phosphopeptides to IMAC Column

8. Digest the phosphoprotein with trypsin (protein:enzyme ratio of 20:1 to 100:1 [w/w]) in 50 mM NH_4HCO_3 buffer (pH 8) for 2 hours at 37°C.

   For these experiments, the protein concentration was 0.1 μg/μl.
9. Load 20 μl (~20 μg) of the trypsin-digested phosphorylated protein with 30 μl of 100 mM acetic acid onto the IMAC column. Incubate the column on a slowly rotating wheel for ~30 minutes at 37°C.

   The column should be checked periodically for leakage from the bottom of the CRC. Leakage can be alleviated by sealing the bottom of the CRC with Parafilm.

10. To remove unbound peptides from the IMAC column, wash the column with 30 μl of H_2O followed by 30 μl of 100 mM acetic acid. Repeat this step three times.

    The IMAC column (with bound phosphopeptides) can be stored in 100 mM acetic acid for at least 2–3 days at 4°C. Even bound phosphopeptides from a complex mixture, such as a cell extract, should be stable, because endogenous phosphatases are digested by the trypsin and/or elute from the column.

    To verify that nonspecific peptides are washed off the column or if the unbound peptides are of interest, analyze the washes by MALDI-MS.

11. Spot a 0.5-μl aliquot of settled beads with 0.5 μl of MALDI matrix on the MALDI target.

Alkaline Phosphatase Digestion of Affinity-bound Phosphopeptides

12. Mix a 5-μl aliquot of the phosphopeptides affinity-bound to the IMAC media (from Step 10) with 5 μl of 50 mM NH_4HCO_3 (pH 7.8) and 1 unit of calf intestine alkaline phosphatase in a 0.7-ml microfuge tube.

13. Incubate the tube on a slowly rotating wheel at 37°C.

14. Monitor the time course of the digestion periodically by spotting a 0.5-μl aliquot of the reaction mixture (supernatant and beads) with 0.5 μl of MALDI matrix onto a MALDI target. Take the first aliquot 30 minutes after beginning Step 13.

    Initial aliquots are typically taken at 30-minute intervals. Depending on the protein, incubation times for the digestion can range from 1 hour to overnight.

Carboxypeptidase Y Digestion of Affinity-bound Phosphopeptides

15. Transfer a 25-μl aliquot of the phosphopeptides affinity-bound to the IMAC media (from Step 10) into another CRC.

16. Wash the column three times each with 30 μl of 50 mM sodium citrate (pH 6).

17. Add 40 μl of 50 mM sodium citrate (pH 6) to the column.

18. Add 1 μl of carboxypeptidase Y solution (1 μg/μl) in H_2O to the column.

19. Incubate the column on a slowly rotating wheel at 37°C.

    Depending on the protein, the carboxypeptidase Y digests can take from a few minutes to overnight to complete. Thus, the time course of the digest (Step 20) should be monitored initially at 30–60-second intervals.

20. To monitor the time course of the reaction:

    a. Wash the beads three times each with 50 μl of 100 mM acetic acid.

    b. Add 30 μl of 100 mM acetic acid.

    c. Spot 0.5 μl of beads with 0.5 μl of MALDI matrix onto the MALDI target for MALDI-MS analysis.

    d. Resume the carboxypeptidase Y reaction by repeating Steps 16–19.

Mass Spectrometry

21. Perform MALDI analyses on the samples from Steps 11, 14, and 20a using a delayed-extraction TOF mass spectrometer.

CASE STUDY

A tryptic digest of a solution containing two α-casein variants (S1 and S2) and trace amounts of β-casein was analyzed by MALDI-MS (see Figure 9.12a). Nine major peaks were observed in the mass range shown. The α-casein tryptic digest was then loaded onto a Fe^{3+} IMAC column. Direct MALDI-MS analysis of the affinity-bound peptides (Figure 9.12b) showed only two major ions, both of which correspond in mass to monophosphorylated tryptic peptides T15 and T14–15. It should be noted that tryptic peptides which contain multiple phosphorylation sites were not observed under these experimental conditions, presumably because the Fe^{3+}-loaded IMAC support binds these phosphopeptides more tightly than does the corresponding Ga^{3+}-IMAC-loaded support (Zhou et al. 2000), and thus are not released during the MALDI process. After phosphatase treatment (1 hour at 37°C) of the α-casein phosphopeptides bound to the Fe^{3+}-IMAC, ions corresponding in mass to dephosphorylated T15 and T14–15 were observed in addition to the phosphorylated T15 and T14–15 (Figure 9.12c). Further digestion with alkaline phosphatase (overnight at 37°C) resulted in nearly complete dephosphorylation of the affinity-bound phosphopeptides (Figure 9.12d). From these data, the number of phosphorylation sites in the affinity-bound peptides can be determined.

To gain sequence information about the affinity-bound phosphopeptides, the immobilized metal ion affinity columns can be subjected to carboxypeptidase Y digestion. The carboxypeptidase Y digest of the tetraphosphorylated peptide T1–2 of β-casein resulted in cleavage of the carboxy-terminal amino acid residues SITR (Zhou et al. 2000). From these data, the location of the four phosphates could be assigned to the remaining four serines on T1–2. It should be noted that carboxypeptidase Y digestion of immobilized metal ion affinity media containing multiple affinity-bound peptides results in complex MALDI spectra, making interpretation of the sequence information difficult. In these cases, sequence information may be obtained by elution of the peptides from the IMAC followed by ESI-MS/MS analysis.

With the development of hybrid quadrupole time-of-flight (QTOF) and TOF-TOF instrumentation with MALDI sources, the MS/MS sequence data can be obtained directly from MALDI desorbed ions. The MALDI-QTOF-MS/MS spectrum of *m/z* 1952, monophosphorylated T14–15 from α-casein, was acquired directly from the Fe^{3+} IMAC resin and is shown in Figure 9.13. The most abundant ion observed, *m/z* 1855, is due to loss of 98 daltons, H_3PO_4, from the protonated molecule. In addition, a nearly complete series of ions containing the carboxyl terminus, *y* ions, and *y* ions plus loss of 98, are observed. *y* ions 1 through 4 are not shifted in mass, indicating that the phospho-group is not located on these residues (A116 through R119). In contrast, the *y* ions y_5 through y_{12} all show an increase in mass of 98 daltons. There is also a series of ions arising from loss of 98 daltons from the *y* ions y_5 (S115) through y_{14} (V106). These data show that the location of the phospho-group is the serine residue S115.

The combination of immobilized metal ion affinity and direct MALDI-MS analysis allows the selective enrichment and detection of phosphopeptides from complex reaction mixtures. The detection of phosphorylation sites in low-abundance phosphoproteins, where typically only a small fraction of the protein is modified, should be enhanced. Subsequent enzymatic digestion or MS/MS of the immobilized phosphopeptides can provide additional information on the number and location of phosphorylation sites in the peptides.

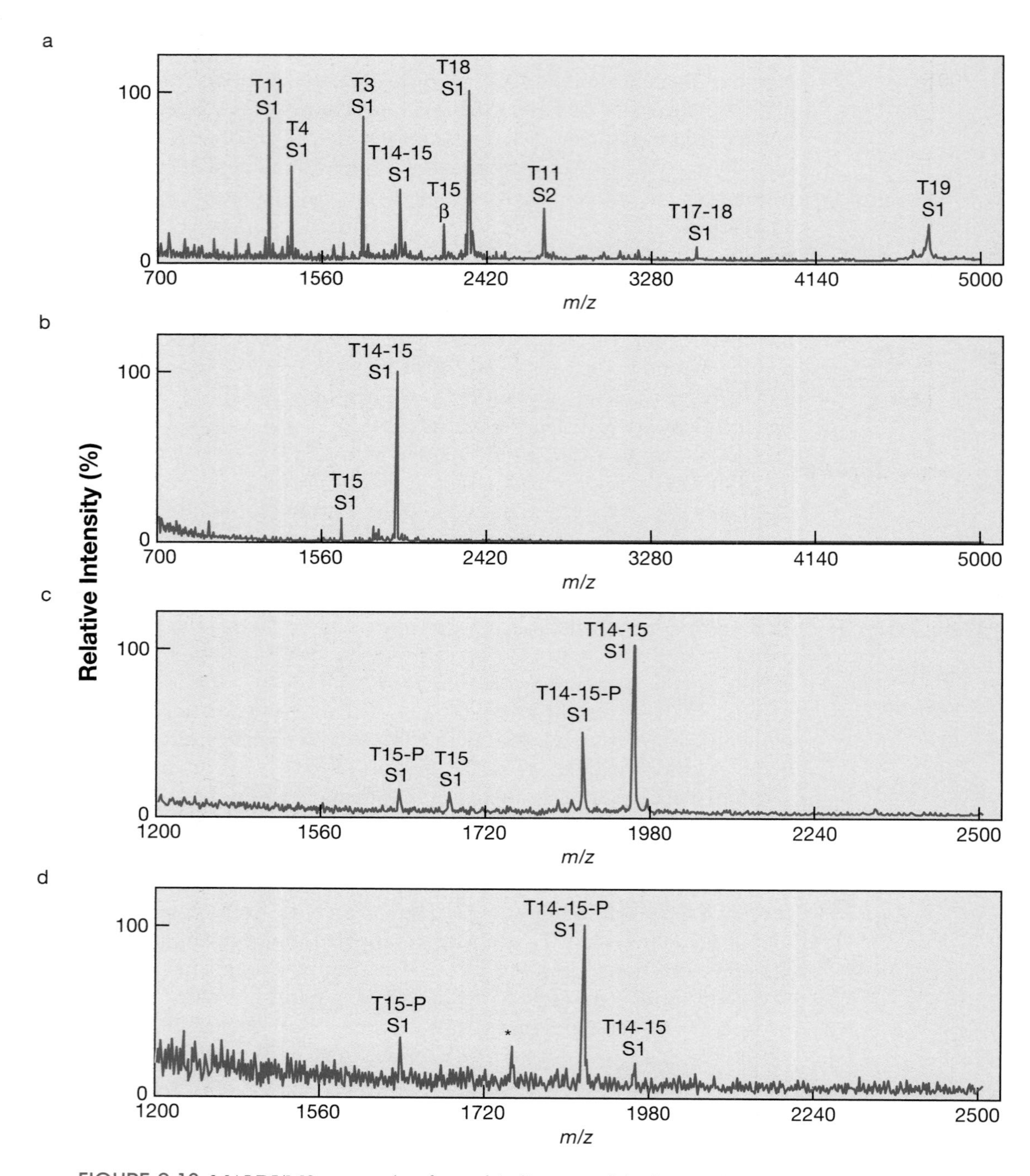

FIGURE 9.12. MALDI/MS spectra (performed in linear mode) of α-casein tryptic digest in solution (*a*); Fe^{3+}-NTA bound phosphopeptides from α-casein tryptic digest (*b*); 1-hour phosphatase treatment of Fe^{3+}-NTA-bound phosphopeptides from α-casein tryptic digest (*c*); and overnight phosphatase treatment of Fe^{3+}-NTA bound phosphopeptides from α-casein tryptic digest (*d*). The ion labeled with an asterisk (*) is a background ion. S1 and S2 are α-casein variants. T is an abbreviation for tryptic peptide.

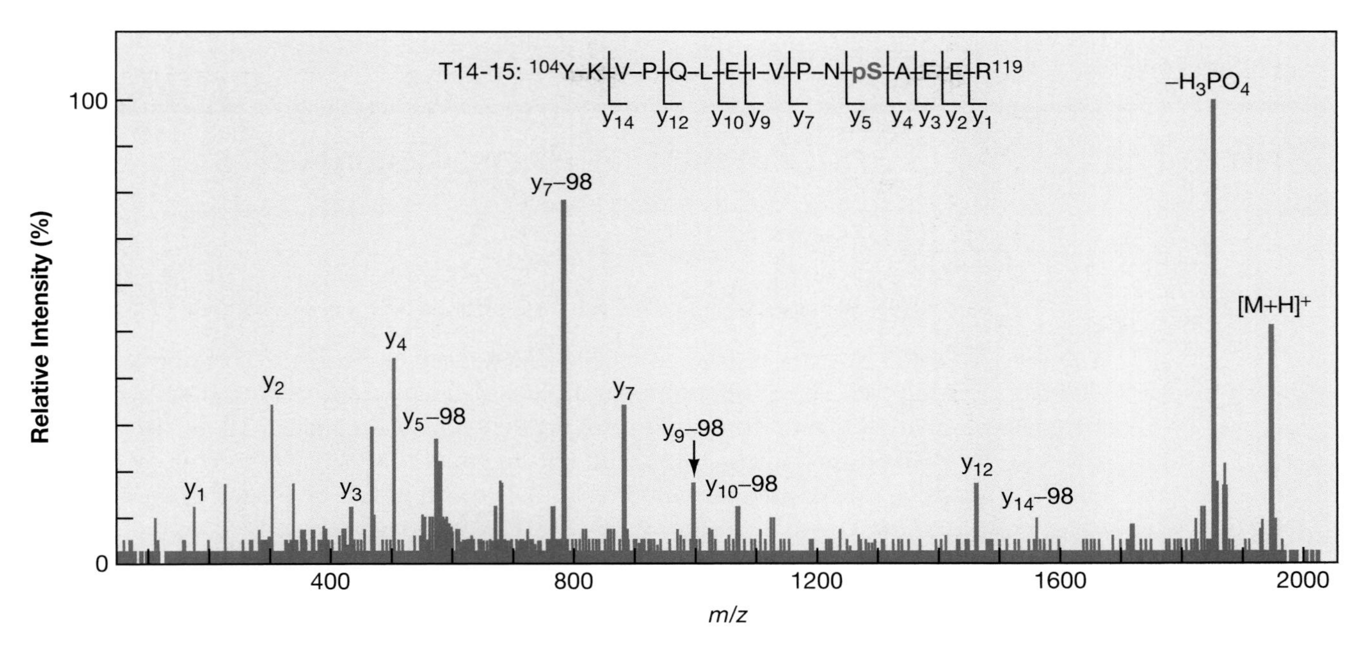

FIGURE 9.13. MALDI-QTOF-MS/MS spectrum of *m/z* 1952, T14–15, from the α-casein tryptic digest acquired directly from the Fe^{3+}-NTA resin. Structurally significant ions are indicated on the spectrum. Matrix was a saturated solution of 2,5-dihydroxybenzoic acid in 70:30 acetonitrile/0.1% TFA.

PROTOCOL 4

Off-line Micro-IMAC Enrichment of Phosphoproteins

IMMOBILIZED METAL AFFINITY CHROMATOGRAPHY (IMAC) can be performed off-line or on-line with μLC-ESI-MS/MS. The on-line configuration has been used successfully despite suboptimal low-level analysis due to the relative disparity between IMAC and μLC flow rates (see the earlier sections on On-line micro-IMAC and Off-line micro-IMAC). With IMAC performed off-line from μLC-ESI-MS, sample volumes from 1 to 100 μl can be loaded using a sample pressure vessel (Goodlett et al. 2000b) or an autosampling system equipped with a sample-trapping cartridge (Hayashi et al. 2001). This protocol describes the construction and use of a micro-IMAC column for the enrichment of phosphoproteins (additional details are available in Corthals et al. [1999] and Figure 9.4). It is similar to the procedure used to generate a micro-capillary HPLC column (Karlsson and Novotny 1988; Kennedy and Jorgenson 1989) (see Chapter 5, Protocol 2 and Chapter 8, Protocol 4). Protocol 5 describes the next step: the analysis of the enriched sample of phosphoproteins by μLC-ESI-MS/MS. This protocol was provided by Garry L. Corthals, Valerie C. Wasinger, and David R. Goodlett (Geneva Proteomics Centre, Geneva University Hospital, Geneva, Switzerland; University of New South Wales, Kensington, Australia; and Institute for Systems Biology, Seattle, Washington, respectively)

MATERIALS

CAUTION: See Appendix 3 for appropriate handling of materials marked with <!>.

Reagents

Ammonium acetate (0.1%) <!> in 50 mM Na_2HPO_4 (pH 8.0) <!>

EDTA (0.1 M)

Ferric chloride ($FeCl_3$) (0.1 M) <!> in 0.1 M acetic acid <!>
See Step 10.

Equipment

Fittings and unions (e.g., Microtight Adapters, Upchurch, Washington)

Helium, pressurized gas <!>

IMAC resin (e.g. Poros-MC, PerSeptive Biosystems)

Polyimide-coated fused-silica capillaries (360 μm O.D. x 50–100 μm I.D.) (e.g., Polymicro Technologies, Tucson, Arizona)

Pressure vessel (contact garry.corthals@dim.hcuge.ch)

Teflon tubing (1/16 inch O.D. x 0.0001 inch I.D.)
See Step 1.

Biological Sample

Peptides from phosphoprotein sample of interest
The phosphoprotein should be digested into peptides according to Chapter 7, Protocol 5.

METHOD

Construction and Preparation of an IMAC Microcolumn

In addition to the method described in this protocol, IMAC columns can also be prepared in constricted gel-loader pipette tips or commercially prepared tips can be purchased (e.g., MC-ZipTip, Millipore). For details on packing and using an IMAC "tip column," see Protocol 1 and Chapter 8, Protocol 2.

1. To one end of a 10-cm-long piece of Teflon tubing (1/16 inch O.D. x 0.0001 inch I.D.), insert a piece of polyimide-coated fused-silica capillary and hold it in place with a union.

2. Before fixing the second of the two polyimide capillaries in place, place the open Teflon end in a slurry of IMAC resin inside a vessel pressurized by helium.

3. Pack the column to a length of ~5 cm at 500 psi pressure.

4. Insert the second piece of fused-silica capillary into the Teflon tubing, and fix it in place with a second union.

5. Place one of the two polyimide capillaries in the pressure vessel. The other capillary serves as an outlet.

   This configuration allows loading of the sample, washing away of contaminants, and eluting of the phosphopeptides from the IMAC column rapidly and at a relatively low pressure.

6. Place a microfuge tube containing H_2O in the pressure vessel, and wash the column with H_2O for 5 minutes at 500 psi. Monitor the solution that elutes from the column with pH paper.

   Monitor the pH during Steps 6–11.

7. Replace the microfuge tube in the pressure vessel with one containing 0.1 M EDTA, and wash the column for ~2.5 minutes (until there is a change in the pH) at 500 psi.

8. Replace the microfuge tube with one containing H_2O and wash the column for 5 minutes at 500 psi.

9. Replace the microfuge tube with one containing 0.1 M acetic acid and wash the column for 5 minutes at 500 psi. Wash until there is a change in the pH.

10. Replace the microfuge tube with one containing 0.1 M $FeCl_3$ in 0.1 M acetic acid and activate the column for 5 minutes at 500 psi.

    Other metals can also be used, such as cupric sulfate, nickel chloride, or gallium nitrate.

11. Replace the microfuge tube with one containing 0.1 M acetic acid and wash the column for 10 minutes at 500 psi to remove any unbound metal ions.

Enrichment of Phosphopeptides

12. Load ≥1 pmole of the protein digest (<5 μl) onto the IMAC column at 200 psi. The peptide mixture should be loaded onto the column in either H_2O or 0.05 M acetic acid.

13. Wash the column with 0.1 M acetic acid for 2.5 minutes at 200 psi.

14. Wash the column with H_2O for 2.5 minutes at 200 psi.

15. Elute the phosphopeptides from the column with 0.1% ammonium acetate in 50 mM Na_2HPO_4 (pH 8.0) at 200 psi. Collect the eluted peptides in microfuge tubes and use directly for MS/MS analysis in Protocol 5.

REGENERATING THE IMAC MEDIA

The IMAC column can be regenerated as follows:

Flush the column with H_2O for 2.5 minutes at 500 psi.
Flush the column with 0.1 M EDTA for 2.5 minutes at 500 psi.
Flush the column with H_2O for 2.5 minutes at 500 psi.
Flush the column with 0.1 M acetic acid for 2.5 minutes at 500 psi.

The column can be stored in 0.1 M acetic acid and reused three times before it should be discarded.

PROTOCOL 5

Analysis of Phosphopeptides by μLC-ESI-MS/MS

WHETHER USING A PRESSURE CELL (Figure 9.4) or gel-loader tips, IMAC-purified peptides (from Protocol 4) can be injected into a μLC separation system and analyzed by μLC-ESI-MS/MS (Becker et al. 1998; Katze et al. 2000). Phosphopeptides behave and elute similarly to nonphosphorylated peptides during RP-HPLC, but because of a decrease in hydrophobicity due to the addition of the phosphate moiety, a phosphopeptide will generally elute before the nonphosphopeptide with the identical amino acid sequence. This difference in elution time is most pronounced for short peptides. Thus, when analyzing phosphopeptides from μLC on-line with ESI-MS/MS, it is wise to acquire data during the wash step (i.e., prior to gradient elution), because some short phosphopeptides may not even bind to the C_{18} resin in 100% aqueous solutions. Otherwise, these very hydrophilic peptides will be missed (however, see Figure 8.24 and accompanying text in Chapter 8).

The following protocol, supplied by Garry L. Corthals, Valerie C. Wasinger, and David R. Goodlett (Geneva Proteomics Centre, Geneva University Hospital, Geneva, Switzerland; University of New South Wales, Kensington, Australia; and Institute for Systems Biology, Seattle, Washington, respectively), provides typical μLC conditions for the separation of IMAC-enriched phosphopeptides when the μLC system is on-line with the ESI-MS/MS.

MATERIALS

CAUTION: See Appendix 3 for appropriate handling of materials marked with <!>.

Reagents

Solvent A

- 0.4% acetic acid <!>
- 0.005% aqueous heptafluoro-butyric acid (HFBA) <!>
- H_2O

Solvent B

- 80% aqueous acetonitrile <!>

  Only acetonitrile is used for Solvent B, as the addition of acid here is unnecessary for good chromatographic separation of peptides (acid is used with systems that monitor separation via UV detection to avoid a rise in the base line with increasing acetonitrile). The addition of acid adds substantially to chemical noise in the mass spectrometer.

Equipment

HPLC system

Mass spectrometer capable of tandem MS analysis of phosphopeptides

Suitable mass spectrometers include a triple-quadrupole MS, an ion-trap MS, or a quadrupole-TOF-MS (e.g., ThermoFinnigan, Micromass, Applied Biosystems, or Bruker Daltonics) (see Chapter 8).

Micro-ESI interface

A micro-ESI interface can be homemade (Goodlett et al. 2000b) or purchased (New Objective, Inc., Woburn, Massachusetts).

Micro-LC column

Choose a 50-µm (I.D.) microcapillary column for optimum sensitivity or a 100-µm (I.D.) microcapillary column for optimum ruggedness, pressure packed, in either case, with, e.g., ODS-AQ (Waters Corp.). Columns can be purchased from various vendors (e.g., LC Packings, The Netherlands or Michrom Bioresources or New Objective) or homemade (see Chapter 8, Protocols 4 and 5).

Biological Sample

Peptides eluted from an IMAC column (Protocol 4, Step 15)

METHOD

1. Set up, wash, and equilibrate the µLC column according to the manufacturer's instructions.

   The column flow rate depends on the column used, but in general, ~200 nl/min and 500 nl/min across 50 and 100 µm (I.D.), respectively, is achieved by restrictive flow splitting using a length (empirically determined) of capillary tubing set in a splitting tee before the capillary column. This setup allows normal-flow HPLC pumps, capable of operating between 0.5 and 4.0 ml/min, to be used for microcapillary column operation.

2. Load the protein sample (from Protocol 4, Step 15) onto the µLC column.

3. Elute the peptides using a linear gradient from 0% to 60% Solvent B over 50 minutes. The eluting peptides from the column are introduced directly into the MS via a micro-ESI interface.

4. Peptides for CID are subsequently selected for fragmentation by the MS using automated data-dependent MS/MS procedures, optional in most MS control software packages.

In laboratories where a tandem mass spectrometer is not available for sequence analysis of phosphopeptides by CID, phosphatase treatment (Zhang et al. 1998) of IMAC-enriched peptide samples has been used in conjunction with MALDI-TOF analysis to identify the phosphopeptides in the sample (see Protocol 2). The general steps are:

- Record the collective masses of all peptides enriched by micro-IMAC in the single-stage MALDI-TOF-MS.
- Treat another aliquot of the peptide sample with a broad-specificity phosphatase (e.g., calf intestinal phosphatase) in solution. Alternatively, treat the sample still on the MALDI probe with the phosphatase.
- After an appropriate incubation time (which is determined by the properties of the phosphatase), record the masses of the treated protein sample.
- Compare the MALDI-TOF mass spectra for the original untreated and phosphatase-treated samples to identify any mass that shows a decrease of 80 amu (or multiples thereof), implying loss of phosphate due to phosphatase treatment.

This experiment is technically simple, and the interpretation of the spectra is straightforward due to the fact that MALDI-TOF-MS produces almost exclusively singly protonated peptide ions.

PROTOCOL 6

Tyrosine Phosphorylation Site Identification by MALDI-MS

THREE DIFFERENT STRATEGIES UTILIZING MS FOR THE IDENTICATION of protein phosphorylation sites are employed in Protocols 6–8. In this first protocol, radiolabeled proteins are separated using SDS-PAGE and digested in-gel. The phosphopeptides are subjected to two-dimensional LC and identified by MALDI-MS and MALDI-PSD. The identification of eight tyrosine phosphorylation sites in the human Gab-1 protein is provided as an example of the method's utility. The expression and phosphorylation of human Gab-1 with the insulin receptor kinase in vitro as described by Lehr et al. (2000) are summarized as an example of this approach to identifying phosphorylated tyrosine residues.

This protocol was provided by Albert Sickmann and Helmut E. Meyer (Medical Proteom Center, University of Bochum, Germany).

MATERIALS

CAUTION: See Appendix 3 for appropriate handling of materials marked with <!>.

Reagents

Acetonitrile <!>

Digestion buffer A (10 mM NH_4HCO_3 at pH 7.8) <!>

Digestion buffer B (10 mM NH_4HCO_3, 50% acetonitrile at pH 7.8)

Ion exchange buffer A (20 mM $H_3CCOONH_4$ at pH 7.0) <!>

Ion exchange buffer B (0.5 M KH_2PO_4 at pH 4.0) <!>

Matrix solution

The matrix solution is a saturated solution of α-cyano-4-hydroxy-cinnamic acid <!> in 0.1% TFA <!>, 50% acetonitrile (v/v).

Protease solution (0.05 µg/µl trypsin in Digestion buffer A)

Use sequencing grade, modified trypsin (e.g., from Promega).

µRP-HPLC buffer A (0.1% TFA)

µRP-HPLC buffer B (0.08% TFA, 84% acetonitrile)

Equipment

Bath sonicator, cooled to 4°C

Centrifugal evaporator (e.g., SpeedVac, Thermo Savant, or equivalent)

HPLC column (Nucleogel SAX 1000-8/46, 4.6 x 50 mm) (Macherey & Nagel, Germany)

HPLC system (e.g., Beckman Gold system)

MALDI-MS (e.g., Ultraflex, Bruker-Daltonic, Bremen, Germany)

MALDI target, stainless steel

Micro-HPLC pump (e.g., ABI 145D or Dionex Ultimate)

The micropump must be capable of generating a reproducible gradient at ≥20 µl/min.

Test tubes (borosilicate, Na^+- and K^+-free)

UV detector (e.g., Shimadzu SPD 10 A, Tokyo, Japan)
Zero death volume T-split

Biological Sample

Radiolabeled phosphorylated (as described by Lehr et al. 2000) protein(s) <!> of interest within a polyacrylamide gel

Separate the phosphorylated protein by 10% SDS-PAGE (see Chapter 2, Protocol 1).

METHOD

Digestion of the SDS-PAGE Separated Gab-1

1. Dissect gel pieces from the polyacrylamide gel that contain phosphorylated protein bands as revealed in the autoradiography.
2. Dice the gel pieces into 1-mm^2 pieces, and place them test tubes.

   Use specific Na^+- and K^+-free test tubes for digestion, otherwise Na^+ and K^+ adducts will be found during mass spectrometry.
3. Wash the gel pieces in Digestion buffer A for 10 minutes at 37ºC.
4. Wash the gel pieces in Digestion buffer B for 10 minutes at 37ºC.
5. Repeat Steps 3 and 4 twice more.
6. Shrink the gel pieces at room temperature by addition of acetonitrile until they are white (~3 minutes).
7. Swell the gel pieces with protease solution for 20 minutes at 37ºC.

   The temperature depends on the protease (e.g., Glu-C requires 25°C and trypsin requires 37°C).
8. Cover the gel pieces with additional Digestion buffer A to prevent drying, and incubate the gel pieces overnight at 37ºC.

Extraction of the Gel Pieces for Anion-exchange Chromatography

9. Add 20 µl of Digestion buffer A to the gel piece and incubate in a sonication bath for 20 minutes at 4ºC.
10. Collect the supernatant.
11. Repeat Steps 1 and 2 twice more.

Anion Exchange Chromatography

12. Adjust the flow rate through the anion exchange column to 0.5 ml/min.
13. Inject the combined supernatants onto the column.

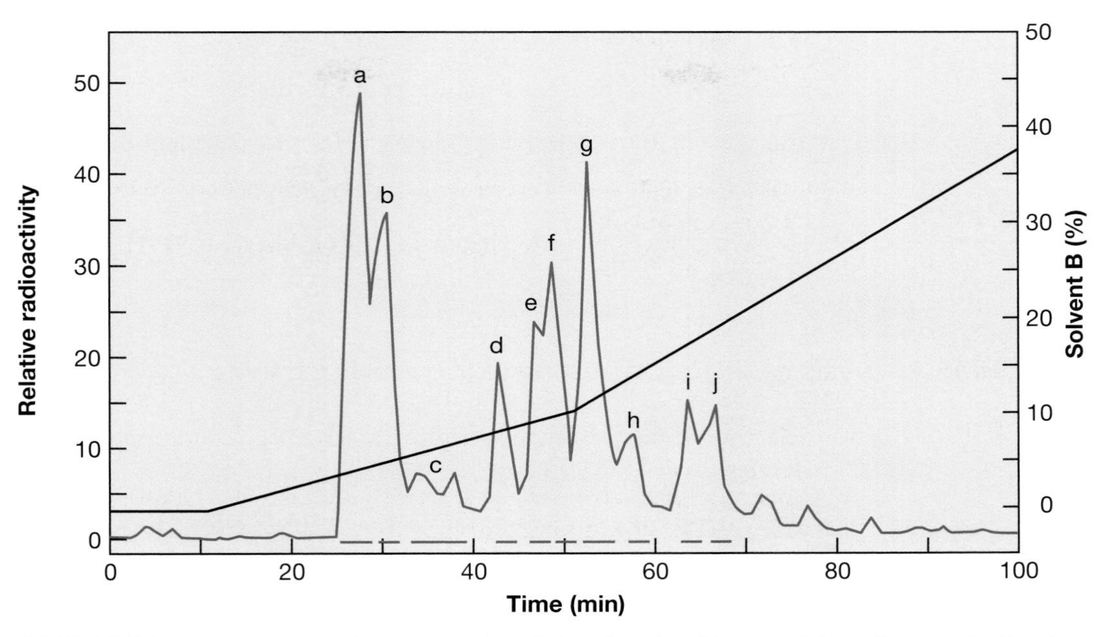

FIGURE 9.14. Anion exchange chromatography of phosphorylated human Gab-1 after tryptic digestion. Fractions of 0.5 ml are collected and the radioactivity is measured and plotted against the retention time. Fractions *a–j* are pooled for further chromatography.

14. Wash the column with Ion exchange buffer A for 20 minutes.
15. Elute peptides from the column by beginning a gradient of Ion exchange buffer B from 0% to 10% over the course of 40 minutes. Collect a 0.5-ml fraction every minute during the elution.
16. Increase the Ion exchange buffer B from 10% to 50% over the course of 75 minutes.
17. Measure the radioactivity in every fraction either using Cerenkov counting or by measuring the radioactivity of fraction aliquots amplified with scintillant.
18. Plot the radioactivity of each fraction versus the fraction number (see Figure 9.14 for an example).

RP-HPLC Chromatography of Radioactive Fractions

19. Select a radioactive fraction from the anion exchange chromatography eluent.
20. Reduce the fraction volume from 500 µl to 80 µl in a centrifugal evaporator.
21. Adjust the µ-HPLC pump flow rate to 70 µl/min.
22. Adjust the µ-HPLC column flow rate to 16 µl/min using a zero death volume T-split.
23. Inject the sample into the 80-µl sample loop.
24. Wash the column with 95% µRP-HPLC buffer A for 30 minutes.
25. Adjust the UV detector to zero. Begin UV detection of the RP-HPLC eluent at 215 nm and 295 nm with a photometer sampling rate of 2 Hz.

26. Begin a gradient of μRP-HPLC buffer B from 5% to 50% over the course of 90 minutes.

27. Collect 8–16-μl fractions from the column eluate.

28. Transfer 0.5 μl of each fraction to a vial for measurement of radioactivity.

    Fraction volume will depend on the size of the peak.

29. Freeze the fractions immediately at –70°C. A typical RP-HPLC chromatogram is shown in Figure 9.15.

MALDI-MS Analysis of Fractions Containing Radioactivity

30. Select a radioactive fraction from the RP-HPLC, and apply 0.3 μl of the fraction to a stainless steel MALDI target.

31. Apply 0.3 μl of matrix solution to the droplet. Allow the sample/matrix to air dry (~1 minute)

32. Record a MALDI-MS spectrum in linear and reflector mode (see Figure 9.16).

33. Compare the in silico peptide masses obtained from a MALDI-MS spectrum with experimentally determined masses obtained from a theoretical digest (e.g., GPMAW, Protein-Prospector) of the phosphorylated protein. Take into consideration:

    - methionine oxidation
    - amino-terminal acetylation
    - phosphorylation of S, T, and Y (in this example, Y phosphorylation)
    - generation of pyroglutamic acid at XXRQXX and XXKQXX
    - incomplete digest

34. Select probable phosphopeptides to be analyzed by MALDI-PSD. Four different MALDI-PSD spectra of phosphotyrosine peptides are shown in Figures 9.17 and 9.18. The common fragmentation pattern of peptides is shown in Figure 9.19 (Biemann 1992).

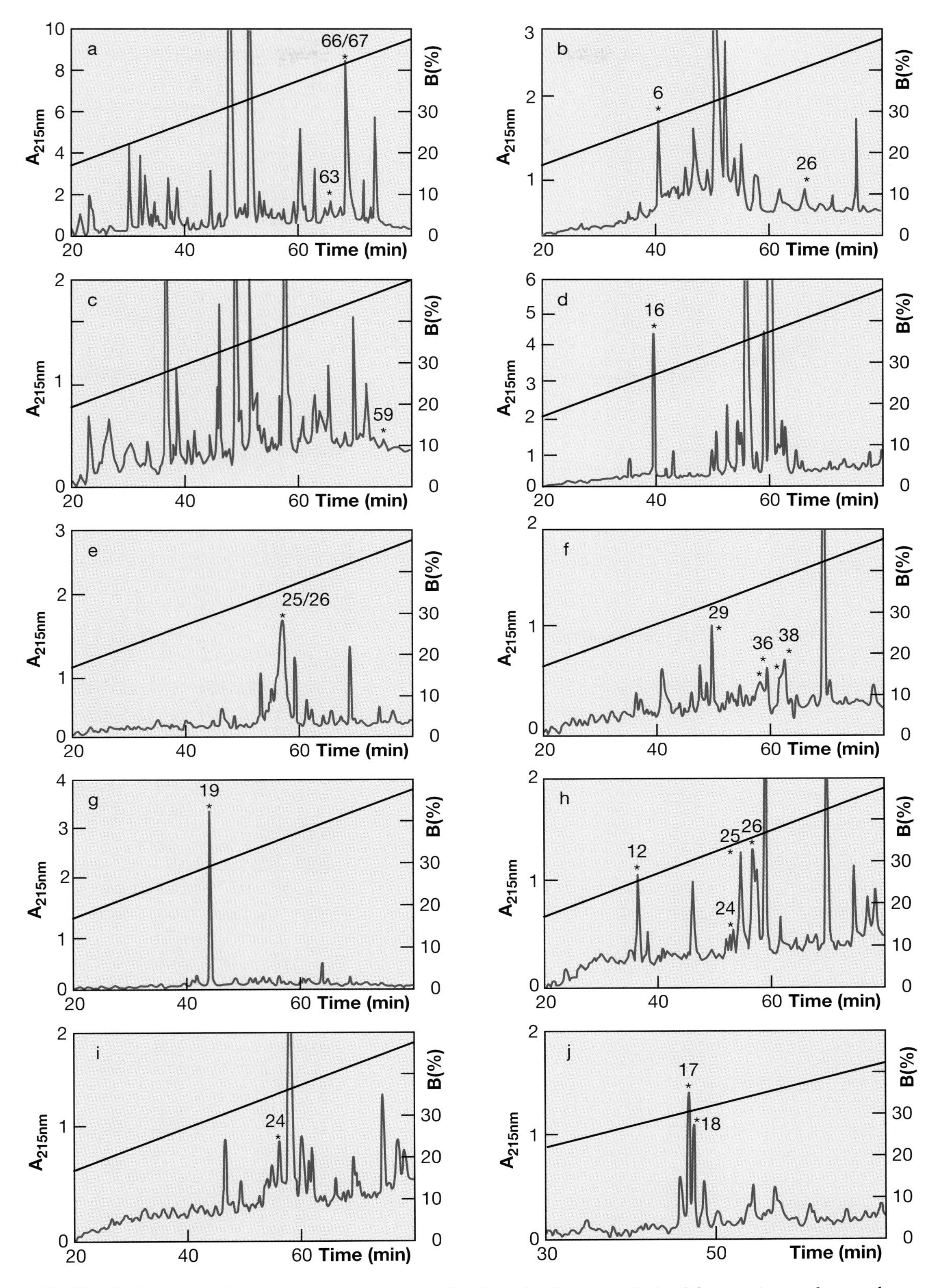

FIGURE 9.15. Reversed-phase chromatography of radioactive fractions derived from anion exchange chromatography (see Figure 9.14). The UV trace at 215 nm (peptide backbone) is plotted against the retention time. Fractions containing radioactivity are marked with an asterisk and subjected to mass spectrometry.

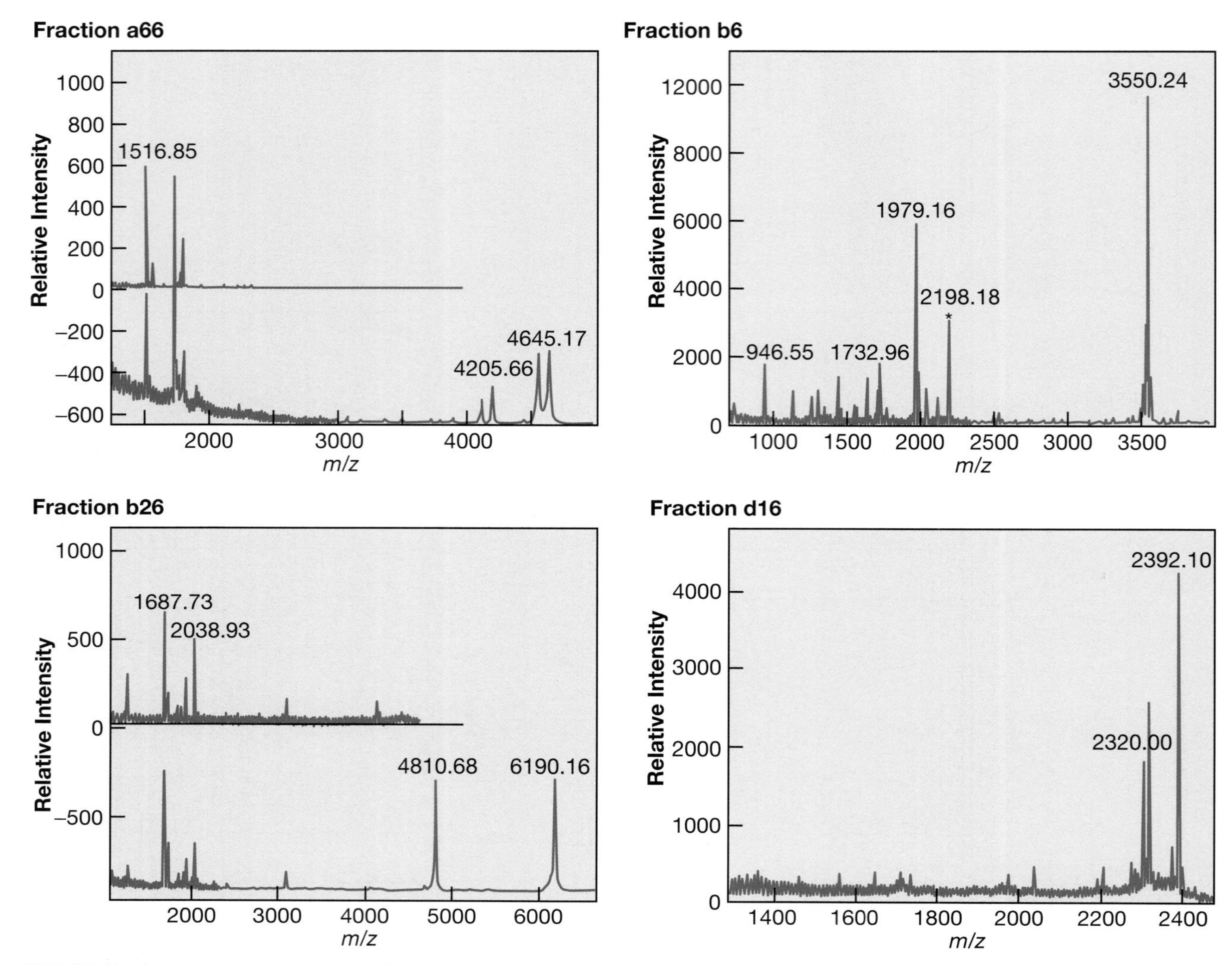

FIGURE 9.16. MALDI-MS spectra of four different phosphopeptide-containing fractions (see Figure 9.15). In fractions a66 and b26, additional measurements in the linear mode are necessary because no potential phosphopeptides are detected in the reflector mode. Especially the large peptides of >4000 daltons can only be detected in the linear mode due to several proline residues in the primary structure of the peptides. Fragment ion spectra of additional peptides are shown in Figures 9.17 and 9.18.

TABLE 9.3. Sequence analysis results of hGab-1 phosphopeptides

Amino acid sequence of identified phosphopeptides	$[M+H]^{+1a}$	Amino acids in hGab-1	Amino acids	Fraction[b]	^{32}P (%)[c]
HGMNGFFQQQMI**Yp**DSPPSR	2320.0	230-248	Y 242	b26, d16	1.8
VSPSSTEADGEL**Yp**VFNTPSGTSSVETQMR	3156,4	273-301	Y 285	f29	1.7
TASDTDSS**Yp**CIPTAGMSPSR	2197.8	365-384	Y 373	b6	2.7
NVLTVGSVSSEELDEN**Yp**PMNPNSPPR	3024.4	431-457	Y 447	h12, g19, c59	27.6
QHSSSFTEPIQEAN**Yp**VPMTPGTFDFSSFGM QVPPPAHMGFR	4367.0	458-498	Y 472	a63, a66-67, f35-38	30.7
FPMSPRPDSVHSTTSSSDSHDSEEN**Yp**VPM NPNLSSEDPNLFGSNSLDGGSSPMIK	6186.6	594-648	Y 619	b26, e25-26, h24-26, i24	17.8
QVE**Yp**LDLDLDSGK	1574.7	654-666	Y 657	j17-18	6.1
SSGSGSSVADERVD**Yp**VVVDQQK	2392.1	675-696	Y 689	d16	11.6

[a]Monoisotopic mass of the single charged ion.
[b]Fraction after anion exchange chromatography (letters)/RP-HPLC (numbers) (see Figures 9.15 and 9.16).
[c]Total amount of radioactivity incorporated into the phosphopeptides corresponds to 100%.

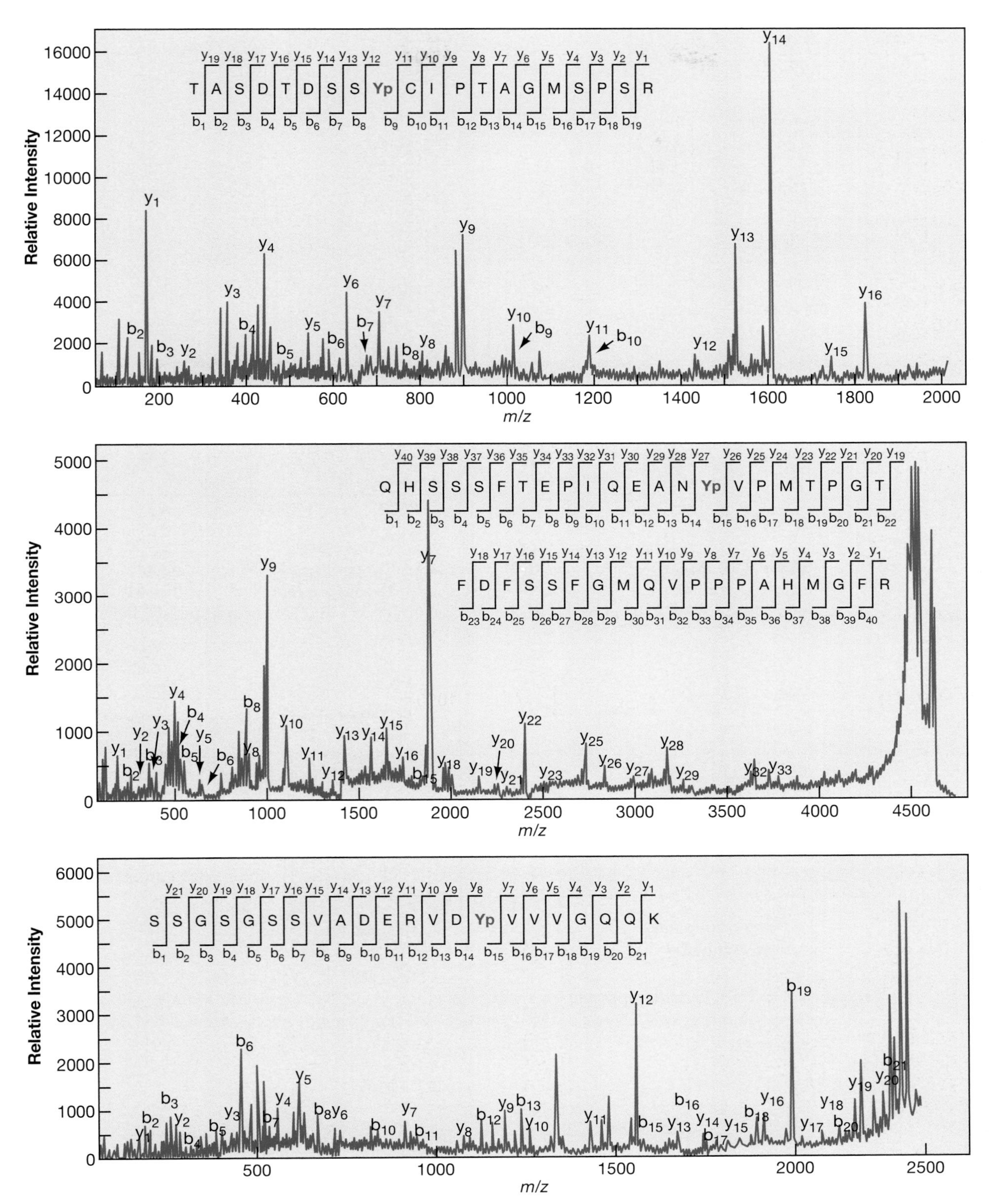

FIGURE 9.17. Three different MALDI-PSD spectra of phosphotyrosine-containing peptides. In each of these spectra, the phosphorylated tyrosine residue is unambiguously identified by b and/or y ions. In the case of the large peptide in the middle of the figure, peptide signals can be detected only in the linear mode. The ion selector was set on the mass 4367.0 (see Table 9.3) and a MALDI-PSD spectrum is recorded in the reflector mode. This shows that PSD spectra can be recorded even when no signal is achieved in the reflector mode.

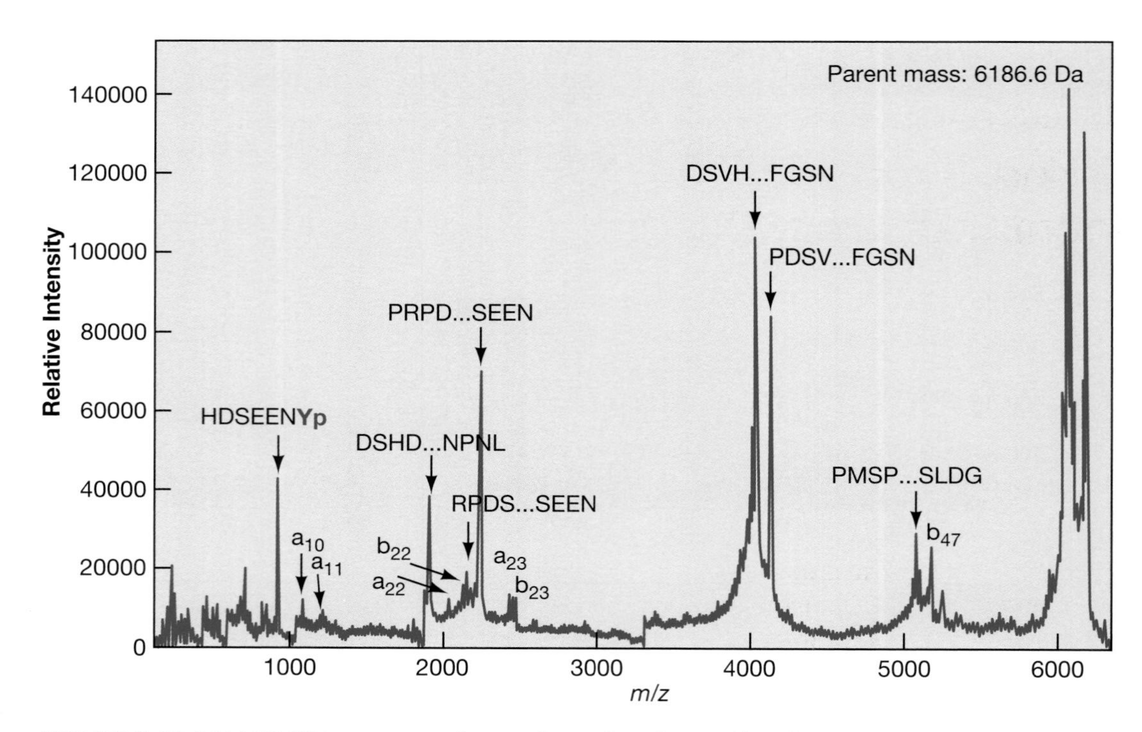

FIGURE 9.18. MALDI-PSD spectrum of a very large phosphopeptide. All major signals in this spectrum are derived from internal cleavages and only some a and b ions are generated. Together with Edman degradation (Edman 1949), the phosphorylation can be localized at the tyrosine residue (see Table 9.3).

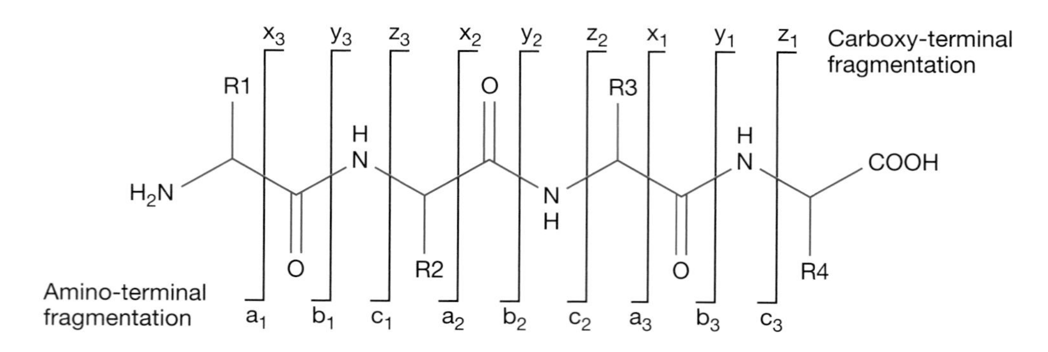

FIGURE 9.19. Common peptide fragmentation. Using MALDI-PSD and ESI-MS/MS, a, b, and y ions are the most common ions. (Modified, with permission, from Biemann 1992 [©Annual Reviews www. Annual Reviews.org].)

PROTOCOL 7

Serine/Threonine Phosphorylation Site Identification by ESI-MS

MUCH OF THE PROTEIN PHOSPHORYLATION WITHIN CELLS is mediated by the activity of independent kinases. However, autophosphorylation of proteins is another way signal transduction pathways are regulated. This protocol is designed to distinguish between autophosphorylation and endogenous phosphorylation. For the localization of all endogenous phosphorylation sites, the protein is digested with trypsin and the peptides are analyzed by on-line coupling of nano-HPLC and ESI-MS. To determine the autophosphorylation sites, the protein is incubated with radioactive ATP prior to digestion with trypsin. Only 20 pmoles of protein is used for each experiment. The stoichiometry of phosphorylated protein is below 5% of the total protein. Prior to analysis, the protein is separated by 10% SDS-PAGE and the protein band is excised. Before digestion, the protein is reduced and alkylated in both cases. This protocol was provided by Albert Sickmann and Helmut E. Meyer (Medical Proteom Center, University of Bochum, Germany).

MATERIALS

CAUTION: See Appendix 3 for appropriate handling of materials marked with <!>.

Reagents

Acetonitrile <!>
Alkylation solution
  40 mM iodoacetamide <!>
  0.1 M Tris-Cl (pH 8.2)
Digestion buffer A (10 mM NH_4HCO_3 at pH 7.8) <!>
Digestion buffer B (10 mM NH_4HCO_3, 50% acetonitrile at pH 7.8)
Formic acid (5%) <!>
β-Mercaptoethanol <!>
Nano-RP-HPLC buffer A (0.1 % formic acid)
Nano-RP-HPLC buffer B (0.1% formic acid, 84% acetonitrile)
Protease solution (0.05 μg/μl trypsin in Digestion buffer A)
  Use sequencing grade, modified trypsin (e.g., from Promega).
Reduction solution
  80 mM dithiothreitol (DTT) <!>
  0.5 mM guanidinium-HCl <!>
  0.8 mM EDTA
  0.1 M Tris-Cl (pH 8.2)
Trifluoroacetic acid (TFA) (0.1%) <!>

Equipment

Bath sonicator, cooled to 4°C
HPLC pump (e.g., ABI 140D, Perkin Elmer)
HPLC system (e.g., Beckman Gold system)
Mass spectrometer, ESI-MS with a nanospray ion source (e.g., LCQ, ThermoFinnigan MAT)
Nano-HPLC column (NAN 75-25-05-PM 75 μm x 250 mm) (LC-Packings, Amsterdam, The Netherlands)
Precolumn (0.5 x 5 mm)
The precolumn and separation column are packed with the same reversed-phase resin (see Tables 5.1 and 5.3).
Precolumn splitter
SEQUEST software
Test tubes (borosilicate, Na^+- and K^+-free)

Biological Molecules

Radiolabeled phosphorylated protein(s) <!> of interest contained within polyacrylamide gel piece(s)
Radiolabel the proteins as described by Suer et al. (2001). Separate the phosphorylated proteins by 10% SDS-PAGE (see Chapter 2, Protocol 1). Excise the gel piece containing the protein of interest as revealed by autoradiography.

METHOD

Reduction, Alkylation, and Digestion of the Phosphoprotein

1. Wash the gel piece in Digestion buffer A in Na^+/K^+-free test tubes for 10 minutes at 37°C.
2. Cover the gel piece with reduction solution for 15 minutes at 37°C.
3. Remove the reduction solution, and cover the gel piece with alkylation solution for 15 minutes at 37°C in the dark.

   Avoid longer incubation times, because methionine is alkylated by iodoacetamide (Sickmann et al. 2000).
4. Add 5 μl of β-mercaptoethanol to the alkylation solution covering the gel piece. Incubate for 5 minutes more at 37°C.
5. Remove the liquid from the container, and wash the gel piece in Digestion buffer A for 10 minutes at 37°C.
6. Discard Digestion buffer A, and wash the gel piece in Digestion buffer B for 10 minutes at 37°C.
7. Repeat Steps 5 and 6 twice more.
8. Shrink the gel piece by addition of acetonitrile until it is white (~3 minutes).
9. Re-swell the gel piece with protease solution for 20 minutes at 37°C.
10. Cover the gel piece with additional Digestion buffer A to prevent drying.
11. Incubate the digest overnight at 37°C.

Sample Concentration

Loading pump
(30 µl/min)
Waste
A
1-4
B
2-3
Precolumn
(71 nl)

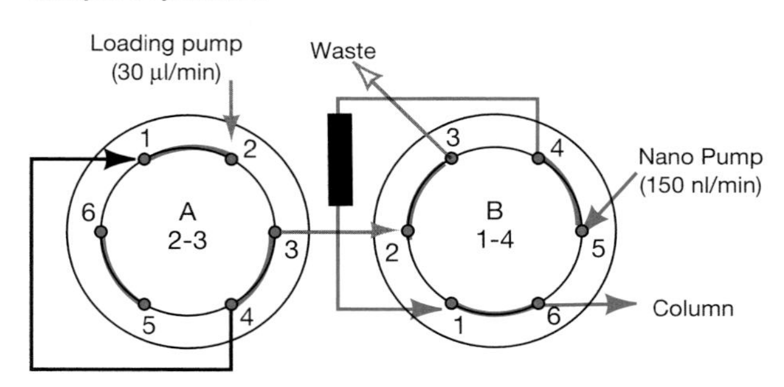

FIGURE 9.20. On-line sample concentration prior to nano-LC-MS. (*Left*) Sample concentration. The sample is injected into the sample loop of valve A and concentrated on the precolumn. Valve B is set to the 2–3 position (waste). Concentration of the sample is performed at a flow rate of 60 µl/min for 10 minutes. The volume of all capillaries and the precolumn is <2 µl (sample loop: additional 30 µl). (*Right*) Separation of the sample. Valve A is in position 2–3 (no loop) and valve B in position 1–4 (nano-column in the flow). The flow rate is reduced from 60 µl/min to 100–200 nl/min.

Extraction of the Gel Piece for Nano-HPLC Separation

12. Discard the digest buffer surrounding the gel piece. Cover the gel piece with 15 µl of 5% formic acid, and incubate it for 20 minutes in a cooled sonication bath.
13. Collect the liquid from around the gel. This contains the digested peptides.
14. Repeat Steps 12 and 13.

Precolumn Concentration of Peptides during Nano-HPLC

Commonly, nano-HPLC is done with flow rates <0.3 µl/min. During the peptide extraction, a volume of ~30 µl is achieved, which is much too large for nano-HPLC. Therefore, a special peptide concentration step as shown in Figure 9.20 is necessary. The major advantage of this method is the on-line concentration of the sample resulting in a low injection volume.

15. Inject the sample into the sample loop. Concentrate the sample onto the μC_{18} precolumn using 0.1% TFA with a flow rate of 60 µl/min for 10 minutes.
16. After concentration, move valve B into the 1-4 position into the pump flow.
17. Elute the peptides as follows:
    a. Adjust the pump flow to 70 µl/min (if using the ABI 140D system).
    b. Adjust the column flow to 100–200 nl/min with a precolumn splitter.
    c. Equilibrate the 75-µm I.D. column with Nano-RP-HPLC buffer A for 10 minutes.
    d. Elute the peptides with a gradient of Nano-RP-HPLC buffer B from 5% to 50% over 90 minutes.

LC-MS/MS Analysis

18. Load the peptides onto an ESI mass spectrometer using a nanospray ion source and a micro-ESI interface.
19. Record the MS spectra with a triple-scan event (full scan, first dependent MS/MS scan, second dependent MS/MS scan) with a scan duration of 2 seconds for each event.

Automatic Data Interpretation with the SEQUEST Algorithm

20. Automatically process the LC-MS/MS raw data file using the SEQUEST algorithm.

Using this protocol to study the phosphorylation of human PI4K92 protein, the first step generated nearly 800 DTA files from the raw data file (settings: grouped scans 2; minimal ion intensity: 1 E-5; mass range: 300–3500 daltons; charge: single-, double-, or triple-charged ion). The search is done against the actual NCBI database (settings: enzyme: trypsin KR/P; missed cleavage sites: 1; modification: C+57 daltons; differential modification: M+16; STY+80 daltons).

In Figure 9.21, the total ion current (TIC, i.e., the intensity of all ions in each scan plotted against the retention time) and the BASE PEAK chromatogram (BASE, i.e., the intensity of the base peak in each scan plotted against the retention time) are shown. The base peak chromatogram is comparable to the UV trace of a photometer. Some automatically interpreted MS/MS spectra of the phosphopeptides RRLSpEQLAHTPTAFK and TASpNPKVENEDEPVR are shown in Figure 9.22.

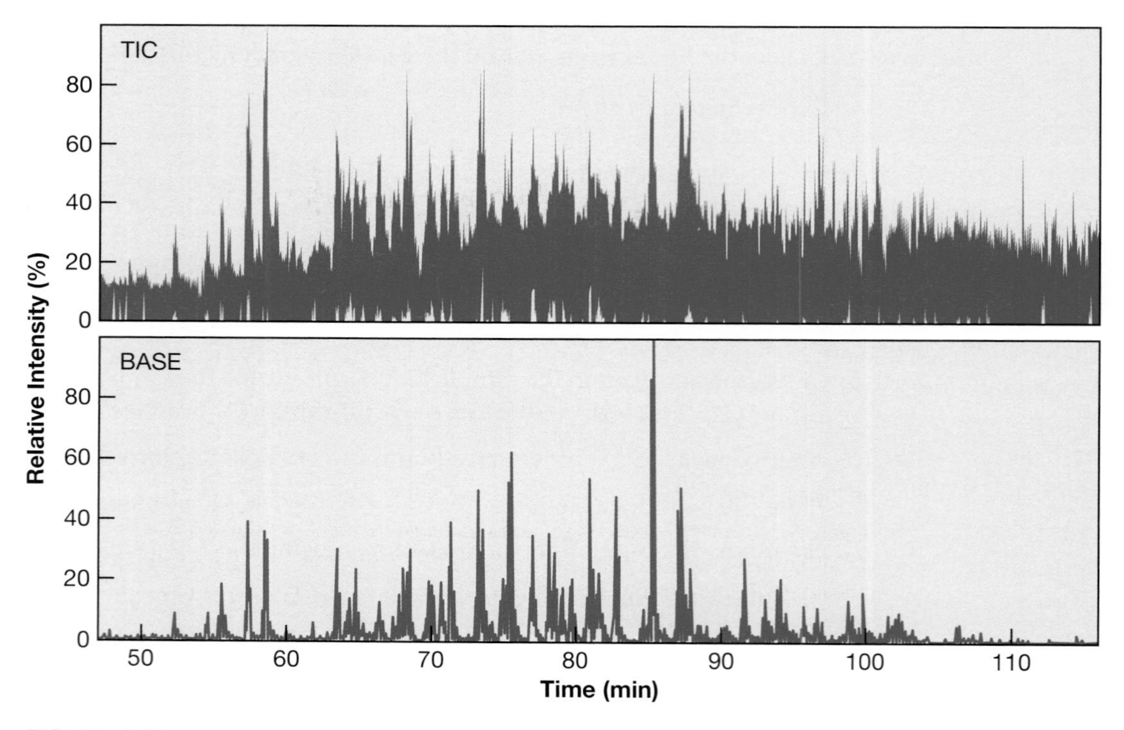

FIGURE 9.21. Nano-LC-MS/MS analysis of PI4K92 phosphopeptides. (*Top*) Total ion chromatogram (TIC). The intensity of all ions in every mass spectrum is plotted against the retention time. (*Bottom*) Base peak chromatogram. The intensity of the base peak in every mass spectrum is plotted against the retention time. This chromatogram is comparable to a UV trace of a HPLC separation, and signals of several peptides with different retention times can be observed.

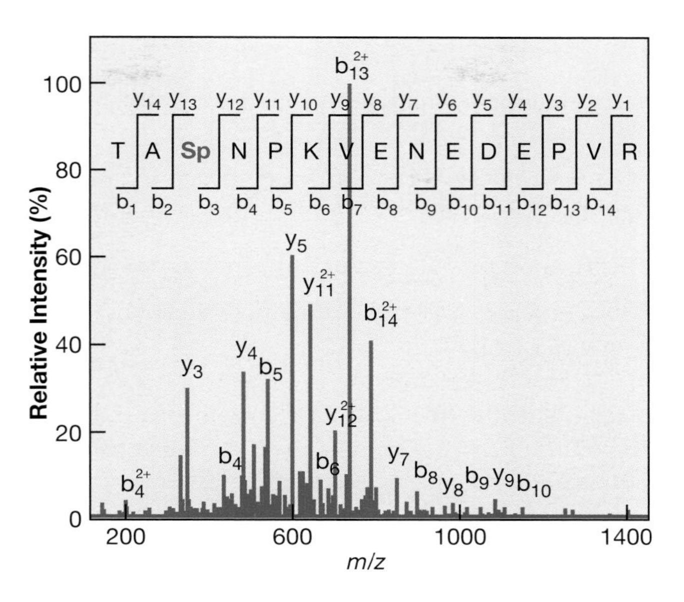

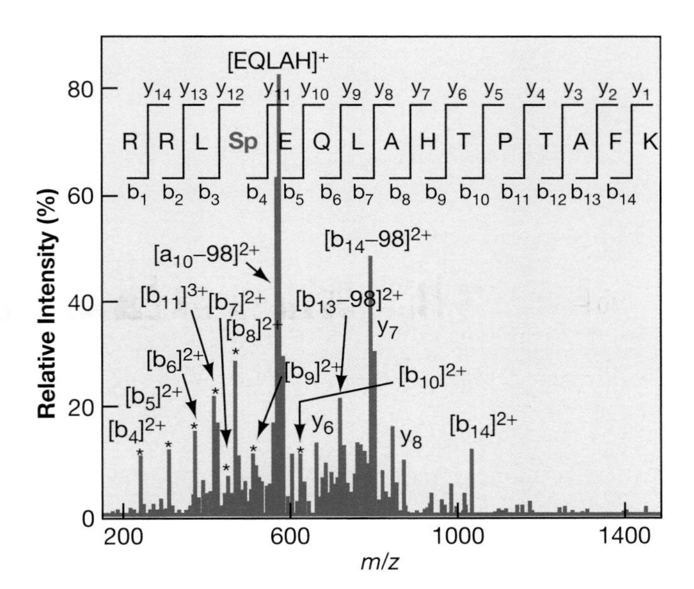

FIGURE 9.22. ESI-MS/MS spectra of two different phosphopeptides derived from human PI4K92 (see Figure 9.21). Surprisingly, no remarkable loss of 80 or 98 daltons is observed in the spectrum of the peptide TASpNPKVENEDEPVR. The phosphorylation site can be localized to the serine residue. The common behavior of phosphopeptides is demonstrated by the peptide RRLSpEQLAHTPTAFK. Several cleavage products are indicated by asterisks (*).

Additional Protocol: Identification of Autophosphorylation Sites Using Radioactive Phosphate

To localize an autophosphorylation site (in this case, of the human PI4K92 protein), incubate the protein with radioactive ATP <!>.

Additional Materials

CAUTION: See Appendix 3 for appropriate handling of materials marked with <!>.

μ-HPLC system (e.g., ABI 140D pump and Dionex Ultimate HPLC system)
MALDI mass spectrometer
μRP-HPLC buffer A (0.1% TFA) <!>
μRP-HPLC buffer B (0.08% TFA, 84% acetonitrile) <!>
RP-HPLC column (0.18 x 150 mm; C_{18}; 3- or 5-μm particle size)
Step 1 requires the reagents and equipment listed in Chapter 2, Protocol 1.

Method

1. Separate the reaction mixture by 10% SDS-PAGE.
2. Reduce, alkylate, and then digest the protein within the gel piece as described in the main protocol.
3. Extract the peptides twice with 10 μl of 1% TFA solution in a sonication bath for 20 minutes at 4°C.
4. Inject the digested peptides into the sample loop of a μ-HPLC system.
5. Separate the peptides as follows:
   a. Adjust the μ-HPLC pump flow to 70 μl/min.
   b. Adjust the μ-HPLC column flow to 1.5 μl/min using a zero death volume T-split.
   c. Inject the sample into the 20-μl sample loop.
   d. Wash the column with 95% μRP-HPLC buffer A for 30 minutes.
   e. Adjust the UV detection to zero. Begin monitoring the eluent at 215 nm and 295 nm with a photometer sampling rate of 2 Hz.
   f. Elute the peptides with a gradient of μRP-HPLC buffer B from 5% to 50% over 90 minutes.
6. Collect 1–1.5-μl fractions of the column eluate, and immediately freeze them at –70°C.
7. Assay for radioactivity in each of the fractions, using the entire sample.

   Prompt freezing of the samples reduces peptide loss by adsorption to the surface of the collection tubes.
8. Analyze the radioactive fractions by MALDI-MS as described in Protocol 6, Steps 31–34.

Figure 9.23 shows the linear and reflector mode spectra together with the recorded PSD spectra. Typical for serine and threonine phosphate is the side-chain cleavage of the phosphate group during mass spectrometry. A summary of all identified phosphorylation sites in the human PI4K92 protein is given in Table 9.4.

TABLE 9.4. Sequence analysis results of PI4K92 phosphopeptides

Amino acid sequence of the identified phosphopeptides	$[M+H]^+$	Amino acid	Kind of phosphorylation
ELPSLSPAPD**T**GLSPSKR	1931.9	T 263	autophosphorylation
ELPSLSPAPDTGL**S**PSKR	1931.9	S 266	endogenous
ELPSL**S**PAPDTGLSPSKR	1931.9	S 258	endogenous
SK**S**DATASISLSSNLKR	1844.9	S 227	endogenous
TA**S**NPKVENEDEPVR	1764.7	S 294	endogenous
SVENLPECGI**T**HEQR	1848.8	T 423	autophosphorylation
RRL**S**EQLHTPTAFK	1763.8	S 496	endogenous
LSEQLHTP**T**AFK	1451.6	T 504	endogenous

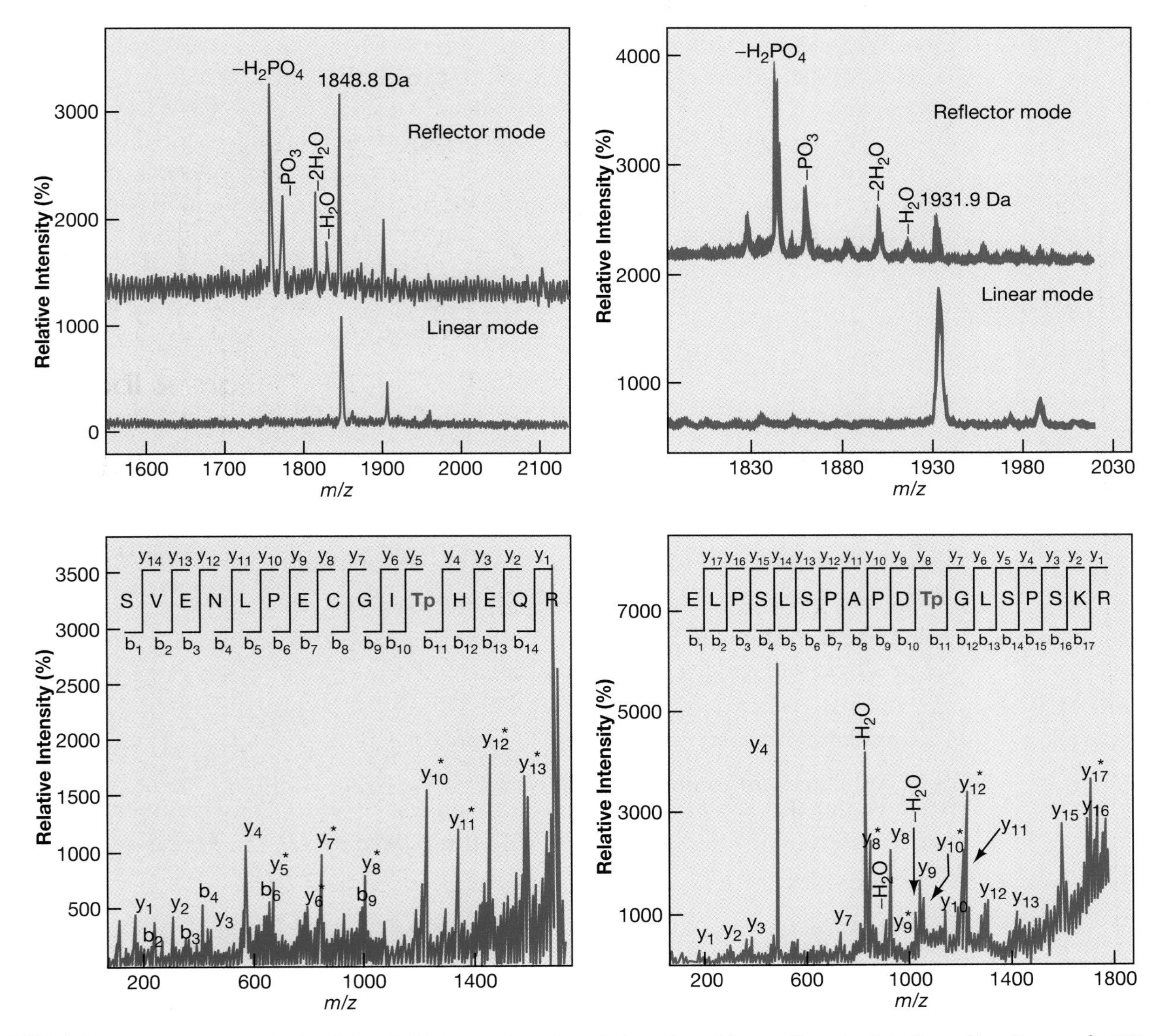

FIGURE 9.23. MALDI-MS analysis of the PI4K92 autophosphorylation sites. After radioactive labeling, digestion, and μLC separation of the peptides, these spectra are recorded using radioactive HPLC fractions. (*Top*) MALDI-MS spectra in reflector and linear mode. Phosphoserine and phosphothreonine undergo specific cleavage in the side chain of the phosphorylated amino acid, which can only be detected in the reflector mode. The linear spectrum does not show these products because they are generated after the ion acceleration. Therefore, the intact peptide and all products reach the ion detector at the same time. (*Bottom*) MALDI-PSD spectra of the ions with 1848.8 daltons and 1931.9 daltons. The specific cleavage products are marked with an asterisk (*).

PROTOCOL 8

Histidine Phosphorylation Site Identification by ESI-MS

HISTIDINE PHOSPHATE IS MODERATELY STABLE under alkaline conditions; however, the half-life of this phospho-amino acid under acidic conditions (pH <5) is less than 20 minutes. Thus, it is better to process proteins containing phosphohistidine rapidly and under alkaline conditions. The use of acidic matrices, such as α-cyano-4-hydroxy-cinnamic acid, for MALDI-MS prevents the detection of phosphohistidine. An alkaline MALDI matrix is a poor alternative because it often decreases the sensitivity of detection of phosphorylated amino acids. Surprisingly, phosphohistidine is stable under acidic conditions when stored on C_{18} column resins. Therefore, HPLC separation on C_{18} resin with an acidic solvent system combined with ESI-MS is possible. This protocol was provided by Albert Sickmann and Helmut E. Meyer (Institut für Physiologische Chemie der Ruhr Universität Bochum, Germany).

MATERIALS

CAUTION: See Appendix 3 for appropriate handling of materials marked with <!>.

Reagents

Nano-RP-HPLC buffer A (0.1% formic acid) <!>
Nano-RP-HPLC buffer B (0.1% formic acid, 84% acetonitrile) <!>
NH_4HCO_3 (50 mM, pH 7.8) <!> with 5% trypsin
Trifluoroacetic acid (TFA) (0.1%) <!>

Equipment

μC_{18} precolumn
HPLC system (e.g., ABI 140D, Dionex Ultimate)
Mass spectrometer, ESI-MS with a nanospray ion source (e.g., LCQ, ThermoFinnigan MAT)
Nano-HPLC column (NAN 75-25-05-PM 75 µm x 250 mm) (LC-Packings, Amsterdam, The Netherlands)
Precolumn splitter
SEQUEST software
UV detector (e.g., Shimadzu SPD 10 A, Tokyo, Japan)

Biological Sample

Phosphoprotein mixture of interest

METHOD

Digestion and Concentration of Peptides

1. Digest the protein mixture in 50 mM NH_4HCO_3 (pH 7.8) with 5% trypsin for 4 hours at 37°C.
2. Inject the digested peptides into the sample loop (see Figure 9.20).
3. Concentrate the sample onto the μC_{18} precolumn using 0.1% TFA with a flow rate of 60 µl/min for 10 minutes.
4. After concentration, move valve B into the 1-4 position into the pump flow.
5. Elute the peptides as follows:
   a. Adjust the pump flow to 70 µl/min (if using the ABI 140D system).
   b. Adjust the column flow to 100–200 nl/min with a precolumn splitter.
   c. Equilibrate the 75-µm I.D. column for 10 minutes with Nano-HPLC buffer A.
   d. Elute the peptides with a gradient of Nano-RP-HPLC buffer B from 5% to 50% over 90 minutes.
6. Load the peptides onto an ESI mass spectrometer using a nanospray ion source.
7. Record the MS spectra with a triple scan event (full scan, first dependent MS/MS scan, second dependent MS/MS scan) with a scan duration of 2 seconds for each event.

Automatic Data Interpretation with the SEQUEST Algorithm

8. Automatically process the LC-MS/MS raw data file using the SEQUEST algorithm.

   The search parameters are explained in the SEQUEST users guide.

Figure 9.24 shows the mass spectra of HPr before digestion. The raw ESI mass spectrum in panel a and the deconvoluted ESI mass spectrum of HPr in panel b show a mass shift of 80 daltons, which indicates monophosphoryated HPr. Finally, Figure 9.25 shows the MS/MS spectrum of the phosphohistidine peptide MEQNSYVIIDETGIHpAR. The database search is done against the NCBI database and leads to this histidine phosphorylated peptide.

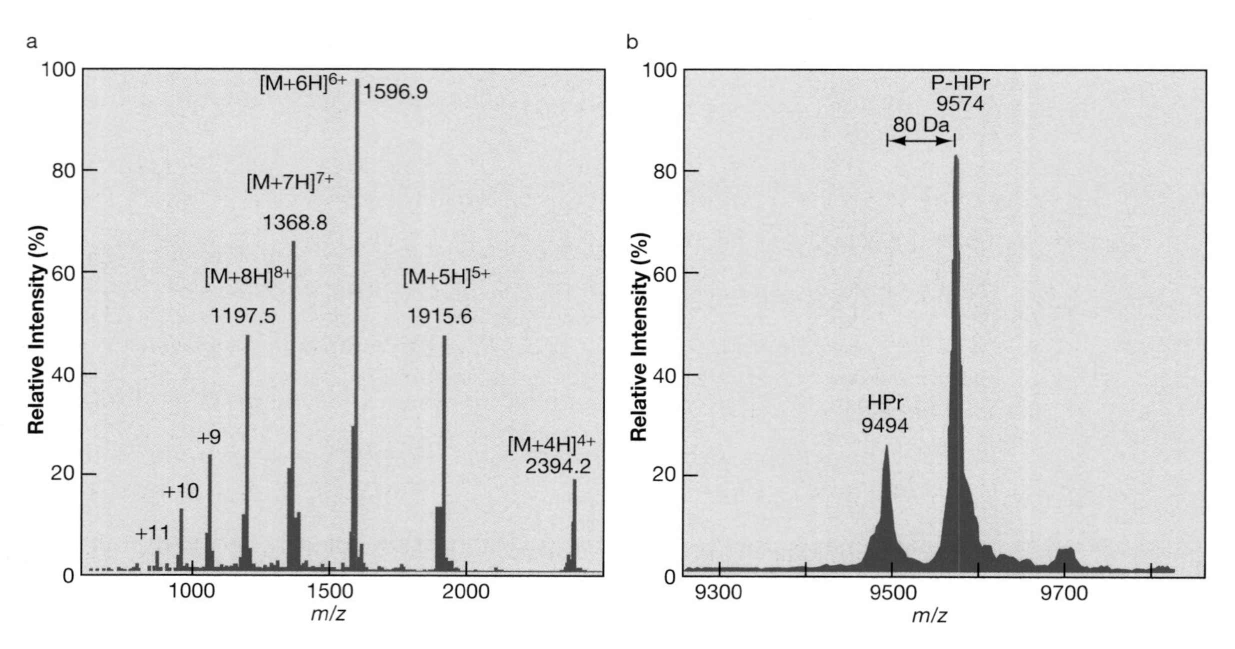

FIGURE 9.24. Nanospray-ESI spectrum of HPr before digestion. Approximately 5 pmoles of freshly phosphorylated HPr is used for low-flow nano-ESI-MS. The charge states 4 to 11 can be observed in the spectra. After deconvolution of the spectrum in *a*, the mass (9494 daltons) for HPr and the mass (9574 daltons) for phosphorylated HPr were calculated. Approximately 25% of the HPr is not phosphorylated.

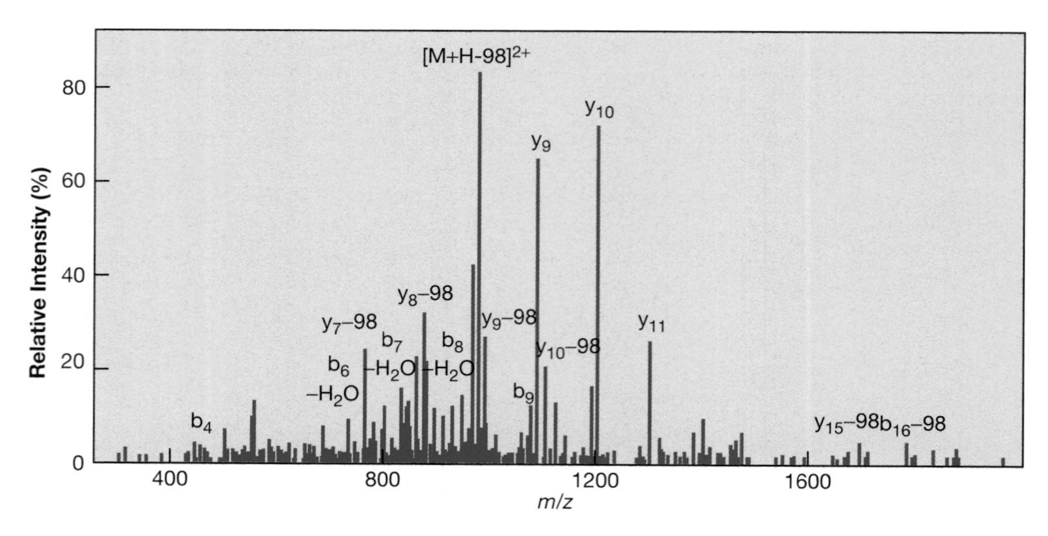

FIGURE 9.25. ESI-MS/MS spectrum of the phosphopeptide MEQNSYVIIDETGIHpAR. Localization of the phosphorylated histidine residue is easily done. The spectrum is recorded ~5 hours later than the spectrum in Figure 9.24, panel *a*. Due to the unstable phosphohistidine residue, ~5% of the HPr is phosphorylated at this point in time.

REFERENCES

Abu-lawi K.I. and Sultzer B.M. 1995. Induction of serine and threonine protein phosphorylation by endotoxin-associated protein in murine resident peritoneal macrophages. *Infect. Immun.* **63:** 498–502.

Adams J.A. 2001. Kinetic and catalytic mechanisms of protein kinases. *Chem. Rev.* **101:** 2271–2290.

Aebersold R. and Goodlett D. 2001. Mass spectrometry in proteomics. *Chem. Rev.* **101:** 269–295.

Aebersold R., Watts J.D., Morrison H.D., and Bures E. 1991. Determination of the site of tyrosine phosphorylation at the low picomole level by automated solid-phase sequence analysis. *Anal. Biochem.* **199:** 51–60.

Aebersold R., Figeys D., Gygi S., Corthals G., Haynes P., Rist B., Zhang Y., and Goodlett D.R. 1998. Towards an integrated analytical technology for the generation of multidimensional protein expression maps. *J. Prot. Chem.* **17:** 533–535.

Affolter M., Watts J.D., Krebs D.L., and Aebersold R. 1994. Evaluation of two-dimensional phosphopeptide maps by electrospray ionization mass spectrometry of recovered peptides. *Anal. Biochem.* **223:** 74–81.

Ahn N.G. and Resing K.A. 2001. Toward the phosphoproteome. *Nat. Biotechnol.* **19:** 317–318.

Amankwa L.N., Harder K., Jirik F., and Aebersold R. 1995. High-sensitivity determination of tyrosine-phosphorylated peptides by on-line enzyme reactor and electrospray ionization mass spectrometry. *Protein Sci.* **4:** 113–125.

Andersson L. and Porath J. 1986. Isolation of phosphoproteins by immobilized metal (Fe^{3+}) affinity chromatography. *Anal. Biochem.* **154:** 250–254.

Annan R.S. and Carr S.A. 1996. Phosphopeptide analysis by matrix-assisted laser desorption time-of-flight mass spectrometry. *Anal. Chem.* **68:** 3413–3421.

———. 1997. The essential role of mass spectrometry in characterizing protein structure: Mapping post-translational modifications. *J. Protein Chem.* **16:** 391–402.

Annan R.S., Huddleston M.J., Verma R., Deshaies R.J., and Carr S.A. 2001. A multidimensional electrospray MS-based approach to phosphopeptide mapping. *Anal. Chem.* **73:** 393–404.

Arad-Dann H., Beller U., Haimovitch R., Gavrieli Y., and Ben-Sasson S.A. 1993. Immunohistochemistry of phosphotyrosine residues: Identification of distinct intracellular patterns in epithelial and steroidogenic tissues. *J. Histochem. Cytochem.* **41:** 513–519.

Asara J.M. and Allison J. 1999. Enhanced detection of phosphopeptides in matrix-assisted laser desorption/ionization mass spectrometry using ammonium salts. *J. Am. Soc. Mass Spectrom.* **10:** 35–44.

Baldwin M.A., Medzihradszky K.F., Lock C.M., Fisher B., Settineri T.A., and Burlingame A.L. 2001. Matrix-assisted laser desorption/ionization coupled with quadrupole/orthogonal acceleration time-of-flight mass spectrometry for protein discovery, identification, and structural analysis. *Anal. Chem.* **73:** 1707–1720.

Bateman R.H., Carruthers R., Hoyes J.B., Jones C., Langridge J.I., Millar A., and Vissers J.P. 2002. A novel precursor ion discovery method on a hybrid quadrupole orthogonal acceleration time-of-flight (Q-TOF) mass spectrometer for studying protein phosphorylation. *J. Am. Soc. Mass Spectrom.* **13:** 792–803.

Becker S., Corthals G.L., Aebersold R., Groner B., and Muller C.W. 1998. Expression of a tyrosine phosphorylated, DNA binding Stat3beta dimer in bacteria. *FEBS Lett.* **441:** 141–147.

Bennett K.L., Stensballe A., Podtelejnikov A., Moniatte M., and Jensen O.N. 2000. Phosphopeptide analysis using a MALDI-QqTOF mass spectrometer. In *Proceedings of the 48th ASMS Conference on Mass Spectrometry and Allied Topics*, pp. 627-628. American Society for Mass Spectrometry, Long Beach, California.

———. 2002. Phosphopeptide detection and sequencing by matrix-assisted laser desorption/ionization quadrupole time-of-flight tandem mass spectrometry. *J. Mass Spectrom.* **37:** 179–190.

Benore-Parsons M., Seidah N.G., and Wennogle L.P. 1989. Substrate phosphorylation can inhibit proteolysis by trypsin-like enzymes. *Arch. Biochem. Biophys.* **272:** 274–280.

Biemann K. 1988. Contributions of mass spectrometry to peptide and protein structure. *Biomed. Environ. Mass Spectrom.* **16:** 99–111.

———. 1992. Mass spectrometry of peptides and proteins. *Annu. Rev. Biochem.* **61:** 977–1010.

Blume-Jensen P. and Hunter T. 2001a. Two-dimensional phosphoamino acid analysis. *Methods Mol. Biol.* **124:** 49–65.

——. 2001b. Oncogenic kinase signalling. *Nature* **411:** 355–365.

Bond J.A., Webley K., Wyllie F.S., Jones C.J., Craig A., Hupp T., and Wynford-Thomas D. 1999. p53-Dependent growth arrest and altered p53-immunoreactivity following metabolic labeling with ^{32}P ortho-phosphate in human fibroblasts. *Oncogene* **18:** 3788–3792.

Boyle W.J., Geer Van der P., and Hunter T. 1991. Phosphopeptide mapping and phosphoamino acid analysis by two-dimensional separation on thin-layer cellulose plates. *Methods Enzymol.* **201**: 110–149.

Cao P. and Stults J.T. 1999. Phosphopeptide analysis by on-line immobilized metal-ion affinity chromatography-capillary electrophoresis electrospray ionization mass spectromtry. *J. Chromatogr. A* **853:** 225–235.

——. 2000. Mapping the phosphorylation sites of proteins using on-line immobilized metal affinity chromatography/capillary electrophoresis/electrospray ionization multiple stage tandem mass spectrometry. *Rapid Commun. Mass Spectrom.* **14:** 1600–1606.

Carr S.A., Huddleston M.J., and Annan R.S. 1996. Selective detection and sequencing of phosphopeptides at the femtomole level by mass spectrometry. *Anal. Biochem.* **239:** 180–192.

Chen J., Qi Y., Zhao R., Zhou G.W., and Zhao Z.J. 2001. Assay of protein tyrosine phosphatases by using matrix-assisted laser desorption ionization time-of-flight mass spectrometry. *Anal. Biochem.* **292:** 51–58.

Chen S.L., Huddleston M.J., Shou W., Deshaies R.J., Annon R., and Carr S.A. 2002. Mass spectrometry-based methods for phosphorylation site mapping of hyperphosphorylated proteins applied to Net1, a regulator of exit from mitosis in yeast. *Mol. Cell. Proteomics* **1:** 186–196.

Cleverley K.E., Betts J.C., Blackstock W.P., Gallo J.-M., and Anderton B. H. 1998. Identification of novel in vitro pka phosphorylation sites on the low and middle molecular mass neurofilament subunits by mass spectrometry. *Biochemistry* **37:** 3917–3930.

Cohen P. 2002. The origins of protein phosphorylation. *Nat. Cell Biol.* **4:** E127–130.

Conrads T.P., Issaq H.J., and Veenstra T.D. 2002. New tools for quantitative phosphoproteome analysis. *Biochem. Biophys. Res. Commun.* **290:** 885–890.

Corthals G.L., Gygi S.P., Aebersold R., and Patterson S.D. 1999. Identification of proteins by mass spectrometry. In *Proteome research: 2D gel electrophoresis and detection methods* (ed. T. Rabilloud), pp. 197–231. Springer, New York.

Covey T.R., Huang E.C., and Henion J.D. 1991a. Structural characterization of protein tryptic peptides via liquid chromatography/mass spectrometry and collision-induced dissociation of their doubly charged molecular ions. *Anal. Chem.* **63:** 1193–2000.

Covey T., Shushan B., Bonner R., Schröder W., and Hucho F. 1991b. LC/MS and LC/MS/MS screening for the sites of post-translational modifications in proteins. In *Methods in protein sequence analysis* (ed. H. Jörnvall et al.), pp. 249–256. Birkhäuser Verlag, Basel, Switzerland.

Craig A.G. 2001. Identification of the sites of phosphorylation in proteins using high performance liquid chromatography and mass spectrometry. *Methods Mol. Biol.* **124:** 87–105.

Edman P. 1949. A method for the determination of the amino acid sequence in peptides. *Arch. Biochem.* **22:** 475–476.

Fenn J.B., Mann M., Meng C.K., Wong S.F., and Whitehouse C.M. 1989. Electrospray ionization for mass spectrometry of large biomolecules. *Science* **246:** 64–71.

Figeys D., Corthals G.L., Gallis B., Goodlett D.R., Ducret A., Corson M.A., and Aebersold R. 1999. Data-dependent modulation of solid-phase extraction capillary electrophoresis for the analysis of complex peptide and phosphopeptide mixtures by tandem mass spectrometry: Application to endothelial nitric oxide synthase. *Anal. Chem.* **71:** 2279–2287.

Flora J.W. and Muddiman D.C. 2001. Selective, sensitive, and rapid phosphopeptide identification in enzymatic digests using ESI-FTICR-MS with infrared multiphoton dissociation. *Anal. Chem.* **73:** 3305–3311.

Francis S.H. and Corbin J.D. 1993. Structure and function of cyclic nucleotide-dependent protein kinases. *Ann. Rev. Physiol.* **56:** 237–272.

Gallis B., Corthals G.L., Goodlett D.R., Ueba H., Kim F., Presnell S.R., Figeys D., Harrison D.G., Berk B.C., Aebersold R., and Corson M. 1999. Identification of flow-dependent endothelial nitric-oxide synthase phosphorylation sites by mass spectrometry and regulation of phosphorylation and nitric oxide production by the phosphatidylinositol 3-kinase inhibitor LY294002. *J. Biol. Chem.* **274:** 30101–30108.

Garnier C., Lafitte D., Jorgensen T.J., Jensen O.N., Briand C., and Peyrot V. 2001. Phosphorylation and oligomerization states of native pig brain HSP90 studied by mass spectrometry. *Eur. J. Biochem.* **268:** 2402–2407.

Gatti A. and Traugh J.A. 1999. A two-dimensional peptide gel electrophoresis system for phosphopeptide mapping and amino acid sequencing. *Anal. Biochem.* **266:** 198–204.

Gibson B.W. and Cohen P. 1990. Liquid secondary ion mass spectrometry of phosphorylated and sulfated peptides and proteins. *Methods Enzymol.* **193:** 480–501.

Gobom J., Nordhoff E., Mirgorodskaya E., Ekman R., and Roepstorff P. 1999. Sample purification and preparation technique based on nano-scale reversed-phase columns for the sensitive analysis of complex peptide mixtures by matrix-assisted laser desorption/ionization mass spectrometry. *J. Mass Spectrom.* **34:** 105–116.

Godovac-Zimmermann J. and Brown L.R. 2001. Perspectives for mass spectrometry and functional proteomics. *Mass Spectrom. Rev.* **20:** 1–57.

Godovac-Zimmermann J., Soskic V., Poznanovic S., and Brianza F. 1999. Functional proteomics of signal transduction by membrane receptors. *Electrophoresis* **20:** 952–961.

Gold M.R., Yungwirth T., Sutherland C.L., Ingham R.J., Vianzon D., Chiu R., Vanoostveen I., Morrison H.D., and Aebersold R. 1994. Purification and identification of tyrosine-phosphorylated proteins from b lymphocytes stimulated through the antigen receptor. *Electrophoresis* **15:** 441–453.

Goodlett D.R., Aebersold R., and Watts J.D. 2000a. Quantitative in vitro kinase reaction as a guide for phosphoprotein analysis by mass spectrometry. *Rapid Commun. Mass Spectrom.* **14:** 344–348.

Goodlett D.R., Bruce J.E., Anderson G.A., Rist B., Pasa-Tolic L., Fiehn O., Smith R.D., and Aebersold R. 2000b. Protein identification with a single accurate mass of a cysteine-containing peptide and constrained database searching. *Anal. Chem.* **72:** 1112–1118.

Goodlett D.R., Wahl J.H., Udseth H.R., and Smith R.D. 1993. Reduced elution speed detection for capillary electrophoresis-mass spectrometry. *J. Microcol. Sep.* **5:** 57–61.

Goshe M.B., Conrads T.P., Panisko E.A., Angell N.H., Veenstra T.D., and Smith R.D. 2001. Phosphoprotein isotope-coded affinity tag approach for isolating and quantitating phosphopeptides in proteome-wide analyses. *Anal. Chem.* **73:** 2578–2586.

Gygi S.P., Rist B., Gerber S.A., Turecek F., Gelb M.H., and Aebersold R. 1999. Quantitative analysis of complex protein mixtures using isotope-coded affinity tags. *Nat. Biotechnol.* **17:** 994–999.

Haebel S., Jensen C., Andersen S.O., and Roepstorff P. 1995. Isoforms of a cuticular protein from larvae of the meal beetle, Tenebrio molitor, studied by mass spectrometry in combination with Edman degradation and two-dimensional polyacrylamide gel electrophoresis. *Protein Sci.* **4:** 394–404.

Hanger D.P., Betts J.C., Loviny T.L.F., Blackstock W.P., and Anderton B.H. 1998. New phosphorylation sites identified in hyperphosphorylated tau (paired helical filament-tau) from alzheimer's disease brain using nanoelectrospray mass spectrometry. *J. Neurochem.* **71:** 2465–2476.

Hart S.R., Waterfield M.D., Burlingame A.L., and Cramer R. 2002. Factors governing the solubilization of phosphopeptides retained on ferric NTA IMAC beads and their analysis by MALDI TOFMS. *J. Am. Soc. Mass Spectrom.* **13:** 1042–1051.

Hayashi F., Underhill D.M., Ozinsky A., Smith K.D., Yi E.C., Eng J.K., Goodlett D.R., and Aderem A. 2001. The innate immune response to bacterial flagellin is mediated by Toll-like receptor 5. *Nature* **410:** 1099–1103.

Huddleston M.J., Annan R.S., Bean M.F., and Carr S.A. 1993. Selective detection of phosphopeptides in complex-mixtures by electrospray liquid-chromatography mass-spectrometry. *J. Am. Soc. Mass Spectrom.* **4:** 710–715.

Huddleston M.J., Annan R.S., Bean M.F., and Carr S.A. 1994. Selective detecton of Thr-, Ser-, and Tyr-phosphopeptides in complex digests by electrospray LC-MS. In *Techniques in protein chemistry V* (ed. J.W. Crabb), pp. 123–130. Academic Press, San Diego, California.

Hunter T. 1995. Protein kinases and phosphatases: The yin and yang of protein phosphorylation and signaling. *Cell* **80:** 225–236.

———. 1998. The Croonian Lecture 1997. The phosphorylation of proteins on tyrosine: Its role in cell growth and disease. *Phil. Trans. R. Soc. Lond. B* **353:** 583–605.

———. 2000. Signaling—2000 and beyond. *Cell* **100:** 113–127.

Hunter A.P. and Games D.E. 1994. Chromatographic and mass spectrometric methods for the identification of phosphorylation sites in phosphoproteins. *Rapid Commun. Mass Spectrom.* **8:** 559–565.

Immler D., Gremm D., Kirsch D., Spengler B., Presek P., and Meyer H.E. 1998. Identification of phosphorylated proteins from thrombin-activated human platelets isolated by two-dimensional gel electrophoresis by electrospray ionization-tandem mass spectrometry (ESI-MS/MS) and liquid chromatography-electrospray ionization-mass spectrometry (LC-ESI-MS). *Electrophoresis* **19:** 1015–1023.

Jaffe H., Veeranna, and Pant H.C. 1998. Characterization of serine and threonine phosphorylation sites in beta-elimination/ethanethiol addition-modified proteins by electrospray tandem mass spectrometry and database searching. *Biochemistry* **37:** 16211–16224.

Janek K., Wenschuh H., Bienert M., and Krause E. 2001. Phosphopeptide analysis by positive and negative ion matrix-assisted laser desorption/ionization. *Rapid Commun. Mass Spectrom.* **15:** 1593–1599.

Jensen O.N. 2000. Modification-specific proteomics: Systematic strategies for analysing post-translationally modified proteins. In *Proteomics: A trends guide* (ed. W. Blackstock and M. Mann), pp. 36–42. Elsevier Science, Amsterdam.

Jensen O.N., Shevchenko A., and Mann M. 1997. Protein analysis by mass spectrometry. In *Protein structure: A practical approach* (ed. T.E. Creighton), pp. 29–57. IRL Press, Oxford, United Kingdom.

Jensen O.N., Wilm M., Shevchenko A., and Mann M. 1999. Peptide sequencing of 2-DE gel-isolated proteins by nanoelectrospray tandem mass spectrometry. *Methods Mol. Biol.* **112:** 571–588.

Ji H., Baldwin G.S., Burgess A.W., Moritz R.L., Ward L.D., and Simpson R.J. 1993. Epidermal growth factor induces serine phosphorylation of stathmin in a human colon carcinoma cell line (LIM 1215). *J. Biol. Chem.* **268:** 13396–13405.

Johnson L.N. and Lewis R.J. 2001. Structural basis for control by phosphorylation. *Chem Rev.* **101:** 2209–2242.

Karas M. and Hillenkamp F. 1988. Laser desorption ionization of proteins with molecular masses exceeding 10,000 daltons. *Anal. Chem.* **60:** 2299–2301.

Karlsson K.E. and Novotny M. 1988. Separation efficiency of slurry-packed liquid chromatography micro-columns with very small inner diameters. *Anal. Chem.* **60:** 1662–1665.

Katta V., Chowdhury S.K., Chait B.T. 1991. Use of a single-quadrupole mass spectrometer for collision-induced dissociation studies of multiply charged peptide ions produced by electrospray ionization. *Anal. Chem.* **63:** 174–179.

Katze M.G., Kwieciszewski B., Goodlett D.R., Blakely C.M., Nedderman P., Tan S.-L, and Aebersold R. 2000. Ser(2194) is a highly conserved major phosphorylation site of the hepatitis C virus nonstructural protein NS5A. *Virology* **278:** 501–513.

Kelleher N.L., Zubarev R.A., Bush K., Furie B., Furie B.C., McLafferty F.W., and Walsh C.T. 1999. Localization of labile posttranslational modifications by electron capture dissociation: The case of gamma-carboxyglutamic acid. *Anal. Chem.* **71:** 4250–4253.

Kennedy R.T. and Jorgenson J. W. 1989. Quantitative analysis of individual neurons by open tubular liquid chromatography with voltammetric detection. *Anal. Chem.* **61:** 1128–1135.

Kolesnikova V.Y., Sklyankina V.A., Baratova L.A., Nazarova T.I., and Avaeva S.W. 1974. Modification of O-phosphoserine residues in phosphoproteins. *Biochemistry* **39:** 235–240.

Krebs E.G. 1994. The growth of research on protein phosphorylation. *Trends Biol. Sci.* **19:** 439–439.

Krishna R.G. and Wold F. 1993.Post-translational modification of proteins. *Adv. Enzymol. Relat. Areas Mol. Biol.* **67:** 265–298.

Lapko V.N., Jiang X.Y., Smith D.L., and Song P.S. 1997. Posttranslational modification of oat phytochrome A: Phosphorylation of a specific serine in a multiple serine cluster. *Biochemistry* **36:** 10595–10599.

Larsen M.R., Sørensen G.L., Fey S.J., Larsen P.M., and Roepstorff P. 2001. Phospho-proteomics: Evaluation of the use of enzymatic de-phosphorylation and differential mass spectrometric peptide mass mapping for site specific phosphorylation assignment in proteins separated by gel electrophoresis. *Proteomics* **1:** 223–238.

Lee C.H., McComb M.E., Bromirski M., Jilkine A., Ens W., Standing K.G., and Perreault H. 2001. On-membrane digestion of beta-casein for determination of phosphorylation sites by matrix-assisted laser desorption/ionization quadrupole/time-of-flight mass spectrometry. *Rapid Commun. Mass Spectrom.* **15:** 191–202.

Lehr S., Herkner A., Sickmann A., Meyer H.E., Krone W., and Müller-Wieland D. 2000. Identification of major tyrosine phosphorylation sites in the human insulin receptor substrate Gab-1 by insulin receptor kinase in vitro. *Biochemistry* **39:** 10898–10907.

Li S. and Dass C. 1999. Iron(III)-immobilized metal ion affinity chromatography and mass spectrometry for the purification and characterization of synthetic phosphopeptides. *Anal. Biochem.* **270:** 9–14.

Liao P.C., Leykam J., Andrews P.C., Cage D.A., and Allison J. 1994. An approach to locate phosphorylation sites in a phosphoprotein: Mass mapping by combining specific enzymatic degradation with matrix-assisted laser desorption/ionization mass spectrometry. *Anal. Biochem.* **219:** 9–20.

Loboda A.V., Krutchinsky A.N., Bromirski M., Ens W., and Standing K.G. 2000. A tandem quadrupole/time-of-flight mass spectrometer with a matrix- assisted laser desorption/ionization source: Design and performance. *Rapid Commun. Mass Spectrom.* **14:** 1047–1057.

Ma Y., Lu Y., Mo W., and Neubert T.A. 2001. Rapid detection of phosphopeptides from protein digests using matrix assisted laser desorption/ionization time-of-flight mass spectrometry and nanoelectrospray quadrupole time-of-flight mass spectrometry. *Rapid Commun. Mass Spectrom.* **15:** 1693–1700.

MacDonald J.A., Mackey A.J., Pearson W.R., and Haystead T.A.J. 2002. A strategy for the rapid identification of phosphorylation sites in the phoshoproteome. *Mol. Cell. Proteomics* **1:** 314–322.

Mann M., Hendrickson R.C., and Pandey A. 2001. Analysis of proteins and proteomes by mass spectrometry. *Annu. Rev. Biochem.* **70:** 437–473.

Mann M., Ong S.E., Gronborg M., Steen H., Jensen O.N., and Pandey A. 2002. Analysis of protein phosphorylation using mass spectrometry: Deciphering the phosphoproteome. *Trends Biotechnol.* **20:** 261–268.

Marcus K., Immler D., Sternberger J., and Meyer H.E. 2000. Identification of platelet proteins separated by two-dimensional gel electrophoresis and analyzed by matrix assisted laser desorption/ionization-time of flight-mass spectrometry and detection of tyrosine-phosphorylated proteins. *Electrophoresis* **21:** 2622–2636.

Merrick B.A., Zhou W., Martin K.J., Jeyarajah S., Parker C.E., Selkirk J.K., Tomer K.B., and Borchers C.H. 2001. Site-specific phosphorylation of human p53 protein determined by mass spectrometry. *Biochemistry* **40:** 4053–4066.

Martensen T.M. 1971. Chemical properties, isolation, and analysis of *O*-phosphates in proteins. *Methods Enzymol.* **107:** 3–23.

Meyer H., Hoffmann-Posorske E., Korte H., and Heilmeyer L.J. 1986. Sequence analysis of phosphoserine-containing peptides. Modification for picomolar sensitivity. *FEBS Lett.* **204:** 61–66.

Meyer H.E., Eisermann B., Heber M., Hoffmann-Posorske E., Korte H., Weigt C., Wegner A., Hutton T., Donella-Deana A., and Perich J.W. 1993. Strategies for nonradioactive methods in the localization of phosphorylated amino acids in proteins. *FASEB J.* **7:** 776–782.

Michel H., Hunt D.F., Shabanowitz J., and Bennett J. 1988. Tandem mass spectrometry reveals that three photosystem II proteins of spinach chloroplasts contain N-acetyl-O-phosphothreonine at their NH_2 termini. *J. Biol. Chem.* **263:** 1123–1130.

Miliotis T., Ericsson P.O., Marko-Varga G., Svensson R., Nilsson J., Laurell T., and Bischoff R. 2001. Analysis of regulatory phosphorylation sites in ZAP-70 by capillary high-performance liquid chromatography coupled to electrospray ionization or matrix-assisted laser desorption ionization time-of- flight mass spectrometry. *J. Chromatogr. B* **752:** 323–334.

Muszynska G., Dobrowolska G., Medin A., Ekamn P., and Porath J.O. 1992. Model studies on iron (III) ion affinity chromatography II. Interaction of immobilized iron (III) ions with phosphorylated amino acids, peptides and proteins. *J. Chromatogr.* **604:** 19–28.

Neubauer G. and Mann M. 1999. Mapping of phosphorylation sites of gel-isolated proteins by nanoelectrospray tandem mass spectrometry: Potentials and limitations. *Anal. Chem.* **71:** 235–242.

Neville D.C.A., Rozanas C.R., Price E.M., Gruis D.B., Verkman A.S., and Townsend R.R. 1997. Evidence for phosphorylation of serine 753 in CFTR using a novel metal-ion affinity resin and matrix-assisted laser desorption mass spectrometry. *Protein Sci.* **6:** 2436–2445.

Nuwaysir L.M. and Stults J.T. 1993. electrospray ionization mass spectrometry of phosphopeptides isolated by on-line immobilized metal-ion affinity chromatography. *J. Am. Soc. Mass Spectrom.* **4:** 662–669.

Oda Y., Nagasu T., and Chait B.T. 2001. Enrichment analysis of phosphorylated proteins as a tool for probing the phosphoproteome. *Nat. Biotechnol.* **19:** 379–382.

Oda Y., Huang K., Cross F.R., Cowburn D., and Chait B.T. 1999. Accurate quantitation of protein expression and site-specific phosphorylation. *Proc. Natl. Acad. Sci.* **96:** 6591–6596.

Olcott M.C., Bradley M.L., and Haley B.E. 1994. Photoaffinity labeling of creatine kinase with 2-azido- and 8-azidoadenosine triphosphate: Identification of two peptides from the ATP-binding domain. *Biochem.* **33:** 11935–11941.

Pandey A., Andersen J.S., and Mann M. 2000a. Use of mass spectrometry to study signaling pathways. *Sci. STKE* **37:** PL1.

Pandey A., Podtelejnikov A.V., Blagoev B., Bustelo X.R., Mann M., and Lodish H.F. 2000b. Analysis of receptor signaling pathways by mass spectrometry: Identification of Vav-2 as a substrate of the epidermal and platelet-derived growth factor receptors. *Proc. Natl. Acad. Sci.* **97:** 179–184.

Papac D.I., Hoyes J., and Tomer K.B. 1994. Direct Analysis of Affinity-Bound Analytes by MALDI/TOF MS. *Anal. Chem.* **66:** 2609–2613.

Pas H.H. and Robillard G.T. 1988. S-phosphocysteine and phosphohistidine are intermediates in the phosphoenolpyruvate-dependent mannitol transport catalyzed by Escherichia coli EIIMtl. *Biochemistry* **27:** 5835–5839.

Porath J., Carlsson J., Olsson I., and Belfrage G. 1975. Metal chelate affinity chromatography, a new approach to protein fractionation. *Nature* **258:** 598–599.

Posada J. and Cooper J.A. 1992. Molecular signal integration. Interplay between serine, threonine, and tyrosine phosphorylation. *Mol. Biol. Cell* **3:** 583–592.

Posewitz M.C. and Tempst P. 1999. Immobilized gallium(III) affinity chromatography of phosphopeptides. *Anal. Chem.* **71:** 2883–2892.

Powell K.A., Valova V.A., Malladi C.S., Jensen O.N., Larsen M.R., and Robinson P.J. 2000. Phosphorylation of dynamin I on Ser-795 by protein kinase C blocks its association with phospholipids. *J. Biol. Chem.* **275:** 11610–11617.

Qian X.-H., Zhou W., Khaledi M.G., and Tomer K.B. 1999. Direct analysis of the products of sequential cleavages of peptides and proteins affinity-bound to immobilized metal ion beads by matrix-assisted laser desorption/ionization mass spectrometry. *Anal. Biochem.* **274:** 174–180.

Qin J. and Chait B.T. 1997. Identification and characterization of posttranslational modifications of proteins by MALDI ion trap mass spectrometry. *Anal. Chem.* **69:** 4002–4009.

Qin J., Fenyo D., Zhao Y.M., Hall W.W., Chao D.M., Wilson C.J., Young R.A., and Chait B.T. 1997. A strategy for rapid, high confidence protein identification. *Anal. Chem.* **69:** 3995–4001.

Raska C.S., Parker C.E., Dominski Z., Marzluff W.F., Glish G.L., Pope R.M., and Borchers C.H. 2002. Direct MALDI-MS/MS of phosphopeptides affinity-bound to immobilized metal ion affinity chromatography beads. *Anal. Chem.* **74:** 3429–3433.

Resing K.A., Johnson R.S., and Walsh K.A. 1995. Mass spectrometric analysis of 21 phosphorylation sites in the internal repeat of rat profilaggrin, precursor of an intermediate filament associated protein. *Biochemistry* **34:** 9477–9487.

Roepstorff P. and Fohlmann J. 1984. Proposal for a common nomenclature for sequence ions in mass spectra of peptides. *Biomed. Mass Spectrom.* **11:** 601.

Schlessinger J. 2000. Cell signaling by receptor tyrosine kinases. *Cell* **103:** 211–225.

Schlosser A., Pipkorn R., Bossemeyer D., and Lehmann W.D. 2000. Analyse der Proteinphosphorylierung durch Kombination von Elastase-Verdau und ESI-Tandem-Massenspektrometrie. In *Proceedings of the 32nd annual meeting of the German Mass Spectrometry Society*, p. 115. German Mass Spectrometry Society, Oldenburg, Germany.

———. 2001. Analysis of protein phosphorylation by a combination of elastase digestion and neutral loss tandem mass spectrometry. *Anal. Chem.* **73:** 170–176.

Schrecker O., Stein R., Hengstenberg W., Gassner M., and Stehlik D. 1975. The staphylococcal PEP dependent phosphotransferase system, proton magnetic resonance (PMR) studies on the phosphoryl carrier protein HPr: Evidence for a phosphohistidine residue in the intact phospho-HPr molecule. *FEBS Lett.* **51:** 309–312.

Shevchenko A., Loboda A., Ens W., and Standing K.G. 2000. MALDI quadrupole time-of-flight mass spectrometry: A powerful tool for proteomic research. *Anal. Chem.* **72:** 2132–2141.

Shevchenko A., Wilm M., Vorm O., and Mann M. 1996. Mass spectrometric sequencing of proteins from silver-stained polyacrylamide gels. *Anal. Chem.* **68:** 850–858.

Shi S.D., Hemling M.E., Carr S.A., Horn D.M., Lindh I., and McLafferty F.W. 2001. Phosphopeptide/phosphoprotein mapping by electron capture dissociation mass spectrometry. *Anal. Chem.* **73:** 19–22.

Sickmann A. and Meyer H.E. 2001. Phosphoamino acid analysis. *Proteomics* **1:** 200–206.

Sickmann A., Dormeyer W., Wortelkamp S., Woitalla D., Kuhn W., and Meyer H.E. 2000. Identification of proteins from human cerebrospinal fluid, separated by two-dimensional polyacrylamide gel electrophoresis. *Electrophoresis* 21: 2721–2728.

Springer M.S., Coy M.F., and Adler J. 1979. Protein methylation in behavioural control mechanisms and in signal transduction. *Nature* **280:** 279–284.

Steen H. and Mann M. 2002. A new derivatization strategy for the analysis of phosphopeptides by precursor ion scanning in positive ion mode. *J. Am. Soc. Mass Spectrom.* **13:** 996–1003.

Steen H., Küster B., and Mann M. 2001a. Quadrupole time-of-flight versus triple-quadrupole mass spectrometry for the determination of phosphopeptides by precursor ion scanning. *J. Mass Spectrom.* **36:** 782–790.

Steen H., Küster B., Fernandez M., Pandey A., and Mann M. 2001b. Detection of tyrosine phosphorylated peptides by precursor ion scanning quadrupole TOF mass spectrometry in positive ion mode. *Anal. Chem.* **73:** 1440–1448.

Stensballe A., Andersen S., and Jensen O.N. 2001. Characterization of phosphoproteins from electrophoretic gels by nanoscale Fe(III) affinity chromatography with off-line mass spectrometry analysis. *Proteomics* **1:** 207–222.

Stensballe A., Jensen O.N., Olsen J.V., Haselmann K.F., and Zubarev R.A. 2000. Electron capture dissociation of singly and multiply phosphorylated peptides. *Rapid Commun. Mass Spectrom.* **14:** 1793–1800.

Storm S.M. and Khawaja X.Z. 1999. Probing for drug-induced multiplex signal transduction pathways using high resolution two-dimensional gel electrophoresis: Application to beta-adrenoceptor stimulation in the rat C6 glioma cell. *Brain. Res. Mol. Brain. Res.* **71:** 50–60.

Vihinen H. and Saarinen J. 2000. Phosphorylation site analysis of semliki forest virus nonstructural protein 3. *J. Biol. Chem.* **275:** 27775–27783.

Wahl J.H., Goodlett D.R., Udseth H.R., and Smith R.D. 1993. Use of small-diameter capillaries for increasing peptide and protein detection sensitivity in capillary electrophoresis-mass spectrometry. *Electrophoresis* **14:** 448–457.

Weckwerth W., Willmitzer L., and Fiehn O. 2000. Comparative quantification and identification of phosphoproteins using stable isotope labeling and liquid chromatography/mass spectrometry. *Rapid Commun. Mass Spectrom.* **14:** 1677–1681.

Weijland A., Williams J.C., Neubauer G., Courtneidge S.A., Wierenga R.K., and Superti-Furga G. 1997. Src regulated by C-terminal phosphorylation is monomeric. *Proc. Natl. Acad. Sci.* **94:** 3590–3595.

Wettenhall R.E.H., Aebersold R., and Hood L.E. 1991. Solid-Phase Sequencing of ^{32}p-labeled phosphopeptides at picomole and subpicomole levels. *Methods Enzymol.* **201:** 186–199.

Wilm M., Neubauer G., and Mann M. 1996. Parent ion scans of unseparated peptide mixtures. *Anal. Chem.* **68:** 527–533.

Wind M., Edler M., Jakubowski N., Linscheid M., Wesch H., and Lehmann W.D. 2001. Analysis of protein phosphorylation by capillary liquid chromatography coupled to element mass spectrometry with 31P detection and to electrospray mass spectrometry. *Anal. Chem.* **73:** 29–35.

Yan J.X., Packer N.H., Gooley A.A., and Williams K.L. 1998. Protein phosphorylation: Technology for the identification of phosphoamino acids. *J. Chromatogr.* **808:** 23–41.

Zhang X.L., Herring C.J., Romano P.R., Szczepanowska J., Brzeska H., Hinnebusch A.G., and Qin J. 1998. Identification of phosphorylation sites in proteins separated by polyacrylamide gel electrophoresis. *Anal. Chem.* **70:** 2050–2059.

Zhao J.Y., Kuang J., Adlakha R.C., and Rao P.N. 1989. Threonine phosphorylation is associated with mitosis in HeLa cells. *FEBS Lett.* **249:** 389–395.

Zhou H., Watts J.D., and Aebersold R. 2001. A systematic approach to the analysis of protein phosphorylation. *Nat. Biotechnol.* **19:** 375–378.

Zhou W., Merrick B.A., Khaledi M.G., and Tomer K.B. 2000. Detection and sequencing of phosphopeptides affinity bound to immobilized metal ion beads by matrix-assisted laser desorption/ionization mass spectrometry. *J. Am. Soc. Mass Spectrom.* **11:** 273–282.

Zubarev R.A., Kelleher N.L., and McLafferty F.W. 1998. Electron capture dissociation of multiply charged protein cations. A nonergodic process. *J. Am. Chem. Soc.* **120:** 3265–3266.

WWW RESOURCES

http://ca.expasy.org/prosite/ PROSITE–Database of protein families and domains

http://pkr.sdsc.edu/html/index.shtml The Protein Kinase Resource homepage. The University of California, San Diego.

http://scansite.mit.edu Scansite homepage, Massachusetts Institute of Technology

http://welcome.to/gpmaw GPMAW (General Protein/Mass Analysis for Windows) homepage

http://www.cbs.dtu.dk/databases/PhosphoBase/ PhosphoBase v2.0–A database of phosphorylation sites. Center for Biological Sequence Analysis (CBS)

http://www.cbs.dtu.dk/services/NetPhos/ Center for Biological Sequence Analysis, NetPhos 2.0 Prediction Server

http://www.lecb.ncifcrf.gov/phosphoDB/ Phosphoprotein Database (PPDB), Laboratory of Experimental and Computational Biology (LECB), NCI-Frederick Cancer Research Facility

CHAPTER 10

Characterization of Protein Complexes*

PROTOCOLS

*The text chapter contains contributions by M. Baca and J.-G. Zhang (The Walter and Eliza Hall Institute of Medical Research, Parkville, Victoria 3050, Australia), M.T. Ryan (La Trobe University, 3086 Melbourne, Australia), J. Rappsilber and M. Mann (University of Southern Denmark, DK-5230 Odense, Denmark), N. Naryshkin, A. Revyakin, and R.H. Ebright (Rutgers University, Piscataway, New Jersey 08854), M. Damelin and P. Silver (Dana Farber Cancer Institute, Boston, Massachusetts), and G. MacBeath (Harvard University, Cambridge, Massachusetts).

ONE OF THE REVELATIONS OF THE HUMAN GENOME PROJECT, thus far, is that the genome harbors only ~30,000 genes (Lander et al. 2001; Venter et al. 2001). However, it should be noted that researchers disagree on the precise number of protein-encoding genes (Hogenesch et al. 2001), with estimates being as high as 60,000 genes. Irregardless of the final number of genes found in the human genome, the number of proteins encoded is likely to be substantially larger than its set of genes due to posttranslational modifications (see Chapter 9) and alternate splicing, which is reported to be highly frequent in humans (Mironov et al. 1999; Black 2000). It is believed that alternative splicing, which occurs in nearly all metazoans and is especially common in vertebrates, provides an economical means for producing functionally diverse polypeptides from a single gene (Lopez 1998; Black 2000; Smith and Valcarcel 2000). By aligning expressed sequence tag (EST) sequences and mapping the resultant mRNA families, it has been estimated that ~35% of human genes yield variably spliced protein products (Croft et al. 2000). Because the published ESTs derive from a limited number of tissues, this value may be an underestimate, and the 30,000 genes could easily produce several hundred thousand different proteins.

ALTERNATIVE mRNA SPLICING HAS SPECIFIC EFFECTS ON PROTEIN PRODUCT ACTIVITY

In the human genome, there are many examples of multiple mRNA transcripts being produced from a single gene (e.g., the *n*-cadherins, calcium-activated potassium channels, and neurexins). Altered mRNA splicing has been shown to have very specific effects on the activity of a protein product. For example, alternative splicing has been shown to influence the ligand binding of growth factors and cell adhesion molecules, determine the subcellular localization of the encoded protein, determine the phosphorylation status of the encoded protein, and affect the activation domains of transcription factors (Lopez 1998; Black 2000; Smith and Valcarcel 2000).

At present, the rules governing splice site choice are poorly understood, which makes it difficult to recognize exons and predict splicing patterns within the genome sequence with any degree of confidence. For example, functional splice sites often do not fully match the recognized consensus sequences. Conversely, there are many splicing site sequence motifs in the genome that are not recognized by the splicing apparatus.

PHYSIOLOGICAL COMPLEXITY IS REPRESENTED BY THE INTERACTOME

For the reasons discussed above, it is not surprising that the physiological complexity of organisms is not merely a consequence of gene numbers (e.g., humans have less than twice as many genes as the 959-cell nematode *Caenorhabditis elegans*). Rather, evolution of the increased physiological complexity of higher-order organisms is undoubtedly due to a number of other mechanisms that include alternative splicing, diversification of gene regulatory networks, and the ability of intracellular signaling pathways to interact with one another. Indeed, it is becoming increasingly apparent that biological signaling pathways can interact with one another to form complex networks comprising a large number of components. Such complexity arises from the overlapping functions of components, from the connections among components, and from the spatial relationship between components (Weng et al. 1999). Additionally, many cellular processes are performed and regulated not by individual proteins, but by proteins acting in large protein assemblies or macromolecular complexes. For instance, the eukaryotic ribosome, which translates RNA into protein, consists of ~80 unique proteins (Wool et al. 1995), and the RNA polymerase II transcription complexes in eukaryotic cells, which are involved in DNA replication, comprise at least 50 different proteins (Pugh 1996).

Only a global view of a cell's mRNA and protein complement, and a detailed knowledge of how these complements change with development and the environment (especially in disease), will allow a full understanding of how a complex organism works. This is the real impetus of proteomics, especially cell-mapping proteomics, which aims to describe all protein-protein interactions (both spatially and temporally) within a given cell. The full complement of protein-protein interactions of a cell's proteome is now referred to as the "interactome." Uncovering protein-protein interaction schemes will have a pivotal role in the post-genome era by shedding light on the molecular mechanisms underlying biological processes (see Weng et al. 1999; Aravind 2000; Chen and Han 2000; Mayer and Hieter 2000; Gerstein et al. 2002).

A number of diverse combinatorial biology methods have proved to be invaluable both in identifying natural protein-protein interactions and also in mapping the specificities and energetics of these interactions (for a review, see Pelletier and Sidhu 2001). In these methods, DNA-encoded libraries of peptides and proteins have been constructed that include a covalent linkage between each peptide/protein and the encoding DNA. Members of the libraries can be selected by virtue of a specific binding interaction, and their protein sequence can be deduced from the sequence of the cognate DNA. These combinatorial biology methods can be categorized into two groups depending on the method for generating DNA-encoded polypeptide libraries. In vitro methods include phage display (for review, see Rodi and

Makowski 1999; Sidhu et al. 2000) and ribosome display technologies (Hanes and Pluckthun 1997). (The principal application of phage display and ribosome display, thus far, has been for the selection and evolution of antibodies.) In contrast, in vivo combinatorial biology methods use polypeptide libraries that are expressed inside living cells along with a target protein of interest. The most widely used in vivo method for the detection of protein-protein interactions is the yeast two-hybrid system (Fields and Song 1989), which is discussed in the next section.

In this chapter, various proteomic approaches for identifying members of protein complexes are discussed. Historically, the biochemical purification of protein complexes in sufficient amounts and purity to enable their characterization has been difficult. Invariably, a different strategy had to be formulated for each complex under consideration, and mild separation techniques that do not destroy protein-protein interactions had to be developed. The following are some of the physical techniques used to detect and purify interacting proteins:

- Blue native polyacrylamide gel electrophoresis (Protocol 4).
- Native agarose gels (Protocol 3).
- Immunoprecipitation.
- Affinity capture (Protocols 1 and 2).

Protein interactions identified using some of these physical techniques are fraught with ambiguities and may not necessarily be a true reflection of biological complexes that occur within the cell. For instance, if the target "bait" used in affinity capture approaches exists in more than one posttranslationally modified form (e.g., phosphorylation state), and each of these forms is involved in a multiprotein complex, then the complex captured by the affinity tag (which is common to all posttranslationally modified forms of the target protein) will represent a composite of a number of complexes. Such posttranslationally modified forms of the target bait may be located in different parts of the cell (in different multiprotein complexes) and may never be in contact in a physiological sense. In addition, to add to the complexity of intracellular networking, such multiprotein complexes will vary temporally (e.g., different stages of the cell cycle). It should also be borne in mind that biological macromolecules have evolved to function in crowded environments such as the inside of cells (where the total concentration of protein and RNA is in the range of 300–400 g/liter [Zimmerman and Minton 1991]) and that protein-protein interactions are protein-concentration-dependent (for reviews on "molecular crowding," see Ellis 2001; Verkman 2002). Hence, if the cellular concentration of expressed recombinant target protein (bait) varies markedly from one experiment to another, the nature of the corresponding multiprotein complexes derived from these experiments may bear no resemblance to each other. For example, the concentration of recombinant bait protein is more likely to resemble the intracellular concentration of the mature protein if the expression is driven off the natural promoter compared to a situation where it is overexpressed constitutively. The latter situation may result in the formation of multiprotein complexes that differ markedly in their protein components. For these reasons, it is important to use more than one physical method to verify any putative protein-protein interaction and, importantly, to verify the biological nature of such interactions using in vivo methods such as fluorescence resonance energy transfer (FRET) (see Protocol 8).

In addition to identifying the components of multiprotein complexes, it is important to understand the spatial organization, or topography, of such complexes. Such information is critical to understanding the function of the complex at the molecular level. This information is also important in the pursuit of small molecules that disrupt protein-protein interactions (e.g., in the rational drug design of a small-molecule antagonist of protein-protein complex formation). In the absence of three-dimensional structure information, chemical cross-linking has been used to reveal closest-neighbor relationships between components within a protein complex. In this chapter, a number of basic protocols for performing chemical cross-linking or label transfer studies are discussed (see Protocols 5–7).

One of the ultimate goals of studying protein-protein interactions is to understand the consequences of the interactions for cell function. This depends in turn on understanding the strength of the interaction in the cell (i.e., the binding constant, K_d), the concentration of the components involved in the interaction, the effect of cellular compartmentation, the influence of competing proteins, and the flux of the individual reactions. (The global nature of such studies, and the ability to predict the effect of external stimuli on cellular pathways, is referred to as total "systems biology" [Ideker et al. 2001; Davidson et al. 2002; Kitano 2002; Noble 2002].) For a discussion of methods for estimating and determining binding constants, see Phizicky and Fields (1995), Schuck (1997a,b), Lakey and Raggett (1998), Medaglia and Fisher (2001, 2002), and Cantor and Schimmel (1980). In this chapter, the fluorescence gel-retardation assay for determining K_d values is discussed (see Protocol 9).

IDENTIFYING PROTEIN-PROTEIN INTERACTIONS USING YEAST TWO-HYBRID ANALYSIS VERSUS PROTEOMIC APPROACHES

Traditional genetic and biochemical methods for identifying protein-protein interactions largely focus on one target gene or protein at a time. Using one of the most popular methods, the yeast two-hybrid system, thousands of new protein-protein interactions have been identified (Phizicky and Fields 1995). Figure 10.1 provides a schematic depiction of the basic principles of the yeast two-hybrid system; for a detailed discussion of the method, including basic protocols, see Sambrook and Russell (2001). In a study of the protein-protein interactions in the yeast *Saccharomyces cerevisiae* (the complete genome sequence is now available), this approach, until recently, largely focused on individual protein complexes (Flores et al. 1999; Mayes et al. 1999). More recently, a high-throughput yeast two-hybrid method was developed for mapping all protein interactions in this organism (Uetz et al. 2000). This study revealed 957 putative protein interactions involving 1004 *S. cerevisiae* proteins; these data are now publicly accessible via the Web Site, http://curatools.curagen.com, thereby enabling individual researchers to manipulate the data themselves. A global network of protein-protein interactions in yeast (2709 putative interactions encompassing 2309 different proteins), generated using a combination of yeast two-hybrid (Ito et al. 2000; Uetz et al. 2000) and biochemical protein interaction data in public databases, has been described recently (Schwikowski et al. 2000).

Using a set of 29 genes that had been previously characterized, Walhout and co-workers have used a similar high-throughput yeast two-hybrid approach to build an interaction matrix among a large complement of proteins encoded by *C. elegans* (Walhout et al. 2000). Interaction maps have also been proposed for a number of viruses such as *Escherichia coli* bacteriophage T4 (Bartel et al. 1996), hepatitis C (Flajolet et al. 2000), and vaccinia virus (McCraith et al. 2000) and more recently for the prokaryote *Helicobacter pylori* (Rain et al. 2001). User-friendly navigable databases, including BIND (http://binddb.org), DIP (http://dip.doe-mbi.ucla.edu), and MINT (http://cbm.bio.uniroma2.it/mint/index.html) (Zanzoni et al. 2002), are being developed to store and represent protein network information in such a way that individual scientists can formulate new interpretations or pose questions for experimental testing (Mayer and Hieter 2000; Xenarios et al. 2000; Bader et al. 2001; Duan et al. 2002).

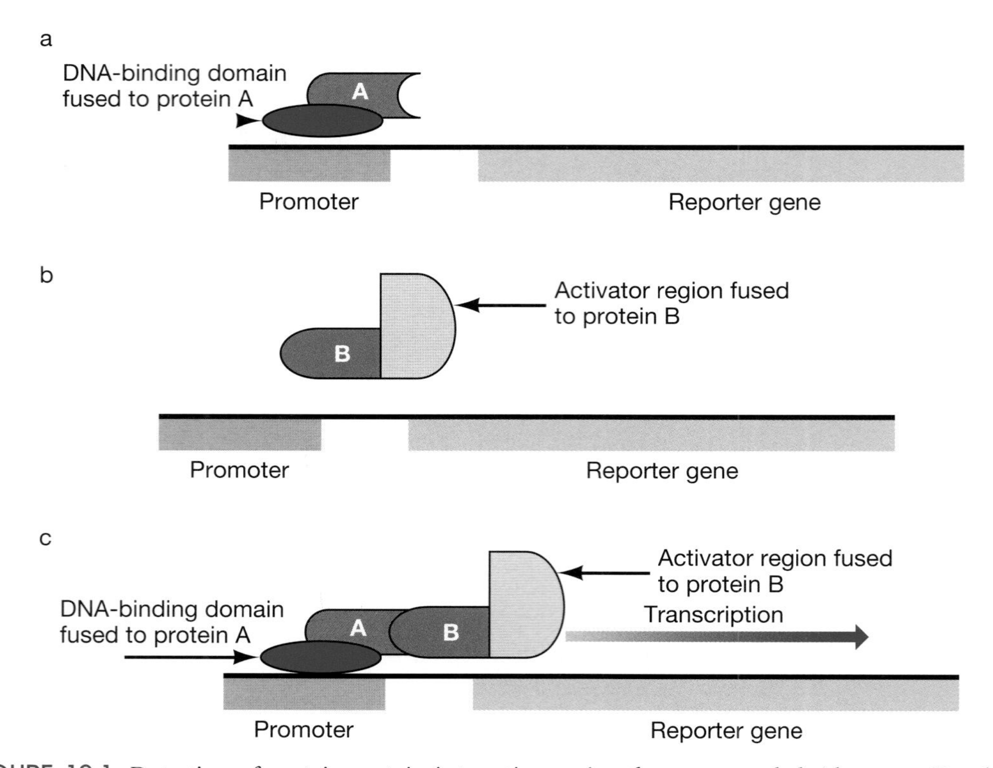

FIGURE 10.1. Detection of protein-protein interactions using the yeast two-hybrid system. Two-hybrid analysis works by separating the coding sequences for the DNA-binding and activation domains of a transcriptional activator and cloning them into separate vector molecules. The coding sequence of a candidate protein whose partners are sought (referred to as "bait") is then fused with the DNA-binding domain. A library of coding sequences for proteins that might interact with the "bait" (called "prey") is made in fusion with the activation domain. Yeast has two sexes, called α and **a**. Therefore, baits and prey can easily be introduced into the same yeast cell by mating. If they physically interact, the DNA-binding and activation domains are closely juxtaposed, and the reconstituted transcriptional activator can mediate the switching on of the gene that effects the color change. (*a*) Fusion of the bait protein and the DNA-binding domain of the transcriptional activator cannot turn on the reporter gene. (*b*) Likewise, fusion of the prey protein and the activating region of the transcriptional activator is also insufficient to switch on the reporter gene. (*c*) When bait and prey associate, however, the DNA-binding domain and activator region are brought close enough together to switch on the reporter gene. The result is gene transcription and a color change that can be monitored. (Redrawn, with permission, from Oliver 2000 [©Macmillan].)

Although yeast two-hybrid analysis has been highly successful in detecting many potential protein-protein interactions, this technique has a number of limitations.

- Because two-hybrid analysis is a transcription-based system, any bait protein that self-activates the reporter genes is unsuitable for screening.
- Proteins that require posttranslational modification(s) for protein-binding activity (e.g., phosphorylation) will be less likely to be isolated by this screen.
- If an additional nonpeptide factor is involved in the protein interaction, the interaction may go undetected. For example, two transcription factors might require a stretch of DNA sequence to form a stable protein complex.
- The interaction may lack biological context. For example, many protein interactions only occur when the cell is in a particular physiological state and thus may go undetected. Other potential interactions, which are detected by this approach, may never occur in vivo, because the proteins are located in separate cellular compartments.

TABLE 10.1. Selected examples of proteomic analyses of multiprotein complexes, organelles, and signaling pathways

Multiprotein complex	
Human spliceosome (HeLa cells)	Neubauer et al. (1998)
Yeast nuclear pore complex	Rout et al. (2000)
E. coli chaperonin GroEL	Houry et al. (1999)
Yeast ribosome	Link et al. (1999)
Yeast spliceosome U1 snRNP[a]	Neubauer et al. (1997)
Yeast 26S proteosome	Verma et al. (2000)
Yeast nuclear pore complex	Allen et al. (2001); Rout et al. (2000)
Yeast protein complexes	Gavin et al. (2002); Ho et al. (2002)
Yeast anaphase-promoting complex	Zachariae et al. (1998)
Organelle/cellular component	
Human nucleolus	Anderson et al. (2002)
Human paraspeckles (subnuclear bodies)	Fox et al. (2002)
Mouse microsomal fraction (HL-60 cells)	Han et al. (2001)
Mouse phagosome (J774 cells)	Garin et al. (2001)
Human epithelial cell membranes (LIM1215)	Simpson et al. (2000)
Signaling pathways/secreted proteins	
EGFR (HeLa cells)[b]	Pandey et al. (2000b)
Secreted proteins during differentiation of mouse 3T3-L1 preadipocytes to adipocytes	Kratchmarova et al. (2002)

[a]snRNP = small nuclear ribonucleoprotein.
[b]EGFR = epidermal growth factor receptor.

PROTEIN-PROTEIN INTERACTIONS CAN BE ANALYZED USING AFFINITY CAPTURE TECHNIQUES

The use of cell-mapping proteomics to identify protein interaction partners is a powerful way to elucidate the biological activity of a protein of unknown function. Important insights into the function of a protein can be gained from knowledge of the biochemical interactions in which the protein is involved, particularly where the function of the interaction partner is well-defined. The task of identifying interacting proteins has usually been tackled with technologies that rely on the expression of gene products from a cDNA library in conjunction with some form of a screen to identify interacting proteins. The most widely used of these methods has been the yeast two-hybrid approach (see the previous section). In contrast to these genetic-based approaches, proteomic identification typically relies on isolating the interacting protein directly from a biological source using:

- ***Immunoprecipitation.*** Coimmunoprecipitation is a classical method for detecting protein-protein interactions and has been used extensively during the past couple of decades (see Phizicky and Fields 1995).
- ***Isolation of multiprotein complexes.*** For a list of multiprotein complexes that have been extensively studied using cell-mapping proteomics, see Table 10.1.
- ***Affinity capture.*** This uses a recombinant fusion protein comprising the target protein and an affinity "tag" (or bait) covalently fused to the amino or carboxyl terminus of the target protein. The tag may be a small peptide epitope (in this case, an antibody is used to capture the fusion protein, as well as proteins interacting with the target protein) or a biospecific protein (e.g., glutathione-*S* thioreductase [GST], which can be affinity-captured using glutathionine) or a synthetic peptide derived from a "structural motif." For a list of commonly used affinity tags, see Table 10.2.

TABLE 10.2. Commonly used fusion partners for studying protein-protein interactions

System	Fusion partner	Size	Fusion point	Ligand (matrix)	Supplier
1. *Enzymes*					
GST	glutathione-*S*-transferse	26 kD	amino terminus	glutathione-agarose	Amersham
pET GST fusion system 41 and 42	glutathione-*S*-transferse	26 kD	amino terminus	glutathione-agarose	Novagen
2. *Carbohydrate-binding proteins*					
pMAL	maltose-binding protein (MBP)	40 kD	amino terminus	cross-linked amylose	Novagen
pET fusion system 34b-38b, CBD.Tag	cellulose-binding domain tag (CBD)	107 aa	amino or carboxyl terminus	polyclonal antibody, cellulose	
3. *Protein-protein interaction*					
pCAL	CBP (carboxy-terminal fragment from muscle myosin light-chain kinase)	4 kD	amino or carboxyl terminus	calmodulin affinity resin	Stratagene
4. *Charged amino acids*					
QIAexpress	6x His tag	6 aa	amino or carboxyl terminus	Ni-NTA (nitrolotriacetic acid Sepharose [Ni^{2+}])	QIAGEN
pET fusion system 14-16, 19-42	6x His tag	6, 8, or 10 aa	amino or carboxyl terminus, internal	Ni-NTA (nitrolotriacetic acid Sepharose [Ni^{2+}])	Novagen Invitrogen
5. *Epitope tags*					
	HAYPYDVPDYA			MAb	
	c-*myc* EQKLISEEDL	10 aa		MAb	Stratgene; Invitrogen
	FLAG DYKDDDDK	8 aa		MAb	Stratagene
	GLU-GLU EEEEYMPME			MAb	
pET fusion system T7.Tag	T7 (gene 10) MASMTGGQQMG	11 or 200 aa	amino terminus, internal	MAb-cross-linked agarose	Novagen
pET fusion system S.Tag	S-peptide KETAAAKFERQHMDS	15 aa		S-protein (104 amino acids) affinity agarose	Novagen
6. *Biotin-binding domain*		8 kD			
PinPoint	biotin carboxylase carrier protein (in vivo biotinylation site)	123 aa	amino terminus	avidin affinity resin	Promega

Protein fusion partners (purification "tags") are typically positioned at the amino-terminal end of the target protein (e.g., where the tag protein is 20–40 kD), whereas peptide epitope tags (e.g., 8–15 amino acids long) may be positioned internally, or at either the amino- or carboxy-terminal end. For epitope tags, this choice is usually made arbitrarily. Although the small size of epitope tags decreases the possibility of functional interference (a common problem with the larger protein tags), they may influence posttranslational processing events such as signal peptide cleavage (Ledent et al. 1997). In addition to helping identify protein-proteins interactions, epitope tags can be used to (1) localize and also track the movement of tagged proteins within a cell and (2) monitor the purification of recombinant proteins during purification by western blot analysis. This is particularly useful because epitope tags obviate the need to create protein-specific antibodies, a time-consuming and expensive undertaking.

In contrast to most of the purification procedures used for the tags listed in this table, only proteins with histidine tags can be purified under denaturing conditions such as guanidinium hydrochloride or 60% isopropanol (Holzinger et al. 1996; Franken et al. 2000).

Fusion proteins are commonly designed so that affinity tags can be removed by specific proteases that recognize cleavage sites engineered between the affinity tag and the protein of interest. Information concerning commercially available kits for constructing affinity capture fusion proteins can be obtained at QIAGEN (http:www.qiagen.com); Novagen (http://www.novagen.com); Stratagene (http://www.stratagene.com); Invitrogen (http://www.invitrogen.com); Promega (http://www.promega.com); and Covance Research Products (CRP) (http://www.CRPinc.com).

Affinity-purified proteins are typically fractionated by gel electrophoresis, using either one-dimensional (1D) or two-dimensional (2D) gels (see Chapters 2 and 4, respectively), and visualized by Coomassie Blue staining. Specific bands are then excised from the gel, proteolytically digested in situ, and identified by mass spectrometry (MS) (see Chapters 7 and 8). Comparison of gel profiles with those obtained from the incubation of a cell lysate with a control resin (i.e., a resin that lacks the immobilized probe protein) or the affinity tag fused to an irrelevant protein is used to identify proteins that interact specifically. Alternatively, an intact complex can be digested, without prior separation of the component proteins, and its component proteins can be identified by direct analysis of the complex peptide mixture (see Chapter 8, Protocol 6).

Affinity purification of interacting proteins from cell lysates is an inherently simple procedure. The main limitation to identifying novel proteins by this technique is in isolating sufficient quantities of protein that enable subsequent characterization by MS-based methods. Success is ultimately dependent on the relative abundance of the interacting protein within the biological source used for the experiment. It is worthwhile to consider the practical aspects of this limitation. Ideally, identification of a protein band on a gel by in situ digestion and MS requires quantities that are detectable by Coomassie staining, i.e., $\geq$0.05 μg of protein. For a 50-kD low-abundance protein present at ten copies per cell, it would take 6×10^{10} cells to provide 0.05 μg of material (see Figure 1.8), and this does not take into account sample losses incurred during isolation of the target protein. For a highly abundant protein present at 10,000 copies per cell, only 6×10^{7} cells would be required to produce the same amount of protein. Although the production and processing of lysate from 10^{8} cells are likely to be straightforward for most mammalian cell lines, scaling this up to 10^{11} cells is a more significant challenge. With this in mind, it is important to address the sensitivity issue in initial pilot experiments. These are best performed using intrinsic [^{35}S]Met/Cys radiolabeling or more sensitive silver staining to detect protein bands on a gel (see Chapter 2, Protocols 5–7 and Chapter 7, Protocol 11). If specific interacting proteins are detected in these studies, then the investigator can proceed with scaling up the amounts of starting material used in the experiment to generate sufficient material for MS-based sequence analysis.

Protocols 1 and 2 provide details for performing two types of affinity capture. Protocol 1 describes the use of FLAG-epitope-tagged proteins for both small-scale analysis and large-scale coimmunoprecipitation of proteins interacting with the human inhibitor of apoptosis protein (IAP) complex using a FLAG-tag IAP fusion protein transiently introduced into mammalian U292 cells (Verhagen et al. 2000). The ectopically expressed (transiently or stably expressed) epitope-tagged protein X (e.g., FLAG-protein X) can be immunoprecipitated using commercially available epitope-tag-specific monoclonal antibodies. The advantage of examining the interactions of endogenously expressed proteins is that these proteins are more likely to be physiological and less likely to be an artifact of overexpression. In scaling up the procedure, however, epitope-tagged overexpressed protein X may be necessary for sufficient coimmunoprecipitating protein to be isolated. Protocol 2 contains a generic affinity capture method using a recombinantly expressed protein that is known or believed to interact with other proteins. Two successful applications are described within Protocol 2, one using the GST recombinant protein system to isolate the interacting partners for the suppressors of cytokine signaling (SOCS) family of proteins, which act as intracellular inhibitors of several cytokine signal transduction pathways (Zhang et al. 1999). DNA fragments encoding a conserved 40-amino-acid region of the SOCS family of proteins (SOCS box) with a 14-amino-acid linker sequence were generated by polymerase chain reaction (PCR) and cloned into the GST bacterial expression vector pGEX-2T (Smith and Johnson 1988). The GST fusion proteins were first purified by affinity chromatography on glutathione Sepharose 4B

Optimization of Variables for Affinity Capture Methods

Given the sensitivity issue, careful selection of conditions that maximize the yield of interacting proteins can make the difference between experimental success and failure. Several variables must be considered:

- ***What is the nature of the probe protein?*** The probe can be either a full-length protein, a relevant interaction domain, or even a smaller bioactive fragment that can be chemically synthesized. Although a full-length protein must, by default, contain all of the regions likely to participate in protein-protein interactions, a smaller active fragment may be easier to produce, may be immobilized at higher density, and may be less susceptible to nonspecific interactions.
- ***How is the probe protein to be immobilized in preparing the affinity matrix?*** Covalent immobilization allows more stringent wash and elution conditions, but may result in partial or even total inactivation of the probe protein. A protein that is noncovalently immobilized via association of a fused affinity tag (e.g., GST) with an immobilized ligand is likely to retain activity; however, if large excesses of the tagged probe protein are eluted from the affinity resin, this may obscure interacting proteins in the SDS-PAGE (SDS-polyacrylamide gel electrophoresis) analysis.
- ***What cell line will be used as a source of the interacting protein present?*** This is a critical consideration. A cell line that is known to express the probe protein naturally is a good place to be looking for interacting partners. However, the investigator should also consider how easy it would be to scale up production of that particular cell line. In this regard, a cell line that grows in suspension is preferable to one which is adherent.
- ***How will the cell lysate be prepared?*** The conditions used for lysing the cells should not inactivate any potential interacting protein, nor interfere with the subsequent affinity capture. If the probe protein is naturally localized to a particular cellular compartment, preparations of that cell compartment may be preferable to using a whole-cell lysate (see Chapter 3).
- ***Do the source cells need to be pretreated prior to their lysis?*** For instance, if the probe protein is a component of a cell signaling pathway, activation of the target cells with the appropriate growth factor may be required prior to lysis.
- ***How will the affinity resin be washed following incubation with cell lysate?*** Some cellular proteins will bind nonspecifically to the affinity matrix; thus, it is important to minimize the capture of these contaminating proteins. Nonspecifically associating proteins are removed by washing the affinity resin after incubation with the cell lysate. Although extensive washing will remove these proteins, too much washing can result in the loss of specifically associated proteins. Consequently, the extent of washing should be adjusted so as to maximize the capture of specifically interacting proteins while removing as much of the background as possible. As nonspecifically associating proteins are never totally removed by washing, it is important to run a parallel control experiment to show that the affinity capture of a cellular protein is due to interaction with the immobilized probe protein and not to other components of the affinity matrix. Comparison of the SDS-PAGE profiles of cellular proteins captured by the probe protein resin versus the control resin is used to distinguish specific from nonspecifically associated proteins (also see below Tandem Affinity Purification).
- ***One- or two-dimensional gel electrophoresis?*** One-dimensional SDS-PAGE is technically less challenging than running two-dimensional gels, but the latter has a far greater capacity to resolve complex mixtures. Determining which of these two techniques should be used will depend on the level of background-contaminating proteins that copurify with the interacting protein(s).

and then covalently coupled to Sepharose resin. This affinity resin was used to isolate SOCS-box-binding proteins from the M1 cell lysate, which were subsequently identified by MS-based methods as elongins B and C.

Tandem Affinity Purification

For stringent purification of protein complexes, a tandem affinity purification (TAP) tag, comprising a fusion cassette encoding a low-affinity tag/enzymatic cleavage site/high-affinity tag, has been described (see Figure 10.2) (Rigaut et al. 1999). In this system, calmodulin-binding peptide (CBP) was chosen as the low-affinity tag because it allows for efficient selection and specific release from the affinity column (e.g., using EGTA under mild conditions), and ProtA (two IgG-binding units of Protein A of *Staphylococcus aureus*) as the high-affinity tag. In contrast to CBP, ProtA can only be released from matrix-bound IgG under denaturing conditions at low pH. With the TAP strategy, a specific TEV protease recognition sequence is inserted between the CBP and ProtA tags to allow specific proteolytic release of the bound complex under native conditions. For additional details and protocols using TAP, see Séraphin et al. (2002).

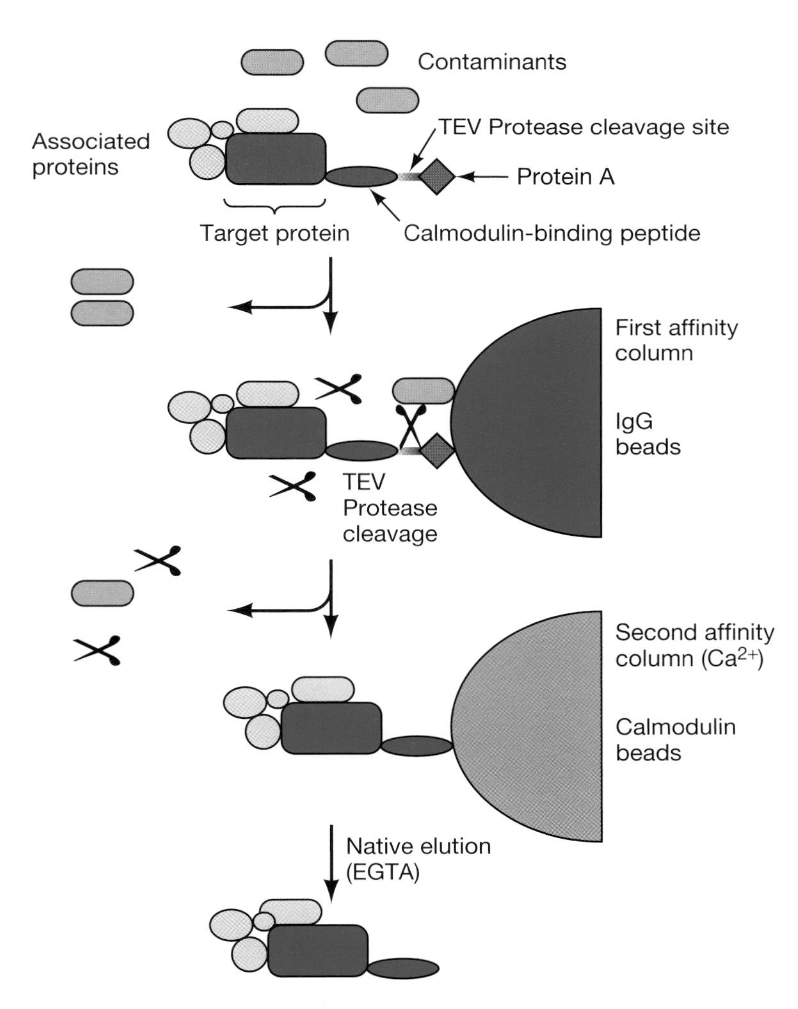

FIGURE 10.2. Scheme of TAP strategy for affinity capture. (Redrawn, in part, with permission, from Rigaut at al. 1999.)

Isolation of Multiprotein Complexes Using Electrophoresis

An important tool for the biochemist is the ability to analyze protein complexes in their native state. Two commonly used methods are native agarose gel electrophoresis and Blue Native–PAGE (BN-PAGE).

Native agarose gel electrophoresis

Native agarose gel electrophoresis is a well-established method for isolating multiprotein complexes (Hames 1990; R. Kim et al. 1999). The method uses a horizontal gel apparatus that is simple to set up, takes a short time to run, and avoids the use of toxic components. This system allows the detection of both positively and negatively charged proteins as well as protein-protein complexes in the same gel (R. Kim et al. 1999). A protein in native agarose can have either a positive or negative charge, depending on its isoelectric point (pI) and the pH of the buffer used to perform the electrophoresis. Proteins with a pI value lower than the buffer pH carry a net negative charge and migrate toward the anode, whereas proteins with a pI higher than the buffer pH carry a positive charge and migrate toward the cathode. The gel is run in a submerged horizontal platform with the wells positioned in the center of the gel. This allows for negatively and positively charged proteins to migrate toward the anode and cathode, respectively. Proteins with different molecular weights and pI values can be tested as well as proteins that form complexes, whether they be two pure proteins forming a complex or a complex formed after incubating a pure protein with a crude extract. Once a complex is formed, it can be cut from the gel and the components isolated using ultrafiltration. This method does not replace isoelectric focusing (IEF) gels because it does not determine the pI of a protein, but it does facilitate the detection of protein-protein complexes and may provide information on the protein's charge at a defined pH. A generic method for performing native agarose gel electrophoresis is given in Protocol 3.

BN-PAGE

BN-PAGE is a technique developed by Schägger and von Jagow (1991) for the separation and analysis of membrane protein complexes, often in their enzymatically active form. Its main advantage over conventional PAGE systems is the use of Coomassie Brilliant Blue G in both the sample loading buffer and cathode buffer. Binding of the dye to native proteins performs two important functions:

- It imparts a slight negative charge on the proteins, thereby enabling them to enter the native gel at neutral pH where the stability of protein complexes is most optimal.
- By binding to hydrophobic regions of proteins, the dye prevents protein aggregation during electrophoresis.

Electrophoresis separates detergent from protein complexes, which often results in the aggregation of hydrophobic membrane proteins. However, the presence of Coomassie dye in BN-PAGE maintains protein solubility, thereby enabling multiprotein complexes to separate one from another according, largely, to their apparent molecular mass (M_r). A further advantage of this technique is that protein bands are visible during electrophoresis and thus subsequent staining of the gel is not always necessary. Although the charge shift on membrane proteins and their complexes exerted by the Coomassie dye can lead to aberrant molecular masses, the M_r of proteins with a pI below 8.6 does not deviate significantly when compared with most soluble protein markers (Schägger et al. 1994).

BN-PAGE has been instrumental in the analysis of protein complexes of the mitochondrial membranes, in particular respiratory complexes (Schägger 1996). It has also been an important tool in the analysis and assembly of mitochondrial protein translocation complexes (Dekker et al. 1998; Ryan et al. 1999; Model et al. 2001). Additionally, the method has been used to study individual or multiple protein complexes from membranes including chloroplasts (Poetsch et al. 2000), endoplasmic reticulum (Wang and Dobberstein 1999), and the plasma membrane (Schamel and Reth 2000). BN-PAGE can also be used for the analysis of soluble protein complexes, as has been observed for the heptameric mitochondrial matrix form of Hsp60.

Not all membrane proteins and their complexes resolve on BN-PAGE. For example, many mitochondrial proteins streak from the high- to the low-M_r range, which may be due to the proteins dissociating from their complexes during electrophoresis or from a change in their solubility. Other complexes resolve extremely well. This variability may be dependent on a number of factors including the detergent employed and the stability of the protein complex, as well as whether Coomassie dye in fact binds to the protein(s) being analyzed. It is important to use a detergent that is efficient at solubilizing membranes but does not disrupt the integrity of the membrane protein complex. Initial studies should determine which detergents are most suitable for maintaining the particular protein complex intact prior to application and for the duration of the run. Most studies employ Triton X-100, *n*-dodecyl maltoside, or digitonin as detergents. In the case of mitochondria, digitonin gives more discernible protein complexes of higher M_r in comparison to dodecyl maltoside or Triton X-100. Indeed, the stable complexes observed on BN-PAGE following Triton X-100 solubilization can instead be seen as supercomplexes when digitonin is used (Schägger and Pfeiffer 2000).

BN-PAGE has also been used to analyze the subunit composition of membrane protein complexes. Total extracts or purified membrane protein complexes can be subjected to BN-PAGE, and individual subunits can be separated using SDS-PAGE in the second dimension. Well-resolved protein spots originating from complexes can be observed and subjected to further downstream processing such as Coomassie staining, immunoblot analysis, or amino acid sequencing. This is a particularly useful technique because two-dimensional gel electrophoresis using IEF in the first dimension may be problematic for resolving membrane proteins. Other applications of BN-PAGE include:

- Purification of membrane protein complexes for crystallization trials for structural analysis (Poetsch et al. 2000).
- Separation of individual mitochondrial respiratory complexes prior to activity determinations using in-gel histochemical staining.
- Examination of cell lines originating from patients with mitochondrial disorders (Dabbeni-Sala et al. 2001).

Protocol 4 describes a generic method for performing BN-PAGE.

UNDERSTANDING THE ARCHITECTURE OF MULTIPROTEIN COMPLEXES

Chemical Cross-linking Provides Information about Protein Proximity and Spatial Organization

Cross-linking fuses proteins that are physically close to each other (Kamp 1988). Once a linked pair is formed, it remains intact even under harsh purification conditions such as SDS-PAGE. The linked proteins can be identified using mass spectrometric methods. This allows one to conclude that these proteins either interact directly or at least reside in close proximity in a protein complex. The strategy can be employed in two scenarios: the identification of weakly or transiently bound complex components and the analysis of nearest-neighbor relations in protein complexes (Rappsilber et al. 2000).

Using cross-linkers, weak interactions can be stabilized throughout the purification of a complex. In this way, novel complex components may be identified by MS that otherwise would have been lost in the washing steps of the purification. This approach has successfully been employed for some membrane complexes (Layh-Schmitt et al. 2000). However, due to the cell membrane, which restricts free access of the cross-linkers to the interior of the cell, and the requirements of cross-linkers to work in specific buffer compositions, in vivo cross-linking is still in its infancy.

Cross-linking can also be used to investigate the spatial organization of proteins in a complex (see, e.g., Wong 1991). For such a topological analysis, a protein complex is incompletely cross-linked and the products are separated using SDS-PAGE (see Chapter 2). The components of each band, i.e., the pairs of neighboring proteins, are then identified using MS, obviating the need for special reagents like antibodies. The cross-linking reaction is best performed on a purified complex as this allows better control of the reaction conditions.

The optimal yield of the reaction cannot be 100%, because this would mean that all proteins of a complex are covalently linked together and pairs cannot be isolated. Instead, the goal should be a distribution of all possible stages of linkage—pairs, trimers, and so on. This means that only a fraction of the complex gives rise to each of the products and the yield is therefore very low. The typical material requirement with standard equipment in a nonspecialized laboratory is ~5 pmoles, allowing the analysis of products that have a yield as low as 2–5%.

Every complex is different with respect to the orientation of functional groups toward each other and within their microenvironment. Therefore, no one ideal cross-linker suits all cases. Cross-linkers with different chemical properties must be tested and the best reaction conditions must be determined empirically.

Cross-linkers are available with a variety of reactive groups, spacer lengths, rigidity of spacers, and hydrophobicity/hydrophilicity properties. Most cross-linkers are composed of two functional groups separated by a spacer. They may have identical (homobifunctional cross-linker) or different (heterobifunctional cross-linker) functional groups on each end. Cross-linkers react only with specific amino acid side chains, most often the side chains of cysteine, lysine, aspartic acid, and glutamic acid, except in the case of photoinducible groups, which are fairly promiscuous. Zero-length cross-linkers function differently by activating an amino acid side chain. It is not known beforehand which functional groups are displayed in a suitable geometry on the surfaces of the proteins in the complex to be linked. Therefore, a set of cross-linkers must be tested. In Protocol 5, the following cross-linkers were chosen: EDC, a zero-length cross-linker that activates carboxyl groups to react with amino groups; SMCC, a heterobifunctional cross-linker that reacts with sulfhydryl groups on one side and amino groups on the other side; and the two homobifunctional cross-linkers BMH and BS^3, the first linking sulfhydryl groups, and the second linking amino groups.

It is not enough that two proteins display the necessary surface groups for cross-linking. The groups must also present the correct geometry to each other. This can be tested by using cross-linkers with tethers of different lengths. Sometimes, the resulting data can even be used for a detailed study on the distances between proteins within the complex. Generally, however, the shorter the linker, the more convincing the conclusion that the linked proteins are direct neighbors. It should be noted that for steric reasons, a longer linker with the identical functional groups will not necessarily yield all of the products given by a shorter one.

To determine the best distribution of products, the reaction time, reaction temperature, and concentration of reactants must be varied. The temperature and the ratio of cross-linker to complex will influence the extent of linkage. The temperature influences the flexibility in the complex, with a higher degree of linkage at elevated temperatures. Too much linker results in complete linkage of the complex, whereas too much of the complex can result in oligomerization and linkage of proteins belonging to separate complexes. As the amount of available complex may actually be too low to allow quantification, the optimal ratio of cross-linker to complex is determined by varying the concentration of the cross-linker. An excess of cross-linker over protein is required as the cross-linker is hydrolyzed. The reaction ends due to this hydrolysis and the modification of the available surface groups of the proteins.

The partially cross-linked complex is denatured and analyzed by SDS-PAGE. To resolve the high-molecular-weight products, a low percentage of acrylamide should be chosen. The products are recognized by comparison to a lane where unlinked complex has been separated. The migration of linked protein species on a acrylamide gel can deviate from the expected behavior on the basis of the calculated molecular mass. This migration pattern may also be influenced by the position of the link due to a different geometry of the resulting complex. Products from different cross-linkers with the same protein composition may therefore have slightly different migration properties. The protein bands may be visualized using Coomassie Brilliant Blue, colloidal Coomassie, fluorescent dyes such as SYPRO stains (i.e., Orange, Red, or Ruby), as well as negative staining using $ZnCl_2$ without restriction with regard to the planned analysis by MS (for staining protocols, see Chapters 2, 4, and 7). Of these, colloidal Coomassie has the highest sensitivity. Although silver staining is even more sensitive, a protocol must be used that is compatible with mass spectrometric analysis (e.g., see Shevchenko et al. 1996; also see Chapter 2, Protocol 7 and Chapter 4, Protocol 12). Standard silver staining protocols employ high concentrations of reagents such as glutaraldehyde, which cross-link the proteins in the gel and impair the subsequent analysis by MS.

For identification by MS, the proteins are enzymatically degraded, with trypsin being the protease of choice. Extensive modification and cross-linking impairs protease activity. The low amount of linkage used in topological studies, however, is tolerated in the standard protocols (Shevchenko et al. 1996). Sensitivity can be severely compromised by losses in sample transfer. In addition, drying down the peptide solution should be avoided (Simpson et al. 1989; Speicher et al. 2000). The mass spectrometric analysis can be done using standard equipment for the identification of proteins such as ion traps, quadrupole time-of-flight (TOF) instruments, or high-sensitivity reflector matrix-assisted laser desorption ionization (MALDI)-TOF instruments (see Chapter 8).

For the initial protein identification, the peptide mass data (see Chapter 8, Peptide Mass Fingerprinting) can be used to search a nonredundant sequence database without restrictions. However, depending on the data quality, the species and protein masses may be used as constraints. The confidence in an identification can be greatly increased if fragmentation data can be obtained in some way, either by post-source decay (PSD) or by tandem MS (MS/MS) experiments in the mass spectrometer (see Chapter 8 for both PSD and MS/MS). For topological analyses, the complex components are known and the searches can therefore be restricted to these proteins. Fragmentation data are then only required if special conclusions are to be drawn from a particular observed mass. This is the case for observed signals that potentially arise from cross-linked peptides.

Degree of difficulty of this technique

As an alternative to determining the identity of cross-linked proteins, the direct identification of the linked peptides is an intriguing way to obtain information about the folding of a protein or the location of interaction surfaces (Young et al. 2000; Pearson et al. 2002). For a number of reasons, however, this technique is also rather arduous. The efficiency of cross-linking is low, resulting in a limited amount of product. The products may be homogeneous in that protein A is linked to protein B, yet the sites of linkage may be heterogeneous.

An additional problem is dynamic range. For example, although 1 pmole of product will give 1 pmole of generated peptide, the amount of cross-linked peptide will be dependent on the extent of cross-linking. For example, if only 10% cross-linking occurs, the yield of cross-linked peptide will be 0.1 pmole.

Furthermore, the sensitivity to detect linked peptides is generally lower, because they may be lost more easily during the sample work up (e.g., less efficiently extracted from the gel plugs when compared to unlinked peptides); MS has a lower sensitivity toward larger peptides, with linked peptides having twice the size. Taken together, these arguments account for the rare observation of linked peptides in a cross-link investigation. This can only be improved if a method is used specifically to enrich or detect linked peptides. In any case, a single mass alone is insufficient to allow conclusions as to what peptides are linked!

Subsecond Photo-cross-linking Using Water-soluble Metal Complexes

Conventional cross-linkers are usually applied to the analysis of interactions of a few proteins in equilibrium, mainly because typical bifunctional reagents, consisting of two electrophilic or UV-light-activated phenones or azides separated by a linker arm, work too slowly to monitor rapid changes in protein environment (Wong 1991). Additionally, the time required for these constitutively reactive reagents can effect large-scale modification of nucleophilic side chains, such as acylation of lysines, on the surface of proteins during the extended incubation times. This raises the concern of artifactual results attributable to structure destabilization.

To circumvent some of these problems, a fundamentally new type of protein cross-linking chemistry (coined PICUP or photo-induced cross-linking of unmodified proteins) based on radical coupling reactions has been developed (Fancy and Kodadek 1999; K. Kim et al. 1999). In this procedure (see Protocol 6), the proteins of interest are mixed with water-soluble metal complexes such as tris(2,2′-bipyridyl)-ruthenium(II) chloride hexahydrate (Ru[II][bpy]$_3^{2+}$) (see Figure 10.3a) (Fancy and Kodadek 1999) or a palladium porphyrin,

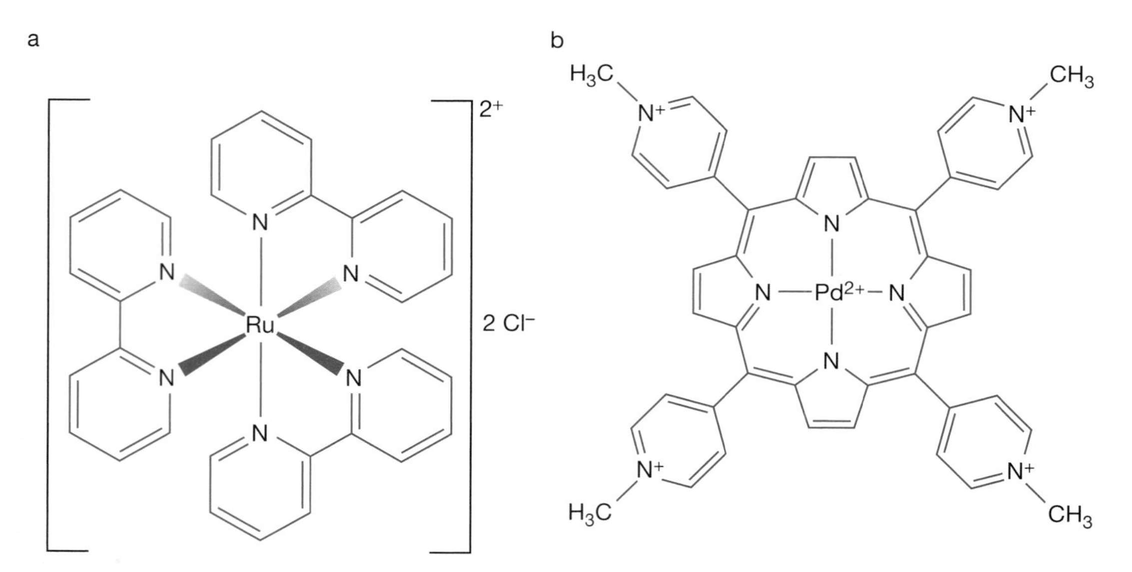

FIGURE 10.3. Water-soluble metal complexes used for sub-second photo-cross-linking: (*a*) (Ru[bpy]$_3$)Cl$_2$; (*b*) Pd(II)tetrakis-(4-*N*-[methyl]-pyridyl-porphyrin). (Courtesy of David A. Fancy, University of Texas Southwestern Medical Center.)

such as Palladium(II) (tetrakis 4-*N*-(methyl)-pyridyl porphyrin) (see Figure 10.3b) (K. Kim et al. 1999) and ammonium persulfate. The mixture is then briefly photolyzed with visible light (>400 nm) to photo-oxidize the metal complex via electron transfer to ammonium persulfate (see Figure 10.4) (Bolletta et al. 1980; Aboul-Enein and Schulte-Frohlinde 1988; Nickel et al. 1994). The now-activated metal complex, either Ru(III) or a Pd(II) porphyrin radical cation, extracts an electron from an aromatic amino acid, such as tryptophan or tyrosine, resulting in a protein-centered radical species that can attack a wide variety of other groups, eventually leading to covalent cross-linking. The speed and ease of using this technique have opened a door for the examination of protein-protein interactions not yet in equilibrium, and thus should aid biochemists in their pursuit to understand cellular function.

The Structural Organization of Protein-DNA and Multiprotein-DNA Complexes Can Be Analyzed by Site-specific Protein-DNA Photo-cross-linking

Site-specific protein-DNA photo-cross-linking is able to define positions of proteins relative to DNA within large multiprotein-DNA complexes, and thus permits analysis of the structural organization of these complexes (Bartholomew et al. 1990; Bell and Stillman 1992; Lagrange et al. 1996). In conjunction with rapid mixing and laser flash photolysis, site-specific protein-DNA photo-cross-linking is also able to define the kinetics of formation of individual protein-DNA interactions within large multiprotein-DNA complexes, and thus also permits analysis of pathways and kinetics of assembly of complexes (S. Druzhinin et al., unpubl.).

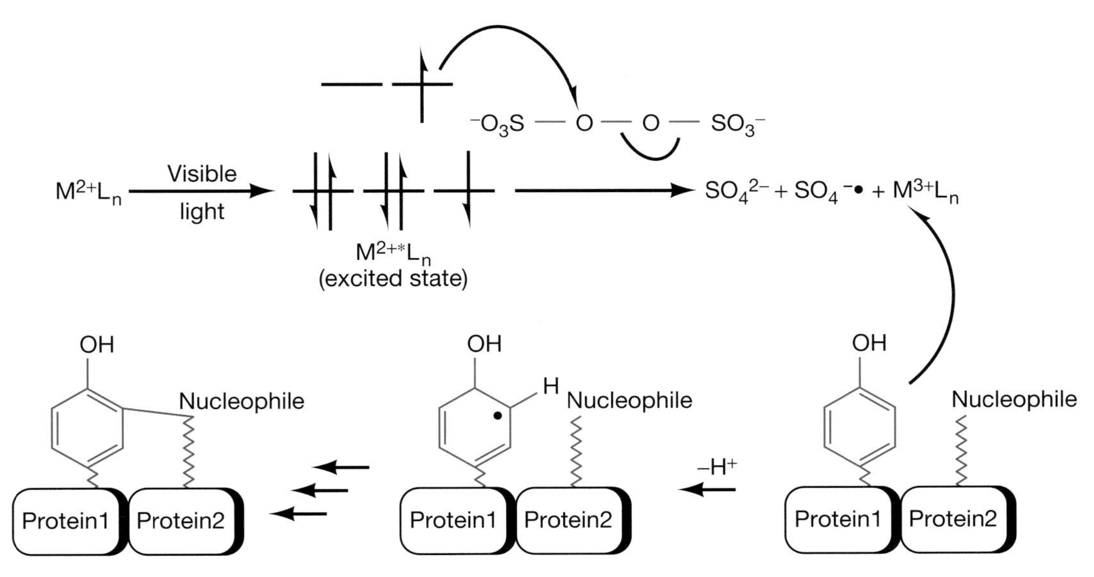

FIGURE 10.4. Scheme depicting photo-oxidation of metal complex (via ammonium persulfate) to form a radical cation that then abstracts an electron from an aromatic amino acid (e.g., tryptophan or tyrosine), resulting in a protein-centered radical species that can then attack a wide variety of other amino acid side chains, resulting in covalent cross-links. After absorption of visible light by a soluble metal complex, the excited electron is trapped by ammonium persulfate, leading to a deficient metal center and sulfate radical anion. The oxidized metal center then abstracts an electron from an aromatic amino acid, such as tyrosine, leaving a radical hole competent for attack by another aromatic side chain or nucleophile. This eventually leads to covalent linking by an undetermined mechanism. (Courtesy of David A. Fancy, University of Texas Southwestern Medical Center.)

Several procedures have been developed for site-specific protein-DNA photo-cross-linking within large multiprotein-DNA complexes (Bartholomew et al. 1990; Bell and Stillman 1992; Lagrange et al. 1996). The most effective procedure involves the following four steps (see Figure 10.5):

1. Chemical (Fidanza et al. 1992; Yang and Nash 1994; Mayer and Barany 1995) and enzymatic (Sambrook and Russell 2001) reactions are used to prepare a DNA fragment containing a phenyl-azide photoactivatible cross-linking agent and an adjacent radiolabel incorporated at a single, defined DNA phosphate (with a 9.7 Å linker between the photoreactive atom of the cross-linking agent and the phosphorus atom of the phosphate, and with an ≈11 Å maximum "reach" between potential cross-linking targets and the phosphorus atom of the phosphate).

2. The multiprotein-DNA complex of interest is formed using the site-specifically derivatized DNA fragment, and the multiprotein-DNA complex is UV-irradiated, initiating covalent cross-linking with proteins in direct physical proximity to the cross-linking agent.

3. Extensive nuclease digestion is performed, eliminating uncross-linked DNA and converting cross-linked DNA to a cross-linked, radiolabeled 3–5-nucleotide "tag."

4. The "tagged" proteins are identified, usually by denaturing PAGE followed by autoradiography.

The procedure is performed in a systematic fashion, with preparation and analysis of at least ten derivatized DNA fragments, each having the photoactivatible cross-linking agent incorporated at a single, defined DNA phosphate (typically, each second DNA phosphate [each 12 Å] on each DNA strand spanning the region of interest; Lagrange et al. 1996, 1998; Kim et al. 1997, 2000a; Wang and Stumph 1998; Bartlett et al. 2000; Naryshkin et al. 2000).

The results of the procedure define the translational positions of proteins relative to the DNA sequence. Plotted on a three-dimensional representation of a DNA helix, the results also define the rotational orientations of proteins relative to the DNA helix axis, and the groove orientations of proteins relative to the DNA major and minor grooves (Lagrange et al. 1996, 1998; Kim et al. 1997, 2000a; Wang and Stumph 1998; Naryshkin et al. 2000).

Protocol 7 presents detailed methods for cross-linking of protein-DNA complexes immobilized on streptavidin-coated paramagnetic beads ("on-bead" cross-linking; see Naryshkin et al. 2000). (For protocols on cross-linking of protein-DNA complexes in solution ["in-solution" cross-linking] or in polyacrylamide gel matrices ["in-gel" cross-linking], see Kim et al. [2000b] and Naryshkin et al. [2001].) On-bead cross-linking offers three important advantages:

- The ability to perform stringent washing to eliminate nonspecific and nonproductive complexes (Kim et al. 2000a; Naryshkin et al. 2000).
- The ability to rapidly change reaction media (e.g., to add substrates, inhibitors, or allosteric effectors; Kim et al. 2000a; Naryshkin et al. 2000).
- The ability to change reaction temperatures rapidly.

In view of these advantages, we now employ on-bead cross-linking in all photo-cross-linking experiments in our laboratory not requiring rapid mixing and laser flash photolysis (Naryshkin et al. 2000; Y. Kim et al., unpubl.).

PROTEIN-PROTEIN INTERACTIONS CAN BE ANALYZED IN VIVO WITH FRET

The ability to study protein complexes in living cells is crucial for understanding the nature of the complexes, how their properties are affected by various conditions, and how their

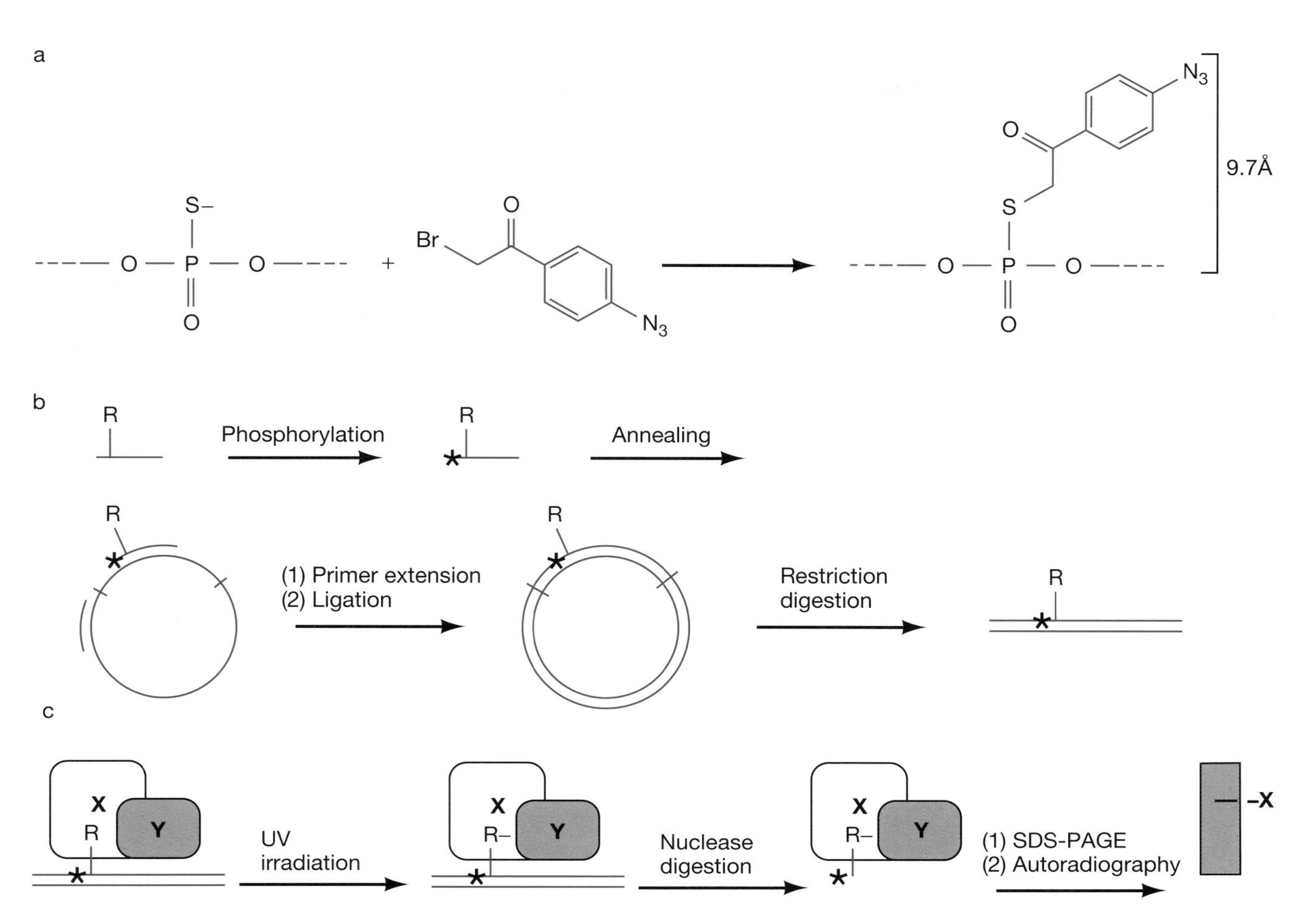

FIGURE 10.5. Site-specific protein-DNA photo-cross-linking (Lagrange et al. 1996, 1998; Kim et al. 1997, 2000a, 2000b; Naryshkin et al. 2000, 2001). (*a,b*) Chemical and enzymatic reactions are used to prepare a DNA fragment containing phenyl-azide photoactivatible cross-linking agent (R) and an adjacent radioactive phosphorus (*) incorporated at a single, defined DNA site. On the basis of the chemistry of incorporation, the maximum distance between the site of incorporation and the photoreactive atom is 9.7 Å; the maximum distance between the site of incorporation and a cross-linked atom is ≈11 Å. (*c*) UV irradiation of the derivatized protein-DNA complex initiates cross-linking. Nuclease digestion eliminates uncross-linked DNA and converts cross-linked, radiolabeled DNA to a cross-linked, radiolabeled three- to five-nucleotide "tag." The "tagged" proteins are identified by denaturing polyacrylamide gel electrophoresis and autoradiography. (Redrawn, with permission, from Lagrange et al. 1996.)

structures and functions are related. The development of green fluorescent protein (GFP) derivatives has provided an opportunity to study protein-protein interactions in living cells with FRET (Heim and Tsien 1996; Miyawaki et al. 1997; Veveer et al. 2002). GFP-based FRET has been used in a wide range of applications (van Roessel and Brand 2002).

FRET involves the indirect excitation of an acceptor, such as yellow fluorescent protein (YFP), by a directly excited donor, such as cyan fluorescent protein (CFP) (Stryer 1978; Clegg 1996). FRET can occur only under certain conditions: When the donor and acceptor are very close in space and have a certain relative orientation. The maximum interfluorophore distance depends on the properties of the specific fluorophores used and is ~52 Å for the CFP-YFP pair (Tsien 1998). Therefore, if two proteins are tagged, one with CFP and the other with YFP, the detection of FRET requires very close association between the target proteins. The use of CFP and YFP as a donor-acceptor pair is possible because of substantial overlap between the emission of CFP and the excitation of YFP (see Figure 10.6). The broad spectra of these fluorophores yields substantial cross-talk (i.e., detection of one fluorophore with fil-

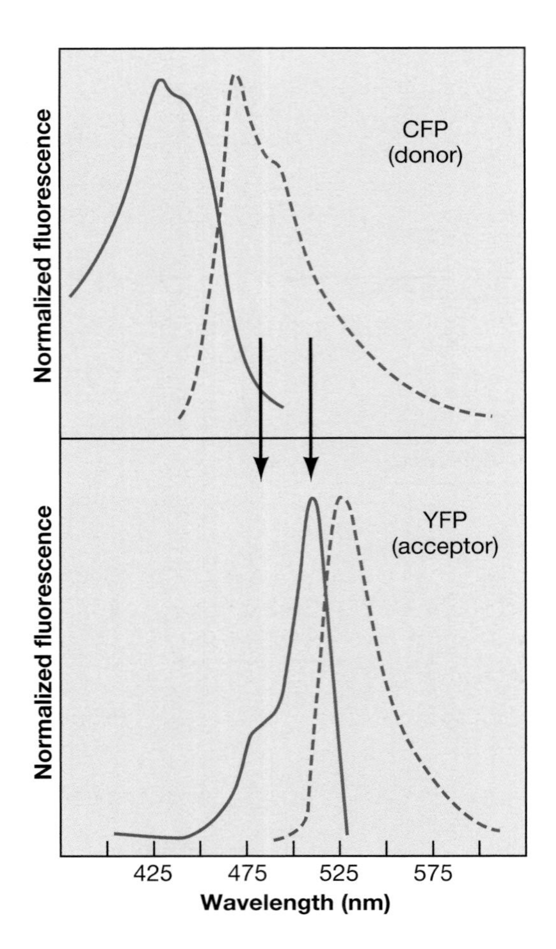

FIGURE 10.6. Excitation and emission spectra of CFP and YFP. The normalized fluorescence intensities for excitation (*solid lines*) and emission (*dashed lines*) as a function of wavelength are shown for CFP (*top*) and YFP (*bottom*). The overlap between CFP emission and YFP excitation allows FRET to occur (*arrows*) under certain conditions. (Courtesy of Marc Damelin and Pamela Silver, Dana Farber Cancer Institute.)

ters used for the other fluorophore), but this problem can be surmounted by appropriate corrections during the data analysis. FRET can be a powerful method for studying protein interactions in living cells for the following reasons:

- Target proteins are studied under physiological conditions while they are carrying out their normal cellular function. This is generally not the case for other in vivo systems, such as the yeast two-hybrid system, in which protein fusions are overexpressed and targeted to the nucleus to allow gene expression readouts.
- A detected FRET signal is more likely to represent a direct interaction than analogous results from yeast two-hybrid or coimmunoprecipitation experiments; the strong dependence of FRET on the interfluorophore distance implies that two target proteins yielding a FRET signal are very close in space.
- Protein interactions can be monitored in vivo over the course of the cell cycle, drug treatments, or other environmental stimuli.
- The structural organization of macromolecular protein complexes, which may contain scores of protein interactions and may be difficult to study in vitro, can be analyzed.

It is also important to realize the limitations of this method. In contrast to in vitro FRET experiments that allow careful calibration of concentrations and other parameters and the use of small fluorophores, in vivo FRET measurements suffer from natural fluctuations among cells and (currently) require very large fluorophores (28-kD GFPs). Thus, the precision of in vitro FRET measurements simply cannot be obtained in vivo. Additionally, the strict requirements for FRET (the distance between CFP and YFP and their relative orientation) imply that negative results are inconclusive. For example, two target proteins may interact, but the nature of the interaction compared to the experimental conditions may preclude energy transfer.

The capabilities of in vivo FRET are enhanced in tractable genetic systems. For example, the budding yeast *S. cerevisiae* offers many advantages for FRET experiments due to the ease of genetic manipulations:

- DNA can be integrated into the genome at a targeted site, such that the fusion protein replaces the endogenous protein in the cell, eliminating competition between the endogenous and fusion proteins. The fusion protein is expressed under its own promoter, and consequently, the expression level of the fusion protein is physiological and consistent from cell to cell.
- The functionality of each fusion protein can be tested genetically by replacing the endogenous protein with the fusion and testing for cell viability under conditions in which the protein is required for growth. This assay ensures that a fusion protein can carry out all essential functions of the endogenous protein.
- FRET experiments can be extended to different genetic backgrounds to study the physiological relevance of specific interactions as well as the organization of protein complexes.
- Many potential protein-protein interactions can be screened with FRET in a relatively short time period, allowing for the analysis of large protein complexes.

One application of FRET in the yeast system has been to study the nuclear pore complex (NPC). The NPC, which is central to the transport of macromolecules between the nucleus and cytoplasm, is a 60-MD membrane-embedded structure that has been difficult to analyze in its entirety in vitro. In one recent study of the NPC's structural organization, more than 100 potential interactions between individual NPC components were screened with FRET and many novel interactions were identified (Damelin and Silver 2002). Another FRET study screened potential interactions between two transport receptors and many of the individual NPC components, thus mapping the translocation pathways of the receptors (Damelin and Silver 2000). Both of these studies demonstrated the potential of GFP-based FRET for analyzing macromolecular protein complexes in vivo. A protocol focusing on a standard microscope-based assay for budding yeast is given in Protocol 8. Many aspects of this method are applicable to other systems and assays.

For measuring FRET between two target proteins, Protocol 8 can be extended to address more complex questions. In particular, the relative ease of manipulations in yeast not only facilitates the actual FRET experiments, as discussed above, but also encourages a wide range of potential follow-up experiments to characterize a given interaction. A FRET interaction between two target proteins can be compared in different genetic backgrounds. The interaction may depend on other proteins in the protein complex or on signaling proteins that regulate the interaction; mutations in the target proteins themselves may also influence the interaction. Note that an added benefit of integrating the fusion constructs into the genome is that mutant and wild-type alleles can be tagged with the same DNA, unless the mutation is very close to the fusion site. For example, the same integration vector was used to fuse YFP to *NUP1* and the *nup1-8* mutant (Damelin and Silver 2000).

The dependence of a FRET interaction on cell cycle phase or on defined environmental conditions can also be investigated. Synchronization of a cell population, drug treatments, and other environmental stimuli can be used to characterize an interaction further. However, because data acquisition can take a long time, dividing a cell culture and staggering the treatments may be necessary.

Ultimately, the study of protein interactions must extend beyond the confines of two Target proteins into the realm of large multiprotein complexes. The FRET protocol is applicable to such studies because the noninteracting proteins within the complex can serve as Control proteins. To this end, as many proteins as possible in a given complex are tagged with CFP and YFP. A recent study on the NPC provides an example of using FRET to study the structural organization of a large complex (Damelin and Silver 2002). Fifteen components of the NPC were tagged with CFP and YFP, and the various combinations were assayed for FRET. Nine novel interactions and four previously documented ones were identified. Interestingly, when the CFP tag on a 120-kD component was moved from the carboxyl to the amino terminus, different FRET interactions were observed. This result demonstrates the specificity of the FRET signal and also illustrates the utility of generating both amino- and carboxy-terminal fusions (see Protocol 8, Step 1).

PROTEIN-PROTEIN INTERACTIONS CAN BE ANALYZED BY DETERMINING BINDING CONSTANTS

Understanding the strength of the interaction between two proteins in a cell is crucial to an understanding of the consequences of that interaction for cell function. To assess the interaction between two proteins, detailed information concerning the following parameters is required:

- ***Binding constant describing the interaction of one protein with another.*** For a simple bimolecular interaction between two proteins P1 and P2, the interaction is governed by the binding constant K_d, according to the equation $K_d = [P1]\,[P2]/[P1{-}P2]$, where P1 and P2 refer to the free (i.e., unbound) concentrations of P1 and P2, respectively. The interaction between proteins P1 and P2 is also expressed in two other ways. First, it is often expressed as an affinity constant, $K_a = [P1{-}P2]/[P1]\,[P2]$, i.e., $K_a = 1/K_d$. Second, it is often expressed as a ratio of two rate constants. The rate of formation (association) of P1–P2 is $k_a\,[P1]\,[P2]$, where k_a is the association rate constant, and the rate of breakdown (dissociation) of P1–P2 is $k_d\,[P1{-}P2]$, where k_d is the dissociation rate constant. At equilibrium, the rate of formation of P1–P2 equals the rate of breakdown of P1–P2 (i.e., $K_d = k_d/k_a$).
- ***Concentrations of the interacting proteins.*** In addition to knowledge of the K_d values for two interacting proteins, the total cellular concentrations of P1 and P2 (i.e., the sum of bound and free concentrations) are needed to evaluate the extent to which P1 and P2 can interact.
- ***Solution conditions.*** Salt concentration, pH, and excipients that cause "molecular crowding" (e.g., polyethylene glycol) (for reviews on molecular crowding, see Ellis 2001 and Verkman 2002) can markedly affect the strength of protein-protein interactions.
- ***Role of cofactors.*** Several factors can influence the nature of interacting proteins, especially small effector molecules and ions such as Ca^{2+}, the cofactors such as the low-molecular-weight metabolites ATP and GTP, and posttranslational modifications (e.g., phosphorylation, glycosylation, acylation, and alkylation). Other macromolecules such as DNA, RNA,

and proteins can influence protein-protein interactions by forming multiprotein complexes involving a large number of components.

- ***Influence of competing proteins.*** Even if two proteins (P1 and P2) have a high affinity for each other, and their total cellular concentrations are sufficiently high for them to interact functionally within the cell, they may not do so in vivo. If a protein can interact with more than one other protein (i.e., P1–P3, P1–P–P4, P1–Pn, etc.), then these other proteins may effectively compete with P2 for P1. For instance, some of the competing proteins (P3, etc.) may be present in concentrations 1000-fold higher than P2, and even if their affinity is much lower (10-fold) than that of P1–P2, they will interact (titrate out) with P1, leaving negligible amounts of P1 available to interact with P2. Of course, this problem is compounded by the likelihood of P3, etc., also interacting with a different subset of proteins. (It is now generally recognized that proteins do not exist in isolation within a cell and that they are involved in more than one intracellular pathway, having several interacting partners.) Moreover, a protein that is capable of interacting with a particular protein, or subset of other proteins, is greatly influenced by its intracellular location. Although two proteins may be capable of interacting in vitro, the question remains as to whether they have the opportunity to interact with each other in vivo, especially if they are located in different cellular compartments. Some proteins are regulated in part by their partitioning between different intracellular compartments and can only interact when they traffic between these compartments (e.g., transcription factors are regulated by their partitioning between the cytoplasm and the nucleus, and they can only interact with the transcription machinery when they partition to the nucleus).

Several methods are available for determining binding constants, such as surface plasmon resonance (Schuck 1997a,b; Medaglia and Fisher 2001, 2002; Natsume et al. 2001, and references therein), gel filtration, sedimentation equilibrium, microcalorimetry (see Appendix 2), and various fluorescence methods (for fluorescence polarization assay and fluorescence gel-retardation assay, see Protocol 9) (for a more detailed understanding of binding constants, see Cantor and Schimmel 1980; Lakey and Raggett 1998).

PROTEIN MICROARRAYS FACILITATE THE LARGE-SCALE STUDY OF PROTEINS

During the past decade, microarray technology has revolutionized the study of gene expression. Arrays of nucleic acids, composed of either single-stranded oligonucleotides or double-stranded PCR products, have been used to measure the abundance of thousands of different nucleic acid species in cell or tissue samples. Because the microarray format has proven to be both convenient and effective for studying nucleic acids, many have anticipated that microarrays of proteins would similarly facilitate the large-scale study of proteins.

Protein Profiling versus Protein Function

There are two primary areas in which microarrays of proteins can prove useful: protein profiling and the determination of protein function. Protein profiling is most analogous to transcription profiling (the large-scale study of gene expression) and involves measuring the abundance, modification, and localization of many different proteins in a cell or tissue sample. This requires arrays of immobilized molecules that can selectively capture specific proteins from complex mixtures and report on their abundance and state of posttranslational modification. The arrayed molecules may be other cellular proteins, but they are more often antibodies, antibody mimics, or even nonproteinaceous receptors such as nucleic acid aptamers. Because antibodies are most often used, however, array-based protein profiling is usually discussed in the context of protein arrays.

Array-based protein profiling is not to be confused with the other application of protein arrays: the high-throughput study of protein function. For this application, the proteins being studied are themselves immobilized in a microarray format. Different assays may then be performed on the arrays to investigate various aspects of the proteins' functions, including their ability to bind to other molecules (including other proteins) or their ability to serve as substrates for protein-modifying enzymes (such as kinases or proteases).

Fabricating Protein Arrays

The substrate

Because the uses for protein arrays are many and varied, no single substrate has been adopted as a standard. However, a number of constraints direct the choice of substrate and continue to drive the field to improve in this area.

- Many applications currently use fluorescence as a detection method. It is therefore best if the substrate is intrinsically nonfluorescent. For this reason, glass is most often used, although plastics and membranes are also used.
- Most applications require the proteins being arrayed to remain folded and functional throughout the experiment. Hydrophilic substrates are therefore most often employed. For the most demanding applications (such as profiling low-abundance proteins in cell lysates), it is also important to use a surface with very low nonspecific binding to minimize background levels on the arrays. Surface chemistry plays an important role because the proteins being arrayed must attach to the substrate irreversibly and yet remain functional.
- It is useful for the substrate to be compatible with standard arrayers and scanners, which typically accommodate 2.5-cm x 7.5-cm x 1-mm microscope slides. Many vendors now sell slides in this format that can be used for a variety of applications. Typically, these slides come with protocols for arraying, blocking, and processing. With that said, two substrates are particularly useful in our hands: aldehyde slides and BSA-NHS slides. Protocol 10 details the preparation of these slides.

Sample preparation and storage

Some proteins require special buffer compositions to retain their activity. In addition, different substrates use different chemistries to mediate attachment of the arrayed proteins. For these reasons, there is no general solution when preparing proteins for arraying. However, for the substrates described above (aldehyde slides and BSA-NHS slides), the parameters discussed below should be considered.

Protein concentration and humectants

The concentration of the protein being spotted determines how much attaches covalently to the surface of the slide within a reasonable period of time. If no humectant is included in the protein sample, the nanodroplet will evaporate within a second or two of spotting, and the concentration of the protein will increase dramatically. Under these conditions, less protein is required for efficient attachment, but the protein will often denature. As a result, we typically include 20–40% glycerol in our protein samples to prevent evaporation.

The inclusion of glycerol has the added benefit that it stabilizes many proteins and enables longer-term storage at –20°C. Assuming the inclusion of glycerol, most assays require the protein being spotted to be present at a concentration of at least 100 µg/ml. Higher concentrations yield more intense spots, but little benefit is derived from exceeding 1 mg/ml. Demanding applications, such as monitoring low-abundance proteins with arrayed antibodies, typically require concentrations of at least 0.5 mg/ml.

If short peptides are being spotted, it is more appropriate to adjust their concentration in units of molarity, rather than mg/ml. This is because peptides will most often attach via their α-amine (one per peptide). With proteins, the contribution from lysine residues becomes more significant (more lysines per molecule) and hence the unit "mg/ml" is more appropriate because it provides a better estimate of the concentration of primary amines in the sample. When spotting peptides, we have found that a concentration of 20–40 µM yields good results, but lower concentrations are also acceptable. (Note that 40 µM corresponds to 0.1 mg/ml for a 2.5-kD peptide.)

Buffer, pH, and ionic strength

Proteins attach to both aldehyde slides and BSA-NHS slides via nucleophilic groups on their surface (the ε-amine of lysine residues or the α-amine at the amino terminus). As such, the buffer cannot contain molecules with strong nucleophiles, such as primary amines. Tris (Tris-[hydroxymethyl]aminomethane) (p*K*a = 8.2), a commonly used buffer, is not compatible with these slides. However, the following are some commonly used buffers that are compatible with both aldehyde and BSA-NHS slides:

MES (2-[*N*-morpholino]-ethanesulfonic acid) (p*K*a = 6.2)
PIPES (piperazine-*N,N′*-bis[ethanesulfonic acid]) (p*K*a = 6.8)
Phosphate (p*K*a2 = 7.2)
MOPS (3-[*N*-morpholino]propanesulfonic acid) (p*K*a = 7.3)
HEPES (*N*-2-hydroxyethylpiperazine-*N′*-2-ethanesulfonic acid) (p*K*a = 7.7)
Borate (p*K*a = 9.2).

The pH of the buffer affects the rate at which the proteins attach to the surface of the slides. High pH favors rapid attachment. In practice, if the protein is printed at a pH of 7 or higher, it attaches efficiently to the slide. If lower-pH buffers are required, it may be necessary to use higher concentrations of proteins or longer reaction times to achieve acceptable results. These parameters depend on the specific application and are best determined empirically.

Ionic strength can have an important role in ensuring that a protein remains folded and functional. Fortunately, ionic strength has little impact on either spotting or attachment and so can be set appropriately depending on the proteins being arrayed.

Reductants

Many proteins require reductants in the buffer to prevent unwanted disulfide bonds from forming. In addition, proteins often suffer from unwanted oxidation (such as at methionine residues) when stored without reductants. Since both dithiothreitol (DTT) and β-mercaptoethanol contain thiols, which are nucleophilic, they cannot be included in the buffer. To circumvent this problem, 1 mM Tris(2-carboxyethylphosphine) hydrochloride (TCEP; from Pierce Endogen, Rockford, Illinois) has been used successfully as a reductant. This is a water-soluble reagent that effectively reduces disulfide bonds but does not interfere with the slide chemistry.

Detergents and denaturants

In general, it is best to avoid the inclusion of detergents in the buffer. Although they do not interfere with the slide chemistry, detergents reduce surface tension and so result in larger, more diffuse spots. If detergents are absolutely necessary, they may be included, but the density of the spots on the arrays must be decreased accordingly. For some applications, it may also be desirable to array proteins or peptides under denaturing conditions and then refold the immobilized molecules on the slide surface. This may be done by including 6 M guanidine hydrochloride in the buffer. Because guanidine hydrochloride also acts as a humectant, it is not necessary to include glycerol under these circumstances. However, the inclusion of 20–40% glycerol results in more uniform spots and so is usually desirable. If guanidine hydrochloride is included in the spotting buffer, it is best not to use BSA-NHS slides since guanidine hydrochloride alters their surface and results in increased nonspecific binding.

Printing the arrays

Protein microarrays can be fabricated using most commercial or homemade arrayers. However, the quality and consistency of the spots vary depending on the arraying method. We typically array proteins in 40% glycerol/60% phosphate-buffered saline (PBS), which is relatively viscous. Split pins can pick up and deliver this buffer; however, the size of the spots tends to decrease on successive slides. To avoid this problem, the Affymetrix 417 arrayer (Affymetrix, Santa Clara, California), which uses a ring-and-pin system to pick up and deliver the sample, can be used. Noncontact methods also work well; for example, PerkinElmer Life Sciences sells the BioChip Arrayer and SpotArray Enterprise, which use piezo-electric dispensing to deliver sub-nanoliter volumes to the surface of the slide and can accommodate solutions with 40% glycerol. Regardless of which arrayer is used, care should be taken in the following areas:

- ***Humidity.*** Although glycerol is very effective at preventing the spots from evaporating, very low humidity still takes its toll. To control the humidity, the arrayer should be enclosed (most commercial arrayers are) and outfitted with a humidifier and humidistat set at 60–70% relative humidity.
- ***Temperature.*** Arrays can be fabricated at room temperature; however, the sample plates should be kept at 4°C when not in use. (Some arrayers feature automated plate handling and keep the plates in a chilled hotel.) Of particular concern is that the interior of some arrayers heats up during a print run due to poorly located motors or excessive friction. It is advisable to monitor the temperature during extended print runs and take appropriate measures to counteract this problem. Some arrayers also tend to heat the pins while drying

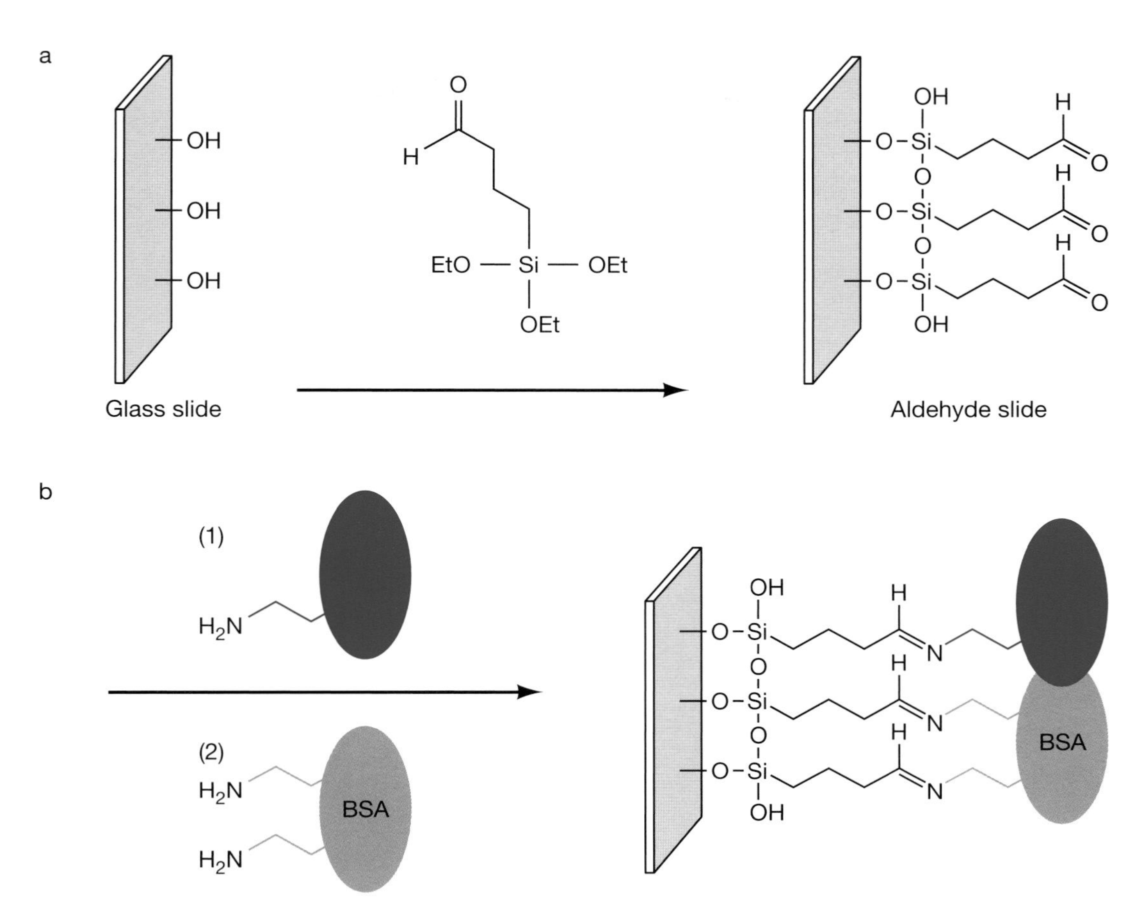

FIGURE 10.7. Aldehyde slide (see text for details). (Courtesy of Gavin MacBeath, Harvard University.)

them. Make sure that the arrayer pauses long enough for the pins to cool down before inserting them in the next sample.

- ***Carryover.*** Because buffers containing 40% glycerol are viscous, extra care should be taken in washing the pins or rings between samples to avoid carryover.

Quenching arrays

After printing the arrays, sufficient time must be allowed for the proteins to attach to the slide surface. On aldehyde slides, amines on the proteins form a Schiff base with the aldehydes on the slide (see Figure 10.7). This reaction is very rapid, so the slides may be quenched as soon as 15 minutes after arraying is complete (longer incubation times are not detrimental). On the BSA-NHS slides, amines on the proteins form amide or urea bonds with the activated carboxylates or activated amines on the BSA, respectively (see Figure 10.8). This reaction is slower than Schiff base formation, and hence, there should be a wait of at least 1 hour before quenching the slides. A method for quenching slides that prevents smearing the arrayed droplets is provided in the panel on ADDITIONAL PROTOCOL: PREVENTING THE FORMATION OF COMETS at the end of Protocol 11.

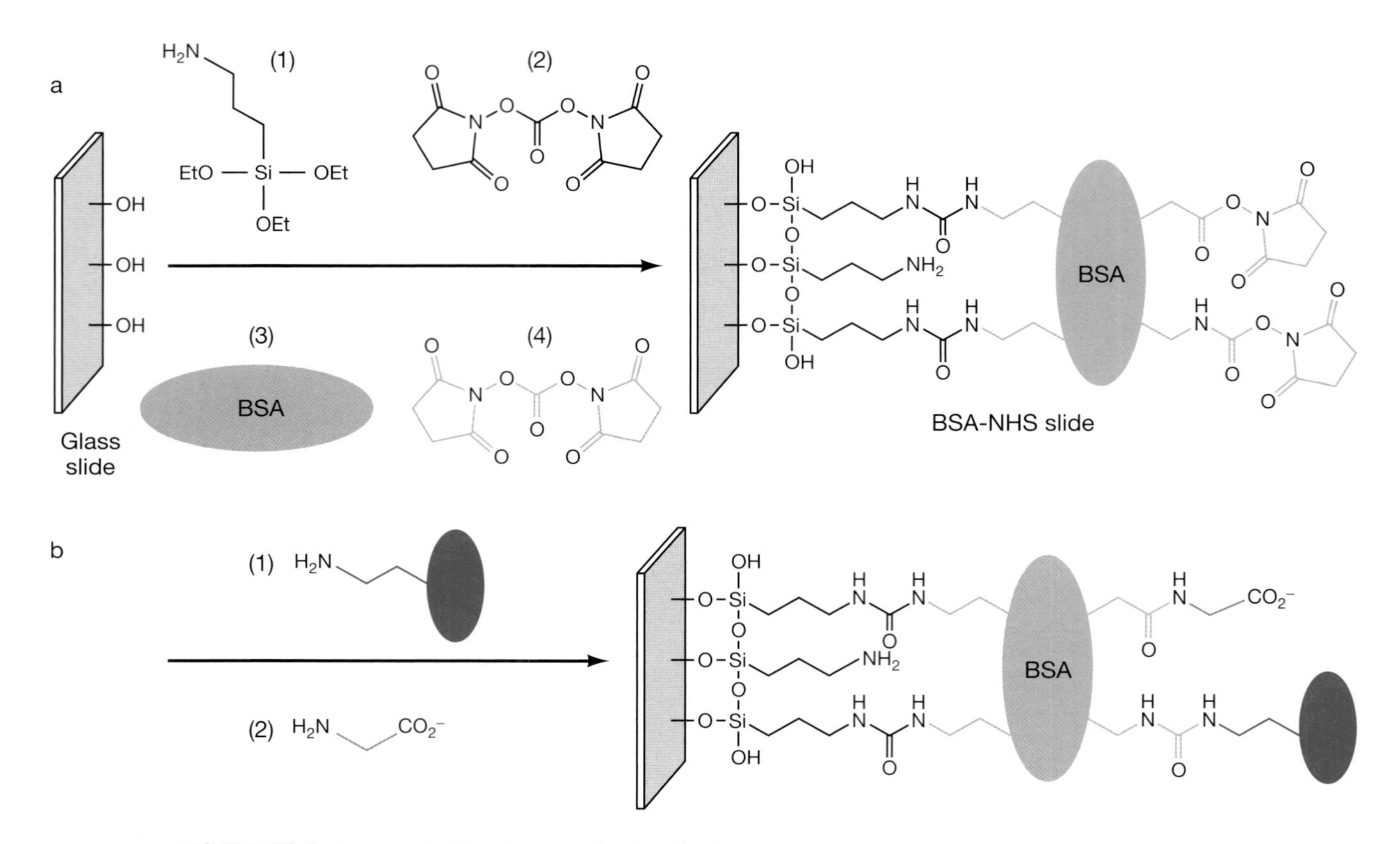

FIGURE 10.8. BSA-NHS slides (see text for details). (Courtesy of Gavin MacBeath, Harvard University.)

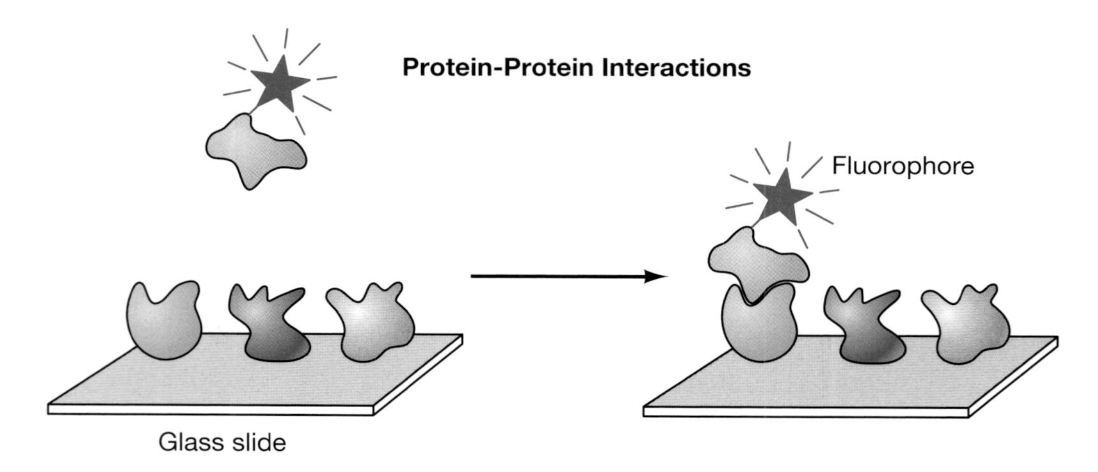

FIGURE 10.9. Protein-protein interactions (see text for details). (Courtesy of Gavin MacBeath, Harvard University.)

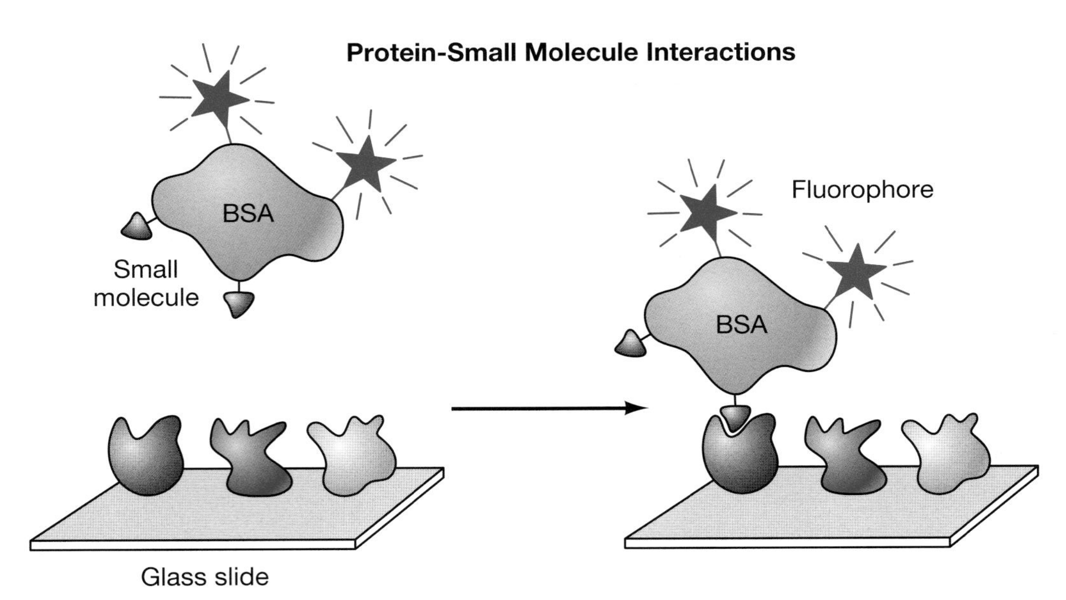

FIGURE 10.10. Protein-small molecule interactions (see text for details). (Courtesy of Gavin MacBeath, Harvard University.)

Assays with Protein Arrays

Protein-protein interactions

Protein arrays can be used to screen for protein-protein interactions. For this application, a protein microarray is incubated with a fluorescently labeled protein, the array is washed, and stable interactions are identified by scanning the slide for fluorescent spots (see Figure 10.9). It is also possible to probe the array with an epitope-, hapten-, or biotin-tagged protein and subsequently visualize bound proteins with a fluorescently labeled antibody or streptavidin conjugate. However, direct labeling of the protein of interest produces the best results. For a labeling and probing protocol for protein-protein interactions, see Protocol 11.

Protein–small molecule interactions

Protein arrays can also be used to screen for protein-small molecule interactions. For this application, a protein microarray is incubated with a fluorescently labeled compound, the array is washed, and stable interactions are identified by scanning the slide for fluorescence (see Figure 10.10 and Protocol 12). It is convenient to label compounds by covalently linking them to a carrier protein, such as bovine serum albumin (BSA), that has previously been labeled with a fluorophore. Not only is it often more convenient to label proteins in this way (rather than by directly coupling them to a fluorescent molecule), but the carrier protein also aids in rendering the labeled compound water-soluble. In addition, the valency of the conjugate can be increased to detect low-affinity interactions if desired. If the compounds cannot easily be immobilized, they may in principle be labeled with tritium, and the interactions identified on the basis of radioactive spots. However, this approach has not yet been tested.

Kinase-substrate interactions

In addition to screening for long-lived interactions, protein arrays can also be used to screen for transient interactions, such as those between an enzyme and its substrate. As long as the

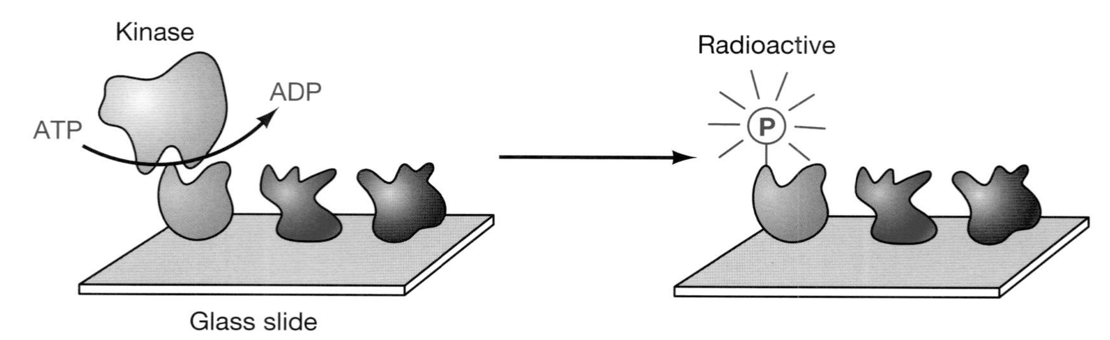

FIGURE 10.11. Enzyme-substrate interactions (see text for details). (Courtesy of Gavin MacBeath, Harvard University.)

enzyme modifies its protein substrate and that modification can be detected, the transient interaction between enzyme and substrate can be inferred. Perhaps the most useful application in this area is the identification of putative substrates for protein kinases. Protein kinases modify their substrates by transferring a phosphate group from adenosine 5′-triphosphate (ATP) to a side chain on the protein (serine, threonine, tyrosine, or histidine). To identify substrates of protein kinases, a protein microarray is incubated with a kinase and [γ-^{33}P]ATP (see Figure 10.11). Following an appropriate incubation, the array is washed, coated with a photographic emulsion, further incubated, and finally developed and imaged with a scanning light microscope (for details, see Protocol 13).

Protein profiling

In addition to screening for interactions, protein arrays can be used to quantitate the abundance and modification states of proteins in complex mixtures (such as serum, cell culture supernatants, or even cell lysates). For this application, it is necessary to array receptor molecules, such as antibodies, on the slides and then use these molecules to capture specifically their cognate antigens from solution. Although various label-free detection methods are under development, it is likely that investigators will continue to rely on fluorescent labeling methods for some time. There are two ways to detect the captured proteins using fluorescent dyes. The direct method (see Figure 10.12) relies on labeling all of the proteins in the complex solution with a fluorescent dye and then capturing the labeled proteins on the antibody array. The advantage of this method is that two different samples can be compared directly in a competitive binding experiment using two different colored dyes. The disadvantage of this approach, however, is that it is often difficult to label low-abundance proteins with high efficiency. Moreover, this approach requires extremely specific reagents to analyze low-abundance proteins (cross-reactivity of the reagents become a problem when the antigen is present at a much lower concentration than other proteins with similar epitopes).

The other way to detect the captured proteins does not require labeling the proteins themselves. Instead, each protein is captured by one reagent and then detected in a second step with a second reagent (see Figure 10.13). The second reagent (antibody) recognizes the protein at a site that does not overlap with the recognition site of the first reagent (antibody). In this sandwich approach, the second reagent is labeled and so provides the signal. The advantage of this approach is that it does not require labeling the proteins themselves. Moreover, additional specificity is gained by using two different reagents for each protein.

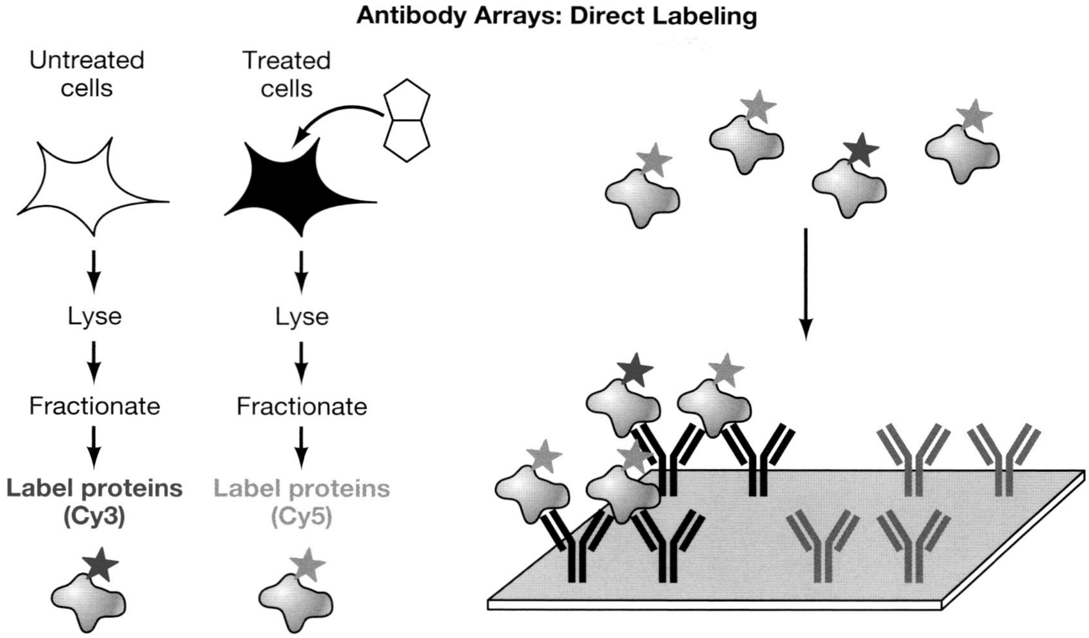

FIGURE 10.12. Antibody arrays: Direct labeling (see text for details). (Courtesy of Gavin MacBeath, Harvard University.)

The disadvantage, however, is that is it more difficult to assemble matched pairs of antibodies for each protein of interest. Although there are reasons to pursue both approaches, the sandwich approach is more appropriate for analyzing very complex solutions (cell lysates) and this method is detailed in Protocol 14.

Reading and Interpreting the Arrays

One of the advantages of constructing protein arrays on glass slides is that they can be scanned and analyzed using commercially available instrumentation. Several companies sell fluorescence slide scanners. In general, there are two types of scanners: those that use lasers to excite the fluorophores on the slide and a photomultiplier tube (PMT) to measure fluorescence and those that use a white-light source coupled with filters to illuminate the slide and a charge-coupled device (CCD) to measure fluorescence. Both types of instruments can be used to read protein arrays. The advantage of the PMT is that it is often faster and less expensive, whereas the advantage of CCD is that it tends to offer a higher signal-to-noise ratio and is more flexible with respect to the range of fluorophores that can be read.

All commercially available scanners come with image-processing software. At a minimum, the software enables the fluorescence of each spot to be quantified and the data to be exported to a spread sheet. The analysis of protein array data may become more sophisticated in the future; however, at present, we have used relatively simple manipulations to interpret our results. For antibody array data, changes in fluorescence from one sample to the next are best expressed as fold-changes. If purified antigens are available, standard curves can be generated, offering absolute, rather than relative, quantification.

Interaction data require more processing. One of the powerful aspects of protein arrays is that they can be used to determine the full $n \times n$ matrix of protein-protein interactions within a set of n proteins. To do this, n arrays of all n proteins are prepared and each array is

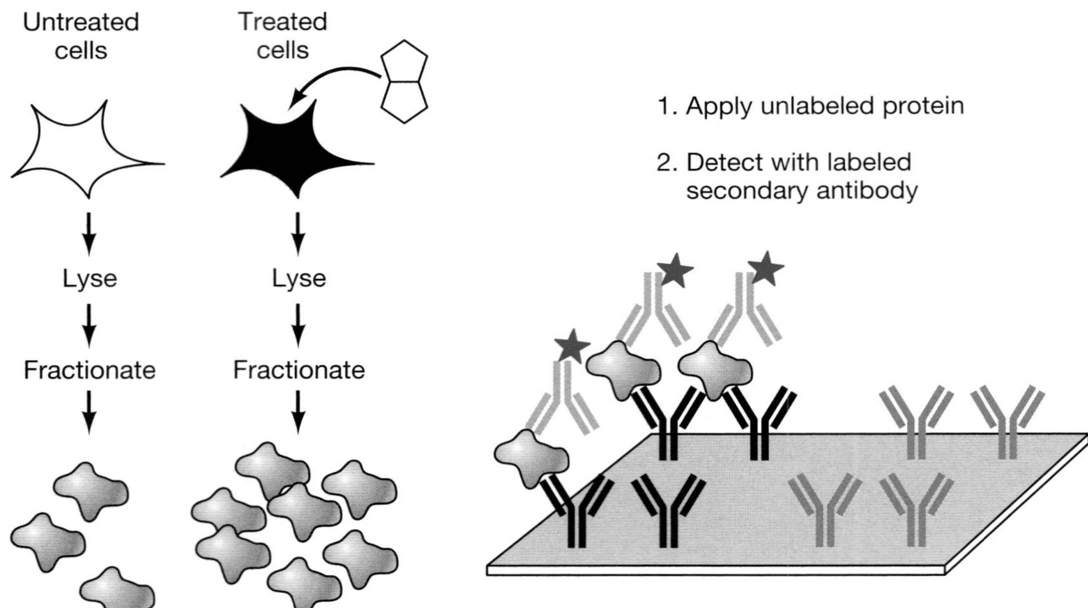

FIGURE 10.13. Antibody arrays: Sandwich labeling (see text for details). (Courtesy of Gavin MacBeath, Harvard University.)

probed with a different member of the set. Each labeled protein, however, will produce a different level of background binding to the arrays and a different maximum signal for specific binding. As a result, it is necessary to apply some sort of normalization to determine which interactions are real and which are likely to arise from nonspecific binding. We have found it useful to quantitate the fluorescence of each spot on an array and then divide that intensity by the mean intensity of all the spots on the array to yield an "x-fold above the mean" value. An arbitrary decision must then be made as to how large an "x-fold above the mean" value is required to score the interaction as real.

As a further complication, some proteins on the array exhibit higher levels of nonspecific binding than others. If only a single array is processed, it is impossible to determine whether the signal at a given spot is due to a specific binding event or arises from nonspecific binding to an intrinsically "sticky" protein. One advantage of the $n \times n$ experiment is that it is possible to see how each immobilized protein interacts with all n labeled proteins. Thus, a second level of confidence can be gained by analyzing each immobilized protein over the n different arrays. Each "x-fold above the mean" value for a given immobilized protein can be divided by the mean of the "x-fold above the mean" values on all n arrays. This effectively removes the nonspecific binding effect and highlights interactions that are more likely to be real.

PROTOCOL 1

Using FLAG Epitope-tagged Proteins for Coimmunoprecipitation of Interacting Proteins

THIS PROTOCOL WAS CONTRIBUTED BY Anne Verhagen (Walter and Eliza Hall Institute of Medical Research, Parkville, Australia).

Protein-protein interactions have a major role in transducing and regulating signaling pathways in cells. One way to detect protein interactions is by coimmunoprecipitation. In the initial analysis, immunoprecipitates can be prepared from [^{35}S]methionine-labeled cell lysates and examined. When an ^{35}S-labeled protein is detected that appears to coimmunoprecipitate specifically with the protein of interest (protein X), the procedure can be scaled up to generate sufficient amounts of the coimmunoprecipitating protein for detection by Coomassie Blue staining. This is normally sufficient for identification by MS technology. When examining protein interactions, it may be possible to immunoprecipitate endogenous protein X directly if antibodies are available. Alternatively, ectopically expressed (transiently or stably expressed) epitope-tagged protein X (e.g., FLAG-protein X) can be immunoprecipitated using commercially available epitope-tag-specific monoclonal antibodies (mAbs). The advantage of examining interactions of endogenously expressed proteins is that these are more likely to be physiological and less likely to be an artifact of overexpression. In scaling up the procedure, however, epitope-tagged overexpressed protein X may be necessary for sufficient coimmunoprecipitating protein to be isolated. This protocol describes the use of FLAG-epitope-tagged proteins for both small-scale analysis and large-scale coimmunoprecipitation of interacting proteins.

MATERIALS

CAUTION: See Appendix 3 for appropriate handling of materials marked with <!>.

Reagents

Anti-FLAG M2 agarose affinity gel (Sigma)

Glycine (100 mM, pH 3) <!>

Horseradish peroxidase (HRP)-conjugated anti-mouse secondary antibody

IEF Buffer

- 9 M urea <!>
- 324 mM dithiothreitol (DTT) <!>
- 2% pharmalytes 3-10 (Amersham Biosciences)
- 0.5% Triton X-100 <!>

Lysis buffer
- 1% Triton X-100
- 10% glycerol
- 150 mM NaCl
- 20 mM Tris (pH 7.5) <!>
- 2 mM EDTA

Supplement lysis buffer with 1 mM PMSF <!>, 10 μg/ml aprotinin <!>, and 10 μg/ml leupeptin, <!> immediately before use.

[^{35}S]Methionine or [^{35}S]methionine/[^{35}S]cysteine <!>

NaCl (saline) (150 mM)

Phosphate-buffered saline (PBS)

Polyethylenimine (PEI) (800 kD) (Fluka) <!>

Prepare a 0.9 mg/ml stock in H_2O, adjust pH to 7, and filter sterilize. This stock can be kept at 4°C wrapped in foil for several months. Always vortex the tube immediately before use.

RF10 medium
- RPMI 1640 media
- 10% fetal calf serum
- 1% penicillin-streptomycin <!>
- 2% glutamine

RPMI 1640 media (serum-free) (Invitrogen Life Technologies)

The use of RPMI and RF10 media is appropriate with 293 T cells. Other cells may require different media.

4x SDS-PAGE loading buffer
- 200 mM Tris (pH 6.8)
- 400 mM DTT
- 8% SDS <!>
- 0.4% bromophenol blue <!>
- 40% glycerol

Sodium azide (1%) <!>

Tris (1 M, pH 8)

Trypsin (cell-culture grade)

Equipment

Centrifuge, large, with fixed-angle rotor (e.g., Sorvall)

Fluorescence microscope with appropriate filters for detection of GFP

Incubator, cell culture at 37°C and 5% CO_2

Microfuge

Microfuge tubes

Needle (23-gauge)

Poly-prep chromatography columns (10-ml) (Bio-Rad)

Rotating wheel or other suitable mixing device at 4°C

SDS-PAGE and 2D-IPG/SDS-PAGE apparatus

Ultrafiltration column (e.g., Centricon-10 for 10,000 m.w. cutoff, Millipore)

See note at Step 40.

Biological Molecules

Anti-FLAG M2 monoclonal antibody (Sigma)

Mammalian expression vector encoding FLAG-epitope-tagged protein X

Vectors for expression of FLAG-epitope-tagged proteins under the control of a cytomegalovirus (CMV) promoter are available from Sigma (pFLAG-CMV series 1-7).

Optional: Mammalian expression vector encoding GFP (CLONTECH)

See the note to Step 3.

293 T cells

METHOD

The following protocols are required for both initial analysis of protein-protein interactions and large-scale purification of protein-X interacting proteins. A summary diagram of the various steps is given in Figure 10.14.

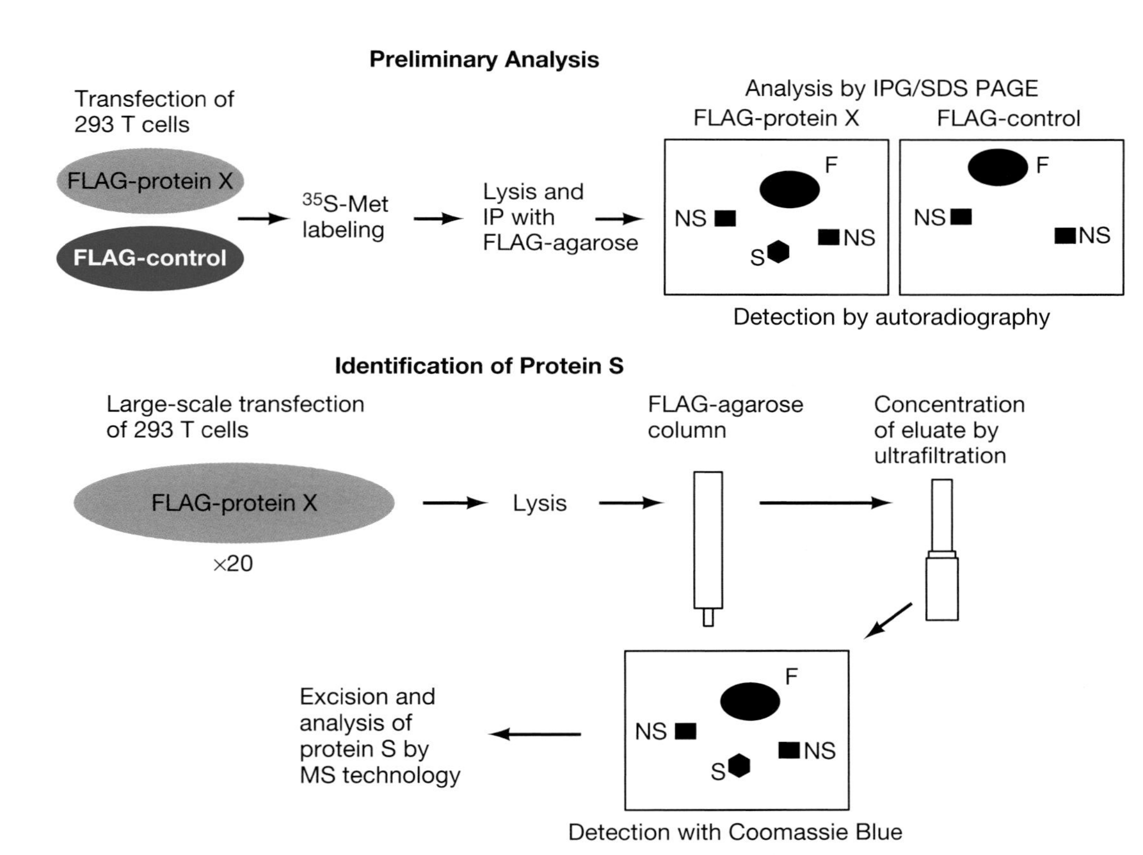

FIGURE 10.14. Use of FLAG-epitope-tagged proteins for coimmunoprecipitation of interacting proteins. In the preliminary analysis, 293 T cells are transfected with FLAG-epitope-tagged protein X and an unrelated FLAG-tagged protein (FLAG-control). Following labeling of the cells with [^{35}S]methionine, the cells are lysed, and an immunoprecipitate is prepared using FLAG-agarose beads. The immunoprecipitate is analyzed by IPG/SDS-PAGE. FLAG-tagged proteins (F), proteins that nonspecifically interact (NS), and a protein that specifically interacts (S) are indicated. For the identification of protein S, large-scale transfection of 293 T cells with Flag-protein X is performed. Protein X and interacting proteins are isolated on a FLAG-agarose column, concentrated on a Centricon ultrafiltration column, separated by 2D gel electrophoresis, and revealed by Coomassie Blue staining. Protein S is then identified by mass spectrometric techniques. (Courtesy of Anne Verhagen, Walter and Eliza Hall Institute of Medical Research, Parkville, Australia.)

Transfection of Cells with Vectors Encoding Epitope-tagged Proteins

Transfection of mammalian cells with plasmid vectors can be performed using many different procedures. The transfection procedure given below has been used successfully for the transfection of 293 T cells. Alternative methods are described by Sambrook et al. (1989) and Sambrook and Russell (2001). Many commercial transfection kits are also available.

Transfection Using Polyethylenimine

This procedure was first described in detail by Boussif et al. (1995). The method below has been adapted for the transfection of 293 T cells (Verhagen et al. 2000).

1. Grow 293 T cells in RF10 medium in a humidified incubator at 5% CO_2 and 37°C.
2. On day 1, harvest the exponentially growing cells by trypsinization and plate out the cells at a density of 1 x 10^6 cells for 10-cm plates or 3 x 10^6 cells per 15-cm plate.
3. On day 2, for a 10-cm plate, mix 12 µg of mammalian expression vector encoding FLAG-epitope-tagged protein-X DNA with 600 µl of PBS. In a separate tube, mix 48 µl of PEI (0.9 mg/ml) with 600 µl of saline. For 15-cm plates, scale up appropriately. Vortex both tubes for 10 seconds and store for 10 minutes at room temperature.

   Optional: It is helpful to have a small amount of GFP-encoding vector (1.2 µg per 10-cm plate) in the DNA mix together with the FLAG-protein X-encoding vector (10.8 µg). This enables transfection efficiency to be assessed by examining the cells under a fluorescent microscope. Green cells are normally evident 24 hours posttransfection.
4. Carefully add the PEI mix to the DNA mix. Vortex again and leave for a further 10 minutes.
5. Rinse the cells once with serum-free RPMI and add 8 ml of fresh serum-free RPMI (supplemented with glutamine and antibiotics) to the 10-cm plates (or 20 ml/15-cm plate).
6. Add the DNA/PEI mix dropwise to the plates and gently mix by tipping the plate backward and forward.
7. Incubate the cells for 4 hours in the incubator.
8. Remove the media, add fresh RF10 medium, and return the cells to the incubator overnight.

Metabolic Labeling for Small-scale Analysis

In the initial analysis of protein interactions, metabolic labeling is used to detect coimmunoprecipitating proteins.

9. One day posttransfection, remove cell culture media and add 3 ml of fresh RF10 medium supplemented with [^{35}S]methionine (final concentration 0.05 mCi/ml).
10. Return the cells to the incubator for 16 hours. The long labeling time allows for the labeling of proteins that do not have a high turnover rate.

    Optional: A higher level of incorporation will occur if RF10 medium is used that is free of unlabeled methionine, although some cells do not survive well in this media for long periods.

Cell Lysis

11. At 48 hours posttransfection, rinse the cells twice with 10 ml of PBS.

    WARNING: Make sure that radioactive waste is disposed of appropriately.

12. Remove all traces of PBS and add 1 ml of ice-cold cell lysis buffer (freshly supplemented with protease inhibitors) per 10-cm plate (or 3 ml per 15-cm plate).
13. Use a cell scraper to dislodge all cellular material from the cell culture plate and transfer the extract to a microfuge tube.
14. Allow lysis to proceed on a rotating wheel for 1 hour at 4°C.
15. Remove nuclear and cellular debris by centrifugation at 13,000*g* for 10 minutes at 4°C and retain the supernatant on ice.

Immunoprecipitation for Small-scale Analysis

16. Transfer the required FLAG-antibody-coupled agarose beads into a microfuge tube (25 µl per immunoprecipitation), and pellet the beads by centrifugation at 13,000*g* for 10 seconds.
17. Remove the supernatant, and wash the beads twice in cell lysis buffer. Centrifuge for 10 seconds after each wash.
18. Resuspend the beads in 10 volumes of lysis buffer and dispense 250 µl (= 25 µl of beads) into microfuge tubes. Centrifuge briefly to pellet the beads and remove excess buffer.
19. Add 1 ml of precleared cell lysate (from Step 15) to the FLAG antibody–agarose pellet. Incubate the beads with cell lysate on a rotating wheel for 1 hour at 4°C.
20. Pellet the beads by centrifugation at 13,000*g* for 10 seconds, and remove the supernatant carefully by suction using a 23-gauge needle.
21. Wash the beads five times by resuspending them each time in 1 ml of fresh lysis buffer. Centrifuge to pellet the beads and aspirate off the supernatant after each wash.
22. Following the final wash, carefully remove all traces of the supernatant by aspiration.
23. Resuspend the beads in 40 µl of 100 mM glycine (pH 3) to elute the proteins. Pellet the beads by centrifugation and transfer the eluate to a fresh tube. Repeat this step.
24. Neutralize the final eluate (80 µl) with 8 µl of 1 M Tris (pH 8).
25. For analysis by one-dimensional SDS-PAGE, add 4x SDS loading buffer to the sample and boil it for 5 minutes. For analysis by two-dimensional immobilized pH gradient/SDS-PAGE, dilute the sample 1:4 with IEF buffer. For details on one-dimensional and two-dimensional PAGE, see Chapters 2 and 4, respectively.

Large-scale Enrichment of the FLAG-tagged Protein and Its Interacting Partners by Antibody Column Chromatography

26. Place 0.8 ml of anti-FLAG antibody–agarose beads in an empty chromatography column.
27. Wash with 3 x 1 ml of 0.1 M glycine, allowing each milliliter to pass through the column before making the next addition.

28. Pass 10 ml of PBS through the column, pipetting the beads up and down to accelerate the process.

29. Pass 2 ml of lysis buffer through the column, allowing the beads to settle and form a bed.

30. Pass 50 ml of precleared cell lysate through the column without disturbing the bed. This is a slow process. Do not allow the beads to dry out at any point. Save 20 μl of precolumn cell lysate for later analysis (S1). Collect the lysate that has been through the column and also collect 20 μl for gel analysis (S2). During collection of the lysate, continue to hydrate the column with lysis buffer.

31. Wash the beads with 2 column volumes of ice-cold lysis buffer. Pipetting the beads up and down will speed up the process, but allow for the beads to settle and drain between each wash.

32. Wash the beads with 2 column volumes of ice-cold PBS.

33. Elute the proteins with 6 x 1 ml of 0.1 M glycine (pH 3), allowing the beads to drain between each addition and collecting 1-ml fractions in separate microfuge tubes.

34. Neutralize the fractions immediately with 50 μl of 1 M Tris (pH 8) and remove 20 μl of each for analysis (S3 to S8).

35. Wash the beads with 2 column volumes of 0.1 M glycine (pH 3), pipetting the beads up and down to accelerate the process.

36. Neutralize the beads with 2 column volumes of PBS.

37. Add sodium azide (to 0.1% v/v final concentration) directly to the PBS in the column. Seal the column and store at 4°C for future use.

38. Analyze the 20-μl samples (S1 to S8) by western blot analysis with anti-FLAG M2 monoclonal antibody and a horseradish peroxidase (HRP)-conjugated anti-mouse secondary antibody.

    The efficiency of the column in removing FLAG-protein X from the cell lysate can be determined by how much remaining FLAG-protein X is in S2 compared with the starting material (S1).

39. Of the eluted fractions S3 to S8, pool those in which FLAG-protein X is detectable by western blot analysis.

Concentration of Proteins for Purification

40. Place the pooled fractions from the column containing FLAG-protein X in a Centricon ultrafiltration tube and concentrate according to the manufacturer's instructions.

    EXPERIMENTAL TIP: A Centricon-10 tube has a molecular-mass cutoff of 10 kD and is suitable if the coimmunoprecipitating protein (i.e., the protein that interacts with protein X) has a molecular mass above this. Other Centricon tubes are available for smaller proteins.

41. When the total pooled fraction has been concentrated to a very small volume (e.g., 50 μl), invert the Centricon tube and recover the concentrate according to the manufacturer's instructions.

42. Dilute the sample 1:4 with IEF buffer.

    EXPERIMENTAL TIP: Some proteins may stick to the ultrafiltration membrane of the Centricon tube. It is possible to wash them off the membrane using IEF buffer.

Purificiation and Identification of the Interacting Proteins

43. Separate the proteins by two-dimensional IPG/SDS-PAGE (see Chapter 4).

44. Stain the gel with Coomassie Blue to visualize proteins that have interacted with protein X.

45. Identify these proteins by nano-electrospray ionization tandem mass spectrometry, as discussed in Chapter 8.

CASE STUDY

DIABLO is a protein identified on the basis of its ability to interact with the cell death inhibitor protein mammalian IAP homolog A (MIHA). For the identification of DIABLO, 293 T cells were transiently transfected with a cDNA encoding MIHA with a carboxy-terminal FLAG epitope tag (FLAG-MIHA). In the initial analysis with [^{35}S]methionine-labeled cell lysates, a protein spot corresponding to DIABLO could be detected that coimmunoprecipitated specifically with FLAG-MIHA and not another unrelated FLAG-tagged protein. To identify DIABLO, cellular lysate prepared from 100 15-cm Petri dishes of 293T cells transiently transfected with FLAG-MIHA was passed through a column of anti-FLAG antibody-coupled agarose beads. After extensive washing, the bound proteins were eluted with acidic glycine and separated by two-dimensional IPG/SDS-PAGE (see Figure 10.14) using a Amersham Biosciences Multiphor system. This involved separation of the proteins in the first dimension on 11-cm (pH 3–10) Immobiline strips, followed by separation in the second dimension on precast, solid-back SDS gradient 8–18% polyacrylamide gels. The gel was stained with Coomassie PhastGel Blue R (following the procedure for staining with Coomassie Blue described in the Multiphor electrophoresis system user's manual, except that for Step 1, fixing solution C was used instead of fixing solution N), and the stained gel spot corresponding to DIABLO was excised for analysis (Verhagen et al. 2000).

PROTOCOL 2

Affinity Purification of Interacting Proteins from Cell Lysates

THIS PROTOCOL WAS CONTRIBUTED BY Manuel Baca and Jian-Guo Zhang (The Walter and Eliza Hall Institute of Medical Research, Parkville, Victoria 3050, Australia).

The identification of protein interaction partners is a powerful method for elucidating the biological activity of a protein. Important insights into the function of a protein can be gained from a knowledge of the biochemical interactions in which the protein is involved, particularly where the function of the interaction partner is well-defined. Proteomic identification relies on isolating the interacting protein directly from a biological source and then identifying that protein through microanalytical methods. Here, recombinant protein or a chemically synthesized bioactive fragment is immobilized on resin and used as a "probe" to capture interacting proteins directly from a cell extract. The affinity-purified proteins are fractionated by gel electrophoresis and visualized by Coomassie staining. Comparison of this gel profile to that obtained from the incubation of cell lysate with a control resin (i.e., resin that lacks the immobilized probe protein) is used to identify proteins that interact specifically. Specific bands are excised from the gel, proteolytically digested in situ, and analyzed by liquid chromatography/tandem mass spectrometry to derive primary sequence information and consequently identify the interacting protein (Figure 10.15). In this protocol, a generic affinity capture procedure is described. This generalized protocol works well in several applications. However, some conditions can be altered, as discussed earlier in the chapter (see the panel on OPTIMIZATION OF VARIABLES FOR AFFINITY CAPTURE METHODS in the introduction to this chapter). A case study is provided at the end of this protocol detailing the identification of elongins B and C as SOCS-box-interacting proteins (Zhang et al. 1999).

MATERIALS

CAUTION: See Appendix 3 for appropriate handling of materials marked with <!>.

Reagents

Coomassie Blue stain
- 50% (v/v) methanol <!>
- 10% (v/v) glacial acetic acid <!>
- 40% (v/v) H_2O
- 0.1% (w/v) Coomassie Brilliant Blue R250 <!>

Ethanolamine-HCL (1 M, pH 8.0) <!>
This can be prepared from ethanolamine solution adjusted to pH 8.0 with HCl.

Immobilization buffer
- phosphate-buffered saline (PBS)
- 0.02% (v/v) Tween-20
- 0.5 mM Tris(2-carboxyethyl)-phosphine hydrochloride (TCEP) (Pierce) <!>

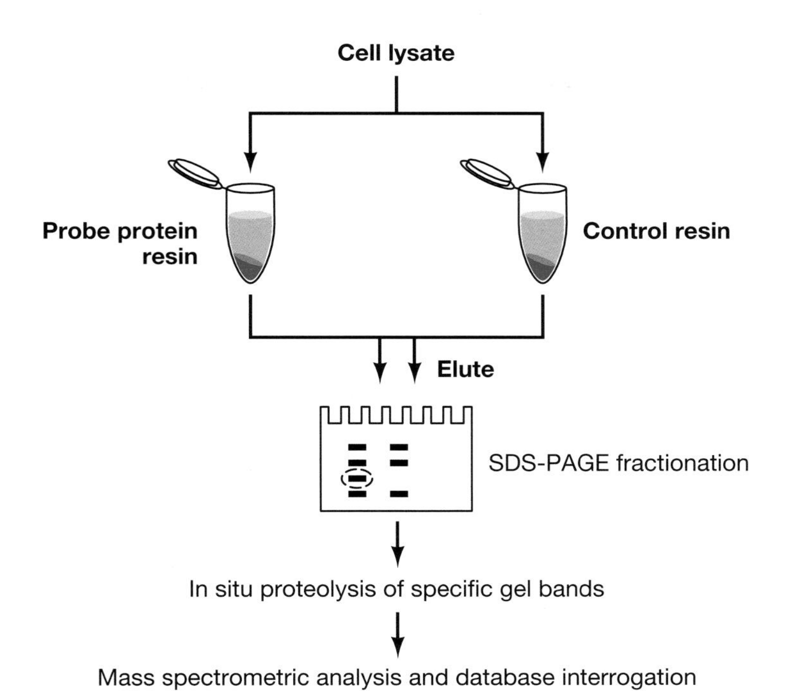

FIGURE 10.15. Outline of affinity capture strategy. (Courtesy of Manuel Baca and Jian-Guo Zhang, The Walter and Eliza Hall Institute of Medical Research.)

N-hydroxysuccinimide (NHS)-activated Sepharose (Amersham Biosciences)
Immediately before use, prepare the resin by washing four times with ice-cold 1 mM HCl <!>.

Immobilization buffer containing 0.02% (v/v) sodium azide <!>.

Lysis buffer
- 10 mM Tris-Cl (pH 7.5)
- 100 mM NaCl
- 0.5% Nonidet P-40

Immediately before use, add the protease inhibitor 1 mM phenylmethylsulfonyl fluoride (PMSF) <!> in addition to Complete Protease Inhibitor Cocktail tablets (Roche Applied Science) and the protein phosphatase inhibitors 1 mM sodium orthovandate (Na_3VO_4) <!> and 1 mM sodium fluoride (NaF) <!>.

2x SDS-PAGE gel-loading buffer
- 2.5% (w/v) SDS <!>
- 25% glycerol
- 125 mM Tris-Cl (pH 6.8)
- 0.01% (w/v) bromophenol blue <!>

Add fresh dithiothreitol (DTT) <!> to 100 mM final concentration immediately before use.

SDS-polyacrylamide gradient gel (4–20%) <!>

Sodium phosphate (0.5 M, pH 7.5)

Equipment

Boiling water bath or heating block
Chromatography columns (disposable 10-ml) (Bio-Rad)
Dialysis membrane
Gel electrophoresis apparatus

Microfuge
Microfuge tubes
Tube mixer
Preferably, use a device that permits end-over-end mixing.
Scalpel
UV spectrophotometer

Biological Molecules

Cell line growing in culture
Probe protein, recombinantly expressed, or chemically synthesized peptide probe

METHOD

Preparation of Affinity Resin

1. Dialyze the recombinantly produced probe protein against immobilization buffer for at least 4 hours. Replace the dialysis buffer and repeat this step twice.

   If the probe is a synthetic peptide that has been HPLC-purified, this may be taken up directly in immobilization buffer.

   IMPORTANT: Completely remove all traces of amine and thiol-containing reagents (e.g., Tris-HCl, DTT, and glutathione) prior to immobilizing a protein covalently on NHS-Sepharose; otherwise, these agents will themselves react with resin-bound NHS groups. TCEP is a nonnucleophilic reducing agent that will keep the protein reduced, but it will not interfere with the subsequence immobilization chemistry.

2. Quantitate the concentration of dialyzed protein by measuring the absorbance at 280 nm.

   EXPERIMENTAL TIP: Use the dialysis buffer as the spectrophotometer blank. The molar extinction coefficient of protein is readily determined by the formula nW(5690) + nY(1280), where nW and nY are the number of tryptophan and tyrosine residues within the fusion protein, respectively.

3. Incubate 0.6 mg of dialyzed protein with 200 μl of washed NHS-activated Sepharose for 3 hours at room temperature, while shaking gently on a rotary mixer.

   Prepare the control resin in tandem with the probe resin according to Steps 3–5. If the probe protein has been expressed as a fusion protein, the control resin should consist of the fusion tag (e.g., GST) immobilized onto NHS-Sepharose. Otherwise, the control resin is simply ethanolamine-blocked NHS-Sepharose.

4. Block any unreacted NHS groups by adding 0.1 volume of buffered 1 M ethanolamine (pH 8.0). Continue shaking gently on a rotary mixer for 3 hours at room temperature or overnight at 4°C.

5. Transfer the resin to a disposable 10-ml chromatography column and wash the resin with 50 ml of immobilization buffer.

   The resin can be stored at 4°C in immobilization buffer containing 0.02% (v/v) sodium azide. The resin shelf-life depends on the stability of the immobilized protein. If the recombinant probe protein is known to degrade quickly, the resin should be used as soon as possible.

   Immediately prior to use, wash the affinity resin with lysis buffer (see Step 9).

Purification of Interacting Proteins

6. Harvest 1 x 10^9 cultured cells and lyse them for 30 minutes on ice using 5 ml of ice-cold lysis buffer.

   This number of cells should be easily attainable for a cell line grown in suspension. For adherent cells, a more feasible cell number is 1 x 10^8 for initial experiments.

7. Transfer the lysate to microfuge tubes and centrifuge at maximum speed for 15 minutes at 4°C.

8. *Optional:* Add the supernatant to 150 µl of control resin freshly washed in lysis buffer. Incubate the mixture for 2 hours at 4°C on a rotary mixer, centrifuge briefly, and then remove the supernatant.

   EXPERIMENTAL TIP: This preclearing step can be used to reduce the capture of proteins that associate nonspecifically with the probe-protein resin. It is better to attempt the procedure without this step initially, but it can be included if a high level of background protein contamination is observed.

9. Transfer 50 µl of each resin to separate microfuge tubes. Wash the resins (several times) with lysis buffer, including a centrifugation step between washes to pellet the resin. Then resuspend the resins to the same initial concentration, and dispense 40 µl of each to fresh tubes for the next step.

10. Split the lysate into two equal portions. Add one half to 40 µl of probe protein resin and the other half to 40 µl of control resin. Incubate on a rotary mixer for 2 hours at 4°C.

    This amount of resin should be suitable for most purposes, but it may need adjustment according to the density of probe protein loaded on the resin. Small proteins or peptides can be immobilized at high density and consequently less resin could be used. Conversely, if the probe protein has a high molecular weight, more resin may have to be used. The greater the volume of resin used, the greater the extent of background protein contamination.

11. Centrifuge the mixtures briefly and remove the supernatants by aspiration. Wash the resins each with 2 ml of cold lysis buffer. Centrifuge the tubes briefly, and remove the supernatant. Repeat this wash step three more times.

    EXPERIMENTAL TIP: This degree of washing is given as a guide. If necessary, increase or decrease the extent of washing according to the level of background protein capture observed in the control resin experiment.

12. Elute the bound proteins by adding 30 µl of 2x SDS-PAGE gel-loading buffer to each resin sample and heating the tubes for 4 minutes at 95°C.

13. Run the samples on a 4–20% SDS-PAGE gradient gel (see Chapter 2, Protocol 1).

14. Visualize the protein bands by staining with Coomassie Blue (see Chapter 2, Protocol 2 or 3).

    If no bands are visible by Coomassie staining, repeat the experiment, but instead silver stain the gel. Note that standard silver staining methods (see Chapter 4, Protocol 11 and Chapter 7, Protocol 11) are not compatible with in situ digestion and mass spectrometric analysis. Although lower in sensitivity, consider using the mass-spectrometry-compatible silver staining procedure in Chapter 2, Protocol 7.

15. Compare the protein profiles on the two gels resulting from the capture of lysate proteins by the immobilized probe protein resin versus those captured by the control resin. Use a clean scalpel blade to excise any bands specifically captured by the probe protein resin for subsequent in situ digestion and mass spectrometric analysis.

Case Study: Identification of Elongin B and Elongin C as SOCS-box-interacting Proteins (Zhang et al. 1999)

The suppressors of cytokine signaling (SOCS) are a family of proteins that contain a central SH2 domain and an ~40-amino-acid motif at the carboxyl terminus known as the SOCS box (Starr et al. 1997; Hilton et al. 1998). In addition, several other non-SH2 protein families also contain a SOCS box at their carboxyl termini (Hilton et al. 1998). At the time of the discovery of the SOCS proteins in 1997, the SOCS box was a newly discovered motif of unknown biological function. The identification of cellular proteins that specifically associate with the SOCS box was used to provide important insights into the possible role of this motif in cell signaling pathways.

Two complementary affinity-capture approaches were used to isolate SOCS-box-interacting proteins. These approaches differed in the nature of the SOCS box probe that was used to capture interacting proteins. In one approach, the SOCS box from SOCS-1 was recombinantly expressed as a GST fusion protein and covalently coupled onto NHS-Sepharose resin. This material was used as an affinity matrix to capture specific interacting proteins from the lysate of the murine monocytic leukemic cell line M1. Alternatively, a 46-residue synthetic peptide corresponding to the SOCS box from SOCS-1 was specifically biotinylated at the amino terminus and immobilized onto streptavidin-agarose resin. This affinity matrix was also used to capture specific interacting proteins from the M1 cell lysate.

When the affinity column eluates from both approaches were analyzed by SDS-PAGE, two specific interacting proteins of 15 kD and 18 kD were detected. Excision of these gel bands and subsequent proteolysis and mass spectrometric analysis were used to identify the proteins as elongin C and elongin B, two proteins known to associate with each other and, in turn, with proteins such as elongin A and the Von Hippel Lindau (VHL) tumor suppressor protein. Interestingly, the elongin B/C heterodimer was known to associate with proteins that contain a (T,S,P)LXXX(C,S)XXX(L,I,V) consensus sequence, and indeed, this motif is a conserved feature of all SOCS boxes. The identification of elongins B and C as SOCS-box-interacting proteins had major implications for the proposed function of the SOCS box. Elongins B and C are believed to target associated proteins to proteasomal destruction; thus, it was suggested that the SOCS box acts as an adaptor for directing SOCS-associated proteins to the protein degradation pathway.

PROTOCOL 3

Native Agarose Gel Electrophoresis of Multiprotein Complexes

THIS PROTOCOL WAS CONTRIBUTED BY Rosalind Kim (University of California, Berkeley, Berkeley, California).

An important tool for the biochemist is the ability to analyze proteins in their native state. Electrophoresis of proteins and protein-protein complexes in native agarose gels using a horizontal gel apparatus is simple to set up, takes a short time to run, and avoids the use of toxic components. This system allows the detection of both positively and negatively charged proteins as well as protein-protein complexes in the same gel (R. Kim et al. 1999). A protein in native agarose can have either a positive or negative charge depending on its pI and the pH of the buffer used to perform the electrophoresis. Proteins with a pI lower than the buffer pH carry a net negative charge and migrate toward the anode, whereas proteins with a pI higher than the buffer pH carry a positive charge and migrate toward the cathode. The gel is run in a submerged horizontal platform with the wells positioned in the center of the gel. This allows for negatively and positively charged proteins to migrate toward the anode and cathode, respectively. Proteins with different molecular weights and pI values can be tested as well as proteins that form complexes, whether they be two pure proteins forming a complex or a complex formed after incubating a pure protein with a crude extract. Once a complex is formed, it can be cut from the gel and the components can be isolated using an Ultrafree-DA centrifugal filter device (Millipore) according to the method of Krowcyznska et al. (1995). This method does not replace IEF gels because it does not determine the pI of a protein, but it does facilitate the detection of protein-protein complexes and may give information on the protein's charge at a defined pH.

MATERIALS

CAUTION: See Appendix 3 for appropriate handling of materials marked with <!>.

Reagents

Agarose (powdered)

Gel destaining solution

- 45% methanol <!>
- 10% acetic acid <!>

Gel staining solution

- 0.12% Coomassie Brilliant Blue R
- 45% methanol
- 10% acetic acid

Glycerol (5%)

Native gel buffer (Buffer A)
- 25 mM Tris-Cl (pH 8.5)
- 19.2 mM glycine <!>

2x Sample-loading buffer
- 20% glycerol
- 0.2% bromophenol blue <!>
- 0.12 M Tris base

SDS-polyacrylamide gel <!>

Prepare an SDS-polyacrylamide gel according to Chapter 2, Protocol 1 or use a precast gel (e.g., Amersham Biosciences). See Step 15.

Equipment

Cellophane (ultraclear) (Idea Scientific Co., Minneapolis, Minnesota)

This cellophane is used for drying the agarose or polyacrylamide gels. Even though the agarose gels can be as thick as 0.7 cm, these gels (after soaking in 5% glycerol) can be placed between two layers of cellophane, stretched using a frame, and dried for 24 hours at 37°C to preserve the data.

Erlenmeyer flask

See Step 1a.

Gel electrophoresis apparatus (horizontal) (e.g., Horizon 58, Whatman Biometra)

Microcon centrifugal filter units (Millipore)

Microfuge

Microwave <!>

Ultrafree-DA unit (Millipore)

See Step 11.

Biological Samples

Cellular extract containing proteins of interest

For the preparation of suitable extracts from animal, yeast, and bacterial cells, see Chapter 3.

Pure protein samples (2–5 µg per lane)

Additional Reagents

Step 15 requires the reagents and equipment listed in Chapter 2, Protocol 1.

METHOD

Preparation of the Native Gel

1. Prepare a horizontal 0.8% agarose gel in native gel buffer. The dimensions of the gel should be ~8 cm x 5.5 cm x 3 mm.

   a. Add powdered agarose and native gel buffer to an Erlenmeyer flask that is three to four times the volume of the native gel buffer used. Weigh the flask.

   b. Cover the flask with an inverted beaker and place it in a microwave until the agarose melts.

   c. Reweigh the flask and add H_2O, if necessary, to bring it back to its original weight.

d. Allow the agarose to cool to 45°C in a water bath.

e. Place the comb in the center of the gel-casting assembly and pour the agarose solution into the assembly.

f. Allow the agarose to harden.

g. Place the gel into a horizontal electrophoresis unit containing native gel buffer. Make sure that the agarose gel is completely submerged. Carefully remove the comb.

Identification of Protein Complexes

2. Using purified proteins, incubate the proteins of interest together under conditions that facilitate protein complex formation. To determine whether proteins in a cell extract may bind to a protein of interest, incubate 5 µg of the pure protein with 45 µg of soluble cell extract under the appropriate conditions for forming protein complexes.

3. Mix 2–5 µg of each protein sample 1:1 (v/v) with 2x sample-loading buffer.

   The protein samples should include each individual protein (and cell extract, if used) and the mixtures incubated as in Step 2.

4. Load the samples into the wells (typically, load ~20 µl per well), and run the gel at a constant voltage of 50 V for 1 hour at room temperature.

5. Stain the gel in staining solution for 20 minutes and destain in destaining solution until the proteins bands can be clearly identified (Figure 10.16).

6. To dry the gel, soak it in 5% glycerol for 5 minutes and dry it between two sheets of cellophane membrane for 24 hours at 37°C.

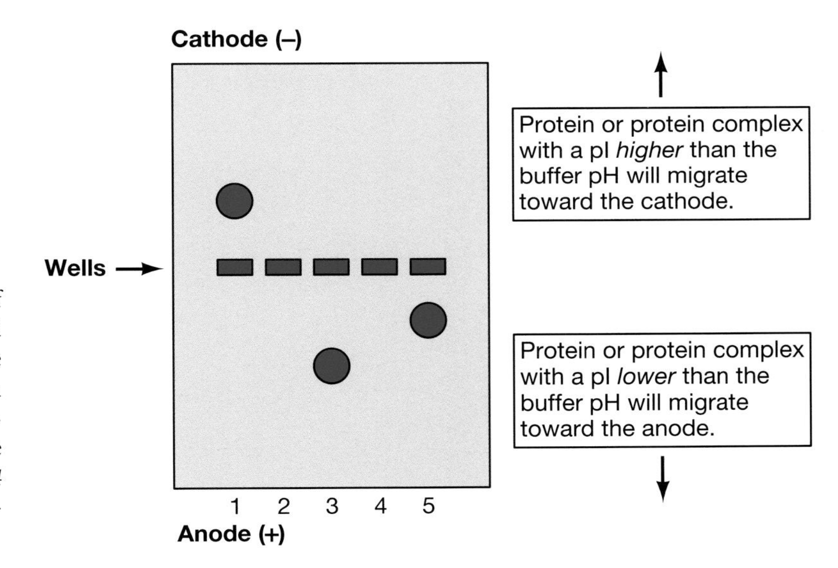

FIGURE 10.16. Schematic representation of migration of negatively and positively charged proteins and a complex of the two. Native agarose gel (0.8%) is shown as described in Protocol 3. (Lane *1*) A positively charged protein; (lane *3*) a negatively charged protein; (lane *5*) a complex of the two proteins (lanes *2* and *4* are blank). (Courtesy of Rosalind Kim, University of California Berkeley.)

Isolation and Identification of the Components of a Protein-Protein Complex

7. Load two identical protein complex samples into adjacent lanes of a 0.8% agarose gel. Run the gel at a constant voltage of 50 V for 1 hour at room temperature.

8. Use a sharp razor blade to separate the two lanes from the gel.

9. Stain one of the two lanes as described in Step 5. Keep the other lane submerged in native gel buffer (Buffer A).

10. Identify the region of the stained gel that contains the protein complex to be recovered. Using the stained gel as a guide, excise the slice of agarose from the unstained gel fragment that corresponds to the complex.

11. Place the unstained gel fragment containing the complex into an Ultrafree-DA unit. The gel nebulizer in this unit is a device that will convert the gel slice to a fine spray upon centrifugation. Place the filter unit into a filtrate vial. Centrifuge the unit at 5000*g* for 10 minutes at either room temperature or 4°C. The gel fragment must pass through the orifice of the nebulizer, thus converting the gel to a fine slurry. The gel particles are captured by the filter unit and discarded (the filter unit is not reusable).

12. Transfer the protein sample in the filtrate vial to a Microcon ultrafiltration concentrator (the cutoff size to use is determined by the size of the proteins under study).

13. Centrifuge the Microcon concentrator at 14,000*g* until the volume of the sample is reduced to 10–20 μl.

    The proteins are retained above the membrane in the Microcon sample reservoir.

14. Recover the concentrated proteins by inverting the Microcon into a new vial. Centrifuge at 1000*g* for 3 minutes.

    For complete details, refer to the manufacturer's instructions.

15. To identify the sizes of the individual components of the protein complex band, analyze the concentrated proteins by SDS-PAGE (see Chapter 2, Protocol 1).

    The percentage of acrylamide in the gel will depend on the sizes of the proteins under study (see Chapter 2, Table 2.2). Alternatively, the proteins (~4 pmoles are required) can be analyzed by ESI- or MALDI-MS (see Chapter 8).

PROTOCOL 4

Protein Analysis Using Blue Native PAGE

THIS PROTOCOL WAS CONTRIBUTED BY Michael T. Ryan (Department of Biochemistry, La Trobe University, 3086 Melbourne, Australia).

BN-PAGE was developed to aid analysis of membrane protein complexes (Schägger and von Jagow 1991). Both the sample-loading buffer and the cathode buffer contain Coomassie Brilliant Blue G. Upon binding to proteins, this anionic dye gives the proteins a slight negative charge, enabling them to enter the native gel at neutral pH, where the stability of protein complexes is optimal. In addition, by binding to hydrophobic regions of proteins, the dye prevents protein aggregation during electrophoresis. This protocol describes a generic method for performing BN-PAGE.

MATERIALS

CAUTION: See Appendix 3 for appropriate handling of materials marked with <!>.

Reagents

Acrylamide stock (49.5%T, 3%C)

Dissolve 48 g of ultrapure acrylamide <!> and 1.5 g of *N,N′*-methylene-bis-acrylamide <!> in H_2O to a total volume of 100 ml. Filter through a Whatman 3MM filter and store the solution at room temperature, away from light.

Ammonium persulfate <!>

Make a fresh 10% (w/v) solution in H_2O.

Anode buffer

Prepare a solution of 50 mM Bis-Tris (bis[2-hydroxyethyl]imino-tris[hydroxymethyl]methane). Adjust the pH to 7.0 with HCl <!>. Store at 4°C.

10x Cathode buffer I

- 500 mM Tricine
- 150 mM Bis-Tris (pH 7.0)
- 0.2% (w/v) Coomassie Brilliant Blue G

Store at 4°C.

10x Cathode buffer II

- 500 mM Tricine
- 150 mM Bis-Tris (pH 7.0)

Store at 4°C.

Destaining solution

- 40% ethanol <!>
- 10% acetic acid <!>

Dithiothreitol (DTT) <!>

3x Gel buffer

200 mM ε-amino-*n*-caproic acid <!>

150 mM Bis-Tris

Adjust the pH to 7.0 with HCl and store the buffer at 4°C. Some investigators use a final concentration of 500 mM ε-amino-*n*-caproic acid in their samples and in the gel to facilitate detergent solubilization of protein samples and to focus complexes during electrophoresis. However, in some cases, this may lead to a reduction in the stability of membrane protein complexes. Reducing the concentration of ε-amino-*n*-caproic acid has not in our hands resulted in any detrimental effects.

Isopropanol <!>

10x Loading dye

5% (w/v) Coomassie Brilliant Blue G

500 mM ε-amino-*n*-caproic acid

100 mM Bis-Tris

Adjust the pH to 7.0 with HCl and store the dye at 4°C.

SDS (2% w/v) <!>

Staining solution

0.25% (w/v) Coomassie Brilliant Blue R250

40% (v/v) ethanol

10% (v/v) acetic acid

Solubilization buffer

Most buffers containing the appropriate nonionic detergent used to solubilize membrane proteins can be applied to BN-PAGE. For mitochondrial translocation (TOM and TIM) complexes, a solubilization buffer consisting of 1% (w/v) digitonin <!> (recrystallized), 20 mM Tris-Cl (pH 7.4), 0.1 mM EDTA, 50 mM NaCl, 10% (w/v) glycerol, and 1 mM PMSF <!> is typically used because this buffer was used in the original immunoprecipitation protocols to determine the interactions between these membrane proteins (Blom et al. 1995). Other detergents (e.g., 0.5% v/v Triton X-100 <!> or 0.2% w/v *n*-dodecyl maltoside) have also been used. However, BN-PAGE analysis of mitochondrial outer membranes solubilized in Triton X-100 has shown that this detergent results in a number of small-molecular-weight TOM proteins dissociating from the main TOM complex (Dekker et al. 1998).

1x SDS-PAGE running buffer

25 mM Tris

192 mM glycine

1% (w/v) SDS

TEMED (*N,N,N′,N′*-tetramethyleneethylenediamine) solution <!>

Transfer buffer

48 mM Tris base

39 mM glycine

0.04% (w/v) SDS

20% (v/v) methanol

To recrystallize digitonin, boil the powder in ethanol and store it for 20 minutes at –20°C. Centrifuge the digitonin precipitate at 5000*g*, and dry the pellet in a vacuum desiccator.

Although concentrations of NaCl above 100 mM are used in the solubilization of membranes with Triton X-100, this salt may be problematic when the proteins are stacked during BN-PAGE. ε-Amino-*n*-caproic acid, which has been reported to enhance the actions of detergents in the solubilization of proteins (Schägger and von Jagow 1991), should be used instead of NaCl. It also acts as a serine protease inhibitor and may therefore maintain the stability and activity of protein complexes. Bis-Tris is used in the gel buffer because this base has a slightly acidic pK (6.5–6.8) and can therefore maintain the pH at 7.0. Tricine, with a pK of 8.15, is used as the trailing ion (Schägger and von Jagow 1991).

Equipment and Conditions

BN-PAGE equipment

Cooled gel system (e.g., SE 600 standard dual cooled vertical unit, Amersham Biosciences)

The SE 600 can accommodate up to four gels at a time, enhancing reproducibility. The large anode buffer chamber with a cooling system enables electrophoresis to run at a constant temperature.

Gel plates (18 x 16 cm)

Spacers (typically 1–1.5-mm thick)

Combs (containing 10–15 teeth)

For preparative BN-PAGE, use a comb containing one normal-sized tooth for loading marker proteins and a single large tooth for the sample.

Mini-gel systems can also be used and a number of different ones are now on the market. Use one with an integral cooling chamber, enabling the apparatus to be connected to a recirculating refrigerated water bath, to maintain a constant temperature under the high-voltage conditions that generate excess heat.

Plates and equipment should be cleaned thoroughly and rinsed well with H_2O to ensure that no traces of detergent (which could otherwise denature proteins) remain. If BN-PAGE is to be performed regularly, a complete PAGE system reserved exclusively for this technique is recommended. A conventional homemade vertical gel apparatus can be also be used and run in a 4°C room. In this case, the temperature should be monitored regularly and the voltage adjusted appropriately to ensure that the gel does not become warm, which may otherwise lead to protein aggregation or disassembly of protein complexes.

Electroblotting apparatus

Gradient maker

This gradient maker should be sufficient to hold a total volume equivalent to that of the gel employed (e.g., 20 ml) and have mini-sized spin bars for mixing of acrylamide solutions.

Magnetic stirrer

Microfuge

Peristaltic pump

Power supply (with maximum 500 V or above)

PVDF membrane

Refrigerated circulating H_2O bath

This bath should be connected to the cooling chamber of the gel apparatus. Because heating during the electrophoresis run does occur, it may be necessary to monitor the temperature of the anode buffer at times, using a thermometer. Positioning the gel chamber on top of a magnetic stirrer and placing a magnetic spin-bar within the anode buffer chamber can maintain the buffer temperature during the electrophoretic run.

Biological Molecules

Marker proteins

The marker proteins must be in their native configurations. We use a set of markers (from Amersham Biosciences) containing thyroglobulin (669 kD), ferritin (440 kD), catalase (232 kD), lactate dehydrogenase (140 kD), and BSA (66 kD). Resuspend 250 µg of protein in 800 µl of 1x loading dye and load 50 µl per lane when using a large gel. This corresponds to 2.5–5 µg of protein per band. These soluble proteins migrate as discrete bands at their appropriate molecular weights.

Protein samples

The protein samples can range from purified complexes to subcellular fractions to total cellular extracts. For mitochondrial extracts, mitochondria are isolated using standard techniques (Spector et al. 1997). Solubilize mitochondrial pellets in an appropriate detergent-containing buffer. Approximately 100 µg of total protein in a volume of 50 µl of solubilization buffer loaded in each well is sufficient to observe most abundant respiratory complexes. To enrich membrane complexes, prepare a total membrane pellet by sonicating a mitochondrial fraction in hypotonic buffer, pellet the intact mitochondria at 10,000*g*, and then centrifuge the supernatant at 100,000*g* to recover membranes.

Additional Reagents

Steps 24–28 require the reagents and equipment listed in Chapter 2, Protocol 1.

METHOD

Preparation of a Gradient Gel

A 4–16.5% acrylamide gradient gel with a 4% acrylamide stacking gel is typically used. This resolves protein complexes from >1000 kD to ~40 kD. However, because the top of the gel can be fragile and if the size of the complex to be analyzed is less than 600 kD, a 6–16.5% (or other) gradient gel may be employed. It may be convenient to make acrylamide mixes sufficient for pouring a number of gels since this will aid in reproducibility (see Table 10.3). The higher-percentage acrylamide mix contains 10% (v/v) glycerol, which aids in the formation of the gradient by preventing unwanted mixing of layers during gel casting. A simple gradient maker, with a small magnetic spin bar placed in the chamber containing the higher-percent acrylamide mix, is centered on top of a magnetic stirrer. The chamber is stirred at considerable speed to aid mixing; however, in some cases, high velocities of stirring can slow the entry of the lower-percent acrylamide mix into the mixing chamber. For additional information on pouring gradient gels, see Figure 10.17 and Chapter 2, Protocol 1.

1. Determine the volume needed to fill the gel, leaving ~2.5 cm for the stacking gel in addition to the lengths of the wells of the comb used. (For 18 x 16-cm gels with 1-mm-thick spacers, use 9 ml of each acrylamide solution, thereby leaving ~5 cm of space above the gradient.) Fill both gradient chambers with exact volumes of each acrylamide mix.

2. To 9 ml of acrylamide, add 4 µl of TEMED and 40 µl of 10% (w/v) ammonium persulfate to each chamber and mix thoroughly.

   Different volumes of acrylamide will require different volumes of TEMED and ammonium persulfate.

3. Turn the pump on and then open the valve between the two gradient chambers to enable the flow of the lower-percent acrylamide mix into the chamber containing the higher-percent acrylamide mix.

   EXPERIMENTAL TIP: The flow rate should be quite fast, because a constant and continuous stream of acrylamide running from the tubing into the gel results in the formation of a more linear gradient.

4. Layer isopropanol onto the surface of the acrylamide between the glass plates, to level the gel and aid in polymerization.

TABLE 10.3. Acrylamide mixes for BN-PAGE

	4%	6%	14%	16.5%	20%
3x Gel buffer	30 ml	30 ml	30 ml	30 ml	30 ml
Acrylamide stock	7.3 ml	11 ml	25.5 ml	30 ml	36.4 ml
Glycerol	–	–	18 ml	18 ml	18 ml
H_2O	62.7 ml	49 ml	26.5 ml	12 ml	5.6 ml
Total 90 ml					

Table courtesy of Michael T. Ryan (LaTrobe University).

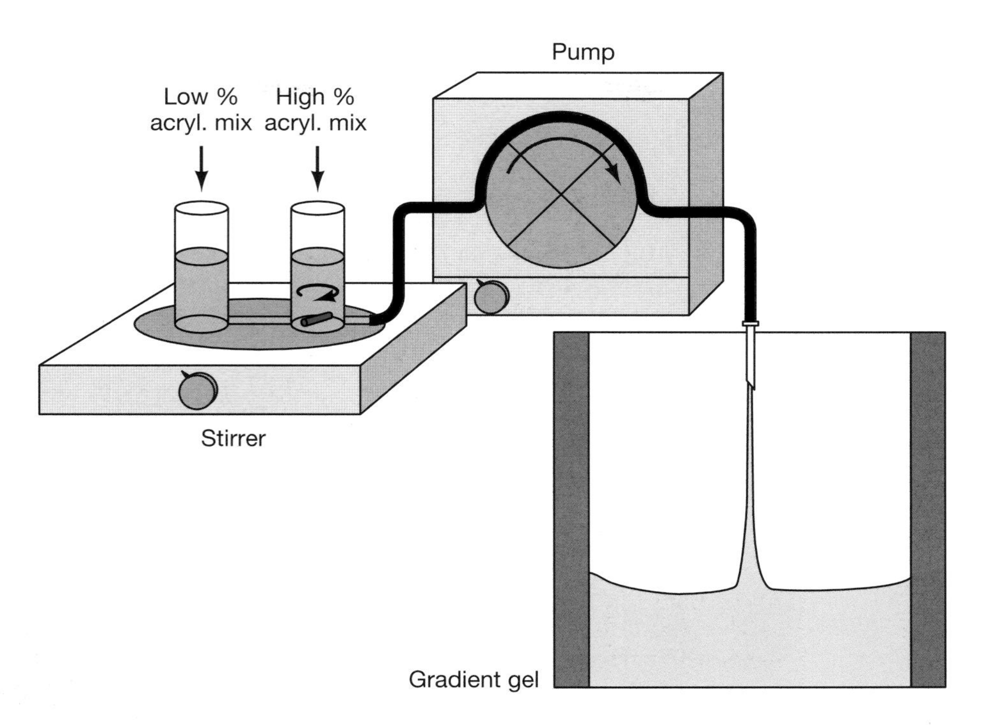

FIGURE 10.17. Schematic representation of pouring a gradient BN-PAGE gel. Low- and high-percent acrylamide mixes (see Table 10.3) are placed in the appropriate gradient chamber. The pump is turned on causing the acrylamide to flow between the gel plates and enabling mixing of the low-percent acrylamide with the high-percent acrylamide. (Courtesy of Michael T. Ryan, LaTrobe University.)

5. Immediately after use, thoroughly wash out the gradient chamber and tubing with H_2O to prevent blockage by polymerized acrylamide.

6. Following polymerization, remove the isopropanol. Wash the top of the gel with H_2O, and blot the inside surfaces of the glass plates dry using filter paper.

7. Prepare a 4% stacking gel solution and layer the solution (see Table 10.3) on top of the polymerized gradient gel.

8. Insert the comb carefully into the stacking gel solution, starting first from one side, to avoiding trapping air bubbles at the gel-comb interface.

9. Once the gel is completely polymerized, store it at 4°C. When kept moist (e.g., surrounded by tissues soaked in H_2O and then wrapped in plastic wrap, and placed in a container), the gel is stable for up to 1 week.

Loading the Protein Samples

10. Remove the comb from the gel and wash the wells with H_2O to remove unpolymerized acrylamide and excess reagents.

11. Resuspend the protein samples in the appropriate solubilization buffer. Centrifuge the samples in a microfuge at maximum speed for 5 minutes at 4°C to pellet any aggregated material.

12. Add loading dye to the protein samples (5 μl is added to 45 μl of sample for large gels or 2 μl to 18 μl of sample for minigels).

13. Load the samples into the wells using a syringe (or gel-loading pipette tips). Make sure to rinse the syringe thoroughly in H_2O between each sample. Gently overlay the samples with Cathode buffer I.

    Alternatively, place Cathode buffer I in the wells first, and then load the samples through the buffer, where they will settle to the bottom of the well.

Electrophoresis Conditions

14. Run large gels at 100 V until the samples have migrated through the stacking gel and then increase the voltage to 500 V for migration of the proteins through the gradient gel (total electrophoresis time ~5 hours).

    Alternatively, when set at 6 mA per gel and a maximum of 180 V, electrophoresis is complete after 16 hours. For minigels, run at 6 mA (maximum 400 V) for 1 hour and 20 minutes.

15. After the Coomassie dye has entered the gel and traveled at least to the top of the gradient gel, replace Cathode buffer I with Cathode buffer II. The lack of Coomassie dye in the latter buffer enables the proteins to become visible in the gel during electrophoresis. Transferring to Cathode buffer II is required if the proteins will be electroblotted.

    Alternatively, if it is known beforehand how long to run the gel and the proteins will not be electroblotted, Cathode buffer I can be used exclusively during electrophoresis. Destain the gel after electrophoresis is complete to visualize protein complexes.

16. Stop the electrophoresis run when the Coomassie Blue dye reaches the bottom of the gel. Proceed to visualize the proteins by staining the gel (Step 17), electroblot the proteins to a PVDF membrane (Step 19), or separate the protein complexes into their component proteins by SDS-PAGE (Step 21).

Staining the Protein Complexes

Protein complexes are often directly visible following electrophoresis. Additional complexes can be observed by staining the gel.

17. Incubate the gel with staining solution for ~1 hour with gentle shaking.

18. Destain the gel by incubating it with multiple changes of destaining solution until clear bands are observed.

Electroblotting the Protein Complexes

19. Soak the gel in 1x SDS-PAGE running buffer for ~20 minutes.

20. Rinse the gel in transfer buffer and electroblot onto a PVDF membrane. Using the semidry method, set transfers of large gels at 200 mA for 1.5 hours.

    EXPERIMENTAL TIP: Nitrocellulose is not recommended, because the Coomassie dye binds irreversibly to this membrane causing problems in downstream processes such as immunodecoration.

Isolating the Protein Complex Components by SDS-PAGE

To analyze the composition of protein complexes, SDS-PAGE can be performed in the second dimension. This is particularly important when analyzing proteins by immunodecoration. Because polyclonal antibodies often cross-react with other proteins in a sample, it can lead to anomalous results when BN-PAGE blots are decorated. To confirm that an immunodetected complex indeed contains the antigen of interest, a lane from the BN-PAGE gel is excised and subjected to SDS-PAGE, thereby resulting in the separation of proteins from the various complexes (Figure 10.18).

Alternatively, the protein profile can be observed by staining (Figure 10.19a) the SDS-PAGE gel. In the case of mitochondria, 150–200 μg of total protein applied to the first dimension is required to observe individual protein spots on the SDS-PAGE gel.

21. Excise lanes from the BN-PAGE gel using a razor blade or the edge of a thin gel spacer.

    EXPERIMENTAL TIP: Gel strips can be wrapped in Parafilm and stored at –80°C for long periods prior to SDS-PAGE.

22. If possible, use a razor blade to remove the blue migration front at the very bottom of the strip (containing dye and detergent) so that the large concentration of Coomassie stain at this end does not trail from the front during SDS-PAGE.

23. Soak the gel strip in 2% (w/v) SDS containing 0.2 M DTT for 20 minutes.

24. Remove excess solution from the gel strip. Use forceps to gently place the strip on one of the gel plates (will be used to prepare the SDS-PAGE gel, see Chapter 2, Protocol 1) at a

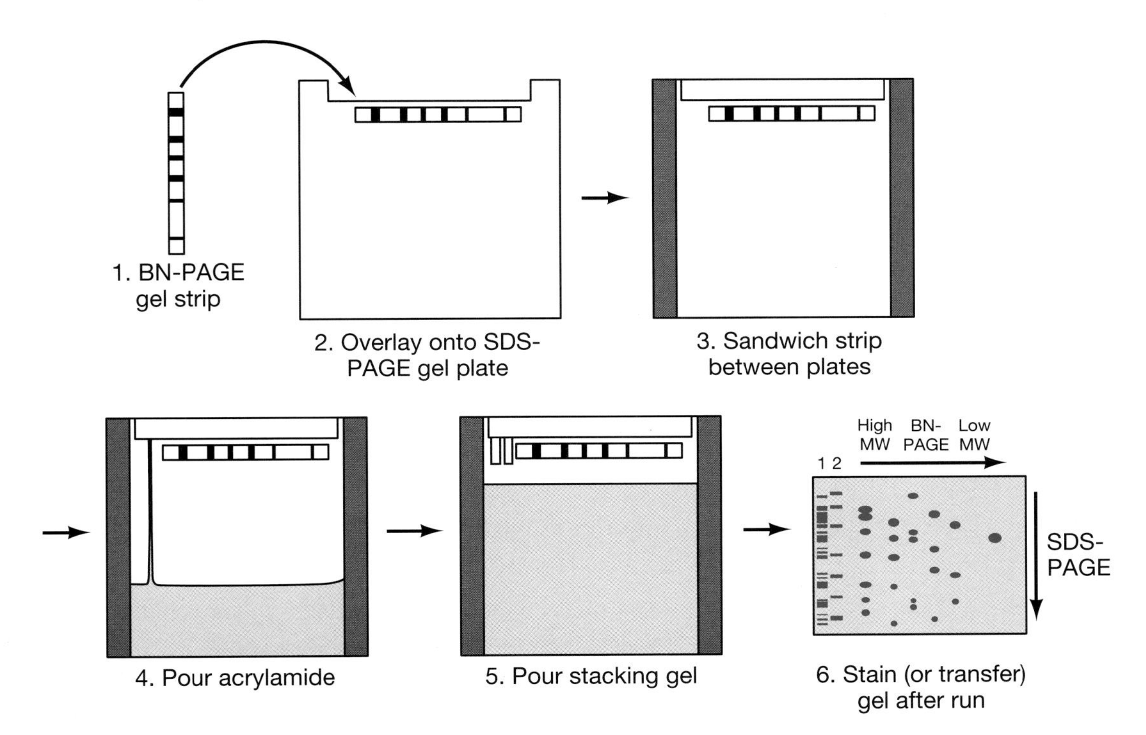

FIGURE 10.18. Schematic representation of performing SDS-PAGE in the second dimension following BN-PAGE. (Courtesy of Michael T. Ryan, LaTrobe University.)

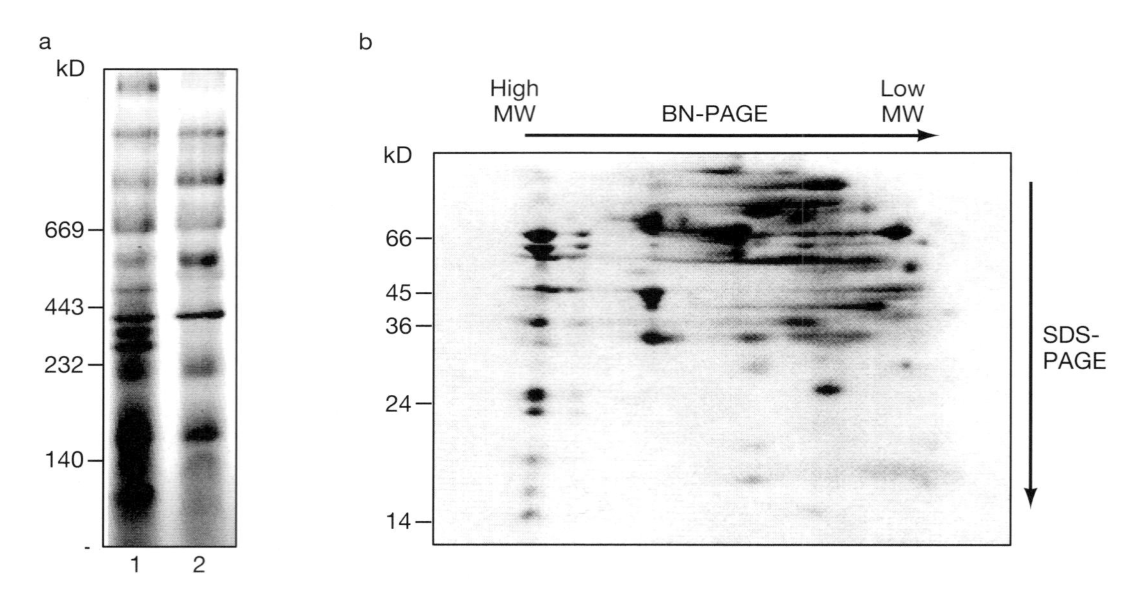

FIGURE 10.19. Coomassie-stained mitochondrial extracts following BN-PAGE (*a*) and 2D-PAGE (*b*). (*a*) Mitochondrial extracts (100 μg) isolated from rat kidney (lane *1*) or rat heart (lane *2*) were solubilized in buffer containing 0.5% (v/v) Triton X-100 and subjected to BN-PAGE (4–16.5% gradient) followed by Coomassie staining. Multiple complexes can be observed. (*b*) Yeast mitochondria (150 μg of protein) were solubilized in digitonin-containing buffer and subjected to BN-PAGE followed by SDS-PAGE in the second dimension and subsequent Coomassie staining. (Courtesy of Michael T. Ryan, LaTrobe University.)

position that the wells would normally occupy. Place the other glass plate on top so that the gel strip is sandwiched between the two plates (Figure 10.18). The spacers should be the same thickness as those used in BN-PAGE.

25. Pour the separation gel solution to ~3 cm below the gel strip and overlay with isopropanol. Avoid wetting the strip by tilting the gel on its side while pouring.

26. After polymerization, remove the isopropanol and pour the stacking gel solution over and around the BN-gel strip. Make sure that bubbles do not become trapped under the gel strip.

27. For comparison, place a comb containing two lanes in the stacking gel at the side of the gel strip to load both SDS-PAGE molecular-weight markers and a sample of the original protein that was applied to BN-PAGE, but in this case, prepared for SDS-PAGE by boiling in Laemmli sample buffer (see Chapter 2, Protocol 1).

28. Run the gel under standard SDS-PAGE conditions (see Chapter 2, Protocol 1).

29. Following electrophoresis, stain the gel, electroblot the proteins, or prepare individual protein spots for mass spectrometry.

    Individual protein spots obtained from 2D-PAGE can be subjected to protein sequence analysis using various techniques. Alternatively, protein complexes can be excised directly from the BN-PAGE gel and subjected to in-gel proteolytic (e.g., tryptic) digestion (Matsudaira 1993, and see Chapter 7, Protocol 13) followed by separation of peptides using HPLC and subsequent sequencing using Edman degradation (see Chapter 6, Protocol 1) or mass spectrometry (see Chapter 8).

PROTOCOL 5

Analysis of the Topology of Protein Complexes Using Cross-linking and Mass Spectrometry

THIS PROTOCOL WAS CONTRIBUTED BY Juri Rappsilber and Matthias Mann (Protein Interaction Laboratory, Department for Biochemistry and Molecular Biology, University of Southern Denmark, Campusvej 55, DK-5230 Odense M, Denmark).

This protocol is designed to allow identification of spatial relationships between proteins within a wide variety of multiprotein complexes. Details for optimizing cross-linking of proteins within multiprotein complexes are provided. Following the cross-linking reaction, the proteins are separated by one-dimensional PAGE, the cross-linked proteins are isolated from the gel, and the individual members of the cross-linked complexes are identified by mass spectrometry (Shevchenko et al. 1996; Pandey et al. 2000a). The protocol can be used to analyze many protein complexes isolated by any purification technique, provided the protein complexes remain in their native configuration.

MATERIALS

CAUTION: See Appendix 3 for appropriate handling of materials marked with <!>.

Reagents

Buffer X
- 20 mM PIPES (pH 7.0)
- Salts according to the requirements of the investigated complex

  The pH of PIPES changes –0.009 units per °C. The buffer must not contain reagents that can react with the cross-linker employed, e.g., dithiothreitol <!>, β-mercaptoethanol <!>, Tris, or phosphate.

Cross-linking reagents
- BS3 (bis[sulfosuccinimidyl]suberate)
- BMH (bismaleimidohexane)
- EDC (1-ethyl-3-[3-dimethylaminopropyl]carbodiimide) <!>
- SMCC (succinimidyl-4-[*N*-maleimidomethyl]cyclohexane-1-carboxylate)

Dimethylsulfoxide (DMSO) (HPLC grade, if possible) <!>

4x SDS-loading buffer
- 200 mM Tris-Cl (pH 6.8)
- 8% SDS <!>
- 0.4% bromophenol blue <!>
- 40% glycerol

Stop solution
- 1 M dithiothreitol (DTT)
- 1 M Tris-Cl (pH 7.8)

Equipment

(*Optional*) Bio-Spin column (Bio-Rad), MicroSpin G-25 column (APBiotech), or NAP column (Amersham Biosciences)
See Step 1.

Mass spectrometer capable of identifying proteins (Ion trap, reflector MALDI-TOF, triple quadrupole, QTOF, or QSTAR mass spectrometer)

Software that calculates peptides from protein sequences (e.g., GPMAW, available at http://welcome.to/gpmaw)

(*Optional*) Ultrafiltration concentrator (e.g., the Centricon Plus-20, Millipore)
See Step 2.

Water bath or heating block preset to 70°C

Biological Sample

Purified protein complex

Purify the protein complex using a method that will keep the complex in its native form. A proven way to isolate larger quantities of a complex is by the recombinant tagging of one of the complex components. The other components can assemble on the tagged protein in the cell or in the cell lysate, and the complex is then isolated by the interaction of the tag with a tag-specific solid matrix, e.g., Protein A tag and immobilized IgG, GST and immobilized glutathione, or affinity epitope (such as Myc or HA) and immobilized antibody. The elution conditions must be mild so as not to destroy the complex. Either competition (e.g., using the peptide epitope) or enzymatic cleavage in the linker region between the tag and the complex member is suitable for elution. The use of salt, detergents, or extreme pH usually result in denaturation of protein complexes and are therefore generally not advisable for elution.

Additional Reagents

Steps 10 and 14 require the reagents and equipment listed in Chapter 2, Protocol 1.

METHOD

The cross-linking reaction can precede the purification. In this case, the elution can be done in SDS-loading buffer. Note that, in general, protein cross-linking in vivo is not a trivial task.

Optimization of the Cross-linking Reaction

Different preparations of a protein complex vary in quality and concentration. It is therefore recommended that all optimization tests be done with a single preparation. However, freezing samples of protein complexes is not advised (unless specifically tested), because it often results in partial denaturation. Rather, plan all work carefully and complete it within a few days, during which the complex is stored at 4°C.

1. (*Optional*) If the buffer components of the protein complex solution are not compatible with cross-linking, exchange the purification buffer with Buffer X. A rapid method for this is gel filtration. Depending on the volume of the solution, use a Bio-Spin column (50–100 µl), MicroSpin G-25 column (10–100 µl), or the gravitation flow-driven NAP columns (0.1–2.5 ml) according to the manufacturers' protocols. The procedure is simple:
   - Equilibrate the column with Buffer X.
   - Apply the sample to the column.
   - Collect the flowthrough.

Gel filtration-based desalting methods have an efficiency of 95–98%. This means, for example, that a solution containing 100 mM Tris before desalting will still contain 2–5 mM Tris in the sample after gel filtration. Thus, to achieve the acceptable concentrations of buffer components, gel filtration may have to be repeated. An alternative desalting method, although it takes longer, is dialysis.

2. (*Optional*) Concentrate the samples. Samples can be rapidly (~10 minutes) and efficiently concentrated using an ultrafiltration device according to the manufacturer's protocol.

   Depending on the concentration of the purified complex, volume reduction may be necessary to ensure that a sufficient amount of protein is loaded onto the gel in Steps 10 and 14. The efficiency of precipitation (done just prior to loading the sample onto the gel) from dilute samples is not as good as ultrafiltration-based methods and is therefore not recommended. To reduce sample losses due to adsorption of protein on the ultrafiltration membrane, the membrane can be coated prior to concentrating the sample, by concentrating a BSA solution to 1 mg/ml in Buffer X and washing the membrane three times with Buffer X.

3. Place 13 microfuge tubes on ice and add 26 µl of protein complex solution to each one.

   IMPORTANT: To distinguish between substrates and the products of the cross-link reaction, include a control to which no cross-linker is added.

4. Weigh out ~1 mg of each cross-linker on a balance and dissolve in a volume of cold DMSO or H_2O that gives a 10 mM solution.

   EDC: 1.9 mg/ml H_2O
   BS^3: 5.7 mg/ml H_2O
   BMH: 2.8 mg/ml DMSO
   SMCC: 3.3 mg/ml DMSO

   IMPORTANT: The cross-linkers hydrolyze rapidly and thus should be handled as quickly as possible. Do not use frozen stocks!

5. Prepare two dilutions, of 1 mM and 0.1 mM in cold water, from the 10 mM stocks of each cross-linker.

6. Add 3 µl of each cross-linker solution (0.1 mM, 1 mM, and 10 mM) to one of the tubes containing protein complex (from Step 3) to test the optimal cross-linker concentration.

7. Incubate the tubes for 1 hour at 2°C on ice.

8. Stop the reaction by adding 1 µl of stop solution. Continue incubating the tubes for a further 10 minutes.

9. Add 10 µl of 4x SDS-loading buffer to each sample. Heat the samples for 10 minutes at 70°C in a water bath or heating block.

10. Separate the components (some will be cross-linked) of the protein complexes on a low-percentage SDS-PAGE gel. Be sure to include protein from the control tube that is absent any cross-linker.

    The percentage of the SDS-PAGE gel is determined by the size of expected cross-linked products and whether it can still be handled; the gel becomes increasingly fragile with lower percentages of acrylamide. A good compromise is a 6% gel, although this may not be required if small proteins are investigated. To increase handling stability of low-percentage gels, add 1/6 volume of a 2% prepolymerized acrylamide solution. A prestained marker allows electrophoresis to be continued until the designated size reaches the bottom of the gel, allowing a better spread of the slower-migrating species.

11. After the best cross-linker and its most suitable concentration have been determined, further optimize the cross-linking reaction conditions by

    - varying the time that the cross-linking reaction is allowed to proceed, ranging from 10 minutes to 2 hours, and
    - varying the temperature from 2°C to room temperature at ~5°C intervals.

Isolation of Cross-linked Products and Further Analysis

12. When the optimum cross-linking conditions have been established, prepare additional sample of protein complex, if necessary, according to Steps 1 and 2.

13. Add the chosen cross-linker to the protein complex sample, and incubate the mixture under the optimized reaction conditions.

    IMPORTANT: Remember to include a control tube containing the protein complex but no cross-linking reagent.

14. Repeat Steps 8–10.

    A number of points must be considered in preparing the SDS-PAGE gel if planning subsequent analysis by mass spectrometry. All solutions must be free of dust to avoid keratin contamination. Ideally, use freshly filtered solutions. Take special care at those steps where the use of gloves is necessary because electrostatic charging increases the risk of contamination. The sensitivity of the mass spectrometric analysis is best if the final protein sample is in as small a volume of acrylamide gel as possible. Measures to reduce the gel volume include the use of 1-mm spacers, employing conditions to obtain well-resolved bands (e.g., possibly precast gels), and cutting the bands very precisely in Step 16.

15. Visualize the proteins by staining (see Chapter 2, Protocols 2–7). Identify the bands containing cross-linked proteins by comparison of the cross-linked material to the untreated complex on the SDS-PAGE gel.

16. Excise the bands of interest from the gel and analyze them by mass spectrometry (see Chapter 8, Protocol 1).

    Although the proteins being analyzed are cross-linked, no special precautions must be taken, other than keeping the volume of buffer used in the digestion (Chapter 7) as small as possible to ensure a sufficient concentration of peptides in the supernatant. This eliminates the need for extraction procedures prior to mass spectrometric analysis, which often result in the loss of some peptide species.

The individual proteins within mixtures of proteins can be difficult to identify unambiguously. Therefore, it is sometimes advisable to perform the database searches with the peptide mass data on a custom database that contains only the known components of the investigated complex. This analysis can also be done manually by comparing the measured peptide masses with a list of calculated peptide masses if sophisticated software is not at hand.

Case Study: Topology of the Nup84p Complex

The Nup84p complex is a six-member subcomplex of the yeast nuclear pore complex (Siniossoglou et al. 1996). To study the topology of the Nup84p complex, it was purified and cross-linked (Rappsilber et al. 2000). The products were separated by SDS-PAGE and identified using MS. For the purification, one of the components (Nup85p) was tagged with the Protein A tag which binds to the heavy chain of IgG. When the Nup84p complex formed, it incorporated the Nup85p–Protein A fusion protein, enabling the entire complex to be purified using IgG immobilized on Sepharose beads. The bound complex was eluted by cleaving the protease site in the linker between Protein A and Nup85p using TEV protease in a buffer that was compatible with cross-linking. The volume of the eluate was sufficiently small that concentration was unnecessary. An aliquot was compared with a control purification of a metabolic enzyme to determine the background of the purification (Figure 10.20). Based on the intensity of the Coomassie-stained bands, the concentration of the complex was estimated to be ~1 pmole in 26 μl. The low concentration of the purified Nup84p complex minimized the likelihood of cross-links between complexes.

To find a suitable cross-linker, five different reagents with reactivity toward sulfhydryl and/or amino groups and one nonspecific linker (BASED) were investigated. The cross-linkers SMCC, BS^3, DSS, and BMH resulted in the appearance of different new bands on the polyacrylamide gel at a concentration of 10 μM (~300-fold excess over the complex). SMCC and BMH gave the best yield of products in both number and amount. BS^3 and DSS resulted in one major product at about the size of band II (Figure 10.21) obtained by SMCC. With BASED, the intensity of nonlinked material on the gel decreased; however, surprisingly, no bands corresponding to cross-linked species were observed. DMA did not result in any apparent reaction.

To refine the reaction conditions, cross-linker concentration, temperature, and reaction time were varied for SMCC and BS^3, the most promising of the cross-linkers that were initially screened. Cross-linking with 100 μM SMCC for 1 hour at 2°C gave the largest diversity of products for this complex, as judged from SDS-PAGE and silver staining. Consequently, these conditions were chosen for the further analysis of the Nup84p complex (Figure 10.21).

The optimal molar ratio of complex to reagent was found to be ~1:3000. A ratio of 1:300 resulted in low yields of cross-linked products, whereas ratios of 1:10,000 or higher practically linked all of the members of Nup84p complex to one another.

Establishing standard conditions for the cross-linking reaction was difficult because the Nup85p complex could only be purified in relatively small amounts (a few hundred fmoles). The quantitation of cross-linking reactions also proved to be difficult because protein complex concentration could only be estimated by comparison of stained bands.

Time-course experiments, performed at 2°C, 24°C, and 30°C, using SMCC, BS^3, and BMH revealed that the reaction was essentially complete after 30–60 minutes. Higher temperatures only increased the staining background on the gel but did not increase cross-linking efficiencies.

The cross-linked products were excised from the final gel and processed under standard conditions to obtain peptides for mass spectrometric analysis. The gel slices were washed, and the proteins were reduced, alkylated, and digested with trypsin (Shevchenko et al. 1996; also see Chapter 7). An aliquot of the supernatant was analyzed on a thin-layer preparation with α-cyano-4-hydroxy-*trans*-cinnamic acid mixed with nitrocellulose as the matrix (Vorm et al. 1994). Mass spectra were obtained as the sum of 200–250 selected measurements on a Bruker REFLEX II, and spectra were internally calibrated using matrix-related signals and trypsin autolysis signals. Peaks were assigned in LaserOne soft-

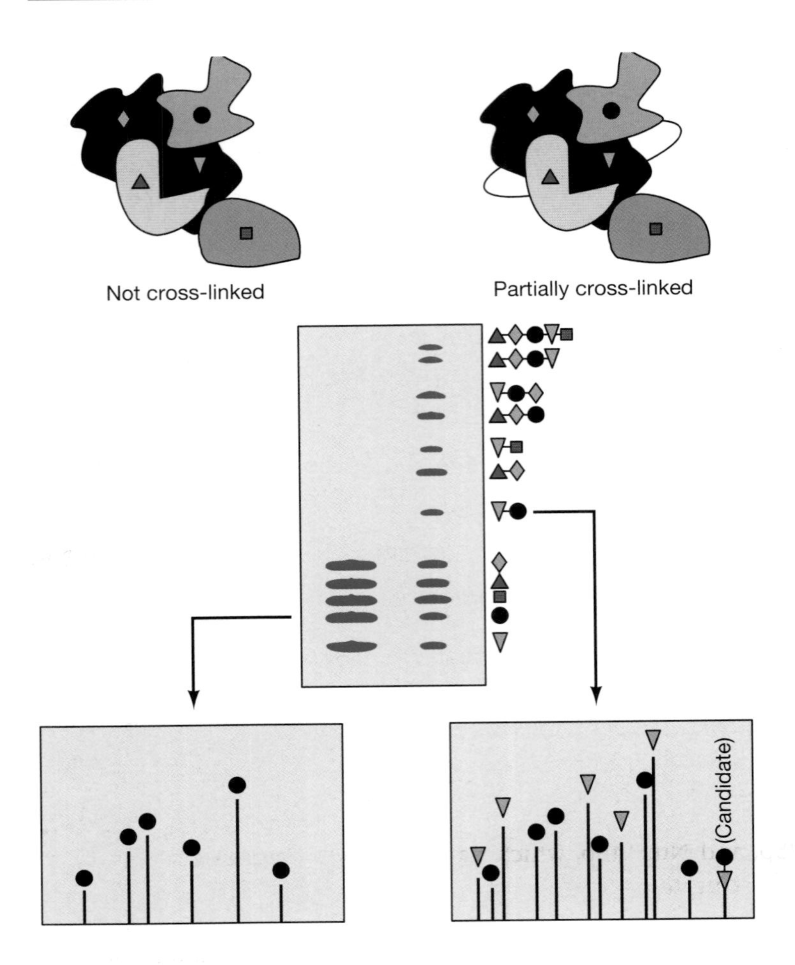

FIGURE 10.20. Schematic outline of the approach used to obtain topological information on a protein complex. A fraction of the purified complex is cross-linked, and the products are separated by 1D SDS-PAGE. In parallel, the noncross-linked complex is subjected to gel electrophoresis in order to recognize the cross-link products. Relevant bands are excised, digested with trypsin, and measured by MALDI-MS. The peptide mass maps obtained are searched against a nonredundant database or a complex-specific database. Based on proteins identified in cross-link products, a model of the complex topology can be constructed. Since there are many possibilities for the masses of cross-linked peptides, the observed masses that correspond to cross-linked peptides are not conclusive even at high mass accuracy. Interpretation of the site of contact based on these peptides requires further analysis, i.e., sequencing. (Courtesy of Juri Rappsilber and Matthias Mann, University of Southern Denmark.)

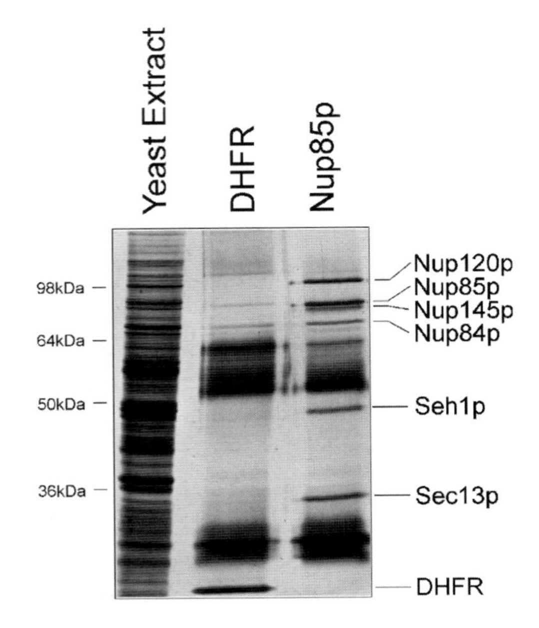

FIGURE 10.21. Silver-stained SDS-PAGE gel (10%) of the Nup84p complex (lane *3*). As a control, DHFR, which has no known protein-binding partners, was purified in parallel to show the background of the purification procedure (lane *2*). (Lane *1*) Total yeast extract. (Courtesy of Juri Rappsilber and Matthias Mann, University of Southern Denmark.)

ware, developed in-house. Using the program PeptideSearch (developed in-house), the peptide masses were searched against a nonredundant database containing all known protein sequences. The identification of a protein by PeptideSearch is based on the comparison of matching peptides with the matches of the other proteins in the database. The four large subunits of the Nup84p complex could be identified in one or the other cross-linked product. The two small subunits, Sec13p and Seh1p, could also be identified when the search was confined to yeast proteins below 120 kD. This shows that, in principle, the analysis of the cross-linked products could have yielded the identities of all six complex components without prior knowledge of the complex. This information could be then used for a topological analysis of the protein complex. The number of matching peptides was determined by comparing the list of measured masses with the tryptic peptides of the six known complex members. This analysis permitted identification of the proteins present in their respective bands (see Figures 10.22 and 10.23 and Table 10.4).

Band III of Figure 10.21 contains Nup85p and Nup145p. The added masses of Nup85p (86 kD) and the carboxy-terminal part of Nup145p (~85 kD, as Nup145p is posttranslationally cleaved into two proteins with different function) are in agreement with the apparent mass of Band III (168 kD). Therefore, we conclude that Nup85p and Nup145p are direct neighbors. Band II contained Nup85p, Nup145p, and Seh1p. The apparent molecular mass of the band is 138 kD, which agrees within the range of uncertainty introduced by the unknown migration behavior of cross-linked protein pairs with the protein pairs Nup85p + Seh1p (86 kD + 40 kD) and Nup145p + Seh1p (85 kD + 40 kD). From these data, we conclude that Seh1p is a direct neighbor of both Nup85p and Nup145p. This is further confirmed by the analysis of Band V which migrates at an apparent mass of 185 kD and contains Seh1p, Nup85p, and Nup145p, which have an added mass of 203 kD. This band therefore contains all three proteins cross-linked to one another. Similar reasoning on the results of the identifications in Band IV shows that Nup120p is a direct neighbor of Nup85p and Nup145p. The higher-molecular-mass Bands VI, VII, and VIII all contain four or more components cross-linked to each other and are therefore less informative. These bands, however, all contain Nup85p and Nup145p, thus supporting a central role for these two proteins. The remaining band (Band I) migrates at an apparent mass of 122 kD and was found to contain all six members of the complex. Considering the apparent mass of the band, each of the four large Nup proteins is potentially linked to either one of the small proteins, Sec13p or Seh1p. Nup85p linked to Seh1p was identified in Band II; therefore, the combination in Band I is likely to be Nup85p + Sec13p. Equally, Nup145p + Seh1p was identified in Band II, suggesting the combination of Nup145p + Sec13p in Band I.

The failure to observe certain protein linkages can also suggest topological information. For example, Sec13p was not found linked to Seh1p (the corresponding mass range is not shown on the gel in Figure 10.21), and Nup84p was not found directly linked to Nup120p. This suggests that Sec13p is not directly adjacent to Seh1p and that Nup84p does not have direct contact to Nup120p. In addition, for each of the two larger proteins, Nup84p and Nup120p, there is only one band indicating a link to one of the smaller proteins, Sec13p and Seh1p (Band I). This is consistent with Nup120p being directly adjacent to either one of the small proteins and Nup84p to the other one, with both pairs being separate from each other.

Taken together, the data suggest a model for the Nup84p complex in which Nup85p and Nup145p form a central pair with two opposing pairs at the side, one composed of Nup84p with either Sec13p or Seh1p and the other being Nup120p with either Seh1p or Sec13p. At present, the orientation of the two opposing pairs, *cis* or *trans*, remains unclear (Figure 10.24). This model satisfies all the constraints derived from the data in Table 10.4 and was validated by deletion analysis (Siniossoglou et al. 2000).

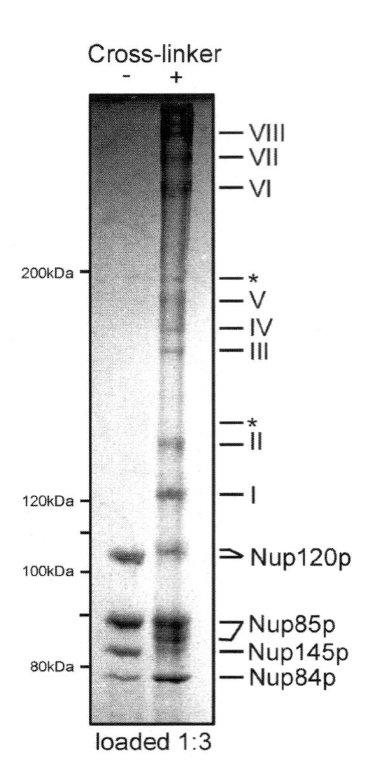

FIGURE 10.22. Silver-stained SDS-PAGE gel (6%) of the purified Nup84p complex either noncross-linked (lane *1*, –) or cross-linked (lane *2*, +) loaded in a ratio of 1:3. At least eight new bands appear after addition of cross-linker (labeled I–VIII). Background bands are denoted with an asterisk. (Courtesy of Juri Rappsilber and Matthias Mann, University of Southern Denmark.)

TABLE 10.4. Identification of cross-linked proteins

Band	$MW_{Exp.}$ (kD)	Sec13p 32 kD	Seh1p 40 kD	Nup84p 84 kD	Nup85p 86 kD	Nup145p 85 kD[a]	Nup120p 121 kD[b]	Linked proteins	$MW_{Calc.}$ (kD)
I	122	6/25%	6/29%	16/25%	11/23%	14/14%	33/35%	Nup120p/(Sec13p or Seh1p)	153/161[c]
								Nup84p/(Sec13p or Seh1p)	116/124
								Nup85p/Sec13p	118
								Nup145p/Sec13p	117
II	138	–/–	8/32%	–/–	11/21%	9/11%	2/–	Nup85p/Seh1p	126
								Nup145p/Seh1p	125
III	168	–/–	1/–	–/–	8/11%	11/8%	–/–	Nup85p/Nup145p	171
IV	175	1/–	–/–	1/–	8/19%	6/7%	6/9%	Nup85p/Nup120p	207
								Nup145p/Nup120p	206
V	185	9/33%	–/–	1/–	7/15%	6/6%	–/–	Nup85p/Nup145p/Sec13p	203
VI	>200	–/–	5/18%	2/–	11/15%	18/19%	11/14%	Nup85p/Nup145p/Nup120p/Seh1p	332
VII	>200	1/–	1/–	9/17%	8/14%	16/15%	10/11%	Nup85p/Nup145p/Nup120p/Nup84p	376
VIII	>200	8/40%	2/–	8/15%	9/23%	23/24%	11/14%	Nup85p/Nup145p/Nup120p/Nup84p/Sec13p	408

Table courtesy of Juri Rappsilber and Matthias Mann (University of Southern Denmark).
x/*y*%, where *x* is the number of peptides and *y* is the sequence coverage observed for the respective proteins.
[a]In mature Nup145p, the amino-terminal third of the protein sequence is cleaved so that Nup145p migrates faster than Nup85p on SDS-PAGE.
[b]Nup120p runs at an apparent molecular mass of 100 kD.

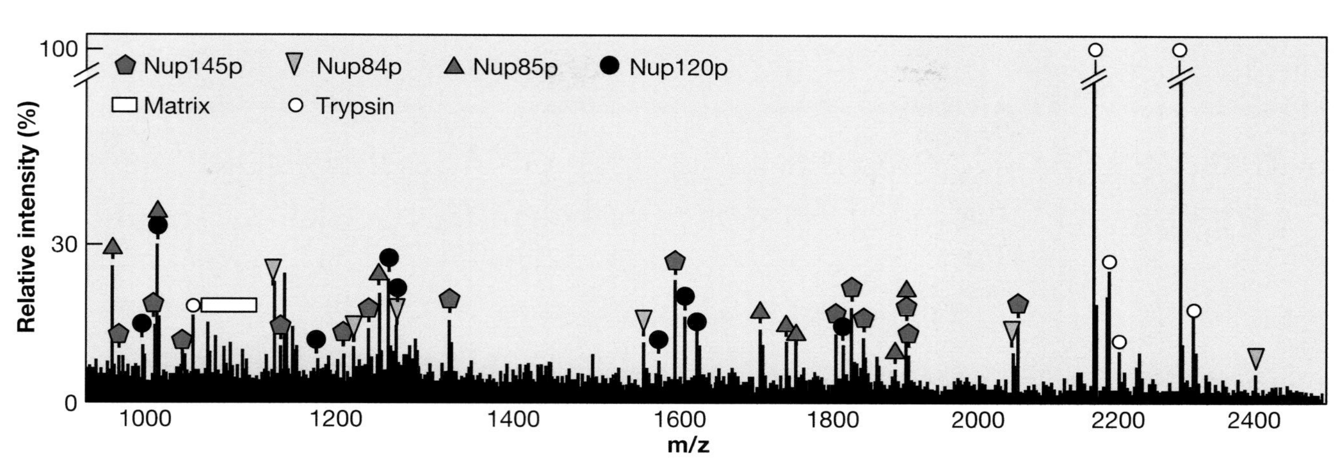

FIGURE 10.23. MALDI peptide mass map obtained after in-gel tryptic digestion of Band VII. Four proteins were identified by iterative searches in the nonredundant database (NRDB) with greater than 50 ppm mass accuracy. The proteins were identified in the order: Nup145p, Nup120p, Nup84p, and Nup85p. (Courtesy of Juri Rappsilber and Matthias Mann, University of Southern Denmark.)

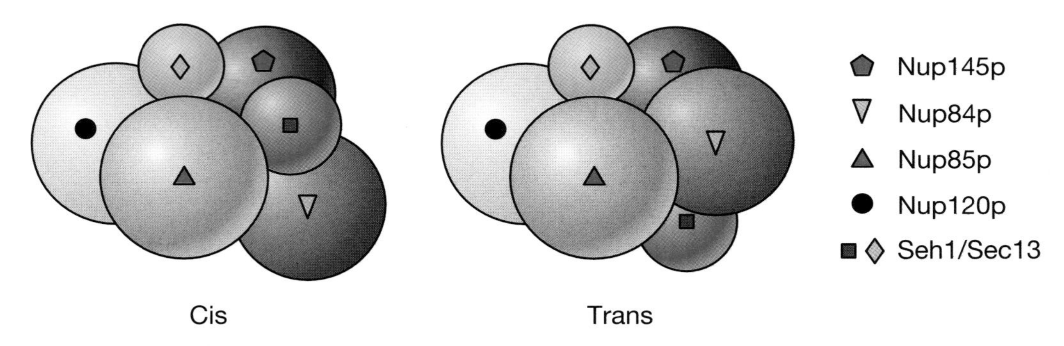

FIGURE 10.24. Proposed topology model for the Nup84p complex. The two small complex members may be on the same side of the complex (*cis*) or on opposing sides (*trans*). Sec13p may be close to Nup120p and Seh1p may be close to Nup84p or vice versa. (Courtesy of Juri Rappsilber and Matthias Mann, University of Southern Denmark.)

PROTOCOL 6

Subsecond Photo-cross-linking: Use of Water-soluble Metal Complexes to Monitor Protein-Protein Interactions

THIS PROTOCOL WAS CONTRIBUTED BY David A. Fancy (University of Texas Southwestern Medical Center Dallas, Texas).

Understanding how the vast arrays of protein-protein interactions are responsible for cellular function is one of the fundamental quests in biology. Over the decades, we have seen increasing complexity in the organization of protein complexes, such as in the proteosome and RNA polymerase II holoenzyme, and we have been challenged to follow the intricate cascades of signal transduction to the cell nucleus. To understand these systems, the investigator is faced with at least two fundamental questions: (1) What is the organization of the proteins in question? (2) How do these interactions change during a catalytic cycle? Of the variety of methods invented to answer these questions, chemical cross-linking has proven to be very useful.

This protocol provides details in using photo-induced cross-linking of unmodified proteins (PICUP), which is based on radical coupling reactions. The proteins of interest are mixed with water-soluble metal complexes, Palladium(II) (tetrakis-4-*N*-[methyl]-pyridyl porphyrin (K. Kim et al. 1999) or Tris-bipyridyl ruthenium (II) dication and ammonium persulfate (Fancy et al. 1999). The mixture is photolyzed with visible light (>400 nm) to photo-oxidize the metal complex via electron transfer to ammonium persulfate (see Figure 10.4) (Bolletta et al. 1980; Aboul-Enein and Schulte-Frohlinde 1988; Nickel et al. 1994). The activated metal-centered radical cation extracts an electron from an aromatic amino acid, resulting in a protein-centered radical species that can attack a wide variety of other groups. This leads to covalent cross-linking between components of the protein complex. Irradiation times can range from subsecond flashes when a 150 W Xe lamp is used, or to multiple seconds when using a hand-held flashlight. Typical yields of cross-linked product range from 30% to 90%.

MATERIALS

CAUTION: See Appendix 3 for appropriate handling of materials marked with <!>.

Reagents

Ammonium persulfate <!>
4x Gel-loading buffer
- 0.2 M Tris
- 8% SDS <!>
- 2.9 M β-mercaptoethanol <!>
- 40% glycerol
- 0.4% xylene cyanol <!>
- 0.4% bromophenol blue <!>

Methanol <!>

$PdCl_2$ <!>
5x Reaction buffer
75 mM sodium phosphate (pH 7.5)
750 mM NaCl
TMPyP (5,10,15,20-tetrakis[1-methyl-4-pyridinio]porphyrin tetra[*p*-toluene sulfonate], Sigma)
Trifluoroacetic acid (TFA) <!>
Tris-bipyridyl ruthenium (II) dication (Ru[bpy]$_3^{2+}$)

Equipment

Heating block preset to 95°C
Photo-cross-linking apparatus
Xenon high-power lamp (150 W "arc lamp," Thermo Oriel) <!>
Cut-on filter (380–2500-nm; Thermo Oriel)
Body of a manual single-lens reflex camera with the lens and back removed
Cardboard barrier
Sample microfuge tube holder
For an illustration of the photo-cross-linking apparatus, see Figure 10.25, and for details, see the panel on GENERAL CONSIDERATIONS on the following page.
SepPak C_{18} cartridge (Waters)
Spectrophotometer, visible
Syringe filter (45 µm) and syringe

Biological Sample

Protein sample of interest
Prepare the protein sample in a total reaction volume of 125 µM TMPyP-Pd(II) or Ru(bpy)$_3^{2+}$ in a 1.7-ml microfuge tube.
See the panel on GENERAL CONSIDERATIONS.

Additional Reagents

Step 5 (of the photo-cross-linking procedure) requires the reagents and equipment listed in Chapter 2, Protocols 1 and 2.

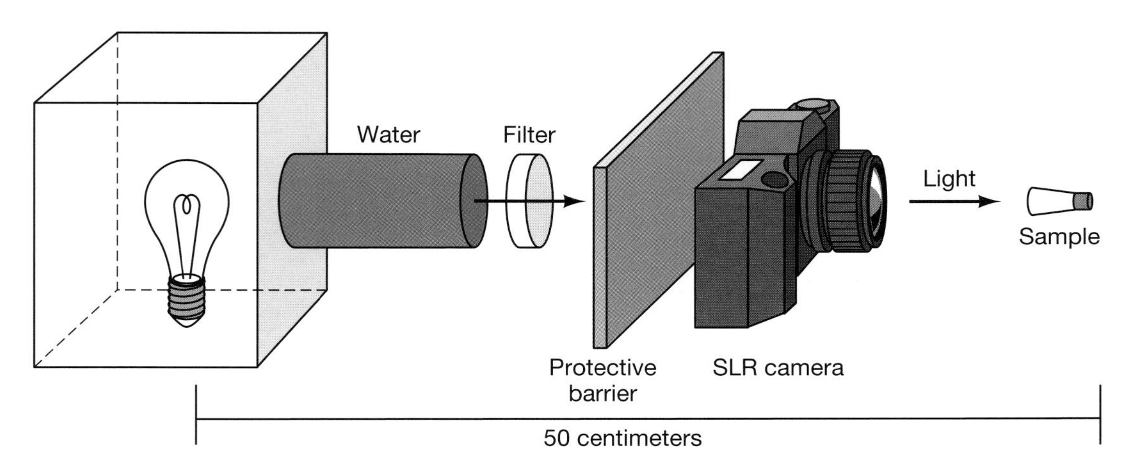

FIGURE 10.25. Typical light source setup for a high-power lamp (150 W Xe). Light is first filtered through 10 cm of distilled H_2O and then through a 380–2500-nm cut-on filter. Exposure time (to 1/1000 second) is controlled using a manual single-lens reflex (SLR) camera with the lens and back removed. (Courtesy of David A. Fancy, University of Texas Southwestern Medical Center.)

METHOD

General Considerations

- Easily oxidized components such as β-mercaptoethanol or dithiothreitol (DTT), which quench cross-linking, should be removed prior to the experiment.
- Protein concentrations can vary depending on the experiment, and thus the amount of metal complex should be determined empirically. For these conditions, 1–10 μM protein is typically used.
- Stock solutions of 30 mM metal complex in H_2O can be stored indefinitely frozen and in the dark. The ammonium persulfate stock should be made fresh before each series of experiments and not used if more than 1 hour old.
- Cross-linking reactions are carried out in 1.7-ml microfuge tubes, which maximizes the irradiation surface, to achieve the highest quantum yield during a flash irradiation.
- After the addition of the metal complex, the protein solution should be kept in the dark because ambient light can cause molecular oxygen-mediated cross-linking of the proteins (Fancy et al. 2000). Ammonium persulfate is added to a final concentration of 2.5 mM, in the dark, just before photolysis.
- It is important to keep a protective barrier between the camera and the light source until just before shutter opening because a continuous direct beam of intense light can burn through the delicate fabric that makes up many camera shutters.
- Open microfuge sample tubes are held horizontally (parallel to the beam of light) in a mounted rack (i.e., the sample microfuge tube holder) placed 50 cm from the 150 W lamp.
- Care should be taken to focus the beam at the point where the liquid sample is to be placed.
- If a hand-held flashlight is used for these experiments, it should be of the highest power conveniently obtainable and focusable. In our experiments, we have successfully used an aluminum body Brinkman flashlight that takes four D-cell batteries. In this case, the distance between the sample and light source should be reduced to no more than 10 cm.

Metallation of TMPyP with Pd(II)

1. Add 0.34 mg (1.9 μmoles) of $PdCl_2$ to a microfuge tube containing 1 mg (1.2 μmoles) of TMPyP in 1.4 ml of H_2O. Heat this mixture for 2 hours in a 95°C heating block.

   EXPERIMENTAL TIP: Metallation can be monitored by the collapse of the four Q band peaks at 518, 554, 584, and 639 nm into two peaks at 525 and 558 nm by visible spectroscopy.

2. Cool the sample and filter it through a 45-μm syringe filter. Add TFA to 1%.
3. Inject the solution onto a SepPak C_{18} cartridge previously equilibrated in 1% TFA. Wash the cartridge with 5 ml of 1% TFA.
4. Elute TMPyP-Pd(II) with 0.5 ml of methanol into a preweighed microfuge tube.

5. After evaporation of the methanol, dissolve TMPyP-Pd(II) in 1 ml of H_2O. Determine its concentration by its extinction coefficient at 417 nm ($\varepsilon = 158 \times 10^3$ $M^{-1}cm^{-1}$).

   This reagent can be stored frozen and in the dark indefinitely.

Photo-cross-linking Proteins

WARNING: Wear safety glasses with tinted lenses to protect the eyes when using the photo-cross-linking apparatus with a xenon lamp.

1. Prepare the protein sample in a total reaction volume of 20 µl, containing protein, 1x reaction buffer, and 125 µM TMPyP-Pd(II) in a 1.7-ml microfuge tube. Protect the reaction tube from light.
2. Add 5 µl of freshly made ammonium persulfate to the solution immediately before photolysis. Secure the sample tube horizontally (so it is parallel to the beam of light) in the sample holder.
3. Remove the barrier between the SLR camera and the light source, and open the camera shutter for 0.5 second by a cable release or, if so equipped, the camera's own autotimer. After the shutter is closed, replace the solid barrier between the camera and the lamp.
4. Immediately add 7 µl of a 4x gel-loading buffer to the sample to quench the reaction.
5. Analyze the photo-cross-linking reaction using SDS-PAGE followed by Coomassie Blue staining or western blotting.

   An example of a typical cross-linking experiment is shown in Figure 10.26. In this experiment, UvsY was cross-linked using the above protocol. One can see in lane 6 that only in the case where the metal complex and ammonium persulfate are present is cross-linking efficient during flash photolysis. Also apparent, in lane 3, is the oxygen-mediated cross-linking when long exposures are taken in the absence of ammonium persulfate.

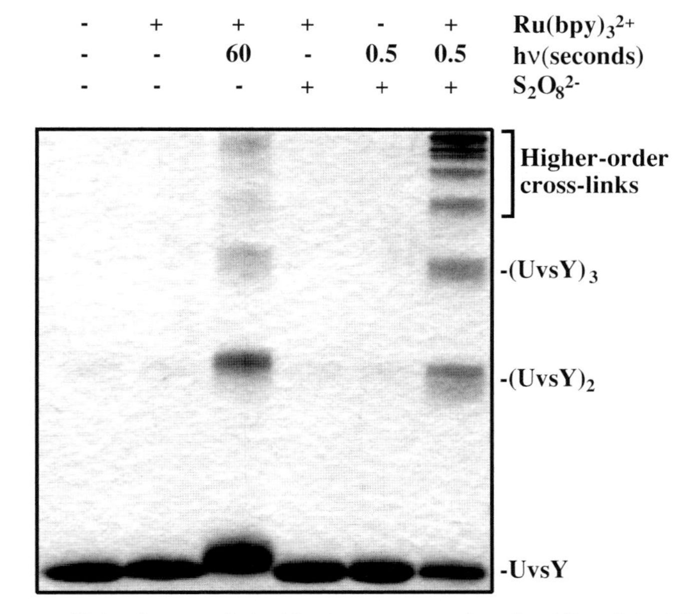

FIGURE 10.26. UvsY is efficiently cross-linked in the presence of $Ru(bpy)_3^{2+}$ and APS (lane 6) during a 0.5-second exposure. A competing pathway is shown in lane 3, but it takes a considerably longer exposure time. Thus, reactions must remain in the dark until flash photolysis is initiated. (Courtesy of David A. Fancy, University of Texas Southwestern Medical Center.)

Optimization and Precautions

As with all experimental procedures, a bit of optimization is required to achieve satisfactory results. Too much cross-linking is often the result of a high concentration of protein, metal complex, or ammonium persulfate, or overexposure of the sample to the light. If decreasing the shutter time or reducing the amount of metal complex and ammonium persulfate does not work, then the protein concentration must be lowered. Interestingly, for some proteins at high concentrations (>1 μM), this technique can pick up protein collisions and thus link proteins that should be strictly monomeric or are unrelated. An example of this is shown in Figure 10.27, lane 5). UvsY (a filamentous protein involved in T4 homologous recombination) was mixed with maltose and binding was nonspecifically cross-linked. Unfortunately, lowering the protein concentration can be problematic if the method of protein detection is not sensitive. Instead, to eliminate nonspecific cross-linking, exogenous amino acid, such as tyrosine or histidine, can be added (to 60 μM) to the solution (Figure 10.27, lane 6). Presumably, under these circumstances, long-lasting radical cations produced on the surface of the oxidized protein would interact with an electron-rich histidine before nonspecifically cross-linking to another protein diffusing through the solution.

If too little or no cross-linking is observed, it is often due to poor detection or a lack of exposed aromatic amino acids on the surface of the protein. At this stage, the investigator may need to prepare a western blot to detect cross-linked products reliably. We have noticed that some monoclonal antibodies detect cross-linked products poorly if irradiation times are considerably longer than 5 seconds (Fancy et al. 2000). This is most likely due to the oxidation of the epitope needed for antibody binding.

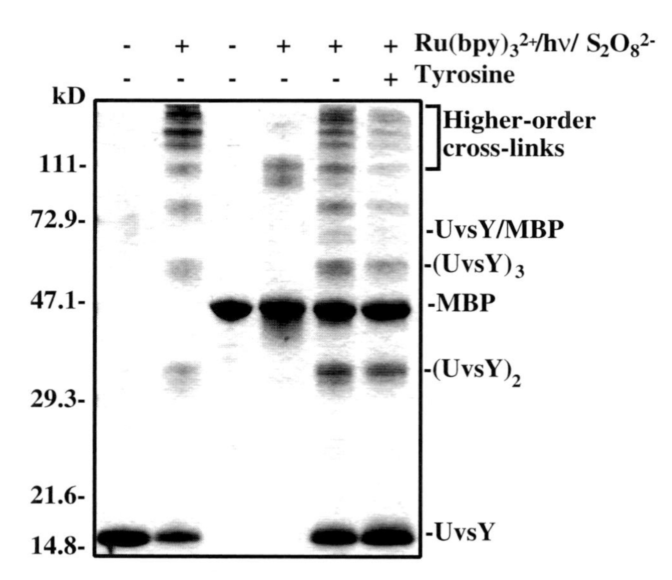

FIGURE 10.27. Adding a small amount of tyrosine to a cross-linking reaction can eliminate some of the nonspecific cross-linking. UvsY and MBP can cross-link to each other (lane 5) reaffirming the suspicion that PICUP can trap transient protein-protein interactions. Adding a threefold excess of tyrosine over protein can decrease nonspecific cross-linking without compromising yield. (Courtesy of David A. Fancy, University of Texas Southwestern Medical Center.)

Case Study: Subsecond Photo-cross-linking

Mapping protein-protein contacts in large multiprotein complexes can be a daunting task, and it is thus important to develop new chemical methods to aid in this pursuit. Affinity modification approaches, in which a reactive tag is appended to a specific residue or agent, have been valuable tools used to probe the immediate environment around the point of attachment (Fancy 2000). Current methods rely heavily on bifunctional cross-linkers or UV-light-activatable photo probes that often result in low yields and protein degradation.

We set out to show that an affinity approach could also be employed using the water-soluble long-wavelength, light-harvesting metal complexes exploited by PICUP. In the following experiment, we show that Pd(II)-metallated Protoporphyrin IX (PPIX-Pd[II]) (Aldrich) can be selectively delivered to the antibody α-HA (or monoclonal antibody 12CA5) (Sigma) by attaching it to the epitope peptide of the influenza hemagglutinin protein NH_2-YPYDVPDYA (HA). PPIX has two carboxyl groups that allow easy coupling to the peptide using 1-ethyl-3-(3-dimethylaminopropyl)carbodiimide (EDC). In this experiment, we incubated various ratios of HA-PPIX-Pd(II) with α-HA or an antibody that it does not bind, α-glutathione *S*-transferase (Sigma). Only if selective delivery of the cross-linking reagent is achieved should one observe a high level of cross-linking between α-HA and HA-PPIX-Pd(II) and not α-GST (Figure 10.28). Protein levels and peptide conjugate levels were kept low (<10 ng) to avoid nonspecific protein cross-linking, which may be attributable to the exogenous addition of HA-PPIX-Pd(II).

Results of this experiment are shown in Figure 10.29. It is clear that upon addition of ammonium persulfate ($S_2O_8^{-2}$) and subsequent photolysis, cross-linking occurs specifically at α-HA. Interestingly, even a 1000:1 ratio of HA-PPIX-Pd(II) to antibody did not induce coupling of α-GST. Also of interest is that addition of ammonium persulfate was not essential to initiate efficient cross-linking of α-HA. This is presumably due to the ability of oxygen to sensitize cross-linking via a potentially different mechanism.

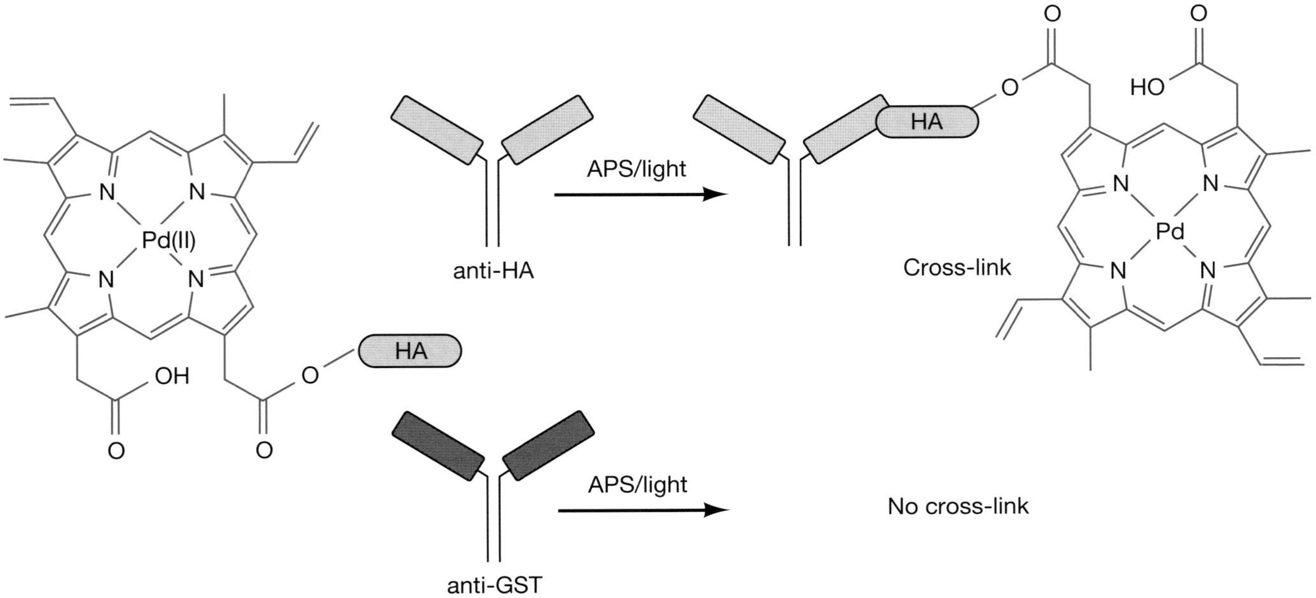

FIGURE 10.28. Selective delivery of HA-PPIX-Pd(II) to α-HA and cross-linking. The epitope for α-HA (NH_2-YPYDVPDYA) can be selectively tethered to PPIX-Pd(II) and then incubated with α-HA (which it recognizes) or with α-GST which it does not. Cross-linking should only efficiently occur when selective recognition is achieved. (Adapted, with permission, from Aboul-Enein and Schulte-Frohlinde 1988 [©American Society for Photobiology].)

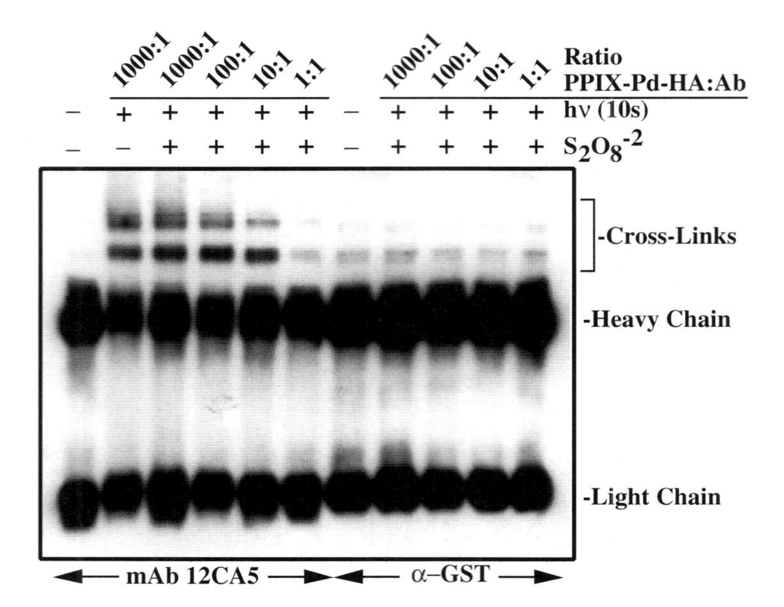

FIGURE 10.29. Selective delivery of HA-PPIX-Pd(II) to α-HA and cross-linking. Cross-linking is only observed when a specific complex between α-HA and HA-PPIX-Pd(II) is formed (lanes *2–5*). When an antibody that does not recognize the HA epitope is employed, there is no cross-linking. For detection, proteins were transferred to PVDF membrane and probed with α-mouse antibodies conjugated to horseradish peroxidase. Detection was achieved through enhanced chemiluminescence (ECL). (Courtesy of David A. Fancy, University of Texas Southwestern Medical Center.)

PROTOCOL 7

Site-specific Protein-DNA Photo-cross-linking: Analysis of Structural Organization of Protein-DNA and Multiprotein-DNA Complexes

THIS PROTOCOL WAS CONTRIBUTED BY Nikolai Naryshkin, Andrey Revyakin, and Richard H. Ebright (Howard Hughes Medical Institute, Waksman Institute, and Department of Chemistry, Rutgers University, Piscataway New Jersey).

Processes of fundamental biological importance, including replication, repair, recombination, and transcription, proceed via assembly of large multiprotein-DNA complexes. Understanding such processes requires information regarding the structural organization of the complexes and information regarding the pathways and kinetics of assembly of the complexes. Site-specific protein-DNA photo-cross-linking is able to define positions of proteins relative to DNA within large multiprotein-DNA complexes, permitting the analysis of structural organization of complexes (Bartholomew et al. 1990; Bell and Stillman 1992; Lagrange et al. 1996). In conjunction with rapid mixing and laser flash photolysis, site-specific protein-DNA photo-cross-linking also is able to define kinetics of formation of individual protein-DNA interactions within large multiprotein-DNA complexes, and thus also allows analysis of pathways and kinetics of assembly of complexes (S. Druzhinin et al., unpubl.).

Several procedures have been developed for site-specific protein-DNA photo-cross-linking within large multiprotein-DNA complexes (Bartholomew et al. 1990; Bell and Stillman 1992; Lagrange et al. 1996). The most effective procedure involves the following four parts (see Figure 10.5):

- ***Part 1:*** Chemical (Fidanza et al. 1992; Yang and Nash 1994; Mayer and Barany 1995) and enzymatic (Sambrook and Russell 2001) reactions are used to prepare a DNA fragment containing a phenyl-azide photoactivatible cross-linking agent and an adjacent radiolabel incorporated at a single, defined DNA phosphate (with a 9.7 Å linker between the photoreactive atom of the cross-linking agent and the phosphorus atom of the phosphate, and with an ≈11 Å maximum "reach" between potential cross-linking targets and the phosphorus atom of the phosphate).
- ***Part 2:*** The multiprotein-DNA complex of interest is formed using the site-specifically derivatized DNA fragment, and the multiprotein-DNA complex is UV-irradiated, initiating covalent cross-linking with proteins in direct physical proximity to the cross-linking agent.
- ***Part 3:*** Extensive nuclease digestion is performed, eliminating uncross-linked DNA and converting cross-linked DNA to a cross-linked, radiolabeled three- to five-nucleotide "tag."
- ***Part 4:*** The "tagged" proteins are identified, usually by denaturing polyacrylamide gel electrophoresis followed by autoradiography.

The procedure is performed in a systematic fashion, with preparation and analysis of at least ten derivatized DNA fragments, each having the photoactivatible cross-linking agent incorporated at a single, defined DNA phosphate (typically, every second DNA phosphate [each 12 Å] on each DNA strand spanning the region of interest) (Lagrange et al. 1996, 1998; Kim et al. 1997, 2000a; Wang and Stumph 1998; Bartlett et al. 2000; Naryshkin et al. 2000). The results of the procedure define the translational positions of proteins relative to the DNA sequence. Plotted on a three-dimensional representation of a DNA helix, the results also define the rotational orientations of proteins relative to the DNA helix axis and the groove orientations of proteins relative to the DNA major and minor grooves (Lagrange et al. 1996, 1998; Kim et al. 1997, 2000a; Wang and Stumph 1998; Naryshkin et al. 2000).

Elsewhere, we have presented detailed protocols for cross-linking of protein-DNA complexes in solution ("in-solution" cross-linking; Kim et al. 2000b; Naryshkin et al. 2001) or in polyacrylamide gel matrices ("in-gel" cross-linking; Kim et al. 2000b; Naryshkin et al. 2001). Here, we present a detailed protocol for the cross-linking of protein-DNA complexes immobilized on streptavidin-coated paramagnetic beads ("on-bead" cross-linking; see Naryshkin et al. 2000). On-bead cross-linking offers three important advantages over the other two methods:

- The ability to perform stringent washing to eliminate nonspecific and nonproductive complexes (Kim et al. 2000a; Naryshkin et al. 2000).
- The ability to change reaction media rapidly (e.g., to add substrates, inhibitors, or allosteric effectors) (Kim et al. 2000a; Naryshkin et al. 2000).
- The ability to change reaction temperatures rapidly.

In view of these advantages, we now employ in our laboratory on-bead cross-linking in all photo-cross-linking experiments not requiring rapid mixing and laser flash photolysis (Naryshkin et al. 2000; Y. Kim et al., unpubl.).

PART 1: Design and Preparation of Biotin-labeled Site-specifically Derivatized DNA Fragment

MATERIALS

CAUTION: See Appendix 3 for appropriate handling of materials marked with <!>.

Reagents

10x Annealing buffer
  400 mM Tris-HCl (pH 7.9)
  500 mM NaCl
  100 mM $MgCl_2$ <!>
ATP (100 mM)
p-Azidophenacyl bromide <!>
Biotin-11-dATP, biotin-11-dCTP, or biotin-16-dUTP (1 mM)
  The choice of biotin-labeled dNTP is determined by the choices of restriction endonucleases (see Biological Molecules, below).
C_{18} HPLC solvent A (50 mM triethylammonium acetate [pH 7.0], 5% acetonitrile <!>)
C_{18} HPLC solvent B (100% acetonitrile)
Chloroform <!>
Controlled-pore glass (CPG) oligonucleotide synthesis supports (1 μmole, 500 Å) (Applied Biosystems)
dA, dC, dG, T β-cyanoethylphosphoramidites (Applied Biosystems) <!>
dNTPs (10 mM)
  Three dNTPs are required to complement the biotin-labeled dNTP in Step 36.
Denaturing loading buffer
  0.3% bromophenol blue <!>
  0.3% xylene cyanol <!>
  12 mM EDTA, in formamide <!>
  The formamide may have to be deionized prior to its addition to the loading buffer. If any yellow color is present, deionize the formamide by stirring on a magnetic stirrer with Dowex XG8 mixed bed resin for 1 hour and filtering it twice through Whatman No. 1 paper. Store the deionized formamide in small aliquots under nitrogen at –70°C.
DNA polymerization mix (25 mM each dATP, dGTP, dCTP, and TTP)
EDTA (0.5 M, pH 8.0)
Elution buffer
  0.5 M ammonium acetate <!>
  10 mM magnesium acetate (pH 7.5)
  1 mM EDTA
Ethanol (70% and absolute) <!>
  Store at –20°C.
Formamide <!>
[γ-^{32}P]ATP (10 mCi/ml, 6000 Ci/mmole) <!>
Low-EDTA TE
  10 mM Tris-HCl (pH 8.0)
  0.1 mM EDTA

Methanol <!>
Nondenaturing polyacrylamide slab gel (5%)
  29:1 acrylamide:bisacrylamide
  0.5x TBE (10 x 7 x 0.15 cm)
Nondenaturing loading buffer
  0.3% bromophenol blue
  0.3% xylene cyanol
  30% glycerol, in H_2O
Oligodeoxyribonucleotide synthesis reagents (e.g., Applied Biosystems)
10x Phosphorylation buffer
  500 mM Tris-HCl (pH 7.6)
  100 mM $MgCl_2$
  15 mM β-mercaptoethanol <!>
PicoGreen dsDNA quantitation kit (Molecular Probes)
Polyacrylamide slab gel (5%)
  29:1 acrylamide:bisacrylamide <!>
  8 M urea <!>
  0.5x TBE (10 x 7 x 0.075 cm)
Polyacrylamide slab gel (5%)
  29:1 acrylamide:bisacrylamide
  0.5x TBE (10 x 7 x 0.15 cm)
Polyacrylamide slab gel (12%)
  19:1 acrylamide:bisacrylamide
  8 M urea <!>
  0.5x TBE (10 x 7 x 0.075 cm)
  TBE (45 mM Tris-borate [pH 8.3] and 1 mM EDTA) is usually prepared as a 5x stock. Dilute the stock buffer just before use and make the gel solution and the electrophoresis buffer from the same stock solution. The 5x stock solution is prepared by combining 54 g of Tris base, 27.5 g of boric acid, and 20 ml of 0.5 M EDTA (pH 8.0) to a final volume of 1 liter with H_2O.
Potassium phosphate (1 M, pH 7.0)
10x Restriction endonuclease digestion buffers
  The choice of digestion buffer(s) is determined by the choices of restriction endonucleases (see Biological Molecules below).
SDS (10%) <!>
Sodium acetate (3 M, pH 5.2)
TE
  10 mM Tris-HCl (pH 7.6)
  1 mM EDTA
Tetraethylthiuram disulfide/acetonitrile (Applied Biosystems)
Triethylammonium acetate (50 mM, pH 7.0) (Prime Synthesis)

Equipment

Autoradiogram markers
Autoradiography intensifying screen
Beaker (600 ml)

Benchtop centrifuge
CHROMA SPIN+TE-10 and SPIN+TE-100 spin columns (BD Biosciences CLONTECH)
DNA/RNA synthesizer (e.g., ABI392, Applied Biosystems)
Fixed-angle microfuge
Germicidal lamp (254 nm) <!>
HPLC system, including
HPLC pump (e.g., L-7100, Hitachi)
HPLC UV detector (e.g., L-3000, Hitachi)
C_{18} RP-HPLC column (5 μm) (e.g., LiChrospher 100 RP-18 column, KGaA)
Intensifying screen
Microfuge tubes, siliconized polypropylene
Oligonucleotide purification cartridge (OPC) (Applied Biosystems)
Phosphoimaging system
Polypropylene round-bottomed tubes (6 ml)
Scalpels (disposable)
Spin-X centrifuge filter (0.22 μm, cellulose acetate)
UV spectrophotometer
Vacuum evaporator (SpeedVac, Thermo Savant, or equivalent)
Water baths or heating blocks preset to 37°C, 55°C, 65°C, and 90°C

Biological Molecules

DNA polymerase I Klenow fragment (10 units/μl)

Restriction endonucleases A and B

Choose restriction endonucleases A and B using the following criteria:

- Endonuclease A should have a unique (or nearly unique) cleavage site 100–150 nucleotides 5′ to the DNA site of interest (Figure 10.5).
- Endonuclease B should have a unique (or nearly unique) cleavage site 100–150 nucleotides 3′ to the DNA site of interest (see Figure 10.5).
- One, and only one, restriction endonuclease (either A or B) should yield cleavage products with 5′ overhanging ends containing dA, dG, or T.

It is important to use restriction endonucleases with cleavage sites relatively distant (100–150 nucleotides) from the DNA site of interest and thus to prepare DNA fragments having relatively long DNA sequences flanking the DNA site of interest. This ensures that in the event a protein binds nonspecifically at the ends of the DNA fragment (see, e.g., Melancon et al. 1983), it will not interfere with binding of protein(s) at the DNA site of interest, and it will not yield cross-links that overlap those of protein(s) at the DNA site of interest (Naryshkin et al. 2000).

Single-stranded DNA-binding protein (Stratagene)

Single-stranded DNA template

The single-stranded DNA template is a single-stranded DNA of M13 derivative (or, less preferably, phagemid derivative) carrying a DNA site for the protein(s) of interest (prepared as in Sambrook and Russell 2001).

T4 DNA ligase (5 units/μl)

T4 DNA polymerase (3 units/μl)

T4 polynucleotide kinase (10 units/μl)

T7 DNA polymerase (10 units/μl)

Upstream primer

The upsteam primer is a 25-nucleotide oligodeoxyribonucleotide complementary to region of single-stranded DNA template ~50 nucleotides 5′ to cleavage site for restriction endonuclease A (see Figure 10.5).

METHOD

Chemical Reactions

1. Perform 24 standard cycles of solid-phase β-cyanoethylphosphoramidite oligodeoxyribonucleotide synthesis to prepare CPG-linked precursor containing residues 3–26 of the desired oligodeoxyribonucleotide. Use the following settings:

   Cycle: 1.0 μmole CE
   DMT: on
   End procedure: manual

2. Replace the iodine/H_2O/pyridine/tetrahydrofuran solution (bottle 15) with a tetraethylthiuram disulfide/acetonitrile solution. Perform one modified cycle of solid-phase β-cyanoethylphosphoramidite oligodeoxyribonucleotide synthesis to add residue 2 and phosphorothioate linkage. Use the following settings:

   Cycle: 1.0 μmole sulfur
   DMT: on
   End procedure: manual

3. Replace the tetraethylthiuram disulfide/acetonitrile solution (bottle 15) with the iodine/H_2O/pyridine/tetrahydrofuran solution. Place a collecting vial on the DNA synthesizer. Perform one standard cycle of solid-phase β-cyanoethylphosphoramidite oligodeoxyribonucleotide synthesis to add residue 1. Use the following settings:

   Cycle: 1.0 μmole CE
   DMT: on
   End procedure: automatic CE

   Steps 1–3 yield an oligodeoxyribonucleotide 26 nucleotides in length with a phosphorothioate linkage between nucleosides 2 and 3. Occasionally (rarely), a low yield is observed in the primer-extension reaction (see Steps 28–31). In this case, prepare an oligodeoxyribonucleotide with 1–10 additional nucleotides at the 3′ end, and preferably with dG or dC at the 3′ end.

4. Remove the collecting vial and screw the cap on tightly. Incubate for 10 hours at 55°C.

5. Transfer the sample to a 6-ml polypropylene round-bottomed tube, place the tube in a vacuum evaporator, and spin for 30 minutes with the vacuum evaporator lid ajar and with no vacuum (to allow evaporation of ammonia).

6. Close the vacuum evaporator lid, apply vacuum, and dry.

7. Remove the trityl group and purify ~0.1 μmole of oligodeoxyribonucleotide on OPC according to the supplier's protocol.

8. Dry in the vacuum evaporator.

9. Resuspend in 100 μl of TE. Remove a 2-μl aliquot and dilute with 748 μl of TE. Determine the concentration from UV absorbance at 260 nm (molar extinction coefficient ~240,000 $M^{-1}cm^{-1}$).

10. To check the purity of the oligodeoxyribonucleotide, mix an aliquot containing 1 nmole of oligodeoxyribonucleotide with an equal volume of formamide. Apply this mixture to

a 12% polyacrylamide TBE slab gel. To monitor the progress of the electrophoresis, load 5 μl of denaturing loading buffer in an adjacent lane. Allow electrophoresis to proceed for 30 minutes at 25 V/cm.

11. Remove the gel, place it on an intensifying screen, and view it in the dark using a 254-nm germicidal lamp. The oligodeoxyribonucleotide should appear as a dark shadow against a green background and should migrate more slowly than bromophenol blue. If the purity is ≥95%, proceed to Step 12.

12. Divide the remainder of the sample into 50-nmole aliquots, transfer them to 1.5-ml siliconized polypropylene microfuge tubes, dry them in a vacuum evaporator, and store them at –20°C. The oligodeoxyribonucleotide is stable for at least 3 years.

13. Dissolve 10 mg (42 μmole) of *p*-azidophenacyl bromide in 1 ml of chloroform. Transfer 100-μl aliquots (4.2 μmole) to 1.5-ml siliconized polypropylene microfuge tubes, and dry them in a vacuum evaporator. Wrap the tubes with aluminum foil, and store desiccated at 4°C (stable indefinitely).

    IMPORTANT: This and all subsequent steps should be carried out under subdued lighting. Fluorescent light and daylight must be excluded. Low-to-moderate levels of incandescent light (e.g., from a single task lamp with 60 W tungsten bulb) are acceptable.

14. Dissolve a 50-nmole aliquot of phosphorothioate oligodeoxyribonucleotide (from Step 12) in 50 μl of H_2O, and resuspend a 42-μmole aliquot of *p*-azidophenacyl bromide (from Step 13) in 220 μl of methanol.

15. In a 1.5-ml siliconized polypropylene microfuge tube, mix:

    50 μl (50 nmoles) of the phosphorothioate oligodeoxyribonucleotide solution
    5 μl of 1 M potassium phosphate (pH 7.0)
    55 μl (1 μmole) of the *p*-azidophenacyl bromide solution

    Incubate the reaction for 3 hours at 37°C.

16. Precipitate the derivatized oligodeoxyribonucleotide by adding 11 μl of 3 M sodium acetate (pH 5.2) and 275 μl of ice-cold absolute ethanol. Invert the tube several times, and store it for 30 minutes at –20°C.

17. Centrifuge the tube at 13,000*g* for 5 minutes at 4°C.

18. Remove the supernatant, and wash the pellet with ice-cold 70% ethanol. Air-dry the pellet for 15 minutes at room temperature. Store at –20°C. The *p*-azidophenacyl-derivatized oligodeoxyribonucleotide is stable for at least 1 year.

19. Resuspend the derivatized oligodeoxyribonucleotide (from Step 18) in 100 μl of 50 mM triethylammonium acetate (pH 7.0).

20. Analyze a 5-μl aliquot by C_{18} RP-HPLC to determine the efficiency of the derivatization reaction.

    a. Set the column flow rate to 1 ml/min.

    b. Equilibrate the column with 10 column volumes of C_{18} HPLC solvent A before loading the sample.

c. After loading the sample, wash the column with 6 column volumes of C_{18} HPLC solvent A.

d. Elute with a 50-minute gradient of 0–70% C_{18} HPLC solvent B in C_{18} HPLC solvent A.

Derivatized and underivatized oligodeoxyribonucleotides elute at ~25% solvent B and ~16% solvent B, respectively.

21. If derivatization efficiency is ≥80%, purify the remainder of the sample using the procedure in Step 20, collecting derivatized oligodeoxyribonucleotide peak fractions.

22. Pool the peak fractions, divide them into 1-ml aliquots, and dry them in a vacuum evaporator. Store the fractions desiccated at –20°C in the dark. The purified derivatized oligodeoxyribonucleotide is stable for at least 1 year.

23. Resuspend one aliquot of the pooled and dried fraction in 100 μl of TE. Remove 5 μl, dilute with 495 μl of H_2O, and determine the concentration from UV absorbance at 260 nm (molar extinction coefficient ~242,000 $M^{-1}cm^{-1}$).

24. Divide the remainder of the derivatized oligodeoxyribonucleotide/TE solution from Step 23 into 20 5-pmole aliquots and one larger aliquot. Dry them in a vacuum evaporator, and store desiccated at –20°C in the dark.

The derivatization procedure yields two diastereomers in an approximate ratio of 1 to1: one in which azidophenacyl is incorporated at the sulfur atom corresponding to the phosphate O1P, and one in which azidophenacyl is incorporated at the sulfur atom corresponding to the phosphate O2P (see Fidanza et al. 1992; Mayer and Barany 1995). Depending on the oligodeoxyribonucleotide sequence and HPLC conditions, the two diastereomers may elute as a single peak, or as two peaks (e.g., at 24% and 25% solvent B). In most cases, no effort is made to resolve the two diastereomers, and experiments are performed using the unresolved diastereomeric mixture. This permits simultaneous probing of protein-DNA interactions in the DNA minor groove (probed by the O1P-derivatized diastereomer) and the DNA major groove (probed by the O2P-derivatized diastereomer) (Lagrange et al. 1996, 1998; Kim et al. 1997, 2000a; Wang and Stumph 1998; Naryshkin et al. 2000).

Enzymatic Reactions

25. Resuspend 5 pmoles of derivatized oligodeoxyribonucleotide in 12 μl of H_2O. To the oligodeoxyribonucleotide add:

    2 μl of 10x phosphorylation buffer
    5 μl of [γ-^{32}P]ATP (50 μCi)
    1 μl (10 units) of T4 polynucleotide kinase

    Incubate for 30 minutes at 37°C.

    IMPORTANT: This and all subsequent steps should be carried out under subdued lighting. Fluorescent light and daylight must be excluded. Low-to-moderate levels of incandescent light (e.g., a single-task lamp with 60 W tungsten bulb) are acceptable.

26. Add 20 μl of H_2O and mix thoroughly.

27. Desalt the radiophosphorylated derivatized oligodeoxyribonucleotide into TE using a CHROMA SPIN+TE-10 spin column according to the manufacturer's protocol.

Immediately proceed to the next step, or, if necessary, store the radiophosphorylated derivatized oligodeoxyribonucleotide solution at –20°C in the dark. The solution is stable for up to 24 hours.

28. In a 1.5-ml siliconized polypropylene microfuge tube, mix:

    40 μl of radiophosphorylated derivatized oligodeoxyribonucleotide
    1 μl of 10 μM upstream primer
    4 μl of 0.5 μM single-stranded DNA template
    5 μl of 10x annealing buffer

    Heat the mixture for 5 minutes at 65°C.

29. Transfer the tube to a 600-ml beaker containing 400 ml of H_2O at 65°C. Incubate the beaker at room temperature to permit slow cooling (65°C to 25°C in ~60 minutes).

30. Add:

    2 μl of DNA polymerization mix (25 mM each dATP, dGTP, dCTP, and TTP)
    1 μl of 100 mM ATP
    0.5 μl (10 units) of T7 DNA polymerase
    1 μl (3 units) of T4 DNA polymerase
    1 μl (5 units) of T4 DNA ligase
    1 μl (1 μg/ml) of single-stranded binding protein

    Incubate the mixture for 5 minutes on ice, followed by 5 minutes at room temperature, followed by 1 hour at 37°C. Perform a parallel reaction without ligase as a "no-ligase" control.

31. Add 1 μl (5 units) of T4 DNA ligase and incubate for an additional 2 hours at 37°C.

32. Terminate the reaction by adding 1 μl of 10% SDS.

33. Desalt into TE using a CHROMA SPIN+TE-100 spin column according to the manufacturer's protocol.

34. Perform a restriction digestion:

    a. Where digestion buffers for restriction endonucleases A and B are compatible: In a 1.5-ml siliconized polypropylene microfuge tube, mix:

        56 μl of site-specifically derivatized DNA/TE solution from Step 33
        6.2 μl of 10x digestion buffer
        1 μl (10 units) of restriction endonuclease A
        1 μl (10 units) of restriction endonuclease B

        Incubate for 1 hour at 37°C. Perform a parallel restriction digestion using 56 μl of "no-ligase" control.

    b. Where digestion buffers for restriction endonucleases A and B are incompatible: In a 1.5-ml siliconized polypropylene microfuge tube, mix:

        56 μl of site-specifically derivatized DNA/TE solution from Step 33
        6.2 μl of 10x digestion buffer having lower salt concentration
        1 μl (10 units) of corresponding restriction endonuclease

Incubate for 1 hour at 37°C. To the digest, add:

7.1 μl of 10x digestion buffer having higher salt concentration
1 μl (10 units) of corresponding restriction endonuclease

Incubate for 1 hour at 37°C. Perform a parallel restriction digestion using 56 μl of "no-ligase" control.

35. Estimate the efficiency of ligation.

 a. Mix 3-μl aliquots of the restriction digest and of the "no-ligase" control restriction digest each with 7 μl of denaturing loading buffer.

 b. Heat the tubes for 5 minutes at 90°C.

 c. Apply the reactions to a 5% polyacrylamide TBE slab gel.

 d. Run the gel for 30 minutes at 25 V/cm.

 e. Dry the gel and subject it to phosphorimaging. Estimate the ligation efficiency by comparing the reaction and "no-ligase" control lanes. If the ligation efficiency is ≥90%, proceed to the next step. If not, repeat the preparation, beginning with Step 25.

36. To the remainder of the restriction digest (from Step 34), add:

 2.5 μl of 1 mM biotin-labeled dNTP
 0.25 μl of 10 mM each of other dNTPs
 0.3 μl (3 units) of DNA polymerase I Klenow fragment

 Incubate for 20 minutes at 25°C.

37. In order, add:

 3 μl of 0.5 M EDTA (pH 8.0)
 10 μl of nondenaturing loading buffer

38. Apply the entire reaction to a *nondenaturing* 5% polyacrylamide TBE slab gel. Run the gel at 25 V/cm until the bromophenol blue has completely run off the gel.

39. Retrieve the derivatized DNA fragment from the gel.

 a. Remove one glass plate and cover the gel with plastic wrap. Attach two autoradiogram markers to the gel.

 b. Expose X-ray film to the gel for 30–60 seconds at room temperature, and process the film.

 c. Cut out the portion of the film corresponding to the derivatized DNA fragment.

 d. Using white paper as background, superimpose the cut-out film on the gel, using the autorad markers as alignment reference points.

 e. Use a disposable scalpel to excise the portion of the gel corresponding to the derivatized DNA fragment.

40. Place the excised gel slice in a 1.5-ml siliconized polypropylene microfuge tube and crush it with a 1-ml pipette tip. Add 400 μl of elution buffer, vortex briefly, and rock the tube for 12 hours at 37°C.

41. Transfer the supernatant to a Spin-X centrifuge filter. Centrifuge in a fixed-angle microfuge at 13,000*g* for 1 minute at room temperature.

42. Transfer the filtrate to a 1.5-ml siliconized polypropylene microfuge tube. Precipitate the derivatized DNA fragment by addition of 1 ml of ice-cold absolute ethanol. Invert the tube several times, and store it for 30 minutes at –20°C.

43. Centrifuge the tube in a fixed-angle microfuge at 13,000*g* for 5 minutes at 4°C.

44. Remove and dispose of the supernatant, wash the pellet with 500 μl of ice-cold 70% ethanol, and air-dry the pellet for 10 minutes at room temperature.

45. Resuspend the pellet in 30 μl of low-EDTA TE.

46. Determine the amount of radioactivity in the sample by Cerenkov counting.

47. Remove a 1-μl aliquot of the sample and determine its DNA concentration using a PicoGreen double-stranded DNA quantitation kit according to the manufacturer's protocol. Calculate the specific activity (expected specific activity ~5000 Ci/mmole).

48. Store the derivatized DNA fragment at –20°C in the dark. The fragment is stable for ~1 week.

PART 2: Preparation and UV Irradiation of Protein-DNA Complex

MATERIALS

CAUTION: See Appendix 3 for appropriate handling of materials marked with <!>.

Reagents

2× Binding and washing buffer

- 10 mM Tris-HCl (pH 7.5)
- 2.0 M NaCl
- 1 mM EDTA

Dynabeads M-280 streptavidin-coated paramagnetic beads (10 mg/ml) (Dynal)

1× Protein-binding buffer

Protein-binding buffer should contain components required for complex formation by protein(s) of interest and also should contain 50 μg/ml of acetylated bovine serum albumin (BSA), which reduces nonspecific adsorption of protein(s) onto beads and surface, and 1× protease inhibitor cocktail (Sigma), which reduces protease degradation of protein(s) of interest.

3× Quench solution

- 3 M urea <!>
- 0.3 M NaI <!>
- 0.3 M β-mercaptoethanol <!>

Equipment

Borosilicate glass culture tube (13 × 100 mm)

Dry bath and heating blocks

Optional: For complex formation at temperatures above ambient, see Step 13.

Magnetic particle concentrator (Dynal)

Microfuge tubes (siliconized polypropylene)

Photochemical reactor (Rayonet RPR-100)

The reactor should be equipped with an RMA400 sample holder and 16 RPR-3500 Å tubes (Southern New England Ultraviolet).

Photochemical reactors should be available at most chemistry departments. With appropriate modifications to the protocol, other photon sources can be substituted, including 305–310-nm UV lasers (S. Druzhinin et al., unpubl.), long-wave cross-linking units (Wang and Stumph 1998), and long-wave germicidal lamps (Bartlett et al. 2000) (the latter two photon sources being only marginally satisfactory, due to the requirement for relatively long UV-irradiation times).

Polystyrene microfuge tubes (04-978-145, Fisher Scientific)

Radioactivity survey meter (for ^{32}P, Ludlum Measurements)

Refrigerated bath

Optional: For complex formation at temperatures below ambient, see Step 13.

Biological Molecules

Biotin-labeled, site-specifically derivatized DNA fragment (60 nM in low-EDTA TE)

From Part 1, Step 48.

Protein(s) of interest

METHOD

1. Resuspend the paramagnetic beads by gently shaking the storage vial. Transfer a 3-μl aliquot into a 1.5-ml siliconized polypropylene microfuge tube, and centrifuge the tube briefly in a microfuge.

2. Place the tube into a magnetic particle concentrator, wait 30 seconds, and remove the supernatant.

3. Resuspend the beads in 30 μl of 1x binding and washing buffer, and mix them gently using a micropipette. Incubate the beads for 2 minutes at room temperature.

4. Centrifuge the tube briefly in a microfuge.

5. Place the tube in a magnetic particle concentrator, wait for 30 seconds, and then remove the supernatant. Repeat Steps 3–5.

6. Resuspend the bead pellet in 10 μl of 2x binding and washing buffer. Add 2 μl of 60 nM biotin-labeled, site-specifically derivatized DNA fragment in low-EDTA TE (from Part 1, Step 48). Incubate the tube for 10 minutes at 25°C, mixing occasionally by gently tapping the tube.

7. Place the tube into a magnetic particle concentrator for 30 seconds. Transfer the supernatant to a new 1.5-ml siliconized polypropylene microfuge tube.

8. Estimate the binding efficiency of the biotin-labeled, site-specifically derivatized DNA fragment by comparing the radioactivity of the bead pellet and supernatant. If the binding efficiency is ≥80%, proceed to the next step. If not, repeat the preparation beginning with Part 1, Step 25.

   IMPORTANT: This and all subsequent steps should be carried out under subdued lighting. Fluorescent light and daylight must be excluded. Low-to-moderate levels of incandescent light (e.g., from single task lamp with 60 W tungsten bulb) are acceptable.

9. Resuspend the bead pellet in 30 μl of 1x binding and washing buffer. Centrifuge the beads briefly in a microfuge.

10. Place the tube into a magnetic particle concentrator, wait for 30 seconds, and then remove the supernatant.

11. Resuspend the bead pellet in 30 μl of 1x protein-binding buffer and centrifuge the beads briefly in a microfuge.

12. Place the tube into a magnetic particle concentrator, wait 30 seconds, and remove the supernatant.

13. Resuspend the bead pellet in 1x protein-binding buffer to the desired DNA concentration. Add the protein(s) to the desired concentration. Incubate the mixture at the desired temperature for the desired time, mixing occasionally by gently tapping the side of the tube.

    In this and subsequent steps, DNA and protein concentrations, reaction temperature, and time are determined empirically and/or are based on previous knowledge about the interactions of the components within the complex.

14. Centrifuge the reaction briefly in a microfuge.

15. Place the tube into a magnetic particle concentrator, wait for 10 seconds, and remove the supernatant.

16. Resuspend the beads in 50 μl of 1x protein-binding buffer preincubated at the desired temperature. Incubate for 2 minutes at the desired temperature.

17. Place the tube into a magnetic particle concentrator, wait 30 seconds, and remove the supernatant.

    Optionally, in this step, add challengers, inhibitors, or effectors (see Kim et al. 2000a; Naryshkin et al. 2000).

18. Resuspend the bead pellet in 20 μl of 1x protein-binding buffer preincubated at the desired temperature and transfer the mixture to a polystyrene microfuge tube. Insert the polystyrene tube inside a 13 x 100-mm borosilicate glass culture tube preincubated at the desired temperature.

19. Place the tube in the photochemical reactor. UV-irradiate the protein-DNA sample for 20 seconds (11 mJ/mm^2 at 350 nm).

    If the desired reaction temperature is at least 10°C higher or lower than the ambient temperature, fill the borosilicate glass culture tube with H_2O, and preincubate it until the H_2O reaches the desired reaction temperature.

20. Transfer the suspension to a 1.5-ml siliconized polypropylene microfuge tube. Add 7 μl of quench solution, vortex briefly, and spin the tube briefly in a microfuge.

21. Place the tube into a magnetic particle concentrator, wait for 10 seconds, and remove 17 μl of the supernatant. Immediately proceed to Part 3.

PART 3: Nuclease Digestion

MATERIALS

CAUTION: See Appendix 3 for appropriate handling of materials marked with <!>.

Reagents

$CaCl_2$ (55 mM) <!>
5x SDS-loading buffer
- 300 mM Tris-HCl (pH 8.0)
- 10% SDS <!>
- 20 mM EDTA
- 25% β-mercaptoethanol <!>
- 0.1% bromophenol blue <!>
- 50% glycerol

Equipment

Heating block preset to 70°C

Biological Molecules

DNase I (240 units/µl)
DNase I should be diluted to 10 units/µl with 20 mM sodium acetate (pH 6.5), 5 mM $CaCl_2$, 0.1 mM PMSF, and 50% glycerol.

Micrococcal nuclease
Micrococcal nuclease should be dissolved and diluted to 10 units/µl with 5 mM $CaCl_2$, 0.1 mM PMSF, and 50% glycerol.

UV-irradiated protein(s)-DNA complex of interest (from Part 2, Step 21)

METHOD

1. To 10 µl of the bead suspension (from Part 2, Step 21), add:
   - 1 µl of 55 mM $CaCl_2$
   - 0.5 µl (5 units) of micrococcal nuclease
   - 0.5 µl (5 units) of DNase I

   Incubate for 5 minutes at 37°C.

2. Terminate the reaction by adding 3 µl of 5x SDS-loading buffer and heating the tube for 3 minutes at 70°C.

   Occasionally (rarely), the digestion reaction yields doublets of bands (resulting from incomplete digestion). In such cases, perform an alternative digestion procedure.

   a. To 10 µl of the bead suspension (from Part 2, Step 21), add 1 µl of 55 mM $CaCl_2$ and 0.5 µl (5 units) of DNase I. Incubate for 10 minutes at 37°C.

   b. Terminate the reaction by adding 1 µl of 10% SDS and heating for 5 minutes at 65°C.

   c. After the sample has cooled to room temperature, add 1 µl of 25 mM $ZnCl_2$, 1 µl of 1 M acetic acid <!>, and 1 µl (30 units) of S1 nuclease. Incubate for 10 minutes at 37°C.

   d. Terminate the reaction by adding 3 µl of 5x SDS-loading buffer, and heating for 3 minutes at 70°C.

PART 4: Identification of Cross-linked Proteins

MATERIALS

CAUTION: See Appendix 3 for appropriate handling of materials marked with <!>.

Reagents

Polyacrylamide (37.5:1 acrylamide:bisacrylamide <!>) slab gel

Use a gel with a polyacrylamide concentration (or polyacrylamide-concentration gradient) that results in maximal resolution in the molecular-weight range of protein(s) of interest.

Prestained protein molecular-mass markers (7–210 kD) (e.g., Bio-Rad)

1x SDS running buffer

25 mM Tris

250 mM glycine (pH 8.3) <!>

0.1% SDS <!>

Equipment

Apparatus for autoradiography or phosphorimaging

Gel dryer

Biological Molecules

Nuclease-digested protein(s)-DNA complex of interest (from Part 3, Step 2)

METHOD

1. Apply the entire sample (16 μl, from Part 3, Step 2) to a polyacrylamide slab gel. Load 5 μl of prestained protein molecular-mass markers into an adjacent lane.
2. Run the gel in SDS running buffer at 25 V/cm until the bromophenol blue runs off the gel.
3. Dry the gel, and perform autoradiography or phosphorimaging.

CASE STUDY: ANALYSIS OF *ESCHERICHIA COLI* RNA POLYMERASE TRANSCRIPTION INITIATION COMPLEXES

The procedure described in this protocol has been successfully applied to analysis of *Escherichia coli* RNA polymerase transcription initiation complexes (Naryshkin et al. 2000, 2001, and unpubl.). The results have permitted analysis of the structural organization of the RNA polymerase-promoter open complex (RPo; Naryshkin et al. 2000), analysis of the structural organization of a putative intermediate in formation of the RNA polymerase-promoter open complex ($RP_{15°C}$; N. Naryshkin et al., unpubl.), and analysis of the effects of a transcriptional activator on the structural organization of RPo and $RP_{15°C}$ (Naryshkin et al. 2000 and unpubl.).

PROTOCOL 8

Analysis of Protein Interactions In Vivo with FRET

THIS PROTOCOL WAS CONTRIBUTED BY Marc Damelin and Pamela Silver (Dana-Farber Cancer Institute, Boston, Massachusetts).

The protocol presented here focuses on a standard microscope-based assay for budding yeast, but many aspects are applicable to other systems and assays. The Experimental Design and Data Analysis procedures can be extended to FRET experiments in any system, including mammalian cell lines and other model organisms. The Strain Construction procedure can be applied to studies in yeast using fluorimetry instead of microscopy (Overton and Blumer 2000). Although the data acquisition and analysis described below can be a daunting task, careful quantitative analysis is absolutely necessary due to a high degree of cross-talk and many unknown parameters that vary from cell to cell.

MATERIALS

Reagents

Molecular biology reagents

Use standard molecular biology reagents as described by Ausubel et al. (1997) and Sambrook and Russell (2001).

Synthetic complete (SC) liquid media with 2% glucose

Use for the growth of yeast cells to be analyzed by FRET. The media should be less than 2 months old.

Yeast manipulation reagents

Use the reagents described by Guthrie and Fink (2002).

Equipment

Filter sets

CFP filter set to visualize CFP (Donor)

YFP filter set to visualize YFP (Acceptor)

FRET filter set to visualize "sensitized emission," which is YFP emission resulting indirectly from CFP excitation.

Each set is composed of an excitation filter, a dichroic mirror, and an emission filter. Filters that have been optimized for CFP and YFP are available from Chroma Technology Corp. (or alternatively from Omega Optical). The filter specifications are shown in Table 10.5. Depending on the microscope being used, the filters can be installed either in three individual filter cubes or on filter wheels (which can eliminate duplicates). It is important that the user be able to switch among filter sets easily and quickly.

TABLE 10.5. Filters required for CFP-YFP FRET

Filter set	Excitation	Dichroic	Emission
CFP	440 nm/20 nm	455 nm LP	480 nm/30 nm
YFP	500 nm/25 nm	525 nm LP	545 nm/35 nm
FRET	440 nm/20 nm	455 nm LP	535 nm/25 nm

Table courtesy of Marc Damelin and Pamela Silver (Dana-Farber Cancer Institute).

Image analysis software

The software must be able to operate the shutter, allowing precise exposures, and to analyze images by calculating the average pixel intensity within a specified region.

Microscope equipment

The microscope must be outfitted with a camera, an electronic shutter, specialized filters, and appropriate light source. See the panel on EQUIPMENT FOR IN VIVO FRET ANALYSIS below.

Standard equipment for molecular biology and yeast manipulations

Biological Molecules

α-GFP antibody

The antibody is used to check expression of the fusion proteins on western blots.

DNA encoding CFP and YFP

It is often most convenient to procure yeast vectors designed for FRET from various laboratories (Damelin and Silver 2000, 2002; Overton and Blumer 2000), but the genes are also available commercially (e.g., CLONTECH). For details on the vectors and their use, see Step 1.

DNA fragments encoding portions of the endogenous proteins of interest

Cells

Yeast strains

A wild-type haploid *ura3 trp1* strain, such as FY23 (Winston et al. 1995), serves as the parent strain. It is strongly preferable that the strain be cream-colored (not pink or red, e.g., in *ade2* mutants), because the pigment obscures FRET signals. Additional yeast strains may be needed to test the functionality of the fusion proteins. For each tagged protein, conditions in which the protein is essential for viability should be identified. If viability cannot be assayed under specific growth conditions, yeast strains with mutations that make the tagged protein essential for viabiltiy (synthetic lethal mutations) should be obtained. Yeast cultures for microscopy are grown on a roller drum at room temperature or at 25°C.

EQUIPMENT FOR IN VIVO FRET ANALYSIS

Listed below is one example of an acceptable setup (Damelin and Silver 2000):

Nikon Diaphot-300 inverted epifluorescence microscope with a 60 x 1.4 NA objective
100-W mercury lamp <!>
Photometrics 200-series liquid-cooled charge-coupled device (CCD) camera
Uniblitz D122 shutter and shutter driver
Appropriate filters

Use of a confocal microscope may be considered but is not necessary for this assay in yeast. Software such as the MetaMorph Imaging System (Universal Imaging Corp.) can handle data acquisition and image analysis.

METHOD

Experimental Design

Due to the complex nature of in vivo FRET experiments, the valid interpretation of the data requires experimental controls that satisfy several criteria. A detailed understanding of the issues surrounding these experiments is extremely useful in the planning stage, as well as during image acquisition and data analysis. For an elaborate and very helpful discussion, see Gordon et al. (1998). The symbols introduced by these authors have been adopted in this protocol (see Table 10.6 and the panel on AN EXPLANATION OF SYMBOLS USED FOR ANALYSIS OF FRET DATA). For applications of this protocol, see Damelin and Silver (2000, 2002).

In its simplest form, the FRET assay is used to investigate a potential interaction between two proteins, Target1 and Target2. Other endogenous proteins, Control1, Control2..., that do not interact with Target1 serve as the negative controls. As described below in Data Analysis and in Gordon et al. (1998), in vivo FRET measurements are not absolute, and they are meaningful only when evaluated statistically relative to similar measurements (i.e., interacting pair vs. noninteracting pair). A negative control is therefore essential to establish a basis for comparison. Using more than one Control protein is preferable, because the additional data create a broader framework for interpretation. Each Control protein must satisfy three requirements:

1. It should not interact with Target1.
2. It should colocalize with Target2.
3. Its expression level should be comparable to that of Target2.

Control proteins are typically other endogenous proteins, but there are possible variations. A mutant Target2 protein may serve as a control, provided its expression level and localization are comparable to wild-type Target2. However, the use of at least one Control protein besides Target1 and Target2 is encouraged.

In the experimental design shown in Figure 10.30, Target1 is fused to CFP, and Target2 and all Control proteins are fused to YFP. This scheme derives from the fact that CFP generates most of the cross-talk in the FRET filter set. Consequently, only pairs with the same CFP fusion protein should be compared directly.

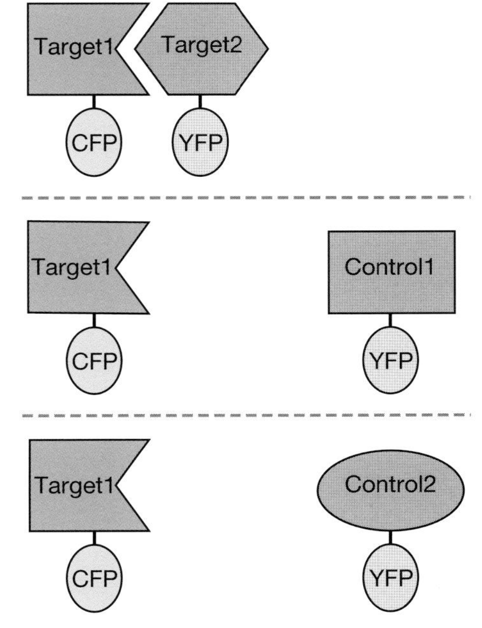

FIGURE 10.30. Standard in vivo FRET experimental design. An endogenous protein (Target1), tagged with CFP, is paired with each of several other endogenous proteins (Target2, Control1, Control2) tagged with YFP. Because in vivo FRET measurements are not absolute, results can be evaluated only by the direct comparison of measurements for an interacting pair (*top panel*) and noninteracting pairs (*middle* and *bottom panels*). (Courtesy of Marc Damelin and Pamela Silver, Dana-Farber Cancer Institute.)

An Explanation of Symbols Used for Analysis of FRET Data

The FRET measurements discussed in this protocol use three filter sets, which are termed the Donor, FRET, and Acceptor (or CFP, FRET, and YFP, respectively). These filter sets are designed to isolate and maximize three signals: the donor fluorescence, the acceptor fluorescence due to FRET, and the directly excited acceptor fluorescence, respectively. The excitation filters for the donor filter set and the FRET filter set are either the same filter or two matched filters. Neutral density filters may be used to match the signals from the three filter sets to the dynamic range of the detector. Defined below are the FRET symbols used in this protocol which represent the signals using the type of filter set (Donor, FRET or Acceptor), the fluorochromes present in the sample (donor only, acceptor only, or both donor and acceptor), and the signal from either just the donor or acceptor when both are present in the sample. Each symbol starts with an uppercase letter representing the filter set:

D for the Donor (CFP) filter set
F for the FRET filter set
A for the Acceptor (YFP) filter set

The second letter is lowercase and indicates which fluorochromes are present in the specimen:

d for donor only (CFP)
a for acceptor only (YFP)
f for both donor and acceptor present (so FRET is possible) (CFP and YFP)

TABLE 10.6. Some FRET symbols and their interpretation

Symbol	Filter set	Fluorochromes present	Meaning
Dd	donor	donor	signal from a donor-only specimen using the donor filter cube
Fd	FRET	donor	signal from a donor-only specimen using the FRET filter set
Ad	acceptor	donor	signal from a donor-only specimen using the acceptor filter set
Da	donor	acceptor	signal from an acceptor-only specimen using the donor filter cube
Fa	FRET	acceptor	signal from an acceptor-only specimen using the FRET filter set
Aa	acceptor	acceptor	signal from an acceptor-only specimen using the acceptor filter set
Df	donor	donor and acceptor	signal from a FRET specimen using the donor filter cube
Ff	FRET	donor and acceptor	signal from a FRET specimen using the FRET filter set
Af	acceptor	donor and acceptor	signal from a FRET specimen using the acceptor filter set

Adapted, with permission, from Gordon et al. (1998).

Strain Construction

1. Construct vectors for tagging the endogenous proteins with CFP or YFP.

   Integrating (nonreplicating) vectors are best suited for this purpose, because the fusion replaces the endogenous copy in the genome and is expressed at comparable levels from cell to cell. The vectors have three primary components: a yeast selection marker such as *TRP1* or *URA3*, an in-frame fusion between the endogenous gene and CFP/YFP, and either a 3′-untranslated region (for carboxy-terminal fusions only) or a promoter (for amino-terminal fusions only). All CFP vectors should have one selection marker (e.g., *TRP1*) and all YFP vectors should have another marker (e.g., *URA3*).

Only a short fragment of the endogenous gene needs to be used because its sole purpose is to target the integration; the genomic copy provides most of the coding sequence. Thus, a short fragment of the gene corresponding to each endogenous protein can be amplified by PCR and inserted into a cassette such as pPS1890 (CFP) or pPS1891 (YFP) (Damelin and Silver 2000). These particular cassettes yield fusions with CFP/YFP at the carboxyl terminus of the endogenous protein. Once integrated into the genome, these fusions are expressed under the endogenous promoter.

EXPERIMENTAL TIP: For some Target proteins, it may be desirable to generate two fusions: one at the amino terminus and one at the carboxyl terminus. Because the FRET signal is highly dependent on the interfluorophore distance, the presence of a fluorophore on one domain of a protein may not yield a FRET signal if another domain is responsible for the protein-protein interaction in question. Performing the experiment with different fusions increases the chance of detecting FRET. Additionally, a fusion to one terminus of a particular target protein may not be functional, in which case, the other fusion should be used.

When designing the integration vectors, it is imperative to ensure that the final vector has a unique restriction site within the DNA encoding the endogenous protein, so that the vector can be linearized prior to the yeast transformation. There should be at least 100 bp of genomic DNA on each side of the restriction site to allow efficient recombination. Ideally, the linearization site will be absent in both *TRP1* and *URA3*, as subsequent experiments may involve swapping the fluorophore tags.

As an alternative to vectors, DNA encoding the fluorescent protein and the selection marker can be amplified by PCR with primers designed for integration at the desired site. This method may be less efficient since there is less genomic DNA to target the recombination.

2. For each fusion protein, transform a wild-type yeast strain (described under Materials) with the linearized vector. Isolate several transformants for further analysis.

3. Grow the transformants in liquid media (SC or drop-out) to log phase, and check expression and localization of the protein fusion under the microscope.

4. For transformants expressing the protein fusion, perform western analysis on whole-cell lysates, probing with α-GFP antibody, to confirm the expression of the full-length fusion.

5. Determine the functionality of each fusion protein by assaying cell viability, expression of the fusion protein, and correct localization of the fusion protein, under conditions in which the fusion protein is required.

   For instance, in one study, Nup82p-YFP was assayed in wild-type cells because *NUP82* is essential for growth, whereas Nup188p-YFP was assayed in *nup170Δ* cells because *NUP188* is essential in that genetic background (Damelin and Silver 2000).

6. If the fusion proteins are functional, generate double-labeled strains for the FRET experiments, for example, by transforming the Target1-CFP strain with each of the YFP constructs. Repeat Steps 3 and 4 to verify coexpression of the fusions.

7. Prepare glycerol stocks of the yeast strains at –80°C selected for further analysis.

Exposure Settings Calibration

The exposure settings for a given filter set must be the same during image acquisition for all strains with the same Target-CFP fusion. The exposure settings for the three filter sets can be different from one another.

8. To prepare the samples, grow the cells to log phase in SC media at room temperature or at 25°C.

9. Acquire sample images of the double-labeled strains in the following order:

   - FRET filter set
   - YFP filter set
   - CFP filter set, to minimize the effects of photobleaching

10. Determine the optimal exposure settings by adjusting the hardware and software parameters, most notably the exposure times.

    Exposures should be as short as possible to minimize effects of photobleaching or photodamage. However, the resulting images must have adequate signal intensity over background and must be amenable to analysis (e.g., check that the desired region of the cell can be selected and analyzed with the software program). Many software applications have a "Scale Image" feature that can be used to change the visualization of images without changing the actual data, and thus may allow shorter exposures.

11. Pipette 2 µl of the sample onto a standard glass slide, gently place a coverslip on the slide, and examine immediately.

    The coverslips must match the objective; typically, use a No. 1.5 coverslip over the sample.

    EXPERIMENTAL TIP: Slides should not be prepared in advance and should not be used for more than 5–10 minutes.

    The exposure settings depend on the equipment used as well as the expression level of the protein fusions. Accordingly, in independent sets of experiments where Target1 represents distinct endogenous proteins, it may be necessary to have different exposure settings. The settings must be the same only for experiments within which the data are directly compared.

Image Acquisition

The most critical aspects of image acquisition are maintaining the same settings and order of exposures for each cell. Acquiring images with all three filter sets records all possible information and permits many different levels of analyses, as described below and in Gordon et al. (1998). If the software allows, it is convenient to use a macro for acquiring three exposures (FRET, then YFP, then CFP) of a cell with the optimal settings determined above in Exposure Settings Calibration. Delays between exposures may be necessary for changing filter sets. For many analysis methods, it is easiest to have images containing only one cell. Therefore, either the region of image acquisition should be limited to a small box in the center of the field or small subregions of larger images should be saved separately.

CFP and YFP are extremely sensitive to photobleaching, and certain precautions should be taken to minimize the effects. Cells should not be visualized by fluorescence before images are acquired, because significant photobleaching occurs even after seconds. Thus, white light (as opposed to the fluorescence lamp) should be used to find and focus on each cell. Similarly, each slide should be used to image only four to five cells, located in different regions of the slide.

The number of cells per strain needed for analysis depends on the specific conditions of the experiment and may be difficult to determine beforehand. A good starting point is to image 20 cells per strain; this includes only cells that coexpress both fusion proteins and provide images amenable to analysis.

12. Acquire images with all filter sets of empty fields to measure background levels.

13. Acquire images with all filter sets of cells expressing only Target1-CFP, to determine *Fd*, *Dd*, and *Ad* (see Gordon et al. 1998, Table 10.6, and the panel above on AN EXPLANATION OF SYMBOLS USED FOR ANALYSIS OF FRET DATA). Prepare the samples as described in Steps 8–11.

    In most cases for CFP, when using the filter sets described above, $Ad = 0$.

14. Acquire images with all filter sets of cells expressing only Target2-YFP, to determine *Fa*, *Da*, and *Aa*. These values depend on the fluorophore and the experimental conditions but not the strain, so these measurements do not need to be repeated for all strains expressing a YFP fusion.

    In most cases for YFP, when using the filter sets described above, $Da = 0$.

15. Acquire images with all filter sets for each of the double-labeled strains.

Data Analysis

IMPORTANT: Images must be analyzed on a cell-by-cell basis.

16. For a given cell, open the three images corresponding to the FRET, YFP, and CFP filter sets.

17. Select a region of the cell for analysis. The regions in all three images should be comparable with respect to region area, as determined by the software and confirmed by eye.

    Several functions available on imaging software applications may be suitable for this purpose. Examples include a function that allows a region to be selected according to pixel intensity ("Threshold") and a function that allows a region to be drawn manually on one image and copied to other images.

18. Log (to a text file or spreadsheet file) the average pixel intensity, *for the selected region,* in each image.

19. Repeat Steps 16–18 for each cell. To facilitate the spreadsheet analysis, log the values for FRET, YFP, and CFP images in the same order for each cell.

20. For each background image (i.e., acquired in Step 12), log the average intensity of a box in the center of the image.

21. Open the data file in a spreadsheet application.

22. Determine the background signal in each filter set by calculating the average value from all background images in that filter set.

23. Subtract the background signal from all the cell measurements (i.e., data from Steps 13–15) in the respective filter sets. *Use these background-corrected cell measurements in all subsequent calculations.*

24. Determine *Fd/Dd* from the CFP strain by calculating the average *Fd/Dd* ratio. This average value can be treated as a constant in Step 26.

25. Determine *Fa/Aa* from the YFP strain by calculating the average *Fa/Aa* ratio. This average value can be treated as a constant in Step 26.

26. For each cell in each of the double-labeled strains, calculate some of the measures of FRET described previously, such as *Ff/Df*, F^c/Df, and *FRET2* (Gordon et al. 1998). Cells with outlying or aberrant (e.g., negative) values should be excluded from further analysis.

    Values obtained from these calculations do not provide an absolute measure of FRET, because they depend on experimental parameters such as exposure time and because in vivo FRET experiments involve many unknowns. The values are meaningful only when compared among the different strains. Moreover, values for the various calculations (*Ff/Df*, F^c/Df, and *FRET2*) for a given cell can be very different from one another.

    F^c represents a calculated FRET value termed "corrected FRET" (Gordon et al. 1998; Youvan et al. 1997).

    $$F^c = Ff - Df(Fd/Dd) - Af(Fa/Aa)$$

    F^c separates the FRET signal from the non-FRET signal if *Ad* and *Da* are effectively zero. However, if *Ad* and *Da* are not zero, then F^c does not correct for the resulting cross-talk, and the FRET versus non-FRET signals are not separated. *Ad* and *Da* can often be made effectively zero by appropriate choice of fluorochromes and filter sets. F^c can be normalized for the concentration of donor by using F^c/Df, which is approximately normalized for the concentration of donor.

    FRET2 is a calculated FRET value, which is beyond the scope of this protocol to derive. For details, see Gordon et al. (1998) and Damelin and Silver (2002).

27. Compare values for the cells in different strains. *Only values for a given calculation (Ff/Df, F^c/Df, or FRET2) can be compared directly.*

    - For a first approximation: Calculate the mean and standard deviation for *Ff/Df*, F^c/Df, and *FRET2* for each strain.

    - For statistical analysis: Compare the values for two given strains using the Rank Sum Test, which compares the data sets about their medians and is appropriate for small samples in which the values are not necessarily distributed evenly about their mean. The standard threshold for the Rank Sum Test is $p < 0.05$, but a higher threshold (e.g., $p < 0.02$) may be used.

        If the values for Target1-CFP/Target2-YFP are significantly greater than those for Target1-CFP/Control-YFP, there is then a FRET interaction between Target1-CFP and Target2-YFP. Such a conclusion is justified only when the proper controls are used (see Experimental Design above). The existence of a FRET interaction implies that the Target proteins are very close to each other in space, because CFP and YFP must be separated by <30 Å. However, the FRET experiment alone does not prove a direct interaction between the Targets.

        There are no "rules" for deciding which calculation (*Ff/Df*, F^c/Df, and *FRET2*) should be ultimately adopted, and the most reasonable choice often depends on the specific experiment. The influence of many aspects of an experiment on the different calculations has been discussed (Gordon et al. 1998). Ideally, the results do not depend on the calculation, as shown explicitly in some cases (Gordon et al. 1998). In other cases, one calculation may be preferable due to the characteristics of those particular experiments; thus, it is often instructive to compare the results for the various calculations. In general, when applicable, the simpler calculations involving fewer measurements are preferable.

        In addition to the controls built into the original experiment, the demonstration that an observed FRET signal is maintained upon swapping fluorophores between the target proteins further substantiates the result. In other words, if Target1-CFP and Target2-YFP yield a FRET interaction, so should Target2-CFP and Target1-YFP in most cases. Different control proteins may be required for this experiment if Target1 and Target2 differ in expression level.

One of the goals of proteomics is the analysis of large multiprotein complexes (e.g., see Damelin and Silver 2002). The FRET protocol is quite useful in this regard, because the noninteracting proteins within the complex can serve as Control proteins. In such a study, as many proteins as possible in a protein complex are tagged with CFP and YFP. The FRET assay is performed on the double-labeled strains, and as in this protocol, only strains with the same CFP fusion are compared to one another. The signals from certain strains may significantly exceed the signals from other strains in the panel, thus constituting FRET interactions. Implicit in this setup is the idea that the identities of the Target2 and Control proteins for each Target-CFP fusion may not be known beforehand, but will be revealed during the course of the experiment. A given protein may serve as Target2 in one case and as a Control in another. Moreover, in some cases, it is possible that more than one FRET interaction will be observed for one Target-CFP fusion.

PROTOCOL 9

Monitoring Protein-Protein Interactions with GFP Chimeras

THESE PROTOCOLS WERE CONTRIBUTED BY Sang-Hyun Park (Department of Cellular and Molecular Pharmacology, University of California, San Francisco, California) and Ronald T. Raines (Department of Biochemistry, University of Wisconsin, Madison, Wisconsin).

Green fluorescent protein (GFP) from the jellyfish *Aequorea victoria* has exceptional physical and chemical properties besides its spontaneous fluorescence. These properties, which include high thermal stability and resistance to detergents, organic solvents, and proteases, endow GFP with enormous potential for biotechnical applications. Since the cDNA of GFP was cloned (Prasher et al. 1992), a variety of GFP variants have been generated that broaden the spectrum of its application (Cubitt et al. 1995; Delagrave et al. 1995; Ehrig et al. 1995; Heim et al. 1995; Crameri et al. 1996; Ward et al. 2000). Among those variants, S65T GFP is unique in having increased fluorescence intensity and faster fluorophore formation than that of the wild-type protein, as well as altered excitation and emission spectra (Heim et al. 1995; Ormö et al. 1996). Moreover, the wavelengths of the excitation and emission maxima of S65T GFP (490 and 510 nm, respectively) are close to those of fluorescein. The fluorescein-like spectral characteristics of S65T GFP enable its use with instrumentation that had been designed specifically for use with fluorescein.

We describe here the use of S65T GFP to probe protein-protein interactions in vitro (Park and Raines 1997, 2000). This method requires fusing S65T GFP to a target protein (X) to create a GFP chimera (X~GFP[S65T]). The target protein is fused to the amino terminus of GFP, rather than to the carboxyl terminus; otherwise, heterogeneity can occur (Park and Raines 1997, 2000). The interaction of the fusion protein with another protein (Y) can be analyzed with common instrumentation by two distinct methods.

The first method is a fluorescence gel-retardation assay. The gel-retardation assay has been used widely to study protein-DNA interactions (Carey 1991). This assay is based on the electrophoretic mobility of a protein-DNA complex being less than that of either molecule alone. In a fluorescence gel-retardation assay, electrophoretic mobility is detected by the fluorescence of S65T GFP. This assay is a rapid method to demonstrate the existence of a protein-protein interaction and to estimate the equilibrium dissociation constant (K_d) of the resulting complex.

The second method is a fluorescence polarization assay, which provides an accurate means to evaluate K_d in a specified homogeneous solution. Fluoresence polarization assays usually rely on fluorescein as an exogenous fluorophore. S65T GFP can likewise serve in this role. Furthermore, the fluorescence polarization assay can be adapted for the high-throughput screening of protein or peptide libraries.

METHOD 1: Fluorescence Gel-retardation Assay

MATERIALS

CAUTION: See Appendix 3 for appropriate handling of materials marked with <!>.

Reagents

Native polyacrylamide gel (6% w/v) <!>
Tris-HCl buffer (10 mM, pH 7.5) containing glycerol (5% v/v)

Equipment

Software for quantifying fluorescence (ImageQuant, Amersham Biosciences)
Fluorescence imaging system (e.g., FluorImager System, Amersham Biosciences)
UV/visible spectrophotometer
Vertical gel electrophoresis system

Biological Molecules

Purified protein of interest (Y)
Purified S65T GFP chimera (X~GFP[S65T]) (CLONTECH)
For details on the construction and purification of the chimera, see Park and Raines (1997, 2000).

METHOD

1. Estimate the concentration of purified X~GFP(S65T) using the extinction coefficient ($\varepsilon = 39.2\ \mathrm{mM^{-1}cm^{-1}}$ at 490 nm) of S65T GFP (Heim et al. 1995).
2. To begin the gel-retardation assay, mix 1.0 µM of X~GFP(S65T) with varying amounts of Y in 10 µl of 10 mM Tris-HCl buffer (pH 7.5) containing 5% glycerol.
3. Incubate the mixtures for 20 minutes at 20°C. Load the mixtures onto a 6% native polyacrylamide gel, and subject the loaded gel to electrophoresis at 10 V/cm for 30 minutes at 4°C.
4. Immediately after electrophoresis, scan the gel with a fluorescence imaging system at 700 V using a built-in filter set (490 nm for excitation and ≥515 nm for emission).
5. Quantify the fluorescence intensities of bound and free X~GFP(S65T) using appropriate software, such as ImageQuant.
6. Determine the value of **R** (= fluorescence intensity of bound X~GFP[S65T]/total fluorescence intensity) for each gel lane from the measured fluorescence intensities.
7. Calculate the value of K_d for each lane with Equation 1, and average those values.

$$K_d = \frac{1-\mathbf{R}}{\mathbf{R}}\left([\mathrm{Y}]_{\mathrm{total}} - \mathbf{R}[\mathrm{X{\sim}GFP}]_{\mathrm{total}}\right) \qquad (1)$$

Case Study: S65T GFP Chimeras

To demonstrate the potential of S65T GFP chimeras in the exploration of protein-protein interactions, we use as a model system the well-characterized interaction of the S peptide and S-protein fragments of bovine pancreatic ribonuclease (RNase A; EC 3.1.27.5) (Raines 1998; Raines et al. 2000). (GFP chimeras have also been used to explore the interaction of [1] the BALF-1 protein and Bax/Bak [Marshall et al. 1999]; [2] CREB and importins [Forwood et al. 2001]; and [3] a cyclophilin and the capsid protein p24 of HIV-1 [Kiessig et al. 2001].) Subtilisin treatment of RNase A yields two tightly associated polypeptide chains: S peptide (residues 1–20) and S protein (residues 21–124) (Richards 1955). Residues 1–15 of S peptide (S15) are necessary and sufficient to form a complex with S protein (Potts et al. 1963).

The fluorescence gel-retardation assay is used to quantify the interaction between S protein and S15. A GFP chimera, S15~GFP(S65T)~His_6, is produced by standard recombinant DNA techniques and purified by Ni^{++}-affinity chromatography (Park and Raines 1997, 2000). A fixed quantity of S15~GFP(S65T)~His_6 is incubated with a varying quantity of S protein prior to electrophoresis in a native polyacrylamide gel. After electrophoresis, the gel is scanned with a fluorescence imaging system, and the fluorescence intensities of bound and free S15~GFP(S65T)~His_6 are quantified (Figure 10.31, top). The value of K_d for the complex formed in the presence of different S-protein concentrations is calculated from the values of **R** and the total concentrations of S protein and S15~GFP(S65T)~His_6 (Equation 1). The average (±SE) value of K_d is $(6 \pm 3) \times 10^{-8}$ M.

A competition assay is used to probe the specificity of the interaction between S15~GFP(S65T)~His_6 and S protein. S15~GFP(S65T)~His_6 and S protein are incubated to allow for complex formation. Varying amounts of S peptide are added, and the resulting mixture is incubated further (20 minutes), and then subjected to native gel electrophoresis. The addition of S peptide converts bound S15~GFP(S65T)~His_6 to the free state (Figure 10.31, bottom). Thus, S15~GFP(S65T)~His_6 and S peptide bind to the same region of S protein.

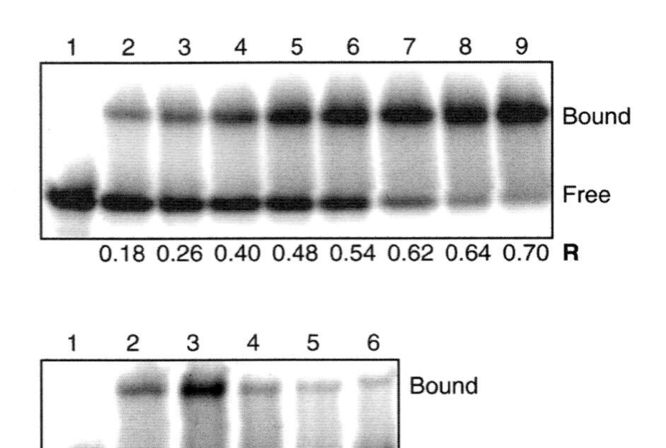

FIGURE 10.31. Gel-retardation assay of a protein-protein interaction. (*Top*, Lanes *1–9*) 1 µM S15~GFP(S65T)~His_6 and 0, 0.2, 0.3, 0.4, 0.5, 0.6, 0.7, 0.8, and 0.9 µM S protein, respectively. The relative mobilities of free and bound S15~GFP(S65T)~His_6 are 0.72 and 0.47, respectively. The value of **R** is obtained for each lane, and values of K_d are calculated by using Equation 1, with the average being $K_d = (6 \pm 3) \times 10^{-8}$ M. (*Bottom*) Gel-retardation assay demonstrating that S peptide competes with S15~GFP(S65T)~His_6 for interaction with S protein. (Lane *1*) 1 µM S15~GFP(S65T)~His_6 and no S protein or S peptide; (lanes *2–6*) 1 µM S15~GFP(S65T)~His_6, 1 µM S protein, and 0, 0.3, 1.0, 3.0, and 10 µM S peptide, respectively.

METHOD 2: Fluorescence Polarization Assay

The fluorescence gel-retardation assay described above is a convenient method to visualize a protein-protein interaction as well as to estimate the value of K_d for the resulting complex. Still, gel-retardation assays have an intrinsic limitation in evaluating equilibrium dissociation constants. In a gel-retardation assay, it is assumed that a receptor-ligand interaction remains at equilibrium during sample loading and electrophoresis. Yet, as samples are loaded and migrate through a gel, complex dissociation is unavoidable and results in an underestimation of the value of K_d. Moreover, if the conditions (e.g., pH, salt type, and concentration) encountered during electrophoresis differ from those in the incubation, then the measured value of K_d may not correspond to the true value.

In a fluorescence polarization assay, the formation of a complex is deduced from an increase in fluorescence polarization, and the equilibrium dissociation constant is determined in a homogeneous aqueous environment. Many fluorescence polarization assays have used fluorescein as the fluorophore (LeTilly and Royer 1993; Heyduk et al. 1996; Malpeli et al. 1996; Fisher et al. 1998). Like a free fluorescein-labeled ligand, a free S65T GFP chimera is likely to rotate more rapidly and therefore to have a lower rotational correlation time than does a bound chimera. An increase in rotational correlation time upon binding results in an increase in fluorescence polarization, which can be used to assess complex formation (Jameson and Sawyer 1995). In contrast to a gel-retardation assay, the fluorescence polarization assay is performed in a homogeneous solution in which the conditions can be dictated precisely.

MATERIALS

Reagents

Tris-HCl buffer (20 mM, pH 8.0) containing NaCl (at various concentrations)

Equipment

Fluorescence spectrometer capable of measuring polarization (e.g., the Beacon fluorescence polarization system, PanVera, Madison, Wisconsin)

Software capable of nonlinear regression analysis (e.g., DeltaGraph, DeltaPoint, Monterey, California)

Biological Molecules

Purified protein of interest (Y)

Purified S65T GFP chimera (X~GFP[S65T]) (CLONTECH)

For details on the construction and purification of the chimera, see Park and Raines (1997, 2000).

METHOD

1. Mix 0.50–1.0 nM X~GFP(S65T) with various concentrations of Y in 1.0 ml of 20 mM Tris-HCl buffer (pH 8.0), with or without NaCl at 20°C.

   Conditions such as buffer, pH, temperature, and salt can be varied to test their effect on the binding interaction.

2. Immediately after mixing, make five to seven polarization measurements at each concentration of Y using a fluorescence spectrometer capable of measuring polarization. For a blank measurement, use a mixture that contains the same components but lacks X~GFP(S65T). Fluorescence polarization (P) is defined as:

$$P = \frac{I_{||} - I_{\perp}}{I_{||} - I_{\perp}} \tag{2}$$

   where $I_{||}$ is the intensity of the emission light parallel to the excitation light plane and $I_{\perp}$ is the intensity of the emission light perpendicular to the excitation light plane. P, being a ratio of light intensities, is a dimensionless number and has a maximum value of 0.5.

3. Determine values of K_d by using software, such as DeltaGraph, to fit the data to the equation:

$$P = \frac{\Delta P \bullet F}{K_d + F} + P_{min} \tag{3}$$

   where P is the measured polarization, ΔP ($= P_{max} - P_{min}$) is the total change in polarization, and F is the concentration of free Y. The fraction of bound protein (f_B) is obtained by using the equation:

$$f_B = \frac{P - P_{min}}{\Delta P} = \frac{F}{K_d + F} \tag{4}$$

   The familiar binding isotherms are obtained by plotting f_B versus F.

CASE STUDY: FLUORESCENCE POLARIZATION

Fluorescence polarization is used to determine the effect of salt concentration on the formation of a complex between S protein and S15~GFP(S65T)~His_6. The value of K_d increases by fourfold when NaCl is added to a final concentration of 0.10 M (see Figure 10.32). A similar salt dependence for the dissociation of RNase S had been observed previously (Schreier and Baldwin 1977). The added salt is likely to disturb the water molecules hydrating the hydrophobic patch in the complex between S peptide and S protein, resulting in a decrease in the binding affinity (Baldwin 1996). Finally, the value of $K_d = 4.2 \times 10^{-8}$ M observed in 20 mM Tris-HCl buffer (pH 8.0) containing NaCl (0.10 M) is similar (i.e., threefold lower) to that obtained by titration calorimetry in 50 mM sodium acetate buffer (pH 6.0) containing NaCl (0.10 M) (Connelly et al. 1990).

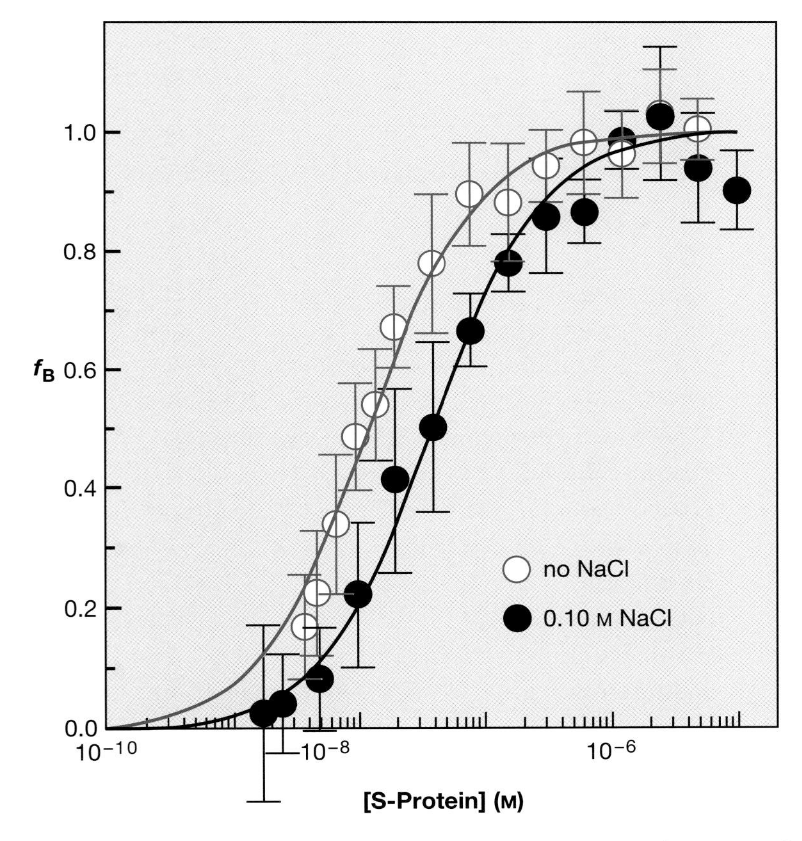

FIGURE 10.32. Fluorescence polarization assay of a protein-protein interaction. S protein (20 µM to 1.0 nM) is added to 20 mM Tris-HCl buffer (pH 8.0) containing S15~GFP(S65T)~His_6 (0.50 nM) at 20°C in a volume of 1.0 ml. Each data point is an average of 5–7 measurements. Binding isotherms are obtained by fitting the data to Equations 3 and 4. The values of K_d in the presence of 0 and 0.10 M NaCl are 1.1×10^{-8} M and 4.2×10^{-8} M, respectively. (Courtesy of Sang-Hyun Park, University of California, San Francisco, and Ronald T. Raines, University of Wisconsin, Madison.)

PROTOCOL 10

Protein Arrays: Preparation of Microscope Slides

THE PROTOCOL WAS CONTRIBUTED BY Gavin MacBeath (Harvard University, Cambridge, Massachusetts).

Many vendors supply slides that can be used for the preparation of protein microarrays. Typically, these slides come with protocols for arraying, blocking, and processing. The following protocols are for two particularly useful substrates.

MATERIALS

CAUTION: See Appendix 3 for appropriate handling of materials marked with <!>.

Reagents

3-Aminopropyltriethoxysilane (United Chemical Technologies, Inc., Bristol, Pennsylvania) <!>
Bovine serum albumin (BSA) (1% w/v) (Fraction V, minimum 98% pure)
N,N′-Disuccinimidyl carbonate
N,N-Diisopropylethylamine (stock is 5.73 M) <!>
N,N-Dimethylformamide (DMF), anhydrous <!>
Ethanol (absolute) <!>
Ethanol (95%)/5% H_2O solution, adjusted to pH 4.5–5.5 with acetic acid <!>
HCl (1% v/v) aqueous solution <!>
Helium gas <!>
Nitrogen gas <!>
Phosphate-buffered saline (PBS) (400 ml, pH 7.5)
Sodium hydroxide (10% w/v), aqueous solution <!>
3-Triethoxysilylbutanal (United Chemical Technologies, Inc.)

Equipment

Centrifuge with swinging bucket rotor and microplate carriers
Microscope aminopropyl-modified slides (Sigma-Aldrich, Corning, TeleChem International)
Microscope glass slides (VWRbrand Micro Slides–plain, VWR Scientific Products)
Slide rack, stainless steel (Wheaton Science Products, Millville, New Jersey)
Staining dish, glass rectangular (Wheaton Science Products)
Vacuum desiccator

METHOD

Fabrication of Aldehyde Slides

Aldehyde slides are glass microscope slides that display an aldehyde functionality on their surface (see Figure 10.7). They may be purchased from commercial vendors such as TeleChem International, Inc. (Sunnyvale, California) under the trade name SuperAldehyde Substrates or fabricated more cost-effectively as described below.

1. Clean plain glass slides by immersing them in an aqueous solution of 10% NaOH (w/v) for 12 hours at room temperature.

   EXPERIMENTAL TIP: Up to 30 slides can be processed conveniently using a 30-slide stainless steel rack and glass rectangular staining dish.

2. Rinse the slides in H_2O, then in an aqueous solution of 1% HCl (v/v), and finally twice in H_2O.

3. Dry the slides by spinning the racks containing the slides in a centrifuge equipped with a swinging bucket rotor and microplate carriers. Centrifuge at 200*g* for 30 seconds at room temperature.

   Drying by centrifugation prevents deposition of spots on the slides.

4. Adjust a 95% ethanol/5% H_2O solution to pH 4.5–5.5 with acetic acid (~0.1% acetic acid).

5. Add 3-triethoxysilylbutanal while stirring to yield a 2% solution of silane (v/v).

6. Stir for 3 minutes at room temperature to allow for hydrolysis and silanol formation and then immerse the slides in the solution. Incubate them for an additional 3 minutes at room temperature.

7. Rinse the slides briefly in absolute ethanol and centrifuge them to dry. Cure them under vacuum for 10 minutes at 110°C or for 24 hours at room temperature.

8. Wash the slides three times for 5 minutes each with 95% ethanol/5% H_2O and dry them by centrifugation. Store the slides at room temperature in a vacuum desiccator or under inert gas.

BSA-NHS Slides

BSA-NHS slides are glass microscope slides that contain a layer of bovine serum albumin (BSA) covalently attached to their surface. In addition, the BSA molecules are chemically activated for covalent attachment of proteins to their surface (see Figure 10.8). BSA-NHS slides are fabricated from aminopropyl-modified slides. These slides can be purchased from a number of vendors (including Sigma-Aldrich, Corning, and TeleChem International) or fabricated from plain glass slides. If purchasing aminopropyl-modified slides, skip to Step 5.

1. Clean plain glass slides as described in the previous section in Steps 1–3.

2. Stir a solution of 95% ethanol/5% H_2O and add 3-aminopropyltriethoxysilane to yield a 3% solution of silane (v/v). (Note: Do not add acetic acid). Stir for 10 minutes at room

temperature to allow for hydrolysis and silanol formation, and then immerse the slides in the solution. Incubate them at room temperature for an additional 10 minutes.

3. Rinse the slides briefly in absolute ethanol and centrifuge them to dry. Cure them under vacuum for 10 minutes at 110°C or for 24 hours at room temperature.

4. Wash the slides three times for 5 minutes each with 95% ethanol/5% H_2O and dry them by centrifugation. Store the slides at room temperature (preferably with a desiccant).

5. Dissolve 10.24 g of *N,N′*-disuccinimidyl carbonate and 6.96 ml of *N,N*-diisopropylethylamine in 400 ml of anhydrous DMF.

   Final concentrations of *N,N′*-disuccinimidyl carbonate and *N,N*-diisopropylethylamine are 100 mM each.

6. Immerse 30 aminopropyl-modified slides (from Step 4 or purchased commercially) in the disuccinimidyl carbonate/diisopropylethylamine solution for 3 hours at room temperature.

   The solution may turn yellow during the 3-hour incubation.

7. Rinse the slides twice with 95% ethanol/5% H_2O.

8. Immerse the slides in 400 ml of PBS (pH 7.5) containing 1% BSA (w/v) for 12 hours at room temperature.

9. Rinse the slides twice with H_2O, twice with 95% ethanol/5% H_2O, and then dry them by centrifugation.

10. Dissolve 10.24 g of 100 mM *N,N′*-disuccinimidyl carbonate and 6.96 ml of 100 mM *N,N*-diisopropylethylamine in 400 ml of anhydrous DMF. Immerse the slides in this solution for 3 hours at room temperature.

11. Rinse the slides three times with 95% ethanol/5% H_2O and dry by centrifugation.

    The slides are now ready for use or can be stored in a desiccator under vacuum for up to 3 months at room temperature without noticeable loss of activity.

PROTOCOL 11

Protein Arrays: Labeling the Protein and Probing the Array for Protein–Protein Interactions

THIS PROTOCOL WAS CONTRIBUTED BY Gavin MacBeath (Harvard University).

To detect protein-protein interactions, a protein microarray can be incubated with a fluorescently labeled protein and the array washed; stable interactions can then be identified by scanning the slide for fluorescent spots (see Figure 10.9). The probe proteins can be labeled with any one of several different commercially available dyes. The method in this protocol uses cyanine-3 (Cy3) or cyanine-5 (Cy5) monofunctionally activated dyes. As with arraying, care must be taken to remove nucleophilic molecules from the protein solution prior to labeling. Common nucleophilic buffer components include Tris buffer, ethanolamine, glycine, dithiothreitol, β-mercaptoethanol, glutathione, and imidazole.

MATERIALS

Reagents

Cy3 or Cy5 monofunctionally activated dyes (Amersham Biosciences)
These dyes come lyophilized in screw-cap tubes with enough dye to label the equivalent of 1 mg of antibody (IgG).

Phosphate-buffered saline (PBS) (pH 7.5)

PBS containing 500 mM glycine (pH 8.0)

PBST (PBS containing 0.1% v/v Tween-20)

Probing buffer
See note to Step 5.

Equipment

Affinity chromatography equipment
See Step 4.

Centrifuge

Dialysis equipment
See Step 4.

Fluorescence slide scanner (e.g., ArrayWoRx fluorescence slide scanner, Applied Precision, Issaquah, Washington)

Humidity chamber

Incubation chambers (CoverWell, Grace Biolabs, Bend, Oregon)
Use these incubation chambers for processing only one array per slide (see Step 6). The CoverWell PC500 accommodates a volume of ~500 µl.

Size-exclusion chromatography equipment
See Step 4.

Slide rotor
See Step 6.

Biological Molecules

Proteins of interest to be fluorescently labeled

Slides printed with protein arrays
For details on printing protein arrrays, see the section of the chapter introduction on Fabricating Protein Arrays. For a summary of highlights, see the panel on GUIDELINES FOR PREPARING PROTEIN ARRAYS below.

GUIDELINES FOR PREPARING PROTEIN ARRAYS

- Prepared proteins should be at a concentration of at least 100 µg/ml (20 µM for peptides), in a buffer with a pH of ≥7, and with 40% glycerol included in the buffer.
- Unacceptable buffers include Tris and ethanolamine.
- Acceptable buffers include MES, PIPES, phosphate, MOPS, HEPES, and borate.
- Avoid the following common additives: dithiothreitol, β-mercaptoethanol, imidazole, glutathione, glycine, aprotinin, and leupeptin.
- Acceptable common additives are TCEP, inorganic ions (Na^+, K^+, Ca^{2+}, Mg^{2+}, Mn^{2+}, NH_4^+, Cl^-, SO_4^{2-}), acetate, sodium azide, EDTA, EGTA, PMSF, and AEBSF.
- For most purposes, use a solution of 40% glycerol/60% PBS (pH 7.5).
- If relatively few proteins are being arrayed, print several arrays per slide (up to 16). While the slide is still dry (after printing but before quenching), draw a grid on the slides with a hydrophobic pen (e.g., PAP pen Liquid Blocker from Newcomer Supply, Middleton, Wisconsin) to physically separate the arrays from each other.
- To avoid smearing the arrayed droplets during quenching, see the panel on ADDITIONAL PROTOCOL: PREVENTING THE FORMATION OF COMETS at the end of this protocol.

METHOD

Labeling the Protein

1. Dissolve the dye in 250 µl of H_2O, and add 5 µl of dye solution for every 10 µg of protein to be labeled. Aim to maintain a protein concentration of at least 100 µg/ml.

   Under these conditions, the extent of labeling is relatively independent of protein concentration. Although the manufacturer recommends labeling at pH 9.3, we routinely label our proteins in PBS at pH 7.5. Low pH favors conjugation of the dye to the α-amine of the protein, whereas high pH results in relatively unselective conjugation to lysine residues.

2. Incubate the dye and protein mixture for 30 minutes at room temperature to allow the dye to react with the protein molecules.

3. Quench the reaction by adding 0.1x volume of PBS containing 500 mM glycine (pH 8.0) and incubate the tube for an additional 30 minutes at room temperature.

4. Remove unincorporated dye either by size exclusion chromatography or by dialysis.

   If the protein contains an affinity tag (such as a His_6-tag or GST-tag), the unincorporated dye is most conveniently removed by affinity chromatography using that tag.

Probing the Array

5. Prepare probing solution by combining probing buffer and fluorescently labeled protein.

   The composition of the probing buffer depends on the labeled protein. In general, the buffer should contain salt, detergent, and BSA to minimize nonspecific binding. In addition, the buffer may contain reductants, metal ions, or other cofactors as required. As a starting point, we recommend PBS (pH 7.5) containing 0.1% Tween-20 (v/v) and 1% BSA (w/v). The optimal concentration of labeled protein varies from experiment to experiment. Begin with 1 µg/ml, which corresponds to 20 nM for a 50-kD protein.

   The volume of buffer to prepare depends on the experiment. Workable protein concentrations range from 1 nM to 1 µM.

6. Probe the protein array with the fluorescently labeled protein.

   Processing single-array slides:

   a. Lay the CoverWell slide incubation chamber on the bench with the *gasket side up*.

   b. Apply 550 µl of probing solution to the CoverWell.

   c. Slowly lower the microarray slide, *array side down*, onto the CoverWell.

   d. Invert the slide and press the slide gently but firmly to seal the CoverWell to the slide.

      If done correctly, the solution should spread over the surface of the slide without trapping air bubbles.

   Processing multi-array slides:

   a. With the slides right-side up, apply 10–30 µl of labeled protein to each array. Do not cover the arrays with coverslips or a CoverWell.

   b. Place the slides in a humid chamber.

7. Incubate the slides for 1 hour at room temperature.

8. Rinse the slides briefly with PBST, and wash them with PBST three times for 2–3 minutes each.

9. Rinse the slides briefly with PBS (without Tween-20), and centrifuge the slides in a slide rotor at 200*g* for 30 seconds to remove excess buffer.

   The slides can be dried under a stream of nitrogen gas if an appropriate rotor is not available.

10. Image the slides with a fluorescence slide scanner (see Reading and Interpreting the Arrays).

Additional Protocol: Preventing the Formation of Comets

Following the attachment of proteins to activated slides, if the arrays are quenched either by pouring quenching solution over the slides or by plunging the slides into the quenching solution, the nanodroplets smear across the surface of the slide and the resulting spots look like comets. Smearing is particularly problematic with aldehyde slides because the reaction between the proteins and the slides is so fast. The following protocol (provided by Gavin MacBeath, Harvard University) is designed to prevent comet formation.

Additional Materials

CAUTION: See Appendix 3 for appropriate handling of materials marked with <!>.

Quenching solutions
For aldehyde slides: PBS (pH 7.5) containing 1% BSA
For BSA-NHS slides: PBS containing 500 mM glycine (pH 8.0) <!>

Method

1. Hold the slide upside down and carefully drop the slide onto the quenching solution from about a centimeter above the solution. The slide will float on the quenching solution due to surface tension.
2. Wait ~10 seconds and then turn the slide right-side up and immerse it in the quenching solution.
3. Shake the slide in quenching solution for at least 1 hour at room temperature or for at least 4 hours at 4°C.
4. Arrays can be stored for several weeks at 4°C by immersing them in PBS containing 1% BSA and 0.02% NaN_3 <!>. For longer-term storage, add 50% glycerol and store at –20°C.

PROTOCOL 12

Protein Arrays: Labeling the Compounds and Probing the Array for Protein–Small Molecule Interactions

THIS PROTOCOL WAS CONTRIBUTED BY Gavin MacBeath (Harvard University).

To detect protein interactions with small molecules, a protein microarray is incubated with a fluorescently labeled compound, the array is washed, and stable interactions are identified by scanning the slide for fluorescence (see Figure 10.10). It is often adventageous to label the small molecules indirectly, by linking them covalently to a previously labeled carrier molecule, such as BSA. As with the labeling of probe proteins (Protocol 11), BSA can be labeled with any one of several commercially available dyes. This protocol describes the labeling of BSA with Cy3 or Cy5, the linkage of the small molecule to the labeled BSA, and the probing of protein microarrays with the fluorescently labeled conjugate.

MATERIALS

CAUTION: See Appendix 3 for appropriate handling of materials marked with <!>.

Reagents

Cy3 and Cy5 monofunctionally activated dyes (Amersham Biosciences)

Disuccinimidyl glutarate (DSG)

N,N′-Disuccinimidyl carbonate

N-[γ-maleimidobutyryloxyl]sulfosuccinimide ester (Sulfo-GMBS, Pierce Biotechnology)

Phosphate-buffered saline (PBS) (pH 7.5)

PBS containing 500 mM glycine (pH 8.0) <!>

PBST supplemented with 1% bovine serum albumin (BSA) (w/v)

Equipment

Fluorescence slide scanner (e.g., ArrayWoRx fluorescence slide scanner, Applied Precision, Issaquah, Washington)

Gel-filtration equipment (e.g., NAP-10 column, Amersham Biosciences)

Humidity chamber

See Step 7.

Incubation chambers (CoverWell, Grace Biolabs, Bend, Oregon)

Use these incubation chambers for processing only one array per slide (see Step 7). The CoverWell PC500 accommodates a volume of ~500 μl.

Biological Molecules

Bovine serum albumin (BSA) to be fluorescently labeled

Small molecules of interest

Slides printed with protein arrays

For details on printing protein arrrays, see the section of the chapter introduction on Fabricating Protein Arrays. For a summary of highlights, see the panel on GUIDELINES FOR PREPARING PROTEIN ARRAYS in Protocol 11.

METHOD

Labeling the BSA

1. Dissolve 1 mg of BSA in 0.5 ml of PBS (pH 7.5). Add this solution to one tube of cyanine monofunctionally activated dye (either Cy3 or Cy5).
2. Incubate the BSA and dye mixture for 30 minutes at room temperature.
3. Quench the reaction by adding 100 μl of PBS containing 500 mM glycine (pH 8.0) and incubate for an additional 30 minutes at room temperature.
4. Remove the unincorporated dye by gel filtration, using PBS (pH 7.5) as the elution buffer.

Coupling Small Molecules to the BSA

5. An appropriate strategy for coupling a small molecule to labeled BSA depends on which functional groups are available on the small molecule.

   It is beyond the scope of this protocol to provide detailed methods on how to deal with every possible situation. For an excellent resource for information and protocols relating to bioconjugate chemistry, see Hermanson (1996). The following recommendations have been provided to serve as general guidelines. Specific details will vary depending on the nature of the molecules under investigation.

   If linking through a carboxylate on the small molecule:

   a. Activate the carboxylate as an *N*-hydroxysuccinimidyl ester using *N,N′*-disuccinimidyl carbonate under basic conditions.

   b. Mix the activated compound with the labeled BSA in PBS (pH 7.5). Incubate for 1 hour at room temperature.

      The appropriate ratio of small molecule to BSA varies depending on the compound, but a good starting point is to use a 10:1 molar excess of small molecule.

   c. Separate the conjugate from the free compound by gel filtration.

   If linking through an amine group on the small molecule:

   a. Activate the small molecule with at least a tenfold molar excess of DSG.

   b. Separate the activated compound from the unreacted DSG and dimeric compound.

   c. Mix the activated compound with the labeled BSA in PBS (pH 7.5) for 1 hour at room temperature.

      The appropriate ratio of small molecule to BSA varies depending on the compound, but a good starting point is to use a 10:1 molar excess of small molecule.

   d. Purify the conjugate by gel filtration.

If linking through a thiol group on the small molecule:

a. Activate the labeled BSA by treating it with sulfo-GMBS.

Dissolve sulfo-GMBS in PBS (pH 7.5) at a concentration of 10 mg/ml. Add 50 equivalents of sulfo-GMBS to an ~1 mg/ml solution of cyanine-labeled BSA in PBS (pH 7.5). For example, add 29 µl of a 10 mg/ml solution of sulfo-GMBS to 1 ml of a 1 mg/ml solution of cyanine-labeled BSA. Incubate for 1 hour at room temperature.

b. Quench the reaction by adding a 1/10 volume of PBS containing 500 mM glycine (pH 8.0) and incubate for 30 minutes at room temperature.

c. Remove unincorporated reagent by gel filtration.

d. Add 100-fold molar excess of the thiol-containing small molecule to the cyanine-BSA-maleimide conjugate in PBS (pH 7.5). Incubate the reaction for 1 hour at room temperature.

e. Purify the final conjugate by gel filtration.

Probing the Array

6. Dilute the doubly labeled BSA conjugates with PBST supplemented with 1% (w/v) BSA.

We recommend starting with a concentration of 10 µg/ml, but higher and lower concentrations (between 0.1 µg/ml and 100 µg/ml) should also be explored.

7. Probe the protein array with the diluted doubly labeled BSA conjugates.

Processing single-array slides:

a. Lay the CoverWell on the bench with the *gasket side up*.

b. Apply 550 µl of probing solution to the CoverWell.

c. Slowly lower the microarray slide, *array side down*, onto the CoverWell.

d. Invert the slide and press it gently but firmly to seal the CoverWell to the slide.

If done correctly, the solution should spread over the surface of the slide without trapping air bubbles.

Processing multi-array slides:

a. With the slides right-side up, apply 10–30 µl of labeled protein to each array. Do not cover the arrays with coverslips or a CoverWell.

b. Place the slides in a humid chamber.

8. Incubate the slides for 1 hour at room temperature.

9. Rinse the slides with PBST, and then wash them with PBST three times for 2–3 minutes each.

10. Rinse the slides with PBS (without Tween-20), and centrifuge the slides in a slide rotor at 200*g* for 30 seconds to remove excess buffer.

11. Image the slides with a fluorescence slide scanner (see Reading and Interpreting the Arrays).

The slides can be dried under a stream of nitrogen gas if an appropriate rotor is not available.

PROTOCOL 13

Protein Arrays: Probing the Array and Detecting Radioactive Spots for Kinase-Substrate Interactions

THIS PROTOCOL WAS CONTRIBUTED BY Gavin MacBeath (Harvard University).

Protein arrays can be probed for potential substrates of protein kinases. To identify substrates of a kinase of interest, a protein microarray is incubated with that kinase and [γ-^{33}P]ATP (see Figure 10.11). Following an appropriate incubation, the array is washed, coated with a photographic emulsion, and incubated further; the emulsion is developed, and then imaged with a scanning light microscope.

MATERIALS

CAUTION: See Appendix 3 for appropriate handling of materials marked with <!>.

Reagents

ATP
Dektol developer (Eastman Kodak Company) <!>
Desiccant
[γ-^{33}P]ATP (20 μCi) <!>
Kinase buffer
  50 mM Tris-Cl (pH 7.5)
  10 mM $MgCl_2$ <!>
  1 mM dithiothreitol (DTT) <!>
Kodak fixer, photographic (Eastman Kodak Company) <!>
NTB-2 autoradiography emulsion (Eastman Kodak Company)
Wash buffer
  20 mM Tris-Cl (pH 7.5)
  150 mM NaCl
  10 mM EDTA
  1 mM EGTA
  0.1% Triton X-100 <!>

Equipment

β-Radiation box
Centrifuge
Incubation chamber (PC200 CoverWell, Grace Biolabs, Bend, Oregon)
Scanning light microscope

Biological Molecules

Appropriate kinase(s)
Proteins of interest

Slides printed with protein arrays

For details on printing protein arrrays, see the section of the chapter introduction on Fabricating Protein Arrays. For a summary of highlights, see the panel on GUIDELINES FOR PREPARING PROTEIN ARRAYS in Protocol 11.

METHOD

1. Wash the arrayed slides with wash buffer three times for 10 minutes each.
2. Wash the slides once with kinase buffer for 10 minutes.
3. Incubate the slides with kinase buffer containing 100 μM ATP for 10 minutes.
4. Wash the slides with kinase buffer for an additional 10 minutes.
5. Dilute each kinase into 200 μl of kinase buffer containing 100 μM ATP and 20 μCi [γ-^{33}P]ATP.
6. Apply the kinase solution to each array using a PC200 CoverWell incubation chamber (see Protocol 11, Step 6) and incubate the slides for 1 hour at room temperature.
7. Wash the slides with wash buffer six times for 5 minutes each.
8. Wash the slides with wash buffer lacking Triton X-100 twice for 5 minutes each.
9. Wash the slides with H_2O three times for 3 minutes each.
10. Centrifuge the slides at 200*g* for 30 seconds to remove excess water.

    The slides can be dried under a stream of nitrogen if an appropriate rotor is not available.

Detecting Radioactive Spots

11. To visualize the radioactive decay, melt NTB-2 autoradiography emulsion for 45 minutes at 45°C in a dark room.
12. Working in the dark, dip the slides in the emulsion for 3 seconds and allow them to dry vertically for 4 hours at room temperature.

    The use of a slide-dipping chamber simplifies this step.

13. Seal the slides in a β-radiation box with a desiccant and incubate in the dark for 4–10 days at 4°C.

    Bring slides to room temperature before developing because the emulsion has a tendency to bubble and/or slide off if they are opened when they are cold.

    EXPERIMENTAL TIP: To avoid contaminating the slides with dust from the desiccant, place them in a permeable container or wrap it in several Kimwipes.

14. To develop the slides, immerse them in:

    Dektol developer for 2 minutes
    H_2O for 10 seconds
    fixer for 5 minutes
    H_2O for 5 minutes

15. Detect the radioactive spots with a scanning light microscope.

PROTOCOL 14

Protein Arrays: Sandwich Approach for Analyzing Complex Solutions

THIS PROTOCOL WAS CONTRIBUTED BY Gavin MacBeath (Harvard University).

Protein arrays can be used to quantitate the abundance and modification states of proteins in complex mixtures. For this application, it is necessary to array capture reagents, such as antibodies, on the slides and then use these molecules to specifically capture their cognate antigens from solution. The investigator has a choice in either labeling the proteins that are under study (see Figure 10.12) or detecting the captured proteins by an indirect labeling method. In the latter case, which is detailed in this protocol, each protein is captured by one antibody (or other capture reagent) and then detected in a second step with a second reagent (see Figure 10.13). The second reagent, which in this case is an antibody, recognizes the protein at a site that does not overlap with the recognition site of the capture reagent. In this sandwich approach, the second reagent is labeled and so provides the signal. The advantage of this approach is that it does not require labeling the proteins themselves. Moreover, additional specificity is gained by using two different reagents for each protein. The disadvantage, however, is that is it more difficult to assemble matched pairs of antibodies for each protein of interest.

MATERIALS

CAUTION: See Appendix 3 for appropriate handling of materials marked with <!>.

Reagents

Lysis buffer
- 20 mM Tris-Cl (pH 7.5)
- 150 mM NaCl
- 1% BSA (w/v)
- 1 mM EDTA
- 1 mM EGTA
- 1% Triton X-100 <!>
- 0.5% Nonidet P-40
- 1 mM β-glycerolphosphate
- 1 mM sodium orthovanadate (Na_3VO_4) <!>
- 1 mM PMSF <!>
- 1 μg/ml leupeptin <!>
- 1 μg/ml pepstatin <!>

Phosphate-buffered saline (PBS) (pH 7.5)

PBS containing 40% glycerol (v/v)

PBS containing 1% (w/v) BSA
PBS containing 500 mM glycine (pH 8.0) <!>
PBS containing 1% BSA (w/v) and 0.02% (w/v) NaN_3 <!>
PBS containing 0.1% Tween (PBST)
PBST containing 1% BSA

Equipment

Affinity chromatography equipment with protein G or protein A beads

BSA-NHS slides
  Prepare as in Protocol 10.

Centrifuge

Fluorescence slide scanner (e.g., ArrayWoRx fluorescence slide scanner, Applied Precision, Issaquah, Washington)

Incubation chambers (CoverWell, Grace Biolabs, Bend, Oregon)
  CoverWell models accommodate the following volumes
    PC200: 200 µl
    PC500: 550 µl

Sonicator

Biological Molecules

Capture antibodies

Detection cocktail
  This cocktail consists of each fluorescently labeled detection antibody present at a concentration of 100–500 ng/ml in PBST containing 1% BSA.

Cells

Mammalian cells

METHOD

Printing Antibody Arrays

For additional details on printing protein arrays, see Fabricating Protein Arrays in the introduction to this chapter. For a summary of highlights, see the panel on GUIDELINES FOR PREPARING PROTEIN ARRAYS in Protocol 11.

1. Prepare the capture antibodies at a concentration of 0.5 mg/ml in PBS (pH 7.5) containing 40% glycerol (v/v). The buffer must be free of nucleophilic components (such as Tris).

   Often, antibodies come formulated with BSA, which will compete for binding to the slide. BSA can be removed by affinity chromatography using protein G or protein A beads.

2. Allow the antibodies to bind to the slides by incubating the slides together for 2 hours at room temperature after arraying the antibodies.

3. Quench the slides with PBS containing 500 mM glycine (pH 8.0) for 1 hour at room temperature.

4. Block the slides for an additional 30–60 minutes with PBS containing 1% BSA (w/v) immediately before use. Alternatively, store the arrays at 4°C immersed in PBS supplemented with 1% BSA (w/v) and 0.02% NaN_3 (w/v).

   Antibody arrays remain stable for 1–2 months under these conditions.

Processing Antibody Arrays

5. Resuspend mammalian cells on ice at a concentration of 10^7 cells/ml in lysis buffer.
6. Lyse the cells by sonication and centrifuge them at 15,000–20,000*g* for 10 minutes at 4°C to remove insoluble material.
7. Wash the antibody-coated slides (from Step 4) twice with PBS.
8. Apply lysate to the slides and incubate them in a humid chamber for 2–3 hours at room temperature.

   The CoverWell PC200 accommodates a volume of 200 μl of lysate and the PC500, 550 μl. Alternatively, 10–30 μl of lysate can be used if the arrays is small, but first draw a border around it with a hydrophobic pen.

   In general, simply cover the array with lysate. The exact volume is not important.
9. Wash the slides three times for 1 minute each with PBST and three times for 1 minute each with PBS.
10. Add detection cocktail to the slides, and incubate them for 1 hour at room temperature.
11. Wash the slides twice for 1 minute each with PBST and twice for 1 minute each with PBS.
12. Centrifuge the slides at 200*g* for 30 seconds to remove excess buffer.
13. Image the slides with a fluorescence slide scanner (see Reading and Interpreting the Arrays in the introduction to this chapter).

REFERENCES

Aboul-Enein A. and Schulte-Frohlinde D. 1988. Biological deactivation and single-strand breakage of plasmid DNA by photosensitization using tris(2,2′-bipyridyl)ruthenium(II) and peroxydisulfate. *Photochem. Photobiol.* **48:** 27–34.

Allen N.P., Huang L., Burlingame A., and Rexach M. 2001. Proteomic analysis of nucleoporin interacting proteins. *J. Biol. Chem.* **276:** 29268–29274.

Andersen J.S., Lyon C.E., Fox A.H., Leung A.K., Lam Y.W., Steen H., Mann M., and Lamond A.I. 2002. Directed proteomic analysis of the human nucleolus. *Curr. Biol.* **12:** 1–11.

Aravind L. 2000. Guilt by association: Contextual information in genome analysis. *Genome Res.* **10:** 1074–1077.

Ausubel F.M., Brent R., Kingston R.E., Moore D.D., Seidman J.G., Smith J.A., and Struhl K., eds. 1997. *Current protocols in molecular biology.* Wiley, New York.

Bader G.D., Donaldson I., Wolting C., Ouellette B.F., Pawson T., and Hogue C.W. 2001. BIND—The Biomolecular Interaction Network Database. *Nucleic Acids Res.* **29:** 252–245.

Baldwin R.L. 1996. How Hofmeister ion interactions affect protein stability. *Biophys. J.* **71:** 2056–2063.

Bartel P.L., Roecklein J.A., SenGupta D., and Fields S.A. 1996. A protein linkage map of *Escherichia coli* bacteriophage T7. *Nat. Genet.* **12:** 72–77.

Bartholomew B., Kassavetis G., Braun B., and Geiduschek E.P. 1990. The subunit structure of *Saccharomyces cerevisiae* transcription factor IIIC probed with a novel photo-cross-linking reagent. *EMBO J.* **9:** 2197–2205.

Bartlett M.S., Thomm M., and Geiduschek E.P. 2000. The orientation of DNA in an archaeal transcription initiation complex. *Nat. Struct. Biol.* **7:** 782–785.

Bell S. and Stillman B. 1992. ATP-dependent recognition of eukaryotic origins of DNA replication by a multiprotein complex. *Nature* **357:** 128–134.

Black D.L. 2000. Protein diversity from alternative splicing: A challenge for bioinformatics and post-genome biology. *Cell* **103:** 367–370.

Blom J., Dekker P.J.T., and Meijer M. 1995. Functional and physical interactions of components of the yeast mitochondrial inner-membrane import machinery (MIM). *Eur. J. Biochem.* **232:** 309–314.

Bolletta F., Juris A., Maestri M., and Sandrini D. 1980. Quantum yield of formation of the lowest excited state of $Ru(bpy)^{2+}_3$ and $Ru(phen)^{2+}_3$. *Inorg. Chim. Acta* **44:** L175–L176.

Boussif O., Lezoualc'h F., Zanta M.A, Mergny M.D., Scherman D., Demeneix B., and Behr J.-P. 1995. A versatile vector for gene and oligonucleotide transfer into cells in culture and *in vivo:* Polyethylenimine. *Proc. Natl. Acad. Sci.* **92:** 7297–7301.

Cantor C.R. and Schimmel P.R. 1980. *Biophysical chemistry, part III. The behavior of biological molecules.* W.H. Freeman, San Francisco, California.

Carey J. 1991. Gel retardation. *Methods Enzymol.* **208:** 103–117.

Chen Z. and Han M. 2000. Building a protein interaction map: Research in the post-genome era. *BioEssays* **22:** 503–506.

Clegg R.M. 1996. Fluorescence resonance energy transfer. In *Fluorescence imaging spectroscopy and microscopy* (ed. X.F. Wang and B. Herman), pp. 179–252. Wiley, New York.

Connelly P.R., Varadarajan R., Sturtevant J.M., and Richards F.M. 1990. Thermodynamics of protein–peptide interactions in the ribonuclease S system studied by titration calorimetry. *Biochemistry* **29:** 6108–6114.

Crameri A., Whitehorn E.A., Tate E., and Stemmer W.P.C. 1996. Improved green fluorescent protein by molecular evolution using DNA shuffling. *Nat. Biotechnol.* **14:** 315–319.

Croft L., Schandorff S., Clark F., Burrage K., Arctander P., and Mattick J.S. 2000. ISIS, the intron information system, reveals the high frequency of alternative splicing in the human genome. *Nat. Genet.* **24:** 340–341.

Cubitt A.B., Heim R., Adams S.R., Boyd A.E., Gross L.A., and Tsien R.Y. 1995. Understanding, improving and using green fluorescent proteins. *Trends Biochem. Sci.* **20:** 448–455.

Dabbeni-Sala F., Di Santo S., Franceschini D., Skaper S.D., and Giusti P. 2001. Melatonin protects against neurotoxicity in rats: A role for mitochondrial complex I activity. *FASEB J.* **15:** 164–170.

Damelin M. and Silver P.A. 2000. Mapping interactions between nuclear transport factors in living cells reveals pathways through the nuclear pore complex. *Mol. Cell* **5:** 133–140.

———. 2002. In situ analysis of spatial relationships between proteins of the nuclear pore complex. *Biophys. J.* (in press).

Davidson E.H., Rast J.P., Oliveri P., Ransick A., Calestani C., Yuh C.H., Minokawa T., Amore G., Hinman V., Arenas-Mena C., et al. 2002. A genomic regulatory network for development. *Science* **295:** 1669–1678.

Dekker P.J.T., Ryan M.T., Brix J., Müller H., Hönlinger A., and Pfanner N. 1998. The preprotein translocase of the outer mitochondrial membrane: Molecular dissection and assembly of the general import pore complex. *Mol. Cell. Biol.* **18:** 6515–6524.

Delagrave S., Hawtin R.E., Silva C.M., Yang M.M., and Youvan D.C. 1995. Red-shifted excitation mutants of the green fluorescent protein. *Bio/Technology* **13:** 151–154.

Duan J.D., Xenarios I., and Eisenberg D. 2002. Describing biological protein interactions in terms of protein states and state transitions. *Mol. Cell Proteomics* **1:** 104–116.

Ehrig T., O'Kane D.J., and Prendergast F.G. 1995. Green-fluorescent protein with altered fluorescence excitation spectra. *FEBS Lett.* **367:** 163–166.

Ellis R.J. 2001. Macromolecular crowding: Obvious but underappreciated. *Trends Biochem. Sci.* **26:** 597–604.

Fancy D.A. 2000. Elucidation of protein–protein interactions using chemical cross-linking or label transfer techniques. *Curr. Opin. Chem.Biol.* **4:** 28–33.

Fancy D.A. and Kodadek T. 1999. New chemistry for the analysis of protein-protein interactions: Rapid and efficient cross-linking triggered by long wavelength light. *Proc. Natl. Acad. Sci.* **96:** 6020–6024.

Fancy D.A., Denison C., Kim K., Xie Y., Holdeman T., Amini F., and Kodadek T. 2000. Scope, limitations and mechanistic aspects of the photo-induced cross-linking of proteins by water-soluble metal complexes. *Chem. Biol.* **7:** 697–708.

Fidanza J., Ozaki H., and McLaughlin L. 1992. Site-specific labeling of DNA sequences containing phosphorothioate diesters. *J. Amer. Chem. Soc.* **114:** 5509–5517.

Fields S. and Song O.K. 1989. A novel genetic system to detect protein–protein interactions. *Nature* **340:** 245–246.

Fisher B.M., Ha J.-H., and Raines R.T. 1998. Coulombic forces in protein-RNA interactions: Binding and cleavage by ribonuclease A and variants at Lys7, Arg10 and Lys66. *Biochemistry* **37:** 12121–12132.

Flajolet M., Rotondo G., Daviet L., Bergametti F., Inchauspe G., Tiollais P., Transy C., and Legrain P. 2000. A genomic approach of the hepatitis C virus generates a protein interaction map. *Gene* **242:** 369–379.

Flores A., Briand J.F., Gadal O., Andrau J.C., Rubbi L., Van Mullem V., Boschiero C., Goussot M., Marck C., Carles C., et al. 1999. A protein-protein interaction map of yeast RNA polymerase III. *Proc. Natl. Acad. Sci.* **96:** 7815–7820.

Forwood J.K., Lam M.H.C., and Jans D.A. 2001. Nuclear import of Creb and AP-1 transcription factors requires importin-β1 and Ran but is independent of importin-α *Biochemistry* **40:** 5208–5217.

Fox A.H., Lam Y.W., Leung A.K., Lyon C.E., Andersen J., Mann M., and Lamond A.I. 2002. Paraspeckles: A novel nuclear domain. *Curr. Biol.* **12:** 13–25.

Franken K.L., Hiemstra H.S., van Meijgaarden K.E., Subronto Y., den Hartigh J., Ottenhoff T.H., and Drijfhout J.W. 2000. Purification of his-tagged proteins by immobilized chelate affinity chromatography: The benefits from the use of organic solvent. *Protein Expr. Purif.* **18:** 95–99.

Garin J., Diez R., Kieffer S., Dermine J.F., Duclos S., Gagnon E., Sadoul R., Rondeau C., and Desjardins M. 2001. The phagosome proteome: Insight into phagosome functions. *J. Cell Biol.* **152:** 165–180.

Gavin A.C., Bosche M., Krause R., Grandi P., Marzioch M., Bauer A., Schultz J., Rick J.M., Michon A.M., Cruciat C.M., Remor M., Hofert C., Schelder M., Brajenovic M., Ruffner H., Merino A., Klein K., Hudak M., Dickson D., Rudi T., Gnau V., Bauch A., Bastuck S., Huhse B., Leutwein C., Heurtier M.A., Copley R.R., Edelmann A., Querfurth E., Rybin V., Drewes G., Raida M., Bouwmeester T., Bork P., Seraphin B., Kuster B., Neubauer G., and Superti-Furga G. 2002. Functional organization of the yeast proteome by systematic analysis of protein complexes. *Nature* **415:** 141–147.

Gerstein M., Lan N., and Jansen R. 2002. Integrating interactomes. *Science* **295:** 284–285.

Gordon G.W., Berry G., Liang X.H., Levine B., and Herman B. 1998. Quantitative fluorescence resonance energy transfer measurements using fluorescence microscopy. *Biophys. J.* **74:** 2702–2713.

Guthrie C. and Fink G. 2002. Guide to yeast genetics and molecular cell biology, parts B, C. *Methods Enzymol.*, vols. 350, 351. Academic Press, San Diego, California.

Hames B.D. 1990. One-dimensional polyacrylamide gel electrophoresis. In *Gel electrophoresis of proteins: A practical approach*, 2nd edition (ed. B.D. Hames and D. Rickwood), pp. 1–148. IRL Press/Oxford University Press, Oxford, United Kingdom.

Han D.K., Eng J., Zhou H., and Aebersold R. 2001. Quantitative profiling of differentiation-induced microsomal proteins using isotope-coded affinity tags and mass spectrometry. *Nat. Biotechnol.* **19:** 946–951.

Hanes J. and Pluckthun A. 1997. In vitro selection and evolution of functional proteins by using ribosome display. *Proc. Natl. Acad. Sci.* **94:** 4937–4042.

Heim R. and Tsien R.Y. 1996. Engineering green fluorescent protein for improved brightness, longer wavelengths and fluorescence resonance energy transfer. *Curr. Biol.* **6:** 178–182.

Heim R., Cubitt A.B., and Tsien R.Y. 1995. Improved green fluorescence. *Nature* **373:** 663–664.

Hermanson G.T. 1996. *Bioconjugate techniques*. Academic Press.

Heyduk T., Ma Y., Tang H., and Ebright R.H. 1996. Fluorescence anisotropy: Rapid, quantitative assay for protein–DNA and protein–protein interaction. *Methods Enzymol.* **274:** 492–503.

Hilton D.J., Richardson R.T., Alexander W.S., Viney E.M., Willson T.A., Sprigg N.S., Starr R., Nicholson S.E., Metcalf D., and Nicola N.A. 1998. Twenty proteins containing a C-terminal SOCS box form five structural classes. *Proc. Natl. Acad. Sci.* **95:** 114–119.

Ho Y., Gruhler A., Heilbut A., Bader G.D., Moore L., Adams S.L., Millar A., Taylor P., Bennett K., Boutilier K., Yang L., Wolting C., Donaldson I., Schandorff S., Shewnarane J., Vo M., Taggart J., Goudreault M., Muskat B., Alfarano C., Dewar D., Lin Z., Michalickova K., Willems A.R., Sassi H., Nielsen P.A., Rasmussen K.J., Andersen J.R., Johansen L.E., Hansen L.H., Jespersen H., Podtelejnikov A., Nielsen E., Crawford J., Poulsen V., Sorensen B.D., Matthiesen J., Hendrickson R.C., Gleeson F., Pawson T., Moran M.F., Durocher D., Mann M., Hogue C.W., Figeys D., and Tyers M. 2002. Systematic identification of protein complexes in *Saccharomyces cerevisiae* by mass spectrometry. *Nature* **415:** 180–183.

Hogenesch J.B., Ching K.A., Batalov S., Su A.I., Walker J.R., Zhou Y., Kay S.A., Schultz P.G., and Cooke M.P. 2001. A comparison of the Celera and Ensembl predicted gene sets reveals little overlap in novel genes. *Cell* **106:** 413–415.

Holzinger A., Phillips K.S., and Weaver T.E. 1996. Single-step purification/solubilization of recombinant proteins: Application to surfactant protein B. *BioTechniques* **20:** 804–808.

Houry W.A., Frishman D., Eckerskorn C., Lottspeich F., and Hartl F.U. 1999. Identification of in vivo substrates of the chaperonin GroEL. *Nature* **402:** 147–154.

Ideker T., Thorsson V., Ranish J.A., Christmas R., Buhler J., Eng J.K., Bumgarner R., Goodlett D.R., Aebersold R., and Hood L. 2001. Integrated genomic and proteomic analyses of a systematically perturbed metabolic network. *Science* **292:** 929–934.

Ito T., Tashiro K., Muta S., Ozawa R., Chiba T., Nishizawa M., Yamamoto K., Kuhara S., and Sakaki Y. 2000. Toward a protein-protein interaction map of the budding yeast: A comprehensive system to examine two-hybrid interactions in all possible combinations between the yeast proteins. *Proc. Natl. Acad. Sci.* **97:** 1143–1147.

Jameson D.M. and Sawyer W.H. 1995. Fluorescence anisotropy applied to biomolecular interactions. *Methods Enzymol.* **246:** 283–300.

Kamp R.M. 1988. Application of bifunctional reagents for topological investigation. In *Modern methods in protein chemistry* (ed. H. Tschesche), vol. 3. pp. 275–298. Walter de Gruyter & Co., Berlin, Germany.

Kiessig S., Reissmann J., Rascher C., Kullertz G., Fischer A., and Thunecke F. 2001. Application of a green fluorescent fusion protein to study protein–protein interactions by electrophoretic methods. *Electrophoresis* **22:** 1428–1435.

Kim K., Fancy D.A., and Kodadek T. 1999. Photo-induced protein cross-linking mediated by palladium porphyrins. *J. Am. Chem. Soc.* **121:** 11896–11897.

Kim R., Yokota H., and Kim S.-H. 1999. Electrophoresis of proteins and protein–protein complexes in a native agarose gel. *Anal. Biochem.* **282:** 147–149.

Kim T.-K., Reinberg D., and Ebright R. 2000a. Mechanism of ATP-dependent promoter melting by transcription factor IIH. *Science* **288:** 1418–1422.

Kim T.-K., Lagrange T., Reinberg D., Naryshkin N., and Ebright R. 2000b. Site-specific protein-DNA photocross-linking. In *DNA-protein interactions* (ed. A. Travers and M. Buckle), pp. 319–335. Oxford University Press, Oxford, United Kingdom.

Kim T.-K., Lagrange T., Wang Y.-H., Griffith J., Reinberg D., and Ebright R. 1997. Trajectory of DNA in the RNA polymerase II transcription preinitiation complex. *Proc. Natl. Acad. Sci.* **94:** 12268–12273.

Kitano H. 2002. Systems biology: A brief overview. *Science* **295:** 1662–1664.

Kodadek T., Gan D.C., and Stemke-Hale K. 1989. The phage T4 uvs recombination protein stabilizes presynaptic filaments. *J. Biol. Chem.* **264:** 16451–16457.

Kratchmarova I., Kalume D.E., Blagoev B., Scherer P.E., Podtelejnikov A.V., Molina H., Bickel P.E., Andersen J.S., Fernandez M.M., Bunkenborg J., Roepstorff P., Kristiansen K., Lodish H.F., Mann M., and Pandey A. 2002. A proteomic approach for identification of secreted proteins during the differentiation of 3T3-L1 preadipocytes to adipocytes. *Mol. Cell. Proteomics* **1:** 213–222.

Krowcyznska A.M., Donoghue K., and Hughes L. 1995. Recovery of DNA, RNA and protein from gels with microconcentrators. *BioTechniques* **18:** 698–703.

Laemmli U.K. 1970. Cleavage of structural proteins during the assembly of the head of bacteriophage T4. *Nature* **227:** 680–685.

Lagrange T., Kapanidis A., Tang H., Reinberg D., and Ebright R. 1998. New core promoter element in RNA polymerase II-dependent transcription: Sequence-specific DNA binding by transcription factor IIB. *Genes Dev.* **12:** 34–44.

Lagrange T., Kim T.K., Orphanides G., Ebright Y., Ebright R., and Reinberg D. 1996. High-resolution mapping of nucleoprotein complexes by site-specific protein-DNA photo-cross-linking: Organization of the human TBP-TFIIA-TFIIB-DNA quaternary complex. *Proc. Natl. Acad. Sci.* **93:** 10620–10625.

Lakey J.H. and Raggett E.M. 1998. Measuring protein-protein interactions. *Curr. Opin. Struct. Biol.* **8:** 119–223.

Lander E.S., Linton L.M., Birren B., Nusbaum C., Zody M.C., Baldwin J., Devon K., Dewar K., Doyle M., FitzHugh W., et al. 2001. Initial sequencing and analysis of the human genome. *Nature* **409:** 860–921.

Layh-Schmitt G., Podtelejnikov A., and Mann M. 2000. Proteins complexed to the P1 adhesin of *Mycoplasma pneumoniae. Microbiology* **146:** 741–747.

Ledent P., Duez C., Vanhove M., Lejeune A., Fonze E., Charlier P., Rhazi-Filali F., Thamm I., Guillaume G., Samyn B., Devreese B., Van Beeumen J., Lamotte-Brasseur J., and Frere J.M. 1997. Unexpected influence of a C-terminal-fused His-tag on the processing of an enzyme and on the kinetic and folding parameters. *FEBS Lett.* **413:** 194–196.

LeTilly V. and Royer C.A. 1993. Fluorescence anisotropy assays implicate protein–protein interactions in regulating *trp* repressor DNA binding. *Biochemistry* **32:** 7753–7758.

Liang X.H., Volkmann M., Klein R., Herman B., and Lockett S.J. 1993. Co-localization of the tumor suppressor protein p53 and human papillomavirus E6 protein in human cervical carcinoma cell lines. *Oncogene* **8:** 2645–2652.

Link A.J., Eng J., Schieltz D.M., Carmack E., Mize G.J., Morris D.R., Garvik B.M., and Yates J.R. III. 1999. Direct analysis of protein complexes using mass spectrometry. *Nat. Biotechnol.* **17:** 676–682.

Lopez A.J. 1998. Alternative splicing of pre-mRNA: Developmental consequences and mechanisms of regulation. *Annu. Rev. Genet.* **32:** 279–305.

Malpeli G., Folli C., and Berni R. 1996. Retinoid binding to retinol-binding protein and the interference with the interaction with transthyretin. *Biochim. Biophys. Acta* **1294:** 48–54.

Marshall W.L., Yim C., Gustafson E., Graf T., Sage D.R., Hanify K., Williams L., Fingeroth J., and Finberg R.W. 1999. Epstein–Barr virus encodes a novel homolog of the *bcl-2* oncogene that inhibits apoptosis and associates with Bax and Bak. *J. Virol.* **73:** 5181–5185.

Matsudaira P. 1993. *A practical guide to protein and peptide purification for microsequencing*. Academic Press, San Diego, California.

Mayer A. and Barany F. 1995. Photoaffinity cross-linking of *Taq*I restriction endonuclease using an aryl azide linked to the phosphate backbone. *Gene* **153:** 1–8.

Mayer M.L. and Hieter P. 2000. Protein networks—Built by association. *Nat. Biotechnol.* **18:** 1242–1243.

Mayes A.E., Verdone L., Legrain P., and Beggs J.D. 1999. Characterization of Sm-like proteins in yeast and their association with U6 snRNA. *EMBO J.* **18:** 4321–4331.

McCraith S., Holtzman T., Moss B., and Fields S. 2000. Genome-wide analysis of vaccinia virus protein–protein interactions. *Proc. Natl. Acad. Sci.* **97:** 4879–4884.

Medaglia M.V. and Fisher R.J. 2001. Analysis of interacting proteins with surface plasmon resonance spectroscopy using BIAcore. In *Molecular cloning: A laboratory manual*, 3rd edition (ed. J. Sambrook and D.W. Russell), pp. 18.96–18.103. Cold Spring Harbor Laboratory Press, Cold Spring Harbor, New York.

———. 2002. Analysis of interacting proteins with surface plasmon resonance using Biacore. In *Protein–protein interactions: A molecular cloning manual* (ed. E. Golemis), pp. 255–272. Cold Spring Harbor

Laboratory Press, Cold Spring Harbor, New York.
Melancon P., Burgess R.R., and Record M.T. Jr. 1983. Direct evidence for the preferential binding of *Escherichia coli* RNA polymerase holoenzyme to the ends of deoxyribonucleic acid restriction fragments. *Biochemistry* **22:** 5169–5176.
Mironov A.A., Fickett J.W., and Gelfand M.S. 1999. Frequent alternative splicing of human genes. *Genome Res.* **9:** 1288–1293.
Mittler R.S., Rankin B.M., and Kiener P.A. 1991. Physical associations between CD45 and CD4 or CD8 occur as late activation events in antigen receptor-stimulated human T cells. *J. Immunol.* **147:** 3434–3440.
Miyawaki A., Llopis J., Heim R., McCaffery J.M., Adams J.A., Ikura M., and Tsien R.Y. 1997. Fluorescent indicators for Ca^{2+} based on green fluorescent proteins and calmodulin. *Nature* **388:** 882–887.
Model K., Meisinger C., Prinz T., Wiedemann N., Truscott K.N., Pfanner N., and Ryan M.T. 2001. Multistep assembly of the protein import channel of the mitochondrial outer membrane. *Nat. Struct. Biol.* **8:** 284–286.
Naryshkin N., Kim Y., Dong Q., and Ebright R.H. 2001. Site-specific protein-DNA photo-cross-linking. In *DNA-protein interactions principles and protocols,* 2nd edition (ed. T. Moss), pp. 337–361. Humana Press, Totowa, New Jersey.
Naryshkin N., Revyakin A., Kim Y., Mekler V., and Ebright R.H. 2000. Structural organization of the RNA polymerase-promoter open complex. *Cell* **101:** 601–611.
Natsume T., Nakayama H., and Isobe T. 2001. BIA-MS-MS: Biomolecular interaction analysis for fuctional interactions. *Trends Biotechnol.* **19:** S28–S33.
Neubauer G., Gottschalk A., Fabrizio P., Seraphin B., Luhrmann R., and Mann M. 1997. Identification of the proteins of the yeast U1 small nuclear ribonucleoprotein complex by mass spectrometry. *Proc. Natl. Acad. Sci.* **94:** 385–390.
Neubauer G., King A., Rappsilber J., Calvio C., Watson M., Ajuh P., Sleeman J., Lamond A., and Mann M. 1998. Mass spectrometry and EST-database searching allows characterization of the multi-protein spliceosome complex. *Nat. Genet.* **20:** 46–50.
Nickel U., Chen Y.H., Schneider S., Silva M.I., Burrows H.D., and Forosinho S.J. 1994. Mechanism and kinetics of the photocatalyzed oxidation of p-phenylenediamines by peroxydisulfate in the presence of tri-2,2´-bipyridyl)ruthenium(II). *J. Phys. Chem.* **98:** 2883–2888.
Noble D. 2002. Modeling the heart—From genes to cells to the whole organ. *Science* **295:** 1678–1682.
Oliver S. 2000. Proteomics: Guilt-by-association goes global. *Nature* **403:** 601–603.
Ormö M., Cubitt A.B., Kallio K., Gross L.A., Tsien R.Y., and Remington S.J. 1996. Crystal structure of the *Aequorea victoria* green fluorescent protein. *Science* **237:** 1392–1395.
Overton M.C. and Blumer K.J. 2000. G-protein-coupled receptors function as oligomers in vivo. *Curr. Biol.* **10:** 341–344.
Pandey A., Andersen J.S., and Mann M. 2000a. Use of mass spectrometry to study signaling pathways. *Science's STKE: http://stke.sciencemag.org/cgi/content/full/OC_sigtrans;2000/37/pl1.*
Pandey A., Podtelejnikov A.V., Blagoev B., Bustelo X.R., Mann M., and Lodish H.F. 2000b. Analysis of receptor signaling pathways by mass spectrometry: Identification of vav-2 as a substrate of the epidermal and platelet-derived growth factor receptors. *Proc. Natl. Acad. Sci.* **97:** 179–184.
Park S.-H. and Raines R.T. 1997. Green fluorescent protein as a signal for protein–protein interactions. *Protein Sci.* **6:** 2344–2349.
———. 2000. Green fluorescent protein chimeras to probe protein–protein interactions. *Methods Enzymol.* **328:** 251–261.
Pearson K.M., Pannell L.K., and Fales H.M. 2002. Intramolecular cross-linking experiments on cytochrome c and ribonuclease A using an isotope multiplet method. *Rapid Commun. Mass Spectrom.* **16:** 149–159.
Pelletier J. and Sidhu S. 2001. Mapping protein–protein interactions with combinatorial biology methods. *Curr. Opin. Biotechnol.* **12:** 340–347.
Phizicky E.M. and Fields S. 1995. Protein-protein interactions: Methods for detection and analysis. *Microbiol. Rev.* **59:** 94–123.
Poetsch A., Neff D., Seelert H., Schägger H., and Dencher N.A. 2000. Dye removal, catalytically active and 2D crystallization of chloroplast H+-ATP synthase purified by blue native electrophoresis. *Biochim. Biophys. Acta* **1466:** 339–349.
Potts J.T., Young D.M., and Anfinsen C.B. 1963. Reconstitution of fully active RNase S by carboxypeptidase-degraded RNase S-peptide. *J. Biol. Chem.* **238:** 2593–2594.

Prasher D.C., Eckenrode V.K., Ward W.W., Prendergast F.G., and Cormier M.J. 1992. Primary structure of the *Aequorea victoria* green-fluorescent protein. *Gene* **111:** 229–233.

Pugh B.F. 1996. Mechanisms of transcription complex assembly. *Curr. Opin. Cell Biol.* **8:** 303–311.

Rain J.-C., Selig L., De Reuse H., Battaglia V., Reverdy C., Simon S., Lenzen G., Petel F., Wojcik J., Schächter V., Chemama Y., Labigne A., and LeGrain P. 2001. The protein–protein interaction map of *Helicobacter pylori. Nature* **409:** 211–215.

Raines R.T. 1998. Ribonuclease A. *Chem. Rev.* **98:** 1045–1065.

Raines R.T., McCormick M., Van Oosbree T.R., and Mierendorf R.C. 2000. The S•Tag fusion system. *Methods Enzymol.* **326:** 362–376.

Rappsilber J., Siniossoglou S., Hurt E.C., and Mann M. 2000. A generic strategy to analyze the spatial organization of multiprotein complexes by cross-linking and mass spectrometry. *Anal. Chem.* **72:** 267–275.

Richards F.M. 1955. Titration of amino groups released during the digestion of ribonuclease by subtilisin. *C.R. Trav. Lab. Carlsberg, Ser. Chim.* **29:** 322–328.

Rigaut G., Shevchenko A., Rutz B., Wilm M., Mann M., and Séraphin B. 1999. A generic protein purification method for protein complex characterization and proteome exploration. *Nat. Biotechnol.* **17:** 1030–1032.

Rodi D.J. and Makowski L. 1999. Phage-display technology—Finding a needle in a vast molecular haystack. *Curr. Opin Biotechnol.* **10:** 87–93.

Rout M.P., Aitchison J.D., Suprapto A., Hjertaas K., Zhao Y., and Chait B.T. 2000. The yeast nuclear pore complex: Composition, architecture, and transport mechanism. *J. Cell Biol.* **148:** 635–651.

Ryan M.T., Müller H., and Pfanner N. 1999. Functional staging of ADP/ATP carrier translocation across the outer mitochondrial membrane. *J. Biol. Chem.* **274:** 20619–20627.

Sambrook J. and Russell D.W. 2001. *Molecular cloning: A laboratory manual,* 3rd edition. Cold Spring Harbor Laboratory Press, Cold Spring Harbor, New York.

Sambrook J., Fritsch E.F., and Maniatis T. 1989. *Molecular cloning: A laboratory manual,* 2nd edition. Cold Spring Harbor Laboratory Press, Cold Spring Harbor, New York.

Schägger H. 1996. Electrophoretic techniques for isolation and quantification of oxidative phosphorylation complexes from human tissues. *Methods Enzymol.* **264:** 555–566.

Schägger H. and Pfeiffer K. 2000. Supercomplexes in the respiratory chains of yeast and mammalian mitochondria. *EMBO J.* **19:** 1777–1783.

Schägger H. and von Jagow G. 1991. Blue native electrophoresis for isolation of membrane protein complexes in enzymatically active form. *Anal. Biochem.* **199:** 223–231.

Schägger H., Cramer W.A., and von Jagow G. 1994. Analysis of molecular masses and oligomeric states of protein complexes by blue native electrophoresis and isolation of membrane protein complexes by two-dimensional native electrophoresis. *Anal. Biochem.* **217:** 220–230.

Schamel W.W. and Reth M. 2000. Monomeric and oligomeric complexes of the B cell antigen receptor. *Immunity* **13:** 5–14.

Schreier A.A. and Baldwin R.L. 1977. Mechanism of dissociation of S-peptide from ribonuclease S. *Biochemistry* **16:** 4203–4209.

Schuck P. 1997a. Reliable determination of binding affinity and kinetics using surface plasmon resonance biosensors. *Curr. Opin. Biotechnol.* **8:** 498–502.

———. 1997b. Use of surface plasmon resonance to probe the equilibrium and dynamic aspects of interactions between biological macromolecules. *Annu. Rev. Biophys. Biomol. Struct.* **26:** 541–566.

Schwikowski B., Uetz P., and Fields S. 2000. A network of protein-protein interactions in yeast. *Nat. Biotechnol.* **18:** 1257–1261.

Séraphin B., Puig O., Bouveret E., Rutz B., and Caspary F. 2002. Tandem affinity purification to enhance interacting protein identification. In *Protein–protein interactions: A molecular cloning manual* (ed. E. Golemis), pp. 313–328. Cold Spring Harbor Laboratory Press, Cold Spring Harbor, New York.

Shevchenko A., Wilm M., Vorm O., and Mann M. 1996. Mass spectrometric sequencing of proteins silver-stained polyacrylamide gels. *Anal. Chem.* **68:** 850–858.

Sidhu S.S., Lowman H.B., Cunnningham B.C., and Wells J.A. 2000. Phage display for selection of novel binding peptides. *Methods Enzymol.* **328:** 333–363.

Simpson R.J., Moritz R.L., Begg G.S., Rubira M.R., and Nice E.C. 1989. Micropreparative procedures for high sensitivity sequencing of peptides and proteins. *Anal. Biochem.* **177:** 221–236.

Simpson R.J., Connolly L.M., Eddes J.S., Pereira J.J., Moritz R.L., and Reid G.E. 2000. Proteomic analysis of the human colon carcinoma cell line (LIM 1215): Development of a membrane protein database. *Electrophoresis* **21:** 1707–1732.

Siniossoglou S., Lutzmann M., Santos-Rosa H., Leonard K., Mueller S., Aebi U., and Hurt E. 2000. Structure and assembly of the Nup84p complex. *J. Cell Biol.* **149:** 41–54.

Siniossoglou S., Wimmer C., Rieger M., Doye V., Tekotte H., Weise C., Emig S., Segref A., and Hurt E.C. 1996. A novel complex of nucleoporins, which includes Sec13p and a Sec13p homolog, is essential for normal nuclear pores. *Cell* **84:** 265–275.

Smith D.B. and Johnson K.S. 1988. Single-step purification of polypeptides expressed in *Escherichia coli* as fusions with glutathione S-transferase. *Gene* **67:** 31–40.

Smith C.W. and Valcarcel J. 2000. Alternative pre-mRNA splicing: The logic of combinatorial control. *Trends Biochem. Sci.* **25:** 381–388.

Spector D.L., Goldman R.D., and Leinwand L.A. 1997. *Cells: A laboratory manual.* Cold Spring Harbor Laboratory Press, Cold Spring Harbor, New York.

Speicher K.D., Kolbas O., Harper S., and Speicher D.W. 2000. Systematic analysis of peptide recoveries from in-gel digestion for protein identification in proteome studies. *J. Biomol. Tech.* **11:** 74–86.

Starr R., Willson T.A., Viney E.M., Murray L.J., Rayner J.R., Jenkins B.J., Gonda T.J., Alexander W.S., Metcalf D., Nicola N.A., and Hilton D.J. 1997. A family of cytokine-inducible inhibitors of signaling. *Nature* **387:** 917–921.

Stryer L. 1978. Fluorescence energy transfer as a spectroscopic ruler. *Annu. Rev. Biochem.* **47:** 819–846.

Szöllösi J., Damjanovich S., Mulhern S.A., and Trón L. 1987. Fluorescence energy transfer and membrane potential measurements monitor dynamic properties of cell membranes: A critical review. *Prog. Biophys. Mol. Biol.* **49:** 65–87.

Trón L., Szöllösi J., Damjanovich S., Helliwell S.H., Arndt-Jovin D.J., and Jovin T.M. 1984. Flow cytometric measurement of fluorescence resonance energy transfer on cell surfaces. Quantitative evaluation of the transfer efficiency on a cell-by-cell basis. *Biophys. J.* **45:** 939–946.

Tsien R. 1998. The green fluorescent protein. *Annu. Rev. Biochem.* **67:** 509–544.

Uetz P., Giot L., Cagney G., Mansfield T.A., Judson R.S., Knight J.R., Lockshon D., Narayan V., Srinisvan M., Pochart P., et al. 2000. A comprehensive analysis of protein-protein interactions in *Saccharomyces cerevisiae. Nature* **403:** 623–627.

van Roessel P. and Brand A.H. 2002. Imaging into the future: Visualizing gene expression and protein interactions with fluorescent proteins. *Nat. Cell Biol.* (suppl. 1) **4:** E15–20.

Venter J.C., Adams M.D., Myers E.W., Li P.W., Mural R.J., Sutton G.G., Smith H.O., Yandell M., Evans C.A., Holt R.A., et al. 2001. The sequence of the human genome. *Science* **291:** 1304–1351.

Verhagen A.M, Ekert P.G., Pakusch M., Silke J., Connolly L.M., Reid G.E., Moritz R.L., Simpson R.J., and Vaux D.L. 2000. Identification of DIABLO, a mammalian protein that promotes apoptosis by binding to and antagonizing IAP proteins. *Cell* **102:** 43–53.

Verkman A.S. 2002. Solute and macromolecular diffusion in cellular aqueous compartments. *Trends Biochem. Sci.* **27:** 27–33.

Verma R., Chen S., Feldman R., Schieltz D., Yates J., Dohmen J., and Deshaies R.J. 2000. Proteasomal proteomics: Identification of nucleotide-sensitive proteasome-interacting proteins by mass spectrometric analysis of affinity-purified proteasomes. *Mol. Biol. Cell* **11:** 3425–3439.

Verveer P.J., Harpur A.G., and Bastiaens P.I.H. 2002. Imaging protein interactions by FRET microscopy. In *Protein–Protein interactions* (ed. E. Golemis), pp. 181–214. Cold Spring Harbor Laboratory Press, Cold Spring Harbor, New York.

Vorm O., Roepstorff P., and Mann M. 1994. Improved resolution and very high sensitivity in MALDI TOF of matrix surfaces made by fast evaporation. *Anal. Chem.* **66:** 3281–3287.

Walhout A.J., Sordella R., Lu X., Hartley J.L., Temple G.F., Brash M.A., Thierry-Mieg N., and Vidal M. 2000. Protein interaction mapping in *C. elegans* using proteins involved in vulval development. *Science* **287:** 116–122.

Wang L. and Dobberstein B. 1999. Oligomeric complexes involved in translocation of proteins across the membrane of the endoplasmic reticulum. *FEBS Lett.* **457:** 316–322.

Wang Y. and Stumph W. 1998. Identification and topological arrangement of *Drosophila* proximal sequence element PSE-binding protein subunits that contact the PSEs of U1 and U6 small nuclear RNA genes. *Mol. Cell. Biol.* **18:** 1570–1579.

Ward W.W., Swiatek G.C., and Gonzalez D.G. 2000. Green fluorescent protein in biotechnology education.

Methods Enzymol. **305:** 672–680.

Weng G., Bhalla U.S., and Iyenger R. 1999. Complexity in biological signaling systems. *Science* **284:** 92–96.

Winston F., Dollard C. and Ricupero-Hovasse S.L. 1995. Construction of a set of convenient *Saccharomyces cerevisiae* strains that are isogenic to S288C. *Yeast* **11:** 53–55.

Wong S.S. 1991. *Chemistry of protein conjugation and cross-linking.* CRC Press, Boca Raton, Florida.

Wool I.G., Chan Y.-L., and Gluck A. 1995. Structure and evolution of mammalian ribosomal proteins. *Biochem. Cell Biol.* **73:** 933–947.

Xenarios I., Rice D.W., Salwinski L., Baron M.K., Marcotte E.M., and Eisenberg D. 2000. DIP: The database of interacting proteins. *Nucleic Acids Res.* **28:** 289–291.

Yang S.-W. and Nash H. 1994. Specific photo-cross-linking of DNA-protein complexes: Identification of contacts between integration host factor and its target DNA. *Proc. Natl. Acad. Sci.* **91:** 12183–12187.

Young M.M., Tang N., Hempel J.C., Oshiro C.M., Taylor E.W., Kuntz I.D., Gibson B.W., and Dollinger G. 2000. High throughput protein fold identification by using experimental constraints derived from intramolecular cross-links and mass spectrometry. *Proc. Natl. Acad. Sci.* **97:** 5802–5806.

Youvan D.C., Coleman W.J., Silva C.M., Petersen J., Bylina E.J., and Yang M.M. 1997. Fluorescence imaging micro-spectrophotometer (FIMS). *Bio/Technology* **1:** 1–16.

Zanzoni A., Montecchi-Palazzi L., Quondam M., Ausiello G., Helmer-Citterich M., and Cesareni G. 2002. MINT: A Molecular INTeraction database. *FEBS Lett.* **513:** 135–140.

Zhang J.-G., Farley A., Nicholson S.E., Willson T.A., Zugaro L.M., Simpson R.J., Moritz R.L., Cary D., Richardson R., Hausmann G., Kile B.J., Kent S.B.H., Alexander W.S., Metcalf D., Hilton D.J., Nicola N.A., and Baca M. 1999. The conserved SOCS box motif in suppressors of cytokine signaling binds to elongins B and C and may couple bound proteins to proteasomal degradation. *Proc. Natl. Acad. Sci.* **96:** 2071–2076.

Zimmerman S.B. and Minton A.P. 1991. Estimation of macromolecular concentrations and excluded volume effects for the cytoplasm of *Escherichia coli. J. Mol. Biol.* **222:** 599–620.

WWW RESOURCES

http://binddb.org BIND—The Biomolecular Interaction Network database

http://cbm.bio.uniroma2.it/mint/index.html Molecular INTeractive database

http://curatools.curagen.com CuraTools for gene sequence and protein data mining and analysis. CuraGen Corporation

http://dip.doe-mbi.ucla.edu Database of Interacting Proteins

http:www.CRPinc.com Covance Research Products Inc. (CRP) home page

http:www.invitrogen.com Invitrogen home page

http:www.novagen.com Novagen home page

http:www.promega.com Promega home page

http://www.qiagen.com Qiagen home page

http:www.stratagene.com Stratagene home page

CHAPTER 11

Making Sense of Proteomics: Using Bioinformatics to Discover a Protein's Structure, Functions, and Interactions

Parag Mallick* and Edward M. Marcotte†

*DOE Center for of Genomics and Proteomics, University of California, Los Angeles; †Department of Chemistry and Biochemistry, and Institute for Cellular and Molecular Biology, University of Texas

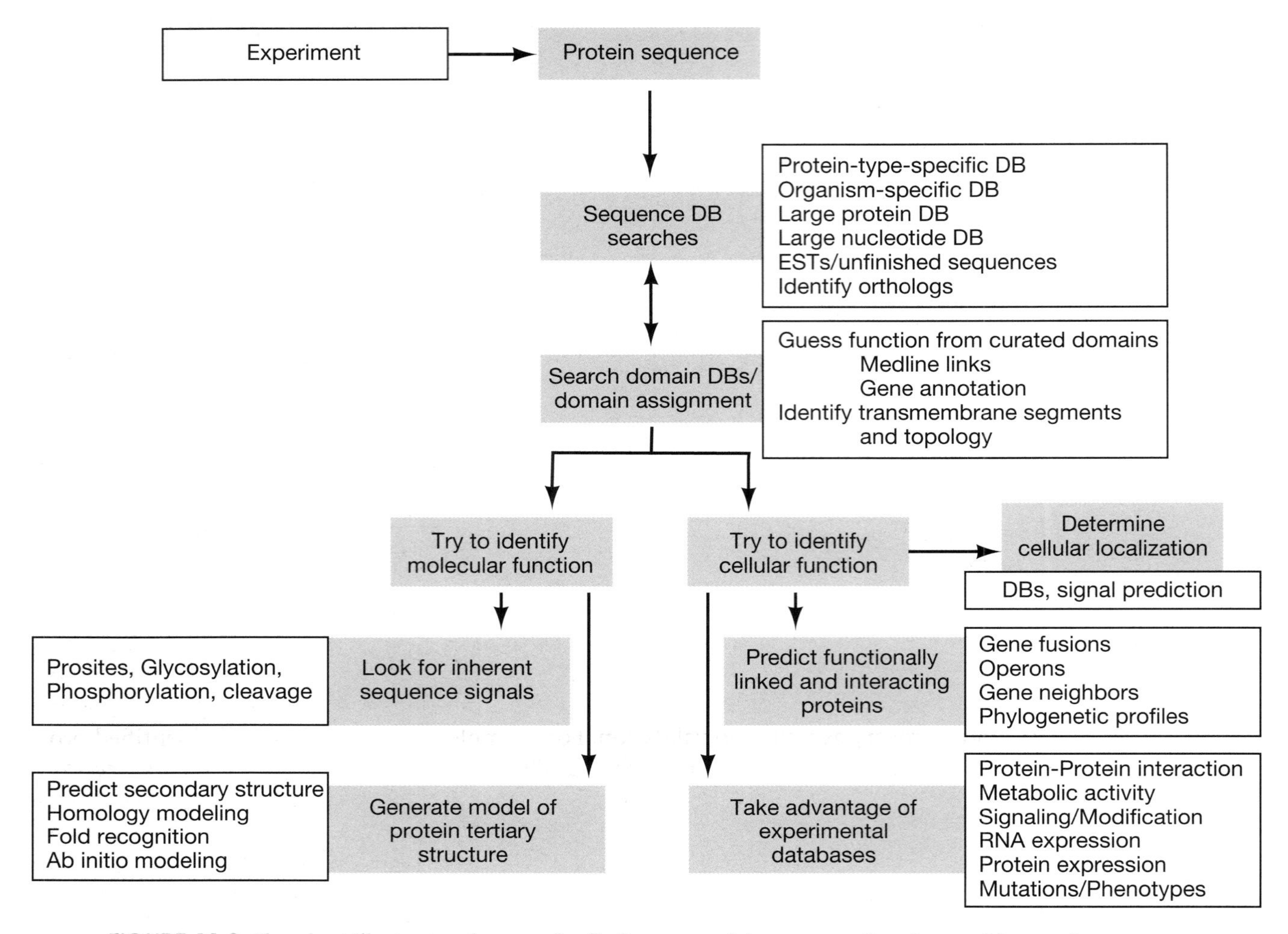

FIGURE 11.1. Flowchart illustrating the steps for finding a protein's structure, function, and interaction partners.

A TYPICAL PROTEOMICS EXPERIMENT MIGHT BEGIN WITH CELLS grown under some specified set of conditions. A subset of the cellular proteins is purified (see Chapter 1) through subcellular fractionation (see Chapter 3) or the use of affinity tags or protein interactions (see Chapter 10). These proteins are then identified or quantified by some combination of one- or two-dimensional gel electrophoresis (Chapters 2 and 4), high-performance liquid chromatography (Chapter 5), amino acid sequencing (Chapter 6), proteolysis (Chapter 7), and/or mass spectrometry (Chapter 8). A diverse set of bioinformatics analyses is required both to perform these experimental steps and to interpret the resulting data. Beyond the initial step of identifying peptides from their sequences or mass spectra, which is discussed in Chapters 6–8, a proteomics investigator would often like to perform analyses to

- identify intact proteins from their constituent peptides,
- find related proteins in multiple species,
- find genes corresponding to the proteins,
- find coexpressed genes or proteins,
- find or test candidate protein interaction partners,

- validate or compare proteomics measurements of posttranslational modifications or protein subcellular localization with computational predictions of the same properties,
- predict structures or domains of proteins, and
- find the functions of proteins discovered in proteomics experiments.

This chapter provides advice and protocols for each of these bioinformatics tasks. More generally, given a protein identified from an experiment, this chapter addresses the essential problem of learning as much as possible, computationally, about the protein's function. The chapter is divided into three sections.

- ***Section 1: Identification of a Gene and Its Homologs.*** Addressed in this section is the problem of identifying the complete gene and its homologs from a peptide or short protein sequence.
- ***Section 2: Predicting Protein Structure and Functions.*** Methods are presented to determine the domain and secondary structure of a protein, and where possible, the tertiary structure and biochemical function.
- ***Section 3: Matching Proteins to Pathways and Identifying the Cellular Roles of Proteins.*** Methods to discover a gene's cellular role are introduced, including protocols for assigning genes to pathways and for identifying potential interaction partners.

In each section, protocols and freely available resources, such as applicable Web servers, are presented as an aid to the analyses. The approaches we suggest are diagrammed in Figure 11.1. It is worth noting that the methods described in this chapter can be applied to prediction and discovery as well as to validation. For example, given experimentally identified protein interactions, investigators have a choice: They can extend the experimental interactions by the addition of computationally identified interactions, as the combined interaction sets are likely to be far more informative, or they can test the experimental interactions by comparison with the computational predictions. Often, computational data can be incorporated intuitively into proteomics experiments, such as by explicitly verifying that experimentally observed membrane proteins are computationally predicted to contain transmembrane segments. The wise investigator will incorporate computational data wherever possible, but set certain classes of bioinformatics analyses aside to provide independent validation of the results. For additional resources on bioinformatics "especially for the novice," see the panel on ADDITIONAL RESOURCES at the end of this chapter.

SECTION 1: Identification of a Gene and Its Homologs

PROGRAMS SUCH AS SEQUEST, DETAILED ELSEWHERE IN THIS VOLUME (Dongre et al. 1997 and Chapter 8), predict a peptide's sequence from a mass spectrum trace. Although these programs often successfully search sequence databases and attach a peptide sequence to a gene, manual searching can often be revealing, as it is able to take into account the individual investigator's experience with a protein and with the experiments in which it was discovered. In addition, it is often of value to know the near and remote homologs of a protein of interest when predicting the function. This section provides suggestions for getting the most information from sequence database searches.

SEQUENCE DATABASE SEARCHING WITH GAPPED-BLAST AND PSI-BLAST

Background and Methodology

Many biologists conduct searches of sequence databases routinely using one of three popular programs: FASTA, Gapped–Basic Local Alignment Search Tool (Gapped-BLAST), and Position-Specific Iterated–BLAST (PSI-BLAST). FASTA is generally considered more sensitive for DNA-DNA comparisons. However, BLAST is usually faster. Needleman-Wunsch (NW) and Smith-Waterman (SW) are two other programs commonly used for sequence comparison. Although NW and SW are more sensitive than either Gapped-BLAST or FASTA, they are not generally used for searching sequence databases because of their prohibitive time requirements. This section focuses on the Gapped-BLAST and PSI-BLAST family of programs. Specific programs in this family are described in Table 11.1. For detailed descriptions of the mechanics of these programs, see Altschul et al. (1990, 1997). In addition, two recent reviews (Altschul and Koonin 1998; Jones and Swindells 2002) provide helpful tips for effectively running these programs.

The Gapped-BLAST suite compares sequences using a substitution matrix, such as BLOSUM62, to mostly identify obvious homologs of a query protein. In contrast, PSI-BLAST identifies remote homologs of a query protein by sitting on top of Gapped-BLAST and defining probable substitutions at *each* position of a query sequence from the results of previous searches. PSI-BLAST is a complex, computation-intensive program and should only be used for finding remote homologs!

TABLE 11.1. The five programs of the Gapped-BLAST family

Program	Query	Database	Comparison method
BLASTP	protein	protein	direct comparison against database
BLASTN	nucleotide	nucleotide	direct comparison against database
BLASTX	nucleotide	protein	translates query in all six reading frames and then performs comparison against database
TBLASTN	protein	nucleotide	translates database in all six reading frames and then performs comparison of query
TBLASTX	nucleotide	nucleotide	translates both the query and the database in all six reading frames into proteins and then compares the generated proteins

PROTOCOL 1

Simple Sequence Search with a Long Peptide Using Gapped-BLAST

TWO PURPOSES OF THIS PROTOCOL ARE:

- ***Identification of the gene from which a peptide fragment is derived.*** Knowing the gene sequence from which a peptide is derived vastly increases the potential to learn about the structure, function, and interactions of the parent protein, its homologs, and its interaction partners.
- ***Identification of genes that code for homologs to the peptide (or its parent protein).*** Knowing a protein's homologs provides insight into structurally and functionally important residues in the protein and enables better identification of remote homologs, secondary structure, tertiary structure, etc.

It is also important to consider from which organism the query sequence was derived. When performing a sequence search, two simple questions must first be answered: What is the appropriate query sequence? What is the appropriate sequence database to search? The possible answers to the first question are discussed in Section 3. For the moment, we assume that the entire known sequence is the query.

Some of the very large number of sequence databases are listed in Table 11.2. A few common databases are described to illustrate the importance of thinking about which database to search, rather than simply choosing the largest.

One popular database, GenPept, is gigantic and nonspecific, containing translated coding regions of all publicly available DNA sequences in addition to proteins that have been specifically deposited. Other databases are smaller and specific to a single genome or a single protein family. For example, the igBLAST database contains exclusively immunoglobulin proteins. The proteins within these specialty databases are not always contained in GenPept. In addition, sequencing centers will often provide users with the ability to search incompletely sequenced genomes. It is generally best to search smaller, more specific databases whenever possible. The searches are less computionally expensive and are often likely to yield useful results. For example, if an investigator were trying to determine which *E. coli* protein a peptide is derived from, it would make the most sense to initially search an *E. coli* database. The NCBI site has a good set of tutorials describing the adjustable input parameters for BLAST. Readers are encouraged to visit:

http://www.ncbi.nlm.nih.gov/Education/BLASTinfo/information3.html

INPUTS

Query sequence (sample): YLGKITRNDAEVLLKKPTVRDGHFLVRQCESSPGEFSISVRFQDSVQHFKVLRDQNGKYY
Database (sample): *C. elegans*
Starting Web Address: http://www.ncbi.nlm.nih.gov/BLAST/

Additional Web Sites for BLAST Searches	Web Sites for sequence searches with small peptides
http://blast.wustl.edu/ http://genome.wustl.edu/Blast/client.pl http://www.sanger.ac.uk/DataSearch	http://prowl.rockefeller.edu/cgi-bin/ProFound http://www.expasy.ch/tools/ http://pir.georgetown.edu/

TABLE 11.2. Databases available on BLAST Web server

Database	Description
A. Peptide sequence databases	
nr	translations of GenBank DNA sequences with redundancies removed, PDB, SwissProt, PIR, and PRF
month	new or revised entries or updates to nr in the previous 30 days
swissprot	latest release of the SwissProt protein sequence database[a]
Drosophila genome	provided by Celera and Berkeley Drosophila genome project
yeast	yeast (*Saccharomyces cerevisiae*) genomic sequences
E. coli	*E. coli* genomic sequences
pdb	sequences of proteins of known three-dimensional structure from the Brookhaven Protein Data Bank
yeast	yeast (*S. cerevisiae*) protein sequences
E. coli	*E. coli* genomic coding sequence translations
pdb	sequences of proteins of known three-dimensional structure from the Brookhaven Protein Data Bank
kabat [kabatpro]	Kabat's database of sequences of immunological interest
alu	translations of select *Alu* repeats from REPBASE, a database of sequence repeats
B. Nucleotide sequence databases	
nt	GenBank, EMBL, DDBJ, and PDB sequences with redundancies removed (EST, STS, GSS, and HTGS sequences excluded)
month	new or revised entries or updates to nr in the previous 30 days
dbest[b]	EST sequences from GenBank, EMBL, and DDBJ with redundancies removed
dbsts[b]	STS sequences from GenBank, EMBL, and DDBJ with redundancies removed
htgs[b]	high-throughput genomic sequences
kabat [kabatnuc]	Kabat's database of sequences of immunological interest
vector	vector subset of GenBank
mito	database of mitochondrial sequences
alu	select *Alu* repeats from REPBASE, a database of sequence repeats; suitable for masking *Alu* repeats from query sequences
epd	eukaryotic promoter database
gss[b]	genome survey sequences, includes single-pass genomic data, exon-trapped sequences, and *Alu* PCR sequences

Reprinted, with permission, from Mount (2001).

[a] The SwissProt database is carefully curated but not always up to date because updates are released after longer intervals. SwissProt and PIR are the preferred protein databases for searches because the nr protein database is a composite of several databases and has duplicates of many sequences. Unfortunately, PIR is not provided as a separate choice on the database menu.

[b] Databases containing sequences that may have been less accurately determined.

METHOD

Inputting the Peptide Sequence

1. Go to http://www.ncbi.nlm.nih.gov/BLAST/.
2. In the section denoted Protein BLAST, click Standard protein-protein BLAST [blastp].
3. Enter the query sequence into the Search box.
4. In the select box denoted Limit by entrez query, choose *Caenorhabditis elegans*.

   Choosing *C. elegans* at this stage specifies the precise database to search. Neglecting this step results in searching GenPept.

5. At the bottom of the page, click the button marked BLAST!
6. Wait a few moments and then click the button marked Format!
7. The results page will appear. A description of how to interpret the results (the output) is given next.

Interpreting BLAST Output

BLAST commonly returns output in several sections, some of which are not always returned (depending on whether the output is to be e-mailed or displayed). The first section of output describes the query just performed, listing the description of the query sequence (if one was provided), and also the database searched against. Regardless of any restrictions placed on the search database, BLAST will often list the database as being the complete database available. For example:

```
Query: = gi|2501594|sp|Q57997|Y577_METJA PROTEIN MJ0577 (162 letters)
Database:  Non-redundant  GenBank  CDS  translations+PDB+SwissProt+SPupdate+PIR
437,713 sequences; 134,605,311 total letters
```

The second section (optionally returned) provides a graphical image map overview of the score of database matches and where they map (align) onto the query sequence. Each line is clickable. Clicking a given line hyperlinks to the alignment between the query protein sequence and the protein identified in the search. A glance at the graphical overview reveals several high-scoring sequences that are highly related to the query. The top bar (pink) is a match to the protein from which the fragment was drawn. The next several full-length bars (blue) are probable distant homologs. Partial length bars reflect similarity between a portion of the query and a database entry. These regions of similarity may be a shared domain or motif. The nature of these lower-scoring hits can be further explored by examining the corresponding alignment.

The third section of output provides descriptions of each significant alignment. The score and E-value or P-value are listed at the end of each line.

WHAT IS THE BLAST SCORE?

P-values and E-values are different ways of representing the significance of the raw score of a sequence alignment. The raw score of an alignment is the sum of substitution and gap scores, which measure imperfect amino acid substitutions and the size of insertions/deletions, respectively, in the sequence alignment. Substitution scores S are given by a look-up table (see PAM, BLOSUM). Gap scores are typically calculated as the sum of a gap opening penalty G and a gap extension penalty L. For a gap of length n, the gap cost would be $G+Ln$. The choice of gap penalties, G and L, is empirical, but it is customary to choose a high value for G (10–15) and a low value for L (1–2).

The E-value, or expectation value, is the *number* of different alignments with scores equal to or better than S that are expected to occur by chance in a database search of this *particular* database. The lower the E-value, the more significant the score. It is important to note that the E-value depends on database size. For example, performing the same database search of the GenPept database over time, even when an identical set of proteins is returned, will result in progressively less significant E-values as the database grows in size.

For comprehensive databases (such as the NCBI GenPept database, with 1,000,000 sequences), a match with an E-value <0.000001 can be trusted as a true homolog. However, a search through all proteins of known structure (e.g., Protein Data Bank, see Table 11.1) with E <0.001 is akin to searching the NCBI NRDB with E approximately <0.1 because the NRDB is nearly 100 times larger. As such, the two results are not comparable unless one corrects for database size. Often, additional remote homologs can be found with higher (less significant) E-values. Manual inspection of the quality of the alignment and of the annotation often enables the user to pick these true homologs from the noise.

Some BLAST implementations return a P-value instead of an E-value. The P-value expresses the *probability* of an alignment occurring with a given raw score or better: This probability score represents the intrinsic quality of the alignment and does not take into account the size of the database searched. The most highly significant P-values will be those close to 0.

The description (also called definition) lines are listed below under the heading "Sequences producing significant alignments." Here, the term "significant" simply refers to all those hits whose E-value was less than some threshold and should not be interpreted as "biological significance."

```
                                                                 Score    E
Sequences producing significant alignments:                      (bits)  Value

gi|17569445|ref|NP_509342.1|  (NM_076941)  C. elegans sem-5 [...   127   1e-30
gi|17508235|ref|NP_493502.1|  (NM_061101)  Src homology domai...    49   6e-07
```

The final section provides the alignments themselves and full descriptions of the sequences as shown below. Pay attention to protein names found in the descriptions. Also note items of the form "`sp|P29355|SEM5_CAEEL`." These are identifiers referencing the protein in other databases. The code `sp` indicates that the entry is in the SwissProt database. The codes `P29355` and `SEM5_CAEEL` are identifiers used to reference that entry within the SwissProt database.

```
>gi|17569445|ref|NP_509342.1|    (NM_076941)  C. elegans sem-5 [Caenorhabditis
elegans]
 gi|134425|sp|P29355|SEM5_CAEEL Sex muscle abnormal protein 5
 gi|283556|pir||S25730 SH2-SH3 protein sem-5 - Caenorhabditis elegans
 gi|247605|gb|AAB21850.1| (S88446) cell-signalling [Caenorhabditis elegans]
 gi|861389|gb|AAA68405.1| (U29082) Hypothetical protein C14F5.5 [Caenorhabditis
elegans]
 gi|228675|prf||1808298A sem-5 gene [Caenorhabditis elegans]
          Length = 228

 Score = 127 bits (319), Expect = 1e-30
 Identities = 60/60 (100%), Positives = 60/60 (100%)

Query:   1   YLGKITRNDAEVLLKKPTVRDGHFLVRQCESSPGEFSISVRFQDSVQHFKVLRDQNGKYY 60
             YLGKITRNDAEVLLKKPTVRDGHFLVRQCESSPGEFSISVRFQDSVQHFKVLRDQNGKYY
Sbjct:  61   YLGKITRNDAEVLLKKPTVRDGHFLVRQCESSPGEFSISVRFQDSVQHFKVLRDQNGKYY 120

>gi|17508235|ref|NP_493502.1| (NM_061101) Src homology domain 2, Src homology
domain 3, protein
          tyrosine kinase (src subfamily) [Caenorhabditis elegans]
 gi|7160701|emb|CAB04427.2| (Z81543) contains similarity to Pfam domain:
PF00017 (Src homology
          domain 2), Score=115.5, E-value=2.6e-39, N=1; PF00018
          (SH3 domain), Score=81.5, E-value=5.5e-21, N=1; PF00069
          (Eukaryotic protein kinase domain), Score=276.2,
          E-value=1.4e-79, N=1~cDNA >
         Length = 507

 Score = 48.5 bits (114), Expect = 6e-07
 Identities = 25/60 (41%), Positives = 35/60 (57%)

Query:   1 YLGKITRNDAEVLLKKPTVRDGHFLVRQCESSPGEFSISVRFQDSVQHFKVLRDQNGKYY 60
           Y GK+ R DAE  L       G FLVR  ES   + S+SVR  DSV+H+++ + +  +G Y+
Sbjct: 125 YFGKMRRIDAEKCLLHTLNEHGAFLVRDSESRQHDLSLSVRENDSVKHYRIRQLDHGGYF 184
```

A high-quality alignment generally contains a small number of gaps and a large number of identical or similar residues. Identical residues are reported along the center line of the alignment. Similar residues are denoted by a + in the center line of the alignment. By looking at the first alignment, it is observed that this protein matches identically with the Sem-5 protein from *C. elegans*, which is actually the gene that the segment was drawn from initially. We can now use the various reference codes (`17569445, NP_509342.1, 134425, P29355,SEM5_CAEEL...`) to find the full-length protein, and also to learn something about the protein as discussed below. Be cautioned at this point not to draw too many conclusions from the GenPept annotation of the protein. Although the annotation may have been drawn from experiment, it was likely drawn instead from homology searches, and hence may not be strictly reliable. In addition, this search should be repeated with the full-length protein in order to collect a diverse collection of homologs to better understand the structure and function of the protein of interest.

What to do with mediocre scores?

BLAST sometimes is unable to locate an identical or even a significant match to a protein segment. This is especially true for organisms that have not been completely sequenced. First try a different database. If after searching all of GenPept, TREMBL, and PIR a quality match has not been discovered, it may be of value to use TBLASTN to search Genbank, or EST and unfinished genome sequence databases. If the exact protein cannot be found, working with a close homolog might still be informative. However, if a homologous gene from a different organism (e.g., Kangaroo, when the query protein is from Human) is chosen, it is best to proceed with caution, as even close homologs often vary in function and may contain quite distinct domains (e.g., see Section 3 of this chapter). It is also a good idea to periodically repeat searches of the organism-specific expressed sequence tag (EST) databases and organism-specific databases available at the sequencing centers.

PSI-BLAST and remote homology searches

A multiple alignment of a sequence family provides much information about which residues are highly conserved and which regions of the sequence can accommodate insertions and deletions. Various methods exist that take into account this position-specific information. The simplest methods use consensus sequences or consensus sequence patterns. More sophisticated methods use *profiles* of the alignment, which consist of scores for allowable amino acid residues, insertions, and deletions specifically calculated for each position in the alignment. Searching with a profile often identifies remote homologs not found with a single sequence and scoring matrix. PSI-BLAST uses results of an initial BLAST search to generate a profile with which to perform additional searches of the same database. At each step, PSI-BLAST "improves" the profile, taking into account the new set of sequences discovered. When using PSI-BLAST, it is therefore important to verify that the sequences included at each stage are actually members of a homologous family. A few falsely identified proteins can badly skew the statistics, dooming further iterations.

PSI-BLAST performs best on large databases, such as SwissProt, PIR, or GenPept. Running the first stage of PSI-BLAST is identical to running Gapped-BLAST. The output of PSI-BLAST, however, presents the user with the option of selecting extra sequences for inclusion in the profile (sequences believed to be part of the family that were initially below threshold) and running extra iterations. Each cycle identifies more sequences, unless PSI-BLAST converges on a set of proteins and cannot find any more potentially homologous sequences within the database. If, at a given iteration, the newly identified sequences diverge greatly from the original query protein, it is likely that a rogue sequence, not truly homologous to the query, has been accidentally included and is biasing the search.

The Importance of Orthologs

For many functional analyses, the user often wants to identify *orthologous* proteins, rather than *paralogous* or simply *homologous* proteins. Typically, genes with a common evolutionary ancestor are denoted homologs. Orthologs are homologous genes generated by a speciation event, whereas paralogs are homologous genes separated by a duplication event (Fitch 1970). The similar sequences identified as statistically significant by algorithms such as BLAST and FastA are typically homologous. Consequently, the term homologous is often used to describe the sequences identified from a BLAST search, although the presence of a shared evolutionary ancestor is not strictly proven. Use of the term "homology" varies by discipline.

In certain biological communities (e.g., biochemistry), the term is used loosely. However, in others (e.g., evolutionary biology), it is used more strictly and cautiously.

Regardless of semantics, for complete transfer of functional annotation from one sequence to another, one typically requires that the sequences be orthologs. However, it is clear that the hits from BLAST sequence searches are often not orthologous. For example, searching a yeast sequence database with the bacterial MutS DNA repair protein will return six classes of homologous DNA repair proteins, all part of the MSH protein family (for *M*ut*S* *H*omologs). However, the functions of these proteins range from DNA mismatch repair, as for MSH2 and MSH6 (Johnson et al. 1996), to reciprocal recombination, as for MSH5 (Hollingsworth et al. 1995). To specifically assign the function of a new DNA repair protein would require knowing to which of these subfamilies the protein belongs.

Assigning a protein to one family out of several paralogous families is best done by calculating a phylogenetic tree for all of the proteins, and then examining which proteins are closest to the query protein in the tree. For an illustration of assigning functions from phylogenetic trees (using the MSH proteins as an example), see Eisen (1998). Programs for calculating phylogenetic trees are widely available, and several such programs are listed in the panel below. For good reviews on phylogenetic analysis, see Thornton and DeSalle (2000) and Mount (2001).

ClustalX (Thompson et al. 1997)
download at ftp://ftp-igbmc.u-strasbg.fr/pub/ClustalX/
available on the Internet at http://www.ebi.ac.uk/clustalw/

Tree View (Rod Page)
download at http://taxonomy.zoology.gla.ac.uk/rod/treeview.html

Phylip (Joe Felsenstein)
http://evolution.genetics.washington.edu/phylip.html
available on the Internet (courtesy of Allison Lim and Louxin Zhang) at
http://sdmc.krdl.org.sg:8080/~lxzhang/phylip/

Paup
http://paup.csit.fsu.edu/

In addition to phylogenetic trees, several rapid methods exist for finding possible orthologous relationships between proteins. One such method is to simply find the protein in the COG database (short for "Clusters of Orthologous Groups" [Tatusov et al. 2001]; http://www.ncbi.nlm.nih.gov/COG/). The curators of COG have organized the known proteins from a number of different genomes into orthologous groups. Within each COG "category," proteins tend to have very closely related function. In fact, they are not strictly orthologs (e.g., the *E. coli* paralogs gyrA and parC are in the same COG category), but they do tend to have very similar sequences and very similar functions. Therefore, if a protein of interest is in the same COG category as a protein of known function, it is reasonable to expect the protein of interest to have a closely related function.

Protocol 2 details a rapid, easily automated method for identifying orthologs based on sequence similarity searches.

PROTOCOL 2

Rapid Identification of Orthologs by the Method of Bidirectional Best Hits

GIVEN A GENE (GENE *A* FROM ORGANISM 1), a rapid sequence-comparison-based method can be used to identify an orthologous gene (gene *B* from organism 2), when both genomes have been completely sequenced. This is an approximate method, not absolutely guaranteed to find orthologs, but often effective. The method, which is called the method of "symmetric best hits" or "bidirectional best hits" (Overbeek et al. 1999), involves searching genome 1 with gene *B*. The best-matching sequence in genome 1 (corresponding to gene *A*, if *A* and *B* are orthologs) would then be used as the query in the reverse sequence search against genome 2. If the best-matching sequence in this search is gene *B*, then genes *A* and *B* are bidirectional best hits and are operationally defined as orthologs. This protocol is best automated if many sequences are to be examined, but the following protocol will serve well for only a few cases.

INPUTS

As an illustration of this protocol's utility, we will find the *E. coli* ortholog of a *D. radiodurans* gene.

Organism 1: *E. coli*
Organism 2: *Deinococcus radiodurans*

Gene *B*: aspartokinase I gene (Genbank identifier 6459119), from *D. radiodurans*

METHOD

1. Find the genome of organism 1 (in this case, *E. coli*) in the Entrez genome database.

   Many of the "prominent" organisms are available at http://www.ncbi.nlm.nih.gov/PMGifs/Genomes/org.html. In this example, click on the name *Escherichia coli* K12, which brings up the Web page dedicated to the *E. coli* K12 genome (http://www.ncbi.nlm.nih.gov/cgi-bin/Entrez/framik?db=G&gi=115).

2. Click to BLAST against protein sequences. Click to new protein sequence. Paste in the amino acid sequence of protein B (the *D. radiodurans* protein 6459119) and press go.

3. The best match from this BLAST search, protein A (in this case, the protein with the Genbank identifier 1786183), has the potential of being an ortholog of protein B. To evaluate protein A by performing the reverse sequence search, begin by clicking on the name of the protein A sequence, which will hyperlink to its sequence entry. Copy the protein sequence from the bottom of the entry.

4. Find the genome of organism 2 (in this case, *D. radiodurans*) in the Entrez genome database at http://www.ncbi.nlm.nih.gov/PMGifs/Genomes/org.html.

5. Click to BLAST against protein sequences. Click to new protein sequence. Paste in the copied sequence protein of protein A (e.g., 1786183) and press go.

6. If the best-matching sequence from this search is protein B—in this case, 6459119—then the two proteins (e.g., 1786183 and 6459119) are bidirectional best hits. In the event that the function of the *D. radiodurans* protein was unknown, a good guess as to its function would be that it has a similar function as its annotated ortholog, the *E. coli* aspartokinase I.

SECTION 2: Predicting Protein Structure and Function

IN THE PREVIOUS SECTION, A GENE OR A SET OF GENES was identified that corresponds to a peptide of interest, allowing many questions to be answered about the function of the gene product. A set of programs and the biological questions they address are shown in Table 11.3. The types of biochemical information that these tools allow investigators to glean about a protein from its sequence are described schematically in Figure 11.2. To facilitate the use of these programs, two "meta" servers have been set up, Gerard Klegyt's ProSal (http://xray.bmc.uu.se/sbnet) and Burkhardt Rost's META-PP ("META Predict Protein Server") (http://maple.bioc.columbia.edu/pp/submit_meta.html). Programs accessed by META-PP and ProSal are indicated by checkmarks in the last two columns under the headings MS and PS, respectively. Other relevant programs of interest are listed in Table 11.3 as well.

Protein Domain Structure Prediction

Before rigorously examining the function of a protein of interest, it is important to assess its domain structure. Although there is no single accepted definition, domains are generally acknowledged to be stable, compact, semi-independent, autonomously folding regions within a protein. Although some proteins consist of a single domain, others consist of several domains. Many protein domains are formed from a single continuous segment of a protein chain, but they can also be composed of discontinuous segments. The size of individual domains varies widely. For example, both E-selectin and lipoxygenase-1 are two-domain proteins; E-selectin is 36 residues long, whereas lipoxygenase-1 is 692 residues long. However, the distribution of domain sizes is narrow; 80% of domains are composed of <200 residues.

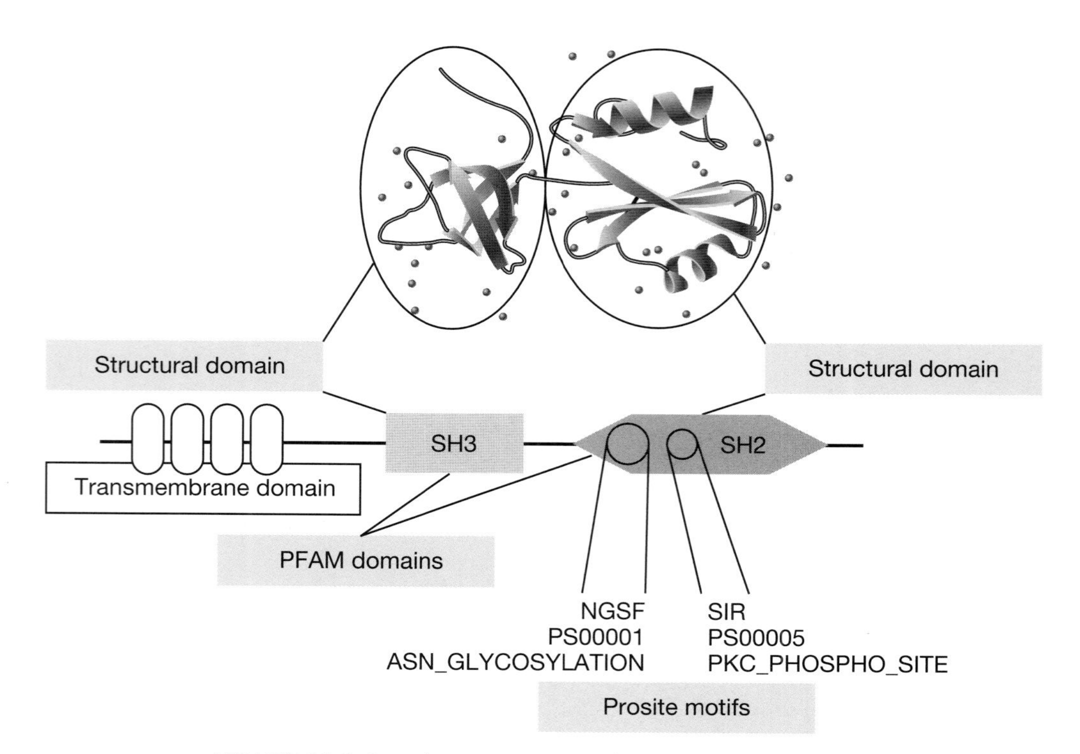

FIGURE 11.2. Protein sequence motifs and domain structure.

TABLE 11.3. Web servers for identifying a variety of protein features

Program	URL	MS	PS
Identifying the secondary structure of a protein			
Jpred	http://jura.ebi.ac.uk:8888/	✓	✓
PSI-pred	http://insulin.brunel.ac.uk/psiform.html	✓	
Phd	http://cubic.bioc.columbia.edu/pp	✓	✓
Sspro	http://promoter.ics.uci.edu/BRNN-PRED/	✓	
Prof	http://www.aber.ac.uk/~phiwww/prof/index.html	✓	
Identifying the tertiary structure of a protein			
SDSC1	http://cl.sdsc.edu/hm.html	✓	
CPHmodels	http://www.cbs.dtu.dk/services/CPHmodels/	✓	✓
SWISS-MODEL	http://www.expasy.ch/swissmod/SWISS-MODEL.html	✓	✓
ModBase	http://pipe.rockefeller.edu/modbase-cgi/index.cgi		
Loopp	http://ser-loopp.tc.cornell.edu/loopp.html	✓	
Superfamily	http://stash.mrc-lmb.cam.ac.uk/SUPERFAMILY/	✓	
Sausage	http://rsc.anu.edu.au/~drsnag/TheSausageMachine.html	✓	
Meta Server	http://bioinfo.pl/Meta/		✓
UCLA-DOE FRSVR	http://fold.doe-mbi.ucla.edu/		
Identifying transmembrane segments/topology			
DAS	http://www.sbc.su.se/~miklos/DAS/	✓	
TMHMM	http://www.cbs.dtu.dk/services/TMHMM-1.0/	✓	✓
TopPred2	http://www.sbc.su.se/~erikw/toppred2/	✓	
MOMENT	http://www.doe-mbi.ucla.edu/Services/moment/	✓	
Identifying protein domains			
Pfam	http://pfam.wustl.edu/hmmsearch.shtml	✓	✓
Interpro	http://www.ebi.ac.uk/interpro/scan.html		✓
Prodom	http://protein.toulouse.inra.fr/prodom/blast_form.html		✓
BLOCKS	http://blocks.fhcrc.org/blocks/blocks_search.html		
SMART	http://smart.embl-heidelberg.de/		✓
DomFISH	http://www.bmm.icnet.uk/~3djigsaw/dom_fish/		
Identifying documented sequence motifs and patterns			
Prosite	http://www.expasy.ch/tools/scnpsit1.html		✓
Emotif	http://elucidate.stanford.edu/emotif/emotif-search.html		✓
GeneQuiz	http://jura.ebi.ac.uk:8765/gqsrv/submit		✓
PRINTS	http://bioinf.man.ac.uk/cgi-bin/dbbrowser/fingerPRINTScan/ muppet/FPScan.cgi		
Identifying subcellular localization			
SignalP	http://www.cbs.dtu.dk/services/SignalP/		
PsortII	http://psort.nibb.ac.jp/form2.html		
ChloroP	http://www.cbs.dtu.dk/services/ChloroP/	✓	
Identifying posttranslational modification or cleavage			
Prosite	http://www.expasy.ch/tools/scnpsit1.html		✓
NetPhos	http://www.cbs.dtu.dk/services/NetPhos/	✓	
NetPico	http://www.cbs.dtu.dk/services/NetPicoRNA/	✓	

MS is META-PP and PS is ProSal.

Knowing domain boundaries is not simply of academic concern, but it is of great practical importance to the functional and structural characterization of a protein. For example, nuclear magnetic resonance (NMR) studies are sometimes not feasible on a full-length protein, but they are tractable on each of the single domains. Furthermore, conformational flexibility from flexible linker regions between domains often impedes protein crystallization. A priori knowledge of domain boundaries allows constructs more suitable for experimental structure elucidation to be built. Experimental methods such as phage display and yeast two-hybrid often also benefit from knowledge of domain structure.

The importance of domain structure is not limited to experiments; computational annotation and prediction benefit from knowing domain structure as well. For example, subsequences corresponding to individual domains are usually better query sequences than whole-protein sequences in similarity searches because individual domains likely correspond to recurring functional and evolutionary units. Consequently, manyfold recognition and comparative modeling methods perform best when applied to the sequences of single domains. Each domain in a multidomain protein often catalyzes a distinct biochemical reaction. Consequently, it is important when inferring function to assign the function to the correct domain. Identifying domains from protein sequences is an active field of study. The majority of available methods rely on either comparative sequence searches, tertiary structure/contact prediction, general properties of domains (e.g., sequence length), or a combination of these methods to infer domain boundaries. Methods relying on sequence searches have been successful at identifying contiguous domains when sequence similarity is readily apparent. However, at low sequence similarity, evolutionary relationships are difficult to discern. Furthermore, if the domain connectivity is conserved within a protein family, domain positions cannot be inferred. Methods relying strictly on general properties of domains are limited in their training by the incompleteness of protein structure databases, which due to experimental complexity, do not contain many proteins with greater than three domains. Methods based on structure/contact prediction are limited by the ability to predict structures with relative accuracy. As no single method is dominant, several servers, each implementing a different approach, are described. In addition, a number of methods produce "sequence domains" that only approximate structural domains, but are still quite useful when performing sequence searches and multiple sequence alignments of protein families.

As mentioned above, the two tools ProSal and META-PP enable users to easily submit sequences for domain analysis via ProDom (Corpet et al. 2000) and Pfam (Bateman et al. 2002). A search page at SBASE (Vlahovicek et al. 2002; http://hydra.icgeb.trieste.it/~kristian/SBASE/) enables users to search for domain matches within their sequence using BLAST. An alternate library of domains/motifs is available from BLOCKS (Henikoff et al. 1999) at http://www.blocks.fhcrc.org/blocks_search.html. Curated domain databases are also a widely used tool for predicting protein function. Discussed in a later section is how these databases and other sequence signals can be used to assess protein function. Sternberg's DomFish server (http://www.bmm.icnet.uk/~3djigsaw/dom_fish/) will conveniently attempt to locate PFAM domains within a query sequence as well.

Alternate methods not strictly based on homology with characterized domain libraries can be found in SnapDragon (George and Heringa 2002; http://mathbio.nimr.mrc.ac.uk/~rgeorge/snapdragon/), which simply requires the input of a sequence and e-mail address, or in DGS (Wheelan et al. 2000; http://www.ncbi.nlm.nih.gov/Structure/dgs/DGSWeb.cgi), which only takes as input the length of the sequence. DGS is generally best used only for sequences of less than 600 amino acids.

Protein Transmembrane Domain Prediction

Approximately 20–30% of the proteins within a typical microbial genome contain transmembrane domains, and many proteomics experiments (see Chapters 1 and 3) may explicitly focus on membrane proteins. Membrane proteins are often difficult to overexpress and have thus far not been readily amenable to structural studies. Consequently, only a handful of structures of membrane proteins are available, in contrast to the tens of thousands of structures of soluble proteins. Therefore, computational methods that can identify membrane proteins and shed light on their structure and topology are quite useful. Much transmembrane helix prediction has focused on predicting the topology of the helices. Topology prediction methods begin with the observations that transmembrane helices are often 20–30 residues long with a high overall hydrophobicity and that short nontranslocated loops contain many positively charged residues, whereas short translocated loops contain few such residues. Knowledge of the hydrophobicity of transmembrane segments enables identification of transmembrane segments from hydrophobicity plots, and knowledge of loop charge enables prediction of the orientation of a protein in the membrane.

Topology prediction methods vary in complexity depending on how the hydrophobicity scale has been derived, what algorithm is used for prediction, and whether or not information from homologous sequences is taken into account. The best current methods claim that >90% of all transmembrane segments can be correctly identified and that the full topology is correctly predicted for >80% of all proteins. Predictions are slightly better for prokaryotic proteins than for eukaryotic proteins. Topology prediction of β-barrel membrane proteins (e.g., porins) is more difficult since the membrane-spanning β-strands are short and thus hard to detect in the sequence. Reasonable predictions can be made when the protein can be aligned with the sequence of a protein of known structure.

A variety of transmembrane helix predictors are available including DAS (Cserzo et al. 1997), TMHMM (Krogh et al. 2001), TopPRED (Claros and von Heijne 1994), and PHDhtm and PHDtopology (Rost et al. 1996) all of which are accessed via META-PP. These methods use a combination of clever statistics or neural networks to fish out potential transmembrane helices. An alternative approach based on simple window hydrophobic moment is available at http://www.doe-mbi.ucla.edu/Services/moment/. TMHMM was recently evaluated to be the most accurate (Moller et al. 2001). In general, we suggest a consensus approach.

Protein Secondary Structure Prediction

Secondary structure prediction lies at the heart of many tertiary structure prediction methods. The use of evolutionary information and patterns of amino acid sequence conservation improved prediction accuracy substantially in the 1990s. Recently, the evolutionary information gathered by improved search tools and larger databases has boosted prediction accuracy to its current 75–85% accuracy for three-state predictions (i.e., predicting α-helices, β-strands, and random coils).

A number of methods that are available employ neural networks, hidden Markov models, or consensus methods. The methods JPRED (Cuff et al. 1998), PHD (Rost and Sander 1994), PROF (Ouali and King 2000), PSIpred (McGuffin et al. 2000), and SSpro (Baldi et al. 1999) are each implemented within META-PP. PSIpred and JPRED are among the most accurate predictors of protein secondary structure (http://cubic.bioc.columbia.edu/eva/sec/common.html).

Protein Tertiary Structure Prediction

The connection between protein structure and function has been well-documented, and, consequently, a number of methods are available to predict a protein's function from its structure. In addition, even preliminary knowledge of a protein's structure can suggest interaction hot spots, functional regions, and potential targets of mutagenesis.

Structure prediction methods are often quite complex, and details of their methodology are not provided here (for detailed information, see Mount 2001). Briefly, structure prediction methods tend to belong to one of three classes: homology modeling, fold recognition, and de novo/ab initio. These classes are roughly defined by the degree of similarity that the method requires between the query sequence and the sequence of a matching structure from the Protein Data Bank (PDB) (http://www.rcsb.org). De novo methods generate structural models of proteins that cannot be readily modeled from existing structures, whereas fold recognition and homology modeling methods generate predictions of structural models using existing structures. Whenever possible (generally when a structure within the PDB shares sequence identity >30% with the query protein), homology modeling methods are preferred to fold recognition methods. Likewise, fold recognition methods are preferred to de novo methods.

More specifically, fold recognition and homology modeling methods find a homologous template structure within the database of experimentally solved structures, determine the equivalent positions between the probe sequence and template structure, and then place the probe sequence onto the template's backbone to generate a model. Consequently, homology modeling and fold recognition methods typically return

- a coordinate file containing the actual three-dimensional model,
- an accession code identifying the template structure from which the model was generated,
- an alignment between the probe sequence and the template showing the equivalent positions, and
- a score describing how well the probe sequence matched the template structure.

If a simple BLAST search of a query protein against the PDB database finds a reliable homolog, then the user can submit a sequence to either ModBase (http://pipe.rockefeller.edu/modbase-cgi/index.cgi) (Pieper et al. 2002) or Swiss-Model (http://www.expasy.ch/swissmod/SWISS-MODEL.html) (Peitsch 1996). Both of these services keep a catalog of generated structural models.

If no obvious homolog emerges, the ProSal server can be used to submit the amino acid sequence to the Structure Prediction Meta Server, which is similar in nature to META-PP; it takes a query sequence and submits it to a variety of fold recognition servers. Unlike META-PP, the Structure Prediction Meta Server collates the top ten structural template predictions from each server into a single page, translating the server results into a uniform format. For each server, the first line of condensed, reformatted output shows the server's name and provides links to the original raw output, the CASP format output, and the PDB model output. Next, the template structure, sequence-structure compatibility score, and alignment between the template structure and probe are given for the ten best predictions ranked by sequence-structure compatibility score. Two additional columns, FSSP and SCOP, contain identifiers describing the fold-type of the template structure. Fold-type and information that can be derived from fold-type are discussed below. It is important to note that each server has a different scoring system. For example, scores from GenThreader range from 0 to 1, whereas scores from BIO INBGO range from $-\infty$ to ∞. The raw output from each server describes how to interpret that server's scores. Predictions with "better" compatibility scores are more reliable.

Several attempts have been made to benchmark structure prediction methods including CASP (http://predictioncenter.llnl.gov/), CAFASP (http://bioinfo.pl/cafasp/), and LiveBench (http://bioinfo.pl/LiveBench/). Readers are encouraged to see Fischer et al. (1999), Bujnicki et al. (2001), and Moult et al. (2001).

Using Structural Classification Databases to Learn about Protein Function

Proteins of similar structure often share functional attributes such as active sites, interaction patches, metal-binding sites, and interacting partners. Consequently, it is valuable to know not only which protein is most similar in structure to the protein of interest, but also the family of structures similar to a protein. Furthermore, if severalfold recognition methods predict proteins that are classified as being structurally similar, the user gains confidence in the prediction. For more details on methods used to classify structures into families, see Mount (2001). The three databases most commonly used to determine structural relationships among proteins are SCOP (http://scop.berkeley.edu), CATH (http://www.biochem.ucl.ac.uk/bsm/cath/_new/index.html), and FSSP (http://www2.ebi.ac.uk/dali/fssp/fssp.html), which classify proteins by methods that are purely manual, a combination of manual and automated, and purely automated, respectively.

Accessing the data in these databases is simple: Accession codes for SCOP and FSSP are given within the Protein Prediction Meta Server output and can be used to directly access a structural family. The PDB code of the template structure can also be used to find structural neighbors and hints about a protein's function. For example, entering the accession code 1mbe into the CATH database (http://www.biochem.ucl.ac.uk/bsm/cath/_new/index.html) returns the information that there is one domain matching that PDB code and that its CATH classification identifier is 1.10.10.60. Clicking on the hyperlink associated with the CATH code returns information about the structure: In this case, 1mbe is a member of a protein family of mainly α-helical proteins, in which the α-helices are arranged as an orthogonal bundle and are homeodomain-like. Consequently, proteins in this family bind DNA. Further analysis would reveal the parts of the protein that directly contact the DNA and are thus functionally important.

Sequence Signals That Reveal Function

In addition to inferring the function of a target protein from its sequence and structural homologs, many signals exist within the protein itself that are indicative of its function. These signals may help in the interpretation of proteomics experiments, as well as in the validation of results from these experiments.

Linking Entrez with Medline

In Protocol 1, the example query protein is matched to several accession numbers, including 17569445 and SEM5_CAEEL. These two accession numbers are, respectively, the identifiers for the NCBI and SwissProt databases. These identifiers are useful to know because precomputed information is often accessed using these identifiers (e.g., access to protein models in ModBase). References to articles potentially describing the sequencing or function of this protein can be accessed via these identification names/numbers at the NCBI (http://www.ncbi.nlm.nih.gov) site or the SwissProt (www.expasy.ch) database. SwissProt is often particularly useful for quickly assimilating information about a protein, as the curated nature of the database implies that the vast amounts of raw data associated with a protein have been conveniently summarized.

In general, a significant problem associated with many proteomics experiments is the conversion of the raw proteomics data into a list of proteins named and annotated in a standard format. Associating each protein with a consistently chosen accession number can ease this process and the resulting interpretation of the experiment.

Annotated domain databases

Some sequence signals are found within extended domains, such as those identified earlier within the Prodom, Pfam, and BLOCKS databases. Domain databases are a better way to get at the function of a protein than the annotation given in full-length protein sequence databases. Often, domain databases are hand-curated and empirically seem to suffer less from problems of misannotation sometimes associated with noncurated databases. Furthermore, the sequence patterns are frequently quite specific, and distinct subtypes of homologous domains may be distinguishable.

As an example of protein domain identification, consider entering the Sem-5 protein sequence (used as an example in Protocol 1) in the sequence box at http://pfam.wustl.edu/hmmsearch.shtml. This search reveals the domain composition of the protein: two SH3 domains and one SH2 domain. The functions of each domain can often be found by clicking on the hyperlinks provided. Other similar information is available at http://smart.embl-heidelberg.de/ and at http://www.blocks.fhcrc.org/blocks_search.html.

Sequence pattern resources

In addition to extended sequence patterns, regular expressions have been derived to describe common motifs indicative of protein functional sites. These simple motifs can provide considerable support for proteomics experiments, by supplying computational evidence for protease sites (such as might be useful in Chapters 7 and 8), glycosylation sites, and phosphorylation sites (see Chapter 9).

Motifs are commonly given in a *regular expression* notation. For example, the motif `[LIVM]-[ST]-A-[STAG]-H-C` can be used to identify serine/threonine proteases. In this notation, x denotes a position where any amino acid is accepted. Often, more than one specific amino acid type may be acceptable for a given position in a motif: These are indicated by listing the acceptable amino acids for a given position between square brackets `[ ]`. For example, `[ALT]` stands for Ala or Leu or Thr. Ambiguities may also be indicated by listing within curly brackets `{ }` the amino acids that are unacceptable at a given position. In this format, `{AM}` stands for any amino acid except Ala and Met. Each element in a pattern is separated from its neighbor by a –. A repeated pattern element is indicated by following an element with a numerical value or a numerical range between parentheses, such as `x(3)` corresponding to `x-x-x` and `x(2,4)` corresponding to `x-x or x-x-x or x-x-x-x`.

Putting these rules together, the pattern `[AC]-x-V-x(4)-{ED}` corresponds to `[Ala or Cys]–any residue-Val-any residue–any residue–any residue–any residue–{any residue except Glu or Asp}`. ProSites (Falquet et al. 2002), PRINTS (Attwood et al. 2002), and Emotifs (Huang and Brutlag 2001) are all examples of short sequence pattern databases. Each of these services can be accessed via either META-PP or ProSal. In addition, META-PP also submits sequences to ChloroP (Emanuelsson et al. 1999), NetOglyc (Gupta et al. 1999), NetPhos (Blom et al. 1999), and NetPico (Blom et al. 1996) which search for sequence determinants of chloroplast transit, *O*-glycosylation, phosphorylation, and picornaviral protease cleavage sites, respectively.

SECTION 3: Matching Proteins to Pathways and Identifying the Cellular Roles of Proteins

BEYOND FINDING THE BIOCHEMICAL FUNCTION of a protein, it is also often possible to discover clues about a protein's cellular role. The clues are derived in part from the protein sequence, in part from the relationship of this sequence to those of other proteins in the same protein family, and in part from additional data, such as results of gene expression experiments, often available from high-throughput functional genomics experiments.

In proteomics experiments, these types of computational analyses of cellular function may help in interpreting results (e.g., when uncharacterized proteins are observed in an experiment, these techniques may suggest functions for the proteins) as well as validating results (e.g., when protein interactions are measured experimentally, these techniques may provide additional computational validation of the observed interactions). Section 3 provides protocols for finding the cellular roles of proteins by computationally identifying a protein's specific pathway (or pathways) and its interaction partners. For experimental determination of cellular roles, see Chapter 3 for information about subcellular localization and Chapter 10 for information about mapping protein interaction partners.

Discovering Functionally Linked Proteins

Three general computational approaches can be used to discover or validate protein interactions, complexes, and pathways:

- Discovery of functionally linked proteins by observation of gene fusion events (Protocol 3).
- Discovery of pathways and functionally linked proteins via the identification of operons (Protocol 4).
- Discovery of functionally linked proteins based on coinheritance of proteins across many organisms (Protocol 5).

The following are two additional approaches using data other than sequence data:

- Identification of interaction partners from databases of experimentally determined interactions.
- Discovery of functionally linked proteins from protein or gene expression patterns (Protocol 6).

In the first approach, illustrated in Figure 11.3, two genes in one organism can be inferred to be functionally linked due to the discovery of a third gene, which is the fusion of the two

FIGURE 11.3. A functional link can often be inferred between two separate proteins after finding a third gene that is a fusion of the first two (Enright et al. 1999; Marcotte et al. 1999). Here, gyrase A and gyrase B of *E. coli* are each homologous to a distinct region of topoisomerase II from *Plasmodium falciparum*, indicated by shaded boxes, suggesting (correctly) a possible functional relationship between *gyrA* and *gyrB*.

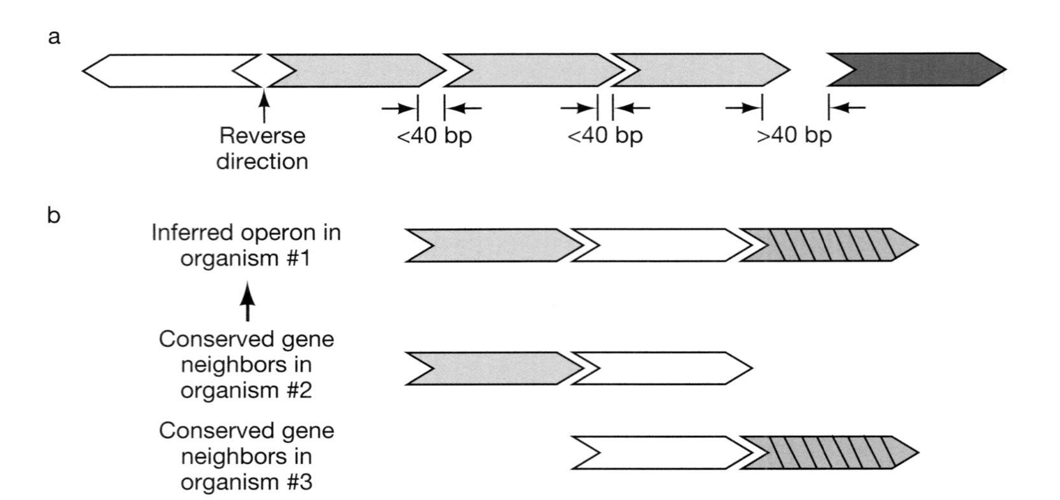

FIGURE 11.4. Computational reconstruction of bacterial operons by two approaches. (*a*) The distances between genes in the same operon (*light blue*) tend to be shorter than distances between adjacent genes in different operons (*dark blue*). In genomes with gene densities similar to those of *E. coli* K12, genes separated by less than 40 nucleotides are more likely to be in the same operon (Salgado et al. 2000); genes separated by longer intergenic distances are more likely to be in different operons. The boundaries of an operon can be estimated by finding the first large intergenic distance or gene encoded in the opposite direction. (*b*) An operon is reconstructed by searching for homologs of neighboring genes conserved as neighbors in other genomes (similar colors) (Bork et al. 1998; Dnadekar et al. 1998; Overbeek et al. 1999).

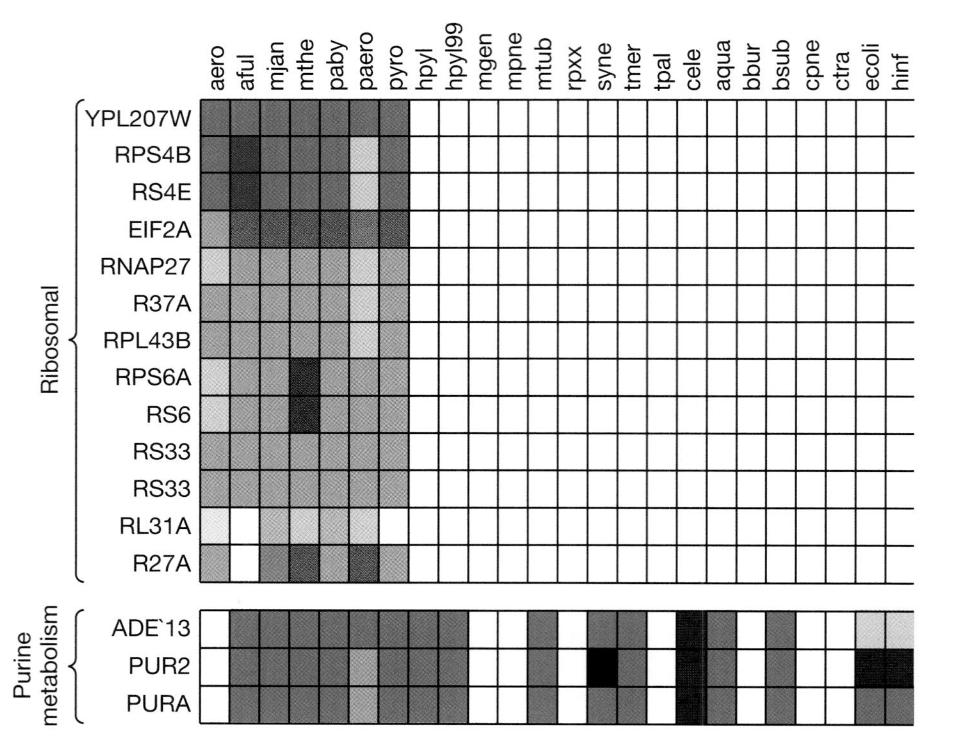

FIGURE 11.5. Proteins in the same pathway are often coinherited across organisms. This trend can be exploited by calculating the phylogenetic profile of a protein, indicated as a row in the figure, recording in which genomes (abbreviated on column headings) a protein has a detectable homolog (*colored* boxes, where darker shading indicates stronger sequence similarity) or lacks detectable homologs (*white* boxes). Here, proteins with similar phylogenetic profiles can be seen to have similar function. (Adapted, with permission, from Marcotte 2000.)

separate genes (Enright et al. 1999; Marcotte et al. 1999). Empirically, it appears that gene fusions typically only occur between genes of related function; consequently, this method can be used to rapidly suggest functional partners for a gene.

The second approach relies upon the tendency for bacterial genes to be organized into operons. Although many operons are known for some organisms (e.g., for known operons of *E. coli*, see the RegulonDB database at http://www.cifn.unam.mx/Computational_Genomics/regulondb/), many more are uncharacterized. Two computational protocols, illustrated in Figure 11.4, are given for reconstructing operons from a genome sequence. Both of these methods ignore the identification of promoters and regulatory sequences, and instead exploit other properties of operons. One protocol exploits short intergenic distances characteristic of operons (Salgado et al. 2000), and the second protocol exploits the tendency for genes in operons to be conserved as neighbors in other genomes (Tamames et al. 1997; Dandekar et al. 1998; Overbeek et al. 1999).

The third approach for identifying pathway and interaction partners exploits the tendency for proteins in the same pathway to be coinherited across organisms (Pellegrini et al. 1999; Huynen et al. 2000), as illustrated in Figure 11.5. A protein sequence is compared with sequences of all known proteins from available fully sequenced genomes. From this analysis, a phylogenetic profile is calculated, describing which genomes contain homologs of the query gene. This phylogenetic profile can then be compared to those of all other genes in a genome to suggest functional partners of the query gene.

Finally, data other than genomic data carry information about gene function and pathways. Protocols and Internet URLs are presented for the computational analysis of existing protein interaction data, gene disruption phenotypes, subcellular localization, and gene and protein expression data. Additional information on these topics is included in Chapters 3, 8, and 10.

PROTOCOL 3

Searching for Functionally Linked Proteins Using Gene Fusions: A Domain-based Search

THIS METHOD PERMITS THE IDENTIFICATION of proteins that are functionally linked to a query protein. Although this process is best automated, the search can be performed manually if desired. The search can be performed by comparing either protein domain structures or protein sequences. This protocol presents the domain-based approach, because it is easiest to implement with existing Web-based tools.

This version of the method takes advantage of a protein domain database, such as those discussed in Section 2 of this chapter. The ProDom database is used as an example here. It illustrates the domain structure of a given protein and allows the display of all other proteins sharing a given domain (Corpet et al. 2000). ProDom can be searched directly at http://prodes.toulouse.inra.fr/prodom/doc/prodom.html or accessed after first identifying a sequence in a participating protein sequence database, such as SwissProt (http://ca.expasy.org/sprot/). SwissProt provides direct hyperlinks between each protein and the corresponding domain entries in ProDom.

Certain domains (called promiscuous domains) are known to be of limited value in this process (Marcotte et al. 1999), as they link large numbers of different proteins together. The SH3 domains, ATP-binding cassettes, ATPase domain, and tyrosine kinase domains are examples of promiscuous domains. Following these promiscuous domain links in Step 4 of the procedure will generally produce spurious linkages, indicated by finding an exceptionally large number of links to distinct proteins inferred from a single fusion protein.

INPUTS

For this example, functional links will be identified for the *E. coli* gyrase A protein.

METHOD

1. Find the entry for the query protein (called, generically, protein A) in the domain database.

   For example, the entry for *E. coli* gyrase A is found at http://protein.toulouse.inra.fr/cgi-bin/ReqProdomII.pl?acc_seq0=P09097.

2. Find all of the proteins sharing a homologous domain with the query protein. In ProDom, click on the icon to the left of the gene name.

3. Search for proteins with additional domains beyond those of the query protein.

Each of these proteins will be evaluated for its potential as a fusion protein (AB) linking the query protein (A) to another protein (B).

4. For each protein with an additional domain, find all of the proteins sharing homologous domains. Again, in ProDom, click on the icon to the left of the gene name.

For example, in ProDom Release 2001.3, 414 genes share domains with the *Plasmodium falciparum* topoisomerase II. Search this list for genes from the same organism as the query protein. Here, of the 414 genes, only 4 are from *E. coli*. Eliminate all genes from this list that are homologous to the query protein (i.e., that share a domain with the query protein). The remaining genes, in this case, the homologous genes ParC and gyrase B, are predicted to be functionally linked to the query protein. In this case, one of these predictions is correct (gyrase B), and one is incorrect (ParC), illustrating that the method does not distinguish among paralogs.

In general, proteins linked by this method have a better than random chance of being in the same cellular pathway or of being interaction partners (Enright et al. 1999; Marcotte et al. 1999).

PROTOCOL 4

Rapid Identification of Operons

THIS PROTOCOL PERMITS THE DISCOVERY OF THE OPERON to which a query gene belongs. Two methods are provided (see Figure 11.4). Method A uses short intergenic distances to identify a gene's operon. For more details about this method and the choice of thresholds, see Figure 4 of Salgado et al. (2000), from which the thresholds are derived. Method B uses the WIT2 Web Site to identify groups of genes that are found clustered together in a variety of organisms. This conservation of gene neighbors is characteristic of operons. An alternative approach for identifying conserved operon structure is to use the STRING Web server (Snel et al. 2000) at http://www.bork.embl-heidelberg.de/STRING/.

METHODS

Method A: Identifying Operons by Short Intergenic Distances

1. Go to the Entrez bacterial genome Internet site http://www.ncbi.nlm.nih.gov/PMGifs/Genomes/eub_g.html and find the genome of interest. Follow the hyperlink to the genome entry.

   For example, *E. coli* K12 is hyperlinked via the file name NC_000913 to http://www.ncbi.nlm.nih.gov/cgi-bin/Entrez/framik?db=Genome&gi=115.

2. Calculate the gene density for this organism as the number of total bases in the genome divided by the number of protein-coding genes. The values for the total bases and the protein-coding genes are located near the top of the Web page.

   The thresholds given below are calculated for *E. coli* K12, which has a gene density of ~1084 nucleotides/gene. Note that genomes with very different gene densities may require recalibration of the thresholds.

3. Select the protein-coding gene feature table.

   For *E. coli* K12, this is found at http://www.ncbi.nlm.nih.gov/cgi-bin/Entrez/altik?gi=115&db=Genome.

4. Find the gene of interest in the table. On the basis of the gene's location in the chromosome, given at the left of the table, calculate the number of nucleotides separating the gene from the genes on either side of it.

   A neighboring gene is likely to be in the same operon if it is encoded on the same strand of the DNA, and it is nearer than ~40 base pairs.

5. Repeat the analysis for each gene included in the operon until the apparent boundaries of the operon are discovered.

Method B: Identifying Operons by Conservation of Gene Neighbors

1. Go to the WIT database Internet site http://wit.mcs.anl.gov/WIT2/.
2. From the menu at the top of the Web page, select General Search.
3. Find the gene of interest via the search menu. For example, to search for *E. coli* ubiquinol oxidase:
   - Select: All ORFs in the Select Data Categories to Search frame
     Bacteria–Escherichia coli in the Select Specific Organism to Search frame
   - Input: ubiquinol oxidase or *cyoA*, searching for an exact match to the term.
   - Click on the Search button.
4. Follow the links to the gene of interest; in this case to ORF 4472.
5. On the frame at the left side of the Web page, select Preserved Operons to get a series of tables showing which neighbors of the gene of interest are conserved as neighbors in another genome. The relationships discovered are summarized at the bottom of the page. In the example of *cyoA*, the four neighboring genes, *cyoB* to *cyoE*, are predicted correctly to be in the same operon as *cyoA*.
6. If desired, the precise locations and orientations of the genes can be examined from the Contig Region link from the main Web page for the query gene (i.e., from the same frame where Preserved Operons is located).

PROTOCOL 5

Finding Functionally Linked Proteins by Comparing Phylogenetic Profiles

THE PURPOSE OF THIS PROTOCOL is to identify proteins functionally linked to a query protein. Two methods are provided: Method A uses a Protein Link Explorer (PLEX; http://apropos.icmb.utexas.edu/function-prediction.html). PLEX will perform a BLASTp search of the query protein against a database of fully sequenced genomes. PLEX reports the proteins with statistically similar sequences, the organisms in which these proteins are found, and the resulting phylogenetic profile of the query protein. From this profile, the investigator can rapidly determine the breadth of organisms containing the query protein, and by extension, the breadth of organisms containing the system in which the query protein operates.

Method B is an alternative approach for identifying proteins with similar inheritance patterns using the COGs database (http://www.ncbi.nlm.nih.gov/COG/).

METHODS

Method A: PLEX

1. Paste a protein sequence into the PLEX.
2. Choose the genome from which the query protein is derived. If the genome is not present, choose one containing a close homolog of the query protein. Choose a threshold for the links to be reported. PLEX will then compare the phylogenetic profile of the query protein with the phylogenetic profile of every protein encoded by the chosen genome.

   PLEX returns those proteins whose phylogenetic profiles match above the threshold selected. These proteins have the most similar inheritance patterns with the query protein and are therefore candidates to operate in the same cellular pathway as the query protein. A quantitative score measuring the degree of coinheritance ranks each hit. The confidence is reported for which each protein in the list is linked to the query protein, based upon calibration of scores against known pathways.

Method B: COG

1. Identify the protein of interest in COG. Note the phylogenetic distribution of the corresponding COG category.
2. Search for other COG categories sharing this distribution using the phylogenetic patterns search feature of the COGs database, or by clicking on the phylogenetic pattern itself, which is hyperlinked to a list of COG categories with similar inheritance patterns. Proteins in these COG categories will be strongly coinherited with the query protein, and are therefore predicted to function in the same pathway.

Protein Interaction Databases

Protein interaction data are increasingly available for human proteins as well as for proteins of several model organisms, especially yeast, *Caenorhabditis elegans*, and the bacteria *Helicobacter pylori*. These interaction data can consist of protein-protein, protein–nucleic acid, and protein–small molecule interactions, which are collected in a number of databases that allow a protein of interest to be visualized in the midst of a network of interaction partners.

Protein-protein interaction data are mainly derived from genome-wide, high-throughput yeast two-hybrid experiments in which interactions are measured between all gene pairs in a genome. More than 4000 unique protein interactions were observed between yeast proteins in three large-scale experiments (Ito et al. 2000, 2001; Uetz et al. 2000). More recently, complexes of yeast proteins have been isolated, and the protein constituents have been identified by mass spectrometric approaches (Rain et al. 2001; Gavin et al. 2002; Ho et al. 2002), thereby identifying hundreds of additional interactions among yeast proteins. A test similar to the high-throughput yeast two-hybrid experiments, performed on bacterial proteins, identified more than 1200 interactions between proteins of the human gastric pathogen *H. pylori* (Rain et al. 2001).

In addition to large-scale experimental approaches, a number of groups have been attempting to cull the previously measured protein-protein interactions from the biological literature (Humphreys et al. 2000; Proux et al. 2000; Thomas et al. 2000; Blaschke et al. 2001; Marcotte et al. 2001). This systematic collection of protein interaction data provides necessary checks on the quality of the large-scale interaction data. Large-scale protein interaction data have varied widely in accuracy (von Mering et al. 2002), but many interactions in the databases have been observed by multiple experimental methods, providing some measure of confidence in the correctness of these interactions.

Protein interaction databases that combine the interactions from these large-scale screens with the interactions extracted from the literature include the BIND database and the Database of Interacting Proteins (DIP). As of this writing, DIP (http://dip.doe-mbi.ucla.edu/) currently contains >17,500 interactions between ~6800 proteins, a majority of which are from yeast and the bacterium *H. pylori* (Xenarios et al. 2002). The BIND database (http://www.bind.ca/) includes >6000 interactions, primarily focusing on yeast proteins. These databases provide practical tools for identifying known interaction partners of a protein of interest.

The interaction partners for yeast proteins, and to a lesser extent bacterial and human proteins, can often be found in the interaction databases listed in the panel below. For proteins from other organisms, it is often possible to identify *interologs* (Matthews et al. 2001), which are proteins presumed to interact on the basis of the known interactions of the proteins' orthologs in another genome, as illustrated in Figure 11.5. To search for interologs (see Figure 11.6), first identify the yeast (or *H. pylori* or human) ortholog of the protein, and then search for interactions for this yeast ortholog. Any interaction partners of the ortholog, themselves yeast proteins, would then be subjected to the reverse process of identifying their orthologs in the genome of interest.

Protein-DNA interaction data are also accumulating, especially with the advent of large-scale assays of transcription-factor-binding specificities (Ren et al. 2000; Bulyk et al. 2001; Iyer et al. 2001). Many protein-DNA interactions and binding specificities are cataloged in the TRANSFAC database (http://transfac.gbf.de/TRANSFAC/) and the RegulonDB database (http://www.cifn.unam.mx/Computational_Genomics/regulondb/).

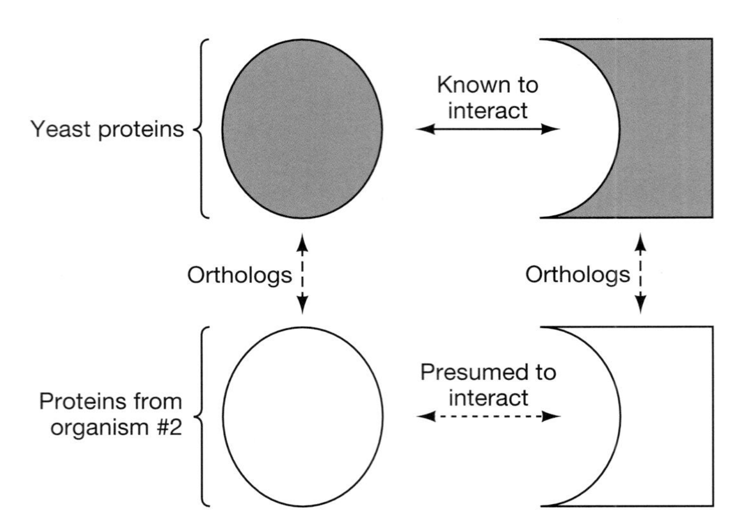

FIGURE 11.6. The experimentally determined protein-protein interactions in one organism can often reveal protein interactions in a second organism. Here, the orthologs of the yeast interaction partners are also likely to interact (Matthews et al. 2001).

URLs of Protein, DNA, and Small Molecule Interaction Databases

BIND	http://www.bind.ca/
BRITE	http://www.genome.ad.jp/brite/
DIP	http://dip.doe-mbi.ucla.edu/
DPInteract	http://arep.med.harvard.edu/dpinteract/
INTERACT	http://www.bioinf.man.ac.uk/resources/interact.shtml
LIGAND	http://www.genome.ad.jp/dbget/ligand.html
MINT	http://cbm.bio.uniroma2.it/mint/
MIPS	http://mips.gsf.de/proj/yeast/CYGD/db/index.html
REBASE	http://rebase.neb.com/rebase/rebase.html
REGULONDB	http://www.cifn.unam.mx/Computational_Genomics/regulondb/
RELIBASE	http://www.ccdc.cam.ac.uk/prods/relibase
TRANSFAC	http://transfac.gbf.de/TRANSFAC/
TRRD	http://wwwmgs.bionet.nsc.ru/mgs/dbases/trrd4/
PIMRider	http://pim.hybrigenics.com/pimriderlobby/current/PimRiderLobby.htm

Metabolic and Signaling Pathway Databases

Beyond cataloging protein interactions, a number of groups are creating databases to store and search known cellular pathways. Such pathway databases include the metabolic pathway databases KEGG (Kanehisa and Goto 2000) and EcoCyc/MetaCyc (Karp et al. 2000), the signal transduction database STKE, and the regulatory database TRANSFAC (Wingender et al. 2001). For proteins whose functions have been characterized in metabolism or signal transduction, these databases allow rapid assessment of the systems in which these proteins operate.

METABOLIC AND SIGNALING PATHWAY DATABASES	
BioCarta	http://www.biocarta.com/
BRITE	http://www.genome.ad.jp/brite/
ECOCYC/METACYC	http://biocyc.org/
EMP	http://emp.mcs.anl.gov/
GeneNet	http://brac.postech.ac.kr/eng/
KEGG	http://www.genome.ad.jp/kegg/
LIGAND	http://www.genome.ad.jp/dbget/ligand.html
MIPS	http://mips.gsf.de/proj/yeast/CYGD/db/index.html
PFBP	http://www.ebi.ac.uk/research/pfbp/
REGULONDB	http://www.cifn.unam.mx/Computational_Genomics/regulondb/
STKE	http://stke.sciencemag.org/
TRRD	http://wwwmgs.bionet.nsc.ru/mgs/dbases/trrd4/
WIT	http://wit.mcs.anl.gov/WIT2/

Protein and Gene Expression Patterns

Due to the prevalence of publicly available large-scale gene expression data sets, quite a bit can be learned about a gene and the conditions under which it is active simply by looking up the gene's expression patterns. These data come from a variety of sources, primarily EST libraries, Serial Analysis of Gene Expression (SAGE) libraries, and DNA microarray data. Currently, more than 11 million sequenced ESTs and SAGE measurements for more than 100 SAGE libraries are available from the dbEST and SAGEmap databases, respectively. These ESTs have been clustered into contigs in the UniGene database, which provides a rapid tool to identify all of the ESTs derived from a given gene. Hundreds of DNA microarray data sets for several different organisms, including yeast and human, are also freely available from the Stanford Microarray Database.

GENE EXPRESSION DATABASES	
dbEST*	http://www.ncbi.nlm.nih.gov/dbEST/index.html
SAGEmap	http://www.ncbi.nlm.nih.gov/SAGE/
UniGene	http://www.ncbi.nlm.nih.gov/UniGene/
Stanford Microarray Database	http://genome-www5.stanford.edu/MicroArray/SMD/

*dbEST can also be searched by BLAST at http://www.ncbi.nlm.nih.gov/BLAST/ and setting the search database to the appropriate EST set.

Beyond simply determining expression patterns of genes, these expression data can often reveal which genes are coexpressed, providing clues about which genes work together in the cell. Searching for coexpressing genes is done most easily with DNA microarray data, which provides measurements of gene expression levels for all genes on a microarray, even low-abundance genes that might be missed from EST or SAGE library-based approaches. Many individual DNA microarray data are available in the public domain, and searching for coexpressing genes requires that these different DNA microarray experiments be analyzed simultaneously. Analyzing for mRNA coexpression is relatively easy to do for all of the genes in a genome as follows:

1. Collect multiple microarray experiments performed on the same organism under differing conditions. Protocol 6 describes one manner in which such data can be gathered.
2. In a spreadsheet, create a table of data with genes along the vertical axis, and the gene's expression levels forming rows in the table (e.g., for the gene expression level measured on spotted microarrays, the investigator might choose the log of the mean value of the normalized ratios of red to green pixels). Write out the data as a tab-delimited text file.
3. Install the freely available Cluster and Tree View software (http://rana.lbl.gov/EisenSoftware.htm).
4. Load the expression data into Cluster, and cluster the data using one of a number of different clustering algorithms, such as hierarchical clustering. View the results in Tree View. Clusters of coexpressing genes can often be identified containing the genes of interest.

PROTOCOL 6

Discovery of Functionally Linked Genes from DNA Microarray Data

THIS PROTOCOL ALLOWS THE INVESTIGATOR TO QUERY the thousands of public DNA microarray experiments in order to find coexpressed genes. To identify the coexpression partners of a small set of genes using publicly available data, the expression data can be downloaded and then analyzed as described above. However, perhaps the easiest approach is to directly query the Stanford Microarray Database, as described in this protocol. The premise of this method is that genes that are coexpressed over a sufficiently large number of experiments are likely to be functionally linked. The precise number of experiments required may vary and is a function of the complexity of the gene's expression patterns, but for yeast genes (and increasingly for human genes), sufficient DNA microarray experiments exist in the public domain to make quite strong inferences.

METHOD

1. Go to the Stanford Microarray Database public advanced search page: http://genome-www5.stanford.edu/cgi-bin/SMD/cluster/QuerySetup.pl.

2. Select the organism of interest (e.g., "*Saccharomyces cerevisiae*"). Under Experimenter, select All. Press the Data retrieval and analysis button.

3. Shift-click to highlight all available experiments, and then press the Data retrieval and analysis button.

4. Activate the radio button labeled All genes. Press the Proceed to data filtering button.

5. Press the Retrieve data button.

   The expression levels for all genes will be retrieved from the microarray experiments selected earlier. At the time of this writing, more than 440 experiments are available for yeast, so this retrieval may be quite a slow process. Although searching all genes is slow, it is required in order to find the most coexpressing genes in a genome.

6. Follow the links to cluster the data.

   The final output will be a hierarchical clustering of all of the genes in the genome by their expression patterns across all available microarray experiments. The raw data files can be downloaded and viewed by the programs cited above, Cluster and Tree View. Given sufficiently complex gene expression data, such as those derived from hundreds of microarray experiments, genes that are clustered with the gene of interest are very likely to operate in the same cellular pathway.

Spatial Expression Patterns

Spatial expression patterns of genes can be quite difficult to collect on a large scale. However, these patterns have been measured for several thousand genes in *Xenopus* embryos (Gawantka et al. 1998; Pollet et al. 2000). Screening with randomly chosen cDNAs, the expression patterns of hundreds of unique mRNAs have been detected by whole-mount in situ hybridization. The expression patterns were photographed and stored in a database (AxelDB). Comparison of the spatial expression patterns of these genes reveals that genes sharing similar spatial expression patterns often have related functional roles. Thus, if a gene of interest occurs in the set of tested genes, searching AxelDB for similar spatial expression patterns may potentially identify functionally linked genes. AxelDB can be searched at http://www.dkfz-heidelberg.de/abt0135/axeldb.htm.

Protein Expression Measurements

Gaining protein expression data is, of course, one of the goals of proteomics. In addition to temporal and spatial gene expression patterns, protein expression data offer the promise of many clues to the functions of proteins. Protein expression data are accumulating more slowly than the interaction and phenotype data. Currently, only a few sets of protein expression data are available outside of the private sector. However, a great deal of research effort is being focused in this direction, and it is likely that these data will rapidly become available. Recent developments in mass spectrometric techniques have allowed the first large-scale measurements of protein expression patterns, using two main approaches described below.

- In the first approach, proteins from two sets of cells are proteolyzed to peptides, and the set of cysteine-containing peptides are purified from each sample with sufhydryl-specific affinity reagents (e.g., ICAT [Gygi et al. 1999b], see Chapter 8, Protocol 9). The specific forms of sulfydryl-specific reagents used for the two samples differ in their isotope distribution, such that the peptides from one set of cells will have a predictable mass difference from those of the other cell sample. These cysteine-containing peptides from the two samples are mixed, and then analyzed by mass spectrometry (MS). Because of the mass differences associated with the peptides, the relative abundance of a given peptide in the two cell samples can be measured as their relative peak heights in the mass spectra. This approach has yielded expression measurements of hundreds of proteins (Gygi et al. 1999b; Conrads et al. 2000); the specific details of this approach and the second approach to follow are described in Chapters 1 and 8.
- In the second general approach using MS, peptides generated from cell extracts have been identified based on capillary electrophoresis elution times and high-resolution mass measurements from a mass spectrometer capable of high mass accuracy, such as Fourier transform-ion cyclotron resonance MS (Conrads et al. 2000; Jensen et al. 2000; see Chapter 8). As with the first technique, isotope differences can be exploited in order to measure protein expression level changes between two samples. In this second approach, isotope labeling has been achieved by growing cells of the two samples in media with distinct isotopic distributions. As in the first technique, expression levels can be measured for thousands of peptides.

In both MS approaches, peptides must be mapped to their parent proteins in a database search. In two large-scale tests of protein expression levels, measured protein expression levels are only poorly correlated with mRNA expression levels, suggesting that protein expression data will be of significant value in complementing available mRNA expression measurements (Gygi et al. 1999a; Ideker et al. 2001).

Despite this progress on mass spectrometric proteomics, the majority of public protein expression currently derives from two-dimensional SDS-PAGE gels (e.g., see Chapter 4). These data are not currently organized for convenient searching of expression levels and patterns across many experiments. However, at least one database, Swiss2DPAGE, reports the measured expression patterns for ~800 proteins derived from a small set of organisms. Swiss2DPAGE can be accessed at http://ca.expasy.org/ch2d/.

Mutant Phenotypes

Just as with interaction and pathway data, phenotypic data are accumulating at a rapid rate for thousands of genes. These data can be of use in proteomics in many different ways, but perhaps the most important will be by providing additional validation for proteomics experiments. For example, one expects that proteins functioning in the same pathway should have generally similar or related phenotypes when their corresponding genes are disrupted. Therefore, this becomes a quantitative test that can be applied to any particular set of proteins linked or coexpressed in a proteomics experiment.

In several model organisms, including yeast and *C. elegans*, genes have been systematically disrupted, and the effects of these gene disruptions have been observed. Recent data of this sort include the measurement of disruption phenotypes of more than half of the genes of yeast (Ross-Macdonald et al. 1999; Winzeler et al. 1999), the transposon mutagenesis of most of the genes of *Mycoplasma genitalium* (Hutchison et al. 1999), and the RNAi inhibition of hundreds of genes of *C. elegans* (Fraser et al. 2000; Gonczy et al. 2000; Maeda et al. 2001).

Gene disruption data can be accessed in a number of different ways. If the gene of interest is derived from one of the organisms with an extensive knockout database, it is a simple matter of finding the gene in the database. As with protein interaction interologs, it is possible to extend gene-knockout phenotypes measured in one organism to the orthologs of the genes in a second organism. The standard cautions apply—this is an unproven approach, and one is taking something of a leap of faith. However, it is possible to generate reasonable hypotheses about the potential phenotypes of a disrupted gene in this manner.

Gene disruption data for yeast is most easily accessible from the SGD database (http://genome-www.stanford.edu/Saccharomyces/) or (through subscription) the YPD database (https://www.incyte.com/proteome/databases.jsp). Gene disruption data for *M. genitalium* are available at http://www.sciencemag.org/feature/data/1042937.shl. Finally, gene disruption data for *C. elegans* are available on the Internet in the supplemental information of Fraser et al. (2000) and Gonczy et al. (2000), in the Nematode Expression Pattern Database (http://nematode.lab.nig.ac.jp/), and (by subscription) in WormPD (https://www.incyte.com/proteome/databases.jsp). These phenotypic data are organized for a number of other organisms, such as *Drosophila*, in FlyBase (http://flybase.bio.indiana.edu/), and mice, in MGD (http://www.informatics.jax.org/).

Predicting the Subcellular Localization of a Protein

Finally, it is often of great interest to know not only the function of a protein, but also where that function is performed. Although the subcellular localization may be directly measurable (e.g., for the preparation of specific subcellular fractions, see Chapter 3), subcellular localization can also be predicted to some extent. Two programs, PSORT (Nakai 2000) and SignalP (Emanuelsson et al. 2000), are commonly used to predict protein cellular localization and secretion. SignalP is specifically trained to identify amino-terminal secretory signals. A sequence can be submitted to SignalP with META-PP (mentioned above). The SignalP output is thoroughly described at http://www.cbs.dtu.dk/services/SignalP/output.html. PSORT is a more general program, predicting localization to many different subcellular compartments. PSORT and PSORTII are available for use at http://psort.nibb.ac.jp/. The Web Site maintains an excellent manual clearly describing the output.

Additional Resources

A number of excellent texts are available for in-depth discussions of the topics in this chapter.

- Books

  For general discussion of bioinformatics approaches, appropriate starting texts include:

  Bioinformatics: Sequence and Genome Analysis by David Mount (Cold Spring Harbor Laboratory Press)

  Bioinformatics: A Practical Guide to the Analysis of Genes and Proteins, edited by Baxevanis and Ouellette (Wiley)

  Introduction to Bioinformatics, edited by Attwood and Parry-Smith (Prentice Hall)

  Proteome research: New Frontiers in Functional Genomics, edited by Wilkins, Williams, Appel, and Hochstrasser (Springer-Verlag)

  For a more detailed look at the algorithms involved, we recommend:

  Biological Sequence Analysis by Durbin and colleagues (Cambridge University Press)

  Computational Molecular Biology by Pavel Pevzner (MIT Press)

  For extensive discussion of protein interactions and experimental protocols, we recommend:

  Protein-Protein Interactions, edited by Erica Golemis (Cold Spring Harbor Laboratory Press)

  For general information about protein structure and function, we recommend:

  Proteins by T.E. Creighton (W.H. Freedman)

- Web Sites

  Several excellent on-line tutorials are available. We recommend starting with:

  http://www.biochem.ucl.ac.uk/bsm/dbbrowser/jj/prefacefrm.html

REFERENCES

Altschul S.F. and Koonin E.V. 1998. Iterated profile searches with PSI-BLAST—A tool for discovery in protein databases. *Trends Biochem. Sci.* **23:** 444–447.

Altschul S.F., Gish W., Miller W., Myers E.W., and Lipman D.J. 1990. Basic local alignment search tool. *J. Mol. Biol.* **215:** 403–410.

Altschul S.F., Madden T.L., Schaffer A.A., Zhang J., Zhang Z., Miller W., and Lipman D.J. 1997. Gapped-BLAST and PSI-BLAST: A new generation of protein database search programs. *Nucleic Acids Res.* **25:** 3389–3402.

Attwood T.K., Blythe M.J., Flower D.R., Gaulton A., Mabey J.E., Maudling N., McGregor L., Mitchell A.L., Moulton G., Paine K., and Scordis P. 2002. PRINTS and PRINTS-S shed light on protein ancestry. *Nucleic Acids Res.* **30:** 239–241.

Baldi P., Brunak S., Frasconi P., Soda G., and Pollastri G. 1999. Exploiting the past and the future in protein secondary structure prediction. *Bioinformatics* **15:** 937–946.

Bateman A., Birney E., Cerruti L., Durbin R., Etwiller L., Eddy S.R., Griffiths-Jones S., Howe K.L., Marshall M., and Sonnhammer E.L. 2002. The Pfam protein families database. *Nucleic Acids Res.* **30:** 276–280.

Blaschke C., Oliveros J.C., and Valencia A. 2001. Mining functional information associated with expression arrays. *Funct. Integr. Genomics* **1:** 256–268.

Blom N., Gammeltoft S., and Brunak S. 1999. Sequence and structure-based prediction of eukaryotic protein phosphorylation sites. *J. Mol. Biol.* **294:** 1351–1362.

Blom N., Hansen J., Blaas D., and Brunak S. 1996. Cleavage site analysis in picornaviral polyproteins: Discovering cellular targets by neural networks. *Protein Sci.* **5:** 2203–2216.

Bork P., Dandekar T., Diaz-Lazcoz Y., Eisenhaber F., Huynen M., and Yuan Y. 1998. Predicting function: From genes to genomes and back. *J. Mol. Biol.* **283:** 707–725.

Bujnicki J.M., Elofsson A., Fischer D., and Rychlewski L. 2001. Structure prediction meta server. *Bioinformatics* **17:** 750–751.

Bulyk M.L., Huang X., Choo Y., and Church G.M. 2001. Exploring the DNA-binding specificities of zinc fingers with DNA microarrays. *Proc. Natl. Acad. Sci.* **98:** 7158–7163.

Claros M.G. and von Heijne G. 1994. TopPred II: An improved software for membrane protein structure predictions. *Comput. Appl. Biosci.* **10:** 685–686.

Conrads T.P., Anderson G.A., Veenstra T.D., Pasa-Tolic L., and Smith R.D. 2000. Utility of accurate mass tags for proteome-wide protein identification. *Anal. Chem.* **72:** 3349–3354.

Corpet F., Servant F., Gouzy J., and Kahn D. 2000. ProDom and ProDom-CG: Tools for protein domain analysis and whole genome comparisons. *Nucleic Acids Res.* **28:** 267–269.

Cserzo M., Wallin E., Simon I., von Heijne G., and Elofsson A. 1997. Prediction of transmembrane alpha-helices in prokaryotic membrane proteins: The dense alignment surface method. *Protein Eng.* **10:** 673–676.

Cuff J.A., Clamp M.E., Siddiqui A.S., Finlay M., and Barton G.J. 1998. JPred: A consensus secondary structure prediction server. *Bioinformatics* **14:** 892–893.

Dandekar T., Snel B., Huynen M., and Bork P. 1998. Conservation of gene order: A fingerprint of proteins that physically interact. *Trends Biochem. Sci.* **23:** 324–328.

Dongre A.R., Eng J.K., and Yates J.R. III. 1997. Emerging tandem-mass-spectrometry techniques for the rapid identification of proteins. *Trends Biotechnol.* **15:** 418–425.

Eisen J.A. 1998. A phylogenomic study of the MutS family of proteins. *Nucleic Acids Res.* **26:** 4291–4300.

Emanuelsson O., Nielsen H., and von Heijne G. 1999. ChloroP, a neural network-based method for predicting chloroplast transit peptides and their cleavage sites. *Protein Sci.* **8:** 978–984.

Emanuelsson O., Nielsen H., Brunak S., and von Heijne G. 2000. Predicting subcellular localization of proteins based on their N-terminal amino acid sequence. *J. Mol. Biol.* **300:** 1005–1016.

Enright A.J., Iliopoulos I., Kyrpides N.C., and Ouzounis C.A. 1999. Protein interaction maps for complete genomes based on gene fusion events. *Nature* **402:** 86–90.

Falquet L., Pagni M., Bucher P., Hulo N., Sigrist C.J., Hofmann K., and Bairoch A. 2002. The PROSITE database, its status in 2002. *Nucleic Acids Res.* **30:** 235–238.

Fischer D., Barret C., Bryson K., Elofsson A., Godzik A., Jones D., Karplus K.J., Kelley L.A., MacCallum R.M.,

Pawowski K., Rost B., Rychlewski L., and Sternberg M. 1999. CAFASP-1: Critical assessment of fully automated structure prediction methods. *Proteins* (suppl.) **37:** 209-217.

Fitch W.M. 1970. Distinguishing homologous from analogous proteins. *Syst. Zool.* **19:** 99–113.

Fraser A.G., Kamath R.S., Zipperlen P., Martinez-Campos M., Sohrmann M., and Ahringer J. 2000. Functional genomic analysis of *C. elegans* chromosome I by systematic RNA interference. *Nature* **408:** 325–330.

Gavin A.C., Bosche M., Krause R., Grandi P., Marzioch M., Bauer A., Schultz J., Rick J.M., Michon A.M., Cruciat C.M., Remor M., Hofert C., Schelder M., Brajenovic M., Ruffner H., Merino A., Klein K., Hudak M., Dickson D., Rudi T., Gnau V., Bauch A., Bastuck S., Huhse B., Leutwein C., Heurtier M.A., Copley R.R., Edelmann A., Querfurth E., Rybin V., Drewes G., Raida M., Bouwmeester T., Bork P., Seraphin B., Kuster B., Neubauer G., and Superti-Furga G. 2002. Functional organization of the yeast proteome by systematic analysis of protein complexes. *Nature* **415:** 141–147.

Gawantka V., Pollet N., Delius H., Vingron M., Pfister R., Nitsch R., Blumenstock C., and Niehrs C. 1998. Gene expression screening in *Xenopus* identifies molecular pathways, predicts gene function and provides a global view of embryonic patterning. *Mech. Dev.* **77:** 95–141.

George R.A. and Heringa J. 2002. SnapDRAGON: A method to delineate protein structural domains from sequence data. *J. Mol. Biol.* **316:** 839–851.

Gonczy P., Echeverri G., Oegema K., Coulson A., Jones S.J., Copley R.R., Duperon J., Oegema J., Brehm M., Cassin E., Hannak E., Kirkham M., Pichler S., Flohrs K., Goessen A., Leidel S., Alleaume A.M., Martin C., Ozlu N., Bork P., and Hyman A.A. 2000. Functional genomic analysis of cell division in *C. elegans* using RNAi of genes on chromosome III. *Nature* **408:** 331–336.

Gupta R., Birch H., Rapacki K., Brunak S., and Hansen J.E. 1999. O-GLYCBASE version 4.0: A revised database of O-glycosylated proteins. *Nucleic Acids Res.* **27:** 370–372.

Gygi S.P., Rochon Y., Franza B.R., and Aebersold R. 1999a. Correlation between protein and mRNA abundance in yeast. *Mol. Cell. Biol.* **19:** 1720–1730.

Gygi S.P., Rist B., Gerber S.A., Turecek F., Gelb M.H., and Aebersold R. 1999b. Quantitative analysis of complex protein mixtures using isotope-coded affinity tags. *Nat. Biotechnol.* **17:** 994–999.

Henikoff S., Henikoff J.G., and Pietrokovski S. 1999. Blocks+: A non-redundant database of protein alignment blocks derived from multiple compilations. *Bioinformatics* **15:** 471–479.

Ho Y., Gruhler A., Heilbut A., Bader G.D., Moore L., Adams S.L., Millar A., Taylor P., Bennett K., Boutilier K., Yang L., Wolting C., Donaldson I., Schandorff S., Shewnarane J., Vo M., Taggart J., Goudreault M., Muskat B., Alfarano C., Dewar D., Lin Z., Michalickova K., Willems A.R., Sassi H., Nielsen P.A., Rasmussen K.J., Andersen J.R., Johansen L.E., Hansen L.H., Jespersen H., Podtelejnikov A., Nielsen E., Crawford J., Poulsen V., Sorensen B.D., Matthiesen J., Hendrickson R.C., Gleeson F., Pawson T., Moran M.F., Durocher D., Mann M., Hogue C.W., Figeys D., and Tyers M. 2002. Systematic identification of protein complexes in *Saccharomyces cerevisiae* by mass spectrometry. *Nature* **415:** 180–183.

Hollingsworth N.M., Ponte L., and Halsey C. 1995. MSH5, a novel MutS homolog, facilitates meiotic reciprocal recombination between homologs in *Saccharomyces cerevisiae* but not mismatch repair. *Genes Dev.* **9:** 1728–1739.

Huang J.Y. and Brutlag D.L. 2001. The EMOTIF database. *Nucleic Acids Res.* **29:** 202–204.

Humphreys K., Demetriou G., and Gaizauskas R. 2000. Two applications of information extraction to biological science journal articles: Enzyme interactions and protein structures. *Pac. Symp. Biocomput.* **5:** 505–516.

Hutchison C.A., Peterson S.N., Gill S.R., Cline R.T., White O., Fraser C.M., Smith H.O., and Venter J.C. 1999. Global transposon mutagenesis and a minimal Mycoplasma genome. *Science* **286:** 2165–2169.

Huynen M., Snel B., Lathe W. III, and Bork P. 2000. Predicting protein function by genomic context: Quantitative evaluation and qualitative inferences. *Genome Res.* **10:** 1204–1210.

Ideker T., Thorsson V., Ranish J.A., Christmas R., Buhler J., Eng J.K., Bumgarner R., Goodlett D.R., Aebersold R., and Hood L. 2001. Integrated genomic and proteomic analyses of a systematically perturbed metabolic network. *Science* **292:** 929–934.

Ito T., Chiba T., Ozawa R., Yoshida M., Hattori M., and Sakaki Y. 2001. A comprehensive two-hybrid analysis to explore the yeast protein interactome. *Proc. Natl. Acad. Sci.* **98:** 4569–4574.

Ito T., Tashiro K., Muta S., Ozawa R., Chiba T., Nishizawa M., Yamamoto K., Kuhara S., and Sakaki Y. 2000. Toward a protein-protein interaction map of the budding yeast: A comprehensive system to examine two-hybrid interactions in all possible combinations between the yeast proteins. *Proc. Natl. Acad. Sci.* **97:** 1143–1147.

Iyer V.R., Horak C.E., Scafe C.S., Botstein D., Snyder M., and Brown P.O. 2001. Genomic binding sites of the yeast cell-cycle transcription factors SBF and MBF. *Nature* **409:** 533–558.

Jensen P.K., Pasa-Tolic L., Peden K.K., Martinovic S., Lipton M.S., Anderson G.A., Tolic N., Wong K.K., and Smith R.D. 2000. Mass spectrometric detection for capillary isoelectric focusing separations of complex protein mixtures. *Electrophoresis* **21:** 1372–1380.

Johnson R.E., Kovvali G.K., Prakash L., and Prakash S. 1996. Requirement of the yeast MSH3 and MSH6 genes for MSH2-dependent genomic stability. *J. Biol. Chem.* **271:** 7285–7288.

Jones D.T. and Swindells M.B. 2002. Getting the most from PSI-BLAST. *Trends Biochem. Sci.* **27:** 161–164.

Kanehisa M. and Goto S. 2000. KEGG: Kyoto encyclopedia of genes and genomes. *Nucleic Acids Res.* **28:** 27–30.

Karp P.D., Riley M., Saier M., Paulsen I.T., Paley S.M., and Pellegrini-Toole A. 2000. The EcoCyc and MetaCyc databases. *Nucleic Acids Res.* **28:** 56–59.

Krogh A., Larsson B., von Heijne G., and Sonnhammer E.L. 2001. Predicting transmembrane protein topology with a hidden Markov model: Application to complete genomes. *J. Mol. Biol.* **305:** 567–580.

Maeda I., Kohara Y., Yamamoto M., and Sugimoto A. 2001. Large-scale analysis of gene function in *Caenorhabditis elegans* by high-throughput RNAi. *Curr. Biol.* **11:** 171–176.

Marcotte E.M. 2000. Computational genetics: Finding protein function by nonhomology methods. *Curr. Opin. Struct. Biol.* **10:** 359–365.

Marcotte E.M., Xenarios I., and Eisenberg D. 2001. Mining literature for protein-protein interactions. *Bioinformatics* **17:** 359–363.

Marcotte E.M., Pellegrini M., Ng H.L., Rice D.W., Yeates T.O., and Eisenberg D. 1999. Detecting protein function and protein-protein interactions from genome sequences. *Science* **285:** 751–753.

Matthews L.R., Vaglio P., Reboul J., Ge H., Davis B.P., Garrels J., Vincent S., and Vidal M. 2001. Identification of potential interaction networks using sequence-based searches for conserved protein-protein interactions or "interologs". *Genome Res.* **11:** 2120–2126.

McGuffin L.J., Bryson K., and Jones D.T. 2000. The PSIPRED protein structure prediction server. *Bioinformatics* **16:** 404–405.

Moller S., Croning M.D., and Apweiler R. 2001. Evaluation of methods for the prediction of membrane spanning regions. *Bioinformatics* **17:** 646-653.

Moult J., Fidelis K., Zemla A., and Hubbard T. 2001. Critical assessment of methods of protein structure prediction (CASP): Round IV. *Proteins* (suppl. 5) **45:** 2-7.

Mount D.W. 2001. *Bioinformatics: Sequencing and genome analysis.* Cold Spring Harbor Laboratory Press, Cold Spring Harbor, New York.

Nakai K. 2000. Protein sorting signals and prediction of subcellular localization. *Adv. Protein Chem.* **54:** 277–344.

Ouali M. and King R.D. 2000. Cascaded multiple classifiers for secondary structure prediction. *Protein Sci.* **9:** 1162–1176.

Overbeek R., Fonstein M., D'Souza M., Pusch G.D., and Maltsev N. 1999. The use of gene clusters to infer functional coupling. *Proc. Natl. Acad. Sci.* **96:** 2896–2901.

Peitsch M.C. 1996. ProMod and Swiss-Model: Internet-based tools for automated comparative protein modelling. *Biochem. Soc. Trans.* **24:** 274–279.

Pellegrini M., Marcotte E.M., Thompson M.J., Eisenberg D., and Yeates T.O. 1999. Assigning protein functions by comparative genome analysis: Protein phylogenetic profiles. *Proc. Natl. Acad. Sci.* **96:** 4285–4288.

Pieper U., Eswar N., Stuart A.C., Ilyin V.A., and Sali A. 2002. MODBASE, a database of annotated comparative protein structure models. *Nucleic Acids Res.* **30:** 255–259.

Pollet N., Schmidt H.A., Gawantka V., Vingron M., and Niehrs C. 2000. Axeldb: A *Xenopus laevis* database focusing on gene expression. *Nucleic Acids Res.* **28:** 139–140.

Proux D., Rechenmann F., and Julliard L. 2000. A pragmatic information extraction strategy for gathering data on genetic interactions. *Proc. Int. Conf. Intell. Syst. Mol. Biol.* **8:** 279–285.

Rain J.C., Selig L., De Reuse H., Battaglia V., Reverdy C., Simon S., Lenzen G., Petel F., Wojcik J., Schachter V., Chemama Y., Labigne A., and Legrain P. 2001. The protein-protein interaction map of *Helicobacter pylori*. *Nature* **409:** 211–215.

Ren B., Robert F., Wyrick J.J., Aparicio O., Jennings E.G., Simon I., Zeitlinger J., Schreiber J., Hannett N., Kanin E., Volkert T.L., Wilson C.J., Bell S.P., and Young R.A. 2000. Genome-wide location and function of DNA binding proteins. *Science* **290:** 2306–2309.

Ross-Macdonald P., Coelho P.S., Roemer T., Agarwal S., Kumar A., Jansen R., Cheung K.H., Sheehan A., Symoniatis D., Umansky L., Heidtman M., Nelson F.K., Iwasaki H., Hager K., Gerstein M., Miller P., Roeder G.S., and Snyder M. 1999. Large-scale analysis of the yeast genome by transposon tagging and gene disruption. *Nature* **402:** 413–418.

Rost B. and Sander C. 1994. Combining evolutionary information and neural networks to predict protein secondary structure. *Proteins* **19:** 55–72.

Rost B., Fariselli P., and Casadio R. 1996. Topology prediction for helical transmembrane proteins at 86% accuracy. *Protein Sci.* **5:** 1704–1718.

Salgado H., Moreno-Hagelsieb G., Smith T.F., and Collado-Vides J. 2000. Operons in *Escherichia coli*: Genomic analyses and predictions. *Proc. Natl. Acad. Sci.* **97:** 6652–6657.

Snel B., Lehmann G., Bork P., and Huynen M.A. 2000. STRING: A web-server to retrieve and display the repeatedly occurring neighbourhood of a gene. *Nucleic Acids Res.* **28:** 3442–3444.

Tamames J., Casari G., Ouzounis C., and Valencia A. 1997. Conserved clusters of functionally related genes in two bacterial genomes. *J. Mol. Evol.* **44:** 66–73.

Tatusov R.L., Natale D.A., Garkavtsev I.V., Tatusova T.A., Shankavaram U.T., Rao B.S., Kiryutin B., Galperin M.Y., Fedorova N.D., and Koonin E.V. 2001. The COG database: New developments in phylogenetic classification of proteins from complete genomes. *Nucleic Acids Res.* **29:** 22–28.

Thomas J., Milward D., Ouzounis C., Pulman S., and Carroll M. 2000. Automatic extraction of protein interactions from scientific abstracts. *Pac. Symp. Biocomput.* **5:** 538–549.

Thompson J.D., Gibson T.J., Plewniak F., Jeanmougin F., and Higgins D.G. 1997. The CLUSTAL_X windows interface: Flexible strategies for multiple sequence alignment aided by quality analysis tools. *Nucleic Acids Res.* **25:** 4876–4882.

Thornton J.W. and DeSalle R. 2000. Gene family evolution and homology: Genomics meets phylogenetics. *Annu. Rev. Genomics Hum. Genet.* **1:** 41–73.

Uetz P., Giot L., Cagney G., Mansfield T.A., Judson R.S., Knight J.R., Lockshon D., Narayan V., Srinivasan M., Pochart P., Qureshi-Emili A., Li Y., Godwin B., Conover D., Kalbfleisch T., Vijayadamodar G., Yang M., Johnston M., Fields S., and Rothberg J.M. 2000. A comprehensive analysis of protein-protein interactions in *Saccharomyces cerevisiae. Nature* **403:** 623–627.

Vlahovicek K., Murvai J., Barta E., and Pongor S. 2002. The SBASE protein domain library, release 9.0: An online resource for protein domain identification. *Nucleic Acids Res.* **30:** 273–275.

von Mering C., Krause R., Snel B., Cornell M., Oliver S.G., Fields S., and Bork P. 2002. Comparative assessment of large-scale data sets of protein-protein interactions. *Nature* **417:** 399–403.

Wheelan S.J., Marchler-Bauer A., and Bryant S.H. 2000. Domain size distributions can predict domain boundaries. *Bioinformatics* **16:** 613–618.

Wingender E., Chen X., Fricke E., Geffers R., Hehl R., Liebich I., Krull M., Matys V., Michael H., Ohnhauser R., Pruss M., Schacherer F., Thiele S., and Urbach S. 2001. The TRANSFAC system on gene expression regulation. *Nucleic Acids Res.* **29:** 281–283.

Winzeler E.A., Shoemaker D.D., Astromoff A., Liang H., Anderson K., Andre B., Bangham R., Benito R., Boeke J.D., Bussey H., Chu A.M., Connelly C., Davis K., Dietrich F., Dow S.W., El Bakkoury M., Foury F., Friend S.H., Gentalen E., Giaever G., Hegemann J.H., Jones T., Laub M., Liao H., Davis R.W., et al. 1999. Functional characterization of the *S. cerevisiae* genome by gene deletion and parallel analysis. *Science* **285:** 901–906.

Xenarios I., Salwinski L., Duan X.J., Higney P., Kim S.M., and Eisenberg D. 2002. DIP, the Database of Interacting Proteins: A research tool for studying cellular networks of protein interactions. *Nucleic Acids Res.* **30:** 303–305.

WWW RESOURCES

http://apropos.icmb.utexas.edu/function-prediction.html Protein Link Explorer (PLEX), University of Texas.

http://bioinfo.pl/cafasp/ CAFASP (Critical Assessment of Fully Automated Structure Prediction).

http://bioinfo.pl/LiveBench/ The Live Bench Project—Continuous benchmarking of protein structure prediction servers.

http://ca.expasy.org/ch2d/ SWISS-2Dpage—Two-dimensional polyacrylamide gel electrophoresis database.

http://ca.expasy.org/sprot/ SWISS-PROT protein knowledgebase.
http://cubic.bioc.columbia.edu/eva/sec/common.html EVA summary
http://dip.doe-mbi.ucla.edu/ The Database of Interacting Proteins.
http://flybase.bio.indiana.edu/ The FlyBase—A Database of the *Drosophila* Genome.
http://genome-www5.stanford.edu/cgi-bin/SMD/cluster/QuerySetup.pl The Stanford Microarray Database.
http://genome-www.stanford.edu/Saccharomyces/ The *Saccharomyces* Genome Database (SGD).
http://maple.bioc.columbia.edu/pp/submit_meta.html The META Server.
http://mathbio.nimr.mrc.ac.uk/~rgeorge/snapdragon/ SnapDRAGON domain boundary prediction.
http://nematode.lab.nig.ac.jp/ NEXTDB (The Nematode Expression Pattern DataBase).
http://pfam.wustl.edu/hmmsearch.shtml Pfam HHM database, Washington University in St. Louis.
http://pipe.rockefeller.edu/modbase-cgi/index.cgi ModBase—Database of comparative protein structure models.
http://predictioncenter.llnl.gov/ Protein Structure Prediction Center, Lawrence Livermore National Laboratory.
http://prodes.toulouse.inra.fr/prodom/doc/prodom.html The Protein Domain Database—ProDom.
http://psort.nibb.ac.jp/ PSORT (Prediction of Protein Sorting Signals and Localization Sites in Amino Acid Sequences).
http://rana.lbl.gov/EisenSoftware.htm The Eisen Lab Software page.
http://scop.berkeley.edu SCOP (Structural Classification of Proteins).
http://smart.embl-heidelberg.de/ SMART sequence database.
http://transfac.gbf.de/TRANSFAC TRANSFAC (The Transcription Factor Database).
http://wit.mcs.anl.gov/WIT2/ WIT: Interactive Metabolic Reconstruction.
http://xray.bmc.uu.se/sbnet/prosal.html ProSal (Protein Sequence Analysis Launcher).
http://www.bind.ca/ The Biomolecular Interaction Network Database (BIND).
http://www.biochem.ucl.ac.uk/bsm/cath/_new/index.html CATH_Protein Structure Classification.
http://www.blocks.fhcrc.org/blocks_search.html BLOCKS searcher.
http://www.bmm.icnet.uk/~3djigsaw/dom_fish/ Domain Fishing 1.0.
http://www.bork.embl-heidelberg.de/STRING/ STRING (Search Tool for Recurring Instances of Neighbouring Genes).
http://www.cbs.dtu.dk/services/SignalP/output.html CBS SignalP V1.1, World Wide Web Prediction Server, Center for Biological Sequence Analysis.
http://www.cifn.unam.mx/Computational_Genomics/regulondb/ RegulonDB v3.2: A Database on Transcriptional Regulation and Genome Organization.
http://www.dkfz-heidelberg.de/abt0135/axeldb.htm The Axeldb database home page.
http://www.doe-mbi.ucla.edu/Services/moment/ MOMENT Transmembrane Helix Prediction.
http://www2.ebi.ac.uk/dali/fssp/fssp.html The FSSP database, EMBL-EBI.
http://www.expasy.ch/swissmod/SWISS-MODEL.html SWISS MODEL.
http://hydra.icgeb.trieste.it/~kristian/SBASE/ BLAST-FTHOM database.
https://www.incyte.com/proteome/databases.jsp The Proteome BioKnowledge® Library, Incyte Genomics.
http://www.informatics.jax.org/ Mouse Genome Informatics.
http://www.ncbi.nlm.nih.gov The National Center for Biotechnology Information (NCBI) home page.
http://www.ncbi.nlm.nih.gov/BLAST/ The NCBI BLAST page.
http://www.ncbi.nlm.nih.gov/cgi-bin/Entrez/altik?gi=115&db=Genome *Escherichia coli* K12, complete genome, NCBI.
http://www.ncbi.nlm.nih.gov/cgi-bin/Entrez/framik?db=G&gi=115 The NCBI Entrez Genome page.
http://www.ncbi.nlm.nih.gov/COG/ The NCBI Clusters of Orthologous Groups of proteins (COGs) page.
http://www.ncbi.nlm.nih.gov/Education/BLASTinfo/information3.html The NCBI BLAST Information Guide.
http://www.ncbi.nlm.nih.gov/PMGifs/Genomes/org.html The NCBI Prominent Organisms Taxonomy/List.
http://www.ncbi.nlm.nih.gov/Structure/dgs/DGSWeb.cgi The NCBI DGS Algorithm page.
http://www.rcsb.org Protein Data Bank (PDB).
http://www.sciencemag.org/feature/data/1042937.shl Hutchison C.A. et al. 1999. Global transposon mutagenesis and a minimal mycoplasma genome.

Reference Tables

REFERENCE TABLES

TABLE A1.1. Metric prefixes

Prefix		Numerical equivalent
yotta	(Y)	10^{24}
zetta	(Z)	10^{21}
exa	(E)	10^{18}
peta	(P)	10^{15}
tera	(T)	10^{12}
giga	(G)	10^{9}
mega	(M)	10^{6}
kilo	(k)	10^{3}
hecto	(h)	10^{2}
deka	(da)	10
deci	(d)	10^{-1}
centi	(c)	10^{-2}
milli	(m)	10^{-3}
micro	(μ)	10^{-6}
nano	(n)	10^{-9}
pico	(p)	10^{-12}
femto	(f)	10^{-15}
atto	(a)	10^{-18}
zepto	(z)	10^{-21}
yocto	(y)	10^{-24}

TABLE A1.2. Pressure conversion table

kPa	psi	bar	MPa
690	100	6.9	0.7
3450	500	34.5	3.4
6890	1000	68.9	6.9
10340	1500	103.4	10.3
13790	2000	137.9	13.8
17240	2500	172.4	17.2
20680	3000	206.8	20.7
24130	3500	241.3	24.1
27580	4000	275.8	27.6
31030	4500	310.3	31.0
34470	5000	344.7	34.5
37920	5500	379.2	37.9
41370	6000	413.7	41.4
44820	6500	448.2	44.8
48260	7000	482.6	48.3
51710	7500	517.1	51.7
55160	8000	551.6	55.2
58610	8500	586.1	58.6
62050	9000	620.5	62.1
65500	9500	655.0	65.5
68950	10000	689.5	68.9

TABLE A1.3. Moles/weight conversion table

Molecular mass (daltons)	1 μg	1 nmole
100	10 nmoles or 6 x 10^{15} molecules	0.1 μg
1,000	1 nmole or 6 x 10^{14} molecules	1 μg
10,000	100 pmoles or 6 x 10^{13} molecules	10 μg
20,000	50 pmoles or 3 x 10^{13} molecules	20 μg
30,000	33 pmoles or 2 x 10^{13} molecules	30 μg
40,000	25 pmoles or 1.5 x 10^{13} molecules	40 μg
50,000	20 pmoles or 1.2 x 10^{13} molecules	50 μg
60,000	17 pmoles or 1.0 x 10^{13} molecules	60 μg
70,000	14 pmoles or 8.4 x 10^{12} molecules	70 μg
80,000	12 pmoles or 7.2 x 10^{12} molecules	80 μg
90,000	11 pmoles or 6.6 x 10^{12} molecules	90 μg
100,000	10 pmoles or 6 x 10^{12} molecules	100 μg
120,000	8.3 pmoles or 5 x 10^{12} molecules	120 μg
140,000	7.1 pmoles or 4.3 x 10^{12} molecules	140 μg
160,000	6.3 pmoles or 3.8 x 10^{12} molecules	160 μg
180,000	5.6 pmoles or 3.3 x 10^{12} molecules	180 μg
200,000	5 pmoles or 3 x 10^{12} molecules	200 μg

TABLE A1.4. Selected buffers and their pK values at 25°C

Trivial name	Buffer name	pK_a	$\Delta pK_a/\Delta t$[a]
Phosphate (pK_1)	–	2.15	0.0044
Malate (pK_1)	–	3.40	–
Formate	–	3.75	0.0
Succinate (pK_1)	–	4.21	–0.0018
Citrate (pK_2)	–	4.76	–0.0016
Acetate	–	4.76	0.0002
Malate	–	5.13	–
Pyridine	–	5.23	–0.014
Succinate (pK_2)	–	5.64	0.0
MES	2-(*N*-Morpholino)ethanesulfonic acid	6.10	–0.011
Cacodylate	Dimethylarsinic acid	6.27	–
Dimethylglutarate	3,3-Dimethylglutarate (pK_2)	6.34	0.0060
Carbonate (pK_1)	–	6.35	–0.0055
Citrate (pK_3)	–	6.40	0.0
BIS-Tris	[Bis-(2-hydroxyethyl)imino]tris(hydroxymethyl) methane	6.46	0.0
ADA	*N*-2-Acetamidoiminodiacetic acid	6.59	–0.011
Pyrophosphate	–	6.60	–
EDPS (pK_1)	*N,N*′-Bis(3-sulfopropyl)ethylenediamine	6.65	–
Bis-Tris propane	1,3-Bis[tris(hydroxymethyl)methylamino]propane	6.80	–
PIPES	Piperazine-*N,N*′-bis(2-ethanesulfonic acid)	6.76	–0.0085
ACES	*N*-2-Acetamido-2-aminoethanesulfonic acid	6.78	–0.020
MOPSO	3-(*N*-Morpholino)-2-hydroxypropanesulfonic acid	6.95	–0.015
Imidazole	–	6.95	–0.020
BES	*N,N*′-Bis-(2-hydroxyethyl)2-aminoethanesulfonic acid	7.09	–0.016
MOPS	3-(*N*-Morpholino)propanesulfonic acid	7.20	–0.015
Phosphate (pK_2)	–	7.20	–0.0028
EMTA	3,6-Endomethylene-1,2,3,6-tetrahydrophthalic acid	7.23	–
TES	2-[Tris(hydroxymethyl)methylamino]ethanesulfonic acid	7.40	–0.020
HEPES	*N*-2-Hydroxyethylpiperazine-*N*′-2-ethanesulfonic acid	7.48	–0.014
DIPSO	3-[*N*-Bis(hydroxyethyl)amino]-2-hydroxypropanesulfonic acid	7.60	–0.015
TEA	Triethanolamine	7.76	–0.020
POPSO	Piperazine-*N,N*′-bis(2-hydroxypropanesulfonic acid)	7.85	–0.013
EPPS, HEPPS	*N*-2-Hydroxyethylpiperazine-*N*′-3-propanesulfonic acid	8.00	–
Tris	Tris(hydroxymethyl)aminomethane	8.06	–0.028
Tricine	*N*-[Tris(hydroxymethyl)methyl]glycine	8.05	–0.021
Glycinamide	–	8.06	–0.029
PIPPS	1,4-Bis(3-sulfopropyl)piperazine	8.10	–
Glycylglycine	–	8.25	–0.025
Bicine	*N,N*-Bis(2-hydroxyethyl)glycine	8.26	–0.018
TAPS	3-{[Tris(hydroxymethyl)methyl]amino}propanesulfonic acid	8.40	0.018
Morpholine	–	8.49	–
PIPBS	1,4-Bis(4-sulfobutyl)piperazine	8.60	–
AES	2-Aminoethylsulfonic acid, taurine	9.06	–0.022
Borate	–	9.23	–0.008
Ammonia	–	9.25	–0.031
Ethanolamine	–	9.50	–0.029
CHES	Cyclohexylaminoethanesulfonic acid	9.55	–0.029
Glycine (pK_2)	–	9.78	–0.025
EDPS	*N,N*′-Bis(3-sulfopropyl)ethylenediamine	9.80	–
APS	3-Aminopropanesulfonic acid	9.89	–
Carbonate (pK_2)	–	10.33	–0.009
CAPS	3-(Cyclohexylamino)propanesulfonic acid	10.40	0.032
Piperidine	–	11.12	–
Phosphate (pK_3)	–	12.33	–0.026

Reprinted, with permission, from Blanchard (1984).

[a]The change in pK per degree centigrade. These values can be used to correct the pK values in the table to higher or lower temperatures.

TABLE A1.5. Approximate molarities and specific gravities of concentrated acids and ammonia

	Percentage by weight	Approximate molarity (M)	Milliliter/liter to prepare M-solution	Specific gravity
Acetic acid	99.6	17.4	57.5	1.05
Ammonia	25	13.3	75.1	0.91
	35	18.1	55.2	0.88
Formic acid	90	23.6	42.4	1.205
	98	25.9	38.5	1.22
Hydrochloric acid	36	11.6	85.9	1.18
Nitric acid	70	15.7	63.7	1.42
Perchloric acid	60	9.2	108.8	1.54
	72	12.2	82.1	1.70
Phosphoric acid	90	16.0	62.4	1.75
		(N = 48.1)	(N = 20.8)	
Sulfuric acid	98	18.3	54.5	1.835
		(N = 36.7)	(N = 27.3)	

Adapted, with permission, from Dawson et al. (1969).

TABLE A1.6. Mass changes due to some posttranslational modifications of peptides and proteins

Modification	Monoisotopic mass change	Average mass change
Homoserine formed from Met by CNBr treatment	–29.9928	–30.0935
Pyroglutamic acid formed from Gln	–17.0265	–17.0306
Disulfide bond formation	–2.0157	–2.0159
Carboxy-terminal amide formed from Gly	–0.9840	–0.9847
Deamidation of Asn and Gln	–0.9840	–0.9847
Methylation	14.0157	14.0269
Hydroxylation	15.9949	15.9994
Oxidation of Met	15.9949	15.9994
Proteolysis of a single peptide bond	18.0106	18.0153
Formylation	27.9949	28.0104
Acetylation	42.0106	42.0373
Carboxylation of Asp and Glu	43.9898	44.0098
Phosphorylation	79.9663	79.9799
Sulfation	79.9568	80.0642
Cysteinylation	119.0041	119.1442
Pentoses (Ara, Rib, Xyl)	132.0423	132.1161
Deoxyhexoses (Fuc, Rha)	146.0579	146.1430
Hexosamines (GalN, GlcN)	161.0688	161.1577
Hexoses (Fru, Gal, Glc, Man)	162.0528	162.1424
Lipoic acid (amide bond to lysine)	188.0330	188.3147
N-acetylhexosamines (GalNAc, GlcNAc)	203.0794	203.1950
Farnesylation	204.1878	204.3556
Myristoylation	210.1984	210.3598
Biotinylation (amide bond to lysine)	226.0776	226.2994
Pyridoxal phosphate (Schiff Base formed to lysine)	231.0297	231.1449
Palmitoylation	238.2297	238.4136
Stearoylation	266.2610	266.4674
Geranylgeranylation	272.2504	272.4741
N-acetylneuraminic acid (sialic acid, NeuAc, NANA, SA)	291.0954	291.2579
Glutathionylation	305.0682	305.3117
N-glycolylneuraminic acid (NeuGc)	307.0903	307.2573
5′-Adenosylation	329.0525	329.2091
4′-Phosphopantetheine	339.0780	339.3294
ADP-ribosylation (from NAD)	541.0611	541.3052
Adventitious modifications		
acrylamide	71.0371	71.0788
glutathione	304.0712	304.3038
β-mercaptoethanol	75.9983	76.1192

Adapted, with permission, from Burlingame and Carr (1996).

TABLE A1.7. Common amino acids *not* encoded in the genetic code

Symbol	Name and composition	Monoisotopic mass	Average mass
Abu	2-aminobutyric acid C_4H_7NO	85.05276	85.1057
AECys	aminoethylcysteine $C_5H_{10}N_2OS$	146.05138	146.2133
Aib	2-aminoisobutyric acid C_4H_7NO	85.05276	85.1057
Cme	carboxymethylcysteine $C_5H_7NO_3S$	161.01466	161.1815
Cys(O_3H)	cysteic acid $C_3H_5NO_4S$	150.99393	151.1430
Dha	dehydroalanine C_3H_3NO	69.02146	69.0630
Dhb	dehydroamino-2-butyric acid —	83.03711	83.0898
Gla	4-carboxyglutamic acid C_4H_5NO	173.03242	173.1253
Hse	homoserine $C_4H_7NO_2$	101.04768	101.1051
Hyl	hydroxylysine $C_6H_{12}N_2O_2$	144.08988	144.1736
Hyp	hydroxyproline $C_5H_7NO_2$	113.04768	113.1161
Iva	isovaline C_5H_9NO	99.06841	99.1326
Nlc	norleucine $C_6H_{11}NO$	113.08406	113.1595
Orn	ornithine $C_5H_{10}N_2O$	114.07931	114.1473
Pip	2-piperidinecarboxylic acid C_6H_9NO	111.06841	111.1436
Pyr	pyroglutamic acid $C_5H_5NO_2$	111.03203	111.1002
Sar	sarcosine C_3H_5NO	71.03711	71.0788

Adapted, with permission, from Burlingame and Carr (1996).

TABLE A1.8. Structures of the 20 common amino acid side chains found in proteins

Amino acid	Three-letter code	One-letter code
Alanine	Ala	A
Arginine	Arg	R
Asparagine	Asn	N
Aspartic Acid	Asp	D
Cysteine	Cys	C
Glutamine	Gln	Q
Glutamic Acid	Glu	E
Glycine	Gly	G
Histidine	His	H
Isoleucine	Ile	I
Leucine	Leu	L
Lysine	Lys	K
Methionine	Met	M
Phenylalanine	Phe	F
Proline	Pro	P
Serine	Ser	S
Threonine	Thr	T
Tryptophan	Trp	W
Tyrosine	Tyr	Y
Valine	Val	V

The full name, three-letter code, and one-letter code are given for each amino acid (e.g., alanine, Ala, A), as well as the chemical formula. Color code: (*red*) oxygen atoms; (*dark blue*) nitrogen atoms; (*light blue*) hydrogen atoms; (*gray*) carbon atoms; (*yellow*) sulfur atoms. For additional amino acid information, see http://prowl.rockefeller.edu/aainfo/contents/htm.

TABLE A1.9. Properties of amino acids

Amino acid residue	pK_a of ionizing side chain[a]	Average residue mass[b] (daltons)	Monoisotopic mass (daltons)[b]	Occurrence in proteins[c] (%)	Percent buried residues[d] (%)	V_r[e] ($Å^3$)	van der Waals volume[f] ($Å^3$)	Accessible surface area[g] ($Å^2$)	Ranking of amino acid polarities[h]
Alanine	–	71.0788	71.03711	7.5	38 (12)	92	67	67	9 (7)
Arginine	12.5 (>12)	156.1876	156.10111	5.2	0	225	148	196	15 (19)
Asparagine	–	114.1039	114.04293	4.6	10 (2)	135	96	113	16 (16)
Aspartic acid	3.9 (4.4–4.6)	115.0886	115.02694	5.2	14.5 (3)	125	91	106	19 (18)
Cysteine	8.3 (8.5–8.8)	103.1448	103.00919	1.8	47 (3)	106	86	104	7 (8)
Glutamine	–	128.1308	128.05858	4.1	6.3 (2.2)	161	114	144	17 (14)
Glutamic acid	4.3 (4.4–4.6)	129.1155	129.04259	6.3	20 (2)	155	109	138	18 (17)
Glycine	–	57.0520	57.02146	7.1	37 (10)	66	48		11 (9)
Histidine	6.0 (6.5–7.0)	137.1412	137.05891	2.2	19 (1.2)	167	118	151	10 (13)
Isoleucine	–	113.1595	113.08406	5.5	65 (12)	169	124	140	1 (2)
Leucine	–	113.1595	113.08406	9.1	41 (10)	168	124	137	3 (1)
Lysine	10.8 (10.0–10.2)	128.1742	128.09496	5.8	4.2 (0.1)	171	135	167	20 (15)
Methionine	–	131.1986	131.04049	2.8	50 (2)	171	124	160	5 (5)
Phenylalanine	–	147.1766	147.06841	3.9	48 (5)	203	135	175	2 (4)
Proline	–	97.1167	97.05276	5.1	24 (3)	129	90	105	13 (–)
Serine	–	87.0782	87.03203	7.4	24 (8)	99	73	80	14 (12)
Threonine	–	101.1051	101.04768	6.0	25 (5.5)	122	93	102	12 (11)
Tryptophan	–	186.2133	186.07931	1.3	23 (1.5)	240	163	217	6 (6)
Tyrosine	10.9 (9.6–10.0)	163.1760	163.06333	3.3	13 (2.2)	203	141	187	8 (10)
Valine	–	99.1326	99.06841	6.5	56 (15)	142	105	117	4 (3)

[a]The pK_a values in most cases are at 25°C. The expected pK_a values in proteins, shown in parentheses, are determined from model compounds in which titration of side chains is decoupled from charge effects of α-substituents. (Data from Cantor and Schimmel 1980.)

[b]Data from Burlingame and Carr (1996).

[c]Frequency of occurrence of each amino acid residue in the primary structures of 105,990 sequences in the nonredundant OWL protein database (release 26.0 e) (Trinquier and Sanejouand 1998).

[d]This column represents the tendency of an amino acid to be buried (defined as <5% of residue available to solvent) in the interior of a protein and is based on the structures of nine proteins (total of ~2000 individual residues studied, with 587 [29%] of these buried). Values indicate how often each amino acid was found buried, relative to the total number of residues of this amino acid found in the proteins (values in parentheses indicate the number of buried residues of this amino acid found relative to all buried residues in the proteins). (Data from Schien 1990; for other calculation methods with similar results, see Janin 1979 and Rose at al. 1985.)

[e]Average volume (V_r) of buried residues, calculated from the surface area of the side chain (Richards 1977; Baumann et al. 1989).

[f]Data from Darby and Creighton (1993).

[g]Total accessible surface area (ASA) of amino acid side chain for residue X in a Gly-X-Gly tripeptide with the main chain in an extended conformation (Miller et al. 1987). The ASA or cavity surface area is defined as the surface traced by the center of a sphere with the radius of a water molecule (0.15 mm) as it is rolled over the surface of a molecular model of the solution (Lee and Richards 1971).

[h]Values shown represent the mean ranking of amino acids according to the frequency of their occurrence at each sequence rank for 38 published hydrophobicity scales (Trinquier and Sanejouand 1998). Although the majority of these hydrophobicity scales are derived from experimental measurements of chemical behavior or physicochemical properties (e.g., solubility in water, partition between water and organic solvent, chromatographic migration, or effects on surface tension) of isolated amino acids, several "operational" hydrophobicity scales based on the known environment characteristics of amino acids in proteins, such as their solvent accessibility or their inclination to occupy the core of proteins (based on the position of residues in the tertiary structures as observed by X-ray crystallography or NMR) are included (Trinquier and Sanejouand 1998). The lower rankings represent the most hydrophobic amino acids, and higher values represent the most hydrophilic amino acids. For comparative purposes, the hydrophobicity scale of Radzicka and Wolfenden is shown in parentheses. This scale was derived from the measured hydration potential of amino acids that is based on their free energies of transfer from the vapor phase to cyclohexane, 1-octanol, and neutral aqueous solution (Radzicka and Wolfenden 1988).

Techniques

TECHNIQUES

TECHNIQUE 1

Quantifying Protein by Bicinchoninic Acid

THIS METHOD IS A VARIATION OF THE LOWRY ASSAY developed by Smith et al. (1985), which uses bicinchoninic acid (BCA) to enhance the detection of Cu^{+} generated under alkaline conditions at sites of complexion between Cu^{2+} and protein. The resulting tetradentate BCA–cuprous ion chromophore absorbs at 562 nm. The speed with which color develops in the BCA assay is influenced by the temperature of the reaction. At 37°C, the Cu^{+}-protein complexes develop chiefly by oxidation of cysteine, tyrosine, and tryptophan residues. At 60°C, the temperature used in this protocol, additional Cu^{+}-protein complexes arise from oxidation of peptide bonds (Smith et al. 1985; Wiechelman et al. 1988). The concentration limits of common reagents on the BCA assay are listed in Table A2.1.

This technique is divided into four parts: Standard Procedure, Construction of a Calibration Curve for the BCA Protein Assay, Microprocedure, and 96-Well Microtiter Plate Procedure.

TABLE A2.1. Concentration limits of common reagents on the BCA and Bradford assays

Compound	Bradford	BCA
Buffers		
acetate	0.6 M[a]	0.2 M,[a] 0.25 M (pH 5.58)
cacodylate-Tris	0.1 M[b]	–
glycine	0.1 M	1 M (pH 11)[c]
HEPES	100 mM	100 mM
MES	700 mM	50 mM[d]
MOPS	200 mM	50 mM[d]
Na^{+}-citrate	50 mM	<1 mM[a]
PIPES	500 mM	50 mM[d]
potassium phosphate	1 M	–
sodium phosphate	1 M	0.1 M[c]
sodium acetate	0.6 M[a]	0.2 M (pH 5.5)
TES	–	50 mM[d]
Tris	2 M	0.1 M, 0.25 M (pH 11.25)[c]
Tricine	OK	–
BES	2.5 M	–
Salts		
$MgCl_2$	1 M	–
ammonium sulfate	1 M	interferes[c]
NaCl	5 M	1 M[a]
urea	6 M	3 M
KCl	1 M	–
guanidine HCl	OK	4 M[c]
glycine	–	1.0 M (pH 11.8)
Detergents and Denaturants		
Lubrol	–	1%[c]
Brij 35	interferes[e]	1%[c]
CHAPS	1%	1%
Na-deoxycholate	0.25%	–
Nonidet P-40	interferes[e]	1%
octyl glucoside	2%[e]	1%

TABLE A2.1. *(Continued)*

Compound	Bradford	BCA
SDS	0.1%	1%[c]
sodium cholate	interferes[e]	–
Triton X-100	0.1%[f]	1%
Tween-20	interferes[e]	–
glucopyranoside	OK[g]	–
Media		
Eagle's MEM	OK	–
Hank's salt solution	OK	–
Earle's salt solution	OK	–
Sugars		
glucose	OK	10 mM[e]
sucrose	1 M[a]	1 M
Chelators		
EDTA	100 mM	10 mM
EGTA	0.05 M	–
Reducing Agents		
2-mercaptoethanol	1 M	50 μM[h]
dithiothreitol	1 M	1 mM
Alcohols, Polar Compounds		
acetone	OK	–
DMSO	–	5%[a]
ethanol	OK	–
glycerol	99%	10%
methanol	OK	–
Miscellaneous		
sodium azide	–	0.2%[c]
DNA	1 mg/ml	–
lipids	–	interferes[b]
$MgCl_2$	1 M	–
RNA	0.3 mg/ml	–
ATP	1 mM	–
ampholytes	0.5%, 1%[i,j]	interferes
HCl	0.1 M[a]	0.1 N
NaOH	0.1 M[a]	0.1 N
NAD	1 mM	–
phenol	5%	–
amino acids	OK	–
polypeptides (<3 kD)	OK	–
3-ethanolamine	interferes[k]	interferes

Adapted, with permission, from Bollag et al. (1996 [©Wiley]).

A dash (–) indicates that this compound has not been tested. "OK" indicates that this compound was used successfully, but no concentration was indicated. Note that there may be some variation in permissible amounts due to differences in the assay protocol.

[a]Stoscheck (1990).

[b]Kessler and Fanestil (1986).

[c]Smith et al. (1985).

[d]Kaushal and Barnes (1986).

[e]Fanger (1987).

[f]Low concentration (0.008%) of Triton X-100 results in both improved sensitivity and protein-to-protein variability for the estimation of low-molecular-weight proteins (Friedenauer and Berlet 1989).

[g]Unlike other detergents, the glucopyranoside detergents, when used to solubilize membrane bound proteins, do not interfere with the Bradford assay (Fanger 1987). Certain ampholytes (e.g., 2D Pharmalyte [pH 3–10], Polybuffer [pH 6–9] from Sigma; and Servaly+ [pH 7–9] from Serva), ammonium sulfate, and reducing agents (glucose, 2-mercaptoethanol, DTT), interfere with the BCA protein assay (Brown et al. 1989). These substances can be removed prior to performing the BCA assay by DOC/TCA precipitation (Brown et al. 1989).

[h]Hill and Straka (1988).

[i]Spector (1978).

[j]Read and Northcote (1981); Peterson (1983).

[k]Concentrations of 150 mM 3-ethanolamine interfere with both the Bradford and BCA protein assays.

TECHNIQUE 1A: Standard Procedure

THE STANDARD BCA ASSAY IS CARRIED OUT in a volume of ~1 ml, using 50 μl of a series of dilutions of the test sample. The concentration of protein in the sample is calculated by interpolation into a plot of concentration versus absorbance obtained using known amounts of a reference protein (in this example, bovine serum albumin; please see Technique 1B).

The *range of sensitivity* for this protocol is 0.02–2.0 mg of protein/ml.

MATERIALS

CAUTION: See Appendix 3 for appropriate handling of materials marked with <!>.

Reagents

Bovine serum albumin (BSA) stock solution

Dissolve 100 mg of BSA in 10 ml of H_2O.

Sample buffer

Standard reagent A

1.0 g of sodium BCA (4,4-dicarboxy-2, 2′-biquinoline, disodium salt) (26 mM final)

2.0 g of $Na_2CO_3 \cdot H_2O$ (0.16 M final)

0.16 g of sodium tartrate·$2H_2O$ (7 mM final)

0.4 g of NaOH (0.1 M final) <!>

0.95 g of $NaHCO_3$ (0.11 M final)

Adjust the volume to 100 ml with H_2O. Stir the solution and, if necessary, adjust the pH to 11.5 with 10 M NaOH.

Standard reagent B

Dissolve 0.4 g of $CuSO_4 \cdot 5H_2O$ in 10 ml of H_2O.

Standard reagents A and B are stable for 1–3 weeks when stored in plastic containers at room temperature, and indefinitely at 4°C. Note that although the disodium salt of BCA is soluble at neutral pH, the free acid is not readily soluble even at basic pH.

Although Standard reagents A and B are simple to prepare, more reproducible results are obtained using commercial reagents, which are available in a variety of kits from Pierce Chemical Company (Rockford, Illinois). Commercial reagents are compatible with commonly used biological buffers, have a higher sensitivity, and display an expanded linear range.

Standard working reagent

Mix 100 volumes of Standard reagent A with 2 volumes of Standard reagent B. This solution is stable for 2–3 days at room temperature and should be green in color.

Equipment

Cuvettes (glass or polystyrene)

Spectrophotometer

Test tubes (small disposable) or microfuge tubes

Water bath or heating block preset to 60°C

Biological Sample

Test sample(s)

When using the BCA method to measure the concentration of protein in solution, it is essential to assay in parallel a set of samples containing known amounts of a reference protein, such as BSA. The results obtained from the reference samples are used to construct a calibration curve (see Technique 1B).

- Assay dilutions of test and reference samples in duplicate (preferably, in triplicate).
- If the same cuvette is to be used throughout the assay, read the samples with the lower protein content *first* in order to minimize errors arising from sample carryover due to incomplete rinsing.
- After the cuvette has been rinsed with H_2O between samples, tap the inverted cuvette (gently) on Kimwipes to remove excess water droplets.
- For optimal results, add the BCA working buffer to the protein standards and "unknowns" sequentially and, following color reaction development, read the absorbance values of the samples in the same order.
- Each assay is only as good as the calibration curve. Do not take short cuts: Assay of unknown protein solutions must be accompanied by the generation of a new calibration curve.

METHOD

1. Dilute aliquots of the test sample(s) with sample buffer to a concentration that will fall within the range of the calibration curve (0.25–2.0 mg/ml). To do this, estimate the protein concentration by measuring the absorbance of the test sample at 280 nm.

   The BCA method can be used with a wide variety of buffers (see Table A2.1), but it is sound practice to include a "buffer blank" in the assay to validate this assumption.

2. Use small disposable test tubes or microfuge tubes to prepare dilutions of the reference protein (BSA) ranging from 0.2 to 2.0 mg/ml.

3. Aliquot 50 μl of each dilution of test sample and reference protein into microfuge tubes. The amount of reference protein in the tubes ranges from 10 to 100 μg. Add 1 ml of standard working reagent to each tube and mix the contents well by vortexing.

4. Incubate the tubes for 30 minutes at 60°C (or for 2 hours at ambient temperature).

   Incubating the samples for longer periods of time can increase the sensitivity of the assay. Conversely, if the color is becoming too intense, terminate the incubation earlier. Take care to treat samples and standards identically. Although the assay may also be performed at room temperature, incubation at temperatures <60°C results in diminished sensitivity (due to the inability of peptide bonds to react with the cupric ions at lower temperatures) and greater sample-to-sample variability. A procedure for developing the color in seconds, using a microwave oven, has been described by Akins and Tuan (1992).

5. Cool the reactions to room temperature and then measure the absorbance at 562 nm, using glass or polystyrene cuvettes. The color is stable for at least 1 hour.

6. Plot A_{562} versus protein concentration to generate the calibration curve, and determine the exact concentration of protein in the test samples by interpolation as shown in Technique 1B.

TECHNIQUE 1B: Construction of a Calibration Curve for the BCA Protein Assay

DILUTIONS OF REFERENCE SAMPLES (PROTEIN STANDARDS) should be assayed in duplicate (preferably, in triplicate). Each assay is only as good as the calibration curve. Do not take short cuts: Assay of unknown protein solutions must be accompanied by the generation of a new calibration curve. It is essential that the calibration curve remains linear (i.e., adheres to Beer's law) over the range of protein concentrations tested; otherwise, a significant margin of error will be associated with the protein estimates of the unknown samples. Table A2.2 illustrates how to arrange absorbance data in a format that facilitates construction of the calibration curve (Figure A2.1).

Assuming that the test samples were assayed in parallel with the protein standards used to generate a calibration curve (see, e.g., Figure A2.1 and Tables A2.2 and A2.3), the protein concentration of the unknown samples can be calculated as follows:

1. For each dilution set of the test protein, calculate the average absorbance at 562 nm for a given volume of test sample.
2. Convert the average absorbances calculated in Step 1 to mass of protein by interpolation of the calibration curve.
3. Calculate the protein concentration in each dilution set by dividing the mass of protein by the sample volume (i.e., the volume of test protein plus the volume of buffer).
4. Disregard average absorbance values that are outside the linear portion of the calibration curve.

TABLE A2.2. Sample data for a calibration curve for the BCA assay

Sample number	Protein (μg)	Standard solution 2 mg/ml (μl)	Experiment buffer (μl)	BCA reagent (ml)	A_{562}	Ave A_{562}
1	0	–	50	1	0	0
2		6.25	43.75	1	0.21	
3	12.5	6.25	43.75	1	0.21	0.20
4		6.25	43.75	1	0.18	
5		12.50	37.5	1	0.53	
6	25.0	12.50	37.5	1	0.45	0.49
7		12.50	37.5	1	0.50	
8		18.75	31.25	1	0.80	
9	37.5	18.75	31.25	1	0.72	0.76
10		18.75	31.25	1	0.77	
11		25.00	25.00	1	1.06	
12	50.0	25.00	25.00	1	1.02	1.05
13		25.00	25.00	1	1.07	
14		50.00	–	1	1.95	
15	100.0	50.00	–	1	2.10	2.02
16		50.00	–	1	2.01	

Adjust the concentration of the protein standard to 20 μg/ml by measuring the A_{280nm} or the A_{260nm}/A_{280nm} to determine the protein content, and dilute with sample buffer as necessary.

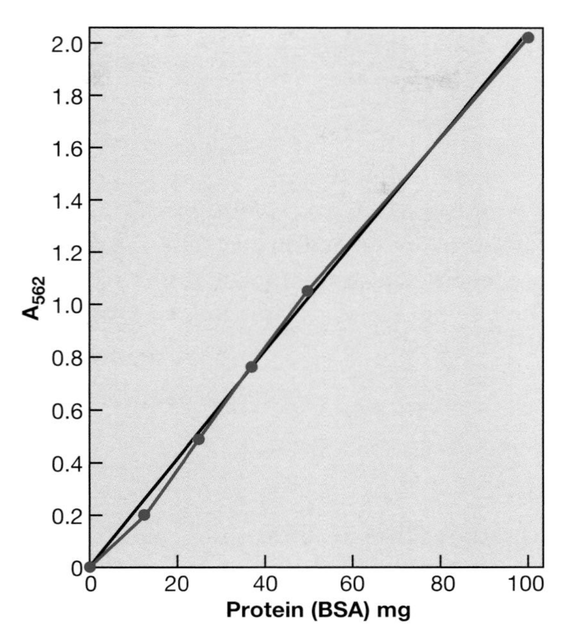

FIGURE A2.1. Calibration curve for the BCA protein assay.

5. Calculate the average of the protein concentrations of appropriate dilution sets (Step 3) to obtain the final estimate of protein concentration of the test sample. For example, using the calibration curve in Figure A2.1 and the data in Table A2.3:

   a. The average absorbance for samples 7, 8, and 9 is (0.88 + 0.87 + 0.88)/3 = 0.88.

   b. Using the calibration curve (Figure A2.1), a horizontal line at A_{562} = 0.88 intersects the calibration curve at a point that corresponds to 43 µg of protein.

   c. The protein concentration is 43 µg/50 µl = 8.75 µg/µl = 8.75 mg/ml.

   d. For the data set shown in Table A2.3, ignore the lowest and highest average readings since they are unreliable (either off scale [i.e., >2] or too low [i.e., <0.1]).

   e. Averaging the remaining two protein concentration values (i.e., 8.3 and 8.75) gives the final estimate of protein concentration of the unknown sample ([8.3 + 8.7]/2 = 8.5 mg/ml).

TABLE A2.3. Sample calculations for determining protein concentration using the BCA assay

Sample number	Sample (µl)	Experiment buffer (µl)	BCA reagent (ml)	A_{562}	Ave A_{562}	Protein µg	Protein mg/ml
1	50	–	1	>2			
2	50	–	1	>2		unreliable	
3	50	–	1	>2			
4	10	40	1	1.66			
5	10	40	1	1.70	1.67	83.0	8.3
6	10	40	1	1.65			
7	5	45	1	0.88			
8	5	45	1	0.87	0.88	43.0	8.7
9	5	45	1	0.87			
10	2.5	47.5	1	0.09			
11	2.5	47.5	1	0.07	0.08	unreliable	
12	2.5	47.5	1	0.09			

Technique 1C: Microprocedure

THE STANDARD BCA ASSAY (TECHNIQUE 1A) USES LARGE VOLUMES of both reagents and samples and cannot easily be automated. The following method is more parsimonious and can be adapted for use with a microplate reader (see Technique 1D and Hinson and Webber 1988).

The *range of sensitivity* for this protocol is 0.5–20 μg protein/ml.

MATERIALS

CAUTION: See Appendix 3 for appropriate handling of materials marked with <!>.

Reagents

Bovine serum albumin (BSA) stock solution

Dissolve 1 mg of BSA in 10 ml of H_2O.

Microreagent A

0.8 g of $Na_2CO_3 \cdot H_2O$

1.6 g of NaOH <!>

1.6 g of sodium tartrate$\cdot 2H_2O$

Adjust the volume to 100 ml with H_2O. Stir the solution and adjust the pH to 11.5 with 10 M NaOH.

Microreagent B

Dissolve 4.0 g of sodium BCA in 100 ml of H_2O.

Microreagents A and B are stable for 1–3 weeks when stored in plastic containers at room temperature, and indefinitely at 4°C. Note that although the disodium salt of BCA is soluble at neutral pH, the free acid is not readily soluble even at basic pH.

Although Microreagents A and B are simple to prepare, more reproducible results are obtained using commercial reagents, which are available in a variety of kits from Pierce Chemical Company (Rockford, Illinois). Commercial reagents are compatible with commonly used biological buffers, have a higher sensitivity, and display an expanded linear range.

Microreagent C

Dissolve 0.4 g of $CuSO_4 \cdot 5H_2O$ in 10 ml of H_2O.

Microworking reagent

Mix 1 volume of Microreagent C with 25 volumes of Microreagent B, and then add 26 volumes of Microreagent A. The microworking reagent is stable for 24 hours at room temperature.

Sample buffer

Equipment

Cuvettes (glass or polystyrene)

Spectrophotometer

Test tubes (small disposable) or microfuge tubes

Water bath or heating block preset to 60°C

Biological Sample

Test sample(s)

METHOD

1. Dilute aliquots of the test sample(s) with sample buffer to a concentration that will fall within the range of the calibration curve (0.2–30 μg/ml). To do this, estimate the protein concentration by measuring the absorbance of the test sample at 280 nm.

   The BCA method can be used with a wide variety of buffers (see Table A2.1), but it is sound practice to include a "buffer blank" in the assay to validate this assumption.

2. Use small disposable test tubes or microfuge tubes to prepare dilutions of a stock solution of BSA, ranging from 0.4 to 2.0 μg/ml of protein, to use as reference standards.

3. Transfer 500 μl of each test sample and each reference standard (the final range of reference standards is 0.2–1.0 μg of protein) into microfuge tubes. Add 500 μl of microworking reagent to each tube and mix well.

4. Incubate the tubes for 1 hour at 60°C.

5. Cool the reactions to room temperature and then measure the absorbance at 562 nm, using glass or polystyrene cuvettes. The color developed is stable for at least 1 hour.

6. Plot A_{562} versus protein concentration to generate the calibration curve, and determine the exact concentration of protein in the test samples by interpolation as shown in Technique 1B.

TECHNIQUE 1D: 96-Well Microtiter Plate Procedure

THE BCA "MICROPROCEDURE" FOR QUANTIFYING PROTEIN (Technique 1C) can be carried out in microtiter plates, and the results read in microplate readers. This modification permits the protein concentration of very large numbers of samples to be determined with ease. If the microplate reader is interfaced with a computer, more than 1000 samples can be read per hour. This protocol has been adapted from Hinson and Webber (1988).

The *range of sensitivity* for this protocol is 0.5–20 μg protein/ml.

MATERIALS

CAUTION: See Appendix 3 for appropriate handling of material marked with <!>.

Reagents

Bovine serum albumin (BSA) stock solution (500 μg/ml)

Dilution buffer

Generally, the dilution buffer is phosphate-buffered saline (PBS). Some detergents can be used if solubility is a problem. For reagents that interfere with the BCA assay, see Table A2.1. The dilution buffer used for preparing samples should also be used for preparing the standard for consistency.

Standard reagent A

5 g of sodium BCA (4,4-dicarboxy-2, 2´-biquinoline, disodium salt)

10 g of $Na_2CO_3 \cdot H_2O$

0.8 g of sodium tartrate

2 g of NaOH <!>

4.75 g of $NaHCO_3$

Adjust the volume to 500 ml with H_2O. Stir the solution and, if necessary, adjust the pH to 11.25 using 10 M NaOH.

Standard reagent B

Dissolve 2 g of $CuSO_4 \cdot 5H_2O$ in 100 ml of H_2O.

Standard reagents A and B are stable for 1–3 weeks when stored in plastic containers at room temperature, and indefinitely at 4°C. Note that although the disodium salt of BCA is soluble at neutral pH, the free acid is not readily soluble even at basic pH. These reagents may be purchased ready prepared as a kit from Pierce Chemical Company (Rockford, Illinois).

Standard working reagent

Mix 50 volumes of Standard reagent A with 1 volume of Standard reagent B. This solution is stable for 2–3 days at room temperature and should be green in color.

Equipment

ELISA plate reader (e.g., Multiscan MCC/340, Pathtech Diagnostics Pty. Ltd.)

Microtiter plates (96-well disposable)

Oven preset to 60°C

Pipette (multichannel 20–200 μl)

Biological Sample

Test sample(s)

METHOD

1. Use the multichannel pipette to add 30 µl of the dilution buffer to each well in the microtiter plate.

   It may not be necessary dilute the sample, but it is useful to prepare a range of concentrations to ensure that at least one will fit on the standard curve to establish protein concentration.

   EXPERIMENTAL TIP: Draw a schematic of the microtiter plate that illustrates the distribution and concentration of the protein samples (see below).

	1	2	3	4	5	6	7	8	9	10	11	12
A	30 µl of 500 µg/ml BSA	30 µl of 500 µg/ml BSA	30 µl of 500 µg/ml BSA	Sample #1	Sample #1	Sample #2	Sample #2	Sample #3	Sample #3	Sample #4	Sample #4	Sample #5
B												
C												
D		Serial dilution										
E												
F												
G												
H												

2. Add 30 µl of 500 µg/ml BSA solution to the top wells in the first three lanes (i.e., positions A1, A2, and A3). Mix thoroughly by pipetting the solution up and down several times. The standard is performed in triplicate to verify reproducibility.

   As there was 30 µl of dilution buffer already present in the well, a 1:1 dilution has been performed; therefore, the starting concentration for the BSA standard is 250 µg/ml.

3. Add 30 µl of test sample (in duplicate) to the top wells in the remaining lanes, i.e., positions A4 to A12. Mix thoroughly by pipetting the solution up and down several times.

4. Use the multichannel pipette to carry out a serial dilution of each test sample and the BSA standard. Transfer 30 µl of sample from a well in row A (which has a total volume of 60 µl per well) to the corresponding well in row B. Pipette the solution up and down in the well to mix thoroughly. Without changing pipette tips, transfer 30 µl of sample from row B to the corresponding well in row C. Continue the serial dilution until all wells in the microtiter dish contain the appropriate sample.

5. Use the multichannel pipette to transfer 20 µl of each sample from the original plate to a fresh microtiter plate.

   To minimize cross-contamination and avoid changing tips between rows, begin the transfers from the bottom row, where the samples are most dilute, and work up the plate row-by-row, to row A.

6. Pour the working reagent into a reservoir that will accommodate the multichannel pipette. Add 250 µl of the working reagent to each well containing 20 µl of sample. Mix thoroughly by pipetting the solutions up and down.

7. Incubate the plate for 30 minutes at 60°C in an oven.

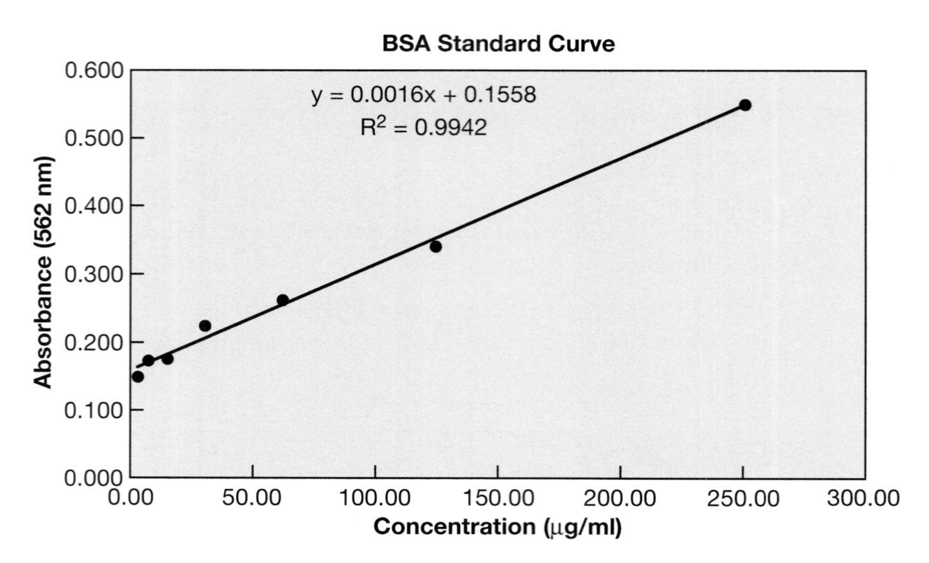

FIGURE A2.2. Calibration curve for the 96-well microtiter plate BCA protein assay. The R^2 value is an indication of how well the trend-line fits the plotted values, and it can be interpreted as the proportion of the variance in y attributable to the variance in x. The closer the R^2 value is to 1.0, the better the fit.

8. Remove the plate from the oven, and place it on the bench top. Allow the microtiter plate to cool to room temperature (~10 minutes).
9. Ensure that the wells are free of air bubbles and that the bottom of the plate is clean. Read the absorbance at 562 nm (or close to) using an ELISA plate reader.
10. Plot the average of the three absorbance readings for the BSA standard to generate a standard curve. Plot the absorbance (562 nm) as a function of BSA concentration (µg/ml). A straight line should result. For a sample standard curve, see Figure A2.2.
11. Calculate the sample concentrations using the straight-line formula. Remember to take into account the dilution factors.

TECHNIQUE 2

Coomassie Blue Dye-binding Assay (Bradford Assay) for the Determination of Protein Concentration

THE EASE, SPEED, AND SENSITIVITY OF THE BRADFORD ASSAY (Bradford 1976) no doubt accounts for its continuing popularity. Binding of the Coomassie Blue dye is generally carried out in a reaction mixture that contains sufficient phosphoric acid to ensure that most of the dye molecules become protonated and absorb light at 465 nm. However, the pH of the assay must be high enough to allow a small amount of dye to survive in the nonprotonated blue form, which absorbs at 595 nm in solution and is available to bind to protein. Equilibrium shifts between free forms of the dye prevent depletion of the unprotonated blue form as a consequence of binding to protein.

The dye-protein complexes are traditionally detected by an increase in absorption at 595 nm, which results from conversion of protonated dye to unprotonated blue molecules bound to protein. However, a small metachromatic shift occurs on binding of Coomassie to protein, and the λ_{max} of the dye-protein complex is 620 nm, where the sensitivity of the method is maximal.

The chief disadvantage of the method is the large variability of color development among different proteins. The change in absorbance per unit mass of protein varies over a three- to fourfold range depending on the nature of the protein assayed (Pierce and Suelter 1977; Van Kley and Hale 1977). Ideally, the reference protein used to construct the standard curve should be the same as the test protein. However, in most cases, this is impractical and the dye response of a sample is generally quantified relative to that of a generic protein (e.g., BSA), which permits comparisons with previous studies using the same standard. Note, however, that BSA is atypical because it exhibits an unusually large response to the Coomassie Blue dye and hence may result in an underestimate of the concentration of a test protein. Ovalbumin and γ-globulin may be better choices for protein standards, because their dye-binding capacity is similar to that of other proteins (Read and Northcote 1981). Because of the uncertainty in relating the absorbance values obtained with the test protein to those obtained with the reference protein, the Bradford assay has been largely displaced by the BCA method as the technique of choice to quantify protein in solution.

The *range of sensitivity* for this protocol is 10–100 µg/ml for the standard assay and 1–10 µg/ml for the microassay.

MATERIALS

CAUTION: See Appendix 3 for appropriate handling of materials marked with <!>.

Reagents

Bradford stock solution

Dissolve 350 mg of Coomassie Blue G250 (Serva Blue G) in 100 ml of 95% ethanol <!>. Mix the resulting solution with 200 ml of 88% phosphoric acid <!> (stable indefinitely at room temperature). Although the amount of soluble dye in various commercial preparations of Coomassie Blue G250 varies markedly (Wilson 1979) (generally, Serva Blue G is regarded as having the greatest dye content), the quality of the dye is not critical for routine protein determination.

Bradford working buffer

Mix together 425 ml of distilled H_2O, 15 ml of 95% ethanol, 30 ml of 88% phosphoric acid, and 30 ml of Bradford stock solution. Pass the reagent through Whatman No. 1 filter paper, and store it in an amber bottle at room temperature. The working buffer remains usable for several weeks. However, during this time, dye may precipitate from the solution; hence, it is recommended that the stored reagent be filtered before use.

Buffer or solvent used to prepare the sample

Reference protein stock solution (1 mg/ml)

Dissolve 10 mg of BSA in 10 ml of distilled H_2O (store the stock solution at –20°C in small 200-μg aliquots). Since the moisture content of lyophilized proteins varies markedly (e.g., 5–15%) depending on the nature of their storage, the precise concentration of protein in the protein standard stock solution should be determined by amino acid analysis or from its absorbance at 280 nm. The absorbance at 280 nm of a 1 mg/ml solution of BSA, in a 1-cm light-path-length cuvette is 0.70. The corresponding values for γ-globulin and ovalbumin are 1.38 and 0.79, respectively.

Equipment

Cuvettes

Plasticware and glassware should be absolutely clean and detergent-free. Disposable plastic cuvettes are recommended because dye accumulates on cuvette walls and is difficult to remove. Traces of dye bound to glassware or plasticware can be removed by a thorough rinsing with methanol <!> or washing with concentrated glassware detergent, followed by H_2O and ethanol. Quartz (silica) cuvettes should not be used, since the dye sticks to them tenaciously.

Microfuge tubes

Spectrophotometer

Biological Sample

Sample of test protein

METHOD

1. Prepare a dilution series of the reference protein in duplicate or triplicate, as follows: Pipette 2.5, 5, 7.5, 10, 15, and 100 μl of a 1 mg/ml solution of the reference protein into microfuge tubes, and adjust each to 100 μl with the buffer used to prepare the test sample. Pipette 100 μl of the buffer into a microfuge tube to provide the reagent blank.

2. Pipette 2.5–20 μg of the test protein solution (maximum volume 100 μl) into microfuge tubes. Adjust the volume in each tube to 100 μl with buffer. If even an estimate of the con-

centration of protein in the test solution is unknown, assay a range of dilutions (e.g., 1, 1/10, 1/100, 1/1000). Prepare duplicates (or triplicates) of each sample.

3. Add 1 ml of Bradford working buffer to each tube of sample and reference protein. Mix the reactions well by inversion or gentle vortexing. Avoid foaming as this will lead to poor reproducibility.

4. Incubate the reactions for 5 minutes at 25°C.

5. Measure the A_{620} of the samples and standards against the reagent blank between 2 minutes and 1 hour after mixing.

   Color development is complete after 5 minutes. Prolonged incubation (10–15 minutes) causes microprecipitation of protein that increases with time and can result in visible aggregates. This phenomenon is especially noticeable at higher protein concentrations and is believed to be due to the acidic conditions of the assay. To minimize any error due to color loss, read the OD of test samples within 10 minutes of the standards. Error due to color loss should then be <2% (Peterson 1983).

6. Prepare a calibration curve and estimate the protein content of the test samples as described in Technique 1B for the BCA assay.

   EXPERIMENTAL TIP: The standard curve for the Bradford assay becomes nonlinear at high concentrations of protein due to the depletion of free dye. For this reason, it is essential to include a calibration curve with each assay. If necessary, the linearity of the assay can be improved by plotting the A_{620}/A_{465} ratios, which corrects for free-dye depletion (Sedmak and Grossberg 1977).

TECHNIQUE 3

Purification of Urea

THIS TECHNIQUE REMOVES CYANATE IONS THAT ARE PRODUCED IN UREA solutions upon standing; the extent of formation varies with the pH of the solution (Hagel et al. 1971). Cyanate ion formation in aqueous urea solutions occurs more rapidly at high pH and high temperatures. Hence, unbuffered urea solutions should be stored under acidic conditions at 4°C. Alternatively, after deionization and filtration, the water from the urea filtrate can be evaporated, and the crystallized urea can be collected and stored in the dry state at 4°C. Under these conditions, crystalline urea can be stored for several weeks without significant cyanate ion formation. The concentration limits to common reagents on the Bradford assay are listed in Table A2.1 in Technique 1.

MATERIALS

CAUTION: See Appendix 3 for appropriate handling of materials marked with <!>.

Reagents

Deionizing resin (e.g., AG 501-X8 mixed-bed ion-exchange resin, Bio-Rad) <!>
AG 501-X8, which consists of equivalent amounts of AG 1-X8 (a strong anion exchange resin), and AG 50W-X8 (a strong cation exchange resin) contains a dye that changes from blue-green to gold when its exchange capacity is exhausted.

Heptane, warmed to 40–50°C <!>
Heptane is extremely flammable.

Urea, analar (or technical) grade <!>

Equipment

Buchner funnel or Scintered glass filtration funnel
Filter papers (e.g., Whatman No. 2 and Whatman cellulose Grade 1 Chr) or equivalent
Ice bath

METHOD

Removal of Cyanate Ions from Urea Solutions

1. Prepare an aqueous solution of urea to the desired molarity.
2. Just prior to use, deionize the resin by either
   - passing the solution through a column of mixed-bed deionizing resin

   or

- swirling the urea solution with deionizing resin for several hours at room temperature. Recover the urea solution by filtering it through cellulose filter paper, Grade 1 Chr (11 μm) or a sintered glass filtration funnel of high porosity.

  The deionized urea solution should have a conductivity of less than 2 μS. Urea solutions containing amine buffers such as Tris have an advantage in that they compete with the protein for the cyanate.

Recrystallization of Urea

3. Dissolve urea (10 mg/ml) in warm heptane (40–50°C) and filter rapidly.
4. Cool the mixture to 5°C in an ice bath, which will result in crystal formation.
5. Use a Pasteur pipette to remove the mother liquor and set it aside at –20°C for further recrystallization.
6. Harvest the crystals by filtration using Whatman No. 2 filter paper. Allow the recovered crystals to dry in the dark in a chemical fume hood.
7. Store the recrystallized urea at –20°C in a dry container in the dark.

TECHNIQUE 4

Separation of Peptides and Proteins by HPCEC

Thomas P. Hennessy,* Marion I. Huber,† Klaus K. Unger,† Reinhard I. Boysen,* and Milton T.W. Hearn*

**Centre for Bioprocess Technology, Department of Biochemistry and Molecular Biology, Monash University, Clayton, Victoria 3800, Australia; †Institut für Anorganische und Analytische Chemie, Johannes Gutenberg Universität, Mainz, FGR*

ALTHOUGH CLOSELY RELATED TO HPLC AND HPCE, high-performance capillary electrochromatography (HPCEC) is now claiming its own unique orthogonal place in the separation sciences, particularly with regard to the nanoscale analysis of peptides and proteins. Moreover, the technique capitalizes on existing mobile-phase technology and knowledge to improve separations. To extensively exploit the hybrid characteristics of CEC, the peptide or protein chemist, as well as the chromatographer, must understand how the background electrolyte simultaneously propagates flow and achieves the selectivity needed for high-resolution separations, particularly with biomolecules (Walhagen et al. 2000a; Zhang et al. 2000).

In packed-capillary CEC, the electro-osmotic flow (EOF) is predominantly caused by the charged surface of the particles in the column bed (Ross et al. 1996). An analogous situation occurs with chemically modified open tubular CEC, where the EOF is generated from the residual silanol groups of the etched and surface-modified capillary (Walhagen et al. 2000b; Matyska et al. 2001). The contributions to the EOF of the capillary wall and inner and outer surface of the particles, respectively, have been discussed elsewhere (Grimes et al. 2000; Liapis and Grimes 2000; Walhagen et al. 2000b). Thus, in CEC, the EOF velocity depends on the surface charge of the sorbent, the temperature, viscosity, and composition, such as dielectric constant, ionic strength, and pH, of the eluent (Walhagen et al. 2001). Moreover, in the implementation of CEC analytical procedures, the unique biophysical properties exhibited by peptides and proteins must be taken into consideration. A particular peptide or protein will display a defined isoelectric point (pI) value whereby the net charge of the molecule reaches zero with a buffer of specific pH value. The nature of the amino acid sequence and the extent of structure/folding that has occurred in the presence of a particular background electrolyte determine the effective charge and its relationship to the pI value (Walhagen et al. 2001). Apart from chromatographic interactions, which can be most conveniently adjusted via the organic modifier content in the background electrolyte, the electrophoretic migration velocity can also be tuned by varying the pH and/or the ionic strength of the buffer. It must be emphasized that the separation in CEC is governed by both chromatographic retention (thermodynamic process) and electrophoretic migration (kinetic process) (Walhagen et al. 2000a, 2001; Zhang et al. 2000). However, this book is not the platform to discuss in theoretical depth all of the relevant variables and their dependencies; instead, some general considerations toward the practical understanding of retention/migration processes are detailed. The major task throughout this technique will be to provide familiarization with instrumental aspects of this separation technique and how it is applied in peptide analysis (for a sketch of the instrumental setup, see Figure A2.3).

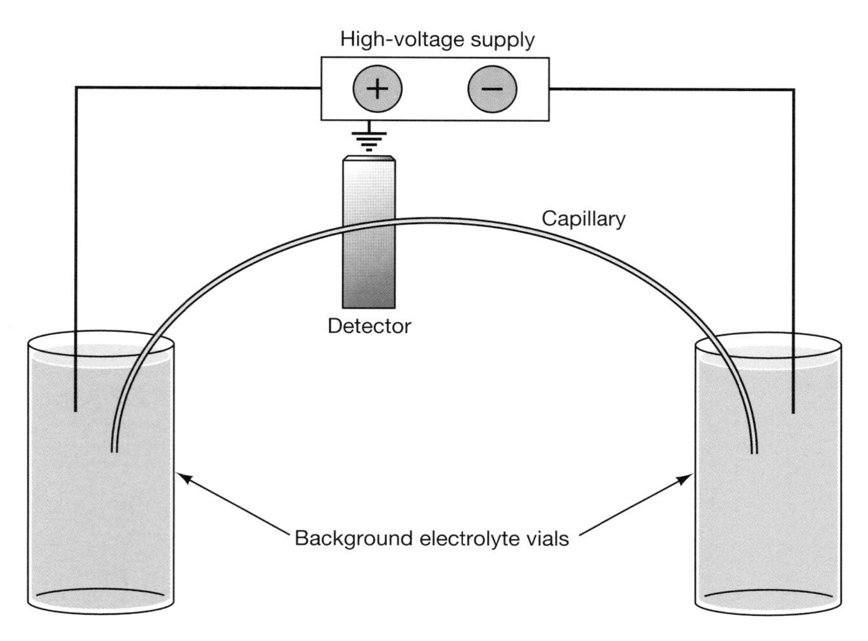

FIGURE A2.3. Schematic showing the instrumental setup for CEC separation of peptides or proteins.

MATERIALS

CAUTION: See Appendix 3 for appropriate handling of materials marked with <!>.

Reagents

H_3PO_4 <!> (or other acids and bases to prepare buffers)

HPLC-grade acetonitrile <!>, 2-propanol <!>, methanol <!>, or other suitable organic solvent with excellent transparency in the UV range down to 210 nm

Na_2HPO_4 (or other suitable salts) <!>

Thiourea <!> or sodium nitrate <!>

Equipment

Analytical C_{18} column

Other *n*-alkylsilica sorbents can be used in place of C_{18}. The average particle diameter of the sorbent should be 0.5–5 µm, porosity 60–300 Å, and a suitable phase ratio should be established. Pack the sorbent in analytical capillaries of the appropriate dimensions, e.g., 3 µm, 80 Å, 340 mm x 100 µm I.D. Sorbents of different surface chemistries can also be employed, such as ion exchange or mixed mode, to achieve alternative selectivities, provided the surface allows generation of a relatively high EOF under the experimental conditions.

CEC apparatus

The CEC equipment consists of a high-voltage supply, spectrophotometer with interface, analytical column, cartridge thermoelement, autosampler, PC, printer and software, CE system with attendant data management systems, and system automation controllers. The equipment system may be constructed in the lab from modular components or a commercially available instrument (e.g., the Agilent 3^D CE system).

Nitrogen, pressurized gas <!>

Nitrogen is used for autosampler operation or pressurization of background electrolyte vials.

Sonicator, benchtop (e.g., Eltrosonic Type 07, Kühn & Bayer)

Biological Sample

Proteins or peptides of interest

PREPARATION OF BACKGROUND ELECTROLYTE

1. Prepare 1 liter of each buffer stock solution at 100 mM. For example, weigh 14.196 g of Na_2HPO_4 and dissolve with 900 ml of H_2O. Adjust the pH with concentrated H_3PO_4 and then add H_2O to the final volume. Store the stock solutions at 4°C.
2. Prepare background electrolyte by diluting an aliquot of the buffer stock solution with the appropriate amount of H_2O and adding the desired amount of organic modifier, e.g., 100 ml of acetonitrile. As an example, 5 ml of stock solution, 45 ml of H_2O, and 50 ml of acetonitrile yield a 5 mM background electrolyte

   The molarity refers to the overall composition. The volumes of organic solvent, stock solution, and H_2O must be measured independently and are then combined because of the volume contraction that occurs upon mixing, which may be up to 30 ml/liter prepared solvent (depending on the nature of the organic solvent). Failure to do so can lead to substantial errors in mobile phase composition.

   Be aware that changes in the organic modifier content will also influence the pH of the eluent. By convention, the definition of the pH value used in the procedures described here only applies for H_2O-only solutions. For the background electrolytes, the pH detailed is really the pH of the buffer used in the background electrolyte.
3. Mix the solvent either by stirring with a Teflon-coated magnetic flea or by shaking in a stoppered cylinder (time depends on volume).

   WARNING: Parafilm should *not* be used under any circumstances to cover flasks or cylinders containing the eluents used in CEC, because the organic solvents in combination with buffer components may dissolve components in the Parafilm, yielding extraneous peaks in the chromatogram.
4. Filter all eluents through a 0.2-μm PTFE filter.

   Filtering of eluents increases the column lifetime and contributes to degassing.
5. Degas the background electrolyte by sonication.

   IMPORTANT: Degas solvents for use as background electrolyte extraordinarily thoroughly.
6. Close unused eluent bottles with a stopper to avoid evaporation of the organic solvent.

METHOD

Preparation of the Sample

1. Dissolve the protein sample in H_2O (weak mobile phase) to achieve the desired concentration. If the sample is insoluble, add small amounts (typically <25% of the total volume) of organic modifier (strong mobile phase).

   IMPORTANT: Do *not* dissolve the sample in background electrolyte, because buffers/additives may be changed frequently especially during method development.
2. Inspect the sample for clearness. Filter the sample through a 0.2-μm PTFE filter if it is insoluble, or opalescence or solid particles are present. Alternatively, centrifuge the sample and use the supernatant for injection.

   Samples that are not fully dissolved should not be injected into the column because they can block the column.
3. Store the sample at 4°C or –20°C (depending on the planned storage time and usage) until ready for use. Peptides and protein samples can degrade at room temperature and during storage.

   Avoid repeated freezing/thawing of the sample. Instead, prepare small aliquots of the peptide or protein sample, store them at –20°C, and use each aliquot only once.

Isocratic Separations of Peptides and Proteins with CEC

The elution window for peptides and proteins is known to be very narrow in comparison to other molecules; thus, it may prove helpful to construct a separation around data acquired in HPLC and/or capillary electrophoresis. However, if such data are not available, the investigator must anticipate the appropriate separation conditions, e.g., the stationary-phase properties, peptide composition, and conformational propensities. In the case of peptide samples of unknown composition or structure, such as would arise from the enzymatic digestion of a new protein, follow the strategy that mimics a course similar to that followed with micro-HPLC with narrow-bore columns.

The ionic strength of the background electrolytes applied most commonly ranges from 5 to 20 mM, and the volume fraction of organic modifier encompasses the wide range of 20–70%, depending on peptide charge and hydrophobicity for isocratic elution. The pH can be adjusted with various buffers from pH 3 to 8. Phosphate, acetate, and Tris buffers are popular for these adjustments, with acetate buffers of relatively low ionic strengths particularly suitable for CEC-MS applications due to their volatility.

A good starting point for finding a set of separation conditions is to utilize a 10 mM ammonium acetate buffer and to vary the organic modifier (e.g., acetonitrile) content in the background electrolyte. Once the elution window is determined, optimize the separation by varying other conditions. If the variation of conditions is performed too drastically in terms of EOF velocity alteration, cavity formation is likely to occur and the capillary must be reconditioned externally on an HPLC pump.

Testing the Instrumental CEC

1. Install the capillary in the instrument as detailed by the manufacturer or as appropriate if a lab-made instrument is utilized.

2. Supply the instrument with background electrolyte and sample(s).

3. Produce a blank run (inject background electrolyte) and run under the same conditions as those intended for the separation of the peptide or protein sample. Repeat if "ghost" peaks occur.

4. Produce a background run (inject the solution in which the sample is dissolved when different from the background electrolyte)

   Since buffers are frequently changed during the optimization procedure, dissolve samples in H_2O if possible or H_2O and an organic modifier to help solvation; *do not* dissolve samples in the buffer/background electrolyte. The importance of the background run in establishing an acceptable separation with the peptide or protein mixture cannot be overemphasized.

5. Measure the dead time of the column with thiourea or sodium nitrate (or any other non-interactive solute).

   The choice of the dead time marker is dictated by the separation conditions. Neutrality must be secured for the conditions in question if accurate values of the EOF are to be determined.

6. Test the column performance with an isocratic run and an appropriate test mixture, e.g., a suitable HPCEC test mixture included 0.15% (w/v) dimethylphthalate, 0.15% (w/v) diethylphthalate, 0.01% (w/v) diphenyl, 0.03% (w/v) *O*-terphenyl, and 0.32% (w/v) dioctylphthalate in methanol (Walhagen et al. 2000a,b, 2001).

   First, this test allows the evaluation of the column bed integrity (low integrity will be associated with split, fronting, or tailing peaks) and column performance (in terms of plate numbers). Second, this test allows, if repeated at regular intervals, the monitoring of the performance during the lifetime of a column and the assessment of batch-to-batch differences of column fillings.

Analytical RPCEC of Peptides and Proteins

1. Prior to sample analysis, conduct at least one blank run under the same conditions as those intended for the peptide sample. Repeat if "ghost" peaks occur.

2. Separate the sample with an analytical RPCEC procedure in order to assess the sample in terms of purity, peak profile, and elution conditions with the experimental conditions as outlined below.

3. Perform peak assignment.

Selection of the Electrochromatographic Conditions for the Analytical Separation of Peptide and Protein Mixtures

The following electrochromatographic conditions are representative of the type of standard analytical protocol that can be employed. Numerous variations in terms of choice of reversed-phase sorbent, applied voltage, run duration, background electrolyte composition (including the choice of different organic solvent modifiers, buffer species, or ion-pairing reagents) and pH, temperature, and sample size (controlled by injection mode/duration) can be contemplated. The following isocratic procedure represents a robust protocol around which these variations can be constructed.

- Capillary column: A C_4, C_8, or C_{18} reversed-phase sorbent packed into a suitable column (e.g., 3 µm, 300 Å, 330 mm x 100 µm I.D.)
- Background electrolyte: Acetonitrile/5 mM Na_2HPO_4 buffer (pH 7.5) (50:50 v/v)
- Isocratic elution
- Injection: 5 kV, 3-second sample, 5 kV, 3-second background electrolyte (see also the panel on ADDITIONAL REMARKS AND CONCLUSIONS below)
- Electric: 15 kV
- Detection: 214 nm
- Temperature: Room temperature

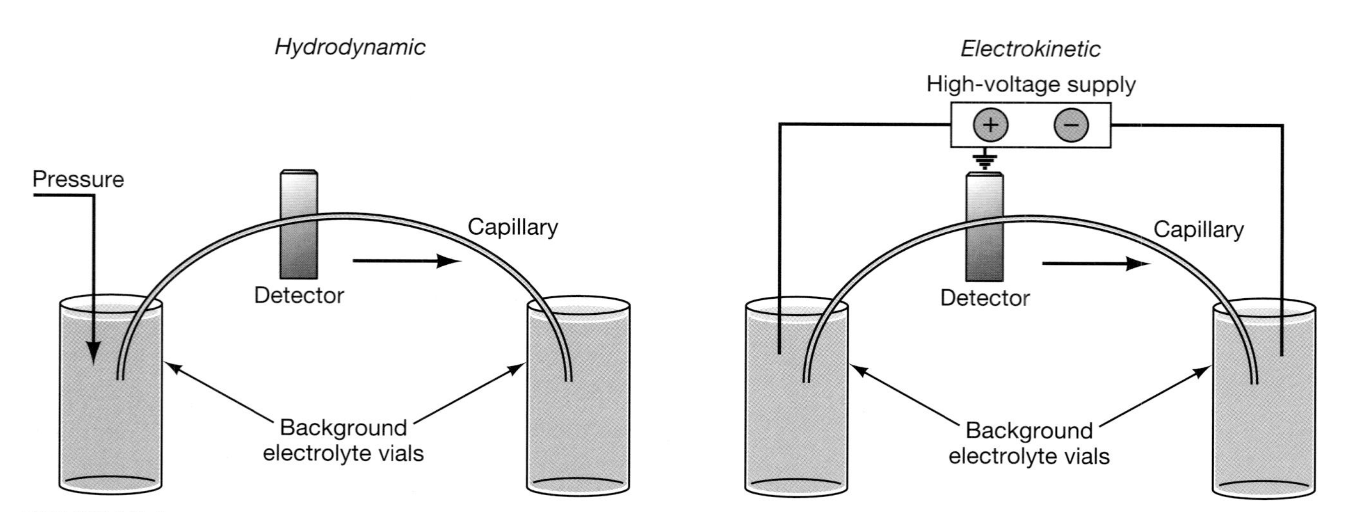

FIGURE A2.4. Schematic illustrating the different injection modes based on pressure-driven hydrodynamic flow as used in µ-HPLC and electrokinetic flow as used in CEC. The direction of liquid flow is highlighted by arrows.

Additional Remarks and Conclusions

- ***Injection:*** There are several choices for CEC injection (see Figure A2.4): hydrodynamic injection (by pressure) and electrokinetic injection (applying voltage, current, or power). As opposed to capillary electrophoresis, injection by voltage is the most frequently used injection technique in CEC, despite the obvious problem with differences in injection concentrations for molecules with different mobilities. Reproducible hydrodynamic injection is not applicable due to the nature of the sorbent immobilized in the capillary. High pressures cannot currently be delivered accurately enough to the capillary to favor this injection mode. In the various examples found in the literature, the injection consists of two individual parts. The first is the injection of the sample itself followed by injection of background electrolyte, thus exploiting a focusing effect if the electrophoretic velocity of the background electrolyte is faster as compared to the components of the sample.
- ***Electrical polarity effects:*** Most instruments allow for reversal of the electric field polarity. This feature can be exploited to reverse the direction of flow. Voltage, current, and power are related to one another by the resistance of the capillary/background electrolyte system; thus, mechanically stable frits may experience disintegration in an electric field applied across the capillary inner diameter. In the constant voltage mode, current can be utilized to indicate the stability of the system. Rapid alteration or decrease of current thus displays cavity formation or a clogged or broken capillary and can thus be used to identify unstable or inappropriate conditions.
- ***Operation mode:*** Most instruments (this applies for lab-made as well as commercially available instruments) require capillaries designed to consist of a packed bed and an open tubular segment. In instruments with fixed locations for the detection capillaries, these instruments can be operated either in the "regular" operation mode or in the mode referred to as "short end" operation mode with the injection and flow direction adapted appropriately (see conditions for CASE STUDY and the legend to Figure A2.5).

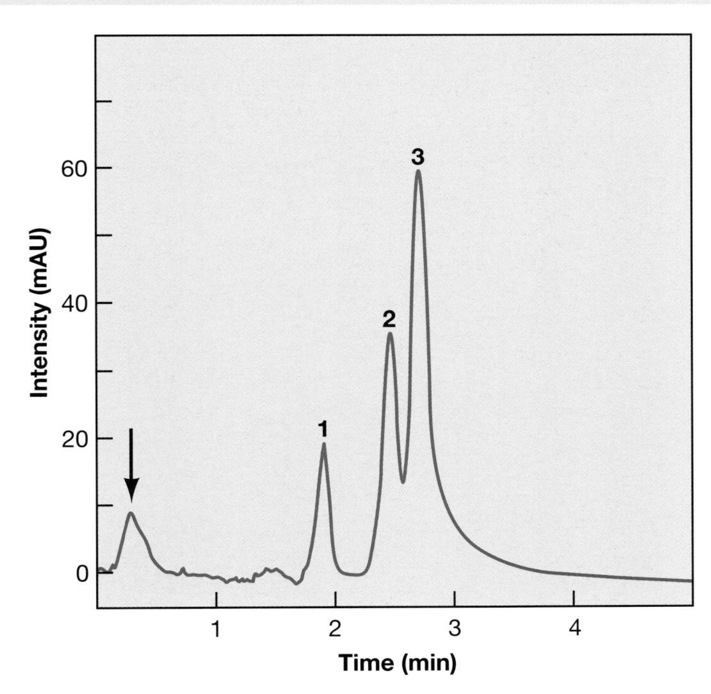

FIGURE A2.5. Electrochromatogram of three synthetic TRAP (thrombin receptor antagonist peptide) analogs. Note that peptides assigned to peaks 2 and 3 only differ in order of an alanine residue substitution of the corresponding amino acid residue in the sequence. Hydrodynamic injection results in the first peak (as an eigen peak not assigned) in the electrochromatogram (for other instrumental details, see text). Sequences:

Peptide peak 1: H-Ser-Ala-Leu-Leu-Arg-Asn-Pro-OH
Peptide peak 2: H-Ser-Phe-Ala-Leu-Arg-Asn-Pro-OH
Peptide peak 3: H-Ser-Phe-Leu-Ala-Arg-Asn-Pro-OH

Case Study

Three synthetic thrombin receptor antagonist peptide analogs (TRAP analogs) were analyzed by CEC. The capillary column employed was operated in the short-end operational mode (see also ADDITIONAL REMARKS AND CONCLUSIONS above). The electrochromatogram is displayed in Figure A2.5. The separation conditions are detailed below:

- Instrument: Agilent Technologies ^{3D}CE modified with external pressure supply
- Stationary phase: Noncommercial reversed-phase (C_{18}) porous silica (3 µm, 80 Å)
- Column: 330 mm x 100 µm I.D.; effective length (packed bed length including frits) 8 cm
- Background electrolyte: Acetonitrile/5 mM Na_2HPO_4 buffer (pH 7.5) (20:80 v/v)
- Elution mode: Isocratic
- Operation mode: Short end
- Injection: 12 bar, 1 second
- Electric: –10 kV
- Detection: 200 nm
- Temperature: 15°C (cartridge), 25°C (sample tray)
- Pressurization: 9 bar to both vials (inlet and outlet)

TECHNIQUE 5

Automated Two-dimensional LC-MS/MS for Large-scale Protein Analysis

Toshiaki Isobe,[*†] Yoshio Yamauchi,[†] Masato Taoka,[*] and Nobuhiro Takahashi[†§]

[*]Department of Chemistry, Graduate School of Science, Tokyo Metropolitan University, Tokyo 192-0397, Japan; [†]Integrated Proteomics System Project, Pioneer Research on Genome the Frontier, MEXT, c/o Department of Chemistry, Graduate School of Science, Tokyo Metropolitan University, Tokyo 192-0397, Japan; [§]Department of Applied Biological Science, and Department of Biotechnology, United Graduate School of Agriculture, Tokyo University of Agriculture and Technology, Tokyo 183-8509, Japan

AMONG THE PROTEIN SEPARATION AND IDENTIFICATION METHODS currently in use, two-dimensional polyacrylamide gel electrophoresis (2D-PAGE) followed by mass spectrometry (MS) is the most widely used. 2D-PAGE resolves thousands of proteins in a single gel. However, the obstacles to high-throughput large-scale protein analysis, a key issue in proteomics, by 2D-PAGE are manifold.

- The identification of these proteins requires time-consuming steps to digest the proteins within the gel, extract individual protein "spots" from the gel, and prepare the extracted samples for MS analysis.
- Experimental 2D-PAGE has not been fully automated.
- Some cellular proteins are rarely isolated in 2D-PAGE studies, such as membrane-associated proteins and those with extremes in either pI or molecular weight.

Nearly two decades ago, we presented an automated 2D high-performance liquid chromatography (2D-HPLC) technique for systematic separation of very complex peptide or protein mixtures (Takahashi et al. 1985, 1991; Isobe et al. 1991b), and applied the technique to the sequencing of very large proteins and protein genetic variants (Takahashi et al. 1984, 1987a,b), and for the profiling of proteins expressed in developing rat cerebella (Isobe et al. 1991a; Taoka et al. 1992, 1994). This technique has now been refined by incorporating new MS technology, such as replacing a conventional UV detector by a mass spectrometer with an electrospray ionization (ESI) source. This technique describes a new platform for proteomics: on-line 2D-LC-MS (MS/MS), which provides a fully integrated analytical system for non-gel-based large-scale protein identification.

THE PRINCIPLE OF 2D-LC IS SIMILAR TO THAT OF 2D-PAGE

2D-PAGE separates proteins according to their intrinsic charge and molecular weight by the combination of isoelectric focusing (IEF) and SDS-PAGE. The two-dimensional chromatographic technique described here employs two columns having different separation specificities:

- An ion-exchange column that separates proteins and peptides according to charge.
- A reversed-phase column that separates molecules on the basis of hydrophobicity.

Chromatography is performed sequentially by the two columns under time-dependent control of the flow system for each column. The resolution power of the chromatography is multiplied by the combination of the two columns. For example, if each column has a peak capacity of 50, then the theoretical separation capacity of the 2D-LC system is 50 x 50 = 2500. Integration of MS adds significant advantages to the 2D-LC technique. With the ability of MS to separate biological samples very precisely according to molecular mass, direct coupling of MS with 2D-LC further increases the resolution of the 2D-LC system. When combined with automated tandem mass spectrometry (MS/MS), the system also provides structural information on protein samples simultaneously upon detection. Thus, ~10,000 MS/MS spectra can be obtained in a typical 2D-LC-MS/MS analysis of extremely complex peptide mixtures, such as a tryptic digest of crude cellular extracts, from which thousands of peptides can be assigned using the available gene and protein databases.

INSTALLATION OF THE 2D-LC SYSTEM

Several installations are possible for performing automated 2D-LC-MS (MS/MS). For instance, Link et al. (1999) have developed a system coupled with ESI-MS, in which a single capillary column is packed with two independent chromatography phases in series (also see Chapter 8, Protocols 4 and 5). The system presented here is composed of two independent, conventional HPLC assemblies with an ion exchange or a reversed-phase column connected in tandem through a six-way column-switching valve, and a mass spectrometer with ESI source. In this system, the elution of solutes from the ion-exchange column (IXC) is stopped during the chromatography in order to perform the reversed-phase chromatography (RPC),

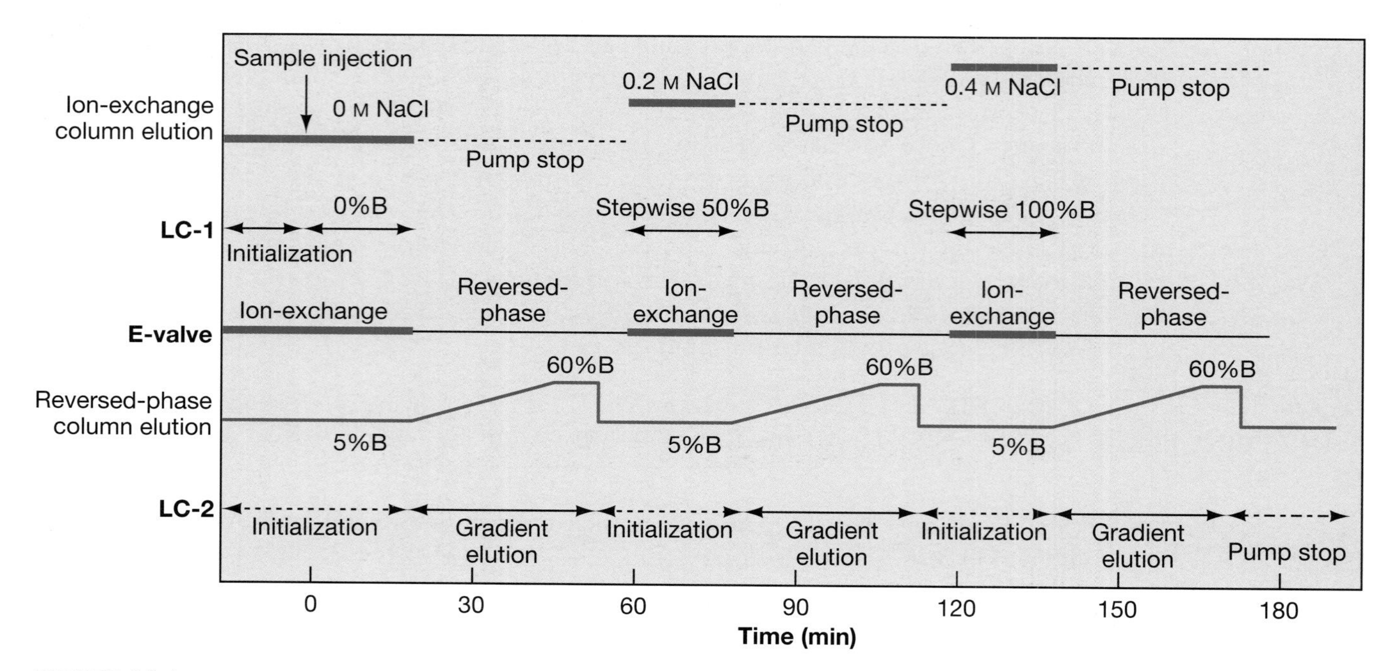

FIGURE A2.6. Time-dependent elution program of 2D-LC system. The program synchronizes ion-exchange and reversed-phase chromatography by controlling two LC assemblies (LC-1 and LC-2) and an electrical six-way column-switching valve (E-valve).

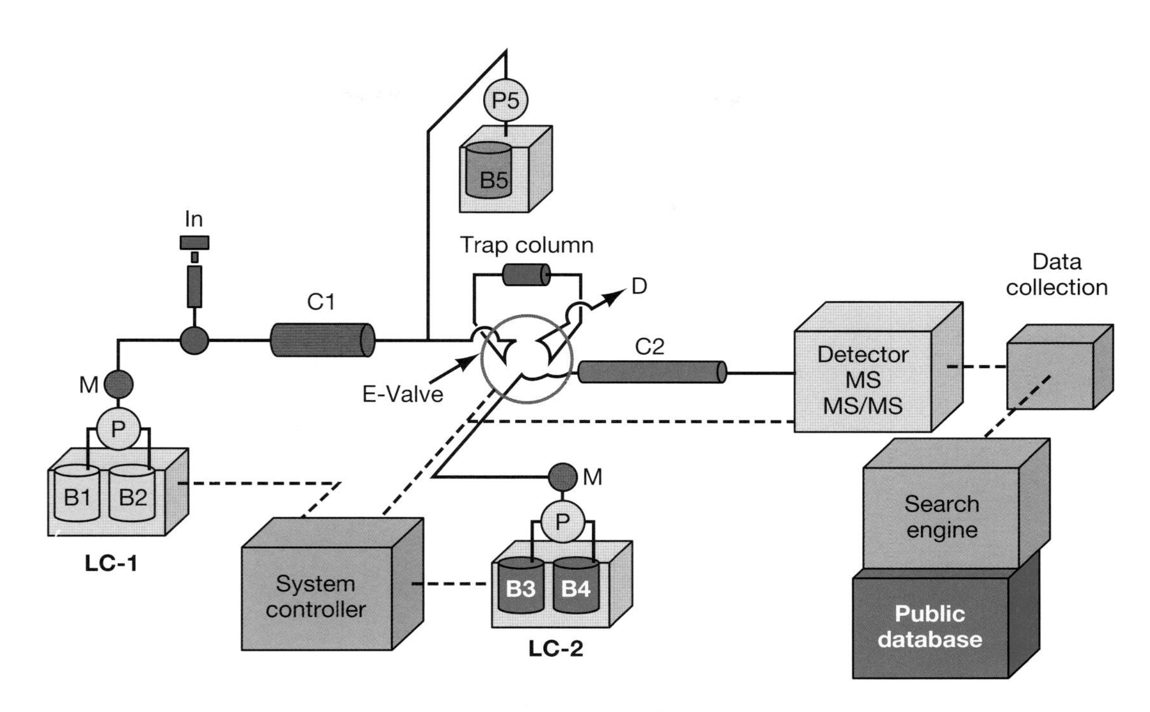

FIGURE A2.7. Schematic diagram of the 2D-LC-MS (MS/MS) system. The assemblies are explained in the text. (M) Manifold; (D) drain.

to which the eluent from the IXC is applied. After the completion of the chromatography of the RPC, chromatography of the IXC is started again; the eluent is applied to the RPC, and the RPC is repeated as in the first step. These separation cycles are repeated over and over again by a PC-based control program (see Figure A2.6). Figure A2.7 shows a flow diagram of the system, together with the multiple column assembly connected to the column-switching valve. This system is equipped with a small reversed-phase "trap" column inserted between the two analytical columns that serves to remove salts from peptides and proteins eluted from the IXC. This installation is important for reproducible analysis as the salts prevent efficient ionization of samples on MS analysis and often cause troubles in the system. A PC program controls the two HPLC assemblies and the column-switching valve to synchronize the ion-exchange and the reversed-phase chromatography. The flow system is followed by a series of PCs for MS data collection and handling.

MATERIALS

CAUTION: See Appendix 3 for appropriate handling of materials marked with <!>.

Reagents

Ammonia, aqueous <!>

Ammonium bicarbonate buffer (10 mM, pH 8.0) <!>

Column buffers and solvents:

Buffer B1: 0.025 M Tris-Cl buffer (pH 8.0)

Buffer B2: 0.025 M Tris-Cl buffer (pH 8.0) containing 0.4 M NaCl

Solvent B3: 0.2% formic acid <!> in H_2O
Solvent B4: 0.2% formic acid in acetonitrile <!>
Solvent B5: 1% formic acid in H_2O
Dithiothreitol (DTT) <!>
HCl (6 M) <!>
Iodoacetamide (IAA) <!>
TPCK-treated trypsin (Worthington Biochemicals)
Tris-Cl (0.5 M, pH 8.5) containing 7 M guanidinium hydrochloride <!> and 10 mM EDTA

Equipment

Columns
First dimension: Bioassist Q (5 µm, 2 mm I.D. x 35 mm) (Tosoh Corp.)
Second dimension: Mighty-sil C_{18} (3 µm, 0.32 mm I.D. x 100 mm) (Kanto Chemicals, Tokyo)
Trap column; Mighty-sil C_{18} (15 µm, 1 mm I.D. x 5 mm) (Kanto Chemicals, Tokyo)
HPLC assemblies (Shimadzu) with an ion exchange or a reversed-phase column connected in tandem through a six-way column-switching valve (VICI)
Mass spectrometer with ESI source (QTOF 2, Micromass)
Nitrogen gas <!>

Biological Sample

Protein mixture, lyophilized

METHOD

Preparation of the Protein Sample

1. Dissolve 200–500 µg of protein in 200 µl of 0.5 M Tris-Cl (pH 8.5) containing 7 M guanidinium hydrochloride and 10 mM EDTA. Bubble the solution with nitrogen gas for 10 minutes.

2. Add 100 µg of DTT (~3 mM final concentration) to the protein solution at room temperature with gentle mixing. Incubate the solution for 2 hours at room temperature.

3. Add 250 µg of IAA (~7 mM final concentration) and leave the solution in the dark for 1 hour at room temperature.

4. Dialyze the solution for 8 hours against four changes of 10 mM ammonium bicarbonate buffer (pH 8.0).

   If the amount of protein in the solution is below 100 µg, the solution can be dialyzed using a small dialysis unit (Pierce Chemical Company) to minimize adsorption to the dialysis membrane. Because the sample solution is often viscous, neither ultrafiltration nor gel filtration is recommended for desalting.

5. Add 1–2 µg of TPCK-treated trypsin into the protein solution and digest overnight at 37°C.

6. Acidify the digest to pH 2 by adding an aliquot of 6 M HCl and, where necessary, remove any precipitates by centrifugation.

7. Neutralize the supernatant with aqueous ammonia to pH 8, dilute the peptide mixture with an equal volume of H_2O, and apply to the 2D-LC-MS/MS system.

Two-dimensional Liquid Chromatography

Refer to Figure A2.7 while carrying out Steps 8–18.

8. Activate the computer program that controls the HPLC systems. Steps 9–18 are performed automatically. The total time required to analyze the protein sample using this protocol is 12 hours.

9. Equilibrate columns C1 (ion-exchange) and C2 (reversed-phase) for 20 minutes with Buffer B1 and Solvent B3, respectively, at flow rates of 100 μl/min (for column C1) and 5 μl/min (for column C2).

10. Apply the protein sample to column C1 through the sample injector (In), and elute with Buffer B1 for 5 minutes at a flow rate of 100 μl/min.

11. Mix the eluent on-line with Solvent B5, which is continuously supplied by pump 5 (P5). The mixture flows into a reversed-phase "trap" column, where salts are separated from peptides and proteins by continuous washing with Solvent B5.

12. LC-1 is automatically stopped, and after 2 minutes, the column-switching valve moves to connect the trap column to column C2.

13. Introduce the salt-free samples adsorbed on the trap column into the reversed-phase column (column C2) and separate them for 40 minutes with a linear gradient formed from Solvents B3 and B4 (5–60% acetonitrile in 0.2% formic acid) at a flow rate of 5 μl/min.

14. Load the eluted samples directly into the ion source of the mass spectrometer. MS and MS/MS data are collected automatically.

15. Equilibrate column C2 again with Solvent B3 for 15 minutes after the linear gradient elution is finished. The first cycle of the 2D-LC is completed.

16. The computer reactivates the LC-1 system, so that proteins are again eluted from the ion-exchange column C1 for 5 minutes at a flow rate of 100 μl/min, this time with a mixture of Buffers B1 and B2 (95% B1:5% B2).

17. Repeat Steps 11–15.

18. Repeat the separation cycles eight more times, each time changing the ratio of Buffers B1 and B2 used to elute proteins from the ion-exchange column. Below is the percentage of Buffer B2 in each of the ten cycles.

Cycle	1	2	3	4	5	6	7	8	9	10
% B2	0	5	10	15	20	25	30	40	50	100

Data Collection and Database Searching

MS/MS data are generally converted to text files listing mass values and intensities of the fragment ions. The files are analyzed using appropriate search engines that scan gene and protein databases and report back with peptide assignments and putative identification of the proteins. Figure A2.8 and Table A2.4 show the results of an analysis of an insoluble protein fraction from *Caenorhabditis elegans*. The collected MS/MS data were converted to text files by MassLynx software (Micromass, Manchester, UK) and analyzed by the Mascot Daemon algorithm (Matrix Science, London, UK) to assign peptides and identify proteins from the *C. elegans* database (National Center for Biotechnology Information). The database search was performed with the following parameters:

- Fixed modification: carbamidemethyl (Cys)
- Variable modification: *N*-acetyl (Protein), oxidation (Met), phospho (Ser, Thr, Tyr), pyro-Glu (amino terminal)
- Maximum missed cleavages: 3
- Peptide molecular weight tolerance: ±500 ppm
- Peptide charge number: +2 and +3
- MS/MS tolerance: ±0.8 dalton

TABLE A2.4. Portions of a small sample of proteins identified by 2D-LC-MS/MS analysis of the insoluble protein fraction of *C. elegans*

Protein name	Accession number	Number of peptides
(U18546)NaK-ATPase alpha subunit[Caenorhabditis elegans]	gi|604515	15
(AF148953)myotactin formB[Caenorhabditis elegans]	gi|5002308	14
hypothetical protein K11D9.2a-Caenorhabditis elegans	gi|7505778	13
hypothetical protein W04D2.1a-Caenorhabditis elegans	gi|7508956	5
hypothetical protein Y57G11C.15-Caenorhabditis elegans	gi|7510283	3
hypothetical protein C47E12.5-Caenorhabditis elegans	gi|7497568	3
hypothetical protein T22D1.4-Caenorhabditis elegans	gi|7508220	3
hypothetical protein D2023.7-Caenorhabditis elegans	gi|7498188	2
(Y14949)SURF-4 protein[Caenorhabditis elegans]	gi|2440040	2
hypothetical protein F33D11.11-Caenorhabditis elegans	gi|7500388	2
(AF036692)Hypothetical protein C44B12.1[Caenorhabditis elegans]	gi|14573941	2
NADH-UBIQUINONEOXIDOREDUCTASECHAIN1	gi|128633	2
(AF040646)Similar to sugar transporter coded for by C. elegans cDNAy	gi|13592445	2
cytochrome b[Caenorhabditis elegans]	gi|5834888	1
hypothetical protein F57F4.3-Caenorhabditis elegans	gi|7504539	1
hypothetical protein C12D12.2-Caenorhabditis elegans	gi|7495879	1
hypothetical protein C14F11.3-Caenorhabditis elegans	gi|7496015	1
PUTATIVE GLUCOSYLTRANSFERASE C08H9.3	gi|1731299	1
INTEGRIN BETA PAT-3 PRECURSOR	gi|13431773	1
synaptobrevin SNB-1-Caenorhabditis elegans	gi|7511661	1

The table lists protein name, gene identification number, and number of peptides assigned to identify each protein. Note that all of the hypothetical proteins were predicted to be membrane-associated proteins by the SOSUI algolithm (http://sosui.proteome.bio.tuat.ac.jp/sosuiframe0.html).

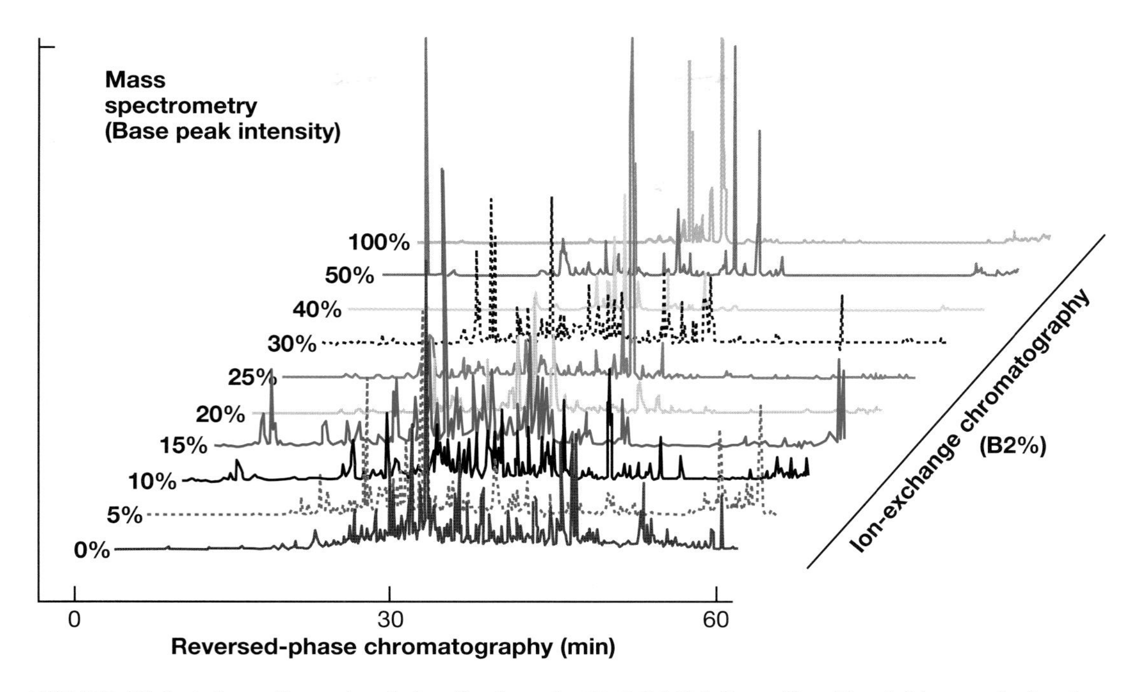

FIGURE A2.8. A three-dimensional visualization of a 2D-LC-MS/MS profile of insoluble protein fraction of *C. elegans*.

CONCLUSIONS

Technique 5 detailed an analytical platform for proteomics, 2D-LC-MS (MS/MS), which provides a fully integrated approach to non-gel-based protein profiling and identification. The high resolution of the system is achieved by sequential multidimensional chromatography on ion-exchange and reversed-phase columns combined with high-resolution MS, while reproducibility is maintained by the automation of the total system. The technique is amenable for analysis of expression profiles of cellular proteins and identification of proteins in highly complex protein mixtures. In particular, it is very powerful for large-scale analysis of protein components in biological complexes, such as functional membrane domains and cellular organelles. We are currently applying this technique for the analysis of functional protein/membrane complexes, rafts, and cellular organelles such as the endoplasmic reticulum. However, several obstacles must be overcome during future development of this LC-based technology.

- The resolution of large proteins is limited when cellular proteins are directly analyzed. This is probably because large proteins bind very tightly to the reversed-phase column and elute with low recovery during the chromatography.
- The peptide-based 2D-LC-MS/MS analysis following protease digestion of protein mixture allows mass identification of proteins regardless of their physicochemical characteristics, molecular size, and solubility, but it makes quantitative information of cellular levels of proteins ambiguous. To analyze quantitative changes in proteins, the method must therefore be combined with an appropriate technology for mass spectrometric quantitation, such as isotope labeling of cells (Gygi et al. 1999).

- The informatics for data handling, processing, and evaluation must be improved. The 2D-LC-MS/MS technology produces a large amount of MS/MS data in 12 hours and typically reports back with a list of more than 1000 peptides and several hundred candidate proteins. At present, this data processing step is not fully integrated and usually requires several days to complete. The evaluation of the final data is extremely time-consuming not only because of the lack of a complete knowledge database, but also because of a lack of a common standard among protein and gene entries in the current genome and protein databases. In many cases, there are more than ten entries for one protein and each has a different accession number in the database. Far more confusing is that one cannot determine if a slight difference in amino acid sequence between two protein entries indicates different proteins or merely artifacts due to experimental error.

A Case Study: Analysis of a Portion of the *C. elegans* Proteome

The system presented here is applicable for protein and peptide separations. In MS mode, proteins and peptides are separated by 2D-LC and detected by MS, giving rise to the elution profile of proteins and peptides with additional information on the molecular mass. As reported earlier on our prototype system, this profile could be used for analysis of expression levels of cellular proteins and their dynamics (Isobe et al. 1991b; Taoka et al. 1992). In general, the 2D-LC-MS system can resolve hundreds of proteins expressed in cells and is therefore a good supplement to 2D-PAGE. For this purpose, a simple mass spectrometer with an electrospray ionization source, such as ESI-TOF-MS without a collision cell, can be used, and an LC-based "protein profiling" system can be constructed that is fully automated from sample injection to data collection. However, this system cannot provide information on protein identification, because the mass data of a protein molecule is insufficient to specify a particular protein, unless candidates are predicted.

The 2D-LC-MS/MS system is extremely powerful for peptide analysis. It can separate peptides to much higher resolution than proteins, with simultaneous collection of the precise molecular mass of each peptide and MS/MS data for internal amino acid sequence. In fact, the system can separate and identify all of the tryptic fragments of human serum albumin in a single chromatographic run, except for several small peptides that are not adsorbed to the reversed-phase column. The most significant application of this system is mass identification of proteins in crude extracts of cells or identification of protein components in functional complex and cellular organelles. Combined with the automated generation of collision-induced dissociation (CID) of peptide fragments and computer-assisted retrieval of the spectra, one can expect to assign thousands of peptides in a highly complex peptide mixture generated with protease digests of crude protein mixture, from which hundreds of proteins are generally identified from the gene and protein databases. This is possible because a very precise peptide mass value with a portion of the amino acid sequence (sequence tag) is often sufficient for unequivocal assignment of one corresponding peptide and identification of one particular protein in the database. Because the identification depends on a part of the protein structure (e.g., peptide mass and sequence tag), this technique is applicable to any protein mixture, even intractable ones, such as those that include membrane-bound proteins. Figure A2.8 shows one such example, in which tryptic digests of an insoluble protein fraction of the nematode *C. elegans* are analyzed directly by this system. Although the figure illustrates total ion chromatogram (TIC), a plot of base peak intensities during total ion monitor of chromatography, the automated MS/MS analysis actually generated almost 8000 MS/MS spectra, which allowed ~1200 peptide and 400 protein identifications to be assigned (Yamauchi et al., unpubl.). Table A2.4 illustrates a portion of the proteins identified by this system. The table includes many "hypothetical" proteins presumably located in the membrane that are identified in the nematode genome but have not yet been characterized as functional proteins.

TECHNIQUE 6

Modification of Phosphoamino Acids to Facilitate Identification

Alastair Aitken

Division of Biomedical and Clinical Laboratory Sciences, University of Edinburgh, Scotland

IN ADDITION TO DIRECT SEQUENCING OF PHOSPHOPEPTIDES, methods have been developed that convert the phosphoamino acid to a more stable derivative. These are generally applicable to phosphoserine and, to a lesser extent, phosphothreonine. Only one of these methods has proved particularly useful in practice when multiple phosphorylation sites are encountered (as is frequently encountered).

- Phosphoserine can be identified after conversion to β-methylaminoalanine by β-elimination and addition of methylamine (Annan et al. 1982).
- The conversion of phosphoserine to alanine by β-elimination followed by reduction with sodium borohydride has also been used (Richardson et al. 1978). *O*-glycosylated serine residues are similarly converted to alanine.
- During these reactions, the corresponding threonine derivatives are converted, with lower efficiency, to 2-aminobutanoic acid. Therefore, independent confirmation of the presence of phosphate in a given peptide should be sought, because *O*-glycosylated serine would also undergo β-elimination and reduction during the derivatization procedure. In this case, however, sequence analysis of the unmodified peptide would not give a PTH-serine at the cycle of appearance of the glycosylated residue, whereas a substantial amount of the dithiothreitol (DTT) adduct of dehydroserine would be expected in the case of a phosphorylated residue.
- A particularly useful method, which is detailed in this technique, exists for the derivatization of phosphoserine to the stable adduct *S*-ethylcysteine (Meyer et al. 1986) by β-elimination of the phosphate followed by addition of ethanethiol. Due to the alkaline conditions required for the conversion to *S*-ethyl-Cys, Asn and Gln may be deamidated and identified as PTH-Asp or PTH-Glu. This procedure may be used for the selective isolation of phosphoseryl peptides (Holmes 1987). A derivatized peptide from a doubly phosphorylated species will elute correspondingly later than the singly derivatized species, indicating the applicability of the method to multiple phosphoseryl peptides. HPLC before and after derivatization should produce highly purified peptides even from a very complex mixture because the elution position of all others will be unaffected. The dehydroalanine intermediate from amino-terminal phosphoserine peptides will rearrange to pyruvate and carboxy-terminal phosphoserine peptides will give ethylamine.
- It has been possible to raise antiphosphoamino acid antibodies to particular phosphopeptide sequences in proteins. Although there are commercial antibodies claimed against phosphoserine or phosphothreonine, these do not appear to be generally successful for identification of these derivatives. In contrast, the use of antiphosphotyrosine antisera is well established.
- The development of metal-ion affinity chromatography to specifically isolate phosphoproteins has also been useful (see Chapter 9, Protocols 1, 3, and 4).

MATERIALS

CAUTION: See Appendix 3 for appropriate handling of materials marked with <!>

Reagents

Dimethylsulfoxide (DMSO) <!>
Ethanethiol (10 M) <!>
Ethanol <!>
Glacial acetic acid <!>
NaOH (5 M) <!>

Equipment

Heat block preset to 50°C

HPLC solvents

The HPLC solvents are used to prepare linear gradients of H_2O and acetonitrile <!> in 0.1% trifluoracetic acid <!> (for additional details, see Chapter 5).

Nitrogen <!>

RP-HPLC column (e.g., Vydac C_{18})

See text preceding Step 1.

Biological Sample

Phosphopeptides of interest

METHOD

Selective Isolation of Phosphoseryl Peptides from a Complex Mixture

Steps 1, 2, 8, and 9 are additional and necessary if the phosphopeptides are part of a mixture of proteins (e.g., a cell extract). If purified phosphopeptides are being modified, only Steps 3–7 need be carried out.

1. (*Optional*) Apply the underivatized peptides to an RP-HPLC column.

2. (*Optional*) Elute the peptides under standard conditions (see Chapter 5). Collect fractions containing the phosphopeptide(s).

Derivatization of Phosphoserine Residues

3. Prepare incubation mixture in a microfuge tube by combining:

H_2O	200 µl
DMSO	200 µl
ethanol	100 µl
5 M NaOH	65 µl
10 M ethanethiol	60 µl

WARNING: Ethanethiol has an extremely noxious odor. Carry out manipulations in a chemical fume hood.

4. Dissolve the peptide in the tube of incubation mixture.

5. Flush the tube with nitrogen and incubate the reaction for 1 hour at 50°C.

   An extended reaction time (18 hours at 50°C) may be used because the β-elimination step of derivatization of a phophoserine adjacent to a proline residue (particularly if the proline is carboxy-terminal) is slow. This can be carried out with acetonitrile, which minimizes subsequent manipulations (Aitken et al. 1995).

6. After cooling the tube on ice to minimize potential hazard, add 10 μl of glacial acetic acid in small aliquots to neutralize the solution.

7. Either apply the derivatized peptide directly to a protein sequencer or concentrate it by vacuum centrifugation.

   The elution position of this PTH derivative on HPLC during automated sequencing is well-characterized.

Separation of *S*-ethylcysteinyl Peptides from Underivatized Peptides

8. Apply the impure *S*-ethylcysteinyl peptides to the same RP-HPLC column used in Step 1.

9. Elute under the same conditions as in Step 2. The derivatized peptide(s) can be expected to emerge at 4–5% acetonitrile later than the native phosphopeptide.

TECHNIQUE 7

Identification of Phosphoamino Acids by Thin-layer Electrophoresis

Alastair Aitken

Division of Biomedical and Clinical Laboratory Sciences, University of Edinburgh, Scotland

THE ANALYSIS OF PHOSPHOAMINO ACIDS is frequently confined to qualitative or semiquantitative determinations based on their separation by 1D or 2D thin-layer or paper electrophoresis. This is still an extremely useful technique because only inexpensive equipment is required, and the spots corresponding to phosphoamino acids may be excised and their radioactivity (^{32}P) quantitated in a scintillation counter.

MATERIALS

CAUTION: See Appendix 3 for appropriate handling of materials marked with <!>.

Reagents

HCl (6 N) <!>
HPLC-grade H_2O
pH 1.9 electrophoresis buffer (acetic acid <!>:formic acid <!>:H_2O 4:1:45 v/v/v)
pH 3.5 electrophoresis buffer (pyridine <!>:acetic acid:H_2O 1:10:89 v/v/v)
Phenol (0.5% v/v) <!>
Phosphoserine, phosphothreonine, and phosphotyrosine (1 µg/µl [i.e., ~5 nmole/µl]) (Sigma)
Ninhydrin spray (0.2% ninhydrin w/v in ethanol) <!>

Equipment

Argon (pure, sequencer grade) <!>
Centrifugal evaporator (e.g., SpeedVac, Thermo Savant, or equivalent)
Electrophoresis tank
Glass test tubes, small
  Number the hard-glass test tubes with a diamond scriber.
Liquid scintillation counter
Oven or block heater preset to 110°C
Plastic-backed thin-layer cellulose sheets (e.g., Camlab, Machery-Nagel)
Screw-top glass or PTFE containers for hydrolysis of peptides (Waters-Milligen or Ciba-Corning)

Biological Sample

Phosphopeptide samples of interest, radiolabeled with ^{32}P <!>

METHOD

1. After desalting in a volatile buffer, transfer the phosphopeptide samples to small test tubes, previously numbered. Remove the volatile sample buffer in vacuo, either in a desiccator or a centrifugal evaporator.
2. Place the numbered tubes in the hydrolysis container(s) with 2 ml of 6 N HCl and 0.5% (v/v) phenol. Flush the container with pure argon for 2 minutes, close the container, and evacuate it if possible.
3. Heat the container in an oven or block heater for 1–4 hours at 110°C to hydrolyze the ^{32}P-labeled phosphopeptides.

   The recovery of phosphoamino acids varies with the primary structure around the phosphorylation site as well as the acid lability of the phosphoamino acid. The order of lability of free phosphoamino acids is phosphotyrosine > phosphoserine > phosphothreonine.
4. Cool the hydrolysis tubes, and then quickly transfer them to a centrifugal evaporator and evacuate the tubes without initially turning on the rotation (to prevent washing down traces of HCl into the tubes). Dissolve the residues in 50 μl of HPLC-grade H_2O and evaporate them in a centrifugal evaporator (in the conventional manner) to dryness to remove traces of HCl.
5. Dissolve the dried residue in 5–10 nmoles each of unlabeled phosphoserine, phosphothreonine, and phosphotyrosine.
6. Subject the samples to electrophoresis at pH 1.9 in the first dimension and pH 3.5 in the second dimension on plastic-backed thin-layer cellulose sheets.
7. Stain the electrophoretogram with ninhydrin to locate the unlabeled phosphoamino acids by spraying or dipping it in the ethanolic ninhydrin solution and allowing the electrophoretogram to develop. Heat in a warm oven to hasten the process (see Figure A2.9).
8. Cut out the regions of the electrophoretogram containing phosphoamino acids and count them in a liquid scintillation counter.

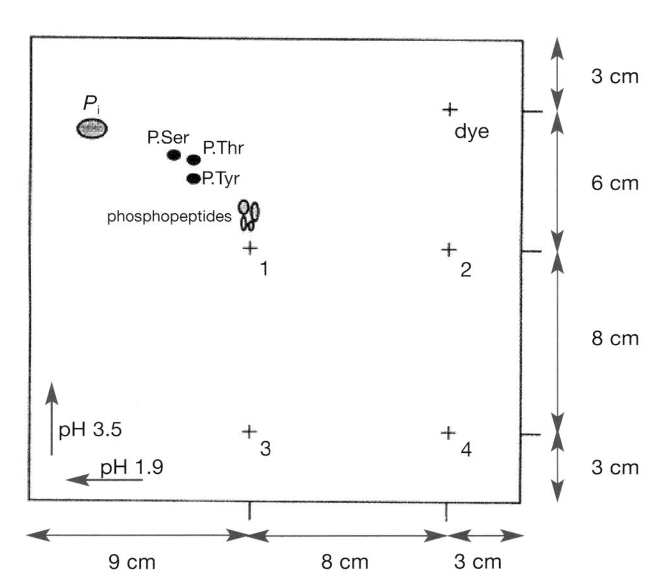

FIGURE A2.9. Cellulose TLC plate indicating sample and dye origins and ^{32}P-labeled products from a sample applied to origin 1. Four samples are typically applied on a TLC plate at the indicated sample origins, shown as 1, 2, 3, and 4. Before electrophoresis in the first dimension, a marker dye mixture of ε-DNP-lysine and xylene cyanol FF is applied to the dye origin depicted in the upper right corner. The plate is rotated 90° counterclockwise for second-dimension electrophoresis at pH 3.5. The positions of phosphoserine (P.Ser), phosphothreonine (P.Thr), phosphotyrosine (P.Tyr), released orthophosphate (P_i), and partially hydrolyzed phosphopeptides are shown for a sample applied to origin 1. (Reproduced, with permission, from Blume-Jensen and Hunter 2001.)

TECHNIQUE 8

Isothermal Titration Calorimetry: Measuring Intermolecular Interactions

Adrian Velazquez-Campoy and Ernesto Freire

Department of Biology and Biocalorimetry Center, The Johns Hopkins University, Baltimore, Maryland 21218

THE ASSOCIATION OF BIOLOGICAL MACROMOLECULES with one another or with their ligands has a central role in the structural assembly and functional regulation of biological systems (e.g., binding of hormones or toxins to receptors, allosteric enzymatic control, and regulation of transcription and replication). In a time when proteomics is considered as the next necessary step toward a more advanced understanding in diverse fields such as cell biology, developmental biology, and genetics, it is imperative to have access to a fast, robust, and reliable technique to measure the energetics of intermolecular interactions (e.g., protein-ligand, protein-protein, or protein-nucleic acid interactions) and to develop a quantitative description of the driving forces that govern molecular associations and regulatory processes. Recent reviews highlight isothermal titration calorimetry (ITC) as a powerful tool to accomplish these goals (Jelesarov and Bosshard 1999; Leavitt and Freire 2001).

The simplest binding process corresponds to a macromolecule with a single ligand-binding site (1:1 stoichiometry) $M + L \leftrightarrow ML$, where M and L are the free reactant species, macromolecule and ligand, and ML represents the binary complex. More complex processes, like the one corresponding to a macromolecule with n-independent and equivalent ligand-binding sites (1:n stoichiometry, with noninteracting identical binding sites), $M + nL \leftrightarrow ML_n$, can be easily generalized from the simple case. This model and other extensions to sets of independent or interacting sites (positive and negative cooperativity) have been discussed in the literature (Cantor and Schimmel 1980; Wyman and Gill 1990; Freire et al. 1990; Van Holde et al. 1998).

The strength of the association complex is given in terms of the association constant, K_a, or the dissociation constant, K_d,

$$K_a = \frac{[ML]}{[M][L]} = \frac{1}{K_d} \quad (1)$$

which are related to the Gibbs energy of binding, ΔG:

$$\Delta G = -RT \ln K_a = RT \ln K_d \quad (2)$$

which, in turn, can be expressed in terms of the enthalpy, ΔH, and entropy, ΔS, of binding:

$$\Delta G = \Delta H - T\Delta S \quad (3)$$

For a binding process to occur, the Gibbs energy of binding must be negative. Basically, ΔH reflects the interactions between macromolecule and ligand (hydrogen bonds, electrostatic and van der Waals interactions) taking as a reference their respective interactions with the solvent. The more negative ΔH, the more favorable the interactions in the binary complex.

The entropy, ΔS, represents the balance between two phenomena with opposite contributions. A large desolvation of hydrophobic surfaces upon binding is accompanied by an extremely large increase in entropy corresponding to the release of water molecules. On the other hand, the binding imposes conformational and roto-translational restrictions, resulting in a loss of degrees of freedom of the individual reactants and, consequently, a loss of conformational and translational entropy.

A variety of experimental techniques and analytical methods exist to study binding processes. Most of the traditional methods used to study binding are based on partitioning the system into free and bound species (e.g., equilibrium dialysis, ultrafiltration, gel exclusion chromatography, ultracentrifugation, and radioligand binding) and are both time- and material-consuming. Spectroscopic methods that detect a binding event by measuring changes in an observable parameter proportional to the extent of binding saturation fail to provide the stoichiometry directly. These methods require special reporter groups (e.g., chromophores or fluorophores) and therefore cannot be used to measure every binding event. All of these techniques offer a partial description of the binding reaction and only allow the determination of the association constant and evaluation of the Gibbs energy of binding, but they do not permit the dissection of the latter into enthalpic and entropic contributions. This analysis has been proven to be fundamental to precisely describe and understand the interactions underlying different binding processes (Gomez and Freire 1995; Parker et al. 1999; Todd et al. 2000; Velazquez-Campoy et al. 2000a).

Among the techniques used to study binding, ITC is unique because, in a single experiment, it provides a complete thermodynamic characterization of the energetics of a binding process, i.e., the determination of all the thermodynamic quantities describing that process: Gibbs energy, association constant, enthalpy and entropy of binding, as well as the stoichiometry of binding. Like any other experimental technique, ITC exhibits very important advantages and some limitations.

Advantages of ITC

Heat is a universal signal. Any binding process is accompanied by a heat effect susceptible to being measured by ITC. Even if under certain experimental conditions the reaction heat is near zero, it is possible to slightly change the experimental setup (with a change in temperature, pH, or buffer) in order to have a measurable heat effect.

Given that the measured signal is the reaction heat, it is possible to determine directly the enthalpy of binding in addition to the binding affinity using ITC. In addition, ITC does not require the presence of reporter groups (very often nonnatural), because the signal proportional to the extent of binding is the heat effect.

ITC provides a complete model-independent thermodynamic characterization of the binding process, because it directly measures the enthalpy of binding (reaction heat at constant pressure) and the association constant. Unlike spectroscopic techniques, there is no need to obtain the temperature dependence of the association constant to estimate the so-called van't Hoff enthalpy. The van't Hoff method lacks accuracy because (1) of the narrow range of temperatures allowed in experiments, (2) the association constant very often appears temperature-independent, due to the so-called enthalpy/entropy compensation phenomenon, (3) poor precision in the association constant determination may be translated into large errors for ΔH and ΔS, and (4) the van't Hoff analysis is model-dependent. ITC is a nondestructive technique. After running the experiment, it is possible to recover and recycle the reactants.

Limitations of ITC

In ITC, *global* heat effects are measured. The identification of specific contributions and interactions in the binding process requires ad hoc experimental designs. A more detailed study, usually involving different environmental conditions, a number of modifications of the reactants (e.g., mutant proteins and/or small library of ligands), and a structure-based thermodynamic dissection (Luque and Freire 1998), might reveal the nature of the specific contributions and individual interactions driving the binding process.

As described below, there is a well-defined range (dependent on the macromolecule concentration in the reaction cell) for the accurate determination of the association constant (roughly, 10^4 M^{-1} $\leq K_a \leq 10^9$ M^{-1}), even though the range can be extended by using displacement titrations (see below).

OUTLINE OF THE PROCEDURE

Principle of the Technique

A calorimetric titration measures the heat released or absorbed by the stepwise addition of a molecule (ligand) into the reaction cell of the calorimeter containing a solution of the interacting macromolecule while maintaining constant temperature and pressure. Changing the chemical composition of the sample as a result of the addition of ligand, using a precision computer-controlled syringe, triggers the reaction. The content of the sample cell is continuously stirred to ensure rapid mixing of the reactants. Thus, the system goes through a sequence of equilibrium states, and the direct thermodynamic observable is the heat associated with every change of state. A second cell, where no binding reaction takes place, is used as a reference.

The measured heat is related to both the enthalpy and the extent of reaction. For each injection, the heat released or absorbed, q_i, is given by:

$$q_i = V\Delta H\,([ML]_i - [ML]_{i-1}) \tag{4}$$

where V is the reaction cell volume, ΔH is the enthalpy of binding, and $[ML]_i$ is the change in complex concentration after the ith injection. A typical ITC experiment is illustrated in Figure A2.10. Modern ITC instruments are designed on the dynamic compensation principle, in which the measured signal is the amount of thermal power necessary to maintain null, at any time, the temperature difference between the reaction and reference cells, i.e., the amount of thermal power generated by the reaction triggered by the ligand injection. The time evolution of the signal exhibits characteristic deflections from the baseline in the form of peaks, each corresponding to one injection. The heat after each injection is obtained by calculating the area under each peak. Because q_i is proportional to the increase in bound macromolecule concentration, its magnitude will decrease as the macromolecule is titrated until reaching complete saturation. After saturation, additional injections produce small identical peaks, due to nonspecific phenomena such as dilution and mixing effects, whose heat effect must be subtracted from all the injection peaks before performing data analysis. Data analysis of the heat effect for each injection leads to the calculation of the thermodynamic binding parameters (n, K_a, ΔH).

Type of Calorimetric Experiments: Information Available by ITC

Simultaneous determination of the association constant and the enthalpy of binding

As in every equilibrium-binding technique, reactant concentrations must be within an appropriate range in order to obtain reliable estimates of the association constant. A practical rule of thumb is given by the parameter c, where $c = K_a \times [M]_T$, which must lie between

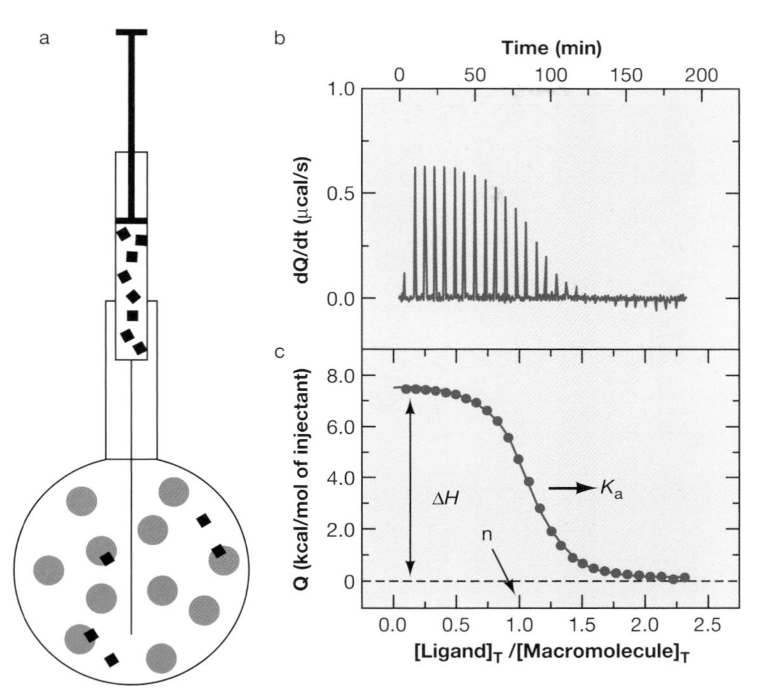

FIGURE A2.10. (*Panel a*) Illustration of the configuration of an ITC reaction cell. The cell volume is 1.4 ml and initially is filled with the macromolecule solution (*blue*). The injection syringe is filled with the ligand solution (*black*). At specified time intervals (400 seconds), a small volume (10 µl) of the ligand solution is injected into the cell, triggering the binding reaction and producing the characteristic peak sequence in the recorded signal (*panel b*). After saturating the macromolecule, the residual heat effects (the so-called "dilution peaks"), if any, are due to mechanical and dilution phenomena. The experiment corresponds to the titration of HIV-1 protease with acetyl-pepstatin (a universal inhibitor of aspartic proteases), performed in 10 mM sodium acetate, 2% DMSO (pH 5.0) at 25°C. The concentrations of reactants are 20 µM HIV-1 protease and 300 µM acetyl-pepstatin. After integration of the area under each peak (and subtraction of the dilution heat effects and normalization per mole of injected ligand), the individual heats are plotted against the molar ratio (*panel c*) from which, through nonlinear regression, it is possible to estimate the thermodynamic parameters n, K_a, and ΔH. The solid line corresponds to the theoretical curve with $n = 1.03$, $K_a = 2.3 \times 10^6\ \text{M}^{-1}$, and $\Delta H = 7.7$ kcal/mole.

0.1 and 1000, thus imposing a limit to the lowest and largest association constant measurable at a given macromolecule concentration (Wiseman et al. 1989). This phenomenon is illustrated in Figure A2.11.

As the c value increases, the transition from low to high ligand concentration is more abrupt. In the case of very high association constants (macromolecule concentration much higher than the dissociation constant), all of the ligand added in any injection will bind to the macromolecule until saturation occurs and all the peaks, except the last ones, therefore exhibit the same heat effect. For low association constants (macromolecule concentration far below the dissociation constant), from the very first injection, only a fraction of the ligand added will bind, producing a less steep titration in which saturation is hardly reached. To obtain accurate estimates of the association constant, an intermediate case is desirable.

To obtain a satisfactory titration curve, the concentration of ligand should be enough to exceed the stoichiometric binding after completion of the injection sequence (i.e., ligand concentration in the syringe should be 10–20 times the concentration of macromolecule in the cell). In case of poor solubility of the ligand, the reverse titration should be attempted.

The constraints dictated by parameter c impose an experimental limitation. For very high association constants, optimal concentrations are too small to be practical. For very low asso-

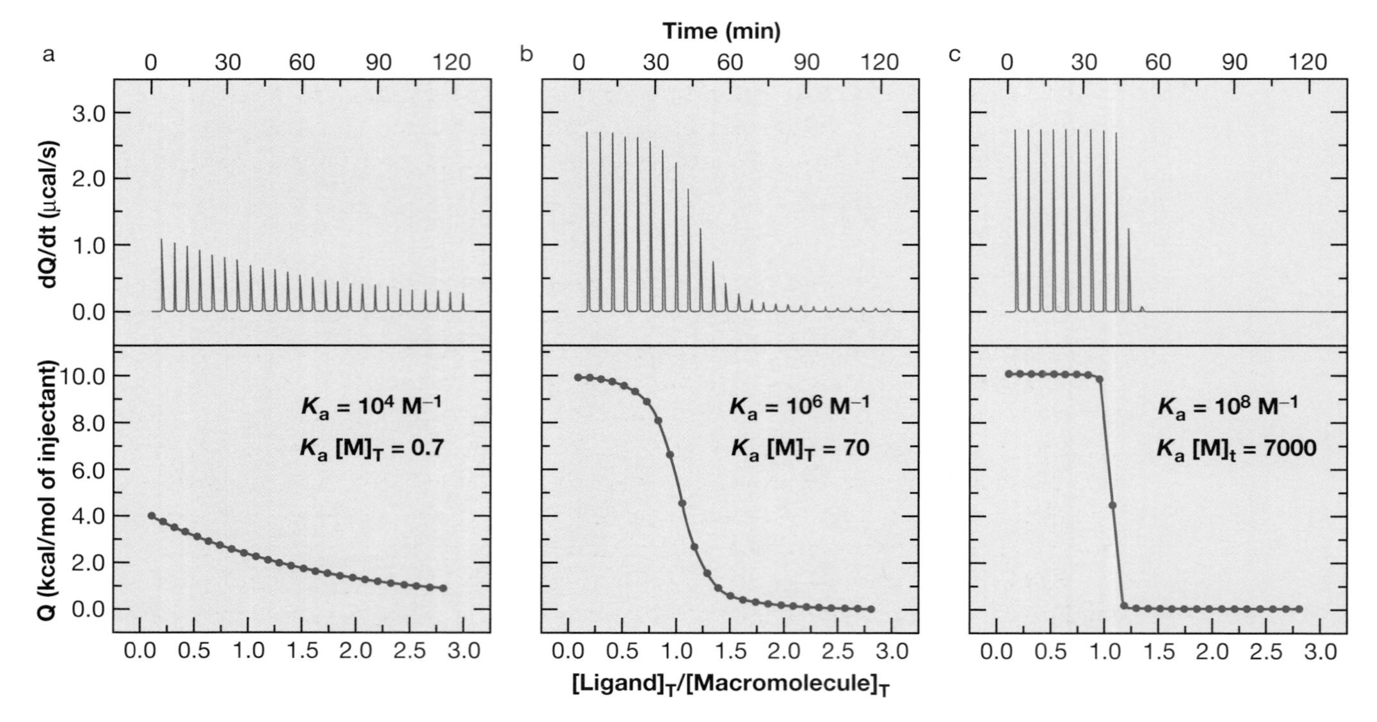

FIGURE A2.11. Illustration of the effect of the association constant value on the shape of a titration curve. The plots represent three titrations simulated using the same parameters (concentrations of reactants and binding enthalpy), but different association constants. Low (*panel A*), moderate (*panel B*), and high affinity (*panel C*) binding processes are shown. To obtain accurate estimates of the association constant, an intermediate case is desirable (*center*). The limit value of the curve when the molar ratio reaches zero, taking as a reference the heat effect associated to the dilution peaks, provides a good estimate of the enthalpy only when the parameter c is large enough, because such limit value is given by $(c/c+1) \times \Delta H$ (Indyk and Fisher 1998).

ciation constants, the concentrations may be prohibitively high, either because there is a possibility of aggregation or the economic/production costs are simply too great.

It is sometimes possible to change the experimental conditions (temperature or pH, without compromising stability against aggregation or unfolding) in order to modify the association constant toward accessible experimental values (Doyle et al. 1995; Baker and Murphy 1996; Doyle and Hensley 1998; Velazquez-Campoy et al. 2000b). In this case, appropriate equations are required to extrapolate to the original conditions. However, there is an extension of the ITC protocol aimed at overcoming such drawbacks. Without changing experimental conditions at all, displacement ITC extends without limits the useful range for the association constant determination (Sigurskjold 2000; Velazquez-Campoy et al. 2001). Basically, a displacement experiment consists of a titration of the high-affinity ligand into a solution of the macromolecule prebound to a weaker ligand, therefore decreasing the apparent affinity of the potent ligand. The thermodynamic parameters for the binding of the high-affinity ligand are calculated from the apparent binding parameters characterizing the displacement titration, and the known binding parameters for the weak ligand. A similar experimental design can be used to measure the binding of weak ligands.

Determination of the enthalpy of binding

Binding enthalpy is usually determined from titrations in the optimal range of reactant concentrations. However, a greater accuracy can be achieved if enthalpy is measured by performing injections of ligand into a large excess of macromolecule (far above the dissociation constant). Therefore, although there will be no saturation, all the ligand injected will be fully associated, giving a good measurement of the binding reaction heat. Blank experiments (see below) must be included in order to eliminate the contribution of different phenomena (dilution of reactant, solution mixing, and other nonspecific effects) occurring simultaneously during the injections and to obtain the heat effect associated with the binding process.

Blank experiment

To estimate the heat effect associated with mechanical and dilution phenomena (where binding does not take place), blank experiments can be performed. In this case, the ligand is injected into a buffer solution without the macromolecule. However, this may not be the ideal way to estimate such heat effects, because the degree of solvation of the ligand and the chemical composition would be different compared to the titration with the macromolecule present. For this reason, some investigators usually consider the average effect of the last injection peaks as the reference heat effect. If, during the experiment, it is not possible to reach complete saturation, precluding such averaging, it is possible to include in the fitting function a constant term accounting for such dilution heat effects.

Blank experiments, also called "heat of dilution experiments," are required to measure binding enthalpies under excess macromolecule concentration, as mentioned previously. In addition, these types of experiments can be used as a test to detect possible self-association of the ligand (especially very hydrophobic ones) at high concentrations. On the basis of this consideration, protein oligomerization processes can be studied calorimetrically by performing dilution experiments in which a concentrated solution of protein is injected into a solution containing buffer. When injected, the complex partially dissociates giving rise to a heat effect proportional to the enthalpy of association. From the data analysis, it is possible to estimate the association constant and the enthalpy for the oligomerization process (Lovatt et al. 1996).

Determination of heat capacity change of binding

The temperature derivative (at constant pressure) of the binding enthalpy, i.e., the heat capacity change upon binding or binding heat capacity, ΔC_P, is defined as

$$\Delta C_P = \left(\frac{\partial \Delta H}{\partial T}\right)_P \qquad (5)$$

and can be determined by repeating the same experiment at several temperatures. The heat capacity is given by the slope of an enthalpy versus temperature plot.

ΔC_p has been shown to originate from surface dehydration upon binding and, to a lesser extent, from the difference in vibrational modes between the complex and the free species. It provides information about the nature of the interactions driving the binding, in addition to the temperature dependence of the thermodynamic functions (Gibbs energy, enthalpy, and entropy of binding).

Determination of coupled proton transfer process

ITC is one of the more suitable techniques for the assessment of the existence of protonation/deprotonation processes coupled to binding. When a binding process is coupled to proton transfer between the bulk solution and the bound complex, the enthalpy of binding will depend on the ionization enthalpy of the buffer molecule, ΔH_{ion}, as shown in Equation 6.

$$\Delta H = \Delta H_0 + n_H \Delta H_{ion} \qquad (6)$$

where ΔH_0 is the buffer-independent enthalpy of binding and n_H is the number of protons being exchanged. Therefore, by repeating the titration under the same conditions, but using several buffers with different ionization enthalpies, there will be a linear dependence of ΔH versus ΔH_{ion}, from which it is possible to estimate n_H (slope) and ΔH_0 (intercept with the axis) (Biltonen and Langerman 1979; Gomez and Freire 1995).

- If n_H is zero, there is no net proton transfer.
- If n_H is positive, there is a protonation, i.e., a proton transfer from solution to complex.
- If n_H is negative, there is a deprotonation.

Amount of Sample Required

Modern calorimeters (compared to old ones) require only small amounts of sample to complete an experiment. Advances in material sciences and electronics, and improvements in the operation design principle, allow measurements of far less than 1 µcal, making it possible to conduct experiments at macromolecule concentrations close to typical dissociation constants in biological processes. Considering that the reaction cell volume is ~1 ml, the amount of macromolecule required is in the range of 1–100 nanomoles (~1 mg for an average protein).

Preparation of the Sample for Calorimetry

To obtain an accurate thermodynamic description of the binding process under study, reactants must be of the highest purity achievable. However, minor traces of contaminants will sometimes not distort the interactions enough to affect the results. When considering purity, it is important to distinguish between chemical and conformational homogeneity. Even if the macromolecule solution is 100% chemically pure, it is essential that it is correctly folded and that none of the protein is denatured.

It is also important that the solutions of macromolecule and ligand are chemically equilibrated, perfectly matched in salt content, pH, and cosolvents (e.g., organic solvents such as ethanol and DMSO). Because ITC detects global heat effects, the thermal effect associated with the mixing of different buffers will be included in every injection. Therefore, if possible, reactants should be dialyzed against the same buffer solution (if the molecular weight of the ligand is too low, at least it should be dissolved in the same buffer solution and then pH-adjusted). Extreme care should be taken when using additional organic cosolvents (e.g., ethanol and DMSO), because their heat of dilution is exceptionally large.

When performing experiments at a given temperature, samples should be cooled slightly below the temperature, because the handling of the sample while loading the calorimetric cell will raise the sample temperature. If the final cell temperature after loading is higher than the experimental temperature, the calorimeter will go through a long cooling process.

Analysis of the Results

Before the widespread use of computers, it was common to utilize mathematical transformations (e.g., Scatchard plots) to linearize binding equations in order to obtain thermodynamic parameters from linear least-square regression. However, such manipulations often incurred a systematic propagation of errors and uneven distribution of statistical weights over the binding curve. Nowadays, it is customary to implement nonlinear fitting procedures to directly analyze the raw calorimetric data. Considering Equation 4, the total cumulative heat released or absorbed, Q_i, is proportional to the total concentration of binary complex:

$$Q_i = V\Delta H[ML]_i \tag{7}$$

The analysis of the calorimetric titration data can be performed by estimating the variable model parameters (n, K_a, and ΔH) by fitting either the cumulative heat, Q_i, or the individual heat, q_i, for each injection. The analysis in terms of the individual heat is more convenient because it eliminates error propagation associated with cumulative data.

Even when assuming a model obeying 1:1 stoichiometry, it is useful to introduce a parameter n, which functions as the number of binding sites, as in the general 1:n model. When estimating the parameters, n should be equal to unity, within the experimental error, otherwise the following statements would hold:

$n > 1$ There is an error in the determination of the reactants concentration (real ligand concentration is lower and/or real macromolecule concentration is higher). There is more than one binding site (specific binding or not) per molecule.

$n < 1$ There is an error in the determination of the reactants concentration (real ligand concentration is higher and/or real macromolecule concentration is lower). There is less than one binding site per molecule, i.e., the sample is not chemically or conformationally homogeneous.

In considering the general binding model 1:n, Equations 3 and 4 can be expressed in terms of the free concentration of ligand:

$$\begin{aligned} q_i &= V\Delta H[M]_T\left(\frac{nK_a[L]_i}{1+K_a[L]_i} - \frac{nK_a[L]_{i-1}}{1+K_a[L]_{i-1}}\right) \\ Q_i &= V\Delta H[M]_T\left(\frac{nK_a[L]_i}{1+K_a[L]_i}\right) \end{aligned} \tag{8}$$

where $[M]_T$ is the total concentration of macromolecule. Although the binding equations, Equation 8, acquire an easy form in terms of the free ligand concentration, the independent known variable in ITC experiments is the total ligand concentration

$$[L]_T = [L] + [ML] \tag{9}$$

and the 1:n model permits an explicit analytical solution (Wiseman et al. 1989; Freire et al. 1990):

$$\begin{aligned} Q_i &= V\Delta H\,\frac{1 + [M]_T nK_a + K_a[L]_T - \sqrt{(1 + [M]_T nK_a + K_a[L])_T{}^2 - 4[M]_T nK_a^2[L]_T}}{2K_a} \\ q &= Q_i - Q_{i-1} \end{aligned} \tag{10}$$

In practice, a modified version of Equation 10, where the dilution effect on the macromolecule concentration due to the addition of the ligand is considered, is used:

$$\begin{aligned} Q_i &= V\Delta H\,\frac{1 + [M]_{T,i} nK_a + K_v[L]_{T,i} - \sqrt{(1 + [M]_{T,i} nK_a + K_a[L]_{T,i})^2 - 4[M]_{T,i} nK_a^2[L]_{T,i}}}{2K_a} \\ q_i &= Q_i - Q_{i-1}\,\frac{[M]_{T,i}}{[M]_{T,i-1}} \end{aligned} \tag{11}$$

and the total concentrations of reactants after any injection i are given by:

$$\begin{aligned} [M]_{T,i} &= [M]_{T,0}\left(1 - \frac{v}{V}\right)^i \\ [L]_{T,i} &= [L]_0\left(1 - \left(1 - \frac{v}{V}\right)^i\right) \end{aligned} \tag{12}$$

where $[M]_{T,0}$ is the initial concentration of macromolecule in the cell, $[L]_0$ is the concentration of ligand in the syringe, v is the injection volume, and V is the cell volume. Nonlinear fitting of q_i as a function of $[L]_{T,i}$ provides n, K_a, and ΔH as adjustable parameters. It is possible to avoid performing blank experiments to estimate the dilution heat effect and include in Equations 10–11 a constant term, q_d, representing the dilution heat effect contribution. Note that in the fitting procedure, q_d floats.

In general, models with different sets of binding sites require a numerical approach in which, via an iterative process, the model parameters (n_j, $K_{a,j}$, and ΔH_j, where j stands for each set of binding sites) are determined.

ITC: MEASURING INTERMOLECULAR INTERACTIONS TECHNIQUE

This technique was designed for experiments performed using the VP-ITC titration calorimeter from MicroCal, LLC. (Northampton, Massachusetts), which, at the present time, is the most sensitive and easy to use ITC instrument. With a VP-ITC titration calorimeter or similar instrument, it is possible to perform an experiment in 1 hour (including data analysis) with 1 mg of sample. If using a calorimeter other than the VP-ITC, the method described here may be of help as a guide.

Pancreatic ribonuclease A (EC 3.1.27.5) and the nucleotides 2′- and 3′-cytidine monophosphate, inhibitors of its endonuclease activity, provide a model system for ITC. The binding of these compounds, extremely exothermic and with moderate affinity, has been extensively characterized (Flogel and Biltonen 1975; Wiseman et al. 1989; Straume and Freire 1992).

MATERIALS

CAUTION: See Appendix 3 for appropriate handling of materials marked with <!>.

Reagents

2′CMP and 3′CMP

Prepare 2′CMP (C 7137, Sigma) and 3′CMP (C 1133, Sigma) at 1.5 mM in 15 mM potassium acetate (pH 5.5) dissolving the compounds directly in the buffer (~1 ml). Measure the concentration of each monophosphate by diluting it 1:20 in 100 mM sodium phosphate (pH 7.0) and using an extinction coefficient of 7400 $M^{-1}cm^{-1}$ at 260 nm.

Potassium acetate (15 mM, pH 5.5) <!>

Ribonuclease A

Prepare ribonuclease A (R 5500, Sigma) at 0.1 mM in 15 mM potassium acetate (pH 5.5) dissolving the protein in the buffer (~5 ml) and dialyzing it overnight at 4°C against 4 liters of the same buffer. Measure the concentration of the enzyme using an extinction coefficient of 9800 $M^{-1}cm^{-1}$ at 278 nm.

Sodium phosphate (100 mM, pH 7.0) <!>

Equipment

Calorimeter

Glass vials (12 x 75 mm and 6 x 50 mm)

Loading syringe (a long-needle syringe)

METHOD

1. Equilibrate the calorimeter at 24.5°C.

2. Prepare 2.2 ml of ribonuclease A, 0.5 ml of nucleotide (in a small vial), and 10 ml of buffer. Degas all solutions with a vacuum pump for 15 minutes.

3. While the solutions are degassing, thoroughly clean the calorimetric reaction cell.

4. Use the loading syringe to fill the reference cell with H_2O or 15 mM potassium acetate (pH 5.5) (the latter may help to reduce noise level).

5. Rinse the reaction cell with 15 mM potassium acetate (pH 5.5). Slowly inject the ribonuclease solution into the reaction cell. Carefully remove bubbles.

6. Fill the 250-µl injection syringe with the nucleotide solution. Rinse the outside of the syringe tip with 15 mM potassium acetate (pH 5.5) or H_2O and dry it.

   Although the syringe nominally indicates 250 µl, the available total injection volume is 309 µl. Filling the syringe incompletely will generate nonuniform injections.

7. Carefully insert the injection syringe into the reaction cell. Avoid bending the needle or mixing the liquids inside and outside the needle.

8. Equilibrate the calorimeter at 25°C.

9. Set the running parameters for the experiment as follows:

   Number of injections: 30
   Reference power[1]: 10 µcal/sec
   Concentration in syringe: ~1.5 mM
   Stirring speed: 490 rpm
   Feedback mode[2]: high
   Comments: ad lib
   Duration: automatically set according to the injection volume
   Filter: 2 seconds
   Temperature: 25°C
   Initial delay: 180 seconds
   Concentration in cell: ~100 µM
   File name: *.itc
   Equilibration options[3]: fast
   Injection volume: first 2 µl, the rest 8 µl
   Spacing between injections: 400 seconds

   Experiments with other molecules may require different parameter values.

   [1]If the reaction is expected to be so exothermic and/or the concentration of ligand is so high that the heat effect in the first injections exceeds –200 µcal, raise the reference power to prevent the signal from going below zero.

   [2]Set the feedback to high feedback gain, otherwise a drop in performance will occur. Given a heat effect, the peak corresponding to the injection will be smaller and broader (i.e., there will be a loss in sensitivity and an increase in the time for running an experiment).

   [3]In the equilibration options, compared to the other choices, fast equilibration is more convenient because it permits simultaneous thermal and mechanical equilibration, and it allows the user to manually start the injection sequence whenever the baseline is good (see Step 10).

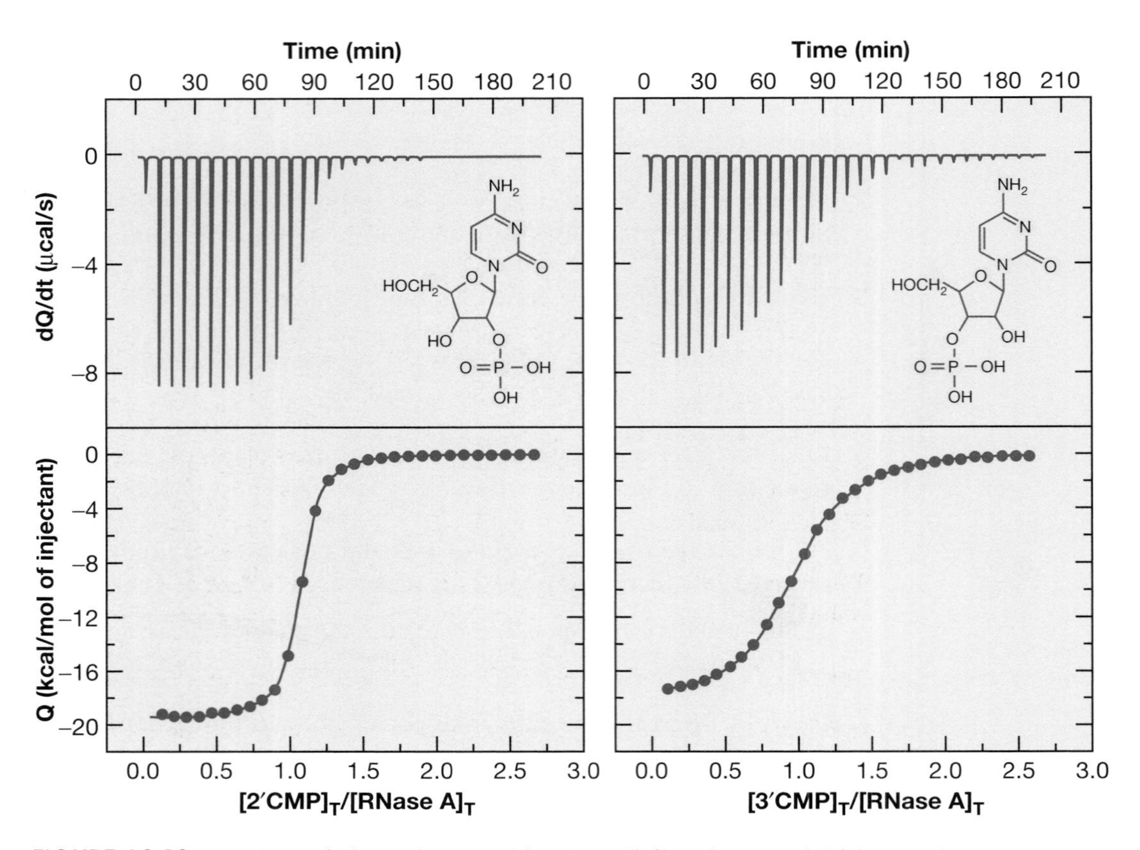

FIGURE A2.12. Titrations of ribonuclease A with 2′CMP (*left*) and 3′CMP (*right*). Experiments were performed in 15 mM potassium acetate (pH 5.5) at 25°C. The concentrations of reactants are 80 μM ribonuclease A and 1.3 mM 2′CMP and 3′CMP. Because the titrations were carried out using the same reactant concentrations and experimental conditions, from the steeper titration for 2′CMP it is possible to conclude that this compound is a more potent ligand than 3′CMP. The solid lines correspond to theoretical curves with $n = 1.02$, $K_a = 2.9 \times 10^6$ M^{-1}, and $\Delta H = -19.3$ kcal/mole for 2′CMP and $n = 0.98$, $K_a = 2.4 \times 10^5$ M^{-1}, and $\Delta H = -18.6$ kcal/mole for 3′CMP.

10. Start the experiment. There will be thermal and mechanical equilibration stages. Initiate the injection sequence after a stable no-drift noise-free baseline occurs (as seen in the 1 μcal/sec scale).

    The first injection is usually erroneous because there is some mixing of the solutions inside and outside of the needle when inserting the injection syringe and/or during the time required to reach thermal and mechanical equilibration. Therefore, set the first injection to be a smaller volume than subsequent injections to minimize the amount of wasted ligand. In the data analysis, the first injection will be taken into consideration for the calculation of the total ligand concentration in the cell. However, its associated heat effect will not be included in the fitting procedure.

11. Use the software provided by MicroCal for data analysis according to the equations described previously. Results should be consistent with the information shown in Figure A2.12.

REFERENCES

Aitken A., Howell S., Jones D., Madrazo J., and Patel Y. 1995. 14-3-3 α and δ are the phosphorylated forms of Raf-activating 14-3-3 β and ζ. *In vivo* stoichiometric phosphorylation in brain at a Ser-Pro-Glu-Lys motif. *J. Biol. Chem.* **270:** 5706–5709.

Akins R.E. and Tuan R.S. 1992. Measurement of protein in 20 seconds using a microwave BCA assay. *Biotechniques* **12:** 496–499.

Annan W.D., Manson W., and Nimmo J.A. 1982. The identification of phosphoseryl residues during the determination of amino acid sequence in phosphoproteins. *Anal. Biochem.* **121:** 62–68.

Baker B.M. and Murphy K.P. 1996. Evaluation of linked protonation effects in protein binding using isothermal titration calorimetry. *Biophys. J.* **71:** 2049–2055.

Baumann G., Frommel C., and Sander C. 1989. Polarity as a criterion in protein design. *Protein Eng.* **2:** 329–334.

Biltonen R.L. and Langerman N. 1979. Microcalorimetry for biological chemistry: Experimental design, data analysis and interpretation. *Methods Enzymol.* **61:** 287–319.

Blanchard J.S. 1984. Buffers for enzymes. *Methods Enzymol.* **104:** 404–414.

Blume-Jensen P. and Hunter T. 2001. Two-dimensional phosphoamino acid analysis. *Methods Mol. Biol.* **124:** 49–65.

Bollag D.M., Rozycki M.D., and Edelstein S.J. 1996. *Protein methods*, 2nd edition, pp. 76–78. Wiley, New York.

Bradford M.M. 1976. A rapid and sensitive method for the quantitation of microgram quantities of protein utilizing the principle of protein-dye ligand binding. *Anal. Biochem.* **72:** 248–254.

Brown R.E., Jarvis K.L., and Hyland K.J. 1989. Protein measurement using bicinchoninic acid: Elimination of interfering substances. *Anal. Biochem.* **180:** 136–139.

Burlingame A.L. and Carr S.A., eds. 1996. *Mass spectrometry in the biological sciences.* Humana Press, Totowa, New Jersey.

Cantor C.R. and Schimmel P.R. 1980. *Biophysical chemistry: The behavior of biological macromolecules.* W.H. Freeman, New York.

Dawson R.M.C., Elliott D.C., Elliott W.H., and Jones K.M., eds. 1969. *Data for biochemical research.* Clarendon Press, Oxford, United Kingdom.

Darby N.J. and Creighton T.E. 1993. Protein structure. In *In focus* (ed. D. Rickwood), p. 4. IRL Press, Oxford, United Kingdom.

Doyle M.L. and Hensley P. 1998. Tight ligand binding affinities determined from thermodynamic linkage to temperature by titration calorimetry. *Methods Enzymol.* **295:** 88–99.

Doyle M.L., Louie G.L., Dal Monte P.R., and Sokoloski T.D. 1995. Tight binding affinities determined from linkage to protons by titration calorimetry. *Methods Enzymol.* **259:** 183–194.

Fanger B.O. 1987. Adaptation of the Bradford Protein Assay to membrane-bound proteins by solubilizing in glucopyranoside detergents. *Anal. Biochem.* **162:** 11–17.

Flogel M. and Biltonen R.L. 1975. Calorimetric and potentiometric characterization of the ionization behavior of ribonuclease A and its complex with 3′-cytosine monophosphate. *Biochemistry* **14:** 2603–2609.

Freire E., Mayorga O.L., and Straume M. 1990. Isothermal titration calorimetry. *Anal. Chem.* **62:** 950A–959A.

Friedenauer S. and Berlet H.H. 1989. Sensitivity and variability of the Bradford Protein Assay in the presence of detergents. *Anal. Biochem.* **178:** 263–268.

Gomez J. and Freire E. 1995. Thermodynamic mapping of the inhibitor site of the aspartic protease endothiapepsin. *J. Mol. Biol.* **252:** 337–350.

Grimes B.A., Meyers J.J., and Liapis A.I. 2000. Determination of the intraparticle electroosmotic volumetric flow-rate, velocity and Peclet number in capillary electrochromatography from pore network theory. *J. Chromatogr.* **890:** 61–72.

Gygi S.P., Rist B., Gerber S.A., Turecek F., Gelb M.H., and Aebersold R. 1999. Quantitative analysis of complex protein mixtures using isotope-coded affinity tags. *Nat. Biotechnol.* **17:** 994–999.

Hagel P., Gerding J.J.T., Fieggen W., and Bloemendal H. 1971. Cyanate formation in solutions of urea. I. Calculations of cyanate concentrations at different temperature and pH. *Biochem. Biophys. Acta* **243:** 366–373.

Hill H.D. and Straka J.G. 1988. Protein determination using bicinchoninic acid in the presence of sulfhydryl reagents. *Anal. Biochem.* **170:** 203–208.

Hinson D.L. and Webber R.J. 1988. Miniaturization of the BCA protein assay. *BioTechniques* **6:** 14–16.

Holmes C.F.B. 1987. A new method for the selective isolation of phosphoserine-containing peptides. *FEBS Lett.* **215:** 21–24.

Indyk L. and Fisher H.F. 1998. Theoretical aspects of isothermal titration calorimetry. *Methods Enzymol.* **295:** 350–364.

Isobe T., Takahashi N., and Putnam F.W. 1991a. Protein and peptide mapping by two-dimensional HPLC. In *HPLC of peptides and proteins: Separation, analysis and conformation* (ed. R.S. Hodges et al.), pp. 835–845. CRC Press, New York.

Isobe T., Uchida K., Taoka M., Shinkai F., Manabe T., and Okuyama T. 1991b. Automated high performance liquid chromatographic system for mapping proteins in highly complex mixtures. *J. Chromatogr.* **588:** 115–123.

Janin J. 1979. Surface and inside volumes in globular proteins. *Nature* **277:** 491–492.

Jelesarov I. and Bosshard H.R. 1999. Isothermal titration calorimetry and differential scanning calorimetry as complementary tools to investigate the energetics of biomolecular recognition. *J. Mol. Recognit.* **12:** 3–18.

Kaushal V. and Barnes L.D. 1986. Effect of zwitterionic buffers on measurement of small masses of protein with bicinchoninic acid. *Anal. Biochem.* **157:** 291–294.

Kessler R.J. and Fanestil D.D. 1986. Interference by lipids in the determination of protein using bicinchoninic acid. *Anal. Biochem.* **159:** 138–142.

Leavitt S. and Freire E. 2001. Direct measurement of protein binding energetics by isothermal titration calorimetry. *Curr. Opin. Struct. Biol.* **11:** 560–566.

Lee B. and Richards F.M. 1971. The interpretation of protein structures: Estimation of static accessibility. *J. Mol. Biol.* **55:** 379–400.

Liapis A.I. and Grimes B.A. 2000. Modeling the velocity field of the electroosmotic flow in charged capillaries and in capillary columns packed with charged particles: Interstitial and intraparticle velocities in capillary electrochromatography systems. *J. Chromatogr.* **877:** 181–215.

Link A.J., Eng J., Schieltz D.M., Carmack E., Mize G.J., Morris D.R., Garvik B.M., and Yates J.R. 1999. Direct analysis of protein complexes using mass spectrometry. *Nat. Biotechnol.* **17:** 676–682.

Lovatt M., Cooper A., and Camilleri P. 1996. Energetics of cyclodextrin-induced dissociation of insulin oligomers. *Eur. Biophys. J.* **24:** 354–357.

Luque I. and Freire E. 1998. A system for the structure-based prediction of binding affinities and molecular design of peptide ligands. *Methods Enzymol.* **295:** 100–127.

Matyska M.T., Pesek J.J., Boysen R.I., and Hearn M.T.W. 2001. Characterization of open tubular capillary electrochromatography columns for the analysis of synthetic peptides using isocratic conditions. *Anal. Chem.* **73:** 5116–5125.

Meyer H.E., Hoffmann-Posorske E., Korte H., and Heilmeyer L.M.G., Jr. 1986. Sequence analysis of phosphoserine-containing peptides. Modification for picomolar sensitivity. *FEBS Lett.* **204:** 61–66.

Miller S., Janin J., Lesk A.M., and Chothia C. 1987. Interior and surface of monomeric proteins. *J. Mol. Biol.* **196:** 641–656.

Parker M.H., Lunney E.A., Ortwine D.F., Pavlovsky A.G., Humblet C., and Brouillette C.G. 1999. Analysis of the binding of hydroxamic acid and carboxylic acid inhibitors to the stromelysin-1 (matrix metalloproteinase-3) catalytic domain by isothermal titration calorimetry. *Biochemistry* **38:** 13592–13601.

Peterson G.L. 1983. Determination of total protein. *Methods Enzymol.* **91:** 95–121.

Pierce J. and Suelter C.H. 1977. An evaluation of the Coomassie brilliant blue G-250 dye-binding method for quantitative protein determination. *Anal. Biochem.* **81:** 478–480.

Radzicka A. and Wolfenden R. 1988. Comparing the polarities of the amino acids: Side-chain distribution coefficients between the vapor phase, cyclohexane, 1-octanol, and neutral aqueous solution. *Biochemistry* **27:** 1664–1670.

Read S.M. and Northcote D.H. 1981. Minimization of variation in the response to different proteins of the Coomassie blue G dye-binding assay for protein. *Anal. Biochem.* **116:** 53–64.

Richards F.M. 1977. Areas, volumes, packing and protein structure. *Annu. Rev. Biophys. Bioeng.* **6:** 151–176.

Richardson W.S., Munksgaard E.C., and Butler W.T. 1978. Rat incisor phosphoprotein. The nature of the phosphate and quantitation of the phosphoserine. *J. Biol. Chem.* **253:** 8042–8046.

Rose G.D., Geselowitz A.R., Lesser G.J., Lee R.H., and Zehfus M.H. 1985. Hydrophobicity of amino acids in globular proteins. *Science* **229:** 834–838.

Ross G., Dittmann M., Bek F., and Rozing G. 1996. Capillary electrochromatography—Enhancement of LC separation in packed capillary columns by means of electrically driven mobile phases. *Am. Lab.* **28:** 3–12.

Schien C.H. 1990. Solubility as a function of protein structure and solvent components. *BioTechnology* **8:** 308–317.

Sedmak J.J. and Grossberg S.E. 1977. A rapid, sensitive and versatile assay for protein using Coomassie brilliant blue G250. *Anal. Biochem.* **79:** 544–552.

Sigurskjold B.W. 2000. Exact analysis of competition ligand binding by displacement isothermal titration calorimetry. *Anal. Biochem.* **277:** 260–266.

Smith P.K., Krohn R.I., Hermanson G.T., Mallia A.K., Gartner F.H., Provenzano M.D., Fujimoto E.K., Cocke N.M., Olson B.J., and Klenk D.C. 1985. Measurement of protein using bicinchoninic acid. *Anal. Biochem.* **150:** 76–85.

Spector T. 1978. Refinement of the Coomassie Blue method of protein quantitation. *Anal. Biochem.* **86:** 142–146.

Stoscheck C.M. 1990. Quantitation of protein. *Methods Enzymol.* **182:** 50–68.

Straume M. and Freire E. 1992. Two-dimensional differential scanning calorimetry: Simultaneous resolution of intrinsic protein structural energetics and ligand binding interactions by global linkage analysis. *Anal. Biochem.* **203:** 259–268.

Takahashi N., Isobe T., and Putnam F.W. 1991. Multidimensional, microscale HPLC technique in protein sequencing. In *HPLC of proteins, peptides, and polynucleotides* (ed. M.T.W. Hearn), pp. 307–330. VCH, New York.

Takahashi N., Ortel T.L., and Putnam F.W. 1984. Single-chain structure of human ceruloplasmin; the complete amino acid sequence of the whole molecule. *Proc. Natl. Acad. Sci.* **81:** 390–394.

Takahashi N., Ishioka N., Takahashi Y., and Putnam F.W. 1985. Automated tandem high-performance liquid chromatographic system for separation of extremely complex peptide mixtures. *J. Chromatogr.* **326:** 407–418.

Takahashi N., Takahashi Y., Blumberg B.S., and Putnam F.W. 1987a. Amino acid substitutions in genetic variants of human serum albumin and in sequences inferred from molecular cloning. *Proc. Natl. Acad. Sci.* **84:** 4413–4417.

Takahashi N., Takahashi Y., Isobe T., Putnam F.W., Fujita M., Satoh C., and Neel J.V. 1987b. Amino acid substitutions in inherited albumin variants from Amerindian and Japanese populations. *Proc. Natl. Acad. Sci.* **84:** 8001–8005.

Taoka M., Yamakuni T., Song S.Y., Yamakawa Y., Seta K., Okuyama T., and Isobe T. 1992. A rat cerebellar protein containing the cdc10/SW16 motif. *Eur. J. Biochem.* **207:** 615–620.

Taoka M., Isobe T., Okuyama T., Watanabe M., Kondo H., Yamakawa Y., Ozawa F., Hishinuma F., Kubota M., Minegishi A., Song S.Y., and Yamakuni T. 1994. Murine cerebellar neurons express a novel gene encoding a protein related to cell cycle control and cell fate determination proteins. *J. Biol. Chem.* **269:** 9946–9951.

Todd M.J., Luque I., Velazquez-Campoy A., and Freire E. 2000. Thermodynamic basis of resistance to HIV-1 protease inhibition: Calorimetric analysis of the V82F/I84V active site resistant mutant. *Biochemistry* **39:** 11876–11883.

Trinquier G. and Sanejouand Y.-H. 1998. Which effective property of amino acids is best preserved by the genetic code? *Protein Eng.* **11:** 153–169.

Van Holde K.E., Johnson W.C., and Ho P.S. 1998. *Principles of physical biochemistry.* Prentice Hall, Upper Saddle River, New Jersey.

Van Kley H. and Hale S.M. 1977. Assay for protein by dye binding. *Anal. Biochem.* **81:** 485–487.

Velazquez-Campoy A., Kiso Y., and Freire E. 2001. The binding energetics of first- and second-generation HIV-1 protease inhibitors: Implications for drug design. *Arch. Biochem. Biophys.* **390:** 169–175.

Velazquez-Campoy A., Todd M.J., and Freire E. 2000a. HIV-1 protease inhibitors: Enthalpic versus entropic optimization of the binding affinity. *Biochemistry* **39:** 2201–2207.

———. 2000b. Thermodynamic dissection of the binding energetics of KNI-272, a potent HIV-1 protease inhibitor. *Protein Sci.* **9:** 1801–1809.

Walhagen K., Unger K.K., and Hearn M.T.W. 2000a. Capillary electroendoosmotic chromatography of peptides. *J. Chromatogr. A* **887:** 165–185.

——. 2000b. Influence of temperature on the behaviour of small linear peptides in capillary electrochromatography. *J. Chromatogr. A.* **893:** 401–409.

Walhagen K., Unger K.K., and Hearn M.T.W. 2001. Capillary electrochromatography analysis of hormonal cyclic and linear peptides. *Anal. Chem.* **73:** 4924–4936.

Wiechelman K.J., Braun R.B., and Fitzpatrick J.D. 1988. Investigation of the bicinchoninic acid protein assay: Identification of the groups responsible for color formation. *Anal. Biochem.* **175:** 231–237.

Wilson C.M. 1979. Studies and critique of Amido black 10B, Coomassie blue R and Fast green FCF as stains for proteins after polyacrylamide gel electrophoresis. *Anal. Biochem.* **96:** 263–278.

Wiseman T., Williston S., Brandts J.F., and Nin L.N. 1989. Rapid measurement of binding constants and heats of binding using a new titration calorimeter. *Anal. Biochem.* **179:** 131–137.

Wyman J. and Gill S.J. 1990. *Binding and linkage: Functional chemistry of biological macromolecules.* University Science Books, Mill Valley, California.

Zhang S.H., Huang X., Zhang J., and Horvath C. 2000. Capillary electrochromatography of proteins and peptides with a cationic acrylic monolith. *J. Chromatogr.* **887:** 465–477.

WWW RESOURCES

http://prowl.rockefeller.edu/aainfo/contents.htm Amino acid information, PROWL, protein chemistry and mass spectrometry resource developed in collaboration between ProteoMetrics and Rockefeller University.

http://sosui.proteome.bio.tuat.ac.jp/sosuiframe0.html SOSUI system, Department of Biotechnology, Tokyo University of Agriculture and Technology.

Cautions

THE FOLLOWING GENERAL CAUTIONS should always be observed.

- **Become completely familiar with the properties of substances used before** beginning the procedure.
- **The absence of a warning** does not necessarily mean that the material is safe, since information may not always be complete or available.
- **If exposed to toxic substances,** contact your local safety office immediately for instructions.
- **Use proper disposal procedures** for all chemical, biological, and radioactive waste.
- **For specific guidelines on appropriate gloves**, consult your local safety office.
- **Handle concentrated acids and bases** with great care. Wear goggles and appropriate gloves. Wear a face shield when handling large quantities.

  Do not mix strong acids with organic solvents as they may react. Sulfuric acid and nitric acid especially may react highly exothermically and cause fires and explosions.

  Do not mix strong bases with halogenated solvent as they may form reactive carbenes which can lead to explosions.
- **Never pipette** solutions using mouth suction. This method is not sterile and can be dangerous. Always use a pipette aid or bulb.
- **Keep halogenated and nonhalogenated solvents separately** (e.g., mixing chloroform and acetone can cause unexpected reactions in the presence of bases). Halogenated solvents are organic solvents such as chloroform, dichloromethane, trichlorotrifluoroethane, and dichloroethane. Some nonhalogenated solvents are pentane, heptane, ethanol, methanol, benzene, toluene, *N,N*-dimethylformamide (DMF), dimethylsulfoxide (DMSO), and acetonitrile.
- **Laser radiation**, visible or invisible, can cause severe damage to the eyes and skin. Take proper precautions to prevent exposure to direct and reflected beams. Always follow the manufacturer's safety guidelines and consult your local safety office. See caution below for more detailed information.
- **Flash lamps**, due to their light intensity, can be harmful to the eyes. They also may explode on occasion. Wear appropriate eye protection and follow the manufacturer's guidelines.
- **Photographic fixatives and developers** also contain chemicals that can be harmful. Handle them with care and follow manufacturer's directions.
- **Power supplies and electrophoresis equipment** pose serious fire hazard and electrical shock hazards if not used properly.
- **Microwave ovens and autoclaves in the lab require certain precautions.** Accidents have occurred involving their use (e.g., to melt agar or bacto-agar stored in bottles or to sterilize). If the screw top is not completely removed and there is not enough space for the steam to vent, the bottles can explode and cause severe injury when the containers are removed from the microwave or autoclave. Always completely remove bottle caps before microwaving or autoclaving. An alternative method for routine agarose gels that do not require sterile agar is to weigh out the agar and place the solution in a flask.
- **Ultra-sonicators use high-frequency sound waves** (16–100 kHz) for cell disruption and other purposes. This "ultrasound," conducted through air, does not pose a direct hazard to humans, but the associated high volumes of audible sound can cause a variety of effects, including headache, nausea,

and tinnitus. Avoid direct contact of the body with high-intensity ultrasound (not medical imaging equipment). Use appropriate ear protection and display signs on the door(s) of laboratories where the units are used.

- **Use extreme caution when handling cutting devices** such as microtome blades scalpels, razor blades, or needles. Microtome blades are extremely sharp! Use care when sectioning. If unfamiliar with their use, have someone demonstrate proper procedures. For proper disposal, use the "sharps" disposal container in the lab. Discard used needles *unshielded*, with the syringe still attached. This prevents injuries (and possible infections; see Biological Safety) while manipulating used needles since many accidents occur while trying to replace the needle shield. Injuries may also be caused by broken Pasteur pipettes, coverslips, or slides.

GENERAL PROPERTIES OF COMMON CHEMICALS

The hazardous materials list can be summarized in the following categories:

- Inorganic acids, such as hydrochloric, sulfuric, nitric, or phosphoric, are colorless liquids with stinging vapors. Avoid spills on skin or clothing. Spills should be diluted with large amounts of water. The concentrated forms of these acids can destroy paper, textiles, and skin as well as cause serious injury to the eyes.
- Inorganic bases such as sodium hydroxide are white solids which dissolve in water and under heat development. Concentrated solutions will slowly dissolve skin and even fingernails.
- Salts of heavy metals are usually colored powdered solids which dissolve in water. Many of them are potent enzyme inhibitors and therefore toxic to humans and to the environment (e.g., fish and algae).
- Most organic solvents are flammable volatile liquids. Avoid breathing the vapors which can cause nausea or dizziness. Also avoid skin contact.
- Other organic compounds, including organosulphur compounds such as mercaptoethanol or organic amines, can have very unpleasant odors. Others are highly reactive and should be handled with appropriate care.
- If improperly handled, dyes and their solutions can stain not only the sample, but also skin and clothing. Some of them are also mutagenic (e.g., ethidium bromide), carcinogenic, and toxic.
- All names ending with "ase" (e.g., catalase, β-glucuronidase, or zymolase) refer to enzymes. There are also other enzymes with nonsystematic names like pepsin. Many of them are provided by manufacturers in preparations containing buffering substances, etc. Be aware of the individual properties of materials contained in these substances.
- Toxic compounds are often used to manipulate cells. They can be dangerous and should be handled appropriately.
- Be aware that several of the compounds listed have not been thoroughly studied with respect to their toxicological properties. Handle each chemical with the appropriate respect. Although the toxic effects of a compound can be quantified (e.g., LD_{50} values), this is not possible for carcinogens or mutagens where one single exposure can have an effect. Also realize that dangers related to a given compound may also depend on its physical state (fine powder vs. large crystals/diethylether vs. glycerol/dry ice vs. carbon dioxide under pressure in a gas bomb). Anticipate under which circumstances during an experiment exposure is most likely to occur and how best to protect yourself and your environment.

HAZARDOUS MATERIALS

In general, proprietary materials are not listed here. Follow the manufacturer's safety guidelines that accompany the product.

Acetic acid (glacial) is highly corrosive and must be handled with great care. Liquid and mist cause severe burns to all body tissues. It may be harmful by inhalation, ingestion, or skin absorption. Wear appropriate gloves and goggles and use in a chemical fume hood. Keep away from heat, sparks, and open flame.

Acetic anhydride is extremely destructive to the skin, eyes, mucous membranes, and upper respiratory tract. It may be harmful by inhalation, ingestion, or skin absorption. Wear appropriate gloves and safety glasses. and use in a chemical fume hood.

Acetone causes eye and skin irritation and is irritating to mucous membranes and upper respiratory tract. Do not breathe the vapors. It is also extremely flammable. Wear appropriate gloves and safety glasses.

Acetonitrile (Methyl cyanide) is very volatile and extremely flammable. It is an irritant and a chemical asphyxiant that can exert its effects by inhalation, ingestion, or skin absorption. Treat cases of severe exposure as cyanide poisoning. Wear appropriate gloves and safety glasses and use only in a chemical fume hood. Keep away from heat, sparks, and open flame.

Acrylamide (unpolymerized) is a potent neurotoxin and is absorbed through the skin (the effects are cumulative). Avoid breathing the dust. Wear appropriate gloves and a face mask when weighing powdered acrylamide and methylene-bisacrylamide. Use in a chemical fume hood. Polyacrylamide is considered to be nontoxic, but it should be handled with care because it might contain small quantities of unpolymerized acrylamide.

$AgNO_3$, *see* Silver nitrate

ε-Amino-*n*-caproic acid (6-Aminocaproic acid; Hexanoic acid) may be harmful by inhalation, ingestion, or skin absorption. Wear appropriate gloves and safety glasses.

ε-Amino-*n*-butyric acid (GABA), *see* Butyric acid

3-Aminopropyltriethoxysilane, *see* Silane

Ammonia, NH_3, is corrosive, toxic, and can be explosive. It may be harmful by inhalation, ingestion, and skin absorption. Use only with mechanical exhaust. Wear appropriate gloves and safety glasses.

Ammonium bicarbonate, NH_4HCO_3, may be harmful by inhalation, ingestion, or skin absorption. Wear appropriate gloves and safety glasses and use in a chemical fume hood.

Ammonium carbonate, $(NH_4)_2CO_3$, may be harmful by inhalation, ingestion, or skin absorption. Wear appropriate gloves and safety glasses and use in a chemical fume hood.

Ammonium hydrogen carbonate, *see* Ammonium bicarbonate

Ammonium hydroxide, NH_4OH, is a solution of ammonia in water. It is caustic and should be handled with great care. As ammonia vapors escape from the solution, they are corrosive, toxic, and can be explosive. Use only with mechanical exhaust. Wear appropriate gloves and use only in a chemical fume hood.

Ammonium molybdate, $(NH_4)_6Mo_7O_{24}{\cdot}4H_2O$, (or its **tetrahydrate**) may be harmful by inhalation, ingestion, or skin absorption. Wear appropriate gloves and safety glasses and use in a chemical fume hood.

Ammonium persulfate, $(NH_4)_2S_2O_8$, is extremely destructive to tissue of the mucous membranes and upper respiratory tract, eyes, and skin. Inhalation may be fatal. Wear appropriate gloves, safety glasses, and protective clothing and use only in a chemical fume hood. Wash thoroughly after handling.

Ammonium sulfate, $(NH_4)_2SO_4$, may be harmful by inhalation, ingestion, or skin absorption. Wear appropriate gloves and safety glasses.

Aprotinin may be harmful by inhalation, ingestion, or skin absorption. It may also cause allergic reactions. Exposure may cause gastrointestinal effects, muscle pain, blood pressure changes, or bronchospasm. Wear appropriate gloves and safety glasses and use only in a chemical fume hood. Do not breathe the dust.

Arc lamps are potentially explosive. Follow manufacturer's guidelines. When turning on arc lamps, make sure nearby computers are turned off to avoid damage from electromagnetic wave components. Computers may be restarted once the arc lamps are in operation.

Argon is a nonflammable high-pressure gas. It may be harmful by inhalation, ingestion, or skin absorption. Wear appropriate gloves and safety goggles. Use with sufficient ventilation and do not breathe the gas.

p-Azidophenacyl bromide is highly flammable and corrosive. It may be harmful by inhalation, ingestion, and skin absorption. Wear appropriate gloves and safety goggles and use only in a chemical fume hood. Do not breathe the dust. Keep away from heat, sparks, and open flame.

Bisacrylamide is a potent neurotoxin and is absorbed through the skin (the effects are cumulative). Avoid breathing the dust. Wear appropriate gloves and a face mask when weighing powdered acrylamide and methylenebisacrylamide.

Bleach (sodium hypochlorite), NaOCl, is poisonous, can be explosive, and may react with organic solvents. It may be fatal by inhalation and is also harmful by ingestion and destructive to the skin. Wear appropriate gloves and safety glasses and use in a chemical fume hood.

Blood (human) and blood products and Epstein-Barr virus. Human blood, blood products, and tissues may contain occult infectious materials such as hepatitis B virus and HIV that may result in laboratory-acquired infections. Investigators working with EBV-transformed lymphoblast cell lines are also at risk of EBV infection. Any human blood, blood products, or tissues should be considered a biohazard and should be handled accordingly. Wear disposable appropriate gloves, use mechanical pipetting devices, work in a biological safety cabinet, protect against the possibility of aerosol generation, and disinfect all waste materials before disposal. Autoclave contaminated plasticware before disposal; autoclave contaminated liquids or treat with bleach (10% [v/v] final concentration) for at least 30 minutes before disposal. Consult the local institutional safety officer for specific handling and disposal procedures.

Bromophenol blue may be harmful by inhalation, ingestion, or skin absorption. Wear appropriate gloves and safety glasses and use in a chemical fume hood.

***n*-Butanol** is irritating to the mucous membranes, upper respiratory tract, skin, and especially the eyes. Avoid breathing the vapors. Wear appropriate gloves and safety glasses and use in a chemical fume hood. *n*-butanol is also highly flammable. Keep away from heat, sparks, and open flame.

***sec*-Butanol** is irritating to the mucous membranes, upper respiratory tract, the skin, and especially the eyes. Avoid breathing the vapors. Wear appropriate gloves and safety glasses and use in a chemical fume hood. *sec*-butanol is also highly flammable. Keep away from heat, sparks, and open flame.

Butyl Chloride is highly flammable. Keep away from heat, sparks, and open flame. It may be harmful by inhalation, ingestion, and skin absorption. Wear appropriate gloves and safety goggles. Do not breathe vapor.

Calcium chloride is hygroscopic and may cause cardiac disturbances. It may be harmful by inhalation, ingestion, or skin absorption. Do not breathe the dust. Wear appropriate gloves and safety goggles.

CAPS, *see* 3-(Cyclohexylamino)-1-propanesulfonic acid

Carbon dioxide, CO_2, in all forms may be fatal by inhalation, ingestion, or skin absorption. In high concentrations, it can paralyze the respiratory center and cause suffocation. Use only in well-ventilated areas. In the form of dry ice, contact with carbon dioxide can also cause frostbite. Do not place large quantities of dry ice in enclosed areas such as cold rooms. Wear appropriate gloves and safety goggles.

Cetyltrimethylammonium bromide (CTAB) is toxic and an irritant and may be harmful by inhalation, ingestion, or skin absorption. Wear appropriate gloves and safety glasses. Avoid breathing the dust.

CHAPS, *see* 3-[(3-Cholamidopropyl)dimethyl-ammonio]-1-propanesulfonate

Chloroform, $CHCl_3$, is irritating to the skin, eyes, mucous membranes, and respiratory tract. It is a carcinogen and may damage the liver and kidneys. It is also volatile. Avoid breathing the vapors. Wear appropriate gloves and safety glasses and always use in a chemical fume hood.

3-[(3-Cholamidopropyl)dimethyl-ammonio]-1-propanesulfonate (CHAPS) is an irritant and may be harmful by inhalation, ingestion, or skin absorption. Wear appropriate gloves and safety glasses.

Citraconic anhydride is highly toxic and may be harmful by inhalation, ingestion, or skin absorption. Wear appropriate gloves and safety goggles. Avoid inhalation.

Citric acid is an irritant and may be harmful by inhalation, ingestion, or skin absorption. It poses a risk of serious damage to the eyes. Wear appropriate gloves and safety goggles. Do not breathe the dust.

Coomassie Brilliant Blue may be harmful by inhalation, ingestion, or skin absorption. Wear appropriate gloves and safety glasses.

***p*-Cresol** may be fatal if inhaled, ingested, or absorbed through the skin. It is corrosive and is extremely destructive to the eyes, skin, mucous membranes, and upper respiratory tract. Wear appropriate gloves and safety glasses and use in a chemical fume hood.

Crystal Violet can cause severe burns. It may be harmful by inhalation, ingestion, and skin absorption. Wear appropriate gloves and safety goggles and use in a chemical fume hood, Do not breathe the dust.

CTAB, *see* Cetyltrimethylammonium bromide

β-Cyanoethyl diisopropylchlorophosphoramidite (2-Cyanoethyl diisopropylchlorophosphoramidite) is highly corrosive and causes burns. Heating may cause an explosion. It is extremely destructive to mucous membranes and upper respiratory track, eyes, and skin. Wear appropriate gloves and safety goggles and use in a chemical fume hood. Do not breathe the vapor. Keep away from heat, sparks, and open flame.

α-Cyano-4-hydroxycinnamic acid may cause cardiac disturbances. Chronic effects may be delayed. It may be harmful by inhalation, ingestion, or skin absorption. Wear appropriate gloves and safety glasses.

Cyanogen bromide is extremely toxic and is volatile. It may be fatal by inhalation, ingestion, or skin absorption. Do not breathe the vapors. Wear appropriate gloves and always use in a chemical fume hood. Keep away from acids.

3-(Cyclohexylamino)-1-propanesulfonic acid (**CAPS**) is an irritant and may be harmful by inhalation, ingestion, or skin absorption. Wear appropriate gloves and safety glasses. Do not breathe the dust.

DCM, ***see*** **Dichloromethane**

DHB, ***see*** **2,5-dihydroxybenzoic acid**

Dichloromethane (**DCM**), **CH_2Cl_2**, (also known as **Methylene chloride**) is toxic if inhaled, ingested, or absorbed through the skin. It is also an irritant and is suspected to be a carcinogen. Wear appropriate gloves and safety glasses and use in a chemical fume hood. Do not breathe the vapors.

Digitonin may be fatal if inhaled, ingested, or absorbed through the skin. Wear appropriate gloves and safety glasses and use in a chemical fume hood.

2,5-Dihydroxybenzoic acid (**DHB**) may be harmful by inhalation, ingestion, or skin absorption. Wear appropriate gloves and safety glasses. Do not breathe the dust.

Diisopropylethylamine (**DIEA**) is extremely destructive to the mucous membranes, upper respiratory tract, skin, and eyes. It may be harmful by ingestion or skin absorption. Inhalation may be fatal. Wear appropriate gloves and safety glasses and always use in a chemical fume hood. Keep away from heat sparks, and open flame.

N,N**-Dimethylformamide** (**DMF**), **$HCON(CH_3)_2$**, is a possibel carcinogen and is irritating to the eyes, skin, and mucous membranes. It can exert its toxic effects through inhalation, ingestion, or skin absorption. Chronic inhalation can cause liver and kidney damage. Wear appropriate gloves and safety glasses and use in a chemical fume hood.

Dimethyl sulfate (**DMS**), **$(CH_3)_2\ SO_4$**, is extremely toxic and is a carcinogen. Avoid breathing the vapors. Wear appropriate gloves and safety glasses and use only in a chemical fume hood. Dispose of solutions containing dimethyl sulfate by pouring them slowly into a solution of sodium hydroxide or ammonium hydroxide and allowing them to sit overnight in the chemical fume hood. Contact your local safety office before reentering the lab to clean up a spill.

Dimethyl sulfoxide (**DMSO**) may be harmful by inhalation or skin absorption. Wear appropriate gloves and safety glasses and use in a chemical fume hood. DMSO is also combustible. Store in a tightly closed container. Keep away from heat, sparks, and open flame.

Dioctylphthalate is toxic and is suspected of being a carcinogen. It may be harmful by inhalation, ingestion, or skin absorption. Wear appropriate gloves and safety goggles.

1,4-Dioxane is highly flammable both in liquid and vapor form. It is a possible carcinogen and is highly toxic by inhalation, ingestion, or skin absorption. Do not breathe the vapor. Wear appropriate gloves and safety glasses. Keep away from heat, sparks, and open flame.

Dithiothreitol (**DTT**) is a strong reducing agent that emits a foul odor. It may be harmful by inhalation, ingestion, or skin absorption. When working with the solid form or highly concentrated stocks, wear appropriate gloves and safety glasses and use in a chemical fume hood.

DMF, ***see N,N*****-Dimethylformamide**

Dry ice, ***see*** **Carbon dioxide**

EDC, ***see N*****-ethyl-*****N′*****-(dimethylaminopropyl)-carbodiimide**

Epoxy and acrylic resins, ***see*** **Resins**

Ethanol (EtOH), CH3CH_2OH, may be harmful by inhalation, ingestion, or skin absorption. Wear appropriate gloves and safety glasses.

Ethanolamine, $HOCH_2CH_2NH_2$, is toxic and harmful by inhalation, ingestion, or skin absorption. Handle with care and avoid any contact with the skin. Wear appropriate gloves and goggles and use in a chemical fume hood. Ethanolamine is highly corrosive and reacts violently with acids.

Ether, ***see*** **1,4-Dioxane**

1-Ethyl-3-[3-dimethylaminopropyl] carbodiimide (**EDC**) is irritating to the mucous membranes and upper respiratory tract. It may be harmful by inhalation, ingestion, or skin absorption. Wear appropriate gloves and safety glasses. Handle with care.

***N*-Ethylmorpholine** may be harmful by inhalation, ingestion, or skin absorption. Wear appropriate gloves and safety glasses.

Ferric chloride, $FeCl_3$, may be harmful by inhalation, ingestion, or skin absorption. Wear appropriate gloves and safety glasses and use only in a chemical fume hood.

Formaldehyde, HCHO, is highly toxic and volatile. It is also a possible carcinogen. It is readily absorbed through the skin and is irritating or destructive to the skin, eyes, mucous membranes, and upper respiratory tract. Avoid breathing the vapors. Wear appropriate gloves and safety glasses and always use in a chemical fume hood. Keep away from heat, sparks, and open flame.

Formamide is teratogenic. The vapor is irritating to the eyes, skin, mucous membranes, and upper respiratory tract. It may be harmful by inhalation, ingestion, or skin absorption. Wear appropriate gloves and safety glasses and always use a chemical fume hood when working with concentrated solutions of formamide. Keep working solutions covered as much as possible.

Formic acid, HCOOH, is highly toxic and extremely destructive to tissue of the mucous membranes, upper respiratory tract, eyes, and skin. It may be harmful by inhalation, ingestion, or skin absorption. Wear appropriate gloves and safety glasses (or face shield) and use in a chemical fume hood.

Glutaraldehyde is toxic. It is readily absorbed through the skin and is irritating or destructive to the skin, eyes, mucous membranes, and upper respiratory tract. Wear appropriate gloves and safety glasses and always use in a chemical fume hood.

Glycine may be harmful by inhalation, ingestion, or skin absorption. Wear gloves and safety glasses. Avoid breathing the dust.

Guanidine hydrochloride is irritating to the mucous membranes, upper respiratory tract, skin, and eyes. It may be harmful by inhalation, ingestion, or skin absorption. Wear appropriate gloves and safety glasses. Avoid breathing the dust.

Guanidinium hydrochloride, *see* Guanidine hydrochloride

HCOOH, *see* Formic acid

Helium (gas) may cause frostbite and may be harmful by inhalation, ingestion, or skin absorption. Wear appropriate gloves and safety goggles. Do not wear contact lenses.

Heptafluorobutyric acid is corrosive and may be harmful by inhalation, ingestion, or skin absorption. Wear appropriate gloves and safety goggles.

Heptane may be harmful by inhalation, ingestion, or skin absorption. Wear appropriate gloves and safety glasses. It is extremely flammable. Keep away from heat, sparks, and open flame.

Hexane is extremely flammable and may be harmful by inhalation, ingestion, or skin absorption. Wear appropriate gloves and safety glasses and use only in a chemical fume hood. Keep away from heat, sparks, and open flame.

HNO_3, *see* Nitric acid

H_2NOH, *see* Hydroxylamine

H_3PO_4, *see* Phosphoric acid

H_2SO_4, *see* Sulfuric acid

Human tissues/fluids , *see* Blood (human) and blood products

Hydrochloric acid, HCl, is volatile and may be fatal if inhaled, ingested, or absorbed through the skin. It is extremely destructive to mucous membranes, upper respiratory tract, eyes, and skin. Wear appropriate gloves and safety glasses and use with great care in a chemical fume hood. Wear goggles when handling large quantities.

Hydrogen fluoride is extremely toxic, corrosive and can cause severe burns. It may be harmful by inhalation, ingestion, and skin absorption. Wear appropriate gloves and safety goggles and use only in a chemical fume hood.

Hydroxylamine, H_2NOH, is corrosive and toxic. It may be harmful by inhalation, ingestion, or skin absorption. Wear appropriate gloves and safety glasses and use only in a chemical fume hood.

Imidazole is corrosive and may be harmful by inhalation, ingestion, or skin absorption. Wear appropriate gloves and safety glasses and use in a chemical fume hood.

Iodoacetamide (IAA), C_2H_4INO, can alkylate amino groups in proteins and can therefore cause problems if the antigen is being purified for amino acid sequencing. It is toxic and harmful by inhalation, ingestion, or skin absorption. Wear appropriate gloves and safety glasses and use only in a chemical fume hood. Do not breathe the dust.

Iodoacetic acid, *see* Acetic acid

Isopropanol is flammable and irritating. It may be harmful by inhalation, ingestion, or skin absorption. Wear appropriate gloves and safety glasses. Do not breathe the vapor. Keep away from heat, sparks, and open flame.

$K_3Fe(CN)_6$, ***see* Potassium ferricyanide**

$KH_2PO_4/K_2HPO_4/K_3PO_4$, ***see* Potassium phosphate**

Leupeptin (or its **hemisulfate**) may be harmful by inhalation, ingestion, or skin absorption. Wear appropriate gloves and safety glasses and use in a chemical fume hood.

Magnesium acetate tetrahydrate may be harmful by inhalation, ingestion, or skin absorption. Wear appropriate gloves and safety glasses.

Magnesium chloride, $MgCl_2$, may be harmful by inhalation, ingestion, or skin absorption. Wear appropriate gloves and safety glasses and use in a chemical fume hood.

β-Mercaptoethanol (2-Mercaptoethanol), $HOCH_2CH_2SH$, may be fatal if inhaled or absorbed through the skin and is harmful if ingested. High concentrations are extremely destructive to the mucous membranes, upper respiratory tract, skin, and eyes. β-Mercaptoethanol has a very foul odor. Wear appropriate gloves and safety glasses and always use in a chemical fume hood.

MES, ***see*** **2-(*N*-Morpholino)ethanesulfonic acid**

Methanol, MeOH or **H_3COH,** is poisonous and can cause blindness. It may be harmful by inhalation, ingestion, or skin absorption. Adequate ventilation is necessary to limit exposure to vapors. Avoid inhaling these vapors. Wear appropriate gloves and goggles and use only in a chemical fume hood.

Methyl acetate is extremely flammable in both liquid and vapor form. Vapor may cause flash fire. It may be harmful by inhalation, ingestion, or skin absorption. Wear appropriate gloves and safety glasses. Keep away from heat, sparks, and open flame.

Methyl cyanide, ***see*** **Acetonitrile**

Methylene blue is irritating to the eyes and skin. It may be harmful by inhalation, ingestion, or skin absorption. Wear appropriate gloves and safety glasses.

Methyl green may be harmful by inhalation, ingestion, or skin absorption. Wear appropriate gloves and safety glasses and use in a chemical fume hood.

Methyl iodide may be fatal and can affect the central nervous system. It is also a carcinogen and an irritant. It is harmful by inhalation, ingestion, or skin absorption. Wear appropriate gloves and safety goggles and use only in a chemical fume hood.

2-(*N*-Morpholino)ethanesulfonic acid (MES) may be harmful by inhalation, ingestion, or skin absorption. Wear appropriate gloves and safety glasses.

3-(*N*-Morpholino)-propanesulfonic acid (MOPS) may be harmful by inhalation, ingestion, or skin absorption. It is irritating to mucous membranes and upper respiratory tract. Wear appropriate gloves and safety glasses and use in a chemical fume hood.

Na_2CO_3, ***see* Sodium carbonate**

NaF, ***see* Sodium fluoride**

$NaH_2PO_4/Na_2HPO_4/Na_3PO_4$, ***see* Sodium phosphate**

NaN_3, ***see* Sodium azide**

Na_3VO_4, ***see* Sodium orthovanadate**

$(NH_4)_2CO_3$, ***see* Ammonium carbonate**

NH_4HCO_3, ***see* Ammonium bicarbonate**

$(NH_4)_6Mo_7O_{24}\cdot 4H_2O$, ***see* Ammonium molybdate**

$(NH_4)_2SO_4$, ***see* Ammonium sulfate**

$(NH_4)_2S_2O_8$, ***see* Ammonium persulfate**

Nitric acid, HNO_3, is volatile and must be handled with great care. It is toxic by inhalation, ingestion, and skin absorption. Wear appropriate gloves and safety goggles and use in a chemical fume hood. Do not breathe the vapors. Keep away from heat, sparks, and open flame.

Nitrogen, liquid or gaseous can cause severe damage due to extreme temperature. Handle frozen samples with extreme caution. Do not breathe the vapors. Seepage of liquid nitrogen into frozen vials can result in an exploding tube upon removal from liquid nitrogen. Use vials with O-rings when possible. Wear cryo-mitts and a face mask.

Pepstatin may be harmful by inhalation, ingestion, or skin absorption. Wear appropriate gloves and safety glasses and use in a chemical fume hood.

Phenyl isocyanate is highly toxic, combustible, and flammable. It may be harmful by inhalation, ingestion, or skin absorption. It can cause burns. Wear appropriate gloves and safety goggles. Keep away from heat, sparks, and open flame.

Phenyl-methyl-sulfonyl fluoride (PMSF), $C_7H_7FO_2S$ or $C_6H_5CH_2SO_2F$, is a highly toxic cholinesterase inhibitor. It is extremely destructive to the mucous membranes of the respiratory tract, eyes, and skin. It may be fatal by inhalation, ingestion, or skin absorption. Wear appropriate gloves and safety glasses and always use in a chemical fume hood. In case of contact, immediately flush eyes or skin with copious amounts of water and discard contaminated clothing.

Phosphoric acid, H_3PO_4, is highly corrosive and is harmful by inhalation, ingestion, or skin absorption. Wear appropriate gloves and safety glasses. Do not inhale the vapor.

α-Picoline is flammable and the effects from exposure are acute. It is harmful by inhalation, ingestion, or skin absorption. Wear appropriate gloves and safety goggles.

β-Picoline is flammable and the effects from exposure are acute. It is harmful by inhalation, ingestion, or skin absorption. Wear appropriate gloves and safety goggles.

PMSF, *see* Phenyl-methyl-sulfonyl fluoride

Polyethylenimine is highly corrosive and causes eye and skin burns. It may be harmful by inhalation, ingestion, or skin absorption. Wear appropriate gloves and safety goggles.

Ponceau S is an irritant and may be harmful by inhalation, ingestion, or skin absorption. Wear appropriate gloves and safety glasses.

Potassium carbonate may be harmful by inhalation, ingestion, or skin absorption. Wear appropriate gloves and safety glasses and use in a chemical fume hood.

Potassium chloride, KCl, may be harmful by inhalation, ingestion, or skin absorption. Wear appropriate gloves and safety glasses.

Potassium ferricyanide, $K_3Fe(CN)_6$, may be fatal by inhalation, ingestion, or skin absorption. Wear appropriate gloves and safety glasses and always use with extreme care in a chemical fume hood. Keep away from strong acids.

Potassium hydroxide, KOH and KOH/methanol, is highly toxic and may be fatal if swallowed. It may be harmful by inhalation, ingestion, or skin absorption. Solutions are corrosive and can cause severe burns. It should be handled with great care. Wear appropriate gloves and safety goggles.

Potassium phosphate, $KH_2PO_4/K_2HPO_4/K_3PO_4$, may be harmful by inhalation, ingestion, or skin absorption. Wear appropriate gloves and safety glasses. Do not breathe the dust. *K_2HPO_4•$3H_2O$ is dibasic and KH_2PO_4 is monobasic.*

Propane is a flammable high-pressure liquid and gas and may form explosive mixtures with air. Vapor may travel a considerable distance to source of ignition and flash back. Keep away from heat, sparks, and flame. Wear appropriate gloves and safety glasses and use a mechanical exhaust.

Pyridine is highly toxic and extremely destructive to the mucous membranes, upper respiratory tract, skin, and eyes. It may be harmful by inhalation, ingestion, or skin absorption. It is a possible mutagen and may cause male infertility. Keep away from heat, sparks, and open flame. Wear appropriate gloves and safety glasses and always use in a chemical fume hood.

Pyronin Y may be mutagenic. It may be harmful by inhalation, ingestion, or skin absorption. Wear appropriate gloves and safety glasses.

Radioactive substances: When planning an experiment that involves the use of radioactivity, include the physicochemical properties of the isotope (half-life, emission type and energy), the chemical form of the radioactivity, its radioactive concentration (specific activity), total amount, and its chemical concentration. Order and use only as much as really needed. Always wear appropriate gloves, lab coat, and safety goggles when handling radioactive material. **X-rays** and **gamma** rays are electromagnetic waves of very short wavelengths either generated by technical devices or emitted by radioactive materials. They may be emitted isotropically from the source or may be focused into a beam. Their potential dangers depend on the time period of exposure, the intensity experienced, and the wavelengths used. Be aware that appropriate shielding is usually of lead or other similar material. The thickness of the shielding is determined by the energy(s) of the X-rays or gamma rays. Consult the local safety office for further guidance in the appropriate use and disposal of radioactive materials. Always monitor thoroughly after using radioisotopes. A convenient calculator to perform routine radioactivity calculations can be found at: http://www.graphpad.com/calculators/radcalc.cfm

Resins are suspected carcinogens. The unpolymerized components and dusts may cause toxic reactions, including contact allergies with long-term exposure. Avoid breathing the vapors and dusts. Wear appropriate gloves and safety glasses and always use in a chemical fume hood. Sensitivity to these chemicals may develop with repeated contact.

SDS, *see* Sodium dodecyl sulfate

Silane is extremely flammable and corrosive. It may be harmful by inhalation, ingestion, or skin absorption. Keep away from heat, sparks, and open flame. The vapor is irritating to the eyes, skin, mucous membranes, and upper respiratory tract. Wear appropriate gloves and safety goggles and always use in a chemical fume hood.

Silica is an irritant and may be harmful by inhalation, ingestion, or skin absorption. Wear appropriate gloves and safety glasses. Do not breathe the dust.

Silver nitrate, $AgNO_3$, is a strong oxidizing agent and should be handled with care. It may be harmful by inhalation, ingestion, or skin absorption. Avoid contact with skin. Wear appropriate gloves and safety glasses. It can cause explosions upon contact with other materials.

Sinapinic acid may be harmful by inhalation, ingestion, or skin absorption. Wear appropriate gloves and safety glasses.

Sodium acetate, *see* Acetic acid

Sodium azide, NaN_3, is highly poisonous. It blocks the cytochrome electron transport system. Solutions containing sodium azide should be clearly marked. It may be harmful by inhalation, ingestion, or skin absorption. Wear appropriate gloves and safety goggles and handle it with great care. Sodium azide is an oxidizing agent and should not be stored near flammable chemicals.

Sodium borodeuteride is a flammable solid, corrosive, and causes burns. It is water-reactive and is harmful by inhalation, ingestion, or skin absorption. Wear appropriate gloves and safety goggles and use in a chemical fume hood.

Sodium carbonate, Na_2CO_3, may be harmful by inhalation, ingestion, or skin absorption. Wear appropriate gloves and safety glasses and use in a chemical fume hood.

Sodium citrate, *see* Citric acid

Sodium dihydrogen phosphate, NaH_2PO_4, (sodium phosphate, monobasic) may be harmful by inhalation, ingestion, or skin absorption. Wear appropriate gloves and safety glasses and use in a chemical fume hood.

Sodium dodecyl sulfate (SDS) is toxic, an irritant, and poses a risk of severe damage to the eyes. It may be harmful by inhalation, ingestion, or skin absorption. Wear appropriate gloves and safety goggles. Do not breathe the dust.

Sodium fluoride, NaF, is highly toxic and causes severe irritation. It may be fatal by inhalation, ingestion, or skin absorption. Wear appropriate gloves and safety glasses and use only in a chemical fume hood.

Sodium hydrogen phosphate, Na_2HPO_4, (sodium phosphate, dibasic) may be harmful by inhalation, ingestion, or skin absorption. Wear appropriate gloves and safety glasses and use in a chemical fume hood.

Sodium hydroxide, NaOH, and solutions containing NaOH are highly toxic and caustic and should be handled with great care. Wear appropriate gloves and a face mask. All other concentrated bases should be handled in a similar manner.

Sodium hypochlorite, NaOCl, *see* Bleach

Sodium iodide, NaI, may be harmful by inhalation, ingestion, or skin absorption. Wear appropriate gloves and safety glasses and use in a chemical fume hood.

Sodium nitrate, $NaNO_3$, may be harmful by inhalation, ingestion, or skin absorption. Wear appropriate gloves and safety glasses and use in a chemical fume hood.

Sodium orthovanadate, Na_3VO_4, may be harmful by inhalation, ingestion, or skin absorption. Wear appropriate gloves and safety glasses and use in a chemical fume hood.

Sodium phosphate, $NaH_2PO_4/Na_2HPO_4/Na_3PO_4$, is an irritant to the eyes and skin. It may be harmful by inhalation, ingestion, or skin absorption. Wear appropriate gloves and safety goggles. Do not breathe the dust.

Sodium toluene sulfinate, *see* Toluenesulfonic acid

Streptomycin is toxic and a suspected carcinogen and mutagen. It may cause allergic reactions. It may be harmful by inhalation, ingestion, or skin absorption. Wear appropriate gloves and safety glasses.

Succinic anhydride is a possible mutagen and is a severe eye irritant. It may be harmful by inhalation, ingestion, or skin absorption. Wear appropriate gloves and safety glasses and use only in a chemical fume hood. Do not breathe the dust.

Sulfosalicylic acid (dihydrate) is extremely destructive to the mucous membranes and respiratory system. Do not breathe the dust. Wear appropriate gloves and safety glasses and use only in a chemical fume hood.

Sulfuric acid, H_2SO_4, is highly toxic and extremely destructive to tissue of the mucous membranes and upper respiratory tract, eyes, and skin. It causes burns, and contact with other materials (e.g., paper) may cause fire. Wear appropriate gloves, safety glasses, and lab coat and use in a chemical fume hood.

SYPRO Orange/Red/Ruby contains **DMSO.** ***See*** **DMSO.**

TAME, ***see N*****-α-Tosyl-L-arginine methyl ester hydrochloride**

TBP, ***see*** **Tributylphosphine**

TCEP, ***see*** **Tris-(carboxyethyl)phosphine hydrochloride**

TEMED, ***see N,N,N′,N*****-Tetramethylethylenediamine**

Tetrabutylammonium thiocyanate may be harmful by inhalation, ingestion, or skin absorption. Wear appropriate gloves and safety glasses.

N,N,N′,N′**-Tetramethylethylenediamine** (**TEMED**) is highly caustic to the eyes and mucus membranes and may be harmful by inhalation, ingestion, or skin absorption. Wear appropriate gloves and tightly sealed safety goggles.

TFA, ***see*** **Trifluoroacetic acid**

Thiourea may be carcinogenic and may be harmful by inhalation, ingestion, or skin absorption. Wear appropriate gloves and safety glasses and use in a chemical fume hood.

TMAO, ***see*** **Trimethylamine N-oxide**

Toluene, $C_6H_5CH_3$, vapors are irritating to the eyes, skin, mucous membranes, and upper respiratory tract. Toluene can exert harmful effects by inhalation, ingestion, or skin absorption. Do not inhale the vapors. Wear appropriate gloves and safety glasses and use in a chemical fume hood. Toluene is extremely flammable. Keep away from heat, sparks, and open flame.

Toluenesulfonic acid is very corrosive, causes burns, and is extremely destructive to the upper respiratory tract. It may be harmful by inhalation, ingestion, or skin absorption. Wear appropriate gloves and safety glasses and use only in a chemical fume hood.

N**-α-Tosyl-L-arginine methyl ester hydrochloride** (**TAME**) may be harmful by inhalation, ingestion, or skin absorption. Wear appropriate gloves and safety glasses and use only in a chemical fume hood.

N**-Tosyl-L-phenylalanine chloromethyl ketone** (**TPCK**) may be harmful by inhalation, ingestion, or skin absorption. Wear appropriate gloves and safety glasses and use only in a chemical fume hood.

Tributylphosphine (**TBP**) may be fatal. It may be harmful by inhalation, ingestion, or skin absorption. Wear appropriate gloves and safety goggles and use in a chemical fume hood. Do not breathe the vapor.

Triethanolamine may be harmful by inhalation, ingestion, or skin absorption. Wear appropriate gloves and safety glasses and use only in a chemical fume hood.

Trifluoroacetic acid (**TFA**) (**concentrated**) may be harmful by inhalation, ingestion, or skin absorption. Concentrated acids must be handled with great care. Decomposition causes toxic fumes. Wear appropriate gloves and a face mask and use in a chemical fume hood.

Trimethylamine N-oxide (**TMAO**) causes eye irritation and may be harmful by inhalation, ingestion, or skin absorption. Wear appropriate gloves and safety goggles.

Tris may be harmful by inhalation, ingestion, or skin absorption. Wear appropriate gloves and safety glasses.

Tris (carboxyethyl) phosphine hydrochloride (**TCEP**) is corrosive to the mucous membranes, upper respiratory tract, eyes, and skin and can cause burns. It may be harmful by inhalation, ingestion, or skin absorption. Wear appropriate gloves and safety glasses and use in a chemical fume hood. Do not breathe the vapor or mist.

Triton X-100 causes severe eye irritation and burns. It may be harmful by inhalation, ingestion, or skin absorption. Wear appropriate gloves and safety goggles.

Urea may be harmful by inhalation, ingestion, or skin absorption. Wear appropriate gloves and safety glasses.

UV light and/or **UV radiation** is dangerous and can damage the retina. Never look at an unshielded UV light source with naked eyes. Examples of UV light sources that are common in the laboratory include hand-held lamps and transilluminators. View only through a filter or safety glasses that absorb harmful wavelengths. UV radiation is also mutagenic and carcinogenic. To minimize exposure, make sure that the UV light source is adequately shielded. Wear protective appropriate gloves when holding materials under the UV light source.

4-Vinylpyridine may be fatal by inhalation, ingestion, or skin absorption. Wear appropriate gloves and safety goggles. Use only in a chemical fume hood. Do not breathe the vapors.

Xylene is flammable and may be narcotic at high concentrations. It may be harmful by inhalation, ingestion, or skin absorption. Wear appropriate gloves and safety glasses and use only in a chemical fume hood. Keep away from heat, sparks, and open flame.

Xylene cyanol, ***see*** **Xylene**

Zinc chloride, $ZnCl_2$, is corrosive and poses possible risk to the unborn child. It may be harmful by inhalation, ingestion, or skin absorption. Wear appropriate gloves and safety glasses. Do not breathe the dust.

Suppliers

With the exception of those suppliers listed in the text with their addresses, all suppliers mentioned in this manual can be found in the BioSupplyNet Source Book and on the Web Site at:

http://www.biosupplynet.com

If a copy of the BioSupplyNet Source Book was not included with this manual, a free copy can be ordered by any of the following methods:

- Complete the Free Source Book Request Form found at the Web Site at:

  http://www.biosupplynet.com

- E-mail a request to info@biosupplynet.com
- Fax a request to 1-919-659-2199.

Index